Heat and Mass Transfer

Rajendra Karwa

Heat and Mass Transfer

Second Edition

Rajendra Karwa
Jodhpur Institute of Engineering
and Technology
Jodhpur, Rajasthan, India

ISBN 978-981-15-3990-9 ISBN 978-981-15-3988-6 (eBook)
https://doi.org/10.1007/978-981-15-3988-6

This Springer imprint is published by the registered company Springer Nature Singapore Pte Ltd.
The registered company address is: 152 Beach Road, #21-01/04 Gateway East, Singapore 189721, Singapore

Preface

This textbook of heat and mass transfer has been written to meet the need that exists in the opinion of the author among the undergraduate students of Mechanical, Automobile and Production Engineering, graduate students of Thermal Engineering, and students appearing in various competitive examinations. The heat transfer part of the book will also be useful for students of Chemical Engineering.

A classical treatment of the fundamentals of heat transfer has been presented in this book. The basic approach of separate discussions of conduction, convection and radiation has been used. The book contains 16 chapters and 3 appendices. Each chapter contains sufficient number of solved problems. Step-wise answers are given to all unsolved problems.

In Chap. 1, an overview of different modes of heat transfer has been presented.

Chapters 2–4 present the conventional treatment of one-dimensional heat conduction through plane wall, cylindrical and spherical systems, fins, and simple systems with volumetric heat generation.

In Chap. 5, analytical treatment of some cases of two-dimensional steady-state heat conduction has been presented followed by a discussion of finite-difference numerical methods which are often used in practice for solving complex problems.

Chapter 6 is devoted to the transient heat conduction where lumped heat capacity analysis and solution of problems based on Heisler charts have been given. Numerical method of solving transient conduction problems has been presented with a number of illustrative examples.

Analytical solutions of some simple convection heat transfer problems, especially the convection with laminar flow, have been presented in Chap. 7. Empirical relations for forced convection and natural or free convection heat transfer have been presented in Chaps. 8 and 9, respectively.

Chapters 10 and 11 deal with the fundamentals of radiation heat transfer and the exchange of thermal radiation between surfaces separated by transparent medium, respectively. The method of radiation network has been used extensively in the analysis of radiation problems. Gaseous radiation problems have been dealt with using the conventional Hottel charts in Chap. 12.

Chapter 13 has been divided into two parts. In the first part, the basic modes of condensation have been presented followed by the presentation of the analytical solution due to Nusselt for laminar film condensation on a vertical surface. The second part discusses the phenomenon of pool boiling followed by discussion on forced boiling in vertical and horizontal pipes.

The conventional thermal analysis of heat exchangers (the log-mean-temperature difference and effectiveness approaches) is presented in Chap. 14 followed by introduction to design methodology of heat exchangers considering the design of double-pipe heat exchanger.

Chapter 15 presents a brief introduction to mass diffusion in a quiescent medium and convective mass transfer. Analogies between heat, mass and momentum transfer have been presented.

In Chap. 16, thermal analysis and discussion of conventional and enhanced performance solar air heaters are presented along with the mathematical model of the solar air heater.

The first edition of the book received an excellent response. In this edition, some minor modifications and corrections have been made, and at the end of each chapter, summary of the chapter has been incorporated. Many new solved and unsolved problems have been added to provide problems of varying complexity.

The students are advised to refer to the reference books, handbooks and journals, some of which are also listed at the end of book, for details beyond the coverage of this textbook and also for the new developments in the field of heat transfer. Computers have made possible the numerical solution of quite complex problems. Readers are advised to refer to advanced references for the computer-aided solution of heat transfer problems.

The author sincerely expresses deep sense of gratitude and indebtedness to the authors and publishers of various advanced books, handbooks, journals and other references which have been consulted and whose material has been used in the preparation of this book.

In spite of the care taken in preparing the manuscript of this book and reading the proofs, there is always a scope for improvement and some errors might have crept in. I will be grateful to the readers if they can suggest ways to improve the contents and bring the errors to my attention, if any, noticed by them.

Jodhpur, India Dr. Rajendra Karwa

Contents

About the Author

Dr. Rajendra Karwa pursued his Bachelor's and Master's degree in Mechanical Engineering from the University of Jodhpur, Jodhpur, India. He obtained his Ph.D. in 1998 from IIT Roorkee, India. He retired from M.B.M. Engineering College, Faculty of Engineering & Architecture, Jai Narain Vyas University, Jodhpur in 2012 as Professor and Head, Department of Mechanical Engineering after serving more than 30 years. He has served as visiting faculty in Addis Ababa University, Addis Ababa (Ethiopia) from 2003-2005. He is actively involved in design of industrial systems and products, and has successfully completed a large number of industrial projects. Currently he is working as Director (Academics) at Jodhpur Institute of Engineering & Technology, Jodhpur. He has supervised 15 research scholars leading to the award of Ph.D. and M.Tech. He has published about 70 research papers and has authored two books. His areas of interest are heat transfer, machine design, solar thermal applications, and exergy analysis.

List of Symbols[1]

A	Frontal area, m^2;
A	Heat transfer surface area, m^2;
A_c	Cross-sectional area, m^2;
b	Width, m;
c	Speed of light, m/s;
c	Specific heat of solids and liquids, J/(kg K);
c_p	Specific heat at constant pressure, J/(kg K);
c_v	Specific heat at constant volume, J/(kg K);
c_{fx}	Local skin friction coefficient = $\tau_{wx}/(1/2\rho U_\infty^2)$;
C_D	Drag coefficient = $(F_D/A)/(1/2\rho U_\infty^2)$;
C	Heat capacity = mc, J/K;
C	Mass concentration, kg/m^3;
C	Molar concentration, $kmol/m^3$;
d, D	Characteristic dimension, m;
d, D	Diameter, m;
D	Diffusion coefficient[a], m^2/s;
D_h, d_h	Hydraulic diameter = $4A/P$, m;
e	Roughness height, m;
e/D_h, e/D	Relative roughness height;
E	Electric potential;
E	Energy, J;
$E_{b\lambda}$	Monochromatic hemispherical emissive power of black body, $W/(m.m^2)$;
f	Fanning friction factor = $\tau_w/(1/2\rho U^2)$;
f_{app}	Apparent Fanning friction factor in the hydrodynamic entrance region;
F_D	Drag force, N;
F_{ij}	View factor;
g	Gravitational acceleration, m/s^2;
G	Mass velocity, $kg/(s\ m^2)$;
G	Irradiation, W/m^2;
h	Heat transfer coefficient, $W/(m^2\ K)$;

[1](at most of the places, small t has been used when temperature is in °C)

h_{fg}	Enthalpy of evaporation (latent heat), J/kg;
h_m	Mass transfer coefficient, m/s;
h_r	Radiation heat transfer coefficient, W/(m^2 K), W/(m^2 °C);
h_x	Local heat transfer coefficient at position x, W/(m^2 K), W/(m^2 °C);
h_w	Wind heat transfer coefficient, W/(m^2 K), W/(m^2 °C);
H	Height, m;
I	Solar radiation intensity (insolation), W/m^2;
I_λ	Monochromatic intensity of radiation;
J	Radiosity, W/m^2;
k	Thermal conductivity, W/(m K), W/(m °C);
l	Mean free path, m;
L	Length, m;
L	Fundamental dimension of length;
L_c	Corrected fin length, m;
L_{hy}	Hydrodynamic entrance length, m;
L_{th}	Thermal entrance length, m;
m	Mass flow rate, kg/s;
m	A fin parameter, 1/m;
M	Molecular weight;
M	Fundamental dimension of mass;
n	Number of radiation shields;
N	Molal diffusion rate;
NTU	Number of transfer units;
p	Pressure, Pa;
p_i	Partial pressure, Pa;
P	Power, W;
P	Perimeter, m;
q	Heat transfer rate, W;
q'', q_w	Heat flux, W/m^2;
Q	Quantity of heat, J;
q_g	Volumetric heat generation rate, W/m^3;
r	Radius (usually variable), m;
R	Radius, m;
R	Gas constant, J/(kg K);
R	Temperature group $(T_1 - T_2)/(t_2 - t_1)$;
R_k	Thermal resistance to heat conduction, K/W;
S	Temperature group $(t_2 - t_1)/(T_1 - t_1)$;
S	Conduction shape factor;
T_a	Ambient temperature, °C, K;
T_b,T_{fm}	Bulk mean air temperature = $(T_o + T_i)/2$, °C;
T_i	Inlet temperature, °C;
T_o	Outlet temperature, °C;
T_s	Surface temperature, °C, K;
T_{sky}	Sky temperature, K;
T_{sat}, T_s	Saturation temperature, °C;
T_w	Wall temperature, °C, K;

T_∞	Free-stream temperature, °C;
u, v, w	Velocity, m/s;
U	Overall heat transfer coefficient, W/(m^2 K), W/(m^2 °C);
U	Velocity, m/s;
U_∞	Free-stream velocity, m/s;
v	Specific volume, m^3/kg;
V	Volume, m^3;
W	Weight, N;
W	Width of the duct, m;
W/H	Duct aspect ratio;
x, y, z	Variable distances in space;
x_i	Mole fraction;

Dimensionless Numbers

Bi	Biot number = hL/k_{solid};
e^+	Roughness Reynolds number, Eq. (8.63);
Ec	Eckert number = $u^2/c_p \Delta t$;
Fo	Fourier number = $\alpha\tau/L^2$;
g	Heat transfer function, Eq. (8.69);
Gr	Grashof number = $g\beta L^3 \Delta t/\nu^2$;
Gz	Graetz number = Re Pr (D/L);
Le	Lewis number[a] = Sc/Pr = α/D;
p/e	Relative roughness pitch;
Nu	Nusselt number = hL/k_{fluid};
Pe	Peclet number = Re Pr;
Pr	Prandtl number = $\mu c_p/k$;
Pr_t	Turbulent Prandtl number = $\varepsilon_M/\varepsilon_H$;
R	Roughness function, Eq. (8.67);
Ra	Rayleigh number = Gr Pr = $g\beta H^3 \Delta t/\alpha\nu$;
Ra^*	Rayleigh number (based on heat flux) = $g\beta q'' H^4/\alpha\nu k$;
Re	Reynolds number = $\rho U d/\mu$ = GL/μ;
Re_x	Reynolds number based on longitudinal length = $U_\infty x/\nu$;
Re_{cr}	Critical Reynolds number;
Sc	Schmidt number[a] = $\nu/D = \mu/\rho D$ = Le Pr;
Sh	Sherwood number[a] = $h_m L/D$;
St_m	Mass transfer Stanton number = h_m/U;
St	Stanton number = $h/(Gc_p)$;
St_x	Local Stanton number = $h_x/(Gc_p)$;

Greek Symbols

α, β, γ, ϕ, ψ	Angle (degree or rad);
α	Thermal diffusivity = $k/\rho c$, m^2/s;
α	Absorptivity (radiation);
β	Coefficient of volumetric expansion, 1/K;
β	Temperature coefficient of thermal conductivity, 1/K;
β	Collector slope (degree);
δ	Velocity boundary layer thickness, m;
δ	Thickness, m;
δ_{md}	Momentum displacement thickness, m;
δ_{vd}	Velocity displacement thickness, m;
δp, Δp	Pressure drop in the duct, Pa;
δ_t	Thermal boundary layer thickness;
Δt, ΔT	Temperature difference, °C, K;
ε	Fin effectiveness;
ε	Heat exchanger effectiveness;
ε	Emissivity;
ε_H	Thermal eddy diffusivity, m^2/s;
ε_M	Momentum eddy diffusivity or viscosity, m^2/s;
ϕ	Relative humidity;
η	Thermal efficiency;
η_f	Efficiency of fin;
λ	Darcy friction factor (= $4f$)
λ	Wavelength, m;
λ_{max}	Wavelength at maximum value of $E_{b\lambda}$;
μ	Dynamic viscosity, Pa s, N s/m^2, kg/(m s);
v	Kinematic viscosity = μ/ρ, m^2/s;
π	Dimensionless group;
θ	Temperature excess, K;
θ	Time, s;
ρ	Density of fluid, kg/m^3;
ρ	Reflectivity;
σ	Stefan–Boltzmann constant;
σ	Surface tension, N/m;
τ	Time, s;
τ	Shear stress between fluid layers, Pa;
$(\tau\alpha)$	Transmittance–absorptance product;
ω	Solid angle, sr;
ω	Specific humidity;
ψ	Stream function;

Superscript and Subscript

a	Ambient
b	Bulk, blackbody
cr	Critical state
f	Fluid
f	Film
fd	Fully developed
g	Gas
hy	Hydrodynamic
i, 1	Inlet or initial
i	Based on the inside surface of a pipe
l	Laminar or liquid
m	Mass transfer quantity
m	Mean
max	Maximum
min	Minimum
o, 2	Outlet
o	Based on the outside surface of a pipe
o	Stagnation
s	Smooth surface
s	Surface
th	Thermal
v	Vapour
w	Wall
x	Based on variable length
∞	Free-stream condition
$\bar{}$ (overbar)	Mean or molar

Space Coordinates

r, θ, z	Cylindrical, m, rad, m;
r, θ, ϕ	Spherical, m, rad, rad;
x, y, z	Cartesian, m, m, m.

[a] Diffusion D is mass diffusivity, m^2/s.

Note: The symbol L in the dimensionless groups stands for a generic length and is defined according to the particular geometry under consideration; it may be diameter, hydraulic diameter, plate length, etc.

1 Introduction

1.1 Introduction

Thermodynamically, heat is a form of energy that is transferred between two systems (or a system and its surroundings) by virtue of a temperature difference.

Classical thermodynamics treats the processes as though only the end states exist, presuming that the system exists in a state of thermodynamic equilibrium. It provides no information about the rates of irreversible flows. The heat flow with temperature difference is one example of irreversible flow. In studies of processes that involve flow of heat or mass, the rate of flow is an important parameter.

Heat can be transferred in three different ways: by *conduction*, *convection* and *radiation*. In most of the engineering applications, it is a combination of the two or three modes. Pure conduction is found only in solids, and convection is possible only in fluids. A detailed study of these modes will be given in this book. Here, a brief description of each mode is presented to familiarize the reader with them.

1.2 Heat Transfer by Conduction

Thermal conduction is a process by which heat is transmitted by the direct contact between particles of a body without any motion of the material as a whole. The phenomenon of conduction heat transfer can be experienced by a simple experiment. Heat one end of a metal rod. The other end of the rod will become hotter and hotter with the passage of time. Heat reaches from the heated end of the rod to the other end by conduction through the material of the rod.

Conduction occurs in all media—solids, liquids, and gases when a temperature gradient exists. In opaque solids, it is the only mechanism by which the heat can flow. In the fluids, the molecules have freedom of motion and energy is also transferred by movement of the fluid. However, if the fluid is at rest, the heat is transferred by conduction.

Heat transfer by conduction requires that the temperature distribution in a medium is non-uniform, that is, a temperature gradient exists in the body. Fourier noted in 1811 that the heat flow in a homogenous solid is directly proportional to the temperature gradient. Consider a plane wall of thickness dx, whose area perpendicular to the direction x is A as shown in Fig. 1.1. If the thickness of the wall is very small as compared to its height and width, it is a

R. Karwa, *Heat and Mass Transfer*,
https://doi.org/10.1007/978-981-15-3988-6_1

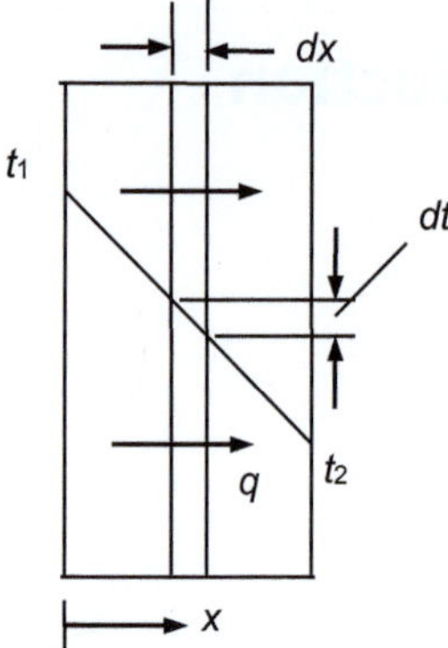

Fig. 1.1 Conduction heat flow through a solid

case of *one-dimensional* (in direction x only) heat flow. Let one of the faces of the wall is at a temperature t_1 and the other at temperature t_2. For the elemental thickness dx, the temperature difference is dt. Then, the conduction heat flux q/A (the heat flux is the heat flow rate per unit area of the surface), according to Fourier, is

$$\frac{q}{A} \propto \frac{dt}{dx} \tag{1.1}$$

When the constant of the proportionality is inserted in Eq. (1.1), we get

$$\frac{q}{A} = -k\frac{dt}{dx} \tag{1.2}$$

The constant of proportionality k is known as *thermal conductivity* of the material. It is a physical property of a substance and characterizes the ability of a material to conduct heat. The negative sign in the equation indicates that the heat flow is in the direction of falling temperature.

From Eq. (1.2), we get

$$k = -\left(\frac{q}{A}\right)\frac{dx}{dt}\ \mathrm{W/(m\,K)} \tag{1.3}$$

Thus thermal conductivity determines the quantity of heat flowing per unit time through the unit area with a temperature drop of 1°C (K) per unit length.

1.3 Heat Transfer by Convection

The term is applied to transport of heat as a volume of liquid or gas moves from a region of one temperature to that of another temperature. Thus, the transport of heat is linked with the movement of the medium itself. The convection can be observed in liquids if we carry out a simple experiment. Consider a pot with water which is placed over a burner. The water at the bottom of the pot is heated and becomes less dense than before due to its thermal expansion. Thus, the water at the bottom, which is less dense than the cold water in the upper portion, rises upwards. It transfers its heat by mixing as it rises. The movement of the water, referred to as the *convection currents*, can be observed by putting a few crystals of potassium permanganate in the bottom of the pot. Density differences and the gravitational force of the earth act to produce a force known as *buoyancy force*, which drives the flow. When the flow is due to the density differences only, it is called *natural or free convection*. Density differences may also be caused by the composition gradients. For example, water vapour rises mainly due to the lower density of water vapour present in the moist air.

Convection is termed as *forced* if the fluid is forced to flow over a surface or in a duct by external means such as a fan, pump or blower, that is, the forced convection implies mechanically induced flow.

Heat transfer processes involving change of phase of a fluid (boiling of liquids or condensation of vapours) are also considered to be convection because of the motion of fluid that is set up due to the rising vapour bubbles during boiling or the falling liquid droplets during the condensation.

The heat transfer by convection is always accompanied by conduction. The combined process of heat transfer by convection and conduction is referred to as convective heat transfer. The heat transfer, between a solid surface and a fluid, is expressed by Newton's law of cooling as

$$q = hA\Delta t \tag{1.4}$$

where Δt is the temperature difference between the fluid and the surface, A is the area of the surface transferring heat and h is known as *heat transfer coefficient or film coefficient*.

In general, the value of the heat transfer coefficient h depends on the fluid-flow conditions, the thermophysical properties of the fluid and the type of flow passage. Order of magnitude of convective heat transfer coefficients is given in Table A6, Appendix.

Consider a heated plate at temperature t_w as shown in Fig. 1.2, which is exposed to a fluid flowing parallel to the plate surface at velocity U_∞ and is at a temperature t_∞. The velocity of the fluid at the plate surface will be zero because of the viscous effect. Since the velocity of the fluid at the wall is zero, we can imagine that there exists a thin film of the fluid quite close to the wall, which is practically stationary. In most of the cases, the temperature gradient is confined to this thin layer where at a greater distance from the wall, only a small temperature difference exists because of the mixing of the fluid. The heat is, thus, transferred across the film by conduction. This heat is then carried away by fluid motion.

Assuming that the temperature variation in the fluid film is linear, the flow of heat across the film can be expressed by Fourier's law as

$$q = k_f A \frac{(t_w - t_\infty)}{\delta_f} \tag{1.5}$$

where δ_f is the thickness of the fluid film and k_f is the thermal conductivity of the fluid.

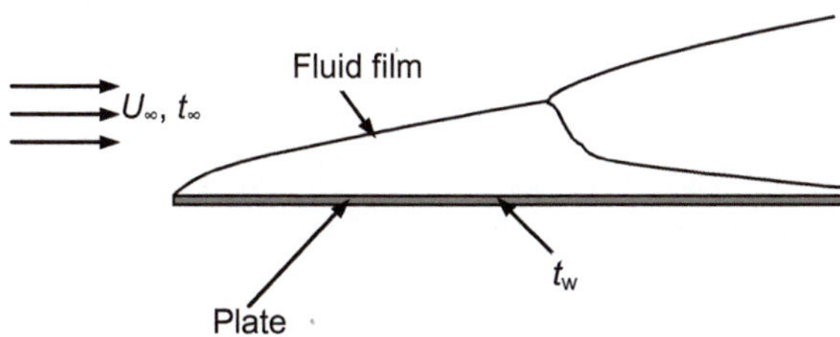

Fig. 1.2 Cooling of a hot plate by flowing fluid

Comparison of Eq. (1.5) with Newton's equation, which is widely used, gives $h = k_f/\delta_f$. This equation relates the convection heat transfer to the conduction. It shows that the value of the heat transfer coefficient is directly proportional to the thermal conductivity of the fluid. It is to be kept in mind that there are other factors also which govern the magnitude of the heat transfer coefficient as mentioned earlier. From the order of magnitude of the heat transfer coefficient in Table A6, it can be seen that the gases provide a lower value of heat transfer coefficient in comparison to the liquids. This can be attributed to their lower thermal conductivity as compared to the liquids.

In most of the heat exchangers, heat is transferred between hot and cold fluid streams across a solid wall. In such cases, it is convenient to combine the two film coefficients (i.e. of the hot and cold fluid streams) to give a single coefficient known as *overall heat transfer coefficient U*, which is defined as

$$U = \frac{q}{A\Delta t} \tag{1.6}$$

Here Δt is the temperature difference between the two fluids.

1.4 Heat Transfer by Radiation

The thermal radiation is the process of heat propagation by means of electromagnetic waves produced by virtue of the temperature of the body. It depends on both the temperature and an optical property known as emissivity ε of the body. In contrast to the conduction and convection heat transfers, radiation can take place through a perfect vacuum. Solids, liquids and gases may radiate energy. For example, water vapour and carbon dioxide are the principal sources of the gaseous radiation in furnaces.

Boltzmann established that the rate at which a body gives out the heat by radiation is proportional to the fourth power of the absolute temperature of the body, that is,

$$q \propto AT^4 \tag{1.7}$$

When the constant of the proportionality is inserted,

$$q = \sigma AT^4 \tag{1.8}$$

where the constant of proportionality σ is known as *Stefan–Boltzmann constant*, and A is the surface area of the body.

A body may absorb, transmit or reflect radiant energy. A *blackbody* absorbs the entire radiation incident on it. Thus, it is a perfect absorber. Technically, the blackbody is a hypothetical body. It does not necessarily refer to the colour of the body, though bodies black in colour usually absorb most. A blackbody is also a perfect or ideal radiator, for which the emissivity $\varepsilon = 1$. For real bodies, ε is less than 1 and they do not emit as much energy as a blackbody.

The net heat radiated between two black bodies 1 and 2 at temperatures T_1 and T_2 that see each other completely (i.e. they exchange heat by radiation between themselves only); the net energy exchange is proportional to the difference in T_1^4 and T_2^4. Thus

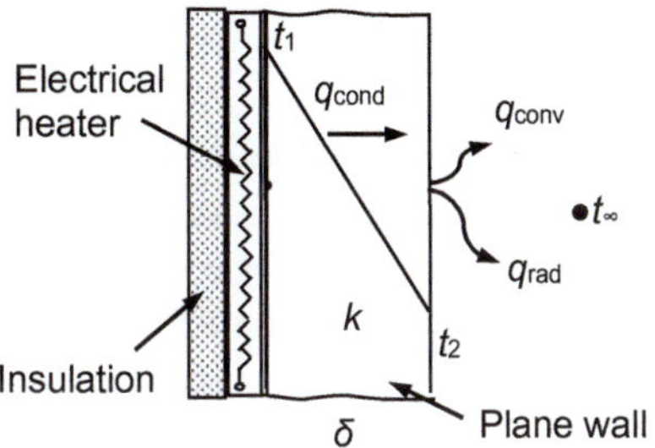

Fig. 1.3 Combination of modes of heat transfer

$$q_{1-2} = \sigma A(T_1^4 - T_2^4) \tag{1.9}$$

The radiation exchange between bodies which are not black is quite complex and will be dealt with in detail later.

1.5 Simultaneous Heat Transfer

Heat can be transferred in three different modes: by conduction, convection and radiation. In most of the engineering applications, it is a combination of the two or three modes. Pure conduction is found only in opaque solids but in the case of semitransparent solids radiation heat transfer is also involved. Convection is possible only in fluids. In vacuum, heat transfer is possible only by radiation. In case of a solid surface exposed to a fluid (gas or liquid), convection heat transfer is always involved. If the gas to which the solid surface is exposed is transparent to radiation, heat transfer also takes place by radiation. It is to note that liquids and some gases (presented in Chap. 12) are absorbers of radiation.

Figure 1.3 presents a case where all the three modes of heat transfer are present. Heat transfers by conduction through the wall from left face to the right face and then is rejected from the right face by convection to the surrounding air and by radiation to the surroundings. The energy balance would give

$$q_{\text{cond}} = q_{\text{conv}} + q_{\text{rad}} \tag{1.10}$$

1.6 Summary

In this chapter, the basic modes of heat transfer have been presented, which will be discussed in greater detail in coming chapters to apply them to various problems of heat transfer.

Newton'slaw of cooling presented for convection heat transfer looks quite simple but it is to be noted that the heat transfer coefficient h is a function which depends on the fluid-flow conditions, the thermophysical properties of the fluid and the geometric configuration of the heated surface or duct, and for its estimate mathematical or experimental approach is used.

A simple case of radiation exchange between two black bodies that see each other completely (i.e. they exchange heat by radiation between themselves only) has been considered in Eq. (1.9). In general, the radiation heat exchange between bodies, which are not black, involves complex interaction because of surface emissive properties and when a system of radiation heat exchange between a number of bodies is involved.

Review Questions

1.1. State and explain in brief the different modes of heat transfer.
1.2. Write Fourier's equation for one-dimensional steady-state conduction heat transfer.
1.3. Define thermal conductivity and heat transfer coefficient.
1.4. Discuss the mechanism of convective heat transfer.
1.5. State Boltzmann's law for the radiation heat transfer. What is the net heat exchange between two blackbodies, which exchange heat by radiation between themselves only?

One-Dimensional Steady-State Heat Conduction

2

2.1 Introduction

We presented Fourier's law of heat conduction in Chap. 1. In this chapter, we shall use the law to calculate the heat flow in systems where one-dimensional heat flow occurs. A plane wall made of isotropic material, whose thickness is much smaller compared to its length and width, is the simplest case of such one-dimensional conduction heat flow. The heat flow will also be one-dimensional in cylindrical and spherical solid when the temperature gradient is only in the radial direction. Some two- or three-dimensional systems can also be approximated as one-dimensional if the temperature variation, represented by the temperature gradient, in one direction is significantly greater than the other directions.

Before attempting the analysis of one-dimensional heat conduction problems, we shall discuss some important terms relating to the conductors and establish the general heat conduction equations in the rectangular, cylindrical and spherical coordinates.

2.2 Temperature Field and Temperature Gradient

Heat flows by conduction when different points in a body are at different temperatures. In general, the temperature may vary both in space (x, y and z) and time (τ). The analytical investigation of the heat conduction is basically a study of the space-time variation of the temperature and determination of the equation:

$$t = t(x, y, z, \tau) \tag{2.1}$$

which is the mathematical expression of the temperature field. Equation (2.1) represents a set of temperatures at all points of the space at any given time.

The temperature field expressed by Eq. (2.1) is referred to as transient or non-steady temperature field. If the temperature in the space does not change with time, the temperature field is a function of the space coordinates only and it is referred to as a steady state. Mathematically, it is expressed as

$$t = t(x, y, z,)$$

R. Karwa, *Heat and Mass Transfer*,
https://doi.org/10.1007/978-981-15-3988-6_2

and

$$\frac{\partial t}{\partial \tau} = 0 \tag{2.2}$$

The temperature fields represented by Eqs. (2.1) and (2.2) are three-dimensional fields since they are function of three coordinates.

A temperature field, which is a function of two coordinates say x and y, is termed as two-dimensional and is described by the following equation:

$$t = t(x, y)$$

and

$$\frac{\partial t}{\partial z} = 0 \tag{2.3}$$

Similarly for a one-dimensional case, the equation of temperature field is

$$t = t(x)$$

and

$$\frac{\partial t}{\partial y} = \frac{\partial t}{\partial z} = 0 \tag{2.4}$$

If different points of the body having the same temperature are joined, we obtain an isothermal surface. Intersection of such isothermal surfaces by a plane gives a family of isotherms on this plane as shown in Fig. 2.1. The figure shows isotherms which differ by temperature $\pm\delta t$.

The temperature gradient is a vector normal to the isothermal surface. Mathematically, it is the derivative in this direction:

$$grad(t) = \vec{n}\frac{\partial t}{\partial n} \tag{2.5}$$

where $\vec{n}$ is the unit vector normal to the isothermal surface and $\frac{\partial t}{\partial n}$ is the temperature derivative along the normal. Projections of the vector grad t can be made on the coordinate

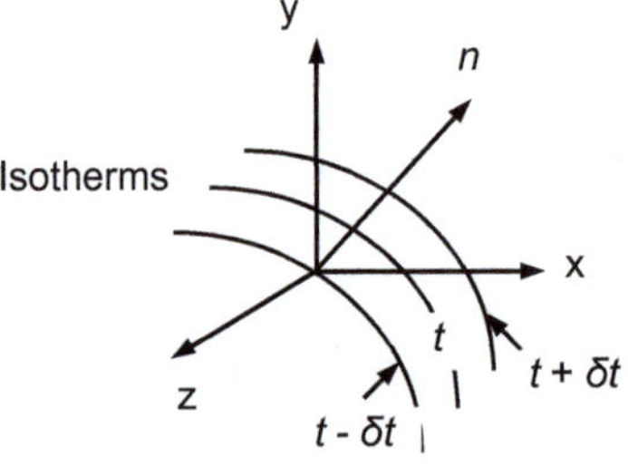

Fig. 2.1 Temperature vectors

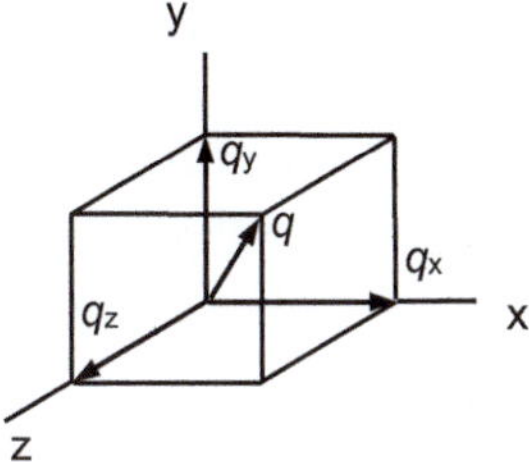

Fig. 2.2 Vector representation of Fourier's law

axes ox, oy and oz. Accordingly, the components of the normal heat flow vector $\vec{q}$ are (Fig. 2.2)

$$
\begin{aligned}
q_x &= -k\frac{\partial t}{\partial x} \\
q_y &= -k\frac{\partial t}{\partial y} \\
q_z &= -k\frac{\partial t}{\partial z}
\end{aligned}
\tag{2.6}
$$

where the negative sign indicates that the heat flow is in the direction of negative temperature gradient.

The vector representing the rate of heat flow is

$$\vec{q} = iq_x + jq_y + kq_z \tag{2.7}$$

The vector q at a point is along a path normal to the isothermal surface at that point.

2.3 Thermal Conductivity

As already mentioned, the thermal conductivity is a physical property of a substance. In general, it depends on temperature, pressure and nature of the substance. Mostly, it is determined experimentally. Typical values of the thermal conductivity of several common engineering materials are given in a tabulated form in Appendix A. The thermal conductivity of the materials of interest differs by many thousand times. The highest values are for the metals, followed by dense ceramics, organic solids and liquids, while the lowest values are for the gases. Super-insulations have been developed for cryogenic applications with thermal conductivity as low as 0.3 mW/(m °C).

2.3.1 Thermal Conductivity of Solids

Heat is transmitted through the solids by the elastic vibrations of the atoms and molecules (crystal lattice vibrations) and by free electrons (electronic thermal conduction). The transfer of heat by free electrons is very effective. The mechanism of electronic thermal conduction is similar to the electric conduction. When the temperature at one end of a rod is raised, the electrons in that region receive a somewhat higher energy and acquire an increased velocity. As these electrons move through the solid, they collide with other particles and transfer their energy to these particles. Thus the heat is transferred from one end to the other. As the electrons can be accelerated to a very high velocity, the electronic thermal conduction is more effective than the molecular conduction.

2.3.2 Thermal Conductivity of Metals and Alloys

Metals have large number of free electrons (referred to as electron gas) and they contribute most to the heat transfer. Hence, the metals have high thermal conductivity. Since the electrons are the carriers of the heat and electricity in metals, the thermal conductivity of a metal is proportional to the electric conductivity. In general, good electrical conductors are almost always good heat conductors. Examples are silver, copper and aluminium. Since the scattering of the electrons intensifies with rising temperature, the thermal and electrical conductivities decline with the increase in the temperature. The thermal conductivity of the metals drops sharply in the presence of impurities. The phenomenon can also be explained by an increase in structure heterogeneity, which causes electron scattering. In contrast to the pure metals, the thermal conductivity of the alloys increases with rise in the temperature.

Electrical insulators are usually bad heat conductors.

2.3.3 Thermal Conductivity of Construction and Heat-Insulating Materials

Insulators and non-metals have lower conductivities because the heat is transferred by the vibrations of the atoms only. Thermal conductivity of the non-metals (solid dielectrics) depends on their density, structure, porosity and moisture content. Their thermal conductivity usually increases with temperature.

Many construction and heat-insulating materials (bricks, concrete, asbestos, etc.) have a porous structure, and it is not possible to consider them continuous. Their thermal conductivity is taken as the thermal conductivity of a solid of the same shape and size, through which the same amount of heat is transmitted under the given temperature and boundary conditions.

The thermal conductivity of porous solids and powders depends on their volumetric density. For example, the thermal conductivity of asbestos powder increases from about 0.1 to 0.25 W/(m K) with the increase in the density from 400 to 800 kg/m^3 because the thermal conductivity of the air filling the pores is much less than that of the solid constituents of the porous material.

The thermal conductivity of the damp porous material is higher than the dry material. The thermal conductivity of a damp material is significantly greater than the thermal conductivities of the dry material and water taken separately. This can be attributed to the capillary movement of water within the material. The thermal conductivity of the construction and building materials range from about 0.03 to 3.0 W/(m K). Materials having thermal conductivity below 0.25 W/(m K) are used for heat insulation.

2.3.3.1 R-Values of Insulating Materials

An insulating material's resistance to conductive heat flow is measured in terms of R-value, which is calculated from $R = \delta/k$ where δ is the thickness of the material in m and k is the thermal conductivity in W/(m K). Thus, the R-value is measured in m^2 K/W in SI units. In FPS system of units (used in the United States), R-values are measured in ft^2.°F. hr/Btu, which are 5.67 times larger than R-values in SI units. The R-value should not be confused with thermal resistivity. It can be interpreted as the thermal resistance of a 1 m^2 cross-section of the material. The higher the R-value of an insulating material, the greater is the thermal resistance the material offers to conduction heat flow.

For calculating the R-value of a multi-layered insulation system, the R-values of the individual layers are added.

It is to note that the R-values take only conduction into account. It does not include convection and radiation.

2.3.4 Thermal Conductivity of Gases

The heat conduction in gases at ordinary pressure and temperature occurs through the transport of the kinetic energy of the molecules because of their random motion and collision. The molecules in the high-temperature region have higher kinetic energy than those in the lower temperature region. When the molecules from the high-temperature region move to the low-temperature region, they give up a part of their kinetic energy through collision with the low-temperature molecules. From the kinetic theory of gases, the thermal conductivity of a gas is given by

$$k = (1/3)v_m l \rho c_v \tag{2.8}$$

where l is the mean free path and v_m is the mean velocity of the molecules.

With the increase in the pressure, the gas density increases but the product $(l\rho)$ remains constant. Therefore, the thermal conductivity does not depend on pressure with the exception of very high (of the order of critical pressure or more) or very low pressures. Since the mean travel velocity of the gas molecules depends on the temperature, and the specific heat of diatomic and polyatomic gases increase with the temperature, their thermal conductivity increase with the temperature. The thermal conductivity of helium and hydrogen is five to ten times greater than that of other gases. Due to the small mass of their molecules, the mean velocity is high. This explains the reason for their high thermal conductivity. The thermal conductivity of steam and other imperfect gases depend on pressure. The thermal conductivity of a mixture of the gases cannot be determined from the additive law and is found experimentally.

2.3.5 Thermal Conductivity of Liquids

The mechanism of heat conduction in liquids is qualitatively the same as in the gases. However, due to the closely spaced molecules, the molecular force exerts a strong influence on the energy transfer by the collision.

Since the density of liquids decreases with the increase in the temperature, their thermal conductivity decreases with the increase in the temperature. However, water and glycerin are important exceptions. In case of water, it increases with temperature in one temperature range and decreases in another. With rising pressure, the thermal conductivity of the liquids increases.

In brief, the gases transfer heat by direct collisions between molecules and their thermal conductivity is low compared to most solids because they are dilute media. Non-metallic solids transfer heat by lattice vibrations so that there is no net motion of the media as the energy propagates through them. Such heat transfer is often described in terms of 'phonons', quanta of lattice vibrations. Metals are much better thermal conductors than non-metals because the mobile electrons which participate in electrical conduction also cause the transfer of heat. The metals which are good electrical conductors are also good heat conductors.

At a given temperature, the thermal and electrical conductivities of metals are proportional. This behaviour is quantified by the Wiedemann–Franz Law (Fraas Arthur 1989). The law (Holman 1992) states that the ratio of the thermal conductivity k to the electrical conductivity σ of a metal is proportional to the temperature T, i.e.

$$\frac{k}{\sigma} = LT \tag{2.9}$$

where L is a constant of proportionality known as the Lorenz number. Its value is 2.44×10^{-8} WΩK^{-2}. However, the experiments have shown that the value of Lorenz number is not exactly the same for all materials. Its value, as reported in the literature, ranges from 2.23×10^{-8} WΩK^{-2} for copper to 3.08×10^{-8} WΩK^{-2} for tungsten at 0°C.

Qualitatively, the relationship of the thermal and electrical conductivities is based on the fact that the both heat and electrical transports involve the transport of free electrons in the metals. Interested readers may refer to https://www.doitpoms.ac.uk/tlplib/thermal_electrical/printall.php for greater details.

2.4 General Heat Conduction Equations

In general, the temperature gradient may exist in all three directions of a solid. There may be internal heat generation. The temperature can also vary with the time (unsteady state). Hence, it is necessary to develop a general heat conduction equation, which can be used to evaluate the heat transfer in any direction under steady or unsteady state and with or without heat generation. First, we consider a rectangular parallelepiped and then a cylinder and spherical system of coordinates.

2.4.1 General Heat Conduction Equation in Cartesian Coordinates

Consider an elementary parallelepiped of a solid as shown in Fig. 2.3. The volume of the element is $\delta V = \delta x \cdot \delta y \cdot \delta z$.

Let δQ_x is the quantity of the heat entering the solid through the face area $(\delta y \cdot \delta z)$ in time $\delta \tau$ as shown in the figure. Using Fourier's heat conduction equation for the unidirectional conduction heat flow, we have

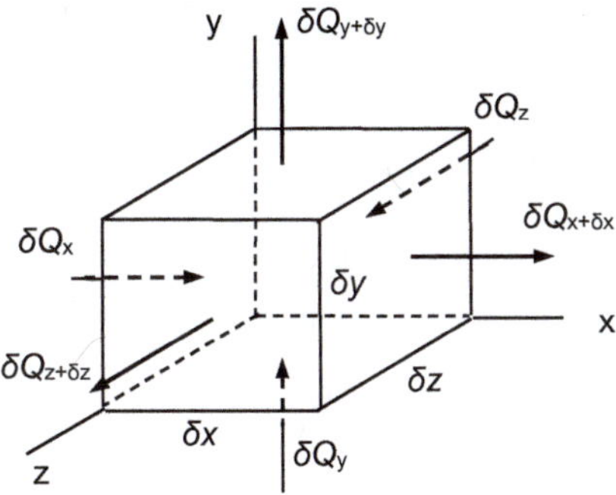

Fig. 2.3 Heat conduction in rectangular parallelepiped system

$$\delta Q_x = -k(\delta y \delta z)\frac{\partial t}{\partial x}\delta\tau \tag{i}$$

where the temperature gradient in the x-direction has been taken as partial derivative because the temperature is function of x, y, z and time τ.

The quantity of heat leaving the element can be obtained by Taylor's expansion of δQ_x:

$$\delta Q_{x+\delta x} = \delta Q_x + \frac{\partial}{\partial x}(\delta Q_x)\delta x + \text{higher order terms} \tag{ii}$$

The net heat inflow into the element due to the difference of heat entering and leaving is (leaving the higher order terms)

$$\begin{aligned}\delta Q_x - \delta Q_{x+\delta x} &= -\frac{\partial}{\partial x}\left[-k_x(\delta y \delta z)\frac{\partial t}{\partial x}\cdot \delta\tau\right]\delta x \\ &= \left[\frac{\partial}{\partial x}\left(k_x\frac{\partial t}{\partial x}\right)\right]\delta x \delta y \delta z \cdot \delta\tau \\ &= \left[\frac{\partial}{\partial x}\left(k_x\frac{\partial t}{\partial x}\right)\right]\delta V \delta\tau\end{aligned} \tag{iii}$$

Similarly in the y- and z-directions,

$$\delta Q_y - \delta Q_{y+\delta y} = \left[\frac{\partial}{\partial y}\left(k_y\frac{\partial t}{\partial y}\right)\right]\delta V \delta\tau \tag{iv}$$

$$\delta Q_z - \delta Q_{z+\delta z} = \left[\frac{\partial}{\partial z}\left(k_z\frac{\partial t}{\partial z}\right)\right]\delta V \delta\tau \tag{v}$$

The net amount of heat stored in the element due to difference in conduction heat flow in and out of the element is the sum of Eqs. (iii), (iv) and (v), i.e.

$$\delta Q = \left[\frac{\partial}{\partial x}\left(k_x\frac{\partial t}{\partial x}\right) + \frac{\partial}{\partial y}\left(k_y\frac{\partial t}{\partial y}\right) + \frac{\partial}{\partial z}\left(k_z\frac{\partial t}{\partial z}\right)\right]\delta V \delta\tau \tag{vi}$$

If there is heat generation in the element at a rate of q_g per unit volume per unit time, then the heat generated in the element in time $\delta\tau$ is

$$\delta Q_g = q_g \delta V \delta\tau \tag{vii}$$

The heat stored in the element due to the difference of conduction heat flow and heat generated within the element increases the internal energy of the element, i.e.

$$\delta E = \delta Q + \delta Q_g \tag{viii}$$

The change in the internal energy is also given by

$$\delta E = mc\delta T = \rho\delta Vc\frac{\partial t}{\partial\tau}\delta\tau \tag{ix}$$

where m is the mass of the element and c is the specific heat of the material. Hence,

$$\delta Q + \delta Q_g = \rho\delta Vc\frac{\partial t}{\partial\tau}\delta\tau \tag{x}$$

Substitution of the values of δQ and δQ_g gives

$$\left[\frac{\partial}{\partial x}\left(k_x\frac{\partial t}{\partial x}\right)+\frac{\partial}{\partial y}\left(k_y\frac{\partial t}{\partial y}\right)+\frac{\partial}{\partial z}\left(k_z\frac{\partial t}{\partial z}\right)\right]\delta V\delta\tau + q_g\delta V\delta\tau = \rho\delta Vc\frac{\partial t}{\partial\tau}\delta\tau$$

or

$$\frac{\partial}{\partial x}\left(k_x\frac{\partial t}{\partial x}\right)+\frac{\partial}{\partial y}\left(k_y\frac{\partial t}{\partial y}\right)+\frac{\partial}{\partial z}\left(k_z\frac{\partial t}{\partial z}\right)+q_g = \rho c\frac{\partial t}{\partial\tau} \tag{2.10}$$

which is the *general heat conduction equation in Cartesian coordinates*.

Case (A) Homogeneous materials ($k_x = k_y = k_z = k$)
If the thermal conductivity is constant, i.e. $k_x = k_y = k_z = k$, which is true for homogeneous materials, then the general heat conduction equation can be written as

$$\frac{\partial^2 t}{\partial x^2}+\frac{\partial^2 t}{\partial y^2}+\frac{\partial^2 t}{\partial z^2}+\frac{q_g}{k} = \frac{\rho c}{k}\frac{\partial t}{\partial\tau} \tag{2.11}$$

The term $(k/\rho c)$ is called the *thermal diffusivity* and is denoted by α, which has been discussed later on. Hence, the above equation can be written as

$$\frac{\partial^2 t}{\partial x^2}+\frac{\partial^2 t}{\partial y^2}+\frac{\partial^2 t}{\partial z^2}+\frac{q_g}{k} = \frac{1}{\alpha}\frac{\partial t}{\partial\tau} \tag{2.12}$$

Case (B) No Heat generation

When there is no heat generation $q_g = 0$ and we have

$$\frac{\partial^2 t}{\partial x^2} + \frac{\partial^2 t}{\partial y^2} + \frac{\partial^2 t}{\partial z^2} = \frac{1}{\alpha}\frac{\partial t}{\partial \tau} \tag{2.13a}$$

or

$$\Delta^2 t = \frac{1}{\alpha}\frac{\partial t}{\partial \tau} \tag{2.13b}$$

The above equation is known as *Fourier's equation.*

Case (C) Steady-state heat conduction

For steady-state heat conduction, the derivative of the temperature with time is zero and the general heat conduction equation yields

$$\frac{\partial^2 t}{\partial x^2} + \frac{\partial^2 t}{\partial y^2} + \frac{\partial^2 t}{\partial z^2} + \frac{q_g}{k} = 0 \tag{2.13c}$$

This equation is known as *Poisson's equation.*

Case (D) Steady-state heat conduction with no heat generation

For this condition, general heat conduction equation yields

$$\frac{\partial^2 t}{\partial x^2} + \frac{\partial^2 t}{\partial y^2} + \frac{\partial^2 t}{\partial z^2} = 0 \tag{2.13d}$$

or

$$\Delta^2 t = 0 \tag{2.13e}$$

This equation is known as the *Laplace equation.*

Case (E) Two- and one-dimensional steady-state heat conduction without heat generation

For two- and one-dimensional steady-state heat conduction without heat generation, the Laplace equation yields the following equations.

For two-dimensional case,

$$\frac{\partial^2 t}{\partial x^2} + \frac{\partial^2 t}{\partial y^2} = 0 \tag{2.14}$$

For one-dimensional case,

$$\frac{\partial^2 t}{\partial x^2} = 0 \tag{2.15}$$

2.4.1.1 Thermal Diffusivity

In steady state, the thermal conductivity is the only property of the substance that determines the temperature distribution across the substance. But in the unsteady state, the temperature distribution is influenced by the thermal conduction as well as the thermal capacity (ρc) of the material. Thus, the thermal diffusivity, which is a combination of these thermophysical properties of the material, controls the temperature distribution in the unsteady state. Thermal diffusivity is a property of a material (being a combination of three physical properties ρ, c and k). The higher the thermal diffusivity of a material, the higher the rate of temperature propagation, i.e. the equalization of the temperature at all points of the space will proceed at a higher rate. It can also be termed as a measure of the thermal inertia of a substance.

In general, the metals have high thermal conductivity, which impart them high thermal diffusivities. The non-metals and insulators have lower heat capacities than the metals but their thermal conductivities are much lower, and this leads to low values of thermal diffusivity of these solids compared to the metals. Typical values of ρ, c, k and α of some substances are given in Table 2.1. More details can be seen in Appendix A.

Temperature gradients by local heating cause localized stress in a material, which may be of great significance in some applications. Local heating is observed in welded joins, parts of internal combustion engines, at furnace inner surfaces, and in massive concrete structures during hydration of the cement. When heat is applied locally to a small area on the surface of a solid, the solid absorbs heat locally at a rate depending on its heat capacity. The heat is conducted away from the heated location at a rate depending on its conductivity. Thus, a high heat capacity means higher temperature gradients, while a high thermal conductivity causes

Table 2.1 Thermophysical properties of some materials

Material	Density ρ, kg/m^3	Specific heat c, kJ/(kg K) at 20°C	Thermal conductivity k, W/(m K)	Thermal diffusivity α, (m^2/s) × 10^6
Metals (at 20°C)				
Aluminium	2710	0.895	204	84.1
Brass (70% Cu, 30% Zn)	8520	0.380	110	34
Copper	8950	0.380	386	113.5
Silver (99.9%)	10,520	0.234	419	170.2
Steel (C = 0.5%)	7830	0.465	55	15.1
18–8 Stainless steel	7820	0.460	16.3	4.53
Mercury(at 0°C)	13,630	0.14	8.2	4.3
Non-metals				
Brick	1600	0.84	0.69	0.51
Concrete	1900	0.88	1.37	0.82
Liquids and gases				
Water (at ≈25°C)	997.4	4.179	0.604	0.145
Air (at 300 K)	1.177	1.006	0.0262	22.1

high rate of carrying away of heat to other parts of the solid that tends to reduce the temperature gradients. These two opposing effects are combined in the thermal diffusivity. A high thermal diffusivity reduces the temperature gradients caused by the local heating and, thus, reduces the local stresses.

2.4.2 General Heat Conduction Equation in Cylindrical Coordinates

The equation in cylindrical coordinates is useful in dealing with the conduction heat transfer in systems with cylindrical geometry such as pipes, wires, rods, etc.

Figure 2.4 shows the primary element with dimensions δr in direction r, $r\delta\theta$ in direction θ and δz in direction z (along the axis of the cylinder). The volume of the element is $(\delta r \cdot r\delta\theta \cdot \delta z)$.

Let δQ_r is the quantity of heat entering the element through the face whose area is $r\delta\theta \cdot \delta z$ in time $\delta\tau$. Using Fourier's equation for unidirectional heat flow, we have

$$\delta Q_r = -k(r\delta\theta\delta z)\frac{\partial t}{\partial r}\delta\tau \tag{i}$$

where the temperature gradient in direction r is the partial derivative in radial direction.

The quantity of heat leaving the element can be obtained by Taylor's expansion of δQ_r:

$$\delta Q_{r+\delta r} = \delta Q_r + \frac{\partial}{\partial r}(\delta Q_r)\delta r + \text{higher order terms} \tag{ii}$$

The net flow into the element due to the difference of the heat entering and leaving the element in r-direction (neglecting higher order terms) is

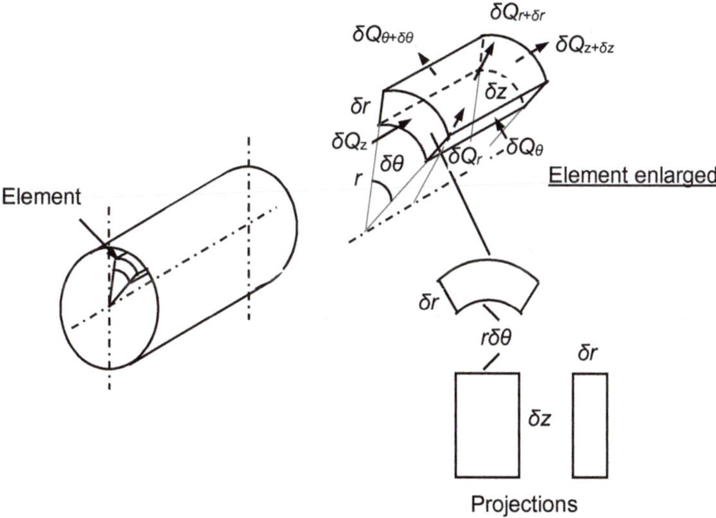

Fig. 2.4 Heat conduction in a cylindrical system

$$\begin{aligned}
\delta Q_r - \delta Q_{r+\delta r} &= -\frac{\partial}{\partial r}\left[-k(r\delta\theta \cdot \delta z)\frac{\partial t}{\partial r}\delta\tau\right]\delta r \\
&= k\frac{\partial}{\partial r}\left(r\frac{\partial t}{\partial r}\right)\frac{(\delta r \cdot r\delta\theta \cdot \delta z)}{r} \cdot \delta\tau \\
&= k\frac{1}{r}\frac{\partial}{\partial r}\left(r\frac{\partial t}{\partial r}\right)\delta V \cdot \delta\tau \\
&= k\frac{1}{r}\left(\frac{\partial t}{\partial r} + r\frac{\partial^2 t}{\partial r^2}\right)\delta V \cdot \delta\tau \\
&= k\left(\frac{1}{r}\frac{\partial t}{\partial r} + \frac{\partial^2 t}{\partial r^2}\right)\delta V\delta\tau \qquad \text{(iii)}
\end{aligned}$$

The area of the face perpendicular to the θ-direction of the heat flow is $(\delta r \cdot \delta z)$, and the net heat flow into the element is

$$\begin{aligned}
\delta Q_\theta - \delta Q_{\theta+\delta\theta} &= -\frac{\partial}{r\partial\theta}\left[-k(\delta r \cdot \delta z)\frac{\partial t}{r\partial\theta}\delta\tau\right]r\delta\theta \\
&= k\frac{1}{r}\frac{\partial}{\partial\theta}\left(\frac{\partial t}{r\partial\theta}\right)(\delta r \cdot r\delta\theta \cdot \delta z)\delta\tau \qquad \text{(iv)} \\
&= k\frac{1}{r^2}\frac{\partial^2 t}{\partial\theta^2}\delta V\delta\tau
\end{aligned}$$

Similarly, the area of the face perpendicular to the z-direction of the heat flow is $(\delta r \cdot r\delta\theta)$ and the net heat flow into the element is

$$\begin{aligned}
\delta Q_z - \delta Q_{z+\delta z} &= -\frac{\partial}{\partial z}\left[-k(\delta r \cdot r\delta\theta)\frac{\partial t}{\partial z}\delta\tau\right]\delta z \\
&= k\frac{\partial^2 t}{\partial z^2}\delta r \cdot r\delta\theta \cdot \delta z \cdot \delta\tau \qquad \text{(v)} \\
&= k\frac{\partial^2 t}{\partial z^2}\delta V\delta\tau
\end{aligned}$$

The net amount of heat stored in the element of Fig. 2.4 is the sum of the net heat flow into the element in r-, θ- and z-directions, i.e. the sum of the right-hand side terms of Eqs. (iii), (iv) and (v), and is

$$\delta Q = k\left(\frac{1}{r}\frac{\partial t}{\partial r} + \frac{\partial^2 t}{\partial r^2} + \frac{1}{r^2}\frac{\partial^2 t}{\partial\theta^2} + \frac{\partial^2 t}{\partial z^2}\right)\delta V\delta\tau \qquad \text{(vi)}$$

If there is internal heat generation at the rate of q_g per unit volume per unit time, then the heat generated in the element in time $\delta\tau$ is

$$\delta Q_g = q_g\delta V\delta\tau \qquad \text{(vii)}$$

The net heat δQ stored in the element due to the difference of in- and outflows, and the heat generation δQ_g within the element increase the internal energy of the element hence

$$\delta E = \delta Q + \delta Q_g \tag{viii}$$

Substitution of the values of δQ and δQ_g gives

$$\delta E = k\left(\frac{1}{r}\frac{\partial t}{\partial r} + \frac{\partial^2 t}{\partial r^2} + \frac{1}{r^2}\frac{\partial^2 t}{\partial \theta^2} + \frac{\partial^2 t}{\partial z^2}\right)\delta V \delta\tau + q_g \delta V \delta\tau \tag{ix}$$

The change in the internal energy is also given by

$$\delta E = \rho \delta V c \frac{\partial t}{\partial \tau}\delta\tau \tag{x}$$

Substitution gives

$$\rho \delta V c \frac{\partial t}{\partial \tau}\delta\tau = k\left(\frac{1}{r}\frac{\partial t}{\partial r} + \frac{\partial^2 t}{\partial r^2} + \frac{1}{r^2}\frac{\partial^2 t}{\partial \theta^2} + \frac{\partial^2 t}{\partial z^2}\right)\delta V \cdot \delta\tau + q_g \delta V \cdot \delta\tau$$

or

$$\left(\frac{1}{r}\frac{\partial t}{\partial r} + \frac{\partial^2 t}{\partial r^2} + \frac{1}{r^2}\frac{\partial^2 t}{\partial \theta^2} + \frac{\partial^2 t}{\partial z^2}\right) + \frac{q_g}{k} = \frac{\rho c}{k}\frac{\partial t}{\partial \tau}$$

or

$$\left(\frac{1}{r}\frac{\partial t}{\partial r} + \frac{\partial^2 t}{\partial r^2} + \frac{1}{r^2}\frac{\partial^2 t}{\partial \theta^2} + \frac{\partial^2 t}{\partial z^2}\right) + \frac{q_g}{k} = \frac{1}{\alpha}\frac{\partial t}{\partial \tau} \tag{2.16}$$

which is the *general heat conduction equation in cylindrical coordinates*, where the thermal conductivity has been assumed to be constant. The equation can be arrived at by the transformation of Eq. (2.12) to cylindrical coordinates, see Example 2.1.

For variable thermal conductivity, the equation in general form can be written as

$$\frac{1}{\rho c}\left[\frac{1}{r}\frac{\partial}{\partial r}\left(k_r r \frac{\partial t}{\partial r}\right) + \frac{1}{r^2}\frac{\partial}{\partial \theta}\left(k_\theta \frac{\partial t}{\partial \theta}\right) + \frac{\partial}{\partial z}\left(k_z \frac{\partial t}{\partial z}\right)\right] + \frac{q_g}{\rho c} = \frac{\partial t}{\partial \tau} \tag{2.17a}$$

If there is no internal heat generation and the thermal conductivity is constant, then $q_g = 0$ and $k_r = k_\theta = k_z = k$. Equation (2.16) reduces to

$$\frac{1}{r}\frac{\partial t}{\partial r} + \frac{\partial^2 t}{\partial r^2} + \frac{1}{r^2}\frac{\partial^2 t}{\partial \theta^2} + \frac{\partial^2 t}{\partial z^2} = \frac{1}{\alpha}\frac{\partial t}{\partial \tau} \tag{2.17b}$$

For the steady-state heat conduction with no heat generation and constant value of the thermal conductivity k, Eq. (2.16) reduces to

$$\left(\frac{1}{r}\frac{\partial t}{\partial r} + \frac{\partial^2 t}{\partial r^2} + \frac{1}{r^2}\frac{\partial^2 t}{\partial \theta^2} + \frac{\partial^2 t}{\partial z^2}\right) = 0 \quad (2.17c)$$

In the case of steady-state and one-dimensional heat conduction (in the radial direction only) without heat generation, the above equation reduces to

$$\frac{1}{r}\frac{dt}{dr} + \frac{d^2 t}{dr^2} = 0 \quad (2.17d)$$

The temperature distribution in this case is a function of radius r only; hence, the partial derivative has been changed to the full derivative.

Example 2.1 Beginning with the three-dimensional heat conduction equation in Cartesian coordinates, obtain the equation in cylindrical coordinates.

Solution

The rectangular and cylindrical coordinates are inter-related by the following relations, refer to Fig. 2.5:

$$x = r\cos\theta$$
$$y = r\sin\theta$$

and

$$z = z \quad (a)$$

The general heat conduction equation in rectangular coordinates is

$$\frac{\partial^2 t}{\partial x^2} + \frac{\partial^2 t}{\partial y^2} + \frac{\partial^2 t}{\partial z^2} + \frac{q_g}{k} = \frac{1}{\alpha}\frac{\partial t}{\partial \tau} \quad (2.12)$$

which is transformed into cylindrical coordinates using Eq. (a), knowing that x and y are functions of r and θ while z is independent of r and θ.

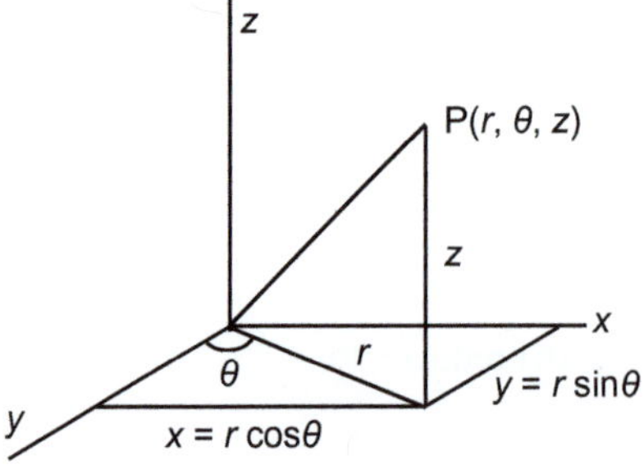

Fig. 2.5 Example 2.1

$$\left(\frac{\partial t}{\partial r}\right) = \frac{\partial t}{\partial x}\frac{\partial x}{\partial r} + \frac{\partial t}{\partial y}\frac{\partial y}{\partial r} = \frac{\partial t}{\partial x}\cos\theta + \frac{\partial t}{\partial y}\sin\theta \tag{i}$$

and

$$\left(\frac{\partial t}{\partial \theta}\right) = \frac{\partial t}{\partial x}\frac{\partial x}{\partial \theta} + \frac{\partial t}{\partial y}\frac{\partial y}{\partial \theta} = -\frac{\partial t}{\partial x} r\sin\theta + \frac{\partial t}{\partial y} r\cos\theta \tag{ii}$$

Solution of Eqs. (i) and (ii) gives

$$\left(\frac{\partial t}{\partial x}\right) = \cos\theta\frac{\partial t}{\partial r} - \left(\frac{\sin\theta}{r}\right)\frac{\partial t}{\partial \theta}$$

and

$$\left(\frac{\partial t}{\partial y}\right) = \sin\theta\frac{\partial t}{\partial r} + \left(\frac{\cos\theta}{r}\right)\frac{\partial t}{\partial \theta}$$

Second derivatives of t with x and y are

$$\frac{\partial}{\partial x}\left(\frac{\partial t}{\partial x}\right) = \cos\theta\frac{\partial}{\partial r}\left(\frac{\partial t}{\partial x}\right) - \left(\frac{\sin\theta}{r}\right)\frac{\partial}{\partial \theta}\left(\frac{\partial t}{\partial x}\right)$$

or

$$\frac{\partial}{\partial x}\left(\frac{\partial t}{\partial x}\right) = \cos\theta\frac{\partial}{\partial r}\left[\cos\theta\frac{\partial t}{\partial r} - \left(\frac{\sin\theta}{r}\right)\frac{\partial t}{\partial \theta}\right] - \left(\frac{\sin\theta}{r}\right)\frac{\partial}{\partial \theta}\left[\cos\theta\frac{\partial t}{\partial r} - \left(\frac{\sin\theta}{r}\right)\frac{\partial t}{\partial \theta}\right]$$

or

$$\frac{\partial^2 t}{\partial x^2} = \cos^2\theta\frac{\partial^2 t}{\partial r^2} + \frac{\sin\theta\cos\theta}{r^2}\frac{\partial t}{\partial \theta} + \left(\frac{\sin^2\theta}{r}\right)\frac{\partial t}{\partial r} + \left(\frac{\sin^2\theta}{r^2}\right)\frac{\partial^2 t}{\partial \theta^2} + \left(\frac{\sin\theta\cos\theta}{r^2}\right)\frac{\partial t}{\partial \theta}$$

Similarly,

$$\frac{\partial}{\partial y}\left(\frac{\partial t}{\partial y}\right) = \sin\theta\frac{\partial}{\partial r}\left(\frac{\partial t}{\partial y}\right) + \left(\frac{\cos\theta}{r}\right)\frac{\partial}{\partial \theta}\left(\frac{\partial t}{\partial y}\right)$$

or

$$\frac{\partial}{\partial y}\left(\frac{\partial t}{\partial y}\right) = \sin\theta\frac{\partial t}{\partial r}\left[\sin\theta\frac{\partial t}{\partial r} + \left(\frac{\cos\theta}{r}\right)\frac{\partial t}{\partial \theta}\right] + \left(\frac{\cos\theta}{r}\right)\frac{\partial}{\partial \theta}\left[\sin\theta\frac{\partial t}{\partial r} + \left(\frac{\cos\theta}{r}\right)\frac{\partial t}{\partial \theta}\right]$$

or

$$\frac{\partial^2 t}{\partial y^2} = \sin^2\theta\frac{\partial^2 t}{\partial r^2} - \left(\frac{\sin\theta\cos\theta}{r^2}\right)\frac{\partial t}{\partial \theta} + \frac{\cos^2\theta}{r}\left(\frac{\partial t}{\partial r}\right) + \frac{\cos^2\theta}{r^2}\frac{\partial^2 t}{\partial \theta^2} - \left(\frac{\sin\theta\cos\theta}{r^2}\right)\frac{\partial t}{\partial \theta}$$

Hence,

$$\frac{\partial^2 t}{\partial x^2} + \frac{\partial^2 t}{\partial y^2} = \left(\cos^2\theta + \sin^2\theta\right)\left[\frac{\partial^2 t}{\partial r^2} + \frac{1}{r}\left(\frac{\partial t}{\partial r}\right) + \frac{1}{r^2}\frac{\partial^2 t}{\partial \theta^2}\right]$$

or

$$\frac{\partial^2 t}{\partial x^2} + \frac{\partial^2 t}{\partial y^2} = \frac{\partial^2 t}{\partial r^2} + \frac{1}{r}\left(\frac{\partial t}{\partial r}\right) + \frac{1}{r^2}\frac{\partial^2 t}{\partial \theta^2}$$

Substitution in Eq. (2.12) gives the desired equation as

$$\left(\frac{1}{r}\frac{\partial t}{\partial r} + \frac{\partial^2 t}{\partial r^2} + \frac{1}{r^2}\frac{\partial^2 t}{\partial \theta^2} + \frac{\partial^2 t}{\partial z^2}\right) + \frac{q_g}{k} = \frac{1}{\alpha}\frac{\partial t}{\partial \tau} \tag{2.16}$$

2.4.3 General Heat Conduction Equation in Spherical Coordinates

Figure 2.6a shows the elemental area on the surface of a sphere of radius r. For the strip ABCD in the figure, the radius is $r\sin\theta$. Also refer to the section of the sphere in projections in Fig. 2.6c. Thus, the sides of the elemental area are $r\delta\theta$ and $r\sin\theta \cdot \delta\phi$. The elemental volume can be made by moving a distance δr in the direction r as shown in Fig. 2.6b. Thus, the sides of the element in Fig. 2.6b are δr, $r\delta\theta$ and $r\sin\theta \cdot \delta\phi$ in the directions r, θ and ϕ, respectively. The volume of the element is

$$\delta V = \delta r \cdot r\delta\theta \cdot r\sin\theta \cdot \delta\phi \tag{i}$$

The heat δQ_r in r-direction enters the element through face of area ($r\delta\theta \cdot r\sin\theta \cdot \delta\phi$), which is perpendicular to r-direction. The heat inflow in time $\delta\tau$ is

$$\delta Q_r = -k(r\delta\theta \cdot r\sin\theta \cdot \delta\phi)\frac{\partial t}{\partial r}\delta\tau \tag{ii}$$

The outflow is

$$\delta Q_{r+\delta r} = \delta Q_r + \frac{\partial}{\partial r}(\delta Q_r)\delta r \tag{iii}$$

The net heat inflow into the element is

$$\begin{aligned}\delta Q_r - \delta Q_{r+\delta r} &= \delta Q_r - \left[\delta Q_r + \frac{\partial}{\partial r}(\delta Q_r)\delta r\right] = -\frac{\partial}{\partial r}(\delta Q_r)\delta r \\ &= k(\delta r \cdot \delta\theta \cdot \sin\theta \cdot \delta\phi)\frac{\partial}{\partial r}\left[r^2\frac{\partial t}{\partial r}\right]\delta\tau \\ &= k\delta V\frac{1}{r^2}\frac{\partial}{\partial r}\left[r^2\frac{\partial t}{\partial r}\right]\delta\tau\end{aligned} \tag{iv}$$

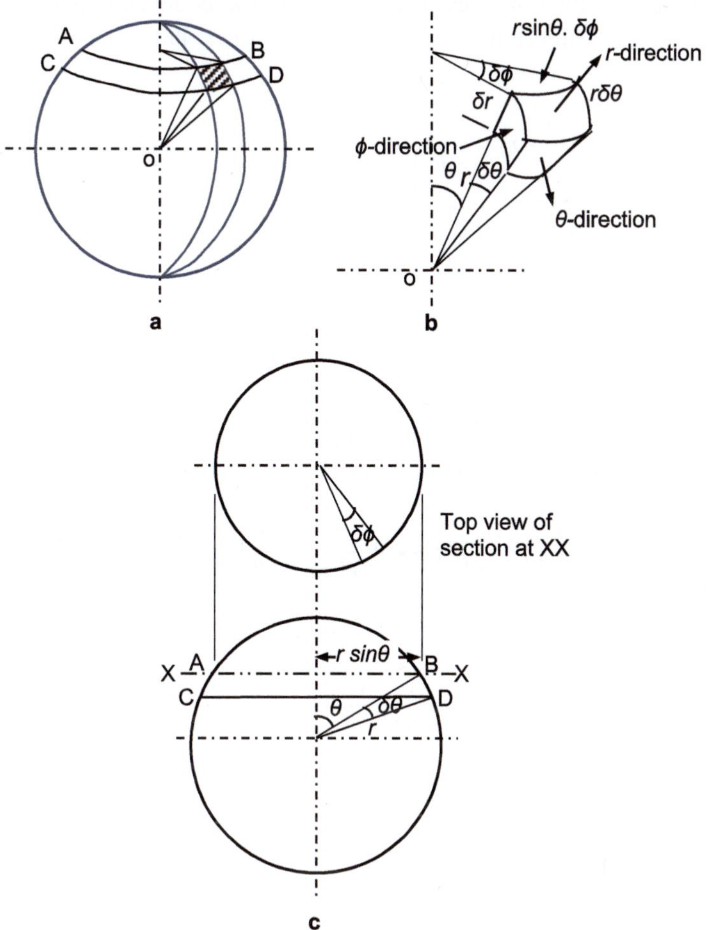

Fig. 2.6 Heat conduction in a spherical system

Flow area perpendicular to the θ direction is $(\delta r \cdot r\sin\theta \cdot \delta\phi)$, and the elemental distance in θ direction is $r\delta\theta$. The net heat inflow into the element in this direction is

$$\begin{aligned}
\delta Q_\theta - \delta Q_{\theta+\delta\theta} &= \delta Q_\theta - \left[\delta Q_\theta + \frac{\partial}{r\partial\theta}(\delta Q_\theta) r\delta\theta\right] = -\frac{\partial}{r\partial\theta}(\delta Q_\theta) r\delta\theta \\
&= -\frac{\partial}{r\partial\theta}\left[-k(\delta r \cdot r\sin\theta \cdot \delta\phi)\frac{\partial t}{r\partial\theta} \cdot \delta\tau\right] r\delta\theta \\
&= k(\delta r \cdot \delta\phi)\delta\theta \frac{\partial}{\partial\theta}\left(\sin\theta \frac{\partial t}{\partial\theta}\right)\delta\tau \\
&= k(\delta r \cdot r\delta\theta \cdot r\sin\theta \cdot \delta\phi)\frac{1}{r^2\sin\theta}\frac{\partial}{\partial\theta}\left(\sin\theta\frac{\partial t}{\partial\theta}\right)\delta\tau \\
&= k\delta V \frac{1}{r^2\sin\theta}\frac{\partial}{\partial\theta}\left(\sin\theta\frac{\partial t}{\partial\theta}\right)\delta\tau \qquad \text{(v)}
\end{aligned}$$

In the similar way, the flow area perpendicular to the ϕ direction is $(\delta r \cdot r\delta\theta)$ and the elemental distance in ϕ direction is $r\sin\theta \cdot \delta\phi$. The net heat inflow into the element in this direction is

$$\begin{aligned}
\delta Q_{\phi} - \delta Q_{\phi+\delta\phi} &= \delta Q_{\phi} - \left[\delta Q_{\phi} + \frac{\partial}{r\sin\theta\partial\phi}(\delta Q_{\phi}) r\sin\delta\phi\right] = -\frac{\partial}{r\sin\theta\partial\phi}(\delta Q_{\phi}) r\sin\theta\delta\phi \\
&= -\frac{\partial}{r\sin\theta\partial\phi}\left[-k(\delta r\cdot r\delta\theta)\frac{\partial t}{r\sin\theta\partial\phi}\cdot\delta\tau\right] r\sin\theta\delta\phi \\
&= k(\delta r\cdot r\delta\theta\cdot r\sin\theta\cdot\delta\phi)\frac{1}{r^2\sin^2\theta}\left(\frac{\partial^2 t}{\partial\phi^2}\right)\delta\tau \\
&= k\delta V\frac{1}{r^2\sin^2\theta}\left(\frac{\partial^2 t}{\partial\phi^2}\right)\delta\tau
\end{aligned} \quad \text{(vi)}$$

The total of the net inflows into the element is the summation of the net of the conduction heat inflow and outflow in all three directions of the heat flow, and is

$$\begin{aligned}
\delta Q &= k\delta V\frac{1}{r^2}\frac{\partial}{\partial r}\left(r^2\frac{\partial t}{\partial r}\right)\delta\tau + k\delta V\frac{1}{r^2\sin\theta}\frac{\partial}{\partial\theta}\left[\sin\theta\frac{\partial t}{\partial\theta}\right]\delta\tau \\
&+ k\delta V\frac{1}{r^2\sin^2\theta}\left(\frac{\partial^2 t}{\partial\phi^2}\right)\delta\tau \\
&= k\delta V\left[\frac{1}{r^2}\frac{\partial}{\partial r}\left(r^2\frac{\partial t}{\partial r}\right) + \frac{1}{r^2\sin\theta}\frac{\partial}{\partial\theta}\left(\sin\theta\frac{\partial t}{\partial\theta}\right) + \frac{1}{r^2\sin^2\theta}\left(\frac{\partial^2 t}{\partial\phi^2}\right)\right]\delta\tau
\end{aligned} \quad \text{(vii)}$$

Heat generated in the element in time $\delta\tau$ is

$$\delta Q_g = q_g \delta V \delta\tau \quad \text{(viii)}$$

Change in the internal energy of the element in time $\delta\tau$ is

$$\delta E = \rho\delta Vc\left(\frac{\partial t}{\partial\tau}\right)\delta\tau \quad \text{(ix)}$$

Energy balance for the element gives

$$\delta E = \delta Q + \delta Q_g \quad \text{(x)}$$

Substitution of the values of δE, δQ and δQ_g gives

$$\begin{aligned}
&\rho\delta V\cdot c\cdot\left(\frac{\partial t}{\partial\tau}\right)\delta\tau \\
&= k\delta V\left[\frac{1}{r^2}\frac{\partial}{\partial r}\left(r^2\frac{\partial t}{\partial r}\right) + \frac{1}{r^2\sin\theta}\frac{\partial}{\partial\theta}\left(\sin\theta\frac{\partial t}{\partial\theta}\right) + \frac{1}{r^2\sin^2\theta}\left(\frac{\partial^2 t}{\partial\phi^2}\right)\right]\delta\tau \\
&+ q_g\delta V\delta\tau
\end{aligned}$$

or

$$\frac{\rho c}{k}\left(\frac{\partial t}{\partial \tau}\right) = \left[\frac{1}{r^2}\frac{\partial}{\partial r}\left(r^2\frac{\partial t}{\partial r}\right) + \frac{1}{r^2 \sin\theta}\frac{\partial}{\partial \theta}\left(\sin\theta\frac{\partial t}{\partial \theta}\right) + \frac{1}{r^2\sin^2\theta}\left(\frac{\partial^2 t}{\partial \phi^2}\right)\right] + \frac{q_g}{k}$$

or

$$\frac{1}{r^2}\frac{\partial}{\partial r}\left(r^2\frac{\partial t}{\partial r}\right) + \frac{1}{r^2 \sin\theta}\frac{\partial}{\partial \theta}\left(\sin\theta\frac{\partial t}{\partial \theta}\right) + \frac{1}{r^2\sin^2\theta}\left(\frac{\partial^2 t}{\partial \phi^2}\right) + \frac{q_g}{k} = \frac{1}{\alpha}\left(\frac{\partial t}{\partial \tau}\right) \tag{2.18}$$

which is the desired equation assuming thermal conductivity as constant.

2.5 One-Dimensional Steady-State Heat Conduction

For the one-dimensional steady-state heat conduction without heat generation,

$$\frac{\partial^2 t}{\partial x^2} = 0 \tag{2.15}$$

Integrating the above equation, we obtain

$$\frac{\partial t}{\partial x} = C_1 \tag{i}$$

and

$$t = C_1 x + C_2 \tag{ii}$$

The boundary conditions in Fig. 2.7 are

$$\text{at } x = x_1, \quad t = t_1$$

and

$$\text{at } x = x_2, \quad t = t_2$$

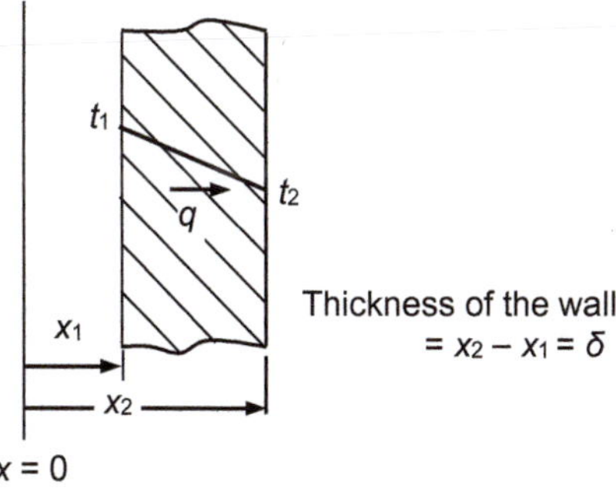

Fig. 2.7 One-dimensional steady-state heat conduction through a plane wall

Substitution in Eq. (ii) gives

$$t_1 = C_1x_1 + C_2$$
$$t_2 = C_1x_2 + C_2$$

Solution of these equations gives

$$C_1 = \frac{t_1 - t_2}{x_1 - x_2} = -\frac{t_1 - t_2}{x_2 - x_1}$$
$$C_2 = t_1 + \frac{t_1 - t_2}{x_2 - x_1}x_1$$

Substitution of values of constants C_1 and C_2 in Eq. (ii) gives

$$t = -\frac{t_1 - t_2}{x_2 - x_1}x + \frac{t_1 - t_2}{x_2 - x_1}x_1 + t_1$$

or

$$\frac{t - t_1}{t_1 - t_2} = \frac{x_1 - x}{x_2 - x_1}$$

which indicates a linear distribution of the temperature, refer to Fig. 2.7.

From Eq. (i),

$$\frac{\partial t}{\partial x} = C_1 = -\frac{t_1 - t_2}{x_2 - x_1}$$

Hence, the rate of heat transfer is

$$q = -kA\frac{\partial t}{\partial x} = kA\frac{t_1 - t_2}{x_2 - x_1} = kA\frac{t_1 - t_2}{\delta}$$
$$= \frac{t_1 - t_2}{\delta/kA} = \frac{\Delta t}{R_k} \tag{2.19}$$

where A is wall area perpendicular to the direction of heat flow and $R_k = \delta/kA$.

We compare the above equation with Ohm's law for an electric conductor, which is

$$I = \frac{E}{R} = \frac{V_1 - V_2}{R}$$

The electric current I corresponds to the heat flow q, the electrical potential E corresponds to the thermal potential and the electrical resistance corresponds to resistance R_k to the heat conduction. Thus, Fourier's equation of heat conduction is exactly analogous to Ohm's law for an electrical conductor. We shall use this electrical analogy frequently as it is quite useful in solving the complex heat conduction problems.

We represent Eq. (2.19) as in Fig. 2.8, where the temperature difference Δt is the driving force for the flow of heat and $R_k = R$ (= δ/kA) is the *thermal resistance*, which the wall offers to the flow of heat by conduction. The reciprocal of the thermal resistance is known as *thermal conductance* of the wall.

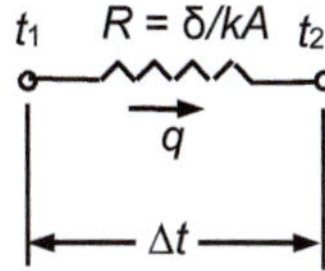

Fig. 2.8 Thermal network

2.5.1 Composite Plane Wall

Walls made of several layers of different materials are called composite walls, refer to Fig. 2.9. Walls of houses, furnaces, boilers, pipes with layer of insulating materials, etc. are some of the examples of composite walls.

The composite wall in Fig. 2.9 consists of three layers of thicknesses δ_1, δ_2 and δ_3. The thermal conductivities of these layers are k_1, k_2 and k_3, respectively. The temperature of the outer layers of the wall is t_1 and t_4 as shown in the figure, with interface temperatures as t_2 and t_3. It is being assumed that different layers are having perfect contact between them, and hence the adjacent surfaces are at the same temperature.

In the steady-state condition, the heat flow q is the same for all the layers and is constant. The equations of heat transfer through these layers are

$$q = k_1 A \frac{t_1 - t_2}{\delta_1} \quad \text{for the first layer} \tag{i}$$

$$q = k_2 A \frac{t_2 - t_3}{\delta_2} \quad \text{for the second layer} \tag{ii}$$

$$q = k_3 A \frac{t_3 - t_4}{\delta_3} \quad \text{for the third layer} \tag{iii}$$

The temperature differences across the layers, from above equations, are

$$t_1 - t_2 = q\left(\frac{\delta_1}{k_1 A}\right) \tag{iv}$$

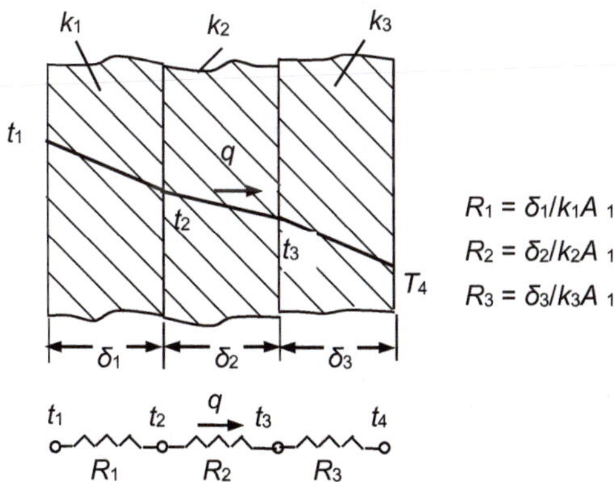

Fig. 2.9 A composite wall

$$t_2 - t_3 = q\left(\frac{\delta_2}{k_2 A}\right) \quad \text{(v)}$$

$$t_3 - t_4 = q\left(\frac{\delta_3}{k_3 A}\right) \quad \text{(vi)}$$

Adding the above equations, we get

$$t_1 - t_4 = q\left(\frac{\delta_1}{k_1 A} + \frac{\delta_2}{k_2 A} + \frac{\delta_3}{k_3 A}\right) \quad \text{(vii)}$$

or

$$q = \frac{t_1 - t_4}{\frac{\delta_1}{k_1 A} + \frac{\delta_2}{k_2 A} + \frac{\delta_3}{k_3 A}}$$

or

$$q = \frac{t_1 - t_4}{R_1 + R_2 + R_3} \quad \text{(viii)}$$

where R_1, R_2 and R_3 are resistances of the layers to the conduction heat flow. For n layers, the equation can be written as

$$q = \frac{t_1 - t_{n+1}}{\sum_{i=1}^{n} R_i} = \frac{t_1 - t_{n+1}}{\frac{1}{A}\sum_{i=1}^{n} \frac{\delta_i}{k_i}} \quad (2.20)$$

2.5.2 One-Dimensional Steady-State Heat Conduction Through a Plane Homogeneous Wall Considering Film Coefficients

In Fig. 2.10, the two faces of the wall are exposed to fluids at temperatures t_i and t_o. The heat is transferred to and from the wall by convection. In the convection heat transfer process, a thin fluid film is formed on the surface, which offers resistance to the flow of heat (refer to Chap. 7 for details).

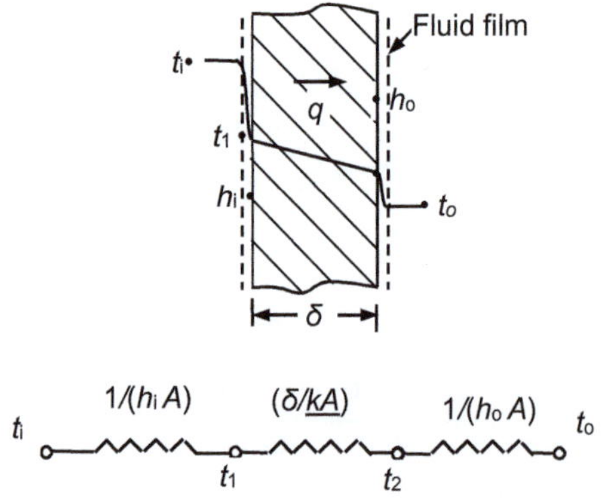

Fig. 2.10 One-dimensional steady-state heat conduction through a plane wall considering film coefficients

Newton's equation for convection heat transfer gives

$$q = hA\Delta t \tag{1.4}$$

where

h is known as convective heat transfer or film coefficient,
A is the area of the surface transferring heat by convection and
Δt is temperature difference between the wall (surface) and the fluid in contact.

Hence, in the present case, at the left face,

$$\begin{aligned} q &= h_i A(t_i - t_1) \\ &= \frac{t_i - t_1}{1/h_i A} = \frac{t_i - t_1}{R_i} \end{aligned} \tag{i}$$

where $R_i = 1/h_i A$ is known as film resistance.

Similarly at the right face

$$\begin{aligned} q &= h_o A(t_2 - t_o) \\ &= \frac{t_2 - t_o}{1/h_o A} = \frac{t_2 - t_o}{R_o} \end{aligned} \tag{ii}$$

From Eqs. (i) and (ii),

$$t_i - t_1 = qR_i$$
$$t_2 - t_o = qR_o$$

For the wall,

$$t_1 - t_2 = q\left(\frac{\delta}{kA}\right) \tag{iii}$$

Combing the temperature equations, we get

$$(t_i - t_1) + (t_1 - t_2) + (t_2 - t_o) = q\left(\frac{1}{h_i A} + \frac{\delta}{kA} + \frac{1}{h_o A}\right)$$

$$q = \frac{t_i - t_o}{\frac{1}{h_i A} + \frac{\delta}{kA} + \frac{1}{h_o A}} \tag{2.21}$$

which is the desired equation for the present case.

2.5.2.1 Conduction Heat Transfer Through a Composite Plane Wall Considering Film Coefficients

Introducing film coefficients in Eq. (2.20), the heat transfer equation takes the form as

$$q = \frac{t_i - t_o}{\frac{1}{h_i A} + \frac{1}{A}\sum_{i=1}^{n}\frac{\delta_i}{k_i} + \frac{1}{h_o A}} \tag{2.22}$$

For example, for a wall with three layers (n = 3), the equation yields

$$q = \frac{t_i - t_o}{\frac{1}{h_i A} + \frac{\delta_1}{k_1 A} + \frac{\delta_2}{k_2 A} + \frac{\delta_3}{k_3 A} + \frac{1}{h_o A}}$$

2.5.3 One-Dimensional Steady-State Conduction Heat Transfer Through a Plane Homogeneous Wall Considering Heat Transfer by Convection and Radiation from the Wall Surface

In general, the wall surface will transfer heat both by convection and radiation to its surroundings. The expression for the heat flow to the wall can be written in the following form:

$$q = (h_{ri} + h_{ci}).A.(t_i - t_1) = \frac{(t_i - t_1)}{1/[(h_{ri} + h_{ci})A]} \tag{i}$$

where h_r is termed as the radiation heat transfer coefficient. Similarly, heat flow from the wall is

$$q = (h_{ro} + h_{co}).A.(t_2 - t_o) = \frac{(t_2 - t_o)}{1/[(h_{ro} + h_{co})A]} \tag{ii}$$

Considering the electrical analogy, the overall resistance to heat flow is

$$\frac{1}{(h_{ri} + h_{ci})A} + \frac{\delta}{kA} + \frac{1}{(h_{ro} + h_{co})A} \tag{iii}$$

Here $1/h_r$ and $1/h_c$ are two resistances R_r and R_c, respectively, in parallel, refer to Fig. 2.11. The heat transfer equation takes the form

$$q = \frac{t_i - t_o}{\frac{1}{(h_{ri} + h_{ci})A} + \frac{\delta}{kA} + \frac{1}{(h_{ro} + h_{co})A}} \tag{2.23}$$

Example 2.2 The heating surface of a boiler is in contact with flue gas (at 1200°C) on one side and boiling water (at 200°C) on the other side. The metal wall thickness is 10 mm, and its thermal conductivity is 45 W/(m K). The local or surface coefficient of heat transfer from the flue gas to the wall is 50 W/(m^2 K) and from wall to the boiling water is 4000 W/(m^2 K). Calculate the rate of heat flow from the flue gas to the water and the surface temperatures.

What will be the effect on heat flow rate if heat transfer coefficients for one or both sides are doubled?

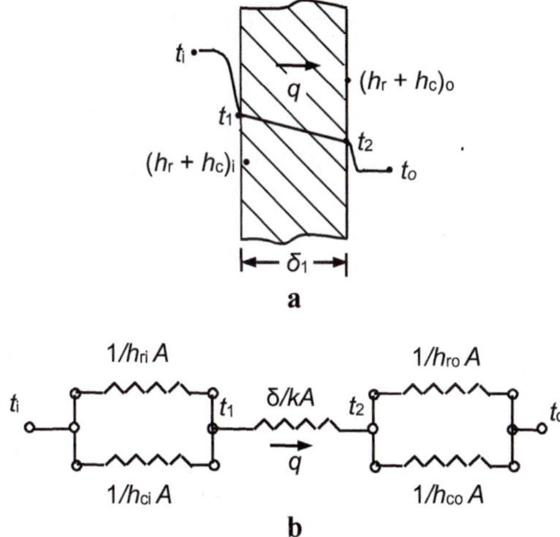

Fig. 2.11 One-dimensional steady-state heat conduction through a plane wall considering heat transfer by convection and radiation from wall surface

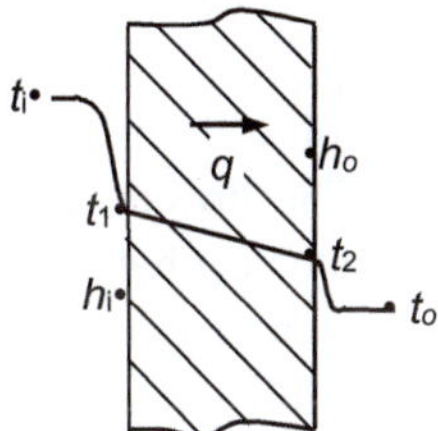

Fig. 2.12 Example 2.2

Solution

(i) The resistance to the heat flow is (refer to Fig. 2.12)

$$
\begin{aligned}
R_k &= R_i + R_{\text{wall}} + R_o \\
&= \frac{1}{Ah_i} + \frac{\delta}{kA} + \frac{1}{Ah_o} \\
&= \frac{1}{1 \times 50} + \frac{10}{1000 \times 1 \times 45} + \frac{1}{1 \times 4000} = 0.02047.
\end{aligned}
$$

The rate of heat transfer per unit area,

$$
q = \frac{t_i - t_o}{R_k} = \frac{1000}{0.02047} = 48{,}852\,\text{W}.
$$

Considering heat flow through the flue gas film at the wall,

$$q = h_i A(t_i - t_1)$$

or

$$48{,}852 = 50 \times 1 \times (1200 - t_1)$$

or

$$t_1 = 223°\text{C}$$

Similarly at the other face,

$$t_2 = \frac{q}{h_o A} + t_o = \frac{48{,}852}{4000} + 200 = 212.2°\text{C}$$

It is to note that the resistance of the wall material (R_{wall}) and that of the film on the waterside (R_o) are negligible compared to the resistance of the film (R_i) on the gas side. Thus, the resistance R_i is the controlling resistance and its change has a greater effect on the heat transfer. By following the above procedure, it can be shown that heat transfer increases to 95,493 W if h_i is doubled as compared to q = 49,147 W when h_o is doubled.

Example 2.3 A furnace wall consists of 200 mm of refractory fireclay brick, 100 mm of kaolin brick and 6 mm of steel plate. The fireside of the refractory is at 1150°C, and the outside of the steel is at 30°C. An accurate heat balance over the furnace shows the heat loss from the wall to be 300 W/m^2. It is known that there may be thin layers of air between the layers of brick and steel. To how many millimetres of kaolin are these air layers equivalent? The thermal conductivities are as follows:

Refractory fireclay bricks, k_f = 1.7 W/(m K)
kaolin brick, k_i = 0.17 W/(m K)
steel, k_s = 17 W/(m K)

Solution

$$q = \frac{A(t_1 - t_4)}{\frac{\delta_f}{k_f} + \frac{\delta_i}{k_i} + \frac{\delta_s}{k_s}}$$

Substitution gives

$$300 = \frac{1 \times (1150 - 30)}{\frac{0.2}{1.7} + \frac{\delta_i}{0.17} + \frac{0.006}{17}}$$

which gives effective thickness of kaolin, δ_i = 0.6146 m = 614.6 mm. Thus, the air layers are equivalent to (614.6 – 100) = 514.6 mm of kaolin.

Example 2.4 The thermal conductivity of a cylindrical specimen 150 mm in diameter was determined using an instrument consisting of two flat plates between which the cylindrical specimen sides were placed. When heat transfer rate was 100 W, the hot and cold end plate temperatures were recorded to be 170°C and 20°C, respectively. The cylindrical specimen is 15 mm in length.

It was found that due to improper matching air clearances of thickness of about 0.15 mm were formed between the cold and hot surfaces of the instrument and the specimen ends. Determine the thermal conductivity of the specimen.

Conductivity of the air can be taken as 2.7×10^{-2} and 3.8×10^{-2} W/(m °C) at the cold and hot surfaces, respectively.

Solution

The resistance to the heat transfer, considering the air resistances, is

$$R = \left(\frac{\delta}{k}\right)_{\text{air,cold}} + \left(\frac{\delta}{k}\right)_{\text{specimen}} + \left(\frac{\delta}{k}\right)_{\text{air,hot}}$$

$$= \left(\frac{0.15}{1000 \times 2.7 \times 10^{-2}}\right) + \left(\frac{15}{1000 \times k}\right) + \left(\frac{0.15}{1000 \times 3.8 \times 10^{-2}}\right)$$

$$= 9.503 \times 10^{-3} + \left(\frac{0.015}{k}\right)$$

The heat transfer from Fourier's equation is

$$q = \frac{A(t_1 - t_4)}{\sum \frac{\delta}{k}}$$

Substitution gives

$$100 = \frac{\pi}{4}(0.15)^2 \times (170 - 20) \times \left(9.503 \times 10^{-3} + \frac{0.015}{k}\right)^{-1}$$

Solution of above equation gives k = 0.882 W/(m K), which is the thermal conductivity of the specimen.

Example 2.5 The walls of a paint drying chamber are built-up of a layer of brick [thickness δ = 250 mm and k = 0.7 W/(m K)]. The temperature in the chamber is estimated to be 115° C. The heat flow from 1 m^2 of the chamber wall is not to exceed 100 W when ambient temperature is 25°C for which a layer of felt [k = 0.045 W/(m K)] is to be applied outside the brick layer. Calculate the thickness of the felt if the surface heat transfer coefficients at inner and outer walls are 30 W/(m^2 °C) and 20 W/(m^2 °C), respectively.

Solution

Total resistance to heat flow

$$R = \frac{1}{A}\left(\frac{1}{h_i} + \frac{\delta_b}{k_b} + \frac{\delta_f}{k_f} + \frac{1}{h_o}\right)$$

where subscripts b and f refer to brick and felt, respectively.

$$R = \frac{1}{1.0}\left(\frac{1}{30} + \frac{0.25}{0.7} + \frac{\delta_f}{0.045} + \frac{1}{20}\right)$$

or

$$= 0.4405 + 22.22\delta_f \tag{i}$$

From heat transfer equation,

$$q = \left(\frac{t_i - t_o}{R}\right)$$

or

$$100 = \left(\frac{115 - 25}{R}\right)$$

or

$$R_k = 0.9$$

Substitution of this value of R_k in Eq. (i) gives

$$\delta_f = 0.0207\,\text{m} = 20.7\,\text{mm, say 21 mm}$$

Example 2.6 The walls of a house have a composite construction of bricks [thickness = 330 mm, k = 0.8 W/(m K)] and 20-mm-thick plaster [k = 1.2 W/(m K)] on both sides. The inner and outer surface heat transfer coefficients are 15 and 30 W/(m^2 K). Determine the heat loss per m^2 of the wall surface if the temperature inside the room is 22°C and outside temperature is –5°C.

When the wind blows fast, the outside film coefficient increases to 200 W/(m^2 °C). Determine the increase in the heat loss.

Identify the controlling resistance that determines the heat flow rate.

Solution

The values of various resistances, for the unit area of the wall, are

$$\text{Convective (inner)}, R_i = \frac{1}{h_i} = \frac{1}{15} = 0.067\,\text{m}^2\,\text{K/W}$$

$$\text{Inner plaster}, R_{pi} = \frac{\delta}{k} = \frac{20}{1000 \times 1.2} = 0.0167\,\text{m}^2\,\text{K/W}$$

$$\text{Bricks}, R_b = \frac{\delta}{k} = \frac{330}{1000 \times 0.8} = 0.4125\,\text{m}^2\,\text{K/W}$$

$$\text{Plaster (outer)}, R_{po} = \frac{\delta}{k} = \frac{20}{1000 \times 1.2} = 0.0167\,\text{m}^2\,\text{K/W}$$

$$\text{Convective (outer)}, R_o = \frac{1}{h_o} = \frac{1}{30} = 0.033\,\text{m}^2\,\text{K/W}$$

The total resistance,

$$\sum R = R_i + R_{pi} + R_b + R_{po} + R_o = 0.5459$$

The heat flow rate,

$$q = \frac{T_i - T_o}{\sum R} = \frac{22 - (-5)}{0.5459} = 49.46\ \mathrm{W/m^2}$$

When the wind blows, the outer convective resistance R_o changes to 1/200 = 0.005 and the total resistance,

$$\sum R = R_i + R_{pi} + R_b + R_{po} + R_o = 0.5179$$

The new heat flow rate,

$$q' = \frac{22 - (-5)}{0.5179} = 52.13\ \mathrm{W/m^2}$$

$$\text{Percentage change in heat flow} = \frac{52.13 - 49.46}{49.46} \times 100 = 5.4\%$$

which is small because the controlling resistance is the resistance of the bricks and is almost 75–80% of the total resistance to the heat flow.

Example 2.7 A house has a multi-layer composite wall constructed as shown in Fig. 2.13. The temperature of the air inside the room is 22°C, and the surface heat transfer coefficient between the room air and the wall is 6 W/(m^2 K). The outer surface heat transfer coefficient is 20 W/(m^2 K). The different wall thicknesses (δ) in mm and the thermal conductivities

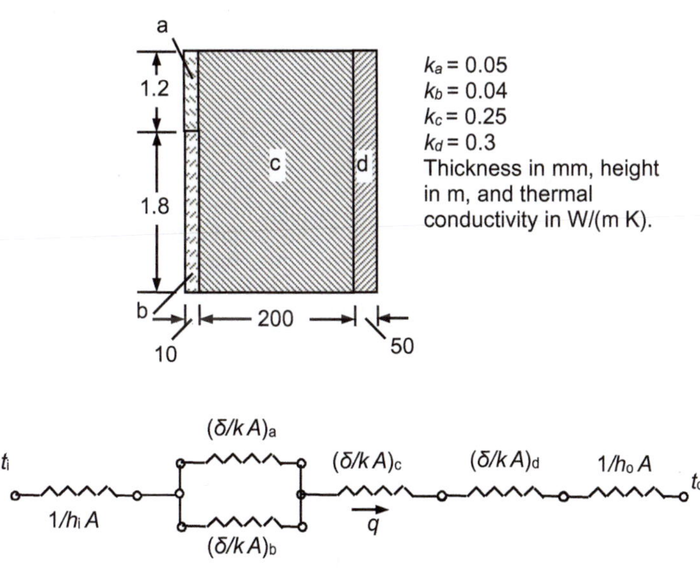

Fig. 2.13 Example 2.7

(k) in W/(m K) are indicted in the figure. Calculate the heat transfer rate across the wall section per m length of the wall. Outside air temperature is 5°C.

Solution

The resistance to heat transfer, refer to the thermal network in Fig. 2.13,

$$\sum R = \frac{1}{h_i A} + \frac{1}{\left(\frac{\delta_a}{k_a A_a}\right)^{-1} + \left(\frac{\delta_b}{k_b A_b}\right)^{-1}} + \left(\frac{\delta_c}{k_c A}\right) + \left(\frac{\delta_d}{k_d A}\right) + \frac{1}{h_o A}$$

$$= \frac{1}{6 \times 3} + \frac{1}{\left(\frac{0.01}{0.05 \times 1.2}\right)^{-1} + \left(\frac{0.01}{0.04 \times 1.8}\right)^{-1}} + \left(\frac{0.2}{0.25 \times 3}\right) + \left(\frac{0.05}{0.3 \times 3}\right) + \frac{1}{20 \times 3} = 0.47$$

Hence, the heat transfer rate is

$$q = \frac{\Delta t}{\sum R} = \frac{22 - 5}{0.47} = 36.17\ \text{W}$$

Example 2.8 Find the heat flow through a wall, per m^2 of wall face area, shown in Fig. 2.14. The thermal conductivities of brick material and steel are 1.0 W/(m K) and 40 W/(m K), respectively.

Calculate the percentage reduction in the heat flow if the bolts are not used.

What will be the temperature at the outer surface of the wall neglecting the effect of bolts?

Solution

Part (i)

Figure 2.15 shows the analogous electrical circuit for various thermal resistances. It is to be noted that the bolts provide a parallel path to the flow of heat. The effect of bolt heads and nuts has been neglected.

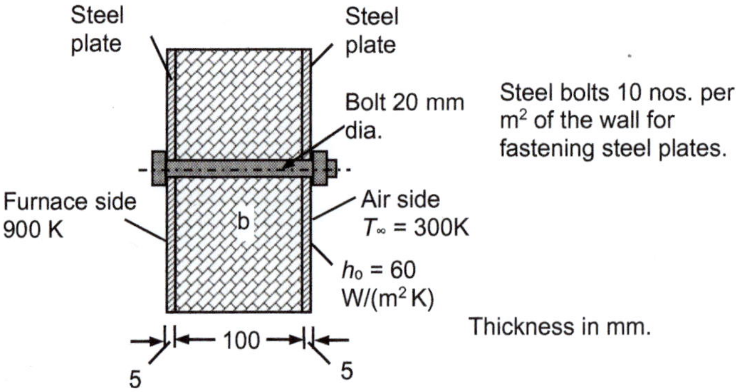

Fig. 2.14 Wall cross-section

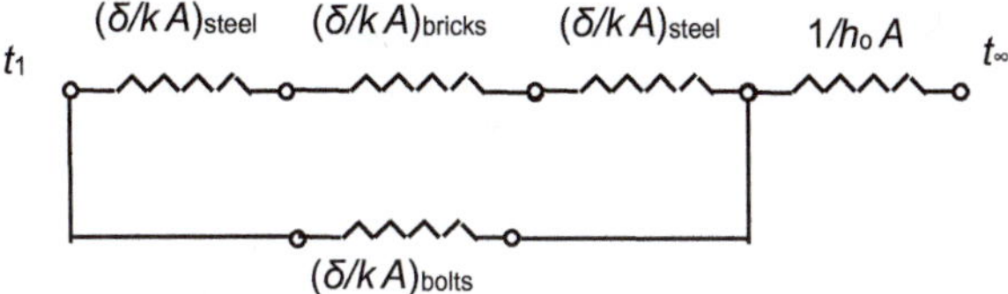

Fig. 2.15 Thermal network with bolts

The heat flow areas and length of heat paths for different layers and bolts are

$$A_{bolts} = 10 \times \frac{\pi}{4}(d_{bolt})^2 = 10 \times \frac{\pi}{4}(0.02)^2 = 3.1416 \times 10^{-3}\,\mathrm{m}^2$$
$$A_{steel} = A_{bricks} = A - A_{bolts} = 1.0 - 3.1416 \times 10^{-3} = 0.99686\,\mathrm{m}^2$$
$$\delta_{steel} = 0.005\,\mathrm{m}, \delta_{brick} = 0.1\,\mathrm{m}, \delta_{bolt} = 0.110\,\mathrm{m}$$

The values of various resistances are

$$R_{steel} = \left(\frac{\delta}{kA}\right)_{steel} = \frac{0.005}{40 \times 0.99686} = 1.254 \times 10^{-4}$$
$$R_{brick} = \left(\frac{\delta}{kA}\right)_{brick} = \frac{0.1}{1 \times 0.99686} = 0.1003$$
$$R_{bolt} = \left(\frac{\delta}{kA}\right)_{bolt} = \frac{0.110}{40 \times 3.1416 \times 10^{-3}} = 0.87535$$
$$R_o = \frac{1}{h_oA} = \frac{1}{60 \times 1} = 0.0167.$$

The total resistance to heat transfer is

$$\sum R = \left(\frac{1}{R_{steel} + R_{brick} + R_{steel}} + \frac{1}{R_{bolt}}\right)^{-1} + R_o = 0.1069$$

The heat transfer,

$$q = \frac{t_1 - t_\infty}{\sum R} = \frac{900 - 300}{0.1069} = 5612.7\,\mathrm{W}$$

Part (ii) Heat transfer without bolts:
The values of various resistances are

$$R_{steel} = \left(\frac{\delta}{kA}\right)_{steel} = \frac{0.005}{40 \times 1.0} = 1.25 \times 10^{-4}$$
$$R_{brick} = \left(\frac{\delta}{kA}\right)_{brick} = \frac{0.1}{1 \times 1.0} = 0.1$$
$$R_o = \frac{1}{h_oA} = \frac{1}{60 \times 1} = 0.0167$$

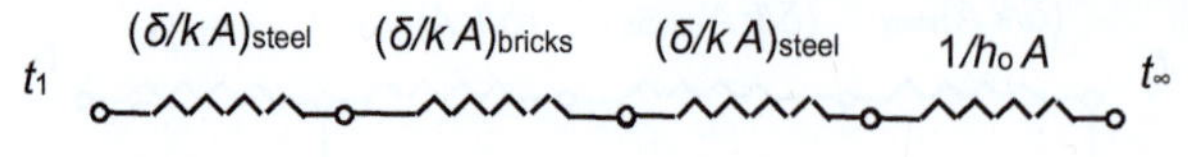

t_1 $(\delta/kA)_{steel}$ $(\delta/kA)_{bricks}$ $(\delta/kA)_{steel}$ $1/h_o A$ t_∞

Fig. 2.16 Thermal network without bolts

The total resistance to heat transfer is (refer to network in Fig. 2.16)

$$\sum R = R_{\text{steel}} + R_{\text{brick}} + R_{\text{steel}} + R_o = 0.11695$$

The heat transfer

$$q = \frac{t_1 - t_\infty}{\sum R} = \frac{900 - 300}{0.11695} = 5130.4\,\text{W}$$

Percentage reduction in heat loss is

$$\Delta q = \frac{5612.7 - 5130.4}{5612.7} \times 100 = 8.6\%.$$

Part (iii) Temperature at the outer surface of the wall:

The temperature T_2 can be found from the heat transfer equation (neglecting the effects of the bolts):

$$q = \frac{t_2 - t_\infty}{R_o}$$

or

$$5130.3 = \frac{T_2 - 300}{0.0167}$$

or

$$t_2 = 385.7\ \text{K}$$

Example 2.9 A masonry wall consists of 100 mm brick outer face with 10 mm mortar joint, a 200 mm concrete wall and a 10 mm insulating board on the inside as shown in Fig. 2.17a. The outside and inside bulk air temperatures are 40 and 25°C, respectively. Determine the heat flux if the outside heat transfer coefficient is 15 W/(m^2 K) and the inside film coefficient is 5 W/(m^2 K). Thermal conductivities given are bricks, k_b = 1.2; mortar, k_m = 0.6; concrete, k_c = 0.8; board, k_b = 0.15 W/(m K).

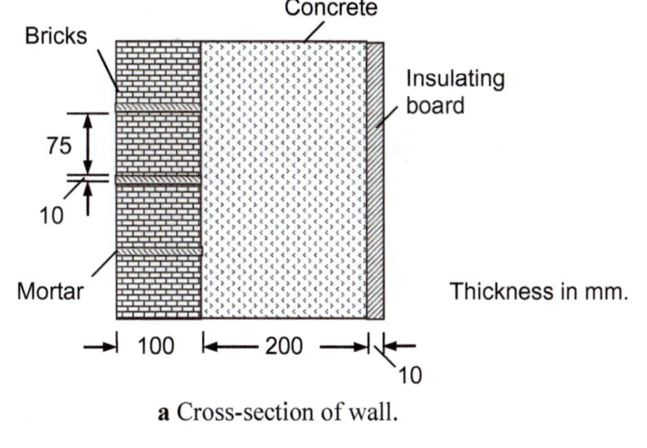

a Cross-section of wall.

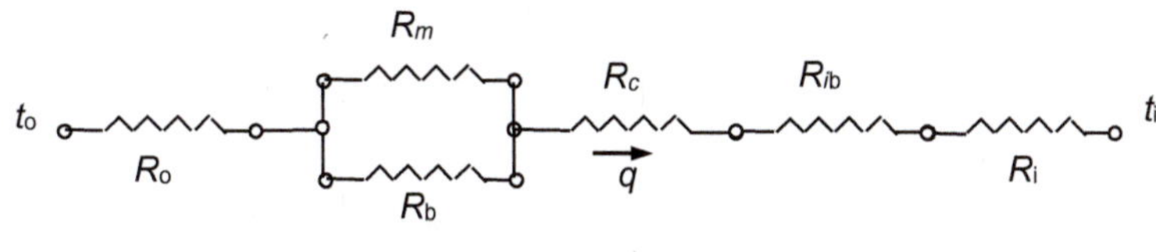

b Thermal network.

Fig. 2.17 Example 2.9

Solution

Resistances for unit height and width of the wall:

(i) Convective (outer)

$$R_o = \frac{1}{h_o A_o} = \frac{1}{15 \times 1} = 0.0666$$

(ii) Brick

$$R_b = \frac{\delta_b}{k_b A_b} = \frac{0.1}{1.2 \times 75/85} = 0.0944$$

(iii) Mortar

$$R_m = \frac{\delta_m}{k_m A_m} = \frac{0.1}{0.6 \times 10/85} = 1.4166$$

(iv) Concrete

$$R_c = \frac{\delta_c}{k_c A_c} = \frac{0.2}{0.8 \times 1} = 0.25$$

(v) Insulating board

$$R_{ib} = \frac{\delta_{ib}}{k_{ib} A_{ib}} = \frac{0.01}{0.15 \times 1} = 0.0666$$

(vi) Convective (inner)

$$R_i = \frac{1}{h_i A_i} = \frac{1}{5 \times 1} = 0.2$$

All resistances are in K/W. The total resistance to heat flow is (refer to the network in Fig. 2.17b)

$$\sum R = R_0 + \left(\frac{1}{R_b} + \frac{1}{R_m}\right)^{-1} + R_c + R_{ib} + R_i = 0.6717\,\mathrm{K/W}$$

Heat flux,

$$q'' = \frac{T_o - T_i}{\sum R} = \frac{40 - 25}{0.6717} = 22.33\,\mathrm{W/m^2}$$

which is from the outside to inside.

Example 2.10 Rods of copper, brass and steel are welded together to form a Y-shaped figure as shown in Fig. 2.18. The cross-sectional area of each rod is 400 mm^2. The end of the copper rod is maintained at 120°C and the ends of the brass and steel rods at 20°C. The lengths of the rods are copper 768 mm, brass 172 mm and steel 45 mm.

(a) What is the temperature of the junction point?
(b) What is the heat flowing in the copper rod?

The thermal conductivities of copper, brass and steel are 384, 86 and 45 W/(m K), respectively. The rods are insulated from outside.

Solution

Heat balance at the junction gives

Heat inflow at the junction through copper rod, q_c

= heat out flow at the junction through brass and steel rods, $(q_b + q_s)$

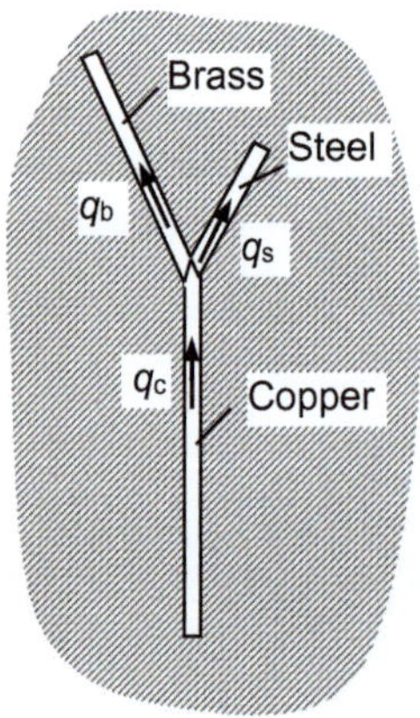

Fig. 2.18 Example 2.10

where, from Fourier's equation,

$$q_c = kA\frac{\Delta t}{L} = 384 \times \frac{400}{(1000)^2} \times \frac{120 - t}{0.768} = 0.2 \times (120 - t)$$

$$q_b = kA\frac{\Delta t}{L} = 86 \times \frac{400}{(1000)^2} \times \frac{t - 20}{0.172} = 0.2 \times (t - 20)$$

$$q_s = kA\frac{\Delta t}{L} = 45 \times \frac{400}{(1000)^2} \times \frac{t - 20}{0.045} = 0.4 \times (t - 20)$$

Substitution in heat balance equation gives

$$0.2 \times (120 - t) = 0.2 \times (t - 20) + 0.4 \times (t - 20)$$

or

$$t = 45°\text{C}$$

This gives

$$q_c = 0.2(120 - 45) = 15\,\text{W}$$

Example 2.11 The walls of a cold storage are proposed to be made of three layers as shown in Fig. 2.19a. The temperature of the inside wall of the storage is likely to be –10°C. The outside surface of the wall is likely to be exposed to surrounding air at 20°C (dew point = 15°C) with film coefficient h_o = 20 W/(m^2 K). Determine the position in the wall

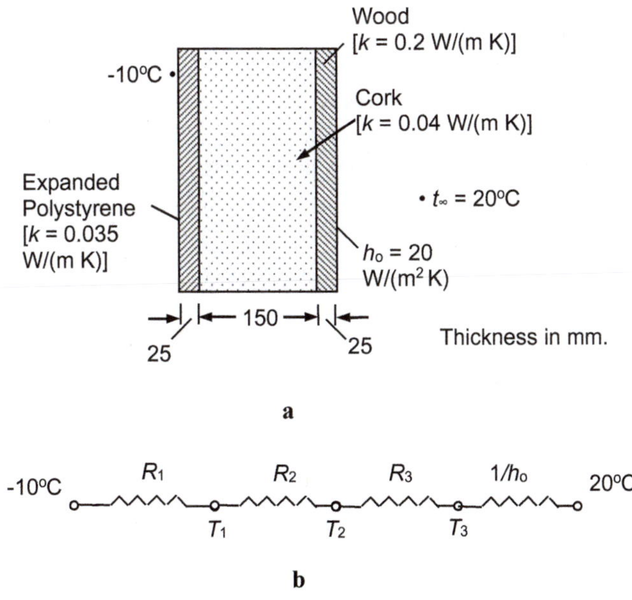

Fig. 2.19 **a** Cross-section of cold storage wall, **b** thermal network

where (i) condensation of the moisture of the air, and (ii) freezing of the condensed moisture may take place if the air diffuses into the wood and cork. Also determine the heat flow rate across the wall.

Solution

Figure 2.19b shows the thermal network for A = 1 m^2. The magnitudes of the resistances are

(i) Polystyrene $R_1 = \frac{\delta}{k} = \frac{0.025}{0.035} = 0.7143$
(ii) Cork $R_2 = \frac{\delta}{k} = \frac{0.15}{0.04} = 3.75$
(iii) Wood $R_3 = \frac{\delta}{k} = \frac{0.025}{0.2} = 0.125$
(iv) Convective $R_o = \frac{1}{h_o} = \frac{1}{20} = 0.05$

The heat flow equation for various layers can be written as

$$q = \frac{t_1 - (-10)}{0.7143} \tag{i}$$

$$q = \frac{t_2 - t_1}{3.75} \tag{ii}$$

$$q = \frac{t_3 - t_2}{0.125} \tag{iii}$$

$$q = \frac{20 - t_3}{0.05} \tag{iv}$$

and heat flow across the wall is

$$q = \frac{20 - (-10)}{0.7143 + 3.75 + 0.125 + 0.05} = 6.47\ \text{W/m}^2$$

Substituting the value of q in Eqs. (i)–(iv) gives t_1 = –5.4°C, t_2 = 18.86°C and t_3 = 19.67°C.

The analysis shows that the temperatures of the two sides of the cork layer are –5.4°C (inside) and 18.86°C (outside). Hence, the condensation of the moisture of the diffusing air and freezing of the condensate will take place in the cork layer.

Let the planes with 0°C and 15°C lie at distances δ_1 and δ_2 from the inner side of the cork as shown in Fig. 2.20. We can write

$$q = \frac{0 - (-5.4)}{\delta_1/0.04}$$

and

$$q = \frac{15 - 0}{\delta_2/0.04}$$

Putting q = 6.47 W/m^2, we get δ_1 = 33.4 mm (for 0°C) and δ_2 = 92.7 mm (for 15°C).

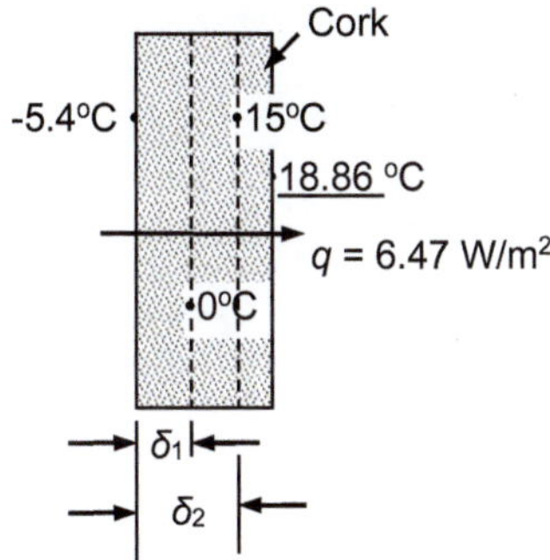

Fig. 2.20 Heat flow across cork layer

2.6 One-Dimensional Steady-State Heat Conduction Through a Cylindrical Shell

For steady-state conduction, $\frac{\partial t}{\partial \tau} = 0$. In case of one-dimensional radial heat flow, the derivatives with respect to θ and z are zero. If there is no heat generation $q_g = 0$ and Eq. (2.17b) reduces to

$$\frac{d^2 t}{dr^2} + \frac{1}{r}\frac{dt}{dr} = 0$$

or

$$r\frac{d^2 t}{dr^2} + \frac{dt}{dr} = 0$$

or

$$\frac{d}{dr}\left(r\frac{dt}{dr}\right) = 0$$

or

$$r\frac{dt}{dr} = C_1$$

or

$$\frac{dt}{dr} = \frac{C_1}{r} \tag{i}$$

or

$$t = C_1 \ln r + C_2 \tag{ii}$$

The boundary conditions are

$$\text{at } r = r_1, \quad t = t_1 \tag{iii}$$

and

$$\text{at } r = r_2, \quad t = t_2 \tag{iv}$$

This gives

$$t_1 = C_1 \ln r_1 + C_2 \tag{v}$$

$$t_2 = C_1 \ln r_2 + C_2 \tag{vi}$$

Solution of Eqs. (v) and (vi) gives

$$C_1 = -\frac{t_1 - t_2}{\ln(r_2/r_1)}$$

$$C_2 = t_1 + \frac{t_1 - t_2}{\ln(r_2/r_1)} \ln r_1$$

Substitution of values of the constants in Eq. (ii) gives the equation of temperature distribution as

$$t = -\frac{t_1 - t_2}{\ln(r_2/r_1)} \ln r + t_1 + \frac{t_1 - t_2}{\ln(r_2/r_1)} \ln r_1$$

Rearrangement of the terms gives

$$\frac{t - t_1}{t_2 - t_1} = \frac{\ln r - \ln r_1}{\ln(r_2/r_1)} = \frac{\ln(r/r_1)}{\ln(r_2/r_1)} \tag{2.24}$$

This is equation of variation of the temperature in non-dimensional form, which is an equation of logarithmic curve. It is to be kept in mind that the wall material has been assumed to be homogeneous and the thermal conductivity is constant.

The conduction heat transfer from Fourier's equation is

$$q = -kA \frac{dt}{dr}$$

or

$$q = -k(2\pi r_2 L) \left(\frac{dt}{dr} \right)_{r=r_2}$$

where $2\pi r_2 L$ is the heat transfer area at $r = r_2$ and $\left(\frac{dt}{dr}\right)_{r=r_2} = \frac{C_1}{r_2}$, refer to Eq. (i). Substitution of the value of C_1 gives derivative as

$$\left(\frac{dt}{dr}\right)_{r=r_2} = -\frac{1}{r_2}\frac{t_1 - t_2}{\ln(r_2/r_1)}$$

Hence,

$$q = -k(2\pi r_2 L)\left(-\frac{1}{r_2}\frac{t_1 - t_2}{\ln(r_2/r_1)}\right)$$

or

$$q = (2\pi kL)\left(\frac{t_1 - t_2}{\ln(r_2/r_1)}\right)$$

or

$$q = \frac{t_1 - t_2}{\frac{1}{2\pi kL}\ln(r_2/r_1)} = \frac{t_1 - t_2}{R} \tag{2.25}$$

where

$$R = \frac{1}{2\pi kL}\ln(r_2/r_1) \tag{2.26}$$

is the resistance to the heat conduction through a cylindrical wall.

Alternative Method

Alternatively, Eq. (2.25) can be derived by considering an elemental cylindrical shell of thickness dr at radius r, refer to Fig. 2.21.

The heat flow, from Fourier's equation, is

$$q = -kA\frac{dt}{dr}$$

or

$$q = -k(2\pi rL)\left(\frac{dt}{dr}\right)$$

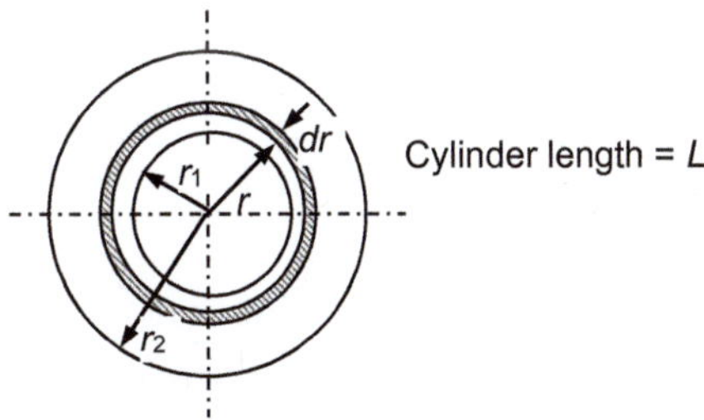

Fig. 2.21 Heat flow across wall of a cylindrical shell

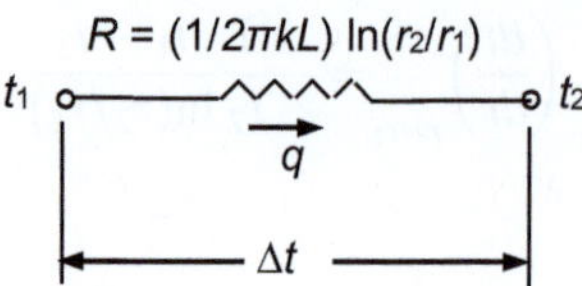

Fig. 2.22 Thermal network for a cylindrical shell

where dt/dr is the temperature gradient. The surface area at radius r is $2\pi rL$. Rearranging

$$q\frac{dr}{r} = -(2\pi kL)dt$$

Integration between the limits $r = r_1$, $t = t_1$ and $r = r_2$, $t = t_2$ gives

$$q\int_{r_1}^{r_2}\frac{dr}{r} = -(2\pi kL)\int_{t_1}^{t_2} dt$$

or

$$q\ln(r_2/r_1) = 2\pi kL(t_1 - t_2)$$

or

$$q = \frac{t_1 - t_2}{\frac{1}{2\pi kL}\ln(r_2/r_1)} = \frac{t_1 - t_2}{R} \tag{2.25}$$

The thermal network in this case can be presented as in Fig. 2.22, where the temperature difference Δt is the driving force for the heat flow and $R = \frac{1}{2\pi kL}\ln(r_2/r_1)$ is the thermal resistance, which the wall offers to the flow of the heat by conduction.

Logarithmic mean radius

Equation (2.25) can be rewritten as

$$q = k\left[\frac{2\pi L(r_2 - r_1)}{\ln(r_2/r_1)}\right]\frac{(t_1 - t_2)}{(r_2 - r_1)} \tag{i}$$

If the heat transfer equation is written in a simple plane wall format, then

$$q = kA'_m\frac{(t_1 - t_2)}{(r_2 - r_1)} \tag{ii}$$

Comparing Eqs. (i) and (ii), we obtain

$$A'_m = \frac{2\pi L(r_2 - r_1)}{\ln(r_2/r_1)} = \frac{(2\pi r_2 L - 2\pi r_1 L)}{\ln(2\pi r_2 L/2\pi r_1 L)} = \frac{(A_2 - A_1)}{\ln(A_2/A_1)}$$

where A_1 and A_2 are the inner and outer surface areas of the cylindrical shell. Area A'_m can be termed as logarithmic mean area. Putting $A'_m = 2\pi r'_m L$, we have

$$2\pi r'_m L = \frac{2\pi L(r_2 - r_1)}{\ln(r_2/r_1)}$$

or

$$r'_m = \frac{r_2 - r_1}{\ln(r_2/r_1)} \tag{2.26}$$

where r'_m can be termed as *logarithmic mean radius*. The arithmetic mean radius is

$$r_m = \frac{r_1 + r_2}{2}$$

and

$$\begin{aligned}\frac{r_m - r'_m}{r_m} &= 1 - \left(\frac{r_2 - r_1}{\ln(r_2/r_1)}\right) \times \frac{2}{r_1 + r_2}\\ &= 1 - \frac{(r_2/r_1 - 1)}{\ln(r_2/r_1)} \times \frac{2}{1 + r_2/r_1}\end{aligned}$$

For $r_2/r_1 = 2$, $\frac{r_m - r'_m}{r_m} = 0.0382$, i.e. for $r_2/r_1 < 2$, the arithmetic mean radius r_m deviates from the logarithmic mean radius by less than 3.8% and can be used without significant error for the calculation of the heat transfer through a cylindrical shell.

2.6.1 One-Dimensional Steady-State Heat Conduction Through a Cylindrical Shell Considering Film Coefficients

When the process of heat conduction is accompanied with convection heat transfer at the inner and outer surfaces of the cylinder (refer to Fig. 2.23), Eq. (2.25) is modified to include the film resistances and is written in terms of inner and outer fluid temperatures t_i and t_o, respectively.

The convection heat transfer at the inner surface gives

$$q = h_i A_i (t_i - t_1) = h_i (2\pi r_1 L)(t_i - t_1)$$

or

$$q = \frac{t_i - t_1}{1/(h_i A_i)} = \frac{t_i - t_1}{1/[h_i(2\pi r_1 L)]} = \frac{t_i - t_1}{R_i} \tag{i}$$

where $R_i = 1/h_i A_i$ is the film resistance at the inner surface of the cylinder.

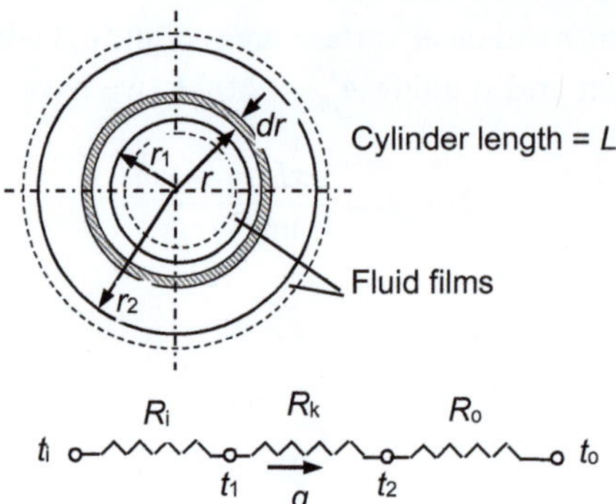

Fig. 2.23 One-dimensional steady-state heat conduction through a cylindrical shell considering film coefficients

Similarly, the convection heat transfer at the outer surface gives

$$q = h_o A_o (t_2 - t_o) = h_o (2\pi r_2 L)(t_2 - t_o)$$

or

$$q = \frac{t_2 - t_o}{1/(h_o A_o)} = \frac{t_2 - t_o}{1/[h_o(2\pi r_2 L)]} = \frac{t_2 - t_o}{R_o} \tag{ii}$$

where $R_o = 1/h_o A_o$ is the film resistance at the outer surface of the cylinder.

The conduction heat transfer through the wall is

$$q = \frac{t_1 - t_2}{[1/(2\pi k L)] \ln(r_2/r_1)} = \frac{t_1 - t_2}{R_k} \tag{iii}$$

From Eqs. (i), (ii) and (iii), we get

$$\begin{aligned} t_i - t_1 &= qR_i \\ t_1 - t_2 &= qR_k \\ t_2 - t_o &= qR_o \end{aligned} \tag{iv}$$

Combining the above equations, we obtain

$$t_i - t_o = q(R_i + R_k + R_o)$$

or

$$q = \frac{t_i - t_o}{R_i + R_k + R_o} = \frac{t_i - t_o}{\sum R}$$

or

$$q = \frac{t_i - t_o}{\frac{1}{h_i A_i} + \frac{1}{2\pi kL}\ln\frac{r_2}{r_1} + \frac{1}{h_o A_o}} \tag{2.27}$$

i.e. the three resistances are in series. The thermal network is shown in Fig. 2.23.

2.6.1.1 Overall Heat Transfer Coefficient

In heat exchangers, the equation of heat exchange between the fluids flowing in and outside the tube is written as

$$q = UA(t_i - t_o) \tag{2.28}$$

where U is termed as *overall heat transfer coefficient.*

Since the heat flow area (surface area) for a cylinder is different at the inner and outer surfaces, the equation of heat exchange must be written as

$$q = U_i A_i (t_i - t_o) \tag{i}$$

or

$$q = U_o A_o (t_i - t_o) \tag{ii}$$

We derived the heat exchange equation as

$$q = \frac{t_i - t_o}{\frac{1}{h_i A_i} + \frac{1}{2\pi kL}\ln\frac{r_2}{r_1} + \frac{1}{h_o A_o}} \tag{2.27}$$

Comparison with Eq. (i) gives

$$\begin{aligned} U_i &= \frac{1}{A_i}\left(\frac{1}{h_i A_i} + \frac{1}{2\pi kL}\ln\frac{r_2}{r_1} + \frac{1}{h_o A_o}\right)^{-1} \\ &= \left(\frac{1}{h_i} + \frac{A_i}{2\pi kL}\ln\frac{r_2}{r_1} + \frac{A_i}{h_o A_o}\right)^{-1} \end{aligned} \tag{2.29a}$$

Since $A_i = 2\pi r_1 L$ and $A_o = 2\pi r_2 L$, the alternative form of the equation of the overall heat transfer coefficient is

$$U_i = \left(\frac{1}{h_i} + \frac{A_i}{2\pi kL}\ln\frac{A_o}{A_i} + \frac{A_i}{h_o A_o}\right)^{-1} \tag{2.29b}$$

Similarly, we can write

$$U_o = \left(\frac{A_o}{h_i A_i} + \frac{A_o}{2\pi kL}\ln\frac{A_o}{A_i} + \frac{1}{h_o}\right)^{-1} \tag{2.30a}$$

where the overall heat transfer coefficient has been referred to the outer surface area.

Defining $h_{io} = h_i A_i / A_o$, which is the film coefficient at the inner surface referred to the outer surface, we obtain

$$U_o = \left(\frac{1}{h_{io}} + \frac{A_o}{2\pi k L} \ln \frac{A_o}{A_i} + \frac{1}{h_o} \right)^{-1} \tag{2.30b}$$

Most of the heat exchangers employ metallic tubes and the tube wall resistance is negligible compared to the two film resistances and is neglected. Then the equation of the overall heat transfer coefficient takes the following form:

$$U_o = \left(\frac{1}{h_{io}} + \frac{1}{h_o} \right)^{-1} = \frac{h_{io} h_o}{h_{io} + h_o} \tag{2.31}$$

2.6.2 Composite Cylindrical Wall

Pipes and tubes carrying high- or low-temperature fluids are generally insulated with one or more layers of insulating materials (termed as lagging of the pipes) to reduce heat loss or gain by the fluid while passing through the tube. The heat transfer in such cases can be determined as follows.

Consider a composite cylinder wall comprising three layers as shown in Fig. 2.24. The radii and thermal conductivities of these layers are mentioned in the figure. In the steady-state condition, the quantity of heat passing through each layer is the same. The adjacent surfaces are assumed to be in perfect contact and hence are at the same temperature. The heat flow equation for the layers can be written as

$$\begin{aligned} q &= \frac{2\pi k_1 L (t_1 - t_2)}{\ln(r_2/r_1)} \\ q &= \frac{2\pi k_2 L (t_2 - t_3)}{\ln(r_3/r_2)} \\ q &= \frac{2\pi k_3 L (t_3 - t_4)}{\ln(r_4/r_3)} \end{aligned} \tag{i}$$

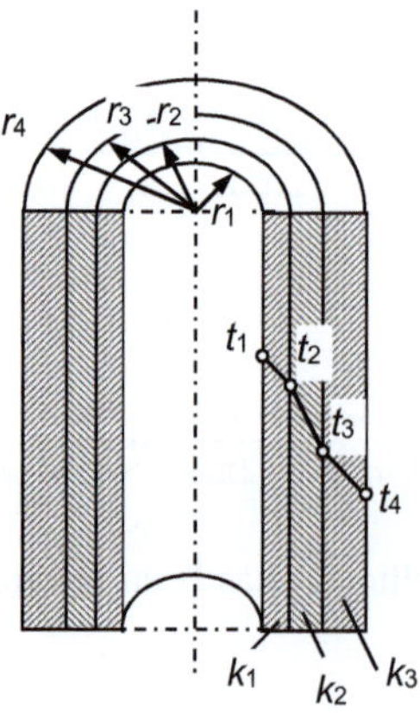

Fig. 2.24 Composite cylindrical wall

From the above equations,

$$\begin{aligned}(t_1 - t_2) &= q\left[\frac{\ln(r_2/r_1)}{2\pi k_1 L}\right]\\ (t_2 - t_3) &= q\left[\frac{\ln(r_3/r_2)}{2\pi k_2 L}\right]\\ (t_3 - t_4) &= q\left[\frac{\ln(r_4/r_3)}{2\pi k_3 L}\right]\end{aligned} \quad \text{(ii)}$$

Adding the equations, we obtain

$$(t_1 - t_4) = q\left[\frac{\ln(r_2/r_1)}{2\pi k_1 L} + \frac{\ln(r_3/r_2)}{2\pi k_2 L} + \frac{\ln(r_4/r_3)}{2\pi k_3 L}\right]$$

or

$$q = \frac{(t_1 - t_4)}{\left[\frac{\ln(r_2/r_1)}{2\pi k_1 L} + \frac{\ln(r_3/r_2)}{2\pi k_2 L} + \frac{\ln(r_4/r_3)}{2\pi k_3 L}\right]} = \frac{(t_1 - t_4)}{R_1 + R_2 + R_3} \tag{2.32}$$

The denominator is the sum of the resistances of the different layers. For n layers, the equation can be written as

$$q = \frac{(t_1 - t_{n+1})}{\frac{1}{2\pi L}\sum_{j=1}^{n} \frac{\ln(r_{j+1}/r_j)}{k_j}} = \frac{(t_1 - t_{n+1})}{\sum_{j=1}^{n} R_j} \tag{2.33}$$

Introducing the film resistances, above equations give

$$\begin{aligned} q &= \frac{(t_i - t_o)}{\left[\frac{1}{h_i A_i} + \frac{\ln(r_2/r_1)}{2\pi k_1 L} + \frac{\ln(r_3/r_2)}{2\pi k_2 L} + \frac{\ln(r_4/r_3)}{2\pi k_3 L} + \frac{1}{h_o A_o}\right]}\\ &= \frac{(t_i - t_o)}{R_i + R_1 + R_2 + R_3 + R_o}\end{aligned}$$

and

$$q = \frac{(t_i - t_o)}{\frac{1}{h_i A_i} + \frac{1}{2\pi L}\sum_{j=1}^{n} \frac{\ln(r_{j+1}/r_j)}{k_j} + \frac{1}{h_o A_o}} \tag{2.34}$$

Example 2.12 Determine steady-state heat transfer rate through the wall of a 5-m-long cylinder having inside and outside radius of 2.7 and 3 m, respectively. The inside and outside wall temperatures are 100°C and 30°C, respectively. The thermal conductivity of the wall material is 45 W/(m K). Also determine the error in estimate when the cylindrical wall is treated as a flat plate.

Solution

Heat flow through a cylindrical wall is given by

$$q = \frac{2\pi kL(t_1 - t_2)}{\ln(r_2/r_1)}$$

or

$$q = \frac{2\pi \times 45 \times 5 \times (100 - 30)}{\ln(3/2.7)} = 939{,}253\ \text{W}$$

Treating the cylindrical shell as a flat plate,

$$A = 2\pi r_m L = 2\pi\left(\frac{r_1 + r_2}{2}\right)L$$
$$= 2\pi\left(\frac{2.7 + 3}{2}\right) \times 5 = 89.53\ \text{m}^2$$

and

$$q' = kA\frac{t_1 - t_o}{r_2 - r_1} = 45 \times 89.53 \times \frac{100 - 30}{3 - 2.7} = 940{,}065\ \text{W}$$

The error in heat transfer estimate is

$$\frac{q' - q}{q} = 0.086\%$$

which is very small.

Comments: Large radii thin cylinders can be treated as flat plates. If the above problem refers to a cylinder with $r_1 = 10$ mm and $r_2 = 20$ mm, then $q = 142{,}769$ W and $q' = 148{,}440$ W. The error, in this case, is 4%. The flat plate equation gives a higher value of the heat transfer rate.

Example 2.13 A 15-mm-diameter copper rod loses heat to the surrounding air with a surface heat transfer coefficient of 10 W/(m^2 K). If the rod is to be uniformly coated with a material of thermal conductivity 0.2 W/(m K) and the coated surface has a heat transfer coefficient of 15 W/(m^2 K), find the thickness of the coating which will keep the state of heat loss unchanged.

Solution

For the rod without insulation, the heat is transferred by convection from its surface and is given by

$$q_1 = \frac{t_w - t_a}{\left(\frac{1}{2\pi r_1 h_1}\right)} \text{ per m length of the rod}$$

where t_w is the temperature of the surface of the rod and t_a is the surrounding air temperature.

For the rod with insulation, the heat flows from the wire surface through the insulation layer and is then transferred by convection from its surface. It is given by

$$q_2 = \frac{t_w - t_a}{\left(\frac{1}{2\pi k_i}\ln\frac{r_2}{r_1} + \frac{1}{2\pi r_2 h_2}\right)} \text{ per m length of the rod}$$

where k_i is the thermal conductivity of the insulating material.

Equating the above equations for an equal rate of heat transfer in both conditions, we have

$$\frac{t_w - t_a}{\left(\frac{1}{2\pi r_1 h_1}\right)} = \frac{t_w - t_a}{\left(\frac{1}{2\pi k_i}\ln\frac{r_2}{r_1} + \frac{1}{2\pi r_2 h_2}\right)}$$

or

$$\frac{1}{r_1 h_1} = \frac{1}{k_i}\ln\frac{r_2}{r_1} + \frac{1}{r_2 h_2}$$

Substituting the values of various terms, we get

$$\frac{1}{0.0075 \times 10} = \frac{1}{0.2}\ln\frac{r_2}{0.0075} + \frac{1}{r_2 \times 15}$$

or

$$-2.2262 = \ln r_2 + \frac{0.0133}{r_2}$$

Solution by trial and error gives $r_2 = 0.094$ m = 94 mm.

Example 2.14 Calculate the overall heat transfer coefficient based on the inner diameter for a steel pipe covered with fibre glass insulation. The following data are given:

ID of pipe, d_i = 20 mm
Thickness of pipe, δ_s = 2 mm
Thickness of insulation, δ_i = 20 mm
Heat transfer coefficient (inside), h_i = 10 W/(m^2 K)
Heat transfer coefficient (outside), h_o = 5 W/(m^2 K)
Conductivity of insulation, k_i = 0.05 W/(m K)
Conductivity of steel, k_s = 46 W/(m K)
Inside fluid temperature, t_i = 200°C
Ambient temperature, t_o = 30°C
Also, find the heat loss from the pipe per m of its length.

Solution

Extending Eq. (2.29b) for the present case, the overall heat transfer coefficient in this case is

$$
\begin{aligned}
U_i &= \left(\frac{1}{h_i} + \frac{A_i}{2\pi k_s L}\ln\frac{r_2}{r_1} + \frac{A_i}{2\pi k_i L}\ln\frac{r_3}{r_2} + \frac{A_i}{h_o A_o}\right)^{-1} \\
&= \left[\frac{1}{h_i} + \frac{r_1}{k_s}\ln\left(\frac{r_2}{r_1}\right) + \frac{r_1}{k_i}\ln\left(\frac{r_3}{r_2}\right) + \left(\frac{r_1}{r_3}\right)\frac{1}{h_o}\right]^{-1} \\
&= \left[\frac{1}{10} + \frac{10}{1000 \times 46}\ln\left(\frac{12}{10}\right) + \frac{10}{1000 \times 0.05}\ln\left(\frac{32}{12}\right) + \left(\frac{10}{32}\right) \times \frac{1}{5}\right]^{-1} \\
&= 2.79\ \text{W/(m}^2\,{}^\circ\text{C)}
\end{aligned}
$$

The heat transfer rate is

$$
\begin{aligned}
q &= U_i A_i (t_i - t_o) \\
&= U_i (2\pi r_1 L) \times (t_i - t_o) \\
&= 2.79 \times (2\pi) \times (10/1000) \times 1 \times (200 - 30) \\
&= 29.80\ \text{W per m length of the pipe}
\end{aligned}
$$

Alternatively, Eq. (2.34) may be used to determine the heat transfer rate:

$$
q = \frac{(t_i - t_o)}{\left[\frac{1}{h_i A_i} + \frac{\ln(r_2/r_1)}{2\pi k_s L} + \frac{\ln(r_3/r_2)}{2\pi k_i L} + \frac{1}{h_o A_o}\right]}
$$

where $A_i = 2\pi r_1 L$ and $A_o = 2\pi r_3 L$.

Example 2.15 Air at 100°C flows through a thin-walled steel tube with inside diameter of 25 mm. The wall thickness is 0.5 mm. Thermal conductivity of the steel is 40 W/(m K). The tube is exposed to an environment at 20°C with heat transfer coefficient of 10 W/(m^2 K). The film coefficient at the inner surface is 50 W/(m^2 K). Calculate

(i) Overall heat transfer coefficient, U_i,
(ii) Heat loss per m length,
(iii) What thickness of insulation having thermal conductivity of 0.035 W/(m K) must be applied to reduce the heat loss by 50%?

Solution

(i) From Eqs. (2.29a), (2.29b), the overall heat transfer coefficient based on the inner surface of pipe can be written as

$$
U_i = \left[\frac{1}{h_i} + \frac{r_1}{k_s}\ln\left(\frac{r_2}{r_1}\right) + \left(\frac{r_1}{r_2}\right)\frac{1}{h_o}\right]^{-1}
$$

Substituting $r_1 = 12.5 \times 10^{-3}$ m, $r_2 = 13 \times 10^{-3}$ m, $k_s = 40$ W/(m K), $h_i = 50$ W/(m^2 K), $h_o = 10$ W/(m^2 K), we get

$$U_i = \left[\frac{1}{50} + \frac{12.5}{1000 \times 40}\ln\left(\frac{13}{12.5}\right) + \left(\frac{12.5}{13}\right) \times \frac{1}{10}\right]^{-1} = 8.608\,\mathrm{W/(m^2K)}$$

(ii) Heat loss per m length,

$$\begin{aligned} q &= U_i A_i (t_i - t_o) \\ &= U_i (2\pi r_1 L) \times (t_i - t_o) \\ &= 8.608 \times (2\pi) \times (12.5/1000) \times 1 \times (100 - 20) \\ &= 54.1\,\mathrm{W}\text{ per m length of the pipe} \end{aligned}$$

(iii) The heat loss after insulation is reduced to the half, i.e. $q' = 54.1/2 = 27.05$ W. The required insulation thickness can be determined from Eq. (2.34):

$$q' = \frac{(t_i - t_o)}{\left[\frac{1}{h_i A_i} + \frac{\ln(r_2/r_1)}{2\pi k_s L} + \frac{\ln(r_3/r_2)}{2\pi k_i L} + \frac{1}{h_o A_o}\right]}$$

or

$$27.05 = \frac{(100 - 20)}{\left[\frac{1}{50 \times 2\pi \times 12.5 \times 10^{-3}} + \frac{\ln(0.013/0.0125)}{2\pi \times 40} + \frac{\ln(r_3/0.013)}{2\pi \times 0.035} + \frac{1}{10 \times 2\pi \times r_3}\right]}$$

By trial and error, $r_3 \approx 18$ mm.

Example 2.16 A steam pipe having an outside diameter of 20 mm is to be covered with two layers of insulation, each of thickness 10 mm. The average thermal conductivity of one material is 5 times that of the other. Assuming that the inner and outer surface temperatures of the composite insulation are fixed, by how much will be the heat transfer be reduced when the better insulating material is placed next to the pipe surface than when it is placed away from the pipe?

Solution

From the given data,

$R_1 = 10$ mm, $R_2 = 20$ mm, $R_3 = 30$ mm and let $k_2 = 5k_1$.

Heat transfer q_1 with better insulating material next to the pipe from Eq. (2.33) is

$$q_1 = \frac{2\pi L(t_1 - t_3)}{\left[\frac{\ln(r_2/r_1)}{k_1} + \frac{\ln(r_3/r_2)}{k_2}\right]} = \frac{2\pi L(t_1 - t_3)}{\left[\frac{\ln(20/10)}{k_1} + \frac{\ln(30/20)}{5k_1}\right]} = 1.2916 \times [2\pi k_1 L(t_1 - t_3)]$$

Heat transfer q_2 with better insulating material placed as outer layer is

$$q_2 = \frac{2\pi L(t_1 - t_3)}{\left[\frac{\ln(r_2/r_1)}{k_1} + \frac{\ln(r_3/r_2)}{k_2}\right]} = \frac{2\pi L(t_1 - t_3)}{\left[\frac{\ln(20/10)}{5k_1} + \frac{\ln(30/20)}{k_1}\right]} = 1.8379 \times [2\pi k_1 L(t_1 - t_3)]$$

Hence,

$$\frac{q_2}{q_1} = \frac{1.8379 \times [2\pi k_1 L(t_1 - t_3)]}{1.2916 \times [2\pi k_1 L(t_1 - t_3)]} = 1.423$$

The analysis shows that for the given surface temperatures, the better insulating material must be placed next to the pipe.

Example 2.17 A conical cylinder is insulated along its tapered surface. The cylinder is of length L with one end of radius R_1 and the other of radius R_2. The surface with radius R_1 is maintained at temperature T_1 and the other at temperature T_2. Find the expression of the heat flow along the length of the cylinder. Consider one-dimensional heat flow along the axis of the cylinder only.

Solution

The heat flow through section xx at distance x is, refer to Fig. 2.25,

$$q = -kA_x \frac{dT}{dx} \tag{i}$$

The radius at distance x is

$$\begin{aligned} R &= R_1 + \frac{(R_2 - R_1)x}{L} \\ &= R_1 + Cx \end{aligned}$$

where $C = \frac{(R_2 - R_1)}{L}$.

The area of the cross-section xx is

$$A_x = \pi(R_1 + Cx)^2$$

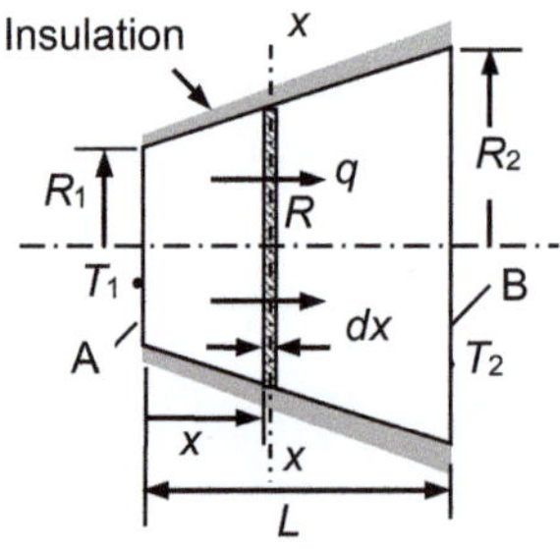

Fig. 2.25 Heat conduction in axial direction of a conical cylinder

Substitution of the value of A_x in Eq. (i) gives

$$q = -k\pi(R_1 + Cx)^2 \frac{dT}{dx} \tag{ii}$$

Integration of Eq. (ii) gives

$$q \int_0^L \frac{dx}{(R_1 + Cx)^2} = -k\pi \int_{T_1}^{T_2} dT \tag{iii}$$

Let $(R_1 + Cx) = a$, then $Cdx = da$. The new boundary conditions are

$$x = 0, \quad a = R_1$$
$$x = L, \quad a = R_1 + CL = R_1 + [(R_2 - R_1)/L]L = R_2$$

Equation (iii) transforms to

$$\int_{R_1}^{R_2} \frac{qda}{Ca^2} = -k\pi \int_{T_1}^{T_2} dT$$

or

$$\frac{q}{C}\left[\frac{-1}{a}\right]_{R_1}^{R_2} = k\pi(T_1 - T_2)$$

or

$$\frac{q}{C}\left(\frac{1}{R_1} - \frac{1}{R_2}\right) = k\pi(T_1 - T_2)$$

or

$$q\left(\frac{L}{R_2 - R_1}\right)\left(\frac{R_2 - R_1}{R_2 R_1}\right) = k\pi(T_1 - T_2)$$

or

$$q = k\pi \frac{(T_1 - T_2)R_1 R_2}{L}.$$

2.7 One-Dimensional Steady-State Heat Conduction Through a Spherical Shell

Three-dimensional general heat conduction equation in spherical coordinates is

$$\frac{1}{r^2}\frac{\partial}{\partial r}\left(r^2\frac{\partial t}{\partial r}\right)+\frac{1}{r^2\sin\theta}\frac{\partial}{\partial\theta}\left(\sin\theta\frac{\partial t}{\partial\theta}\right)+\frac{1}{r^2\sin^2\theta}\left(\frac{\partial^2 t}{\partial\phi^2}\right)+\frac{q_g}{k}=\frac{1}{\alpha}\left(\frac{\partial t}{\partial\tau}\right) \tag{2.18}$$

For steady-state condition, $\left(\frac{\partial t}{\partial \tau}\right) = 0$ and, for one-dimensional radial heat flow, the derivatives $\frac{\partial}{\partial\theta}$ and $\frac{\partial^2}{\partial\phi^2}$ are also zero. Further if heat generation is not present, the equation reduces to

$$\frac{1}{r^2}\frac{d}{dr}\left(r^2\frac{dt}{dr}\right)=0$$

or

$$r^2\frac{dt}{dr}=C_1$$

or

$$\frac{dt}{dr}=\frac{C_1}{r^2} \tag{i}$$

or

$$t=-\frac{C_1}{r}+C_2 \tag{ii}$$

The boundary conditions are

$$t=t_1 \quad \text{at } r=r_1$$

and

$$t=t_2 \quad \text{at } r=r_2$$

Applying these boundary conditions, we get

$$t_1=-\frac{C_1}{r_1}+C_2 \tag{iii}$$

and

$$t_2=-\frac{C_1}{r_2}+C_2 \tag{iv}$$

From Eqs. (iii) and (iv), we have

$$C_1 = \frac{t_1 - t_2}{1/r_2 - 1/r_1}$$

and

$$C_2 = t_1 + \frac{t_1 - t_2}{r_1} \frac{1}{1/r_2 - 1/r_1}$$

Substitution of the values of constants C_1 and C_2 in Eq. (ii) gives

$$\frac{t - t_1}{t_2 - t_1} = \frac{r_1 r_2}{r_2 - r_1}\left(\frac{1}{r_1} - \frac{1}{r}\right)$$

which is the temperature distribution equation in non-dimensional form. It is an equation of hyperbola. Therefore, the temperature distribution across the wall of a spherical shell is hyperbolic.

The heat transfer by conduction through the wall of the shell can be determined by substituting value of dt/dr from Eq. (i) in Fourier's equation. Hence,

$$q = -kA\frac{dt}{dr} = -kA\frac{C_1}{r^2}$$

Substituting value of C_1 and $A = 4\pi r^2$ for the spherical surface, we obtain

$$q = -k(4\pi r^2)\left(\frac{t_1 - t_2}{1/r_2 - 1/r_1} \times \frac{1}{r^2}\right)$$

or

$$q = 4\pi k\left(\frac{t_1 - t_2}{1/r_1 - 1/r_2}\right)$$

or

$$q = 4\pi k\frac{r_1 r_2 (t_1 - t_2)}{r_2 - r_1} \tag{2.35a}$$

or

$$q = \frac{(t_1 - t_2)}{\frac{(r_2 - r_1)}{4\pi k r_1 r_2}} = \frac{(t_1 - t_2)}{R} \tag{2.35b}$$

where

$$R = \frac{(r_2 - r_1)}{4\pi k r_1 r_2} \tag{2.35c}$$

is the resistance of the wall to the conduction heat transfer. The thermal network is shown in Fig. 2.26.

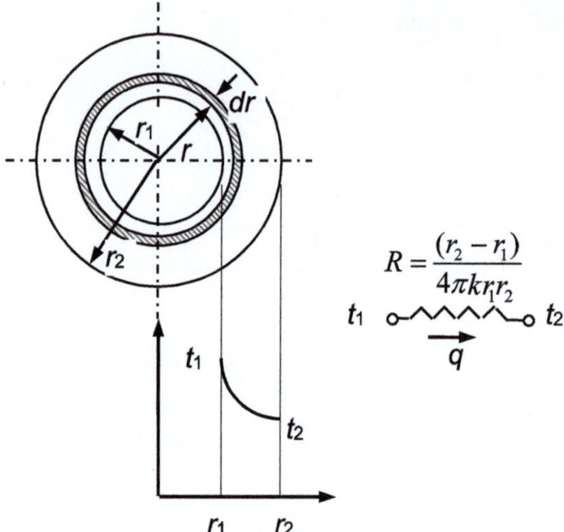

Fig. 2.26 One-dimensional steady-state heat conduction through a spherical shell

Alternative method

Equations (2.35a)–(2.35c), (2.35a) and (2.35b) can be obtained by considering an elementary shell of thickness dr at radius r, refer to Fig. 2.26. The surface area of the elemental shell is $4\pi r^2$. Hence, from Fourier's equation,

$$q = -kA\frac{dt}{dr} = -k(4\pi r^2)\frac{dt}{dr}$$

or

$$q\frac{dr}{r^2} = -4\pi k dt$$

Integration between the limits, $r = r_1$, $t = t_1$, and $r = r_2$, $t = t_2$, gives

$$q\int_{r_1}^{r_2}\frac{dr}{r^2} = -4\pi k\int_{t_1}^{t_2} dt$$

or

$$q\left(\frac{1}{r_1} - \frac{1}{r_2}\right) = -4\pi k(t_2 - t_1)$$

or

$$q = -4\pi k\frac{(t_2 - t_1)}{\left(\frac{1}{r_1} - \frac{1}{r_2}\right)}$$

or

$$q = 4\pi k(t_1 - t_2)\frac{r_1 r_2}{(r_2 - r_1)} \tag{2.35a}$$

or

$$q = \frac{(t_1 - t_2)}{\frac{(r_2 - r_1)}{4\pi k r_1 r_2}} = \frac{(t_1 - t_2)}{R} \tag{2.35b}$$

which is the desired equation.

Effective mean radius
Equation (2.35a) can be rewritten as

$$q = k(4\pi r_1 r_2)\frac{(t_1 - t_2)}{(r_2 - r_1)}$$

If the heat transfer equation is written in a simple plane wall format, then

$$q = kA'_m \frac{(t_1 - t_2)}{(r_2 - r_1)}$$

Comparing the equations, we obtain

$$A'_m = 4\pi r_1 r_2 = \sqrt{A_1 A_2} \tag{2.36a}$$

where A_1 and A_2 are the inner and outer surface areas of the shell. Putting $A'_m = 4\pi r_m^2$, we have

$$A'_m = 4\pi r_m^2 = 4\pi r_1 r_2$$

or

$$r_m = \sqrt{r_1 r_2} \tag{2.36b}$$

Thus the mean effective radius is the geometric mean of the radii r_1 and r_2.

2.7.1 One-Dimensional Steady-State Heat Conduction Through a Spherical Shell Considering the Film Coefficients

Following the procedure adopted for the cylindrical shell, we have

$$q = h_i A_i (t_i - t_1) = h_i (4\pi r_1^2)(t_i - t_1)$$

or

$$q = \frac{t_i - t_1}{1/(h_i A_i)} = \frac{t_i - t_1}{1/[h_i(4\pi r_1^2)]} = \frac{t_i - t_1}{R_i} \tag{i}$$

where $R_i = 1/h_i A_i$ is the film resistance at the inner surface of the shell.

Similarly, the convection heat transfer at the outer surface gives

$$q = h_o A_o (t_2 - t_o) = h_o(4\pi r_2^2)(t_2 - t_o)$$

or

$$q = \frac{t_2 - t_o}{1/(h_o A_o)} = \frac{t_2 - t_o}{1/[h_o(4\pi r_2^2)]} = \frac{t_2 - t_o}{R_o} \tag{ii}$$

where $R_o = 1/h_o A_o$ is the film resistance at the inner surface of the shell.

The conduction heat transfer through the wall is

$$q = \frac{(t_1 - t_2)}{\frac{(r_2 - r_1)}{4\pi k r_1 r_2}} = \frac{(t_1 - t_2)}{R_k} \tag{iii}$$

From Eqs. (i), (ii) and (iii), we get

$$\begin{aligned} t_i - t_1 &= qR_i \\ t_1 - t_2 &= qR_k \\ t_2 - t_o &= qR_o \end{aligned} \tag{iv}$$

Combining the above equations, we obtain

$$t_i - t_o = q(R_i + R_k + R_o)$$

or

$$q = \frac{t_i - t_o}{R_i + R_k + R_o} = \frac{t_i - t_o}{\sum R}$$

or

$$q = \frac{t_i - t_o}{\frac{1}{h_i A_i} + \frac{(r_2 - r_1)}{4\pi k r_1 r_2} + \frac{1}{h_o A_o}} \tag{2.37}$$

i.e. the three resistances are in series. The thermal network is shown in Fig. 2.27.

2.7.2 Composite Spherical Shell

Spherical shells containing high- or low-temperature fluids are generally insulated with one or more layers of insulating materials to reduce heat loss or gain by the fluid. The heat transfer in such cases can be determined as follows.

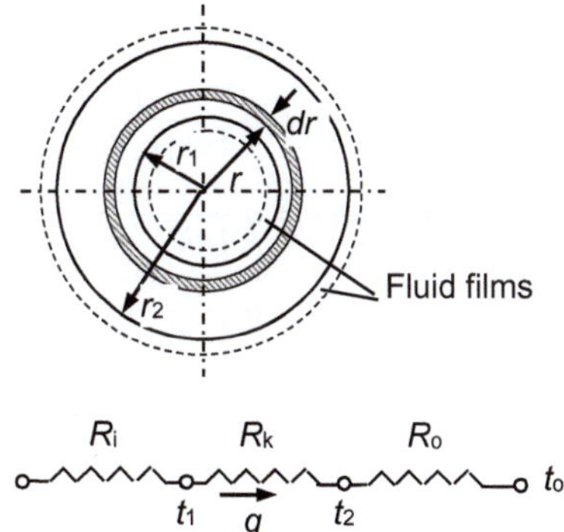

Fig. 2.27 One-dimensional steady-state heat conduction through a spherical shell considering the film coefficients

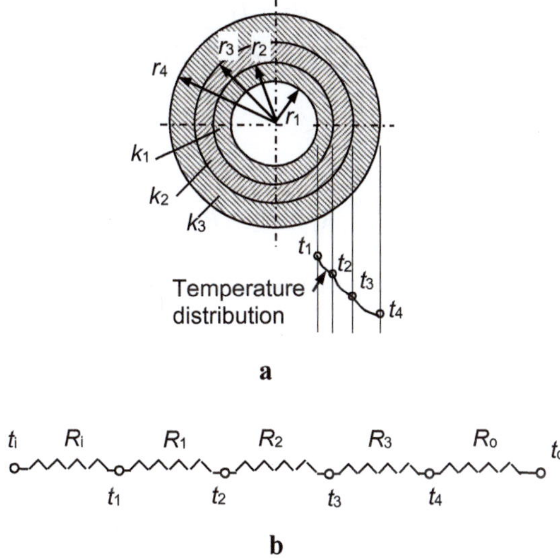

Fig. 2.28 **a** A composite shell, **b** thermal network

Consider a spherical shell with the wall comprising three layers as shown in Fig. 2.28a. The radii and thermal conductivities of these layers are mentioned in the figure. In the steady-state condition, the quantity of heat passing through each layer is the same. The adjacent surfaces are assumed to be in perfect contact and hence are at the same temperature. The heat flow equation for the layers can be written as

$$q = \frac{4\pi k_1 r_1 r_2 (t_1 - t_2)}{r_2 - r_1}$$

$$q = \frac{4\pi k_2 r_2 r_3 (t_2 - t_3)}{r_3 - r_2}$$

$$q = \frac{4\pi k_3 r_3 r_4 (t_3 - t_4)}{r_4 - r_3}$$

From the above equations,

$$(t_1 - t_2) = q\frac{r_2 - r_1}{4\pi k_1 r_1 r_2}$$
$$(t_2 - t_3) = q\frac{r_3 - r_2}{4\pi k_2 r_2 r_3}$$
$$(t_3 - t_4) = q\frac{r_4 - r_3}{4\pi k_3 r_3 r_4}$$

Adding the equations, we obtain

$$(t_1 - t_4) = q\left[\frac{r_2 - r_1}{4\pi k_1 r_1 r_2} + \frac{r_3 - r_2}{4\pi k_2 r_2 r_3} + \frac{r_4 - r_3}{4\pi k_3 r_3 r_4}\right]$$

or

$$q = \frac{(t_1 - t_4)}{\frac{r_2 - r_1}{4\pi k_1 r_1 r_2} + \frac{r_3 - r_2}{4\pi k_2 r_2 r_3} + \frac{r_4 - r_3}{4\pi k_3 r_3 r_4}} \tag{2.38a}$$

or

$$q = \frac{(t_1 - t_4)}{R_1 + R_2 + R_3}$$

The denominator is the sum of the resistances of the different layers. For n layers, the equation can be written as

$$q = \frac{(t_1 - t_{n+1})}{\frac{1}{4\pi}\sum_{j=1}^{n}\frac{r_{j+1} - r_j}{k_j r_j r_{j+1}}} = \frac{(t_1 - t_{n+1})}{\sum_{j=1}^{n} R_j} \tag{2.39}$$

Introducing the film resistances,

$$q = \frac{(t_i - t_o)}{\left[\frac{1}{h_i A_i} + \frac{r_2 - r_1}{4\pi k_1 r_1 r_2} + \frac{r_3 - r_2}{4\pi k_2 r_2 r_3} + \frac{r_4 - r_3}{4\pi k_3 r_3 r_4} + \frac{1}{h_o A_o}\right]} \tag{2.40a}$$
$$= \frac{(t_i - t_o)}{R_i + R_1 + R_2 + R_3 + R_o}$$

The thermal network is shown in Fig. 2.28b. The equation can be written in a general form as

$$q = \frac{(t_i - t_o)}{\frac{1}{h_i A_i} + \frac{1}{4\pi}\sum_{j=1}^{n}\frac{r_{j+1} - r_j}{k_j r_j r_{j+1}} + \frac{1}{h_o A_o}} \tag{2.40b}$$

Example 2.18 A hemispherical electric heated oven of 1.0 m internal diameter is made of 225 mm thick layer of fire bricks. On the outer surface, a magnesia 85 layer of 100 mm thickness is applied for insulation. The temperature at the inner surface of the oven is 800°C. If the ambient temperature is 20°C, and outside convection film coefficient is 10 W/(m^2 K), calculate the heat loss from the oven through the hemispherical shell with and without insulation. Also find out the temperature of the outer surface of the fire bricks. The thermal conductivities of fire bricks and magnesia 85 are 0.3 W/(m K) and 0.05 W/(m K), respectively.

Solution

(i) **Heat loss**

Heat loss through a spherical shell with composite wall is given by

$$q = \frac{(t_1 - t_o)}{\frac{r_2 - r_1}{4\pi k_1 r_1 r_2} + \frac{r_3 - r_2}{4\pi k_2 r_2 r_3} + \frac{1}{4\pi r_3^2 h_o}} = \frac{(t_1 - t_o)}{R_1 + R_2 + R_o}$$

Here r_1 = 500 mm, r_2 = 500 + 225 = 725 mm, r_3 = 725 + 100 = 825 mm, t_1 = 800°C, t_o = 20°C, k_1 = 0.3 W/(m K), k_2 = 0.05 W/(m K) and h_o = 10 W/(m^2 K).

Substitution of the values of various parameters gives

$$R_1 = \frac{r_2 - r_1}{4\pi k_1 r_1 r_2} = \frac{(725 - 500) \times 1000}{4\pi \times 0.3 \times 500 \times 725} = 0.1646$$

$$R_2 = \frac{r_3 - r_2}{4\pi k_2 r_2 r_3} = \frac{(825 - 725) \times 1000}{4\pi \times 0.05 \times 725 \times 825} = 0.266$$

$$R_o = \frac{1}{4\pi r_3^3 h_o} = \frac{(1000)^2}{4\pi \times (825)^2 \times 10} = 0.0117$$

The heat loss from the hemispherical shell will be half of the spherical shell, i.e.

$$q = \frac{1}{2}\left(\frac{t_1 - t_o}{R_1 + R_2 + R_o}\right) = \frac{1}{2}\left(\frac{800 - 20}{0.1646 + 0.266 + 0.0117}\right) = 881.75\,\text{W}$$

Heat loss without magnesia-85 layer will be

$$q = \frac{1}{2}\left(\frac{t_1 - t_o}{R_1 + R_o}\right) = \frac{1}{2}\left(\frac{800 - 20}{0.1646 + 0.0117}\right) = 2212.1\,\text{W}$$

(ii) **Temperature of the outer surface of the firebrick layer**

Heat flow through the brick layer equals the heat loss. Hence,

(i) with magenia-85 layer:

$$881.75 \times 2 = \left(\frac{t_1 - t_b}{R_1}\right) = \left(\frac{800 - t_b}{0.1646}\right)$$

or

$$t_b = 509.73°\text{C}$$

(ii) without magenia-85 layer:

$$2212.1 \times 2 = \left(\frac{t_1 - t_b}{R_1}\right) = \left(\frac{800 - t_b}{0.1646}\right)$$

or

$$t_b = 71.78\,°\mathrm{C}.$$

2.8 Measurement of Thermal Conductivity

2.8.1 Thermal Conductivity Measurement of Solids

Experimental determination of thermal conductivity of a material, based on Eq. (1.3), involves measurement of heat flow rate through the material and the temperature difference across the material specimen under steady-state condition. Various methods used are known as the plate, tube and sphere methods.

The *plate method* is based on one-dimensional heat conduction through a plane wall. Figure 2.29 shows the schematic layout of a guarded hot plate apparatus for measurement of thermal conductivity of solids. The material, whose thermal conductivity is to be determined, is in the form of a circular disc of area A and thickness δ. Heat is supplied to the lower side of the disc by a flat electric heater M. Heat is removed from the top of the disc by arranging a cooler above the disc. Compensation or guard heater G surrounds the main heater and is installed to prevent escape of heat of the main heater from the edge. Additional compensation heater C_1 is provided below the back insulation, under the main heater M, to compensate the loss of heat in the downward direction. The installation of guard heater G and compensation heater C_1 ensures one-dimensional heat flow through the disc, i.e. from the lower heated face to upper cooled face when there is no temperature difference across the side and back insulations (e.g. $t_3 = t_4$). The whole assembly is further adequately insulated to prevent loss of heat to the surrounding.

When steady state is reached, the electric power input (q) to the main heater and temperatures of the hot and cold faces of the disc t_1 and t_2, respectively, are measured with the help of thermocouples installed there. These measurements must be made only when there is no temperature difference across the side and back insulations (e.g. $t_3 = t_4$), as explained earlier. In this situation, all the heat input to the main heater passes through the specimen

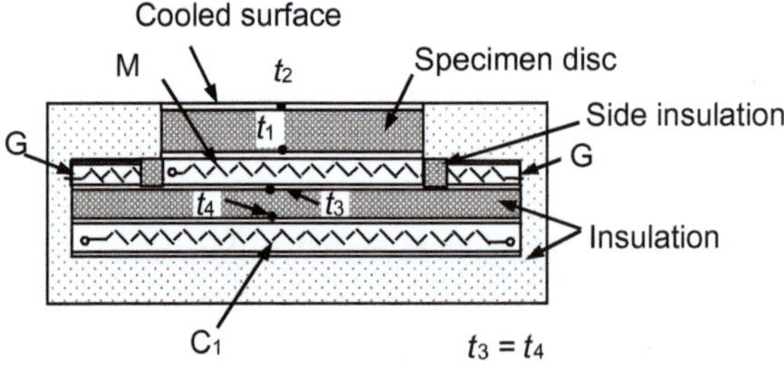

Fig. 2.29 Schematic layout of experimental setup to determine thermal conductivity by plate method

disc. Knowing q, t_1 and t_2, the thermal conductivity is determined from the following relation:

$$k = \frac{q\delta}{A(t_1 - t_2)} = \frac{q\delta}{\pi r^2 (t_1 - t_2)} \tag{2.41}$$

The equipment, described above, is widely used to determine the thermal conductivity of non-metals, i.e. solid of low thermal conductivity. In the case of high thermal conductivity such as metals, the temperature difference $(t_1 - t_2)$ is small and their precise measurement will be required.

Figure 2.30 shows a simple instrument which has been used to measure the thermal conductivity of metals. Here, the rod 1 whose thermal conductivity is to be measured is joined axially with another metal rod 2 whose thermal conductivity is known. Thermocouples are affixed axially to both the rods. The rods are properly insulated from outside to minimize heat loss and ensure one-dimensional heat flow in the axial direction only. Knowing the temperature gradients through the two rods, the thermal conductivity of the rod 1 can be determined using the following relation:

$$q = k_1 A\left(\frac{\Delta t}{\Delta x}\right)_1 = k_2 A\left(\frac{\Delta t}{\Delta x}\right)_2$$

or

$$k_2 = k_1 \left(\frac{\Delta t}{\Delta x}\right)_1 \Big/ \left(\frac{\Delta t}{\Delta x}\right)_2 \tag{2.42}$$

The *tube method* of determining thermal conductivity of insulating materials is based on law of heat conduction through a cylindrical wall, i.e.

$$k = \frac{q \ln(r_2/r_1)}{2\pi L(t_1 - t_2)} \tag{2.43}$$

The instrument, shown schematically in Fig. 2.31a, consists of a central electric heater which is surrounded by a copper tube of external diameter d_1. The material whose thermal

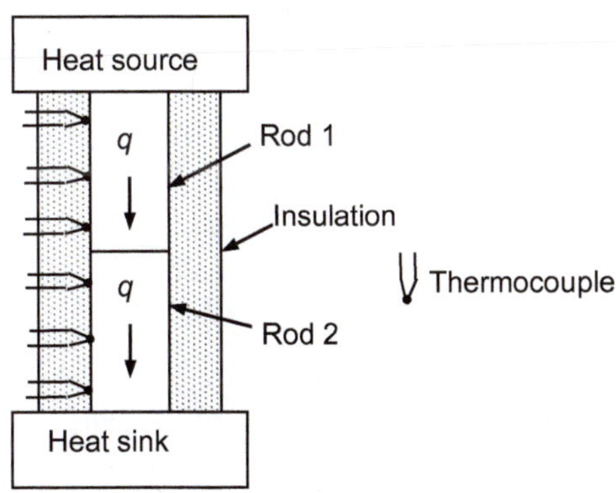

Fig. 2.30 Schematic layout of experimental setup to determine thermal conductivity by rod method

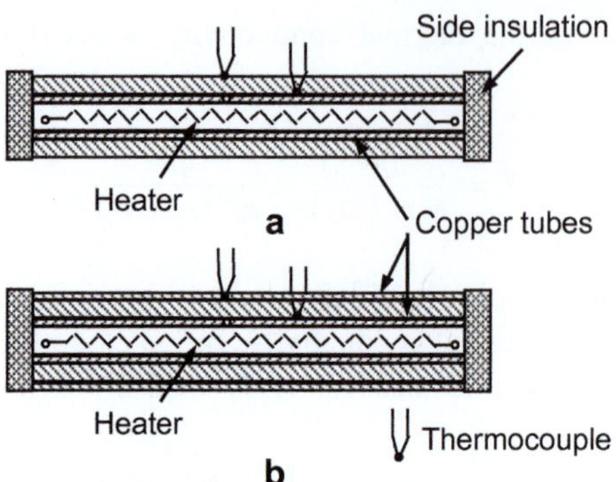

Fig. 2.31 Schematic layout of experimental setup to determine thermal conductivity by tube method

conductivity is to be determined envelops the copper tube. The temperatures t_1 and t_2 of the outer surface of the copper tube and that of the outer surface of the material, respectively, are measured by affixing thermocouples as shown in the figure. Temperatures t_1 and t_2 used in Eq. (2.43) are the average temperatures of these surfaces. Knowing the heat input (i.e. the power supplied to the electric heater), dimensions r_1, r_2, and length L of the tube, the thermal conductivity is calculated from the relation given above.

In the case of materials in powder form, another copper tube of inside radius r_2 is used. The powder is filled in the space between these tubes, see Fig. 2.31b. Temperature t_2, in this case, is the temperature of the inner surface of the outer copper tube. However, the thermal conductivity of the powder insulating material is strong function of the density of the material packed into the space between tubes. Loosely filled material shows a lower thermal conductivity.

In the tube-type instruments described above, the length of the tube must be large (say 2–3 m) so that the end loss is negligible and heat flows only in radial direction as desired.

In the *sphere method*, the instrument consists of two spheres of diameter d_1 and d_2, where $d_2 > d_1$, refer to Fig. 2.32. The test material fills the space between the spheres. The inner sphere houses an electric heater. Measuring the heat input (power to the heater), temperatures t_1 and t_2 in the steady-state condition and knowing the sphere diameters, the thermal conductivity of the filled material is determined from the following relation:

$$k = \frac{q}{4\pi(t_1 - t_2)}\left(\frac{1}{r_1} - \frac{1}{r_2}\right) \tag{2.44}$$

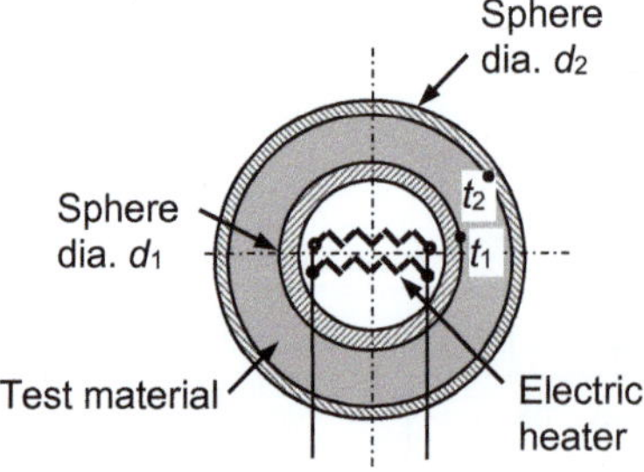

Fig. 2.32 Schematic layout of experimental setup to determine thermal conductivity by sphere method

2.8.2 Thermal Conductivity Measurements of Liquids and Gases

To determine thermal conductivity of liquids and gases, it is an essential requirement that the convection currents are suppressed.

Figure 2.33a depicts one of the methods, wherein the heat from the heated plate flows downwards through a very thin layer of liquid δ (of the order of 0.5 mm) to the cold plate at the bottom. The liquid layer is very thin and is heated from above; hence, the convection currents are not setup.

An alternative method of concentric cylinders (Fig. 2.33b) may be used wherein the heat flows in radial direction through a very thin layer of liquid between two concentric cylinders. This arrangement has also been used for gases. The thermal conductivity is calculated from Eq. (2.43).

Example 2.19 In an instrument used to determine the thermal conductivity, the specimen to be tested is placed between the hot and cold surfaces. The specimen is a circular disc of 150 mm diameter and 15 mm thickness. The temperatures of the hot and cold surfaces of the instrument are recorded to be 170°C and 20°C, respectively. Under steady-state condition, the heat flow through the specimen is 105 W. Calculate the thermal conductivity of the material of the specimen if due to the presence of the guard heaters there is no heat flow in the radial direction.

Solution

$$\begin{aligned} k &= \frac{q\delta}{A(t_1 - t_2)} = \frac{q\delta}{\pi r^2 (t_1 - t_2)} \\ &= \frac{105 \times 0.015}{\pi \times (0.075)^2 \times (170 - 20)} = 0.594\,\text{W/(m K)}. \end{aligned}$$

Example 2.20 A hollow aluminium sphere, with an electric heater at its centre, is used to determine the thermal conductivity of an insulating material. The inner and outer radii of the sphere are 0.1 and 0.125 m, respectively. In a particular test, a spherical shell of an insulating material of 100 mm thickness was cast on the outer surface of the sphere. In the steady state,

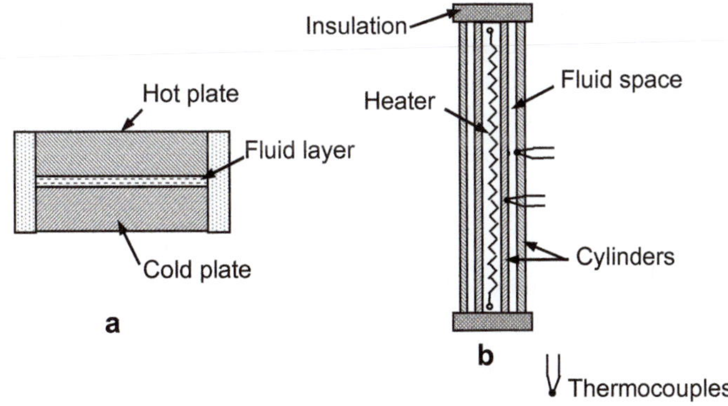

Fig. 2.33 Schematic layout of experimental setups to determine thermal conductivity of liquid or gases

the temperature at the outer surface of the aluminium wall was recorded as 200°C. The system is exposed to room air at 30°C, and the convective heat transfer coefficient is estimated to be 20 W/(m^2 K). If the energy supplied to the electric heater is at the rate of 100 W, what is the thermal conductivity of the insulation? Neglect heat loss by radiation.

Solution

From the convective heat transfer equation,

$$q = hA(t_2 - t_a)$$

where t_2 is the temperature at the outer surface of the shell and t_a is the temperature of the surrounding air.

The above equation gives

$$100 = 20 \times 4\pi \times (0.225)^2(t_2 - 30)$$

The solution gives

$$t_2 = 37.86°\text{C}$$

The heat flow rate by conduction through the wall of the spherical shell is

$$q = \frac{4\pi k r_1 r_2 (t_1 - t_2)}{r_2 - r_1}$$

Substitution of values of various terms gives

$$100 = \frac{4\pi \times k \times 0.125 \times 0.225 \times (200 - 37.86)}{0.225 - 0.125}$$

Solution gives k = 0.1745 W/(m K).

2.9 Effect of Variable Thermal Conductivity

Thermal conductivity of a dry solid substance depends on temperature and when the temperature difference across the wall of such substance is large, proper account of the variation of thermal conductivity with the temperature must be taken. For most of the materials, this dependence is linear if the difference between boundary temperatures of the substance is not large. Mathematically, it can be expressed as

$$k = k_o(1 + \beta t) \tag{2.45}$$

where k_o is the thermal conductivity at 0°C, and β is a constant which is negative for metals and positive for insulating materials. The constant β is usually determined experimentally.

When variation of thermal conductivity with temperature is considered linear, the heat flow and temperature distribution through the plane, cylindrical and spherical walls can be determined as under.

2.9.1 Plane Wall

Fourier's equation for conductive heat transfer is

$$q = -kA\frac{dt}{dx}$$

Introducing value of k from Eq. (2.45),

$$q = -k_o(1+\beta t)A\frac{dt}{dx}$$

or

$$qdx = -k_oA(1+\beta t)dt \tag{i}$$

Integrating between the limits of boundary conditions of the plane wall, refer to Fig. 2.34, we get

$$q\int_0^\delta dx = -Ak_o\int_{t_1}^{t_2}(1+\beta t)\,dt$$

or

$$q\delta = -Ak_o\left(t+\beta\frac{t^2}{2}\right)_{t_1}^{t_2}$$

or

$$q = -\frac{Ak_o}{\delta}\left[(t_2-t_1)+\frac{\beta}{2}(t_2^2-t_1^2)\right] \tag{ii}$$

or

$$q = -\frac{Ak_o}{\delta}(t_2-t_1)\left[1+\frac{\beta}{2}(t_2+t_1)\right]$$

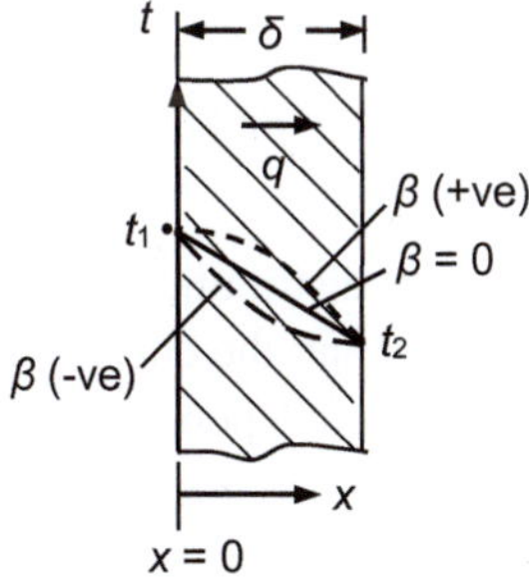

Fig. 2.34 Temperature distribution across a plane wall with temperature-dependent thermal conductivity

Hence, heat flow rate is

$$q = -Ak_o\frac{(t_2 - t_1)}{\delta}\left[1 + \beta\frac{(t_2 + t_1)}{2}\right]$$
$$= -Ak_o(1 + \beta t_m)\frac{(t_2 - t_1)}{\delta}$$
$$= Ak_m\frac{(t_1 - t_2)}{\delta} \tag{2.46}$$

where k_m is the thermal conductivity at mean temperature $t_m = \frac{(t_1 + t_2)}{2}$ of the wall.

The equation establishes that in the case of linear variation of thermal conductivity with temperature, a constant value of the thermal conductivity k_m corresponding to the mean temperature may be used. However, the variation of the thermal conductivity with temperature is not linear in all cases and in such cases the above simplification is not applicable (see Example 2.25).

For the expression of the temperature distribution, integrate Eq. (i) between the limits 0 to x and t_1 to t, which gives

$$q\int_0^x dx = -k_oA\int_{t_1}^{t}(1 + \beta t)dt$$

or

$$qx = -Ak_o\left(t + \beta\frac{t^2}{2}\right)_{t_1}^{t}$$

or

$$q = -\frac{Ak_o}{x}\left[(t - t_1) + \frac{\beta}{2}(t^2 - t_1^2)\right]$$

Substitution of value of q from Eq. (ii) gives

$$-\frac{Ak_o}{\delta}\left[(t_2 - t_1) + \frac{\beta}{2}(t_2^2 - t_1^2)\right] = -\frac{Ak_o}{x}\left[(t - t_1) + \frac{\beta}{2}(t^2 - t_1^2)\right]$$

Simplifying and rearranging the terms, we get

$$t = \frac{1}{\beta}\left\{(1 + \beta t_1)^2 - \left[(1 + \beta t_1)^2 - (1 + \beta t_2)^2\right]\frac{x}{\delta}\right\}^{1/2} - \frac{1}{\beta} \tag{2.47}$$

The equation gives the temperature distribution through the wall. In general, for the positive values of β, the curve is convex and the curve is concave for the negative values of β, see Fig. 2.34. For $\beta = 0$, i.e. when k is independent of the temperature, the distribution is linear.

2.9.2 Cylindrical Shell

Fourier's equation for conductive heat transfer, in this case, is

$$q = -kA\frac{dt}{dr} = -k(2\pi rL)\frac{dt}{dr}$$

Introducing value of k from Eq. (2.45),

$$q = -k_o(1+\beta t)(2\pi rL)\frac{dt}{dr}$$

or

$$q\frac{dr}{r} = -2\pi Lk_o(1+\beta t)dt \tag{i}$$

Integrating between the limits of boundary conditions of the cylindrical wall, we get

$$q\int_{r_1}^{r_2}\frac{dr}{r} = -2\pi Lk_o\int_{t_1}^{t_2}(1+\beta t)dt$$

or

$$q\ln(r_2/r_1) = -2\pi Lk_o\left(t+\beta\frac{t^2}{2}\right)_{t_1}^{t_2}$$

or

$$q\ln(r_2/r_1) = -2\pi Lk_o\left[(t_2-t_1)+\frac{\beta}{2}\left(t_2^2-t_1^2\right)\right] \tag{ii}$$

or

$$q = -2\pi Lk_o\left[1+\frac{\beta}{2}(t_2+t_1)\right]\frac{(t_2-t_1)}{\ln(r_2/r_1)}$$

or

$$q = -2\pi Lk_o[1+\beta t_m]\frac{(t_2-t_1)}{\ln(r_2/r_1)}$$

or

$$q = -2\pi Lk_m\frac{(t_2-t_1)}{\ln(r_2/r_1)} \tag{2.48}$$

where k_m is the thermal conductivity at mean temperature $t_m = \frac{(t_1 + t_2)}{2}$ of the wall, i.e. for the linear variation of thermal conductivity with temperature, a constant value of the thermal conductivity k_m corresponding to the mean temperature may be used.

2.9.3 Spherical Shell

Fourier's equation for conductive heat transfer, in this case, is

$$q = -kA\frac{dt}{dr} = -k(4\pi r^2)\frac{dt}{dr}$$

Introducing value of k from Eq. (2.45),

$$q = -k_o(1 + \beta t)(4\pi r^2)\frac{dt}{dr}$$

or

$$q\frac{dr}{r^2} = -4\pi k_o(1 + \beta t)dt \tag{i}$$

Integrating between the limits of boundary conditions of the spherical wall, we get

$$q\int_{r_1}^{r_2}\frac{dr}{r^2} = -4\pi k_o\int_{t_1}^{t_2}(1 + \beta t)dt$$

or

$$q\left(-\frac{1}{r}\right)_{r1}^{r_2} = -4\pi k_o\left(t + \beta\frac{t^2}{2}\right)_{t_1}^{t_2}$$

or

$$q\left(\frac{1}{r_1} - \frac{1}{r_2}\right) = -4\pi k_o\left[(t_2 - t_1) + \frac{\beta}{2}(t_2^2 - t_1^2)\right] \tag{ii}$$

or

$$q\left(\frac{r_2 - r_1}{r_1 r_2}\right) = -4\pi k_o\left[1 + \frac{\beta}{2}(t_2 + t_1)\right](t_2 - t_1)$$

or

$$q = -4\pi k_o[1 + \beta t_m]\frac{r_1 r_2(t_2 - t_1)}{r_2 - r_1}$$

or

$$q = 4\pi k_m \frac{r_1 r_2 (t_1 - t_2)}{r_2 - r_1}, \tag{2.49}$$

where k_m is the thermal conductivity at mean temperature $t_m = \frac{(t_1 + t_2)}{2}$ of the wall. In this case also, for the linear variation of thermal conductivity with temperature, a constant value of the thermal conductivity k_m corresponding to the mean temperature may be used.

Example 2.21 Derive the equation of temperature distribution through the wall of a cylindrical shell if the thermal conductivity of the wall material is a linear function of temperature.

Solution

Fourier's equation for conductive heat transfer, in this case, is

$$q = -kA\frac{dt}{dr} = -k_o(1 + \beta t)(2\pi rL)\frac{dt}{dr}$$

or

$$q\frac{dr}{r} = -2\pi Lk_o(1 + \beta t)dt$$

Integrating between the limits of boundary conditions of the cylindrical wall, we get

$$q\ln(r_2/r_1) = -2\pi Lk_o\left[(t_2 - t_1) + \frac{\beta}{2}\left(t_2^2 - t_1^2\right)\right] \tag{i}$$

Integrating between the limits of t_1 at r_1 and t at r

$$q\ln(r/r_1) = -2\pi Lk_o\left[(t - t_1) + \frac{\beta}{2}\left(t^2 - t_1^2\right)\right] \tag{ii}$$

From Eqs. (i) and (ii),

$$\frac{\ln(r_2/r_1)}{\ln(r/r_1)} = \frac{(t_2 - t_1) + \frac{\beta}{2}\left(t_2^2 - t_1^2\right)}{(t - t_1) + \frac{\beta}{2}\left(t^2 - t_1^2\right)}$$

or

$$(t - t_1) + \frac{\beta}{2}\left(t^2 - t_1^2\right) - \frac{\ln(r/r_1)}{\ln(r_2/r_1)}\left[(t_2 - t_1) + \frac{\beta}{2}\left(t_2^2 - t_1^2\right)\right] = 0$$

Putting $\left\{\frac{\ln(r/r_1)}{\ln(r_2/r_1)}\left[(t_2 - t_1) + \frac{\beta}{2}\left(t_2^2 - t_1^2\right)\right]\right\} = C$ in the above equation, we get

$$(t - t_1) + \frac{\beta}{2}(t^2 - t_1^2) - C = 0$$

or

$$t^2 + \frac{2}{\beta}t - t_1^2 - \frac{2}{\beta}t_1 - \frac{2}{\beta}C = 0$$

or

$$t^2 + \frac{2}{\beta}t - C_1 = 0 \tag{iii}$$

where $C_1 = t_1^2 + \frac{2}{\beta}t_1 + \frac{2}{\beta}C$.

Solution of Eq. (iii) gives

$$t = \frac{1}{2}\left\{-\frac{2}{\beta} \pm \left[\left(\frac{2}{\beta}\right)^2 + 4C_1\right]^{1/2}\right\}$$

or

$$t = -\frac{1}{\beta} \pm \left[\left(\frac{1}{\beta}\right)^2 + C_1\right]^{1/2}$$

Substituting value of C_1, we have

$$t = -\frac{1}{\beta} \pm \left[\left(\frac{1}{\beta}\right)^2 + t_1^2 + \frac{2}{\beta}t_1 + \frac{2}{\beta}C\right]^{1/2}$$

The equation of C may be rearranged as follows:

$$C = \left\{\frac{\ln(r/r_1)}{\ln(r_2/r_1)}\left[(t_2 - t_1) + \frac{\beta}{2}(t_2^2 - t_1^2)\right]\right\}$$

or

$$C = -\frac{\ln(r/r_1)}{\ln(r_2/r_1)}\left[\frac{(1 + \beta t_1)^2 - (1 + \beta t_2)^2}{2\beta}\right]$$

Substituting value of C, we get

$$t = -\frac{1}{\beta} \pm \left[\left(\frac{1}{\beta}\right)^2 + t_1^2 + \frac{2}{\beta}t_1 - \frac{2}{\beta}\frac{\ln(r/r_1)}{\ln(r_2/r_1)}\left[\frac{(1 + \beta t_1)^2 - (1 + \beta t_2)^2}{2\beta}\right]\right]^{1/2}$$

or

$$t = -\frac{1}{\beta} \pm \frac{1}{\beta}\left[1 + \beta^2 t_1^2 + 2\beta t_1 - \frac{\ln(r/r_1)}{\ln(r_2/r_1)}\left[(1+\beta t_1)^2 - (1+\beta t_2)^2\right]\right]^{1/2}$$

or

$$t = -\frac{1}{\beta} \pm \frac{1}{\beta}\left[(1+\beta t_1)^2 - \frac{\ln(r/r_1)}{\ln(r_2/r_1)}\left[(1+\beta t_1)^2 - (1+\beta t_2)^2\right]\right]^{1/2}$$

At $r = r_2$, $t = t_2$. The condition is satisfied when positive sign of square root is used. This gives the required temperature distribution equation as

$$t = -\frac{1}{\beta} + \frac{1}{\beta}\left[(1+\beta t_1)^2 - \frac{\ln(r/r_1)}{\ln(r_2/r_1)}\left[(1+\beta t_1)^2 - (1+\beta t_2)^2\right]\right]^{1/2}$$

Example 2.22 Following the procedure of the preceding example, derive the equation of temperature distribution through the wall of a spherical shell if the thermal conductivity of the wall material is a linear function of temperature.

Solution

Fourier's equation for conductive heat transfer, in this case, is

$$q = -kA\frac{dt}{dr} = -k(4\pi r^2)\frac{dt}{dr}$$

Introducing value of k from Eq. (2.45),

$$q = -k_o(1+\beta t)(4\pi r^2)\frac{dt}{dr}$$

or

$$q\frac{dr}{r^2} = -4\pi k_o(1+\beta t)dt$$

Integrating between the limits of boundary of t_1 at r_1 and t_2 at r_2, we get

$$q\int_{r_1}^{r_2}\frac{dr}{r^2} = -4\pi k_o\int_{t_1}^{t_2}(1+\beta t)dt$$

or

$$q\left(-\frac{1}{r}\right)_{r1}^{r_2} = -4\pi k_o\left(t + \beta\frac{t^2}{2}\right)_{t_1}^{t_2}$$

or

$$q\left(\frac{1}{r_1}-\frac{1}{r_2}\right)=-4\pi k_o\left[(t_2-t_1)+\frac{\beta}{2}\left(t_2^2-t_1^2\right)\right]$$

or

$$q\left(\frac{r_2-r_1}{r_1r_2}\right)=-4\pi k_o\left[1+\frac{\beta}{2}(t_2+t_1)\right](t_2-t_1) \qquad \text{(i)}$$

Similarly integrating between the limits of boundary of t_1 at r_1 and t at r, we get

$$q\left(\frac{r-r_1}{r_1r}\right)=-4\pi k_o\left[1+\frac{\beta}{2}(t+t_1)\right](t-t_1) \qquad \text{(ii)}$$

From Eqs. (i) and (ii),

$$\left(\frac{r-r_1}{r_1r}\right)\left(\frac{r_1r_2}{r_2-r_1}\right)=\frac{-4\pi k_o\left[1+\frac{\beta}{2}(t+t_1)\right](t-t_1)}{-4\pi k_o\left[1+\frac{\beta}{2}(t_2+t_1)\right](t_2-t_1)}$$

$$\left(\frac{r-r_1}{r_2-r_1}\right)\left(\frac{r_2}{r}\right)=\frac{\left[(t-t_1)+\frac{\beta}{2}\left(t^2-t_1^2\right)\right]}{\left[(t_2-t_1)+\frac{\beta}{2}\left(t_2^2-t_1^2\right)\right]}$$

or

$$\left[(t-t_1)+\frac{\beta}{2}\left(t^2-t_1^2\right)\right]-\left(\frac{r_2}{r}\right)\left(\frac{r-r_1}{r_2-r_1}\right)\left[(t_2-t_1)+\frac{\beta}{2}\left(t_2^2-t_1^2\right)\right]=0$$

or

$$t^2+\frac{2}{\beta}t-t_1^2-\frac{2}{\beta}t_1-\frac{2}{\beta}\left(\frac{r_2}{r}\right)\left(\frac{r-r_1}{r_2-r_1}\right)\left[(t_2-t_1)+\frac{\beta}{2}\left(t_2^2-t_1^2\right)\right]=0$$

or

$$t^2+\frac{2}{\beta}t-t_1^2-\frac{2}{\beta}t_1+\left(\frac{r_2}{r}\right)\left(\frac{r-r_1}{r_2-r_1}\right)\frac{(1+\beta t_1)^2-(1+\beta t_2)^2}{\beta^2}=0$$

or

$$t^2+\frac{2}{\beta}t+C=0 \qquad \text{(iii)}$$

where $-t_1^2-\frac{2}{\beta}t_1+\left(\frac{r_2}{r}\right)\left(\frac{r-r_1}{r_2-r_1}\right)\frac{(1+\beta t_1)^2-(1+\beta t_2)^2}{\beta^2}=C$ in the above equation.

Solution of Eq. (iii) gives

$$t = \frac{1}{2}\left\{-\frac{2}{\beta} \pm \left[\left(\frac{2}{\beta}\right)^2 - 4C\right]^{1/2}\right\}$$

or

$$t = -\frac{1}{\beta} \pm \left[\left(\frac{1}{\beta}\right)^2 - C\right]^{1/2}$$

Substituting value of C, we have

$$t = -\frac{1}{\beta} \pm \left[\left(\frac{1}{\beta}\right)^2 + t_1^2 + \frac{2}{\beta}t_1 - \left(\frac{r_2}{r}\right)\left(\frac{r - r_1}{r_2 - r_1}\right)\frac{(1+\beta t_1)^2 - (1+\beta t_2)^2}{\beta^2}\right]^{1/2}$$

or

$$t = -\frac{1}{\beta} \pm \frac{1}{\beta}\left\{1 + \beta^2 t_1^2 + 2\beta t_1 - \left(\frac{r_2}{r}\right)\left(\frac{r - r_1}{r_2 - r_1}\right)\left[(1+\beta t_1)^2 - (1+\beta t_2)^2\right]\right\}^{1/2}$$

or

$$t = -\frac{1}{\beta} \pm \frac{1}{\beta}\left[(1+\beta t_1)^2 - \left(\frac{r_2}{r}\right)\left(\frac{r - r_1}{r_2 - r_1}\right)\left[(1+\beta t_1)^2 - (1+\beta t_2)^2\right]\right]^{1/2}$$

At $r = r_2$, $t = t_2$. The condition is satisfied when positive sign of square root is used. This gives the required temperature distribution equation as

$$t = -\frac{1}{\beta} + \frac{1}{\beta}\left[(1+\beta t_1)^2 - \left(\frac{r_2}{r}\right)\left(\frac{r - r_1}{r_2 - r_1}\right)\left[(1+\beta t_1)^2 - (1+\beta t_2)^2\right]\right]^{1/2}$$

Example 2.23 The thermal conductivity of the material of a plane wall varies linearly with temperature as $k = k_o(1 + \beta t)$. Verify that for $\beta > 0$, the temperature profile is concave downwards as shown in Fig. 2.34.

Solution

Fourier's equation for a plane wall is

$$q = -k_o(1 + \beta t)A\frac{dt}{dx}$$

or

$$\frac{q}{k_o A} = -(1+\beta t)\frac{dt}{dx}$$

or

$$-\frac{q}{k_o A} = \frac{dt}{dx} + \beta t\frac{dt}{dx}$$

Differentiating with respect to x,

$$0 = \frac{d^2t}{dx^2} + \beta\left[\left(\frac{dt}{dx}\right)^2 + t\frac{d^2t}{dx^2}\right]$$

or

$$\frac{d^2t}{dx^2} = -\frac{\beta}{1+\beta t}\left(\frac{dt}{dx}\right)^2 < 0$$

The second derivative is negative; hence, the temperature profile is concave downwards.

Example 2.24 The surface temperatures of a 200-mm-thick fireclay wall are 1000°C and 200°C. The thermal conductivity of the fireclay is a linear function of the temperature given by the following equation:

$$k = 0.8(1+0.0007t)\ \text{W}/(\text{m}^\circ\text{C})$$

where t is in °C.

Calculate the rate of heat flow, in W/m^2, through the wall and determine the temperature distribution in the wall.

Solution

(i) For linear variation of the thermal conductivity, the mean value of the thermal conductivity may be used, which is

$$k_m = 0.8(1+0.0007t_m) = 0.8\left[1+0.0007\left(\frac{t_1+t_2}{2}\right)\right] = 1.136\ \text{W}/(\text{m}^\circ\text{C})$$

Heat transfer

$$q = k_m A\frac{t_1-t_2}{\delta} = 1.136 \times 1 \times \frac{1000-200}{0.2} = 4544\ \text{W}$$

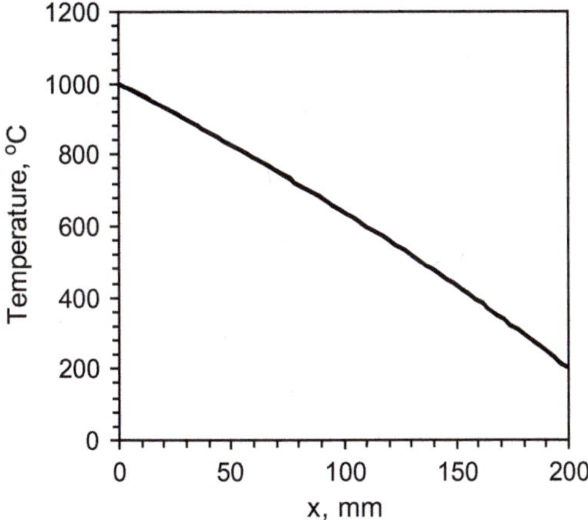

Fig. 2.35 Temperature distribution across the wall of Example 2.24

(ii) **Temperature distribution**

$$t = \frac{1}{\beta}\left\{(1+\beta t_1)^2 - \left[(1+\beta t_1)^2 - (1+\beta t_2)^2\right]\frac{x}{\delta}\right\}^{1/2} - \frac{1}{\beta}$$

Here β = 0.0007. Hence,

$$(1+\beta t_1)^2 = (1+0.0007 \times 1000)^2 = 2.89$$
$$(1+\beta t_2)^2 = (1+0.0007 \times 200)^2 = 1.2996$$

Hence,

$$t = 1428.6 \times (2.89 - 7.952x)^{1/2} - 1428.6$$

Figure 2.35 shows the temperature distribution.

Example 2.25 The thermal conductivity of a plane wall varies as

$$k = k_o(1 + Bt + Ct^2)$$

where k_o is the thermal conductivity at temperature t = 0 and B and C are constants.

If the wall thickness is δ and the surface temperatures are t_1 and t_2, show that the steady heat flux q'' through the wall is given by

$$q'' = -k_o\frac{(t_2 - t_1)}{\delta}\left[1 + \frac{B}{2}(t_2 + t_1) + \frac{C}{3}(t_2^2 + t_1 t_2 + t_1^2)\right]$$

Solution

From Fourier's equation, the heat flow is given by

$$q\int_{0}^{\delta} dx = -A\int_{t_1}^{t_2} k dt$$

or

$$q\delta = -A\int_{t_1}^{t_2} k_o(1 + Bt + Ct^2)\, dt$$

or

$$= -Ak_o\left(t + B\frac{t^2}{2} + C\frac{t^3}{3}\right)_{t_1}^{t_2}$$

or

$$= -Ak_o\left[(t_2 - t_1) + \frac{B}{2}(t_2^2 - t_1^2) + \frac{C}{3}(t_2^3 - t_1^3)\right]$$

or

$$q\delta = -Ak_o(t_2 - t_1)\left[1 + \frac{B}{2}(t_2 + t_1) + \frac{C}{3}(t_2^2 + t_1t_2 + t_1^2)\right]$$

Hence, heat flux is

$$q'' = \frac{q}{A} = -k_o\frac{(t_2 - t_1)}{\delta}\left[1 + \frac{B}{2}(t_2 + t_1) + \frac{C}{3}(t_2^2 + t_1t_2 + t_1^2)\right]$$

Example 2.26 For $k_o = 50$, $B = -0.0005$ and $C = -1 \times 10^{-6}$, show that for small temperatures $t_1 = 200°C$ and $t_2 = 100°C$, the mean value of thermal conductivity can be used.

Solution

(i) The effective thermal conductivity, from the previous example, is

$$k = k_o\left[1 + \frac{B}{2}(t_2 + t_1) + \frac{C}{3}(t_2^2 + t_1t_2 + t_1^2)\right]$$

Substitution gives $k = 45.08$ W/(m K).

(ii) From equation, $k = k_o(1 + Bt + Ct^2) = 47.0$ W/(m K) at $t_1 = 100$°C

$$= 43.0\,\mathrm{W/(m\,K)} \text{ at } t_1 = 200^\circ\mathrm{C}$$

Hence, mean value $k_m = (k_1 + k_2)/2 = 45.0$ W/(m K).

(iii) The mean temperature, $t_m = (t_1 + t_2)/2 = 150$°C

$$\text{At } 150^\circ\mathrm{C}, k = 50 \times (1 - 0.0005 \times 150 - 1 \times 10^6 \times 150^2) = 45.12$$

The analysis shows that if the effect of the temperature is not significant and temperature difference is not large, value of k corresponding to the mean temperature can be used.

Example 2.27 Show that in case of a spherical shell, the heat conducted with variable thermal conductivity $k = k_o(1 + \beta T)$ is higher by $50\beta(T_1 + T_2)$ per cent over the shell with constant value of thermal conductivity k_o, and T_1 and T_2 are wall temperatures.

Solution

With the constant value of the thermal conductivity,

$$q = \frac{4\pi k_o r_1 r_2 (T_1 - T_2)}{r_2 - r_1}$$

The heat conducted with the variable thermal conductivity is

$$q' = \frac{4\pi k_o \left(1 + \beta \frac{T_1 + T_2}{2}\right) r_1 r_2 (T_1 - T_2)}{r_2 - r_1}$$

Hence, the percentage increase in the heat transfer is

$$\Delta q = \frac{q' - q}{q} \times 100$$

Substitution of the values of q and q' gives

$$\Delta q = 50\beta(T_1 + T_2) \text{ percent}$$

It can be shown, using appropriate equations, that the above result is also valid for a plane wall and tube.

Since the value of β is positive for insulating materials and negative for metals, the effect of the rise in temperature is to increase the heat loss through the insulation and decrease the conduction heat transmission through the metal walls.

Example 2.28 A spherical shell is made of a material with thermal conductivity $k = k_o T^2$. Derive the expression for the conduction heat transfer rate, if the inner and outer walls are held at temperatures T_1 and T_2, respectively.

Solution

Substituting $k = k_o T^2$ and $A = 4\pi r^2$ in Fourier's equation, we get

$$q = -k_o T^2 \left(4\pi r^2\right) \frac{dT}{dr}$$

Separating the variables and integrating,

$$q \int_{R_i}^{R_o} \frac{dr}{r^2} = -4\pi k_o \int_{T_1}^{T_2} T^2 dT$$

or

$$\frac{q}{4\pi k_o}\left(\frac{1}{R_i} - \frac{1}{R_o}\right) = -\frac{1}{3}\left(T_2^3 - T_1^3\right)$$

or

$$q = \frac{4}{3}\pi k_o \left(\frac{R_o R_i}{R_o - R_i}\right)\left(T_1^3 - T_2^3\right)$$

Example 2.29 The wall of a steam boiler furnace is made of layer of firebricks of thickness 125 mm (k_b = 0.28 + 0.002T) and insulation bricks layer of 250 mm thickness (k_i = 0.7 W/(m K)). The inside surface temperature of the firebrick wall is 1000°C, and outside insulation brick temperature is 60°C.

(i) Calculate the amount of heat lost per m^2 of the furnace wall and the temperature at the interface of the two layers.
(ii) If the layer of the insulation brick wall is reduced to half of its original thickness and an insulating material with k_m = 0.113(1 + 0.002T) is to be filled between the fire brick and insulation brick layers, what must be the thickness of this insulating material so that the heat loss remains the same. Assume that the inside and outside wall temperatures remain as before.

Temperature T in thermal conductivity equations is the mean temperature in K.

Solution

(i) Given data are k_b = 0.28 + 0.002T, k_i = 0.7 W/(m K), $A_b = A_i = A$ = 1 m^2.
Let temperature at the interface of fire and insulation bricks is T_2.
The heat transfer, refer to Fig. 2.36a,

$$q = k_b A_b \frac{1273 - T_2}{\delta_b} = k_i A_i \frac{T_2 - 333}{\delta_i}$$

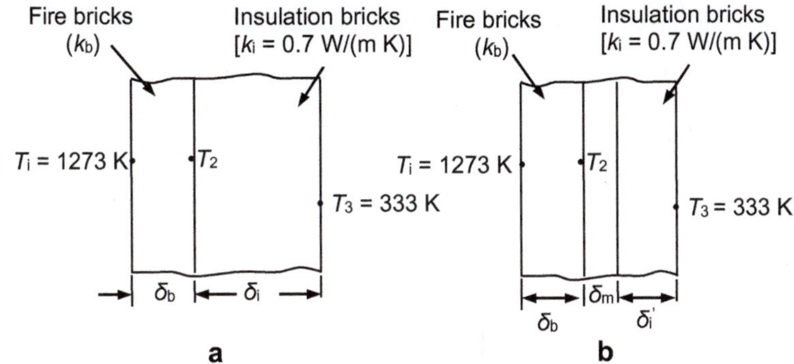

Fig. 2.36 Example 2.29

or

$$\left[0.28 + 0.002 \times \left(\frac{1273 + T_2}{2}\right)\right] \times 1 \times \frac{1273 - T_2}{125} = 0.7 \times 1 \times \frac{T_2 - 333}{250}$$

or

$$T_2 = 1165.6\,\text{K}$$

And the heat loss is

$$q = k_i A \frac{T_2 - 333}{\delta_i} = 0.7 \times 1 \times \frac{1165.6 - 333}{0.25} = 2331.3\,\text{W}$$

(ii) With three layers, the heat flow equation for the reduced thickness δ_i' of the insulation bricks and the same heat loss rate,

$$q = k_i A \frac{T_3 - 333}{\delta_i'}$$

or

$$2331.3 = 0.7 \times 1 \times \frac{T_3 - 333}{0.125}$$

or

$$T_3 = 749.3\,\text{K}$$

Heat flow equation for the firebrick layer is

$$q = k_b A_b \frac{1273 - T_2}{\delta_b} = 2331.3\,\text{W}$$

As there is no change as far as first layer is concerned, hence, T_2 is the same as in Part (i), i.e. T_2 = 1165.6 K.

For the second layer of the insulating material filled between layers of fire bricks and insulation bricks,

$$2331.3 = k_m A_m \frac{T_2 - T_3}{\delta_m}$$

where

$$\begin{aligned} A_m &= A_b = A_i = 1\,\text{m}^2, \\ k_m &= 0.113 \times [1 + 0.002 \times (T_2 + T_3)/2] \\ &= 0.113 \times [1 + 0.002 \times (1165.6 + 749.3)/2] = 0.3294\,\text{W/(m K)} \end{aligned}$$

Substitution gives

$$2331.3 = 0.3294 \times 1 \times \frac{1165.6 - 749.3}{\delta_m}$$

Solution gives δ_m = 0.0588 m = 59 mm.

Example 2.30 The wall of a furnace is built-up of a 250 mm thick layer of fireclay bricks whose thermal conductivity is given by

$$k = 0.83 \times (1 + 0.0007 t_m)\,\text{W/(m°C)}$$

Calculate the rate of heat loss per m^2 of the wall if the temperature of the gas in the furnace is 1200°C and the ambient temperature is 20°C. The film coefficients from the inner and outer wall surfaces are 30 W/(m^2 K) and 20 W/(m^2 K), respectively.

Solution

The total heat resistance to the heat transfer per m^2 of the wall is

$$\begin{aligned} R &= \frac{1}{h_i} + \frac{\delta}{k} + \frac{1}{h_O} \\ &= \frac{1}{30} + \frac{0.25}{k} + \frac{1}{20} \end{aligned}$$

As the thermal conductivity of the wall material is function of its temperature, the problem will be solved by following an iterative method.

We assume mean wall temperature t_m as 600°C for the first trial. This gives

$$k = 0.83(1 + 0.0007 \times 600) = 1.1786\,\text{W/(m°C)}$$

and

$$R = 0.296$$

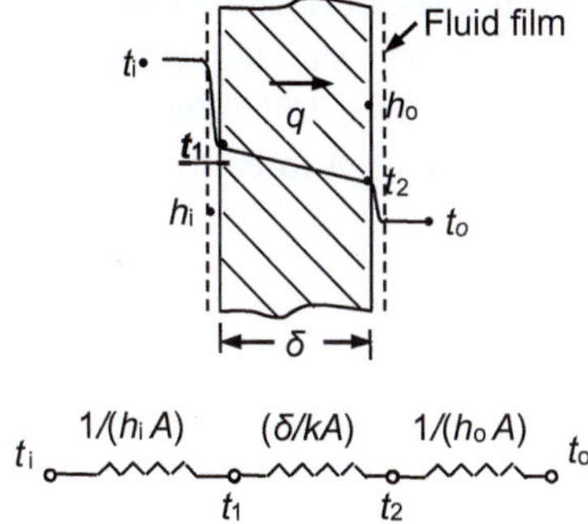

Fig. 2.37 System of Example 2.30 and thermal network

Knowing the total wall resistance, the heat transfer rate is

$$\frac{q}{A} = \frac{t_i - t_o}{R} = \frac{1200 - 20}{0.296} = 3986.5\,\text{W/m}^2$$

Let the wall surface temperatures are t_1 and t_2, refer to Fig. 2.37, then

$$\frac{q}{A} = \frac{t_i - t_1}{1/h_i} = \frac{t_1 - t_2}{\delta/k} = \frac{t_2 - t_o}{1/h_o} = 3986.5\,\text{W/m}^2$$

This gives

$$t_1 = t_i - \frac{q}{A} \times \frac{1}{h_i} = 1200 - 3986.5 \times \frac{1}{30} = 1067.1°\text{C}$$

$$t_2 = t_0 + \frac{q}{A} \times \frac{1}{h_o} = 20 + 3986.5 \times \frac{1}{20} = 219.3°\text{C}$$

This gives first estimate of the mean temperature as

$$t_m = \frac{t_1 + t_2}{2} = 643.2°\text{C}$$

The second trial with the estimated values of t_m gives

$$k = 0.83(1 + 0.0007 \times 643.2) = 1.2037\,\text{W/(m °C)}$$

For k = 1.2037 W/(m °C),

$$R = 0.291$$

and

$$\frac{q}{A} = 4055\,\text{W/m}^2$$

Third trial is not needed as the value of q/A from the second trial is not in much difference from that obtained from the first trial.

Example 2.31 A 50-mm-thick magnesia insulation has its faces at 25°C and 45°C, and the heat loss is found to be 25 W/m^2. When the face temperatures are changed to 240°C and 300°C, respectively, the heat loss becomes 100 W/m^2. Find the thermal conductivity–temperature relation of the material assuming it to be linear.

Solution

The heat loss for a plane wall is given by

$$q = kA\frac{t_1 - t_2}{\delta} \tag{i}$$

Let

k	$k_o(1 + at)$, i.e. a linear relation as specified
t	mean temperature = $(t_1 + t_2)/2$ = (25 + 45)/2 = 35°C in the first case= (240 + 300)/2 = 270°C in the second case
a	a constant
k_o	thermal conductivity at 0°C
δ	thickness of the wall = 0.05 m
A	heat flow area = 1 m^2

Case (i) q = 25 W/m^2 and from Eq. (i),

$$25 = k_o(1 + 35a) \times 1 \times \frac{45 - 25}{0.05}$$

or

$$k_o(1 + 35a) = 0.0625 \tag{ii}$$

Case (ii) q = 100 W/m^2 and from Eq. (i),

$$100 = k_o(1 + 270a) \times 1 \times \frac{300 - 240}{0.05}$$

or

$$k_o(1 + 270a) = 0.0833 \tag{iii}$$

Solution of Eqs. (ii) and (iii) gives k_o = 0.0594 and $a = 1.491 \times 10^{-3}$. Hence, the thermal conductivity–temperature relation is

$$k = 0.0594(1 + 1.491 \times 10^{-3}t)$$

where t is the mean temperature of the material in °C.

2.10 Critical Thickness of Insulation

2.10.1 Critical Thickness of Insulation for Cylinders

The thermal resistance to heat transfer for a hollow cylinder with a layer of insulating material applied to the cylindrical surface is given by, refer to Fig. 2.38a,

$$\begin{aligned} R_t &= \frac{1}{2\pi r_i l h_i} + \frac{\ln\left(\frac{r_1}{r_i}\right)}{2\pi k_m l} + \frac{\ln\left(\frac{r_2}{r_1}\right)}{2\pi k l} + \frac{1}{2\pi r_2 l h_o} \\ &= R_i + R_w + R_k + R_o \end{aligned}$$

The first term R_i in the equation of R_t is the resistance due to the inner film, second term R_w is the metal wall resistance, the third term R_k represents the resistance of the insulation material and the last term R_o is the outer film resistance. The first two terms are constant, while R_k and R_o vary with the change in the thickness of the insulation. Neglecting the first two terms, we have

$$R_t' = \frac{\ln\left(\frac{r_2}{r_1}\right)}{2\pi k l} + \frac{1}{2\pi r_2 l h_o} \tag{i}$$

As we increase the thickness of the insulation, i.e. the radius r_2, keeping inside radius of insulation r_1 constant, the resistance R_k to conduction heat flow through the layer of insulation increases while the resistance R_o to the heat transfer by convection from the surface of insulation decreases, see Fig. 2.38. The sum of the resistances, $R_t' = R_k + R_c$, first decreases to a certain minimum at $r = r_c$ and then again increases. It means that as the radius r_2 increases, the heat flow q first increases to a certain maximum value at $r_2 = r_c$ and then decreases. The value of r_2 at the maximum of heat flow q (or the minimum of the resistance R) is called the *critical radius of insulation.*

The critical radius of the insulation can be determined as under.

The heat transfer rate through the wall and insulation of thickness $(r_2 - r_1)$ with film coefficients h_i and h_o at the inner and outer surfaces, respectively, is given by

$$q = \frac{(t_i - t_o)}{R_t}$$

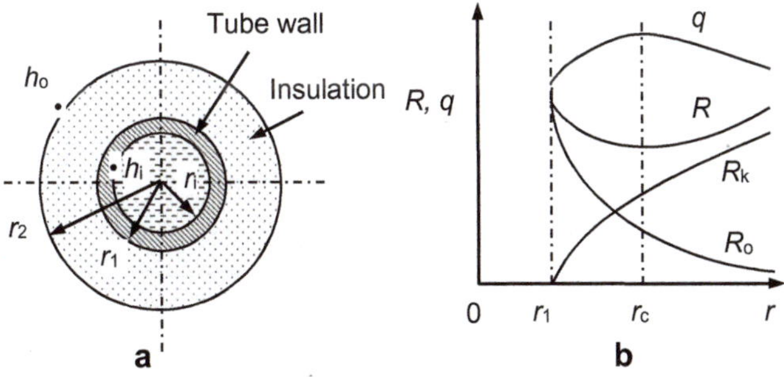

Fig. 2.38 Critical radius of insulation

where

$$R_t = \frac{1}{2\pi r_i l h_i} + \frac{\ln\left(\frac{r_1}{r_i}\right)}{2\pi k_m l} + \frac{\ln\left(\frac{r_2}{r_1}\right)}{2\pi k l} + \frac{1}{2\pi r_2 l h_o}$$

The condition for the maximum or minimum heat transfer is obtained by differentiating the equation of the total resistance R_t with respect to r_2 and equating to zero, i.e.

$$\frac{\partial R_t}{\partial r_2} = 0 = \frac{\partial}{\partial r_2}\left[\frac{1}{2\pi r_i l h_i} + \frac{\ln\left(\frac{r_1}{r_i}\right)}{2\pi k_m l} + \frac{\ln\left(\frac{r_2}{r_1}\right)}{2\pi k l} + \frac{1}{2\pi r_2 l h_o}\right]$$

or

$$\frac{\partial R_t}{\partial r_2} = \frac{1}{2\pi l}\left[\frac{1}{k}\left(\frac{r_1}{r_2}\right)\frac{1}{r_1} - \frac{1}{h_o r_2^2}\right] = 0$$

or

$$r_2 = r_c = \frac{k}{h_o} \tag{2.50}$$

This is the equation of the critical radius of insulation for a cylinder, which is independent of the tube radius r_1. It can be shown that the value of the second derivative of the resistance R_t at $r_2 = r_c$ refers to the minimum value of the resistance, i.e.

$$\frac{\partial^2 R_t}{\partial r_2^2} = \frac{1}{2\pi l}\frac{\partial}{\partial r_2}\left(\frac{1}{k} \times \frac{1}{r_2} - \frac{1}{r_2^2} \times \frac{1}{h_o}\right)$$

or

$$= \frac{1}{2\pi l}\left(\frac{1}{k} \times \frac{-1}{r_2^2} + \frac{2}{r_2^3} \times \frac{1}{h_o}\right)$$

Putting $r_2 = r_c = k/h_o$,

$$\frac{\partial^2 R_t}{\partial r_2^2} = \frac{1}{2\pi l}\left(-\frac{h_o^2}{k^3} + \frac{2h_o^2}{k^3}\right)$$

or

$$\frac{\partial^2 R_t}{\partial r_2^2} = \frac{h_o^2}{2\pi l k^3}$$

which is a positive quantity.

Thus, the condition $r_2 = r_c$ refers to the minimum resistance to the heat transfer or to the maximum heat transfer.

We can conclude that, for the heat insulation of a pipe of radius r_1, the insulating material must be selected such that the critical radius r_c is less than r_1 under the given conditions while in the case of the electric current carrying cables, the insulation radius r_2 must be equal to r_c to dissipate maximum heat. Hence, the insulating material for the cables must be selected such that the critical radius r_c is greater than the bare cable radius r_1.

2.10.2 Critical Thickness of Insulation for Spherical Vessel

The critical radius of the insulation for this can be determined as under.

The heat transfer rate through the wall of the sphere and insulation of thickness $(r_2 - r_1)$ with film coefficients h_i and h_o at the inner and outer surfaces, respectively, is given by

$$q = \frac{(t_i - t_o)}{R_t}.$$

where

$$R_t = \frac{1}{4\pi r_i^2 h_i} + \frac{1}{4\pi k_m}\left(\frac{1}{r_i} - \frac{1}{r_1}\right) + \frac{1}{4\pi k}\left(\frac{1}{r_1} - \frac{1}{r_2}\right) + \frac{1}{4\pi r_2^2 h_o}$$

Again the condition for the maximum or minimum heat transfer is obtained by differentiating the equation of the total resistance R_t with respect to r_2 and equating to zero, i.e.

$$\frac{\partial R_t}{\partial r_2} = 0 = \frac{1}{4\pi}\frac{\partial}{\partial r_2}\left[\frac{1}{r_i^2 h_i} + \frac{1}{k_m}\left(\frac{1}{r_i} - \frac{1}{r_1}\right) + \frac{1}{k}\left(\frac{1}{r_1} - \frac{1}{r_2}\right) + \frac{1}{r_2^2 h_o}\right]$$

or

$$\frac{1}{4\pi}\left[\frac{1}{k}\left(-\frac{-1}{r_2^2}\right) + \frac{-2}{r_2^3 h_o}\right] = 0$$

or

$$r_2 = r_c = \frac{2k}{h_o} \tag{2.51}$$

This is the equation of the critical radius of insulation for a sphere, which is independent of the sphere radius r_1. Again it can be shown that the value of the second derivative of the resistance R_t at $r_2 = r_c$ refers to the minimum value of the resistance.

$$\begin{aligned}&\frac{1}{4\pi}\frac{\partial}{\partial r_2}\left[\frac{1}{k}\left(-\frac{-1}{r_2^2}\right) + \frac{-2}{r_2^3 h_o}\right]\\ &= \frac{1}{4\pi}\left\{\left[-\frac{-1}{k} \times \frac{(-2)}{r_2^3}\right] + \frac{-2 \times (-3)}{r_2^4 h_o}\right\}\end{aligned}$$

Putting $r_2 = r_c = 2\ k/h_o$,

$$\frac{\partial^2 R_t}{\partial r_2^2} = \frac{1}{4\pi}\left(\frac{h_o^3}{8k^4}\right)$$

which is a positive quantity.

Thus, the condition $r_2 = r_c$ refers to the minimum resistance to the heat transfer or to the maximum heat transfer. The discussion presented for cylindrical cell regarding the selection of the insulation material also applies here.

Example 2.32 An electric current carrying cable of radius r_1 is covered with a layer of insulation of thickness $r_2 - r_1$. If the resistance heating rate is q per m length of the cable, show that in the steady-state condition, the cable surface temperature is given by

$$\frac{q}{2\pi}\left[\frac{1}{k}\ln\frac{r_2}{r_1} + \frac{1}{r_2 h}\right] + t_\infty$$

where k is the thermal conductivity of the insulation material, h is the outside film coefficient and t_∞ is the surrounding temperature.

Hence, show that $r_2 = k/h$ gives the minimum cable surface temperature.

Solution

The rate of heat transfer per unit length of the insulated cable at surface temperature t_s is given by

$$q = \frac{t_s - t_\infty}{\left[\frac{1}{2\pi k}\ln\frac{r_2}{r_1} + \frac{1}{2\pi r_2 h}\right]}$$

or

$$t_s = \frac{q}{2\pi}\left[\frac{1}{k}\ln\frac{r_2}{r_1} + \frac{1}{r_2 h}\right] + t_\infty$$

For the cable surface temperature to be minimum or maximum, the condition is

$$\frac{\partial t_s}{\partial r_2} = 0$$

or

$$\frac{\partial}{\partial r_2}\left[\frac{q}{2\pi}\left(\frac{1}{k}\ln\frac{r_2}{r_1} + \frac{1}{r_2 h}\right) + t_\infty\right] = 0$$

or

$$\left[\frac{q}{2\pi}\left(\frac{1}{k}\times\frac{r_1}{r_2}\times\frac{1}{r_1}+\frac{-1}{r_2^2 h}\right)\right]=0$$

or

$$\left[\frac{q}{2\pi}\left(\frac{1}{k}\times\frac{1}{r_2}-\frac{1}{r_2^2 h}\right)\right]=0$$

or

$$r_2=\frac{k}{h}$$

The second derivative is

$$\frac{\partial}{\partial r_2}\left[\frac{q}{2\pi}\left(\frac{1}{k}\times\frac{1}{r_2}-\frac{1}{r_2^2 h}\right)\right] = \frac{q}{2\pi}\left(\frac{1}{k}\times\frac{-1}{r_2^2}-\frac{-2}{r_2^3 h}\right)$$

Substituting $r_2 = k/h$, the second derivate $\frac{\partial^2 t_s}{\partial r_2^2}=\frac{q}{2\pi}\left(\frac{h^2}{k^3}\right)$ which is a positive quantity. Hence, the condition $r_2 = k/h$ refers to the minimum temperature of the cable surface.

Example 2.33 For the cable of the previous example, show that the condition for keeping the insulated cable at the same temperature as if it were not insulated is

$$\ln\frac{r_2}{r_1}=\frac{k}{h}\left(\frac{1}{r_1}-\frac{1}{r_2}\right)$$

Solution

The cable surface temperature with insulation of thickness $(r_2 - r_1)$ from the previous example is

$$t_s=\frac{q}{2\pi}\left[\frac{1}{k}\ln\frac{r_2}{r_1}+\frac{1}{r_2 h}\right]+t_\infty$$

When insulation is not present, the cable surface temperature t_s' can be obtained from the above equation by excluding the insulation resistance and putting $r_2 = r_1$. This gives

$$t_s'=\frac{q}{2\pi}\left[\frac{1}{r_1 h}\right]+t_\infty$$

Equating the two temperature equations for the given condition, we have

$$\frac{q}{2\pi}\left[\frac{1}{k}\ln\frac{r_2}{r_1}+\frac{1}{r_2 h}\right]+t_\infty=\frac{q}{2\pi}\left[\frac{1}{r_1 h}\right]+t_\infty$$

Simplification gives

$$\ln\frac{r_2}{r_1}=\frac{k}{h}\left(\frac{1}{r_1}-\frac{1}{r_2}\right)$$

Example 2.34 An electric current cable is 5 mm in radius. It is to be covered by a uniform sheathing of plastic (k = 0.2 W/(m K)). The convective film coefficient between the surface of the plastic sheathing and surrounding air is 15 W/(m^2 K). Calculate

(i) thickness of insulation which will not alter the temperature of the wire surface, and
(ii) the insulation thickness for minimum wire temperature.

Comment on the result.

Solution

(i) The condition for keeping the insulated cable at the same temperature as if it were not insulated has been obtained in previous example, which is

$$\ln\frac{r_2}{r_1}=\frac{k}{h}\left(\frac{1}{r_1}-\frac{1}{r_2}\right)$$

From the given data, r_1 = 5 mm, k = 0.2 W/(m K) and h = 15 W/(m^2 K). Substitution gives, by trial and error, r_2 = 0.057 m = 57 mm.

(ii) For the minimum wire temperature, the condition is

$$r_2=r_c=\frac{k}{h_o}=\frac{0.2}{15}\times 1000=13.33\,\text{mm}$$

Since r_2 calculated in Part (i) of the example is greater than r_c, any insulation thickness which is less than r_2 calculated in Part (i) will cause decrease in the wire surface temperature for the same heat generation rate in the wire due to Joule heating.
The wire surface temperature is minimum when r_2 = r_c. The wire surface temperature for r_2 = r_1 = 5 mm (i.e. the bare wire) and r_2 = 57 mm is the same.

Example 2.35 Show that heat dissipation from an electric cable provided with critical insulation thickness is given by

$$q=\frac{2\pi kL(t_s-t_\infty)}{1+\ln\frac{k}{h_o r_1}}$$

where the notations have the usual meaning.

Solution

The heat dissipation from an electric cable of radius r_1 with insulation thickness $(r_o - r_1)$ is given by

$$q = \frac{t_s - t_\infty}{\frac{1}{2\pi kL}\ln\frac{r_o}{r_1} + \frac{1}{2\pi r_o L h_o}} = \frac{2\pi kL(t_s - t_\infty)}{\ln\frac{r_o}{r_1} + \frac{k}{r_o h_o}}$$

For the critical insulation thickness, substitute r_o by k/h_o, which gives the desired result.

Example 2.36 An electric wire of 3 mm radius is to be provided with plastic sheathing [k = 0.15 W/(m K)]. The convective film coefficient on the surface of the bare cable as well as insulated cable can be assumed to be 10 W/(m^2 K). For a particular value of current flow I_1, the wire surface temperature is found to be 60°C when the surrounding temperature was 20°C. Find

(i) The insulation thickness so that the wire surface temperature will be minimum when current flow is unchanged. What will be wire surface temperature in this case? Assume that the wire resistance does not change with change in its temperature.
(ii) The maximum possible increase in the current carrying capacity if the wire surface temperature is to be limited to the same value after the insulation of critical thickness.

Solution

(i) To keep the wire as cool as possible, the thickness of the insulation must be equal to the critical thickness, which is

$$r_c = \frac{k}{h_o} = \frac{0.15}{10} = 0.015\,\text{m}$$

i.e. the thickness of the insulation is 15 – 3 = 12 mm.
For the bare wire, the heat dissipation rate is

$$q = h_o(2\pi r_1 L) \times (t_1 - t_\infty) = 10 \times (2\pi \times 0.003 \times 1) \times (60 - 20) = 7.54\,\text{W per m length of the cable}$$

Heat dissipation rate from the insulated cable is given by, refer to Example 2.35,

$$q' = \frac{2\pi kL(t_1 - t_\infty)}{1 + \ln\left(\frac{k}{h_o r_1}\right)} = \frac{2\pi \times 0.15 \times 1 \times (T_1' - 20)}{1 + \ln\left(\frac{0.15}{10\times 0.003}\right)} = 0.3612(T_1' - 20)$$

When the current flow is unchanged, the heat generation rate will remain the same, i.e. $q' = q$. Hence,

$$7.54 = 0.3612(t_1' - 20)$$

or

$$t_1' = 40.87°\text{C}$$

(ii) The heat dissipation rate and hence the current carrying capacity of the cable will be maximum when the critical thickness of the insulation is provided. For $t_1 = t_1' = 60\,°\text{C}$, the heat dissipation rate for $r_o = r_c$ will be, from Part (i),

$$q = 0.3612(60 - 20) = 14.45\,\text{W}$$

For the given wire resistance,

$$q\,(\text{without insulation}) = I_1^2R = 7.54\,\text{W}$$
$$q'\,(\text{with insulation}) = I_2^2R = 14.45\,\text{W}$$

Thus,

$$\frac{I_2}{I_1} = \sqrt{\frac{14.45}{7.54}} = 1.384$$

i.e. the current can be increased by 38.4%.

Example 2.37 A vertical non-insulated steam pipe of 4 m length and 80 mm diameter has surface temperature t_w = 160°C. The surrounding air is at 20°C. Find the loss of the heat from the pipe by convection if the surface heat transfer coefficient is 9 W/(m^2 K).

What will be the heat loss from the pipe length if an asbestos-based insulation [k = 0.16 W/(m K)] of 30 mm thickness is applied? Give your comments.

Solution

Heat loss without insulation

$$q = h(\pi DL)(t_w - t_\infty) = 9 \times \pi \times 0.08 \times 4 \times (160 - 20) = 1266.7\,\text{W}$$

Heat loss when insulation is provided (r_1 = 40 mm, r_2 = 70 mm)

$$q' = \frac{t_w - t_\infty}{\frac{1}{2\pi kL}\ln\frac{r_2}{r_1} + \frac{1}{2\pi r_2 L h_o}} = \frac{160 - 20}{\frac{1}{2\pi\times 0.16\times 4}\ln\frac{70}{40} + \frac{1}{2\pi\times 0.07\times 4\times 9}} = 691.97\,\text{W}$$

Comments: The critical radius of insulation, from the given data, is

$$r_c = \frac{k}{h_o} = \frac{0.16}{9} = 0.0178\,\text{m} = 17.8\,\text{mm}$$

As the critical radius is less than the tube radius, the given insulation will be effective whatever may be thickness of the insulation applied. The loss will decrease with the increase in the thickness of insulation.

Note: In the above analysis, it has been assumed that the surface temperature of the tube and the convective heat transfer coefficient remains the same even after the layer of the insulation is applied.

Example 2.38 If the pipe in Example 2.37 is covered with insulation having thermal conductivity of 0.5 W/(m K), discuss the effect of variation in insulation thickness.

Solution

Heat loss without insulation

$$q = h(\pi DL)(t_w - t_\infty) = 9 \times \pi \times 0.08 \times 4 \times (160 - 20) = 1266.7\,\text{W}$$

Heat loss with insulation provided for r_1 = 40 mm

$$q' = \frac{t_w - t_\infty}{\frac{1}{2\pi kL}\ln\frac{r_2}{r_1} + \frac{1}{2\pi r_2 L h_o}} = \frac{160 - 20}{\frac{1}{2\pi\times 0.5\times 4}\ln\frac{r_2}{40} + \frac{1}{2\pi\times r_2\times 4\times 9}}$$

The critical thickness of insulation

$$r_c = \frac{k}{h_o} = \frac{0.5}{9} = 0.0555\,\text{m}$$

The variation of the heat loss with r_2 is shown in table below and depicted in Fig. 2.39.

r_2 (mm)	Heat loss, Q (W)
$r_2 = r_1 = 40$	1266.70
45	1300.96
50	1318.68
$r_c = 55.5$	1324.36
60	1321.44
70	1300.0
80	1267.92
90	1231.8

Example 2.39 An electric wire is 6 mm in diameter. The wire is to be covered by at least 4 mm rubber insulation [k = 0.1 W/(m K)] from electric consideration. How the cooling conditions will be changed? The heat transfer coefficient from the surface is 10 W/(m^2 K).

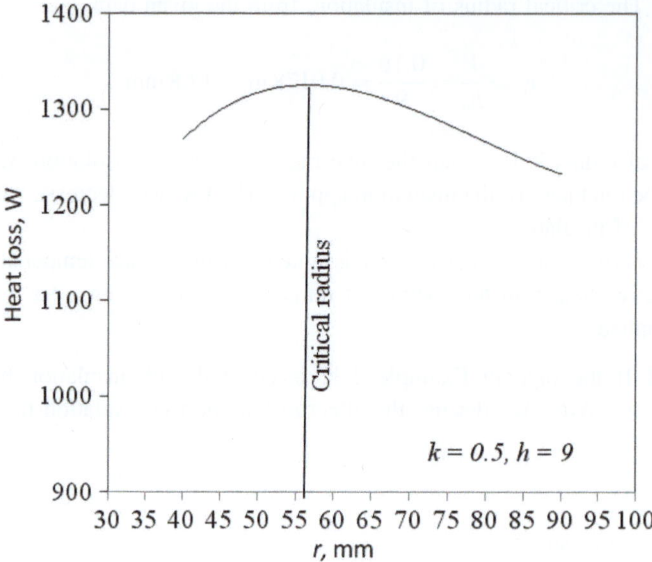

Fig. 2.39 Heat loss versus insulation radius

Solution

The critical radius of insulation is

$$r_c = \frac{k}{h_o} = \frac{0.1}{10} = 0.01\,\text{m} = 10\,\text{mm}$$

As $r_c > r_{\text{wire}}$ and $r_{\text{wire}} < r_2 < r_c$, the insulated wire will be cooled better than the bare wire. At insulation thickness of 7 mm, the cooling effect will be at maximum. The following calculations can be carried out for clear understanding:

Bare wire:

$$q = h_o(\pi DL)(t_{\text{wire}} - t_\infty)$$

or

$$t_{\text{wire}} - t_\infty = \frac{q}{h_o \pi DL} = \frac{q \times 1000}{10 \times \pi \times 6 \times 1} = 5.305q \text{ for 1 m length}$$

Insulated wire:

$$q = \frac{t_{wire} - t_\infty}{\frac{1}{2\pi kL}\ln\frac{r_2}{r_1} + \frac{1}{2\pi r_2 L h_o}}$$

or

$$t_{\text{wire}} - t_\infty = q\left(\frac{1}{2\pi \times 0.1 \times 1}\ln\frac{7}{3} + \frac{1}{2\pi \times 0.007 \times 1 \times 10}\right) = 3.622q$$

Similarly, for the insulation thickness of 7 mm,

$$t_{\text{wire}} - t_\infty = 3.508q$$

and for the insulation thickness of 10 mm,

$$t_{\text{wire}} - t_\infty = 3.557q.$$

From the above calculations, it can be seen that the wire temperature, over the surrounding temperature, ($t_{\text{wire}} - t_\infty$) decreases with the increase in the insulation thickness up to $r_2 = r_c$ and then increases with the further increase in the thickness. Again the wire temperature is minimum at $r_2 = r_c$.

Example 2.40 A cylindrical pipe of 20 mm outer diameter is to be insulated so that the heat loss from the pipe surface is not more than 65 W per m length. The pipe surface is 280°C, and it can be assumed that the surface temperature remains the same after application of the insulation layer. The surrounding temperature is 30°C. The heat transfer coefficient from the surface is 10 W/(m^2 K). Available for the service is asbestos [k = 0.14 W/(m K)] or slag wool [k = 0.08 W/(m K)]. Determine the thicknesses of these insulating materials and comment on the result.

Solution

The critical radius of insulation,

$$r_c = \frac{k}{h_o} = \frac{0.14}{10} = 0.014\,\text{m} = 14\,\text{mm for asbestos}$$

and

$$r_c = \frac{k}{h_o} = \frac{0.08}{10} = 0.008\,\text{m} = 8\,\text{mm for slag wool}$$

Thus the critical radius of insulation r_c is less than the pipe radius r_1 for the slag wool. Therefore, the use of the slag wool as insulation is recommended. If asbestos is used as insulation, the thickness of it should be such that r_2 is well beyond r_c, refer to Fig. 2.40. For slag wool, the curve below $r = r_1$ is shown dashed because it is hypothetical only.

Heat loss from the bare pipe:

$$q = h(\pi DL)(t_w - t_\infty) = 10 \times \pi \times 0.02 \times 1 \times (280 - 30) = 157.1\,\text{W}$$

Heat loss with insulation:

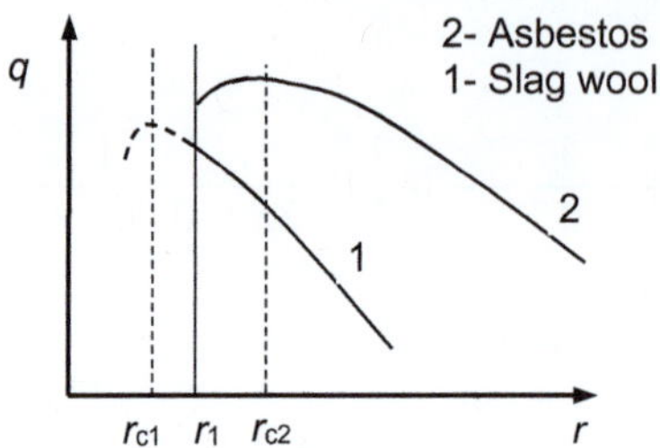

Fig. 2.40 Heat loss versus radius for alternative insulations of Example 2.40

$$q = \frac{t_w - t_\infty}{\frac{1}{2\pi kL}\ln\frac{r_2}{r_1} + \frac{1}{2\pi r_2 L h_o}}$$

For asbestos $r_2 = r_c = 14$ mm ($r_1 = 10$ mm). Heat loss from the above equation is

$$q_a = \frac{280 - 30}{\frac{1}{2\pi \times 0.14 \times 1}\ln\frac{14}{10} + \frac{1}{2\pi \times 0.014 \times 1 \times 10}} = 164.54\,\text{W}.$$

Asbestos insulation thickness for heat loss of 65 W per m length:

$$65 = \frac{280 - 30}{\frac{1}{2\pi \times 0.14 \times 1}\ln\frac{r_2}{10} + \frac{1}{2\pi \times r_2 \times 1 \times 10}}$$

Solution gives $r_2 = 280$ mm, by trial and error.
Similarly for the slag wool ($r_c = 8$ mm),

$$65 = \frac{280 - 30}{\frac{1}{2\pi \times 0.08 \times 1}\ln\frac{r_2}{10} + \frac{1}{2\pi \times r_2 \times 1 \times 10}}$$

which gives $r_2 = 61$ mm by trial and error.

Example 2.41 Show that for a small diameter tube, when critical radius is greater than the tube radius r_1, the heat loss can be reduced by applying an insulating layer of thickness ($r_2 - r_1$) such that

$$\frac{1}{k}\ln\frac{r_2}{r_1} + \frac{1}{h_o r_2} > \frac{1}{h_o r_1}$$

Solution

Resistance to heat flow when insulation layer is applied,

$$R' = \frac{1}{2\pi lk}\ln\left(\frac{r_2}{r_1}\right) + \frac{1}{2\pi r_2 l h_o}$$

Resistance to heat flow from a bare tube,

$$R = \frac{1}{2\pi r_1 l h_o}$$

For reduction in the heat loss R' must be greater than R, i.e.

$$\frac{1}{2\pi l k} \ln\left(\frac{r_2}{r_1}\right) + \frac{1}{2\pi r_2 l h_o} > \frac{1}{2\pi r_1 l h_o}$$

Simplification of the above equation gives the desired result. For the cooling of the electric cables, the condition will be $R' < R$.

2.11 Thermal Contact Resistance

In many engineering applications, solid surfaces are not having metallurgical bond. At the interface of such surfaces, a resistance to the heat flow has been observed. This resistance is called thermal contact resistance. It can be the dominant resistance when high conductivity metals are involved. A drop in temperature at the interface of the two materials due to the resistance of the interface can be seen in Fig. 2.41.

Real surfaces are not perfectly smooth. Hence, when such surfaces are pressed against each other, contact between them occurs only at a limited number of spots as shown in Fig. 2.41b, which has been exaggerated for a better understanding. The voids are filled with the surrounding fluid.

The heat transfer across the interface takes place through two parallel paths:

1. The conduction at the high spots, which are in direct contact
2. The conduction through the fluid entrapped in the voids.

Since the fluid layer is very thin, there is no convection. At high temperatures, radiation across the gap may also contribute to the heat transfer.

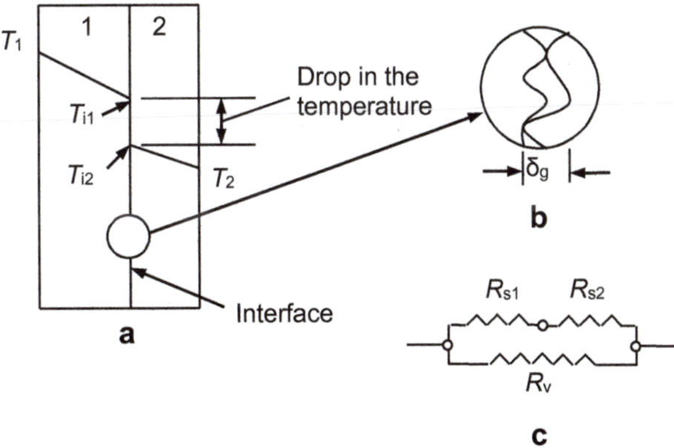

Fig. 2.41 **a** Effect of the thermal contact resistance on temperature profile, **b** contact surfaces (enlarged), **c** the thermal network at the interface

The heat flow equation can be written as

$$q = \frac{T_{i1} - T_{i2}}{\left(\frac{1}{R_s} + \frac{1}{R_v}\right)^{-1}} \tag{i}$$

where

R_s resistance of the spots in direct contact,
R_v resistance of the fluid layer,
$(T_{i1} - T_{i2})$ temperature drop across the interface.

The resistance R_s consists of two resistances R_{s1} and R_{s2} in series; the resistances R_s and R_v are in parallel as shown in Fig. 2.41c.

If it is assumed that the void thickness δ_g, refer to Fig. 2.41b, is equally divided between the solid surfaces 1 and 2, then the resistance R_s can be expressed as

$$R_s = \frac{\delta_g/2}{k_1 A_s} + \frac{\delta_g/2}{k_2 A_s} \tag{ii}$$

The fluid resistance can be expressed as

$$R_v = \frac{\delta_g}{k_f A_v} \tag{iii}$$

where

k_1 thermal conductivity of solid 1,
k_2 thermal conductivity of solid 2,
A_s area of the spots of solid-to-solid contact,
A_v void area and
k_f thermal conductivity of the entrapped fluid.

Substitution in Eq. (i) gives

$$q = (T_{i1} - T_{i2})\left[\left(\frac{\delta_g}{2k_1 A_s} + \frac{\delta_g}{2k_2 A_s}\right)^{-1} + \left(\frac{\delta_g}{k_f A_v}\right)^{-1}\right] \tag{iv}$$

The interface or the contact conductance is defined as

$$q = hA(T_{i1} - T_{i2}) \tag{v}$$

where A is the total contact area of the surfaces.

Comparison of Eqs. (iv) and (v) gives the expression of the contact conductance as

$$h = \frac{1}{\delta_g}\left[\left(\frac{A_s}{A}\right)\frac{2k_1 k_2}{k_1 + k_2} + \left(\frac{A_v}{A}\right)k_f\right] \tag{2.52}$$

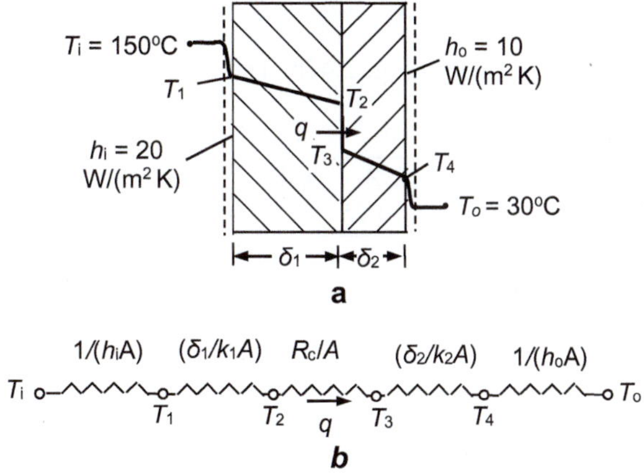

Fig. 2.42 Example 2.42

It is not easy to use this equation because of the difficulty in determination of the values of A_s, A_v and δ_g. However, the equation shows that h is a function of void thickness δ_g, thermal conductivity k_f of the entrapped fluid, and area A_s.

The void thickness δ_g depends on the roughness of the surfaces in contact. Thus, the roughness of the surfaces plays an important role in determining the contact resistance. The resistance increases with the increase in the roughness.

Increase in the joint pressure for the surfaces in contact decreases the resistance. With the increase in the joint pressure, the high spots are deformed (crushed). This increases the contact area A_s and reduces δ_g. Experimental studies have shown that the conductance varies almost directly with the joint pressure.

At very low gas pressures, when the free path of the molecules is large compared to the dimension of the void space, the effective thermal conductivity of the entrapped gas decreases. This leads to an increase in the contact resistance.

Studies have shown that the contact conductance ranges from 3120 to 71,000 W/(m^2 K) for steel, brass and aluminium surfaces ground to various degrees of roughness for pressures from 1.31 to 55.1 bar with air, oil or glycol between the surfaces (Frass 1989).

Thin foils made of materials of good thermal conductivity when inserted between the surfaces reduce the resistance if the foil is softer than the materials of the surfaces in contact.

The resistance is reported to reduce by as much as 75% when thermal grease like Dow 340[1] is used (Holman 1992). The values of the contact resistance of some surfaces are also tabulated in Holman (1992) and Mills (1995). However, the data is scarce and not reliable (Mills 1995).

Example 2.42 The composite wall (A = 2 m^2) shown in Fig. 2.42 consists of layers of bricks (k_1 = 0.6 W/(m K), δ_1 = 0.1 m) and mineral wool (k_2 = 0.045 W/(m K), δ_2 = 0.05 m). The contact resistance is estimated to be 0.25 m^2 K/W. Determine (a) the rate of heat transfer through the wall and (b) the temperature distribution.

[1]A grease like silicone fluid thickened with metal oxide filler [k = 0.4 W/(m K)].

Solution

(a) The electric network is shown in Fig. 2.42b. The total resistance to heat transfer is

$$R_t = \frac{1}{h_i A} + \frac{\delta_1}{k_1 A} + \frac{R_c}{A} + \frac{\delta_2}{k_2 A} + \frac{1}{h_o A}$$
$$= \frac{1}{20 \times 2} + \frac{0.1}{0.6 \times 2} + \frac{0.25}{2} + \frac{0.05}{0.045 \times 2} + \frac{1}{10 \times 2} = 0.839\,\text{K/W}$$

Heat transfer rate,

$$q = \frac{T_i - T_o}{R_t} = \frac{150 - 30}{0.839} = 143.03\,\text{W}$$

(b) We can write

$$q = \frac{T_i - T_1}{1/h_i A}$$

which gives

$$T_1 = T_i - \frac{q}{h_i A}$$
$$= 150 - \frac{143.03}{20 \times 2} = 146.42^\circ\text{C}$$

Similarly,

$$T_2 = T_1 - \frac{q\delta_1}{k_1 A}$$
$$= 146.42 - \frac{143.03 \times 0.1}{0.6 \times 2} = 134.5\,^\circ\text{C},$$
$$T_3 = T_2 - \frac{qR_c}{A}$$
$$= 134.5 - \frac{143.03 \times 0.25}{2} = 116.62\,^\circ\text{C},$$
$$T_4 = T_3 - \frac{q\delta_2}{k_2 A}$$
$$= 116.62 - \frac{143.03 \times 0.05}{0.045 \times 2} = 37.15\,^\circ\text{C}$$

and

$$T_0 = T_4 - \frac{q}{h_o A}$$
$$= 37.15 - \frac{143.03}{10 \times 2} = 30^\circ\text{C}$$

The temperature distribution is shown in the figure.

2.12 Summary

In this chapter, basically Fourier's law of heat conduction has been applied to systems where one-dimensional heat flow occurs.

Mechanism of heat conduction in metals and alloys, construction and heat-insulating materials, and liquid and gases has been discussed. In general, the thermal conductivity depends on temperature, pressure and nature of the substance. It differs by many thousand times for the materials of interest. The highest values are for the metals, followed by dense ceramics, organic solids and liquids, while the lowest values are for the gases. Super-insulation materials have been developed for cryogenic applications. Typical values of the thermal conductivity of several common engineering materials are given in a tabulated form in Appendix A.

In general, the temperature gradient may exist in all three directions of a solid. There may be internal heat generation. The temperature can also vary with the time (unsteady state). Hence, a general heat conduction equation must be developed, which can be used to evaluate the heat transfer in any direction under steady or unsteady state and with or without heat generation. The general heat conduction equations in the rectangular, cylindrical and spherical coordinates have been developed.

In the unsteady state, a material property termed as thermal diffusivity, which is ratio of thermal conductivity (k) and thermal capacity (ρc) of the material, controls the temperature distribution. The higher the thermal diffusivity of a material, the higher the rate of temperature propagation, i.e. the equalization of the temperature at all points of the space will proceed at a higher rate.

Fourier's equation of heat conduction is exactly analogous to Ohm's law for an electrical conductor. We use this electrical analogy frequently as it is quite useful in solving the complex heat conduction problems. The temperature difference Δt is the driving force for the flow of heat and δ/kA is the thermal resistance in case of a plane wall.

Where the wall face is exposed to a fluid, the heat is transferred to and from the wall by convection. In such cases, film resistance $1/hA$ is considered along with conduction resistance of the wall. If heat also transfers by radiation from the surface of the solid, the resistance is $1/[(h_r + h_c)A]$, where h_r is termed as the radiation heat transfer coefficient, which has been discussed in Chap. 11.

In case of one-dimensional steady-state heat conduction through a cylindrical shell, the wall resistance is $\ln(r_2/r_1)/(2\pi kL)$ and film resistances at the inner and outer surface of the shell are $1/h_iA_i$ and $1/h_oA_o$, respectively. In heat exchangers, the equation of heat exchange between the fluids flowing in and outside the tube is written as $q = UA(t_i - t_o)$ where U is termed as overall heat transfer coefficient, refer to Eqs. (2.29a), (2.29b), (2.30a) and (2.30a).

In case of one-dimensional steady-state heat conduction through a spherical shell, $(r_2 - r_1)/(4\pi k r_1 r_2)$ is the resistance of the wall to the conduction heat transfer and film resistances are $1/h_iA_i$ and $1/h_oA_o$.

In Sect. 2.8, experimental schemes of measurement of thermal conductivity of solids, liquids and gases have been presented.

Thermal conductivity of a dry solid substance depends on temperature and when the temperature difference across the wall of such substance is large, the effect of the variation of thermal conductivity with the temperature must be considered. This has been considered in Sect. 2.9.

In case of a hollow cylinder or sphere, with the increase in the thickness of the insulating material, the conduction resistance R_k of the insulation increases while the film resistance R_o decreases. Mathematical analysis shows that the sum of the resistances $(R_k + R_o)$ first decreases to a certain minimum value at radius $r = r_c$ and then again increases. It means that as the insulation radius r_2 increases, the heat flow q first increases to a certain maximum value at $r_2 = r_c$ and then decreases. The value of r_2 at the maximum of heat flow q (or the minimum of the resistance) is called the critical radius of insulation. The knowledge of the concept of critical radius of insulation helps in the selection of type of insulation material. For the heat insulation of a pipe of radius r_1, the insulating material must be selected such that the critical radius r_c is less than r_1 while in the case of the electric current carrying cables, the insulation radius r_2 must be equal to r_c to dissipate maximum heat. Hence, the insulating material for the cables must be selected such that the critical radius r_c is greater than the bare cable radius r_1. Mathematical expressions for the critical radius of insulation have been deduced.

In many engineering applications, solid surfaces are not having metallurgical bond. At the interface of such surfaces, a resistance to the heat flow has been observed, which is termed as thermal contact resistance. It can be a dominant resistance when high conductivity metals are involved. A basic mathematical treatment for estimate of the thermal contact resistance has been given.

A significant number of illustrative examples have been included in this chapter on one-dimensional steady-state heat conduction.

Review Questions

2.1. Define one- and two-dimensional heat transfer when applied to conduction problems.

2.2. Discuss the mechanism of heat conduction in metals and non-metals. What is the effect of alloying on thermal conductivity of metals?

2.3. Write a short note on insulating materials.

2.4. Discuss the factors that affect thermal conductivity of metals, insulating materials, liquids and gases.

2.5. (i) Derive three-dimensional general heat conduction equation in
   (a) Cartesian coordinates
   (b) Cylindrical coordinates
   (c) Spherical coordinates
   for transient condition considering uniform volumetric heat generation.

   (ii) Deduce one-dimensional steady-state conduction equations from the equations derived in Part (i) if there is no heat generation. The thermal conductivity of the material can be assumed to be constant.

2.6. Define thermal diffusivity and explain its physical significance giving suitable example.

2.7. What do you mean by critical radius of insulation? Show that it is given by (i) (k/h_o) for a cylindrical configuration and (ii) $(2\, k/\, h_o)$ for a spherical vessel, where k is the thermal conductivity of the insulating material and h_o is the convection coefficient of the heat loss to the surrounding.

2.8. What is the significance of critical radius of insulation with reference to (a) insulation of pipes to reduce the heat flow, and (b) insulation of the electric cables?

2.9. Define overall heat transfer coefficient and develop its expression for heat transfer in the tube flow.

2.10. How would you measure the thermal conductivity of an insulating material in the form of powder? Draw a neat-labelled sketch of the equipment suggested.

2.11. Present a scheme for experimental measurement of thermal conductivity of material of a flat plate.

2.12. What is thermal contact resistance? Discuss its mechanism and the parameters on which it depends. What are the engineering applications where it is relevant?

2.13. Show that the geometric mean radius R_m of a spherical shell of outside radius R_2 and inside radius R_1 is given by

$$R_m = (R_1 R_2)^{0.5}$$

2.14. Show that the temperature distribution across the wall of a hollow sphere operating with steady and uniform surface temperature T_1 on the inner surface radius R_1 and T_2 on the outer surface radius R_2 is given by

$$\frac{T - T_1}{T_2 - T_1} = \left(\frac{1/R_1 - 1/r}{1/R_1 - 1/R_2}\right)$$

where $R_1 < r < R_2$.

2.15. The thermal conductivity of a certain material varies with temperature as

$$k = k_o\left(1 + aT + bT^2\right)$$

where a and b are constants.

Derive from the first principles an expression for the heat flow rate for a hollow sphere of the inner and outer radii R_1 and R_2, respectively. The temperatures of the inner and outer surfaces are T_1 and T_2, respectively.

[**Ans**. Following the method of Sect. 2.9.3, $q = -k_o(1 + aT + bT^2)4\pi r^2 \frac{dt}{dr}$; Integration gives

$$q \int_{R_1}^{R_2} \frac{dr}{r^2} = -4\pi k_o \int_{T_1}^{T_2} (1 + aT + bT^2)dt;$$

$$q\left(\frac{R_2 - R_1}{R_1 R_2}\right) = -4\pi k_o\left[(T_2 - T_1) + \frac{a}{2}\left(T_2^2 - T_1^2\right) + \frac{b}{3}\left(T_2^3 - T_1^3\right)\right]$$

or

$$q = -4\pi k_o\left(\frac{R_1 R_2}{R_2 - R_1}\right)(T_2 - T_1)\left[1 + \frac{a}{2}(T_2 + T_1) + \frac{b}{3}\left(T_2^2 + T_1 T_2 + T_1^2\right)\right]$$

Problems

2.1 Draw the thermal network (electrical analog) and determine the heat flow rate through the wall shown in Fig. 2.43a. Assume one-dimensional heat flow and uniform temperature at the interface of different layers.
[**Ans**. The thermal network is shown in Fig. 2.43b; $R_a = 0.0375$; $R_b = 0.25$; $R_c = 0.25$; $R_d = 0.2$; $R_e = 2.86$; $R_f = 2.5$; $R_g = 2.143$; Resistances R_b, R_c and R_d are in parallel, and R_e and R_f are also in parallel; The total resistance is 3.59 and $q = 125.35$ W/m^2.]

2.2 The wall of a house may be approximated by 225 mm layer of ordinary bricks [$k = 0.5$ W/(m K)] with both sides of it plastered with 20-mm-thick layer of cement plaster [$k = 1.2$ W/(m K)]. What thickness of plywood sheet [$k = 0.15$ W/(m K)] be added to the inside of the wall to reduce the heat loss (or gain) through the wall by 20 percent? Neglect the contact resistance.
If there is an air space of average 0.25 mm thickness [$k = 0.0257$ W/(m K)] between the wall and the plywood, determine the effect on the heat flow rate. Assume only conduction heat flow across the air layer.
[**Ans**. $\sum R = 0.483$ without plywood sheet; Since $q \propto 1/\sum R$, after application of plywood desired $\sum R' = 0.483/0.8 = (0.483 + \delta_{ply}/0.15)$; This gives $\delta_{ply} = 18.1$ mm. With the air layer $\sum R_a' = 0.483/0.8 + 0.25/(1000 \times 0.0257) = 0.6135$. Reduction in heat flow $= (1/\sum R - 1/\sum R')/(1/\sum R) = 21.3\%$.]

2.3 A vessel with flat bottom, refer to Fig. 2.44, is made of 2-mm-thick nickel sheet [$k = 60$ W/(m K)]. The outside of its bottom has a plating of 0.5-mm-thick copper [$k = 350$ W/(m K)]. Calculate the reduction in the heat transfer rate due to the deposition of scale on the lower side of the bottom. $R_{scale} = 0.001$ K/W. The film

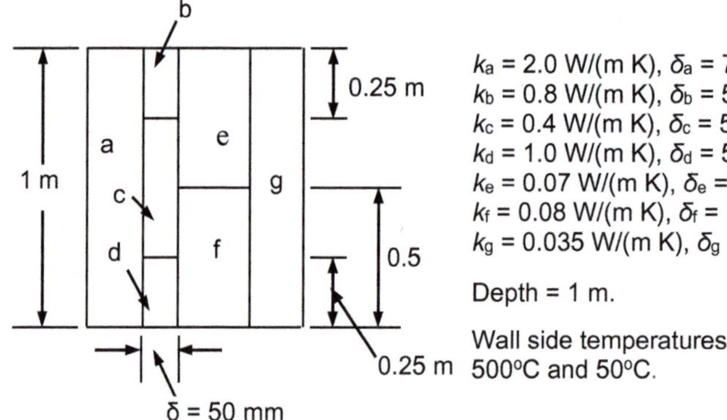

a Wall cross-section.

b Thermal network.

Fig. 2.43 Problem 2.1

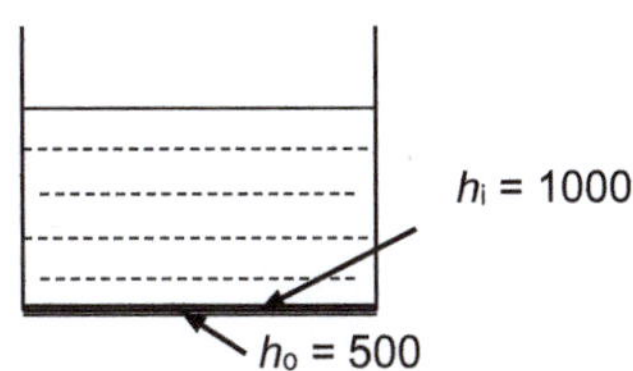

Fig. 2.44 Problem 2.3

coefficients on the two sides of the bottom are 1000 W/(m^2 K) and 500 W/(m^2 K), respectively.
[**Ans**. Without scale, $\sum R = 3.035 \times 10^{-3}$; $q \propto 1/\sum R = 329.5$; With the scale, $\sum R = 4.035 \times 10^{-3}$; $q \propto 1/\sum R = 247.8$; Reduction in the heat transfer rate due to the deposition of scale = (329.5 – 247.8)/329.5 = 24.8%.]

2.4 Two layers of glass (each 4 mm thick) in a window panel are 3 mm apart. The air [k = 0.025 W/(m K)] in between these layers can be assumed to be dry and stagnant. Find the heat loss through the panel for a temperature difference across the panel of 20°C. If the two layers of glass are replaced by a single glass of 10 mm thickness, how the loss will change? [k_{glass} = 0.8 W/(m K)].
[**Ans**. q_1/A = 153.85 W/m^2; q_2/A = 1600 W/m^2.]

2.5 Derive an expression for the heat transfer rate from a cylindrical tank of length L with hemispherical covers. The wall thickness throughout is (R_2 - R_1). The temperatures at the inner and outer surfaces are T_1 and T_2, respectively ($T_1 > T_2$). Assume one-dimensional heat flow.
[**Ans**. $q = 2\pi kL\frac{(T_1-T_2)}{\ln\left(\frac{R_2}{R_1}\right)} + 2\times 2\pi k\frac{(T_1-T_2)}{\left(\frac{1}{R_1}-\frac{1}{R_2}\right)}$]

2.6 An aluminium pipe carries steam at 110°C. The pipe [k = 185 W/(m K)] has an inside diameter of 100 mm and outside diameter of 120 mm. The pipe is located in a room where the ambient air temperature is 30°C and convective heat transfer coefficient between the pipe and air is 15 W/(m^2 K). Determine the heat transfer rate per unit length of the pipe.
To reduce the heat loss from the pipe, it is covered with a 50-mm-thick layer of insulation [k = 0.20 W/(m K)]. Determine the heat transfer rate per unit length from the insulated pipe. Assume that the convective resistance of the steam is negligible.
[**Ans**. Here R_1 = 50 mm; R_2 = 60 mm; R_3 = 110 mm; k_{al} = 185 W/(m K); k_i = 0.2 W/(m K); h_o = 15 W/(m^2 K); This gives

(i) $q = 2\pi \times 1 \times \frac{(110-30)}{\frac{1}{k_{al}}\ln\left(\frac{R_2}{R_1}\right) + \frac{1}{h_o R_2}} = 451.98\,\text{W}$

(ii) $q = 2\pi \times 1 \times \frac{(110-30)}{\frac{1}{k_{al}}\ln\left(\frac{R_2}{R_1}\right) + \frac{1}{k_i}\ln\left(\frac{R_3}{R_2}\right) + \frac{1}{h_o R_3}} = 138.18\,\text{W}$]

2.7 A steel pipe carrying a hot fluid has inside diameter of 120 mm and an outside diameter of 160 mm. It is insulated at the outside with asbestos. The fluid temperature is 150°C, and the surrounding air temperature is 20°C, h (fluid side) = 100 W/(m^2 K), h (air side) = 8 W/(m^2 K), $k_{asbestos}$ = 0.18 W/(m K) and k_{steel} = 42 W/(m K). How thick should the asbestos layer be provided in order to limit the heat loss to 150 W per metre length of the pipe?

[**Ans**. $150 = 2\pi \times 1 \times \frac{(150-20)}{\frac{1}{h_i R_1} + \frac{1}{k_{\text{steel}}} \ln\left(\frac{R_2}{R_1}\right) + \frac{1}{k_{\text{asbestos}}} \ln\left(\frac{R_3}{R_2}\right) + \frac{1}{h_o R_3}}$ This gives R_3 = 182.7 mm and thickness of insulation = R_3 – 80 = 102.7 mm.]

2.8 Determine overall heat transfer coefficient (based on the inner diameter) for the data of Problem 2.7 and calculate the heat transfer rate.
[**Ans**. U_i = 3.0606 W/(m^2 K); $q = UA_i(T_i - T_o)$ = 150 W.]

2.9 Determine the heat flow rate for the data of Problems 2.6 and 2.7 neglecting the pipe wall resistance and comment on the result.
[**Ans**. Neglect $(1/k)\ \ln(R_2/R_1)$ term. The heat transfer rates are 138.22 W and 150.18 W as against 138.18 W and 150 W, respectively. The effect of the wall resistance in both cases is negligible because the resistances offered by the insulation are much larger in magnitude.]

2.10 A steel pipe d_1/d_2 = 100/110 mm is covered with 80 mm thick layer of an insulating material. The thermal conductivity of the insulation material depends on temperature and is given by $k = 0.06\,(1 + 0.4 \times 10^{-2} T)\ \text{W/(m °C)}$. Determine the heat loss per m length of the pipe, if the temperature of the outer surface of the pipe is 200°C and the temperature of the outer surface of the insulation is 40°C.
[**Ans**. For insulation, R_1 = 110/2 = 55 mm, R_2 = 55 + 80 = 135 mm; $k = 0.06 \times \left[1 + 0.4 \times 10^{-2} \times \frac{(T_1 + T_2)}{2}\right] = 0.0888$; $\frac{q}{L} = 2\pi k \frac{(T_1 - T_2)}{\ln\left(\frac{R_2}{R_1}\right)} = 99.4\ \text{W}$]

2.11 A steam pipeline of d_1/d_2 = 100/110 mm is covered with two layers of insulation of equal thickness $\delta_1 = \delta_2$ = 50 mm. The average thermal conductivity of the first layer of insulation material is 0.05 W/(m K) and that of the second layer is 0.1 W/(m K). If the layer of greater thermal conductivity is placed next to the pipe, what will be the effect on the heat loss? Neglect thermal resistance of the pipe material.
[**Ans**. $R_t = \frac{1}{2\pi k_1} \ln\left(\frac{R_2}{R_1}\right) + \frac{1}{2\pi k_2} \ln\left(\frac{R_3}{R_2}\right)$; R_1 = 55 mm, R_2 = 105 mm, R_3 = 155 mm. For the first part of the problem, k_1 = 0.05, k_2 = 0.1 W/(m K) and L = 1 m, which gives R_t = 2.678 K/W. In the second part, k_1 = 0.1 and k_2 = 0.05. W/(m K). This gives R_t = 2.2688 K/W. Thus the resistance decreases by 18%. Hence, the heat loss will increase by 18%. The results show that the better insulation must be put next to the pipe.]

2.12 An electronic instrument for a probe is contained in a spherical shell made of 30 mm thick layer of mild steel (inside radius of 150 mm) and a 10 mm thick outside layer of stainless steel. The two layers can be assumed to be in perfect thermal contact. The instrument inside the shell generates heat at a rate of 6 kW. Estimate the inside surface temperature, if the outer surface is estimated to be at 5°C [k_{ms} = 40 W/(m K) and k_{ss} = 15 W/(m K)].
[**Ans**. T_1 = 27.57°C.]

2.13 The energy loss from a 16 mm outer diameter pipe, which carries a hot fluid, is to be reduced. Available for the service are magnesia 85 [k = 0.071 W/(m K)] and glass wool [k = 0.038 W/(m K)]. Would you help in making the correct selection if heat transfer coefficient h is likely to be 5 W/(m^2 K)?
[**Ans**. The critical radius, $(r_c)_{\text{mag}}$ = 14.2 mm, $(r_c)_{\text{glass}}$ = 7.6 mm; $(r_c)_{\text{mag}}$ is greater than outer radius of the pipe; hence, it should not be used. For the glass wool, any thickness will reduce the heat loss.]

2.14 A thin-walled-long cylindrical vessel of 0.8 m outer diameter contains a chemical undergoing exothermic reaction. The outer surface temperature of the vessel is 180°C when the surrounding air is at 30°C. Determine the thickness of polyurethane foam (rigid) insulation required so that the outer surface temperature of the insulation is not more than 40°C. The convection heat transfer coefficient on outer surface is estimated to be 20 W/(m^2 K). Neglect radiation heat loss.

[**Ans.** Heat flow by conduction through the insulation of thickness (r_2 - r_1) equals the convective heat transfer from the outer surface, i.e. $q = \frac{2\pi k_i l (T_1 - T_2)}{\ln\left(\frac{r_2}{r_1}\right)} = h \times 2\pi r_2 l \times (T_2 - T_\infty)$. Substituting T_1 = 180°C, T_2 = 40°C, T_∞ = 30°C, r_1 = 0.4 m, h = 20 W/(m^2 K) and k_i = 0.025 W/(m K) from Table A3.2 of Appendix A, we get r_2 = 0.417 m; Insulation thickness (r_2 - r_1) = 0.017 m = 17 mm.]

2.15 A hollow aluminum sphere [k_a = 204 W/(m K)] with a 100 W electric heater at its centre is used to determine the thermal conductivity of an insulating material. The inner and outer radii of the sphere are 0.1 and 0.125 m, respectively. In a particular test, a spherical shell of an insulating material of 75 mm thickness was applied on the outer surface of the sphere. In the steady state, the temperature at the inner surface of the aluminum wall was recorded as 200°C. The system is exposed to room air at 30°C and the convective heat transfer coefficient is estimated to be 20 W/(m^2 K). What is the thermal conductivity of the insulation? Neglect heat loss by radiation and contact resistance.

[**Ans.** r_1 = 0.1 m, r_2 = 0.125 m, r_3 = 0.2 m, T_1 = 200°C, T_∞ = 30°C, k_a = 204 W/(m K), h = 20 W/(m^2 K), q = 100 W; Heat flow rate, $q = \frac{(T_1 - T_\infty)}{\frac{1}{4\pi}\left[\frac{1}{k_a}\left(\frac{1}{r_1} - \frac{1}{r_2}\right) + \frac{1}{k_i}\left(\frac{1}{r_2} - \frac{1}{r_3}\right) + \frac{1}{h r_3^2}\right]}$, solution after substitution of various known values gives k_i = 0.15 W/(m K).]

2.16 A fluid is stored in a thin metallic spherical shell (outside diameter of 2 m) with 200 mm thick insulation [k_i = 0.05 W/(m K)]. The shell outer surface temperature is –20°C. Determine the thickness of the insulation which experiences temperature $\leq$ 0°C when the surrounding air temperature is 30°C and convective heat transfer coefficient is 10 W/(m^2 K).

[**Ans.** r_1 = 1 m, r_2 = 1.2 m, T_1 = −20°C, T_∞ = 30°C, k_i = 0.05 W/(m K), h = 10 W/(m^2 K); Heat flow rate, $q = \frac{(T_\infty - T_1)}{\frac{1}{4\pi}\left[\frac{1}{k_i}\left(\frac{1}{r_1} - \frac{1}{r_2}\right) + \frac{1}{h r_2^2}\right]}$.Substitution gives q = 184.65 W; Using heat flow equation between shell radius r_1 and radius r_0 where the temperature is 0°C, $184.65 = \frac{(0 - T_1)}{\frac{1}{4\pi k_i}\left(\frac{1}{r_1} - \frac{1}{r_0}\right)}$, which gives r_0 = 1.073 m, i.e. the thickness of the insulation which experiences temperature $\leq$ 0°C is 73 mm.]

2.17 A cylinder is covered with three different insulation materials A, B and C as shown in Fig. 2.45. Determine the heat flow rate per unit length of the cylinder. Assume radial heat flow only.

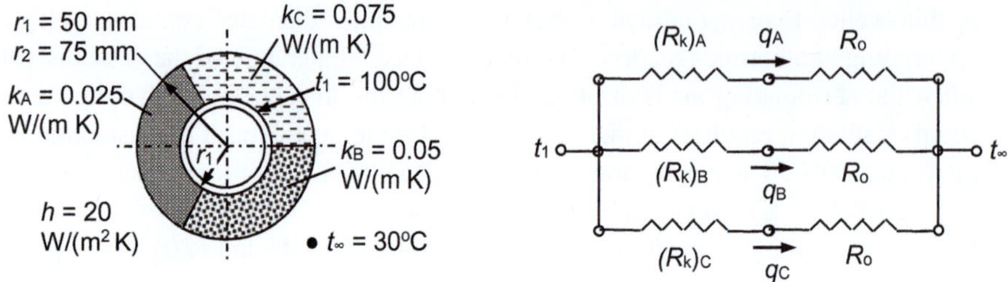

Fig. 2.45 Problem 2.16

[**Ans**. Heat flows through three paths comprising the insulation materials *A*, *B* and *C* (see the thermal network).

$$\frac{q}{L} = \frac{q_A}{L} + \frac{q_B}{L} + \frac{q_C}{L} = \frac{1}{3}\left\{\frac{(t_1 - t_\infty)}{\frac{1}{2\pi}\left[\frac{1}{k_A}\ln\left(\frac{r_2}{r_1}\right) + \frac{1}{hr_2}\right]} + \frac{(t_1 - t_\infty)}{\frac{1}{2\pi}\left[\frac{1}{k_B}\ln\left(\frac{r_2}{r_1}\right) + \frac{1}{hr_2}\right]} + \frac{(t_1 - t_\infty)}{\frac{1}{2\pi}\left[\frac{1}{k_C}\ln\left(\frac{r_2}{r_1}\right) + \frac{1}{hr_2}\right]}\right\}$$

$$= 49.53\,\text{W/m}$$

Note: The insulations will have different temperatures but circumferential heat flow has not been considered.]

2.18 Cross-sectional area of a rod varies with x as $A_x = A_0e^{bx}$ where A_0 and b are constants. Determine temperature distribution equation for one-dimensional steady-state axial conduction heat flow in the rod. The thermal conductivity of the material is constant, and lateral surface of the rod is insulated.

[**Ans**. q_x = constant gives $-kA_x\frac{dT}{dx} = C$. Hence, $dT = \frac{-C}{kA_x}dx = \frac{-C}{kA_0}e^{-bx}dx$. Integration gives $T = \frac{C}{kA_0b}e^{-bx} + C_2 = C_1e^{-bx} + C_2$ where $C_1 = \frac{C}{kA_0b}$; Constants C_1 and C_2 can be determined from temperature boundary conditions.]

References

Fraas Arthur P (1989) Heat exchanger design, 2nd edn. Wiley, New York
Holman JP (1992) Adapted for SI units by White PRS. Heat transfer. McGraw-Hill Book Co., New York
Mills AF (1995) Heat and mass transfer. Irwin, USA

Extended Surfaces (Fins) 3

3.1 Introduction

The heat flow from or to a surface in contact with a fluid is given by

$$q = hA\Delta t \tag{3.1}$$

The heat transfer rate q in this case can be increased by increasing the surface area. This can be achieved by attaching metal pieces to the heat-transferring surface. These extended surfaces are known as fins. Fins are widely used in various engineering equipments. They are provided on the cylinders of air-cooled internal combustion engines. Small diameter metal rods attached vertically to the condenser tubes at the back of a domestic refrigerator also work as fins. Electrical appliances such as transformers and motors are provided with the fins for efficient dissipation of heat generated in these electrical equipments.

Fins of various shapes and in different arrangements have been employed. Some of them are shown in Fig. 3.1. The *longitudinal* fins are long metal strips attached to the surface as shown in Fig. 3.1a. These fins are commonly employed with tubes where the fluid moves along the axis of the tube. *Transverse fins*, Fig. 3.1b, are employed in cross-flow (flow perpendicular to the axis of the tube). *Disc-type transverse fins* are either welded to the tube surface or shrunk on to the tube. These fins may be attached to the tube in the form of helix (such as welding a metal strip to the tube continuously in the form of a helix). *Discontinuous and star fins* are other types of transverse fins. A majority of fins are of rectangular cross-section, Fig. 3.1d. Thick fins and fins made by casting are generally of the trapezoidal cross-section with rounded edges. *Spine- or stud-type fins* are basically metal cones, cylinders or pipes which extend from the pipe or tube surface. They are employed both in case of longitudinal and cross-flows. These fins are mechanically rugged and hence have longer life in corrosive atmosphere than thin plate fins.

Let us consider fins attached to the outer surface of a cylinder, as shown in Fig. 3.1, to increase the heat transfer rate from the cylinder surface to the surrounding. For the heat to be conducted to the fin, the fin temperature t_f must be lower than the cylinder surface temperature t_s. Hence, the difference between the fin and surrounding temperatures $(t_f - t_\infty)$ is less than the difference between the cylinder surface and surrounding temperatures $(t_s - t_\infty)$. Further, the temperature of the fin continuously decreases from the fin base (i.e. the cylinder wall) to the fin tip. Thus, each unit area of the fin surface is less effective in heat transfer (heat

R. Karwa, *Heat and Mass Transfer*,
https://doi.org/10.1007/978-981-15-3988-6_3

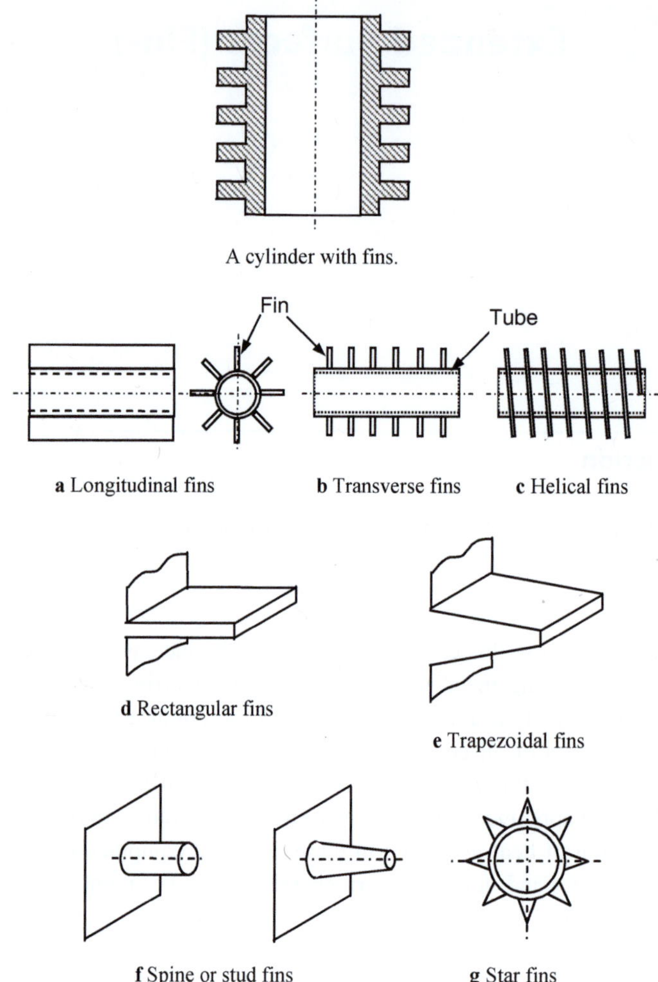

Fig. 3.1 Fin shapes and arrangements

transfer is proportional to the area × temperature difference) than the area of the cylinder surface to which the fins have been attached. However, the effectiveness with which fins transfer heat also depends on the fin profile (shape), its length and the number of fins provided per m^2 of the surface.

In the end, it may be noted that the final choice of the type of fin and their arrangement depends not only on the heat transfer performance but also on the resistance offered to the flow of the surrounding fluid, cost and the ease of fabrication.

3.2 Heat Transfer from a Fin of Uniform Cross-Section

Figure 3.2 shows a fin of uniform cross-section projecting from a wall, which is to be cooled. Let the fin is of uniform cross-section A_c throughout its length and the wall is at temperature t_s. The heat, which is conducted to the fin, is rejected by convection from the fin surface to the surrounding fluid at temperature t_∞.

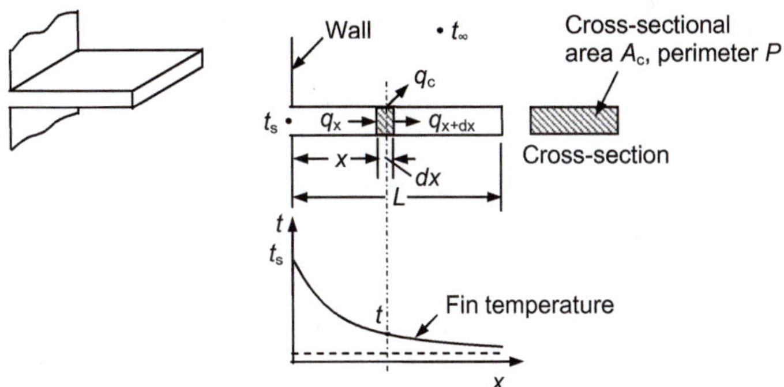

Fig. 3.2 Heat balance on a fin of uniform cross-section

An exact analytical solution of the heat propagation through a fin involves considerable difficulties. However, an approximate solution can be made using the following assumptions:

(i) Temperature at any cross-section of the fin is uniform. The assumption is valid for fins of small cross-section and made of high thermal conductivity material. This assumption reduces the problem to that of one-dimensional heat conduction along the axis of the fin only.
(ii) Steady-state condition.
(iii) Thermal conductivity of the fin material is constant, and
(iv) The heat transfer coefficient h from the fin surface to the surrounding is constant for the entire fin surface.

Most of the fins used in practice are plate type, Fig. 3.1a, with small thickness of the plate and the error due to the first approximation is less than 1%. The greatest uncertainty is in the value of the heat transfer coefficient, which is seldom uniform over the entire fin surface. In the case of severe non-uniformity of the heat transfer coefficient and two- or three-dimensional conduction heat flow, numerical techniques are used to solve the problem.

Consider a very small elemental length dx of the fin at a distance x from the wall surface, Fig. 3.2. The rate of heat flow into the element by conduction is

$$q_x = -kA_c \frac{dt}{dx} \tag{i}$$

The rate of heat flow out of the element by conduction is

$$q_{x+dx} = q_x + \frac{d}{dx}(q_x)dx$$

Heat transfer to the surrounding fluid by convection from the surface area (perimeter × dx) of the element from Newton's law is

$$q_c = hA(t - t_\infty) = h(Pdx)(t - t_\infty)$$

where t is the temperature of the element and P is the perimeter of the fin.

In the steady state, heat inflow by conduction must equal the heat outflow by conduction and convection, i.e.

$$q_x = q_{x+\delta x} + q_c$$

or

$$q_x = q_x + \frac{d}{dx}(q_x)dx + h(Pdx)(t - t_\infty)$$

or

$$\frac{d}{dx}(q_x)dx + h(Pdx)(t - t_\infty) = 0$$

Substitution of the value of q_x from Eq. (i) gives

$$\frac{d}{dx}\left(-kA_c\frac{dt}{dx}\right)dx + h(Pdx)(t - t_\infty) = 0$$

or

$$\frac{d^2t}{dx^2} = \frac{hP}{kA_c}(t - t_\infty) \tag{ii}$$

Let $(t - t_\infty) = \theta$, then

$$\frac{d^2t}{dx^2} = \frac{d^2\theta}{dx^2}$$

Substitution in Eq. (ii) gives

$$\frac{d^2\theta}{dx^2} = \frac{hP}{kA_c}\theta$$

Putting $\frac{hP}{kA_c} = m^2$, the equation is transformed into

$$\frac{d^2\theta}{dx^2} = m^2\theta$$

which is a second-order differential equation. Its general solution is

$$\theta = C_1e^{mx} + C_2e^{-mx} \tag{3.2}$$

The constants C_1 and C_2 are to be determined from the boundary conditions of the problem. For this purpose, we shall discuss three different cases.

3.2.1 A Very Long Fin

If the fin is very long compared to its cross-section dimensions, then the fin end approaches the surrounding fluid temperature. The boundary conditions for this case are

(i) at $x = 0$, $t = t_s$ and
(ii) as $x \rightarrow \infty$, $t \rightarrow t_\infty$

The first condition gives

$$t_s - t_\infty = C_1 + C_2 \tag{i}$$

and the second condition gives

$$t_\infty - t_\infty = 0 = C_1 e^{m\infty} + C_2 e^{-m\infty}$$

or

$$0 = C_1 e^{m\infty}$$

This gives $C_1 = 0$.
Hence, from Eq. (i),

$$C_2 = t_s - t_\infty$$

Substitution in Eq. (3.2) gives

$$\theta = (t - t_\infty) = (t_s - t_\infty)e^{-mx}$$

or

$$\frac{(t - t_\infty)}{(t_s - t_\infty)} = e^{-mx} \tag{3.3}$$

The equation reveals that the temperature along the fin length varies exponentially as depicted in Fig. 3.2.

Heat rejection rate from the entire surface area of the fin q_{fin} equals the heat flow by conduction at the fin base, i.e.

$$q_{fin} = -kA_c \left(\frac{dt}{dx}\right)_{x=o}$$

Using Eq. (3.3), we have

$$q_{fin} = -kA_c \left\{\frac{d}{dx}[(t_s - t_\infty)e^{-mx} + t_\infty]\right\}_{x=o}$$

$$q_{fin} = mkA_c(t_s - t_\infty) = \sqrt{\frac{hP}{kA_c}} \times kA_c(t_s - t_\infty)$$

or

$$q_{fin} = \sqrt{hPkA_c}(t_s - t_\infty) \tag{3.4}$$

3.2.1.1 Comments on Parameter *M*

From the equation of $m = \sqrt{(hP/kA_c)}$, it follows that m is proportional to $\sqrt{h}$, inversely proportional to $\sqrt{k}$ and is a function of fin geometry presented by $\sqrt{(P/A_c)}$. Figure 3.3 shows the dependence of the temperature distribution in the fin along its length for different values of parameter m. For a fin made of high thermal conductivity material, which will give a low value of m for a given fin and heat transfer coefficient h, a large excess temperature $(t - t_\infty)$ is obtained along the length of the fin. Similar effect will be seen for low values of the heat transfer coefficient and ratio of fin perimeter to its area of cross-section.

3.2.2 Negligible Heat Transfer from the Fin End as Compared to the Heat Transferred from the Fin Surface ($A_c \ll PL$)

This situation can occur when the area of the fin end is very small in comparison to the surface area of the entire fin ($A_c \ll PL$) as in the case of a thin long fin. Mathematically, this can be stated by equating the heat reaching the fin end by conduction to zero, which is equivalent to an *insulated tip fin*. Hence,

$$q_{\text{end}} = -kA_c\left(\frac{dt}{dx}\right)_{x=L} = 0$$

or

$$\left(\frac{dt}{dx}\right)_{x=L} = 0$$

The differentiation of Eq. (3.2) with respect to x gives

$$\frac{dt}{dx} = C_1 m e^{mx} + C_2(-m)e^{-mx}$$

Hence,

$$\left(\frac{dt}{dx}\right)_{x=L} = 0 = C_1 m e^{mL} - mC_2 e^{-mL}$$

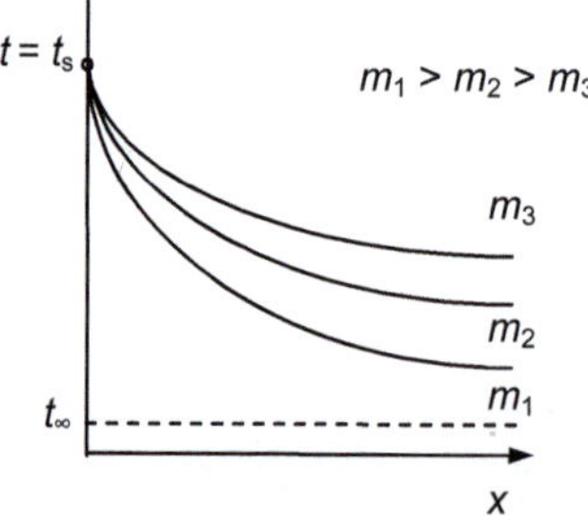

Fig. 3.3 Effect of fin parameter m on temperature distribution along the fin

or

$$C_1 e^{mL} - C_2 e^{-mL} = 0 \tag{i}$$

At the fin base ($x = 0$), the temperature is t_s, i.e.

$$t_s - t_\infty = C_1 + C_2 \tag{ii}$$

From Eqs. (i) and (ii),

$$C_1 = \frac{t_s - t_\infty}{1 + e^{2mL}}$$
$$C_2 = \frac{t_s - t_\infty}{1 + e^{-2mL}}$$

Substitution of the values of constant C_1 and C_2 in Eq. (3.2) gives

$$\theta = \frac{t_s - t_\infty}{1 + e^{2mL}} e^{mx} + \frac{t_s - t_\infty}{1 + e^{-2mL}} e^{-mx}$$

or

$$\frac{t - t_\infty}{t_s - t_\infty} = \frac{e^{mx}}{1 + e^{2mL}} + \frac{e^{-mx}}{1 + e^{-2mL}}$$

Using the relation $\cosh mL = \frac{e^{mL} + e^{-mL}}{2}$, the above equation can be transformed into

$$\frac{t - t_\infty}{t_s - t_\infty} = \frac{\cosh m(L - x)}{\cosh mL} \tag{3.5}$$

The rate of heat transfer from the fin equals the heat conducted into the fin base, i.e.

$$q_{fin} = -kA_c \left(\frac{dt}{dx}\right)_{x=o}$$

Using Eq. (3.5), we have

$$q_{fin} = -kA_c \left\{ \frac{d}{dx} \left[(t_s - t_\infty) \frac{\cosh m(L - x)}{\cosh mL} + t_\infty \right] \right\}_{x=o}$$

or

$$q_{fin} = -kA_c (t_s - t_\infty) \frac{1}{\cosh mL} \left\{ \frac{d}{dx} [\cosh m(L - x)] \right\}_{x=o}$$

or

$$q_{fin} = -kA_c (t_s - t_\infty) \frac{1}{\cosh mL} [-m \sinh m(L - x)]_{x=o}$$

or

$$q_{fin} = kA_c m(t_s - t_\infty)\frac{\sinh mL}{\cosh mL}$$

Substituting the value of m, we obtain the heat transfer equation as

$$q_{fin} = \sqrt{hPkA_c}(t_s - t_\infty)\tanh mL \tag{3.6}$$

The values of exponential and hyperbolic functions are tabulated in Table 3.1 and Fig. 3.4 shows variation of tanh(mL) with mL. It can be seen that for mL = 5.0, tanh(mL) = 0.9999 ≈ 1

Table 3.1 Exponential and hyperbolic functions

mL	e^{mL}	e^{-mL}	sinh(mL)	cosh(mL)	tanh(mL)
0.0	1.00	1.00	0.00	1.00	0.00
0.1	1.105	0.905	0.10	1.005	0.0997
0.2	1.221	0.819	0.20	1.020	0.1973
0.3	1.350	0.741	0.305	1.045	0.2913
0.4	1.492	0.670	0.411	1.0311	0.380
0.5	1.649	0.607	0.521	1.128	0.462
0.6	1.822	0.549	0.637	1.186	0.537
0.7	2.014	0.497	0.759	1.255	0.6044
0.8	2.226	0.449	0.888	1.337	0.664
0.9	2.460	0.407	1.027	1.433	0.7163
1.0	2.718	0.368	1.175	1.543	0.7616
1.1	3.004	0.333	1.336	1.669	0.8005
1.2	3.320	0.301	1.509	1.811	0.8337
1.3	3.670	0.272	1.698	1.971	0.862
1.4	4.055	0.247	1.904	2.151	0.8854
1.5	4.482	0.223	2.129	2.352	0.905
1.6	4.953	0.202	2.376	2.577	0.922
1.7	5.474	0.1827	2.646	2.828	0.9354
1.8	6.050	0.1653	2.942	3.107	0.947
1.9	6.686	0.150	3.268	3.418	0.956
2.0	7.389	0.1353	3.627	3.762	**0.964**
2.1	8.166	0.1224	4.022	4.144	0.9705
2.2	9.025	0.111	4.457	4.568	0.976
2.3	9.974	0.100	4.937	5.037	0.980
2.4	11.02	0.0907	5.466	5.557	0.984
2.5	12.18	0.0821	6.05	6.132	0.987
2.6	13.46	0.074	6.695	6.770	0.989
2.7	14.88	0.067	7.406	7.473	0.991
2.8	16.445	0.061	8.192	8.253	0.9926
2.9	18.174	0.055	9.060	9.115	0.994
3.0	20.09	0.050	10.018	10.068	**0.995**
4.0	54.60	0.0183	27.29	27.31	0.9993
5.0	148.4	6.74×10^{-3}	74.20	74.21	**0.9999**

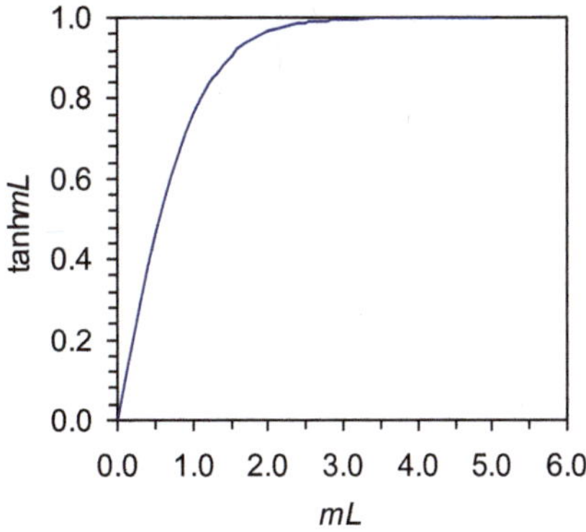

Fig. 3.4 Variation of tanhmL with mL

and Eqs. (3.4) and (3.6) give the same values of rate of heat transfer. Since the advantage gained by using a fin length greater than given by mL = 2 is negligible (tanhmL = 0.964 for mL = 2), this must be considered the maximum useful limit of the fin length.

3.2.3 Short Fins (Fin with Heat Loss from the Fin End)

In the case of short fins, the heat loss from the fin end cannot be neglected. The heat loss from the fin end equals the heat reaching the end by conduction, i.e.

$$h_L A_c (t_L - t_\infty) = -kA_c \left(\frac{dt}{dx}\right)_{x=L} \tag{i}$$

where h_L is the heat transfer coefficient at the fin end.

Equation (3.2) can be written as

$$\theta = C_1(\sinh mx + \cosh mx) + C_2(\cosh mx - \sinh mx)$$

or

$$(t - t_\infty) = (C_1 - C_2)\sinh mx + (C_1 + C_2)\cosh mx$$

or

$$(t - t_\infty) = C_3 \sinh mx + C_4 \cosh mx$$

At x = 0, t = t_s. This gives

$$(t_s - t_\infty) = C_4$$

Hence,

$$(t - t_\infty) = C_3 \sinh mx + (t_s - t_\infty)\cosh mx \tag{ii}$$

Using this temperature relation in Eq. (i), we get

$$-kA_c\left[\frac{d}{dx}[C_3\sinh mx+(t_s-t_\infty)\cosh mx+t_\infty]\right]_{x=L}=h_LA_c[C_3\sinh mL+(t_s-t_\infty)\cosh mL]$$

or

$$-k[C_3m\cosh mL+m(t_s-t_\infty)\sinh mL]=h_L[C_3\sinh mL+(t_s-t_\infty)\cosh mL]$$

Simplification gives

$$C_3=-(t_s-t_\infty)\frac{h_L\cosh mL+mk\sinh mL}{h_L\sinh mL+mk\cosh mL}$$

Substitution of the value of C_3 in Eq. (ii) gives

$$(t-t_\infty)=(t_s-t_\infty)\cosh mx-(t_s-t_\infty)\frac{h_L\cosh mL+mk\sinh mL}{h_L\sinh mL+mk\cosh mL}\sinh mx$$

Simplification gives

$$\frac{t-t_\infty}{t_s-t_\infty}=\frac{\cosh m(L-x)+(h_L/mk)\sinh m(L-x)}{(h_L/mk)\sinh mL+\cosh mL} \tag{3.7}$$

which is the temperature distribution equation.

The rate of heat transfer from the fin equals the heat conducted into the fin base, i.e.

$$q_{fin}=-kA_c\left(\frac{dt}{dx}\right)_{x=o}$$

Using Eq. (3.7), we have

$$q_{fin}=-kA_c\left\{\frac{d}{dx}\left[(t_s-t_\infty)\frac{\cosh m(L-x)+(h_L/mk)\sinh m(L-x)}{(h_L/mk)\sinh mL+\cosh mL}+t_\infty\right]\right\}_{x=o}$$

or

$$q_{fin}=-kA_c(t_s-t_\infty)m\left[\frac{-\sinh m(L-x)-(h_L/mk)\cosh m(L-x)}{(h_L/mk)\sinh mL+\cosh mL}\right]_{x=o}$$

or

$$q_{fin}=mkA_c(t_s-t_\infty)\left[\frac{\sinh m(L-x)+(h_L/mk)\cosh m(L-x)}{(h_L/mk)\sinh mL+\cosh mL}\right]_{x=o}$$

or

$$q_{fin}=mkA_c(t_s-t_\infty)\left[\frac{\sinh mL+(h_L/mk)\cosh mL}{(h_L/mk)\sinh mL+\cosh mL}\right]$$

or

$$q_{fin} = kmA_c(t_s - t_\infty)\frac{(h_L/mk) + \tanh mL}{(h_L/mk)\tanh mL + 1}$$

Substituting the value of m, we get the heat transfer equation as

$$q_{fin} = \sqrt{hPkA_c}(t_s - t_\infty)\frac{(h_L/mk) + \tanh mL}{(h_L/mk)\tanh mL + 1} \tag{3.8}$$

Equations (3.7) and (3.8) reduce to Eqs. (3.5) and (3.6) when $h_L = 0$, i.e. when there is no heat rejection from the fin end. Further value of factor $mL > 5$ reduces Eq. (3.8) to Eq. (3.4) of the very long fin.

The heat transfer equation of the short fin can be expressed in the same form as that of fin of Sect. 3.2.2 if a corrected length L_c, as defined below, is used in all equations of this case.

The heat lost, in case of a short fin, from the fin end is

$$h_L A_c(t_L - t_\infty) \tag{iii}$$

We extend the fin length L by ΔL, see Fig. 3.5, such that the heat lost by the lateral surface of the extended fin length equals the heat loss from the fin end given by Eq. (iii). The heat transfer from the lateral surface of the extended length is

$$h_L(P\Delta L)(t_L - t_\infty) \tag{iv}$$

assuming that the fin end temperature does not change when extended by length ΔL. Equating Eqs. (iii) and (iv), we get

$$h_L(P\Delta L)(t_L - t_\infty) = h_L A_c(t_L - t_\infty)$$

or

$$\Delta L = \frac{A_c}{P}$$

and the corrected length is

$$L_c = L + \Delta L$$

For a circular cross-section fin,

$$\Delta L = \frac{A_c}{P} = \frac{(\pi/4)d^2}{\pi d} = \frac{d}{4}$$

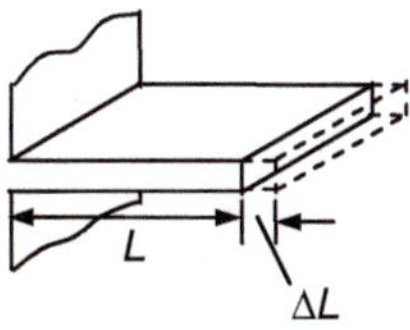

Fig. 3.5 Corrected fin length

For a rectangular cross-section fin,

$$\Delta L = \frac{A_c}{P} \approx \frac{W\delta}{2W} = \frac{\delta}{2}$$

Replacing L by L_c in Eqs. (3.5) and (3.6), we get the equations for the short fin as

$$\frac{t - t_\infty}{t_s - t_\infty} = \frac{\cosh m(L_c - x)}{\cosh mL_c} \tag{3.9}$$

and

$$q_{fin} = \sqrt{hPkA_c}(t_s - t_\infty)\tanh mL_c \tag{3.10}$$

The error which results from this approximation is very small if $(h\delta/2\ k)^{1/2} \leq 0.5$.

3.3 Hollow Fins

It is important to note that the equations developed here can be applied to hollow fins. For such fins, the perimeter will be the sum of the inside and outside if the heat transfer takes place from the inside surface also. The cross-sectional area is the solid area. The equations will not apply to these conditions unless the heat transfer coefficient at the inside and outside surfaces is equal. This is evident from the examination of the derivation.

3.4 Composite Fins

Such fins usually have a core of high thermal conductivity metal such as copper and a sheathing of steel. They are used in high-temperature applications where the core metal may get oxidized. The sheathing protects the core metal. Figure 3.6 shows the cross-section of such a fin.

The resistance to heat flow from the core caused by the sheath is usually very small compared with the external convection coefficient, and hence the fin cross-section at any axial position can be assumed to be at a uniform temperature. With this approximation, the conduction heat transfer equation at any section is

$$q = -(k_1A_1 + k_2A_2 + \ldots) \times \left(\frac{dt}{dx}\right)_x \tag{3.11}$$

and

$$m = \sqrt{\frac{hP}{(k_1A_1 + k_2A_2 + \ldots)}} \tag{3.12}$$

where P is the perimeter of the composite fin, i.e. the fin taken as a whole.

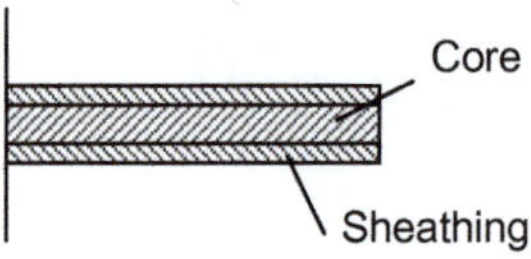

Fig. 3.6 A composite fin

Example 3.1 Heat generated in a bearing by friction causes the temperature at the shaft end to increase to 60°C above the ambient temperature. How is the temperature distributed along the shaft as we move away from the bearing? Calculate the amount of heat transferred through the shaft if the convective heat transfer coefficient for the shaft surface is 7 W/(m^2 K) and the thermal conductivity of the shaft material is 60 W/(m K). The shaft diameter is 60 mm and may be assumed a rod of infinite length.

Solution

The temperature distribution is given by

$$\frac{(t - t_\infty)}{(t_s - t_\infty)} = e^{-mx} \tag{3.3}$$

Here,

$$m = \sqrt{\frac{hP}{kA_c}} = \sqrt{\frac{h\pi d}{k(\pi/4)d^2}} = \sqrt{\frac{4h}{kd}} = \sqrt{\frac{4 \times 7}{60 \times 0.06}} = 2.79$$

and

$$(t_s - t_\infty) = 60^\circ\text{C}$$

Hence,

$$(t - t_\infty) = 60e^{-2.79x}$$

Heat rejection rate from the entire surface area of the shaft is

$$q_{fin} = \sqrt{hPkA_c}(t_s - t_\infty) \tag{3.4}$$

or

$$q_{fin} = \sqrt{7 \times \pi \times 0.06 \times 60 \times (\pi/4) \times (0.06)^2} \times 60 = 28.39\,\text{W}$$

Example 3.2 A 600-mm-long pump shaft of stainless steel [k = 20 W/(m K)] is 30 mm in diameter. The heat transfer coefficient for the outside surface of the shaft to the cooling air at 30°C is 30 W/(m^2 K). The impeller end of the shaft is immersed in a hot fluid which may raise the temperature of the shaft end to 250°C. At what distance the bearing must be placed so that the maximum temperature at the bearing is not greater than 80°C?

If the shaft is hollow with d_i = 20 mm, what will be the temperature at the bearing location determined above? The shaft is not cooled from the inner surface.

Solution

(i) **Solid Shaft**

Here

$$m = \sqrt{\frac{hP}{kA_c}} = \sqrt{\frac{h\pi d}{k(\pi/4)d^2}} = \sqrt{\frac{4h}{kd}} = \sqrt{\frac{4 \times 30}{20 \times 0.03}} = 14.14 \text{ m}^{-1}$$

and

$$mL = 8.485 > 5$$

The temperature distribution for this case is

$$\frac{(t - t_\infty)}{(t_s - t_\infty)} = e^{-mx} \tag{3.3}$$

Substitution gives

$$\frac{80 - 30}{250 - 30} = e^{-14.14x}$$

or

$$x = 0.1048 \text{ m}$$

(ii) **Hollow Shaft**

$$\begin{aligned} A_c &= (\pi/4)(d_o^2 - d_i^2) = (\pi/4)(0.03^2 - 0.02^2), \\ P &= \pi d_o = \pi \times 0.03, \\ m &= \sqrt{\frac{hP}{kA_c}} = 18.97 \text{ m}^{-1} \end{aligned}$$

and the product mL is again greater than 5.

The temperature distribution for this case is also given by

$$\frac{(t - t_\infty)}{(t_s - t_\infty)} = e^{-mx} \tag{3.3}$$

or

$$t = (t_s - t_\infty)e^{-mx} + t_\infty$$

At $x = 0.1048$ m it is

$$t = (250 - 30)e^{-18.97 \times 0.1048} + 30 = 60.12°\text{C}$$

The temperature is lower due to the reduced area for conduction heat transfer along the shaft.

Example 3.3 Three solid rods, made of silver [k_s = 420 W/(m K)], aluminium [k_a = 205 W/(m K)] and iron [k_i = 70 W/(m K)], are coated with a uniform layer of wax all around. The rods are placed vertically in boiling water bath with 250 mm length of each rod projecting outside. If all rods are 5 mm in diameter, 300 mm in length and have the same surface heat transfer coefficient of 40 W/(m^2 K), work out the ratio of lengths up to which the coated material will melt on each rod if the melting temperature of the wax is 40°C. The surrounding temperature is 20°C.

Solution

The rods work as fins of length 250 mm each. The corrected length for each rod is

$$L_c = L + d/4 = 250 + 5/4 = 251.25\,\text{mm}.$$

The value of parameter m is

$$m = \sqrt{\frac{hP}{kA_c}} = \sqrt{\frac{h\pi d}{k(\pi/4)d^2}} = \sqrt{\frac{4h}{kd}} = \sqrt{\frac{4\times 40}{k\times 0.005}} = \frac{178.9}{\sqrt{k}}$$

Hence, parameters m and mL_c for the three rods are

$$m_s = \frac{178.9}{\sqrt{k_s}} = \frac{178.9}{\sqrt{420}} = 8.73 \text{ and } m_sL_c = 8.73\times 0.25125 = 2.19$$
$$m_a = \frac{178.9}{\sqrt{k_a}} = \frac{178.9}{\sqrt{205}} = 12.5 \text{ and } m_aL_c = 12.5\times 0.25125 = 3.14$$
$$m_i = \frac{178.9}{\sqrt{k_i}} = \frac{178.9}{\sqrt{70}} = 21.38 \text{ and } m_iL_c = 21.38\times 0.25125 = 5.37$$

Let the lengths up to which wax will melt are L_1, L_2 and L_3 on the first, second and third rods, respectively. It means that at lengths L_1, L_2 and L_3, the three rods are having the same temperature, i.e. t = 40°C.

The equation of temperature distribution for a short fin is

$$\frac{t-t_\infty}{t_s-t_\infty} = \frac{\cosh m(L_c-x)}{\cosh mL_c} \tag{3.9}$$

or

$$0.25 = \frac{\cosh m(L_c-x)}{\cosh mL_c}$$

or

$$x = L_c - \frac{\cosh^{-1}(0.25\cosh mL_c)}{m} \tag{i}$$

For silver rod, let the length $x = L_1$ at which $t = 40°C$, then from Eq. (i),

$$L_1 = 0.25125 - \frac{\cosh^{-1}[0.25\cosh(2.19)]}{8.73} = 0.193\,\text{m}$$

Similarly for the aluminium and steel rods, the lengths L_2 and L_3, respectively, are

$$L_2 = 0.25125 - \frac{\cosh^{-1}[0.25\cosh(3.14)]}{12.5} = 0.113\,\text{m}$$

and

$$L_3 = 0.25125 - \frac{\cosh^{-1}[0.25\cosh(5.37)]}{21.38} = 0.065\,\text{m}$$

Hence,

$$L_1 : L_2 : L_3 :: 2.97 : 1.74 : 1$$

If we assume the fins to be very long, then the temperature distribution is given by

$$\frac{(t - t_\infty)}{(t_s - t_\infty)} = e^{-mx} \tag{3.3}$$

and the calculated lengths using this equation are

$$L_1 = 0.159\,\text{m},\ L_2 = 0.111\ \text{m},\ L_3 = 0.065\ \text{m}$$

Recalling that the product mL for these cases is 2.18, 3.13 and 5.36, respectively, based on the length L. The assumption of a very long fin ($mL \geq 5$) for the above-calculated values leads to the greatest error for the first case ($mL = 2.18$), a small error for the second case ($mL = 3.13$) and no error for the third case ($mL = 5.36$). Thus, it can be concluded that the assumption of a very long fin can lead to a significant error in the results if product mL is not ≥ 3.

Example 3.4 An aluminium fin [k = 200 W/(m K)] is 3.0 mm in thickness and 100 mm in width. It protrudes 75 mm from a wall at temperature of 200°C. The surrounding temperature is 20°C. Calculate heat loss from the fin if the surface heat transfer coefficient is 10 W/(m^2 K).

Solution

$$m = \sqrt{\frac{hP}{kA_c}} = \sqrt{\frac{h(2W + 2\delta)}{kW\delta}} = \sqrt{\frac{10 \times 206 \times 1000}{200 \times 100 \times 3}} = 5.86$$

$$mL = 5.86 \times 0.075 = 0.4395$$

$$h/mk = 10/(5.86 \times 200) = 8.532 \times 10^{-3}.$$

(a) From Eq. (3.8) of the short fin,

$$q_{fin} = \sqrt{hPkA_c}(t_s - t_\infty)\frac{(h_L/mk) + \tanh mL}{(h_L/mk)\tanh mL + 1}$$

$$q_{fin} = \left(\sqrt{10 \times 206 \times 10^{-3} \times 200 \times 100 \times 3 \times 10^{-6}}\right) \times (200 - 20)\frac{8.532\times10^{-3} + \tanh(0.4395)}{8.532\times10^{-3}\times\tanh(0.4395) + 1}$$

$$= 26.6\,\text{W}.$$

(b) Using corrected length L_c in Eq. (3.6), we have

$$q_{fin} = \sqrt{hPkA_c}(t_s - t_\infty)\tanh mL_c$$

where $L_c = L + \delta/2 = 0.075 + 0.0015 = 0.0765$ m. This gives

$$mL_c = 5.86 \times 0.0765 = 0.44829$$

and

$$q_{fin} = \left(\sqrt{10 \times 206 \times 10^{-3} \times 200 \times 100 \times 3 \times 10^{-6}}\right) \times (200 - 20)\tanh(0.44829)$$
$$= 26.61\,\text{W}.$$

The analysis clearly shows that even for $mL = 0.44$, the equation with corrected length can be used.

Example 3.5 A 20-mm-diameter and 500-mm-long steel rod [k = 50 W/(m K)] is projecting through a furnace wall as shown in Fig. 3.7. The rod portion in wall may be assumed insulated. Determine the temperature of the rod just outside the furnace wall. Comment on the result.

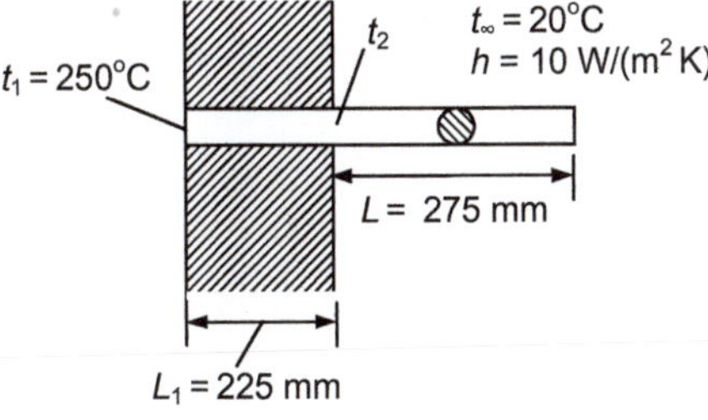

Fig. 3.7 Example 3.5

Solution

Rod projecting outside the wall acts as a fin of length L = 275 mm with base temperature t_2.

Fin parameter,

$$m = \sqrt{\frac{hP}{kA_c}}$$

$$= \sqrt{\frac{10 \times \pi \times 0.02}{50 \times (\pi/4) \times 0.02^2}} = 6.324$$

$$mL = 6.324 \times 0.275 = 1.739 < 3.$$

Since product mL is less than 3, the heat transfer from fin will be calculated from the following equation:

$$q_{fin} = \sqrt{hPkA_c}(t_2 - t_\infty)\tanh(mL_c)$$

where L_c is the corrected length = $L + d/4$ for circular cross-section fin. Hence,

$$q_{fin} = \sqrt{h \times \pi \times d \times k \times (\pi/4) \times d^2}(t_2 - t_\infty)\tanh[m(L + d/4)]$$

$$= \sqrt{10 \times \pi \times 0.02 \times 50 \times (\pi/4) \times 0.02^2}(t_2 - 20)\tanh[6.324 \times (0.275 + 0.02/4)]$$

$$= 0.09375 \times (t_2 - 20)$$

The heat flow by conduction through the rod length L_1 equals the heat rejected by the fin. Hence,

$$kA_c\frac{t_1 - t_2}{L_1} = 0.09375 \times (t_2 - 20)$$

or

$$50 \times (\pi/4) \times 0.02^2 \times \frac{250 - t_2}{0.225} = 0.09375 \times (t_2 - 20)$$

or

$$t_2 = 118.17\,^\circ\mathrm{C}.$$

Example 3.6 Obtain an expression for the optimum thickness of a straight rectangular fin. Use the equation of heat transfer for the fin of Sect. 3.2.2.

Solution

The weight of the fin shown in Fig. 3.8 is

$$W = \rho bL\delta$$

where ρ is the density of the fin material and b is the fin width.

Optimization here is to find a combination of δ and L for a given fin width b and fin area $A_1 = L\delta$.

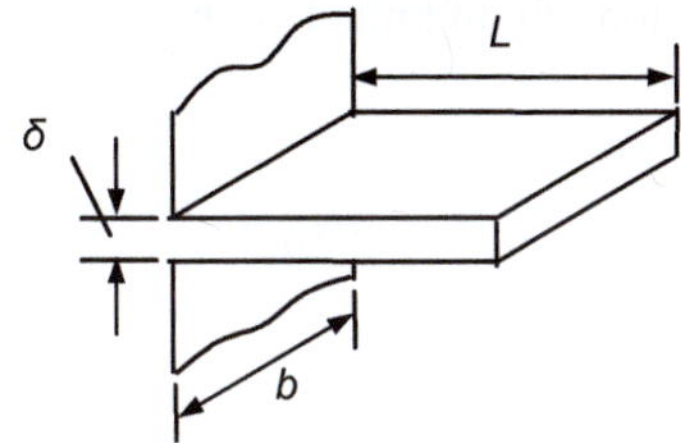

Fig. 3.8 Example 3.6

The heat flow through the fin in consideration is given by

$$q_{fin} = \sqrt{hPkA_c}(t_s - t_\infty)\tanh mL \tag{3.6}$$

Putting $L = A_1/\delta$, parameter $P \approx 2b$ (for $\delta \ll b$) and $m = \sqrt{(hP/kA_c)} = \sqrt{(2\,hb/kb\delta)} = \sqrt{(2\,h/k\delta)}$, the heat flow is

$$q_{fin} = b\sqrt{2hk\delta} \times \tanh\left(\sqrt{\frac{2h}{k\delta}} \times \frac{A_1}{\delta}\right)(t_s - t_\infty)$$

The heat flow will be maximum for $dq_{fin}/d\delta = 0$. Differentiating the equation of q_{fin} with respect to δ and equating to zero, we get

$$\frac{dq_{fin}}{d\delta} = 0$$

$$= \left[\begin{array}{c} b\sqrt{2hk} \times \dfrac{\delta^{-1/2}}{2} \times \tanh\left(\sqrt{\dfrac{2h}{k\delta}} \times \dfrac{A_1}{\delta}\right) + \\ b\sqrt{2hk\delta} \times \dfrac{1}{\cosh^2\left(\sqrt{\frac{2h}{k\delta}} \times \frac{A_1}{\delta}\right)} \sqrt{\dfrac{2h}{k}} \times A_1 \times \left(-\dfrac{3}{2}\delta^{-5/2}\right) \end{array}\right](t_s - t_\infty)$$

Simplification gives

$$\tanh\left(\sqrt{\frac{2h}{k\delta}} \times \frac{A_1}{\delta}\right) = \left[\frac{3\left(\sqrt{\frac{2h}{k\delta}} \times \frac{A_1}{\delta}\right)}{\cosh^2\left(\sqrt{\frac{2h}{k\delta}} \times \frac{A_1}{\delta}\right)}\right]$$

Let $\sqrt{\frac{2h}{k\delta}} \times \frac{A_1}{\delta} = a$, then

$$\tanh(a) = \frac{3a}{\cosh^2(a)}$$

Solution gives $a = 1.4192$. This gives

$$\frac{L}{\delta} = 0.71\sqrt{\frac{2k}{h\delta}} \tag{3.13}$$

Example 3.7 Show that for a long fin of finite length L the heat transfer from the fin surface to the surrounding is given by

$$q_{fin} = \sqrt{hPkA_c}(t_s - t_\infty)\coth mL$$

if at the fin end $t \approx t_\infty$.

Solution

From Eq. (3.2),

$$\theta = C_1 e^{mx} + C_2 e^{-mx} \tag{3.2}$$

The constants C_1 and C_2 are to be determined from the boundary conditions of the problem. The boundary conditions for this case are

(a) at $x = 0$, $t = t_s$, and
(b) at $x = L$, $t = t_\infty$.

The first condition gives

$$t_s - t_\infty = C_1 + C_2 \tag{i}$$

and the second condition gives

$$t_\infty - t_\infty = 0 = C_1 e^{mL} + C_2 e^{-mL}$$

or

$$C_1 e^{mL} + C_2 e^{-mL} = 0 \tag{ii}$$

Solution of Eqs. (i) and (ii) gives

$$C_1 = -\frac{e^{-mL}}{e^{mL} - e^{-mL}}(t_s - t_\infty)$$
$$C_2 = \frac{e^{mL}}{e^{mL} - e^{-mL}}(t_s - t_\infty)$$

Substitution of values of C_1 and C_2 in Eq. (3.2) gives

$$\frac{(t - t_\infty)}{(t_s - t_\infty)} = \frac{e^{m(L-x)} - e^{-m(L-x)}}{e^{mL} - e^{-mL}} = \frac{\sinh[m(L-x)]}{\sinh(mL)} \tag{3.14}$$

Heat rejection rate from the entire surface area of the fin q_{fin} equals the heat flow by conduction at the fin base, i.e.

$$q_{fin} = -kA_c\left(\frac{dt}{dx}\right)_{x=o}$$

Using Eq. (3.3), we have

$$q_{fin} = -kA_c\left\{\frac{d}{dx}\left[(t_s - t_\infty)\frac{\sinh[m(L-x)]}{\sinh(mL)} + t_\infty\right]\right\}_{x=o}$$

or

$$q_{fin} = -kA_c\frac{1}{\sinh(mL)}\{\cosh[m(L-x)]\}_{x=o}(-m)(t_s - t_\infty)$$

or

$$q_{fin} = kA_c m\frac{\cosh(mL)}{\sinh(mL)}(t_s - t_\infty)$$

or

$$q_{fin} = \sqrt{hPkA_c}(t_s - t_\infty)\coth mL \tag{3.15}$$

Equation (3.4) can be obtained by putting $L = \infty$.

Example 3.8 Show that for a fin of finite length L the heat transfer from the fin surface to the surrounding is given by

$$q_{fin} = \sqrt{hPkA_c}(t_s - t_\infty)\frac{\cosh(mL) - \frac{t_L - t_\infty}{t_s - t_\infty}}{\sinh(mL)}$$

if at the fin end temperature is t_L. Other terms have usual meaning.

Solution

From Eq. (3.2),

$$t_s - t_\infty = \theta = C_1 e^{mx} + C_2 e^{-mx}$$

The constants C_1 and C_2 are to be determined from the boundary conditions of the problem. The boundary conditions for this case are

(c) at $x = 0$, $t = t_s$, and
(d) at $x = L$, $t = t_L$.

The first condition gives

$$t_s - t_\infty = C_1 + C_2$$

or

$$\theta_s = C_1 + C_2 \tag{i}$$

and the second condition gives

$$t_L - t_\infty = C_1 e^{mL} + C_2 e^{-mL}$$

or

$$\theta_L = C_1 e^{mL} + C_2 e^{-mL} \tag{ii}$$

Solution of Eqs. (i) and (ii) gives

$$C_1 = \theta_s \frac{\theta_L/\theta_s - e^{-mL}}{e^{mL} - e^{-mL}}$$

$$C_2 = \theta_s \left(1 - \frac{\theta_L/\theta_s - e^{-mL}}{e^{mL} - e^{-mL}}\right)$$

Substitution of values of C_1 and C_2 in Eq. (3.2) gives

$$\frac{\theta}{\theta_s} = \left(\frac{\theta_L/\theta_s - e^{-mL}}{e^{mL} - e^{-mL}}\right) e^{mx} + \left(1 - \frac{\theta_L/\theta_s - e^{-mL}}{e^{mL} - e^{-mL}}\right) e^{-mx}$$

$$= \frac{(\theta_L/\theta_s)(e^{mx} - e^{-mx}) + \left[e^{m(L-x)} - e^{-m(L-x)}\right]}{e^{mL} - e^{-mL}}$$

or

$$\frac{t - t_\infty}{t_s - t_\infty} = \frac{\frac{t_L - t_\infty}{t_s - t_\infty} \sinh mx + \sinh m(L - x)}{\sinh mL} \tag{3.1}$$

Heat rejection rate from the entire surface area of the fin q_{fin} equals the heat flow by conduction at the fin base, i.e.

$$q_{fin} = -kA_c \left(\frac{dt}{dx}\right)_{x=o}$$

Using Eq. (3.1), we have

$$q_{fin} = -kA_c \left\{\frac{d}{dx}\left[(t_s - t_\infty)\frac{\frac{t_L - t_\infty}{t_s - t_\infty}\sinh mx + \sinh m(L - x)}{\sinh mL} + t_\infty\right]\right\}_{x=o}$$

or

$$q_{fin} = -kA_c \frac{1}{\sinh(mL)} \left\{\frac{t_L - t_\infty}{t_s - t_\infty}\cosh mx - \cosh[m(L - x)]\right\}_{x=o} (m)(t_s - t_\infty)$$

or

$$q_{fin} = kA_c m \frac{\cosh(mL) - \frac{t_L - t_\infty}{t_s - t_\infty}}{\sinh(mL)}(t_s - t_\infty)$$

or

$$q_{fin} = \sqrt{hPkA_c}(t_s - t_\infty)\frac{\cosh(mL) - \frac{t_L - t_\infty}{t_s - t_\infty}}{\sinh(mL)} \tag{3.2}$$

Check: If $t_L = t_\infty$, Eq. (3.2) gives

$$q_{fin} = \sqrt{hPkA_c}(t_s - t_\infty)\coth mL$$

which is Eq. (3.15).

3.5 Effectiveness and Efficiency of Fins

To study the performance of a fin in transferring the heat, a term called *fin effectiveness* ε_{fin} is used. It is a ratio of heat transfer with the fin to the heat that would be transferred without fin. Effectiveness of a fin must be always greater than 1. Mathematically, it can be expressed as

$$\varepsilon_{fin} = \frac{q_{fin}}{hA_c(t_s - t_\infty)} \tag{3.16}$$

Thus for a very long fin (Sect. 3.2.1), it is

$$\varepsilon_{fin} = \frac{\sqrt{hPkA_c}(t_s - t_\infty)}{hA_c(t_s - t_\infty)} = \sqrt{\frac{Pk}{hA_c}} \tag{3.17}$$

From the above result, it can be concluded that the fin effectiveness is high when

(i) the fin is made of high thermal conductivity material,
(ii) the fin has high ratio of parameter to the base area (P/A_c) and
(iii) the convective heat transfer coefficient is low.

The third conclusion indicates that the fins are very effective in transferring heat from surfaces with low heat transfer coefficient.

Another term that is used to assess the performance of a fin in transferring heat is *fin efficiency* η_{fin}, which is defined as

$$\eta_{fin} = \frac{q_{fin}}{q_{ideal}} \tag{i}$$

where q_{ideal} is the heat that would be transferred if the entire fin surface were at the fin base temperature. That is,

$$\eta_{fin} = \frac{q_{fin}}{hA_f(t_s - t_\infty)} \tag{ii}$$

where A_f ($=PL$) is the fin surface area.

Considering the fin of Sect. 3.2.2 (a fin with negligible heat transfer from its end as compared to its surface), the fin efficiency is

$$\eta_{fin} = \frac{\sqrt{hPkA_c}(t_s - t_\infty)\tanh mL}{hPL(t_s - t_\infty)}$$
$$= \sqrt{\frac{kA_c}{hP}}\frac{\tanh mL}{L}$$
$$= \frac{\tanh mL}{mL}$$

Mathematically, for small values of *mL*,

$$\frac{\tanh mL}{mL} = 1 - \frac{(mL)^2}{3} \quad \text{(iii)}$$

while for large values of *mL*,

$$\frac{\tanh mL}{mL} \approx \frac{1}{mL} \quad \text{(iv)}$$

From Eq. (iii), it can be seen that as mL tends to zero, $\frac{\tanh mL}{mL}$ tends to the limiting value of unity. From Eq. (iv), and also from the variation of tanh(mL) with mL presented in Table 3.1, it can be seen that the fin efficiency $\eta_{\text{fin}}\left(= \frac{\tanh mL}{mL}\right)$ decreases rapidly as mL increases. Typically at $mL = 1.0$, $\eta_{\text{fin}} = 71.63\%$; for $mL = 2.0$, it is 48.2% and it reduces to 25% at $mL = 4.0$.

For the trivial case of $mL = 0$, when $L = 0$ (i.e. no fin at all), the fin efficiency reaches its maximum value of unity. This situation is meaningless. Thus, it may be concluded that the fin efficiency cannot be maximized with respect to its length. Generally, the fin performance is maximized with respect to its mass, volume or cost. This maximization process has economic meaning also.

As regards the parameter m, a small value of this parameter leads to a high fin efficiency, which is achieved when

(i) fin is made of high thermal conductivity material,
(ii) fin has low P/A_c ratio, i.e. a thick rectangular, square or circular cross-section fin, and
(iii) the convection heat transfer coefficient is low.

The first and third conclusions are the same for both fin effectiveness and efficiency. It means that for high effectiveness and efficiency, the fins must be made of high thermal conductivity materials and must be provided on surfaces with low value of heat transfer coefficient. It will be shown by taking an illustrative example that the fins may reduce the heat transfer rate in applications where very high values of heat transfer coefficient are encountered.

The second conclusion for high fin efficiency, regarding the ratio of fin parameter to cross-sectional area, is opposite to that for high effectiveness.

Since the basic purpose of the installation of fins on a surface is to enhance the heat transfer rate, the fin effectiveness is a more meaningful parameter for assessing the usefulness of the fins.

The fin effectiveness can be related to the fin efficiency. From Eq. (ii),

$$q_{fin} = \eta_{fin} hA_f(t_s - t_\infty)$$

Substitution in Eq. (3.16) gives

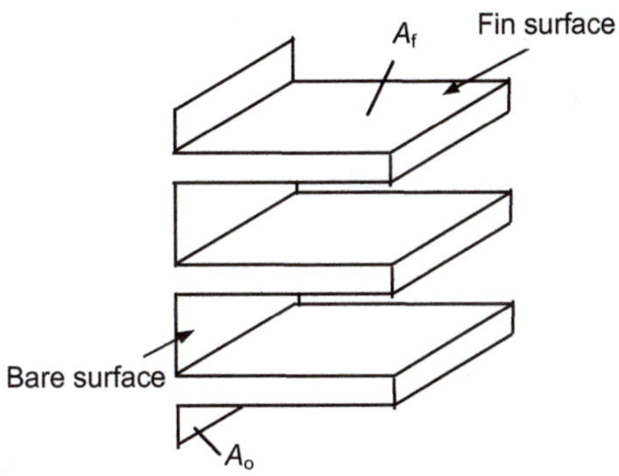

Fig. 3.9 Surface with multiple fins

$$\varepsilon_{fin} = \frac{\eta_{fin} h A_f (t_s - t_\infty)}{h A_c (t_s - t_\infty)}$$

or

$$\varepsilon_{fin} = \eta_{fin}\left(\frac{A_f}{A_c}\right) \tag{3.18}$$

It means that the fin effectiveness equals the fin efficiency multiplied by the ratio of fin surface and cross-sectional areas.

The analysis given above applies only to a single fin. Usually, a surface with a number of fins consists of bare portion between the fins, refer to Fig. 3.9. In such cases, the total heat transfer from the surface is calculated by combining the heat flow from the bare portion of the surface with the heat flow from the fins. Accordingly, the efficiency of the finned surface can be defined as the ratio of the total heat transfer rate of the combined area of the bare surface A_o and that of the fin A_f to the heat which would be transferred if this total area were maintained at the fin base temperature t_s.

The total heat flow can be expressed for a surface with multiple fins as

$$\begin{aligned} q_{total} &= q_{fin} + q_o \\ &= hA_f(t_s - t_\infty)\eta_{fin} + hA_o(t_s - t_\infty) \\ &= h(\eta_{fin}A_f + A_o)(t_s - t_\infty) \end{aligned} \tag{3.19a}$$

The total heat flow can also be found in terms of the effectiveness. In this case

$$\begin{aligned} q_{total} &= q_{fin} + q_o \\ &= \varepsilon_{fin} hA_b(t_s - t_\infty) + hA_o(t_s - t_\infty) \\ &= h(\varepsilon_{fin}A_b + A_o)(t_s - t_\infty) \end{aligned} \tag{3.19b}$$

where A_b is the sum of the base area of fins = *number of fins* × A_c.

The overall efficiency of the finned surface can be expressed as

$$\eta_{total} = \frac{hA_f(t_s - t_\infty)\eta_{fin} + hA_o(t_s - t_\infty)}{h(A_f + A_o)(t_s - t_\infty)}$$

or

$$\eta_{total} = \frac{A_f\eta_{fin} + A_o}{(A_f + A_o)} = \frac{A_f\eta_{fin} + A_o}{A_{fw}}$$

where A_{fw} = total surface area of the finned wall = $A_f + A_o$. Putting $A_o = A_{fw} - A_f$, we get

$$\eta_{total} = \frac{A_f\eta_{fin} + A_{fw} - A_f}{A_{fw}}$$

or

$$\eta_{total} = \frac{A_{fw} - (1 - \eta_{fin})A_f}{A_{fw}}$$

or

$$\eta_{total} = 1 - \frac{A_f}{A_{fw}}(1 - \eta_{fin}) \tag{3.19c}$$

In Eq. (3.19a, 3.19b, 3.19c), the values of the heat transfer coefficient from the fins and bare portion of the wall have been assumed to be equal. In fact they differ. Let the values of these coefficients are h_w and h_f for the base portion and the fins, respectively. Then the heat flow rate from the wall is

$$\begin{aligned} q_{total} &= q_{fin} + q_o \\ &= h_fA_f(t_s - t_\infty)\eta_{fin} + h_wA_o(t_s - t_\infty) \end{aligned} \tag{i}$$

Let

$$q_{total} = h_{red}A_{fW}(t_s - t_\infty) \tag{ii}$$

where $A_{fw} = A_f + A_o$ is the total area of the surface transferring the heat.

Comparing Eqs. (i) and (ii), we get

$$h_{red} = h_f\eta_{fin}\frac{A_f}{A_{fw}} + h_w\frac{A_o}{A_{fw}} \tag{3.20}$$

The coefficient of heat transfer h_{red} is called the *reduced heat transfer coefficient* and is a mean or effective heat transfer coefficient for a finned wall, which accounts for the heat removal from the bare portion of the wall between the fins, fin surface and fin efficiency.

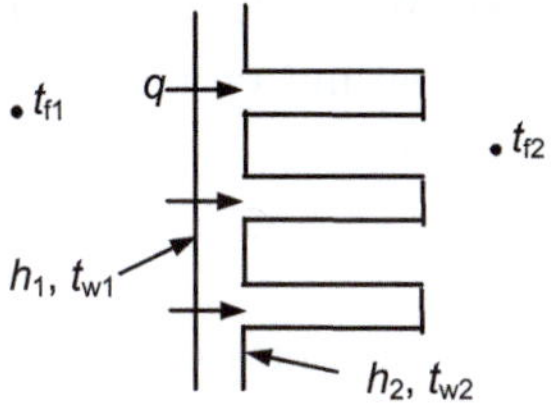

Fig. 3.10 A finned wall

3.6 Heat Transfer from a Finned Wall

The heat flow through a finned wall, with plane side at temperature t_{w1} and the finned side at temperature t_{w2}, can be described by the following set of equations (Fig. 3.10):

$$q = h_1 A_1 (t_{f1} - t_{w1})$$
$$q = \frac{k}{\delta} A_1 (t_{w1} - t_{w2})$$
$$q = h_{red} A_{fw} (t_{w2} - t_{f2})$$

Reducing the equation to

$$\frac{q}{A_1} = \frac{(t_{f1} - t_{f2})}{\frac{1}{h_1} + \frac{\delta}{k} + \frac{1}{h_{red}}\left(\frac{A_1}{A_{fw}}\right)} = U(t_{f1} - t_{f2}) \tag{3.21}$$

where U is the overall heat transfer coefficient for heat flow through a finned wall and is defined as

$$U = \frac{1}{\frac{1}{h_1} + \frac{\delta}{k} + \frac{1}{h_{red}}\left(\frac{A_1}{A_{fw}}\right)} \tag{3.22}$$

The ratio of the area of the fins' surface A_{fw} to the base surface area A_1 is called as *finning factor*.

3.7 Intensification of Heat Transfer by Finning

Use of fin on a heat transfer surface is an effective method of heat transfer enhancement when the heat transfer coefficient for this wall is much smaller compared to the other side. This is illustrated by a numerical problem given below.

For heat transfer through any wall, the film resistance is $1/hA$. Thus, the film resistance depends not only on the heat transfer coefficients but also on the size of the surfaces represented by area A. Hence, the film resistance can be decreased by increasing the surface area by finning.

Example 3.9 For a wall, h_1 = 1000 W/(m^2 K) and h_2 = 10 W/(m^2 K). Determine the enhancement of heat transfer by providing fins on the wall with the lower heat transfer coefficient. The finning factor is 3 and $h_{red}= h_2$.

Solution

The overall heat transfer coefficient for the surface without fins is (neglecting wall resistance)

$$U = \frac{1}{\frac{1}{1000} + \frac{1}{10}} = 9.90\,\text{W}/\left(\text{m}^2\text{K}\right)$$

When surface is finned,

$$U = \frac{1}{\frac{1}{1000} + \frac{1}{10}\left(\frac{1}{3}\right)} = 29.13\,\text{W}/\left(\text{m}^2\text{K}\right)$$

Example 3.10 The following data refer to a heat exchanger, which transfers heat from a heated liquid to air:

Liquid-side surface area, A_l = 2 m^2
Air-side finned surface area, A_f = 7 m^2
Air-side total area, A_{fw} = 10 m^2
Fins:
Length = 15 mm
Thickness = 0.5 mm
Material: steel [k = 45 W/(m K)]
Heat transfer coefficients:
Liquid side, h_l = 250 W/(m^2 K)
Air side, h_a = 50 W/(m^2 K).

Determine the overall heat transfer coefficient U based on the liquid-side area. Neglect the wall resistance to conduction heat flow. Compare the value with surface without fins.

Solution

The overall heat transfer coefficient for a finned wall is given by (Eq. 3.22)

$$U = \frac{1}{\frac{1}{h_l} + \frac{\delta}{k} + \frac{1}{h_{red}}\left(\frac{A_l}{A_{fw}}\right)}$$

Neglecting wall resistance, we have

$$U = \frac{1}{\frac{1}{h_l} + \frac{1}{h_{red}}\left(\frac{A_l}{A_{fw}}\right)} \tag{i}$$

where

$$h_{red} = h_f \eta_{fin} \frac{A_f}{A_{fw}} + h_w \frac{A_o}{A_{fw}} \tag{ii}$$

Here $h_f = h_w = 50$, $A_f = 7$ m^2, $A_{fw} = 10$ m^2, $A_o = A_{fw} - A_f = 10 - 7 = 3$ m^2 and let

$$\eta_{fin} = \frac{\tanh mL}{mL} \tag{3.18}$$

Putting $m = \sqrt{(2h/k\delta)} = \sqrt{(2 \times 50 \times 1000/45 \times 0.5)} = 66.67$, we have

$$\eta_{fin} = \frac{\tanh(66.67 \times 0.015)}{66.67 \times 0.015} = 0.7616$$

Substituting the values of various terms in Eq. (ii) gives

$$h_{red} = 50 \times 0.7616 \times \frac{7}{10} + 50 \times \frac{3}{10} = 41.65\,\text{W}/(\text{m}^2\text{K})$$

And this gives

$$U = \frac{1}{\frac{1}{250} + \frac{1}{41.65}\left(\frac{2}{10}\right)} = 113.61\,\text{W}/(\text{m}^2\text{K})$$

For the wall without fins, the overall heat transfer coefficient is

$$U = \left(\frac{1}{h_l} + \frac{1}{h_a}\right)^{-1} = \left(\frac{1}{250} + \frac{1}{50}\right)^{-1} = 41.67\,\text{W}/(\text{m}^2\text{K})$$

Example 3.11 A finned tube of heat exchanger, shown in Fig. 3.11, carries hot water which is cooled by blowing air over the fins. Determine the overall heat transfer coefficient.

Solution

From Eqs. (3.22) and (2.29a), overall heat transfer coefficient for the present case is

$$U_i = \frac{1}{\frac{1}{h_i} + \frac{A_i}{2\pi kL}\ln\left(\frac{R_0}{R_i}\right) + \frac{1}{h_{red}}\left(\frac{A_i}{A_{fw}}\right)}$$

where h_{red} is reduced heat transfer coefficient given by

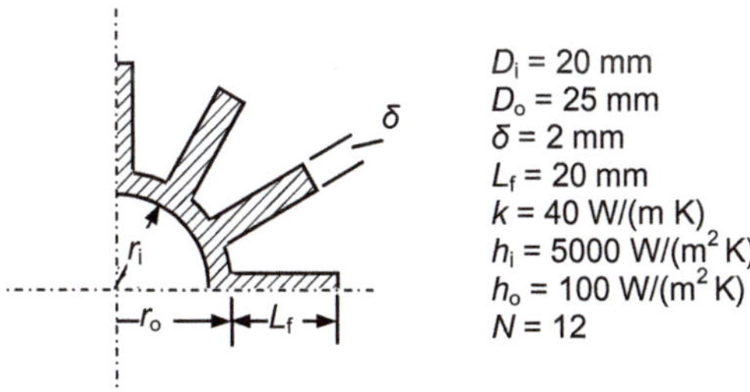

Fig. 3.11 A tube with longitudinal fins

$$h_{red} = h_f \eta_{fin} \frac{A_f}{A_{fw}} + h_w \frac{A_o}{A_{fw}} \tag{3.20}$$

For the given thin rectangular section fins ($P \approx 2b$),

$$m = \sqrt{\frac{hP}{kA_c}} = \sqrt{\frac{h \times 2b}{kb\delta}} = \sqrt{\frac{2h}{k\delta}} = \sqrt{\frac{2 \times 100}{40 \times 2/1000}} = 50$$

$$mL_f = 50 \times \frac{20}{1000} = 1.0.$$

Fin efficiency, neglecting heat transfer from fin end,

$$\eta_{fin} = \frac{\tanh mL_f}{mL_f} = \frac{\tanh 1.0}{1.0} = 0.762$$

Fin surface area for unit length of tube,

$$A_f = N(2L_f + \delta) = 12 \times (2 \times 20 + 2)/1000 = 0.504\,\text{m}^2$$

Total outside heat transfer area for unit length of tube,

$$A_{fw} = A_f + (\pi D_o - N\delta) = 0.504 + (\pi \times 25 - 12 \times 2)/1000 = 0.5585\,\text{m}^2$$

Hence, for $h_f = h_w = h_o = 100$ W/(m^2 K), the reduced heat transfer coefficient

$$h_{red} = 100 \times 0.762 \times \frac{0.504}{0.5585} + 100 \times \frac{0.5585 - 0.504}{0.5585} = 78.52$$

Substitution gives the overall heat transfer coefficient as

$$U_i = \frac{1}{\frac{1}{5000} + \frac{\pi \times (20/1000) \times 1}{2 \times \pi \times 40 \times 1} \ln\left(\frac{25}{20}\right) + \frac{1}{78.52}\left(\frac{\pi \times 20/1000 \times 1}{0.5585}\right)} = 592.2\,\text{W}/\left(\text{m}^2\text{K}\right)$$

If fins are not provided,

$$\begin{aligned} U_i &= \frac{1}{\frac{1}{h_i} + \frac{A_i}{2\pi kL} \ln\left(\frac{R_0}{R_i}\right) + \frac{1}{h_o}\left(\frac{A_i}{A_o}\right)} \\ &= \frac{1}{\frac{1}{5000} + \frac{\pi \times (20/1000) \times 1}{2 \times \pi \times 40 \times 1} \ln\left(\frac{25}{20}\right) + \frac{1}{100}\left(\frac{20}{25}\right)} \\ &= 121.13\,\text{W}/\left(\text{m}^2\text{K}\right) \end{aligned}$$

The effect of fins on heat transfer intensification can be seen.

Example 3.12 One end of a very long aluminium rod [k = 205 W/(m^2 K)] is in contact with a heated wall. Its surface is in contact with a cold fluid. Neglect the contact resistance between the rod and the wall.

(i) By what percentage the rate of heat removal will increase if the rod diameter is doubled?
(ii) If the rod is made of stainless steel [k = 15 W/(m^2 K)], by what percentage the heat transfer rate will change?

Solution

Using the condition of infinite length fin,

$$q_{fin} = \sqrt{hPkA_c}(t_s - t_\infty)$$
$$\propto P^{1/2}k^{1/2}A_c^{1/2} \propto d^{1/2}k^{1/2}(d^2)^{1/2} = d^{3/2}k^{1/2}$$

(i) When the diameter of the rod is doubled,

$$q'_{fin} \propto (2d)^{3/2}k^{1/2}$$

Increase in the heat transfer rate is

$$\Delta q'_{fin} = \frac{q'_{fin} - q_{fin}}{q_{fin}} = \frac{(2d)^{3/2} - d^{3/2}}{d^{3/2}} \times 100 = 182.8\%$$

(ii) When the material of the rod is changed to stainless steel,

$$q'_{fin} \propto d^{3/2}k_1^{1/2}$$

Change in the heat transfer rate is

$$\Delta q'_{fin} = \frac{q'_{fin} - q_{fin}}{q_{fin}} = \frac{(k_1)^{1/2} - k^{1/2}}{k^{1/2}} \times 100 = \frac{(15)^{1/2} - (205)^{1/2}}{(205)^{1/2}} \times 100 = -72.95\%$$

Example 3.13 Two long 10-mm-diameter copper rods [k = 360 W/(m K)] are to be soldered together end to end. The melting point of the solder is 650°C. If the heat transfer coefficient between the copper rod surface and air is 5 W/(m^2 K), and the surrounding air temperature is 25°C, determine the minimum heat input rate to keep the soldered surfaces at 650°C.

Solution

Treating the junction of the two rods as x = 0 plane (refer to Fig. 3.12), the problem is that of a very long fin with T_s = 650°C. The required heat input rate will be twice the heat transfer rate by conduction into the fin, i.e.

$$q = 2q_{fin} = 2\sqrt{hPkA_c}(t_s - t_\infty)$$
$$= 2 \times \sqrt{5 \times \pi \times 0.01 \times 360 \times (\pi/4) \times 0.01^2} \times (650 - 25)$$
$$= 83.3\,\text{W}$$

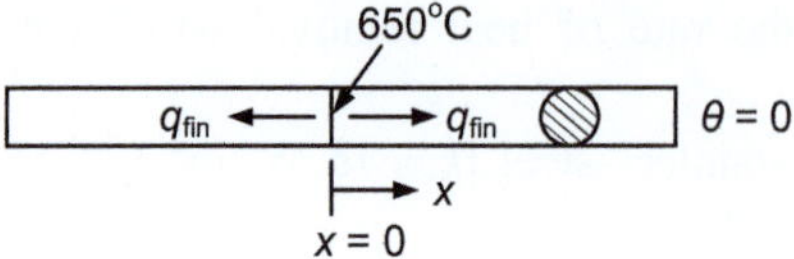

Fig. 3.12 Example 3.13

Example 3.14 Two long rods of the same diameter, one made of brass [k = 85 W/(m K)] and the other of copper [k = 375 W/(m K)], have one of their ends inserted into a furnace, refer to Fig. 3.13. Both rods are exposed to the same environment. At a plane 105 mm away from the furnace wall, the temperature of the brass rod is 120°C. At what distance from the furnace end, the same temperature would be reached in the copper rod?

Solution

For a long fin, the equation of temperature distribution is

$$t = (t_s - t_\infty)e^{-mx} + t_\infty \tag{i}$$

For the brass rod, t = 120°C at x = 0.105 m, i.e.

$$120 = (t_s - t_\infty)e^{-0.105m_1} + t_\infty \tag{ii}$$

For the copper rod, Eq. (i) yields

$$120 = (t_s - t_\infty)e^{-m_2 a} + t_\infty \tag{iii}$$

Equating Eqs. (ii) and (iii), we get

$$(t_s - t_\infty)e^{-0.105m_1} + t_\infty = (t_s - t_\infty)e^{-m_2 a} + t_\infty$$

or

$$0.105m_1 = am_2$$

or

$$a = 0.105\frac{m_1}{m_2} = 0.105\left(\sqrt{\frac{hP}{kA_c}}\right)_1\left(\sqrt{\frac{kA_c}{hP}}\right)_2$$

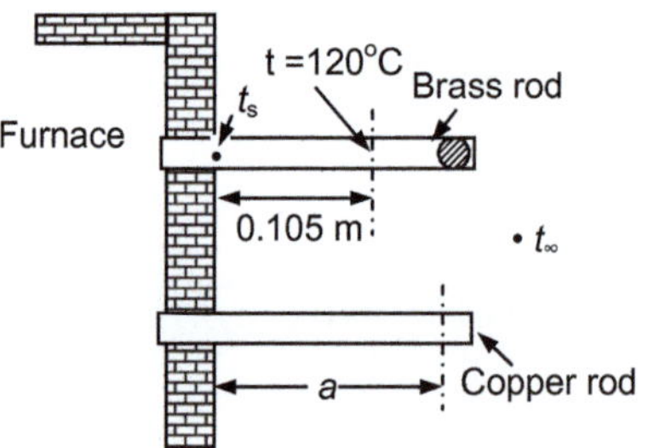

Fig. 3.13 Example 3.14

or

$$a = 0.105\left(\sqrt{\frac{hP}{kA_c}}\right)_1\left(\sqrt{\frac{kA_c}{hP}}\right)_2$$

Since $(hP/A_c)_1 = (hP/A_c)_2$,

$$a = 0.105\sqrt{\frac{k_2}{k_1}} = 0.105\sqrt{\frac{375}{85}} = 0.2205\,\text{m}$$

The desired answer is 220.5 mm.

Example 3.15 Heat dissipation from a surface is to be enhanced by providing wide rectangular section fins. Total of 200 fins are to be provided in 1 m height. Each fin is 1 mm thick and 50 mm long. The convection heat transfer coefficient for the surface without fins is $h = 20$ W/(m^2 K), which drops to $h_f = 15$ W/(m^2 K) when fins are installed on the surface. Calculate the enhancement in heat transfer if the thermal conductivity of the fin material is 210 W/(m K).

Solution

Heat dissipation from the wall without fins,

$$q_o = h_o A_c(t_s - t_\infty) = 20 \times 1 \times b \times (t_s - t_\infty) = 20(b\theta_s)\,\text{W}$$

where $\theta_s = t_s - t_\infty$ and b = width of the fin.

Heat dissipation from finned wall:

Surface area without fins (bare surface area), $A_b = 1 \times b - b \times 1/1000 \times 200 = 0.8b$ m^2.

Contribution of the bare surface,

$$q_b = h_b A_b \theta_s = 15 \times 0.8b \times \theta_s = 12(b\theta_s)\,\text{W}$$

From the given data for the fin,

$$m = \sqrt{\frac{hP}{kA_c}} = \sqrt{\frac{h(2b)}{k(b\delta)}} = \sqrt{\frac{2h}{k\delta}} = \sqrt{\frac{2 \times 15 \times 1000}{210 \times 1}} = 11.95$$

Assuming parameter $P \approx 2b$ for rectangular section fins,

$$mL = 11.95 \times 0.05 = 0.5975$$

Using corrected length L_c in Eq. (3.6), we have

$$q_{fin} = \sqrt{hPkA_c}(t_s - t_\infty)\tanh mL_c$$

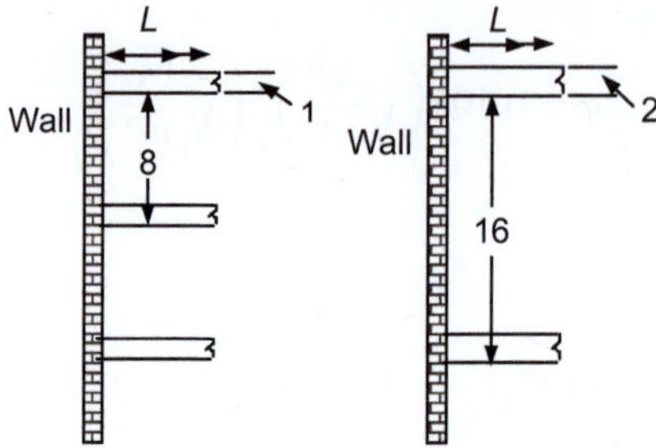

Fig. 3.14 Example 3.16

where $L_c = L + \delta/2 = 0.05 + 0.001/2 = 0.0505$ m. This gives

$$q_{fin} = \left(\sqrt{15 \times 2b \times 210 \times b \times 1 \times 10^{-3}}\right) \times \theta_s \tanh(11.95 \times 0.0505) = 1.354(b\theta_s)\,\mathrm{W}$$

Total heat transfer q_{total} from the fined wall is

$$q_{total} = N \times q_{fin} + q_b = 200 \times 1.354(b\theta_s) + 12(b\theta_s) = 282.8(b\theta_s)\,\mathrm{W}$$

Heat transfer enhancement,

$$\frac{q_{total} - q_o}{q_o} \times 100 = \frac{282.8(b\theta_s) - 20(b\theta_s)}{20(b\theta_s)} \times 100 = 1314\%$$

Example 3.16 In order to enhance heat transfer from a heat exchanger surface at 90°C, 40 mm long fins are to be installed. There are two alternative schemes from equal weight consideration (Fig. 3.14):

(i) 1-mm-thick plate fins at 8 mm pitch, or
(ii) 2-mm-thick plate fins at 16 mm pitch.

The fins are to be made of aluminium [k = 205 W/(m K)]. Assuming surface heat transfer coefficient h = 10 W/(m^2 K) and t_∞ = 20°C, select the better arrangement based on the heat transfer consideration.

Solution

Heat transfer from a fin of finite length is given by

$$q_{fin} = \sqrt{hPkA_c}(t_s - t_\infty)\tanh mL_c$$

For both the arrangements,

$$h = 10\,\mathrm{W}/\left(\mathrm{m^2K}\right), P \approx 2b, k = 205\,\mathrm{W}/\left(\mathrm{m^2K}\right), A_c = b \times \underline{\delta}, \text{ and } (t_s - t_\infty) = 70^\circ\mathrm{C}$$

Substituting the above values, heat transfer equation for unit width becomes

$$q_{fin} = \sqrt{10 \times 2 \times 1 \times 205 \times 1 \times \delta} \times (70) \times \tanh mL_c$$

or

$$q_{fin} = 4482 \times \sqrt{\delta} \times \tanh mL_c \tag{i}$$

Case (i)

$$m = \sqrt{\frac{hP}{kA_c}} = \sqrt{\frac{h(2b)}{k(\delta b)}} = \sqrt{\frac{2h}{k\delta}} = \sqrt{\frac{2 \times 10 \times 1000}{205 \times 1}} = 9.877$$
$$L_c = L + \delta/2 = 40 + 0.5 = 40.5\ \text{mm}$$
$$mL_c = 9.877 \times 40.5/1000 = 0.4$$

Substitution in Eq. (i) gives

$$q_{fin} = 4482 \times \sqrt{1/1000} \times \tanh(0.4) = 53.86\ \text{W per m width.}$$

Case (ii)

$$m = \sqrt{\frac{hP}{kA_c}} = \sqrt{\frac{h(2b)}{k(\delta b)}} = \sqrt{\frac{2h}{k\delta}} = \sqrt{\frac{2 \times 10 \times 1000}{205 \times 2}} = 6.984$$
$$L_c = L + \delta/2 = 40 + 1 = 41\ \text{mm}$$
$$mL_c = 6.984 \times 41/1000 = 0.2864.$$

Substitution in Eq. (i) gives

$$q_{fin} = 4482 \times \sqrt{2/1000} \times \tanh(0.2864) = 55.9\ \text{W per m width.}$$

Since two 1 mm thick fins can be installed in one pitch space for 2 mm thick fins, the first arrangement of 1 mm fins will be better.

Example 3.17 In the previous example, what is the overall effectiveness of the first arrangement?

Solution

In 1.0 m height of the surface, there will be 1000/8 = 125 fins. The bare surface in between the fins will be 125 × 7/1000 = 0.875 m in height. The total heat transfer from the surface would be

$$\begin{aligned} q_{total} &= q_{fin} + q_o \\ &= 125 \times 53.86 + hA_o(t_s - t_\infty) \\ &= 6732.5 + 10 \times 0.875 \times 70 = 7345\ \text{W/m}^2 \end{aligned}$$

The overall effectiveness of the finned surface is the ratio of the heat transfer from the surface with fins to the heat transfer from the surface without fins, i.e.

$$\varepsilon = \frac{q_{total}}{hA(t_s - t_\infty)} = \frac{7345}{10 \times 1 \times 70} = 10.5$$

Example 3.18 A 12.5-mm-diameter aluminium rod [k = 200 W/(m K)] is exposed to an environment at 30°C, while the base temperature is 100°C. The heat transfer coefficient h = 10 W/(m^2 K) for the surface as well as end of the fin. Determine the total heat transfer rate from the fin if its length is 25, 50, 100, 150 or 200 mm. Also determine the effectiveness of the fin for all these lengths and comment on the result.

Solution

The value of parameter m is

$$m = \sqrt{\frac{hP}{kA_c}} = \sqrt{\frac{h\pi d}{k(\pi/4)d^2}} = \sqrt{\frac{4h}{kd}} = \sqrt{\frac{4 \times 10}{200 \times 0.0125}} = 4.0\ \text{m}^{-1}$$

The maximum value of product mL = 0.8, which is less than 3.0. Hence, the heat transfer rate will be calculated considering it to be a short fin, i.e.

$$q_{fin} = \sqrt{hPkA_c}(t_s - t_\infty)\frac{h_L/mk + \tanh mL}{h_L/mk \tanh mL + 1}$$

Here,

$$\frac{h_L}{mk} = \frac{10}{4 \times 200} = 0.0125$$

$$\sqrt{hPkA_c} \times (t_s - t_\infty) = \sqrt{10 \times \pi \times 0.0125 \times 200 \times (\pi/4) \times 0.0125^2} \times 70 = 6.871$$

Substitution gives

$$q_{fin} = 6.871 \times \frac{0.0125 + \tanh(4L)}{0.0125 \tanh(4L) + 1} \quad \text{(i)}$$

and

$$\varepsilon = \frac{q_{fin}}{hA_c(t_s - t_\infty)} = \frac{q_{fin}}{h \times \pi/4 \times d^2 \times (t_s - t_\infty)} = \frac{q_{fin}}{10 \times \pi/4 \times 0.0125^2 \times 70} = 11.64 q_{fin} \quad \text{(ii)}$$

Length of fin, L (m)	mL	tanhmL	q_{fin} (W)	ε
0.025	0.1	0.0997	0.77	8.96
0.05	0.2	0.1974	1.438	16.74
0.1	0.4	0.38	2.684	31.24
0.15	0.6	0.537	3.75	43.65
0.2	0.8	0.664	4.61	53.66

The values of heat transfer rate and effectiveness of the fin as calculated from Eqs. (i) and (ii) for different fin lengths are tabulated.

The effectiveness increases with the length of the fin. It is to be noted that while the length of the fin and hence its weight for the given cross-section of the fin increases in the present example by 8 times, the effectiveness increases by 6 times only. Thus the shorter fins are more economical.

Note: For an infinitely long fin,

$$q_{fin} = \sqrt{hPkA_c}(t_s - t_\infty)$$

Substitution of values of various terms gives

$$q_{fin} = 6.87\,\text{W}$$

For mL = 3, we get

$$q_{fin} = \sqrt{hPkA_c}(t_s - t_\infty)\tanh mL = 6.84\,W$$

The heat transfer rate for a fin with mL = 3 is only 0.5% lower than for a fin of infinite length and 8.9 times of the fin with 25 mm length. For mL = 3, L is 3/m = ¾ = 0.75 m, i.e. the length is 30 times of the minimum length of 25 mm in this example while the effectiveness is only 8.9 times. Thus, it can be seen that the short fins are economical.

3.8 Error in Temperature Measurement with Thermometer Well

The arrangement shown in Fig. 3.15 is used to measure the temperature of fluid flowing through a pipe or duct. The thermometer well is a thin hollow metallic or ceramic cylinder. The metallic cylinder may be welded to the pipe wall. Thermometer to measure the temperature is dipped into the oil filled in the well and thus reads the temperature at the end of the well. It will be shown by the fin analysis that the thermometer does not indicate the actual temperature of the fluid flowing through the duct.

The thermometer well can be regarded as a hollow fin with the end at temperature t_L and the base at temperature t_s. The actual temperature of the fluid in the duct is t_∞. Assuming negligible heat transfer from the fin end as compared to the heat transferred from the fin surface, we have (refer to Sect. 3.2.2)

$$\frac{t - t_\infty}{t_s - t_\infty} = \frac{\cosh m(L - x)}{\cosh mL} \tag{3.5}$$

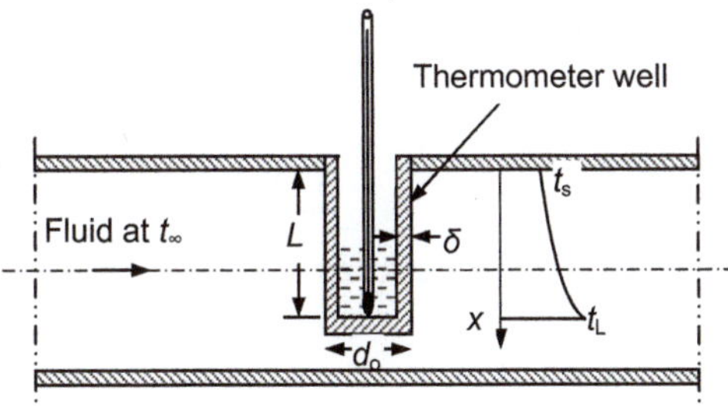

Fig. 3.15 A thermometer well

Putting $t = t_L$ at $x = L$, we get

$$t_L - t_\infty = (t_s - t_\infty)\frac{1}{\cosh mL} \tag{3.23}$$

For a hollow fin,

$$m = \sqrt{\frac{hP}{kA_c}} = \sqrt{\frac{h\pi d_o}{k\pi d_o \delta}} = \sqrt{\frac{h}{k\delta}} \tag{3.24}$$

For the error $(t_\infty - t_L)$ to be minimum, cosh(mL), i.e. m and L must be large. This can be achieved by the following measures:

(i) The well must be made of low thermal conductivity material, such as ceramic;
(ii) The well should be as thin as possible. However, it must have sufficient structural strength and must be able to withstand the fluid pressure and corrosion during the service.
(iii) The well should be as large as possible in length. In a given pipe diameter, the length can be increased by installing the well in an inclined position.
(iv) The heat transfer rate from the fluid to the well can be effectively increased by providing fins on the outer surface of the well, refer to Fig. 3.16.

The thermometer indicates a temperature lower than the gas temperature firstly because of the temperature drop in the film formed at the well surface and secondly due to the conduction of the heat through the wall of the well towards the pipe surface which is at a lower temperature. The resistance to the heat flow from the fluid to the well can be reduced by providing fins or increasing the heat transfer coefficient. The heat flow by conduction to the pipe wall can be reduced by increasing the conduction resistance, i.e. by increasing the length of the heat flow path, reducing the heat flow area and using a low thermal conductivity material for the well. The conduction heat flow can also be reduced by reducing the temperature difference between the well end and the pipe surface. This can be achieved by insulating the pipe from the outside, which will increase the pipe surface temperature.

Note: In the above-presented analysis, we have not considered radiation heat exchange between the surface of the well and the pipe wall.

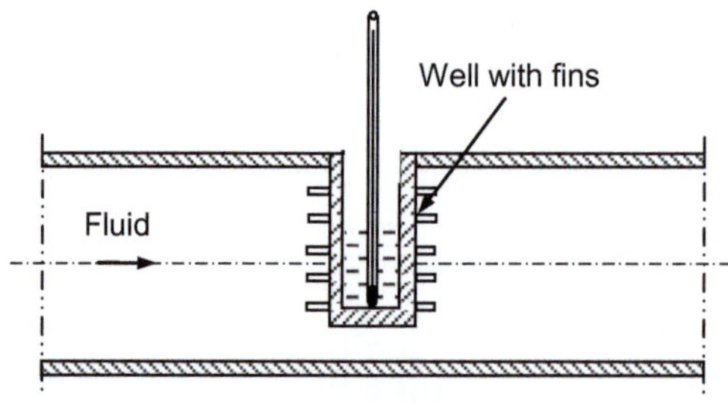

Fig. 3.16 A thermometer well with fins

Example 3.19 Temperature of air flowing through a pipe is measured with a mercury-in-glass thermometer placed in a steel well filled with oil. How great is the error in the measurement if the thermometer indicated a temperature of 100°C? The temperature of the base of the well is 60°C. The well is 120 mm long, and the thickness of the wall of the well is 1.0 mm. The thermal conductivity of the well material is 45 W/(m K), and air to the well surface heat transfer coefficient is 30 W/(m^2 K).

Solution

From Eq. (3.23),

$$t_L - t_\infty = (t_s - t_\infty)\frac{1}{\cosh mL}$$

Here,

$$m = \sqrt{\frac{h}{k\delta}} = \sqrt{\frac{30}{45 \times 1/1000}} = 25.82$$

L = 0.12 m
t_L = 100°C
t_s = 60°C.

Substitution in the equation gives

$$100 - t_\infty = (60 - t_\infty)\frac{1}{\cosh(25.82 \times 0.12)}$$

or

$$t_\infty = 103.96\,^\circ\text{C}$$

Thus the error in measurement of the temperature is $(t_\infty - t_L)$ = 3.96°C.

Note: Radiation heat exchange between the well surface and the pipe inner surface may introduce additional error.

3.9 When Fins Are to Be Used?

The installation of fins on a heat-transferring surface does not necessarily increase the heat transfer rate. In fact, there are certain conditions when the fins on a surface may cause a reduction in the heat transfer. Here, we shall discuss the conditions when the fins are useful.

In Sect. 3.5, fin effectiveness has been defined. The equation of effectiveness can be written for a very long fin (Sect. 3.2.1) as

$$\varepsilon_{fin} = \frac{\sqrt{hPkA_c}(t_s - t_\infty)}{hA_c(t_s - t_\infty)} = \sqrt{\frac{Pk}{hA_c}} = \sqrt{\frac{hP}{kA_c}} \times \frac{k}{h} = \frac{mk}{h} = \frac{1}{(h/mk)} \tag{i}$$

When the effectiveness is unity, the heat transfer rate with or without fins is the same. From the above equation, we can see that the combined term (*h*/*mk*) is unity in this case. It is to note that this conclusion is also applicable to the short fin of Sect. 3.2.3. The combined term (h/mk) has special significance for the fins. We can write

$$\frac{h}{mk} = \frac{h}{k} \times \sqrt{\frac{kA_c}{hP}} = \sqrt{\frac{h}{k}\left(\frac{A_c}{P}\right)} = \sqrt{\frac{h\delta'}{k}} \tag{ii}$$

where $\delta' = A_c/P =$ is a linear dimension. It equals $D/4$ for a circular cross-section fin and $\delta/2$ for a thin rectangular cross-section fin.

The combined term ($h\delta'/k$) is a non-dimensional term and is known as Biot number (Bi). That is,

$$Bi = \frac{h\delta'}{k} \tag{3.25}$$

The Biot number compares the relative magnitudes of internal resistance to conduction heat flow and the external (surface) resistance to convection heat transfer. A low value of the Biot number means that the internal resistance to conduction heat transfer is small in comparison to the external one to the convection heat transfer.

In terms of the Biot number, the heat transfer rate from Eq. (3.8) for the short fin is

$$q_{fin} = \sqrt{hPkA_c}(t_s - t_\infty)\frac{\sqrt{Bi} + \tanh mL}{\sqrt{Bi}\tanh mL + 1} \tag{3.26}$$

and, from Eq. (3.4) of Sect. 3.2.1,

$$\begin{aligned} q_{fin} &= \sqrt{hPkA_c}(t_s - t_\infty) \\ &= hA_c(t_s - t_\infty)/(h/mk) \\ &= hA_c(t_s - t_\infty)/\sqrt{Bi} \end{aligned} \tag{3.27}$$

Now we discuss the effect of the Biot number on the heat transfer rate.

(a) When Bi = 1,

$$q_{fin} = hA_c(t_s - t_\infty)$$

which means the heat transfer with fins q_{fin} is the same as heat transfer without fins. It means that when Bi = 1 the fins do not contribute to heat transfer from the surface, and hence there is no advantage of fins.

(b) When Bi > 1,

$$q_{fin} < hA_c(t_s - t_\infty)$$

which means that finned surface heat transfer q_{fin} is less than the heat transfer without fins, i.e. the fins will act as insulator and heat transfer rate is reduced when the fins are installed.

(c) When Bi < 1,

$$q_{fin} > hA_c(t_s - t_\infty)$$

In this case, heat transfer rate increases when the fins are provided on the surface.

From the above discussion of desirability of Bi < 1, the following can be concluded:

(i) For a given value of the heat transfer coefficient h, the fin should be made of high conductivity material.
(ii) For the given value of the heat transfer coefficient and thermal conductivity of fin material k, the fin geometrical dimensions must be selected to give a high value of (P/A_c). One example of such a fin is a thin and wide fin.
(iii) For the given value of thermal conductivity of fin material k and fin shape, as the value of heat transfer coefficient h increases the fin effectiveness or utility decreases and when hA_c/P equals k, the fin does not enhance the heat transfer rate. If h is so high that the Biot number is greater than unity, the fins act as heat insulator and reduce the heat transfer rate.

For surfaces in contact with condensing vapours or boiling liquids and liquid metals, the magnitude of heat transfer coefficient h is very high and the fins are not useful.

In general, the heat transfer coefficient is quite low for surfaces in contact with a gas at low velocities or in the condition of natural convection heat transfer. The fins are very effective for such surfaces to enhance the heat transfer rate.

The magnitude of the heat transfer coefficient for the surfaces in contact with liquids is moderate to high and due attention must be given to the selection of material and shape of the fin so that the Biot number is sufficiently less than unity. In the case of short and thick fins, the assumption of one-dimensional heat flow made for the analysis presented in the previous sections is not valid. The heat flow in such fins becomes two-dimensional as shown in Fig. 3.17 (note that the heat flow lines are not parallel to the axis of the fin). This will have effect on the conclusion regarding the minimum value of the Biot number. To be sure, that the use of fin in such cases is advantageous, it is advised to fulfil the following condition (Eckert and Drake 1972):

$$Bi \leq 0.2 \tag{3.28}$$

The effectiveness and efficiency defined in Sect. 3.5 can be related to the Biot number as shown below.

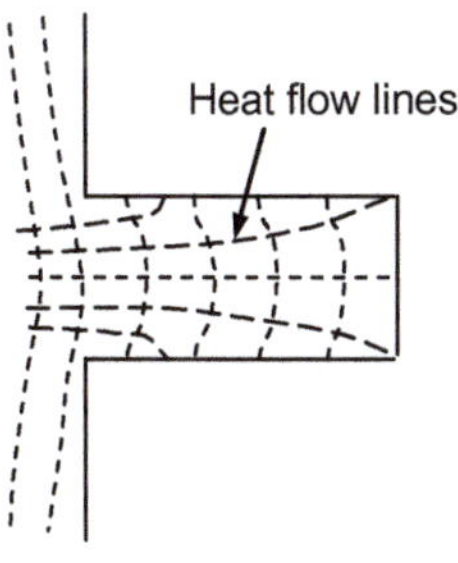

Fig. 3.17 Temperature distribution and heat flow paths in a short rectangular fin

For a very long fin, the effectiveness of the fin, from Eq. (3.17), can be expressed as

$$\varepsilon = \sqrt{\frac{kP}{hA_c}} = \sqrt{\frac{k}{h\delta'}} = \sqrt{\frac{1}{Bi}} \tag{3.29}$$

Thus the effectiveness increases as the Biot number decreases.

Similarly, Eq. (3.18) of efficiency for the fin of Sect. 3.2.2 (a fin with negligible heat transfer from its end as compared to its surface) in the terms of the Biot number is

$$\eta_{fin} = \frac{\tanh mL}{mL} = \frac{\tanh\left(L\sqrt{Bi}/\delta'\right)}{\left(L\sqrt{Bi}/\delta'\right)} \tag{3.30}$$

The efficiency of the fin in this case tends to its maximum value of unity as the Biot number tends to zero.

Example 3.20 Three rods, one made of glass, one of pure aluminium and the third of steel, each having diameter of 10 mm and a length of 250 mm, are used as fins. When the base temperature is 180°C for each fin and ambient temperature 30°C, find the distributions of temperature in the rods and their heat dissipations. The convective heat transfer coefficient is 25 W/(m^2 K). Determine the effectiveness and efficiency of these fins. The thermal conductivities of the three materials are 0.8, 200 and 50 W/(m K), respectively.

Solution

The three rods projecting from the surface work as fins. The value of the parameter m for the circular section rod is

$$m = \sqrt{\frac{hP}{kA_c}} = \sqrt{\frac{4h}{kd}} = \sqrt{\frac{4 \times 25}{k \times 0.01}} = \frac{100}{\sqrt{k}}$$

Therefore, the value of parameter m for the glass fin m_g, aluminium fin m_a and steel fin m_s is 111.8, 7.07 and 14.14, respectively. The product mL for these fins is 27.95, 1.768 and 3.535, respectively.

For the glass and steel fins, $mL \geq 3$; hence, equation of Case (A) may be used. For the aluminium fin, equations of Case (B) with end correction may be used.

The equations of temperature distribution are

Glass fin:

$$\frac{t - t_\infty}{t_s - t_\infty} = e^{-mx}$$

$$\frac{\theta}{\theta_o} = e^{-111.8x}$$

Aluminium fin:

$$\frac{\theta}{\theta_o} = \frac{\cosh m(L_c - x)}{\cosh m(L_c)}$$

Here $L_c = L + d/4 = 0.2525$ m, $\cosh(mL_c) = 3.064$. This gives

$$\frac{\theta}{\theta_o} = 0.3263\cosh(1.785 - 7.07x)$$

Steel fin:

$$\frac{\theta}{\theta_o} = e^{-14.14x}$$

Heat dissipation:

Glass fin:

$$\begin{aligned} q &= \sqrt{hPkA_c}(t_s - t_\infty) \\ &= \sqrt{25 \times (\pi \times 0.01) \times 0.8 \times (\pi/4) \times (0.01)^2}(180 - 30) \\ &= 1.054 \text{ W}. \end{aligned}$$

Aluminium fin:

$$\begin{aligned} q &= \sqrt{hPkA_c}(t_s - t_\infty)\tanh(mL_c) \\ &= \sqrt{25 \times (\pi \times 0.01) \times 200 \times (\pi/4) \times (0.01)^2} \times (180 - 30) \times \tanh(7.07 \times 0.2525) \\ &= 15.75 \text{ W} \end{aligned}$$

Steel fin:

$$\begin{aligned} q &= \sqrt{hPkA_c}(t_s - t_\infty) \\ &= \sqrt{25 \times (\pi \times 0.01) \times 50 \times (\pi/4) \times (0.01)^2}(180 - 30) \\ &= 8.33 \text{ W} \end{aligned}$$

Heat dissipation without fin,

$$q = hA(t_s - t_\infty) = 25 \times (\pi/4) \times (0.01)^2 \times (180 - 30) = 0.2945 \text{ W}$$

Effectiveness

Glass fin:

$$\varepsilon = \frac{1.054}{0.2945} = 3.58$$

Aluminium fin:

$$\varepsilon = \frac{15.75}{0.2945} = 53.48$$

Steel fin:

$$\varepsilon = \frac{8.33}{0.2945} = 28.28$$

Efficiency

The heat dissipation, if whole fin surface is at the base temperature,

$$q = hA_f(t_s - t_\infty) = 25 \times [\pi \times 0.01 \times 0.25 + (\pi/4) \times (0.01)^2] \times (180 - 30) = 29.75\,\mathrm{W}$$

This gives efficiency of the three fins as

$$\eta_g = \frac{1.054}{29.75} \times 100 = 3.54\%$$
$$\eta_{Al} = \frac{15.75}{29.75} \times 100 = 52.94\%$$
$$\eta_s = \frac{8.33}{29.75} = 28\%$$

Despite the same length, cross-section and the base temperature, it can be seen that the effectiveness and efficiency of the aluminum fin (a high thermal conductivity material) is very high.

Example 3.21 Typical heat transfer coefficient data is given in column 2 of Table 3.2. We have to investigate the utility of fins for surfaces in contact with fluids given in column 1 of the table. Fins are to be made either of copper [k = 385 W/(m K)] or steel [k = 70 W/(m K)]. Fin geometric data is as follows: thickness δ = 2 mm, width W = 100 mm and length L = 60 mm.

Table 3.2 Example 3.21

	h, W/(m^2 K)	Copper fins, k = 385 W/(m K)		Steel fins, k = 70 W/(m K)	
		ε	Bi = $h\delta'/k$	ε	Bi = $h\delta'/k$
1. Free Convection					
Gases	5–20	60.1–57.4	(1.3–5.2) × 10^{-5}	55.2–45.8	(7.14–28.5) × 10^{-5}
Liquids	50–500	52.7– 27.1	(1.3–13) × 10^{-4}	34.68–11.8	(7.14–71.4) × 10^{-4}
2. Forced Convection					
Gases	5–100	60.1-47.14	(1.3 − 26) × 10^{-5}	55.2–26.2	7.14 × 10^{-5}-1.4 × 10^{-3}
Liquids	50–10000	52.7–6.2	1.3 × 10^{-4} - 0.026	34.68–2.64	7.14 × 10^{-4}- 0.143
3. Condensation					
Gases	5000–15000	8.77–5.06	0.013–0.039	3.75–2.16	0.071–0.214
Liquids	5 × 10^4 – 15 × 10^4	2.77–1.6	0.13–0.39	1.19–0.68	0.71–2.14
4. Boiling	500–50000	27.73–2.77	1.3 × 10^{-3} - 0.13	11.9–1.19	7.1 × 10^{-3} - 0.715

Solution

The fin effectiveness ε is given by

Case (A): A very long fin:

$$\varepsilon = \frac{1}{\sqrt{Bi}}$$

Case (B): A fin with negligible heat rejection from its end:

$$\varepsilon = \frac{\tanh mL}{\sqrt{Bi}} \approx \frac{1}{\sqrt{Bi}} \quad \text{if } mL > 3$$

Case (C): A short fin:

$$\varepsilon = \frac{1 + (\tanh mL)/\sqrt{Bi}}{1 + \sqrt{Bi}(\tanh mL)}$$

From the given data, fin cross-sectional area $A = W\delta = 200$ mm^2 and fin surface area $A_f = PL = 2(W + \delta)L = 12240$ mm^2. Thus $A << PL$, i.e. the conditions of Case (A) and (B) may be applied.

Further, the product mL for a rectangular section duct and given data is

Copper fin:

$$mL = L\sqrt{\frac{2h}{k\delta}} = 0.0967\sqrt{h}$$

Steel fin:

$$mL = 0.2268\sqrt{h}$$

From the above values of product mL, we can see that when $h \geq 960$ for the copper fin and $h \geq 175$ for the steel fin, the value of product $mL \geq 3$ and the effectiveness equation of Case (A) can be used.

Calculated values of effectiveness ε and the Biot number for the different conditions are given along with the corresponding values of h in Table 3.2. It can be readily seen that the effectiveness decreased with the increase in h. For the given geometric dimensions of the fin in the present example, the effectiveness $\varepsilon = 82.4$ for $h = 5$ W/(m^2 K) and is only 1.6 for the copper fin when dropwise condensation occurs ($h = 15 \times 10^4$) and even less than unity for the steel fin in contact with the condensing vapour.

The Biot number is a quite useful parameter to decide the utility of a fin. It must be kept in mind that the Biot number depends on the thermal conductivity of the fin material, heat transfer coefficient and the fin geometry represented by δ'.

3.10 Heat Transfer from a Bar Connected to Two Heat Sources at Different Temperatures

In Fig. 3.18, thermal reservoirs 1 and 2 are connected with a solid bar of uniform cross-section A and parameter P. The length of the bar is L. For the small element of length dx, shown in the figure as shaded, heat q_x entering the element by conduction is

$$q_x = -kA_c \frac{dt}{dx}$$

Heat leaving the element by conduction is

$$q_{x+dx} = q_x + \frac{d}{dx}(q_x)dx$$

Heat transferred by convection is

$$q_c = hPdx(t - t_\infty)$$

Heat balance for the element gives

$$q_x - q_{x+dx} - q_c = 0$$

or

$$-\frac{d}{dx}(q_x)dx - q_c = 0$$

or

$$\frac{d}{dx}\left(-kA_c \frac{dt}{dx}\right)dx + hPdx(t - t_\infty) = 0$$

or

$$\frac{d^2t}{dx^2} - \frac{hP}{kA_c}(t - t_\infty) = 0$$

The above equation is the same as Eq. (ii) of Sect. 3.2. Its solution is

$$(t - t_\infty) = C_1 e^{mx} + C_2 e^{-mx}$$

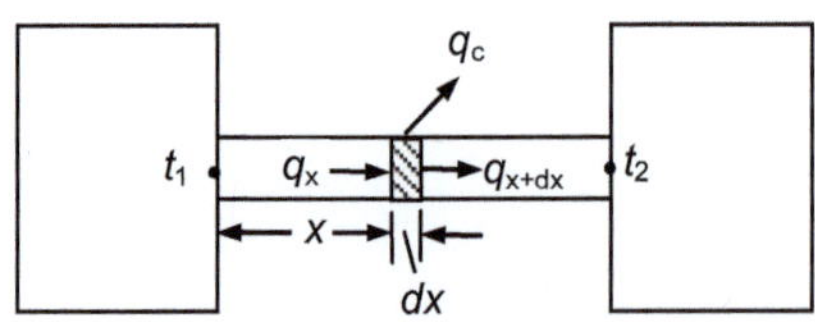

Fig. 3.18 Heat transfer from a bar connected to two heat sources

where

$$m = \sqrt{\frac{hP}{kA_c}}$$

The boundary conditions for the present problem are

$$\text{At}\, x = 0, t = t_1$$
$$\text{At}\, x = L, t = t_2$$

Applying the boundary conditions,

$$t_1 - t_\infty = C_1 + C_2$$
$$(t_2 - t_\infty) = C_1 e^{mL} + C_2 e^{-mL}$$

Solving the above equations, we get

$$C_1 = \frac{\theta_2 - \theta_1 e^{-mL}}{e^{mL} - e^{-mL}}$$
$$C_2 = \frac{\theta_1 e^{mL} - \theta_2}{e^{mL} - e^{-mL}}$$

where $\theta_1 = t_1 - t_\infty$ and $\theta_2 = t_2 - t_\infty$

Substitution of the values of C_1 and C_2 in Eq. (i) gives

$$t - t_\infty = \frac{\theta_2 - \theta_1 e^{-mL}}{e^{mL} - e^{-mL}} e^{mx} + \frac{\theta_1 e^{mL} - \theta_2}{e^{mL} - e^{-mL}} e^{-mx}$$

which can be transformed into the following form:

$$\begin{aligned} t - t_\infty &= \frac{\theta_1 \sinh m(L - x)}{\sinh mL} + \frac{\theta_2 \sinh mx}{\sinh mL} \\ &= \frac{\theta_1 \sinh m(L - x) + \theta_2 \sinh mx}{\sinh mL} \end{aligned} \tag{3.31}$$

The rate of heat loss from the bar can be found from the relation

$$\begin{aligned} (q_c)_{total} &= \int_0^L hPdx(t - t_\infty) \\ &= hP \int_0^L (t - t_\infty)dx \\ &= hP \int_0^L \frac{\theta_1 \sinh m(L - x) + \theta_2 \sinh mx}{\sinh mL} dx \\ &= \frac{hP}{\sinh mL} \int_0^L [\theta_1 \sinh m(L - x) + \theta_2 \sinh mx] dx \\ &= \frac{hP}{\sinh mL} \left[\frac{-\theta_1(1 - \cosh mL) + \theta_2(\cosh mL - 1)}{m} \right] \end{aligned}$$

or

$$q_c = \frac{hP}{m \sinh mL}(\theta_1 + \theta_2)(\cosh mL - 1) \tag{3.32}$$

The position of maximum temperature in the bar can be determined by differentiating Eq. (3.31) with respect to x and equating to zero. Thus

$$\frac{d}{dx}(t - t_\infty) = \frac{1}{\sinh mL}\left\{\frac{d}{dx}[\theta_1 \sinh m(L - x) + \theta_2 \sinh mx]\right\} = 0$$

This gives

$$-\theta_1 m \cosh m(L - x) + m\theta_2 \cosh mx = 0$$

Thus

$$\frac{\theta_1}{\theta_2} = \frac{\cosh mx}{\cosh m(L - x)} \tag{3.33}$$

When both the reservoirs are at the same temperature, i.e. $\theta_1 = \theta_2$, then the above equation gives

$$\cosh mx = \cosh m(L - x)$$

or

$$x = L/2$$

i.e. the maximum temperature is at midplane of the rod, which is a logical result.

Example 3.22 Can we obtain the temperature distribution equation of the bar connected to two heat sources at different temperatures from the equation of the fin?

Solution

From Example 3.7, the temperature distribution equation for a bar with one end at temperature t_s and the other at temperature t_∞ is

$$\frac{t - t_\infty}{t_s - t_\infty} = \frac{\sinh m(L - x)}{\sinh mL} \tag{3.14}$$

Consider the sub-problem (a) and (b) shown in Fig. 3.19. Their temperature distribution equations will be

$$t - t_\infty = (t_1 - t_\infty)\frac{\sinh m(L - x)}{\sinh mL}$$

and

$$t - t_\infty = (t_2 - t_\infty)\frac{\sinh mx}{\sinh mL}$$

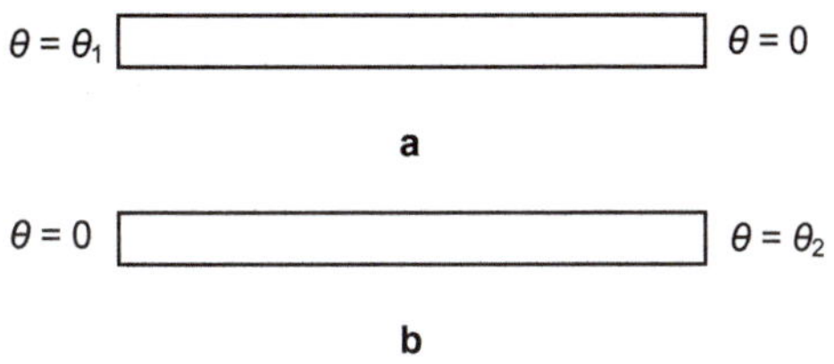

Fig. 3.19 Examples 3.22

Superposition of sub-problems (a) and (b) gives the original problem. So the temperature distribution for the bar with ends at temperature excess θ_1 and θ_2 is

$$t - t_\infty = \theta_1 \frac{\sinh m(L - x)}{\sinh mL} + \theta_2 \frac{\sinh mx}{\sinh mL}$$

which is the desired result.

Example 3.23 One end of a long rod is inserted into a furnace while the other end projects into outside air. Under steady state, the temperature of the rod is measured at two points 75 mm apart and found to be 125°C and 88.5°C when the ambient temperature is 20°C. The rod diameter is 250 mm, and the convective heat transfer coefficient is 6 W/(m^2 K). Find the thermal conductivity of the rod material.

Solution

Given data is h = 6 W/(m^2 K), $P = \pi D = \pi \times 0.25$ m and $A_c = (\pi/4)\, D^2 = (\pi/4) \times (0.25)^2$ m^2, and k is unknown.

For a very long fin,

$$\frac{t - t_\infty}{t_s - t_\infty} = e^{-mx}$$

$$m = \sqrt{\frac{hP}{kA_c}} = \sqrt{\frac{6 \times \pi \times 0.25 \times 4}{\pi \times (0.25)^2 k}} = \frac{9.8}{\sqrt{k}}$$

Let given temperatures are at $x = L_1$ (t_1 = 125°C) and $x = L_1 + 0.075$ (t_2 = 88.5°C), then

$$\frac{125 - 20}{t_s - 20} = e^{-\left(\frac{9.8 L_1}{\sqrt{k}}\right)}$$

and

$$\frac{88.5 - 20}{t_s - 20} = e^{-\left[\frac{9.8(L_1 + 0.075)}{\sqrt{k}}\right]}$$

Dividing the first equation by the second, we get

$$\frac{125 - 20}{88.5 - 20} = e^{\left(\frac{9.8}{\sqrt{k}}\times 0.075\right)}$$

Simplification gives $k = 2.96$ W/(m K).

Alternatively, x_1 may be considered to be zero, then $t_1 = t_s = 125$°C and the substitution gives

$$\frac{88.5 - 20}{125 - 20} = e^{-\left(\frac{9.8}{\sqrt{k}}\times 0.075\right)}$$

which gives the same result.

Example 3.24 An iron rod 10 mm in diameter and 500 mm long is giving up heat to its surrounding at an average rate of 10 W/(m^2 K) by combined radiation and conduction. The ends of the rod are firmly connected to two heat sources at 120°C. The thermal conductivity of the iron is 40 W/(m K). The temperature of the surrounding is 20°C.

(i) What is the temperature at the middle of the rod?
(ii) How much heat is flowing out of each source?
(iii) How much heat is flowing at the middle of the rod?

Solution

The parameter m is, from the given data,

$$m = \sqrt{\frac{hP}{kA_c}} = \sqrt{\frac{10 \times \pi \times 0.01 \times 4}{\pi \times (0.01)^2 \times 40}} = 10$$

Product $mL = 10 \times 0.5 = 5$, $e^{mL} = 148.41$, $e^{-mL} = 6.74 \times 10^{-3}$ and $e^{mL/2} = 12.182$. The temperature excess at each end, $\theta_1 = \theta_2 = 120–20 = 100$°C.

(i) From Eq. (3.31), we have

$$t - t_\infty = \frac{\theta_1 \sinh m(L - x) + \theta_2 \sinh mx}{\sinh mL}$$

Temperature at the middle of the bar

$$t_{L/2} - t_\infty = 2\theta_1 \frac{\sinh(mL/2)}{\sinh mL}$$

or

$$t_{L/2} - t_\infty = 2\theta_1 \frac{e^{mL/2} - e^{-mL/2}}{e^{mL} - e^{-mL}} = 2\theta_1 \left(\frac{e^{mL} - 1}{e^{2mL} - 1}\right) e^{mL/2} = 2\theta_1 \left(\frac{e^{mL/2}}{e^{mL} + 1}\right)$$

or

$$t_{L/2} = 2\theta_1 \frac{e^{mL/2}}{e^{mL}+1} + t_\infty$$
$$= 2 \times 100 \times \frac{12.182}{148.41+1} + 20 = 36.3^\circ\text{C}$$

(ii) The heat transfer rate

$$q = \frac{hP}{m \sinh mL}(\theta_1 + \theta_2)(\cosh mL - 1) \tag{3.32}$$

where

$$\sinh mL = \frac{e^{mL} - e^{-mL}}{2} = 74.2$$
$$\cosh mL = \frac{e^{mL} + e^{-mL}}{2} = 74.208$$

Hence,

$$q = \frac{10 \times (\pi \times 0.01)}{10 \times 74.2}(100+100) \times (74.208 - 1) = 6.2\,\text{W}$$

(iii) The maximum of the temperature occurs at the middle of the rod when $\theta_1 = \theta_2$, and hence at this plane $d\theta/dx = 0$. Thus, the heat does not flow at the middle of the rod.

Example 3.25 The ends of a 30-mm-diameter and 1-m-long rod are maintained at a constant temperature of 100°C. The rod transfers heat to the surrounding air at 25°C. The heat transfer coefficient is estimated to be 32 kW/(m^2 °C). If the centre of the rod is maintained at 50°C, find the thermal conductivity of the rod material.

Solution

Temperature at the middle of the bar (refer to Example 3.24)

$$t_{L/2} = 2\theta_1 \frac{e^{mL/2}}{e^{mL}+1} + t_\infty$$

Substituting θ_1 = 100–25 = 75°C, and $t_{L/2} - t_\infty$ = 50–25 = 25°C, the equation transforms to

$$e^{mL} + 1 = 6e^{mL/2}$$

Let $e^{mL/2} = x$, then

$$x^2 - 6x + 1 = 0$$

The equation gives $x = e^{mL/2} = 5.8284$ and 0.1716. The first value gives $m = 3.5254$ for $L = 1$ m. Using the equation of parameter m, we get

$$m = \sqrt{\frac{hP}{kA_c}}$$

or

$$3.5284 = \sqrt{\frac{32 \times \pi \times 0.03 \times 4}{k \times \pi \times (0.03)^2}}$$

or

$$k = 343.3 \text{ W/(m K)}$$

The second value of x gives negative value of k, and hence is not applicable.

Example 3.26 Derive an expression for the temperature distribution along a uniform cross-section fin with internal heat generation at the rate of q_g W/m^3 along its length. The fin end is insulated. The base temperature of the fin is t_s while the surrounding air is at temperature t_∞. Can the heat flow from the base towards the fin be equal to zero? Find out the condition.

Solution

We make an energy balance on an element of the fin of thickness dx as shown in Fig. 3.20. In the steady state, heat generated in the element and heat inflow by conduction must equal to the heat outflow by conduction and convection, i.e.

$$\begin{aligned}&\text{Energy inflow into left face} + \text{heat generated in the element}\\&\quad = \text{energy outflow from the right face} + \text{energy lost by Convection}\end{aligned}$$

or

$$q_x + q_g = q_{x+dx} + q_c$$

or

$$-kA_c\frac{dt}{dx} + q_g(A_c dx) = -kA_c\frac{dt}{dx} + \frac{d}{dx}\left(-kA_c\frac{dt}{dx}\right)dx + h(Pdx)(t - t_\infty)$$

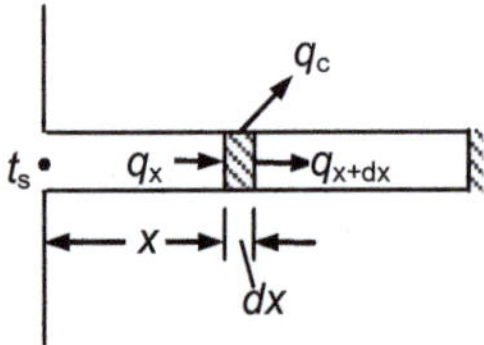

Fig. 3.20 Example 3.26

where t is the temperature of the element and P is the perimeter. Simplification gives

$$\frac{d^2t}{dx^2} - \frac{hP}{kA_c}(t - t_\infty) + \frac{q_g}{k} = 0$$

Let $(t - t_\infty) = \theta$, then $\frac{d^2t}{dx^2} = \frac{d^2\theta}{dx^2}$. Putting $\frac{hP}{kA_c} = m^2$, the equation is transformed to

$$\frac{d^2\theta}{dx^2} - m^2\theta + \frac{q_g}{k} = 0$$

The above equation can be further transformed by using $\theta' = \theta - \frac{q_g}{km^2}$ to give

$$\frac{d^2\theta'}{dx^2} - m^2\theta' = 0$$

which is a second-order differential equation. Its general solution is

$$\theta' = C_1 e^{mx} + C_2 e^{-mx} \quad \text{(i)}$$

The constants C_1 and C_2 are to be determined from the boundary conditions of the problem.

(i) at $x = 0$, $\theta' = \theta'_1 = t_s - t_\infty - \frac{q_g}{km^2}$ and
(ii) at $x = L$, $\frac{d\theta}{dx} = \frac{d\theta'}{dx} = 0$

The first condition gives

$$C_1 + C_2 = t_s - t_\infty - \frac{q_g}{km^2} \quad \text{(ii)}$$

and the second condition gives

$$\frac{d\theta'}{dx} = 0 = mC_1 e^{mL} - mC_2 e^{-mL}$$

or

$$C_1 e^{mL} - C_2 e^{-mL} = 0 \quad \text{(iii)}$$

Solving Eqs. (ii) and (iii) for the constants C_1 and C_2, we get

$$C_1 = \frac{t_s - t_\infty - \frac{q_g}{km^2}}{1 + e^{2mL}}$$

$$C_2 = \frac{t_s - t_\infty - \frac{q_g}{km^2}}{1 + e^{-2mL}}$$

Substitution in Eq. (i) gives

$$\theta' = \frac{t_s - t_\infty - \frac{q_g}{km^2}}{1 + e^{2mL}} e^{mx} + \frac{t_s - t_\infty - \frac{q_g}{km^2}}{1 + e^{-2mL}} e^{-mx}$$

or

$$\left(t - t_\infty - \frac{q_g}{km^2}\right) = \left(t_s - t_\infty - \frac{q_g}{km^2}\right)\left(\frac{e^{mx}}{1+e^{2mL}} + \frac{e^{-mx}}{1+e^{-2mL}}\right)$$

or

$$\frac{t - t_\infty - \frac{q_g}{km^2}}{t_s - t_\infty - \frac{q_g}{km^2}} = \frac{e^{mx}}{1+e^{2mL}} + \frac{e^{-mx}}{1+e^{-2mL}}$$

or

$$\frac{t - t_\infty - \frac{q_g}{km^2}}{t_s - t_\infty - \frac{q_g}{km^2}} = \frac{\cosh m(L-x)}{\cosh mL} \tag{3.34}$$

The condition of zero heat flow at the base is

$$q_{fin} = -kA_c\left(\frac{dt}{dx}\right)_{x=o} = 0$$

or

$$\left(\frac{dt}{dx}\right)_{x=o} = 0$$

From Eq. (3.34), we have

$$\begin{aligned}\left(\frac{dt}{dx}\right)_{x=0} &= \left(t_s - t_\infty - \frac{q_g}{km^2}\right)\frac{1}{\cosh mL}\left\{\frac{d}{dx}[\cosh m(L-x)]\right\}_{x=0} \\ &= \left(t_s - t_\infty - \frac{q_g}{km^2}\right)\frac{1}{\cosh mL}(-m \sinh mL)\end{aligned}$$

From the above equation, it can be seen that the heat flow at the base of the fin is zero when $t_s - t_\infty = \frac{q_g}{km^2}$.

Example 3.27 A composite fin consists of a cylindrical rod 3 mm diameter and 100 mm long of one material. It is uniformly covered with another material forming outside diameter of 10 mm and length of 100 mm. Thermal conductivity of inner material is 15 W/(m K) and that of outer is 45 W/(m K). Convective heat transfer coefficient is 12 W/(m^2 K). Determine (i) the effectiveness of the composite fin neglecting heat rejection from the fin end, and (ii) the efficiency of the fin.

Solution

We make the energy balance on an element of the composite fin of thickness dx as shown in Fig. 3.21.

Fig. 3.21 A composite fin

Energy inflow into left face = Energy outflow from the right face + energy lost by convection

or

$$q_x = q_{x+dx} + q_c$$

or

$$q_x = q_x + \frac{d}{dx}(q_x)dx + h(Pdx)(t - t_\infty)$$

or

$$\frac{d}{dx}(q_x)dx + h(Pdx)(t - t_\infty) = 0 \tag{i}$$

where $q_x = \left(-k_i A_i \frac{dt}{dx}\right) + \left(-k_o A_o \frac{dt}{dx}\right)$ for the composite fin.

The above equation is based on the assumption that there is no temperature gradient in radial direction of the composite fin, i.e. the temperature at any plane parallel to the wall is uniform throughout the cross-section.

Substituting the value of q_x, we get

$$\frac{d}{dx}\left[\left(-k_i A_i \frac{dt}{dx}\right) + \left(-k_o A_o \frac{dt}{dx}\right)\right]dx + h(Pdx)(t - t_\infty) = 0$$

or

$$(k_i A_i + k_o A_o)\frac{d^2 t}{dx^2} - hP(t - t_\infty) = 0$$

or

$$\frac{d^2 t}{dx^2} - \frac{hP}{(k_i A_i + k_o A_o)}(t - t_\infty) = 0$$

Let $(t - t_\infty) = \theta$, then $\frac{d^2 t}{dx^2} = \frac{d^2\theta}{dx^2}$. Putting $\frac{hP}{(k_i A_i + k_o A_o)} = m^2$, the equation is transformed to

$$\frac{d^2\theta}{dx^2} - m^2\theta = 0$$

Heat transfer from the fin is

$$q_{fin} = \left[-k_i A_i \left(\frac{dt}{dx}\right)_{x=0}\right] + \left[-k_o A_o \left(\frac{dt}{dx}\right)_{x=0}\right]$$

Following the procedure used earlier

$$q_{fin} = (k_i A_i + k_o A_o) m (t_s - t_\infty) \tanh mL \tag{3.35}$$

where

$$m = \sqrt{\frac{hP}{(k_i A_i + k_o A_o)}} \tag{3.36}$$

Heat transfer rate from the surface when there is no fin,

$$q = h(A_i + A_o)(t_s - t_\infty)$$

Thus the effectiveness of the fin is

$$\varepsilon = \frac{q_{fin}}{q} = \frac{(k_i A_i + k_o A_o)}{h(A_i + A_o)} m \tanh mL$$

where

k_i	= 15 W/(m K)
k_o	= 45 W/(m K)
A_i	= $(\pi/4)\ (3/1000)^2$ m^2
A_o	= $(\pi/4) \times [(10/1000)^2 - (3/1000)^2]$ m^2
h	= 12 W/(m^2 K)
P	= $\pi d_o = \pi \times (10/1000)$ m
$m = \sqrt{\frac{hP}{(k_i A_i + k_o A_o)}}$	= 10.652
L	= 0.1 m.

Substitution gives ε = 29.85.
Fin efficiency for case B gives

$$\eta_{fin} = \frac{\tanh mL}{mL} = 0.74$$

Example 3.28 Heat dissipation from a surface (1 m × 1 m) is to be increased by 50% by providing rectangular section fins. The fins are to be 5 mm thick, and 20 fins are to be provided. The surface temperature is 100°C. It is expected that the surface temperature will drop to 90°C when the fins will be installed but the convection heat transfer coefficient can be assumed to remain unchanged at 20 W/(m^2 K). Calculate the length of fins if the thermal conductivity of the fin material is 200 W/(m K) and the surrounding temperature is 20°C.

Solution

Heat dissipation without fins,

$$q = hA_c(t_s - t_\infty) = 20 \times 1 \times (100 - 20) = 1600\,\text{W}$$

Desired heat dissipation rate, q' = 1.5 × 1600 = 2400 W.
The surface area without fins (bare surface area) = 1 × 1 – 1 × 5/1000 × 20 = 0.9 m^2.
Contribution of the bare surface, q_b = 20 × 0.9 × (90 − 20) = 1260 W.
Heat to be dissipated by the fins, q_f' = 2400 – 1260 = 1140 W.
Heat to be dissipated by one fin, $q_f = q_f'$ /(number of fins) = 1140/20 = 57 W.
The fin is quite thin. Hence, we take account of the heat rejection by the fin end by using corrected fin length. Thus,

$$q_{fin} = \sqrt{PhkA_c}(t_s - t_\infty)\tanh mL_c \quad \text{(i)}$$

where P = 2 × (1000 + 5)/1000 = 2.01 m, A_c = 1000 × 5/1000^2 = 5 × 10^{-3}. Value of the parameter m from the given data is

$$m = \sqrt{\frac{hP}{kA_c}} = \sqrt{\frac{20 \times 2.01}{200 \times 5 \times 10^{-3}}} = 6.34$$

Substitution of the values of various terms in Eq. (i) gives

$$57 = \sqrt{2.01 \times 20 \times 200 \times 5 \times 10^{-3}} \times (90 - 20)\tanh mL_c$$

Simplification gives

$$\tanh mL_c = 0.12843$$

or

$$mL_c = 0.12915$$

or

$$L_c = \frac{0.12915}{m} = \frac{0.12915}{6.34} = 0.02037\,\text{m}$$
$$= 20.4\,\text{mm}.$$

Hence, the fin length,

$$L = L_c - \frac{\delta}{2} = 20.4 - 2.5 = 17.9\,\text{mm}$$

Check: Considering the heat loss from fin end, the heat dissipation equation is

$$q_{fin} = \sqrt{PhkA_c}(t_s - t_\infty)\frac{h/mk + \tanh mL}{(h/mk)\tanh mL + 1}$$

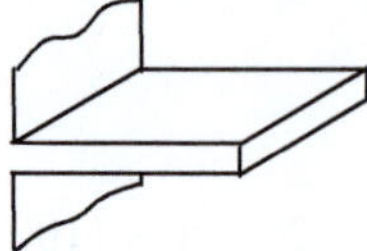

Fig. 3.22 A rectangular section fin

Substituting $mL = (6.34 \times 0.0179) = 0.11349$ and values of other parameters, we obtain

$$q_{fin} = \sqrt{2.01 \times 20 \times 200 \times 0.005} \times (90 - 20) \times \frac{20/(6.34 \times 200) + \tanh(0.11349)}{[20/(6.34 \times 200)] \times \tanh(0.11349) + 1}$$
$$= 57.13\,\mathrm{W}$$

Example 3.29 Rectangular section (800 × 3 mm^2) aluminium fins of 30 mm length, as shown in Fig. 3.22, are provided on a flat wall. The heat transfer coefficient from the fin surface is 10 W/(m^2 K). The thermal conductivity of the fin material is 200 W/(m K). The temperature at the base of the fin is 300°C, and the temperature of the surrounding air is 20° C. Calculate the temperature at the end of the fin.

Solution

From Eq. (3.7), we have

$$\frac{t - t_\infty}{t_s - t_\infty} = \frac{\cosh m(L - x) + (h_L/mk)\sinh m(L - x)}{(h_L/mk)\sinh mL + \cosh mL}$$

Let the temperature at the fin end ($x = L$) is t_L, then

$$\frac{t_L - t_\infty}{t_s - t_\infty} = \frac{1}{(h_L/mk)\sinh mL + \cosh mL}$$

or

$$t_L = (t_s - t_\infty)\frac{1}{(h_L/mk)\sinh mL + \cosh mL} + t_\infty$$

From the given data,

$$m = \sqrt{\frac{hP}{kA_c}} = \sqrt{\frac{10 \times 2 \times (800 + 3) \times 1000}{200 \times 800 \times 3}} = 5.784$$

This gives

$$mL = 5.784 \times (30/1000) = 0.17353$$

Hence,

$$t_L = (300 - 20)\frac{1}{[10/(5.784 \times 200)] \times \sinh(0.17353) + \cosh(0.17353)} + 20 = 295.35^{\circ}\mathrm{C}$$

Note: The area of the fin end is 800 × 3 = 2400 mm^2, while the total surface area is 50580 mm^2. Thus the fin end area is only 4.75%. If we neglect the heat rejection from the fin end, then we have (Eq. 3.23)

$$\frac{t_L - t_\infty}{t_s - t_\infty} = \frac{1}{\cosh mL}$$

which gives t_L = 295.84°C, which is only marginally different from the earlier calculated value.

Example 3.30 A tube consists of integral longitudinal fins as shown in Fig. 3.23. The outside diameter of the tube is 100 mm. Length of the tube is 1 m. Each fin is 5 mm in thickness and 50 mm in height. There are 20 such fins. The temperature at the base of the fins is 100°C, and the surrounding temperature is 20°C. The heat transfer coefficient from the fin surface and also from the surface between the fins can be assumed to be 10 W/(m^2 K). The thermal conductivity of the fin and tube material is 50 W/(m K). Calculate the amount of heat transferred from the finned wall to the surrounding. Also calculate the amount of heat that would be transferred from the tube wall without fins under the same conditions.

Solution

From Eq. (3.8), the heat transfer from a fin is

$$q_{fin} = \sqrt{hPkA_c}(t_s - t_\infty)\frac{(h_L/mk) + \tanh mL}{(h_L/mk)\tanh mL + 1}$$

Here h = 10 W/(m^2 K), P = 2(b + δ) = 2(1000 + 5) × 1/1000 m, A_c = 1 × 5/1000 m^2, k = 50 W/(m K), t_s = 100°C, t_∞ = 20°C and L = 50 mm.

The values of various parameters give

$$m = \sqrt{\frac{hP}{kA_c}} = \sqrt{\frac{10 \times 2 \times (1000 + 5) \times 1/1000}{50 \times 1 \times 5/1000}} = 8.97$$

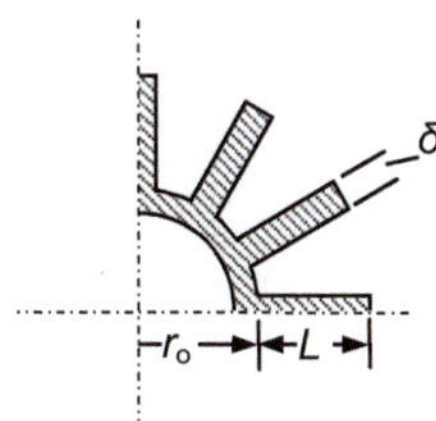

Fig. 3.23 A tube with longitudinal fins

This gives

$$mL = 8.97 \times 50/1000 = 0.4483$$
$$\frac{h}{mk} = \frac{10}{8.97 \times 50} = 0.0223$$
$$\sqrt{hPkA_c} = \sqrt{10 \times (2 \times 1005/1000) \times 50 \times 1 \times 5/1000} = 2.2417$$

Hence,

$$q_{fin} = 2.2417 \times (100 - 20)\frac{0.0223 + \tanh 0.4483}{0.0223 \times \tanh 0.4483 + 1} = 78.67\,\text{W}$$

Total heat transfer from 20 fins = 20 × 78.67 = 1573.4 W.
The bare surface of the tube = (πd_o − 20 × fin thickness δ) × length of the tube

$$= \frac{(\pi \times 100 - 20 \times 5)}{1000} \times 1 = 0.2142\,\text{m}^2$$

Heat transfer from the tube surface

$$= hA(t_s - t_\infty) = 10 \times 0.2142 \times (100 - 20) = 171.3\,\text{W}$$

Total heat transfer from the tube = 1573.4 + 171.3 = 1744.7 W.
Heat transfer from the tube without fins,

$$= h(\pi d_o L) \times (t_s - t_\infty)$$
$$= 10 \times (\pi \times 100 \times 1)/1000 \times (120 - 20) = 251.33\,\text{W}$$

Hence, effectiveness = 1744.7/251.33 = 6.94.

3.11 Generalized Equation of Fin

We have studied the performance of uniform cross-section fins protruding from a flat wall. In practical applications, the profile or cross-section of the fin may vary along its length as shown in Fig. 3.24 and may be attached to circular surfaces.

The calculation for fins with variable cross-section is more complicated than that for the straight fins of uniform cross-section.

Let the shape of the fin is such that

$$A(x) = f(x)$$

and

$$P(x) = \varphi(x)$$

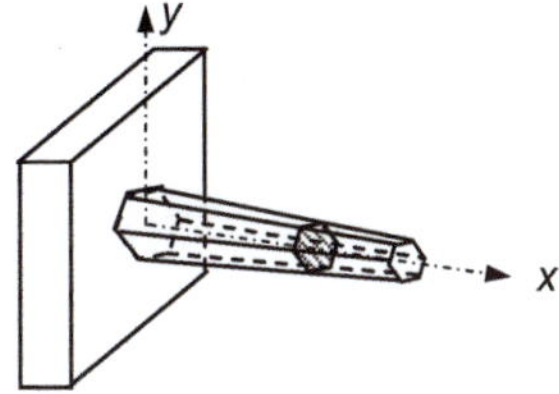

Fig. 3.24 A fin of variable cross-section

Taking infinitesimal element of thickness dx of the fin at a distance x from the wall, refer to Fig. 3.24, the heat balance equation can be written as

$$-kA_x\frac{dt}{dx} = -kA_x\frac{dt}{dx} + \frac{d}{dx}\left(-kA_x\frac{dt}{dx}\right)dx + hP_xdx(t - t_\infty)$$

or

$$\frac{d}{dx}\left(A_x\frac{dt}{dx}\right) = \frac{h}{k}P_x(t - t_\infty)$$

or

$$A_x\frac{d^2t}{dx^2} + \frac{dA_x}{dx} \times \frac{dt}{dx} - \frac{h}{k}P_x(t - t_\infty) = 0$$

or

$$\frac{d^2t}{dx^2} + \frac{dA_x/dx}{A_x} \times \frac{dt}{dx} - \frac{h}{k} \times \frac{P_x}{A_x}(t - t_\infty) = 0 \quad (3.37)$$

This is the generalized equation applicable to fins of any profile or cross-section where the cross-section is some function of x.

Equation (3.37) is a modified Bessel's equation, and its solution is obtained with the help of Bessel's function.

Note: For a fin of uniform cross-section, A_x and P_x are not functions of x but are constants. Hence, their derivatives are zero and the above equation reduces to

$$\frac{d^2t}{dx^2} - \frac{hP}{kA}(t - t_\infty) = 0$$

which is the equation for the uniform cross-section fin.

3.12 Fin of Minimum Weight (Isachenko et al.1977)

The design of fin for maximum heat transfer at minimum weight is of special significance in some applications. This can be achieved if the specific rate of heat flow (heat flow through the unit area of cross-section of the fin) remains constant over the entire length of the fin.

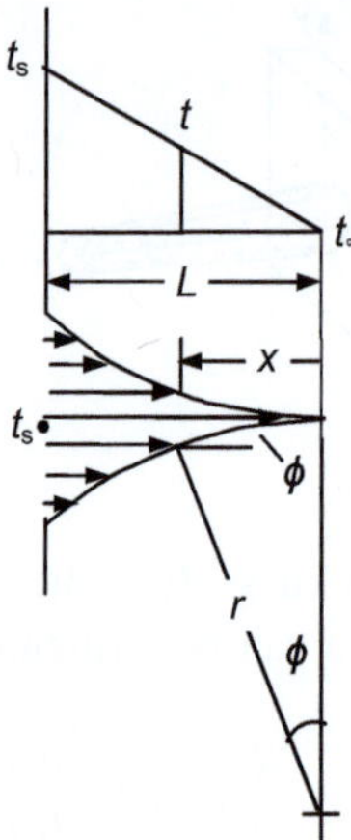

Fig. 3.25 Fin of minimum weight

Under such condition, the lines of heat flow will be parallel to the axis of the fin and the temperature distribution along the fin length will be linear, Fig. 3.25. Therefore, we may write for any cross-section of the fin:

$$t - t_\infty = \frac{x}{L}(t_s - t_\infty) \tag{i}$$

where x is the distance measured from the tip and L is the total length of the fin.

Consider an elementary surface of the fin at distance x. If the rate of heat flow along the fin axis is q, the heat flow through the elemental surface will be q sinϕ, where ϕ is angle of the elemental surface with the axis of the fin. The heat balance at the elemental surface gives

$$q \sin\phi = h(t - t_\infty)$$

or

$$q \sin\phi = \frac{hx}{L}(t_s - t_\infty)$$

This gives

$$\sin\phi = \frac{hx}{Lq}(t_s - t_\infty) = \frac{x}{Lq/[h(t_s - t_\infty)]} \tag{3.38}$$

The contour of this fin is a circular arc of radius r, since sinϕ = x/r. From Eq. (3.38), $r = Lq/[h(t_s - t_\infty)$. This fin possesses the least weight. However, it is difficult to manufacture such fins. A triangular fin differs only slightly from the circular arc fin and can be manufactured with ease. Such fins are commonly used in applications where the minimum weight is the design consideration.

3.13 Straight Fin of Triangular Section

The area normal to the heat flow is a function of the distance along the fin. With origin at the vertex of the triangle, the area of cross-section at distance x is

$$A_x = \left(\delta \frac{x}{L}\right) b \tag{i}$$

Following the procedure outlined for uniform cross-section fins, the heat balance for the infinitesimal element of thickness dx gives, refer to Fig. 3.26,

$$\frac{d}{dx}\left(-kA_x \frac{dt}{dx}\right) + hP_x(t - t_\infty) = 0$$

When $b >> \delta_x$, $P_x = 2(b + \delta_x) \approx 2b$ and is constant. Assuming constant thermal conductivity and substituting the value of A_x from Eq. (i), the above equation transforms to

$$-k\frac{\delta b}{L}\frac{d}{dx}\left(x\frac{dt}{dx}\right) + h(2b)(t - t_\infty) = 0$$

or

$$\frac{d}{dx}\left(x\frac{dt}{dx}\right) - \frac{2hL}{k\delta}(t - t_\infty) = 0$$

or

$$x\frac{d^2t}{dx^2} + \frac{dt}{dx} - \beta(t - t_\infty) = 0$$

where $\beta = \dfrac{2hL}{k\delta}$.

Let $\theta = t - t_\infty$, then $\dfrac{dt}{dx} = \dfrac{d\theta}{dx}$, and $\dfrac{d^2t}{dx^2} = \dfrac{d^2\theta}{dx^2}$. Hence, we obtain

$$\frac{d^2\theta}{dx^2} + \frac{1}{x}\frac{d\theta}{dx} - \frac{\beta}{x}\theta = 0 \tag{3.39}$$

Equation (3.39) is a modified Bessel equation and its solution is

$$\theta = C_1 I_0(2\sqrt{\beta x}) + C_2 K_0(2\sqrt{\beta x}) \tag{3.40}$$

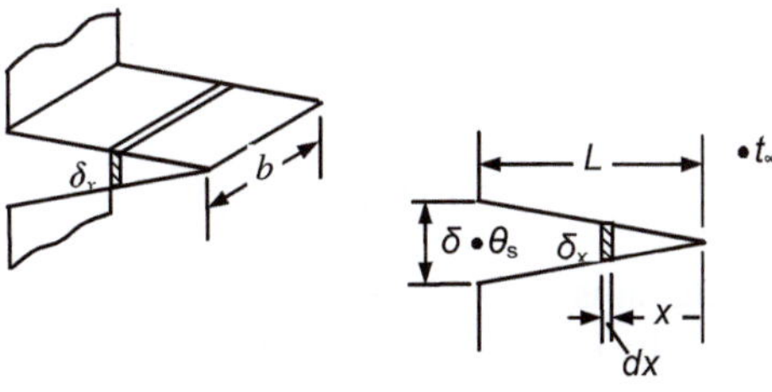

Fig. 3.26 Atriangular fin

Table 3.3 Modified Bessel functions of first and second kinds

α	$I_0(\alpha)$	$I_1(\alpha)$	$K_0(\alpha)$	$K_1(\alpha)$
0.0	1.0000	0.0000	∞	∞
0.2	1.0100	0.1005	1.753	4.775
0.4	1.0404	0.2040	1.114	2.185
0.6	1.0921	0.3137	0.777	1.303
0.8	1.1665	0.4329	0.565	0.862
1.0	1.2661	0.5652	0.421	0.602
1.2	1.394	0.715	0.319	0.435
1.4	1.553	0.886	0.244	0.321
1.6	1.750	1.085	0.188	0.241
1.8	1.990	1.371	0.146	0.183
2.0	2.280	1.591	0.114	0.140
2.2	2.629	1.914	0.0893	0.1079
2.4	3.049	2.298	0.0702	0.0837
2.6	3.553	2.755	0.0554	0.0653
2.8	4.157	3.301	0.0438	0.0511
3.0	4.881	3.953	0.0347	0.0402
3.2	5.747	4.734	0.02760	0.0316
3.4	6.785	5.670	0.02196	0.0250
3.6	8.028	6.793	0.01750	0.0198
3.8	9.517	8.140	0.01397	0.0157
4.0	11.302	9.760	0.0112	0.0125
4.2	13.442	11.706	0.00893	0.00994
4.4	16.010	14.046	0.00715	0.00792
4.6	19.023	16.863	0.00573	0.00633
4.8	22.794	20.253	0.00460	0.00505
5.0	27.240	24.336	0.00369	0.00404
5.2	32.584	29.254	0.00297	0.00324
5.4	39.009	35.182	0.00238	0.00260
5.6	46.738	42.328	0.00192	0.00208
5.8	56.038	50.946	0.00154	0.00167
6.0	67.234	61.342	0.00125	0.00134
6.2	80.72	73.89	0.001005	0.00108
6.4	96.98	89.03	0.00081	0.00087
6.6	116.54	107.30	0.00065	0.00067
6.8	140.14	129.38	0.000526	0.000563
7.0	168.6	156.04	0.000425	0.000454

where I_0 is modified Bessel function of I kind, and K_0 is modified Bessel function of II kind.

The values of functions $I_0(\alpha)$ and $K_0(\alpha)$ are tabulated in Table 3.3 as function of α. In the present case, $\alpha = 2\sqrt{(\beta x)}$.

At $x = 0$, $\alpha = 2\sqrt{(\beta x)} = 0$. For $\alpha = 0$, $K_o(\alpha) = \infty$ from the table. It means that at the fin tip ($x = 0$), the temperature excess θ is infinity from Eq. (3.40), which is physically not possible. Hence, the constant C_2 must be zero. This gives

$$\theta = C_1 I_0(2\sqrt{\beta x})$$

The other boundary condition is
$\theta = \theta_s$ (the temperature at the base of the fin) at $x = L$.

This gives

$$\theta_s = C_1 I_0(2\sqrt{\beta L})$$

or

$$C_1 = \frac{\theta_s}{I_0(2\sqrt{\beta L})}$$

and hence the temperature distribution equation applicable to the present case is

$$\theta = \theta_s \frac{I_0(2\sqrt{\beta x})}{I_0(2\sqrt{\beta L})} \tag{3.41}$$

The temperature θ_o of the fin end is obtained by putting $x = 0$ in the above equation

$$\theta_o = \theta_s \frac{1}{I_0(2\sqrt{\beta L})} \tag{3.42}$$

The heat flow rate q_{fin} equals the heat flow by conduction at the fin base, i.e.

$$\begin{aligned} q_{fin} &= \left(kA\frac{dt}{dx}\right)_{x=L} \\ &= k(b\delta)\left[\frac{d}{dx}\left(\theta_s \frac{I_0(2\sqrt{\beta x})}{I_0(2\sqrt{\beta L})}\right)\right]_{x=L} \\ &= k(b\delta)\theta_s \frac{1}{I_0(2\sqrt{\beta L})}\left\{\frac{d}{dx}\left[I_0(2\sqrt{\beta x})\right]\right\}_{x=L} \end{aligned}$$

Since $\left\{\frac{d}{dx}[I_n(\alpha)]\right\}_{x=L} = I_{n+1}(\alpha)\frac{d}{dx}(\alpha)$, we get

$$\begin{aligned} q_{fin} &= k(b\delta)\theta_s \frac{1}{I_0(2\sqrt{\beta L})}\left\{I_1(2\sqrt{\beta x}) \times 2\sqrt{\beta}\frac{1}{2\sqrt{x}}\right\}_{x=L} \\ &= k(b\delta)\theta_s \frac{I_1(2\sqrt{\beta L})}{I_0(2\sqrt{\beta L})} \times \frac{\sqrt{\beta}}{\sqrt{L}} \end{aligned}$$

Substituting $\beta = \frac{2hL}{k\delta}$ and simplifying, we get

$$q_{fin} = \sqrt{2hk\delta} \times b\theta_s \frac{I_1(2\sqrt{\beta L})}{I_0(2\sqrt{\beta L})} \tag{3.43}$$

Example 3.31 The heat dissipation from an air-cooled flat surface is to be enhanced by installing either triangular or rectangular fins. Either fin is to be 25 mm thick, 100 mm long and made from aluminium [k = 230 W/(m K)]. The wall temperature is 630°C, and the heat transfer coefficient between the surface and the ambient air at 30°C is 72 W/(m^2 K). Compare the effectiveness of the two fins based on the heat flow per unit weight.

Solution

(i) **Triangular fin**:

From Eq. (3.43),

$$q_{fin} = \sqrt{2hk\delta} \times b\theta_s \frac{I_1(2\sqrt{\beta L})}{I_0(2\sqrt{\beta L})} \tag{i}$$

Here h = 72 W/(m^2 K), k = 230 W/(m K), δ = 0.025 m, b = 1 m (assumed), $\theta_s = t_s - t_\infty$ = 630 – 30 = 600°C and L = 0.1 m.

The factor $\beta = 2\ hL/k\delta = 2 \times 72 \times 0.1/(230 \times 0.025) = 2.504$ and $2\sqrt{(\beta L)} = 2\sqrt{(2.504 \times 0.1)} = 1$.

From Table 3.3, $I_1(1) = 0.565$ and $I_0(1) = 1.266$.

Substitution of values of various parameters in Eq. (i) gives

$$q_{fin} = \sqrt{2 \times 72 \times 230 \times 0.025} \times \frac{0.565}{1.266} \times 1 \times 600$$
$$= 7705\ \text{W per m width of the fin}$$

(ii) **Rectangular fin**:

From the given data,

$$P = 2(1 + 0.025) = 2.05\ \text{m}^2$$
$$A_c = b\delta = 0.025\ \text{m}^2$$
$$m = \sqrt{\frac{hP}{kA_c}} = 5.07$$
$$\frac{h}{mk} = 0.062$$

Hence, from Eq. (3.8),

$$q_{fin} = \sqrt{PhkA_c}(t_s - t_\infty)\frac{h/mk + \tanh mL}{(h/mk)\tanh mL + 1}$$
$$= \sqrt{2.05 \times 72 \times 230 \times 0.025} \times 600 \times \frac{0.062 + \tanh(0.507)}{0.062 \times \tanh(0.507) + 1}$$
$$= 8990\ \text{W per m width}$$

The weight of the rectangular section fin per unit width is twice that of the triangular fin, while the heat transfer capacity is only 16.7 percent higher. Thus, the triangular fin is more effective per unit weight.

3.14 Straight Fin of Trapezoidal Section

Consider a fin of trapezoidal cross-section as shown in Fig. 3.27. The fin operates in conditions specified for the forgoing case. The dimensions and notations are given in the figure. The general solution given in Eq. (3.40) is

$$\theta = C_1 I_0(2\sqrt{\beta x}) + C_2 K_0(2\sqrt{\beta x}) \tag{i}$$

where I_0 is modified Bessel function of I kind, and K_0 is modified Bessel function of II kind. Here $\beta = \frac{2hL_1}{k\delta_1}$.

One of the boundary conditions is

$\theta = \theta_s$ (the temperature at the base of the fin) at $x = L_1$.

This gives

$$\theta_s = C_1 I_0(2\sqrt{\beta L_1}) + C_2 K_0(2\sqrt{\beta L_1}) \tag{ii}$$

For simplification, we assume that heat transfer from the fin end is negligible, i.e.

$$\left(\frac{d\theta}{dx}\right)_{x=L_2} = 0 \text{ at x} = L_2$$

This gives

$$\begin{aligned} 0 &= C_1 I_1(2\sqrt{\beta L_2}) 2\sqrt{\beta}\tfrac{1}{2\sqrt{L_2}} - C_2 K_1(2\sqrt{\beta L_2}) 2\sqrt{\beta}\tfrac{1}{2\sqrt{L_2}} \\ 0 &= C_1 I_1(2\sqrt{\beta L_2}) - C_2 K_1(2\sqrt{\beta L_2}) \end{aligned} \tag{iii}$$

where

$$\left\{\frac{d}{dx}[K_n(\alpha)]\right\}_{x=L} = -K_{n+1}(\alpha)\frac{d}{dx}(\alpha)$$

Solving Eqs. (ii) and (iii), we get values of constants C_1 and C_2. Finally substitution of their values in Eq. (i) gives

$$\frac{\theta}{\theta_s} = \frac{I_o(2\sqrt{\beta x})K_1(2\sqrt{\beta L_2}) + I_1(2\sqrt{\beta L_2})K_o(2\sqrt{\beta x})}{I_o(2\sqrt{\beta L_1})K_1(2\sqrt{\beta L_2}) + I_1(2\sqrt{\beta L_2})K_o(2\sqrt{\beta L_1})} \tag{3.44}$$

The temperature of the fin end θ_2 can be found from the above equation by putting $x = L_2$.

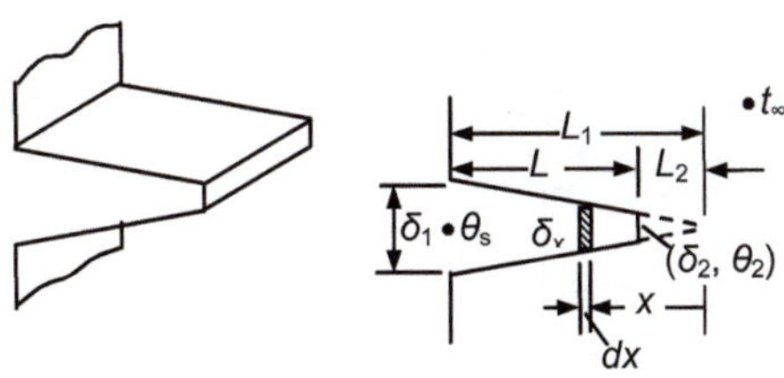

Fig. 3.27 A trapezoidal fin

The heat flow rate q_{fin} equals the heat flow by conduction at the fin base, i.e.

$$q_{fin} = \left(kA_{base}\frac{dt}{dx}\right)_{x=L_1}$$
$$= \left[\frac{I_1(2\sqrt{\beta L_1})K_1(2\sqrt{\beta L_2}) - I_1(2\sqrt{\beta L_2})K_1(2\sqrt{\beta L_1})}{I_0(2\sqrt{\beta L_1})K_1(2\sqrt{\beta L_2}) + I_1(2\sqrt{\beta L_2})K_0(2\sqrt{\beta L_1})}\right]\frac{k\delta_1 b\sqrt{\beta}}{\sqrt{L_1}}\theta_s$$

Substituting $\beta = \frac{2hL_1}{k\delta_1}$ and simplifying, we get

$$q_{fin} = \left[\frac{I_1(2\sqrt{\beta L_1})K_1(2\sqrt{\beta L_2}) - I_1(2\sqrt{\beta L_2})K_1(2\sqrt{\beta L_1})}{I_0(2\sqrt{\beta L_1})K_1(2\sqrt{\beta L_2}) + I_1(2\sqrt{\beta L_2})K_0(2\sqrt{\beta L_1})}\right]\sqrt{2hk\delta_1} \times b\theta_s \tag{3.45}$$

The heat transfer from the tip of the fin can be approximately accounted for by an increase in the fin length L by half of its thickness.

3.15 Annular Fin

Figure 3.28 shows an annular fin. Consider an elemental ring of width dr at radius r. The surface area of the ring rejecting heat to the surrounding is $2(2\pi r dr)$ and the cross-sectional area is $(2\pi r\delta)$. Heat entering the elemental ring by conduction is

$$q_r = -kA_c\left(\frac{dt}{dr}\right) = -k(2\pi r\delta) \times \left(\frac{dt}{dr}\right)$$

Heat leaving the element is

$$q_{r+dr} = q_r + \left(\frac{dq_r}{dr}\right)dr$$

Heat rejected by convection to the surrounding from the elemental surface area is

$$q_c = hA(t - t_s)$$
$$= h[2(2\pi r dr)](t - t_\infty)$$

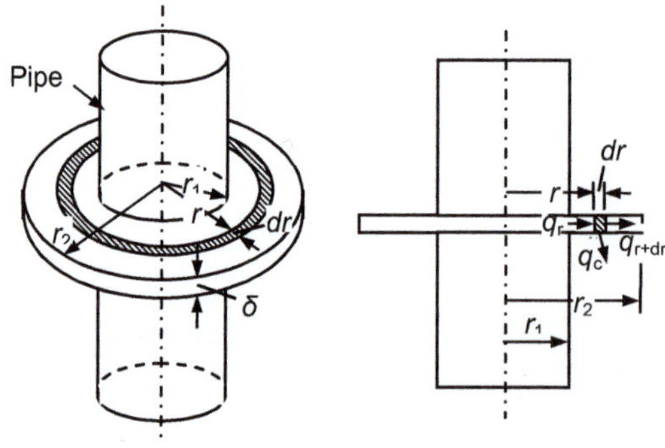

Fig. 3.28 An annular fin

In equilibrium, the heat entering the element equals the heat leaving the element, i.e.

$$q_r = q_{r+dr} + q_c$$
$$= q_r + \left(\frac{dq_r}{dr}\right)dr + q_c$$

or

$$-\left(\frac{dq_r}{dr}\right)dr = q_c$$

or

$$-\frac{d}{dr}\left(-k(2\pi r\delta) \times \frac{dt}{dr}\right)dr = 4\pi r dr \times h \times (t - t_\infty)$$

or

$$k(2\pi\delta)\frac{d}{dr}\left(r\frac{dt}{dr}\right) = 4\pi r \times h \times (t - t_\infty)$$

or

$$\frac{d}{dr}\left(r\frac{dt}{dr}\right) = \left(\frac{2h}{k\delta}\right) \times r \times (t - t_\infty)$$

Putting $\theta = t - t_\infty$,

$$\frac{d}{dr}\left(r\frac{d\theta}{dr}\right) = \left(\frac{2h}{k\delta}\right) \times r\theta$$

or

$$\frac{d\theta}{dr} + r\frac{d^2\theta}{dr^2} - \left(\frac{2h}{k\delta}\right) \times r\theta = 0$$

or

$$\frac{d^2\theta}{dr^2} + \frac{1}{r}\frac{d\theta}{dr} - \frac{2h\theta}{k\delta} = 0 \tag{i}$$

This is a form of Bessel's equation of zero order, and its solution is

$$\theta = C_1 I_0(\beta r) + C_2 K_0(\beta r) \tag{ii}$$

where $\beta = \sqrt{(2\ h/k\delta)}$ and I_0 and K_0 are as defined earlier. Constants C_1 and C_2 are to be determined from the boundary conditions.

One of the boundary conditions is

$\theta = \theta_s$, the temperature at the base of the fin, i.e. at $r = r_1$.

This gives

$$\theta_s = C_1 I_0(\beta r_1) + C_2 K_0(\beta r_1) \tag{iii}$$

For simplification, we assume that heat transfer from the fin end is negligible. This is true when $\delta << (r_2 - r_1)$. This gives

$$\left(\frac{d\theta}{dx}\right)_{r=r_2} = 0$$

Hence,

$$0 = C_1 I_1(\beta r_2)\beta - C_2 K_1(\beta r_2)\beta \tag{iv}$$

Solving Eqs. (iii) and (iv), we get values of constants C_1 and C_2 as

$$C_1 = \theta_s \frac{K_1(\beta r_2)}{I_0(\beta r_1)K_1(\beta r_2) + I_1(\beta r_2)K_0(\beta r_1)}$$

$$C_2 = \theta_s \frac{I_1(\beta r_2)}{I_0(\beta r_1)K_1(\beta r_2) + I_1(\beta r_2)K_0(\beta r_1)}$$

Substitution of their values in Eq. (ii) gives

$$\frac{\theta}{\theta_s} = \frac{I_0(\beta r)K_1(\beta r_2) + I_1(\beta r_2)K_0(\beta r)}{I_0(\beta r_1)K_1(\beta r_2) + I_1(\beta r_2)K_0(\beta r_1)} \tag{3.46}$$

The temperature of the fin end θ_2 can be found from the above equation by putting $r = r_2$. The heat flow rate q_{fin} equals the heat flow by conduction at the fin base, i.e.

$$\begin{aligned} q_{fin} &= \left(-kA_{base}\frac{dt}{dr}\right)_{r=r_1} \\ &= 2\pi k\delta\beta r_1 \left[\frac{I_1(\beta r_2)K_1(\beta r_1) - I_1(\beta r_1)K_1(\beta r_2)}{I_0(\beta r_1)K_1(\beta r_2) + I_1(\beta r_2)K_0(\beta r_1)}\right]\theta_s \end{aligned} \tag{3.47}$$

The heat transfer from the tip of the fin can be approximately accounted for by increasing r_2 by half of the fin thickness δ.

Example 3.32 An annular aluminium alloy fin [k = 160 W/(m K)] is installed on a 25 mm OD tube. The fin has uniform thickness of 0.4 mm and has outer radius of 37.5 mm. If the tube wall temperature is 150°C, the surrounding air temperature is 20°C and the average convective heat transfer coefficient is 30 W/(m^2 K), calculate the heat loss from the fin.

Solution

Neglecting the heat loss from the fin end,

$$q_{fin} = 2\pi k\delta\beta r_1 \left[\frac{I_1(\beta r_2)K_1(\beta r_1) - I_1(\beta r_1)K_1(\beta r_2)}{I_0(\beta r_1)K_1(\beta r_2) + I_1(\beta r_2)K_0(\beta r_1)}\right]\theta_s \tag{3.47}$$

Here k = 160 W/(m K), δ = 0.4/1000 m, θ_s = (150–20) = 130°C, h = 30 W/(m^2 K). Hence,

$$\beta r_1 = \sqrt{\frac{2h}{k\delta}} r_1 = \sqrt{\frac{2\times 30\times 1000}{160\times 0.4}} \times \frac{12.5}{1000} = 0.3827,$$
$$\beta r_2 = \beta r_1 \frac{r_2}{r_1} = 1.1482$$

This gives

$$q_{fin} = 2\pi \times 160 \times 0.0004 \times 0.3827 \times \left[\frac{I_1(1.1482)K_1(0.3827) - I_1(0.3827)K_1(1.1482)}{I_0(0.3827)K_1(1.1482) + I_1(1.1482)K_0(0.3827)}\right] \times 130$$

Substitution of the values of Bessel's function from Table 3.3 gives

$$q_{fin} = 23.1 \text{ W}$$

3.16 Fin Efficiency Plots

The equations derived in previous sections for rectangular, triangular and circumferential fins are very inconvenient for calculations. The heat flow rate can be determined approximately for most of the engineering applications using the following equation:

$$q_{fin} = \eta_f h A_s \theta_s \tag{3.48}$$

where η_f is the fin efficiency, which can be read from Fig. 3.29 for rectangular, triangular and circumferential fins of rectangular cross-section, and A_s is surface area of the fin. The abscissa parameter in Fig. 3.29 is $L_c^{3/2}(h/kA_m)^{1/2}$ where L_c (the corrected length) and A_m are defined in the figure. For other fin shapes, the readers can refer to Gardner (1945) and advanced texts on the subject.

Example 3.33 Repeat Example 3.31 using the efficiency approach.

Solution

(i) **Triangular fin**:

$$L_c = L = 0.1, A_m = (t/2)L = (0.025/2) \times 0.1 = 0.00125$$

$$\text{Parameter}\, L_c^{3/2}\sqrt{\frac{h}{kA_m}} = 0.1^{3/2} \times \sqrt{\frac{72}{230 \times 0.00125}} = 0.5$$

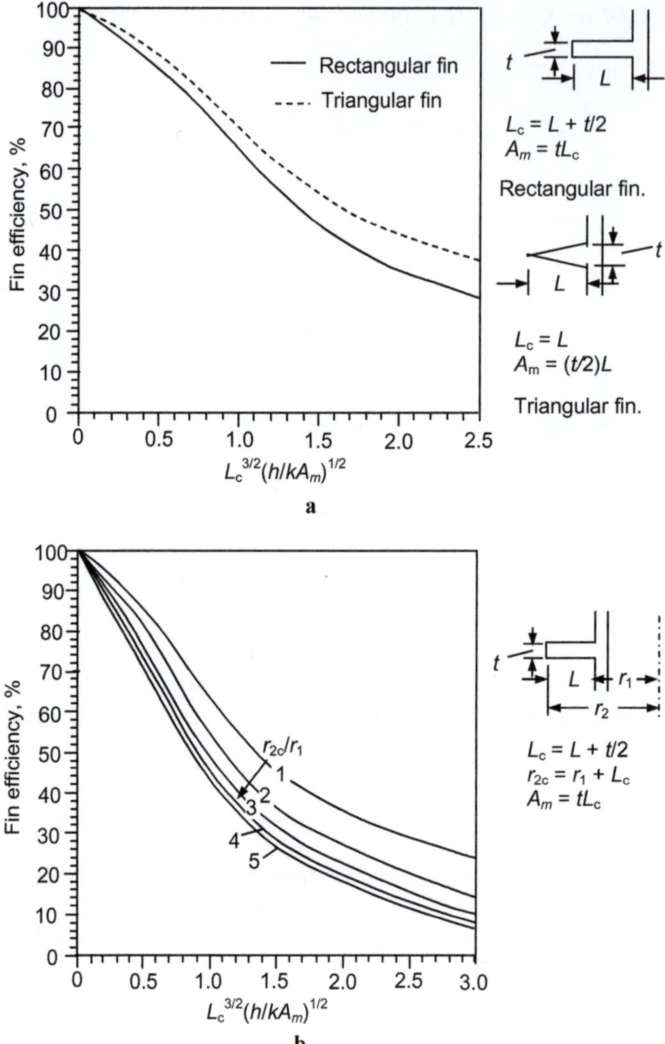

Fig. 3.29 **a** Efficiencies of rectangular and triangular fins, **b** Efficiency of circumferential or annular fins of rectangular section. Holman JP, adapted for SI units by White PRS, Heat Transfer, McGraw-Hill Book Co, New York, Copyright 1992. The material is reproduced with permission of McGraw-Hill Education (Asia)

From Fig. 3.29a, the fin efficiency, $\eta_f = 0.88$ for the calculated value of the above parameter. Hence,

$$\begin{aligned} q_{fin} &= hA_s\theta_s\eta_f = h \times \left[(2b) \times \sqrt{L^2 + (t/2)^2}\right]\theta_s\eta_f \\ &= 72\left[(2 \times 1) \times \sqrt{0.1^2 + (0.025/2)^2}\right] \times (630 - 30) \times 0.88 \\ &= 7662\ \mathrm{W/m\ width\ of\ the\ fin} \end{aligned}$$

(ii) **Rectangular fin**:

$$L_c = L + t/2 = 0.1125, t = 0.025, A_m = tL_c = 0.025 \times 0.1125 = 0.0028125$$

$$\text{Parameter}\, L_c^{3/2}\sqrt{\frac{h}{kA_m}} = 0.1125^{3/2} \times \sqrt{\frac{72}{230 \times 0.0028125}} \approx 0.4$$

From Fig. 3.29a, the fin efficiency $\eta_f \approx 0.91$ for the calculated value of above parameter. Hence,

$$\begin{aligned} q_{fin} &= hA_s\theta_s\eta_f = h \times (2bL_c) \times \theta_s\eta_f \\ &= 72 \times (2 \times 1 \times 0.1125) \times (630 - 30) \times 0.91 \\ &= 8845\ \text{W/m width of the fin} \end{aligned}$$

The results are quite close to the solution in Example 3.31.

Example 3.34 Repeat above example for the triangular fin with L = 0.2 m and 0.05 m keeping fin base thickness fixed at 0.025 m.

Solution

(i) ***L* = 0.2 m:**

$$L_c = L = 0.2, A_m = (t/2)L = (0.025/2) \times 0.2 = 0.0025$$

$$\text{Parameter}\, L_c^{3/2}\sqrt{\frac{h}{kA_m}} = (0.2)^{3/2} \times \sqrt{\frac{72}{230 \times 0.0025}} = 1.0$$

From Fig. 3.29, the fin efficiency, η_f = 0.7 and

$$\begin{aligned} q_{fin} &= hA_s\theta_s\eta_f = h\left[(2b) \times \sqrt{L^2 + (t/2)^2}\right]\theta_s\eta_f \\ &= 72\left[(2 \times 1) \times \sqrt{0.2^2 + (0.025/2)^2}\right] \times (630 - 30) \times 0.7 \\ &= 12120\ \text{W/m width of the fin} \end{aligned}$$

i.e. 1.55 times of that for fin of length 0.1 m while the weight of the fin increased to 2 times.

(ii) ***L* = 0.05 m:**

$$L_c = L = 0.05, A_m = (t/2)L = (0.025/2) \times 0.05 = 0.000625$$

$$\text{Parameter}\, L_c^{3/2}\sqrt{\frac{h}{kA_m}} = (0.05)^{3/2} \times \sqrt{\frac{72}{230 \times 0.000625}} = 0.25$$

From Fig. 3.29, the fin efficiency, η_f = 0.95 and

$$q_{fin} = hA_s\theta_s\eta_f = h\left[(2b) \times \sqrt{L^2 + (t/2)^2}\right]\theta_s\eta_f$$

$$= 72 \times \left[(2 \times 1) \times \sqrt{0.05^2 + (0.025/2)^2}\right] \times (630 - 30) \times 0.95$$

$$= 4229 \text{ W/m width of the fin}$$

i.e. 0.55 times of that for fin of length 0.1 m while the weight of the fin is reduced to half. We can conclude that the heat transfer per unit weight is greater for fins of smaller heights.

Example 3.35 Repeat Example 3.32 using the fin efficiency approach with no correction for the fin end loss (as fin thickness is very small).

Solution

For $L_c = L = 0.025$, $A_m = tL = 0.0004 \times 0.025 = 0.00001$, $h = 30$ W/(m^2 K) and $k = 160$ W/(m K),

$$\text{Parameter}\, L_c^{3/2}\sqrt{\frac{h}{kA_m}} = (0.025)^{3/2} \times \sqrt{\frac{30}{160 \times 0.00001}} = 0.54$$

and

$$\frac{r_{2c}}{r_1} = 3$$

From Fig. 3.29b, fin efficiency $\eta_f \approx 0.73$. Hence,

$$\begin{aligned} q_{fin} &= \eta_f hA_s\theta_s \\ &= 0.73 \times 30 \times [2\pi(0.0375^2 - 0.0125^2)] \times (150 - 20) \\ &= 22.36\,\text{W} \end{aligned}$$

which is quite close to the solution given in Example 3.32.

Example 3.36 A steel pipe carries hot gas at 800°C. In order to increase heat transfer rate from the pipe, a finned aluminum cylinder is slipped over the steel pipe as shown in Fig. 3.30a. The contact resistance to heat transfer at the pipe and cylinder interface is estimated to be 10^{-4} m^2 K/W. For the data given in the figure, determine heat transfer rate per unit length of pipe.

Solution

The thermal network for the system is shown in Fig. 3.30b. The heat transfer rate is

$$q = \frac{t_{gas} - t_\infty}{R_{total}}$$

where $R_{total} = R_i + R_{wall} + R_c + R_2 + R_{eq}$. The various resistances for unit length of pipe are calculated below.

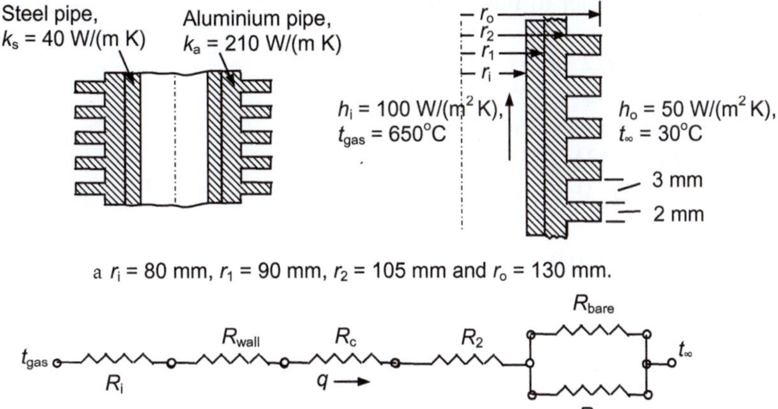

Fig. 3.30 Example 3.36

Inner surface film resistance:

$$R_i = \frac{1}{h_i A_i} = \frac{1}{h_i(2\pi r_i L)} = \frac{1}{100 \times 2 \times \pi \times 0.08 \times 1} = 0.0199$$

Pipe wall resistance:

$$R_{wall} = \frac{1}{2\pi k_s L}\ln\left(\frac{r_1}{r_i}\right) = \frac{1}{2 \times \pi \times 40 \times 1}\ln\left(\frac{90}{80}\right) = 0.468 \times 10^{-3}$$

Contact resistance:

$$R_c = \frac{10^{-4}}{2\pi r_1 L} = \frac{10^{-4}}{2 \times \pi \times 0.09 \times 1} = 0.176 \times 10^{-3}$$

Aluminium wall resistance:

$$R_2 = \frac{1}{2\pi k_a L}\ln\left(\frac{r_2}{r_1}\right) = \frac{1}{2 \times \pi \times 210 \times 1}\ln\left(\frac{105}{90}\right) = 0.117 \times 10^{-3}$$

Film resistance of surface of aluminium cylinder without fins:

$$R_{bare} = \frac{1}{h_o A_{bare}} = \frac{1}{h_o(2\pi r_2 \times L_{bare})} = \frac{1}{50 \times 2 \times \pi \times 0.105 \times 0.6} = 0.0505$$

Fin resistance using Eq. (3.48):

$$R_{fin} = \frac{t_s - t_\infty}{q_{fin}} = \frac{1}{\eta_{fin} h_o A_{fin}}$$

Efficiency of the fin (refer to Fig. 3.29b):

$$L_c = L + t/2 = 25 + 2/2 = 26\,\text{mm}$$
$$r_{2c} = r_1 + L_c = 105 + 26 = 131\,\text{mm}$$
$$A_m = tL_c = 2 \times 26 = 52\,\text{mm}^2$$
$$r_{2c} / r_1 = 131/105 = 1.248$$
$$L_c^{3/2}(h/kA_m)^{1/2} = 0.026^{3/2}\left[50/\left(210 \times 52 \times 10^{-6}\right)\right]^{1/2} = 0.284$$

From Fig. 3.29b, $\eta_{fin} = 0.91$.
Fin surface area per unit length of pipe:

$$A_{fin} = \text{no. of fins} \times \pi(r_{2c}^2 - r_2^2) \times 2 = 200 \times \pi \times (0.131^2 - 0.105^2) \times 2 = 7.71\,\text{m}^2$$

Hence,

$$R_{fin} = \frac{1}{\eta_{fin} h_o A_{fin}} = \frac{1}{0.91 \times 50 \times 7.71} = 2.85 \times 10^{-3}$$
$$R_{eq} = \left[\frac{1}{R_{bare}} + \frac{1}{R_{fin}}\right]^{-1} = \left[\frac{1}{0.0505} + \frac{1}{2.85 \times 10^{-3}}\right]^{-1} = 2.69 \times 10^{-3}$$

Total resistance,

$$\begin{aligned} R_{total} &= R_i + R_{wall} + R_c + R_2 + R_{eq} \\ &= 0.0199 + 0.468 \times 10^{-3} + 0.176 \times 10^{-3} + 0.117 \times 10^{-3} + 2.69 \times 10^{-3} = 0.02335 \end{aligned}$$

Heat transfer rate,

$$q = \frac{650 - 30}{0.02335} = 26552\,\text{W/m length}$$

3.17 Summary

The heat transfer rate from or to a surface in contact with a fluid can be enhanced by providing fins, which basically increase the surface area for heat transfer. Fins are widely used in various engineering equipments. Fins of various shapes and in different arrangements have been employed. A majority of fins are of rectangular cross section. Thick fins and fins made by casting are generally of the trapezoidal cross-section with rounded edges. Spine- or stud-type fins are mechanically rugged, and hence have longer life in corrosive atmosphere than thin plate fins. It may be noted that the final choice of the type of fin and their arrangement depends not only on the heat transfer performance but also on the resistance offered to the flow of the surrounding fluid, cost and the ease of fabrication.

In this chapter, mathematical treatments of various types of uniform cross-section, triangular, trapezoidal and annular fins have been presented to determine the temperature variation along the fin length and heat transfer rate from the fin surface assuming one-dimensional steady-state heat flow condition.

It is to note that an exact analytical solution of the heat propagation through a fin involves considerable difficulties. Mathematical solutions presented here are approximate because of the following major simplifying assumptions.

(i) Temperature at any cross-section of the fin is uniform. The assumption is valid for fins of small cross-section and made of high thermal conductivity material. This assumption reduces the problem to that of one-dimensional heat conduction along the axis of the fin only.
(ii) The heat transfer coefficient h from the fin surface to the surrounding is constant for the entire fin surface.

The greatest uncertainty is in the value of the heat transfer coefficient, which is seldom uniform over the entire fin surface. In the case of severe non-uniformity of the heat transfer coefficient and two- or three-dimensional conduction heat flow, numerical techniques are used to solve the problem. A discussion on the same has been presented in Chap. 5.

In order to do mathematical analysis of the uniform cross-section fins, the fins have been divided into three classes: (i) a very long fin (mathematically a fin of infinite length), (ii) a fin with negligible heat transfer from the fin end as compared to the heat transferred from the fin surface (treated as a fin with insulated end) and (iii) a short fin (fin with heat loss from the fin end).

The heat transfer equation of the short fin can be expressed in the same form as that of fin with insulated end if a corrected length L_c, defined as $L_c = L + \Delta L$ where $\Delta L = A_c/P$, is used. In fact, we can use heat transfer equation as $q_{fin} = \sqrt{hPkA_c}(t_s - t_\infty)\tanh mL_c$ for all cases because in case of a very long fin product mL will be greater than 5 and $\tanh(mL)$ can be taken as unity, refer to Table 3.1.

To study the performance of a fin in transferring the heat, a term called fin effectiveness has been used, which is a ratio of heat transfer with the fin to the heat that would be transferred without fin. Effectiveness of a fin must be always greater than 1 and as high as possible. In general, the fin effectiveness is high when (i) the fin is made of high thermal conductivity material, (ii) it has high ratio of parameter to the base area (P/A_c) and (iii) the convective heat transfer coefficient is low. Condition for enhancement of heat transfer from a fin has been deduced in terms of a non-dimensional number termed as Biot number Bi.

Since the basic purpose of the installation of fins on a surface is to enhance the heat transfer rate, the fin effectiveness is a meaningful parameter for assessing the usefulness of the fins. However, another term that has been used to assess the performance of a fin in transferring heat is fin efficiency η_{fin}, which is defined as ratio of the heat transfer with the fin to the heat that would be transferred if the entire fin surface were at the fin base temperature. Since most of the heat-transferring surfaces consist of a number of fins with bare portion of the surface between the fins, the total heat transfer from the surface is obtained by combining the heat flow from the fins q_{fin} with the heat flow from the bare portion of the surface q_{o}. Using the fin efficiency, the heat flow rate from the wall can be readily expressed as $q_{total} = q_{fin} + q_o = h_f A_f (t_s - t_\infty)\eta_{fin} + h_w A_o (t_s - t_\infty)$, where h_{w} and h_{f} are heat transfer coefficients for the bare portion and the fins, respectively.

In Sect. 3.12, design of fin for maximum heat transfer at minimum weight has been developed. The contour of this fin is a circular arc. However, it is difficult to manufacture such fins. A triangular fin differs only slightly from the circular arc fin and can be manufactured with ease. Such fins are commonly used in applications where minimum weight is design consideration. A modification of the triangular fin is the fin of trapezoidal cross-section. Mathematical equations for temperature distribution and heat transfer from triangular and trapezoidal fins have been derived in Sects. 3.13 and 3.14, respectively. Analysis of annular or circumferential fins for cylindrical surfaces is given in Sect. 3.15. The equations derived for triangular, trapezoidal and circumferential fins are very inconvenient for calculations. The heat flow rate can be determined approximately for most of the engineering applications from Eq. (3.48) knowing the fin efficiency from efficiency plots in Fig. 3.29 for rectangular, triangular and circumferential fins of rectangular cross-section, surface area of the fin and the temperature difference between the wall and the ambient air.

A significant number of illustrative examples have been included in the chapter.

Review Questions

3.1 Write a short note on heat transfer enhancement from a surface.

3.2 Why fins are used? Discuss various types of fins used in engineering applications.

3.3 Derive the expression for temperature distribution and total heat flow rate under steady-state conditions for a fin of uniform cross-section which is so long that the temperature of the end of the fin can be assumed to be equal to the surrounding temperature.

3.4 Derive the expression for temperature distribution and total heat flow rate under steady-state conditions for a uniform cross-section fin of length *L*. The heat transfer from the free end of the fin may be neglected. Explain the different parameters that form a basis for assessing the utility of a fin.

3.5 Derive the equation for heat dissipation by a fin of uniform cross-section if the area of the fin end is a very small proportion of the total fin surface area and its contribution to the heat dissipation can be neglected.

3.6 Show that the temperature distribution for a fin of finite length, when the heat loss from the fin end by convection is taken into account, is given by

$$\frac{t - t_\infty}{t_s - t_\infty} = \frac{\cosh m(L - x) + (h_L/mk)\sinh m(L - x)}{(h_L/mk)\sinh mL + \cosh mL}$$

Hence, show that the heat transfer from this fin is given by

$$q_{fin} = \sqrt{hPkA_c}(t_s - t_\infty)\frac{(h_L/mk) + \tanh mL}{(h_L/mk)\tanh mL + 1}$$

where h_L is the heat transfer coefficient from the fin end.

3.7 How would you introduce correction in the fin length to take account of the heat transfer from the fin end? What will be the form of equation of the heat transfer from the fin?

3.8 Define the efficiency and effectiveness of a fin. Discuss the parameters that affect the fin efficiency and effectiveness.

3.9 Show that the total efficiency of a finned wall is given by

$$\eta_{total} = 1 - \frac{A_f}{A_{fw}}(1 - \eta_{fin})$$

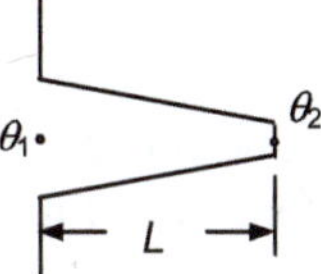

Fig. 3.31 Q. 3.16

where

A_f = surface area of all fins attached to the wall
A_{fw} = total surface area of the finned wall rejecting heat
η_{fin} = fin efficiency.

3.10 Discuss the design of thermometer well to minimize the error in the measurement of the temperature of fluids flowing through a duct.

3.11 'Fins are more effective for surfaces with low value of surface heat transfer coefficient'. Justify the statement.

3.12 Show that fins provided on a surface will increase the heat transfer rate from the surface only if the Biot number is less than 1.

3.13 A thin rod of length L has its ends connected to two walls, which are maintained at temperatures T_1 and T_2, respectively. The rod loses heat to the surrounding fluid at T_∞ by convection. Derive the equations of temperature distribution in the rod and total heat rejection rate from the entire surface of the rod.

3.14 Present a scheme of experimental setup to measure heat transfer coefficient from a fin surface.

3.15 Show that for the triangular profile fin shown in Fig. 3.26, the temperature variation equation is

$$\theta = \theta_s \frac{I_0\left(2\sqrt{\beta x}\right)}{I_0\left(2\sqrt{\beta L}\right)}$$

where $I_o(\alpha)$ is Bessel's function, $\beta = (2hL/k\delta)$, θ_s = temperature excess at the fin base and δ is thickness of the fin at the base.
Determine the total heat rejected by the fin.

3.16 Figure 3.31 shows a fin of trapezoidal profile. Derive the equation for the temperature distribution. Neglect the heat transfer from the tip of the fin. θ_1 and θ_2 are the temperature excesses above the ambient. The width of the fin (perpendicular to the plane of the paper) is very large compared to the height L of the fin. Fin thicknesses at the base and tip are δ_1 and δ_2, respectively.

Problems

3.1 A part of a very long 25-mm-diameter copper rod [k = 370 W/(m K)] is inserted into a furnace. Temperatures at two points 100 mm apart along its length are measured to be 150°C and 120°C when the steady state was achieved. If the ambient air is at 30°C, estimate the effective heat transfer coefficient from the rod surface.
[Hint: Apply the fin equation for temperature distribution, i.e.

$$\frac{t_1 - t_\infty}{t_2 - t_\infty} = \exp[m(x_2 - x_1)]$$

[**Ans.** m = 2.88 m^{-1}; h = 19.2 W/(m^2 K)]

3.2 One end of a long rod is inserted into a furnace, while the other end projects into outside air. Under steady state, the temperature of the rod is measured at two points 75 mm apart and found to be 125°C and 88.5°C when the ambient temperature is 20°C. The rod diameter is 250 mm, and the thermal conductivity of the rod material is 3.0 W/(m K). Find the convective heat transfer coefficient.
[**Ans.** Given k = 3.0 W/(m K), $P = \pi D = \pi \times 0.25$ m, $A_c = (\pi/4)\ D^2 = (\pi/4) \times (0.25)^2$ m^2 and h is unknown. For a very long fin, $\frac{t - t_\infty}{t_s - t_\infty} = e^{-mx}$, where $m = \sqrt{\frac{hP}{kA_c}} = \sqrt{\frac{h\times\pi\times 0.25\times 4}{\pi\times(0.25)^2\times 3}} = 2.31\sqrt{h}$; Distance x_1 may be considered to be zero (refer to Example 3.23), then t_s = 125°C and t = 88.5°C at x = 0.075 m. Substitution gives $\frac{88.5-20}{125-20} = e^{-(2.31\sqrt{h}\times 0.075)}$ or h = 6.08 W/(m^2 K).]

3.3 One end of a long rod is inserted into a furnace, while the other end projects into outside air. Under steady-state condition, the temperatures of the rod at 50 mm and 125 mm from the furnace wall measured with the help of thermocouples indicate 125°C and 88.5°C when the ambient temperature is 20°C. Find the temperature of the rod at the furnace wall.
[**Ans.** For a very long fin, we have $\frac{t-t_\infty}{t_s-t_\infty} = e^{-mx}$; for the given temperatures of t_1 = 125°C at x_1 = 0.05 m and t_2 = 88.5°C at x_2 = 0.125 m, we have $\frac{125 - 20}{t_s - 20} = e^{-0.05m}$ and $\frac{88.5 - 20}{t_s - 20} = e^{-0.125m}$, solution gives m =5.695 m^{-1}. With this value of m, $\frac{125 - 20}{t_s - 20} = e^{-0.05m}$ gives t_s = 159.6°C.]

3.4 Heat generated in a bearing causes the temperature of the end of a shaft to rise to 50°C above the ambient. The shaft is 50 mm in diameter and 800 mm long. Determine the equation of temperature distribution along the shaft if the heat transfer coefficient for the shaft surface is 10 W/(m^2 K) and the thermal conductivity of the shaft material is 45 W/(m K). What is the shaft temperature 100 mm away from the heated end?
[**Ans.** m = 4.216 m^{-1}, mL = 3.373 > 3, hence, the equation of a very long fin can be used, which gives $\theta = 50e^{-4.216x}$. Temperature at 100 mm from the heated end, that is, at x = 100 mm is 32.8°C above the ambient.]

3.5 The head of a solid steel valve is at 600°C. The stem of the valve has a diameter of 10 mm and is cooled by water at 60°C. The heat transfer coefficient is 60 W/(m^2 K). The length of the stem is 200 mm. Determine temperature of the stem 50 mm from the heated end. k = 40 W/(m K).

[**Ans.** $m = 24.495$, $mL = 4.9 > 3$. The temperature distribution equation of a very long fin gives $T = 218.67°C$. Equation (3.9) with end correction gives 218.75°C.]

3.6 Heat dissipation by convection from a metal tank containing cooling oil is to be increased by 50 per cent by addition of fins to the wall. The fins will be 10 mm thick and spaced 100 mm apart between the centres. The surface temperature of the tank is 100°C, and the surrounding air temperature is 20°C. Determine the height of each fin on the assumption that the convective heat transfer coefficient remains unchanged and the surface temperature of the tank is expected to drop to 95°C when fins are fitted. The thermal conductivity of the fins and tank material is 200 W/(m K). Take for simplicity 1 m × 1 m surface area of the tank. $h = 20$ W/(m^2 K).

[**Ans.** Heat dissipation without fins, $q = hA_c(t_s - t_\infty) = 1600$ W/m^2. Desired heat dissipation rate = 1.5 × 1600 = 2400 W. Fin width = width of the tank = 1 m; number of fins, n = (height of tank/fin pitch) = 1000/100 = 10; fin thickness, $\delta = 10$ mm. Surface area not containing fins = 1 × 1 − 10 × 10/1000 = 0.9 m^2; heat dissipation by this surface area = 20 × 0.9 × (95 – 20) = 1350 W; Heat to be dissipated by the fins = 2400–1350 = 1050 W; Heat to be dissipated by one fin = 1050/10 = 105 W; Fin perimeter, $P = 2(W + \delta) \approx 2.0$ m; $A_c = 1 \times 10/1000 = 1$ 10^{-2}; $m = 4.47$. Considering fin end correction, $q_{fin} = \sqrt{(hPkA_c)} \times (t_s - t_\infty) \tanh(mL_c)$ gives $\tanh(mL_c) = 0.1565$ and $mL_c = 0.1578$; $L_c = 0.1578/4.47 = 35.3$ mm; Required length of fin = $L_c - \Delta L = 35.3 - \delta/2 = 30.3$, say 31 mm.]

3.7 A steel well of 12 mm ID is placed in a duct of 75 mm ID for a thermometer. The fluid temperature in the duct is 300°C. Determine the length of the well so that the error in the temperature measurement is less than 1.5°C. The heat transfer coefficient is 100 W/(m^2 K). Wall thickness of the well is 1 mm. The temperature at the well base = 270°C. $k = 50$ W/(m K).

[**Ans.** $m = 44.72$; $(t_L - t_\infty)/(t_s - t_\infty) = 1/\cosh mL$ gives $L = 82.5$ mm for $t_L - t_\infty = 1.5°C$ and $t_s - t_\infty = 30°C$. As the length of the well is greater than the diameter of the duct, the well is to be located inclined.]

3.8 For a rectangular section fin, the thickness and length are 5 mm and 100 mm, respectively. The width of the fin is 300 mm. The thermal conductivity of the fin material is 20 W/(m K). How will the effectiveness of the fin change as the heat transfer coefficient changes from 10 to 100 W/(m^2 K)?

[**Ans.** First case, $h = 10$ W/(m^2 K): $mL = 1.426$, Bi = 1.25×10^{-3}, and $\varepsilon = \tanh mL_c/\sqrt{Bi} = 25.4$; Second case, $h = 100$ W/(m^2 K): $mL = 4.509 > 3$, Bi = 0.0125 and $\varepsilon = 1/\sqrt{Bi} = 8.94$; Fins are more effective when the heat transfer coefficient is low.]

3.9 20-mm-diameter steel rods [$k = 50$ W/(m K)] are to be used as fins on one side of a wall. The heat transfer coefficients for the two sides of the wall are 900 W/(m^2 K) and 25 W/(m^2 K), respectively. The length of the fins is to be restricted to 100 mm. On which side of the wall, would you install the fins? Also determine the heat transfer enhancement or the overall effectiveness if one fin is installed per 2000 mm^2 of the wall area.

[**Ans.** For the high h side, Biot number $Bi_1 = h_1D/4k = 900 \times (0.02/4)/50 = 0.09$; for the other side, $Bi_2 = 0.0025$, as $Bi_2 < Bi_1$, the fins must be installed on the low heat transfer coefficient side. $m = \sqrt{(hP/kA_c)} = 10$ and $L_c = L + D/4 = 105$ mm. $mL_c = 1.05$, $q_{fin} = \sqrt{(hPkA_c)}\ (t_s - t_\infty) \tanh mL_c$. Heat transfer from finned surface = $h \times (2000 \times 10^{-6} - \pi/4\ D^2)\ (t_s - t_\infty) + \sqrt{(hPkA_c)}\ (t_s - t_\infty) \tanh mL_c = 0.16495(t_s - t_\infty)$. Heat transfer from 2000 mm^2 area without fins is

h (2000 × 10^{-6}) × (t_s − t_∞) = 0.05(t_s − t_∞). The ratio of these heat transfer rates = 3.3, which is the enhancement.]

3.10 The following data refer to a uniform section fin.
Thickness = 6 mm; length = 60 mm; width = 1 m; temperature excess at the base = 100°C; h = 10 W/(m^2 K); thermal conductivity of the fin material = 60 W/(m K).
Determine the temperature at the fin end and the efficiency of the fin.
[**Ans.** m = 7.45 m^{-1}; mL_c = 0.47; q_{fin} = $\sqrt{(hPkA_c)}$ (t_s − t_∞) tanhmL_c = 117.93 W; θ_2/θ_s = 1/coshmL_c gives θ_2 = 89.89°C; Fin efficiency = 117.93/[hPL_c(t_s − t_∞)] = 93.04%.]

3.11 Determine the end temperature and heat flow from a straight trapezoidal fin of length L = 75 mm, width = 1 m, δ_1 = 2 mm, δ_2 = 0.5 mm, h = 250 W/(m^2 K), k = 400 W/(m K) and θ_1= 100°C.
[**Ans.** L_1 = 0.1 m, L_2 = 0.025 m, $\beta = 2hL_1/(k\delta_1)$ = 62.5, $2\sqrt{(\beta L_2)}$ = 2.5, $2\sqrt{(\beta L_1)}$ = 5, $I_0[2\sqrt{(\beta L_1)}]$ = 27.24, $I_0[2\sqrt{(\beta L_2)}]$ = 3.301, $I_1[2\sqrt{(\beta L_1)}]$ = 24.336, $I_1[2\sqrt{(\beta L_2)}]$ = 2.5265, $K_0[2\sqrt{(\beta L_1)}]$ = 0.00369, $K_0[2\sqrt{(\beta L_2)}]$ = 0.0628, $K_1[2\sqrt{(\beta L_1)}]$ = 0.00404, $K_1[2\sqrt{(\beta L_2)}]$ = 0.0745; From Eq. (3.44) for L_2 = 0.025 m, θ_2= 19.85°C; From Eq. (3.45), q = 1768.6 W.]

3.12 In order to increase the heat transfer rate from a cylindrical pipe (OD = 100 mm), circumferential fins of 4 mm thickness and 50 mm height are added at a pitch of 10 mm. If the heat transfer coefficient h = 40 W/(m^2 K) and the thermal conductivity of the fin and pipe material is 50 W/(m K), determine the heat transfer enhancement.
[**Ans.** Without fins, q = 12.566 (t_s − t_∞); For the circumferential fins, efficiency ≈ 69% from Fig. 3.29b for $L_c^{3/2}(h/kA_m)^{1/2}$ ≈ 0.73 and r_{2c}/r_1 ≈ 2; Fin surface area = $2\pi(r_2^2 - r_1^2)$, q_{fin} = fin efficiency × fin surface area × h × (t_s − t_∞) = 1.3, 4.0 (t_s − t_∞); From Eq. (3.47), q_{fin} = 1.3(t_s − t_∞) for $\beta = \sqrt{(2\ h/k\delta)}$ = 20 (where $\delta = t$), βr_1 = 1, βr_2 = 2, $I_o(\beta r_1)$ = 1.2661, $I_1(\beta r_1)$ = 0.5652, $I_1(\beta r_2)$ = 1.591, $K_o(\beta r_1)$ = 0.421, $K_1(\beta r_1)$ = 0.602, $K_1(\beta r_2)$ = 0.14. Number of fins/m = 100; Heat transfer from bare surface = 0.6 × 12.566(t_s − t_∞); Heat rejection from the finned tube = number of fins × 1.3 (t_s − t_∞) + 0.6 × 12.566(t_s − t_∞) = 137.54(t_s − t_∞); Enhancement = 137.54/12.566 = 10.95.]

References

Eckert ERG, Drake RM Jr (1972) Analysis of heat and mass transfer. International student edition, McGraw-Hill Kogakusha, Tokya
Gardner KA (1945) Efficiency of extended surfaces. Trans ASME. J Heat Transfer 67:621–631
Holman JP (1992) Adapted for SI units by White PRS. Heat transfer, McGraw-Hill Book Co, New York
Isachenko VP, Osipova VA, Sukomel AS (1977) Heat transfer, 2nd edn. MIR Publishers, Moscow

4 Conduction with Heat Generation

4.1 Plane Wall with Uniform Heat Generation

Figure 4.1 shows a plane wall of thickness δ. Heat is generated within the wall at uniform rate q_g per unit volume and is liberated over the entire volume. The generated heat is transferred to the wall surfaces by conduction and is rejected to the surrounding from one or both faces of the wall. In equilibrium, when the heat generated equals the heat rejected, the wall surface temperatures are T_1 and T_2.

For the analysis, the following assumptions are being made:

1. steady-state conditions,
2. one-dimensional heat flow,
3. constant thermal conductivity of the wall material (homogeneous material).

Consider an elemental strip of thickness dx at distance x from the left face of the wall. From the Fourier's conduction equation, the heat entering the left face of the elemental strip is

$$q_x = -kA\frac{dt}{dx} \tag{1.2}$$

The heat leaving the face at distance $x + dx$ is

$$q_{x+dx} = q_x + \frac{d}{dx}(q_x)dx$$

Net heat leaving the elemental strip is

$$\begin{aligned} q_{x+dx} - q_x &= \frac{d}{dx}(q_x)dx \\ &= \frac{d}{dx}\left(-kA\frac{dt}{dx}\right)dx \\ &= -kA\left(\frac{d^2t}{dx^2}\right)dx \end{aligned}$$

R. Karwa, *Heat and Mass Transfer*,
https://doi.org/10.1007/978-981-15-3988-6_4

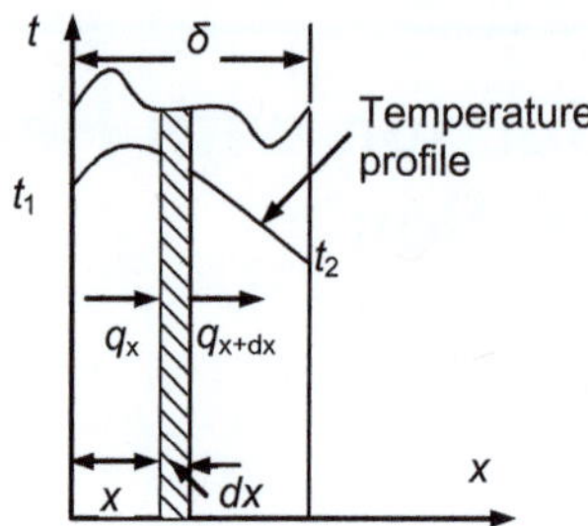

Fig. 4.1 Conduction with heat generation in a plane wall

The heat generated in the elemental strip is

$$q'_g = (Adx)q_g$$

where Adx is the volume of the elemental strip and q_g is the rate of heat generation per unit volume.

In the equilibrium, the total heat generated in the elemental strip must equal the net heat leaving the strip, i.e.

$$-kA\left(\frac{d^2t}{dx^2}\right)dx = (Adx)q_g$$

or

$$\frac{d^2t}{dx^2} + \frac{q_g}{k} = 0 \tag{4.1}$$

Equation (4.1) can also be obtained from the general heat conduction equation by assuming one-dimensional steady-state heat conduction with constant thermal conductivity.

Integrating Eq. (4.1), we get

$$\frac{dt}{dx} = -\frac{q_g}{k}x + C_1$$

and

$$t = -\frac{q_g}{2k}x^2 + C_1x + C_2 \tag{4.2}$$

The constants of the integration in Eq. (4.2) are determined from the boundary conditions. We consider two different cases.

4.1.1 Case (A) Surfaces at Different Temperatures

Heat is rejected from both the surfaces and the surfaces are at different temperatures t_1 and t_2. This may happen when the heat rejection rates from the two faces are not the same due to the different surface heat transfer coefficients at the two faces. This gives the following boundary conditions:

$$\text{at}\, x = 0, \quad t = t_1$$
$$\text{and, at}\, x = \delta, \quad t = t_2$$

Applying the boundary conditions to Eq. (4.2), we get

$$\text{at}\, x = 0, \quad t_1 = C_2$$
$$\text{and, at}\, x = \delta,$$
$$t_2 = -\frac{q_g}{2k}\delta^2 + C_1\delta + C_2 = -\frac{q_g}{2k}\delta^2 + C_1\delta + t_1$$

This gives

$$C_1 = \frac{t_2 - t_1}{\delta} + \frac{q_g}{2k}\delta$$

Inserting the value of C_1 and C_2 in Eq. (4.2), we get

$$t = -\frac{q_g}{2k}x^2 + \left(\frac{t_2 - t_1}{\delta} + \frac{q_g}{2k}\delta\right)x + t_1 = \frac{q_g}{2k}(\delta - x)x + \left(\frac{t_2 - t_1}{\delta}\right)x + t_1 \tag{4.3}$$

Equation (4.3) is the equation of temperature distribution through the wall. For a given value of heat generation rate q_g, the temperature distribution is a function of the distance x only and the equation can be written as

$$t(x) = \frac{q_g}{2k}(\delta - x)x + \left(\frac{t_2 - t_1}{\delta}\right)x + t_1 \tag{4.4}$$

Equation (4.4) can be transformed into non-dimensional form as follows:

$$t(x) - t_2 = \frac{q_g}{2k}(\delta - x)x + \left(\frac{t_2 - t_1}{\delta}\right)x + t_1 - t_2$$

or

$$\begin{aligned}\frac{t(x) - t_2}{t_1 - t_2} &= \left(\frac{q_g}{2k}\right)\frac{(\delta - x)x}{t_1 - t_2} - \frac{x}{\delta} + 1\\ &= \left[\frac{q_g\delta^2}{2k(t_1 - t_2)}\right]\left(1 - \frac{x}{\delta}\right)\frac{x}{\delta} + \left(1 - \frac{x}{\delta}\right)\\ &= B\left(1 - \frac{x}{\delta}\right)\frac{x}{\delta} + \left(1 - \frac{x}{\delta}\right)\\ &= \left(1 - \frac{x}{\delta}\right)\left(\frac{Bx}{\delta} + 1\right)\end{aligned}$$

or

$$\frac{t(x) - t_2}{t_1 - t_2} = \left(1 - \frac{x}{\delta}\right)\left(\frac{Bx}{\delta} + 1\right) \tag{4.5}$$

where

$$B = \left[\frac{q_g \delta^2}{2k(t_1 - t_2)}\right]. \tag{4.6}$$

Parameter B is a dimensionless term.

4.1.1.1 The Maximum Temperature and Its Location Within the Wall

It can be determined by differentiating Eq. (4.5) with respect to the dimensionless distance (x/δ) and setting the derivative to zero.

$$\begin{aligned}\frac{d}{d(x/\delta)}\left[\frac{t(x) - t_2}{t_1 - t_2}\right] &= 0 = -\left(\frac{Bx}{\delta} + 1\right) + B\left(1 - \frac{x}{\delta}\right)\\ &= -\frac{Bx}{\delta} - 1 + B - \frac{x}{\delta}B\\ &= B\left(1 - \frac{2x}{\delta}\right) - 1\end{aligned}$$

This gives the condition as

$$\frac{x}{\delta} = \frac{B - 1}{2B} \tag{4.7}$$

Equation (4.7) gives the value of dimensionless distance x/δ, where the maximum value of the temperature occurs. By substituting this value in Eq. (4.5), we get the value of maximum temperature as

$$\left[\frac{t_{max} - t_2}{t_1 - t_2}\right] = \frac{(B+1)^2}{4B} \tag{4.8}$$

Effect of factor *B* on the temperature distribution:

The factor *B* has been defined as

$$B = \left[\frac{q_g \delta^2}{2k(t_1 - t_2)}\right] \tag{4.6}$$

(i) When there is no heat generation, i.e. $q_g = 0$, factor *B* is zero and Eq. (4.5) reduces to

$$\frac{t(x) - t_2}{t_1 - t_2} = \left(1 - \frac{x}{\delta}\right), \tag{4.9a}$$

which gives the temperature distribution as

$$t(x) = t_1 + \frac{x}{\delta}(t_2 - t_1) \tag{4.9b}$$

The temperature distribution is linear in this case as shown in Fig. 4.2.

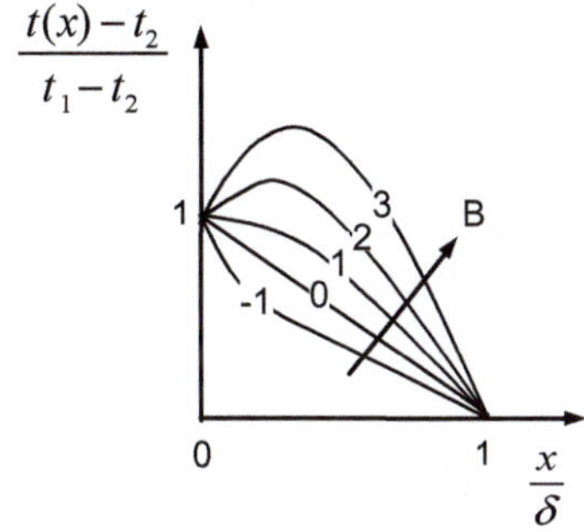

Fig. 4.2 Dimensionless temperature distribution as a function of factor B

For the case of heat generation, the positive value of q_g, factor B is positive. For B = 1, Eq. (4.7) gives (x/δ) = 0, i.e. the maximum value of the temperature occurs at x = 0. With the increase in the value of B, the value of (x/δ) given by (4.7) increases and in the limit approaches 0.5 when B has a very high value, refer Fig. 4.2. In general with heat generation, the maximum temperature is located within the wall (x/δ = 0–0.5). When the wall absorbs heat, q_g is negative and factor B will have a negative value.

The rate of heat transfer from any face of the wall is

$$q = -kA\frac{dt}{dx}$$

The value of the derivative (dt/dx) from Eq. (4.4) is

$$\begin{aligned}\frac{dt}{dx} &= \frac{q_g}{2k}\delta - \frac{q_g}{k}x + \left(\frac{t_2 - t_1}{\delta}\right)\\ &= \frac{q_g}{2k}(\delta - 2x) + \left(\frac{t_2 - t_1}{\delta}\right)\end{aligned}$$

Substituting the value of dt/dx, we get

$$q = kA\left[\left(\frac{t_1 - t_2}{\delta}\right) - \frac{q_g}{2k}(\delta - 2x)\right] \tag{4.10}$$

When there is no heat generation,

$$q = kA\left(\frac{t_1 - t_2}{\delta}\right)$$

With the heat generation, the heat transfer from the left face, q_L, can be obtained by putting x = 0 in Eq. (4.10) to give

$$\begin{aligned}q_L &= kA\left[\left(\frac{t_1 - t_2}{\delta}\right) - \frac{q_g}{2k}\delta\right]\\ &= kA\left(\frac{t_1 - t_2}{\delta}\right)(1 - B)\end{aligned}$$

Similarly, the heat transfer from the right face, q_R, can be obtained by putting x = δ in Eq. (4.10) to give

$$q_R = kA\left[\left(\frac{t_1 - t_2}{\delta}\right) + \frac{q_g}{2k}\delta\right]$$
$$= kA\left(\frac{t_1 - t_2}{\delta}\right)(1 + B)$$

The total heat loss from the wall is

$$q_t = -q_L + q_R \tag{4.11}$$

where the negative sign with q_L indicates heat flow from the left face is in the negative x-direction.

Substitution of values of q_L and q_R in Eq. (4.10) gives

$$q_t = A\delta q_g$$

i.e. the heat from the wall equals the heat generated in the wall.

It can be seen that when $B = 1$, i.e. t_{max} at $x = 0$, $q_L = 0$, which means that there is no heat loss from the left-hand face and the total heat is rejected from the right-hand face.

Example 4.1 Two steel plates are separated by a steel rod of 25 mm diameter and 300 mm long. The rod is welded to each plate. The space between the plates is filled with insulation material as shown in Fig. 4.3. Voltage difference between the two plates causes the current to flow through the steel rod. Due to the flow of current, heat is generated in the rod at a rate of 15 W. Under steady-state condition, the temperatures at the ends are 100°C and 70°C. Find the maximum temperature in the rod. Also, calculate the heat flow from each end. The thermal conductivity of the steel rod is 40 W/(m K).

Solution

Heat generated per unit volume

$$q_g = \frac{15}{(\pi/4) \times (0.025)^2 \times 0.3} = 101.86 \times 10^3$$

$$B = \left[\frac{q_g \delta^2}{2k(t_1 - t_2)}\right] = \left[\frac{101.86 \times 10^3 \times 0.3^2}{2 \times 40 \times (100 - 70)}\right] = 3.82$$

$$\frac{t_{max} - t_2}{t_1 - t_2} = \frac{(3.82 + 1)^2}{4 \times 3.82} = 1.52$$

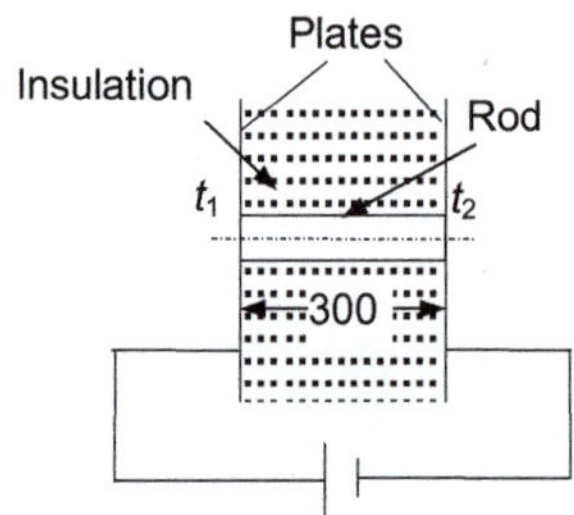

Fig. 4.3 Example 4.1

or

$$\begin{aligned} t_{\max} &= 1.52(t_1 - t_2) + t_2 \\ &= 1.52(100 - 70) + 70 = 115.6 \end{aligned}$$

Heat transfer from the left face is

$$\begin{aligned} q_L &= -kA\left[\left(\frac{t_1 - t_2}{\delta}\right) - \frac{q_g}{2k}\delta\right] \\ &= -40 \times (\pi/4) \times (0.025)^2\left[\left(\frac{100 - 70}{0.3}\right) - \frac{101.86 \times 10^3}{2 \times 40} \times 0.3\right] \\ &= 5.535\,\text{W}. \end{aligned}$$

Heat transfer from the right face is

$$\begin{aligned} q_R &= kA\left[\left(\frac{t_1 - t_2}{\delta}\right) + \frac{q_g}{2k}\delta\right] \\ &= 40 \times (\pi/4) \times (0.025)^2\left[\left(\frac{100 - 70}{0.3}\right) + \frac{101.86 \times 10^3}{2 \times 40} \times 0.3\right] \\ &= 9.461\,\text{W}. \end{aligned}$$

Total heat flow from the rod ends q_L and q_R equals the heat generation.

Example 4.2 Heat is generated at an interface between two slabs. One is steel [k = 35 W/(m K)] of 50 mm thickness and the other is brass [k = 70 W/(m K)] of 50 mm thickness. The temperatures of the outer surfaces of the steel and brass slabs are maintained at 100°C and 60°C, respectively. The heat generation rate at the contact surface of the slabs is 1.5×10^5 W/m^2. Determine the heat flow through the steel slab.

Solution

Let the temperature at the interface of the slabs is t_i. From Fourier's heat conduction equation, the heat flow through the left slab is (refer Fig. 4.4)

$$\begin{aligned} q_L &= k_1 A\left(\frac{t_i - t_1}{\delta_1}\right) \\ &= 35 \times 1 \times \left(\frac{t_i - 100}{0.05}\right) = 700 \times (t_i - 100). \end{aligned}$$

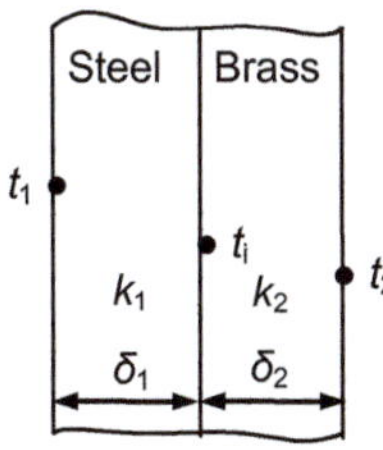

Fig. 4.4 Example 4.2

and through the right slab is

$$q_R = k_2 A\left(\frac{t_i - t_2}{\delta_2}\right)$$
$$= 70 \times 1 \times \left(\frac{t_i - 60}{0.05}\right) = 1400 \times (t_i - 60).$$

The total heat flow must equal the heat generated. Thus

$$q_L + q_R = q_g$$

or

$$700 \times (t_i - 100) + 1400 \times (t_i - 60) = 1.5 \times 10^5$$

or

$$t_i = 144.76°\text{C}.$$

Knowing the interface temperature t_i, the values of q_L and q_R through the steel and brass, respectively, are

$$q_L = 3.13 \times 10^4\ \text{W}$$
$$q_R = 1.18 \times 10^5\ \text{W}.$$

4.1.2 Case (B) Surfaces at the Same Temperature

This gives the following boundary conditions:

$$\text{at}\, x = 0, \quad t = t_1$$
$$\text{and, at}\, x = \delta, \quad t = t_1$$

Applying the boundary conditions to Eq. (4.2), we get

$$\text{at}\, x = 0, \quad t_1 = C_2$$
$$\text{and at}\, x = \delta,$$

$$t_1 = -\frac{q_g}{2k}\delta^2 + C_1\delta + C_2 = -\frac{q_g}{2k}\delta^2 + C_1\delta + t_1$$

This gives

$$C_1 = \frac{q_g}{2k}\delta$$

Inserting the value of C_1 and C_2 in Eq. (4.2) we get

$$t = -\frac{q_g}{2k}x^2 + \left(\frac{q_g}{2k}\delta\right)x + t_1 = \frac{q_g}{2k}(\delta - x)x + t_1 \tag{4.12}$$

Equation (4.12) is the equation of temperature distribution through the wall. For a given value of heat generation rate q_g, the temperature distribution is a function of the distance x only and the equation can be written as

$$t(x) = \frac{q_g}{2k}(\delta - x)x + t_1 \tag{4.13}$$

The above equation can also be obtained by putting $t_2 = t_1$ in Eq. (4.4).

It is obvious that in the present case, the maximum temperature must be at $x = \delta/2$ and

$$\begin{aligned} t_{\max} &= \frac{q_g}{2k}\left(\delta - \frac{\delta}{2}\right)\frac{\delta}{2} + t_1 \\ &= \frac{q_g}{8k}\delta^2 + t_1 \end{aligned} \tag{4.14}$$

The heat flow from the right face is

$$q_R = -kA\left(\frac{dt}{dx}\right)_{x=\delta}$$

or

$$q_R = kA\left(\frac{q_g}{2k}\right)\delta = \frac{1}{2}(A\delta) \times q_g$$

From the symmetry, the total heat transfer rate q_t is

$$q_t = 2q_R = (A\delta) \times q_g$$

The heat conducted to the face q_R is rejected to the surrounding by convection. If the surrounding fluid is at temperature t_∞ and h is the heat transfer coefficient at the wall, then

$$\frac{(A\delta) \times q_g}{2} = hA(t_1 - t_\infty)$$

or

$$t_1 = \frac{q_g}{2h}\delta + t_\infty$$

Substituting the value of t_1 in Eq. (4.13), we get

$$t = \frac{q_g}{2k}(\delta - x)x + \frac{q_g}{2h}\delta + t_\infty. \tag{4.15}$$

The maximum temperature is at the midpláne of the wall, and its value can be obtained by putting $x = \delta/2$.

Example 4.3 An electric current I = 10A flows through a metal strip of rectangular cross-sections (8 mm × 1 mm) and 1 m long. The voltage difference between the strip ends is 230 V. Find the temperature in the mid of strip thickness if the temperature of the surface of the strip is 200°C. The thermal conductivity of the strip material is 25 W/(m K).

Solution

Heat generated per unit volume,

$$q_g = \frac{V \times I}{(\text{cross-section}) \times \text{length}} = \frac{230 \times 10}{8 \times 1 \times 10^{-6}} = 2.875 \times 10^8 \text{ W/m}^3.$$

Temperature at the midplane of the strip is

$$\begin{aligned} t_{\max} &= \frac{q_g}{8k}\delta^2 + t_s \\ &= \frac{2.875 \times 10^8}{8 \times 25} \times 0.001^2 + 200 = 201.44°\text{C}. \end{aligned}$$

Example 4.4 A 100 mm thick plane wall generates heat at the rate of 1×10^5 W/m^3. One side of the wall is exposed to the ambient air at 30°C while the other side of the wall is insulated. The convective heat transfer coefficient on the airside is 400 W/(m^2 K). Proceeding from the basic principles, determine the location and value of the maximum temperature to which the wall is subjected. The thermal conductivity of the wall material is 2 W/(m K).

Solution

The one-dimensional steady-state conduction equation with heat generation is

$$\frac{d^2t}{dx^2} + \frac{q_g}{k} = 0 \tag{4.1}$$

Integrating the above equation, we get

$$\frac{dt}{dx} = -\frac{q_g}{k}x + C_1 \tag{i}$$

and

$$t = -\frac{q_g}{2k}x^2 + C_1 x + C_2 \tag{ii}$$

The constants C_1 and C_2 are to be found from the boundary conditions of the problem. At $x = 0$, the wall is insulated and hence

$$\frac{dt}{dx} = 0,$$

which gives $C_1 = 0$.

At $x = \delta$, the heat reaching the wall by conduction is rejected to the surrounding by convection (refer Fig. 4.5). Thus

$$-kA\left(\frac{dt}{dx}\right)_{x=\delta} = hA[t(\delta) - t_\infty]$$

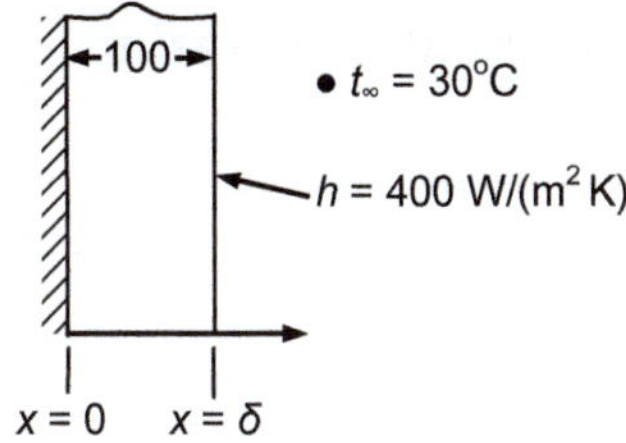

Fig. 4.5 Example 4.4

or

$$-\left(\frac{dt}{dx}\right)_{x=\delta} = \frac{h}{k}[t(\delta) - t_\infty] \tag{iii}$$

From Eq. (i), with $C_1 = 0$, dt/dx at $x = \delta$ is

$$\left(\frac{dt}{dx}\right)_{x=\delta} = -\frac{q_g}{k}\delta$$

and from Eq. (ii), $t(\delta)$ is

$$t(\delta) = -\frac{q_g}{2k}\delta^2 + C_2 \tag{iv}$$

Substitution of the values of dt/dx and $t(\delta)$ in Eq. (iii), we have

$$\frac{q_g}{k}\delta = \frac{h}{k}\left[-\frac{q_g}{2k}\delta^2 + C_2 - t_\infty\right]$$

which gives

$$C_2 = \frac{q_g}{h}\delta + \frac{q_g}{2k}\delta^2 + t_\infty$$

Putting the value of C_2 in Eq. (ii), we get the equation of the temperature distribution through the wall as

$$t = -\frac{q_g}{2k}x^2 + \frac{q_g}{h}\delta + \frac{q_g}{2k}\delta^2 + t_\infty$$

At the insulated wall side, $dt/dx = 0$, hence the temperature is maximum. Putting $x = 0$, we get the maximum temperature

$$\begin{aligned} t_{\max} &= \frac{q_g}{h}\delta + \frac{q_g}{2k}\delta^2 + t_\infty \\ &= \frac{1 \times 10^5}{400} \times 0.1 + \frac{1 \times 10^5}{2 \times 2} \times 0.1^2 + 30 = 305°\text{C} \end{aligned}$$

4.2 Cylinder with Uniform Heat Generation

Consider an elemental cylinder of thickness *dr* at distance *r* from the central axis in Fig. 4.6. From the Fourier's conduction equation, the heat entering the shell surface at radius *r* is

$$q_r = -k2\pi rL\frac{dt}{dr} \tag{i}$$

The heat leaving the face at distance *r* + *dr* is

$$q_{r+dr} = q_r + \frac{d}{dr}(q_r)dr \tag{ii}$$

Net heat leaving the elemental cylinder is

$$\begin{aligned} q_{r+dr} - q_r &= \frac{d}{dr}(q_r)dr \\ &= \frac{d}{dr}\left(-k2\pi rL\frac{dt}{dr}\right)dr \\ &= -k2\pi L\frac{d}{dr}\left(r\frac{dt}{dr}\right)dr \end{aligned} \tag{iii}$$

The heat generated in the elemental cylindrical cell is

$$q_g' = (2\pi rLdr)q_g \tag{iv}$$

where $2rLdr$ is the volume of the elemental cylinder and q_g is the rate of heat generation per unit volume.

In the equilibrium, the total heat generated in the elemental cylinder must equal the net heat leaving the elemental cylinder, i.e.

$$-k2\pi L\frac{d}{dr}\left(r\frac{dt}{dr}\right)dr = (2\pi rLdr)q_g$$

or

$$\frac{d}{dr}\left(r\frac{dt}{dr}\right) + \frac{q_g}{k}r = 0 \tag{4.16}$$

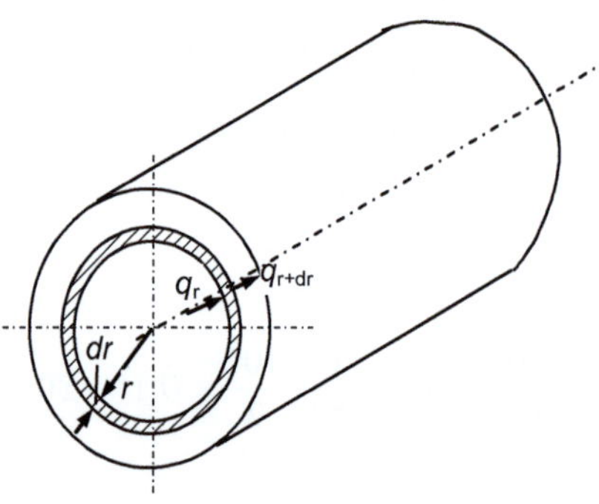

Fig. 4.6 Conduction with heat generation in a cylinder

Equation (4.16) can also be obtained from the general heat conduction equation by assuming one-dimensional steady state heat conduction with constant thermal conductivity.

Integrating Eq. (4.16), we get

$$r\frac{dt}{dr} + \frac{q_g}{k}\frac{r^2}{2} + C_1 = 0$$

or

$$\frac{dt}{dr} + \frac{q_g}{2k}r + \frac{C_1}{r} = 0 \tag{v}$$

Further integration gives

$$t + \frac{q_g}{4k}r^2 + C_1 \ln r + C_2 = 0 \tag{vi}$$

The constants of the integration in Eq. (vi) are determined from the boundary conditions, which are

$$\begin{aligned} \text{at}\, r &= 0, & dt/dr &= 0 \\ \text{and at}\, r &= R, \text{i.e.at the surface} & t &= t_1 \end{aligned}$$

Applying the first boundary condition to Eq. (v), we get $C_1 = 0$

Inserting $C_1 = 0$ and applying the second boundary condition to Eq. (vi) gives

$$t_1 + \frac{q_g}{4k}R^2 + 0 + C_2 = 0$$

or

$$C_2 = -\left(\frac{q_g}{4k}R^2 + t_1\right)$$

Inserting the value of C_2 in Eq. (vi) we get

$$t + \frac{q_g}{4k}r^2 - \left(\frac{q_g}{4k}R^2 + t_1\right) = 0$$

or

$$t = \frac{q_g}{4k}(R^2 - r^2) + t_1 \tag{4.17}$$

which is the equation of temperature distribution through the cylindrical wall.

The heat conducted to the surface of the cylinder is transferred by convection to the surrounding fluid, hence

$$\begin{aligned} q &= hA(t_1 - t_\infty) \\ \text{or} \quad -kA\left(\frac{dt}{dr}\right)_{r=R} &= hA(t_1 - t_\infty) \\ \text{or} \quad -k\left(-\frac{q_g}{2k}R\right) &= h(t_1 - t_\infty) \end{aligned}$$

or

$$t_1 = t_\infty + \frac{q_g}{2h} R$$

Substituting the value of t_1 in Eq. (4.17), we get

$$t = \frac{q_g}{4k}(R^2 - r^2) + \frac{q_g}{2h} R + t_\infty \tag{4.18}$$

The maximum temperature is at the centre $r = 0$ where $dt/dr = 0$ and is

$$t_{\max} = \frac{q_g}{4k} R^2 + \frac{q_g}{2h} R + t_\infty \tag{4.19}$$

Example 4.5 A current of 100 A flows through a stainless steel wire of 5 mm diameter and 1 m long. The wire is submerged in a liquid bath maintained at 10°C. The heat transfer coefficient for the wire surface is 500 W/(m^2 K). If the thermal conductivity of the wire material is 20 W/(m K) and resistivity is 70 × 10^{-6} Ω-cm, calculate the centreline temperature of the wire.

Solution

Resistance of the wire

$$\begin{aligned} R_e &= \frac{\rho L}{A} \\ &= \frac{70 \times 10^{-6} \times 100}{\pi \times (0.25)^2} = 0.03565\,\Omega.. \end{aligned}$$

Heat generated due to ohmic heating of wire

$$I^2 R_e = 100^2 \times 0.03565 = 356.5\,\text{W}.$$

Heat generation rate per unit volume

$$q_g = \frac{356.5}{\pi \times (2.5/1000)^2 \times 1} = 1.8156 \times 10^7\,\text{W/m}^3.$$

Maximum temperature from Eq. (4.19) is

$$\begin{aligned} t_{\max} &= \frac{q_g}{4k} R^2 + \frac{q_g}{2h} R + t_\infty \\ &= \frac{1.8156 \times 10^7}{4 \times 20}\left(\frac{2.5}{1000}\right)^2 + \frac{1.8156 \times 10^7}{2 \times 500}\left(\frac{2.5}{1000}\right) + 10 \\ &= 56.81^\circ\text{C}. \end{aligned}$$

Example 4.6 A plate with uniformly distributed inner heat source with a volumetric heat generation rate of q_g is exposed on both sides to a flowing fluid. The local coefficients of heat transfer from the surfaces of the plate to the surrounding fluid and the bulk fluid temperature on the two sides of the plate are h_1 and h_2 and t_{f1} and t_{f2}, respectively. The thickness of the plate is δ and thermal conductivity is k.

(a) Show that when the maximum temperature is at the middle of the plate

$$t_{f2} - t_{f1} = q_g\delta\left(\frac{1}{h_1} - \frac{1}{h_2}\right)\frac{1}{2}$$

(b) Show that when $t_{f1} = t_{f2}$ and $h_2 = 2h_1$, location of the maximum temperature is given by

$$\frac{x_o}{\delta} = \frac{h_1\delta + k}{2h_1\delta + 3k}$$

Solution

(a) For the maximum temperature, with uniform volumetric heat generation rate, to occur at the middle

$$t_1 = t_2$$

The heat lost from the left face is given by, refer Fig. 4.7,

$$q_L = kA\left(-\frac{t_1 - t_2}{\delta} + \frac{q_g}{2k}\delta\right) \tag{i}$$

Hence, for $t_1 = t_2$,

$$q_L = A\delta\frac{q_g}{2} \tag{ii}$$

The heat lost from the left face is also given by

$$q_L = h_1A(t_1 - t_{f1}) \tag{iii}$$

Equating Eqs. (ii) and (iii),

$$(t_1 - t_{f1}) = \frac{q_g}{2h_1}\delta \tag{iv}$$

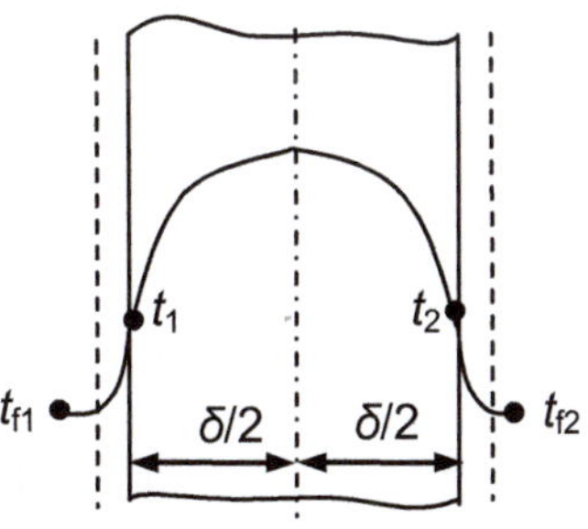

Fig. 4.7 Example 4.6

Similarly for the right-hand face,

$$(t_2 - t_{f2}) = \frac{q_g}{2h_2}\delta \tag{v}$$

From Eqs. (iv) and (v) for $t_1 = t_2$,

$$(t_{f2} - t_{f1}) = \frac{q_g\delta}{2}\left(\frac{1}{h_1} - \frac{1}{h_2}\right).$$

(b) From Eqs. (i) and (iii),

$$kA\left(-\frac{t_1 - t_2}{\delta} + \frac{q_g}{2k}\delta\right) = h_1 A(t_1 - t_{f1})$$

or

$$\left(-\frac{t_1 - t_2}{\delta} + \frac{q_g}{2k}\delta\right) = \frac{h_1}{k}(t_1 - t_{f1}) \tag{vi}$$

Similarly at the right face,

$$\left(\frac{t_1 - t_2}{\delta} + \frac{q_g}{2k}\delta\right) = \frac{h_2}{k}(t_2 - t_{f2}) \tag{vii}$$

Putting $h_2 = 2h_1$ and $t_{f2} = t_{f1}$ in the above equation, we obtain

$$\left(\frac{t_1 - t_2}{\delta} + \frac{q_g}{2k}\delta\right) = \frac{2h_1}{k}(t_2 - t_{f1}) \tag{viii}$$

Multiplying both sides of Eq. (vi) by 2, we get

$$-2\frac{t_1 - t_2}{\delta} + \frac{q_g}{k}\delta = \frac{2h_1}{k}(t_1 - t_{f1}) \tag{ix}$$

Subtracting Eq. (viii) from Eq. (ix), we get

$$\begin{aligned} -3\frac{t_1 - t_2}{\delta} + \frac{q_g}{2k}\delta &= \frac{2h_1}{k}(t_1 - t_{f1} - t_2 + t_{f1}) \\ &= \frac{2h_1}{k}(t_1 - t_2) \end{aligned}$$

or

$$-3 + \frac{q_g}{2k(t_1 - t_2)}\delta^2 = \frac{2h_1}{k}\delta$$

Putting $\left(\frac{q_g\delta^2}{2k(t_1 - t_2)} = B\right)$ we get

$$-3 + B = \frac{2h_1}{k}\delta$$

The simplification gives the value of parameter B as

$$B = \frac{2h_1}{k}\delta + 3$$

The position of maximum temperature from Eq. (4.7),

$$\frac{x}{\delta} = \frac{B-1}{2B}$$

Substituting the value of B, we get

$$\frac{x}{\delta} = \frac{\frac{2h_1}{k}\delta + 3 - 1}{2\left(\frac{2h_1}{k}\delta + 3\right)}$$

or

$$\frac{x_o}{\delta} = \frac{h_1\delta + k}{2h_1\delta + 3k}.$$

Example 4.7 A 5 mm thick plate with uniformly distributed internal heat source having a volumetric rate of heat libration q_g = 3.0 × 10^8 W/m^3, is exposed on both sides to cooling fluids. The local coefficients of heat transfer from the two sides of the plate to the fluid to which it is exposed are 3000 W/(m^2 K) and 1800 W/(m^2 K) and the temperatures of the fluid are 140°C and 150°C, respectively. If the thermal conductivity of the plate material is 30 W/(m K), determine position and magnitude of the maximum temperature in the plate and temperature of both sides of the plate.

Solution

From Example 4.6,

$$-\frac{t_1 - t_2}{\delta} + \frac{q_g}{2k}\delta = \frac{h_1}{k}(t_1 - t_{f1})$$

or

$$-\frac{t_1 - t_2}{0.005} + \frac{3.0 \times 10^8}{2 \times 30} \times 0.005 = \frac{3000}{30}(t_1 - 140)$$

or

$$-1.5t_1 + t_2 = -195 \tag{i}$$

Similarly for the right-hand face

$$\frac{t_1 - t_2}{\delta} + \frac{q_g}{2k}\delta = \frac{h_2}{k}(t_2 - t_{f1})$$

or

$$\frac{t_1 - t_2}{0.005} + \frac{3.0 \times 10^8}{2 \times 30} \times 0.005 = \frac{1800}{30}(t_2 - 150)$$

or

$$t_1 - 1.3t_2 = -170 \tag{ii}$$

Solving Eqs. (i) and (ii) for t_1 and t_2, we get

$$t_1 = 445.8\,°\text{C}$$
$$t_2 = 473.7\,°\text{C}.$$

Value of parameter B is

$$B = \frac{q_g\delta^2}{2k(t_1 - t_2)} = \frac{3.0 \times 10^8 \times 0.005^2}{2 \times 30 \times (445.8 - 473.7)} = -4.48$$

The position of maximum temperature from Eq. (4.7),

$$\frac{x}{\delta} = \frac{B - 1}{2B} = \frac{-4.48 - 1}{2 \times (-4.48)} = 0.6116$$

or

$$x = 3.06\,\text{mm}.$$

Maximum temperature from Eq. (4.8) is

$$\left[\frac{t_{max} - t_2}{t_1 - t_2}\right] = \frac{(B + 1)^2}{4B}$$

or

$$t_{max} = \frac{(B + 1)^2}{4B}(t_1 - t_2) + t_2 = \frac{(-4.48 + 1)^2}{4 \times (-4.48)}(445.8 - 473.7) + 473.7 = 492.55°\text{C}.$$

Example 4.8 A plate, with a uniform rate of volumetric heat generation q_g, is exposed on both sides to flowing fluids. The thickness of the plate is δ and the thermal conductivity of the material is k. The temperature of the fluid on one side of the plate is t_{f1} with heat transfer coefficient h_1. Determine the temperature of the fluid on the opposite side of the plate so that the rate of heat flow from this surface is equal to zero.

Solution

The temperature of the opposite surface of the plate must be equal to the fluid temperature as there is no heat transfer from this surface, i.e. $t_2 = t_{f2}$ since $q = h_2 A (t_2 - t_{f2}) = 0$.

The total heat generated in the plate,

$$q = q_g A\delta$$

The heat is rejected from only one face, i.e.

$$h_1 A(t_1 - t_{f1}) = q_g A\delta$$

or

$$t_1 = \frac{q_g A\delta}{h_1 A} + t_{f1}$$

The conduction heat flow to this face is

$$kA\left(-\frac{t_1 - t_2}{\delta} + \frac{q_g}{2k}\delta\right) = q_g A\delta$$

or

$$-(t_1 - t_2) = \frac{q_g}{2k}\delta^2$$

Substitution of the values of t_2 and t_1, we get

$$-\frac{q_g\delta}{h_1} - t_{f1} + t_{f2} = \frac{q_g\delta^2}{2k}$$

or

$$t_{f2} = t_{f1} + q_g\delta\left(\frac{\delta}{2k} + \frac{1}{h_1}\right).$$

Example 4.9 A hollow cylindrical copper bar, having inner and outer diameters of 13 mm and 50 mm, respectively, carries a current density of 5000 A/cm^2. When the outer surface temperature is maintained at 40°C and no heat is removed through the inner surface, find the position and value of the maximum temperature. For copper, assume electrical resistivity $\rho = 2 \times 10^{-6}$ Ω-cm and thermal conductivity $k = 381$ W/(m K).

Solution

Heat generation rate per unit volume,

$$q_g = \frac{I^2 R}{AL} = \frac{(5000 \times 10^4 \times A)^2}{AL} \times \frac{\rho L}{A} = (5000 \times 10^4)^2 \times 2 \times 10^{-8} = 50 \times 10^6\ \text{W/m}^3.$$

The differential equation for one-dimensional heat flow with heat generation in cylindrical coordinates is

$$\frac{d}{dr}\left(r\frac{dt}{dr}\right)+\frac{q_g}{k}r=0 \tag{4.16}$$

Integration of Eq. (4.16) gives

$$r\frac{dt}{dr}+\frac{q_g}{k}\frac{r^2}{2}+C_1=0$$

or

$$\frac{dt}{dr}+\frac{q_g}{2k}r+\frac{C_1}{r}=0$$

Further integration gives the temperature distribution equation as

$$t+\frac{q_g}{4k}r^2+C_1\ln r+C_2=0 \tag{i}$$

The constants of the integration in Eq. (i) are determined from the boundary conditions, which are

at $r=r_1$, $dt/dr=0$ as no heat is removed from the inner surface,

and at $r=r_2$, $t=t_2$

The first boundary condition gives

$$0+\frac{q_g}{2k}r_1+\frac{C_1}{r_1}=0$$

or

$$C_1=-\frac{q_g}{2k}r_1^2$$

Inserting value of constant C_1 in the temperature distribution equation, we get

$$t+\frac{q_g}{4k}r^2-\frac{q_g}{2k}r_1^2\ln r+C_2=0$$

The second boundary condition gives

$$t_2+\frac{q_g}{4k}r_2^2-\frac{q_g}{2k}r_1^2\ln r_2+C_2=0$$

or

$$C_2=\frac{q_g}{2k}r_1^2\ln r_2-\frac{q_g}{4k}r_2^2-t_2$$

Using the numerical values given in the problem,

$$C_1 = -\frac{q_g}{2k}r_1^2 = -\frac{50\times 10^6}{2\times 381}\times\left(\frac{13}{2000}\right)^2 = -2.772$$

$$C_2 = \frac{50\times 10^6}{2\times 381}\times\left(\frac{13}{2000}\right)^2 \ln\left(\frac{50}{2000}\right) - \frac{50\times 10^6}{4\times 381}\times\left(\frac{50}{2000}\right)^2 - 40 = -70.82$$

Substitution of the values of the constants and other parameters in Eq. (i) gives

$$t + \frac{q_g}{4k}r^2 + C_1 \ln r + C_2 = 0$$

or

$$t + \frac{50\times 10^6}{4\times 381}r^2 - 2.772 \ln r - 70.82 = 0$$

The maximum temperature is at the inner surface (r = 13/2000 m) and is

$$t = -\frac{50\times 10^6}{4\times 381}\left(\frac{13}{2000}\right)^2 + 2.772\times\ln\left(\frac{13}{2000}\right) + 70.82 = 55.47°\text{C}.$$

The heat transfer at the outer surface of the bar is

$$q = -kA\left(\frac{dt}{dr}\right)_{r=r_2}$$

where

$$\left(\frac{dt}{dr}\right)_{r=r_2} = -\frac{q_g}{2k}r_2 - \frac{C_1}{r_2}$$
$$= -\frac{50\times 10^6}{2\times 381}\times\frac{50}{2000} - \frac{-2.772\times 2000}{50} = -1529.6$$

Hence,

$$q = -381\times 2\pi\times\frac{50}{2000}\times 1\times(-1529.6) = 91542\,\text{W}.$$

Check: The heat transfer must equal the heat generation rate, which is

$$q = q_g\times\frac{\pi}{4}(r_2 - r_1)^2\times L = 50\times 10^6\times\pi\times\left[\left(\frac{50}{2000}\right)^2 - \left(\frac{13}{2000}\right)^2\right]\times 1.0 = 91539\,\text{W}.$$

Example 4.10 A fuel element is made in the form of a long hollow cylinder having an inner diameter ($2r_1$) of 20 mm and outer diameter ($2r_2$) = 30 mm. The element is made of uranium with a thermal conductivity k_1 = 30 W/(m K). The inner and outer surfaces of the fuel element carry tightly fitting stainless steel cladding of 1 mm thickness. The thermal conductivity of steel k_2 is 20 W/(m K). The volumetric rate of heat generation is 6×10^7 W/m^3 and is assumed to be uniform over the cross-section.

The fuel element is being cooled with carbon dioxide flowing along an outer channel. The bulk temperature of the coolant in the outer channel is 200°C. The local coefficient of heat transfer from the outer cladding to the coolant is 500 W/(m^2 K). Determine the maximum temperature for the fuel element.

Solution

The heat generated in unit length of the fuel element is

$$\begin{aligned} q &= q_g V = q_g \times \pi \times (r_2^2 - r_1^2) \times 1 \\ &= 6 \times 10^7 \times \pi \times (15^2 - 10^2) \times 10^{-6} \times 1 = 23561.94\,\text{W/m length.} \end{aligned}$$

If the temperature of the outer surface of the fuel element is t_2, then

$$q = \frac{t_2 - t_{f2}}{\frac{1}{2\pi k_2 L}\ln\frac{r_3}{r_2} + \frac{1}{2\pi r_3 L} \times \frac{1}{h_2}}$$

Substituting L = 1 m, k_2 = 20 W/(m K), r_3 = (r_2 + 1) = 0.016 m, r_3/r_2 = 16/15, h_2 = 500 W/(m^2 K) and t_{f2} = 200°C, we get

$$23561.94 = \frac{t_2 - 200}{\frac{1}{2\pi \times 20 \times 1}\ln\frac{16}{15} + \frac{1}{2\pi \times 0.016 \times 1} \times \frac{1}{500}}$$

or

$$t_2 = 680.85°\text{C}.$$

From Eq. (4.16), by integration, we have

$$\frac{dt}{dr} + \frac{q_g}{2k} r + \frac{C_1}{r} = 0 \quad \text{(i)}$$

Further integration gives

$$t + \frac{q_g}{4k} r^2 + C_1 \ln r + C_2 = 0 \quad \text{(ii)}$$

The boundary conditions are

at $r = r_1 = 10\,\text{mm}$, $dt/dr = 0$ as no heat is removed from the inner surface,

and at $r = r_2 = 15\text{mm}$, $t = t_2 = 680.85°\text{C}$.

Applying the boundary conditions to Eqs. (i) and (ii), we have, for the given data of the problem,

$$C_1 = -\frac{q_g}{2k} r_1^2 = -\frac{6 \times 10^7}{2 \times 30} \times (0.01)^2 = -100$$

and

$$680.85 + \frac{6 \times 10^7}{4 \times 30} \times (0.015)^2 - 100 \ \ln\ (0.015) + C_2 = 0$$

or

$$C_2 = -1213.32$$

Substitution of the values of C_1 and C_2 in Eq. (ii) we get the temperature distribution equation as

$$t + \frac{q_g}{4k} r^2 - 100 \ln r - 1213.32 = 0$$

The maximum temperature $t = t_{max}$ will occur at the inner surface where $dt/dr = 0$. Hence, from the above equation,

$$\begin{aligned} t_{max} &= -\frac{q_g}{4k} r_1^2 + 100 \ln r_1 + 1213.32 \\ &= -\frac{6 \times 10^7}{4 \times 30} (0.01)^2 + 100 \ln(0.01) + 1213.32 = 702.8\,°\text{C}. \end{aligned}$$

Example 4.11 A stainless steel tube with an inner diameter of 9.5 mm and an outer diameter of 10 mm is heated by passing an electric current. The tube carries a current of 300 A. The specific resistance and the thermal conductivity of the steel are 0.8 $\Omega.mm^2/m$ and 18.0 W/(m K), respectively. Calculate the volumetric rate of heat liberation from the tube and the temperature drop across the wall of the tube if all the heat generated in the wall of the tube is transferred from its inner surface. Consider one-meter length of the tube.

Solution

Electric resistance of the tube,

$$R = \frac{\rho L}{A} = \frac{0.8 \times 1}{(\pi/4) \times (10^2 - 9.5^2)} = 0.1044\,\Omega/\text{m}.$$

The rate of heat generation per m length of the tube,

$$q = I^2 R = 300^2 \times 0.1044 = 9396\,\text{W/m}.$$

Volumetric rate of heat generation,

$$q_g = \frac{I^2 R}{AL} = \frac{9396 \times 10^6}{(\pi/4) \times (10^2 - 9.5^2) \times 1} = 1.227 \times 10^9\,\text{W/m}^3.$$

Following the procedure of the previous example, we have

$$\frac{dt}{dr} + \frac{q_g}{2k} r + \frac{C_1}{r} = 0$$

and

$$t + \frac{q_g}{4k} r^2 + C_1 \ln r + C_2 = 0$$

The boundary conditions are

$$\text{at } r = r_1, \quad t = t_1,$$

$$\text{and} \quad \text{at } r = r_2, \quad dt/dr = 0 \text{ as no heat is removed from the outer surface}$$

Applying the second boundary condition, we have

$$C_1 = -\frac{q_g}{2k} r_2^2$$

This gives the temperature distribution equation as

$$t + \frac{q_g}{4k} r^2 - \frac{q_g}{2k} r_2^2 \ln r + C_2 = 0$$

The condition $t = t_1$ at $r = r_1$ gives

$$C_2 = -t_1 - \frac{q_g}{4k} r_1^2 + \frac{q_g}{2k} r_2^2 \ln r_1$$

Substituting the value of constant C_2 gives the temperature distribution equation as

$$t + \frac{q_g}{4k} r^2 - \frac{q_g}{2k} r_2^2 \ln r - t_1 - \frac{q_g}{4k} r_1^2 + \frac{q_g}{2k} r_2^2 \ln r_1 = 0$$

The temperature drop across the wall $(t_2 - t_1)$, where t_2 is the temperature at radius r_2,

$$\begin{aligned} t_2 - t_1 &= \frac{q_g}{2k}\left(-\frac{r_2^2}{2} + r_2^2 \ln r_2 - r_2^2 \ln r_1 + \frac{r_1^2}{2}\right) \\ &= \frac{1.227 \times 10^9}{2 \times 18}\left[-\frac{5^2}{2} + 5^2 \ln 5 - 5^2 \ln\left(\frac{9.5}{2}\right) + \left(\frac{9.5}{2}\right)^2 \times \frac{1}{2}\right] \times 10^{-6} \\ &= 2.25^\circ\text{C}. \end{aligned}$$

Example 4.12 Show that the temperature distribution relation for a long cylinder with uniform heat generation can be expressed in non-dimensional form as

$$\frac{t - t_w}{t_{\max} - t_w} = 1 - \left(\frac{r}{R}\right)^2$$

where t_w is the temperature at the outer surface of the cylinder.

Solution

From Eq. (4.17), we have

$$t = \frac{q_g}{4k}(R^2 - r^2) + t_w$$

or

$$t - t_w = \frac{q_g}{4k}(R^2 - r^2) \tag{i}$$

The maximum temperature is at $r = 0$, thus

$$t_{\max} - t_w = \frac{q_g}{4k}R^2 \tag{ii}$$

From Eqs. (i) and (ii), we get

$$\frac{t - t_w}{t_{\max} - t_w} = 1 - \left(\frac{r}{R}\right)^2 \tag{4.20}$$

Example 4.13 Show that for an electric current carrying wire, the maximum temperature at the axis of the wire is given by

$$t_{\max} = t_w + \frac{i^2}{4kk_e}R^2$$

where k and k_e are the thermal and electric conductivities of the wire material, respectively, and i is current density (electrical conductivity k_e is reciprocal of the electrical resistivity ρ).

Solution

From Example 4.12, we have

$$t_{\max} = t_w + \frac{q_g}{4k}R^2$$

For an electric current carrying wire, $q_g = I^2R_e/AL) = I^2\,(\rho L/A)/(AL) = (I/A)^2/k_e = i^2/k_e$, where R_e is the electrical resistance (the electrical conductivity k_e is reciprocal of the resistivity). Hence,

$$t_{\max} = t_w + \frac{i^2}{4kk_e}R^2$$

4.3 Solid Sphere with Uniform Heat Generation

Consider an elemental spherical shell of thickness dr at distance r from the centre, see Fig. 4.8. From the Fourier's conduction equation, the heat entering the shell surface at radius r is

$$q_r = -kA\frac{dt}{dr} = -k(4\pi r^2)\frac{dt}{dr}$$

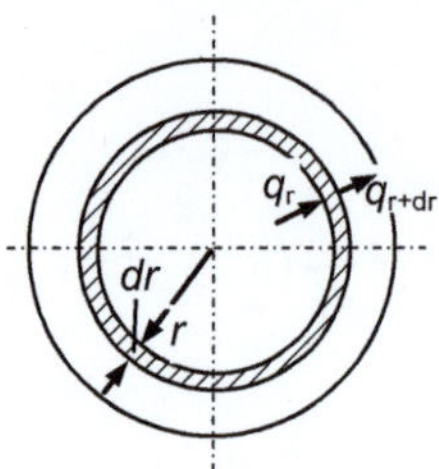

Fig. 4.8 Conduction with heat generation in a sphere

The heat leaving the face at distance (r + dr) is

$$q_{r+dr} = q_r + \frac{d}{dr}(q_r)dr$$

Net heat leaving the elemental shell is

$$\begin{aligned} q_{r+dr} - q_r &= \frac{d}{dr}(q_r)dr \\ &= \frac{d}{dr}\left(-4k\pi r^2 \frac{dt}{dr}\right)dr \\ &= -4k\pi \frac{d}{dr}\left(r^2 \frac{dt}{dr}\right)dr \end{aligned}$$

The heat generated in the elemental shell is

$$q'_g = (4\pi r^2 dr)q_g$$

where $4\pi r^2 dr$ is the volume of the elemental cell and q_g is the rate of heat generation per unit volume.

In the equilibrium, the total heat generated in the elemental spherical cell must equal the net heat leaving the elemental sphere, i.e.

$$-4k\pi \frac{d}{dr}\left(r^2 \frac{dt}{dr}\right)dr = (4\pi r^2 dr)q_g$$

or

$$\frac{d}{dr}\left(r^2 \frac{dt}{dr}\right) + \frac{q_g}{k}r^2 = 0 \tag{4.21}$$

Equation (4.21) can also be obtained from the general heat conduction equation by assuming one-dimensional steady-state heat conduction with constant thermal conductivity.

Integrating Eq. (4.21), we get

$$r^2 \frac{dt}{dr} = -\frac{q_g}{k}\frac{r^3}{3} + C_1 \tag{i}$$

At $r = 0$, $dt/dr = 0$ for a solid sphere. Applying this boundary condition, we get

$$C_1 = 0$$

Hence,

$$\frac{dt}{dr} = -\frac{r}{3}\frac{q_g}{k}$$

Its integration gives

$$t = -\frac{q_g}{3k}\frac{r^2}{2} + C_2 \tag{ii}$$

The constant of the integration C_2 can be determined from the following boundary condition.

At $r = R, t = t_2$.

Applying the second boundary condition, we get

$$t_2 = -\frac{q_g}{6k}R^2 + C_2$$

This gives

$$C_2 = \frac{q_g}{6k}R^2 + t_2$$

Inserting the value of C_2 in Eq. (ii), we get

$$t = -\frac{q_g}{6k}r^2 + \frac{q_g}{6k}R^2 + t_2$$

or

$$t = \frac{q_g}{6k}(R^2 - r^2) + t_2 \tag{4.22}$$

which is the equation of temperature distribution through the spherical wall.

The maximum temperature is at the centre ($r = 0$), where $dt/dr = 0$. Hence,

$$t_{\max} = \frac{q_g}{6k}R^2 + t_2 \tag{4.23}$$

The heat conducted to the surface of the sphere is transferred by convection to the surrounding fluid, hence

$$q = hA(t_2 - t_\infty)$$

or

$$-kA\left(\frac{dt}{dr}\right)_{r=R} = hA(t_2 - t_\infty)$$

or

$$-k\left(-\frac{q_g}{3k}R\right) = h(t_2 - t_\infty)$$

or

$$t_2 = t_\infty + \frac{q_g}{3h}R$$

Substituting the value of t_2 in Eq. (4.22), we get the temperature distribution equation in terms of the surrounding temperature t_∞ as

$$t = \frac{q_g}{6k}(R^2 - r^2) + \frac{q_g}{3h}R + t_\infty \quad (4.24)$$

Example 4.14 A 90 mm diameter orange generates 5.25 × 10³ W/m³ heat while it undergoes the ripening process. If the external surface of the orange is at 5°C, determine the temperature at the centre of the orange. Also, determine the heat flow from the outer surface of the orange. The orange may be treated as a spherical body. The average thermal conductivity may be taken as 0.25 W/(m K) for the orange material.

Solution

The temperature at the centre of the orange from Eq. (4.23) is

$$t_{max} = \frac{q_g}{6k}R^2 + t_2$$
$$= \frac{5.25 \times 10^3}{6 \times 0.25}\left(\frac{0.09}{2}\right)^2 + 5 = 12.1\,°\text{C}.$$

The heat flow from the outer surface of the orange equals the heat generated in the orange and is

$$= \frac{4}{3}\pi R^3 q_g = \frac{4}{3}\pi\left(\frac{0.09}{2}\right)^3 \times 5.25 \times 10^3 = 2.0\,\text{W}.$$

4.4 Heat Transfer Through Piston Crown

In internal combustion engines, the piston head (crown) is subjected to a very high gas temperature during the combustion. The heat flowing to the piston crown from the hot gases is transferred by conduction through the crown to its edge and then to the cylinder wall. The centre of the crown is at the farthest point from the cylinder wall and is at the highest temperature. A very high temperature at the crown centre may cause piston failure and may also lead to combustion problems in the case of petrol engines. An approximate analysis is presented below to estimate this temperature.

Piston head is a thin disc of thickness b which is subjected to uniform heat flux q_{gas} due to convection and radiation from the gases, refer Fig. 4.9. It is assumed in the presented

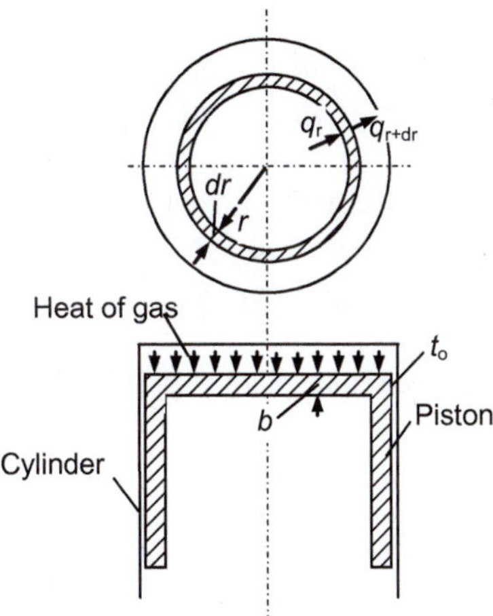

Fig. 4.9 Heat transfer through piston crown

analysis that the heat transfer from the lower part of the head to the surrounding air is negligible. Thus, there is only radial heat transfer through the piston crown by conduction.

Consider an elemental ring of radial thickness *dr* at distance *r* from the centre. From the Fourier's conduction equation, the heat entering the ring at radius *r* is

$$q_r = -kA\frac{dt}{dr} = -k(2\pi rb)\frac{dt}{dr}$$

The heat leaving the face at distance (*r* + *dr*) is

$$q_{r+dr} = q_r + \frac{d}{dr}(q_r)dr$$

Net heat leaving the elemental ring is

$$\begin{aligned} q_{r+dr} - q_r &= \frac{d}{dr}(q_r)dr \\ &= \frac{d}{dr}\left(-k2\pi rb\frac{dt}{dr}\right)dr \\ &= -2\pi bk\frac{d}{dr}\left(r\frac{dt}{dr}\right)dr \end{aligned} \tag{i}$$

The heat incident at the surface of the elemental ring is

$$q'_g = (2\pi rdr)q_{gas} \tag{ii}$$

where $2\pi rdr$ is the surface area of the elemental ring and q_{gas} is the heat flux due to the gas.

In the equilibrium, the heat incident must equal the net heat leaving the elemental ring, i.e.

$$-2\pi bk\frac{d}{dr}\left(r\frac{dt}{dr}\right)dr = (2\pi rdr)q_{gas}$$

or

$$\frac{d}{dr}\left(r\frac{dt}{dr}\right)+\frac{q_{\text{gas}}}{bk}r=0 \tag{4.25}$$

Integrating the above equation, we get

$$r\frac{dt}{dr}+\frac{q_{\text{gas}}}{bk}\frac{r^2}{2}=C_1$$

At r = 0, dt/dr = 0 for a solid circular plate. Applying this boundary condition, we get

$$C_1=0$$

Hence,

$$\frac{dt}{dr}+\frac{r}{2}\frac{q_{\text{gas}}}{bk}=0$$

Further integration gives

$$t+\frac{q_{\text{gas}}}{2bk}\frac{r^2}{2}=C_2 \tag{iii}$$

The constant of the integration C_2 can be determined from the following boundary condition.

At $r=R, t=t_o$.

Applying the second boundary condition, we get

$$t_o+\frac{q_{\text{gas}}}{4bk}R^2=C_2$$

Inserting the value of C_2 in Eq. (iii), we get the equation of temperature distribution as

$$t=\frac{q_{\text{gas}}}{4bk}(R^2-r^2)+t_0 \tag{4.26}$$

The maximum temperature will occur at the centre (r = 0) and is

$$t_{\max}=\frac{q_{\text{gas}}}{4bk}R^2+t_0 \tag{4.27}$$

The total heat transferred to the piston crown is

$$q_{\text{total}}=\pi R^2 q_{\text{gas}}$$

Example 4.15 The base of an electric iron is made of aluminium. The weight of the base is 1.5 kg. The surface area of the base is 0.04 m^2. An electric heater of 1000 W capacity is placed over the base. When heating is turned on, the base is placed in a vertical position. How long the iron will take to reach a temperature of 120°C if at the start of the heating the base is at the room temperature? The room temperature is 20°C and the heat transfer coefficient from the surface of the base to the surrounding is 20 W/(m^2 K). Take following data for aluminium ρ = 2700 kg/m^3, c = 900 J/(kg °C), and k = 200 W/(m K).

Solution

The energy balance gives

$$q_{\text{heater}} = \text{increase in heat content of the base} + \text{heat lost to the surroundings}$$

or

$$q_{\text{heater}} = mc\frac{d\theta}{d\tau} + hA\theta \tag{i}$$

where

θ $t_{\text{base}} - t_{\text{ambient}}$
$\frac{d\theta}{d\tau}$ rate of increase of the temperature of the base,
m mass of the base,
c specific heat of the base material,
A area of base transferring heat by convection.

Rearranging Eq. (i), we have

$$\frac{mc}{hA}\frac{d\theta}{d\tau} = \frac{q_{\text{heater}}}{hA} - \theta \tag{ii}$$

Let $mc/hA = a$ and $q_{\text{heater}}/hA = b$, then

$$a\frac{d\theta}{d\tau} = b - \theta$$

or

$$\frac{d\theta}{b-\theta} = \frac{1}{a}d\tau$$

Integration between $\theta = 0$ at $\tau = 0$ to $\theta = \theta_1$ at $\tau = \tau_1$ gives

$$[-\ln(b-\theta)]_0^{\theta_1} = \frac{1}{a}\tau_1$$

or

$$\ln\left(\frac{b}{b-\theta_1}\right) = \frac{1}{a}\tau_1$$

or

$$\tau_1 = a\ln\left(\frac{b}{b-\theta_1}\right)$$

From the given data

$$a = \frac{mc}{hA} = \frac{1.5 \times 900}{20 \times 0.04} = 1687.5$$

$$b = \frac{q_{\text{heater}}}{hA} = \frac{1000}{20 \times 0.04} = 1250$$

Substitution gives

$$\tau_1 = 1687.5 \times \ln\left(\frac{1250}{1250 - 100}\right) = 140.7\,\text{s}.$$

Example 4.16 Prove that the temperature distribution in a plane wall shown in Fig. 4.10 with uniform heat generation q_g is given by

$$t = -\frac{1}{\beta} + \sqrt{\left(t_c + \frac{1}{\beta}\right)^2 - \frac{q_g}{\beta k_o}x^2}$$

where the thermal conductivity is given by $k = k_o(1 + \beta t)$.

Solution

Heat balance for the elemental strip gives

$$q_{x+dx} - q_x = (Adx)q_g$$

or

$$q_x + \frac{d}{dx}(q_x)dx - q_x = Adxq_g$$

or

$$\frac{d}{dx}(q_x) = Aq_g$$

or

$$\frac{d}{dx}\left(-kA\frac{dt}{dx}\right) = Aq_g$$

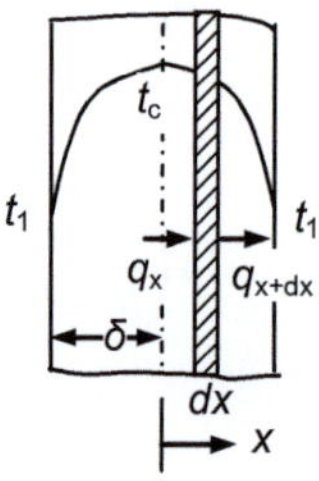

Fig. 4.10 Example 4.16

or

$$\frac{d}{dx}\left(-k\frac{dt}{dx}\right) = q_g$$

Integration gives

$$k\frac{dt}{dx} + q_g x = C_1 \qquad \text{(i)}$$

At $x = 0$, $dt/dx = 0$, hence $C_1 = 0$. Putting $k = k_o(1 + \beta t)$, we have

$$k_o(1+\beta t)\frac{dt}{dx} + q_g x = 0$$

or

$$\frac{dt}{dx} + \beta t\frac{dt}{dx} + \frac{q_g}{k_o}x = 0$$

Further integration gives

$$t + \frac{\beta t^2}{2} + \frac{q_g}{k_o} \times \frac{x^2}{2} = C_2 \qquad \text{(ii)}$$

At $x = 0$, $t = t_c$, hence

$$C_2 = t_c + \frac{\beta t_c^2}{2}$$

Substitution of the value of C_2 in Eq. (ii) gives

$$t + \frac{\beta t^2}{2} + \frac{q_g}{k_o} \times \frac{x^2}{2} - t_c - \frac{\beta t_c^2}{2} = 0$$

Solution of this quadratic equation is

$$t = \frac{-\frac{2}{\beta} + \sqrt{\left(\frac{2}{\beta}\right)^2 - 4\left(\frac{q_g}{\beta k_o}x^2 - 2\frac{t_c}{\beta} - t_c^2\right)}}{2}$$

$$= -\frac{1}{\beta} + \sqrt{\left(t_c + \frac{1}{\beta}\right)^2 - \frac{q_g}{\beta k_o}x^2},$$

which is the desired result.

Example 4.17 Derive the expression of the distribution of temperature in a cylindrical object with uniform heat generation. The material thermal conductivity is given by $k_o(1 + \beta t)$.

Solution

The differential equation for a cylinder is

$$\frac{d}{dr}\left(-kr\frac{dt}{dr}\right) = q_g r$$

Integration gives

$$kr\frac{dt}{dr} + q_g\frac{r^2}{2} = C_1 \tag{i}$$

At the centre ($r = 0$), $dt/dr = 0$, hence $C_1 = 0$. Putting $k = k_o(1 + \beta t)$, we have

$$k_o(1+\beta t)r\frac{dt}{dr} + q_g\frac{r^2}{2} = 0$$

or

$$\frac{dt}{dr} + \beta t\frac{dt}{dr} + \frac{q_g}{2k_o}r = 0$$

Further integration gives

$$t + \frac{\beta t^2}{2} + \frac{q_g}{2k_o} \times \frac{r^2}{2} = C_2 \tag{ii}$$

At $r = 0$, $t = t_c$ hence

$$C_2 = t_c + \frac{1}{2}\beta t_c^2$$

Substitution of the value of C_2 in Eq. (ii) gives

$$t + \frac{\beta t^2}{2} + \frac{q_g}{4k_o}r^2 - t_c - \frac{1}{2}\beta t_c^2 = 0$$

Solution of this quadratic equation is

$$\begin{aligned} t &= \frac{-\frac{2}{\beta} + \sqrt{\left(\frac{2}{\beta}\right)^2 - 4\left(\frac{q_g}{2\beta k_o}r^2 - 2\frac{t_c}{\beta} - t_c^2\right)}}{2} \\ &= -\frac{1}{\beta} + \sqrt{\left(t_c + \frac{1}{\beta}\right)^2 - \frac{q_g}{2\beta k_o}r^2}, \end{aligned}$$

which is the desired result.

Example 4.18 The left face of a plane wall of thickness L and thermal conductivity k is exposed to a microwave radiation which causes volumetric heat generation in the wall, refer Fig. 4.11. The heat generation varies as

$$q_g = q_o\left(1 - \frac{x}{L}\right)$$

where q_o (W/m^3) is a constant.

The other side of the wall is insulated. If the exposed surface of the wall is at a constant temperature T_s, obtain an expression of temperature distribution.

Solution

The differential equation for the elemental strip in the wall is

$$k\frac{d^2t}{dx^2} + q_g = 0$$

Substituting the given equation of heat generation, we get

$$k\frac{d^2t}{dx^2} + q_0\left(1 - \frac{x}{L}\right) = 0$$

Integrating once,

$$\frac{dt}{dx} + \frac{q_0}{k}x - \frac{q_0}{kL}\frac{x^2}{2} = C_1 \quad \text{(i)}$$

At $x = L$, $dt/dx = 0$ since there is no heat transfer at this face. This gives

$$C_1 = \frac{q_o}{k}\left(L - \frac{L^2}{2L}\right) = \frac{q_o L}{2k}$$

Equation (i) takes the form

$$\frac{dt}{dx} + \frac{q_o}{k}x - \frac{q_o}{k}\left(\frac{x^2}{2L}\right) - \frac{q_o L}{2k} = 0$$

Integrating again, we get

$$t + \frac{q_o}{k}\frac{x^2}{2} - \frac{q_o}{k}\left(\frac{x^3}{6L}\right) - \frac{q_o L}{2k}x = C_2$$

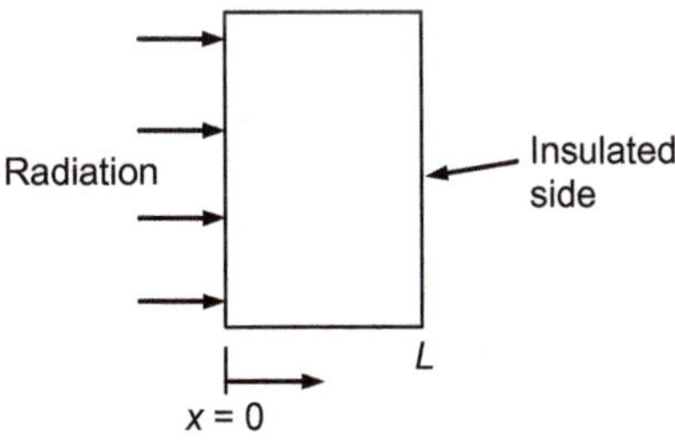

Fig. 4.11 Example 4.18

Applying the second boundary condition of $t = t_s$ at $x = 0$, we get the value of constant $C_2 = t_s$. Substitution gives the expression of the temperature distribution as

$$t = t_s + \frac{q_o}{6kL}\left(x^3 - 3x^2L + 3L^2x\right)$$

At $x = L$, the temperature is maximum. Its value is obtained by putting $x = L$ in the above equation as

$$t_{\max} = t_s + \frac{q_o L^2}{6k}$$

Example 4.19 The outer surfaces of the composite wall shown in Fig. 4.12 are exposed to a fluid at 20°C. The convective heat transfer coefficients are 500 W/(m^2 K) and 1000 W/(m^2 K) on the left and right faces, respectively. The middle wall B experiences uniform volumetric heat generation rate q_g, while there is no heat generation in walls A and C. The temperature at the interfaces are T_1 = 220°C and T_2 = 200°C. Assuming one-dimensional heat flow and negligible contact resistance at the two interfaces, determine

(i) The heat generation rate, q_g (W/m^3),
(ii) The thermal conductivity k_b, and
(iii) If the coolant flow on the left side of the wall fails causing practically no loss of heat from this face ($h_1 = 0$), determine the temperatures T_1, T_2 and $T_{\max}$ of both wall sides.

Solution

(i) **Heat generation rate, q_g**
Heat loss from the left face (A = 1 m^2),

$$q_L = \frac{t_1 - t_\infty}{\frac{L_a}{k_a} + \frac{1}{h_1}} = \frac{220 - 20}{\frac{60}{1000 \times 30} + \frac{1}{500}} = 5 \times 10^4 \text{ W}.$$

Similarly, heat loss from the right face,

$$q_R = \frac{t_2 - t_\infty}{\frac{L_c}{k_c} + \frac{1}{h_2}} = \frac{200 - 20}{\frac{30}{1000 \times 30} + \frac{1}{1000}} = 9 \times 10^4 \text{ W}.$$

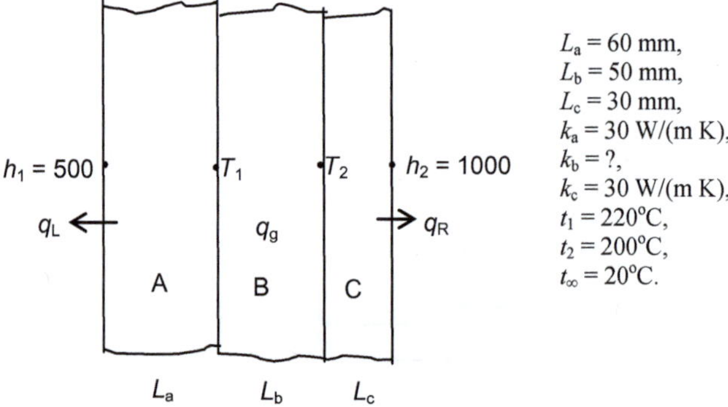

Fig. 4.12 Example 4.19

Heat generated in the walls equals the sum of q_L and q_R. Hence, the heat generation rate is

$$q_g = \frac{q_L + q_R}{AL_b} = \frac{5 \times 10^4 + 9 \times 10^4}{1 \times 0.05} = 2.8 \times 10^6 \text{ W/m}^3.$$

(ii) For the middle wall with uniform heat generation rate and surface temperatures T_1 and T_2, the heat flow q_R from Eq. (4.10) by putting $x = L_b$ is

$$q_R = k_b\left(\frac{t_1 - t_2}{L_b}\right) + \frac{q_g L_b}{2}$$

or

$$9 \times 10^4 = k_b\left(\frac{220 - 200}{0.05}\right) + \frac{2.8 \times 10^6 \times 0.05}{2}$$

This gives k_b = 50 W/(m K).

(iii) Temperature distribution through the middle wall is given by Eq. (4.4):

$$\begin{aligned} t &= \frac{q_g}{2k_b}(L_b - x)x + \left(\frac{t_2 - t_1}{L_b}\right)x + t_1 \\ &= \frac{2.8 \times 10^6}{2 \times 50}(0.05 - x)x + \left(\frac{200 - 220}{0.05}\right)x + 220 \end{aligned}$$

or

$$t = 1000x - 2.8 \times 10^4 x^2 + 220$$

Position of T_{max} from Eq. (4.7) is

$$\frac{x}{L_b} = \frac{B - 1}{2B}$$

where

$$B = \frac{q_g L_b^2}{2k_b(t_1 - t_2)} = \frac{2.8 \times 10^6 \times (0.05)^2}{2 \times 50 \times (220 - 200)} = 3.5$$

Substitution gives

$$x = \frac{3.5 - 1}{2 \times 3.5} \times 0.05 = 0.0179 \text{ m from left side of the middle wall,}$$

and

$$t_{max} = 1000 \times 0.0179 - 2.8 \times 10^4 \times 0.0179^2 + 220 = 228.9^\circ\text{C}.$$

Alternatively, Eq. (4.8) may be used.

The temperatures of sides of the wall can be found from the convection heat transfer equations as follows.

$$q_L = h_1 A(t_1' - t_\infty)$$

$$t_1' = \frac{q_L}{h_1 A} + t_\infty = \frac{5 \times 10^4}{500 \times 1} + 20 = 120^\circ\text{C}$$

Similarly,

$$t_2' = \frac{q_R}{h_2 A} + t_\infty = \frac{9 \times 10^4}{1000 \times 1} + 20 = 110^\circ\text{C}.$$

The temperature distribution through the composite wall is shown in Fig. 4.13a.

(iv) When $h_1 = 0$,

$$t_1' = t_1$$

and

$$q_R = (9 + 5) \times 10^4\ \text{W}.$$

Therefore,

$$t_2' = \frac{q_R}{h_2 A} + t_\infty = \frac{14 \times 10^4}{1000 \times 1} + 20 = 160^\circ\text{C}.$$

From conduction heat transfer equation,

$$q_R = k_c \frac{t_2 - t_2'}{L_c}$$

or

$$t_2 = \frac{q_R L_c}{k_c} + t_2' = \frac{14 \times 10^4 \times 0.03}{30} + 160 = 300^\circ\text{C}.$$

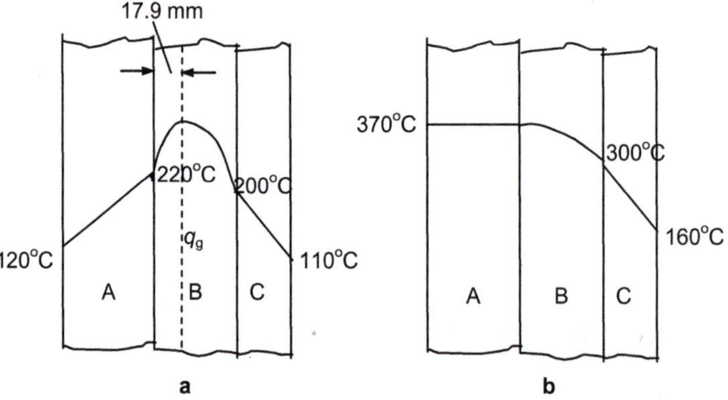

Fig. 4.13 Example 4.19

The maximum temperature (at the left interface),

$$t_{\max} = t_2 + \frac{q_g}{2k_b} L_b^2 = 300 + \frac{2.8 \times 10^6}{2 \times 50} \times (0.05)^2 = 370^\circ\text{C}.$$

The temperature distribution is shown in Fig. 4.13b.

Example 4.20 Heat is uniformly generated in the wall of a very long hollow cylindrical shell of thickness ($R_2 - R_1$). The heat is removed through the inner surface. The thermal conductivity of the wall material varies with temperature as $k = k_o(1 + \beta t)$. Show that the temperature distribution in the wall is given by

$$t = -\frac{1}{\beta} + \sqrt{\left(t_2 + \frac{1}{\beta}\right)^2 - \frac{q_g R_2^2}{2k_o \beta}\left[\left(\frac{r}{R_2}\right)^2 - 2\ln\left(\frac{r}{R_2}\right) - 1\right]}$$

where t_2 is the temperature of the outer surface of the shell.

Solution

The differential equation for one-dimensional heat flow in cylindrical coordinates is

$$\frac{d}{dr}\left(kr\frac{dt}{dr}\right) + q_g r = 0$$

Integration of the above equation gives

$$kr\frac{dt}{dr} + q_g\frac{r^2}{2} = C_1 \quad \text{(i)}$$

At $r = R_2$, $dt/dr = 0$ since no heat is removed from the outer surface. This boundary condition gives

$$C_1 = \frac{q_g R_2^2}{2}$$

Equation (i) takes the form

$$kr\frac{dt}{dr} + q_g\frac{r^2}{2} - \frac{q_g R_2^2}{2} = 0$$

or

$$k_o(1 + \beta t)\frac{dt}{dr} + q_g\frac{r}{2} - \frac{q_g R_2^2}{2r} = 0$$

or

$$\frac{dt}{dr} + \beta t\frac{dt}{dr} + \frac{q_g}{2k_o} r - \frac{q_g R_2^2}{2k_o}\frac{1}{r} = 0$$

Integration of the above equation gives

$$t + \frac{\beta t^2}{2} + \frac{q_g}{2k_o} \cdot \frac{r^2}{2} - \frac{q_g R_2^2}{2k_o} \ln r = C_2 \quad \text{(ii)}$$

The second boundary condition is $t = t_2$ at $r = R_2$. This gives

$$C_2 = t_2 + \frac{\beta T_2^2}{2} + \frac{q_g}{2k_o} \cdot \frac{R_2^2}{2} - \frac{q_g R_2^2}{2k_o} \ln R_2$$

Substitution of the value of C_2 in Eq. (ii) gives

$$t + \frac{\beta t^2}{2} + \frac{q_g R_2^2}{4k_o}\left[\left(\frac{r}{R_2}\right)^2 - 1\right] - \frac{q_g R_2^2}{2k_o} \ln\left(\frac{r}{R_2}\right) - t_2 - \frac{\beta t_2^2}{2} = 0 \quad \text{(iii)}$$

or

$$t^2 + \frac{2t}{\beta} + \frac{q_g R_2^2}{2\beta k_o}\left[\left(\frac{r}{R_2}\right)^2 - 1\right] - \frac{q_g R_2^2}{\beta k_o} \ln\left(\frac{r}{R_2}\right) - \frac{2t_2}{\beta} - t_2^2 = 0$$

This is a quadratic equation whose solution is

$$t = -\frac{1}{\beta} + \sqrt{\left(t_2 + \frac{1}{\beta}\right)^2 - \frac{q_g R_2^2}{2k_o \beta}\left[\left(\frac{r}{R_2}\right)^2 - 2\ln\left(\frac{r}{R_2}\right) - 1\right]} \quad \text{(iv)}$$

If thermal conductivity is constant $\beta = 0$ and putting k_o by k in Eq. (iii), we get

$$t + \frac{q_g R_2^2}{4k}\left[\left(\frac{r}{R_2}\right)^2 - 1\right] - \frac{q_g R_2^2}{2k} \ln\left(\frac{r}{R_2}\right) - t_2 = 0$$

or

$$t = t_2 - \frac{q_g R_2^2}{4k}\left[\left(\frac{r}{R_2}\right)^2 - 2\ln\left(\frac{r}{R_2}\right) - 1\right]$$

Example 4.21 How the above-derived expression will change if the shell is cooled from the outer surface instead of the inner one?

Solution

We can get the desired result by replacing R_2 by R_1 and t_2 by t_1. This gives

$$t = -\frac{1}{\beta} + \sqrt{\left(t_1 + \frac{1}{\beta}\right)^2 - \frac{q_g R_1^2}{2k_o \beta}\left[\left(\frac{r}{R_1}\right)^2 - 2\ln\left(\frac{r}{R_1}\right) - 1\right]}$$

and

$$t = t_1 - \frac{q_g R_1^2}{4k}\left[\left(\frac{r}{R_1}\right)^2 - 2\ln\left(\frac{r}{R_1}\right) - 1\right]$$

when k is constant.

The expression can be derived by following the procedure of the above example.

Example 4.22 Inner and outer surfaces of a cylindrical shell are maintained at uniform temperatures T_1 and T_2, respectively. If there is a uniform rate of heat generation q_g within the shell, obtain the following expressions for the steady state, one-dimensional radial distribution of temperature, heat flux and heat rate.

$$T(r) = T_2 + \frac{q_g R_2^2}{4k}.\left[1 - \left(\frac{r}{R_2}\right)^2\right] - \left\{\frac{q_g R_2^2}{4k}\left[1 - \left(\frac{R_1}{R_2}\right)^2\right] + (T_2 - T_1)\right\}\left(\frac{\ln r - \ln R_2}{\ln R_1 - \ln R_2}\right)$$

Heat rate,

$$q(r) = \pi L q_g r^2 - \frac{2\pi L k}{\ln(R_2/R_1)}\left[\frac{q_g R_2^2}{4k}\left(1 - \frac{R_1^2}{R_2^2}\right) + (T_2 - T_1)\right]$$

Heat flux,

$$q''(r) = \frac{q_g r}{2} - \frac{k}{r}\left[\frac{q_g R_2^2}{4k}\left(1 - \frac{R_1^2}{R_2^2}\right) + (T_2 - T_1)\right]\frac{1}{\ln(R_2/R_1)}$$

Solution

(i) **Temperature distribution**

From Eq. (vi), Sect. 4.2,

$$t + \frac{q_g}{4k}.r^2 + C_1 \ln r + C_2 = 0 \tag{i}$$

The boundary conditions are

$$t = T_1 \text{ at } r = R_1$$
$$\text{and} \quad t = T_2 \text{ at } r = R_2.$$

Applying the boundary conditions,

$$T_1 + \frac{q_g}{4k} R_1^2 + C_1 \ln R_1 + C_2 = 0$$

and

$$T_2 + \frac{q_g}{4k} R_2^2 + C_1 \ln R_2 + C_2 = 0$$

Solving for C_1 and C_2, we get

$$C_1 = \left[\frac{q_g}{4k}(R_2^2 - R_1^2) + (T_2 - T_1)\right]\left(\frac{1}{\ln R_1 - \ln R_2}\right)$$

$$C_2 = -T_2 - \frac{q_g R_2^2}{4k} - \left[\frac{q_g}{4k}(R_2^2 - R_1^2) + (T_2 - T_1)\right]\left(\frac{\ln R_2}{\ln R_1 - \ln R_2}\right)$$

Substitution of values of C_1 and C_2 in Eq. (i) gives

$$\begin{aligned} T(r) = T_2 + \frac{q_g R_2^2}{4k}.\left[1 - \left(\frac{r}{R_2}\right)^2\right] \\ - \left\{\frac{q_g R_2^2}{4k}\left[1 - \left(\frac{R_1}{R_2}\right)^2\right] + (T_2 - T_1)\right\}\left(\frac{\ln r - \ln R_2}{\ln R_1 - \ln R_2}\right) \end{aligned} \quad \text{(ii)}$$

(ii) **Heat rate**

$$q(r) = -k(2\pi rL)\frac{dT}{dr} \quad \text{(iii)}$$

Differentiating Eq. (ii),

$$\frac{dT}{dr} = -\frac{q_g r}{2k} - \left\{\frac{q_g R_2^2}{4k}\left[1 - \left(\frac{R_1}{R_2}\right)^2\right] + (T_2 - T_1)\right\}\left(\frac{1/r}{\ln R_1 - \ln R_2}\right)$$

Substitution in Eq. (iii) gives

$$q(r) = \pi L q_g r^2 - \frac{2\pi Lk}{\ln(R_2/R_1)}\left[\frac{q_g R_2^2}{4k}\left(1 - \frac{R_1^2}{R_2^2}\right) + (T_2 - T_1)\right]$$

(iii) **Heat flux**

$$\begin{aligned} q''(r) &= \frac{q(r)}{A(r)} = \frac{q(r)}{2\pi rL} \\ &= \frac{q_g r}{2} - \frac{k}{r}\left[\frac{q_g R_2^2}{4k}\left(1 - \frac{R_1^2}{R_2^2}\right) + (T_2 - T_1)\right]\frac{1}{\ln(R_2/R_1)} \end{aligned}$$

Example 4.23 Show that the heat generation rate in the cylinder of Fig. 4.14 for a linear radial temperature distribution varies inversely with radius.

Solution

For steady-state radial heat conduction with heat generation for a cylinder, Eq. (2.16) gives

$$k\left(\frac{1}{r}\frac{\partial t}{\partial r}\right) + q_g = 0. \quad \text{(i)}$$

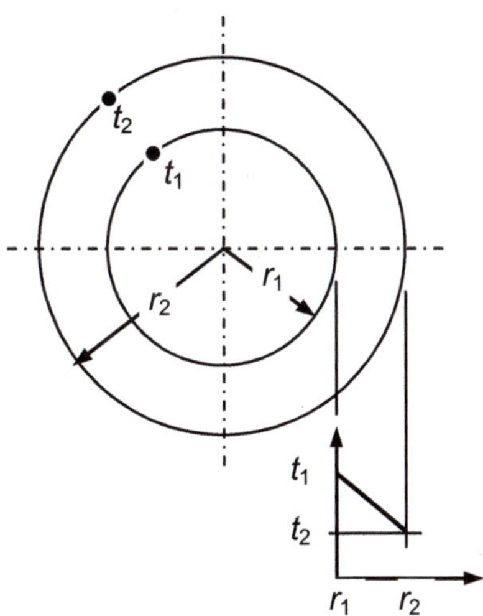

Fig. 4.14 Hollow cylinder with heat generation

For linear temperature distribution with radius, $\frac{\partial t}{\partial r}$ = constant, say C. Integration of relation $\frac{\partial t}{\partial r} = C$ gives $C = \frac{t_2 - t_1}{r_2 - r_1}$. Substitution in Eq. (i) gives

$$q_g = \left(\frac{t_1 - t_2}{r_2 - r_1}\right)\frac{k}{r} \propto \frac{1}{r},$$

i.e. the heat generation rate q_g varies inversely with radius r.

Example 4.24 Heat flows axially through a rod of variable cross-sectional area given by $A_x(x) = A_o e^{bx}$ where A_o and b are constants. Considering volumetric heat generation rate $q_g = q_o e^{-bx}$, obtain the expression for q_x when the left face and lateral surface are insulated. Thermal conductivity of the rod material is constant.

Solution

For the control volume, refer Fig. 4.15,

$$q_x + (A_x dx)q_g = q_{x+dx} \tag{i}$$

Substituting $q_{x+dx} = q_x + \frac{d}{dx}(q_x)dx$, we have, from Eq. (i),

$$A_x dx q_g = \frac{d}{dx}(q_x)dx$$

or

$$\begin{aligned}\frac{d}{dx}(q_x) &= A_x q_g \\ &= A_0 e^{bx} q_0 e^{-bx} \\ &= A_0 q_0\end{aligned}$$

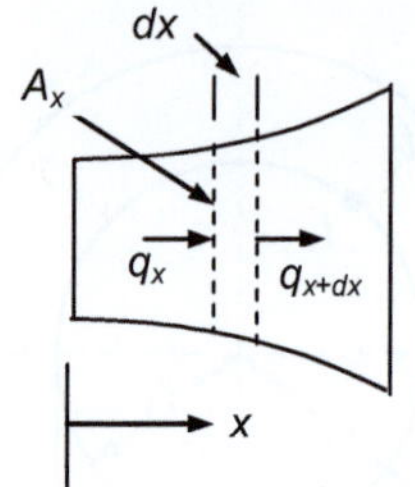

Fig. 4.15 Example 4.24

or

$$q_x = A_0 q_0 x + C_1 \quad \text{(ii)}$$

Left face is insulated, i.e. at $x = 0$, $q_x = 0$. This condition gives $C_1 = 0$. Hence, Eq. (ii) reduces to

$$q_x = A_0 q_0 x.$$

Example 4.25 Heat generation in a cylindrical container of radioactive waste varies as

$$q_o\left[1-\left(\frac{r}{r_o}\right)^2\right]$$

If the heat is rejected from the surface by convection to fluid at temperature t_∞, determine the surface temperature of the container.

Solution

The rate of heat generation in the cylinder is

$$\begin{aligned} Q_g &= \int_0^{r_o} q_g(2\pi rLdr) \\ &= \int_0^{r_o} q_o\left[1-\left(\frac{r}{r_o}\right)^2\right](2\pi rLdr) \\ &= 2\pi Lq_o \int_0^{r_o}\left[1-\left(\frac{r}{r_o}\right)^2\right] rdr \\ &= 2\pi Lq_o\left[\frac{r^2}{2}-\frac{r^4}{r_o^2}\right]_0^{r_o} = \frac{\pi Lq_o r_o^2}{2}. \end{aligned} \quad \text{(i)}$$

The rate of heat rejection by convection is

$$Q_c = h(2\pi r_o L)(t_s - t_\infty) \quad \text{(ii)}$$

In equilibrium, $Q_g = Q_c$, which gives

$$\frac{\pi L q_o r_o^2}{2} = h(2\pi r_o L)(t_s - t_\infty)$$

or

$$t_s = t_\infty + \frac{q_o r_o}{4h}.$$

Example 4.26 For the heat generation rate given in Example 4.25, determine the equation of radial temperature distribution.

Solution

From the general heat conduction equation in cylindrical coordinates, we have for one-dimensional heat conduction

$$\frac{1}{r}\frac{d}{dr}\left(r\frac{dt}{dr}\right) = -\frac{q_g}{k}$$

$$= -\frac{q_o}{k}\left[1 - \left(\frac{r}{r_o}\right)^2\right]$$

or

$$\frac{d}{dr}\left(r\frac{dt}{dr}\right) = -\frac{q_o}{k}\left(r - \frac{r^3}{r_o^2}\right)$$

Integration of the above equation gives

$$r\frac{dt}{dr} = -\frac{q_o}{k}\left(\frac{r^2}{2} - \frac{r^4}{4r_o^2}\right) + C_1$$

At $r = 0$, $dt/dr = 0$, which gives $C_1 = 0$. Hence,

$$r\frac{dt}{dr} = -\frac{q_o}{k}\left(\frac{r^2}{2} - \frac{r^4}{4r_o^2}\right)$$

or

$$\frac{dt}{dr} = -\frac{q_o}{k}\left(\frac{r}{2} - \frac{r^3}{4r_o^2}\right) \quad \text{(i)}$$

Further integration gives

$$t = -\frac{q_o}{k}\left(\frac{r^2}{4} - \frac{r^4}{16r_o^2}\right) + C_2 \quad \text{(ii)}$$

Hence, the surface temperature is

$$t(r_o) = -\frac{q_o}{k}\left(\frac{r_o^2}{4} - \frac{r_o^4}{16r_o^2}\right) + C_2$$

or

$$t(r_o) = -\frac{3q_o r_o^2}{16k} + C_2 \quad \text{(iii)}$$

At the surface of the cylinder, heat reaching by conduction equals the heat rejected by convection hence

$$-k\frac{dt}{dr} = h[t(r_o) - t_\infty]$$

Substitution from Eqs. (i) to (iii) in the above equation gives

$$-k\left[-\frac{q_o}{k}\left(\frac{r}{2} - \frac{r^3}{4r_o^2}\right)\right]_{r=r_o} = h\left(-\frac{3q_o r_o^2}{16k} + C_2 - t_\infty\right)$$

or

$$\frac{q_o r_o}{4h} = -\frac{3q_o r_o^2}{16k} + C_2 - t_\infty$$

$$C_2 = \frac{q_o r_o}{4h} + \frac{3q_o r_o^2}{16k} + t_\infty.$$

Substitution of the value of C_2 in Eq. (ii) gives

$$t = -\frac{q_o}{k}\left(\frac{r^2}{4} - \frac{r^4}{16r_o^2}\right) + \frac{q_o r_o}{4h} + \frac{3q_o r_o^2}{16k} + t_\infty$$

or

$$t = \frac{q_o r_o}{4h} + \frac{q_o r_o^2}{k}\left[\frac{3}{16} - \frac{1}{4}\left(\frac{r}{r_o}\right)^2 + \frac{1}{16}\left(\frac{r}{r_o}\right)^4\right] + t_\infty,$$

which is the desired radial temperature distribution equation.

Check: At $r = r_o$, the temperature from the above equation is

$$t(r_o) = \frac{q_o r_o}{4h} + t_\infty,$$

which is the same as determined in the previous example.

Example 4.27 The composite wall shown in Fig. 4.16a consists of two layers ($k_1 = 50$ W/(m K), $\delta_1 = 0.05$ m and $k_2 = 210$ W/(m K), $\delta_2 = 0.025$ m). There is a uniform generation of heat at the rate of 1×10^6 W/m^3 in layer 1. The contact resistance between the layers is estimated to be 0.001 m^2 K/W. Left face of layer 1 is insulated. The open face of layer 2 is subjected to convection environment. Determine the temperature distribution through the wall.

Solution

Starting from the fundamental equation of conduction heat transfer through a plane wall, we have

$$\frac{d^2t}{dx^2} + \frac{q_g}{k} = 0$$

Integration gives

$$\frac{dt}{dx} + \frac{q_g}{k}x - C_1 = 0 \tag{i}$$

At $x = 0$, $dt/dx = 0$ as the wall is insulated. This give $C_1 = 0$. Equation (i) reduces to

$$\frac{dt}{dx} + \frac{q_g}{k}x = 0$$

Further integration gives

$$t = -\frac{q_g x^2}{2k} + C_2 \tag{ii}$$

At $x = \delta_1$, $t = t_2$ hence

$$t_2 = -\frac{q_g \delta_1^2}{2k} + C_2$$

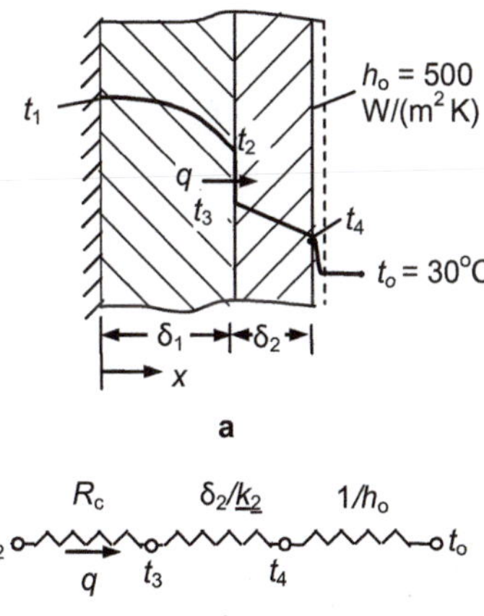

Fig. 4.16 Example 4.27

or

$$C_2 = \frac{q_g \delta_1^2}{2k} + t_2$$

Substitution of the value of C_2 in Eq. (ii) gives

$$t = \frac{q_g}{2k}(\delta_1^2 - x^2) + t_2 \quad \text{(iii)}$$

Equation (iii) gives temperature distribution in layer 1 with heat generation, which is parabolic. Temperature distribution through layer 2 is a straight line because there is no heat generation in layer 2 and the thermal conductivity is constant.

The temperature t_1–t_4 can be determined as under.

Heat flow rate through layer 2 equals the heat generated in the wall, i.e. $q_g \delta_1$ for unit face area of the wall. Heat flow rate equation gives

$$q = q_g \delta_1 = \frac{t_2 - t_o}{R_t}, \quad \text{(iv)}$$

where refer Fig. 4.16b,

$$R_t = R_c + \frac{\delta_2}{k_2} + \frac{1}{h_o} = 0.001 + \frac{0.025}{210} + \frac{1}{500} = 0.00312.$$

Equation (iv) gives

$$\begin{aligned} t_2 &= q_g \delta_1 R_t + t_o \\ &= 1 \times 10^6 \times 0.05 \times 0.00312 + 30 = 186\,°\text{C}. \end{aligned}$$

From Eq. (iii), at $x = 0$

$$\begin{aligned} t_1 &= \frac{q_g}{2k}\delta_1^2 + t_2 \\ &= \frac{1 \times 10^6}{2 \times 50} \times (0.05)^2 + 186 = 211\,°\text{C}. \end{aligned}$$

We can write

$$q = \frac{t_2 - t_3}{R_c}$$

which gives

$$\begin{aligned} t_3 &= t_2 - qR_c \\ &= 186 - (1 \times 10^6 \times 0.05) \times 0.001 = 136\,°\text{C}. \end{aligned}$$

Similarly,

$$\begin{aligned} t_4 &= t_3 - \frac{q\delta_2}{k_2} \\ &= 136 - \frac{(1 \times 10^6 \times 0.05) \times 0.025}{210} = 130.05\,°\text{C}. \end{aligned}$$

and

$$t_0 = t_4 - \frac{q}{h_o}$$
$$= 130.05 - \frac{(1 \times 10^6 \times 0.05)}{500} = 30.05°\text{C},$$

which is approximately equal to the given value.

Example 4.28 Heat generation in a spherical container of radioactive waste varies as

$$q_o \left[1 - \left(\frac{r}{r_o} \right)^2 \right].$$

If the heat is rejected from the surface by convection to fluid at temperature t_∞, determine the equation of radial temperature distribution.

Solution

From the general heat conduction equation in spherical coordinates, we have for one-dimensional heat conduction

$$\frac{1}{r^2} \frac{\partial}{\partial r} \left(r^2 \frac{\partial t}{\partial r} \right) = -\frac{q_g}{k}$$
$$= -\frac{q_o}{k} \left[1 - \left(\frac{r}{r_o} \right)^2 \right]$$

or

$$\frac{\partial}{\partial r} \left(r^2 \frac{\partial t}{\partial r} \right) = -\frac{q_o}{k} \left(r^2 - \frac{r^4}{r_o^2} \right)$$

Integration of the above equation gives

$$r^2 \frac{dt}{dr} = -\frac{q_o}{k} \left(\frac{r^3}{3} - \frac{r^5}{5r_o^2} \right) + C_1$$

At $r = 0$, $dt/dr = 0$, which gives $C_1 = 0$. Hence,

$$r^2 \frac{dt}{dr} = -\frac{q_o}{k} \left(\frac{r^3}{3} - \frac{r^5}{5r_o^2} \right)$$

or

$$\frac{dt}{dr} = -\frac{q_o}{k} \left(\frac{r}{3} - \frac{r^3}{5r_o^2} \right) \quad \text{(i)}$$

Further integration gives

$$t = -\frac{q_o}{k}\left(\frac{r^2}{6} - \frac{r^4}{20r_o^2}\right) + C_2 \tag{ii}$$

Hence, the surface temperature is

$$t(r_o) = -\frac{q_o}{k}\left(\frac{r_o^2}{6} - \frac{r_o^4}{20r_o^2}\right) + C_2$$

or

$$t(r_o) = -\frac{7q_o r_o^2}{60k} + C_2 \tag{iii}$$

At the surface of the sphere, heat reaching by conduction equals the heat rejected by convection hence

$$-k\frac{dt}{dr} = h[t(r_o) - t_\infty]$$

Substitution from Eqs. (i) and (iii) in the above equation gives

$$-k\left[-\frac{q_o}{k}\left(\frac{r}{3} - \frac{r^3}{5r_o^2}\right)\right]_{r=r_o} = h\left(-\frac{7q_o r_o^2}{60k} + C_2 - t_\infty\right)$$

or

$$\frac{2q_o r_o}{15h} = -\frac{7q_o r_o^2}{60k} + C_2 - t_\infty$$

$$C_2 = \frac{2q_o r_o}{15h} + \frac{7q_o r_o^2}{60k} + t_\infty.$$

Substitution of the value of C_2 in Eq. (ii) gives

$$t = -\frac{q_o}{k}\left(\frac{r^2}{6} - \frac{r^4}{20r_o^2}\right) + \frac{2q_o r_o}{15h} + \frac{7q_o r_o^2}{60k} + t_\infty$$

or

$$t = \frac{2q_o r_o}{15h} + \frac{q_o r_o^2}{k}\left[\frac{7}{60} - \frac{1}{6}\left(\frac{r}{r_o}\right)^2 + \frac{1}{20}\left(\frac{r}{r_o}\right)^4\right] + t_\infty,$$

which is the desired radial temperature distribution equation.

Example 4.29 A cylindrical reactor fuel element with uniform heat generation rate q_g has steel cladding as shown in Fig. 4.17. Heat is rejected from the outer surface by convection to a coolant at temperature t_∞. Determine the expressions for temperature distributions in fuel element and cladding.

Solution

For the cylindrical fuel element,

$$\frac{1}{r}\frac{d}{dr}\left(r\frac{dt_f}{dr}\right) = -\frac{q_g}{k_f},$$

or

$$\frac{d}{dr}\left(r\frac{dt_f}{dr}\right) = -\frac{q_g r}{k_f}$$

Integration gives

$$r\frac{dt_f}{dr} = -\frac{q_g r^2}{2k_f} + C_1$$

At $r = 0$, $dt_f/dr = 0$, therefore $C_1 = 0$. Hence, the equation is

$$r\frac{dt_f}{dr} = -\frac{q_g r^2}{2k_f}$$

or

$$\frac{dt_f}{dr} = -\frac{q_g r}{2k_f}$$

Further integration gives

$$t_f = -\frac{q_g r^2}{4k_f} + C_2 \tag{i}$$

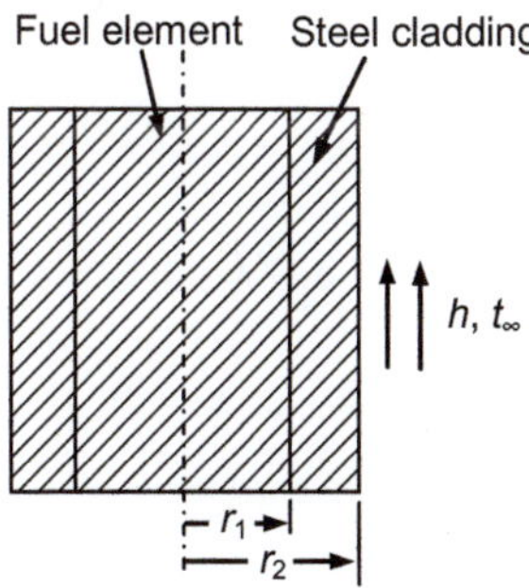

Fig. 4.17 Example 4.29

For the cylindrical cladding,

$$\frac{1}{r}\frac{d}{dr}\left(r\frac{dt_c}{dr}\right) = 0,$$

or

$$\frac{d}{dr}\left(r\frac{dt_c}{dr}\right) = 0$$

Integration gives

$$r\frac{dt_c}{dr} = C_3$$

or

$$\frac{dt_c}{dr} = \frac{C_3}{r}$$

Further integration gives

$$t_c = C_3 \ln r + C_4 \tag{ii}$$

The constants C_2, C_3 and C_4 can be found from the following boundary conditions.

At radius r_1,

$$t_f(r_1) = t_c(r_1) \tag{iii}$$

Conduction heat flow at r_1 gives

$$-k_f\left(\frac{dt_f}{dr}\right)_{r=r_1} = -k_c\left(\frac{dt_c}{dr}\right)_{r=r_1} \tag{iv}$$

At the cladding surface, heat is rejected by convection, hence

$$-k_c\left(\frac{dt_c}{dr}\right)_{r=r_2} = h[t_c(r_2) - t_\infty] \tag{v}$$

From Eq. (iii),

$$-\frac{q_g r_1^2}{4k_f} + C_2 = C_3 \ln r_1 + C_4 \tag{vi}$$

From Eq. (iv),

$$-k_f\left(\frac{dt_f}{dr}\right)_{r=r_1} = -k_c\left(\frac{dt_c}{dr}\right)_{r=r_1}$$

or

$$-k_f\left(-\frac{q_g r_1}{2k_f}\right) = -k_c\frac{C_3}{r_1}$$

or

$$C_3 = -\frac{q_g r_1^2}{2k_c}$$

From Eq. (v),

$$-k_c\frac{C_3}{r_2} = h[C_3 \ln r_2 + C_4 - t_\infty]$$

or

$$\frac{q_g r_1^2}{2r_2 h} = -\frac{q_g r_1^2}{2k_c}\ln r_2 + C_4 - t_\infty$$

or

$$C_4 = \frac{q_g r_1^2}{2r_2 h} + \frac{q_g r_1^2}{2k_c}\ln r_2 + t_\infty$$

Substituting values of C_3 and C_4 in Eq. (vi), we obtain

$$C_2 = -\frac{q_g r_1^2}{2k_c}\ln r_1 + \frac{q_g r_1^2}{4k_f} + \frac{q_g r_1^2}{2r_2 h} + \frac{q_g r_1^2}{2k_c}\ln r_2 + t_\infty$$

or

$$C_2 = \frac{q_g r_1^2}{2k_c}\ln\left(\frac{r_2}{r_1}\right) + \frac{q_g r_1^2}{4k_f} + \frac{q_g r_1^2}{2r_2 h} + t_\infty$$

Substituting values of constants C_2–C_4 in Eqs. (i) and (ii), we obtain the expressions for temperature distributions in fuel element and cladding as

$$t_f = -\frac{q_g r^2}{4k_f} + \frac{q_g r_1^2}{2k_c}\ln\left(\frac{r_2}{r_1}\right) + \frac{q_g r_1^2}{4k_f} + \frac{q_g r_1^2}{2r_2 h} + t_\infty$$

or

$$t_f = \frac{q_g}{4k_f}(r_1^2 - r^2) + \frac{q_g r_1^2}{2k_c}\ln\left(\frac{r_2}{r_1}\right) + \frac{q_g r_1^2}{2r_2 h} + t_\infty \quad (1)$$

and

$$t_c = -\frac{q_g r_1^2}{2k_c}\ln r + \frac{q_g r_1^2}{2r_2 h} + \frac{q_g r_1^2}{2k_c}\ln r_2 + t_\infty$$

or

$$t_c = \frac{q_g r_1^2}{2k_c}\ln\left(\frac{r_2}{r}\right) + \frac{q_g r_1^2}{2r_2 h} + t_\infty \tag{2}$$

4.5 Summary

This chapter is devoted to the heat conduction processes when there is heat generation in the solid itself. Firstly the treatment has been presented for plane wall, cylindrical and spherical solids with a uniform rate of heat generation per unit volume with constant thermal conductivity of the solid material, and for the steady-state one-dimensional heat conduction. The following basic differential equations for these cases have been developed in Sects. 4.1–4.3.

Plane wall with uniform heat generation,

$$\frac{d^2t}{dx^2} + \frac{q_g}{k} = 0 \tag{4.1}$$

Cylinder with uniform heat generation,

$$\frac{d}{dr}\left(r\frac{dt}{dr}\right) + \frac{q_g}{k}r = 0 \tag{4.16}$$

Solid sphere with uniform heat generation,

$$\frac{d}{dr}\left(r^2\frac{dt}{dr}\right) + \frac{q_g}{k}r^2 = 0 \tag{4.21}$$

The students are advised to start from these basic differential equations for solution to problems. Integration of the above equations and the determination of the constants of the integration from the given boundary conditions will provide the applicable equation of temperature distribution for different cases of interest, which can be utilized for calculation of the rate of heat transfer using the Fourier's equation or to locate the position and magnitude of maximum temperature in the solid.

For cases with variable thermal conductivity, the students may start from equations obtained from the general heat conduction equations in Chap. 1; for example, from equation $\frac{d}{dx}\left(-k\frac{dt}{dx}\right) = q_g$ for a plane wall (refer Example 4.16), or from equation $\frac{d}{dr}\left(kr\frac{dt}{dr}\right) + q_g r = 0$ for a cylindrical element (refer Examples 4.17 and 4.20). Alternatively, one may start from fundamentals.

For the problems involving the heat generation, which is variable as in Example 4.18, it is advised to start from fundamentals only.

An approximate treatment of the heat transfer through piston crown has been presented in Sect. 4.4 to determine temperature distribution and maximum temperature in the crown.

A number of solved examples have been included to illustrate the application of the above-discussed approaches for a variety of problems.

Review Questions

4.1 Show that for a plane wall of thickness 2 l with a uniformly distributed heat generation q_g per unit volume, the temperature t_o at the midplane is given by

$$t_o = \frac{q_g l^2}{2k} + t_w$$

where t_w is the temperature on either side of the wall and k is the thermal conductivity of the wall material.

4.2 Develop an expression of temperature distribution in a sphere of radius R_o made of a homogeneous material in which energy is released at a uniform rate per unit volume. The surface temperature of the sphere is T_o and the thermal conductivity of the sphere material may be assumed to be constant. Also, give the expression of maximum temperature in the sphere.

4.3 Inner and outer surfaces of a spherical shell are maintained at uniform temperatures t_1 and t_2, respectively. If there is a uniform rate of heat generation q_g within the shell, obtain the following expressions for the steady-state, one-dimensional radial distribution of temperature, heat flux and heat rate.

$$t(r) = t_2 + \frac{q_g R_2^2}{6k} \cdot \left[1 - \left(\frac{r}{R_2}\right)^2\right] - \left\{\frac{q_g R_2^2}{6k}\left[1 - \left(\frac{R_1}{R_2}\right)^2\right] + (t_2 - t_1)\right\}\left(\frac{1/r - 1/R_2}{1/R_1 - 1/R_2}\right)$$

Heat rate,

$$q(r) = \frac{4\pi q_g r^3}{3} - 4\pi k\left[\frac{q_g R_2^2}{6k}\left(1 - \frac{R_1^2}{R_2^2}\right) + (t_2 - t_1)\right]\left(\frac{1}{1/R_1 - 1/R_2}\right)$$

Heat flux,

$$q''(r) = \frac{q_g r}{3} - \frac{k}{r^2}\left[\frac{q_g R_2^2}{6k}\left(1 - \frac{R_1^2}{R_2^2}\right) + (t_2 - t_1)\right]\left(\frac{1}{1/R_1 - 1/R_2}\right)$$

[**Hint**: Follow the procedure of Example 4.22]

Problems

4.1 Heat is generated at the interface between two slabs each 50 mm thick. One of the slabs is of steel [$k = 35$ W/(m K)] and other is of brass [$k = 75$ W/(m K)]. The temperature at the outside of the steel slab is 95°C, while at the outside of the brass is 47°C. Calculate the heat flow rate at the outer surfaces and the interface temperature when the rate of heat generation is 175 kW/m^2 of the contact area.
[**Ans.** $q_L = k_1 A\left(\frac{t_i - t_1}{\delta_1}\right) = 700 \times (t_i - 95)$; $q_R = k_2 A\left(\frac{t_i - t_2}{\delta_2}\right) = 1500 \times (t_i - 47)$; $q_L + q_R = q_g$; $t_i = 141.82$°C; $q_L = 32.77\,\text{kW/m}^2$; $q_R = 142.23\,\text{kW/m}^2$].

4.2 A long cylindrical rod of 100 mm radius experiences uniform heat generation rate of 2 × 10^4 W/m^3. It carries a tightly fitting sleeve having an outer radius of 150 mm and a thermal conductivity of 5 W/(m K). The surface of the sleeve is exposed to cross-flow of air at 30°C. If the thermal conductivity of the rod material is 0.5 W/(m K) and the convective heat transfer coefficient is 40 W/(m^2 K), determine

(i) Temperature at the interface of the rod and the sleeve, and
(ii) The maximum temperature and its location.

[**Ans.** $\frac{q}{L} = \frac{t_1 - t_\infty}{\frac{1}{2\pi k_2}\ln\left(\frac{R_2}{R_1}\right) + \frac{1}{2\pi R_2 h}}$, where q/L equals the heat generated per unit length of the rod, i.e. $q/L = q_g \times \pi R_1^2 \times 1$. This gives interface temperature, $t_1 = 54.77$°C. $t_{\max} = \frac{q_g}{4k_1} R_1^2 + t_1 = 154.77$°C.]

4.3 A large concrete platform 1.0 m high is to be constructed by pouring concrete on a brick base. The top surface of the concrete is exposed to the atmosphere at 300 K and the unit surface conductance is estimated to be 10 W/(m^2 K). The concrete sets by an exothermal chemical reaction, which produces heat at the rate of 60 W/m^3. Assuming one-dimensional steady-state conduction with a uniform rate of heat generation per unit volume, find the maximum temperature within the concrete. Take thermal conductivity of the concrete as 1.0 W/(m K) and neglect the heat loss from the base. (Hint: The bottom surface is insulated, hence at the bottom dt/dx is zero and the temperature is the maximum. Use the following equation: $T_{\max} = \frac{q_g \delta}{h} + \frac{q_g \delta^2}{2k} + T_\infty$)
[**Ans.** $t_{\max} = 63$°C]

4.4 The outer freshly plastered wall surface is held at 35°C during curing, when the rate of generation of heat due to the setting is 5 × 10^4 W/m^3. The plaster is 15 mm thick. Determine the temperature at inner surface of the plaster. The inner surface can be assumed to be insulated. The thermal conductivity of the plaster material is 1 W/(m K).
[**Ans.** The temperature at the wall is maximum, hence the boundary conditions are: $x = 0$, $dt/dx = 0$; at $x = \delta = 0.015$ m, $t = 35$°C, refer Fig. 4.18. Integrate Eq. (4.1) and applying boundary condition to obtain temperature distribution equation $t = \frac{q_g}{2k}(\delta^2 - x^2) + t_1$; Putting $x = 0$ in the temperature distribution equation, we get $t_{\max} = 40.6$°C.]

4.5 A proposed nuclear fuel element 60 mm in diameter is to be clad in a 5 mm thick aluminium sheathing. The fuel element generates heat at a rate of 4 × 10^5 kW/m^3. The outside surface of the aluminium cover is likely to be at 80°C because of the contact with

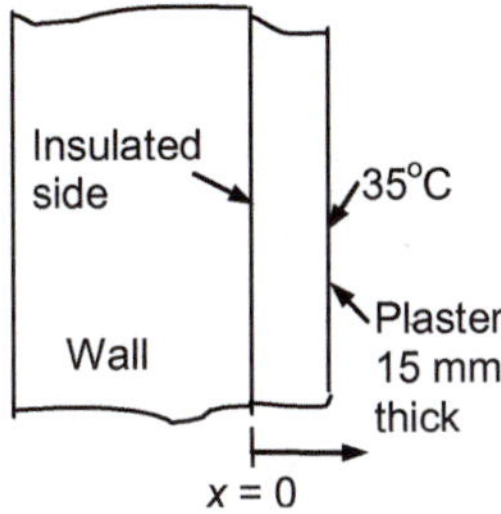

Fig. 4.18 Problem 4.4

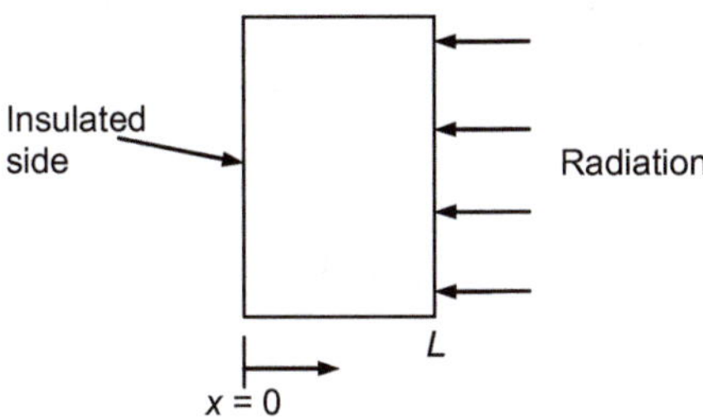

Fig. 4.19 Problem 4.7

a cooling fluid. Estimate the maximum temperature of the aluminium cladding if its thermal conductivity can be expressed as $k = 210(1 - t/1000)$, where t is in °C.
[**Ans.** $q_g(\pi R^2 L) = 2\pi L k_o[1 - \beta(t_o + t_i)/2]\frac{(t_i - t_o)}{\ln(R_o/R_i)}$, where $\beta = 1/1000$; $t_i = 237$°C.]

4.6 A current of 80 A is passed through a 2 mm diameter 500 mm long stainless steel wire [$k = 20$ W/(m K) and electrical resistivity $\rho = 76$ μΩ-cm]. If the wire is dipped in a liquid bath at 80°C and $h = 1500$ W/(m^2 K), calculate the maximum temperature of the wire.
[**Ans.** Resistance of wire $R = \rho L/A = 0.121\Omega$, heat generated, $H = I^2R = 774.4$ W, heat generation rate, $q_g = H$/Volume of wire $= 493 \times 10^6$ W/m^3. Convective heat transfer equation $hA(t_s - t_\infty) = H$ gives $t_s = 244.33$°C. Maximum temperature at the centre, $t_{\max} = \frac{q_g R_o^2}{4k} + t_s = 250.5$°C.]

4.7 The plane wall shown in Fig. 4.19 is subjected to a radiation which causes volumetric heat generation in the wall. The heat generation varies as

$$q_g = q_0\left(\frac{x}{L}\right)$$

Derive the equation of distribution and determine the maximum temperature in the wall if the temperature at the wall surface (at $x = L$) is t_s.
[**Ans.** $t = t_s + \frac{q_o}{6kL}(L^3 - x^3)$; $t_{\max} = (t)_{x=0} = t_s + \frac{q_o}{6k}L^2$.]

Steady-State Two-Dimensional Heat Conduction

5

5.1 Introduction

In the preceding chapters, the cases of one-dimensional steady-state conduction heat flow were analysed. We consider now two-dimensional steady-state conduction heat flow through solids without heat sources.

The Laplace equation that governs the temperature distribution for two dimensional heat conduction system is

$$\frac{\partial^2 t}{\partial x^2} + \frac{\partial^2 t}{\partial y^2} = 0 \tag{5.1}$$

Equation (5.1) is based on the assumption of constant and equal thermal conductivities in both x and y space coordinates. Solution of this equation gives the temperature in the body as a function of coordinates x and y.

Knowing the temperature distribution in the body, the heat flow in the x- and y-directions at a point can be determined from

$$q_x = -kA_x \frac{\partial t}{\partial x} \tag{5.2a}$$

$$q_y = -kA_y \frac{\partial t}{\partial y} \tag{5.2b}$$

The rate of heat flow at the point is the resultant of the components q_x and q_y, i.e.

$$q = iq_x + jq_y \tag{5.3}$$

It follows that in order to determine the heat flow, the temperature field must be known. Thus, the problem reduces to the solution of Eq. (5.1). The common techniques available for the solution of Eq. (5.1) are

(i) Mathematical analysis (analytical solution),
(ii) Graphical analysis,
(iii) Method of analogy,
(iv) Numerical solutions using either a finite-difference or finite-element method.

R. Karwa, *Heat and Mass Transfer*,
https://doi.org/10.1007/978-981-15-3988-6_5

5.2 Analytical Solution of Two-Dimensional Heat Conduction Problems

Consider a rectangular section bar, as shown in Fig. 5.1, which is very long in z-direction. Three lateral sides of the bar are maintained at a constant temperature T_o. For the fourth side we consider different conditions, case (i)–(iii), as outlined below.

Case (i)

The fourth side ($y = H$) has a sinusoidal temperature distribution, $T = T_m \sin(\pi x/W)$, imposed on it.

Using $\theta = T - T_o$, the Laplace equation, Eq. (5.1), is transformed to

$$\frac{\partial^2 \theta}{\partial x^2} + \frac{\partial^2 \theta}{\partial y^2} = 0 \tag{5.4}$$

The boundary conditions are

$$\begin{array}{llll} \text{(i)}\ \theta = 0 & \text{at}\, y = 0 & \text{(ii)}\theta = 0 & \text{at}\, x = 0; \\ \text{(iii)}\ \theta = 0 & \text{at}\, x = W; & \text{(iv)}\ \theta = \theta_m \sin(\pi x/W) & \text{at}\, y = H \end{array} \tag{5.5}$$

where $\theta_m = T_m - T_o$ is amplitude of the sine function. $T = T_m \sin(\pi x/W)$ is transformed to $\theta = \theta_m \sin(\pi x/W)$.

The solution of Eq. (5.4), using the separation of variable technique (based on the assumption that the solution to the differential equation takes a product form), is

$$\theta(x, y) = XY \tag{5.6}$$

where $X = X(x)$ and $Y = Y(y)$.

Substitution in Eq. (5.4) gives

$$-\left(\frac{1}{X}\right)\frac{d^2X}{dx^2} = \left(\frac{1}{Y}\right)\frac{d^2Y}{dy^2} \tag{5.7}$$

Each side of Eq. (5.7) is independent of the other, since x and y are independent variables. Hence, the left side of Eq. (5.7) can equal the right side only if both the sides have a constant value, greater than zero, say λ^2. That is,

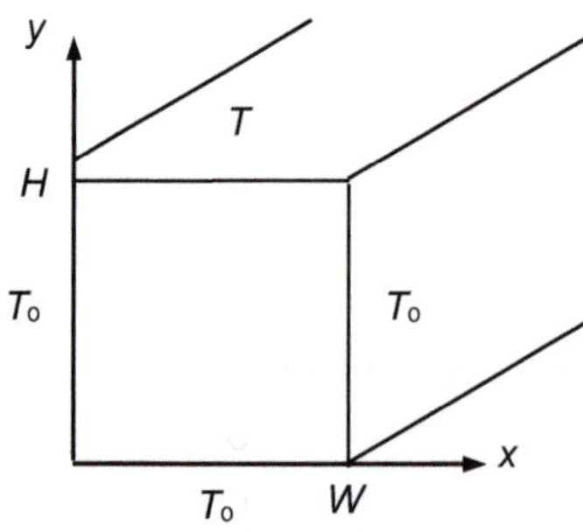

Fig. 5.1 A rectangular section bar with given thermal boundary conditions

$$-\left(\frac{1}{X}\right)\frac{\mathrm{d}^2X}{\mathrm{d}x^2} = \left(\frac{1}{Y}\right)\frac{\mathrm{d}^2Y}{\mathrm{d}y^2} = \lambda^2 \quad (5.8)$$

Thus, we get two ordinary differential equations as

$$\frac{\mathrm{d}^2X}{\mathrm{d}x^2} + \lambda^2 X = 0 \quad (5.9)$$

and

$$\frac{\mathrm{d}^2Y}{\mathrm{d}y^2} - \lambda^2 Y = 0 \quad (5.10)$$

The value of the constant λ^2 is to be determined from the given boundary conditions. The general solution of Eq. (5.9) is

$$X = C_1 \cos\lambda x + C_2 \sin\lambda x$$

and that of Eq. (5.10) is

$$Y = C_3 \exp(-\lambda y) + C_4 \exp(\lambda y)$$

Substitution in Eq. (5.6) gives

$$\theta(x, y) = XY = (C_1 \cos\lambda x + C_2 \sin\lambda x)[C_3 \exp(-\lambda y) + C_4 \exp(\lambda y)] \quad (5.11)$$

Applying the boundary conditions, we have

$$0 = (C_1 \cos\lambda x + C_2 \sin\lambda x)(C_3 + C_4) \quad \text{(i)}$$

$$0 = C_1[C_3 \exp(-\lambda y) + C_4 \exp(\lambda y)] \quad \text{(ii)}$$

$$0 = (C_1 \cos\lambda W + C_2 \sin\lambda W)[C_3 \exp(-\lambda y) + C_4 \exp(\lambda y)] \quad \text{(iii)}$$

$$\theta_m \sin(\pi x/W) = (C_1 \cos\lambda x + C_2 \sin\lambda x)[C_3 \exp(-\lambda H) + C_4 \exp(\lambda H)] \quad \text{(iv)}$$

Equations (i) and (ii) give

$$C_3 = -C_4$$

and

$$C_1 = 0.$$

Substitution in Eq. (iii) gives

$$0 = C_4 C_2 \sin(\lambda W)[\exp(\lambda y) - \exp(-\lambda y)] \quad (5.12)$$

Equation (5.12) requires that

$$\sin(\lambda W) = 0$$

or

$$\lambda = n\pi/W \tag{5.13}$$

where n is a positive integer.

By substituting values of constants C_1 to C_4 and λ in Eq. (5.11), we get

$$\theta(x,y) = T - T_o = \sum_{n=1}^{\infty} C_n \sin\left(\frac{n\pi x}{W}\right) \sinh\left(\frac{n\pi y}{W}\right) \tag{5.14}$$

where the term $\sinh(n\lambda y/W)$ replaces the exponential term and constants C_2 and C_4 have been combined. The solution of the differential equation is a sum of the solutions for each value of n, up to infinity.

Using the fourth boundary condition of $\theta = \theta_m \sin(\pi x/W)$ at $y = H$, we get

$$\theta_m \sin\left(\frac{\pi x}{W}\right) = \sum_{n=1}^{\infty} C_n \sin\left(\frac{n\pi x}{W}\right) \sinh\left(\frac{n\pi H}{W}\right) \tag{5.15}$$

This holds only for $C_n = 0$ when $n > 1$. For $n = 1$, we get

$$\theta_m \sin\left(\frac{\pi x}{W}\right) = C_1 \sin\left(\frac{\pi x}{W}\right) \sinh\left(\frac{\pi H}{W}\right),$$

which gives

$$C_1 = \frac{\theta_m}{\sinh\left(\frac{\pi H}{W}\right)}.$$

The final solution of the differential equation is

$$\theta(x,y) = T - T_o = \theta_m \left[\frac{\sin\left(\frac{\pi x}{W}\right) \sinh\left(\frac{\pi y}{W}\right)}{\sinh\left(\frac{\pi H}{W}\right)}\right] \tag{5.16}$$

Case (ii)

If the temperature along $y = H$ side is given by an arbitrary function $f(x)$, the boundary condition (iv) in Eq. (5.5) changes. The remaining conditions (i) to (iii) remain valid and the solution is as determined above, refer Eq. (5.14), i.e.

$$\theta(x,y) = T - T_o = \sum_{n=1}^{\infty} C_n \sin\left(\frac{n\pi x}{W}\right) \sinh\left(\frac{n\pi y}{W}\right) \tag{5.14}$$

The fourth boundary condition gives

$$f(x) = \sum_{n=1}^{\infty} C_n \sin\left(\frac{n\pi x}{W}\right) \sinh\left(\frac{n\pi H}{W}\right) \tag{5.17}$$

The quantities $C_n \sinh(n\pi H/W)$ must be the coefficients of the Fourier sine series for $f(x)$ in the interval $(0 < x < L)$, i.e.

$$C_n \sinh\left(\frac{n\pi H}{W}\right) = \frac{2}{W}\int_0^W f(x) \sin\left(\frac{n\pi x}{W}\right) dx \tag{5.18}$$

and thus,

$$\theta(x,y) = T - T_o = \frac{2}{W}\sum_{n=1}^{\infty}\left[\frac{1}{\sinh\left(\frac{n\pi H}{W}\right)}\int_0^W f(x)\sin\left(\frac{n\pi x}{W}\right)dx\right]\sin\left(\frac{n\pi x}{W}\right)\sinh\left(\frac{n\pi y}{W}\right) \tag{5.19}$$

The temperature field is shown in Fig. 5.2. The heat flow lines (dashed lines) are drawn perpendicular to the isotherms (firm lines).

Case (iii)

The three sides of the bar are held at temperatures T_o, while the fourth side ($y = H$) is at a constant temperature T_b. The fourth boundary condition ($\theta = T_b - T_o = \theta_b$ at $y = H$) gives

$$\theta_b = \sum_{n=1}^{\infty} C_n \sin\left(\frac{n\pi x}{W}\right) \sinh\left(\frac{n\pi H}{W}\right) \tag{5.20}$$

The value of C_n is determined by expanding the constant temperature difference θ_b in a Fourier series over the interval $(0 < x < W)$, which is

$$\theta_b = \theta_b \frac{2}{\pi}\sum_{n=1}^{\infty}\frac{(-1)^{n+1}+1}{n}\sin\left(\frac{n\pi x}{W}\right) \tag{5.21}$$

Comparison of Eqs. (5.20) and (5.21) gives

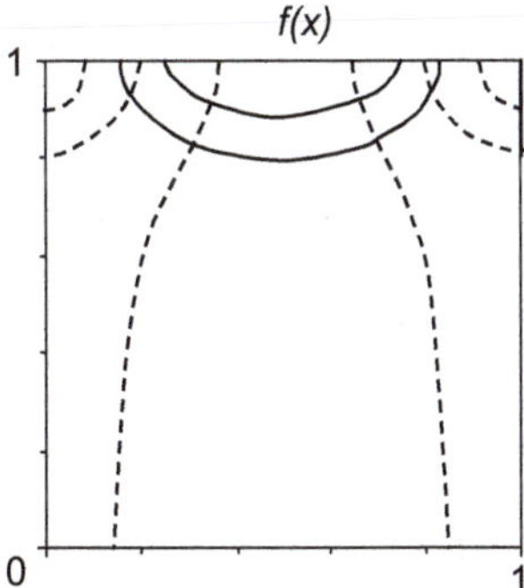

Fig. 5.2 Isotherms and heat flow lines in a rectangular plate

$$C_n = \theta_b \left(\frac{2}{\pi}\right) \frac{(-1)^{n+1} + 1}{n \sinh\left(\frac{n\pi H}{W}\right)} \tag{5.22}$$

The final solution is

$$\frac{\theta}{\theta_b} = \frac{T - T_o}{T_b - T_o} = \frac{2}{\pi} \sum_{n=1}^{\infty} \frac{(-1)^{n+1} + 1}{n \sinh\left(\frac{n\pi H}{W}\right)} \times \sin\left(\frac{n\pi x}{W}\right) \times \sinh\left(\frac{n\pi y}{W}\right) \tag{5.23}$$

$$= \frac{4}{\pi} \sum_{j=1}^{\infty} \frac{1}{(2j-1) \sinh\left[\frac{(2j-1)\pi H}{W}\right]} \times \sin\left[\frac{(2j-1)\pi x}{W}\right] \times \sinh\left[\frac{(2j-1)\pi y}{W}\right] \tag{5.23b}$$

In Eq. (5.23b), only the odd-order terms appear. The even-order terms vanish because of the bracketed term $[(-1)^{n+1} + 1]$.

5.3 Conduction Through a Flat Semi-infinite Homogeneous Plate

Consider the flat plate shown in Fig. 5.3, whose length in *y*-direction is infinite. It is assumed that the plate is relatively thin in *z*-direction and *xoy* surfaces are insulated so that there is no temperature gradient in *z*-direction and the temperature field is two dimensional.

The general solution, Eq. (5.11), is

$$\theta(x, y) = XY = (C_1 \cos \lambda x + C_2 \sin \lambda x)[C_3 \exp(-\lambda y) + C_4 \exp(\lambda y)] \tag{5.11}$$

The boundary conditions are

(i) $\theta = 0$ at $x = 0$; (ii) $\theta = 0$ at $y \to \infty$;
(iii) $\theta = 0$ at $x = W$; (iv) $\theta = F(x)$ at $y = 0$ where $\theta = T - T_o$ and $f(x) - T_o = F(x)$.

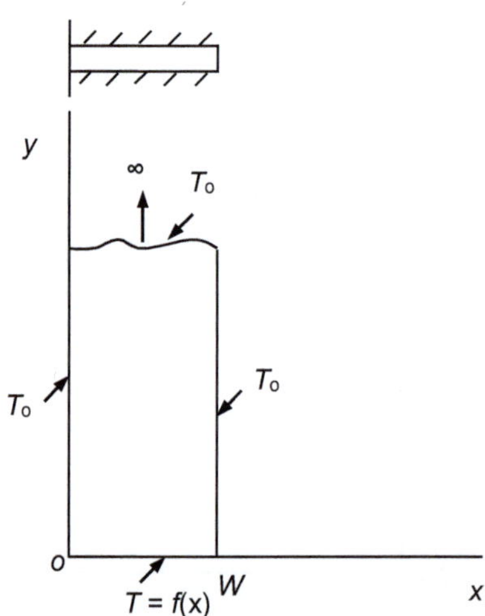

Fig. 5.3 A semi-infinite flat plate with given thermal boundary conditions

(i) $X = \phi(x) = (C_1 \cos \lambda x + C_2 \sin \lambda x)$ will give $\phi(x) = 0$ at $x = 0$ when $C_1 = 0$

(ii) $Y = \psi(y) = C_3 \exp(-\lambda y) + C_4 \exp(\lambda y)$ satisfies the condition $\theta = 0$ as $y \to \infty$. Function $\psi(y) = 0$ at $y \to \infty$ and this is possible when $C_4 = 0$.
Using $C_1 = 0$ and $C_4 = 0$, we obtain from Eq. (5.11)

$$\begin{aligned}\theta(x, y) &= C_2 \sin(\lambda x)[C_3 \exp(-\lambda y)] \\ &= C \sin(\lambda x)[\exp(-\lambda y)]\end{aligned} \tag{5.24}$$

(iii) To satisfy the third boundary condition of $\theta = 0$ at $x = W$,

$$\sin(\lambda x) = 0$$

or

$$\lambda = n\pi/W$$

where $n = 1, 2, 3, \ldots\ldots$

The general solution is, therefore, a sum of solutions for the each value of n up to infinity, i.e.

$$\theta(x, y) = T - T_o = \sum_{n=1}^{\infty} C_n \sin\left(\frac{n\pi x}{W}\right) \exp\left(\frac{-n\pi y}{W}\right) \tag{5.25}$$

The integral constant C_n is determined from the fourth boundary condition, $\theta = F(x)$ at $y = 0$. Thus

$$F(x) = \sum_{n=1}^{\infty} C_n \sin\left(\frac{n\pi x}{W}\right) \tag{5.26}$$

The expression is sine expansion of function $F(x)$ into Fourier series. Hence,

$$C_n = \frac{2}{W} \int_0^W F(x) \sin\left(\frac{n\pi x}{W}\right) \mathrm{d}x \tag{5.27}$$

and the final expression of temperature distribution is

$$\theta(x, y) = \frac{2}{W} \sum_{n=1}^{\infty} \sin\left(\frac{n\pi x}{W}\right) \exp\left(\frac{-n\pi y}{W}\right) \int_0^W F(x) \sin\left(\frac{n\pi x}{W}\right) \mathrm{d}x \tag{5.28}$$

For the special case of $T = T_1$ = constant at $y = 0$, function $f(x) = T_1$ and $F(x) = T_1 - T_o = \theta_1$. The integral of Eq. (5.27) is

$$\int_0^W F(x)\sin\left(\frac{n\pi x}{W}\right)dx = \left\{-\left(\frac{W}{n\pi}\right)\theta_1\left[-\cos\left(\frac{n\pi x}{W}\right)\right]\right\}_0^W = \left(\frac{2W}{n\pi}\right)\theta_1$$

Substitution in Eq. (5.28) gives

$$\frac{\theta}{\theta_1} = \frac{4}{\pi}\sum_{n=1}^{\infty}\frac{1}{n}\exp\left(\frac{-n\pi y}{W}\right)\sin\left(\frac{n\pi x}{W}\right) \tag{5.29a}$$

$$= \frac{4}{\pi}\sum_{j=1}^{\infty}\frac{1}{(2j-1)}\exp\left[\frac{-(2j-1)\pi y}{W}\right]\sin\left[\frac{(2j-1)\pi x}{W}\right] \tag{5.29b}$$

In Eq. (5.29b), only odd-order terms appear.

Example 5.1 Consider the bar shown in Fig. 5.1 with $W = H$. The boundary conditions are: $\theta = 100°\mathrm{C}$ for the $y = H$ surface while for the remaining three sides of the bar, $\theta = 0°\mathrm{C}$. Determine the temperature along the centreline of the bar.

Solution

Substituting $H = W$, $x = W/2$ and $y = H/2 = W/2$ in the right-hand side of Eq. (5.23), we obtain

$$\frac{\theta_c}{\theta_b} = \frac{2}{\pi}\sum_{n=1}^{\infty}\frac{(-1)^{n+1}+1}{n\sinh(n\pi)}\times\sin\left(\frac{n\pi}{2}\right)\times\sinh\left(\frac{n\pi}{2}\right)$$

Retaining the first three odd-order terms (even-order terms are zero), we obtain

$$\frac{\theta_c}{\theta_b} = \frac{2}{\pi}\left[\frac{2}{\sinh(\pi)}\times\sin\left(\frac{\pi}{2}\right)\times\sinh\left(\frac{\pi}{2}\right)+\frac{2}{3\sinh(3\pi)}\times\sin\left(\frac{3\pi}{2}\right)\times\sinh\left(\frac{3\pi}{2}\right)\right.$$
$$\left.+\frac{2}{5\sinh(5\pi)}\times\sin\left(\frac{5\pi}{2}\right)\times\sinh\left(\frac{5\pi}{2}\right)\right]$$
$$= (2/\pi)[0.3985 - 0.00599 + 0.000155] = 0.24998^*$$

or

$$\theta_c = 24.998°\mathrm{C}.$$

*The series will converge to 0.25.

Example 5.2 If in the above example $H/W = 2$, determine the centreline temperature.

Solution

Substituting $H = 2\,W$, $x = W/2$ and $y = H/2 = W$ in the right-hand side of Eq. (5.23), we obtain

$$\frac{\theta_c}{\theta_b} = \frac{2}{\pi}\sum_{n=1}^{\infty}\frac{(-1)^{n+1}+1}{n\sinh(2n\pi)}\times\sin\left(\frac{n\pi}{2}\right)\times\sinh(n\pi)$$

Retaining the first two odd-order terms (even order terms are zero), we obtain

$$\frac{\theta_c}{\theta_b} = \frac{2}{\pi}\left[\frac{2}{\sinh(2\pi)} \times \sin\left(\frac{\pi}{2}\right) \times \sinh(\pi) + \frac{2}{3\sinh(6\pi)} \times \sin\left(\frac{3\pi}{2}\right) \times \sinh(3\pi)\right]$$

$$= 0.054919 - 0.000034 = 0.054885^*$$

or

$$\theta_c = 5.49°\text{C}.$$

*The series will converge to 0.0549.

Note: From the results of the above two examples, it can be seen that (i) θ_c/θ_b is a strong function of ratio H/W, and (ii) the convergence of the series improves as H/W increases. When $H/W \geq 2$, the first term practically equals the sum of the series.

Example 5.3 The material of the long square cross-section bar, shown in Fig. 5.4a, has homogeneous composition. The top side of the bar is held at t_1 = 200°C and the temperature of the remaining three sides is 100°C. Determine the temperature along the centreline of the bar.

Solution

The problem can be solved by the *method of superposition* explained below.

Consider the problem on the left in Fig. 5.5, which has uniform temperature on all sides. Its centreline temperature will be 100°C. The centreline temperature of all the sub-problems on the right will be the same by symmetry. The sub-problems on the right can be superimposed to yield the problem on the left.

Let θ_c is the temperature at the centreline of the sub-problems on the right. Then superposition will give

$$4\theta_c = 100°\text{C}$$

or

$$\theta_c = 25°\text{C}$$

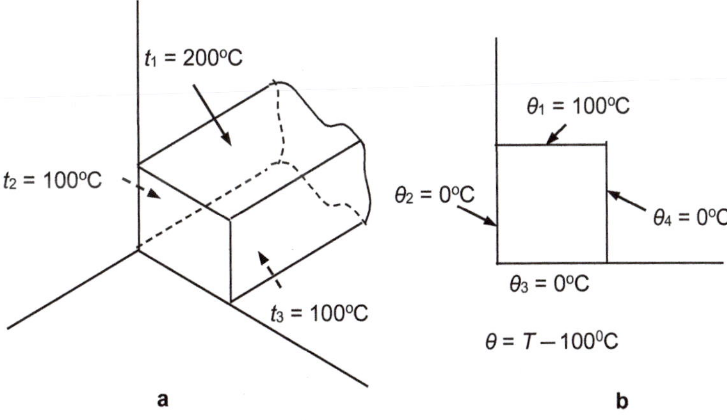

Fig. 5.4 Example 5.3

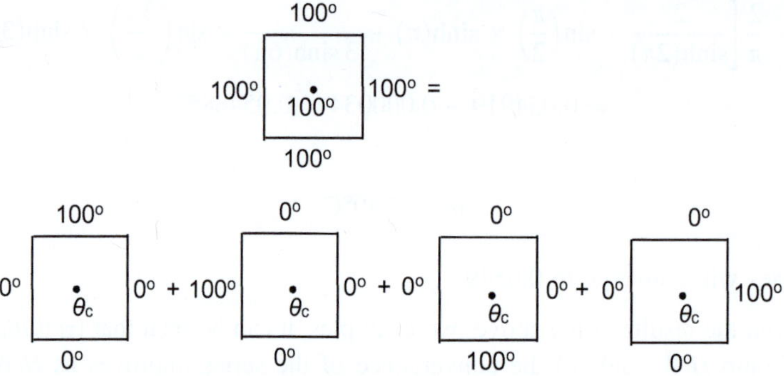

Fig. 5.5 Problem of Fig. 5.4 and its sub-problems

which is the desired centreline temperature of the original problem and hence $t_c = \theta_c + 100 = 125°\text{C}$.

Example 5.4 Using the method of superposition, determine the temperature at point (1, 1) in Fig. 5.6.

Solution

This problem has more than one non-homogeneous boundary condition. It can be reduced into a set of simpler problems with the geometry of the original problem and each having a non-homogeneous boundary condition. Then the solution of these simpler problems can be superimposed at the point (1, 1) to give the solution of the original problem as explained through Fig. 5.7.

Refer Example (5.3),

$\theta_{c1} = 75°\text{C}$ and $\theta_{c2} = 25°\text{C}$. The superposition gives

$\theta_c = 75°\text{C} + 25°\text{C} = 100°\text{C}$.

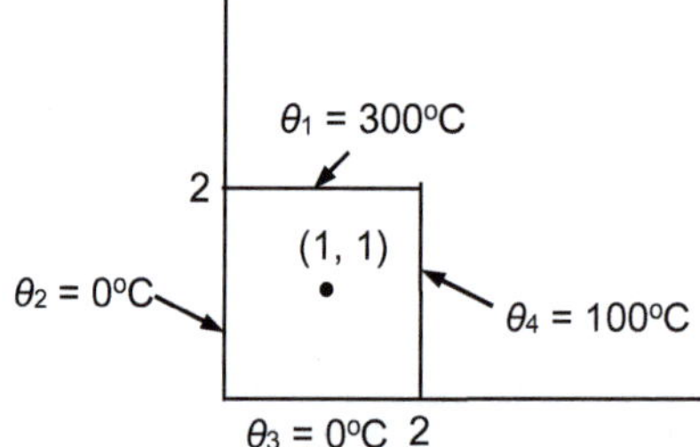

Fig. 5.6 Example 5.4

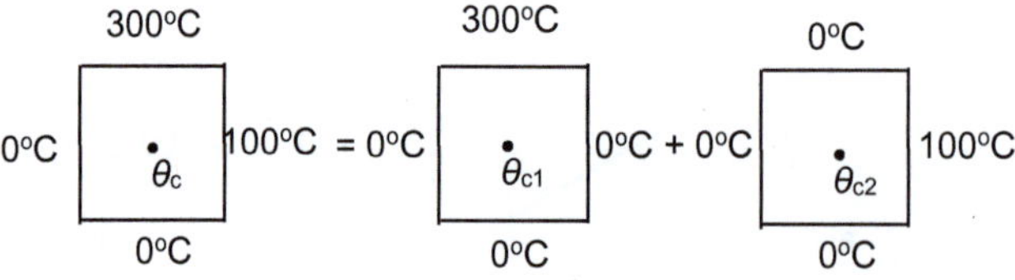

Fig. 5.7 Problem of Fig. 5.6 and its sub-problems

5.4 Mean Value Theorem

The results of the previous two examples can be extended to a homogeneous bar having a regular n-sided polygon, where side i is at temperature T_i ($i = 1, 2, 3, \ldots n$). Its centreline temperature will be the arithmetic mean of the temperatures around its perimeter, i.e.

$$T_c = (1/n)(T_1 + T_2 + \cdots T_n) \tag{5.30a}$$

When n tends to infinity, the polygon becomes a circle and

$$T_c = \frac{1}{2\pi}\int_0^{2\pi} T(\phi)\mathrm{d}\phi \tag{5.30b}$$

where $T(\phi)$ is the equation of angular distribution of temperature around the circumference of the circle. This result is termed the *mean value theorem.*

Example 5.5 Using the method of superposition, obtain the equation of temperature distribution for the long bar in Fig. 5.8.

Solution

As explained earlier, this problem has more than one non-homogeneous boundary condition. It can be reduced into a set of simpler problems with the geometry of the original problem and each having non-homogeneous boundary condition as shown in Fig. 5.9. Then the equations of the temperature distribution of these simpler problems can be superimposed to give the solution of the original problem as explained below.

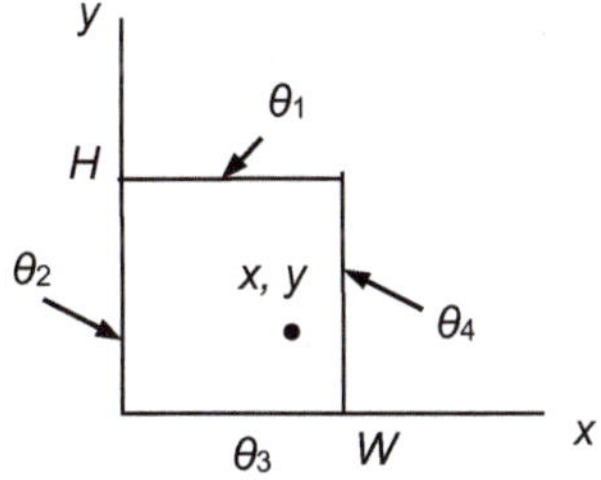

Fig. 5.8 A long bar with non-homogeneous thermal boundary conditions

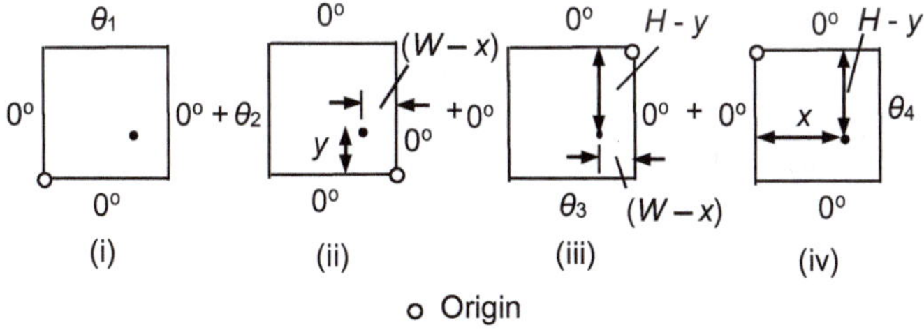

Fig. 5.9 A set of sub-problems for the problem in Fig. 5.8

Part (i) Referring Fig. 5.1, the temperature distribution equation is given by Eq. (5.23b) as

$$\frac{\theta(x,y)}{\theta_1} = \frac{4}{\pi}\sum_{j=1}^{\infty} \frac{\sinh\left[\frac{(2j-1)\pi y}{W}\right]}{(2j-1)\sinh\left[\frac{(2j-1)\pi H}{W}\right]} \times \sin\left[\frac{(2j-1)\pi x}{W}\right] \tag{i}$$

We can denote this function as $t(x, y/\theta_1, W, H)$. Then the functions for the parts (ii) to (iv) of Fig. 5.9 and the relevant temperature distribution equations can be written as below.

Part (ii) The function of this case is $t(y, W - x/\theta_2, H, W)$, see the new position of the origin in the Fig. 5.9b for the measurement of distances instead of distances x and y of Case (i), and

$$\frac{\theta(x,y)}{\theta_2} = \frac{4}{\pi}\sum_{j=1}^{\infty} \frac{\sinh\left[\frac{(2j-1)\pi (W-x)}{H}\right]}{(2j-1)\sinh\left[\frac{(2j-1)\pi W}{H}\right]} \times \sin\left[\frac{(2j-1)\pi y}{H}\right] \tag{ii}$$

Part (iii) The function of this case is $t(W–x, H–y/\theta_3, W, H)$ and

$$\frac{\theta(x,y)}{\theta_3} = \frac{4}{\pi}\sum_{j=1}^{\infty} \frac{\sinh\left[\frac{(2j-1)\pi (H-y)}{W}\right]}{(2j-1)\sinh\left[\frac{(2j-1)\pi H}{W}\right]} \times \sin\left[\frac{(2j-1)\pi (W-x)}{W}\right] \tag{iii}$$

Part (iv) The function of this case is $t(H–y, x/\theta_4, H, W)$ and

$$\frac{\theta(x,y)}{\theta_4} = \frac{4}{\pi}\sum_{j=1}^{\infty} \frac{\sinh\left[\frac{(2j-1)\pi x}{H}\right]}{(2j-1)\sinh\left[\frac{(2j-1)\pi W}{H}\right]} \times \sin\left[\frac{(2j-1)\pi (H-y)}{H}\right] \tag{iv}$$

Superposition, i.e., the summation of Eqs. (i)–(iv), gives the desired result.

Example 5.6 Does the result of Example 5.5 confirm the mean value theorem for the square?

Solution

Substitution of $W = H$ and $x = y = H/2$ in the result gives

$$\begin{aligned}
\theta\left(\frac{H}{2},\frac{H}{2}\right) &= (\theta_1+\theta_2+\theta_3+\theta_4) \times \frac{4}{\pi}\sum_{J=1}^{\infty} \frac{\sinh\left[\frac{(2j-1)\pi}{2}\right]}{(2j-1)\sinh(2j-1)\pi} \times \sin\left[\frac{(2j-1)\pi}{2}\right] \\
&= (\theta_1+\theta_2+\theta_3+\theta_4)\frac{4}{\pi}\left[\frac{\sinh\left(\frac{\pi}{2}\right)}{\sinh(\pi)} \times \sin\left(\frac{\pi}{2}\right) + \frac{\sinh\left(\frac{3\pi}{2}\right)}{3\sinh(3\pi)} \times \sin\left(\frac{3\pi}{2}\right) + \frac{\sinh\left(\frac{5\pi}{2}\right)}{5\sinh(5\pi)} \times \sin\left(\frac{5\pi}{2}\right) + \cdots\right] \\
&= (\theta_1+\theta_2+\theta_3+\theta_4) \times (4/\pi) \times (0.19926 - 0.002994 + 0.00007764) \\
&= 0.24999 \times (\theta_1+\theta_2+\theta_3+\theta_4), \text{ for first three terms} \\
&= (\theta_1+\theta_2+\theta_3+\theta_4)/4.001,
\end{aligned}$$

which confirms the mean value theorem for a square.

Example 5.7 Solve Example 5.4 using the equations developed in Example 5.5.

Solution

The applicable equation is, refer Example 5.5,

$$\theta(x,y) = \theta_1 \times \frac{4}{\pi} \times \sum_{j=1}^{\infty} \frac{\sinh\left[\frac{(2j-1)\pi y}{W}\right]}{(2j-1)\sinh\left[\frac{(2j-1)\pi H}{W}\right]} \times \sin\left[\frac{(2j-1)\pi x}{W}\right]$$
$$+\theta_4 \frac{4}{\pi} \sum_{J=1}^{\infty} \frac{\sinh\left[\frac{(2j-1)\pi x}{H}\right]}{(2j-1)\sinh\left[\frac{(2j-1)\pi W}{H}\right]} \times \sin\left[\frac{(2j-1)\pi(H-y)}{H}\right] \quad ((i))$$

Here $W = H = 2$ and $x = y = 1$. This gives

$$\theta(1,1) = (\theta_1 + \theta_4) \times \frac{4}{\pi} \times \sum_{j=1}^{\infty} \frac{\sinh\left[\frac{(2j-1)\pi}{2}\right]}{(2j-1)\sinh[(2j-1)\pi]} \times \sin\left[\frac{(2j-1)\pi}{2}\right]$$

For the first three terms,

$$\theta(1,1) = (300+100) \times \frac{4}{\pi} \times \left[\frac{\sinh\left(\frac{\pi}{2}\right)}{\sinh(\pi)} \times \sin\left(\frac{\pi}{2}\right) + \frac{\sinh\left(\frac{3\pi}{2}\right)}{3\sinh(3\pi)} \times \sin\left(\frac{3\pi}{2}\right) + \frac{\sinh\left(\frac{5\pi}{2}\right)}{5\sinh(5\pi)} \times \sin\left(\frac{5\pi}{2}\right)\right]$$
$$= 1600 \times \left(0.19927 - 0.002994 + 7.764 \times 10^{-5}\right)/\pi = 100°\text{C}.$$

and hence

$$t_c = 100 + 100 = 200°\text{C}.$$

Example 5.8 Determine temperature at point (3/2, ½) for the configuration of Example 5.4.

Solution

Eq. (i) of Example 5.7 applies. We get (refer Fig. 5.10)

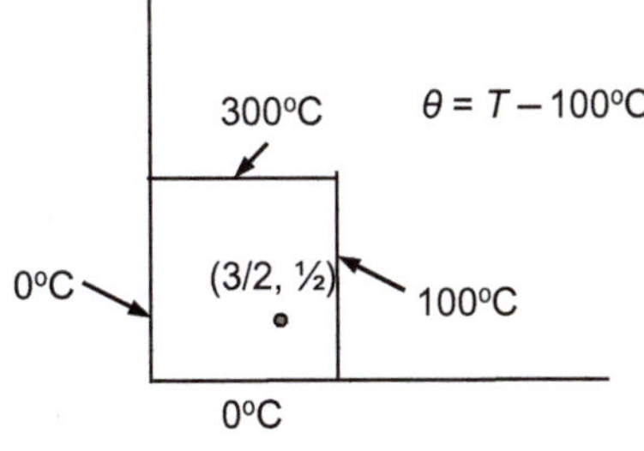

Fig. 5.10 Example 5.8

$$\theta\left(\frac{3}{2},\frac{1}{2}\right) = \theta_1 \times \frac{4}{\pi} \times \sum_{j=1}^{\infty} \frac{\sinh\left[\frac{(2j-1)\pi}{4}\right]}{(2j-1)\sinh[(2j-1)\pi]} \times \sin\left[\frac{(2j-1)3\pi}{4}\right]$$
$$+\,\theta_4 \frac{4}{\pi} \sum_{j=1}^{\infty} \frac{\sinh\left[\frac{(2j-1)3\pi}{4}\right]}{(2j-1)\sinh[(2j-1)\pi]} \times \sin\left[\frac{(2j-1)3\pi}{4}\right]$$

For the first three terms, we have

$$\theta\left(\frac{3}{2},\frac{1}{2}\right) = 300 \times \frac{4}{\pi} \times \left[\frac{\sin\left(\frac{3\pi}{4}\right)\sinh\left(\frac{\pi}{4}\right)}{\sinh \pi} + \frac{\sin\left(\frac{9\pi}{4}\right)\sinh\left(\frac{3\pi}{4}\right)}{3\sinh 3\pi} + \frac{\sin\left(\frac{15\pi}{4}\right)\sinh\left(\frac{5\pi}{4}\right)}{5\sinh 5\pi}\right]$$
$$+\,100 \times \frac{4}{\pi} \times \left[\frac{\sin\left(\frac{3\pi}{4}\right)\sinh\left(\frac{3\pi}{4}\right)}{\sinh \pi} + \frac{\sin\left(\frac{9\pi}{4}\right)\sinh\left(\frac{9\pi}{4}\right)}{3\sinh 3\pi} + \frac{\sin\left(\frac{15\pi}{4}\right)\sinh\left(\frac{15\pi}{4}\right)}{5\sinh 5\pi}\right] = 63.64^{\circ}\text{C}.$$

and

$$t\left(\frac{3}{2},\frac{1}{2}\right) = 63.64 + 100 = 163.64^{\circ}\text{C}.$$

Alternative Method

The problem can be reduced into a set of simple sub-problems as shown in Fig. 5.11 and then equations of Example 5.5 can be applied.

Example 5.9 Determine temperature at point (3/2, ½) for the long rectangular section bar in Fig. 5.12.

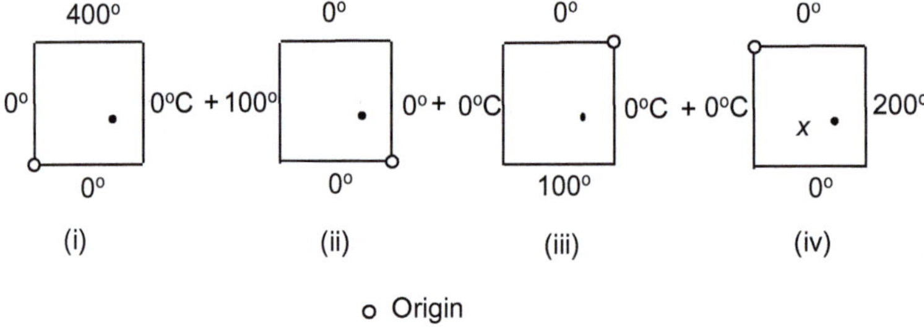

Fig. 5.11 A set of sub-problems of Fig. 5.10

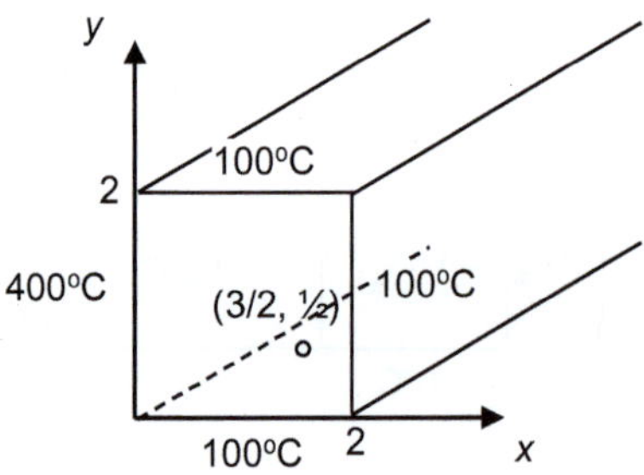

Fig. 5.12 Example 5.9

Solution

The temperature from Eq. (5.23b) is

$$\theta(x,y) = \theta \times \frac{4}{\pi} \times \sum_{j=1}^{\infty} \frac{1}{(2j-1)\sinh\left[\frac{(2j-1)\pi H}{W}\right]} \times \sin\left[\frac{(2j-1)\pi x}{W}\right] \times \sinh\left[\frac{(2j-1)\pi y}{W}\right] \quad \text{(i)}$$

The given problem can be transformed into the problem shown in Fig. 5.13 for which θ = 300°C, x = 3/2, y = ½, W = H = 2.

The calculated values of various terms in the above equation for j = 1–3 have been listed in the Table 5.1.

Substitution in Eq. (i) gives

$$\theta(x,y) = (4/\pi) \times 300 \times [0.8687 \times 0.707/11.5488 + 5.227 \times 0.707/(6195.8 \times 3) + 25.367 \times (-0.707)/(5 \times 3317812) = 20.4°\text{C}.$$

Hence,

$$t\left(\frac{3}{2},\frac{1}{2}\right) = 20.4 + 100 = 120.4°\text{C}.$$

Example 5.10 A rectangular section bar ($W \times H$) is very long in the z-direction. Its three sides are held at temperatures T_o while the fourth side (y = H) is held at a constant temperature T_b as shown in Fig. 5.14 Determine the equation of heat flow rate from the face y = 0.

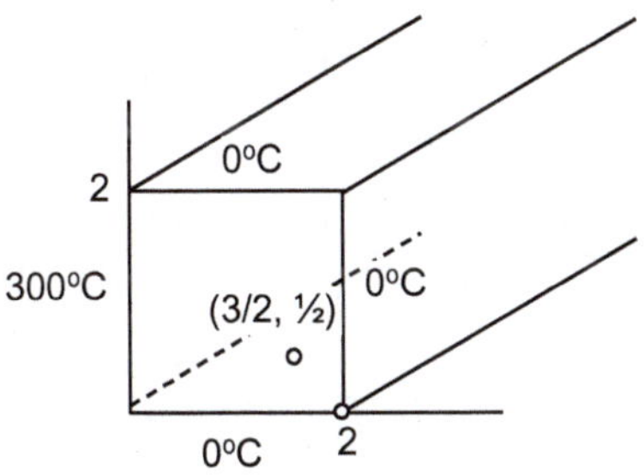

Fig. 5.13 Equivalent system of Fig. 5.12

Table 5.1 Example 5.9

	sinh[(2j – 1) πy/W]	sin[(2j – 1) πx/W]	sinh[(2j–1) πH/W]
j = 1	0.8687	0.707	11.5488
j = 2	5.227	0.707	6195.82
j = 3	25.367	–0.707	3317812

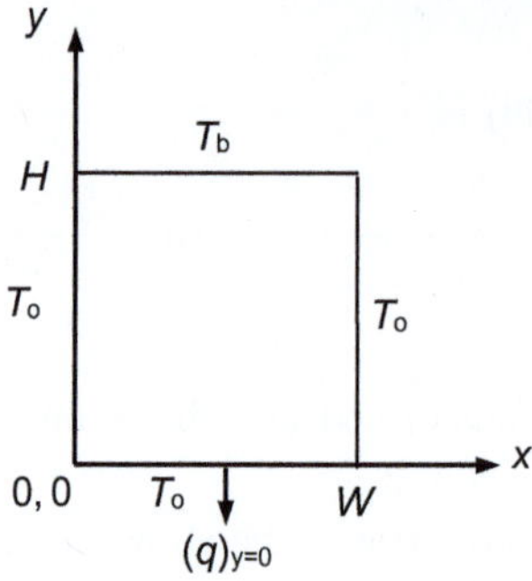

Fig. 5.14 Example 5.10

Solution

For the given bar, the heat flow from the face $y = 0$ for unit length in z-direction is

$$q_{y=0} = -\int_{x=0}^{x=W} \mathrm{d}q_y(x,0)\mathrm{d}x = -\int_{x=0}^{x=W} \left(-k\frac{\partial T}{\partial y}\bigg|_{y=0}\right)\mathrm{d}x = \int_{x=0}^{x=W} k\frac{\partial \theta}{\partial y}\bigg|_{y=0}\mathrm{d}x \quad \text{(i)}$$

where $\theta = T\text{–}T_o$.

From Eq. (5.23), we have

$$\theta = (T_b - T_o)\frac{2}{\pi}\sum_{n=1}^{\infty}\frac{(-1)^{n+1}+1}{n\sinh\left(\frac{n\pi H}{W}\right)} \times \sin\left(\frac{n\pi x}{W}\right) \times \sinh\left(\frac{n\pi y}{W}\right)$$

Hence,

$$\begin{aligned}\frac{\partial \theta}{\partial y}\bigg|_{y=0} &= (T_b - T_o)\frac{2}{\pi}\sum_{n=1}^{\infty}\frac{(-1)^{n+1}+1}{n\sinh\left(\frac{n\pi H}{W}\right)} \times \sin\left(\frac{n\pi x}{W}\right) \times \left\{\frac{\partial}{\partial y}\left[\sinh\left(\frac{n\pi y}{W}\right)\right]\right\}_{y=o} \\ &= (T_b - T_o)\frac{2}{\pi}\sum_{n=1}^{\infty}\frac{(-1)^{n+1}+1}{n\sinh\left(\frac{n\pi H}{W}\right)} \times \sin\left(\frac{n\pi x}{W}\right) \times \left\{\frac{n\pi}{W}\cosh\left(\frac{n\pi y}{W}\right)\right\}_{y=o} \\ &= (T_b - T_o)\frac{2}{\pi}\sum_{n=1}^{\infty}\frac{(-1)^{n+1}+1}{n\sinh\left(\frac{n\pi H}{W}\right)} \times \sin\left(\frac{n\pi x}{W}\right) \times \left(\frac{n\pi}{W}\right)\end{aligned}$$

Substitution in Eq. (i) gives

$$\begin{aligned}q_{y=0} &= k(T_b - T_o)\int_{x=0}^{x=W}\frac{2}{\pi}\sum_{n=1}^{\infty}\frac{(-1)^{n+1}+1}{n\sinh\left(\frac{n\pi H}{W}\right)} \times \sin\left(\frac{n\pi x}{W}\right) \times \left(\frac{n\pi}{W}\right)\mathrm{d}x \\ &= k(T_b - T_o)\frac{2}{\pi}\sum_{n=1}^{\infty}\frac{(-1)^{n+1}+1}{n\sinh\left(\frac{n\pi H}{W}\right)} \times \left(\frac{n\pi}{W}\right)\int_{x=0}^{x=W}\sin\left(\frac{n\pi x}{W}\right)\mathrm{d}x \\ &= k(T_b - T_o)\frac{2}{\pi}\sum_{n=1}^{\infty}\frac{(-1)^{n+1}+1}{n\sinh\left(\frac{n\pi H}{W}\right)} \times \left[-\cos\left(\frac{n\pi x}{W}\right)\right]_{x=0}^{x=W}\end{aligned}$$

or

$$q_{y=0} = k(T_b - T_o)\frac{2}{\pi}\sum_{n=1}^{\infty}\frac{(-1)^{n+1}+1}{n\sinh\left(\frac{n\pi H}{W}\right)}\times[1-\cos(n\pi)]$$

5.5 Graphical Analysis of Two Dimensional Steady-State Conduction: Thermal Flux Plotting and Shape Factor

Consider a quarter, Fig. 5.15b, of a chimney cross-section, Fig. 5.15a, where the inner surface is at some uniform temperature T_1 and the outer surface is at uniform temperature T_2. It is a two-dimensional system. Isotherms (constant temperature) and heat flow lines (adiabats) have been sketched therein following the rule that the heat flow lines cut the isotherms at right angles.

To determine the heat flow, let us consider an elemental curvilinear square[1] a-b-c-d in a heat flow channel. The heat flow through this channel is

$$\Delta q = -k(\Delta x \times 1)\frac{\Delta T}{\Delta y}$$

where the depth of the element (perpendicular to the plane of the paper) has been taken as unity.

If the curvilinear squares are drawn such that $\Delta x = \Delta y$ for all such elements, then

$$\Delta q = -k\Delta T.$$

Let there be N number of temperature increments (uniformly spaced isotherms) between inner and outer surfaces, then the drop in temperature across any two adjacent isotherms is

$$-\Delta T = (T_1 - T_2)/N.$$

and

$$\Delta q = k(T_1 - T_2)/N.$$

If there are M number of heat flow channels, the total heat flow through the quarter section will be the sum of heat flow through all channels, i.e.

$$q = M\Delta q = k\cdot(M/N)(T_1 - T_2) = kS(T_1 - T_2) \tag{5.31}$$

where $S = M/N$ is termed as *conduction shape factor*.

Shape factors for some common geometrical systems and thermal conditions of practical utility have been worked out by various researchers. They are summarized in Table 5.2.

From the above analysis, it can be seen that to determine the total heat flow through a given configuration, it is required to draw curvilinear square elements and count the number

[1]Curvilinear squares closely approximate the true squares only in the limit when their number approaches infinity. In finite form, they must be drawn such that average lengths of the opposite sides of the curvilinear squares are nearly equal and keeping internal angles close to 90°.

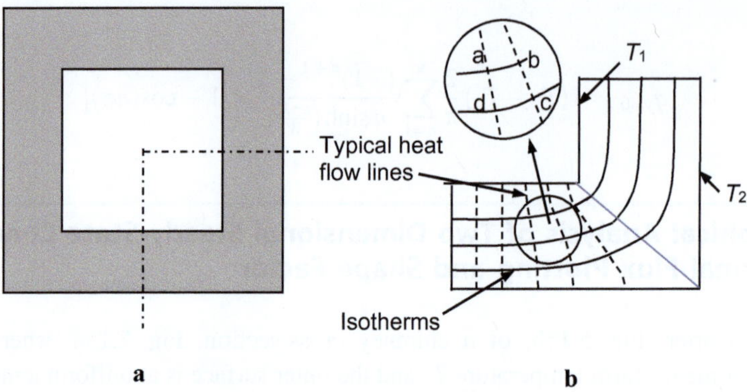

Fig. 5.15 **a** Chimney cross-section **b** Flux plotting in one quarter

Table 5.2 Conduction shape factors

S. No.	Physical system	Schematic	Shape factor	Restrictions
1	Plane wall of thickness δ	A, δ	A/δ	One-dimensional heat flow
2	Hollow cylinder, length L (radial heat flow)	L	$\dfrac{2\pi L}{\ln\left(\frac{R_2}{R_1}\right)}$	$L >> R$
3	Hollow sphere	R1, R2	$\dfrac{4\pi R_1 R_2}{R_2 - R_1}$	-
4	Isothermal sphere of radius R buried in semi-infinite medium having an isothermal surface	Isothermal, z, R	$\dfrac{4\pi R}{1 - \frac{R}{2Z}}$	$z > D/2$

(continued)

Table 5.2 (continued)

S. No.	Physical system	Schematic	Shape factor	Restrictions
5	Isothermal sphere of radius R buried in an infinite medium	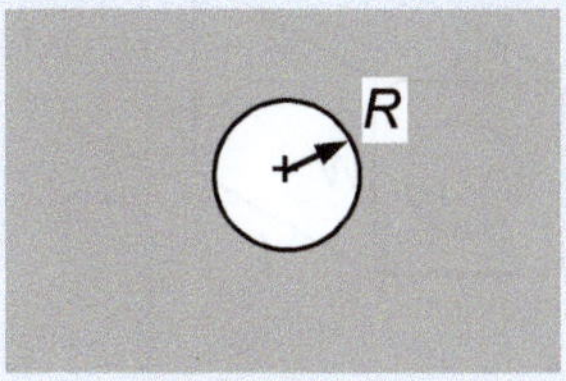	$4\pi R$ (This value can be obtained by putting $R_2 = \infty$ in equation of Case 3, or by putting $z = \infty$ in that of Case 4).	-
6	Isothermal sphere of radius R buried in semi-infinite medium with an insulated surface	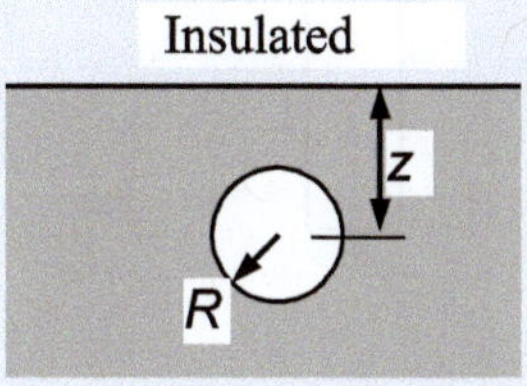	$\frac{4\pi R}{1 + \frac{R}{2Z}}$	-
7	Two isothermal spheres buried in infinite medium	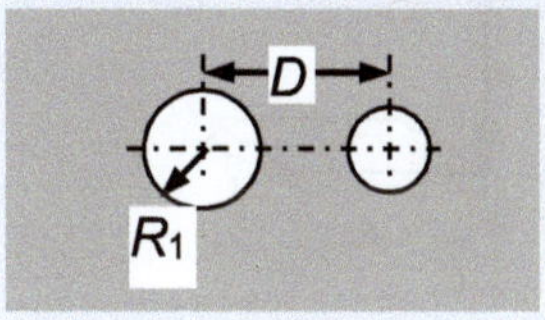	$\frac{4\pi D}{\frac{R_2}{R_1}\left(1 - \frac{C_1^4}{1 - C_2^2}\right) - 2C_2}$ where $C_1 = R_1/D$, $C_2 = R_2/D$ (R_2 refers to smaller sphere)	$D > 5R_{max}$
8	Isothermal horizontal cylinder of radius R buried in semi-infinite medium having isothermal surface		$\frac{2\pi L}{\cosh^{-1}\left(\frac{Z}{R}\right)}$	$L >> R$
			$\frac{2\pi L}{\ln\left(\frac{2Z}{R}\right)}$	$L >> R$ $Z > 3R$
			$\frac{2\pi L}{\ln\left(\frac{L}{R}\right)\left[\frac{\ln(L/2D)}{1 - \ln(L/R)}\right]}$	$Z >> R$ $L >> z$
9	Two isothermal cylinders buried in infinite medium	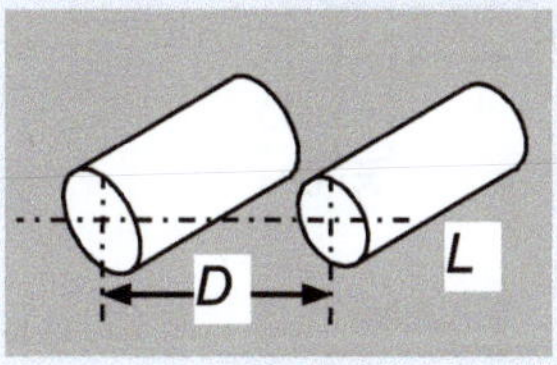	$\frac{2\pi L}{\cosh^{-1}\left(\frac{D^2 - R_1^2 - R_2^2}{2R_1R_2}\right)}$	$L >> R_1, R_2$ $L >> D$
10	Eccentric cylinders of equal lengths	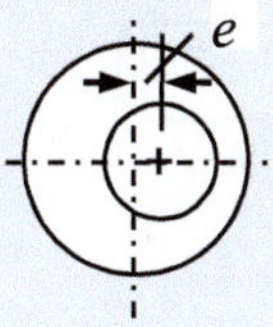	$\frac{2\pi L}{\cosh^{-1}\left(\frac{R_1^2 + R_2^2 - e^2}{2R_1R_2}\right)}$	$L >> 2R_2$ $R_2 > R_1$

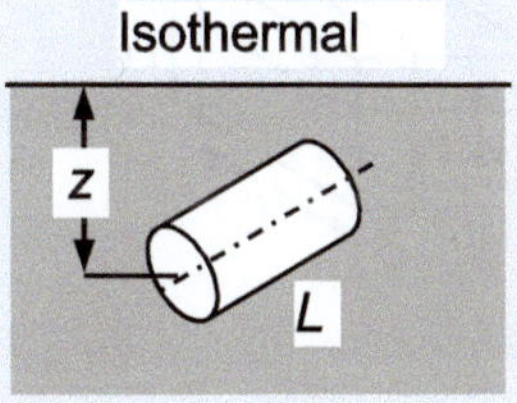

(continued)

Table 5.2 (continued)

S. No.	Physical system	Schematic	Shape factor	Restrictions
11	Cylinder of length L centred in a square duct of equal length	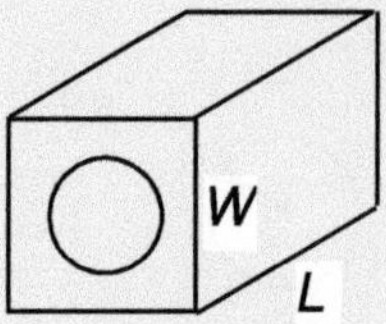	$\frac{2\pi L}{\ln\left(\frac{0.54W}{R}\right)}$	$L >> W$ $W > 2R$
12	Isothermal cylinder of radius R placed vertically in semi-infinite medium having isothermal surface	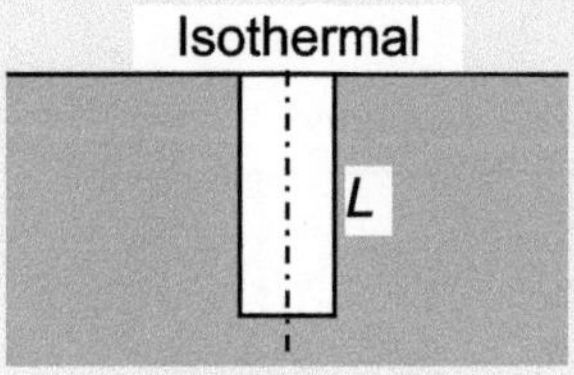	$\frac{2\pi L}{\ln\left(\frac{2L}{R}\right)}$	$L >> 2R$
13	Thin plate buried in semi-infinite medium	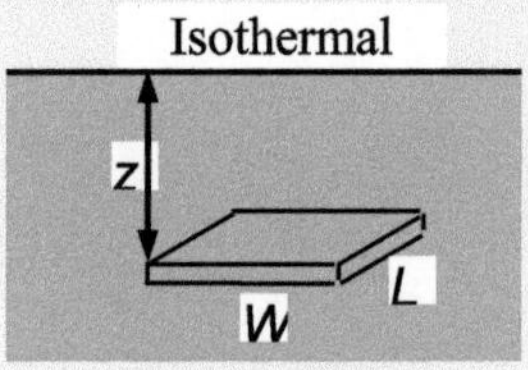	$\frac{\pi W}{\ln\left(\frac{4W}{L}\right)}$ $\frac{2\pi W}{\ln\left(\frac{4W}{L}\right)}$	$z = 0$ $W > L$ $z >> W$ $W > L$
14	An edge formed by the intersection of two plane walls; temperatures of inner and outer walls are T_1 and T_2, respectively	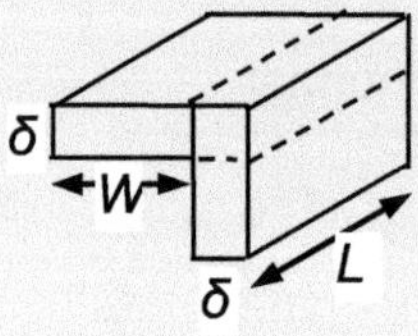	$0.54L$	$W > \delta /5$ $H > \delta /5$
15	Corner at the intersection of three plane walls, each of thickness δ; temperatures of inner and outer walls are T_1 and T_2, respectively	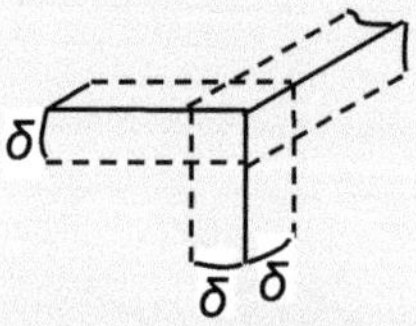	0.15δ	Inside dimension > $\delta/5$
16	Isothermal rectangular parallelepiped buried in semi-infinite medium having isothermal surface	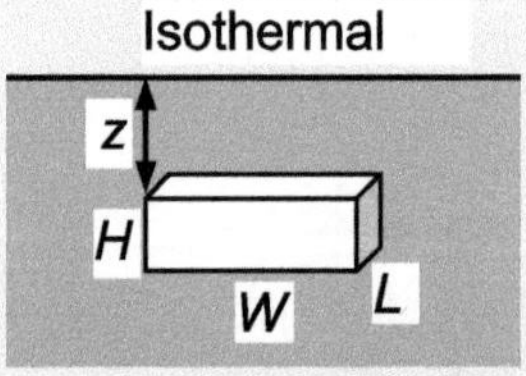	$1.685L \left[\log\left(1+\frac{Z}{W}\right)\right]^{-0.59} \times \left(\frac{Z}{H}\right)^{-0.078}$	$L >> W, H, z$

of heat flow channels and the temperature increments. However, the thermal flux plotting depends on the skill of the person plotting the lines. Electrical analogy, discussed in Sect. 5.6, may be used to sketch the isotherms and heat flow lines.

It is to note that this technique is now not much used but a rough sketching can sometimes help in making a fairly good estimate of the temperatures and the basic concepts of the plots can help in a quick check of the result from other methods. Hence, it must not be totally overlooked.

Example 5.11 Determine the heat transfer rate per unit length by flux plotting through the 300 mm thick insulation on a 200 mm outer diameter pipe. The temperature of the inner surface of the insulation is 150°C and outer surface is at 50°C. The thermal conductivity of the insulating material is 0.06 W/(m K). Check the result against the analytical solution.

Solution

Concentric circles of 200 and 800 mm are drawn accurately. The network of the curvilinear squares can be constructed by freehand plotting in a quarter only, refer Fig. 5.16. There are approximately 5.25 squares in each heat flow lane, and there are a total of 24 flow lanes (6 in each quarter), i.e. $N \approx 5.25$ and $M = 24$. This gives

$$S \approx \frac{24}{5.25} = 4.57$$

and

$$\frac{q}{L} = kS(T_1 - T_2) = 4.57 \times 0.06 \times (150 - 50) = 27.42 \text{ W/m}.$$

From the analytical solution,

$$\frac{q}{L} = 2\pi k \frac{T_1 - T_2}{\ln\left(\frac{R_2}{R_1}\right)} = 2\pi \times 0.06 \times \frac{150 - 50}{\ln\left(\frac{400}{100}\right)} = 27.19 \text{ W/m}.$$

The result of the freehand plotting is in a reasonable agreement with the analytical solution (about 0.8% higher than that of the analytical solution).

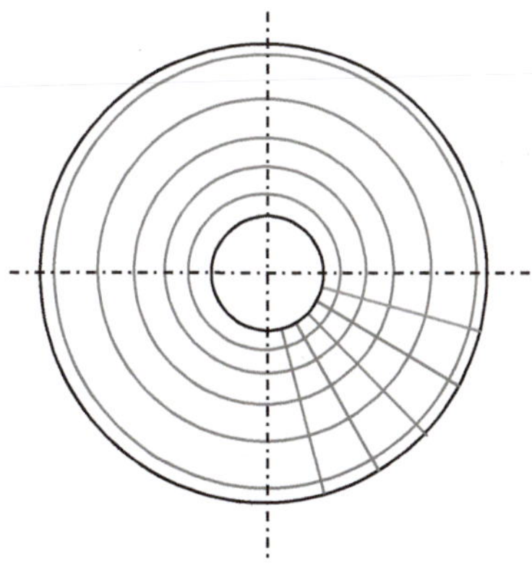

Fig. 5.16 Flux plotting Example 5.11

Example 5.12 A 25 mm OD pipe carries a hot fluid (pipe inner surface temperature = 80° C). It is placed centrally in a thin-walled square duct (85 × 85 mm^2). The space between the pipe and the duct is completely filled with glass wool insulation [k = 0.055 W/(m K)]. Calculate the heat loss for one meter length of the pipe if T_2 = 20°C.

Solution

From the flux plotting, refer Fig. 5.17,

$$N_{av} \approx (6.3 + 6.4 + 6.8 + 7.3)/4 = 6.7.$$

For the 4 flow lanes in one-eighth region of the configuration as shown in the figure, the shape factor

$$\frac{S}{L} \approx (8 \times 4)/6.7 = 4.78$$

and the heat loss per meter length of the pipe is

$$\frac{q}{L} = k\frac{S}{L}(T_1 - T_2) = 0.055 \times 4.78 \times 60 = 15.77 \text{ W/m}.$$

From Case 11, Table 5.2,
$\frac{S}{L} = \frac{2\pi}{\ln\left(\frac{0.54W}{R}\right)} = 4.83$, which is very close to that found from flux plotting.

Example 5.13 Determine the shape factor for the configuration shown in Fig. 5.18. If the configuration refers to a square-section duct ($a \times a$) covered with fireclay of thickness 1.5a to give an outer shape of 4a × 4a square, determine the heat loss for a = 20 mm, T_1 = 100°C and T_2 = 10°C. Thermal conductivity of fireclay is 1.05 W/(m K).

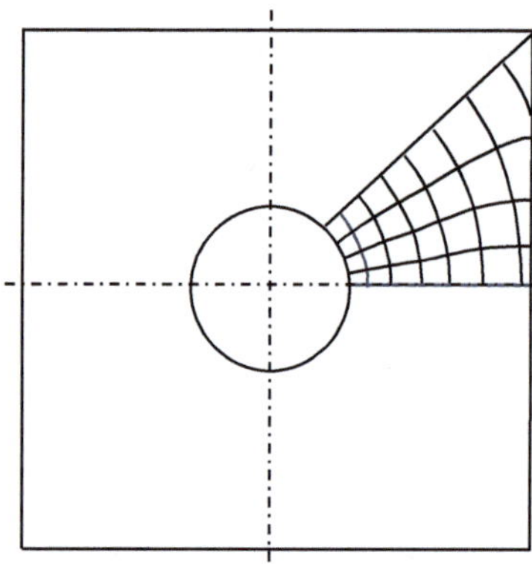

Fig. 5.17 Flux plotting Example 5.12

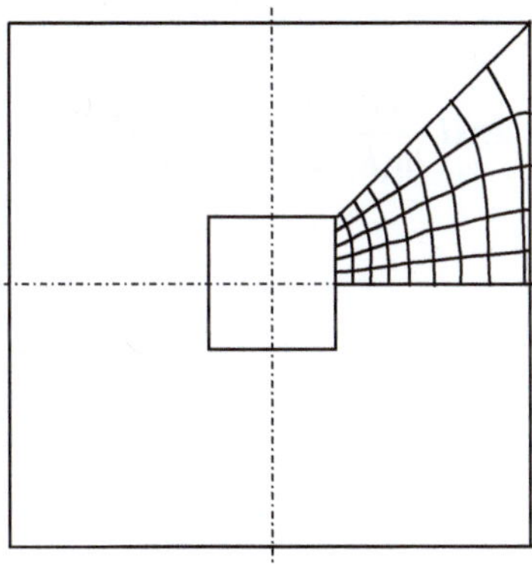

Fig. 5.18 Flux plotting Example 5.13

Solution

Flux plot on one-eight region of the configuration has been shown in Fig. 5.18. Freehand plotting for this region gives

$$N_{av} \approx (8.1 + 8.15 + 8.25 + 8.4 + 8.7)/5 = 8.32.$$

The shape factor,

$$\frac{S}{L} = \frac{M}{N_{av}} = (8 \times 5)/8.32 = 4.81.$$

Heat loss through the insulation, per meter length of the duct, is

$$\frac{q}{L} = k\left(\frac{S}{L}\right)(T_1 - T_2) = 1.05 \times 4.81 \times (100 - 10) = 454.54 \text{ W/m}.$$

Example 5.14 Determine the shape factor for the following systems.

(i) a plane wall,
(ii) a hollow cylinder of length L ($L >> r$),
(iii) a hollow sphere,
(iv) an isothermal sphere of radius R buried in an infinite medium.

Solution

(i) **A plane wall** (thickness δ)

$$q = kA(T_1 - T_2)/\delta = k(A/\delta).(T_1 - T_2)$$

Hence, $S = (A/\delta)$.

(ii) **A hollow cylinder**

$$q = 2\pi kL\frac{T_1 - T_2}{\ln(R_2/R_1)} = k\left[\frac{2\pi L}{\ln(R_2/R_1)}\right](T_1 - T_2)$$

Hence,

$$S = \frac{2\pi L}{\ln(R_2/R_1)}.$$

(iii) **A hollow sphere**

$$q = 4\pi kR_2R_1\frac{T_1 - T_2}{(R_2 - R_1)} = k\left[\frac{4\pi R_2R_1}{(R_2 - R_1)}\right](T_1 - T_2)$$

Hence,

$$S = \frac{4\pi R_2R_1}{(R_2 - R_1)} = 4\pi\left[\frac{1}{R_1} - \frac{1}{R_2}\right]^{-1}.$$

(iv) **An isothermal sphere of radius *r* buried in infinite medium**
For the infinite medium, put $R_2 = \infty$ and $R_1 = R$ in shape factor equation of case (iii). This gives

$$S = 4\pi R.$$

Example 5.15 A 1 m diameter sphere of radioactive material generates heat at the rate of 5 kW. It is buried 1 m in the earth [k = 1 W/(m K)], which has surface temperature of 30°C. What is the surface temperature of the sphere?

Solution

The shape factor from Table 5.2 (Case 4)

$$S = \frac{4\pi R}{\left(1 - \frac{R}{2z}\right)} = \frac{4\pi \times 0.5}{\left(1 - \frac{0.5}{2\times 1}\right)} = 8.378$$
$$q = kS(T_1 - T_2)$$

or

$$5000 = 1 \times 8.378 \times (T_s - 30)$$

or

$$T_s = 626.8^\circ\ \mathrm{C}.$$

Example 5.16 A 50 mm one-dimensional thin-walled pipe carries a cold fluid at 5°C. It is enclosed by a 200 mm pipe, whose surface is at 40°C. The space between the two pipes is filled with dry sawdust [k = 0.06 W/(m K)]. What is the heat transfer rate per m length of the pipe?

If the inner pipe is placed eccentric with 25 mm eccentricity, what will be the heat transfer rate?

Solution

(i) **Concentric cylinders**

$$\frac{q}{L} = kS(T_1 - T_2)$$
$$= k\frac{2\pi}{\ln\left(\frac{R_1}{R_2}\right)}(T_1 - T_2)$$
$$= 0.06 \times \frac{2\pi}{\ln\left(\frac{100}{25}\right)} \times (40 - 5) = 9.52\ \text{W/m}.$$

(ii) **Eccentric cylinders**

$$\frac{S}{L} = \frac{2\pi}{\cosh^{-1}\left(\frac{R_1^2 + R_2^2 - e^2}{2R_1R_2}\right)} = \frac{2\pi}{\cosh^{-1}\left(\frac{25^2 + 100^2 - 25^2}{2\times 25\times 100}\right)} = 4.77$$
$$\frac{q}{L} = k\frac{S}{L}(T_1 - T_2) = 0.06 \times 4.77 \times (40 - 5) = 10.02\ \text{W/m}.$$

Example 5.17 A 500 mm OD pipeline transports crude oil at 100°C. In order to reduce the heat loss from the pipeline, two schemes have been proposed: (i) one of the schemes is to bury it 3.0 m below the earth's surface [k_e = 1.0 W/(m K)], (ii) the second scheme is to cover it with 50 mm thick glass wool insulation [k_g = 0.05 W/(m K)]. If the earth's surface temperature is 0°C and the temperature of the outer surface of the insulation is also estimated to be 0°C, compare the schemes.

Solution

(i) Heat loss per unit length of the pipeline buried in the earth, Case 8, Table 5.2

$$\frac{S}{L} = \frac{2\pi}{\ln\left(\frac{2Z}{R}\right)} = \frac{2\pi}{\ln\left(\frac{2\times 3}{0.25}\right)} = 1.977$$
$$\frac{q}{L} = k_e\frac{S}{L}(T_1 - T_2) = 1.0 \times 1.977 \times (100 - 0) = 197.7\ \text{W/m}.$$

(ii) Insulation covering:

$$\frac{q}{L} = k_g \frac{S}{L}(T_1 - T_2)$$
$$= k_g \times \frac{2\pi}{\ln \frac{R_2}{R_1}} \times (T_1 - T_2)$$
$$= 0.05 \times \frac{2\pi}{\ln\left(\frac{300}{250}\right)} \times (100 - 0) = 172.3 \text{ W/m}.$$

Scheme 2 is better. However, the final choice of the scheme is based on economics.

Example 5.18 A small cubical furnace is of 0.75 m sides on the inside. It is made of 150 mm thick fireclay bricks [k = 1.3 W/(m K)]. If the inside and outside surfaces are at 500°C and 100°C, respectively, determine the rate of heat loss from the furnace through its walls.

Solution

(i) **Walls**: There are six 0.75 × 0.75 m^2 surface areas, which can be treated as one-dimensional conduction cases. Thus, the heat loss through these walls is

$$q_w = 6[kA(T_1 - T_2)]/\delta = 6 \times [1.3 \times (0.75 \times 0.75) \times (500 - 100)/0.15]/1000$$
$$= 11.70 \text{ kW}.$$

(ii) **Edges**: There are 12 edges, each 0.75 m long. These are two-dimensional problems for which the total shape factor from Table 5.2 is

$$S_e = 12 \times (0.54L) = 12 \times 0.54 \times 0.75 = 4.86.$$

(iii) **Corners**: The 8 three-dimensional corners have the total shape factor, refer Table 5.2,

$$S_c = 8 \times (0.15\delta) = 8 \times 0.15 \times 0.15 = 0.18.$$

The heat loss through the edges and corners is

$$q_e + q_c = k(S_e + S_c)(T_1 - T_2) = 1.3 \times (4.86 + 0.18) \times 400\,/1000 = 2.62 \text{ kW}.$$

Total heat loss = 11.70 + 2.62 = 14.32 kW approx.

Example 5.19 Combustion products at an average temperature of 800°C flow through a 20 m high chimney whose cross-section is shown in Fig. 5.19. The ambient temperature is 30°C. Determine the heat loss from the chimney if the inside and outside film coefficients are 100 and 10 W/(m^2 K), respectively.

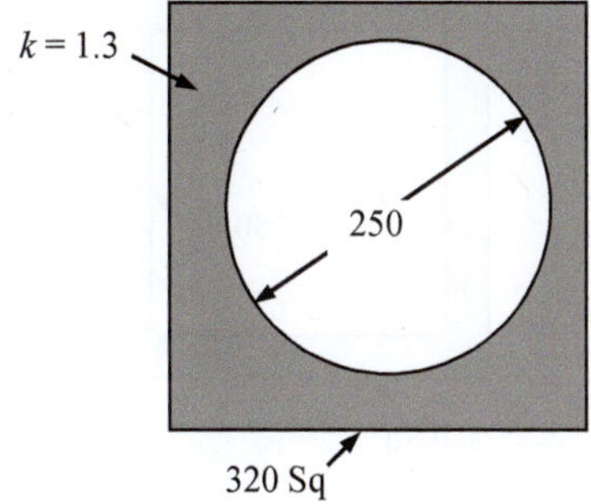

Fig. 5.19 Example 5.19

Solution

From Table 5.2, Case 11, conduction shape factor,

$$S = \frac{2\pi L}{\ln\left(\frac{0.54W}{R}\right)} = \frac{2\pi \times 20}{\ln\left(\frac{0.54 \times 0.32}{0.125}\right)} = 388.06.$$

The thermal resistances are:
Inside convective,

$$R_i = \frac{1}{2\pi R_1 L h_i} = \frac{1}{2\pi \times 0.125 \times 20 \times 100} = 6.366 \times 10^{-4}.$$

Conductive,

$$R_k = \frac{1}{kS} = \frac{1}{1.3 \times 388.06} = 1.982 \times 10^{-3}.$$

Outside convective,

$$R_i = \frac{1}{4WLh_o} = \frac{1}{4 \times 0.32 \times 20 \times 10} = 3.906 \times 10^{-3}.$$

The heat loss,

$$q = \frac{(T_1 - T_2)}{R_i + R_k + R_o} = \frac{(800 - 30)}{6.366 \times 10^{-4} + 1.982 \times 10^{-3} + 3.906 \times 10^{-3}} = 118\text{kW}.$$

Example 5.20 Repeat the above example for the 250 × 250 mm^2 flue passage, Fig. 5.20. Comment on the result.

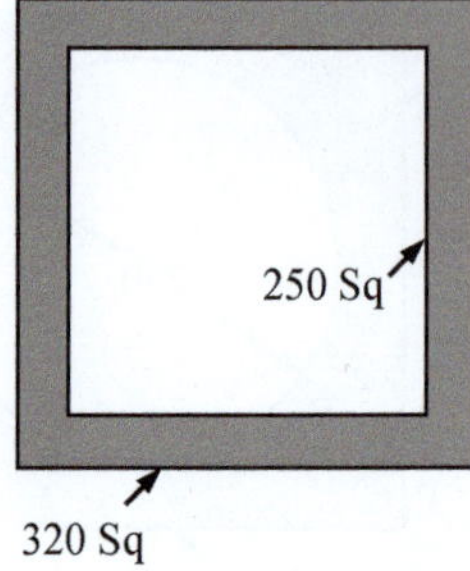

Fig. 5.20 Example 5.20

Solution

From Table 5.2,

$$S_{\text{walls}} = 4(A/\delta) = 4 \times (0.25 \times 20)/0.035 = 571.43.$$

Shape factor for edges:

$$S_e = 4 \times (0.54L) = 4 \times 0.54 \times 20 = 43.2.$$

Hence,

$$S_{\text{total}} = 571.43 + 43.2 = 614.63.$$

$$q = \frac{(T_1 - T_2)}{\frac{1}{h_i A_i} + \frac{1}{kS} + \frac{1}{h_o A_o}} = \frac{(800 - 30)}{\frac{1}{100 \times 1 \times 20} + \frac{1}{1.3 \times 614.63} + \frac{1}{10 \times 1.28 \times 20}} = 136.1 \text{ kW}.$$

The heat flow in this case is more because the mass of the wall resisting conduction heat flow is less.

Example 5.21 A radioactive brick (50 mm × 100 mm × 200 mm) at 800°C is buried 750 mm deep in the earth [k = 0.5 W/(m K)]. Determine the heat transfer rate when the surface temperature of the earth is 30°C.

Solution

From Table 5.2, Case 16,

$$S = 1.685L\left[\log\left(1 + \frac{z}{W}\right)\right]^{-0.59}\left(\frac{z}{H}\right)^{-0.078}$$

$$= 1.685 \times 0.2 \times \left[\log\left(1 + \frac{0.75}{0.1}\right)\right]^{-0.59}\left(\frac{0.75}{0.05}\right)^{-0.078} = 0.2849$$

Heat transfer rate,

$$q = kS\Delta T = 0.5 \times 0.2849 \times (800 - 30) = 109.7 \text{ W}.$$

Example 5.22 A furnace is 1.2 m by 0.9 m by 0.8 m. It is made of a brick wall of 150 mm thickness [for bricks, k = 1.3 W/(m K)]. Inside and outside temperatures are 500°C and 50°C, respectively. Determine the heat loss through the walls.

Solution

Inside dimensions are $0.9 \times 0.6 \times 0.5 \text{ m}^3$.

Plane walls: 2 nos. of $0.9 \times 0.6 \text{ m}^2$, 2 nos. of $0.9 \times 0.5 \text{ m}^2$ and 2 nos. of $0.6 \times 0.5 \text{ m}^2$. The total area is

$$A_{\text{wall}} = 2(0.9 \times 0.6 + 0.9 \times 0.5 + 0.6 \times 0.5) = 2.58 \text{ m}^2.$$

Heat transfer through the walls,

$$q_{\text{wall}} = kA\Delta T/\delta = 1.3 \times 2.58 \times (500 - 50)/0.15 = 10.06 \text{ kW}.$$

Shape factors

Edges: Four edges of 0.9 m, four of 0.6 m and four of 0.5 m.

Total shape factor for the edges,

$$S_e = 4 \times 0.54 \times (0.9 + 0.6 + 0.5) = 4.32\,.$$

Corners: There are 8 similar corners.

Total shape factor for these corners,

$$S_c = 8 \times 0.15 \times 0.15 = 0.18.$$

Heat loss through the edges and corners is

$$q_e + q_c = kS\Delta T = 1.3 \times (4.32 + 0.18) \times (500 - 50) = 2.63 \text{ kW}.$$

Total heat loss, q = 10.06 + 2.63 = 12.69 kW.

Example 5.23 In order to recover oil, steam is supplied through a 250 mm diameter and 250 m long pipe laid down into the earth [k = 1.0 W/(m K)]. Determine the heat loss from the pipe if the steam is at 150°C and the earth is at 20°C.

Solution

From Case 12 of Table 5.2, the shape factor,

$$S = \frac{2\pi L}{\ln\left(\frac{2L}{R}\right)} = \frac{2\pi \times 250}{\ln\left(\frac{2\times 250}{0.125}\right)} = 189.4$$

The heat loss,

$$q = kS\Delta T = 1.0 \times 189.4 \times (150 - 20) = 24.62 \text{ kW}.$$

Example 5.24 Determine the equation of shape factor for Case 8 of Table 5.2 from that of Case 9.

Solution

For Case 9,

$$\frac{S}{L} = \frac{2\pi}{\cosh^{-1}\left(\frac{D^2 - R_1^2 - R_2^2}{2R_1R_2}\right)}$$

Putting $D = z + R_1$, the denominator becomes

$$\cosh^{-1}\left(\frac{z^2 + 2R_1z - R_2^2}{2R_1R_2}\right) = \cosh^{-1}\left(\frac{z^2}{2R_1R_2} + \frac{z}{R_2} - \frac{R_2}{2R_1}\right)$$

When R_1 tends to infinity, the cylinder coverts into plane isothermal surface, i.e. to Case 8, and

$$\cosh^{-1}\left(\frac{z^2}{2R_1R_2} + \frac{z}{R_2} - \frac{R_2}{2R_1}\right) = \cosh^{-1}\left(\frac{z}{R}\right)$$

where $R = R_2$.

Thus

$$\frac{S}{L} = \frac{2\pi}{\cosh^{-1}\left(\frac{z}{R}\right)},$$

which is the desired result.

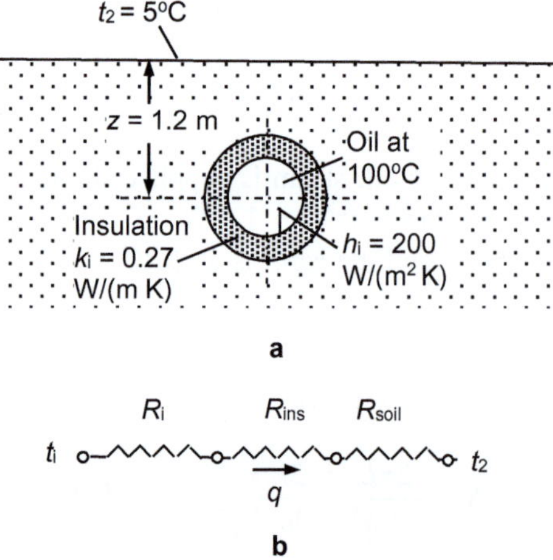

Fig. 5.21 Example 5.25

Example 5.25 A 500 mm diameter long pipe carrying oil at 100°C is covered with 250 mm thick insulation. It is buried in soil [k_{soil} = 0.27 W/(m K)] at a depth of 1.2 m as shown in Fig. 5.21a. Determine the rate of heat transfer from the oil if the convection heat transfer coefficient from oil to pipe surface is 500 W/(m^2 K). Soil surface temperature is 5°C. Neglect pipe wall resistance.

Solution

The electric network is shown in Fig, 5.21b. The total resistance to heat transfer is

$$R_t = R_i + R_{\text{ins}} + R_{\text{soil}}$$

For unit length of pipe,

$$\begin{aligned} R_i &= \frac{1}{h_i A_i} = \frac{1}{h_i \times \pi d_i L} \\ &= \frac{1}{200 \times \pi \times 0.5 \times 1} = 0.00318. \end{aligned}$$

Resistance of insulation,

$$\begin{aligned} R_i &= \frac{1}{2\pi k_i L} \ln\left(\frac{r_1}{r_i}\right) \\ &= \frac{1}{2\pi \times 0.06 \times 1} \ln\left(\frac{0.5}{0.25}\right) = 1.84. \end{aligned}$$

From Table 5.2, Case 8,

$$\begin{aligned} R_{soil} &= \frac{1}{Sk_{\text{soil}}} = \left[\frac{\cosh^{-1}(z/R)}{2\pi L}\right] \times \frac{1}{k_{\text{soil}}} \\ &= \frac{\cosh^{-1}(1.2/0.5)}{2\pi \times 1} \times \frac{1}{0.27} = 0.897. \end{aligned}$$

Hence, $R_t = R_i + R_{\text{ins}} + R_{\text{soil}} = 0.00318 + 1.84 + 0.897 = 2.74.$
Heat transfer rate,

$$q = \frac{t_i - t_2}{R_t} = \frac{100 - 5}{2.74} = 34.67\,\text{W/m length}.$$

5.6 Experimental Investigation of Conduction Process by Method of Analogy: Electro-Thermal Analogy

The electro-thermal analogy has been employed to sketch the curvilinear squares experimentally. It is based on the similarity of the mathematical equations of the steady-state heat conduction and electric conduction presented by the Laplace equations given below.

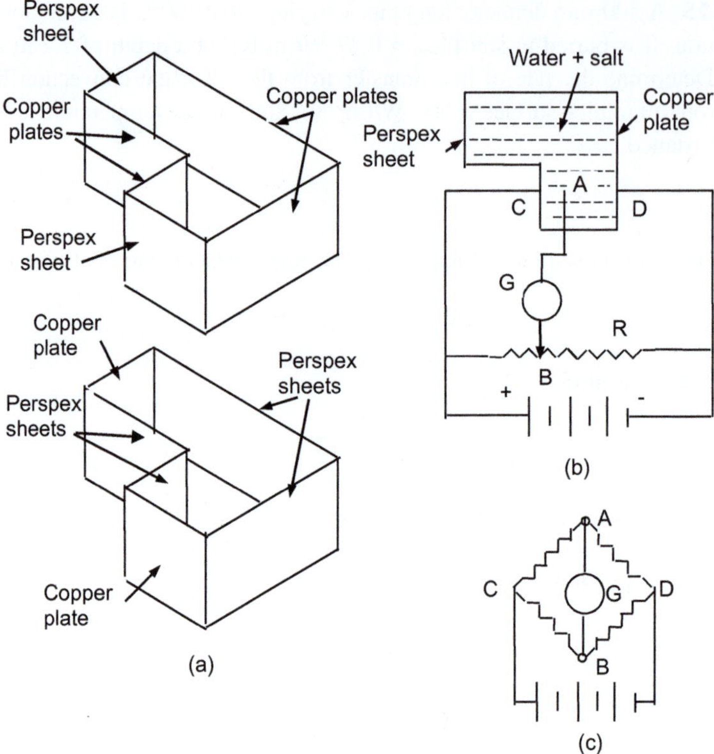

Fig. 5.22 Experimental setup for flux plotting

$$\frac{\partial^2 t}{\partial x^2} + \frac{\partial^2 t}{\partial y^2} = 0 \tag{5.1}$$

$$\frac{\partial^2 E}{\partial x^2} + \frac{\partial^2 E}{\partial y^2} = 0 \tag{5.32}$$

where E is the electrical potential.

Thus, the thermal phenomenon can be investigated by the study of the electrical one. The experimental investigation of the later is simpler than the thermal process. The method involves the construction of an electric analog and experimentally drawing the isothermal (equipotential) and heat flow (current flow) lines. An experimental setup based on this analogy is discussed below.

The setup, shown in Fig. 5.22a, uses two similar hollow boxes '1' and '2' made with two parallel sides of copper plates and the other two of an insulating material (say perspex sheet). The bottom of the boxes is also made of the insulating material. The boxes represent a scaled model of the chimney cross-section discussed earlier.

A metallic tracer at 'A' in Fig. 5.22b can trace the whole area of the box at a certain depth. The tracer is connected to the moving point 'B' of a rheostat 'R' through a galvanometer 'G'. Some electrical potential difference (say 12 V) is applied to the two ends of the rheostat and the same is applied across the two parallel copper plates of the hollow box. The box water, made conductive by adding some salt, is filled to such a level that the point of the metallic tracer touches the water surface. Thus the rheostat and the box form a Wheatstone bridge, see Fig. 5.22c. Tracer 'A' is moved, touching the water surface, such that the galvanometer 'G' always shows zero deflection. The path of the tracer is thus a uniform potential line and this path can be plotted on a paper with the help of a traversing mechanism. Setting the movable point of the rheostat at different positions, constant potential lines can be plotted for the whole of the section. Repeating the above experiment with the second box, similar lines can be plotted which will be found to be nearly perpendicular to the lines drawn with the first box.

The method leads to a better plot of the curvilinear squares than by freehand plotting and, thus, gives a better estimate of the value of the shape factor.

5.7 Numerical Solution Methods

The direct integration of the differential equations has been used to solve simple problems of two-dimensional, three-dimensional and transient heat conduction problems but success in solving complex problems, involving non-linear boundary conditions and temperature- or position-depending thermal properties, is limited. Such problems have been solved using numerical methods. Commonly used numerical methods are: *finite-difference* and *finite-element methods.*

The *finite-difference method* has been used extensively in solving heat conduction problems because of its simplicity in implementation. *The finite-element method* is being widely used to solve problems in structural mechanics. It requires much greater mathematical efforts and has been used for solving heat conduction problems involving complicated geometries.

Here only the basic principles of the finite-difference method are presented. The application of the method has been illustrated with some examples.

5.7.1 Finite-Difference Method

The first step in the finite-difference method is to discretize the spatial and time coordinates to form a mesh of nodes. Then by applying the energy balance to the volume elements surrounding the nodes, a set of linear algebraic equations is obtained. These equations, consisting of as many unknowns as the number of nodes in the mesh, are solved by metric inversion or by iteration. The accuracy of the finite-difference approximation increases with number of nodes hence computers are used to obtain finite-difference solutions. Standard computer programs are available for this purpose. A coarse mesh with few nodes has been used in this section (suitable for calculation by hand) to clarify the basic principles of the method.

Figure 5.23 shows section of a body which has been divided into a number of small volumes by using equal divisions in x- and y-directions. The nodal points (nodes) of these volumes are designated for a two-dimensional system as $(m + 1, n)$, (m, n), $(m-1, n)$, etc., where m and n indicate the x and y increments, respectively. The thermal properties of each volume are assumed to be concentrated at the nodal points. In order to establish temperatures at the nodal points, we use finite differences to approximate the differentials in the Laplace equation, Eq. (5.1). We shall consider the case of constant thermal conductivity of the material.

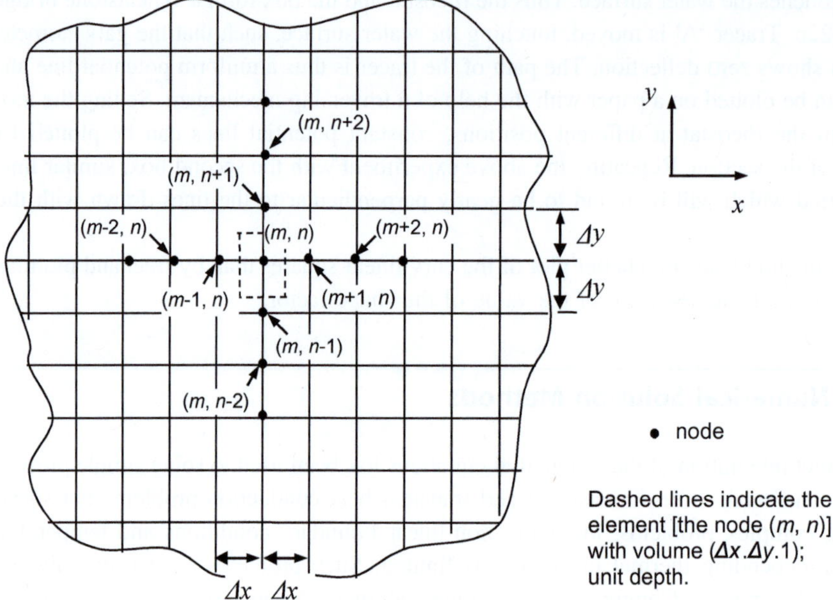

Fig. 5.23 Nomenclature for two-dimensional numerical analysis

The temperature gradients can be written as follows with reference to Fig. 5.23.

$$\left(\frac{\partial t}{\partial x}\right)_{m+1/2,n} \approx \frac{T_{m+1,n} - T_{m,n}}{\Delta x}$$

$$\left(\frac{\partial t}{\partial x}\right)_{m-1/2,n} \approx \frac{T_{m,n} - T_{m-1,n}}{\Delta x}$$

$$\left(\frac{\partial t}{\partial y}\right)_{m,n+1/2} \approx \frac{T_{m,n+1} - T_{m,n}}{\Delta y}$$

$$\left(\frac{\partial t}{\partial y}\right)_{m,n-1/2} \approx \frac{T_{m,n} - T_{m,n-1}}{\Delta y}$$

$$\left(\frac{\partial^2 t}{\partial x^2}\right)_{m,n} \approx \frac{\left(\frac{\partial t}{\partial x}\right)_{m+1/2,n} - \left(\frac{\partial t}{\partial x}\right)_{m-1/2,n}}{\Delta x} = \frac{T_{(m+1,n)} + T_{(m-1,n)} - 2T_{(m,n)}}{(\Delta x)^2}$$

$$\left(\frac{\partial^2 t}{\partial y^2}\right)_{m,n} \approx \frac{\left(\frac{\partial t}{\partial y}\right)_{m,n+1/2} - \left(\frac{\partial t}{\partial y}\right)_{m,n-1/2}}{\Delta y} = \frac{T_{(m,n+1)} + T_{(m,n-1)} - 2T_{(m,n)}}{(\Delta y)^2}$$

Substitution in Eq. (5.1) yields

$$\frac{T_{(m+1,n)} + T_{(m-1,n)} - 2T_{(m,n)}}{(\Delta x)^2} + \frac{T_{(m,n+1)} + T_{(m,n-1)} - 2T_{(m,n)}}{(\Delta y)^2} = 0$$

If $\Delta x = \Delta y$, then

$$T_{(m+1,n)} + T_{(m-1,n)} + T_{(m,n+1)} + T_{(m,n-1)} - 4T_{(m,n)} = 0 \tag{5.33}$$

This is known as *nodal* or *temperature equation.*

Equation (5.33) can also be obtained making use of the *method of heat balances*. Consider node (m, n) in Fig. 5.23. The rates of heat conduction from the nodes (m–1, n), (m +1, n), (m, n–1) and (m, n +1) to node (m, n) are given by the following equations.

$$k(\Delta y.1)\frac{T_{m-1,n} - T_{m,n}}{\Delta x}$$
$$k(\Delta y.1)\frac{T_{m+1,n} - T_{m,n}}{\Delta x}$$
$$k(\Delta x.1)\frac{T_{m,n-1} - T_{m,n}}{\Delta y}$$
$$k(\Delta x.1)\frac{T_{m,n+1} - T_{m,n}}{\Delta y}$$

In the steady state, the net heat transfer to the node (m, n) is zero. For $\Delta x = \Delta y$, it gives Eq. (5.33). The use of the method of heat balance is illustrated in the examples to follow.

Example 5.26 Write down the nodal equation for the node situated at the corner as shown in Fig. 5.24 with one side insulated and the adjacent side subjected to convective heat transfer.

Solution

The rate of heat conducted between 2 and 1, and 3 and 1, respectively, is

$$k\left(\frac{b\Delta y}{2}\right)\frac{(T_2 - T_1)}{\Delta x}$$

and

$$k\left(\frac{b\Delta x}{2}\right)\frac{(T_3 - T_1)}{\Delta y}$$

where b is the depth perpendicular to the plane of the paper.

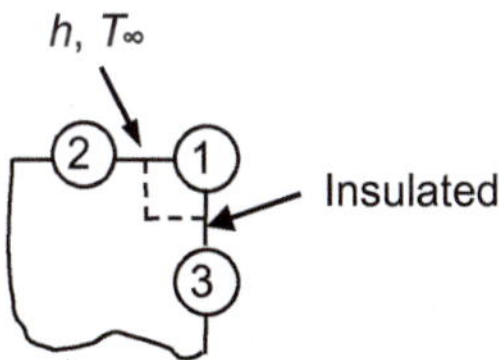

Fig. 5.24 Example 5.26

The rate of heat convected into node 1 is

$$\left(\frac{hb\Delta x}{2}\right)(T_\infty - T_1) = 0$$

In the steady state, the summation of energy transfer rate into node 1 is zero. Thus

$$k\left(\frac{b\Delta y}{2}\right)\frac{(T_2 - T_1)}{\Delta x} + k\left(\frac{b\Delta x}{2}\right)\frac{(T_3 - T_1)}{\Delta y} + \left(\frac{hb\Delta x}{2}\right)(T_\infty - T_1) = 0$$

Assuming a square grid ($\Delta x = \Delta y$), the equation simplifies to

$$T_2 + T_3 + \left(\frac{h\Delta x}{k}\right)T_\infty - \left[\frac{h\Delta x}{k} + 2\right]T_1 = 0$$

which is the desired result.

Example 5.27 Considering one-dimensional heat conduction through a thin fin, show that the nodal equation for the node shown in Fig. 5.25 can be expressed as

$$\left[\frac{hP(\Delta x)^2}{kA} + 2\right]T_m - \left[\frac{hP(\Delta x)^2}{kA}\right]T_\infty - (T_{m+1} + T_{m-1}) = 0$$

Solution

The rate of heat conducted between nodes (m–1) and (m), and between nodes (m + 1) and (m) are, respectively,

$$kA\frac{(T_{m-1} - T_m)}{\Delta x}$$

and

$$kA\frac{(T_{m+1} - T_m)}{\Delta x}$$

The rate at which the heat is convected from the fin surface is

$$hP\Delta x(T_\infty - T_m)$$

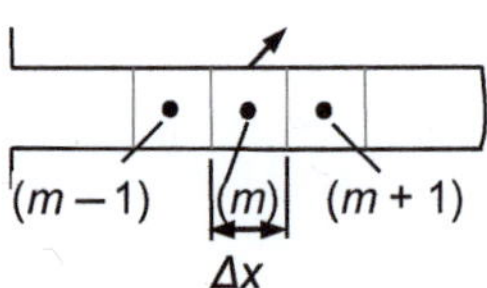

Fig. 5.25 Example 5.27

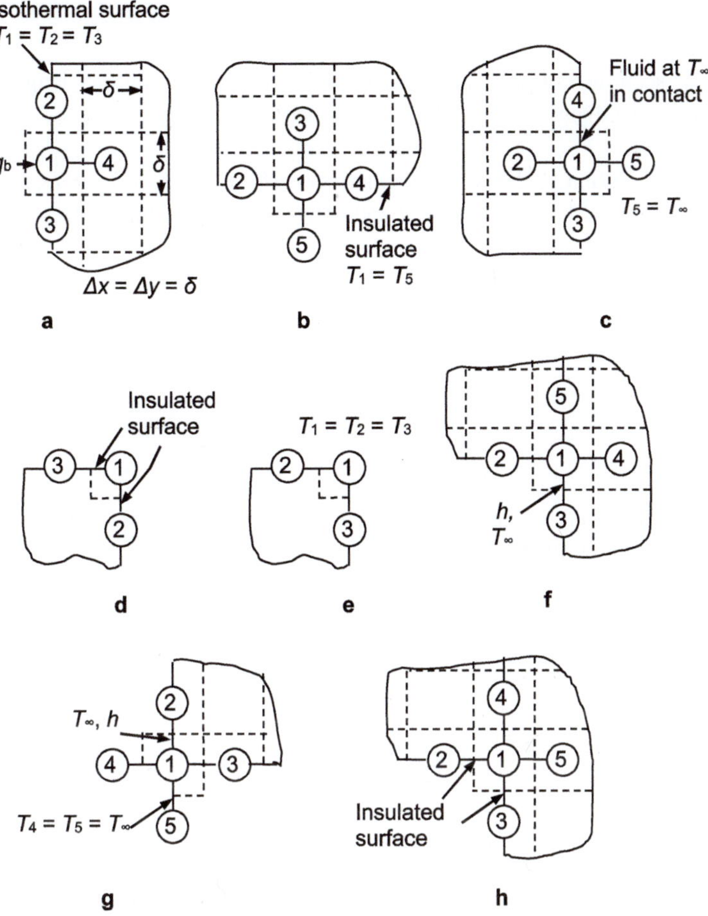

Fig. 5.26 Example 5.28

For the steady state, the summation of the heat transfer rates into the node (m) must be zero, and thus

$$kA\frac{(T_{m-1}-T_m)}{\Delta x}+kA\frac{(T_{m+1}-T_m)}{\Delta x}+hP\Delta x(T_\infty-T_m)=0$$

which on simplification gives

$$\left[\frac{hP(\Delta x)^2}{kA}+2\right]T_m-\left[\frac{hP(\Delta x)^2}{kA}\right]T_\infty-(T_{m+1}+T_{m-1})=0$$

Example 5.28 Write down nodal (temperature) equations for the cases shown in Fig. 5.26a–h referring to the numerical solution of two-dimensional steady-state conduction problems. Consider the dimension perpendicular to the plane of paper as b.

Solution

The solid cross-sections in Fig. 5.26a–h have been divided into arbitrary lengths and widths of Δx and Δy, respectively. Let $\Delta x = \Delta y = \delta$ as depicted in Fig. 5.26a.

Case (a): The nodal point '1' lies on an isothermal surface, i.e. $T_1 = T_2 = T_3$. Hence, there is no heat flow between nodal points 1 and 2 and between 1 and 3. Hence, the nodal equation is

$$q_b(b\delta) + k(b\delta)\frac{(T_4 - T_1)}{\delta} = 0$$

or

$$q_b\frac{\delta}{k} + (T_4 - T_1) = 0.$$

Case (b): The nodal point '1' lies on an insulated surface (as $T_1 = T_5$). Hence $q_{5\text{-}1} = 0$. The nodal equation can be written as

$$q_{2-1} + q_{3-1} + q_{4-1} = 0$$

or

$$k\left(\frac{b\delta}{2}\right)\frac{(T_2 - T_1)}{\delta} + k(b\delta)\frac{(T_3 - T_1)}{\delta} + k\left(\frac{b\delta}{2}\right)\frac{(T_4 - T_1)}{\delta} = 0$$

or

$$T_2 + T_4 + 2T_3 - 4T_1 = 0$$

It is to be noted that area of the heat flow between 2-1 and 4-1 is half of that between 3-1.

Case (c): The surface is in contact with a fluid at bulk temperature T_∞. The film coefficient is h at the surface. The nodal equation is

$$q_{5-1} + q_{4-1} + q_{3-1} + q_{2-1} = 0$$

or

$$h(b\delta)(T_\infty - T_1) + k\left(\frac{b\delta}{2}\right)\frac{(T_4 - T_1)}{\delta} + k\left(\frac{b\delta}{2}\right)\frac{(T_3 - T_1)}{\delta} + k(b\delta)\frac{(T_2 - T_1)}{\delta} = 0$$

or

$$\left(\frac{h\delta}{k}\right)(T_\infty - T_1) + \frac{(T_4 - T_1)}{2} + \frac{(T_3 - T_1)}{2} + (T_2 - T_1) = 0$$

or

$$\frac{(T_3 + T_4)}{2} + T_2 + \left(\frac{h\delta}{k}\right)T_\infty - T_1\left(2 + \frac{h\delta}{k}\right) = 0$$

Case (d): The nodal point '1' lies at the corner of two insulated surfaces. Hence, heat flows only between nodal points 2-1 and 3-1. The nodal equation for this case is

$$q_{2-1} + q_{3-1} = 0$$

or

$$k\left(\frac{b\delta}{2}\right)\frac{(T_2 - T_1)}{\delta} + k\left(\frac{b\delta}{2}\right)\frac{(T_3 - T_1)}{\delta} = 0$$

or

$$T_2 + T_3 - 2T_1 = 0$$

Case (e): The nodal point '1' lies at the corner of two isothermal surfaces. Hence, there is no heat flow to the nodal point 1 and there is no nodal equation for this case.

Case (f): The nodal point '1' lies in a corner where the surfaces are in contact with a fluid at bulk temperature T_∞. Hence, the nodal equation is

$$h(b\delta)(T_\infty - T_1) + k\left(\frac{b\delta}{2}\right)\frac{(T_2 - T_1)}{\delta} + k\left(\frac{b\delta}{2}\right)\frac{(T_3 - T_1)}{\delta} + k(b\delta)\frac{(T_4 - T_1)}{\delta} + k(b\delta)\frac{(T_5 - T_1)}{\delta} = 0$$

or

$$\left(\frac{h\delta}{k}\right)(T_\infty - T_1) + \frac{1}{2}(T_2 - T_1) + \frac{1}{2}(T_3 - T_1) + (T_4 - T_1) + (T_5 - T_1) = 0$$

or

$$T_2 + T_3 + 2T_4 + 2T_5 + \left(\frac{2h\delta}{k}\right)T_\infty - T_1\left(6 + \frac{2h\delta}{k}\right) = 0$$

Case (g): Following the procedure applied to the above cases, the nodal equation can be written as

$$q_{\infty-1} + q_{2-1} + q_{3-1} = 0$$

or

$$h(b\delta)(T_\infty - T_1) + k\left(\frac{b\delta}{2}\right)\frac{(T_2 - T_1)}{\delta} + k\left(\frac{b\delta}{2}\right)\frac{(T_3 - T_1)}{\delta} = 0$$

or

$$2\left(\frac{h\delta}{k}\right)(T_\infty - T_1) + (T_2 - T_1) + (T_3 - T_1) = 0$$

or

$$T_2 + T_3 + 2\left(\frac{h\delta}{k}\right)T_\infty - 2T_1\left(1 + \frac{h\delta}{k}\right) = 0$$

Case (h): For this case,

$$q_{2-1} + q_{3-1} + q_{4-1} + q_{5-1} = 0$$

or

$$k\left(\frac{b\delta}{2}\right)\frac{(T_2 - T_1)}{\delta} + k\left(\frac{b\delta}{2}\right)\frac{(T_3 - T_1)}{\delta} + k(b\delta)\frac{(T_4 - T_1)}{\delta} + k(b\delta)\frac{(T_5 - T_1)}{\delta} = 0$$

or

$$T_2 + T_3 + 2T_4 + 2T_5 - 6T_1 = 0$$

Example 5.29 Write the nodal equation for the interior node near a curved boundary as shown in Fig. 5.27.

Solution

The rate of heat conduction from 1, 2, (m, n–1) and (m–1, n) to the node (m, n) are

$$k\left(\frac{a\Delta x + \Delta x}{2}\right)\frac{T_1 - T_{m,n}}{b\Delta y}$$

$$k\left(\frac{b\Delta y + \Delta y}{2}\right)\frac{T_2 - T_{m,n}}{a\Delta x}$$

$$k\left(\frac{a\Delta x + \Delta x}{2}\right)\frac{T_{m,n-1} - T_{m,n}}{\Delta y}$$

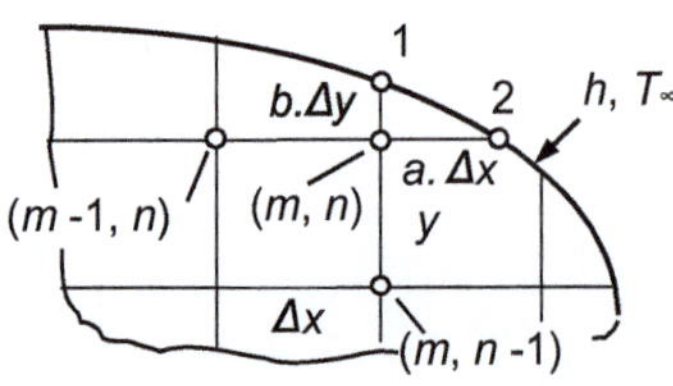

Fig. 5.27 Example 5.29

and

$$k\left(\frac{b\Delta y+\Delta y}{2}\right)\frac{T_{m-1,n}-T_{m,n}}{\Delta y}.$$

In the steady state, the summation of the conducted heat into the node (m, n) is zero. For $\Delta x = \Delta y$, this gives

$$k\left(\frac{a+1}{b}\right)(T_1-T_{m,n})+k\left(\frac{b+1}{a}\right)(T_2-T_{m,n})+k(a+1)(T_{m,n-1}-T_{m,n})+k(b+1)(T_{m-1,n}-T_{m,n})=0$$

or

$$\frac{(a+1)T_1}{b}+\frac{(b+1)T_2}{a}+(a+1)T_{m,n-1}+(b+1)T_{m-1,n} - \left[\frac{(a+1)}{b}+\frac{(b+1)}{a}+(a+1)+(b+1)\right]T_{m,n} = 0$$

Dividing by (a + 1) (b + 1), we have

$$\frac{T_1}{b(b+1)}+\frac{T_2}{a(a+1)}+\frac{T_{m,n-1}}{(b+1)}+\frac{T_{m-1,n}}{(a+1)} - \left[\frac{1}{b(b+1)}+\frac{1}{a(a+1)}+\frac{1}{(b+1)}+\frac{1}{(a+1)}\right]T_{m,n} = 0$$

Simplification gives the final equation as

$$\frac{T_1}{b(b+1)}+\frac{T_2}{a(a+1)}+\frac{T_{m,n-1}}{(b+1)}+\frac{T_{m-1,n}}{(a+1)}-\left(\frac{1}{a}+\frac{1}{b}\right)T_{m,n}=0,$$

which is the desired result.

Nodal equations for some configurations of interest are given in Table 5.3.

Example 5.30 Write down the nodal equation for a nodal point on the boundary between two materials as shown in Fig. 5.28.

Solution

The heat balance equation is, refer Fig. 5.28,

$$(q_{2-1})_A+(q_{2-1})_B+(q_{3-1})_A+(q_{3-1})_B+q_{4-1}+q_{5-1}=0$$

Using Fourier's law for conduction heat transfer, we obtain for unit depth

$$k_A\left(\frac{\Delta x}{2}.1\right)\frac{T_2-T_1}{\Delta y}+k_B\left(\frac{\Delta x}{2}.1\right)\frac{T_2-T_1}{\Delta y}+k_A\left(\frac{\Delta x}{2}.1\right)\frac{T_3-T_1}{\Delta y}+k_B\left(\frac{\Delta x}{2}.1\right)\frac{T_3-T_1}{\Delta y} + k_A(\Delta y.1)\frac{T_4-T_1}{\Delta x}+k_B(\Delta y.1)\frac{T_5-T_1}{\Delta x}=0$$

Table 5.3 Nodal equation for different configurations

Configuration	Nodal equation
Internal node	$T_{(m-1,n)} + T_{(m+1,n)} + T_{(m,n-1)} + T_{(m,n+1)} - 4T_{(m,n)} = 0$
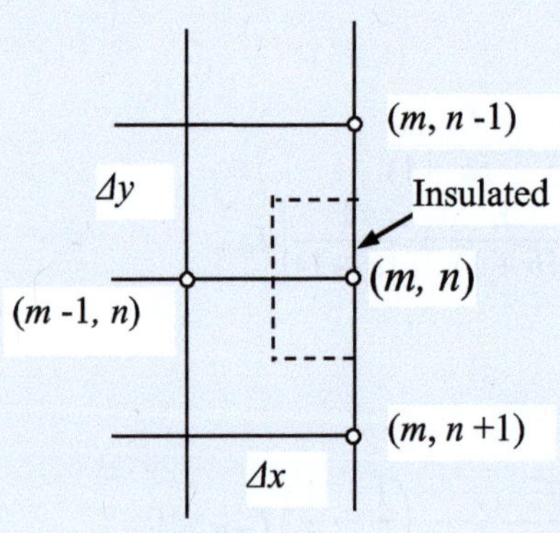 **Node on an insulated surface**	$T_{(m,n-1)} + T_{(m,n+1)} + 2T_{(m-1,n)} - 4T_{(m,n)} = 0$
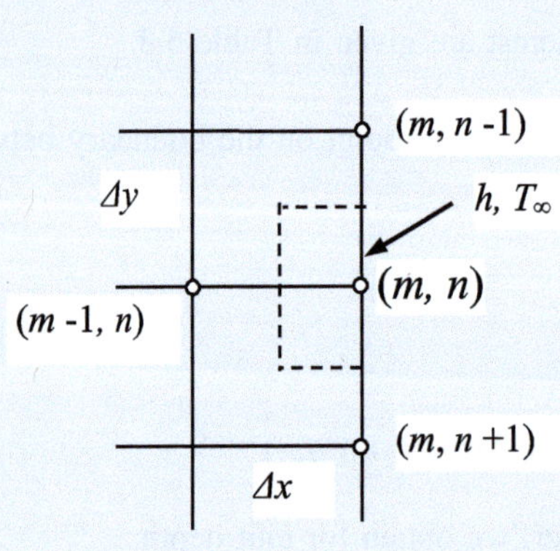 **Node on a surface in contact with fluid**	$(1/2)[T_{(m,n-1)} + T_{(m,n+1)} + 2T_{(m-1,n)}] + h(\Delta x/k)T_\infty - [2 + h(\Delta x/k)]T_{(m,n)} = 0$

(continued)

Table 5.3 (continued)

Configuration	Nodal equation
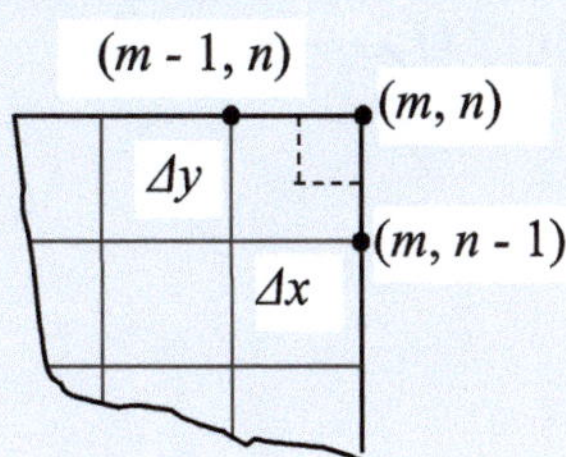 **Insulated external corner**	$T_{(m,n-1)} + T_{(m-1,n)} - 2T_{(m,n)} = 0$
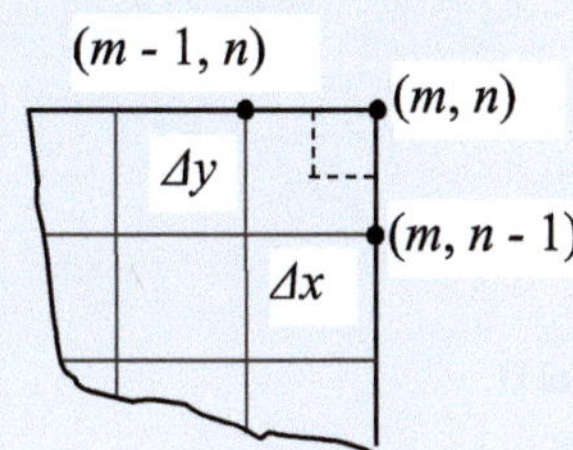 **External corner on isothermal surfaces**	The node lies on the corner of two isothermal surfaces hence there is no nodal equation
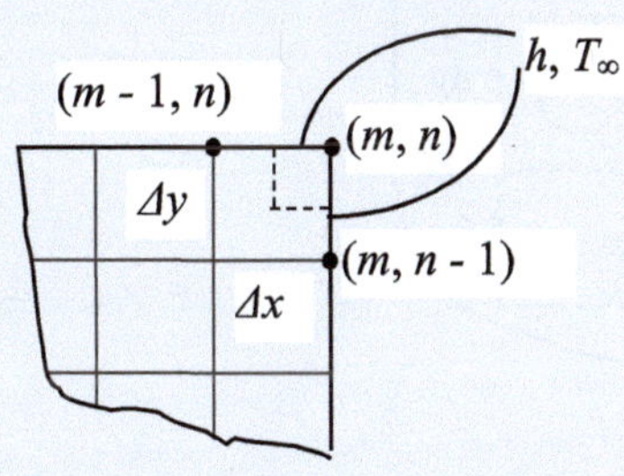 **External corner in contact with a fluid**	$T_{(m,n-1)} + T_{(m-1,n)} + 2h(\Delta x/k)T_\infty$ $-[2 + 2h(\Delta x/k)]T_{(m,n)} = 0$
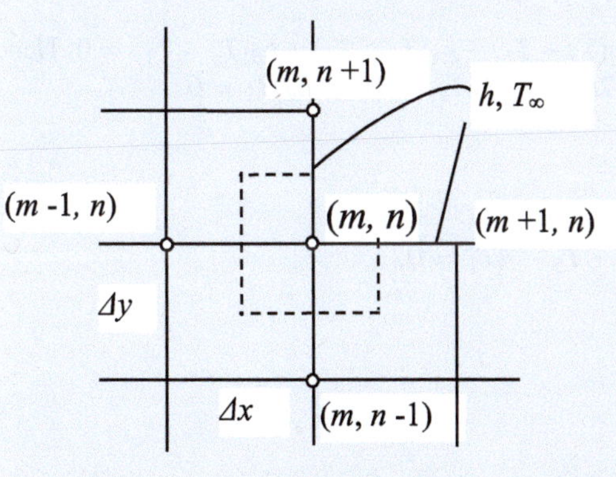 **Interior corner in contact with a fluid**	$T_{(m,n+1)} + T_{(m+1,n)} + 2(T_{(m-1,n)} + T_{(m,n-1)}) +$ $2h(\Delta x/k)T_\infty - [6 + 2h(\Delta x/k)]T_{(m,n)} = 0$

(continued)

Table 5.3 (continued)

Configuration	Nodal equation
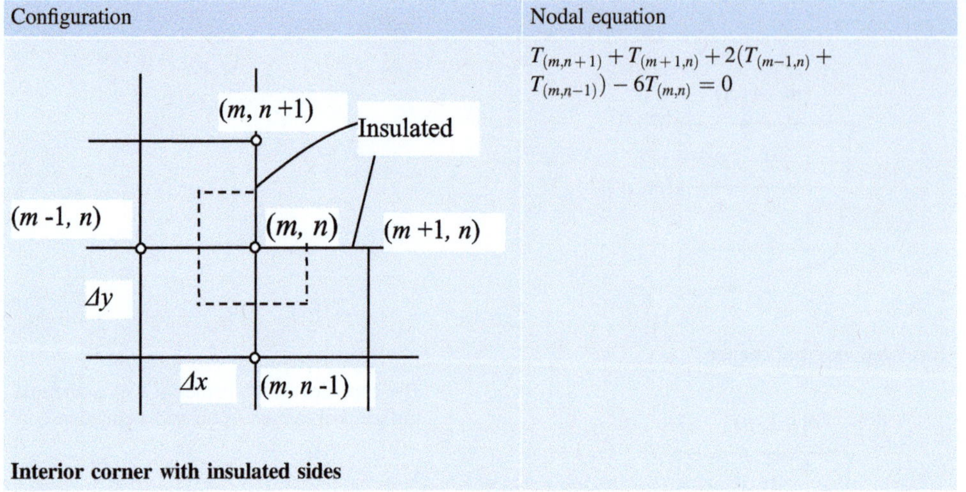 **Interior corner with insulated sides**	$T_{(m,n+1)} + T_{(m+1,n)} + 2(T_{(m-1,n)} + T_{(m,n-1)}) - 6T_{(m,n)} = 0$

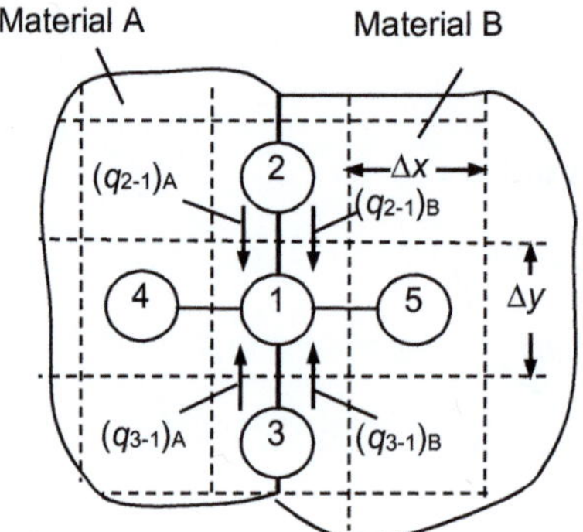

Fig. 5.28 Example 5.30

For $\Delta x = \Delta y$, we have

$$\frac{k_A}{2}(T_2 - T_1) + \frac{k_B}{2}(T_2 - T_1) + \frac{k_A}{2}(T_3 - T_1) + \frac{k_B}{2}(T_3 - T_1) + k_A(T_4 - T_1) + k_B(T_5 - T_1) = 0. \text{ Hence,}$$

$$(k_A + k_B)T_2 + (k_A + k_B)T_3 + 2k_AT_4 + 2k_BT_5 - 4(k_A + k_B)T_1 = 0$$

If $k_A = k_B$, we get

$$T_2 + T_3 + T_4 + T_5 - 4T_1 = 0,$$

which is the same as Eq. (5.33).

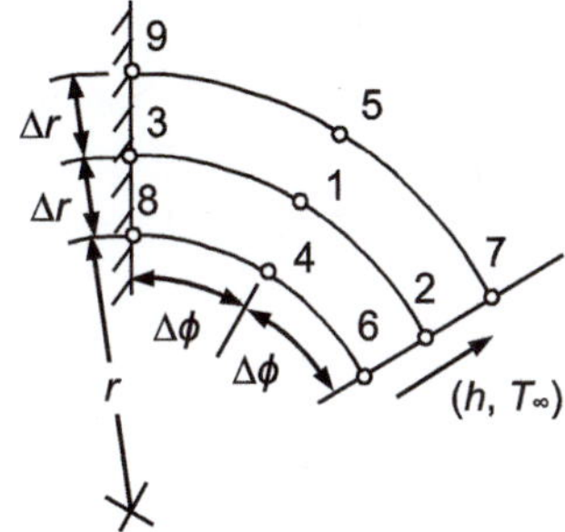

Fig. 5.29 Example 5.31

Example 5.31 For the cylindrical segment shown in Fig. 5.29, write down nodal equations for node 1 to 3. The radial and angular spacings of nodes are shown in the figure.

Solution

Node 1

The heat balance equation for the two-dimensional conduction system gives

$$q_{2-1}+q_{3-1}+q_{4-1}+q_{5-1}=0$$

$$k(\Delta r.1)\frac{T_2-T_1}{(r+\Delta r)\Delta\phi}+k(\Delta r.1)\frac{T_3-T_1}{(r+\Delta r)\Delta\phi}+k\left[\left(r+\frac{\Delta r}{2}\right)\Delta\phi.1\right]\frac{T_4-T_1}{\Delta r}$$
$$+k\left[\left(r+\frac{3\Delta r}{2}\right)\Delta\phi.1\right]\frac{T_5-T_1}{\Delta r}=0$$

Simplification gives the nodal equation as

$$\left(\frac{\Delta r}{\Delta\phi}\right)^2\frac{1}{(r+\Delta r)}(T_2+T_3)+\left(r+\frac{\Delta r}{2}\right)T_4+\left(r+\frac{3\Delta r}{2}\right)T_5$$
$$-2\left[\left(\frac{\Delta r}{\Delta\phi}\right)^2\frac{1}{(r+\Delta r)}+(r+\Delta r)\right]T_1$$
$$=0$$

Node 2

The heat balance equation for this two-dimensional system gives

$$k(\Delta r.1)\frac{T_1-T_2}{(r+\Delta r)\Delta\phi}+k\left[\frac{1}{2}\left(r+\frac{\Delta r}{2}\right)\Delta\phi.1\right]\frac{T_6-T_2}{\Delta r}+k\left[\frac{1}{2}\left(r+\frac{3\Delta r}{2}\right)\Delta\phi.1\right]\frac{T_7-T_2}{\Delta r}+h(\Delta r.1)(T_\infty-T_2)=0$$

Simplification gives the nodal equation as

$$2\left(\frac{\Delta r}{\Delta\phi}\right)^2\frac{1}{(r+\Delta r)}T_1+\left(r+\frac{\Delta r}{2}\right)T_6+\left(r+\frac{3\Delta r}{2}\right)T_7+2h\frac{(\Delta r)^2}{k\Delta\phi}T_\infty$$
$$-2\left[\left(\frac{\Delta r}{\Delta\phi}\right)^2\frac{1}{(r+\Delta r)}+(r+\Delta r)+2h\frac{(\Delta r)^2}{k\Delta\phi}\right]T_2$$
$$=0$$

Node 3

The heat balance equation for this two-dimensional system gives

$$k(\Delta r.1)\frac{T_1 - T_3}{(r+\Delta r)\Delta\phi} + k\left[\frac{1}{2}\left(r+\frac{\Delta r}{2}\right)\Delta\phi.1\right]\frac{T_8 - T_3}{\Delta r} + k\left[\frac{1}{2}\left(r+\frac{3\Delta r}{2}\right)\Delta\phi.1\right]\frac{T_9 - T_3}{\Delta r} = 0$$

Simplification gives the nodal equation as

$$2\left(\frac{\Delta r}{\Delta\phi}\right)^2\frac{1}{(r+\Delta r)}T_1 + \left(r+\frac{\Delta r}{2}\right)T_8 + \left(r+\frac{3\Delta r}{2}\right)T_9 - 2\left[\left(\frac{\Delta r}{\Delta\phi}\right)^2\frac{1}{(r+\Delta r)} + (r+\Delta r)\right]T_3 = 0$$

The above equation can also be obtained by putting $h = 0$ in the nodal equation of node 2.

5.7.2 Solution of Nodal Equations

The use of the numerical method involves writing nodal equations for each node within the body. The resulting set of equations is solved to determine the temperature at each node. It must be noted that smaller the elemental volumes, greater are the number of nodes and more closely the true temperature is approximated. If the number of nodes is very large, the simultaneous solution of the resulting equations might not be possible if carried out by hand. For a small number of nodes, the hand calculation can be carried out. The techniques, which have been used, are being explained in the sections to follow.

5.7.2.1 Relaxation Method

This method can be used to determine the steady-state temperatures at nodal points. The process in this method of solution is explained below with the help of an example.

Let the nodal equations are

$$\text{Node 1}: \quad 300 + 400 + T_2 + T_4 - 4T_1 = 0 \tag{a}$$

$$\text{Node 2}: \quad 400 + 100 + T_1 + T_3 - 4T_2 = 0 \tag{b}$$

$$\text{Node 3}: \quad 100 + 200 + T_2 + T_4 - 4T_3 = 0 \tag{c}$$

$$\text{Node 4}: \quad 200 + 300 + T_1 + T_3 - 4T_4 = 0. \tag{d}$$

The relaxation method proceeds as follows.

1. It starts with an initial guess of unknown temperatures. Since the initially assumed values will usually be in error, the right-hand side of the above nodal equations will differ from zero. So replace the zeros in Eqs. (a)–(d) with R_1, R_2, R_3 and R_4, respectively, where R_1, R_2, etc. are known as *residuals*.

$$700 + T_2 + T_4 - 4T_1 = R_1 \tag{e}$$

$$500 + T_1 + T_3 - 4T_2 = R_2 \quad \text{(f)}$$

$$300 + T_2 + T_4 - 4T_3 = R_3 \quad \text{(g)}$$

$$500 + T_1 + T_3 - 4T_4 = R_4. \quad \text{(h)}$$

2. Next set up a table known as *unit change table*. This table shows the effect of 1° change of temperatures at a node on the residuals. Also, enter the effect on all residuals of block (overall) unit change (see Table 5.4).
3. Set up the *relaxation table*. Calculate the initial residuals from the initial gauss of the temperatures T_1 through T_4 using the residual equations (e)–(h) as illustrated in Table 5.5.

The initially guessed temperature values are now changed such that the residuals are reduced to zero. This procedure should begin with relaxing the largest initial residual. If all residuals are of the same sign, make a block change. Note that a good initial guess of temperatures helps to minimize the effort required. In the left-hand column of the table are entered the change in temperatures from the initially assumed values.

In the present example, the initial residuals are of the same sign, hence a block change has been made. Next relax the largest residuals, which are R_1 and R_2 taking the help of unit change table. First, choose any one, say R_1 and over-relax it but not too much. The change in T_1 has increased R_2. Now, relax R_2. Change in ΔT_2 in the table has relaxed all the residuals. Check the result, so obtained, by substituting the temperature values so obtained in Eqs. (e)–(h).

Example 5.32 Derive an expression for the residual of the node at the end of a fin (refer Fig. 5.30). Assume one-dimensional heat conduction.

Table 5.4 Unit change

	ΔR_1	ΔR_2	ΔR_3	ΔR_4
$\Delta T_1 = +1$	–4	+1	0	+1
$\Delta T_2 = +1$	+1	–4	+1	0
$\Delta T_3 = +1$	0	+1	–4	+1
$\Delta T_4 = +1$	+1	0	+1	–4
Block change = + 1	–2	–2	–2	–2

Table 5.5 Relaxation table

	T_1	R_1	T_2	R_2	T_3	R_3	T_4	R_4
Initial guess	250	125	200	125	175	25	225	25
Block change = + 25	275	75	225	75	200	–25	250	–25
$\Delta T_1 = +25$	300	–25	225	100	200	–25	250	0
$\Delta T_2 = +25$	300	0	250	0	200	0	250	0
Check		0		0		0		0
Solution	300		250		200		250	

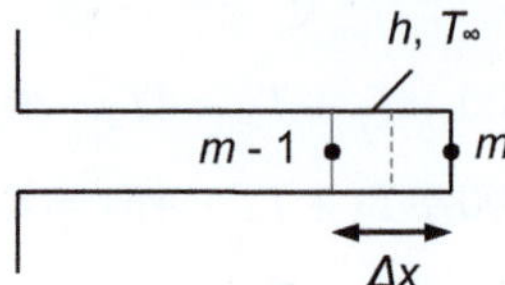

Fig. 5.30 Example 5.32

Solution

The rate of the heat conducted between nodes (*m*–1) and *m* is

$$kA_c\frac{(T_{m-1}-T_m)}{\Delta x} \tag{i}$$

The rate of heat convected is the sum of the heat convected from the end and the lateral surface of the volume:

$$hA_c(T_\infty - T_m)+h\left(\frac{P\Delta x}{2}\right)(T_\infty - T_m) \tag{ii}$$

In the steady state, summation of Eqs. (i) and (ii) must be zero, i.e.

$$kA_c\frac{(T_{m-1}-T_m)}{\Delta x}+hA_c(T_\infty - T_m)+h\left(\frac{P\Delta x}{2}\right)(T_\infty - T_m)$$

Rearranging the terms, we get the nodal equation as

$$T_m\left[1+h\left(\frac{\Delta x}{k}\right)+\frac{hP(\Delta x)^2}{2kA_c}\right]-T_\infty\left[\frac{h\Delta x}{k}+\frac{hP(\Delta x)^2}{2kA_c}\right]-T_{m-1}=0$$

The residual equation can be written as

$$T_m\left[1+h\left(\frac{\Delta x}{k}\right)+\frac{hP(\Delta x)^2}{2kA_c}\right]-T_\infty\left[\frac{h\Delta x}{k}+\frac{hP(\Delta x)^2}{2kA_c}\right]-T_{m-1}=R_m$$

Example 5.33 Write down the residual equation for the insulated corner section shown in Fig. 5.31.

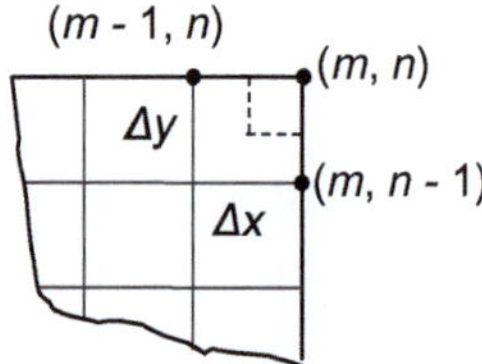

Fig. 5.31 Example 5.33

Solution

The heat flow rates from the nodes (m, n–1) and (m–1, n) to the node (m, n) are

$$k\left(\frac{\Delta x.1}{2}\right)\frac{(T_{m,n-1}-T_{m,n})}{\Delta y} \tag{1}$$

$$k\left(\frac{\Delta y.1}{2}\right)\frac{(T_{m-1,n}-T_{m,n})}{\Delta x} \tag{2}$$

The nodal equation is the summation of the above equations and for $\Delta x = \Delta y$, the simplification gives

$$T_{m,n-1}+T_{m-1,n}-2T_{m,n}=0$$

Hence, the residual equation is

$$T_{m,n-1}+T_{m-1,n}-2T_{m,n}=R_{m,n}$$

Example 5.34 A square-section fin (5 × 5 mm^2), see Fig. 5.32, is attached to a wall at 200° C. The surface of the fin is exposed to air at 20°C. The convective heat transfer coefficient is 100 W/(m^2 K). The fin is made of stainless steel with thermal conductivity of 20 W/(m K). Set up the residual equations in terms of temperature excess $\theta = T - T_\infty$ and determine the heat transfer rate. Length of the fin is 25 mm.

Solution

Node 1 (at the base):

$$q_b = kA_c\frac{(\theta_1-\theta_2)}{\Delta x}+h\left(\frac{P\Delta x}{2}\right)\theta_1$$

where A_c = 25 × 10^{-6} m^2, Δx = 5 × 10^{-3} m, P = 20 × 10^{-3} m, and θ_1 = 180°C. Hence,

$$\begin{aligned} q_b &= 20\times\left(25\times10^{-6}\right)\frac{(180-\theta_2)}{5\times10^{-3}}+100\times\left(\frac{20\times10^{-3}\times5\times10^{-3}}{2}\right)\times180 \\ &= 18.9-0.1\theta_2 \end{aligned} \tag{1}$$

The residual equation is

$$R_1 = q_b+0.1\theta_2-18.9$$

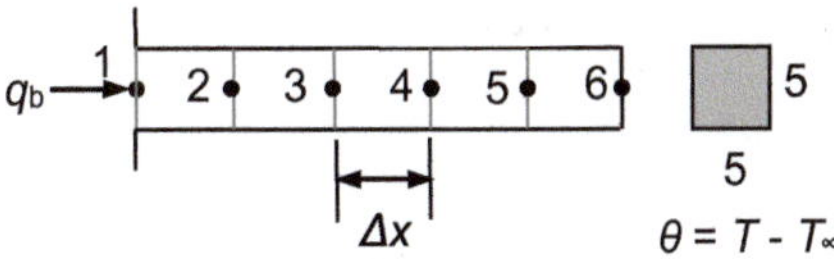

Fig. 5.32 Example 5.34

Node 2:

$$kA_c\frac{(\theta_1-\theta_2)}{\Delta x}+kA_c\frac{(\theta_3-\theta_2)}{\Delta x}-h(P\Delta x)\theta_2=0$$

The substitution of various values gives the residual equation as

$$R_2=180+\theta_3-2.1\theta_2$$

Similarly, the residual equations for the nodes 3 to 5 can be written as

$$R_3=\theta_2+\theta_4-2.1\theta_3$$
$$R_4=\theta_3+\theta_5-2.1\theta_4$$
$$R_5=\theta_4+\theta_6-2.1\theta_5$$

Node 6:

$$kA_c\frac{(\theta_5-\theta_6)}{\Delta x}-h\left(\frac{P\Delta x}{2}+A_c\right)\theta_6=0$$

Substitution gives

$$R_6=\theta_5-1.075\theta_6$$

The set of residual equations is

$$R_1=q_b+0.1\theta_2-18.9$$
$$R_2=180+\theta_3-2.1\theta_2$$
$$R_3=\theta_2+\theta_4-2.1\theta_3$$
$$R_4=\theta_3+\theta_5-2.1\theta_4$$
$$R_5=\theta_4+\theta_6-2.1\theta_5$$
$$R_6=\theta_5-1.075\theta_6$$

The unit change table and relaxation tables are given as Tables 5.6 and 5.7, respectively. Since the residuals are reasonably low after step 8, the process can be terminated.

Table 5.6 Unit change

	ΔR_2	ΔR_3	ΔR_4	ΔR_5	ΔR_6
$\Delta\theta_2=+1$	-2.1	+1	0	0	0
$\Delta\theta_3=+1$	+1	-2.1	+1	0	0
$\Delta\theta_4=+1$	0	+1	-2.1	+1	0
$\Delta\theta_5=+1$	0	0	+1	-2.1	+1
$\Delta\theta_6=+1$	0	0	0	+1	-1.075
Block change = + 1	-1.1	-0.1	-0.1	-0.1	-0.075

Table 5.7 Relaxation table

	θ_2	R_2	θ_3	R_3	θ_4	R_4	θ_5	R_5	θ_6	R_6
Initial guess	130	7	100	–5	75	7.5	65	–1.5	60	0.5
$\Delta\theta_2 = +5$	135	–3.5	100	0	75	7.5	65	–1.5	60	0.5
$\Delta\theta_3 = +3.5$	135	0	103.5	–7.35	75	7.5	65	–1.5	60	0.5
$\Delta\theta_4 = +7$	135	0	103.5	–0.35	82	-3.7	65	5.5	60	0.5
$\Delta\theta_5 = +3$	135	0	103.5	–0.35	82	-0.7	68	–0.8	60	3.5
$\Delta\theta_6 = +3$	135	0	103.5	–0.35	82	-0.7	68	2.2	63	0.275
$\Delta\theta_5 = +1$	135	0	103.5	–0.35	82	0.3	69	0.1	63	1.275
$\Delta\theta_6 = +0.5$	135	0	103.5	–0.35	82	0.3	69	0.6	63.5	0.7375
Approximate solution	135		103.5		82		69		63.5	

Now from Eq. (1),

$$q_b = 18.9 - 0.1 \times 135 = 5.4 \text{ W}.$$

Check:
From the equation of the fin,

$$q_b = \sqrt{hPkA_c}(T_s - T_\infty)\tanh(mL_c)$$

where $m = \sqrt{\frac{hP}{kA_c}} = \sqrt{\frac{100\times(20\times10^{-3})}{20\times(25\times10^{-6})}} = 63.25$

$$L_c = L + \frac{A_c}{P} = 25 \times 10^{-3} + \frac{25 \times 10^{-6}}{20 \times 10^{-3}} = 0.02625 \text{ m}$$
$$\tanh mL_c = \tanh(63.25 \times 0.02625) = 0.9298$$

Hence,

$$q_b = \sqrt{100 \times (20 \times 10^{-3}) \times 20 \times (25 \times 10^{-6})} \times 180 \times 0.9298 = 5.29 \text{ W}.$$

Thus the solution is in reasonable agreement.

Example 5.35 Figure 5.33 shows a rectangular section fin extending from the wall of a furnace. It is exposed to a convection environment (h = 400 W/(m^2 K) and T_∞ = 25°C). The fin is very wide in the z-direction. Calculate the steady-state temperatures of the nodes shown in the figure. The thermal conductivity of the material is 4 W/(m K).

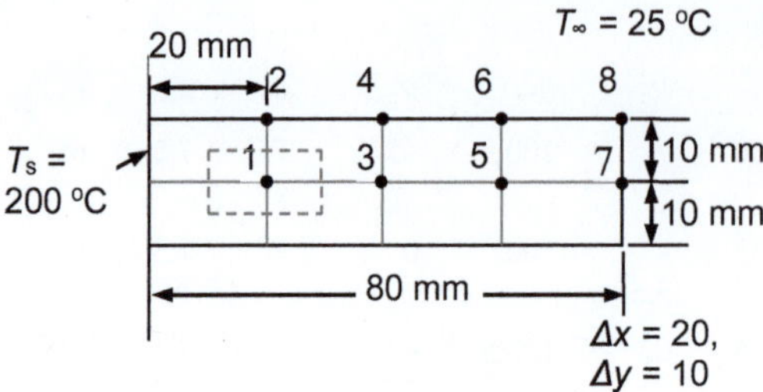

Fig. 5.33 Example 5.35

Solution

Due to the symmetry about the horizontal centreline, there are only 8 different nodal conditions. It is to note that $\Delta x \neq \Delta y$ in this problem.

Interior nodes (1, 3 and 5):
At node 1, the heat balance gives

$$kA_y\frac{(T_s - T_1)}{\Delta x} + kA_y\frac{(T_3 - T_1)}{\Delta x} + 2kA_x\frac{(T_2 - T_1)}{\Delta y} = 0$$

where

$$A_x = (20 \times 1) \times 10^{-3}\text{m}^2 \text{ for depth} = 1\text{m}$$
$$A_y = (10 \times 1) \times 10^{-3}\text{m}^2$$
$$\Delta x = 20 \times 10^{-3}\text{m}$$
$$\Delta y = 10 \times 10^{-3}\text{m}$$

Substitution of various values gives

$$k \times 10 \times \frac{(T_s - T_1)}{20} + k \times 10 \times \frac{(T_3 - T_1)}{20} + 2k \times 20 \times \frac{(T_2 - T_1)}{10} = 0$$

or

$$\frac{(T_s - T_1)}{2} + \frac{(T_3 - T_1)}{2} + 4(T_2 - T_1) = 0$$

or

$$10T_1 - 8T_2 - T_3 - T_s = 0$$

So the nodal equation is

$$10T_1 - 8T_2 - T_3 - 200 = 0 \quad \text{(i)}$$

Similarly the equation for the nodes 3 and 5 can be written as

$$10T_3 - 8T_4 - T_5 - T_1 = 0 \tag{ii}$$

$$10T_5 - 8T_6 - T_7 - T_3 = 0 \tag{iii}$$

Nodes 2, 4, and 6:
These nodes are on the convective boundary. At node 2, the heat balance equation is

$$kA_x\frac{(T_1 - T_2)}{\Delta y} + k\frac{A_y}{2}\frac{(T_4 - T_2)}{\Delta x} + k\frac{A_y}{2}\frac{(T_s - T_2)}{\Delta x} + h(\Delta x.1)(T_\infty - T_2) = 0$$

Putting vales of Δx, Δy, A_x, A_y, k and h, we obtain

$$4 \times 20 \times 10^{-3} \times \frac{(T_1 - T_2)}{10 \times 10^{-3}} + 4 \times 10 \times 10^{-3} \times \frac{(T_4 - T_2)}{2 \times 20 \times 10^{-3}}$$
$$+ 4 \times 10 \times 10^{-3}\frac{(T_s - T_2)}{2 \times 20 \times 10^{-3}} + 400 \times 20 \times 10^{-3}(T_\infty - T_2) = 0$$

or

$$8(T_1 - T_2) + (T_4 - T_2) + (T_s - T_2) + 8(T_\infty - T_2) = 0$$

or

$$18T_2 - 8T_1 - T_4 - T_s - 8T_\infty = 0$$

Substituting values of T_s and T_∞, the nodal equation is

$$18T_2 - 8T_1 - T_4 - 400 = 0$$

Similarly, at node 4, we have

$$18T_4 - 8T_3 - T_6 - T_2 - 200 = 0$$

And at node 6,

$$18T_6 - 8T_5 - T_8 - T_4 - 200 = 0$$

Node 8 (the corner node):

$$k\frac{A_y}{2}\frac{(T_6 - T_8)}{\Delta x} + k\frac{A_x}{2}\frac{(T_7 - T_8)}{\Delta y} + h\frac{(\Delta x + \Delta y).1}{2}(T_\infty - T_8) = 0$$

or

$$4 \times 10 \times 10^{-3} \times \frac{(T_6 - T_8)}{2 \times 20 \times 10^{-3}} + 4 \times 20 \times 10^{-3}\frac{(T_7 - T_8)}{2 \times 10 \times 10^{-3}} + 400$$
$$\times \frac{(20 + 10) \times 10^{-3}}{2}(T_\infty - T_8)$$
$$= 0$$

or

$$(T_6 - T_8) + 4(T_7 - T_8) + 6(T_\infty - T_8) = 0$$

or

$$11T_8 - 4T_7 - T_6 - 150 = 0$$

Node 7 at the fin end:

$$kA_y \frac{(T_5 - T_7)}{\Delta x} + 2k\frac{A_x}{2}\frac{(T_8 - T_7)}{\Delta y} + h(\Delta y.1)(T_\infty - T_7) = 0$$

Substitution of various values gives

$$7T_7 - 4T_8 - T_5 - 50 = 0$$

The complete set of nodal equations is

$$\text{Node 1}: \quad 10T_1 - 8T_2 - T_3 - 200 = 0 \qquad \text{(i)}$$

$$\text{Node 2}: \quad 18T_2 - 8T_1 - T_4 - 400 = 0 \qquad \text{(ii)}$$

$$\text{Node 3}: \quad 10T_3 - 8T_4 - T_5 - T_1 = 0 \qquad \text{(iii)}$$

$$\text{Node 4}: \quad 18T_4 - 8T_3 - T_6 - T_2 - 200 = 0 \qquad \text{(iv)}$$

$$\text{Node 5}: \quad 10T_5 - 8T_6 - T_7 - T_3 = 0 \qquad \text{(v)}$$

$$\text{Node 6}: \quad 18T_6 - 8T_5 - T_8 - T_4 - 200 = 0 \qquad \text{(vi)}$$

$$\text{Node 7}: \quad 7T_7 - 4T_8 - T_5 - 50 = 0 \qquad \text{(vii)}$$

$$\text{Node 8}: \quad 11T_8 - 4T_7 - T_6 - 150 = 0 \qquad \text{(viii)}$$

The above set of equations can be solved by Gaussian elimination or relaxation method. The relaxation method yields

T_1 = 65.91°C, $T_2 = T_9$ = 53.2°C, T_3 = 33.69°C, $T_4 = T_{10}$ = 30.5°C, T_5 = 26.77°C, $T_6 = T_{11}$ = 26.1°C, T_7 = 25.36°C, $T_8 = T_{12}$ = 25.22°C
with the residuals as

$$R_1 = -0.19, R_2 = -0.18, R_3 = 0.22, R_4 = 0.18, R_5 = -0.15, R_6 = -0.08, R_7 = -0.13, R_8 = -0.12.$$

Example 5.36 From the temperatures determined in Example 5.35, determine the heat flow rate from the fin base.

Solution

From Example 5.35, h = 400 W/(m^2 K), k = 4 W/(m K), T_∞ = 25°C, T_s = 200°C, T_1 = 65.91°C, $T_2 = T_9$ = 53.2°C. The heat flow rate from the fin base, refer Fig. 5.34, is

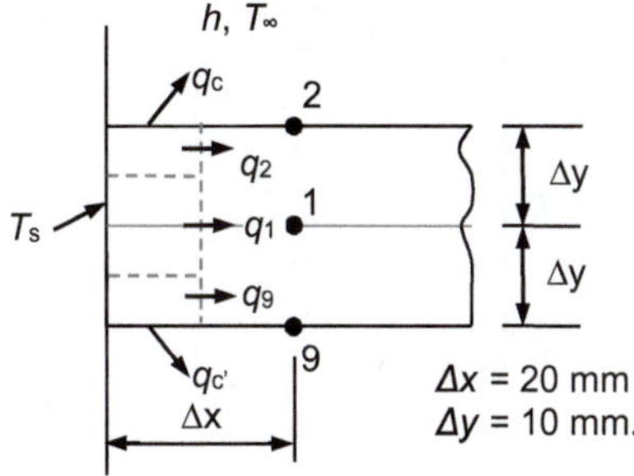

Fig. 5.34 Example 5.36

$$q_{\text{fin}} = q_c + q_1 + q_2 + q_9 + q_{c'},$$

where

$$q_c = q_{c'} = h\left(\frac{\Delta x}{2}.1\right)(T_s - T_\infty) = 400 \times \frac{20}{2000} \times 1 \times (200 - 25) = 700\,\text{W/m}$$

$$q_1 = k(\Delta y.1)\frac{(T_s - T_1)}{\Delta x} = 4 \times \frac{10}{1000} \times 1 \times \frac{(200 - 65.91) \times 1000}{20} = 268.18\,\text{W/m}$$

$$q_2 = k\left(\frac{\Delta y}{2}.1\right)\frac{(T_s - T_2)}{\Delta x} = 4 \times \frac{10}{2000} \times 1 \times \frac{(200 - 53.2) \times 1000}{20} = 146.8\,\text{W/m}$$

$$q_9 = k\left(\frac{\Delta y}{2}.1\right)\frac{(T_s - T_9)}{\Delta x} = 4 \times \frac{10}{2000} \times 1 \times \frac{(200 - 53.2) \times 1000}{20} = 146.8\,\text{W/m}.$$

Hence,

$$q_{\text{fin}} = 700 + 268.18 + 146.8 + 146.8 + 700 = 1961.78\,\text{W/m}.$$

Example 5.37 Display the nodal equations of Example 5.35 in matrix form.

Solution

The matrix is as given below.

$$\begin{vmatrix} 10 & -8 & -1 & & & & & \\ -8 & 18 & & -1 & & & & \\ -1 & & 10 & -8 & -1 & & & \\ & -1 & -8 & 18 & & -1 & & \\ & & -1 & & 10 & -8 & -1 & \\ & & & -1 & -8 & 18 & & -1 \\ & & & & -1 & & 7 & -4 \\ & & & & & -1 & -4 & 11 \end{vmatrix} \begin{vmatrix} T_1 \\ T_2 \\ T_3 \\ T_4 \\ T_5 \\ T_6 \\ T_7 \\ T_8 \end{vmatrix} = \begin{vmatrix} 200 \\ 400 \\ 0 \\ 200 \\ 0 \\ 200 \\ 50 \\ 150 \end{vmatrix}$$

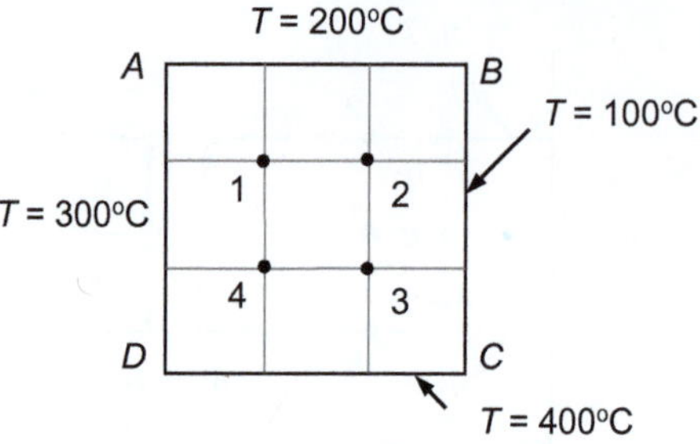

Fig. 5.35 Example 5.38

5.7.2.2 Gaussian Elimination

The relaxation method involves trial and error, hence it cannot be applied to the solution of a set of a large number of equations. Such a set of nodal equations can be solved by Gaussian elimination method. This method is systematic, hence easy to understand and apply. It is being explained below with the help of an example.

Now the computers have made it possible to solve practical problems, which were earlier believed to be not solvable due to a large number of nodal equations involved in such problems.

Example 5.38 Write the nodal equations at the four interior nodal points of the square grid of the section ABCD of a bar shown in Fig. 5.35. The temperature at the boundaries AB, BC, CD and DA are indicated in the figure.

Solution

The nodal equations obtained are

$$\text{Node 1}: \quad 200 + 300 + T_2 + T_4 - 4T_1 = 0 \tag{a}$$

$$\text{Node 2}: \quad 200 + 100 + T_3 + T_1 - 4T_2 = 0 \tag{b}$$

$$\text{Node 3}: \quad 100 + 400 + T_4 + T_2 - 4T_3 = 0 \tag{c}$$

$$\text{Node 4}: \quad 300 + 400 + T_1 + T_3 - 4T_4 = 0 \tag{d}$$

Example 5.39 Explain and use the *Gaussian elimination method* to solve the nodal equations of the previous example.

Solution

The nodal equations are

$$500 + T_2 + T_4 - 4T_1 = 0 \tag{a}$$

$$300 + T_3 + T_1 - 4T_2 = 0 \tag{b}$$

$$500 + T_4 + T_2 - 4T_3 = 0 \tag{c}$$

$$700 + T_1 + T_3 - 4T_4 = 0 \tag{d}$$

The equations may be rewritten as

$$T_1 - T_2 \qquad -T_4 = 500 \tag{e}$$

$$-T_1 + 4T_2 - T_3 \qquad = 300 \tag{f}$$

$$-T_2 + 4T_3 - T_4 = 500 \tag{g}$$

$$-T_1 \qquad - T_3 + 4T_4 = 700 \tag{h}$$

(i) The Gaussian elimination is to first *triangularize* the given set of equations. This is accomplished by some basic operations as explained below.

Eliminate T_1 from Eqs. (f) and (h) by multiplying these equations by 4 and adding Eq. (e) to them. This operation gives

$$4T_1 - T_2 \qquad - T_4 = 500 \tag{i}$$

$$15T_2 - 4T_3 - T_4 = 1700 \tag{j}$$

$$-T_2 + 4T_3 - T_4 = 500 \tag{k}$$

$$-T_2 - 4T_3 + 15T_4 = 3300 \tag{l}$$

Next, eliminate T_2 from Eqs. (k) and (l) by multiplying them by 15 and adding Eq. (j) to them. The result of this operation is

$$4T_1 - T_2 \qquad - T_4 = 500 \tag{m}$$

$$15T_2 - 4T_3 - T_4 = 1700 \tag{n}$$

$$56T_3 - 16T_4 = 9200 \tag{o}$$

$$-64T_3 + 224T_4 = 51200 \tag{p}$$

Rewriting the equations in simplified form, we obtain

$$4T_1 - T_2 \qquad - T_4 = 500 \tag{m}$$

$$15T_2 - 4T_3 - T_4 = 1700 \tag{n}$$

$$7T_3 - 2T_4 = 1150 \tag{o'}$$

$$-2T_3 + 7T_4 = 1600 \tag{p'}$$

Now eliminate T_3 from Eq. (p') by adding to it 2/7 times of Eq. (o'), to obtain

$$4T_1 - T_2 \qquad - T_4 = 500 \tag{m}$$

$$15T_2 - 4T_3 - T_4 = 1700 \tag{n}$$

$$7T_3 - 2T_4 = 1150 \tag{o'}$$

$$45/7T_4 = (1150\text{x}2/7) + 1600 \tag{p''}$$

This is the triangularized set of equations.

(ii) From Eq. (p''), T_4 = 300°C.

Now *back substitute* beginning with the bottom equation and working upwards. The process gives

$$T_3 = 250°\text{C}, T_2 = 200°\text{C}, \text{ and } T_1 = 250°\text{C}.$$

Note: When the number of equations in a set is large, the hand calculation may be a time-consuming process. For this purpose, a standard computer program is available.

The given set of equations can be put in a matrix form as given below.

$$\begin{vmatrix} +4 & -1 & 0 & -1 \\ -1 & +4 & -1 & 0 \\ 0 & -1 & +4 & -1 \\ -1 & 0 & -1 & +4 \end{vmatrix} \begin{vmatrix} T_1 \\ T_2 \\ T_3 \\ T_4 \end{vmatrix} = \begin{vmatrix} 500 \\ 300 \\ 500 \\ 700 \end{vmatrix}$$

In general, the matrix is

$$\begin{vmatrix} \text{A}(1,1) & \text{A}(1,2) & \ldots & \text{A}(1,\text{N}) \\ \text{A}(2,1) & \text{A}(2,2) & \ldots & \text{A}(2,\text{N}) \\ \ldots & \ldots & \ldots & \ldots \\ \text{A}(\text{N},1) & \text{A}(\text{N},2) & \ldots & \text{A}(\text{N},\text{N}) \end{vmatrix} \begin{vmatrix} T(1) \\ T(2) \\ \ldots \\ T(N) \end{vmatrix} = \begin{vmatrix} \text{B}(1) \\ \text{B}(2) \\ \ldots \\ \text{B}(\text{N}) \end{vmatrix}$$

2. The addition of the nodal equations of this example yields

$$T_1 + T_2 + T_3 + T_4 = 1000.$$

This equals the sum of the four boundary temperatures. This result is exact for any four points that are symmetrically placed about the centre of a square.

Example 5.40 Combustion products flow through a chimney whose cross-section is shown in Fig. 5.36. If at any location, the inside surface temperature is 600°C and outside surface temperature is 100°C, determine the heat loss rate from the chimney for its unit length.

Solution

Refer Fig. 5.36b.

Node 1

The heat balance equation for the two-dimensional conduction system gives

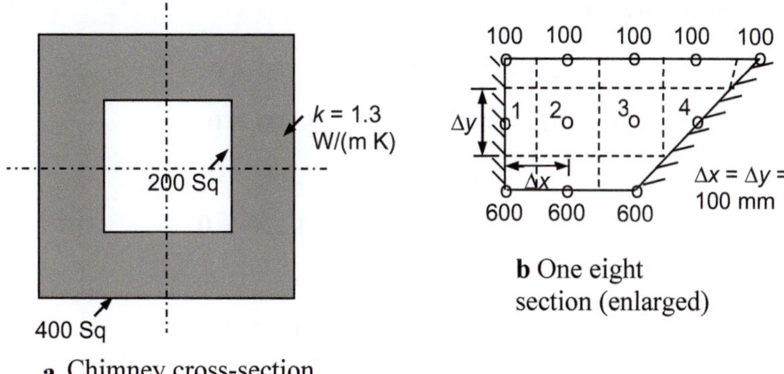

a Chimney cross-section

b One eight section (enlarged)

Fig. 5.36 Example 5.40

$$k\left(\frac{\Delta x}{2}.1\right)\frac{100-T_1}{\Delta y}+k\left(\frac{\Delta x}{2}.1\right)\frac{600-T_1}{\Delta y}+k(\Delta y.1)\frac{T_2-T_1}{\Delta x}=0.$$

Simplification gives the nodal equation for $\Delta x = \Delta y$ as

$$2T_2+700-4T_1=0. \tag{i}$$

Node 2

The heat balance equation for the two-dimensional conduction system gives

$$k(\Delta x.1)\frac{100-T_2}{\Delta y}+k(\Delta x.1)\frac{600-T_2}{\Delta y}+k(\Delta y.1)\frac{T_1-T_2}{\Delta x}+k(\Delta y.1)\frac{T_3-T_2}{\Delta x}=0.$$

Simplification gives the nodal equation as

$$T_1+T_3+700-4T_2=0. \tag{ii}$$

Node 3

The equation is similar to node 2 and is

$$T_2+T_4+700-4T_3=0. \tag{iii}$$

Node 4

The heat balance equation for the two-dimensional conduction system gives

$$k(\Delta x.1)\frac{100-T_4}{\Delta y}+k(\Delta y.1)\frac{T_3-T_4}{\Delta x}=0.$$

Simplification gives the nodal equation as

$$T_3+100-2T_4=0. \tag{iv}$$

Rewriting the nodal equations as

$$4T_1 - 2T_2 \qquad - 700 = 0 \quad \text{(a)}$$

$$T_1 - 4T_2 + T_3 \qquad + 700 = 0 \quad \text{(b)}$$

$$T_2 - 4T_3 + T_4 + 700 = 0 \quad \text{(c)}$$

$$T_3 - 2T_4 + 100 = 0 \quad \text{(d)}$$

The equation can be solved by the Gaussian elimination method.

We eliminate T_1 from Eq. (b) by multiplying this equation by –4 and adding Eq. (a) to it. This operation gives

$$4T_1 - 2T_2 \qquad - 700 = 0 \quad \text{(e)}$$

$$14T_2 - 4T_3 \qquad - 3500 = 0 \quad \text{(f)}$$

$$T_2 - 4T_3 + T_4 + 700 = 0 \quad \text{(g)}$$

$$T_3 - 2T_4 + 100 = 0 \quad \text{(h)}$$

Next eliminate T_2 from Eq. (g) by multiplying this equation by –14 and adding Eq. (f) to it. The result of this operation is

$$4T_1 - 2T_2 \qquad - 700 = 0 \quad \text{(i)}$$

$$14T_2 - 4T_3 \qquad - 3500 = 0 \quad \text{(j)}$$

$$52T_3 - 14T_4 - 13300 = 0 \quad \text{(k)}$$

$$T_3 - 2T_4 + 100 = 0 \quad \text{(l)}$$

Rewriting the equations in simplified form, we obtain

$$4T_1 - 2T_2 \qquad - 700 = 0 \quad \text{(m)}$$

$$7T_2 - 2T_3 \qquad - 1750 = 0 \quad \text{(n)}$$

$$26T_3 - 7T_4 - 6650 = 0 \quad \text{(o)}$$

$$T_3 - 2T_4 + 100 = 0 \quad \text{(p)}$$

Now eliminate T_3 from Eq. (p) by multiplying this equation by - 26 and adding Eq. (o) to it. This gives

$$4T_1 - 2T_2 \qquad - 700 = 0 \quad \text{(q)}$$

$$7T_2 - 2T_3 \qquad - 1750 = 0 \quad \text{(r)}$$

$$26T_3 - 7T_4 - 6650 = 0 \quad \text{(s)}$$

$$45T_4 - 9250 = 0 \quad \text{(t)}$$

This is the triangularized set of equations. From Eq. (t), $T_4 = 205.55°C$.
Now back substitute beginning with Eq. (s) and working upwards. The process gives

$$T_3 = 311.11°C, T_2 = 338.88°C, \text{ and } T_1 = 344.44°C.$$

The heat leaving the outer surface of this chimney section is

$$\begin{aligned} q_{1/8} &= k\left(\frac{\Delta x}{2}.1\right)\frac{T_1 - 100}{\Delta y} + k(\Delta x.1)\frac{T_2 - 100}{\Delta y} + k(\Delta x.1)\frac{T_3 - 100}{\Delta y} + k(\Delta x.1)\frac{T_4 - 100}{\Delta y} \\ &= k(0.5T_1 + T_2 + T_3 + T_4 - 350) \\ &= 1.3 \times (0.5 \times 344.44 + 338.88 + 311.11 + 205.55 - 350) = 881.1\ \text{W/m length.} \end{aligned}$$

The heat flow rate can also be calculated from the heat entering the chimney section, which is

$$\begin{aligned} q_{1/8} &= k\left(\frac{\Delta x}{2}.1\right)\frac{600 - T_1}{\Delta y} + k(\Delta x.1)\frac{600 - T_2}{\Delta y} + k(\Delta x.1)\frac{600 - T_3}{\Delta y} \\ &= k(1500 - 0.5T_1 - T_2 - T_3) \\ &= 1.3 \times (1500 - 0.5 \times 344.44 - 338.88 - 311.11) = 881.1\ \text{W/m length.} \end{aligned}$$

Since the considered chimney section is 1/8 of the total cross-section, the heat loss from the chimney is

$$q = 881.1 \times 8 = 7048.8\ \text{W/m length.}$$

Example 5.41 If in the above problem inner and outer surfaces of the chimney are subjected to convective conditions write the nodal equations.

Solution Refer Fig. 5.37b.

Node 1
The heat balance equation for the two-dimensional conduction system gives

$$k\left(\frac{\Delta x}{2}.1\right)\frac{T_5 - T_1}{\Delta y} + k\left(\frac{\Delta x}{2}.1\right)\frac{T_8 - T_1}{\Delta y} + k(\Delta y.1)\frac{T_2 - T_1}{\Delta x} = 0.$$

Simplification gives the nodal equation for $\Delta x = \Delta y$ as

$$2T_2 + T_5 + T_8 - 4T_1 = 0 \tag{i}$$

Node 2
The heat balance equation for the two-dimensional conduction system gives

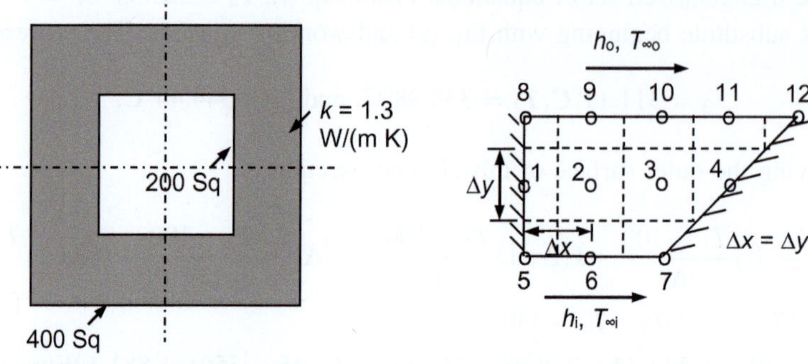

Fig. 5.37 Example 5.41

$$k(\Delta x.1)\frac{T_6 - T_2}{\Delta y} + k(\Delta x.1)\frac{T_9 - T_2}{\Delta y} + k(\Delta y.1)\frac{T_1 - T_2}{\Delta x} + k(\Delta y.1)\frac{T_3 - T_2}{\Delta x} = 0.$$

Simplification gives the nodal equation as

$$T_1 + T_3 + T_6 + T_9 - 4T_2 = 0. \quad \text{(ii)}$$

Node 3

The equation is similar to node 2 and is

$$T_2 + T_4 + T_7 + T_{10} - 4T_3 = 0. \quad \text{(iii)}$$

Node 4

The heat balance equation for the two-dimensional conduction system gives

$$k(\Delta x.1)\frac{T_{11} - T_4}{\Delta y} + k(\Delta y.1)\frac{T_3 - T_4}{\Delta x} = 0.$$

Simplification gives the nodal equation as

$$T_3 + T_{11} - 2T_4 = 0. \quad \text{(iv)}$$

Node 5

The heat balance equation for the two-dimensional conduction system gives

$$k\left(\frac{\Delta x}{2}.1\right)\frac{T_1 - T_5}{\Delta y} + k\left(\frac{\Delta y}{2}.1\right)\frac{T_6 - T_5}{\Delta x} + h_i\left(\frac{\Delta x}{2}.1\right)(T_{\infty i} - T_5) = 0.$$

Simplification gives the nodal equation for $\Delta x = \Delta y$ as

$$T_1 + T_6 + \frac{h_i \Delta x}{k} T_{\infty i} - \left(2 + \frac{h_i \Delta x}{k}\right) T_5 = 0. \quad \text{(v)}$$

Node 6

The heat balance equation for the two-dimensional conduction system gives

$$k(\Delta x.1)\frac{T_2-T_6}{\Delta y}+k\left(\frac{\Delta y}{2}.1\right)\frac{T_5-T_6}{\Delta x}+k\left(\frac{\Delta y}{2}.1\right)\frac{T_7-T_6}{\Delta x}+h_i(\Delta x.1)(T_{\infty i}-T_6)=0.$$

Simplification gives the nodal equation for $\Delta x = \Delta y$ as

$$2T_2+T_5+T_7+\frac{2h_i\Delta x}{k}T_{\infty i}-2\left(\frac{h_i\Delta x}{k}+2\right)T_6=0. \qquad \text{(vi)}$$

Node 7

The heat balance equation for the two-dimensional conduction system gives

$$k(\Delta x.1)\frac{T_3-T_7}{\Delta y}+k\left(\frac{\Delta y}{2}.1\right)\frac{T_6-T_7}{\Delta x}+h_i\left(\frac{\Delta x}{2}.1\right)(T_{\infty i}-T_7)=0.$$

Simplification gives the nodal equation for $\Delta x = \Delta y$ as

$$2T_3+T_6+\frac{h_i\Delta x}{k}T_{\infty i}-\left(3+\frac{h_i\Delta x}{k}\right)T_7=0. \qquad \text{(vii)}$$

Node 8

The heat balance equation for the two-dimensional conduction system gives

$$k\left(\frac{\Delta x}{2}.1\right)\frac{T_1-T_8}{\Delta y}+k\left(\frac{\Delta y}{2}.1\right)\frac{T_9-T_8}{\Delta x}+h_0\left(\frac{\Delta x}{2}.1\right)(T_{\infty o}-T_8)=0.$$

Simplification gives the nodal equation for $\Delta x = \Delta y$ as

$$T_1+T_9+\frac{h_o\Delta x}{k}T_{\infty o}-\left(2+\frac{h_o\Delta x}{k}\right)T_8=0. \qquad \text{(viii)}$$

Node 9

The heat balance equation for the two-dimensional conduction system gives

$$k(\Delta x.1)\frac{T_2-T_9}{\Delta y}+k\left(\frac{\Delta y}{2}.1\right)\frac{T_8-T_9}{\Delta x}+k\left(\frac{\Delta y}{2}.1\right)\frac{T_{10}-T_9}{\Delta x}+h_o(\Delta x.1)(T_{\infty o}-T_9)=0.$$

Simplification gives the nodal equation for $\Delta x = \Delta y$ as

$$2T_2+T_8+T_{10}+\frac{2h_o\Delta x}{k}T_{\infty o}-2\left(\frac{h_o\Delta x}{k}+2\right)T_9=0. \qquad \text{(ix)}$$

Node 10

The heat balance equation for the two-dimensional conduction system gives

$$k(\Delta x.1)\frac{T_3 - T_{10}}{\Delta y} + k\left(\frac{\Delta y}{2}.1\right)\frac{T_9 - T_{10}}{\Delta x} + k\left(\frac{\Delta y}{2}.1\right)\frac{T_{11} - T_{10}}{\Delta x} + h_o(\Delta x.1)(T_{\infty o} - T_{10}) = 0.$$

Simplification gives the nodal equation for $\Delta x = \Delta y$ as

$$2T_3 + T_9 + T_{11} + \frac{2h_o\Delta x}{k}T_{\infty o} - 2\left(\frac{h_o\Delta x}{k} + 2\right)T_{10} = 0. \tag{x}$$

Node 11

The heat balance equation for the two-dimensional conduction system gives

$$k(\Delta x.1)\frac{T_4 - T_{11}}{\Delta y} + k\left(\frac{\Delta y}{2}.1\right)\frac{T_{10} - T_{11}}{\Delta x} + k\left(\frac{\Delta y}{2}.1\right)\frac{T_{12} - T_{11}}{\Delta x} + h_o(\Delta x.1)(T_{\infty o} - T_{11}) = 0.$$

Simplification gives the nodal equation for $\Delta x = \Delta y$ as

$$2T_4 + T_{10} + T_{12} + \frac{2h_o\Delta x}{k}T_{\infty o} - 2\left(\frac{h_o\Delta x}{k} + 2\right)T_{11} = 0. \tag{xi}$$

Node 12

The heat balance equation for the two-dimensional conduction system gives

$$k\left(\frac{\Delta y}{2}.1\right)\frac{T_{11} - T_{12}}{\Delta x} + h_o\left(\frac{\Delta x}{2}.1\right)(T_{\infty o} - T_{12}) = 0.$$

Simplification gives the nodal equation for $\Delta x = \Delta y$ as

$$T_{11} + \frac{h_o\Delta x}{k}T_{\infty o} - \left(\frac{h_o\Delta x}{k} + 1\right)T_{12} = 0. \tag{xii}$$

5.7.2.3 The Gauss–Seidel Iteration Method

The iterative method may yield quicker results when the number of nodal equations is large. One of such methods, the Gauss–Seidel method, is being explained below using the nodal equations of Example 5.35.

Step 1: Rearrange the nodal equations so that the unknown temperature of the node for which the nodal equation was written is on the left as follows.

Table 5.8 The Gauss–Seidel iteration table

n[a]	T_1	T_2	T_3	T_4	T_5	T_6	T_7	T_8
Initial guess	130	80	90	55	50	40	35	30
1	93	65.6	58.3	42.9	41.33	33.53	30.19	27.68
2	78.31	59.41	46.28	35.84	34.47	29.96	27.88	26.50
3	72.16	56.28	39.33	33.38	30.69	28.08	26.67	25.89
4	68.96	54.73	36.67	32.01	28.80	27.13	26.06	25.58
5	67.45	53.98	35.23	31.28	27.83	26.64	25.74	25.42
6	66.71	53.61	34.48	30.89	27.33	26.39	25.57	25.33
7	66.34	53.42	34.08	30.69	27.08	26.26	25.49	25.29
8	66.14	53.32	33.87	30.59	26.94	26.19	25.44	25.27
9	66.04	53.27	33.77	30.53	26.87	26.15	25.42	25.26
10	65.99	53.25	33.71	30.5	26.83	26.13	25.41	25.25

[a]n indicates the number of iterations

$$
\begin{aligned}
T_1 &= (1/10)(8T_2 + T_3 + 200)\\
T_2 &= (1/18)(8T_1 + T_4 + 400)\\
T_3 &= (1/10)(8T_4 + T_5 + T_1)\\
T_4 &= (1/18)(8T_3 + T_6 + T_2 + 200)\\
T_5 &= (1/10)(8T_6 + T_7 + T_3)\\
T_6 &= (1/18)(8T_5 + T_8 + T_4 + 200)\\
T_7 &= (1/7)(4T_8 + T_5 + 50)\\
T_8 &= (1/11)(4T_7 + T_6 + 150)
\end{aligned}
$$

Step 2: Assume an initial set of values for the temperatures T_1 to T_8 as shown in Table 5.8. Using each of the equations of T_1 to T_8, calculate new values of the temperatures. For example, the first equation gives a new value of temperature T_1 = 93°C. Now using the new value of T_1 and still not updated value of T_4 (= 55°C), obtain the new value of T_2 (= 65.6°C). In this manner, sweep through the set of eight equations. This gives a new set of revised or updated estimates of T_1 to T_8 (refer row 3 of the table). This completes one cycle of iteration. In essence, it is the calculation of new values of nodal temperatures using the most recent values of the temperatures.

Step 3: The process explained in Step 2 is repeated until the calculated nodal temperatures converge. This is assumed to be achieved when successive temperatures differ by very small values, i.e. when

$$(T_i)_{n+1} - (T_i)_n \le \delta \quad \text{for all } T_i$$

where $(T_i)_{n+1}$ is the calculated value of the temperature T_i after (n +1) iterations and $(T_i)_n$ is the temperature T_i after nth iteration. Value of δ depends on the desired accuracy of the result.

It is to note that with the increase in the number of iterations, both the accuracy and the cost (time) of the computation increase. The decision regarding the number of the iterations is the tradeoff between the two.

After 10th iteration,

$$(T_i)_{10}-(T_i)_9 \leq (-0.06°\text{C}) \quad \text{for all } T_i$$

Comparison of the results of last iteration with those obtained with the relaxation method shows that they differ by a very small amount.

Example 5.42 Solve Example 5.9 using the finite-difference method.

Solution

A grid with $\Delta x = \Delta y = 0.5$ is being used, see Fig. 5.38. The temperature distribution will be symmetrical about the horizontal axis.

The nodal equations are

$$\text{Node 1}: \quad -4T_1 + 2T_2 + T_3 + 400 = 0 \tag{i}$$

$$\text{Node 2}: \quad -4T_2 + T_1 + T_4 + 500 = 0 \tag{ii}$$

$$\text{Node 3}: \quad -4T_3 + 2T_4 + T_1 + T_5 = 0 \tag{iii}$$

$$\text{Node 4}: \quad -4T_4 + T_3 + T_2 + T_6 + 100 = 0 \tag{iv}$$

$$\text{Node 5}: \quad -4T_5 + T_3 + 2T_6 + 100 = 0 \tag{v}$$

$$\text{Node 6}: \quad -4T_6 + T_5 + T_4 + 200 = 0. \tag{vi}$$

The solution of the equations gives T_1 = 258.03°C, T_2 = 228.57°C, T_3 = 175°C, T_4 = 156.25°C, T_5 = 129.47°C and T_6 = 121.43°C.

Note: 1. A finer grid will give a better approximation.

2. From the mean value theorem, T_3 = [(400 + 100 + 100 + 100)/4] = 175°C.

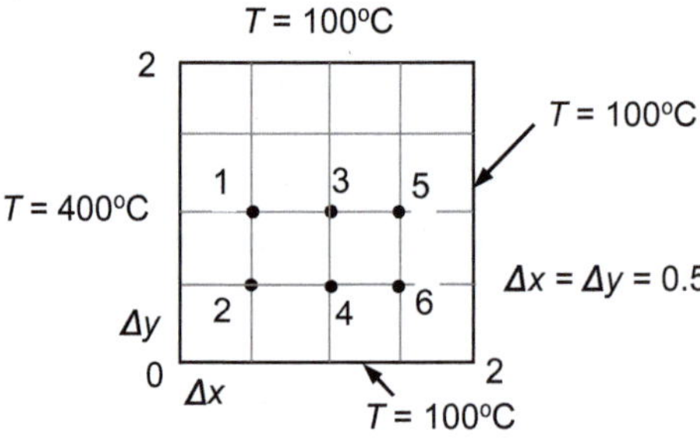

Fig. 5.38 Example 5.42

5.8 Two-Dimensional Steady-State Heat Conduction with Heat Generation

(a) **Internal node, refer Fig.** 5.39a,

In the steady state, the summation of heat flow to the node (m, n) from the surrounding nodes is zero.

$$k(\Delta y.1)\frac{(T_{m-1,n}-T_{m,n})}{\Delta x}+k(\Delta y.1)\frac{(T_{m+1,n}-T_{m,n})}{\Delta x}+k(\Delta x.1)\frac{(T_{m,n-1}-T_{m,n})}{\Delta y}$$
$$+k(\Delta x.1)\frac{(T_{m,n+1}-T_{m,n})}{\Delta y}+q_g(\Delta x.\Delta y.1)=0$$

for $\Delta z = 1$.

Putting $\Delta x = \Delta y$, and on simplification, we obtain

$$T_{m,n}=\frac{1}{4}\left[T_{m-1,n}+T_{m+1,n}+T_{m,n-1}+T_{m,n+1}\right]+q_g\frac{(\Delta x)^2}{4k}$$

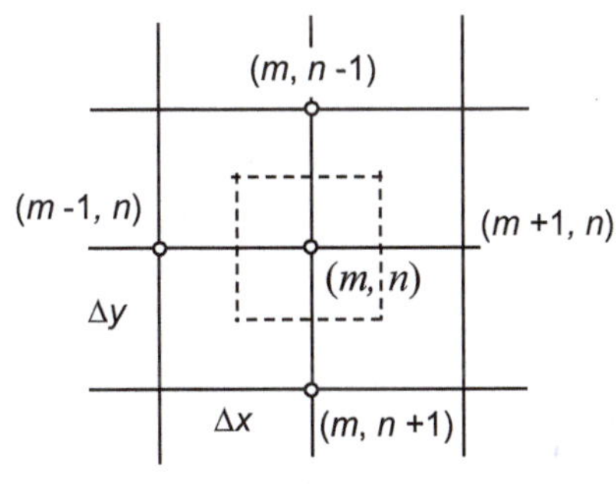

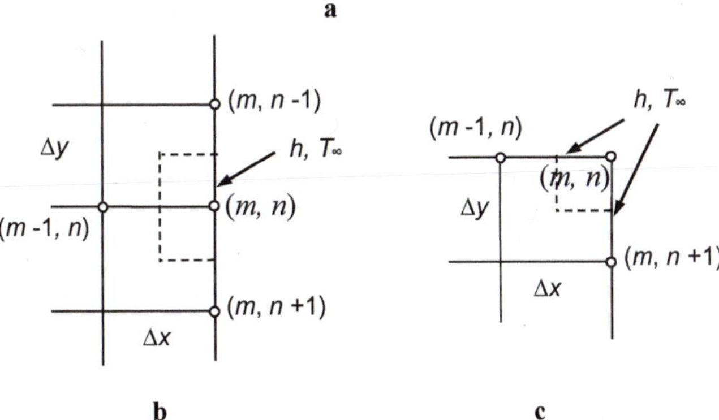

Fig. 5.39 Two-dimensional steady-state heat conduction with heat generation

(b) **Surface node, refer Fig.** 5.39(b),

Summation of the heat conduction and convection into the node gives

$$k(\Delta y.1)\frac{(T_{m-1,n}-T_{m,n})}{\Delta x}+k\left(\frac{\Delta x.1}{2}\right)\frac{(T_{m,n-1}-T_{m,n})}{\Delta y}+k\left(\frac{\Delta x.1}{2}\right)\frac{(T_{m,n+1}-T_{m,n})}{\Delta y}$$
$$+h(\Delta y.1)(T_\infty-T_{m,n})+q_g\frac{(\Delta x.\Delta y.1)}{2}=0$$

Simplification gives

$$T_{m,n-1}+T_{m,n+1}+2T_{m-1,n}-2\left[2+h\left(\frac{\Delta x}{k}\right)\right]T_{m,n}+2h\left(\frac{\Delta x}{k}\right)T_\infty+q_g\frac{(\Delta x)^2}{k}=0$$

for $\Delta x=\Delta y$.

(c) **External corner node**

Heat balance equation in this case is

$$k\frac{(\Delta y.1)}{2}\frac{(T_{m-1,n}-T_{m,n})}{\Delta x}+k\frac{(\Delta x.1)}{2}\frac{(T_{m,n+1}-T_{m,n})}{\Delta y}+h\frac{(\Delta x+\Delta y).1}{2}(T_\infty-T_{m,n})$$
$$+q_g\left(\frac{\Delta x}{2}\cdot\frac{\Delta y}{2}.1\right)=0$$

For $\Delta x=\Delta y$, simplification gives

$$T_{m-1,n}+T_{m,n+1}-2\left[1+h\left(\frac{\Delta x}{k}\right)\right]T_{m,n}+2h\left(\frac{\Delta x}{k}\right)T_\infty+q_g\frac{(\Delta x)^2}{2k}=0$$

Example 5.43 Write down nodal equations for an interior node and a surface node with convection in a cylindrical system for one-dimensional radial heat conduction.

Solution

In Fig. 5.40, the nodes are spaced at equal radial distance Δr with node $m = 0$ at the centre. Node m, situated at a radial distance $m\Delta r$, corresponds to annular ring of radial width Δr and radius $m\Delta r$. Similarly we can define the area corresponding to other nodal points. Node n is surface node.

The heat balance equation for node m can be written for unit length of cylinder as

$$k\left[2\pi\left(m\Delta r-\frac{\Delta r}{2}\right).1\right]\frac{T_{m-1}-T_m}{\Delta r}+\left[2\pi\left(m\Delta r+\frac{\Delta r}{2}\right).1\right]\frac{T_{m+1}-T_m}{\Delta r}=0$$

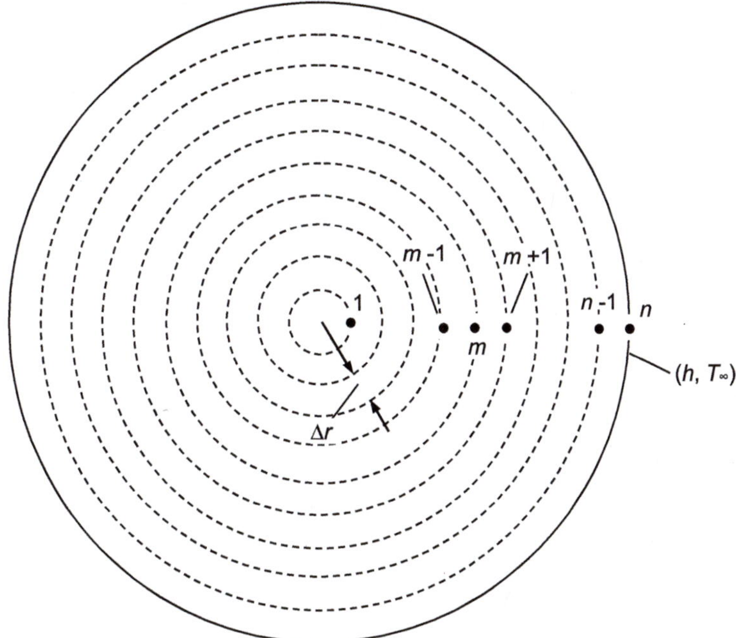

Fig. 5.40 Example 5.43

or

$$k\left[2\pi\Delta r\left(m-\frac{1}{2}\right).1\right]\frac{T_{m-1}-T_m}{\Delta r}+\left[2\pi\Delta r\left(m+\frac{1}{2}\right).1\right]\frac{T_{m+1}-T_m}{\Delta r}=0$$

Simplification gives

$$\left(m-\frac{1}{2}\right)T_{m-1}+\left(m+\frac{1}{2}\right)T_{m+1}-2mT_m=0$$

The heat balance equation for node n for unit length of cylinder is

$$k\left[2\pi\left(n\Delta r-\frac{\Delta r}{2}\right).1\right]\frac{T_{n-1}-T_n}{\Delta r}+h(2\pi n\Delta r.1)(T_\infty-T_n)=0$$

or

$$k\left(n-\frac{1}{2}\right)(T_{n-1}-T_n)+hn\Delta r(T_\infty-T_n)=0$$

or

$$\left(n-\frac{1}{2}\right)T_{n-1}+\frac{hn\Delta r}{k}T_\infty-\left(n-\frac{1}{2}+\frac{hn\Delta r}{k}\right)T_n=0$$

Example 5.44 If the radial heat conduction in Example 5.43 is accompanied with uniform volumetric heat generation, write the nodal equations for node m and n.

Solution

The heat balance equation is considering the heat generation

$$k\left[2\pi\left(m\Delta r-\frac{\Delta r}{2}\right).1\right]\frac{T_{m-1}-T_m}{\Delta r}+\left[2\pi\left(m\Delta r+\frac{\Delta r}{2}\right).1\right]\frac{T_{m+1}-T_m}{\Delta r}+q_g\Delta V=0$$

where $\Delta V = 2\pi m\Delta r.\ \Delta r\ .1$. Substitution gives

$$k\left[2\pi\left(m\Delta r-\frac{\Delta r}{2}\right).1\right]\frac{T_{m-1}-T_m}{\Delta r}+\left[2\pi\left(m\Delta r+\frac{\Delta r}{2}\right).1\right]\frac{T_{m+1}-T_m}{\Delta r}+q_g(2\pi m\Delta r.\Delta r.1)=0$$

or

$$\left(m-\frac{1}{2}\right).(T_{m-1}-T_m)+\left(m+\frac{1}{2}\right).(T_{m+1}-T_m)+\frac{q_g m(\Delta r)^2}{k}=0$$

or

$$\left(m-\frac{1}{2}\right)T_{m-1}+\left(m+\frac{1}{2}\right)T_{m+1}-2mT_m++\frac{q_g m(\Delta r)^2}{k}=0$$

The heat balance equation for node n for unit length of cylinder considering heat generation is

$$k\left[2\pi\left(n\Delta r-\frac{\Delta r}{2}\right).1\right]\frac{T_{n-1}-T_n}{\Delta r}+h(2\pi n\Delta r.1)(T_\infty-T_n)+q_g\Delta V=0$$

where $\Delta V = 2\pi n\Delta r.(\Delta r/2)\ .1$ for surface node. Substitution gives

$$k\left[2\pi\left(n\Delta r-\frac{\Delta r}{2}\right).1\right]\frac{T_{n-1}-T_n}{\Delta r}+h(2\pi n\Delta r.1)(T_\infty-T_n)+q_g[2\pi n\Delta r.(\Delta r/2).1]=0$$

or

$$k\left(n-\frac{1}{2}\right)(T_{n-1}-T_n)+\frac{hn\Delta r}{k}(T_\infty-T_n)+\frac{q_g n(\Delta r)^2}{2k}=0$$

or

$$\left(n-\frac{1}{2}\right)T_{n-1}+\frac{hn\Delta r}{k}T_\infty-\left(n-\frac{1}{2}+\frac{hn\Delta r}{k}\right)T_n+\frac{q_g n(\Delta r)^2}{2k}=0$$

5.9 Summary

In this chapter, analytical solution, graphical analysis, method of analogy and numerical solutions have been presented for the study of problems of two-dimensional steady-state conduction heat flow through solids without heat sources.

The Laplace equation that governs the temperature distribution for two-dimensional heat conduction system is $\partial^2 t/\partial x^2 + \partial^2 t/\partial y^2 = 0$. Since to determine the conduction heat flow, the temperature field must be known, the problem reduces to the solution of the Laplace equation. The common techniques used for the solution are: (i) analytical, (ii) graphical, (iii) method of analogy, and (iv) numerical methods.

In Sect. 5.2, the analytical solution for a rectangular section bar ($W \times H$), which is very long in the z-direction, has been presented whose three lateral sides ($x = 0$ and W, and $y = 0$) are maintained at a constant temperature T_o and for the fourth side we considered three different boundary conditions: (i) the fourth side ($y = H$) has a sinusoidal temperature distribution, $T = T_m \sin(\pi x/W)$, imposed on it, (ii) temperature along the fourth side is given by an arbitrary function $f(x)$, and (iii) the fourth side is at a constant temperature T_b.

In Sect. 5.3, mathematical treatment of the problem of conduction through a flat semi-infinite homogeneous plate has been presented, whose length in y-direction is infinite, it is relatively thin in the z-direction and xoy surfaces are insulated so that there is no temperature gradient in the z-direction and the temperature field is two dimensional.

In Sect. 5.5, the basic concept of the graphical method of thermal flux plotting has been explained and conduction shape factor has been defined. The shape factor presented in Table 5.2 can be readily used for the solution of various physical systems of interest as illustrated by solved problems. However, the graphical method involves drawing of curvilinear square elements and count of the number of heat flow channels and the temperature increments, which depends on the skill of the person plotting the lines. Electrical analogy, discussed in Sect. 5.6, may be used to sketch the isotherms and heat flow lines. It is to note that this technique is now not much used but rough sketching and the concept of the plots can help in a quick check of the result from other methods. Hence, it must not be totally overlooked.

The direct integration of the differential equations has been used to solve simple problems of two-dimensional, three-dimensional and transient heat conduction problems but success in solving complex problems, involving non-linear boundary conditions and temperature- or position-depending thermal properties, is limited. Such problems have been solved using numerical methods. Commonly used numerical methods are: finite difference and finite element methods.

The finite-difference method has been used extensively in solving heat conduction problems because of its simplicity in implementation. The finite element method is being widely used to solve problems in structural mechanics. It requires much greater mathematical efforts and has been used for solving heat conduction problems involving complicated geometries. Here only the basic principles and application of the finite-difference method have been presented. The first step in the finite-difference method is to discretize the spatial and time coordinates to form a mesh of nodes. Then by applying the energy balance to the volume elements surrounding the nodes, a set of linear algebraic equations (termed as nodal equations) is obtained. Various techniques, namely the relaxation method, Gaussian elimination, and the Gauss–Seidel iteration methods for the solution of nodal equations developed by the finite-difference method have been presented.

It is to note that the accuracy of the finite-difference approximation increases with the number of nodes hence computers are used to obtain finite-difference solutions. Standard computer programs are available for this purpose.

Review Questions

5.1 What are the various methods of solving two-dimensional heat conduction problems? Discuss their basics and limitations.
5.2 What are curvilinear squares?
5.3 Define conduction shape factor. Support your answer by considering a configuration where two-dimensional heat conduction is encountered.
5.4 Explain the basic principle of electrothermal analogy?
5.5 Explain the basic method of finite difference. What are the various methods available to obtain a solution after writing the finite-difference equations?
5.6 Compare relaxation and Gaussian elimination methods.

Problems

5.1 Using the equations of temperature distribution obtained in the analytical method for the two-dimensional heat conduction, determine the temperature at (3/4, ¼) of the long square-section rod shown in Fig. 5.41.
[**Ans**. Using the equations presented in Example 5.5, $t(x, y)$ = 127.2°C.]

5.2 A cubical furnace (0.6 m side) is covered with 0.1 m thick layer of insulation [k = 0.035 W/(m K)]. If the temperature difference across the insulation is 80°C, calculate the heat loss through the layer.
[**Ans**. Shape factors: $S_{\text{plane wall}}$ = 3.6, S_{edge} = 0.324, S_{corner} = 0.015, Total shape factor, S = 25.61 m; q = 71.71 W]

5.3 Combustion gases at an average temperature of 1200°C flow through a 3 m long, 100 mm zero-dimensional circular section duct. In order to reduce the heat loss, the duct is covered with insulation [k = 0.05 W/(m^2 K)]. The insulated duct measures 250 mm × 250 mm (square in shape). Determine the heat loss for the duct if the heat transfer coefficients are h_i = 150 W/(m^2 K) and h_0 = 5 W/(m^2 K) for the duct inner and outer surfaces, respectively. T_∞ = 30°C.
[**Ans**. From Case (11), Table 5.2, S = 18.98; Resistances are: $R_i = 1/(2\pi R L h_i)$ = 7.074 × 10^{-3}, $R_k = 1/kS$ = 1.054; and $R_o = 1/(4WLh_o)$ = 0.0666, $q = \Delta t/\sum R$ = 1037.5 W.]

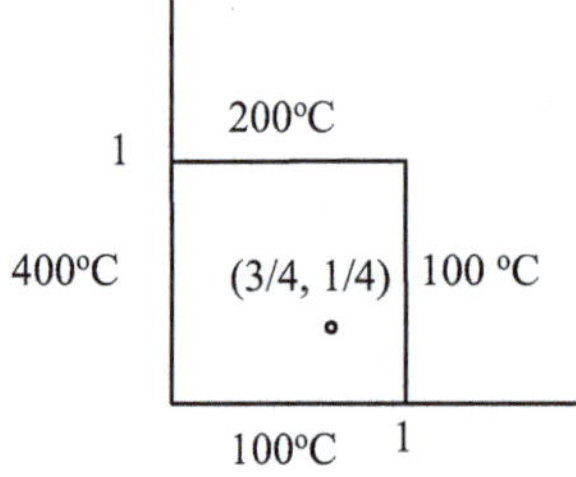

Fig. 5.41 Problem 5.1

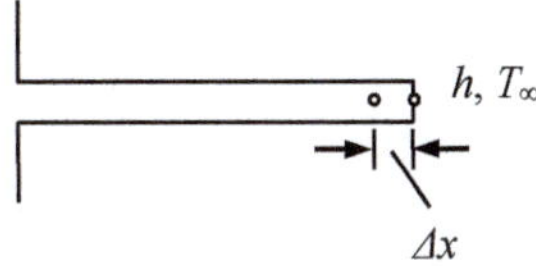

Fig. 5.42 **Problem 5.5**

5.4 A 75 mm diameter hot fluid pipeline (surface temperature = 80°C) and a cold water pipeline 50 mm in diameter (surface temperature = 20°C) are 200 mm apart on centres in a large duct packed with glass wool insulation [k = 0.038 W/(m K)]. Calculate the heat transfer to the cold fluid for 5 m length of the pipes.
[**Ans.** From Case (9), Table 5.2, S = 8.49, $q = k\,S$ (80–20) = 19.36 W].

5.5 For a plate fin of uniform cross-section A_c along its length, show that the temperature for the tip node shown in Fig. 5.42 is given by

$$\left[h\left(\frac{\Delta x}{k}\right)+1+\frac{1}{2}(m\Delta x)^2\right]T_m=\left[h\left(\frac{\Delta x}{k}\right)+\frac{1}{2}(m\Delta x)^2\right]T_\infty+T_{m-1}$$

where $m^2 = hP/kA_c$.
[**Hint**: $hA_c(T_\infty-T_m)+kA_c\left(\frac{T_{m-1}-T_m}{\Delta x}\right)+P\frac{\Delta x}{2}h(T_\infty-T_m)=0$. On simplification and putting $\frac{hP}{kA_c}=m^2$, the result is obtained.]

5.6 If the end of the plate fin of Problem 5.5 is insulated, show that the temperature for the tip node is given by

$$T_{m-1}+\left[m^2.\frac{(\Delta x)^2}{2}\right]T_\infty-\left[1+m^2.\frac{(\Delta x)^2}{2}\right]T_m=0$$

where $m^2 = hP/kA_c$.
[**Hint**: Refer Fig. 5.43. Heat balance equation is $kA_c\left(\frac{T_{m-1}-T_m}{\Delta x}\right)+hP\frac{\Delta x}{2}(T_\infty-T_m)=0$. On simplification and putting $\frac{hP}{kA_c}=m^2$, the result is obtained.]

5.7 The node (m, n) in Fig. 5.44 is situated on a boundary along which uniform heat flux q'' is specified. Show that in the steady-state, the node temperature is given by

$$T_{m,n}=(1/4)(T_{m,n+1}+T_{m,n-1}+2T_{m-1,n})+(q''/2k)\Delta x$$

Verify that the adiabatic boundary limit result deduced from the equation agrees with the corresponding result in Table 5.3.

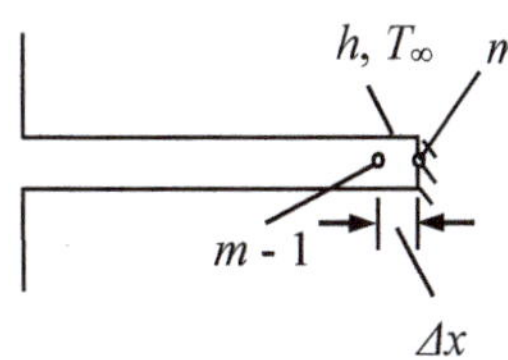

Fig. 5.43 Problem 5.6

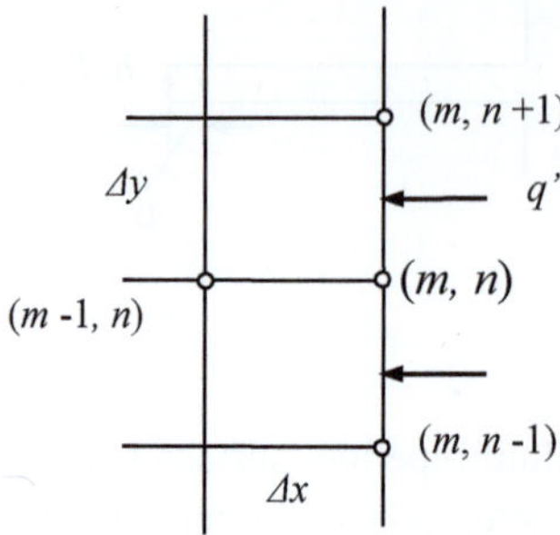

Fig. 5.44 Problem 5.7

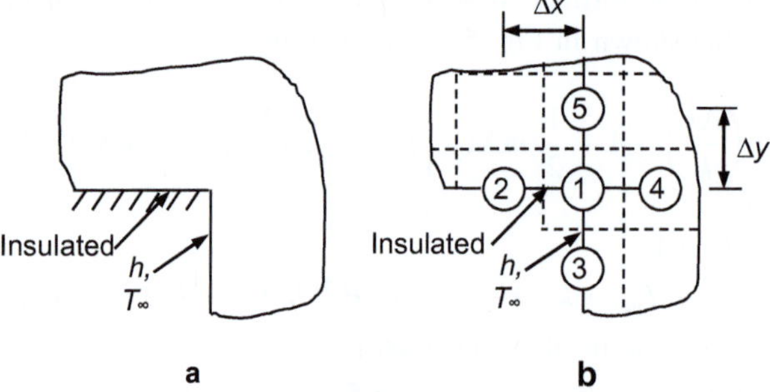

Fig. 5.45 Problem 5.8

[**Ans**. The heat balance equation is

$$k\left(\frac{\Delta x}{2}\right)\frac{(T_{m,n+1}-T_{m,n})}{\Delta y}+k\left(\frac{\Delta x}{2}\right)\frac{(T_{m,n-1}-T_{m,n})}{\Delta y}+k(\Delta y)\frac{(T_{m-1,n}-T_{m,n})}{\Delta x}+q''\Delta y$$
$$=0$$

Substitution of $\Delta x = \Delta y$ gives the desired result. For adiabatic boundary, $q'' = 0$]

5.8 Write down the nodal equation for internal corner of a two-dimensional system with horizontal boundary insulated and vertical boundary subjected to convection heat transfer as shown in Fig. 5.45a.

Verify that the adiabatic boundary limit result deduced from the equation agrees with the corresponding result in Table 5.3.

[**Ans**. The heat balance equation for node 1 gives for unit depth, refer Fig. 5.45b,

$$h\left(\frac{1.\Delta y}{2}\right)(T_\infty-T_1)+k\left(\frac{1.\Delta y}{2}\right)\frac{(T_2-T_1)}{\Delta x}+k\left(\frac{1.\Delta x}{2}\right)\frac{(T_3-T_1)}{\Delta y}+k(1.\Delta y)\frac{(T_4-T_1)}{\Delta x}$$
$$+k(1.\Delta x)\frac{(T_5-T_1)}{\Delta y}=0$$

Putting Δx = Δy, we have $T_2+T_3+2T_4+2T_5+\left(\frac{h\Delta y}{k}\right)T_\infty-\left(6+\frac{h\Delta y}{k}\right)T_1=0.$

When vertical boundary is also adiabatic, put h = 0 and the result is $T_2+T_3+2T_4+2T_5-6T_1=0$, which is the same as for case (h) Example 5.28.]

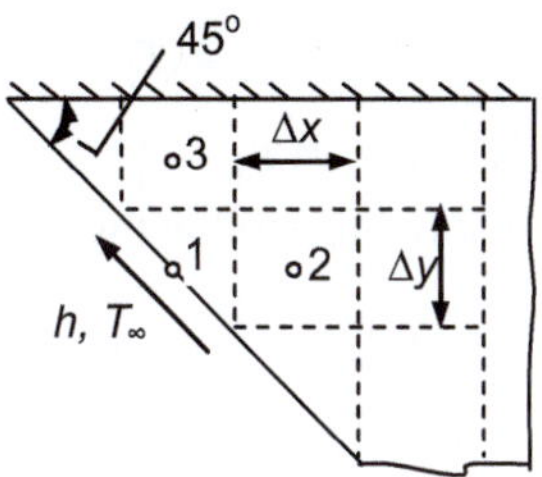

Fig. 5.46 Problem 5.10

5.9 Write down nodal equation for the two-dimensional steady-state conduction problem of Case (g) shown in Fig. 5.26 of Example 5.28 when the vertical boundary is insulated. Consider the dimension perpendicular to the plane of paper as b.
[**Ans**. Referring to the heat balance equation of Case (g) of Example 5.28, the convective heat transfer will reduce to half and the heat balance equation for node 1 will be $h\left(\frac{b\delta}{2}\right)(T_\infty - T_1) + k\left(\frac{b\delta}{2}\right)\frac{(T_2-T_1)}{\delta} + k\left(\frac{b\delta}{2}\right)\frac{(T_3-T_1)}{\delta} = 0$. Simplification gives $T_2 + T_3 + \left(\frac{h\delta}{k}\right)T_\infty - \left(2 + \frac{h\delta}{k}\right)T_1 = 0$. When horizontal boundary is also insulated we get result of Case (d) by putting $h = 0$, i.e. $T_2 + T_3 - 2T_1 = 0$.]

5.10 Write nodal equation for the nodal point 1 on an inclined surface of a system shown in Fig. 5.46. The inclined surface is subjected to a convective heat transfer.
[**Ans**. The heat balance equation for unit depth is $k(\Delta y.1)\left(\frac{T_2-T_1}{\Delta x}\right) + k(\Delta x.1)\left(\frac{T_3-T_1}{\Delta y}\right) + h\left(\sqrt{\Delta x^2 + \Delta y^2}.1\right)(T_\infty - T_1) = 0$. For $\Delta x = \Delta y$ equation transforms to nodal equation as $T_2 + T_3 + \sqrt{2}\frac{h\Delta x}{k}T_\infty - \left(2 + \sqrt{2}\frac{h\Delta x}{k}\right)T_1 = 0$.]

5.11 Write nodal equation for the nodal point 1 on tip of the system shown in Fig. 5.47 The inclined surface is subjected to a convective heat transfer while the horizontal surface is insulated.
[**Ans**. The heat balance equation for unit depth is

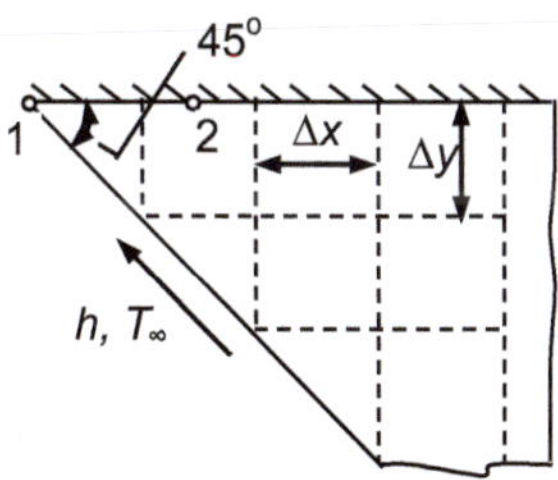

Fig. 5.47 Problem 5.11

$k\left(\frac{\Delta y}{2}.1\right)\left(\frac{T_2-T_1}{\Delta x}\right)+h\left(\frac{1}{2}\sqrt{\Delta x^2+\Delta y^2}.1\right)(T_\infty-T_1)=0$. For $\Delta x = \Delta y$ equation transforms to nodal equation as $T_2+\sqrt{2}\frac{h\Delta x}{k}T_\infty-\left(1+\sqrt{2}\frac{h\Delta x}{k}\right)T_1=0$.]

5.12 Use the relaxation method to solve the following set of equations.

$$500+T_2+T_4-4T_1=R_1 \tag{a}$$

$$300+T_3+T_1-4T_2=R_2 \tag{b}$$

$$500+T_4+T_2-4T_3=R_3 \tag{c}$$

$$700+T_1+T_3-4T_4=R_4 \tag{d}$$

[**Ans.** T_1 = 250°C, T_2 = 200°C, T_3 = 250°C, T_4 = 300°C.]

6 Unsteady or Transient Heat Conduction

6.1 Introduction

When a solid body is suddenly exposed to an environment at different temperatures, the equilibrium temperature condition in the body is established only after some time. The equilibrium condition is referred to as the steady state. We discussed the temperature distribution and heat transfer in the steady state in Chaps. 2–5.

This chapter is devoted to the transient state of conduction, i.e. the heating or cooling of the bodies which takes place in the period before the equilibrium is established. In the transient or unsteady-state heat conduction, the temperature of the solid body varies with time as well as in the space. Such problems are of interest because they are encountered in many industrial processes. The practical problems of the transient heat conduction can be divided into two groups: (i) when the body tends to thermal equilibrium, and (ii) when the temperature of the body is subjected to periodic variation.

6.2 Lumped Heat Capacity Analysis

Consider a body which may be regarded as having uniform temperature throughout at any instant. The assumption of uniform temperature throughout is approximately valid for bodies with a very high thermal conductivity combined with a low value of the convective heat transfer coefficient. Smaller bodies with lower values of the thermal conductivity may also satisfy this condition. Mathematically, we shall assume that the thermal conductivity of the body is infinite. This analysis is called *lumped heat capacity* method.

From the heat balance for the small body shown in Fig. 6.1, which is in interaction with the environment at temperature T_∞, the convection heat loss from the surface of the body at any instant will equal the rate of change of the internal energy of the body. Thus

R. Karwa, *Heat and Mass Transfer*,
https://doi.org/10.1007/978-981-15-3988-6_6

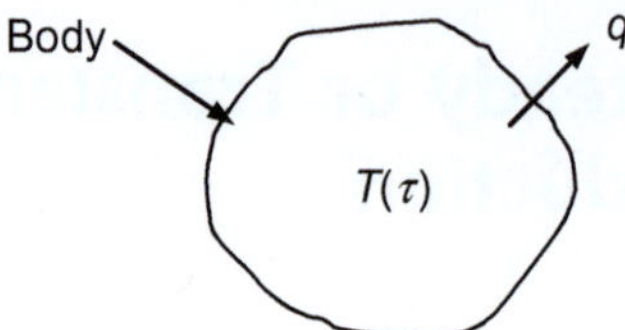

Fig. 6.1 Cooling of a lump

$$q = hA_s(t - t_\infty) = -c(\rho V)\frac{dt}{d\tau} \tag{6.1}$$

where

h is the convection heat transfer coefficient,
A_s is the surface area of the body,
t is the temperature of the body at any instant,
t_∞ is the temperature of the surrounding fluid,
c is the specific heat of the body,
ρ is the density of the material of the body,
(ρV) is the mass of the body of volume V,
$dt/d\tau$ is the rate of change of temperature of the body.

If the body is initially (i.e. at time $\tau = 0$) at a uniform temperature t_i, the temperature of the body after time τ can be determined by integrating Eq. (6.1).

$$\int \frac{dt}{t - t_\infty} = -\left(\frac{hA_s}{c\rho V}\right)\int d\tau$$

or

$$\ln(t - t_\infty) = -\left(\frac{hA_s}{c\rho V}\right)\tau + C_1$$

The constant of integration C_1 can be found from the condition that $t = t_i$ at $\tau = 0$. This gives

$$\ln(t_i - t_\infty) = C_1$$

Hence, we obtain

$$\ln(t - t_\infty) = -\left(\frac{hA_s}{c\rho V}\right)\tau + \ln(t_i - t_\infty)$$

or

$$\ln\frac{t - t_\infty}{t_i - t_\infty} = -\left(\frac{hA_s}{c\rho V}\right)\tau$$
$$\frac{t - t_\infty}{t_i - t_\infty} = \exp\left[-\left(\frac{hA_s}{c\rho V}\right)\tau\right] \tag{6.2}$$

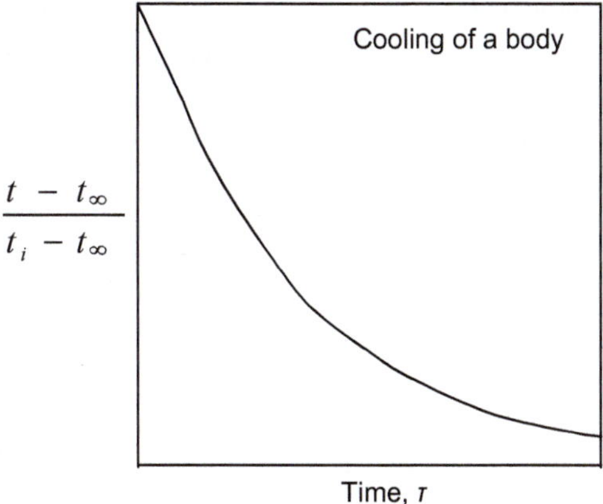

Fig. 6.2 Cooling curve of a lumped heat capacity system

From Eq. (6.2), it is evident that the temperature of the body falls or rises exponentially with time τ as shown in Fig. 6.2.

The exponent of Eq. (6.2) can be transformed into a product of two dimensionless parameters:

$$\begin{aligned}\frac{hA_s}{c\rho V}\tau &= \left(\frac{hV}{kA_s}\right)\left(\frac{A_s^2 k}{c\rho V^2}\right)\tau \\ &= \left(\frac{hL}{k}\right)\left(\frac{k}{c\rho L^2}\right)\tau \\ &= \left(\frac{hL}{k}\right)\left(\frac{\alpha\tau}{L^2}\right)\end{aligned} \tag{6.3}$$

where L (= V/A_s) is referred to as a characteristic dimension of the solid, and α (= $k/c\rho$) is the thermal diffusivity of the solid.

The values of the characteristic dimension L, in Eq. (6.3), for some simple geometric shapes are given in Table 6.1.

The first dimensionless term (hL/k), in Eq. (6.3), is called the *Biot number* Bi. The second dimensionless term is dimensionless time and is referred as *Fourier number* Fo. That is,

$$\mathrm{Bi} = \frac{hL}{k} \tag{6.4}$$

$$\mathrm{Fo} = \frac{\alpha\tau}{L^2} \tag{6.5}$$

Equation (6.2), thus, becomes

$$\frac{t - t_\infty}{t_i - t_\infty} = \exp(-\mathrm{BiFo}) \tag{6.6}$$

The Biot number compares the relative magnitudes of the internal resistance to conduction heat transfer and surface resistance to convection heat transfer. A low value of the Biot number means that the internal resistance is small in comparison to the surface resistance and due to the low internal resistance the temperature throughout the body will be uniform. In

Table 6.1 Characteristic dimension L

Type of solid	Volume, V	Surface area, A_s	Characteristic dimension, $L = V/A_s$
Cylinder (diameter D and length L; $D << L$))	$(\pi/4)D^2L$	πDL	$D/4$
Sphere (diameter D)	$(\pi/6)D^3$	πD^2	$D/6$
Cube (side L)	L^3	$6L^2$	$L/6$
Plate (width W, length L, and thickness δ; $\delta << L$)	$WL\delta$	$2WL$	$\delta/2$

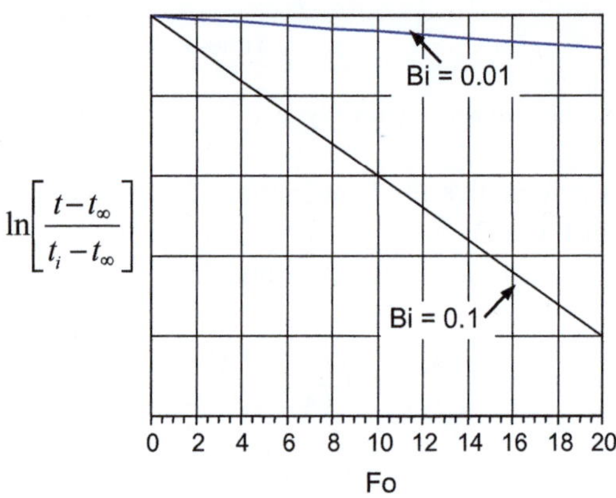

Fig. 6.3 Temperature–time curves

this situation, the transient behaviour of the body is controlled by the convection heat transfer coefficient.

Using the dimensionless numbers Bi and Fo, the temperature–time curves for the bodies covered by the lumped capacity analysis can be reduced to a single universal plot for any value of the heat transfer coefficient h as shown in Fig. 6.3.

An electric capacitor discharges in a circuit with a pure resistance according to the relation:

$$\frac{E}{E_i} = \exp[-\tau/(RC)_e] \tag{6.7}$$

where

R electrical resistance,
C capacitance,
E_i potential at $\tau = 0$,
E potential at $\tau > 0$, that is, $E = E(\tau)$.

The product $(RC)_e$ has the units of time and is known as the time constant.

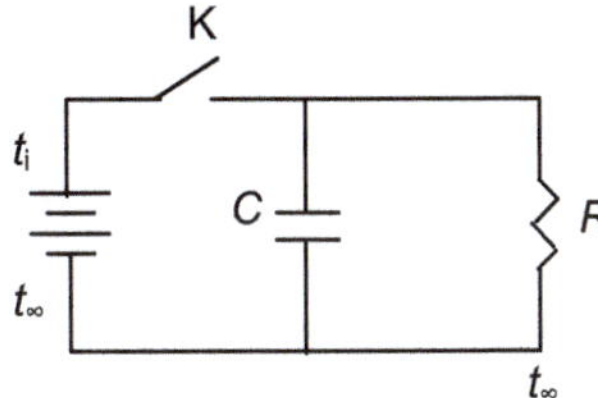

Fig. 6.4 The thermal network for a single-lump heat capacity system

For a thermal system, the thermal capacitance is $C = mc = (\rho V)c$ and the thermal resistance is $1/hA_s$. Using these, Eq. (6.2) can be written as

$$\frac{t - t_\infty}{t_i - t_\infty} = \frac{\theta}{\theta_i} = \exp[-\tau/(RC)_{th}] \tag{6.8}$$

This equation is analogous to Eq. (6.7). In a thermal system, heat is stored while in an electric system, the electric charge is stored and the flow of heat from the thermal system is equivalent to the flow of electric current. This analogy can be used for the analysis of the lumped heat capacity systems by designing an electric system which satisfies the following condition:

$$(RC)_e = (RC)_{th} = \frac{c\rho V}{hA_s} \tag{6.9}$$

Figure 6.4 shows the thermal network for a single-lump heat capacity system. In this network, the thermal capacitor is charged initially to potential (temperature) t_i when the switch K is closed. The stored energy is dissipated through the resistance $R = 1/hA$ when the switch is opened.

The *thermal time constant* is the time required for the temperature difference $(t–t_\infty)$ between the body and the surrounding to reach 0.368 of its initial value $(t_i–t_\infty)$, i.e.

$$\frac{t - t_\infty}{t_i - t_\infty} = 0.368 \tag{6.10}$$

Corresponding to the temperature ratio $(t–t_\infty)/(t_i–t_\infty) = 0.368$, the value of $hA_s\tau/(c\rho V)$ in the exponent in Eq. (6.2) is unity, i.e.

$$\tau = (RC)_{th} = \frac{c\rho V}{hA_s} = \frac{c\rho L}{h}$$

If a thermometer is used to measure an unsteady temperature, it is important to understand the speed with which it follows the change. Generally, a term known as *half-value time* is used to measure this characteristic of the thermometer. It is the time within which the initial difference between the true temperature and the indicated temperature of the thermometer is reduced to half after a sudden change of the true temperature. Corresponding to the temperature ratio $(t–t_\infty)/(t_i–t_\infty) = 0.5$, the value of the exponent in Eq. (6.2) is 0.693. Thus, the half-value time τ_H is given by

$$\tau_H \frac{h}{c\rho L} = 0.693 \tag{6.11}$$

6.2.1 Instantaneous and Total Heat Flow

The instantaneous heat flow from the surface of the lump by convection is

$$q = hA_s(t - t_\infty)$$

Substitution of the value of (t–t_∞) from Eq. (6.2) gives

$$q = hA_s(t_i - t_\infty)\exp\left[-\left(\frac{hA_s}{c\rho V}\right)\tau\right] \tag{6.12}$$

The total heat flow q_t in time τ can be determined by integrating Eq. (6.12):

$$\begin{aligned} q_t &= \int_0^\tau hA_s(t_i - t_\infty)\exp\left[-\left(\frac{hA_s}{c\rho V}\right)\tau\right]d\tau \\ &= \left\{hA_s(t_i - t_\infty)\exp\left[-\left(\frac{hA_s}{c\rho V}\right)\tau\right] \times \frac{1}{-\left(\frac{hA_s}{c\rho V}\right)}\right\}_0^\tau \\ &= -c\rho V(t_i - t_\infty)\left\{\exp\left[-\left(\frac{hA_s}{c\rho V}\right)\tau\right] - 1\right\} \\ &= -c\rho V(t_i - t_\infty)[\exp(-\text{Bi Fo}) - 1] \end{aligned} \tag{6.13}$$

Using Eq. (6.6),

$$q = -c\rho V(t_i - t_\infty)\left(\frac{t - t_\infty}{t_i - t_\infty} - 1\right)$$

or

$$q = c\rho V(t_i - t) \tag{6.14}$$

6.2.2 Applicability of the Lumped Heat Capacity Analysis

The temperature variation in a body is a function of the Biot number. Let us consider three cases applied to a plate of thickness $2L$.

(1) **Bi → ∞ (practically Bi > 100)**
From the equation of Bi = hL/k, it follows that when $h \to \infty$ at the given physical parameters and size of the body, Bi → ∞, i.e. when heat is removed at a very high rate from the surface of the body. Under these conditions, the cooling of the body is governed by the physical properties and the dimensions of the body. The temperature distribution, for this case, is shown in Fig. 6.5 for a plane wall.

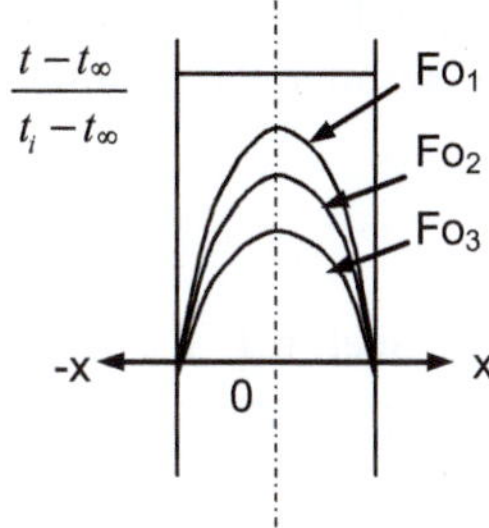

Fig. 6.5 Temperature distribution in a plane wall cooled at Bi $\to \infty$; $Fo_1 < Fo_2 < Fo_3$

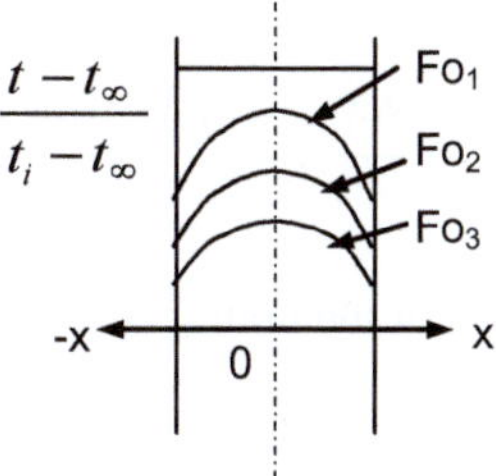

Fig. 6.6 Temperature distribution in a plane wall cooled at 100 > Bi > 0.1; $Fo_1 < Fo_2 < Fo_3$

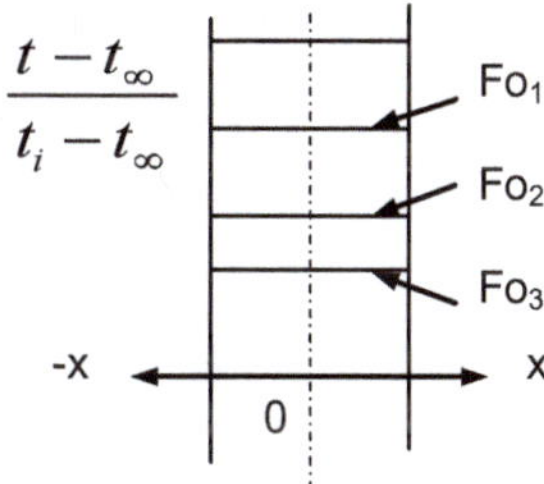

Fig. 6.7 Temperature distribution in a plane wall cooled at Bi < 0.1; $Fo_1 < Fo_2 < Fo_3$

(2) **100 > Bi > 0.1**
The rate of cooling in this case depends both on the internal and external resistances. The temperature distribution curve at any moment of time will appear as shown in Fig. 6.6.

(3) **Very small Biot number (practically Bi < 0.1)**
The Biot number will be small for a thin plate with a large value of thermal conductivity and small heat transfer coefficient. At the small values of the Biot number, the surface temperature differs little from the temperature at the axis of the plate, that is, the temperature distribution across the plate can be assumed to be uniform at any moment of time, Fig. 6.7. Thus, the lumped heat capacity analysis is applicable to the condition when Bi $\leq$ 0.1. In this case, the rate of cooling depends on the value of the heat transfer coefficient only.

Example 6.1 In a lumped heat capacity system, if there is internal heat generation at a constant volumetric rate q_g W/m^3, determine the relationship between time and temperature.

Solution

Following the procedure outlined in Sect. 6.1, the heat balance for the body, which is in interaction with the environment at temperature t_∞, the convection heat loss from the surface of the body at any instant will equal the rate of change of the internal energy of the body and the internal heat generation. Thus

$$q = hA_s(t - t_\infty) = -c(\rho V)\frac{dt}{d\tau} + q_g V$$

or

$$\frac{dt}{d\tau} + \frac{hA_s}{\rho c V}t = \frac{q_g V + hA_s t_\infty}{\rho c V}$$

Its solution is the general solution of the homogeneous equation and the particular integral $t = t_\infty + q_g V/hA_s$. Hence,

$$t = C_1 \exp\left(-\frac{hA_s}{\rho c V}\tau\right) + \left(t_\infty + \frac{q_g V}{hA_s}\right)$$

or

$$t - t_\infty = C_1 \exp\left(-\frac{hA_s}{\rho c V}\tau\right) + \frac{q_g V}{hA_s} \tag{i}$$

Applying the initial condition of t_i at time $\tau = 0$, we obtain

$$t_i - t_\infty = C_1 + \frac{q_g V}{hA_s}$$

or

$$C_1 = (t_i - t_\infty) - \frac{q_g V}{hA_s}.$$

Substitution of the value of C_1 in Eq. (i) gives

$$t - t_\infty = \left[(t_i - t_\infty) - \frac{q_g V}{hA_s}\right]\exp\left(-\frac{hA_s}{\rho c V}\tau\right) + \frac{q_g V}{hA_s}.$$

Rearranging the terms and introducing $hA_s\tau/\rho c V$ = Bi Fo, we obtain

$$t - t_\infty = (t_i - t_\infty)\exp(-\text{Bi Fo}) + \frac{q_g V}{hA_s}[1 - \exp(-\text{Bi Fo})],$$

which is the desired relation.

or

$$\frac{t - t_\infty - \frac{q_g V}{h A_s}}{t_i - t_\infty - \frac{q_g V}{h A_s}} = \exp(-\text{Bi Fo}),$$

Example 6.2 A 0.8 mm diameter Nichrome wire (ρ = 8400 kg/m^3, c = 420 J/(kg K), k = 12 W/(m K) and electric resistivity $\rho_e = 1.2 \times 10^{-6}$ Ωm) is dipped in oil at bulk temperature of 25°C. When 10 A current is passed through the wire, determine the time for the wire to come within 1°C of its steady-state temperature. The convection heat transfer coefficient is 500 W/(m^2 K).

Solution

The Biot number for a cylindrical configuration,

$$\text{Bi} = \frac{h(d/4)}{k} = \frac{500 \times (0.8/4000)}{12} = 0.0083.$$

Since Biot number is less than 0.1, lumped heat capacity analysis can be applied. For a lumped heat capacity system with internal heat generation at a constant volumetric rate q_g W/m^3, temperature–time relationship is given by, refer Example 6.1,

$$\frac{t - t_\infty - \frac{q_g V}{h A_s}}{t_i - t_\infty - \frac{q_g V}{h A_s}} = \exp(-\text{Bi Fo})$$

or

$$\frac{t - t_\infty - \frac{q_g d}{4h}}{t_i - t_\infty - \frac{q_g d}{4h}} = \exp(-\text{Bi Fo}) \tag{i}$$

as

$$\frac{V}{A_s} = \frac{(\pi/4)d^2 L}{\pi d L} = \frac{d}{4}.$$

Here $q_g = I^2 R_e / [(\pi/4)d^2 L]$, where electric resistance,

$$R_e = \frac{\rho_e L}{A_c} = \frac{\rho_e L}{(\pi/4)d^2} = \frac{1.2 \times 10^{-6} \times 1}{(\pi/4) \times (0.8/1000)^2} = 2.39\,\Omega.$$

Hence,

$$q_g = \frac{10^2 \times 2.39}{\frac{\pi}{4} \times \left(\frac{0.8}{1000}\right)^2 \times 1.0} = 4.75 \times 10^8\ \text{W/m}^3.$$

$$\text{Fo} = \frac{\alpha\tau}{L^2} = \left(\frac{k}{\rho c}\right)\frac{\tau}{L^2} = \left(\frac{k}{\rho c}\right)\frac{\tau}{(d/4)^2}$$
$$= \left(\frac{12}{8400 \times 420}\right) \times \frac{\tau}{(0.8/4000)^2} = 85.03\,\tau.$$

In steady state, the heat generation rate equals the heat transfer rate from the wire surface, hence

$$q_g V = hA_s(t_w - t_\infty)$$

or

$$t_w = t_\infty + \frac{q_g V}{hA_s} = t_\infty + \frac{q_g d}{4h} = 25 + \frac{4.75 \times 10^8 \times (0.8/1000)}{4 \times 500} = 215\,°\text{C},$$

where t_w is the wire temperature in steady-state condition. Hence, t = 215–1 = 214°C.
Substitution of the values of various terms in Eq. (i) gives

$$\frac{214 - 25 - \frac{4.75\times10^8\times(0.8/1000)}{4\times 500}}{25 - 25 - \frac{4.75\times10^8\times(0.8/1000)}{4\times500}} = \exp(-0.0083 \times 85.03\tau)$$

Solution gives τ = 7.4 s.

Example 6.3 A 72 mm diameter orange is subjected to a cold environment. The heat transfer coefficient h is estimated to be 10 W/(m^2 K). Could a lumped thermal capacity analysis be applied? Thermal conductivity of apple is 0.6 W/(m K) at 20°C.

Solution

Assuming apple to be spherical, the characteristic dimension, $L = D/6$ = 72/6 = 12 mm
Biot number,

$$\text{Bi} = \frac{hL}{k} = \frac{10 \times 0.012}{0.6} = 0.2,$$

which is greater than 0.1, hence the lumped heat capacity analysis must not be used.

Example 6.4 A right circular cone has a base diameter of 50 mm and a height of 75 mm. The thermal conductivity of the cone material is 15 W/(m K). The heat transfer coefficient is 50 W/(m^2 K). Can we apply the lumped capacity analysis? Neglect the heat transfer from the base.

Solution

The characteristic dimension is

$$L = \frac{V}{A_s} = \frac{\frac{1}{3}\pi R^2 H}{\pi R(R^2 + H^2)^{1/2}} = \frac{RH}{3(R^2 + H^2)^{1/2}},$$

where

R radius of the base = 25 mm,
H height of the cone = 75 mm.

Substitution gives

$$L = \frac{25 \times 75}{3(25^2 + 75^2)^{1/2}} \approx 8 \text{ mm}.$$

Biot number,

$$\text{Bi} = \frac{hL}{k} = \frac{50 \times 0.008}{15} = 0.0267 < 0.1.$$

Lumped capacity analysis can be applied.

Example 6.5 What can be the maximum diameter of a steel ball [k = 40 W/(m K)] subjected to a convective heat transfer coefficient h = 25 W/(m^2 K) for the applicability of the lumped capacity analysis?

Solution

The characteristic dimension is

$$L = \frac{V}{A_s} = \frac{D}{6}$$

For the lump capacity analysis to be applicable, Bi must be less than 0.1.

$$\text{Bi} = 0.1 = \frac{hL_{\text{max}}}{k} = \frac{25 \times D_{\text{max}}/6}{40}$$

or

$$D_{\text{max}} = 0.96 \text{ m},$$

which is the desired result.

Example 6.6 Steel balls 10 mm in diameter are annealed by heating to 880°C and then slowly cooling to 100°C in an air environment at 25°C. The convection heat transfer coefficient is 15 W/(m^2 K). Estimate the time required for this cooling process. For steel take k = 40 W/(m K), c = 450 J/(kg K) and ρ = 7900 kg/m^3.

Solution

For a sphere, characteristic length $L = R/3$ = 0.005/3 = 1/600 m.

Biot number = hL/k = 15/(40 × 600) = 6.25 × 10^{-4} < 0.1, hence the lumped heat capacity solution is applicable. For this condition

$$\frac{t - t_\infty}{t_i - t_\infty} = \exp[-\mathrm{Bi\ Fo}]$$

where

$$\mathrm{Fo} = \frac{\alpha\tau}{L^2} = \frac{k}{\rho c}\frac{\tau}{L^2} = \frac{40}{7900 \times 450 \times (1/600)^2}\tau = 4.051\tau.$$

This gives

$$\frac{100 - 25}{880 - 25} = \exp[-6.25 \times 10^{-4} \times 4.051\tau]$$

or

$$\tau = 962\,\mathrm{s}.$$

Example 6.7 A 1 mm diameter copper wire initially at a temperature of 140°C is suddenly placed in the atmosphere at 40°C. The convective heat transfer coefficient is 12 W/(m^2 K). Calculate the time required for the wire to reach a temperature of 90°C. For copper, ρ = 8954 kg/m^3, c = 0.3831 kJ/(kg K), and k = 386 W/(m K).

Solution

For a cylindrical body, characteristic length $L = V/A_s = r/2$ = 0.25 mm.

Biot number = hL/K = 12 × 0.25/(386 × 1000) = 7.77 × 10^{-6} < 0.1, hence the lumped heat capacity analysis is applicable.

For the given data,

$$\frac{t - t_\infty}{t_i - t_\infty} = \frac{90 - 40}{140 - 40} = 0.5$$

$$\mathrm{Fo} = \frac{\alpha\tau}{L^2} = \frac{k}{\rho c}\frac{\tau}{L^2} = \frac{386}{8954 \times 383.1 \times (0.25/1000)^2}\tau = 1800.4\tau$$

From Eq. (6.6),

$$-\mathrm{Bi\ Fo} = \ln\left(\frac{t - t_\infty}{t_i - t_\infty}\right)$$

Substitution of the values of various terms gives

$$-7.77 \times 10^{-6} \times 1800.4\tau = \ln(0.5)$$

or

$$\tau = 49.55\,\mathrm{s}.$$

Example 6.8 How much heat has been removed from the steel balls of Example 6.6 in 962 s?

Solution

From Eq. (6.13),

$$q = c\rho V(t_i - t_\infty)[\exp(-\text{Bi Fo}) - 1]$$

$$= 450 \times 7900 \times \frac{4}{3} \times \pi \times \left(\frac{5}{1000}\right)^3 \times (880 - 25)$$

$$\times (0.08772 - 1)$$

$$= -1452\,\text{W}.$$

Alternatively using Eq. (6.14)

$$q = c\rho V(t_i - t_1)$$

$$= 450 \times 7900 \times \frac{4}{3} \times \pi \times \left(\frac{5}{1000}\right)^3$$

$$\times (880 - 100)$$

$$= 1452\,\text{W}.$$

Example 6.9 A cubical piece of aluminium 10 mm on a side is to be heated from 50°C to 300°C directly by flame. How long should the piece remain in the flame, if the flame temperature is 800°C and the heat transfer coefficient between the flame and the aluminium piece is 190 W/(m^2 K)? For aluminium, ρ = 2719 kg/m^3, c = 0.871 kJ/(kg K) and k = 215 W/(m K).

Solution

The characteristic dimension for a cube with side L is

$$L = \frac{V}{A_s} = \frac{L^3}{6L^2} = \frac{L}{6}$$

From Eq. (6.4),

$$\text{Bi} = \frac{hL}{k} = \frac{190 \times 1 \times 10^{-2}}{6 \times 215} = 1.473 \times 10^{-3} < 0.1.$$

As the Biot number is less than 0.1, the lumped heat capacity analysis can be used, which gives

$$\frac{t - t_\infty}{t_i - t_\infty} = \exp\left[-\left(\frac{hA_s}{c\rho V}\right)\tau\right] \tag{6.2}$$

Inserting the given data, we have

$$\frac{300 - 800}{50 - 800} = \exp\left[-\left(\frac{190 \times 6 \times 10^2 \times 1000}{0.871 \times 1000 \times 2719 \times 10^3}\right)\tau\right]$$

Solution of the above equation gives

$$\tau = 8.41 \text{ s.}$$

Example 6.10 A chromel–alumel thermocouple (diameter = 0.7 mm) is used to measure the temperature of a gas stream for which the heat transfer coefficient is 600 W/(m^2 K). Estimate the time constant of the thermocouple. Given: ρ = 8500 kg/m^3; c = 400 J/(kg K).

Solution

The thermocouple bead is very small in diameter, hence the lumped heat capacity analysis can be applied.

$$\frac{t - t_\infty}{t_i - t_\infty} = 0.368 = \exp\left[-\left(\frac{hA_s}{c\rho V}\right)\tau\right]$$

or

$$\left(\frac{hA_s}{c\rho V}\right)\tau = 1$$

Substitution of the values of various parameters gives

$$\left[\frac{600 \times \pi(0.7)^2 \times 1000}{400 \times 8500 \times (\pi/6) \times (0.7)^3}\right]\tau = 1$$

or

$$\tau = 0.66 \text{ s.}$$

Example 6.11 The bead of a copper constant thermocouple is 0.5 mm in diameter. If its initial temperature is 30°C and the surrounding air temperature is 80°C, how long will it take the thermocouple bead to attain (i) 79°C (ii) 79.5°C and (iii) 79.9°C? Given: k = 23 W/(m K), c = 410 J/(kg K), ρ = 8900 kg/m^3, and h = 50 W/(m^2 K).

Solution

From Eq. (6.2),

$$\tau = -\left(\frac{c\rho V}{hA_s}\right)\ln\left(\frac{t - t_\infty}{t_i - t_\infty}\right)$$

Substituting values of various terms and $V/A_s = D/6$ for a spherical body, we get

$$\tau = -6.08 \ln\left(\frac{t - 80}{-50}\right)$$

For t = 79.0°C, τ = 23.8 s, for t = 79.5°C, τ = 28.0 s and for t = 79.9°C, τ = 37.8 s.

Note: If the bead diameter is doubled, the above-calculated values will be doubled as time τ is proportional to $L = D/6$. It is evident that the measurement of the temperature of an unsteady thermal system, the diameter of the bead of the thermocouple should be as small as possible.

Example 6.12 Consider a mercury thermometer (k = 10 W/(m K), c = 140 J/(kg K), ρ = 13.6 g/cm^3) for the measurement of the temperature of the air stream in Example 6.11. The mercury bulb can be assumed to be cylindrical in shape (D = 5 mm). The thermal resistance of the thin glass wall may be neglected. How long will it take the bulb to attain 79°C? Given h = 50 W/(m^2 K).

Solution

Here,

$$\tau = -\left(\frac{c\rho}{h} \times \frac{V}{A_s}\right) \times \ln\left(\frac{t - t_\infty}{t_i - t_\infty}\right) = -\left(\frac{140 \times 13600}{50} \times \frac{0.005}{4}\right) \ln\left(\frac{-1}{-50}\right) = 186.2\,\text{s}.$$

Comparison of the result with that of Example 6.11 clearly shows the disadvantage of the mercury thermometer for recording unsteady temperatures.

Example 6.13 If a steel spherical ball of 100 mm diameter is coated with a 1 mm thick layer of dielectric material of thermal conductivity 0.05 W/(m K), estimate the time required to cool it from 500°C to 100°C in an oil bath at 25°C. The convection heat transfer coefficient is 500 W/(m^2 K). For steel, k = 40 W/(m K), c = 450 J/(kg K) and ρ = 7900 kg/m^3.

Solution

The total resistance to heat transfer is

$$\begin{aligned} R_t &= \frac{r_2 - r_1}{4\pi k_i r_1 r_2} + \frac{1}{h(4\pi r_2^2)} \\ &= \frac{1/1000}{4\pi \times 0.05 \times 0.05 \times 0.051} + \frac{1}{500 \times (4\pi \times 0.051^2)} \\ &= 0.6853\,\text{K/W} \end{aligned}$$

The overall heat transfer coefficient,

$$U = \frac{1}{R_t A_2} = \frac{1}{R_t 4\pi r_2^2} = \frac{1}{0.6853 \times 4\pi \times 0.051^2} = 44.64\,\text{W}/(\text{m}^2\,\text{K}).$$

Note: Since the difference in r_2 and r_1 is very small, the overall heat transfer coefficient can also be approximated from $U = (R_t')^{-1} = \left(\frac{r_2 - r_1}{k} + \frac{1}{h}\right)^{-1} = \left(\frac{1/1000}{0.05} + \frac{1}{500}\right)^{-1}$ $= 45.45\,\text{W}/(\text{m}^2\,\text{K})$.

For a sphere, characteristic length $L = D/6 = 0.102/6 = 0.017$ m.

Biot number

$$\mathrm{Bi} = \frac{UL}{k} = \frac{44.64 \times 0.017}{40} = 0.019 < 0.1$$

hence the lumped heat capacity method can be used.

From Eq. (6.2),

$$\left(\frac{UA_s}{c\rho V}\right)\tau = -\ln\left(\frac{t - t_\infty}{t_i - t_\infty}\right)$$

or

$$\tau = -\ln\left(\frac{t - t_\infty}{t_i - t_\infty}\right) \times \left(\frac{c\rho V}{UA_s}\right) = -\ln\left(\frac{t - t_\infty}{t_i - t_\infty}\right) \times \left(\frac{c\rho D}{6U}\right)$$

where $\frac{V}{A_s} = \frac{\pi D^3/6}{\pi D^2} = \frac{D}{6}$. Hence, time for cooling on substitution of values of various terms

$$\tau = -\ln\left(\frac{100 - 25}{500 - 25}\right) \times \left(\frac{450 \times 7900 \times 0.102}{6 \times 44.64}\right) = 2499\,\mathrm{s} = 0.694\,\mathrm{h}.$$

Example 6.14 A rectangular plate of thickness δ = 50 mm (width W, length L; $\delta \ll L$ and $\delta \ll W$), which is initially at a uniform temperature of 400°C, is exposed to a fluid at 25°C. The convective heat transfer coefficient is 50 W/(m^2 K). Determine the time required to lose 50% of its stored heat and temperature of the plate at that time. Neglect radiation heat exchange. For the plate material, k = 210 W/(m K), c = 900 J/(kg K) and ρ = 2710 kg/m^3.

Solution

Characteristic dimension L of a plate from Table 6.1 is $\delta/2$ = 0.025 m.

Biot number,

$$\mathrm{Bi} = \frac{hL}{k} = \frac{50 \times 0.025}{210} = 0.00595 < 0.1.$$

Hence, lumped heat capacity method can be applied.

From Eq. (6.13), the heat flow q_t in time τ is

$$q_t = -c\rho V(t_i - t_\infty)\left\{\exp\left[-\left(\frac{hA_s}{c\rho V}\right)\tau\right] - 1\right\} \tag{6.13}$$

Maximum possible total heat flow is (when body acquires surrounding temperature)

$$q_{\max} = c\rho V(t_i - t_\infty)$$

Hence, 50% of the maximum possible heat flow in time τ is

$$q_t = 0.5c\rho V(t_i - t_\infty)$$

Substitution in Eq. (6.13) gives

$$0.5c\rho V(t_i - t_\infty) = -c\rho V(t_i - t_\infty)\left\{\exp\left[-\left(\frac{hA_s}{c\rho V}\right)\tau\right] - 1\right\}$$

or

$$0.5 = -\left\{\exp\left[-\left(\frac{hA_s}{c\rho V}\right)\tau\right] - 1\right\}$$

or

$$0.5 = \exp\left[-\left(\frac{hA_s}{c\rho V}\right)\tau\right]$$

or

$$-\left(\frac{hA_s}{c\rho V}\right)\tau = \ln 0.5$$

or

$$\tau = -\ln 0.5 \times \left(\frac{c\rho V}{hA_s}\right)$$

From Table 6.1, $V/A_s = \delta/2$ for a plate. Hence,

$$\tau = -\ln 0.5 \times \left(\frac{c\rho\delta}{2h}\right) = -\ln 0.5 \times \left(\frac{900 \times 2710 \times 0.05}{2 \times 50}\right) = 845\,\text{s}.$$

From Eq. (6.2), temperature after time τ is

$$t = t_\infty + (t_i - t_\infty)\exp\left[-\left(\frac{hA_s}{c\rho V}\right)\tau\right]$$

or

$$= t_\infty + (t_i - t_\infty)\exp\left[-\left(\frac{2h}{c\rho\delta}\right)\tau\right]$$

or

$$= 25 + (400 - 25)\exp\left[-\left(\frac{2 \times 50}{900 \times 2710 \times 0.05}\right) \times 845\right] = 212.5\,^\circ\text{C}.$$

Alternatively, simply use $\frac{q_t}{q_{\max}} = \frac{c\rho V(t - t_\infty)}{c\rho V(t_i - t_\infty)} = 0.5$ to calculate t.

Note: When radiation heat exchange is present, radiation heat transfer coefficient is calculated from

$$h_r = \varepsilon\sigma\left(T^2 + T_{\text{sur}}^2\right)(T + T_{\text{sur}})$$

when the body at temperature T is in a large space having surface temperature T_{sur}, for details about radiation heat transfer coefficient refer Chap. 11. It is to note that h_r varies as the body temperature changes. The radiation heat transfer coefficient h_r has its maximum value when body temperature T is maximum, which is at the start of cooling when the body is cooled or is the temperature attained after heating if the body is heated. If the Biot number based on the sum of convection heat transfer coefficient and the maximum radiation heat transfer coefficient is less than 0.1, lumped heat capacity analysis may be applied.

When both convection and radiation are present, the heat balance for the body gives

$$-c(\rho V)\frac{\mathrm{d}T}{\mathrm{d}\tau} = hA_s(T - T_\infty) + \varepsilon A_s\sigma\left(T^4 - T_{sur}^4\right)$$

Thus the variation of the body temperature is given by

$$\int_{T_i}^{T} \mathrm{d}T = T(\tau) - T_i = -\frac{A_s}{\rho c V}\int_0^{\tau}\left[h(T - T_\infty) + \varepsilon\sigma\left(T^4 - T_{sur}^4\right)\right]\mathrm{d}\tau,$$

which is solved numerically.

If only radiation is present (body in an enclosure with vacuum), the heat balance equation is

$$-c(\rho V)\frac{\mathrm{d}T}{\mathrm{d}\tau} = \varepsilon A_s\sigma\left(T^4 - T_{sur}^4\right)$$

Separating variables and integrating from initial condition to time τ, we have

$$\int_0^{\tau} \mathrm{d}\tau = -\frac{\rho c V}{\varepsilon A_s \sigma}\int_{T_i}^{T}\frac{\mathrm{d}T}{T^4 - T_{\text{sur}}^4},$$

which gives

$$\tau = \frac{\rho c V}{4\varepsilon A_s \sigma T_{\text{sur}}^3}\left\{\left[\ln\left(\frac{T_{\text{sur}} + T}{T_{\text{sur}} - T}\right) - \ln\left(\frac{T_{\text{sur}} + T_i}{T_{\text{sur}} - T_i}\right)\right] + 2\left[\tan^{-1}\left(\frac{T}{T_{\text{sur}}}\right) - \tan^{-1}\left(\frac{T_i}{T_{\text{sur}}}\right)\right]\right\}.$$

6.3 Lumped Capacitance, Varying Fluid Temperature

Consider the body of Fig. 6.1, which is in contact with a fluid whose temperature increases with time, i.e.

$$t_\infty = a\tau \tag{i}$$

where a is a constant.

From Eq. (6.1), we write

$$q = hA_s(t - t_\infty) = -c(\rho V)\frac{\mathrm{d}t}{\mathrm{d}\tau}$$

$$\frac{\mathrm{d}t}{\mathrm{d}\tau} + \frac{hA_s}{\rho cV}t - \frac{hA_s}{\rho cV}t_\infty = 0$$

or

$$\frac{\mathrm{d}t}{\mathrm{d}\tau} + \frac{hA_s}{\rho cV}t = \frac{hA_s}{\rho cV}a\tau$$

It is a differential equation whose solution is

$$t = b\exp\left(-\frac{hA_s}{\rho cV}\tau\right) + a\left(\tau - \frac{\rho cV}{hA_s}\right) \tag{ii}$$

Initially, the body and the surrounding fluid are in thermal equilibrium, hence from Eq. (i), $t = 0$ at $\tau = 0$. Applying the condition in the above equation, we get

$$b = a\frac{\rho cV}{hA_s}$$

and the solution, Eq. (ii), becomes

$$t = a\tau - a\frac{\rho cV}{hA_s}\left[1 - \exp\left(-\frac{hA_s}{\rho cV}\tau\right)\right]$$

The variation of the temperature of the body given by this equation is shown in Fig. 6.8. The temperature of the body lags behind the fluid temperature. After the initial transition period, this time lag becomes constant and the temperature of the body varies linearly with time according to the equation

$$t = a\left(\tau - \frac{\rho cV}{hA_s}\right) \tag{6.15}$$

where $\frac{\rho cV}{hA_s}$ is the time lag by which the linear temperature variation of the body is delayed with respect to the fluid temperature.

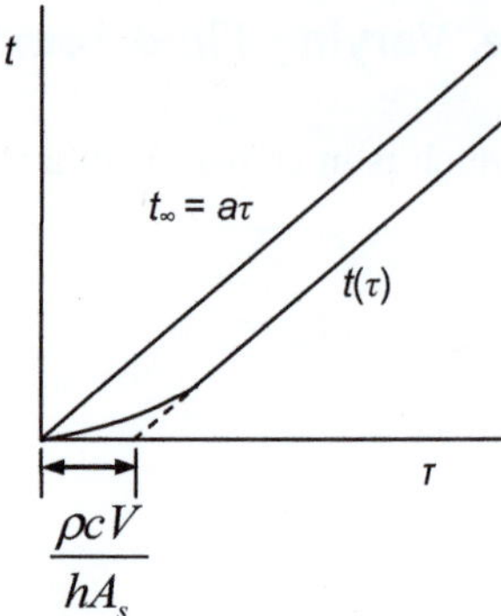

Fig. 6.8 Temperature-time history: linearly varying fluid temperature

Example 6.15 The thermometer of Example 6.12 is to be used to measure the temperature of an electric oven. Calculate the temperature lag of the thermometer if the oven is heated at the rate of 200°C/h. The convective heat transfer coefficient may be taken as 10 W/(m^2 K).

Solution

From Eq. (6.15), the temperature recorded by the thermometer at any moment after the transition period is

$$t = a\left(\tau - \frac{\rho c V}{h A_s}\right)$$

Hence, the temperature lag is

$$\Delta t = a\tau - t = a\frac{\rho c V}{h A_s}$$
$$= \frac{200}{3600} \times \frac{13600 \times 140 \times 0.005}{10 \times 4} = 13.2^\circ\text{C}.$$

6.4 Multiple-Lumped Capacity Systems

Such systems can also be analysed by following the procedure given for a single-lump heat-capacity system.

Let us consider a two-lump system consisting of electrically heated liquid in a container, Fig. 6.9. There is a convective heat transfer from the liquid to the container and from the container to the surrounding fluid at t_∞. The energy balance at any moment on the two lumps (liquid *1* and container 2) gives

$$h_1 A_1 (t_1 - t_2) = -c_1(\rho_1 V_1)\frac{\mathrm{d}t_1}{\mathrm{d}\tau} \tag{i}$$

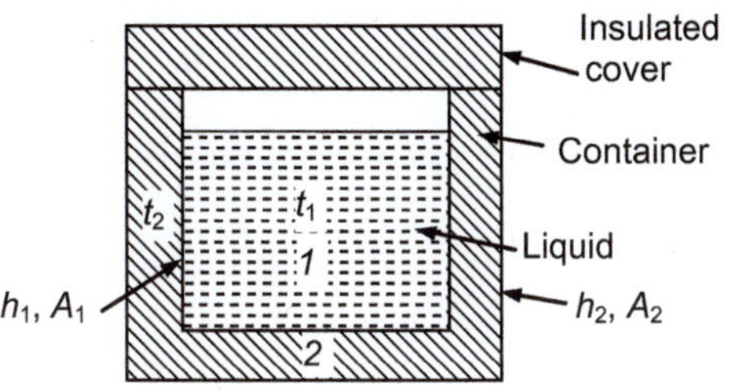

Fig. 6.9 Two-lump system

$$h_1A_1(t_2 - t_1) + h_2A_2(t_2 - t_\infty) = -c_2(\rho_2 V_2)\frac{dt_2}{d\tau} \tag{ii}$$

for the lumps liquid *1* and container *2*, respectively

These are simultaneous linear differential equations for $t_1(\tau)$ and $t_2(\tau)$.

Differentiating Eq. (i) with respect to time τ_1, we get

$$h_1A_1\left(\frac{dt_1}{d\tau} - \frac{dt_2}{d\tau}\right) = -c_1(\rho_1 V_1)\frac{d^2t_1}{d\tau^2}$$

Substitution of the value of $dt_2/d\tau$ from Eq. (ii) and elimination of t_2 from the resulting equation gives

$$\frac{d^2t_1}{d\tau^2} + C_1'\frac{dt_1}{d\tau} + C_2' t_1 = C_2' t_\infty \tag{iii}$$

where

$$C_1' = \frac{h_1A_1}{\rho_1 c_1 V_1} + \frac{h_1A_1}{\rho_2 c_2 V_2} + \frac{h_2A_2}{\rho_2 c_2 V_2}$$
$$C_2' = \left(\frac{h_1A_1}{\rho_1 c_1 V_1}\right)\left(\frac{h_2A_2}{\rho_2 c_2 V_2}\right)$$

The solution of Eq. (iii) is

$$t_1 = Ae^{m_1\tau} + Be^{m_2\tau} + t_\infty \tag{iv}$$

where

$$m_1 = \frac{-C_1' - \left[(C_1')^2 - 4(C_2')\right]^{1/2}}{2}$$
$$m_2 = \frac{-C_1' + \left[(C_1')^2 - 4(C_2')\right]^{1/2}}{2}$$

The constants A and B can be determined from the initial boundary conditions, which are

(i) At the beginning ($\tau = 0$), the two lumps are at the same temperature t_i, i.e.

$$t_1(0) = t_2(0) = t_i.$$

(ii) This condition of $t_1(0) = t_2(0) = t_i$ gives, from Eq. (i),

$$\frac{dt_1}{d\tau} = 0.$$

These conditions give

$$t_i = A + B + t_\infty \tag{v}$$

and

$$\left(\frac{dt_1}{d\tau}\right)_{\tau=0} = Am_1 + Bm_2 + t_\infty = 0 \tag{vi}$$

From Eqs. (v) and (vi),

$$A = \frac{m_2}{m_2 - m_1}(t_i - t_\infty)$$
$$B = -\frac{m_1}{m_2 - m_1}(t_i - t_\infty)$$

Substituting values of A and B in Eq. (iv) gives

$$\frac{t_1 - t_\infty}{t_i - t_\infty} = \frac{m_2}{m_2 - m_1} e^{m_1 \tau} - \frac{m_1}{m_2 - m_1} e^{m_2 \tau} \tag{6.16}$$

The temperature of the container $t_2(\tau)$ can be obtained by substituting the value of $t_1(\tau)$ from Eq. (6.16) and its derivative $dt_1/d\tau$ into Eq. (i).

It is evident that there will be three simultaneous differential equations for a three-lump system, four equations for a four-lump system and so on.

To draw the thermal network, the thermal capacitance of each lump of the system can be determined from the known values of their volume, density and specific heats. The thermal resistance connecting the lump must also be determined.

The thermal network of the two-lump system shown in Fig. 6.9 is presented in Fig. 6.10. Here the connecting resistance is due to convection between the container and liquid. Temperatures t_1 and t_2 reach t_i when the container is insulated from outside an electric heater is run (this is equivalent to opening of switch K_2 and closing of switch K_1, and charging the capacitors to potential t_i). Thereafter, switch K_1 is opened and K_2 is closed, which causes the dissipation of stored energy in the capacitors through the resistances.

We take another example of a two-lump-system as shown in Fig. 6.11. The system consists of two solids (lumps) in perfect thermal contact (negligible contact resistance). Block 2 is exposed to a fluid at temperature t_∞. Initially, both the lumps are at the same temperature, i.e. $t_1(0) = t_2(0) = t_i$ at $\tau = 0$.

$$C_1 = \rho_1 c_1 V_1,\ R_1 = 1/h_1 A_1,\ C_2 = \rho_2 c_2 V_2,\ R_2 = 1/h_2 A_2$$

Fig. 6.10 The thermal network of two-lump system

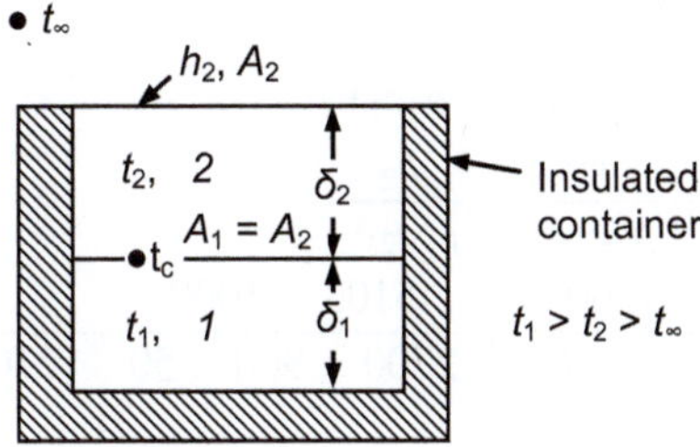

Fig. 6.11 Example 6.16

The heat transfer from lump *1* to *2* is

$$\frac{q}{A} = k_1 \left(\frac{t_1 - t_c}{\delta_1/2} \right) = k_2 \left(\frac{t_c - t_2}{\delta_2/2} \right)$$

where t_c is the temperature at the interface. From the two equations,

$$t_c = \frac{k_1 \delta_2 t_1 + k_2 \delta_1 t_2}{k_2 \delta_1 + k_1 \delta_2}$$

Using the first equation, we get

$$\frac{q}{A} = \frac{q_1}{A_1} = \frac{2k_1}{\delta_1} \left(t_1 - \frac{k_1 \delta_2 t_1 + k_2 \delta_1 t_2}{k_2 \delta_1 + k_1 \delta_2} \right)$$

Rearranging the equation, we get

$$q_1 = A_1 k' (t_1 - t_2)$$

where

$$k' = \frac{2k_1 k_2}{k_2 \delta_1 + k_1 \delta_2}$$

By replacing h_1 of the two-lump system of Fig. 6.9 by k', temperature $t_1(\tau)$ can be determined from Eq. (6.16). The application of the results obtained here is illustrated in the example that follows.

Example 6.16 For the system shown in Fig. 6.11, the following data is given:

For solid *1*: $c_1 = 385$ J/(kg K), $k_1 = 390$ W/(m K) and $\delta_1 = 60$ mm, $\rho_1 = 8950$ kg/m^3; solid *2*: $c_2 = 900$ J/(kg K), $k_2 = 210$ W/(m K) and $\delta_2 = 50$ mm, $\rho_2 = 2700$ kg/m^3; $t_i = 200$°C, $t_\infty = 40$°C and $h_2 = 20$ W/(m^2 K).

Determine the expression for the temperature of solid *1*, assuming applicability of the lumped analysis.

Solution

$$k' = \frac{2k_1k_2}{k_2\delta_1 + k_1\delta_2} = \frac{2 \times 390 \times 210}{(210 \times 60 + 390 \times 50)/1000} = 5103$$

$$\begin{aligned} C_1' &= \frac{k'A_1}{\rho_1 c_1 V_1} + \frac{k'A_1}{\rho_2 c_2 V_2} + \frac{h_2 A_2}{\rho_2 c_2 V_2} \\ &= \frac{k'}{\rho_1 c_1 \delta_1} + \frac{k'}{\rho_2 c_2 \delta_2} + \frac{h_2}{\rho_2 c_2 \delta_2} \\ &= \frac{5103 \times 1000}{8950 \times 385 \times 60} + \frac{5103 \times 1000}{2700 \times 900 \times 50} + \frac{20 \times 1000}{2700 \times 900 \times 50} \\ &= 0.06685 \end{aligned}$$

$$\begin{aligned} C_2' &= \left(\frac{k'A_1}{\rho_1 c_1 V_1}\right)\left(\frac{h_2 A_2}{\rho_2 c_2 V_2}\right) \\ &= \left(\frac{5103 \times 1000}{8950 \times 385 \times 60}\right)\left(\frac{20 \times 1000}{2700 \times 900 \times 50}\right) = 4.06 \times 10^{-6} \end{aligned}$$

$$\begin{aligned} m_1 &= \frac{-C_1' - \left[(C_1')^2 - 4(C_2')\right]^{1/2}}{2} \\ &= \frac{-0.06685 - \left[(0.06685)^2 - 4(4.06 \times 10^{-6})\right]^{1/2}}{2} = -0.06678 \end{aligned}$$

$$\begin{aligned} m_2 &= \frac{-C_1' + \left[(C_1')^2 - 4(C_2')\right]^{1/2}}{2} \\ &= \frac{-0.06685 + \left[(0.06685)^2 - 4(4.06 \times 10^{-6})\right]^{1/2}}{2} = -0.6 \times 10^{-4} \end{aligned}$$

Then

$$\frac{t_1 - t_\infty}{t_i - t_\infty} = \frac{m_2}{m_2 - m_1} e^{m_1 \tau} - \frac{m_1}{m_2 - m_1} e^{m_2 \tau}$$

or

$$\begin{aligned} \frac{t_1 - 40}{200 - 40} &= \frac{-0.6 \times 10^{-4}}{-0.6 \times 10^{-4} - (-0.06678)} e^{-0.06678\tau} \\ &\quad - \frac{(-0.06678)}{-0.6 \times 10^{-4} - (-0.06678)} e^{-0.6 \times 10^{-4}\tau} \end{aligned}$$

or

$$t_1 = 40 - 0.144e^{-0.06678\tau} + 160.14e^{-0.6\times 10^{-4}\tau}$$

6.5 Transient Heat Flow in Semi-infinite Solids

A semi-infinite slab is shown in Fig. 6.12, which is bounded by plane $x = 0$ and extends to infinity in the positive x-direction. The solid also extends to infinity in y- and z-directions. Examples of such bodies are a thick or large block of steel, the earth, etc. When such a body is heated or cooled for a relatively short period, the temperature will change only for a short distance from the surface.

Case A: Surface Temperature of Semi-infinite Body Suddenly Lowered and Maintained at Constant Temperature

First, we consider the case when the surface temperature of the semi-infinite body is suddenly lowered and maintained at temperature t_s.

For a one-dimensional conduction system without heat generation, temperature distribution equation can be deduced from Eq. (2.13a) as

$$\frac{\partial^2 t}{\partial x^2} = \frac{1}{\alpha}\frac{\partial t}{\partial \tau}$$

The boundary conditions are

(i) The body is initially at a uniform temperature,

$$t(x, 0) = t_i \quad \text{at} \quad \tau = 0$$

(ii) At the surface, the temperature is suddenly changed and then maintained at t_s, hence

$$t(0, \tau) = t_s \quad \text{at} \quad \tau > 0$$

(iii) At distance sufficiently away from the surface, the temperature in the body does not change, hence

$$t(\infty, \tau) = t_i \quad \text{at} \quad \tau > 0$$

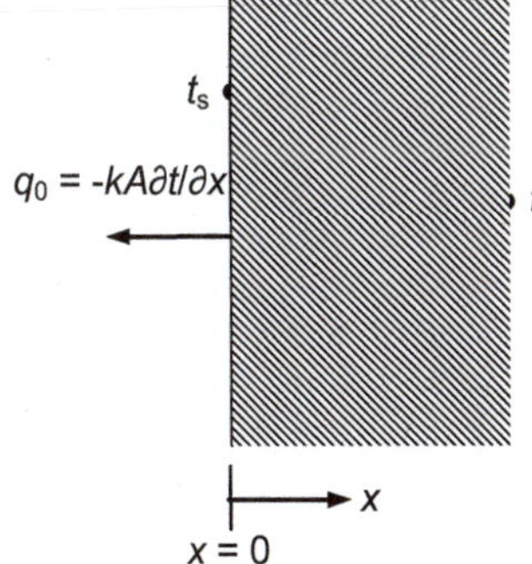

Fig. 6.12 A semi-infinite solid

The differential equation of this problem is solved by the Laplace-transform technique. The solution for temperature distribution, i.e. the temperature of a parallel plane at distance x at any time τ, is

$$\frac{t(x,\tau)-t_s}{t_i-t_s} = \operatorname{erf}\left(\frac{x}{2\sqrt{\alpha\tau}}\right) \tag{6.17}$$

where $\operatorname{erf}\left(\frac{x}{2\sqrt{\alpha\tau}}\right)$ is the Gauss error function. It is defined as

$$\operatorname{erf}\left(\frac{x}{2\sqrt{\alpha\tau}}\right) = \frac{2}{\sqrt{\pi}}\int_0^{\frac{x}{2\sqrt{\alpha\tau}}} e^{-\eta^2}\,d\eta \tag{6.18}$$

where η is a dummy variable and the integral is a function of the upper limit. We can express the non-dimensional temperature distribution as

$$\frac{t(x,\tau)-t_s}{t_i-t_s} = \frac{2}{\sqrt{\pi}}\int_0^{\frac{x}{2\sqrt{\alpha\tau}}} e^{-\eta^2}\,d\eta \tag{6.19}$$

The temperature distribution given by Eq. (6.19) is plotted in Fig. 6.13. The values of the error function are tabulated in Table 6.2.

The heat flow at any plane at distance x from the surface is

$$q_x = -kA\frac{\partial t}{\partial x}$$

The differentiation of Eq. (6.19) gives

$$\begin{aligned}\frac{\partial t}{\partial x} &= (t_i-t_s)\frac{2}{\sqrt{\pi}}\times e^{-\frac{x^2}{4\alpha\tau}}\times\frac{\partial}{\partial x}\left(\frac{x}{2\sqrt{\alpha\tau}}\right)\\ &= \frac{(t_i-t_s)}{\sqrt{\pi\alpha\tau}}\times e^{-\left(\frac{x^2}{4\alpha\tau}\right)}\end{aligned}$$

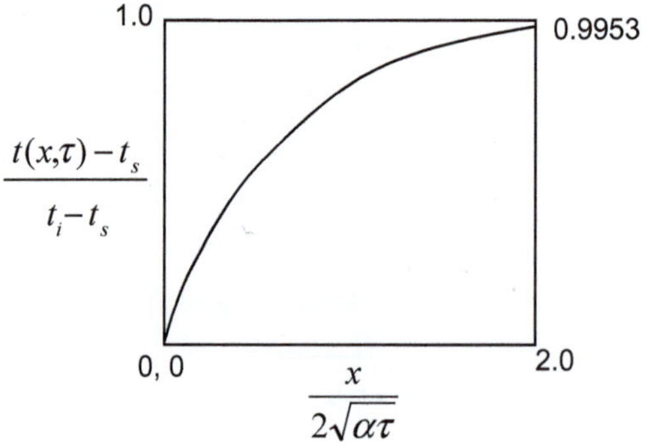

Fig. 6.13 Temperature distribution given by Eq. (6.19)

Table 6.2(a) Error function

$\frac{x}{2\sqrt{\alpha\tau}}$	erf $\frac{x}{2\sqrt{\alpha\tau}}$	$\frac{x}{2\sqrt{\alpha\tau}}$	erf $\frac{x}{2\sqrt{\alpha\tau}}$	$\frac{x}{2\sqrt{\alpha\tau}}$	erf $\frac{x}{2\sqrt{\alpha\tau}}$	$\frac{x}{2\sqrt{\alpha\tau}}$	erf $\frac{x}{2\sqrt{\alpha\tau}}$
0.00	0.0000	0.65	0.6420	1.30	0.9340	1.95	0.9942
0.05	0.05636	0.70	0.6778	1.35	0.9435	2.00	0.9953
0.10	0.1125	0.75	0.7111	1.40	0.9523	2.10	0.9970
0.15	0.1680	0.80	0.7421	1.45	0.9593	2.20	0.9981
0.20	0.2227	0.85	0.7706	1.50	0.9661	2.30	0.9989
0.25	0.2764	0.90	0.7969	1.55	0.9713	2.40	0.9993
0.30	0.3286	0.95	0.8208	1.60	0.9764	2.50	0.9996
0.35	0.3794	1.00	0.8427	1.65	0.9802	2.60	0.9998
0.40	0.4284	1.05	0.8625	1.70	0.9838	2.80	0.99993
0.45	0.4755	1.10	0.8802	1.75	0.9865	3.0	0.99998
0.50	0.5205	1.15	0.8962	1.80	0.9891	3.2	0.999994
0.55	0.5633	1.20	0.9103	1.85	0.9910	3.4	0.999998
0.60	0.6039	1.25	0.9229	1.90	0.9928	3.6	1.000000

Table 6.2(b) Inverted table

erf $\frac{x}{2\sqrt{\alpha\tau}}$	$\frac{x}{2\sqrt{\alpha\tau}}$
0.0	0
0.2	0.1791
0.25	0.2253
0.3	0.2725
0.4	0.3708
0.5	0.4769
0.6	0.5951
0.7	0.7329
0.75	0.8134
0.8	0.9062
0.9	1.1631
1.0	3.6000

Hence,

$$q_x = -kA\frac{(t_i - t_s)}{\sqrt{\pi\alpha\tau}} \times e^{-\left(\frac{x^2}{4\alpha\tau}\right)} \tag{6.20a}$$

The heat flow at the surface is obtained by putting $x = 0$,

$$q_{x=0} = -kA\frac{(t_i - t_s)}{\sqrt{\pi\alpha\tau}} \tag{6.20b}$$

The heat flux at the surface is a function of $(1/\sqrt{\tau})$, hence it diminishes with an increase in time.

The total heat flow in time τ can be obtained by integrating Eq. (6.20b) over the time interval 0 to τ, i.e.

$$\begin{aligned} Q_0 &= -\frac{kA}{\sqrt{\pi\alpha}}(t_i - t_s)\int_0^{\tau}\frac{1}{\sqrt{\tau}}\mathrm{d}\tau \\ &= -\frac{kA}{\sqrt{\pi\alpha}}(t_i - t_s)(2\sqrt{\tau}) \\ &= -1.128kA(t_i - t_s)\sqrt{\frac{\tau}{\alpha}} \end{aligned} \tag{6.21}$$

Case B: Constant Heat Flux on Semi-infinite Solid

The surface of the body is suddenly exposed to uniform heat flux. The body is initially at a uniform temperature, i.e.

$$t(x,0) = t_{\mathrm{i}} \quad \text{at} \quad \tau = 0,$$

and, for $\tau > 0$,

$$\frac{q_o}{A} = -k\left(\frac{\partial t}{\partial x}\right)_{x=0}$$

The temperature distribution, in this case, is expressed as

$$t - t_i = \frac{2q_0\sqrt{\alpha\tau/\pi}}{kA} \times e^{-\left(\frac{x^2}{4\alpha\tau}\right)} - \frac{q_0 x}{kA}\left[1 - \mathrm{erf}\left(\frac{x}{2\sqrt{\alpha\tau}}\right)\right] \tag{6.22}$$

Case C: Convective Boundary Condition

In most of the cases of interest, the heat is transferred to or from the surface of the solid by convection. Hence, at $x = 0$,

$$hA(t_\infty - t)_{x=0} = -kA\left(\frac{\partial t}{\partial x}\right)_{x=0}$$

where t_∞ is the temperature of the fluid to which the surface of the solid is exposed.

The temperature distribution, in this case, is given by

$$\frac{t - t_i}{t_\infty - t_i} = 1 - \mathrm{erf}\left(\frac{x}{2\sqrt{\alpha\tau}}\right) - \exp\left(\frac{hx}{k} + \frac{h^2\alpha\tau}{k^2}\right) \times \left[1 - \mathrm{erf}\left(\frac{x}{2\sqrt{\alpha\tau}} + \frac{h\sqrt{\alpha\tau}}{k}\right)\right] \tag{6.23}$$

The above equation can be written as

$$\frac{t - t_\infty}{t_i - t_\infty} = \mathrm{erf}\left(\frac{x}{2\sqrt{\alpha\tau}}\right) + \exp\left(\frac{hx}{k} + \frac{h^2\alpha\tau}{k^2}\right) \times \left[1 - \mathrm{erf}\left(\frac{x}{2\sqrt{\alpha\tau}} + \frac{h\sqrt{\alpha\tau}}{k}\right)\right] \tag{6.24}$$

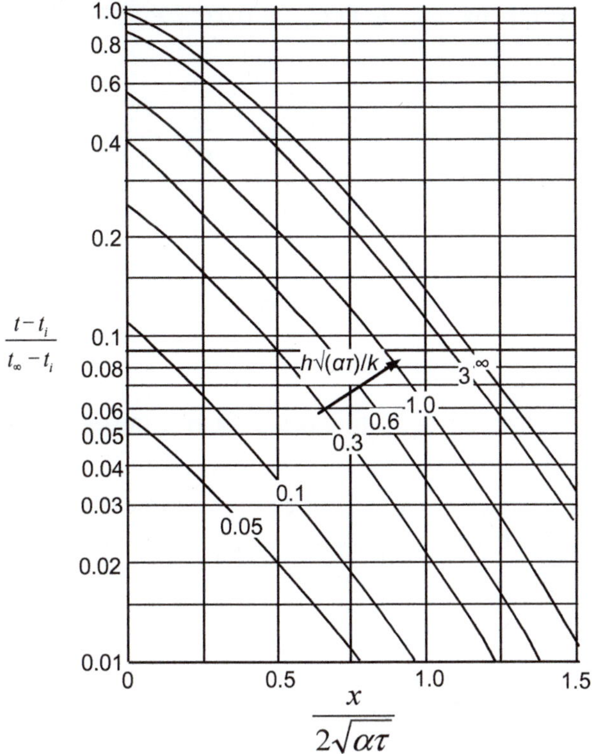

Fig. 6.14 Temperature distribution in semi-infinite solid with convection boundary condition. Holman JP, adapted for SI units by White PRS, Heat Transfer, McGraw-Hill Book Co, New York, Copyright 1992. The material is reproduced with permission of McGraw-Hill Education (Asia)

If h tends to ∞, the surface temperature of the body approaches the temperature of the fluid to which the surface is exposed. For this case, Eq. (6.24) transforms to

$$\frac{t - t_\infty}{t_i - t_\infty} = \text{erf}\left(\frac{x}{2\sqrt{\alpha\tau}}\right)$$

which is the result presented by Eq. (6.17). The solution given by Eq. (6.23) or (6.24) is also available in graphical form in Fig. 6.14.

Example 6.17 A large block of steel, which is initially at a uniform temperature of 30°C, is suddenly changed to and held at 200°C. Calculate the temperature at a depth of 20 mm after 60 s. For steel k = 45 W/(m K) and $\alpha = 1.2 \times 10^{-5}$ m²/s.

Solution

For transient heat conduction in infinite thick solids, Eq. (6.17) applies

$$\frac{t(x,\tau) - t_s}{t_i - t_s} = \text{erf}\left(\frac{x}{2\sqrt{\alpha\tau}}\right)$$

The given data is: $t_s = 200°C$, $t_i = 30°C$, $x = 0.02$, $\alpha = 1.2 \times 10^{-5}$ m²/s and $\tau = 60$ s. Hence,

$$t(x,\tau) - 200 = (30 - 200) \times \text{erf}\left(\frac{0.02}{2\sqrt{1.2 \times 10^{-5} \times 60}}\right)$$

or

$$t(0.02, 60) = -170 \times \text{erf}(0.3727) + 200$$

or

$$t(0.02, 60) = -170 \times 0.406 + 200 = 130.98°\text{C}.$$

Example 6.18 If the surface of the steel block of the previous example is subjected to a uniform heat flux of 200 kW/m^2, calculate the temperature at the same depth after 60 s.

Solution

Equation (6.22) applies, which gives

$$t - t_i = \frac{2q_0\sqrt{\alpha\tau/\pi}}{kA} \times e^{-\left(\frac{x^2}{4\alpha\tau}\right)} - \frac{q_0 x}{kA}\left[1 - \text{erf}\left(\frac{x}{2\sqrt{\alpha\tau}}\right)\right] \tag{6.22}$$

From given data:

$$\frac{x^2}{4\alpha\tau} = \frac{0.02^2}{4 \times 1.2 \times 10^{-5} \times 60} = 0.1389$$
$$\text{erf}\left(\frac{x}{2\sqrt{\alpha\tau}}\right) = 0.406$$
$$q_0 = 200 \times 10^3 \text{ W/m}^2$$
$$A = 1\,\text{m}^2$$
$$k = 45\,\text{W/(m K)}$$
$$\sqrt{\alpha\tau/\pi} = \sqrt{1.2 \times 10^{-5} \times 60/\pi} = 0.01514$$

Inserting the values of various parameters, we obtain

$$t = \frac{2 \times 200 \times 10^3 \times 0.01514}{45 \times 1} \times e^{-(0.1389)} - \frac{200 \times 10^3 \times 0.02}{45 \times 1}[1 - 0.406] + 30 = 94.3°\text{C}$$

Example 6.19 A water pipe is to be buried underground in wet soil ($\alpha = 1.8 \times 10^{-3}$ m²/h). The night temperature can fall to –5°C, which can remain at this value for 9 h. Calculate the minimum depth at which the pipe is to be laid down so that the surrounding soil temperature does not reach 5°C. The soil temperature at the onset of the night is estimated to be 10°C.

Solution

Let at a depth of L, the temperature t just approaches 5°C in 9 h. Assuming the soil as semi-infinite solid, Eq. (6.17) applies. Hence,

$$\frac{t(x,\tau) - t_s}{t_i - t_s} = erf\left(\frac{x}{2\sqrt{\alpha\tau}}\right)$$

or

$$\frac{5-(-5)}{10-(-5)} = \text{erf}\left(\frac{L}{2\sqrt{1.8\times 10^{-3}\times 9}}\right)$$

or

$$0.667 = \text{erf}\left(\frac{L}{0.2546}\right)$$

or

$$\frac{L}{0.2546} = 0.6494$$

or

$$L = 0.165\,\text{m},$$

i.e. the pipe must be laid down at a depth greater than 0.165 m.

Example 6.20 A semi-infinite aluminium slab is initially at a uniform temperature of 400°C. Its surface is suddenly changed to 50°C. How much heat will be removed per unit area of the slab if the temperature at a depth of 100 mm from the surface drops to 100°C? Given $k = 225$ W/(m K), $\alpha = 8.6 \times 10^{-5}$ m^2/s.

Solution

For the given data,

$$\frac{t(x,\tau) - t_s}{t_i - t_s} = \frac{100-50}{400-50} = 0.1429$$

Thus, from Eq. (6.17),

$$erf\left(\frac{x}{2\sqrt{\alpha\tau}}\right) = 0.1429$$

From Table 6.2,

$$\left(\frac{x}{2\sqrt{\alpha\tau}}\right) \approx 0.126$$

or

$$\tau = \left(\frac{0.1}{2\sqrt{8.6 \times 10^{-5}} \times 0.126} \right)^2 = 1831 \text{ s}.$$

Heat transfer at the surface, Eq. (6.21),

$$\frac{q}{A} = 1.128k(t_s - t_i)\sqrt{\frac{\tau}{\alpha}}$$

$$= 1.128 \times 225 \times (50 - 400)\sqrt{\frac{1831}{8.6 \times 10^{-5}}} = -41 \times 10^7 \text{ J/m}^2.$$

The negative sign indicates heat loss from the slab.

Example 6.21 If the ground in Example 6.19 is suddenly exposed to a cold wave at –10°C with h = 50 W/(m^2 K), will the water in the pipe remain above 4°C in the period of 9 h? The thermal conductivity of the wet soil is 2 W/(m K).

Solution

For the given condition, Fig. 6.14 applies. The required parameters are

$$\frac{h\sqrt{\alpha\tau}}{k} = \frac{50 \times \sqrt{1.8 \times 10^{-3} \times 9}}{2.0} = 3.18$$

$$\frac{x}{2\sqrt{\alpha\tau}} = \frac{0.165}{2\sqrt{1.8 \times 10^{-3} \times 9}} = 0.648$$

For these values of the parameters, the figure gives

$$\frac{t(x,\tau) - t_i}{t_\infty - t_i} \approx 0.3$$

Substitution of known temperature data gives

$$\frac{t(x,\tau) - 10}{-10 - 10} \approx 0.3$$

or

$$t = 4°\text{C}.$$

6.6 Transient Heat Conduction in Infinite Plate

Figure 6.15 shows an infinite plate of thickness $2L$. Physically the plate is so large in length and width that end effects may be ignored. This is a case of one-dimensional transient heat conduction. Initially, the plate is at a uniform temperature t_i. At the time zero, both sides of the plate are suddenly reduced to temperature t_1. The differential equation of one-dimensional heat conduction for the unsteady state is

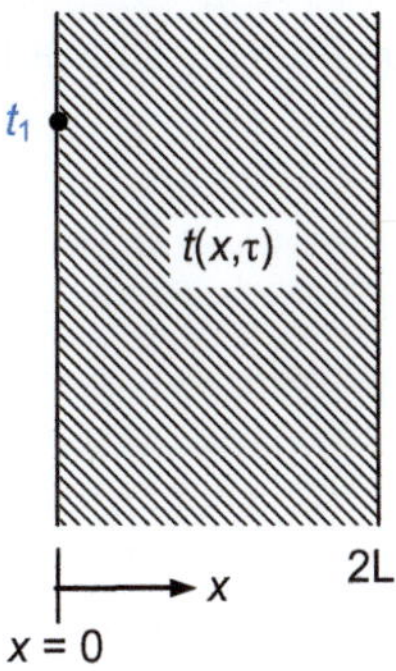

Fig. 6.15 Infinite plate

$$\frac{\partial^2 t}{\partial x^2} = \frac{1}{\alpha}\left(\frac{\partial t}{\partial \tau}\right)$$

Introducing $\theta = t\text{–}t_1$, the equation becomes

$$\frac{\partial^2 \theta}{\partial x^2} = \frac{1}{\alpha}\left(\frac{\partial \theta}{\partial \tau}\right)$$

The initial boundary conditions are

$$\theta = t_{\mathrm{i}} - t_1 = \theta_{\mathrm{i}} \quad \text{at} \quad \tau = 0, \quad 0 \leq x \leq 2L \tag{a}$$

$$\theta = 0 \quad \text{at} \quad x = 0, \quad \tau > 0 \tag{b}$$

$$\theta = 0 \quad \text{at} \quad x = 2L, \quad \tau > 0 \tag{c}$$

Using the technique of separation-of-variables, similar to the one used for the two-dimensional steady-state problem, the solution is of the form

$$\theta(x, \tau) = X(x)H(\tau)$$

It produces two ordinary differential equations

$$\frac{\partial^2 X}{\partial x^2} + \lambda^2 X = 0$$

$$\frac{1}{\alpha}\frac{\partial H}{\partial \tau} + \lambda^2 H = 0$$

where λ^2 is the separation constant.

The form of solution is

$$\theta(x, \tau) = (C_1 \cos \lambda x + C_2 \sin \lambda x)e^{-\lambda^2 \alpha \tau} \tag{i}$$

In order to satisfy the boundary conditions, λ^2 must be > 0. From the boundary condition (b),

$$C_1 = 0.$$

Using boundary condition (c), we obtain

$$0 = (C_2 \sin 2\lambda L)e^{-\lambda^2 \alpha\tau}$$

Since $C_1 = 0$, C_2 cannot be zero. Hence, $\sin(2\lambda L) = 0$ gives

$$\sin(2\lambda L) = \sin(n\pi)$$

or

$$\lambda = \frac{n\pi}{2L}$$

where $n = 1, 2, 3$ .....

Substitution in Eq. (i) gives

$$\theta = \sum_{n=1}^{\infty} C_n e^{-\left(\frac{n\pi}{2L}\right)^2 \alpha\tau} \sin\left(\frac{n\pi x}{2L}\right) \tag{6.25}$$

The initial condition ($\theta = \theta_i$ at $\tau = 0$), when applied to Eq. (6.25), gives

$$\theta_i = \sum_{n=1}^{\infty} C_n \sin\left(\frac{n\pi x}{2L}\right) \tag{6.26}$$

Equation (6.26) is simply a Fourier series expansion of the initial temperature, so that

$$C_n = \frac{1}{L}\int_0^{2L} \theta_i \sin\left(\frac{n\pi x}{2L}\right) dx$$
$$= \frac{4}{n\pi}\theta_i$$

The final solution is

$$\frac{\theta}{\theta_i} = \frac{t - t_1}{t_i - t_1} = \frac{4}{\pi}\sum_{n=1}^{\infty} \frac{1}{n} e^{-\left(\frac{n\pi}{2L}\right)^2 \alpha\tau} \sin\left(\frac{n\pi x}{2L}\right) \tag{6.27}$$

where $n = 1, 3, 5$ .....

Equation (6.27) involves the dimensionless number $\alpha\tau/L^2$, which is the Fourier number based on the half-width L of the plate.

The heat flow, using the Fourier equation, can be obtained as

$$\begin{aligned} q &= -kA\frac{dt}{dx} \\ &= -\frac{4kA}{2L}(\Delta t)_{max}\sum_{n=1}^{\infty} e^{-\left(\frac{n\pi}{2}\right)^2 Fo}\cos\left(\frac{n\pi x}{2L}\right) \end{aligned} \tag{6.28}$$

where Δt_{max} is the difference between the initial and the new face temperature.

6.7 Heisler and Grober Charts

We considered one geometry of unsteady heat conduction in the above section. Solutions have been presented by the researchers for other geometries. The analytical solutions are not useful for rapid calculation, hence the solutions are presented in the graphical form by Heisler. Temperatures as a function of time and spatial position are given in Figs. 6.17, 6.18, 6.19, 6.20, 6.21 and 6.22 for

(i) plates whose thickness is small compared to the other dimensions,
(ii) cylinders whose diameter is small compared to its length,
(iii) Spheres.

All parameters in these figures are dimensionless. The nomenclature used is shown in Fig. 6.16 Temperature used in charts are defined as

$$\theta = t(x,\tau) - t_\infty$$
$$\theta_i = t_i - t_\infty$$
$$\theta_o = t_o - t_\infty$$

where

t_∞ surrounding fluid temperature,
t_o centreline or centre temperature.

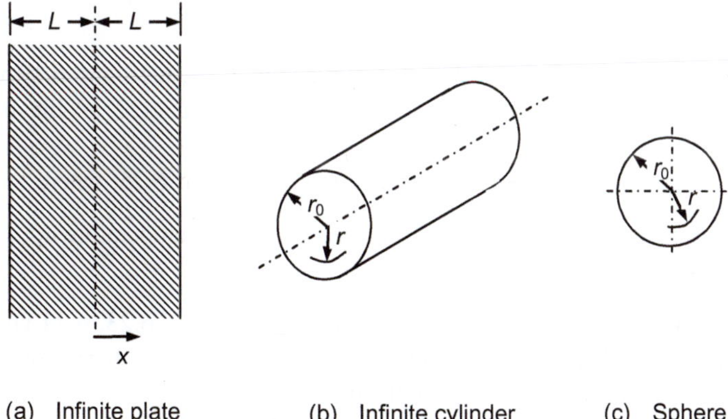

Fig. 6.16 Nomenclature for the Heisler charts

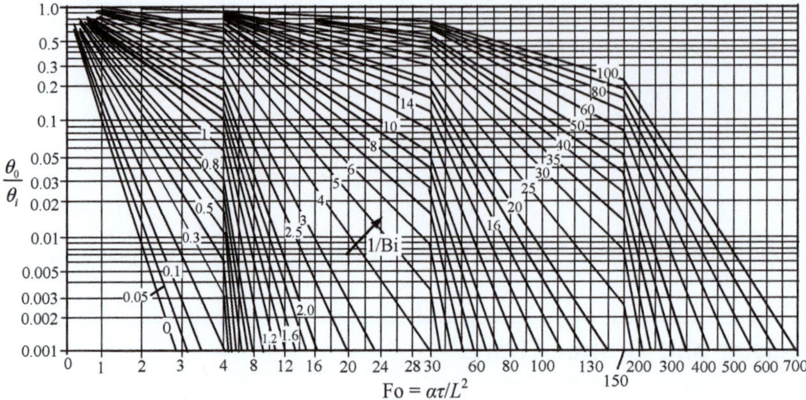

Fig. 6.17 Midplane temperature of an infinite plate of thickness 2*L*. Heisler (1947) Temperature charts for induction and constant temperature heating. Trans ASME 69: 227–236. With permission of ASME

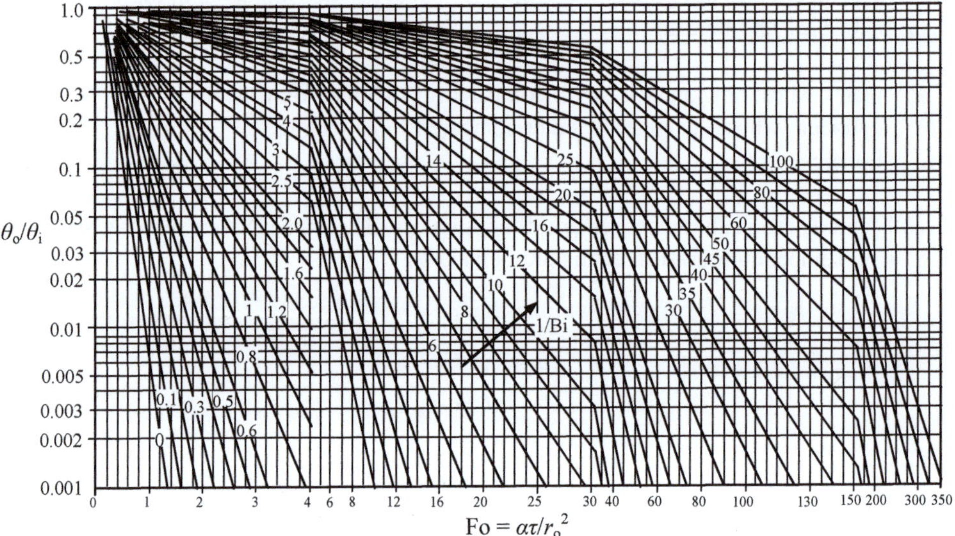

Fig. 6.18 Axis temperature of an infinite cylinder of radius r_o. Heisler (1947) Temperature charts for induction and constant temperature heating. Trans ASME 69: 227–236. With permission of ASME

The centreline temperature θ_0 is obtained using Figs. 6.17, 6.18 and 6.19 while the off-centre temperatures are obtained using Figs. 6.20, 6.21 and 6.22 as illustrated through worked examples.

The heat loss can be determined using Figs. 6.23, 6.24 and 6.25, where

Q_o initial internal energy of the solid, with reference to the temperature t_∞, = $\rho Vc\ (t_i - t_\infty)$,
Q heat lost in time τ.

Note: Inspection of Figs. 6.17, 6.18, 6.19, 6.20, 6.21, 6.22, 6.23, 6.24 and 6.25 shows that the dimensionless temperature distribution and heat flows have been expressed in terms of dimensionless parameters Bi and Fo defined as

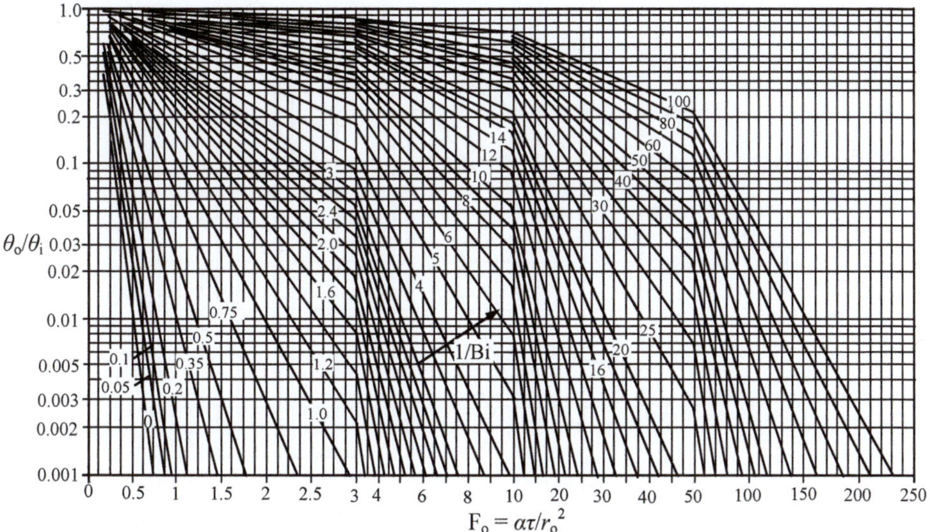

Fig. 6.19 Centre temperature for a sphere of radius r_o. Heisler (1947) Temperature charts for induction and constant temperature heating. Trans ASME 69: 227–236. With permission of ASME

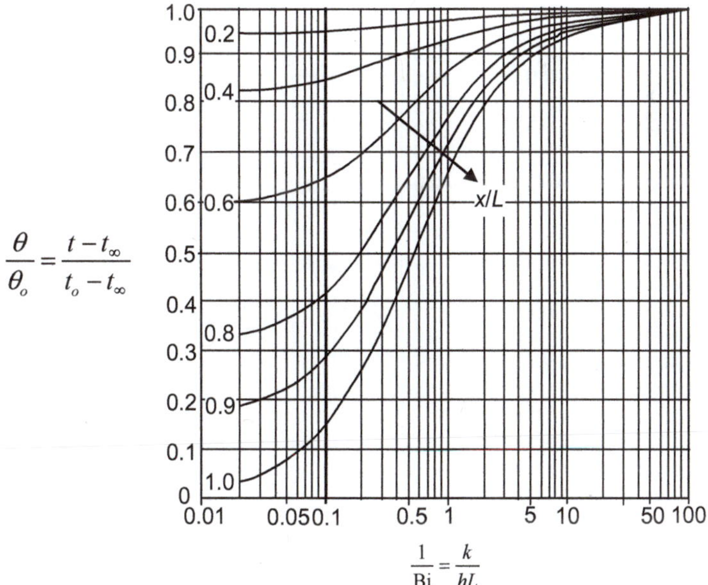

Fig. 6.20 Temperature as a function of centre temperature in an infinite plate of thickness $2L$. Heisler (1947) Temperature charts for induction and constant temperature heating. Trans ASME 69: 227–236. With permission of ASME

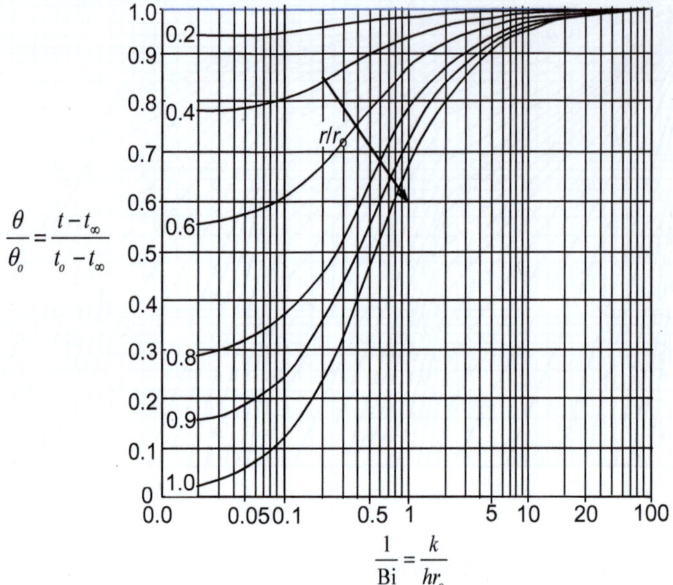

Fig. 6.21 Temperature as a function of axis temperature in an infinite cylinder of radius r_o. Heisler (1947) Temperature charts for induction and constant temperature heating. Trans ASME 69: 227–236. With permission of ASME

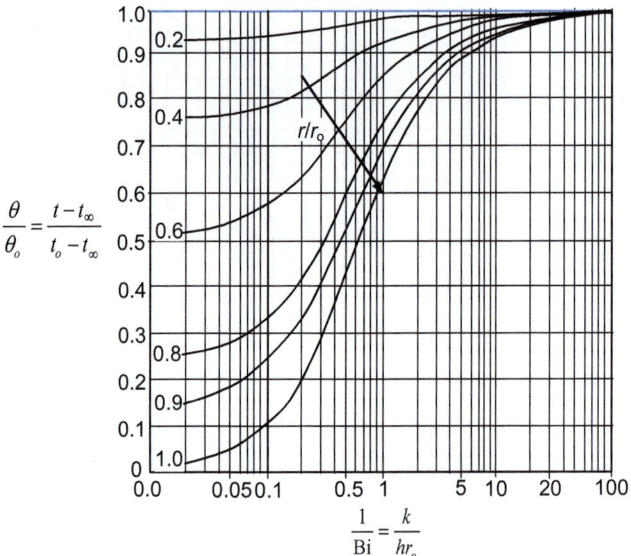

Fig. 6.22 Temperature as a function of centre temperature in a sphere of radius r_o. Heisler (1947) Temperature charts for induction and constant temperature heating. Trans ASME 69: 227–236. With permission of ASME

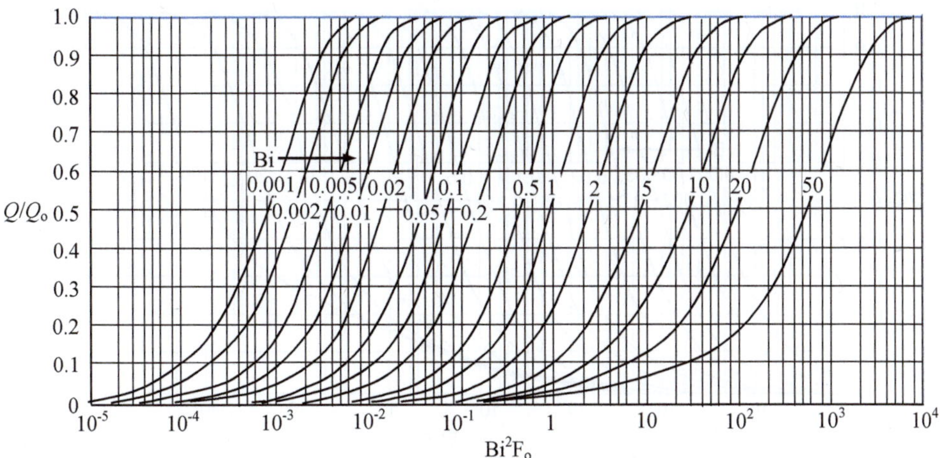

Fig. 6.23 Dimensionless heat loss Q/Q_o with time for a plane wall of thickness $2L$ (Grober et al. 1961)

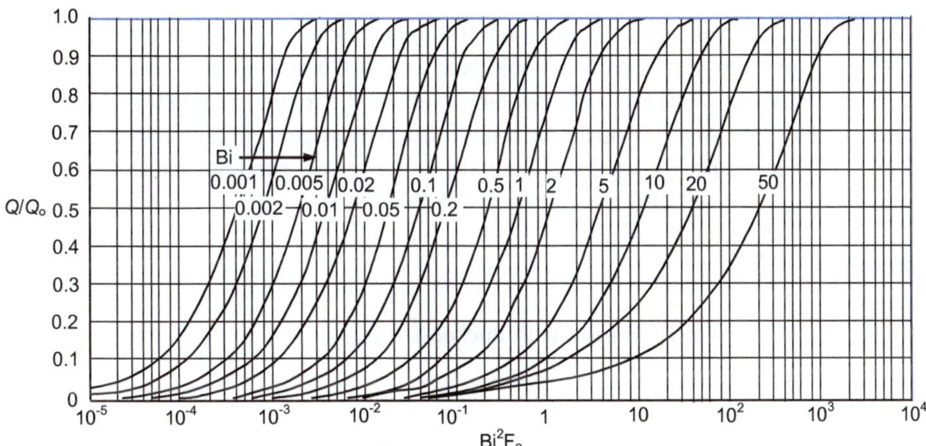

Fig. 6.24 Dimensionless heat loss Q/Q_o with time for an infinite cylinder of radius r_o (Grober et al. 1961)

$$\mathrm{Bi} = \frac{hL}{k} \tag{6.4}$$

$$\mathrm{Fo} = \frac{\alpha\tau}{L^2} \tag{6.5}$$

where L is a characteristic dimension of the body. For a plate, it is the half-thickness L, and for the cylinder and sphere it is the radius r_o, refer Fig. 6.16.

Note that the Heisler charts are applicable for values of Fourier number greater than 0.2.

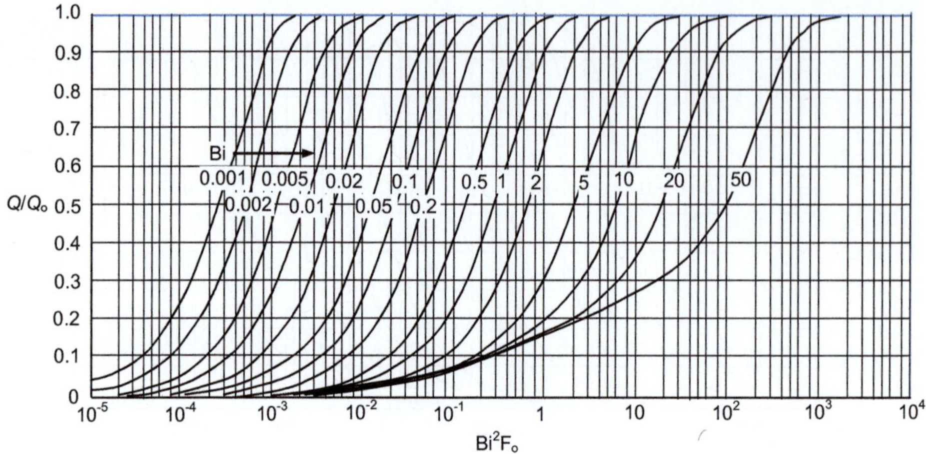

Fig. 6.25 Dimensionless heat loss Q/Q_o with time for a sphere of radius r_o (Grober et al. 1961)

Example 6.22 A large slab of aluminium 100 mm thick is originally at a temperature of 500°C. It is suddenly immersed in a liquid bath at 100°C resulting in a heat transfer coefficient of 1200 W/(m^2 K). Determine the temperature at the centre line and the surface 1 min after the immersion. Also, calculate the total heat removed per unit area of the slab during this period. The properties of the aluminium at the given conditions may be taken as ρ = 2700 kg/m^3, c = 0.9 kJ/(kg K) and k = 215 W/(m K).

Solution

The characteristic length for a plate with thickness δ is

$$L = \frac{\delta}{2} = \frac{100}{2} = 50\,\text{mm}.$$

The Biot number from Eq. (6.4),

$$\text{Bi} = \frac{hL}{k} = \frac{1200 \times 50 \times 10^{-3}}{215} = 0.279 > 0.1.$$

Since the Biot number is greater than 0.1, the lumped heat capacity analysis is not applicable. The problem may be solved using Heisler charts.

The Fourier number,

$$\text{Fo} = \frac{\alpha\tau}{L^2} = \frac{k}{\rho c} \times \frac{\tau}{L^2} = \frac{215}{2700 \times 900} \times \frac{60}{0.05^2} = 2.12.$$

(i) **Centreline temperature, t_o**

From the Heisler chart (Fig. 6.17) at Fo = 2.1 and 1/Bi = 3.6, for the temperature variation at the centre of the plate, we have

$$\frac{t_0 - t_\infty}{t_i - t_\infty} \approx 0.6.$$

Hence, the temperature at the centre of the plate,

$$\begin{aligned} t_0 &= 0.6(t_i - t_\infty) + t_\infty \\ &= 0.6(500 - 100) + 100 = 340°\text{C}. \end{aligned}$$

(ii) **Plate surface temperature, t_s**

The correction factor from Fig. 6.20 for 1/Bi = 3.6 and x/L = 1.0 (where x is the distance from the midplane) is 0.86. Hence, the plate surface temperature t_s is given by

$$\frac{t - t_\infty}{t_i - t_\infty} = \frac{t_0 - t_\infty}{t_i - t_\infty} \times \frac{t - t_\infty}{t_0 - t_\infty} = 0.6 \times 0.86 = 0.516.$$

Simplification gives the surface temperature as

$$\begin{aligned} t &= 0.516 \times (t_i - t_\infty) + t_\infty \\ &= 0.516 \times (500 - 100) + 100 = 306.4°\text{C}. \end{aligned}$$

(iii) **Total heat removal**

The heat transferred from the slab in time τ can be calculated using Fig. 6.23. From the given data,

$$\text{FoBi}^2 = 2.12 \times (0.279)^2 = 0.165,$$

and

$$\frac{hL}{k} = \text{Bi} = 0.279.$$

From Fig. 6.23,

$$\frac{Q}{Q_0} \approx 0.4$$

where Q_0 is the initial internal energy content of the slab in reference to the surrounding temperature. Thus, for the unit surface area of the plate,

$$Q_o = (\rho V)c(t_i - t_\infty) = [2700 \times (100/1000) \times 1] \times 0.9 \times (500 - 100) = 97200\,\text{kJ/m}^2.$$

Hence, the heat lost per unit area of the slab in one minute is

$$Q = 0.4 \times 97200 = 38880\,\text{kJ/m}^2.$$

Example 6.23 For the slab of the previous example, how long it will take for the midplane temperature to drop to 300°C? What will be the temperature at a depth of 20 mm from one of the faces when the midplane temperature will be 300°C?

Solution

From the previous example,

$$\frac{k}{hL} \approx 3.6.$$

For a midplane temperature of 300°C,

$$\frac{t_0 - t_\infty}{t_i - t_\infty} = \frac{300 - 100}{500 - 100} = 0.5.$$

Corresponding to this value of temperature ratio, Fo = 2.7. This gives

$$\text{Fo} = \frac{\alpha\tau}{L^2} = \frac{k}{\rho c} \times \frac{\tau}{L^2} = \frac{215}{2700 \times 900} \times \frac{\tau}{0.05^2} = 2.7$$

or

$$\tau = 76.3\ \text{s for midplane temperature to drop to 300°C.}$$

Since distance x is measured from the midplane, x/L for the plane 20 mm from one of the face is

$$\frac{x}{L} = \frac{0.05 - 0.02}{0.05} = 0.6$$

Correction factor for x/L = 0.6 and 1/Bi = 3.6,

$$\frac{t - t_\infty}{t_0 - t_\infty} = 0.95.$$

Hence,

$$\frac{t - t_\infty}{t_i - t_\infty} = \frac{t_0 - t_\infty}{t_i - t_\infty} \times \frac{t - t_\infty}{t_0 - t_\infty} = 0.5 \times 0.95 = 0.475$$

which gives

$$\begin{aligned} t &= 0.475 \times (t_i - t_\infty) + t_\infty \\ &= 0.475 \times (500 - 100) + 100 = 290°\text{C} \end{aligned}$$

where t is the temperature after 76.3 s at a depth of 20 mm from the surface of the slab when the midplane temperature is 300°C.

Example 6.24 A water pipeline is laid along the centreline of a 300 mm thick, very long and wide, brick masonry wall (k = 0.66 W/(m K), α = 0.0046 cm²/s). The water is initially at a temperature of 10°C. The wall is suddenly exposed to a cold wave [average heat transfer coefficient, h_{av} = 20 W/(m² K)] at –2°C. How long will it take for the water to reach a temperature of 4°C.

Solution

Given data: Midplane temperature t_0 = 4°C, surrounding fluid temperature t_∞ = –2°C, t_i = 10°C and L = 0.15 m. Hence,

$$\frac{t_0 - t_\infty}{t_i - t_\infty} = \frac{4 - (-2)}{10 - (-2)} = 0.5$$

$$\frac{1}{Bi} = \frac{k}{hL} = \frac{0.66}{20 \times 0.15} = 0.22.$$

From Fig. 6.17, Fo $\approx$ 0.6. This gives

$$\tau = \text{Fo} \times \frac{L^2}{\alpha} = 0.6 \times \frac{0.15^2}{0.0046 \times 10^{-4}} = 29347\,\text{s} = 8\,\text{h}\,9\,\text{min}.$$

Example 6.25 A long copper cylinder 100 mm in diameter and initially at 20°C is suddenly exposed to a fluid at 80°C. The average value of the heat transfer coefficient is 500 W/(m² K). Calculate the temperature at a radius of 20 mm and the heat gained per m length of the cylinder in 60 s. For the cylinder material k = 380 W/(m K), ρ = 8950 kg/m³ and c = 380 J/(kg K).

Solution

Thermal diffusivity, $\alpha = \frac{k}{\rho c} = \frac{380}{8950 \times 380} = 11.2 \times 10^{-5} \text{m}^2/\text{s}$.

Fourier number, $\text{Fo} = \frac{\alpha \tau}{r_0^2} = \frac{11.2 \times 10^{-5} \times 60}{0.05^2} = 2.688$

Biot number, $\text{Bi} = \frac{h r_0}{k} = \frac{500 \times 0.05}{380} = 0.0658$

Radius ratio, $\frac{r}{r_0} = \frac{0.02}{0.05} = 0.4$.

From Figs. 6.18 and 6.21,

$$\frac{\theta_o}{\theta_i} \approx 0.7$$

$$\frac{\theta}{\theta_0} \approx 0.98.$$

Hence,

$$\theta = t - t_\infty = \frac{\theta}{\theta_0} \times \frac{\theta_0}{\theta_i} \times \theta_i = 0.98 \times 0.7 \times (20 - 80) = -41.16°\text{C}$$

and

$$t = \theta + t_\infty = -41.16 + 80 = 38.84°\text{C}.$$

Heat gained

$$\text{FoBi}^2 = \frac{h^2 \alpha \tau}{k^2} = 0.0116$$
$$\frac{hr_0}{k} = \frac{500 \times 0.05}{380} = 0.0658.$$

From Fig. 6.24,

$$\frac{Q}{Q_0} \approx 0.3.$$

The internal energy at time $\tau = 0$ is

$$\frac{Q_0}{L} = \rho c(\pi r_o^2)\theta_i$$
$$= 8950 \times 380 \times \pi \times 0.05^2 \times 60 = 1.603 \times 10^6\ \text{J/m}.$$

Hence, heat gained,

$$Q = 0.3Q_0 = 0.3 \times 1.603 \times 10^6 = 4.808 \times 10^5\ \text{J/m}.$$

Example 6.26 For the orange of Example 6.3, estimate the time required for the centre of the orange to reach 10°C if the initial temperature of the orange is 30°C. The temperature of the cold environment is 5°C. The thermal diffusivity of the orange is nearly equal to that of water since it mainly consists of water.

Solution

At the average temperature of ½ (30 + 10) = 20°C, $\alpha = 1.4 \times 10^{-7}$ m^2/s and $k = 0.6$ for water.

$$\frac{1}{\text{Bi}} = \frac{k}{hr_0} = \frac{0.6}{10 \times 36/1000} = 1.67$$

and

$$\frac{\theta_0}{\theta_i} = \frac{t_0 - t_\infty}{t_i - t_\infty} = \frac{10 - 5}{30 - 5} = 0.2.$$

From Fig. 6.19, Fo ≈1.2.

$$\tau = \text{Fo}\frac{r_0^2}{\alpha} = \frac{1.2 \times 0.036^2}{1.4 \times 10^{-7}} = 11108\,\text{s} = 3\,\text{h}\,5\,\text{min}.$$

Example 6.27 A long 50 mm diameter aluminium cylinder initially at 300°C is suddenly exposed to a convective environment at 100°C. The average value of the heat transfer coefficient is 200 W/(m^2 K). Determine the heat lost per m length of the cylinder in 30 s. Given:

$$k = 215\ \text{W}/(\text{m}\,\text{K}), c = 0.9 \times 10^3\,\text{J}/(\text{kg}\,\text{K}), \rho = 2700\,\text{kg}/\text{m}^3, \alpha = 8.85 \times 10^{-5}\text{m}^2/\text{s}.$$

Solution

Fourier number, $\text{Fo} = \frac{\alpha\tau}{r_0^2} = \frac{8.85\times10^{-5}\times30}{0.025^2} = 4.248$

Biot number, $\text{Bi} = \frac{hr_0}{k} = \frac{200\times0.025}{215} = 0.02326.$
$\text{FoBi}^2 = 0.0023.$
From Fig. 6.24,

$$\frac{Q}{Q_i} \approx 0.18.$$

Hence,

$$\begin{aligned} Q &= 0.18Q_i = 0.18 \times \rho c(\pi r_o^2 \times 1)(t_i - t_\infty) \\ &= 0.18 \times 2700 \times 0.9 \times 10^3 \times \pi \times 0.025^2 \times 1 \times (300 - 100) \\ &= 171.8\ \times 10^3\,\text{J}/\text{m}. \end{aligned}$$

Lumped-capacity analysis
Fourier number, $\text{Fo} = \frac{\alpha\tau}{(r_0/2)^2} = \frac{8.85\times10^{-5}\times30}{(0.025/2)^2} = 16.99$
Biot number, $\text{Bi} = \frac{hr_0/2}{k} = \frac{200\times0.025/2}{215} = 0.01163.$
From Eq. (6.13),

$$\begin{aligned} \frac{Q}{Q_i} &= 1 - \exp(-\text{Bi} \times \text{Fo}) \\ &= 0.1793, \end{aligned}$$

which is in reasonable agreement with the above result.

Example 6.28 It was assumed that when Bi $\leq$ 0.1, the body is having uniform temperature and the lumped heat capacity analysis is applicable. Based on the Heisler charts, determine the error in this approximation.

Solution

(i) **Plate**
From Fig. 6.20,
The temperature ratio $\frac{\theta}{\theta_0} = 0.95$, for $\frac{1}{\text{Bi}} = 10$ and $\frac{x}{L} = 1.0$.
This means that the surface temperature $\theta_{x\,=\,L}$ differs by about 5% from the centreline temperature θ_o.

(ii) **Cylinder**

For the Heisler chart, the Biot number is defined as $\mathrm{Bi} = \frac{hr_0}{k}$ while, for the lumped heat capacity analysis, it is defined as $\mathrm{Bi} = \frac{hr_0/2}{k}$.

Hence, Bi = 0.1 refers to Bi = 0.2 for the Heisler chart.

From Fig. 6.21,

$\frac{\theta}{\theta_0} \approx 0.91$ for 1/Bi = 5 and $r/r_o = 1.0$. Thus, the error can be 9%.

(iii) **Sphere**

Following the explanation given for the cylinder,

$$(\mathrm{Bi})_{\mathrm{Heisler}} = (\mathrm{Bi})_{\mathrm{Lumped}} \times \frac{r}{r/3} = 0.3.$$

From Fig. 6.22, $\frac{\theta}{\theta_0} \approx 0.85$ for 1/Bi = 3.33 and $r/r_o = 1.0$. Thus, the error can be 15%.

It can be seen that with the increase in 1/Bi or decrease in the value of the Biot number, the above-approximated errors will reduce. When the value of the Biot number will approach 0.01 (or 1/Bi approaches 100), the error will approach zero. Hence, the above-approximated error of 5–15%, with respect to the Heisler chart, can be regarded as the maximum for the lumped heat capacity analysis. However, it must be noted that the Heisler charts too are approximate.

Example 6.29 A rectangular-shaped sandstone block of thickness $\delta = 150$ mm (width $W \gg \delta$ and length $L \gg \delta$), which is initially at a uniform temperature of 30°C, is exposed on both sides to hot gases at 500°C. The convective heat transfer coefficient is 80 W/(m^2 K). Determine the time required by the block to gain 50% of its maximum storage capacity of heat, and minimum and maximum temperatures of the stone at that time. Neglect radiation heat exchange. For the masonry, $k = 2.9$ W/(m K), $c = 750$ J/(kg K) and $\rho = 2200$ kg/m^3.

Solution

For the given system, characteristic dimension L from Table 6.1 is $\delta/2 = 0.075$ m.

Biot number,

$$Bi = \frac{hL}{k} = \frac{80 \times 0.075}{2.9} = 2.07 > 0.1.$$

Hence, the lumped heat capacity method is not applicable. Grober and Heisler charts may be used to solve the problem.

Time required by the block to gain 50% of its maximum storage capacity of heat

From Grober chart (Fig. 6.23), $\mathrm{Bi}^2\mathrm{Fo} \approx 2.9$ for $Q/Q_o = 0.5$ and Bi = 2.07.

Fourier number,

$$\mathrm{Fo} = \frac{\alpha\tau}{L^2} = \left(\frac{k}{\rho c}\right)\frac{\tau}{L^2} = \left(\frac{2.9}{2200 \times 750}\right)\frac{\tau}{0.075^2} = 3.12 \times 10^{-4}\tau.$$

Hence,

$$2.9 = Bi^2Fo = Bi^2\frac{\alpha\tau}{L^2} = Bi^2\left(\frac{k}{\rho c}\right)\frac{\tau}{L^2}$$
$$= (2.07)^2 \times \left(\frac{2.9}{2200 \times 750}\right)\frac{\tau}{0.075^2} = 13.37 \times 10^{-4}\tau$$

or $\tau = 2169$ s.

For τ = 2169 s,

$$Fo = 3.12 \times 10^{-4} \times 2169 = 0.676.$$

Minimum and maximum temperatures

The minimum temperature is at $x = 0$, i.e. at midplane and the maximum is at the surface ($x = L$). From Heisler chart (Fig. 6.17), with Fo = 0.676 and 1/Bi = 1/2.07 = 0.48, $\theta_o/\theta_i \approx 0.47$. This gives the minimum temperature,

$$t_{x=0} = t_\infty + (\theta_o/\theta_i) \times (t_i - t_\infty) = 500 + 0.47 \times (-470) = 279.1°C.$$

From Heisler chart (Fig. 6.20), with 1/Bi = 1/2.07 = 0.48 and x/L = 1, $\theta/\theta_o \approx 0.46$. This gives the maximum temperature,

$$t_{x=L} = t_\infty + (\theta/\theta_i) \times (t_o - t_\infty) = 500 + 0.46 \times (-220.9) = 398.4°C.$$

Example 6.30 The wall of a furnace is 150 mm thick and is adequately insulated from outside. It is initially at a uniform temperature of 20°C. The wall is exposed to hot gases at 900°C. The convective heat transfer coefficient is 80 W/(m^2 K). Determine the time required for the outer surface to reach a temperature of 725°C. Neglect radiation heat exchange. For the furnace wall, k = 1.2 W/(m K), c = 900 J/(kg K) and ρ = 2500 kg/m^3.

Solution

The wall is equivalent to one-half of a wall of thickness δ (= 2L) with symmetric convection conditions as shown in Fig. 6.26b. Hence, characteristic dimension L from Table 6.1 is $\delta/2$ = 0.3/2 = 0.15 m.

For the wall in Fig. 6.26b, Biot number,

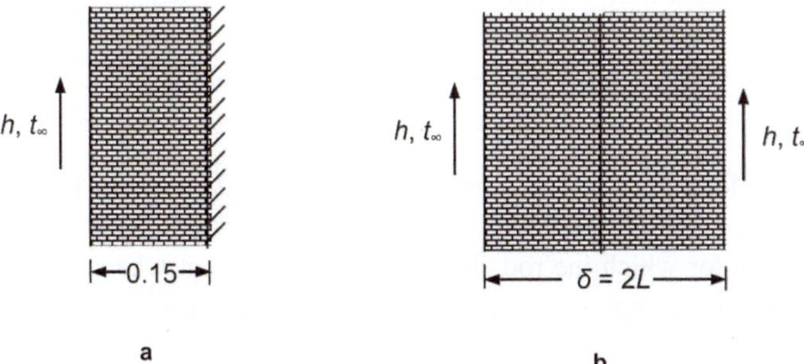

Fig. 6.26 Example 6.30

$$\text{Bi} = \frac{hL}{k} = \frac{80 \times 0.15}{1.2} = 10 > 0.1.$$

Hence, the lumped heat capacity method is not applicable. Heisler graphs will be used.

$$\frac{1}{\text{Bi}} = 0.1$$

$$\frac{\theta_o}{\theta_i} = \frac{t_o - t_\infty}{t_i - t_\infty} = \frac{725 - 900}{20 - 900} = 0.2.$$

For 1/Bi = 0.1 and $\theta_o/\theta_i = 0.2$, Fo ≈ 0.85 from Fig. 6.17. Equation $\text{Fo} = \frac{\alpha\tau}{L^2} = \left(\frac{k}{\rho c}\right)\frac{\tau}{L^2}$ gives

$$\tau = \text{Fo}\left(\frac{\rho c L^2}{k}\right) = 0.85 \times \left(\frac{2500 \times 900 \times 0.15^2}{1.2}\right) = 35859\,\text{s} = 9.96\,\text{h}.$$

Example 6.31 A 2.5 mm thick plastic sheet (k = 0.2 W/(m K), $\alpha = 1 \times 10^{-7}$ m^2/s), initially at 180°C, lies on an insulated surface and is cooled by airflow at 20°C. If the convection heat transfer coefficient is 100 W/(m^2 K), determine the time required for the surface of the sheet to reach 40°C.

Solution

The sheet is equivalent to one-half of a sheet of thickness δ (= 2L) with symmetric convection conditions. Hence, the characteristic length L = 0.0025 m.

The Biot number,

$$\text{Bi} = \frac{hL}{k} = \frac{100 \times 0.0025}{0.2} = 1.25.$$

Figure 6.20 gives $\theta_L/\theta_o = 0.6$ for x/L = 1 and 1/Bi = 0.8.

From given data,

$$\frac{\theta_L}{\theta_i} = \frac{t_L - t_\infty}{t_i - t_\infty} = \frac{40 - 20}{180 - 20} = 0.125,$$

$$\frac{\theta_o}{\theta_i} = \frac{\theta_L}{\theta_i} \times \frac{\theta_o}{\theta_L} = 0.125 \times \frac{1}{0.6} = 0.208.$$

For $\theta_o/\theta_i = 0.208$ and 1/Bi = 0.8, Fo ≈ 1.9 from Fig. 6.17. Knowing that $\text{Fo} = \frac{\alpha\tau}{L^2}$, we get

$$\tau = \frac{\text{Fo}L^2}{\alpha} = \frac{1.9 \times 0.0025^2}{1 \times 10^{-7}} = 118.75\,\text{s}.$$

Example 6.32 A long glass rod of 40 mm diameter (k = 0.78 W/(m K), c = 840 J/(kg K) and ρ = 2700 kg/m^3), initially at 80°C, is cooled in air [t_∞ = 30°C, h = 20 W/(m^2 K)]. Determine the time for which the rod should be cooled so that when removed and isolated from the surroundings it attains an equilibrium temperature of 45°C.

Solution

Heat removed, Q, in cooling to the equilibrium temperature of $t_f = 45°C$ is the difference of the initial heat content and heat content at 45°C with reference to temperature t_∞ of air, i.e.

$$Q = Q_o - Q_f$$

or

$$\frac{Q}{Q_o} = 1 - \frac{Q_f}{Q_o} = 1 - \frac{mc(t_f - t_\infty)}{mc(t_i - t_\infty)} = 1 - \frac{t_f - t_\infty}{t_i - t_\infty}$$

or

$$\frac{Q}{Q_o} = 1 - \frac{45 - 30}{80 - 30} = 0.7.$$

The Biot number,

$$\text{Bi} = \frac{hr_o}{k} = \frac{20 \times 0.02}{0.78} = 0.51,$$

where $L = r_o = 0.02$ m.

For $Q/Q_o = 0.7$ and Bi = 0.51, $\text{Bi}^2\text{F}_o = 0.48$ from Fig. 6.24, which gives

$$\text{Fo} = \frac{(\text{Bi}^2\text{Fo})}{\text{Bi}^2} = \frac{0.48}{0.51^2} = 1.845.$$

Equation $\text{Fo} = \frac{\alpha\tau}{r_o^2} = \left(\frac{k}{\rho c}\right)\frac{\tau}{r_o^2}$ gives

$$\tau = \text{Fo}\left(\frac{\rho c r_o^2}{k}\right) = 1.845 \times \left(\frac{2700 \times 840 \times 0.02^2}{0.78}\right) = 2146\,\text{s}.$$

Example 6.33 A 50 mm diameter long rod ($c = 1000$ J/(kg K) and $\rho = 1200$ kg/m^3), initially at 110°C, is cooled in air [$t_\infty = 30°C$, $h = 60$ W/(m^2 K)]. If the axis temperature reaches 50°C after cooling for 10 min, determine the thermal conductivity of the rod material.

Solution

From given data,

$$\frac{\theta_o}{\theta_i} = \frac{t_{r=0} - t_\infty}{t_i - t_\infty} = \frac{50 - 30}{110 - 30} = 0.25,$$

$$\frac{1}{\text{Bi}} = \frac{k}{hr_o} = \frac{k}{60 \times 0.025} = 0.667k,$$

$$\text{Fo} = \frac{\alpha\tau}{r_o^2} = \left(\frac{k}{\rho c}\right)\frac{\tau}{r_o^2} = \frac{k \times 10 \times 60}{1200 \times 1000 \times 0.025^2} = 0.8k.$$

We have to determine that value of k, which will give $\theta_o/\theta_i = 0.25$ in Fig. 6.18 at values of Fo and 1/Bi from the above equations. By trial-and-error procedure, we find $k = 1.0$ W/(m K).

The Biot number Bi = 1.5 for k = 1.0 W/(m K), hence lumped heat capacity analysis is not applicable.

Example 6.34 A 10 mm diameter steel ball, initially at a temperature of 30°C, is to be heated by immersing in a salt bath at 1000°C and then quenched for hardening. Hardening occurs at locations where the temperature is greater than 800°C. The heat transfer coefficient is 4000 W/(m^2 K). (a) Determine the time required for heating to harden up to a depth of 1 mm and (b) heat flux when t(4 mm, τ) = 800°C,. Given for ball material ρ = 7860 kg/m^3, c = 460 J/(kg K) and k = 40 W/(m K).

Solution

(a) **Time required for heating**

The temperature up to the depth of 1 mm, i.e. from surface to r = 4 mm must be $\geq$ 800°C. From the given data,

$$\frac{r}{r_o} = \frac{4}{5} = 0.8,$$

$$\frac{1}{\text{Bi}} = \frac{k}{hr_o} = \frac{40}{4000 \times 0.005} = 2.$$

For applicability of lumped heat capacity analysis, Biot number Bi = hr_o/(3 k) = 0.166 > 0.1. Hence, the lumped heat capacity analysis is not applicable.

From Fig. 6.22, $\theta/\theta_o \approx 0.84$ for 1/Bi = 2 and r/r_o = 0.8. Hence,

$$\theta_o = \frac{\theta}{0.84} = \frac{800 - 1000}{0.84} = -238.1\,^{\circ}\text{C}.$$

This gives

$$\frac{\theta_o}{\theta_i} = \frac{\theta_o}{t_i - t_\infty} = \frac{-238.1}{30 - 1000} = 0.245.$$

From Fig. 6.19, Fo $\approx$ 1.13 for θ_o/θ_i = 0.245 and 1/Bi = 2. Equation $\text{Fo} = \frac{\alpha\tau}{r_o^2} = \left(\frac{k}{\rho c}\right)\frac{\tau}{r_o^2}$ gives

$$\tau = \text{Fo}\left(\frac{\rho c r_o^2}{k}\right) = 1.13 \times \left(\frac{7860 \times 460 \times 0.005^2}{40}\right) = 2.55\,\text{s}.$$

(b) **Heat flux**

From Fig. 6.22, $\theta/\theta_o \approx 0.77$ for 1/Bi = 2 and r/r_o = 1.0. Hence,

$$\theta = 0.77\theta_o = 0.77 \times (-238.1) = -183.3\,^{\circ}\text{C}$$

or

$$t(r = r_o) = 1000 - 183.3 = 816.7\,^{\circ}\text{C}.$$

$$\text{Heat flux} = h[t(r_o) - t_\infty] = 4000 \times 183.3 = 733.2\,\text{kW/m}^2.$$

6.8 Two- and Three-Dimensional Transient Heat Conduction Systems

The Heisler charts presented in the previous sections are applicable to infinite plates of $2L$ thickness, a long cylinder or a sphere. When the length and width of the plate are not large compared to its thickness, or the length of the cylinder is not large compared to its diameter, the above charts are not applicable. The temperature distribution in such cases is a function of additional space coordinates.

6.8.1 Two-Dimensional Systems

The differential equation for the semi-infinite bar in Fig. 6.27 is

$$\frac{\partial^2 t}{\partial x^2}+\frac{\partial^2 t}{\partial y^2}=\frac{1}{\alpha}\frac{\partial t}{\partial \tau} \tag{6.29}$$

The solution, based on the separation of the variables technique, is of the form

$$t(x,y,\tau)=X(x,\tau)Y(y,\tau) \tag{6.30}$$

It will be shown[1] that the bar in Fig. 6.27 can be formed from two infinite plates of thickness $2L_1$ and $2L_2$, respectively, and the temperature distribution may be expressed as a product of the solution for these plates.

---

[1]For two infinite plates, the differential equations are

$$\frac{\partial^2 t_1}{\partial x^2}=\frac{1}{\alpha}\frac{\partial t_1}{\partial \tau} \tag{i}$$

$$\frac{\partial^2 t_2}{\partial y^2}=\frac{1}{\alpha}\frac{\partial t_2}{\partial \tau} \tag{ii}$$

and their temperature distributions are

$$t_1=t_1(x,\tau) \tag{iii}$$

$$t_2=t_2(y,\tau) \tag{iv}$$

Let the solution to Eq. (6.29) is a simple product solution of above functions, i.e.

$$t=t_1(x,\tau)t_2(y,\tau) \tag{v}$$

Then the derivations are

$$\frac{\partial^2 t}{\partial x^2}=t_2\frac{\partial^2 t_1}{\partial x^2} \tag{vi}$$

$$\frac{\partial^2 t}{\partial y^2}=t_1\frac{\partial^2 t_2}{\partial y^2} \tag{vii}$$

$$\frac{\partial t}{\partial \tau}=t_2\frac{\partial t_1}{\partial \tau}+t_1\frac{\partial t_2}{\partial \tau} \tag{viii}$$

Using Eqs. (i) and (ii), Eq. (viii) transforms to

$$\frac{\partial t}{\partial \tau}=\alpha t_2\frac{\partial^2 t_1}{\partial x^2}+\alpha t_1\frac{\partial^2 t_2}{\partial y^2}$$

Substitution of the value of $\partial t/\partial \tau$ satisfies Eq. (6.29):

$$t_2\frac{\partial^2 t_1}{\partial x^2}+t_1\frac{\partial^2 t_2}{\partial y^2}=\frac{1}{\alpha}\left(\alpha t_2\frac{\partial^2 t_1}{\partial x^2}+\alpha t_1\frac{\partial^2 t_2}{\partial y^2}\right)$$

Therefore, the assumed product solution, Eq. (v), is correct and the dimensionless temperature distribution for the rectangular bar can be given by Eq. (6.31).

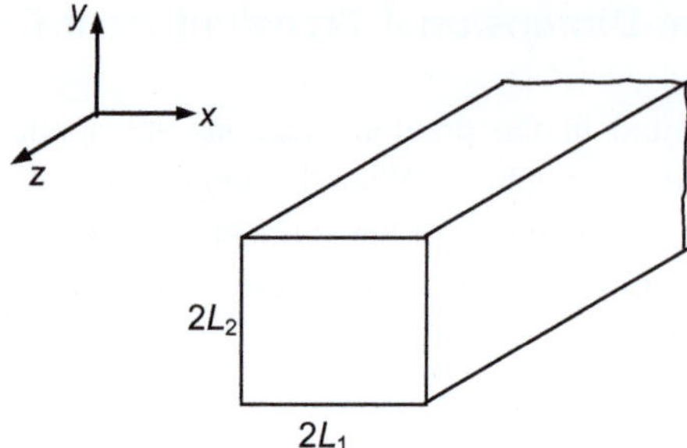

Fig. 6.27 Two-dimensional system

$$\frac{\theta}{\theta_i} = \frac{t(x, y, \tau) - t_\infty}{t_i - t_\infty} = \left(\frac{t(x, \tau) - t_\infty}{t_i - t_\infty}\right)_{2L_1\text{Plate}} \left(\frac{t(y, \tau) - t_\infty}{t_i - t_\infty}\right)_{2L_2\text{Plate}} \tag{6.31}$$

where t_i and t_∞ are defined as earlier.

Similarly, the temperature distribution in semi-infinite plates and cylinders can be expressed.

6.8.2 Three-Dimensional Systems

The results of two-dimensional systems can be extended to a three-dimensional block ($2L_1 \times 2L_2 \times 2L_3$) to give

$$\frac{\theta}{\theta_i} = \frac{t(x, y, z, \tau) - t_\infty}{t_i - t_\infty} = \left(\frac{t(x, \tau) - t_\infty}{t_i - t_\infty}\right)_{2L_1\text{Plate}} \left(\frac{t(y, \tau) - t_\infty}{t_i - t_\infty}\right)_{2L_2\text{Plate}} \left(\frac{t(z, \tau) - t_\infty}{t_i - t_\infty}\right)_{2L_3\text{Plate}} \tag{6.32}$$

The product solutions for the temperature distributions in two- or three-dimensional systems are presented in Table 6.3

Example 6.35 A long rectangular section steel bar (50 mm × 80 mm) is initially at a uniform temperature of 800 °C. It is heat treated by quenching in an oil tank at 80°C. The average heat transfer coefficient is 400 W/(m^2 K). What is the temperature at the centreline after 120 s of quenching? $\alpha = 1.4 \times 10^{-5}$ m^2/s and $k = 50$ W/(m K).

Solution

The problem is a combination of two infinite plates of thicknesses $2L_1$ and $2L_2$, respectively. Hence, equation of Case (2) of Table 6.3 applies

$2L_1$. Plate ($2L_1$ = 50 mm)

$$\frac{1}{\text{Bi}} = \frac{k}{hL_1} = \frac{50}{400 \times 0.025} = 5$$

$$\text{Fo} = \frac{\alpha\tau}{L_1^2} = \frac{1.4 \times 10^{-5} \times 120}{0.025^2} = 2.69.$$

Table 6.3 Product solutions for transient heat conduction in two- and three-dimensional systems

System	Solution
1. Semi-infinite plate 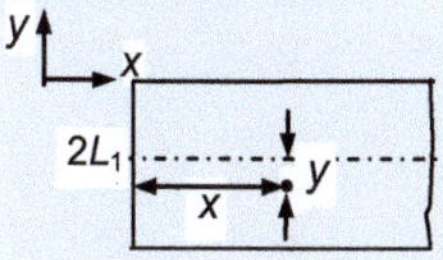	$\frac{\theta}{\theta_i} = \frac{t(y,\tau) - t_\infty}{t_i - t_\infty} = \left(\frac{t(y,\tau) - t_\infty}{t_i - t_\infty}\right)_{2L_1\text{Plate}} \left(\frac{t(x,\tau) - t_\infty}{t_i - t_\infty}\right)_{\substack{\text{semi-infinite}\\ \text{slab}}}$
2. Infinite rectangular bar 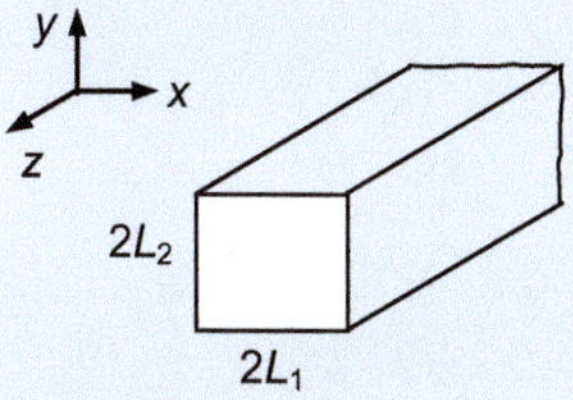	$\frac{\theta}{\theta_i} = \frac{t(x,y,\tau) - t_\infty}{t_i - t_\infty} = \left(\frac{t(x,\tau) - t_\infty}{t_i - t_\infty}\right)_{2L_1\text{Plate}} \left(\frac{t(y,\tau) - t_\infty}{t_i - t_\infty}\right)_{2L_2\text{Plate}}$
3. Semi-infinite cylinder 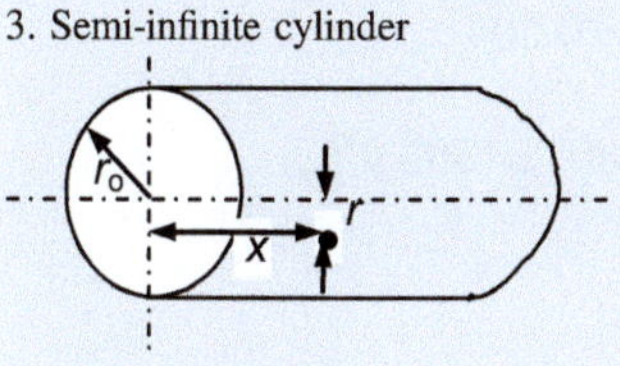	$\frac{\theta}{\theta_i} = \frac{t(r,x,\tau) - t_\infty}{t_i - t_\infty} = \left(\frac{t(r,\tau) - t_\infty}{t_i - t_\infty}\right)_{\substack{\text{Infinite}\\ \text{Cylinder}}} \left(\frac{t(x,\tau) - t_\infty}{t_i - t_\infty}\right)_{\substack{\text{Semi-infinite}\\ \text{slab}}}$
4. A short cylinder 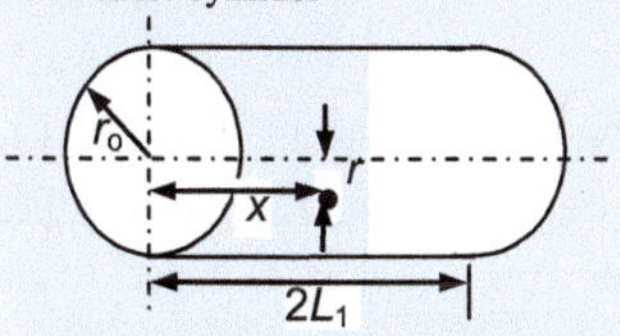	$\frac{\theta}{\theta_i} = \frac{t(r,x,\tau) - t_\infty}{t_i - t_\infty} = \left(\frac{t(r,\tau) - t_\infty}{t_i - t_\infty}\right)_{\substack{\text{Infinite}\\ \text{Cylinder}}} \left(\frac{t(x,\tau) - t_\infty}{t_i - t_\infty}\right)_{2L_1-\text{Plate}}$
5. Semi-infinite rectangular bar 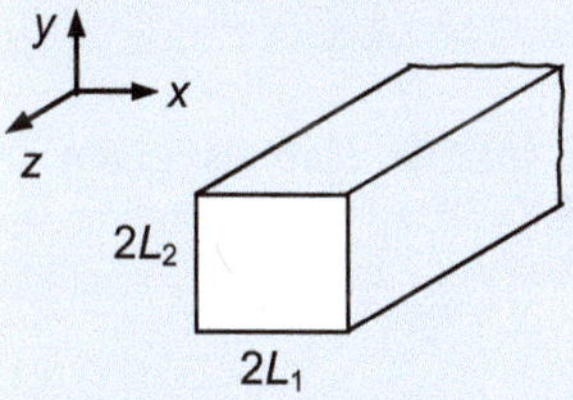	$\frac{\theta}{\theta_i} = \frac{t(x,y,z,\tau) - t_\infty}{t_i - t_\infty} = \left(\frac{t(x,\tau) - t_\infty}{t_i - t_\infty}\right)_{2L_1\text{Plate}} \left(\frac{t(y,\tau) - t_\infty}{t_i - t_\infty}\right)_{2L_2\text{Plate}} \left(\frac{t(z,\tau) - t_\infty}{t_i - t_\infty}\right)_{\substack{\text{Semi-infinite}\\ \text{solid}}}$
6. Rectangular parallelepiped 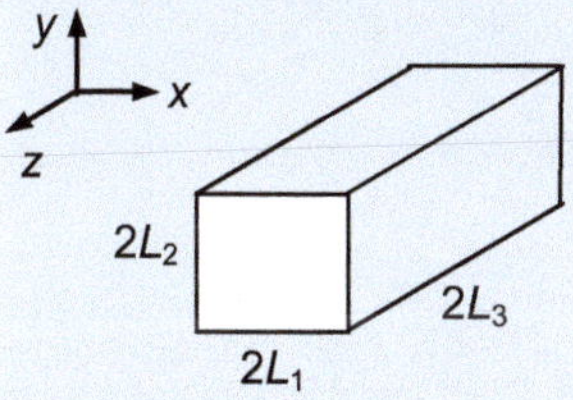	$\frac{\theta}{\theta_i} = \frac{t(x,y,z,\tau) - t_\infty}{t_i - t_\infty} = \left(\frac{t(x,\tau) - t_\infty}{t_i - t_\infty}\right)_{2L_1\text{Plate}} \left(\frac{t(y,\tau) - t_\infty}{t_i - t_\infty}\right)_{2L_2\text{Plate}} \left(\frac{t(z,\tau) - t_\infty}{t_i - t_\infty}\right)_{2L_3\text{Plate}}$

From Fig. 6.17, for $\frac{1}{\text{Bi}} = 5$ and Fo $= 2.69$,

$$\left(\frac{\theta_0}{\theta_i}\right)_{2L_1\text{plate}} \approx 0.63.$$

$2L_2$- Plate ($2L_2$ = 80 mm)

$$\frac{1}{\text{Bi}} = \frac{k}{hL_2} = \frac{50}{400 \times 0.04} = 3.125$$

$$\text{Fo} = \frac{\alpha\tau}{L_2^2} = \frac{1.4 \times 10^{-5} \times 120}{0.04^2} = 1.05.$$

From Fig. 6.17,

$$\left(\frac{\theta_0}{\theta_i}\right)_{2L_2\text{plate}} \approx 0.8.$$

For the given bar, the solution is

$$\left(\frac{\theta_0}{\theta_i}\right) = \left(\frac{\theta_0}{\theta_i}\right)_{2L_1\text{plate}} \left(\frac{\theta_0}{\theta_i}\right)_{2L_2\text{plate}} \approx 0.63 \times 0.8 = 0.504$$

$$t_0 - t_\infty = 0.504 \times (t_i - t_\infty) = 0.504 \times (800 - 80) = 362.9\,^\circ\text{C}.$$

Centreline temperature after 120 s,

$$t_0 = (t_0 - t_\infty) + t_\infty = 362.9 + 80 = 442.9\,^\circ\text{C}.$$

Example 6.36 A semi-infinite steel cylinder [α = 1.4 × 10^{-5} m^2/s and k = 45 W/(m K)] is 50 mm in diameter. It is initially at a uniform temperature of 30°C. Calculate the temperature at the axis of the cylinder 40 mm away from its one end after 100 s if it is exposed to an environment of 200°C with heat transfer coefficient of 350 W/(m^2 K).

Solution

The problem is a combination of infinite cylinder and semi-infinite slab as given in Table 6.3, Case 3. The solution for the centreline temperature is

$$\left(\frac{\theta_0}{\theta_i}\right) = \left(\frac{\theta_0}{\theta_i}\right)_{\substack{\text{infinite}\\\text{cylinder}}} \left(\frac{\theta}{\theta_i}\right)_{\substack{\text{semi-infinite}\\\text{slab}}} \tag{i}$$

The required temperature ratios in Eq. (i) are calculated below.

Infinite Cylinder (r_o = 0.025 m)

$$\frac{1}{\text{Bi}} = \frac{k}{hr_0} = \frac{45}{350 \times 0.025} = 5.143$$

$$\text{Fo} = \frac{\alpha\tau}{r_o^2} = \frac{1.4 \times 10^{-5} \times 100}{0.025^2} = 2.24.$$

From Fig. 6.18,

$$\left(\frac{\theta_0}{\theta_i}\right)_{\substack{\text{infinite}\\ \text{cylinder}}} \approx 0.45.$$

Semi-infinite Slab (x = 0.04 m)

The required parameters for this case are

$$\frac{h\sqrt{\alpha\tau}}{k} = \frac{350 \times \sqrt{1.4 \times 10^{-5} \times 100}}{45} = 0.291$$
$$\frac{x}{2\sqrt{\alpha\tau}} = \frac{0.04}{2 \times \sqrt{1.4 \times 10^{-5} \times 100}} = 0.534.$$

From Fig. 6.14,

$$1 - \left(\frac{\theta}{\theta_i}\right) \approx 0.085$$

or

$$\frac{\theta}{\theta_i} \approx 1 - 0.085 = 0.915.$$

The required solution from Eq. (i) is

$$\left(\frac{\theta_0}{\theta_i}\right) = \left(\frac{\theta_0}{\theta_i}\right)_{\substack{\text{infinite}\\ \text{cylinder}}} \left(\frac{\theta}{\theta_i}\right)_{\substack{\text{semi-infinite}\\ \text{slab}}} = 0.45 \times 0.915 = 0.412$$

or

$$t_0 = 0.412 \times (t_i - t_\infty) + t_\infty = 0.412 \times (30 - 200) + 200 = 129.96\,^{\circ}\text{C}.$$

Example 6.37 A steel cylinder 50 mm in diameter and 80 mm long is initially at a uniform temperature of 650°C. It is dipped in a cooling medium at 100°C. If the heat transfer coefficient is 1000 W/(m^2 K), determine the temperature at 20 mm away from one of its flat faces and at radius 10 mm after 150 s. Given: α = 1.2 × 10^{-5} m^2/s and k = 40 W/(m K).

Solution

The problem is a combination of an infinite cylinder and a 2L plate, refer Case 4, Table 6.3.

Infinite Cylinder (r_o = 0.025 m)

$$\frac{1}{\text{Bi}} = \frac{k}{hr_0} = \frac{40}{1000 \times 0.025} = 1.6$$
$$\text{Fo} = \frac{\alpha\tau}{r_o^2} = \frac{1.2 \times 10^{-5} \times 150}{0.025^2} = 2.88.$$

From Fig. 6.18, $\left(\frac{\theta_0}{\theta_i}\right) \approx 0.05.$

From Fig. 6.21, at $r/r_o = 10/25 = 0.4$, $\left(\frac{\theta}{\theta_o}\right) \approx 0.95.$

Hence,

$$\left(\frac{\theta}{\theta_i}\right)_{\substack{\text{infinite}\\ \text{cylinder}}} = \left(\frac{\theta_0}{\theta_i}\right) \times \left(\frac{\theta}{\theta_0}\right) = 0.05 \times 0.95 = 0.0475.$$

2*L*- Plate (2*L* = 0.08 m)

The required parameters for this case are

$$\frac{1}{\text{Bi}} = \frac{k}{hL} = \frac{40}{1000 \times 0.04} = 1.0$$

$$\text{Fo} = \frac{\alpha\tau}{L^2} = \frac{1.2 \times 10^{-5} \times 150}{0.04^2} = 1.125.$$

From Fig. 6.17, $\left(\frac{\theta_0}{\theta_i}\right) \approx 0.5.$

From Fig. 6.20, at $x/L = (40 - 20)/40 = 0.5$, $\left(\frac{\theta}{\theta_o}\right) \approx 0.9.$

Hence,

$$\left(\frac{\theta}{\theta_i}\right)_{2L-\text{plate}} = \left(\frac{\theta_0}{\theta_i}\right) \times \left(\frac{\theta}{\theta_0}\right) = 0.5 \times 0.9 = 0.45.$$

The solution for the given short cylinder is

$$\left(\frac{\theta}{\theta_i}\right) = \left(\frac{t - t_\infty}{t_i - t_\infty}\right) = \left(\frac{\theta}{\theta_i}\right)_{\substack{\text{infinite}\\ \text{cylinder}}} \left(\frac{\theta}{\theta_i}\right)_{2L-\text{plate}} = 0.0475 \times 0.45 = 0.02138$$

$$t = 0.02138 \times (t_i - t_\infty) + t_\infty = 0.02138 \times (650 - 100) + 100 = 111.76^\circ\text{C}.$$

6.9 Numerical Method of Solving Transient Conduction Problems

In the previous sections, the results of mathematical analysis for regular-shaped solids under transient conduction have been presented in the form of charts. However, in many problems of practical interest, the geometrical shapes do not conform to these regular shapes, and in some cases, the boundary conditions may also vary with time. It is not possible to provide a mathematical solution to such problems. The *finite-difference method* (a numerical technique, which was presented in Chap. 5 to solve two-dimensional steady-state conduction problems), can also be used to solve transient conduction problems.

The differential equation which governs the heat flow equation within a solid under two-dimensional transient conduction is

$$\frac{\partial^2 t}{\partial x^2} + \frac{\partial^2 t}{\partial y^2} = \frac{1}{\alpha}\frac{\partial t}{\partial \tau} \tag{i}$$

Following the procedure presented in Chap. 5, the derivatives in Eq. (i) may be expressed as (refer Fig. 6.28 for the nomenclature of the nodes):

$$\left(\frac{\partial^2 t}{\partial x^2}\right)_{m,n} \approx \frac{T^p_{m+1,n} + T^p_{m-1,n} - 2T^p_{m,n}}{(\Delta x)^2} \tag{ii}$$

$$\left(\frac{\partial^2 t}{\partial y^2}\right)_{m,n} \approx \frac{T^p_{m,n+1} + T^p_{m,n-1} - 2T^p_{m,n}}{(\Delta y)^2} \tag{iii}$$

$$\frac{\partial t}{\partial \tau} \approx \frac{T^{p+1}_{m,n} - T^p_{m,n}}{\Delta \tau} \tag{iv}$$

where the superscripts designate the time increments. Using the above finite-difference relations in Eq. (i), we obtain

$$\frac{T^p_{m+1,n} + T^p_{m-1,n} - 2T^p_{m,n}}{(\Delta x)^2} + \frac{T^p_{m,n+1} + T^p_{m,n-1} - 2T^p_{m,n}}{(\Delta y)^2} = \frac{1}{\alpha}\frac{T^{p+1}_{m,n} - T^p_{m,n}}{\Delta \tau} \tag{v}$$

For $\Delta x = \Delta y$, the equation of $T^{p+1}_{m,n}$ by simplification of Eq. (v) is

$$T^{p+1}_{m,n} = \frac{\alpha \Delta \tau}{(\Delta x)^2}\left(T^p_{m+1,n} + T^p_{m-1,n} + T^p_{m,n+1} + T^p_{m,n-1}\right) + \left(1 - \frac{4\alpha\Delta\tau}{(\Delta x)^2}\right)T^p_{m,n}$$

Introducing $\frac{\alpha\Delta\tau}{(\Delta x)^2}$ = Fo, the Fourier number, we get

$$T^{p+1}_{m,n} = \text{Fo}\left(T^p_{m+1,n} + T^p_{m-1,n} + T^p_{m,n+1} + T^p_{m,n-1}\right) + (1 - 4\text{Fo})T^p_{m,n} \tag{6.33}$$

Thus, the temperature $T^{p+1}_{m,n}$ of the node (m, n) after a time increment of $\Delta\tau$ may be determined if the temperatures of the surrounding nodes at any particular time are known. Such equations are written for each node leading to a set of equations.

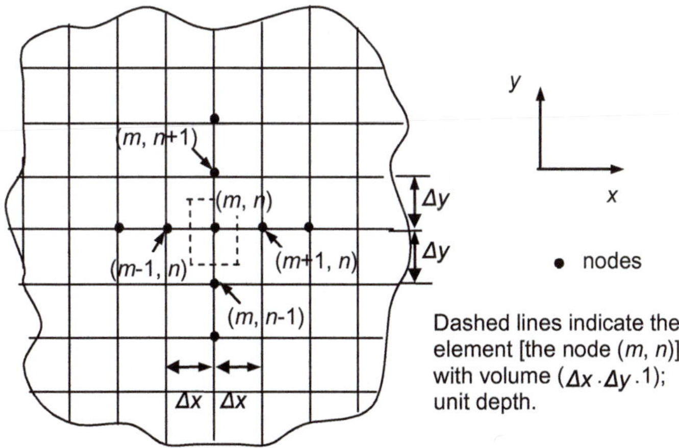

Fig. 6.28 Nomenclature for finite-difference approximation (two-dimensional transient conduction)

If the distance and time increments in Eq. (6.33) are chosen such that

$$\frac{\alpha\Delta\tau}{(\Delta x)^2} = \mathrm{Fo} = \frac{1}{4} \tag{6.34}$$

Then, the temperature of node (m, n) after a time increment $\Delta\tau$ is

$$T_{m,n}^{p+1} = \frac{T_{m+1,n}^{p} + T_{m-1,n}^{p} + T_{m,n+1}^{p} + T_{m,n-1}^{p}}{4} \tag{6.35}$$

which is simply the arithmetic mean of the temperature of the four surrounding nodes at the beginning of the time increment.

For a *one-dimensional system*, Eq. (6.33) becomes

$$T_{m}^{p+1} = \mathrm{Fo}\left(T_{m+1}^{p} + T_{m-1}^{p}\right) + (1 - 2\mathrm{Fo})T_{m}^{p} \tag{6.36}$$

and for

$$\frac{\alpha\Delta\tau}{(\Delta x)^2} = \mathrm{Fo} = \frac{1}{2}, \tag{6.37}$$

$$T_{m}^{p+1} = \frac{T_{m+1}^{p} + T_{m-1}^{p}}{2} \tag{6.38}$$

Equation (6.33) can also be obtained by following the method of heat balances. Consider node (m, n) in Fig. 6.28. The rates of heat conduction from the neighbouring nodes are

$$k(\Delta y \cdot 1)\frac{T_{m+1,n}^{p} - T_{m,n}^{p}}{\Delta x}$$

$$k(\Delta y \cdot 1)\frac{T_{m-1,n}^{p} - T_{m,n}^{p}}{\Delta x}$$

$$k(\Delta x \cdot 1)\frac{T_{m,n+1}^{p} - T_{m,n}^{p}}{\Delta y}$$

$$k(\Delta x \cdot 1)\frac{T_{m,n-1}^{p} - T_{m,n}^{p}}{\Delta y}$$

The change in the internal energy of the elemental volume due to the conduction heat flow into the node (m, n) is

$$\rho c(\Delta x.\Delta y.1)\frac{T_{m,n}^{p+1} - T_{m,n}^{p}}{\Delta\tau}$$

Heat balance at the node (m, n) gives

$$k(\Delta y.1)\frac{T_{m+1,n}^{p} - T_{m,n}^{p}}{\Delta x} + k(\Delta y.1)\frac{T_{m-1,n}^{p} - T_{m,n}^{p}}{\Delta x} + k(\Delta x.1)\frac{T_{m,n+1}^{p} - T_{m,n}^{p}}{\Delta y}$$

$$+ k(\Delta x.1)\frac{T_{m,n-1}^{p} - T_{m,n}^{p}}{\Delta y} = \rho c(\Delta x.\Delta y.1)\frac{T_{m,n}^{p+1} - T_{m,n}^{p}}{\Delta\tau}$$

For $\Delta x = \Delta y$, we get

$$T^p_{m+1,n} + T^p_{m-1,n} + T^p_{m,n+1} + T^p_{m,n-1} - 4T^p_{m,n} = \frac{\rho c}{k\Delta\tau}(\Delta x)^2\left(T^{p+1}_{m,n} - T^p_{m,n}\right)$$

Introducing $\frac{k\Delta\tau}{\rho c(\Delta x)^2} = \text{Fo}$ and solving for $T^{p+1}_{m,n}$, we get

$$T^{p+1}_{m,n} = \text{Fo}\left(T^p_{m+1,n} + T^p_{m-1,n} + T^p_{m,n+1} + T^p_{m,n-1}\right) + (1 - 4\text{Fo})T^p_{m,n} \tag{6.33}$$

The calculation of the temperature $T^{p+1}_{m,n}$ using the above equations requires the selection of certain values of the Fourier number. This choice is limited by the following conditions, known as *stability criteria*.

$$\text{For a two-dimensional system, Fo} \leq 1/4 \tag{6.34}$$

and

$$\text{for a one-dimensional system, Fo} \leq 1/2 \tag{6.37}$$

For a three-dimensional system, the stability criterion can be shown to be Fo $\leq$ 1/6.

By simple analysis, we can show that the rejection of these conditions will lead to unrealistic results. Let us consider a one-dimensional system, with the adjoining nodes of node (m, n) at equal temperatures, say 30°C. Let the temperature of node (m, n) be 40°C. If we select Fo = 1, ignoring the condition of Eq. (6.37), then after the time increment $\Delta\tau$, T^{p+1}_m would be 20°C, which is lower than the temperature of the adjoining nodes and, hence it is an unrealistic result. This conclusion can be shown to be valid for all values of Fo > 1/2 for the one-dimensional case.

Let us select some value of Fo $\leq$ ½, say 1/3. Then from Eq. (6.36), we have

$$T^{p+1}_m = \frac{T^p_{m+1} + T^p_{m-1} + T^p_m}{3} \tag{i}$$

and for Fo = ¼, we get

$$T^{p+1}_m = \frac{T^p_{m+1} + T^p_{m-1} + 2T^p_m}{4} \tag{ii}$$

A comparison of Eqs. (i), (ii) and (6.38) shows that a choice of Fo = ½ in case of one-dimensional system gives a simple equation.

The above conclusions can be extended to the two- and three-dimensional systems also.

In general, smaller the value of increments Δx and $\Delta\tau$, greater is the accuracy of the result but slower will be the convergence to the solution. One may think that a small value of Δx with a large value of $\Delta\tau$ might give a solution with greater accuracy with speed, but the values of the increments cannot be selected arbitrarily. Once the value of Fourier number Fo and the distance increment Δx are selected, the upper limit of the time increment $\Delta\tau$ is fixed from the stability criteria.

We developed the finite-difference equations for a node lying within a solid. Following the method outlined above, we can develop the equations for other physical situations as illustrated in Example 6.38.

Example 6.38 Write the finite difference expression for the temperature of a node on a convective boundary in one-dimensional heat flow.

Solution

Figure 6.29 shows the node (m) on the convective boundary.

The rate of heat conduction from node (m – 1) to node (m) under consideration

$$k(\Delta y.1)\frac{T_{m-1}^{p}-T_{m}^{p}}{\Delta x}$$

Heat convected into the node (m) is

$$h(\Delta y.1)(T_{\infty}^{p}-T_{m}^{p})$$

The change in the internal energy of node (m) due to conduction and convection heat flow is

$$\rho c\left(\frac{\Delta x}{2}.\Delta y.1\right)\frac{T_{m}^{p+1}-T_{m}^{p}}{\Delta \tau}$$

Heat balance at the node (m) gives

$$\rho c\left(\frac{\Delta x}{2}.\Delta y.1\right)\frac{T_{m}^{p+1}-T_{m}^{p}}{\Delta \tau}=k(\Delta y.1)\frac{T_{m-1}^{p}-T_{m}^{p}}{\Delta x}+h(\Delta y.1)(T_{\infty}^{p}-T_{m}^{p})$$

Using $\Delta x = \Delta y$ and rearranging the terms, we get

$$T_{m}^{p+1}-T_{m}^{p}=\frac{2k\Delta \tau}{\rho c(\Delta x)^{2}}(T_{m-1}^{p}-T_{m}^{p})+\frac{h\Delta x}{k}\frac{2k\Delta \tau}{\rho c(\Delta x)^{2}}(T_{\infty}^{p}-T_{m}^{p}).$$

Putting $\frac{k\Delta\tau}{\rho c(\Delta x)^2}$ = Fo and $\frac{h\Delta x}{k}=$ Bi and solving for T_m^{p+1}, we obtain

$$T_{m}^{p+1}=2\text{Fo}(T_{m-1}^{p}+\text{Bi}\times T_{\infty}^{p})+[1-2\text{Fo}-(2\text{Bi}\times \text{Fo})]T_{m}^{p},$$

which is the desired result.

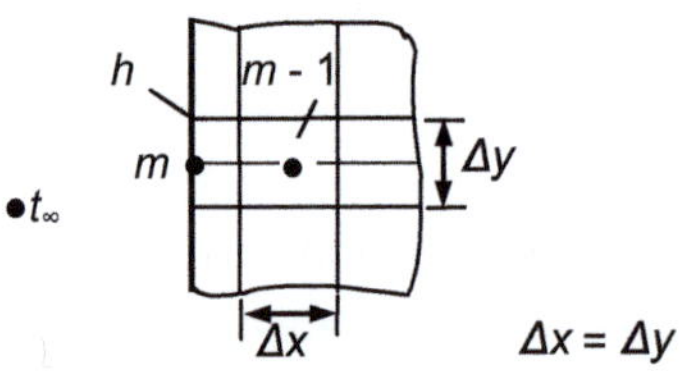

Fig. 6.29 Example 6.38

6.9.1 The Explicit and Implicit Formulations

We developed equations in the previous section following the technique of forward-difference, wherein the temperature of a node after the time interval $\Delta\tau$, that is, $T_{m,n}^{p+1}$ or T_{m}^{p+1} is expressed in terms of the temperatures of the surrounding nodes at the beginning of the time interval. Using this technique, we calculate the temperature $T_{m,n}^{p+1}$ in terms of previous nodal temperatures. Such formulation is termed as *explicit formulation* because the new temperature at the node under consideration appears explicitly on the left hand of the equation. The solution of the problem, thus, proceeds from one time increment to the next till we obtain the desired state. As already discussed, the stability of the solution in this method is governed by the selection of the values of distance and time increments Δx and $\Delta\tau$ according to the stability criteria.

Alternatively, the finite difference equation may be written by using the backward-difference in which the time derivative moves backward (into the past) as explained below. For the interior node (m, n) of Fig. 6.28, we shall write the equations of the rate of conduction heat flow as

$$k(\Delta y.1)\frac{T_{m+1,n}^{p+1}-T_{m,n}^{p+1}}{\Delta x}$$

$$k(\Delta y.1)\frac{T_{m-1,n}^{p+1}-T_{m,n}^{p+1}}{\Delta x}$$

$$k(\Delta x.1)\frac{T_{m,n+1}^{p+1}-T_{m,n}^{p+1}}{\Delta y}$$

$$k(\Delta x.1)\frac{T_{m,n-1}^{p+1}-T_{m,n}^{p+1}}{\Delta y}$$

The change in the internal energy of the elemental volume represented by the node (m, n) is expressed as earlier, that is, by

$$\rho c(\Delta x.\Delta y.1)\frac{T_{m,n}^{p+1}-T_{m,n}^{p}}{\Delta\tau}$$

Heat balance equation is

$$k(\Delta y.1)\frac{T_{m+1,n}^{p+1}-T_{m,n}^{p+1}}{\Delta x}+k(\Delta y.1)\frac{T_{m-1,n}^{p+1}-T_{m,n}^{p+1}}{\Delta x}+k(\Delta x.1)\frac{T_{m,n+1}^{p+1}-T_{m,n}^{p+1}}{\Delta y}$$
$$+k(\Delta x.1)\frac{T_{m,n-1}^{p+1}-T_{m,n}^{p+1}}{\Delta y}=\rho c(\Delta x.\Delta y.1)\frac{T_{m,n}^{p+1}-T_{m,n}^{p}}{\Delta\tau}$$

For $\Delta x = \Delta y$, we get

$$T_{m+1,n}^{p+1}+T_{m-1,n}^{p+1}+T_{m,n+1}^{p+1}+T_{m,n-1}^{p+1}-4T_{m,n}^{p+1}=\frac{\rho c}{k\Delta\tau}(\Delta x)^2\left(T_{m,n}^{p+1}-T_{m,n}^{p}\right)$$

Introducing $\frac{k\Delta\tau}{\rho c(\Delta x)^2}=\mathrm{Fo}$ and solving for $T_{m,n}^{p}$, we get

$$T_{m,n}^{p}=(1+4\mathrm{Fo})T_{m,n}^{p+1}-\mathrm{Fo}\left(T_{m+1,n}^{p+1}+T_{m-1,n}^{p+1}+T_{m,n+1}^{p+1}+T_{m,n-1}^{p+1}\right) \quad (6.38)$$

This equation is called *implicit formulation*. This formulation expresses temperature $T^p_{m,n}$ in terms of $T^{p+1}_{m,n}$. We get a set of equations when all the nodes are considered, which are to be solved simultaneously by using the method discussed in Chap. 5. The greatest advantage of this formulation is that no restriction is imposed on selection of time and distance increments. Thus much larger time increments can be used for the rapid solution of the problem. This actual selection of $\Delta\tau$ and Δx is now based on the trade-off between the accuracy and the cost of computation.

Example 6.39 Write down the nodal equations using explicit and implicit formulations for the cases shown in Fig. 6.30a–c.

Solution

Following the procedure outlined in the previous sections, the nodal equations can be easily written as illustrated below.

A. **Explicit Formulation**

(a) **Interior node with convection**

The summation of heat flow equations by conduction is

$$k\left(\frac{\Delta y}{2}.1\right)\frac{T^p_{m+1,n}-T^p_{m,n}}{\Delta x}+k(\Delta y.1)\frac{T^p_{m-1,n}-T^p_{m,n}}{\Delta x}$$
$$+k(\Delta x.1)\frac{T^p_{m,n+1}-T^p_{m,n}}{\Delta y}+k\left(\frac{\Delta x}{2}.1\right)\frac{T^p_{m,n-1}-T^p_{m,n}}{\Delta y}$$

Heat flow by convection is

$$h\left[\left(\frac{\Delta x+\Delta y}{2}\right).1\right](T_\infty-T^p_{m,n})$$

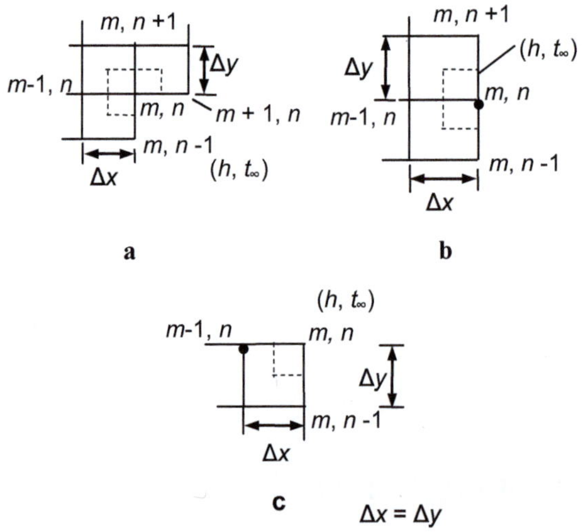

Fig. 6.30 Example 6.39

The summation of the conduction and convection terms given above will be equal to the increase in the internal energy of the elemental volume. That is,

$$k\left(\frac{\Delta y}{2}.1\right)\frac{T^p_{m+1,n}-T^p_{m,n}}{\Delta x}+k(\Delta y.1)\frac{T^p_{m-1,n}-T^p_{m,n}}{\Delta x}$$
$$+k(\Delta x.1)\frac{T^p_{m,n+1}-T^p_{m,n}}{\Delta y}+k\left(\frac{\Delta x}{2}.1\right)\frac{T^p_{m,n-1}-T^p_{m,n}}{\Delta y}$$
$$+h\left[\left(\frac{\Delta x+\Delta y}{2}\right).1\right](T_\infty-T^p_{m,n})=\rho c\left[\left(\Delta x.\Delta y-\frac{\Delta x.\Delta y}{4}\right).1\right]\frac{T^{p+1}_{m,n}-T^p_{m,n}}{\Delta\tau}$$

For $\Delta x = \Delta y$, we get

$$\frac{1}{2}T^p_{m+1,n}+T^p_{m-1,n}+T^p_{m,n+1}+\frac{1}{2}T^p_{m,n-1}-3T^p_{m,n}+\frac{h\Delta x}{k}(T_\infty-T^p_{m,n})$$
$$=\frac{3}{4}\frac{\rho c}{k\Delta\tau}(\Delta x)^2\left(T^{p+1}_{m,n}-T^p_{m,n}\right)$$

Introducing the Fourier number and Biot number, and solving for $T^{p+1}_{m,n}$, we get

$$T^{p+1}_{m,n}=\frac{2}{3}\text{Fo}(T^p_{m+1,n}+2T^p_{m-1,n}+2T^p_{m,n+1}+T^p_{m,n-1})$$
$$+\frac{4}{3}\text{Fo}\times\text{Bi}T_\infty+(1-4\text{Fo}-\frac{4}{3}\text{Fo}\times\text{Bi})T^p_{m,n}$$

The stability condition is

$$1-4\text{Fo}-\frac{4}{3}\text{Fo}\times\text{Bi}\geq 0$$

or

$$\text{Fo}(3+\text{Bi})\leq\frac{3}{4}$$

(b) **Node at plain surface with convection**

The heat balance gives Conduction heat inflow + convection heat inflow = increase in internal energy

or

$$k\left(\frac{\Delta x}{2}.1\right)\frac{T^p_{m,n+1}-T^p_{m,n}}{\Delta y}+k(\Delta y.1)\frac{T^p_{m-1,n}-T^p_{m,n}}{\Delta x}+k\left(\frac{\Delta x}{2}.1\right)\frac{T^p_{m,n-1}-T^p_{m,n}}{\Delta y}$$
$$+h(\Delta y.1)(T_\infty-T^p_{m,n})=\rho c\left[\left(\frac{\Delta x.\Delta y}{2}\right).1\right]\frac{T^{p+1}_{m,n}-T^p_{m,n}}{\Delta\tau}$$

For $\Delta x = \Delta y$, and introducing Fo and Bi, we get

$$T^{p+1}_{m,n}=\text{Fo}(2T^p_{m-1,n}+T^p_{m,n+1}+T^p_{m,n-1}+2\text{Bi}\times T_\infty)+(1-4\text{Fo}-2\text{Fo}\times\text{Bi})T^p_{m,n}$$

The stability condition is

$$\mathrm{Fo}(2+\mathrm{Bi}) \leq \frac{1}{2}.$$

(c) **Node at the exterior corner with convection**

The heat balance equation in this case is
Conduction heat inflow + convection heat inflow = increase in internal energy
or

$$k\left(\frac{\Delta y}{2}.1\right)\frac{T^p_{m-1,n}-T^p_{m,n}}{\Delta x}+k\left(\frac{\Delta x}{2}.1\right)\frac{T^p_{m,n-1}-T^p_{m,n}}{\Delta y}+h\left(\frac{\Delta x+\Delta y}{2}.1\right)(T_\infty-T^p_{m,n})$$
$$=\rho c\left[\left(\frac{\Delta x.\Delta y}{4}\right).1\right]\frac{T^{p+1}_{m,n}-T^p_{m,n}}{\Delta\tau}$$

Using $\Delta x = \Delta y$, and introducing Fo and Bi and simplifying, we get

$$T^{p+1}_{m,n}=2\mathrm{Fo}(T^p_{m-1,n}+T^p_{m,n-1}+2\mathrm{Bi}\times T_\infty)+(1-4\mathrm{Fo}-4\mathrm{Fo}\times\mathrm{Bi})T^p_{m,n}$$

The stability condition is

$$(1-4\mathrm{Fo}-4\mathrm{Fo}\times\mathrm{Bi})\geq 0$$

or

$$\mathrm{Fo}(1+\mathrm{Bi})\leq\frac{1}{4}.$$

B. **Implicit Formulations**

(a) **Interior node with convection**

The summation of conduction heat flow terms gives

$$k\left(\frac{\Delta y}{2}.1\right)\frac{T^{p+1}_{m+1,n}-T^{p+1}_{m,n}}{\Delta x}+k(\Delta y.1)\frac{T^{p+1}_{m-1,n}-T^{p+1}_{m,n}}{\Delta x}$$
$$+k(\Delta x.1)\frac{T^{p+1}_{m,n+1}-T^{p+1}_{m,n}}{\Delta y}+k\left(\frac{\Delta x}{2}.1\right)\frac{T^{p+1}_{m,n-1}-T^{p+1}_{m,n}}{\Delta y}$$

Heat flow by convection is

$$h\left[\left(\frac{\Delta x+\Delta y}{2}\right).1\right](T_\infty-T^{p+1}_{m,n})$$

The summation of the conduction and convection terms given above will be equal to the increase in the internal energy of the elemental volume. That is,

$$k\left(\frac{\Delta y}{2}.1\right)\frac{T^{p+1}_{m+1,n}-T^{p+1}_{m,n}}{\Delta x}+k(\Delta y.1)\frac{T^{p+1}_{m-1,n}-T^{p+1}_{m,n}}{\Delta x}$$
$$+k(\Delta x.1)\frac{T^{p+1}_{m,n+1}-T^{p+1}_{m,n}}{\Delta y}+k\left(\frac{\Delta x}{2}.1\right)\frac{T^{p+1}_{m,n-1}-T^{p+1}_{m,n}}{\Delta y}$$
$$+h\left[\left(\frac{\Delta x+\Delta y}{2}\right).1\right](T_\infty-T^{p+1}_{m,n})=\rho c\left[\left(\Delta x.\Delta y-\frac{\Delta x.\Delta y}{4}\right).1\right]\frac{T^{p+1}_{m,n}-T^{p}_{m,n}}{\Delta\tau}$$

For $\Delta x = \Delta y$, and introducing the Fourier and Biot numbers, we get the equation of present temperature $T^{p}_{m,n}$ of the node (m, n) as

$$T^{p}_{m,n}=\left(\frac{4}{3}\mathrm{Fo}\times\mathrm{Bi}+4\mathrm{Fo}+1\right)T^{p+1}_{m,n}-\frac{2}{3}\mathrm{Fo}(T^{p+1}_{m+1,n}$$
$$+2T^{p+1}_{m-1,n}+2T^{p+1}_{m,n+1}+T^{p+1}_{m,n-1})-\frac{4}{3}\mathrm{Fo}\times\mathrm{Bi}\times T_\infty$$

(b) **Node at plain surface with convection**

The heat balance gives
Conduction heat inflow + convection heat inflow = increase in internal energy
or

$$k(\Delta y.1)\frac{T^{p+1}_{m-1,n}-T^{p+1}_{m,n}}{\Delta x}+k(\frac{\Delta x}{2}.1)\frac{T^{p+1}_{m,n+1}-T^{p+1}_{m,n}}{\Delta y}+k\left(\frac{\Delta x}{2}.1\right)\frac{T^{p+1}_{m,n-1}-T^{p+1}_{m,n}}{\Delta y}$$
$$+h(\Delta y.1)(T_\infty-T^{p+1}_{m,n})=\rho c\left[\left(\frac{\Delta x.\Delta y}{2}\right).1\right]\frac{T^{p+1}_{m,n}-T^{p}_{m,n}}{\Delta\tau}$$

Using $\Delta x = \Delta y$, $\dfrac{k\Delta\tau}{\rho c(\Delta x)^2}=\mathrm{Fo}$, $\dfrac{h\Delta x}{k}=\mathrm{Bi}$ and simplifying, we get

$$T^{p}_{m,n}=(2\mathrm{Fo}\times\mathrm{Bi}+4\mathrm{Fo}+1)T^{p+1}_{m,n}-\mathrm{Fo}(2T^{p+1}_{m-1,n}$$
$$+T^{p+1}_{m,n+1}+T^{p+1}_{m,n-1})-2\mathrm{Fo}\times\mathrm{Bi}\times T_\infty,$$

which is the desired result.

(c) **Node at the exterior corner with convection**

The heat balance equation in this case is

$$k\left(\frac{\Delta y}{2}.1\right)\frac{T^{p+1}_{m-1,n}-T^{p+1}_{m,n}}{\Delta x}+k\left(\frac{\Delta x}{2}.1\right)\frac{T^{p+1}_{m,n-1}-T^{p+1}_{m,n}}{\Delta y}+h\left(\frac{\Delta x+\Delta y}{2}.1\right)(T_\infty-T^{p+1}_{m,n})$$
$$=\rho c\left[\left(\frac{\Delta x.\Delta y}{4}\right).1\right]\frac{T^{p+1}_{m,n}-T^{p}_{m,n}}{\Delta\tau}$$

For $\Delta x = \Delta y$, $\frac{k\Delta\tau}{\rho c(\Delta x)^2}=\mathrm{Fo}$, $\frac{h\Delta x}{k}=\mathrm{Bi}$, the equation gives

Table 6.4 Summary of transient, two-dimensional finite-difference equations ($\Delta x = \Delta y$)

Configuration	Explicit formulation	Implicit formulation
1. Interior node 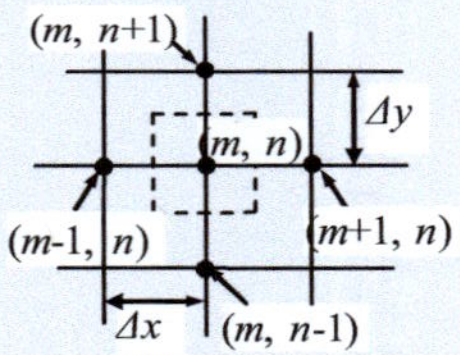	$T_{m,n}^{p+1} = Fo\left(T_{m+1,n}^{p} + T_{m-1,n}^{p} + T_{m,n+1}^{p} + T_{m,n-1}^{p}\right) + (1 - 4Fo)T_{m,n}^{p}$ The stability condition is $Fo \le \frac{1}{4}$	$T_{m,n}^{p} = (1 + 4Fo)T_{m,n}^{p+1} - Fo\left(T_{m+1,n}^{p+1} + T_{m-1,n}^{p+1} + T_{m,n+1}^{p+1} + T_{m,n-1}^{p+1}\right)$
2. Node at interior corner with convection 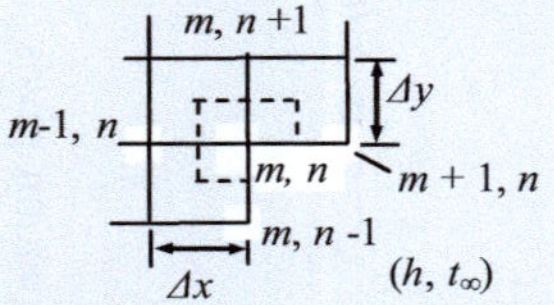	$T_{m,n}^{p+1} = \frac{2}{3}Fo(T_{m+1,n}^{p} + 2T_{m-1,n}^{p} + 2T_{m,n+1}^{p} + T_{m,n-1}^{p}) + \frac{4}{3}Fo \times BiT_{\infty} + (1 - 4Fo - \frac{4}{3}Fo \times Bi)T_{m,n}^{p}$ The stability condition is $Fo(3 + Bi) \le \frac{3}{4}$	$T_{m,n}^{p} = \left(\frac{4}{3}Fo \times Bi + 4Fo + 1\right)T_{m,n}^{p+1} - \frac{2}{3}Fo(T_{m+1,n}^{p+1} + 2T_{m-1,n}^{p+1} + 2T_{m,n+1}^{p+1} + T_{m,n-1}^{p+1}) - \frac{4}{3}Fo \times Bi \times T_{\infty}$
3. Node at plain surface with convection 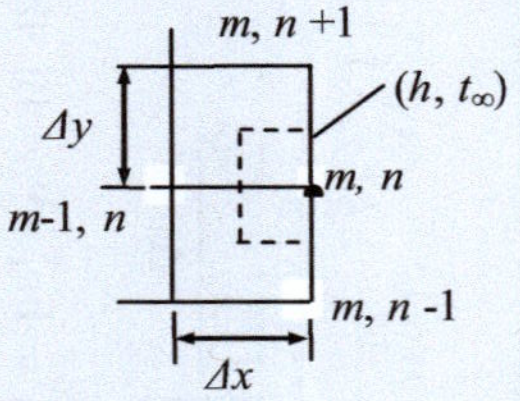	$T_{m,n}^{p+1} = Fo(2T_{m-1,n}^{p} + T_{m,n+1}^{p} + T_{m,n-1}^{p} + 2Bi \times T_{\infty}) + (1 - 4Fo - 2Fo \times Bi)T_{m,n}^{p}$ The stability condition is $Fo(2 + Bi) \le \frac{1}{2}$	$T_{m,n}^{p} = (2Fo \times Bi + 4Fo + 1)T_{m,n}^{p+1} - Fo(2T_{m-1,n}^{p+1} + T_{m,n+1}^{p+1} + T_{m,n-1}^{p+1}) - 2Fo \times Bi \times T_{\infty}$
4. Node at exterior corner with convection 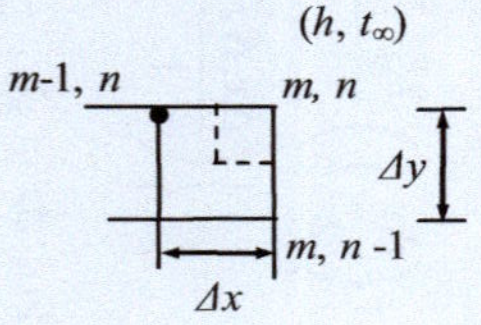	$T_{m,n}^{p+1} = 2Fo(T_{m-1,n}^{p} + T_{m,n-1}^{p} + 2Bi \times T_{\infty}) + (1 - 4Fo - 4Fo \times Bi)T_{m,n}^{p}$ The stability condition is $Fo(1 + Bi) \le \frac{1}{4}$	$T_{m,n}^{p} = (1 + 4Fo + 4Fo \times Bi)T_{m,n}^{p+1}$ $4 - 2Fo(T_{m-1,n}^{p+1} + T_{m,n-1}^{p+1}) - 4Fo \times Bi \times T_{\infty}$

$$T^p_{m,n} = (1 + 4\text{Fo} + 4\text{Fo} \times \text{Bi})T^{p+1}_{m,n} - 2\text{Fo}(T^{p+1}_{m-1,n} + T^{p+1}_{m,n-1}) - 4\text{Fo} \times \text{Bi} \times T_\infty,$$

which is the desired equation.

Summary of transient, two-dimensional finite-difference equations is given in Table 6.4.

Example 6.40 Figure 6.31 shows a plain wall with convection heat transfer on both sides. Write nodal equations for nodes on both sides exposed to convection and for a node lying in the wall following the method of backward differences.

Solution

It is a one-dimensional unsteady-state conduction problem. The wall has been divided by equal distance increments of Δx. The nodes under consideration are (1), (n) and (i) as shown in the figure.

The nodal equation can be written by following the procedure discussed in Sect. 6.9 and Example 6.39.

(i) **Interior surface node (n)**

The heat balance equation is

$$kA\frac{T^{p+1}_{n-1} - T^{p+1}_n}{\Delta x} + h_iA(T^{p+1}_i - T^{p+1}_n) = \rho c\left(\frac{A\Delta x}{2}\right)\frac{T^{p+1}_n - T^p_n}{\Delta\tau}$$

Simplification of the equation gives

$$T^p_n + 2\text{Fo} \times \text{Bi} \times T^{p+1}_i = (2\text{Fo} \times \text{Bi} + 2\text{Fo} + 1)T^{p+1}_n - 2\text{Fo} \times T^{p+1}_{n-1} \tag{i}$$

(ii) **Node (i) inside the wall**

Differential equation of one-dimensional heat conduction is

$$\frac{\partial T}{\partial\tau} = \left(\frac{k}{\rho c}\right)\frac{\partial^2 T}{\partial x^2},$$

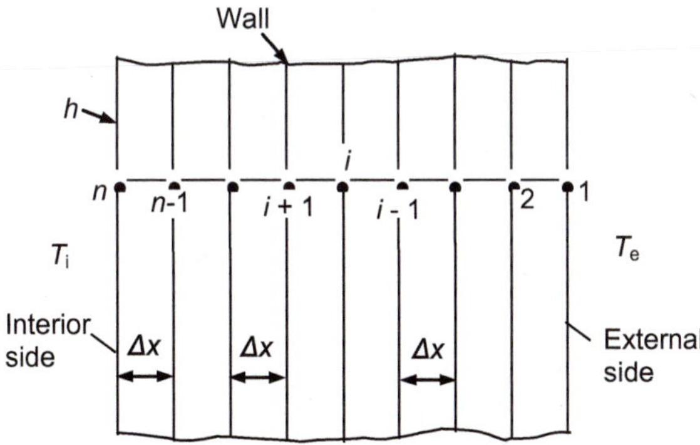

Fig. 6.31 Example 6.40

which gives

$$\frac{T_i^{p+1} - T_i^p}{\Delta\tau} = \alpha\left(\frac{T_{i+1}^{p+1} + T_{i-1}^{p+1} - 2T_i^{p+1}}{(\Delta x)^2}\right).$$

Alternatively, following the method of the heat balance, we have

$$kA\frac{T_{i+1}^{p+1} - T_i^{p+1}}{\Delta x} + kA\frac{T_{i-1}^{p+1} - T_i^{p+1}}{\Delta x} = \rho c(A\Delta x)\frac{T_i^{p+1} - T_i^p}{\Delta\tau}$$

or

$$\frac{T_i^{p+1} - T_i^p}{\Delta\tau} = \frac{k}{\rho c(\Delta x)^2}\left(T_{i+1}^{p+1} + T_{i-1}^{p+1} - 2T_i^{p+1}\right)$$

The required equation is

$$T_i^p = (2\text{Fo} + 1)T_i^{p+1} - \text{Fo}\left(T_{i+1}^{p+1} + T_{i-1}^{p+1}\right) \tag{ii}$$

(iii) **Exterior surface node (1) with convection**

$$kA\frac{T_2^{p+1} - T_1^{p+1}}{\Delta x} + h_o A(T_e^{p+1} - T_1^{p+1}) = \rho c\left(\frac{A\Delta x}{2}\right)\frac{T_1^{p+1} - T_1^p}{\Delta\tau}$$

Simplification gives

$$T_1^p + 2\text{Fo} \times \text{Bi} \times T_e^{p+1} = (2\text{Fo} \times \text{Bi} + 2\text{Fo} + 1)T_1^{p+1} - 2\text{Fo}T_2^{p+1} \tag{iii}$$

For the solution of this problem, the set of implication equations will consist of Eqs. (i), (iii) and equations similar to Eq. (ii) for each node inside the wall.

Example 6.41 Write the finite difference expression for the temperature of a node on a convective boundary of a one-dimensional wall which is suddenly subjected to uniform volumetricheating.

Solution

Figure 6.32 shows the node (m) on the convective boundary.

The rate of heat conduction from node (m – 1) to node (m) under consideration

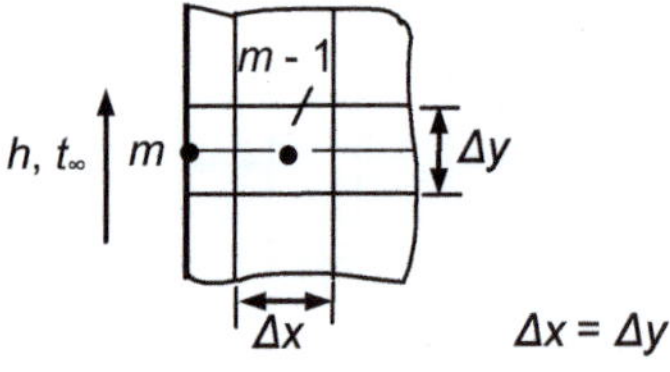

Fig. 6.32 Example 6.41

$$k(\Delta y.1)\frac{T_{m-1}^{p}-T_{m}^{p}}{\Delta x}$$

Heat convected from the node (m) is

$$h(\Delta y.1)(T_{\infty}-T_{m}^{p})$$

Heat generation in the control volume

$$q_g\left(\frac{\Delta x}{2}.\Delta y.1\right)$$

The change in the internal energy of node (m) is

$$\rho c\left(\frac{\Delta x}{2}.\Delta y.1\right)\frac{T_{m}^{p+1}-T_{m}^{p}}{\Delta \tau}$$

Heat balance at the node (m) gives

$$\rho c\left(\frac{\Delta x}{2}.\Delta y.1\right)\frac{T_{m}^{p+1}-T_{m}^{p}}{\Delta \tau}=k(\Delta y.1)\frac{T_{m-1}^{p}-T_{m}^{p}}{\Delta x}+h(\Delta y.1)(T_{\infty}-T_{m}^{p})+q_g\left(\frac{\Delta x}{2}.\Delta y.1\right).$$

Using $\Delta x = \Delta y$ and rearranging the terms, we get

$$T_{m}^{p+1}-T_{m}^{p}=\frac{2k\Delta\tau}{\rho c(\Delta x)^2}(T_{m-1}^{p}-T_{m}^{p})+\frac{h\Delta x}{k}\frac{2k\Delta\tau}{\rho c(\Delta x)^2}(T_{\infty}-T_{m}^{p})+\frac{\Delta\tau}{\rho c}q_g$$

or

$$T_{m}^{p+1}-T_{m}^{p}=\frac{2k\Delta\tau}{\rho c(\Delta x)^2}(T_{m-1}^{p}-T_{m}^{p})+\frac{h\Delta x}{k}\frac{2k\Delta\tau}{\rho c(\Delta x)^2}(T_{\infty}-T_{m}^{p})+\frac{k\Delta\tau}{\rho c(\Delta x)^2}\frac{(\Delta x)^2}{k}q_g$$

Putting $\frac{k\Delta\tau}{\rho c(\Delta x)^2}=\text{Fo}$ and $\frac{h\Delta x}{k}=\text{Bi}$ and solving for T_{m}^{p+1}, we obtain

$$T_{m}^{p+1}=2\text{Fo}\left(T_{m-1}^{p}+\text{Bi}T_{\infty}+\frac{(\Delta x)^2 q_g}{2k}\right)+(1-2\text{Fo}-2\text{Bi.Fo})T_{m}^{p},$$

which is the desired result.

For stability, the coefficient of T_{m}^{p} must be positive, hence

$$(1-2\text{Fo}-2\text{Bi.Fo})\geq 0$$

or

$$\text{Fo}\leq\frac{1}{2(1+\text{Bi})}.$$

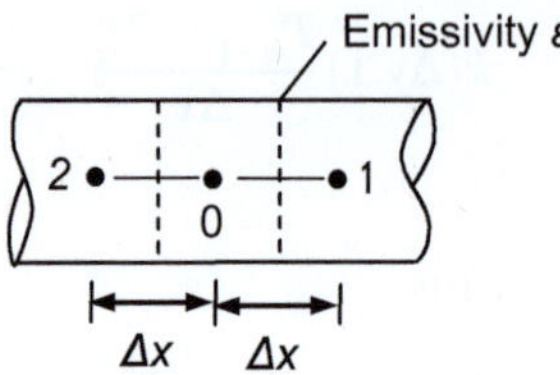

Fig. 6.33 Example 6.42

Example 6.42 Write the finite difference expression for the temperature of node 0 (Fig. 6.33) of a thin rod of the area of cross-section A_c and electrical resistivity ρ_e, which is suddenly subjected to heating due to passing of current I. The rod having surface emissivity ε is located in a large enclosure with vacuum. The enclosure surface temperature is T_{sur}.

Solution

Figure 6.33 shows the node (0) along with nodes 1 and 2.

The rate of heat conduction from node (1) to node (0) is

$$kA_c\frac{T_1^p - T_0^p}{\Delta x}.$$

Similarly the rate of heat conduction from node (2) to node (0) is

$$kA_c\frac{T_2^p - T_0^p}{\Delta x}$$

Heat loss by radiation to the surface of the large space is

$$\varepsilon\sigma(P\Delta x)(T_{sur}^4 - T_0^{4,p})$$

Heat generated in the control volume due to the flow of current is

$$I^2\left(\frac{\rho_e \Delta x}{A_c}\right)$$

The change in the internal energy of node (0) is

$$\rho c(A_c\Delta x)\frac{T_0^{p+1} - T_0^p}{\Delta\tau}$$

Heat balance at the node (0) gives

$$\rho c(A_c\Delta x)\frac{T_0^{p+1} - T_0^p}{\Delta\tau} = kA_c\frac{T_1^p - T_0^p}{\Delta x} + kA_c\frac{T_2^p - T_0^p}{\Delta x} + \varepsilon\sigma P\Delta x(T_{sur}^4 - T_0^{4,p}) + I^2\left(\frac{\rho_e \Delta x}{A_c}\right).$$

Rearranging the terms, we get

$$T_0^{p+1} = \frac{\Delta\tau k}{\rho c(\Delta x)^2}(T_1^p + T_2^p) - \left[2\frac{\Delta\tau k}{\rho c(\Delta x)^2} - 1\right]T_0^p + \frac{\Delta\tau\varepsilon\sigma P}{\rho c A_c}(T_{sur}^4 - T_0^{4,p}) + \left(\frac{I^2\rho_e\Delta\tau}{\rho c A_c^2}\right)$$

$$T_0^{p+1} = \frac{\Delta\tau k}{\rho c(\Delta x)^2}(T_1^p + T_2^p) - \left[2\frac{\Delta\tau k}{\rho c(\Delta x)^2} - 1\right]T_0^p + \frac{k\Delta\tau}{\rho c(\Delta x)^2}\cdot\frac{\varepsilon\sigma P(\Delta x)^2}{kA_c}(T_{sur}^4 - T_0^{4,p})$$
$$+ \frac{k\Delta\tau}{\rho c(\Delta x)^2}\cdot\frac{I^2\rho_e(\Delta x)^2}{kA_c^2}$$

Putting $\frac{k\Delta\tau}{\rho c(\Delta x)^2} = \text{Fo}$, we obtain

$$T_0^{p+1} = \text{Fo}(T_1^p + T_2^p) + (1 - 2\text{Fo})T_0^p + \text{Fo}.\frac{\varepsilon\sigma P(\Delta x)^2}{kA_c}(T_{\text{sur}}^4 - T_0^{4,p}) + \text{Fo}.\frac{I^2\rho_e(\Delta x)^2}{kA_c^2},$$

which is the desired result.

For stability, the coefficient of T_0^p must be positive, hence

$$\text{Fo} \le \frac{1}{2}.$$

6.10 The Schmidt Graphical Method for One-Dimensional Problems

In lieu of the numerical calculations, we can employ the Schmidt graphical technique to solve the one-dimensional transient problems. Differential equation governing this case is

$$\frac{\partial T}{\partial \tau} = \alpha\frac{\partial^2 T}{\partial x^2} \tag{i}$$

The corresponding finite difference equation is

$$T_m^{p+1} - T_m^p = \frac{\alpha\Delta\tau}{(\Delta x)^2}\left(T_{m+1}^p + T_{m-1}^p - 2T_m^p\right) \tag{ii}$$

If we choose the increments to give

$$\frac{\alpha\Delta\tau}{(\Delta x)^2} = \frac{1}{2} \tag{6.39}$$

then,

$$T_m^{p+1} = \frac{1}{2}\left(T_{m+1}^p + T_{m-1}^p\right) \tag{6.40}$$

From the above equation, it is clear that the temperature at node m after the time increment $\Delta\tau$ is the arithmetic mean of the temperatures of the adjacent nodes $(m + 1)$ and $(m - 1)$ at the beginning of the time increment. This arithmetic mean can be done graphically as shown in

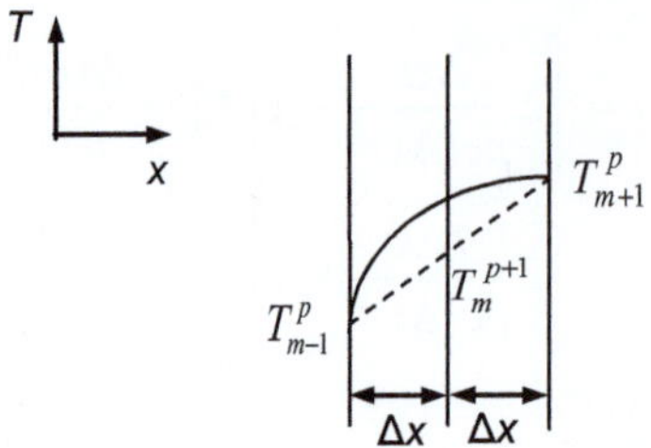

Fig. 6.34 Temperature at node m after the time increment $\Delta\tau$

Fig. 6.34, wherein the value of T_m^{p+1} has been obtained by drawing a line joining the points T_{m+1}^p and T_{m-1}^p.

By continued application of the above procedure, the development of the temperature field with time can be determined from a known initial temperature distribution. Thus to find the temperature distribution after some specified time, the solid is divided into equal distance increment Δx. Then using equation $\Delta\tau = (\Delta x)^2/2\alpha$ from Eq. (6.39), we can find the number of time increments necessary by dividing the total time by $\Delta\tau$.

It is to be noted that to use this graphical solution the temperature distribution in the solid must be known at some time and the chosen value of the time increment $\Delta\tau$ must satisfy Eq. (6.39). The following example illustrates the application of the method when the boundary (surface) temperatures are maintained at constant values.

Example 6.43 A very long 60 mm thick metal plate ($\alpha = 2 \times 10^{-4}$ m^2/min) initially at 50°C is immersed in a fluid ($\bar{h} = \infty$) of temperature 400°C. By the graphical method, determine the temperature distribution in the plate after 1.5 min. Confirm the result by numerical method.

Solution

Let Δx = 10 mm. The corresponding nodes are marked in Fig. 6.35. The time increment is

$$\Delta\tau = \frac{1}{2\alpha}(\Delta x)^2 = \frac{1}{2 \times 2 \times 10^{-4}}(10/1000)^2 = 0.25\,\text{min}.$$

Hence, the number of necessary time increments = 1.5/0.25 = 6.

Figure 6.36 presents the graphical solution. It is to note that due to $\bar{h} = \infty$, the boundary temperatures (at nodes 0 and 6) are maintained at constant values of 400°C throughout the heating process. Start the process by connecting point 0 with 2, 1 with 3, etc. by straight lines. Thus, a new temperature distribution is obtained. Repeat the process. It is to note that the temperature distribution is symmetrical about node 3.

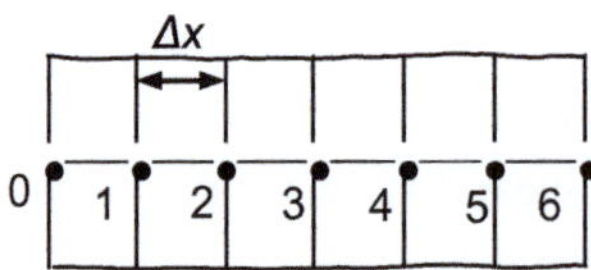

Fig. 6.35 Example 6.43

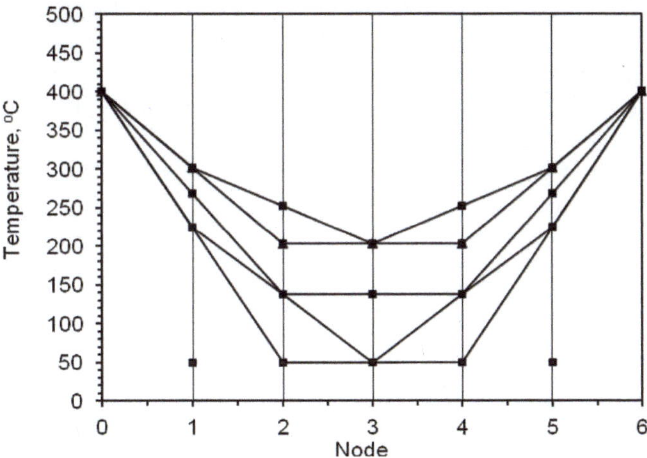

Fig. 6.36 Graphical solution, Schmidt plot

Table 6.5 Numerical solution

Time (min)	Nodes						
	0	1	2	3	4	5	6
0	400	50	50	50	50	50	400
0.25	400	225	50	50	50	225	400
0.5	400	225	137.5	50	137.5	225	400
0.75	400	268.75	137.5	137.5	137.5	268.75	400
1.0	400	268.75	203.125	137.5	203.125	268.75	400
1.25	400	301.563	203.125	203.125	203.125	301.563	400
1.5	400	301.563	252.343	203.125	252.343	301.563	400

The numerical solution can be obtained by using Eq. (6.40). Thus, the equations are

$$T_1^{p+1} = \frac{1}{2}\left(T_0^p + T_2^p\right)$$

$$T_2^{p+1} = \frac{1}{2}\left(T_1^p + T_3^p\right)$$

$$T_3^{p+1} = \frac{1}{2}\left(T_2^p + T_4^p\right)$$

The solution steps are given in Table 6.5.

Example 6.44 A semi-infinite solid (k = 30 W/(m K), $\alpha = 3 \times 10^{-6}$ m^2/s), initially at a uniform temperature of 300°C, is suddenly exposed to a convection environment (h = 200 W/(m^2 K), T_∞ = 30°C). Determine the temperature at the surface and at 30 mm depth after 150 s.

Solution

(i) **Node 0 on the convective boundary**

The rate of heat conduction from node (1) to node (0), refer Fig. 6.37,

$$k(\Delta y.1)\frac{T_1^p - T_0^p}{\Delta x}$$

Heat convected to the node 0 is

$$h(\Delta y.1)(T_\infty - T_0^p)$$

The change in internal energy of node 0 is

$$\rho c\left(\frac{\Delta x}{2}.\Delta y.1\right)\frac{T_0^{p+1} - T_0^p}{\Delta \tau}$$

Heat balance at the node 0 gives

$$\rho c\left(\frac{\Delta x}{2}.\Delta y.1\right)\frac{T_0^{p+1} - T_0^p}{\Delta \tau} = k(\Delta y.1)\frac{T_1^p - T_0^p}{\Delta x} + h(\Delta y.1)(T_\infty - T_0^p).$$

Using $\Delta x = \Delta y$ and rearranging the terms, we get

$$T_0^{p+1} - T_0^p = \frac{2k\Delta \tau}{\rho c(\Delta x)^2}(T_1^p - T_0^p) + h\Delta x\frac{2\Delta \tau}{\rho c(\Delta x)^2}(T_\infty - T_0^p)$$

or

$$T_0^{p+1} = \frac{2k\Delta \tau}{\rho c(\Delta x)^2}(T_1^p - T_0^p) + \frac{h\Delta x}{k}\frac{2k\Delta \tau}{\rho c(\Delta x)^2}(T_\infty - T_0^p) + T_0^p$$

Putting $\frac{k\Delta \tau}{\rho c(\Delta x)^2} = \text{Fo}$ and $\frac{h\Delta x}{k} = \text{Bi}$ and rearranging the terms, we obtain

$$T_0^{p+1} = 2\text{Fo}(T_1^p + \text{Bi}T_\infty) + (1 - 2\text{Fo} - 2\text{Bi.Fo})T_0^p, \quad (1)$$

which is the desired equation.

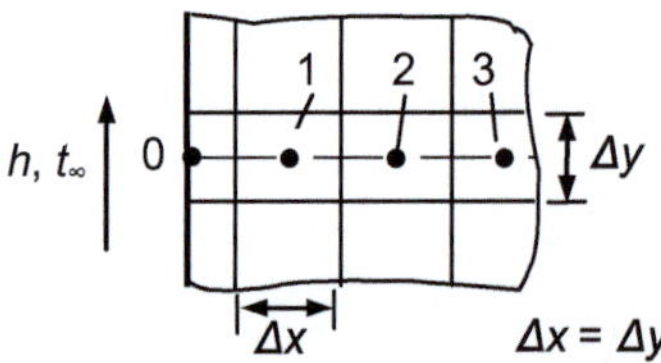

Fig. 6.37 Example 6.44

For stability, the coefficient of T_0^p must be positive, hence

$$(1 - 2\text{Fo} - 2\text{Bi.Fo}) \geq 0$$

or

$$\text{Fo} \leq \frac{1}{2(1 + \text{Bi})}.$$

(ii) **Interior nodes (m = 1, 2, 3….)**

The rate of heat conduction from node 0 to node 1 is

$$k(\Delta y.1)\frac{T_0^p - T_1^p}{\Delta x}$$

The rate of heat conduction from node 2 to node 1

$$k(\Delta y.1)\frac{T_2^p - T_1^p}{\Delta x}$$

The change in the internal energy of node 1 is

$$\rho c(\Delta x.\Delta y.1)\frac{T_1^{p+1} - T_1^p}{\Delta \tau}$$

Heat balance at the node 1 gives

$$\rho c(\Delta x.\Delta y.1)\frac{T_1^{p+1} - T_1^p}{\Delta \tau} = k(\Delta y.1)\frac{T_0^p - T_1^p}{\Delta x} + k(\Delta y.1)\frac{T_2^p - T_1^p}{\Delta x}.$$

Using $\Delta x = \Delta y$ and rearranging the terms, we get

$$T_1^{p+1} = \frac{k\Delta \tau}{\rho c(\Delta x)^2}(T_0^p - T_1^p) + \frac{k\Delta \tau}{\rho c(\Delta x)^2}(T_2^p - T_1^p) + T_1^p$$

Putting $\frac{k\Delta \tau}{\rho c(\Delta x)^2} = \text{Fo}$ and rearranging we obtain

$$T_1^{p+1} = \text{Fo}\left(T_0^p + T_2^p\right) + (1 - 2\text{Fo})T_1^p, \tag{2}$$

Similarly, equations for node 2 to 4 can be written.
For stability, the coefficient of T_1^p must be positive, hence

$$(1 - 2\text{Fo}) \geq 0$$

or

$$\text{Fo} \le \frac{1}{2}.$$

Assuming Δx = 15 mm, Biot number is

$$\text{Bi} = \frac{h\Delta x}{k} = \frac{200 \times 0.015}{30} = 0.1.$$

From stability condition for equation of node 0,

$$\begin{aligned}\text{Fo} &\le \frac{1}{2(1+\text{Bi})}\\ &\le \frac{1}{2(1+0.1)}\\ &\le 0.45.\end{aligned}$$

Let τ = 30 s, then

$$\text{Fo} = \frac{\alpha\tau}{(\Delta x)^2} = \frac{3 \times 10^{-6} \times 30}{0.015^2} = 0.4 < 0.45.$$

Condition is satisfied. Fo = 0.4 also satisfies the stability condition for equations of node 1 to 4.

For Bi = 0.1, Fo = 0.4 and T_∞ = 30, Eq. (1) reduces to

$$T_0^{p+1} = 0.8 \times \left(T_1^p + 3\right) + 0.12T_0^p \tag{3}$$

Equation (2) reduces to

$$T_1^{p+1} = 0.4 \times \left(T_0^p + T_2^p\right) + 0.2T_1^p \tag{4}$$

Similarly, for nodes 2, 3 and 4, we have

$$T_2^{p+1} = 0.4 \times \left(T_1^p + T_3^p\right) + 0.2T_2^p \tag{5}$$

$$T_3^{p+1} = 0.4 \times \left(T_2^p + T_4^p\right) + 0.2T_3^p \tag{6}$$

$$T_4^{p+1} = 0.4 \times \left(T_3^p + T_5^p\right) + 0.2T_4^p \tag{7}$$

For 5 time steps, each of 30 s giving total time 150 s, i.e. p = 0 to 5, solution of Eqs. (3)–(7) is presented in the table below. (Table 6.6)

From the table, the surface temperature T_0 is 261.3°C and the temperature at a depth of 30 mm T_2 is 291.9°C after 150 s. The accuracy of the results from the numerical analysis will increase with a decrease in $\Delta\tau$ and Δx.

Table 6.6 Numerical solution

p	Time (s)	T_0	T_1	T_2	T_3	T_4	T_5
0	0	300	300	300	300	300	300
1	30	278.4	300	300	300	300	300
2	60	275.8	291.4	300	300	300	300
3	90	268.6	288.6	296.6	300	300	
4	120	265.5	283.8	294.8	298.6		
5	150	261.3	280.9	291.9			

The problem can also be solved using Fig. 6.14. The required parameters are

$$\frac{x}{2\sqrt{\alpha\tau}} = \frac{0.00}{2\sqrt{3.0\times10^{-6}\times150}} = 0$$

and

$$\frac{x}{2\sqrt{\alpha\tau}} = \frac{0.03}{2\sqrt{3.0\times10^{-6}\times150}} = 0.71,$$

$$\frac{h\sqrt{\alpha\tau}}{k} = \frac{200\times\sqrt{3.0\times10^{-6}\times150}}{30} = 0.14.$$

From Fig. 6.14, $\frac{T(x)-T_i}{T_\infty-T_i} \approx 0.135$ for $x = 0$ and ≈ 0.026 for $x = 0.03$, which gives $T(x = 0) = 263.5$°C and $T(x = 0.3) = 293.0$°C for $T_\infty = 30$°C and $T_i = 300$°C, which are in reasonable agreement.

Example 6.45 The distribution of temperature t across a large concrete wall 500 mm thick, which is heated from one side, is measured by thermocouples inserted in holes in the wall. It has been found that at a certain instant τ the temperature can be represented approximately by the equation

$$t = 90 - 80x + 16x^2 + 32x^3 - 25.6x^4$$

where x is in meter and temperature in°C. Assuming that the area of the wall is 2.5 m^2, estimate

(i) Heat entering and leaving the wall in unit time,
(ii) Heat energy stored in the wall in unit time,
(iii) Temperature change per unit time at surface ($x = 0$).

Take $c = 0.84$ kJ/(kg K), $\rho = 2175$ kg/m^3, and $k = 2.7$ kJ/(m h K) for the concrete.

Solution

Temperature distribution across the wall is given as

$$t = 90 - 80x + 16x^2 + 32x^3 - 25.6x^4$$

Its derivative is

$$\frac{dt}{dx} = -80 + 32x + 96x^2 - 102.4x^3$$

The second derivative is

$$\frac{d^2t}{dx^2} = 32 + 192x - 307.2x^2$$

(i) The heat entering the wall at $x = 0$,

$$q_{in} = -kA\left(\frac{dt}{dx}\right)_{x=0}$$
$$= -2.7 \times 2.5 \times (-80) = 540\,\text{kJ/h}.$$

The heat leaving the wall at $x = 0.5$ m,

$$q_{out} = -kA\left(\frac{dt}{dx}\right)_{x=0.5}$$
$$= -2.7 \times 2.5 \times (-80 + 32 \times 0.5 + 96 \times 0.5^2 - 102.4 \times 0.5^3)$$
$$= 356.4\,\text{kJ/hr}.$$

(ii) Heat energy stored in the wall = q_{in}–q_{out} = 540–356.4 = 183.6 kJ/h.
(iii) Rate of change of temperature for one-dimensional heat flow, is given by

$$\frac{dt}{d\tau} = \alpha\left(\frac{d^2t}{dx^2}\right)$$

where $\alpha = k/\rho c = 2.7/(2175 \times 0.84) = 1.48 \times 10^{-3}$. Substituting the value of d^2t/dx^2, we have

$$\frac{dt}{d\tau} = 1.48 \times 10^{-3}(32 + 192x - 307.2x^2)$$

At $x = 0$,

$$\frac{dt}{d\tau} = 1.48 \times 10^{-3} \times 32 = 0.04736\,°\text{C/h}$$

and at $x = 0.5$,

$$\frac{dt}{d\tau} = 1.48 \times 10^{-3}(32 + 192 \times 0.5 - 307.2 \times 0.5^2) = 0.0758\,°\text{C/hr}$$

6.11 Summary

This chapter has been devoted to the problems of transient state of heat conduction, i.e. the heating or cooling where the temperature of the solid body varies with time as well as in the space. Such problems are of interest because they are encountered in many industrial processes.

For bodies with a very high thermal conductivity combined with a low value of the convective heat transfer coefficient, uniform temperature throughout the body at any instant can be assumed. Smaller bodies with lower values of the thermal conductivity may also satisfy this condition. The method of lumped heat capacity analysis has been presented for such cases. The analysis shows that the temperature of the body falls or rises exponentially with time and can be presented in terms of two non-dimensional numbers Biot number Bi and Fourier number Fo as $(t-t_\infty)/(t_i-t_\infty)$ = exp(-BiFo). The condition for the applicability of the lumped heat capacity analysis is Bi $\leq$ 0.1.

In Sect. 6.5, the transient behaviour of semi-infinite solids (which is bounded by plane $x = 0$ and extends to infinity in the positive x-direction and also in y- and z-directions) such as a thick or large block of steel, the earth, etc. has been considered when the solid is heated or cooled for a relatively short period. The temperature of such solids changes only for a short distance from the surface. Treatments for three different cases, namely (a) surface temperature of the body is suddenly lowered and maintained at constant temperature, (b) constant heat flux at the surface and (c) the convective boundary condition, have been presented.

For determination of temperature variation with time and spatial position in plates (whose thickness is small compared to the other dimensions), cylinders (whose diameter is small compared to its length), and spheres, solution based on Heisler and Grober charts has been given in Sect. 6.7.

In Sect. 6.8, product solutions for the temperature distributions in two- or three-dimensional transient heat conduction systems are presented.

In many problems of practical interest, the geometrical shapes do not conform to regular shapes discussed above. It is not possible to provide a mathematical solution to such problems. The *finite-difference method* (the numerical technique) can be used to solve all types of transient conduction problems. The method has been presented in Sect. 6.9 followed by a number of illustrative examples.

Using the finite-difference equation, Schmidt gave a graphical method for the solution of one-dimensional problems, which is presented in Sect. 6.10.

Review Questions

6.1 What is the main assumption in the lumped heat capacitance analysis?

6.2. Prove that for a body, whose internal resistance is negligible, the time required for cooling or heating can be obtained from the relation

$$\frac{t - t_\infty}{t_i - t_\infty} = \exp(-\text{BiFo})$$

where Bi and Fo are the Biot and Fourier numbers, respectively, T_i is initial temperature of the body and T_∞ is the surrounding fluid temperature.

6.3 Define Biot and Fourier numbers.
6.4 Draw temperature profile for different values of Biot number in a plane wall symmetrically cooled by convection.
6.5 Write nodal equations for a node at the surface of a plane wall and the node lying inside the wall if the wall is suddenly exposed to a convective atmosphere (h, T_∞).
6.6 Explain explicit and implicit formulations for transient heat conduction problems.
6.7 What is the Schmidt graphical technique to solve the one-dimensional transient problems?

Problems

6.1 Calculate the Biot number for the following.

(a) A triangular fin is 25 mm thick at the base and 100 mm long. It is 75 mm wide and is made of aluminium. The average heat transfer coefficient is 50 W/(m^2 K) along the sides.
(b) A steel sheet [k = 42 W/(m K)], which is in the form of a rectangular parallelepiped (500 mm × 100 mm × 3 m), is at a uniform temperature of 800°C. It is cooled to a centreline temperature of 200°C. The heat transfer coefficient is 50 W/(m^2 K).

Can lump capacity analysis be applied?
[**Ans.** (a) V = ½(25 × 100) × 75 = 93750 mm^3; A_s = 2[√(100^2 + 12.5^2) × 75] = 15116.7 mm^2; $L = V/A_s$ = 6.20 mm; Bi = hL/k = 50 × 6.2/(210 × 1000) = 0.00148. (b) t_{av} = ½ (800 + 200) = 500°C; k = 42 W/(m K); V = 0.5 × 0.1 × 3 = 0.15 m^3; A_s = 2(0.1 × 0.5 + 0.1 × 3 + 3 × 0.5) = 3.70 m^2; $L = V/A_s$ = 0.0405 m; Bi = hL/k = 50 × 0.0405 /42 = 0.0483. Since both values of Bi are less than 0.1, the lumped heat capacity analysis is applicable.]

6.2 A solid copper ball of mass 50 g is quenched in a water bath at 20°C. It cools from an initial temperature of 520°C to 320°C in 10 s. What will be the temperature of the ball after 20 s?

[**Ans.** Lumped heat capacity analysis is applicable. Hence, equation $\frac{t - t_\infty}{t_0 - t_\infty} = e^{\left(-\frac{hA_s}{\rho cV}\tau\right)}$ gives $\left[\frac{320-20}{520-20} = e^{\left(-\frac{hA_s}{\rho cV}\times 10\right)}\right]_{\tau=10}$ and $\left[\frac{t-20}{520-20} = e^{\left(-\frac{hA_s}{\rho cV}\times 20\right)}\right]_{\tau=20}$. Solution of the equations gives t = 200°C.]

6.3 A small copper ball of 5 mm diameter at 550 K is dropped into an oil bath whose temperature is 250 K. The thermal conductivity k of the copper is 375 W/(m K), density is 9000 kg/m^3 and specific heat is 400 J/(kg K). If the heat transfer coefficient is 200 W/(m^2 K), determine the rate of cooling at the beginning of the cooling.

[**Ans.** Differentiation of equation $\frac{t - t_\infty}{t_i - t_\infty} = e^{\left(-\frac{hA_s}{\rho cV}\tau\right)}$ gives $\frac{dt}{d\tau} = -\left(\frac{hA_s}{\rho cV}\right)e^{\left(-\frac{hA_s}{\rho cV}\tau\right)}$ $(t_i - t_\infty)$. At the beginning of the cooling (τ = 0), $\left(\frac{dt}{d\tau}\right)_{\tau=0} = -\left(\frac{hA_s}{\rho cV}\right)(t_i - t_\infty)$, where $A_s = \pi d^2$ and $V = \pi d^3/6$. Substitution gives $\left(\frac{dt}{d\tau}\right)_{\tau=0} = 20\,\text{K/s}$.]

6.4 Steel balls 10 mm in diameter are annealed by heating to 880°C and then slowly cooling to 100°C in an air environment at 25°C. If the time required for this cooling process is 16 min, estimate the heat transfer coefficient. For steel, k = 40 W/(m K), c = 450 J/(kg K) and ρ = 7900 kg/m^3.
[**Ans**. We assume that the lumped heat capacity analysis is applicable. Then from Eq. (6.2) $\left(\frac{hA_s}{c\rho V}\right)\tau = -\ln\left(\frac{t - t_\infty}{t_i - t_\infty}\right)$, or $h = -\ln\left(\frac{t - t_\infty}{t_i - t_\infty}\right) \times \left(\frac{c\rho V}{A_s\tau}\right)$; $\frac{V}{A_s} = \frac{\pi D^3/6}{\pi D^2} = \frac{D}{6}$; Hence, $h = -\ln\left(\frac{t - t_\infty}{t_i - t_\infty}\right) \times \left(\frac{c\rho D}{6\tau}\right)$; On substitution, $h = -\ln\left(\frac{100 - 25}{880 - 25}\right) \times \left(\frac{450 \times 7900 \times 0.01}{6 \times 16 \times 60}\right) = 15.0\,\text{W}/(\text{m}^2\text{K})$; For a sphere, characteristic length $L = D/6$ = 0.01/6 = 1/600 m, Biot number = hL/k = 15/(40 × 600) = 6.25 × 10^{-4} < 0.1, hence the lumped heat capacity analysis is applicable.]

6.5 A very thick concrete wall is initially at a uniform temperature of 20°C. The surface of the wall is suddenly exposed to uniform heat flux of 150 W/m^2. Calculate the temperature of the wall at a depth of 110 mm after 24 h. Take α = 2.4 × 10^{-3} m^2/hr and k = 1.2 W/(m K).

[**Ans**. Equation (6.22) applies; From given data: $\frac{x^2}{4\alpha\tau} = \frac{0.11^2}{4 \times 2.4 \times 10^{-3} \times 24} = 0.0525$, $q_0 = 150\,\text{W/m}^2$, $\text{erf}\left(\frac{x}{2\sqrt{\alpha\tau}}\right) = \text{erf}\left(\frac{0.11}{2\sqrt{2.4 \times 10^{-3} \times 24}}\right) = 0.254$, $\sqrt{\alpha\tau/\pi} = 0.1354$, A = 1 m^2, k = 1.2 W/(m K), Using Eq. (6.22), $t = \frac{2 \times 150 \times 0.1354}{1.2 \times 1} \times e^{-(0.0525)} - \frac{150 \times 0.11}{1.2 \times 1}[1 - 0.254] + 20 = 41.85°\text{C}$]

6.6 A water pipeline is to be placed in the soil. The initial temperature of the soil is 15°C and the surface temperature at the night drops rapidly causing the earth's surface temperature to fall to –20°C. At what depth the pipeline must be located so that the temperature of the water does not drop below 5°C for 10 h period? The thermal diffusivity of the soil is 1.8 × 10^{-3} m^2/h.

[**Ans**. $\frac{t - t_s}{t_i - t_s} = 0.7143 = \text{erf}\left(\frac{x}{2\sqrt{\alpha\tau}}\right)$ gives x = 0.203 m.]

6.7 A large block of steel initially at a uniform temperature of 150°C suddenly has its surface temperature lowered to 50°C. What is the amount of heat removed per unit area of the block when the temperature at a depth of 50 mm has dropped to 100°C? Given: α = 1.2 × 10^{-5} m^2/s, k = 40 W/(m K).

[**Ans**. For $erf\left(\frac{x}{2\sqrt{\alpha\tau}}\right) = \frac{t - t_s}{t_i - t_s} = 0.5$, $\frac{x}{2\sqrt{\alpha\tau}} \approx 0.48$, τ = 226 s for x = 0.05 m; Heat removed, $\frac{Q_0}{A} = 1.128k(t_s - t_i)\sqrt{\frac{\tau}{\alpha}} = 19.58 \times 10^6\,\text{J/m}^2$.]

6.8 If the slab of Problem 6.7 is suddenly exposed to a fluid at 50°C with heat transfer coefficient of 500 W/(m^2 K), calculate the time required for the temperature to reach 100°C at the depth of 50 mm.

[**Ans**. Assume a trial value of time τ and calculate $\frac{h\sqrt{\alpha\tau}}{k}$ and $\frac{x}{2\sqrt{\alpha\tau}}$. Read temperature ratio $(t - t_i)/(t_\infty - t_i)$ from Fig. 6.14. Vary value of τ till the desired temperature ratio of (100–150)/(50–150) = 0.5 is achieved. The time τ = 1650 s corresponding to the temperature ratio of 0.5.]

6.9 Two large blocks of cast-iron (k = 52 W/(m K), c = 420 J/(kg K) and ρ = 7260 kg/m^3) and copper (k = 360 W/(m K), c = 380 J/(kg K) and ρ = 8950 kg/m^3), initially at the same uniform temperature, are subjected to a sudden change in surface temperature. Compare the heat fluxes for the two blocks.
[**Ans**. Blocks can be treated as semi-infinite solids. From Eq. (6.20b), heat flux q'' is $\frac{q_{x=0}}{A} = -k\frac{(t_i - t_s)}{\sqrt{\pi\alpha\tau}}$; Hence, for a given time and temperature difference, $q'' \propto \frac{k}{\sqrt{\alpha}} = \sqrt{k\rho c}$ since $\alpha = k/\rho c$; This gives $\frac{(q'')_{Cu}}{(q'')_{CI}} = \frac{(\sqrt{k\rho c})_{Cu}}{(\sqrt{k\rho c})_{CI}} = \frac{\sqrt{360 \times 8950 \times 380}}{\sqrt{52 \times 7260 \times 420}} = 2.78.$
The heat flux for the copper block is 2.78 times larger than the heat flux for the cast-iron block.]

6.10 A large block of steel initially at a uniform temperature of 150°C is suddenly exposed to a fluid at 50°C with heat transfer coefficient of 500 W/(m^2 K), calculate temperature at a depth of 50 mm after 1650 s. Given: $\alpha = 1.2 \times 10^{-5}$ m^2/s, k = 40 W/(m K).
[**Ans.** $\frac{h^2(\alpha\tau)}{k^2} = \frac{500^2 \times (1.2 \times 10^{-5} \times 1650)}{40^2} = 3.0937$; $\frac{h\sqrt{(\alpha\tau)}}{k} = 1.76$; $\frac{hx}{k} = \frac{500 \times 0.05}{40} = 0.625$; $\frac{x}{2\sqrt{\alpha\tau}} = \frac{0.05}{2\sqrt{1.2 \times 10^{-5} \times 1650}} = 0.1777$; From Eq. (6.23),
$\frac{t - t_i}{t_\infty - t_i} = 1 - \text{erf}\left(\frac{x}{2\sqrt{\alpha\tau}}\right) - \exp\left(\frac{hx}{k} + \frac{h^2\alpha\tau}{k^2}\right) \times \left[1 - \text{erf}\left(\frac{x}{2\sqrt{\alpha\tau}} + \frac{h\sqrt{\alpha\tau}}{k}\right)\right]$; Substitution gives $\frac{t - 150}{50 - 150} = 1 - \text{erf}(0.1777) - \exp(3.7187) \times [1 - \text{erf}(1.9377)]$;
t = 95.18 °C. Alternatively Fig. 6.14 may be used for approximate solution.]

6.11 A large block of metal initially at a uniform temperature of 150°C is suddenly has its surface temperature lowered to 50°C. In 225 s, the temperature at a depth of 50 mm has recorded to be 100°C. Determine thermal conductivity of the slab material. Given: ρ = 7800 kg/m^3 and c = 470 J/(kg K).
[**Ans**. For $erf\left(\frac{x}{2\sqrt{\alpha\tau}}\right) = \frac{t - t_s}{t_i - t_s} = 0.5$, $\frac{x}{2\sqrt{\alpha\tau}} \approx 0.48$, The gives $\alpha = 1.2 \times 10^{-5}$ m^2/s for x = 0.05 m and τ = 225 s; $k = \alpha\rho c = 1.2 \times 10^{-5} \times 7800 \times 470$ = 44 W/(m K).]

6.12 A large plate of stainless steel 40 mm thick and initially at a uniform temperature of 100°C is suddenly exposed to a cooling fluid at 30°C with h = 160 W/(m^2 K). Calculate the temperature at a depth of 10 mm from one of the faces 225 s after the exposure. How much heat is removed per unit area of the plate? Given k = 16 W/(m K), ρ = 7800 kg/m^3 and c = 460 J/(kg K).
[**Ans**. Bi = hL/k = 0.2 > 0.1, Heisler charts may be used. Fo = 2.509, 1/Bi = 5, x/L = 0.5, $\theta_o/\theta_i \approx 0.7$, $\theta/\theta_o = 0.975$; t = 77.78°C, Bi^{2}Fo = 0.10036, $Q/Q_o \approx 0.38$, $Q_o/A = \rho c(2L)\theta_i = 10.046 \times 10^6$ J/m^2 and $Q/A = 3.82 \times 10^6$ J/m^2.]

6.13 A large slab of aluminium 100 mm thick is originally at a temperature of 500°C. It is suddenly immersed in a liquid bath at 100°C resulting in a heat transfer coefficient of 1200 W/(m^2 K). Determine the time elapsed after the immersion when the surface

temperature will be 300°C. The properties of the aluminium at the given conditions may be taken as ρ = 2700 kg/m^3, c = 900 J/(kg K) and k = 215 W/(m K).
[**Ans.** $L = \delta/2 = 50\,\text{mm}$; Bi = hL/k = 1200 × 50 × 10^{-3}/215 = 0.279 > 0.1; Heisler charts may be used; Correction factor from Fig. 6.20 for 1/Bi = 3.6 and x/L = 1.0 is 0.86. Hence, $\dfrac{t - t_\infty}{t_o - t_\infty} = 0.86$ gives $t_o = \dfrac{300 - 100}{0.86} + 100 = 332.6\,^\circ\text{C}$; $\dfrac{\theta_o}{\theta_i} = \dfrac{t_o - t_\infty}{t_i - t_\infty} = \dfrac{332.6 - 100}{500 - 100} = 0.58$; For θ_o/θ_i = 0.58 and 1/Bi = 3.6, Fo ≈ 2.1 from Fig. 6.17; We have Fo $= \dfrac{\alpha\tau}{L^2} = \dfrac{k}{\rho c} \times \dfrac{\tau}{L^2}$, which gives $2.1 = \dfrac{215}{2700 \times 900} \times \dfrac{\tau}{0.05^2}$ or $\tau = 59.3\,\text{s}$.]

6.14 The wall of a furnace is 150 mm thick and is adequately insulated from outside. It is initially at a uniform temperature of 20°C. The wall is exposed to hot gases at 900°C. Determine the convective heat transfer coefficient if the outer surface reaches a temperature of 725°C in 10 h. Neglect radiation heat exchange. For the furnace wall, k = 1.2 W/(m K), c = 900 J/(kg K) and ρ = 2500 kg/m^3.
[**Ans.** $L = 0.15$ m as one side is insulated; $\dfrac{\theta_o}{\theta_i} = \dfrac{t_o - t_\infty}{t_i - t_\infty} = \dfrac{725 - 900}{20 - 900} = 0.2$, $\text{Fo} = \dfrac{\alpha\tau}{L^2} = \left(\dfrac{k}{\rho c}\right)\frac{\tau}{L^2} = \left(\dfrac{1.2}{2500 \times 900}\right)\dfrac{10 \times 3600}{0.15^2} = 0.85$; For Fo = 0.85 and θ_o/θ_i = 0.2, 1/Bi = 0.1 from Fig. 6.17; $h = \dfrac{\text{Bi}k}{L} = \dfrac{10 \times 1.2}{0.15} = 80\,\text{W}/(\text{m}^2\,\text{K})$.]

6.15 A long glass rod of 40 mm diameter (k = 0.78 W/(m K), c = 840 J/(kg K) and ρ = 2700 kg/m^3), initially at 80°C, is cooled in air [t_∞ = 30°C, h = 20 W/(m^2 K)] for 2145 s. Then it is removed and isolated from the surroundings. Determine the equilibrium temperature it attains.
[**Ans.** The Biot number, $\text{Bi} = \dfrac{hL}{k} = \dfrac{20 \times 0.02}{0.78} = 0.51$; Fourier number, $\text{Fo} = \dfrac{\alpha\tau}{L^2} = \left(\dfrac{k}{\rho c}\right)\frac{\tau}{L^2} = \left(\dfrac{0.78}{2700 \times 840 \times 0.02^2}\right) \times 2145 = 1.844$; $\text{Bi}^2\text{Fo} = 1.844 \times 0.51^2 = 0.48$; For Bi = 0.51 and Bi^{2}F$_o$ = 0.48, Q/Q_o = 0.7 from Fig. 6.24; $\frac{Q}{Q_o} = 1 - \dfrac{t_f - t_\infty}{t_i - t_\infty}$ gives equilibrium temperature t_f = 45°C for t_∞ = 30°C and t_i = 80°C.]

6.16 A 2.5 mm thick heated plastic sheet (k = 0.2 W/(m K), α = 1 × 10^{-7} m^2/s) lies on an insulated surface and is cooled by air flow at 20°C. The convection heat transfer coefficient is 100 W/(m^2 K). If the surface of the sheet is found to be 40°C after 2 min, determine the initial temperature of the sheet.

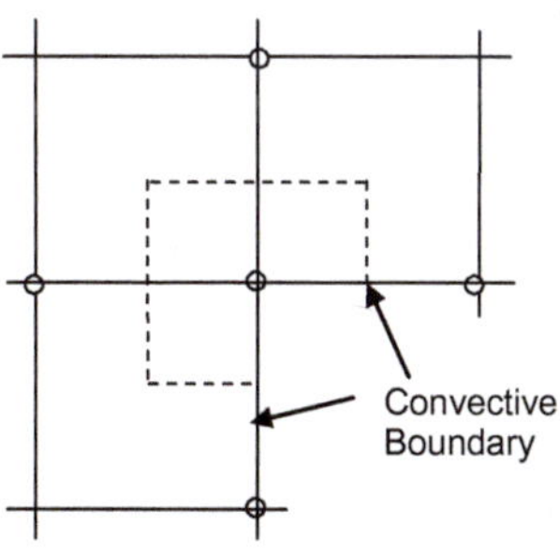

Fig. 6.38 Problem 6.18

[**Ans**. The sheet is equivalent to one-half of a sheet of thickness δ (= $2L$) with symmetric convection conditions. The characteristic length $L = 0.0025$ m; Biot number $\mathrm{Bi} = \frac{hL}{k} = \frac{100 \times 0.0025}{0.2} = 1.25$; Fourier number $\mathrm{Fo} = \frac{\alpha\tau}{L^2} = \frac{1 \times 10^{-7} \times 120}{0.0025^2} = 1.92$; For 1/Bi = 0.8, and Fo = 1.92, $\theta_o/\theta_i \approx 0.205$ from Fig. 6.17; $\theta_L/\theta_o \approx 0.6$ for $x/L = 1$ and 1/Bi = 0.8 from Fig. 6.20; $\theta_i = \frac{\theta_i}{\theta_0} \times \frac{\theta_0}{\theta_L} \times \theta_L = \frac{\theta_i}{\theta_0} \times \frac{\theta_0}{\theta_L} \times (t_L - t_\infty) = \frac{(40 - 20)}{0.205 \times 0.6} = 162.6$; $t_i = \theta_i + t_\infty = 162.6 + 20 = 182.6\,°\mathrm{C}$.]

6.17 A long square section steel bar (50 mm × 50 mm) is initially at a uniform temperature of 800°C. It is heat treated by quenching in an oil tank at 80°C. The average heat transfer coefficient is 400 W/(m^2 K). If the temperature at the centreline is to be 300° C, determine the time of quenching. $\alpha = 1.4 \times 10^{-5}$ m^2/s and $k = 50$ W/(m K).

[**Ans**. The problem is a combination of two infinite plates of equal thicknesses. Hence, $\left(\frac{\theta_0}{\theta_i}\right) = \left[\left(\frac{\theta_0}{\theta_i}\right)_{\mathrm{2Lplate}}\right]^2$; From the given data, $\frac{\theta_0}{\theta_i} = \frac{t_0 - t_\infty}{t_i - t_\infty} = \frac{300 - 80}{800 - 80} = 0.306$;

Hence, $\left(\frac{\theta_0}{\theta_i}\right)_{\mathrm{2Lplate}} = \sqrt{\left(\frac{\theta_0}{\theta_i}\right)} = 0.55$; $\frac{1}{\mathrm{Bi}} = \frac{k}{hL} = \frac{50}{400 \times 0.025} = 5$; From Fig. 6.17, Fo ≈ 3.3, which gives $\tau = \frac{L^2\mathrm{Fo}}{\alpha} = \frac{0.025^2 \times 3.3}{1.4 \times 10^{-5}} = 147.3$ s.]

6.18 Formulate the finite-difference equation for the node at an interior corner with convection, as shown in Fig. 6.38, using the explicit and implicit methods. Obtain the stability criterion for the explicit method. Assume equal grid sizes in both x- and y-directions.

[Hint: Refer Example 6.39.]

6.19 A 250 mm thick concrete wall ($\alpha = 0.0066$ cm^2/s) is split into 5 layers. Determine the time interval $\Delta\tau$ for explicit formulation. Assume one-dimensional heat flow.

[**Ans**. $\Delta x = 250/5 = 50$ mm, Fo = ½ = $\alpha\Delta\tau/(\Delta x)^2$ gives $\Delta\tau = 1894$ s = 31.6 min.]

References

Grober H, Erk S, Grigull U (1961) Fundamentals of heat transfer. McGraw-Hill Book Co, New York

Heisler MP (1947) Temperature charts for induction and constant temperature heating. Trans ASME 69:227–236

Holman JP (1992) Adapted for SI units by White PRS, heat transfer. McGraw-Hill Book Co, New York

7 Convective Heat Transfer

7.1 Introduction

Heat conduction is a process in which the heat is transferred through solids and stagnant layers of fluids (liquids and gases) due to the vibration at the molecular level and electronic effect when a temperature gradient exists. In the case of fluids, the molecules have freedom of motion and heat can also be transferred by the movement of the fluid. This complex process of heat transfer is termed as convection. In convection, there is an observable bulk motion of the fluid and hence can occur only in the case of liquids, gases and multiphase mixtures.

The convective heat transfer can be classified into two categories: (i) *natural or free convection* and (ii) *forced convection.*

7.1.1 Natural Convective Heat Transfer

In this mode of convection, the fluid motion is entirely because of the density differences in the fluid caused due to the local heating in the gravity field. The fluid surrounding the heat source becomes less dense due to the heating and hence moves upwards (known as buoyancy effect). The surrounding colder fluid moves to replace it. This colder fluid is then heated and the process continues, forming the convection currents (the movement of the fluid). Thus, the heat transfers from the bottom of the convection cell to the top. Heating of water in a pot over a stove causing movement of water from the bottom to the top is a typical example of the natural convection. In the nature, convection currents are set up in the air as the solar radiation heats land or water.

7.1.2 Forced Convection Heat Transfer

If the motion of the fluid is caused by some external means such as a fan, blower, pump or wind, the mode of heat transfer is termed as the forced convection. Cooling water through a heat exchanger, air propelled through a solar air heater duct and wind blowing over a heated surface are a few examples where forced convective heat transfer takes place.

R. Karwa, *Heat and Mass Transfer*,
https://doi.org/10.1007/978-981-15-3988-6_7

However, in any forced convection situation, some amount of natural convection is always present. When the natural convection is not negligible, such flows are typically known as *mixed convection.*

The convection is a combined problem of heat and fluid flow. Hence, the rate of heat transfer is affected by the fluid properties such as viscosity, specific heat, density, thermal conductivity, coefficient of thermal expansion, etc. The analytical solution of a convection problem involves application of the equation of motion, energy and continuity, and the Fourier's law of heat conduction. The resulting differential equations that govern the convection are complicated and their exact solution can be given only in a few simple problems of steady flow.

There are three basic methods of determining the rate of heat transfer between fluid and a solid by convection. It will be shown in the sections to follow that the fluid is at rest in the immediate vicinity of the solid surface due to the viscous effect. Therefore, the heat flow at the wall is by conduction and not by convection. The *first method* makes use of this observation and the heat transfer rate can be calculated from the Fourier's law

$$q = -kA\left(\frac{dt}{dy}\right)_{y=0} \tag{7.1}$$

where $(dt/dy)_{y=0}$ is temperature gradient in the fluid at the wall, refer Fig. 7.1, and k is the thermal conductivity of the fluid.

The *second method* is based on analogy between the mechanisms of transfer of fluid momentum to the wall and the transfer of heat by convection. Using the analogy, the rate of heat transfer by convection can be predicted from the measurement of shear stress between the fluid and the wall.

The *third method* is to experimentally determine the heat transfer coefficient, defined by the Newton's equation, because the determination of the temperature gradient at the wall and its variation over the entire heat-transferring surface is very difficult. The Newton's equation for the convective heat transfer coefficient is

$$q = hA\left(t_w - t_f\right) \tag{7.2}$$

According to this equation, the heat transfer is proportional to the area of the surface A and the temperature difference between the wall and the fluid. The factor of proportionality h in the equation is called the convection heat transfer coefficient. It is the quantity of heat transferred in unit time from unit surface area for a unit temperature difference between the wall and the fluid. Studies have shown that the heat transfer coefficient is a function of many parameters.

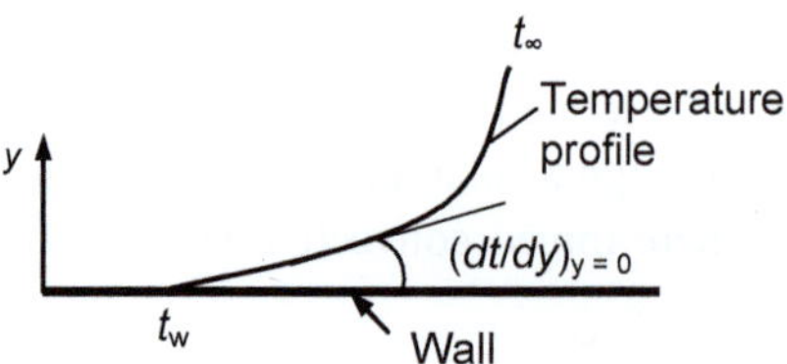

Fig. 7.1 Heat flow at wall

The method of dimensional analysis, based on the dynamic similarity, is used to express the experimental results in the form of a correlation of dimensionless numbers (refer Chap. 8).

In this chapter, analytical treatment of simple cases of forced and natural convection will be presented followed by the discussion of analogy between fluid friction and heat transfer.

7.2 Flow of Fluid Past a Flat Plate

Consider a simple case of steady-state flow of fluid with free-stream velocity U_∞ past a plate of sufficient length as shown in Fig. 7.2. The plate is aligned with the flow so that the x-axis coincides with the flow direction. When the fluid meets the leading edge of the plate, the viscous effect causes it to adhere to the surface. Assuming that there is no slip between the fluid and the surface, the fluid velocity is zero at the plate surface. The fluid velocity increases in the y-direction and approaches the free-stream velocity U_∞ in a layer of thickness δ. This layer is termed as *hydrodynamic boundary layer*. The reason for the drop of velocity through the boundary layer is the viscosity of the fluid. The fluid flow region may be divided into two regions: the boundary layer (region adjacent to the wall), which exhibits the effect of the viscosity and the main flow or free-stream region where the viscosity effect can be neglected.

The thickness of the boundary layer increases in x-direction as more and more fluid is included in the boundary layer and is retarded by the friction. In order to accommodate the included fluid, the streamlines must diverge away from the plate so that the flow area is increased. Due to the divergence of the streamlines, the velocity vector is no longer parallel to the plate and a small velocity component v perpendicular to the plate will exist. Thus, the flow in the boundary layer is two-dimensional.

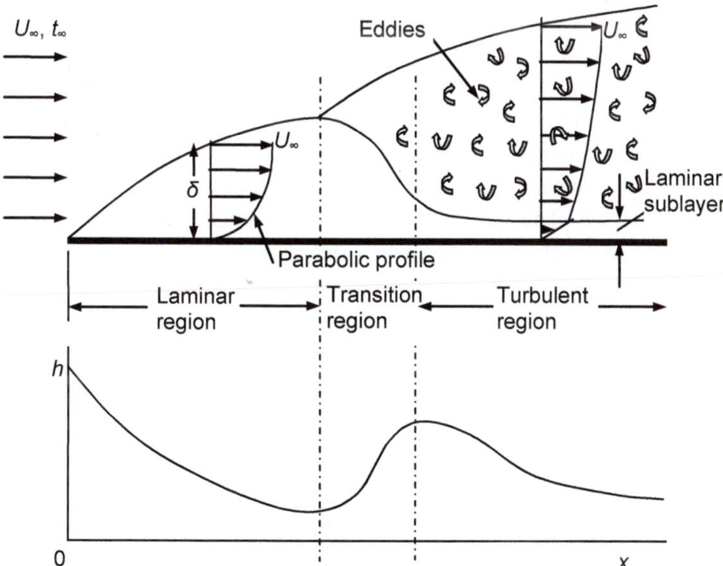

Fig. 7.2 Hydrodynamic boundary layer over a flat plate (the boundary layer thickness is exaggerated for clarity)

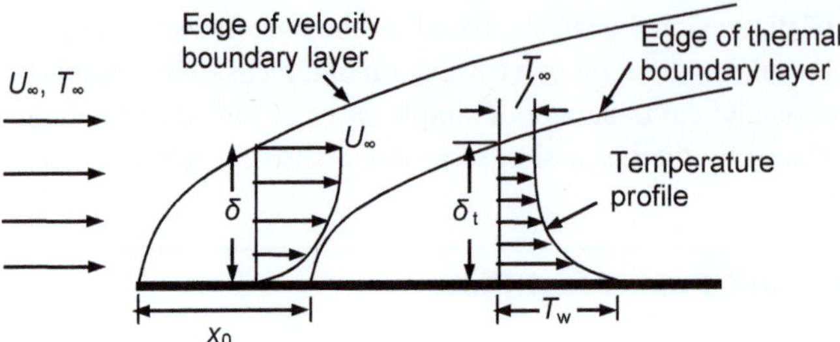

Fig. 7.3 Hydrodynamic (velocity) and thermal boundary layers over a flat plate

The flow along the plate surface (x-direction) is divided into laminar (streamline flow—the fluid flows in distinct fluid layers nearly parallel to the surface) and eddying or turbulent flow (random motion of the fluid in all directions). Initially, the boundary layer development is laminar and the boundary layer thickness increases continuously with the increase in distance x along the plate surface starting from zero at the leading edge of the plate. At some critical distance x_c from the leading edge, transition from laminar to turbulent layer with a thin laminar sublayer takes place. Experiments with fluid at different velocities have revealed that the transition occurs at

$$\frac{U_\infty x}{\nu} > 5 \times 10^5 \tag{7.3}$$

The group $(U_\infty x/\upsilon)$ is a dimensionless number and is called the *Reynolds number*. The Reynolds number corresponding to the critical distance x_c is known as the *critical Reynolds number* ($\mathrm{Re}_c = U_\infty x_c/\upsilon$). The value of the critical Reynolds number is affected by the level of the disturbances present in the free stream approaching the plate and also the plate surface roughness ($\mathrm{Re}_c = 3 \times 10^5 - 5 \times 10^5$ depending on the roughness of the plate surface). For strong disturbances, the critical Reynolds number Re_c has been reported to be as low as 8×10^4. With exceptionally free of disturbances, the flow may remain laminar up to Re_c 10^6 or higher (Schlichting 1979).

If the plate is heated or cooled to a uniform temperature T_w, starting at a distance x_o from the leading edge of the plate, a *thermal boundary layer* will also develop as shown in Fig. 7.3. It is the region where temperature gradient is present. The temperature of the fluid varies from $T = T_w$ at the plate surface to $T = T_\infty$ at the edge of the thermal boundary layer. The thickness of the thermal boundary layer is designated as δ_t. The concept of the thermal boundary layer is analogous to that of the velocity or hydrodynamic boundary layer. The relationship between the two boundary layers will be discussed later.

7.3 Flow in Tubes

Consider the fluid flow through a tube. The velocity distribution is uniform at the inlet cross-section if the tube inlet is rounded and the fluid comes from a large space, Fig. 7.4. Because of the friction, the velocity diminishes at the wall and increases at the centre of the tube. The flow does not become fully developed at once but at a certain distance from the tube inlet. Thus, the boundary layer gradually builds up until it reaches the centre of the tube. The velocity distribution curve, now, acquires a stable form and does not vary down the tube. The

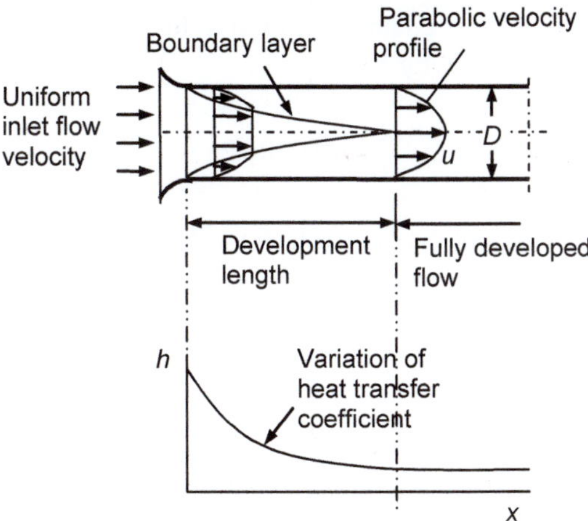

Fig. 7.4 Laminar flow in a tube (Re < 2300)

flow is said to be fully developed. The distance for the velocity profile to be fully developed is termed as *hydrodynamic development length* or entrance length.

The development of the thermal boundary layer in a fluid which is heated or cooled in a duct is qualitatively similar to that of the hydrodynamic boundary layer. As the fluid flows along the duct, the heated or the cooled layer increases in the thickness until the heat is transferred to or from the fluid in the centre of the duct. The temperature profile now acquires the fully developed form (known as thermally fully developed) if the velocity profile is fully developed. The shapes of the velocity and temperature profiles depend on whether the flow is laminar or turbulent.

The hydrodynamic development length for the laminar flow in a circular pipe with a uniform velocity profile at the inlet is approximately given by (Kays and Crawford 1980)

$$\frac{L}{D} \approx 0.05\text{Re} \tag{7.4}$$

where Re is the Reynolds number defined as

$$\text{Re} = \frac{\rho U_{\text{m}} D}{\mu} \tag{7.5}$$

i.e. it is calculated on the basis of the mean velocity U_{m} and the tube diameter D. The mean velocity is calculated by dividing the volumetric flow rate of the fluid by the cross-sectional area of the tube. Thus

$$U_{\text{m}} = \frac{V}{A} = \frac{1}{A}\int_0^A u dA$$

For turbulent flow in tubes, the hydrodynamic development length is short (entrance region is typically less than 10–15 tube diameters) and it depends on the geometric character of the entrance (Kays and Crawford 1980).

The thermal entry length for the laminar flow in a tube may be approximated from the following relation (Kays and Crawford 1980):

$$\frac{L}{D} \approx 0.05\text{Re}\,\text{Pr} \tag{7.6}$$

where Pr is the Prandtl number of the fluid, which is a fluid property and equals $\mu c_p/k$. Thus, the thermal entry length is also dependent on the Prandtl number. For example, for air (Pr = 0.7) flowing with Re = 1000, the development length is 35 tube diameters while for oil (Pr = 100), the same will be 2500 even when flowing with Re = 500. Thus, in an oil exchanger, the fully developed temperature profile is rarely attained.

7.3.1 Laminar Flow Through a Tube

When the flow through a straight tube is laminar, the fluid particles move along the path parallel to the tube axis with no rotation of the particles. The velocity profile of the fully developed laminar flow is of parabolic form as shown in Fig. 7.4.

In the laminar flow, the velocity profile is given by

$$\frac{u}{U_{\max}} = 2\left(\frac{y}{R}\right) - \left(\frac{y}{R}\right)^2 \tag{7.7}$$

where u is the velocity at a distance y from the wall and $U_{\max}$ is the velocity at the axis of the tube.

Volume flow through annulus of radial width dy at radius $(R - y)$ from the centreline, refer Fig. 7.5, is

$$dV = [2\pi(R - y)dy]u$$

Hence, the volume flow rate is

$$\begin{aligned} V &= \int dV = \int_0^R 2\pi u(R - y)dy \\ &= \int_0^R 2\pi\left[2\left(\frac{y}{R}\right) - \left(\frac{y}{R}\right)^2\right] U_{\max}(R - y)dy \\ &= \int_0^R 2\pi U_{\max}\left(2y - 3\frac{y^2}{R} + \frac{y^3}{R^2}\right)dy \\ &= \frac{1}{2}\pi R^2 U_{\max}. \end{aligned}$$

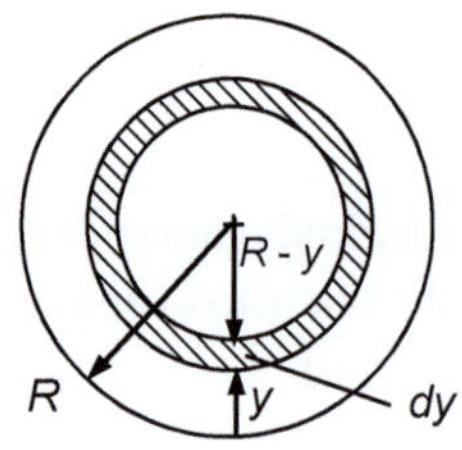

Fig. 7.5 Coordinate system for Eq. (7.7)

The mean velocity is defined as

$$U_m = \frac{V}{A} = \frac{V}{\pi R^2} = \frac{1}{2}\pi R^2 \left(\frac{U_{max}}{\pi R^2}\right)$$

or

$$\left(\frac{U_m}{U_{max}}\right) = 0.5 \tag{7.8}$$

i.e. in the case of laminar flow through a tube, the ratio of the mean velocity to the maximum velocity at the centreline is 0.5.

7.3.2 Turbulent Flow Through a Tube

If the flow Reynolds number exceeds a certain critical value ($Re_c \approx 2300$[1]), the flow becomes turbulent. The fully developed velocity profile in this case is shown in Fig. 7.6. In the case of turbulent flow, three distinct regions in the flow have been found. These regions are a laminar sublayer in the immediate vicinity of the wall, a buffer layer (not shown in the figure) and a prominent turbulent core. The velocity changes abruptly near the wall and takes a somewhat blunter profile in the middle of the tube. The velocity profile becomes nearly flat and the velocity distribution becomes more uniform at very high Reynolds number.

The turbulent flows are of great importance because of greater heat transfer rates in this region.

The ratio of the mean and the maximum velocities is a function of the Reynolds number, i.e.

$$\left(\frac{U_m}{U_{max}}\right)_{turbulent} = f(Re) \tag{7.9}$$

When the fluid is in a turbulent state inside the tube, there is always a layer of fluid at the wall in which the flow is laminar. This layer is known as *laminar sublayer*. The thickness of this layer increases along the development length from $\delta = 0$ at the inlet to its maximum value δ. The thickness of the laminar sublayer is very small and can be visualized from the relation of Schlichting (1979), deduced in terms of Reynolds number using Blasius correlation of friction factor for fully developed turbulent flow in smooth circular duct, as

$$\frac{\delta}{D} = \frac{25}{Re^{7/8}} \tag{7.10}$$

[1]It is to note that the laminar-to-turbulent transition is not sudden but occurs over a range of Reynolds number. The numerical value of the critical Reynolds number Re_c depends on the duct inlet conditions as well as on the surface roughness of the duct. Disturbances such as vibrations on the exterior of the duct wall and flow pulsation also influence the value of the critical Reynolds number. In a circular duct, the lower limit for the critical Reynolds number is accepted to be 2000 below which the flow remains laminar even in the presence of strong disturbances (Potter et al. 2012). The upper limit of Re_c is undefined (Bhatti and Shah 1987), for most practical purposes it is taken as 10^4. The flow in the range $2300 \le Re < 10^4$ is termed as transition flow. The velocity profile in the transition regime is neither parabolic nor the usual turbulent profile.

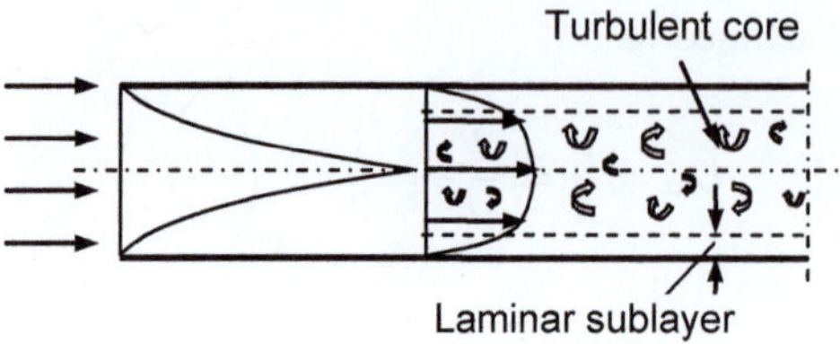

Fig. 7.6 Turbulent flow in a tube

From the relation, it can also be seen that the thickness of the laminar sublayer decreases with the increase in the Reynolds number.

7.4 Equation of Continuity

The equation of continuity is based on the law of conservation of mass. Figure 7.7 shows elemental control volume with dimensions *dx*, *dy* and *dz* in the fluid flow field, where the mass can flow in or out of the element. Let us consider the mass balance for this element.

The mass of the fluid entering the element through the face ABCD, which is perpendicular to the velocity vector *u* in the *x*-direction, in time $d\tau$ is

$$m_x = \rho u(dy.dz)d\tau$$

Mass leaving the element through face EFGH at distance *dx* in *x*-direction is

$$m_{x+dx} = m_x + \frac{\partial m_x}{\partial x}dx$$

The net mass leaving the element in *x*-direction is

$$dm_x = m_{x+dx} - m_x = \left[\frac{\partial}{\partial x}(\rho u)\right](dx.dy.dz)d\tau$$

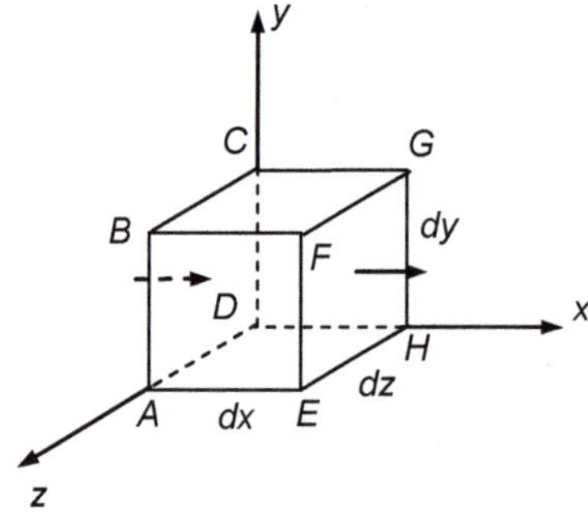

Fig. 7.7 Elemental control volume for deriving equation of continuity

or

$$dm_x = \left[\frac{\partial}{\partial x}(\rho u)\right] dv' d\tau \tag{i}$$

where $dv' = dx.dy.dz$ is the volume of the element.

Similarly, the net fluid amounts leaving in y- and z-directions with velocity vectors v and w, respectively, are

$$dm_y = \left[\frac{\partial}{\partial y}(\rho v)\right] dv' d\tau \tag{ii}$$

$$dm_z = \left[\frac{\partial}{\partial z}(\rho w)\right] dv' d\tau \tag{iii}$$

The total quantity of excess fluid mass leaving the element is the sum of expressions (i), (ii) and (iii), i.e.

$$dm = \left[\frac{\partial}{\partial x}(\rho u) + \frac{\partial}{\partial y}(\rho v) + \frac{\partial}{\partial z}(\rho w)\right] dv' d\tau \tag{iv}$$

The net outflow from the element is due to the decrease in the density of the fluid within the element and is

$$dm = \frac{\partial \rho}{\partial \tau} dv' d\tau \tag{v}$$

From Eqs. (iv) and (v), we get

$$\begin{gathered} \left[\tfrac{\partial}{\partial x}(\rho u) + \tfrac{\partial}{\partial y}(\rho v) + \tfrac{\partial}{\partial z}(\rho w)\right] dv' d\tau + \tfrac{\partial \rho}{\partial \tau} dv' d\tau = 0 \\ \tfrac{\partial}{\partial x}(\rho u) + \tfrac{\partial}{\partial y}(\rho v) + \tfrac{\partial}{\partial z}(\rho w) + \tfrac{\partial \rho}{\partial \tau} = 0, \end{gathered} \tag{7.11}$$

which is the differential equation of continuity in general form.

For an incompressible fluid (liquids), ρ is constant hence

$$\frac{\partial u}{\partial x} + \frac{\partial v}{\partial y} + \frac{\partial w}{\partial z} = 0. \tag{7.12a}$$

For the two-dimensional incompressible fluid flow, the equation is

$$\frac{\partial u}{\partial x} + \frac{\partial v}{\partial y} = 0. \tag{7.12b}$$

7.4.1 The Displacement and Momentum Thickness

The hydrodynamic or momentum boundary layer is the region in which the fluid velocity changes from zero at the plate surface to its free-stream velocity. In fact, there is no precise thickness of boundary layer if this definition is used because, mathematically, the boundary layer extends indefinitely in the *y*-direction. However, the boundary layer thickness is taken to be the distance from the plate surface in which most of the velocity change (from zero to 99 per cent of the free-stream velocity) takes place. To avoid the ambiguity in defining the boundary layer thickness, two boundary layer thickness parameters are defined. These parameters are known as velocity displacement thickness δ_{vd} and momentum displacement thickness δ_{md}. The displacement thickness is a measure of the displacement of the free stream due to the formation of the boundary layer over the plate while the momentum thickness is a measure of the momentum flux displacement caused by the boundary layer.

The *velocity displacement thickness* δ_{vd} can be obtained as follows.

The mass flow rate though the boundary layer (for the unit width of the plate) is

$$\int_{0}^{\infty} \rho.u(dy.1) = \int_{0}^{\infty} \rho.udy$$

where u is the velocity of the fluid at distance y from the plate surface and ρ is the density of the fluid.

For a frictionless (non-viscous) fluid, the velocity of the fluid will be U_∞ and the mass flow rate will be

$$\int_{0}^{\infty} \rho U_\infty dy$$

Thus, the loss in the fluid flow rate due to the formation of the boundary layer is

$$\int_{0}^{\infty} \rho(U_\infty - u)dy$$

If we represent this loss in the fluid flow rate by the flow at U_∞ passing through a thickness δ_{vd}, then

$$\begin{aligned} (\delta_{vd}.1)U_\infty\rho &= \int_{0}^{\infty} \rho(U_\infty - u)dy \\ \delta_{vd} &= \int_{0}^{\infty} \left(1 - \frac{u}{U_\infty}\right)dy \end{aligned} \tag{7.13}$$

A geometrical interpretation of the velocity displacement is shown in Fig. 7.8. An infinite boundary layer ($\delta \to \infty$) has been substituted by a finite layer of thickness δ_{vd} so that the area under the curve would equal the area of rectangle of height δ_{vd} and width U_∞.

The velocity displacement thickness is a measure of the displacement of the main stream resulting from the presence of the flat plate and its boundary (Kays and Crawford 1980).

Similarly, the *momentum displacement thickness* δ_{md} may be determined as follows.

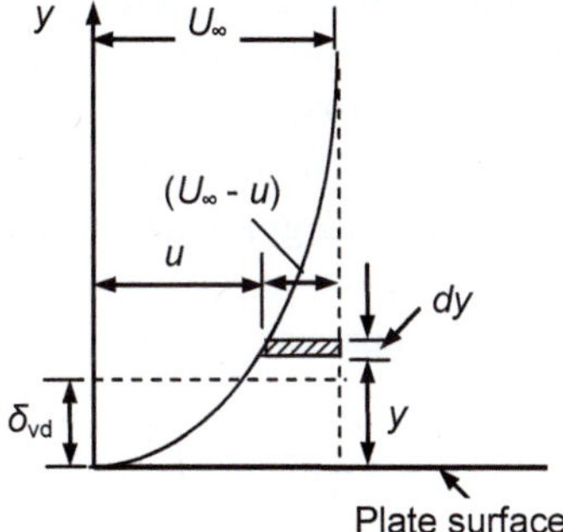

Fig. 7.8 Graphical interpretation of the velocity displacement thickness

The mass flux through the boundary layer strip of height dy (for the unit width of the plate) is

$$= \rho u dy$$

Momentum of this mass is

$$= \rho u dy.u$$

The momentum of the same mass before entering the boundary layer is

$$= (\rho u dy)U_\infty$$

Thus, the loss of momentum of this fluid is

$$= \rho u dy.(U_\infty - u)$$

Total loss of the momentum in the complete boundary layer

$$= \int_0^\infty \rho u(U_\infty - u)dy$$

If we represent this momentum loss of fluid by the momentum of the fluid passing through a thickness δ_{md}, then

$$\delta_{\text{md}}.\rho U_\infty^2 = \int_0^\infty \rho u(U_\infty - u)dy$$

$$\delta_{\text{md}} = \frac{1}{\rho U_\infty^2} \int_0^\infty \rho u(U_\infty - u)dy$$

$$\delta_{\text{md}} = \int_0^\infty \frac{u}{U_\infty}\left(1 - \frac{u}{U_\infty}\right)dy \tag{7.14}$$

The momentum displacement thickness is a measure of the momentum flux decrement caused by the boundary layer, which is proportional to the drag of the plate according to the momentum theorem (Kays and Crawford 1980).

We can define *energy thickness* δ_{ke} as 'the thickness of flow moving at the free-stream velocity and having energy (the kinetic energy) equal to the deficiency of energy in the boundary layer'.

Kinetic energy of the mass of the fluid passing through the elemental strip of area $(dy.1)$ is

$$[\rho u.(dy.1)] \times \left(\frac{1}{2}u^2\right) = \left(\frac{1}{2}\rho u^3\right)dy,$$

If the fluid is non-viscous, the fluid mass $(dy.u.\rho)$ will have free-stream velocity U_∞. The total energy of the mass passing through the boundary layer will be

$$\int \left(\frac{1}{2}\rho u\right)U_\infty^2 dy$$

Thus, the loss of kinetic energy due to the boundary layer is

$$\frac{1}{2}\int_0^\infty \rho u\left(U_\infty^2 - u^2\right)dy$$

If we represent this energy loss by the energy loss of fluid passing through a thickness δ_{ke} at the free-stream velocity. Then

$$\frac{1}{2}\rho U_\infty^3 \delta_{ke} = \frac{1}{2}\int_0^\infty \rho u\left(U_\infty^2 - u^2\right)dy$$

$$\delta_{ke} = \int_0^\infty \frac{u}{U_\infty}\left[1 - \left(\frac{u}{U_\infty}\right)^2\right]dy \tag{7.15}$$

Example 7.1 Determine velocity displacement thickness and momentum thickness if the velocity distribution is linear ($u/U_\infty = y/\delta$, where $u = U_\infty$ at $y = \delta$).

Solution

(i) The velocity displacement thickness, when the limit ∞ is replaced by δ, is given by

$$\delta_{vd} = \int_0^\delta \left(1 - \frac{u}{U_\infty}\right)dy$$

Now substituting the equation of the velocity profile, we have

$$\delta_{vd} = \int_0^{\delta} \left(1 - \frac{y}{\delta}\right) dy = \int_0^{\delta} dy - \int_0^{\delta} \frac{y dy}{\delta}$$
$$= \delta/2$$

(ii) The momentum displacement thickness is given by

$$\delta_{\text{md}} = \int_0^{\delta} \frac{u}{U_\infty}\left(1 - \frac{u}{U_\infty}\right) dy$$
$$= \int_0^{\delta} \frac{y}{\delta}\left(1 - \frac{y}{\delta}\right) dy = \frac{\delta}{6}.$$

7.4.2 The Enthalpy and Conduction Thickness

For a thermal boundary layer, we can define integral thickness parameter in an unambiguous way. These parameters are enthalpy thickness and conduction thickness.

The enthalpy thickness is defined by the relation

$$\delta_e = \frac{\int_0^\infty \rho u e dy}{\rho_\infty U_\infty e_w}$$

where the enthalpy $e = c\ (T - T_\infty)$ and $e_w = c_w\ (T_w - T_\infty)$; c is specific heat.

Substituting the values of e and e_w, the thickness δ_e can be expressed in terms of temperature for constant property fluids ($\rho = \rho_\infty$ and $c = c_w$) as follows:

$$\delta_e = \frac{\int_0^\infty u(T - T_\infty) dy}{U_\infty (T_w - T_\infty)} \tag{7.16}$$

The conduction thickness is defined as

$$\delta_k = \frac{k(T_w - T_\infty)}{q_w} \tag{i}$$

The convection heat transfer coefficient is given by

$$h = \frac{q_w}{(T_w - T_\infty)} \tag{ii}$$

Combining Eqs. (i) and (ii), we get

$$\delta_k = \frac{k}{h} \tag{7.17}$$

Both the thicknesses have been shown in Fig. 7.9.

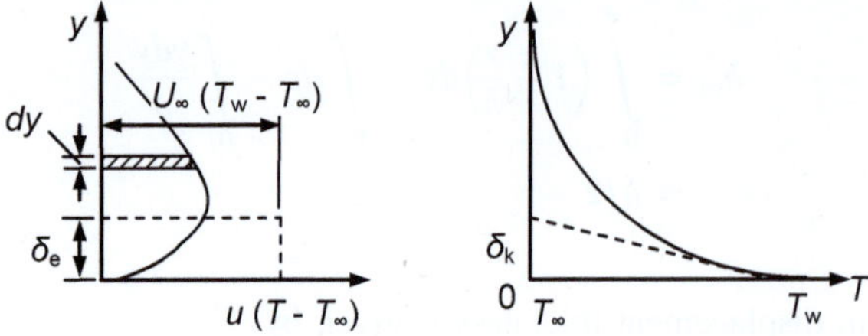

Fig. 7.9 Graphical illustrations of the enthalpy and conduction thicknesses of the boundary layer

7.5 Momentum Equation of Laminar Boundary Layer Over a Flat Plate

The momentum equation of the laminar boundary layer over a flat plate can be derived by making a force-and-momentum balance on the elemental control volume located in the boundary layer as shown in Fig. 7.10.

The following assumptions are being made to simplify the analysis:

1. The fluid is incompressible and the flow is steady.
2. There are no pressure variations in the direction perpendicular to the flow, i.e. in the y-direction.
3. The viscosity of the fluid is constant.
4. Viscous shear forces in y-direction are negligible and hence neglected.
5. The effect of gravitational forces is negligible and hence it is neglected.

Momentum balance in x-direction for the control volume of Fig. 7.10 gives

$$\begin{aligned}&\text{Viscous-shear force} + \text{pressure force}\\&\qquad = \text{rate of momentum transfer in } x\text{-direction}\end{aligned} \tag{7.18}$$

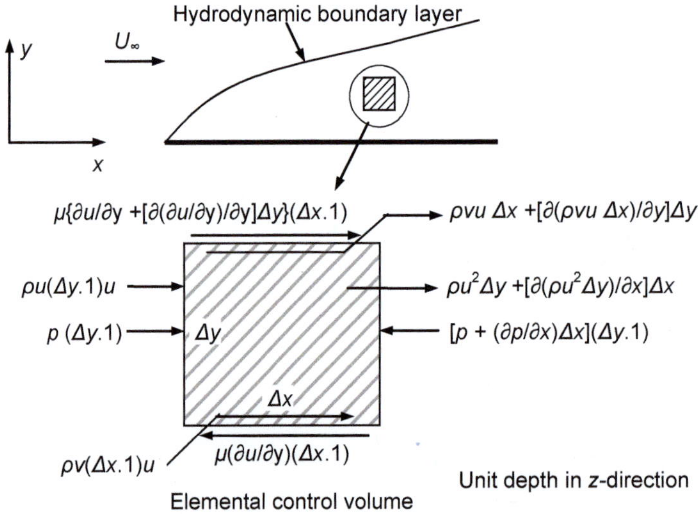

Fig. 7.10 Elemental control volume for force balance on laminar boundary layer

The rate of momentum transfer in x-direction is the product of mass flow rate through a side of the control volume and the x-component of velocity at that point.

The mass entering the left face of the element for unit depth in z-direction

$$m_x = \rho u(\Delta y.1)$$

Hence, the momentum entering this face per unit time is

$$= \rho u(\Delta y.1)u$$

The momentum leaving the right face is

$$= \rho u^2 \Delta y + \frac{\partial}{\partial x}(\rho u^2 \Delta y)\Delta x$$
$$= \rho u^2 \Delta y + 2\rho u \frac{\partial u}{\partial x}\Delta x.\Delta y$$

Similarly, the mass entering the bottom face is

$$= \rho v(\Delta x.1)$$

Hence, the momentum entering the bottom face is

$$= \rho v(\Delta x.1)u$$

The momentum leaving the top face is

$$= \rho vu.\Delta x + \left[\frac{\partial}{\partial y}(\rho vu.\Delta x)\right]\Delta y$$
$$= \rho vu.\Delta x + \left[\frac{\partial}{\partial y}(\rho vu)\right].\Delta x.\Delta y$$

The pressure force on the left face is

$$= p(\Delta y.1)$$

The pressure force on the right face is

$$= \left[p + \frac{\partial p}{\partial x}\Delta x\right](\Delta y.1)$$

Net pressure force in x-direction is

$$= p(\Delta y.1) - \left[p + \frac{\partial p}{\partial x}\Delta x\right](\Delta y.1)$$
$$= -\frac{\partial p}{\partial x}(\Delta x.\Delta y)$$

The viscous shear force on the bottom face is

$$= -\mu \frac{\partial u}{\partial y}(\Delta x.1)$$

and the viscous shear force on the top face is

$$= \mu \left[\frac{\partial u}{\partial y} + \frac{\partial}{\partial y}\left(\frac{\partial u}{\partial y}\right)\Delta y \right](\Delta x.1)$$

The net viscous shear force in x-direction is

$$\mu \frac{\partial}{\partial y}\left(\frac{\partial u}{\partial y}\right)\Delta y.(\Delta x.1) = \mu \frac{\partial^2 u}{\partial y^2}(\Delta x.\Delta y)$$

Substitution of various terms in Eq. (7.18) gives

$$\mu \frac{\partial^2 u}{\partial y^2}(\Delta x.\Delta y) - \frac{\partial p}{\partial x}(\Delta x.\Delta y) =$$
$$\rho u^2 \Delta y + 2\rho u \frac{\partial u}{\partial x}\Delta x.\Delta y - \rho u^2 \Delta y + \rho v u.\Delta x + \left[\frac{\partial}{\partial y}(\rho v u)\right].\Delta x.\Delta y - \rho v u.\Delta x$$

or

$$\begin{aligned} \mu \frac{\partial^2 u}{\partial y^2} - \frac{\partial p}{\partial x} &= 2\rho u \frac{\partial u}{\partial x} + \left[\frac{\partial}{\partial y}(\rho v u)\right] \\ &= 2\rho u \frac{\partial u}{\partial x} + \rho u \frac{\partial v}{\partial y} + \rho v \frac{\partial u}{\partial y} \\ &= \rho u \left(\frac{\partial u}{\partial x} + \frac{\partial v}{\partial y}\right) + \rho u \frac{\partial u}{\partial x} + \rho v \frac{\partial u}{\partial y} \end{aligned}$$

From the continuity equation,

$$\frac{\partial u}{\partial x} + \frac{\partial v}{\partial y} = 0 \tag{7.12}$$

Hence,

$$\mu \frac{\partial^2 u}{\partial y^2} - \frac{\partial p}{\partial x} = \rho u \frac{\partial u}{\partial x} + \rho v \frac{\partial u}{\partial y}$$

or

$$\rho\left(u \frac{\partial u}{\partial x} + v \frac{\partial u}{\partial y}\right) = \mu \frac{\partial^2 u}{\partial y^2} - \frac{\partial p}{\partial x} \tag{7.19}$$

This is the *momentum equation of the laminar boundary layer* over a flat plate.

In the absence of any pressure forces, $\partial p/\partial x = 0$ and the equation becomes

$$\rho\left(u\frac{\partial u}{\partial x}+v\frac{\partial u}{\partial y}\right)=\mu\frac{\partial^2 u}{\partial y^2} \tag{7.20}$$

7.5.1 Solution of Momentum Equation (Blasius Solution)

The momentum and the continuity equations for the laminar flow over a flat plate have been developed as

$$u\frac{\partial u}{\partial x}+v\frac{\partial u}{\partial y}=\nu\frac{\partial^2 u}{\partial y^2} \tag{7.20}$$

and

$$\frac{\partial u}{\partial x}+\frac{\partial v}{\partial y}=0, \tag{7.12}$$

respectively.

We introduce a parameter ψ defined as under, called the stream function, which can replace the components u and v of the velocity by a single function and satisfies the continuity equation:

$$u=(\partial\psi/\partial y) \text{ and } v=-(\partial\psi/\partial x)$$

The boundary conditions of the flow are

$$\begin{aligned}
u &= 0 \text{ at } y = 0\\
v &= 0 \text{ at } y = 0\\
u &= U_\infty \text{ at } x = 0\\
u &\to U_\infty \text{ as } y \to \infty\\
\text{and } (\partial u/\partial y) &\to 0 \text{ as } y \to \infty.
\end{aligned}$$

It is assumed that the shape of the velocity profiles (Fig. 7.11) is geometrically similar at various x-positions starting from the leading edge of the plate. These velocity profiles differ in the y-direction and hence the significant variable is the distance variable (y/δ), in the non-dimensional form. Thus, the velocity profile at any x-location can be expressed as function of (y/δ), i.e.

$$\frac{u}{U_\infty}=g(y/\delta), \tag{7.21}$$

where δ is the thickness of the boundary layer.

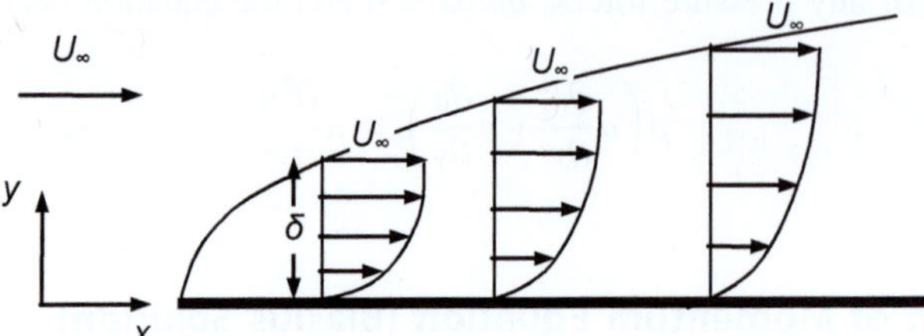

Fig. 7.11 Development of laminar boundary layer over a flat plate

Except very close to the wall, the velocity u is of the order of the free-stream velocity U_∞ and the distance y from the wall is of the order of boundary layer thickness δ, i.e.

$$u \sim U_\infty \text{ and } y \sim \delta$$

Using the above order-of-magnitude terms, the continuity equation can be written in an approximate form as

$$\frac{U_\infty}{x} + \frac{v}{\delta} \approx 0$$

or

$$v \sim U_\infty \delta / x$$

Using the order of magnitude for u, v and y in the momentum equation,

$$u\frac{\partial u}{\partial x} + v\frac{\partial u}{\partial y} = \nu\frac{\partial^2 u}{\partial y^2}, \tag{7.20}$$

we get

$$U_\infty \frac{U_\infty}{x} + \left(\frac{U_\infty \delta}{x}\right) \times \left(\frac{U_\infty}{\delta}\right) \approx \nu\left(\frac{U_\infty}{\delta^2}\right)$$

Simplification gives

$$\delta \sim \sqrt{\frac{\nu . x}{U_\infty}}$$

It can be expressed in the non-dimensional form as

$$\frac{\delta}{x} \sim \sqrt{\frac{\nu}{U_\infty x}} = \frac{1}{\sqrt{\mathrm{Re}_x}} \tag{7.22}$$

where $\mathrm{Re}_x = (U_\infty x/\upsilon)$ is the local Reynolds number, which is a non-dimensional number. The equation indicates the variation of the boundary layer thickness with the free-stream velocity, the viscosity of the fluid and distance x along the plate in the fluid flow direction or with the Reynolds number.

Using order-of-magnitude estimate of δ, Eq. (7.21) can be expressed as

$$\frac{u}{U_\infty} = g\left(\frac{y}{\delta}\right) = g\left[\frac{y}{\sqrt{\nu x/U_\infty}}\right] = g(\eta) \tag{7.23}$$

where

$$\eta = \left[\frac{y}{\sqrt{\nu x/U_\infty}}\right] = y\sqrt{\frac{U_\infty}{\nu x}}. \tag{7.24}$$

The variable η is called the *similarity variable* and $g(\eta)$ is the function we require as a solution.

The stream function ψ, defined earlier, can be expressed in terms of η as

$$\begin{aligned}\psi &= \int u dy = \int U_\infty g(\eta) dy = \int U_\infty g(\eta)\frac{dy}{d\eta} d\eta \\ &= \int U_\infty g(\eta)\sqrt{\frac{\nu x}{U_\infty}} d\eta \\ &= \sqrt{U_\infty \nu x}\int g(\eta) d\eta \\ &= \sqrt{U_\infty \nu x} f\end{aligned} \tag{7.25}$$

where $f = \int g(\eta)\, d\eta$.

The velocity components u and v can be expressed in terms of η and f as

$$\begin{aligned}u &= \frac{\partial \psi}{\partial y} = \frac{\partial \psi}{\partial \eta}\frac{\partial \eta}{\partial y} \\ &= \frac{\partial}{\partial \eta}\left[\left(\sqrt{U_\infty \nu x}\right) f\right] \times \frac{\partial}{\partial y}\left(y\sqrt{\frac{U_\infty}{\nu x}}\right) \\ &= \left(\sqrt{U_\infty \nu x}\right)\frac{\partial f}{\partial \eta} \times \sqrt{\frac{U_\infty}{\nu x}} \\ &= U_\infty \frac{\partial f}{\partial \eta}\end{aligned}$$

Thus,

$$u = U_\infty f' \tag{7.26}$$

where $(\partial f/\partial \eta) = f'$

Similarly,

$$
\begin{aligned}
v &= -\frac{\partial \psi}{\partial x} \\
&= -\frac{\partial}{\partial x}\left[\left(\sqrt{U_\infty v x}\right) f\right] \\
&= -\sqrt{U_\infty v}\frac{\partial}{\partial x}\left[\left(\sqrt{x}\right) f\right] \\
&= -\sqrt{U_\infty v}\left[\sqrt{x}\frac{\partial f}{\partial x} + f\frac{\partial}{\partial x}\left(\sqrt{x}\right)\right] \\
&= -\sqrt{U_\infty v}\left[\sqrt{x}\frac{\partial f}{\partial \eta}\frac{\partial \eta}{\partial x} + \frac{f}{2\sqrt{x}}\right] \\
&= -\sqrt{U_\infty v}\left[\sqrt{x} f' \frac{\partial}{\partial x}\left(y\sqrt{\frac{U_\infty}{v x}}\right) + \frac{f}{2\sqrt{x}}\right] \\
&= -\sqrt{U_\infty v}\left[\sqrt{x} f' \left(y\sqrt{\frac{U_\infty}{v}}\right)\frac{\partial}{\partial x}\left(\frac{1}{\sqrt{x}}\right) + \frac{f}{2\sqrt{x}}\right] \\
&= -\sqrt{U_\infty v}\left[\sqrt{x} f' \left(y\sqrt{\frac{U_\infty}{v}}\right) \times \left(-\frac{1}{2}x^{-3/2}\right) + \frac{f}{2\sqrt{x}}\right] \\
&= \frac{1}{2}\sqrt{\frac{U_\infty v}{x}}\left(f'\eta - f\right)
\end{aligned} \tag{7.27}
$$

Now the derivatives $\partial u/\partial x$, $\partial v/\partial x$ and $\partial^2 u/\partial y^2$ can be determined as follows:

$$
\begin{aligned}
\frac{\partial u}{\partial x} &= \frac{\partial}{\partial x}\left(U_\infty f'\right) \\
&= U_\infty \frac{\partial f'}{\partial x} = U_\infty \frac{\partial f'}{\partial \eta}\frac{\partial \eta}{\partial x} \\
&= U_\infty f'' \frac{\partial}{\partial x}\left(y\sqrt{\frac{U_\infty}{v x}}\right) \\
&= U_\infty f'' \left(y\sqrt{\frac{U_\infty}{v}}\right) \times \left(-\frac{1}{2}x^{-3/2}\right) \\
&= -\frac{U_\infty}{2x}\left(y\sqrt{\frac{U_\infty}{v x}}\right) f'' \\
&= -\frac{U_\infty}{2x}\eta . f''
\end{aligned} \tag{7.28}
$$

where $(\partial f'/\partial \eta) = f''$.

$$
\begin{aligned}
\frac{\partial u}{\partial y} &= \frac{\partial}{\partial y}(U_\infty f') \\
&= U_\infty \frac{\partial f'}{\partial y} = U_\infty \frac{\partial f'}{\partial \eta}\frac{\partial \eta}{\partial y} \\
&= U_\infty f'' \frac{\partial}{\partial y}\left(y\sqrt{\frac{U_\infty}{\nu x}}\right) \\
&= U_\infty f'' \left(\sqrt{\frac{U_\infty}{\nu x}}\right)
\end{aligned}
\tag{7.29}
$$

$$
\begin{aligned}
\frac{\partial^2 u}{\partial y^2} &= \frac{\partial}{\partial y}\left(\frac{\partial u}{\partial y}\right) \\
&= \frac{\partial}{\partial y}\left[U_\infty f''\left(\sqrt{\frac{U_\infty}{\nu x}}\right)\right] \\
&= U_\infty\left(\sqrt{\frac{U_\infty}{\nu x}}\right)\frac{\partial f''}{\partial y} \\
&= U_\infty\left(\sqrt{\frac{U_\infty}{\nu x}}\right)\frac{\partial f''}{\partial \eta}\frac{\partial \eta}{\partial y} \\
&= U_\infty\left(\sqrt{\frac{U_\infty}{\nu x}}\right)f'''\frac{\partial}{\partial y}\left(y\sqrt{\frac{U_\infty}{\nu x}}\right) \\
&= U_\infty\left(\sqrt{\frac{U_\infty}{\nu x}}\right)f'''\left(\sqrt{\frac{U_\infty}{\nu x}}\right) \\
&= U_\infty\left(\frac{U_\infty}{\nu x}\right)f'''
\end{aligned}
\tag{7.30}
$$

where $(\partial f''/\partial\eta) = f'''$.

Substitution of various terms in the momentum equation, Eq. (7.20), gives

$$
U_\infty f'\left(-\frac{U_\infty}{2x}\eta f''\right) + \left[\frac{1}{2}\sqrt{\frac{U_\infty \nu}{x}}(f'\eta - f)\right] \times \left[U_\infty f''\left(\sqrt{\frac{U_\infty}{\nu x}}\right)\right] = \nu\left[U_\infty\left(\frac{U_\infty}{\nu x}\right)f'''\right]
$$

Simplification of the above equation gives

$$
-\frac{1}{2}f'f''\eta + \frac{1}{2}f''(f'\eta - f) = f'''
$$

or

$$
2f''' + ff'' = 0. \tag{7.31}
$$

Table 7.1 Physical and similarity coordinates

Physical coordinates	Similarity coordinates
$u = 0$ at $y = 0$	$f' = 0$ at $\eta = 0$
$v = 0$ at $y = 0$	$f = 0$ at $\eta = 0$
$u \to U_\infty$ as $y \to \infty$	$f' = 1$ at $\eta \to \infty$
$(\partial u/\partial y) \to 0$ as $y \to \infty$	$f'' = 0$ at $\eta \to \infty$

where

$$f = f(\eta) = \int g(\eta) d\eta, \eta = y\sqrt{[U_\infty/(vx)]}, f' = (\partial f/\partial \eta) = u/U_\infty, f'' = (\partial f'/\partial \eta) \text{ and } f''' = (\partial f''/\partial \eta).$$

This is an ordinary but non-linear differential equation of third order for which the boundary conditions are given in Table 7.1.

Equation (7.31) has solved numerically for the function $f(\eta)$. The first solution was obtained by Blasius (1908) in the form of convergent series for small values of η and the asymptotic approximation for the larger values of η. The solution is of the form

$$f(\eta) = \frac{\alpha\eta^2}{2!} - \frac{1}{2}\frac{\alpha^2\eta^5}{5!} + \frac{11}{4}\frac{\alpha^3\eta^8}{8!} - \frac{375}{8}\frac{\alpha^4\eta^{11}}{11!} + \ldots\ldots\ldots \tag{7.32}$$

where $\alpha = 0.33206$.

Differentiation of the equation gives

$$\begin{aligned} f' &= \frac{\partial f}{\partial \eta} = \alpha\eta - \ldots\ldots\ldots \\ f'' &= \frac{\partial f'}{\partial \eta} = \alpha - \ldots\ldots\ldots \\ (f'')_{\eta=0} &= \alpha = 0.33206 \end{aligned} \tag{7.33}$$

The results (values of f, f' and f'' as function of η) are presented in Table 7.2. The distribution of the axial velocity u in the y-direction and that of the transverse velocity v along y-axis is presented in Fig. 7.12.

From the table and the plots, it is evident that at $\eta = 5$, the velocity u is practically equal to the free-stream velocity U_∞ ($f' = u/U_\infty = 0.99$).

If we assume that the thickness of the boundary layer δ corresponds to $u/U_\infty = 0.99$, then at $y = \delta$, $\eta = 5.0$ and we have

$$\eta = \left(y\sqrt{\frac{U_\infty}{vx}}\right)_{y=\delta} = \delta\sqrt{\frac{U_\infty}{vx}} = 5.0$$

Table 7.2 Values of functions $f(\eta)$, etc. for a flat plate

$\eta = y[U_\infty/(\nu x)]^{1/2}$	$f(\eta)$	$f' = u/U_\infty$	f''	$\frac{1}{2}(\eta f' - f)$
0	0	0	0.33206	0
0.2	0.00664	0.06641	0.33199	0.00332
0.4	0.02656	0.13277	0.33147	0.01327
0.6	0.05974	0.19894	0.33008	0.02981
0.8	0.10611	0.26471	0.32739	0.05283
1.0	0.16557	0.32979	0.32301	0.08211
1.2	0.23795	0.39378	0.31659	0.11729
1.6	0.42032	0.51676	0.29667	0.20325
2.0	0.65003	0.62977	0.26675	0.30476
2.4	0.92230	0.72899	0.22800	0.41364
2.8	1.23099	0.81152	0.18401	0.52063
3.2	1.56911	0.87609	0.13913	0.61719
3.6	1.92954	0.92333	0.09809	0.69722
4.0	2.30576	0.95552	0.06424	0.75816
4.4	2.69238	0.97587	0.03897	0.80072
4.8	3.08534	0.98779	0.02187	0.82803
5.0	3.28329	0.99155	0.01591	0.83723
5.2	3.48189	0.99425	0.01134	0.84410
5.6	3.88031	0.99748	0.00543	0.85279
6.0	4.27964	0.99898	0.00240	0.85712
6.4	4.67938	0.99961	0.00098	0.85906
6.8	5.07928	0.99987	0.00037	0.85992
7.2	5.47025	0.99996	0.00013	0.86013
7.6	5.87924	0.99999	0.00004	0.86034
8.0	6.27923	1.00000	0.00001	0.86038
8.4	6.67923	1.00000	0.00000	0.86038
8.8	7.07923	1.00000	0.00000	0.86038

or

$$\delta = 5.0\sqrt{\frac{\nu x}{U_\infty}} \tag{7.34}$$

or

$$\frac{\delta}{x} = 5.0\sqrt{\frac{\nu}{U_\infty x}}$$

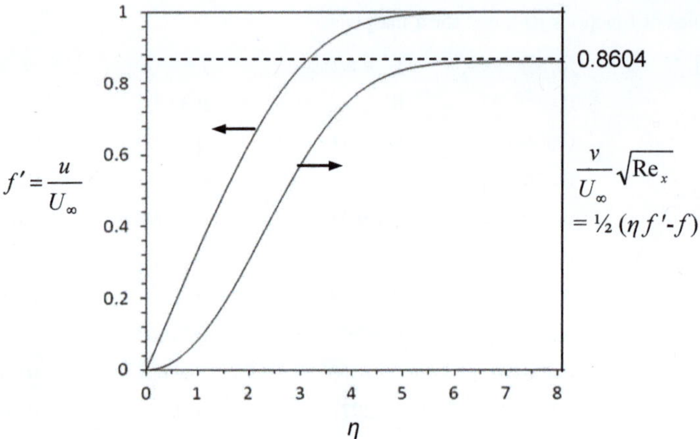

Fig. 7.12 Velocity profiles in laminar boundary layer

or

$$\frac{\delta}{x} = \frac{5.0}{\sqrt{\mathrm{Re}_x}} \tag{7.35}$$

The skin friction coefficient is defined as

$$C_{fx} = \frac{\tau_w}{\frac{1}{2}\rho U_\infty^2}$$

The shear stress τ at the wall is

$$\tau_w = \mu\left(\frac{\partial u}{\partial y}\right)_{y=0}$$

where

$$\begin{aligned}
\left(\frac{\partial u}{\partial y}\right)_{y=0} &= \left[\frac{\partial}{\partial y}\left(\frac{u}{U_\infty}\right)\right]_{y=0} U_\infty \\
&= \left(\frac{\partial f'}{\partial \eta}\right)_{\eta=0}\left(\frac{\partial \eta}{\partial y}\right).U_\infty \\
&= (f'')_{\eta=0}\left(\sqrt{\frac{U_\infty}{\nu x}}\right).U_\infty \\
&= \alpha\left(\sqrt{\frac{U_\infty}{\nu x}}\right).U_\infty \\
&= 0.33206\left(\sqrt{\frac{U_\infty}{\nu x}}\right).U_\infty
\end{aligned}$$

Hence,

$$\tau_w = \mu\left[0.33206\left(\sqrt{\frac{U_\infty}{\nu x}}\right).U_\infty\right] \tag{7.36}$$

Thus,

$$C_{fx} = \mu\left[0.33206\left(\sqrt{\frac{U_\infty}{\nu x}}\right).U_\infty\right] \times \frac{1}{\frac{1}{2}\rho U_\infty^2} = \frac{0.6641}{\sqrt{\frac{U_\infty x}{\nu}}}$$

or

$$C_{fx} = \frac{0.6641}{\sqrt{\mathrm{Re}_x}} \tag{7.37}$$

The average value of the skin friction coefficient for the plate length L can be determined by integrating the local value of the coefficient from $x = 0$ to $x = L$, and then dividing by the plate length L, i.e.

$$\begin{aligned}
\overline{C_f} &= \frac{1}{L}\int_0^L C_{fx}dx \\
&= \frac{1}{L}\int_0^L \frac{0.6641}{\sqrt{\mathrm{Re}_x}}dx \\
&= \frac{1}{L}\int_0^L 0.6641\left(\sqrt{\frac{\nu}{U_\infty x}}\right)dx \\
&= \frac{1}{L}0.6641\left(\sqrt{\frac{\nu}{U_\infty}}\right)\int_0^L \frac{1}{\sqrt{x}}dx \\
&= \frac{1.3282}{L}\sqrt{\frac{\nu}{U_\infty}}.\sqrt{L} = \frac{1.3282}{\sqrt{\frac{U_\infty L}{\nu}}}
\end{aligned}$$

or

$$\overline{C_f} = \frac{1.3282}{\sqrt{\mathrm{Re}_L}} \tag{7.38}$$

where Re_L is based on the total length L of the plate in direction of flow.

Equations (7.35)–(7.38), valid for the laminar region, indicate that

(i) the boundary layer thickness δ increases as square root of the distance x from the leading edge and inversely as square root of the free-stream velocity U_∞,

(ii) the wall shear stress τ_w is inversely proportional to the square root of the distance x and directly proportional to 3/2 power of the free-stream velocity U_∞,

(iii) the local and average skin friction coefficient C_{fx} and $\overline{C_f}$, respectively, vary inversely as square root of both distance and free-stream velocity.

Example 7.2 A fluid (kinematic viscosity $\nu = 2.8 \times 10^{-3}$ m^2/s) flows over a flat plate at 15 m/s. Determine the velocity components u and v at a point located at $x = 2$ m, $y = 40$ mm from the leading edge.

Solution

Flow Reynolds number,

$$\mathrm{Re}_x = \frac{U_\infty x}{\nu} = \frac{15 \times 2}{2.8 \times 10^{-3}} = 10714$$

Boundary layer thickness at $x = 2$ m,

$$\delta = \frac{5x}{\sqrt{\mathrm{Re}_x}} = \frac{5 \times 2}{\sqrt{10714}} = 0.0966 > y$$

i.e. the point lies within the boundary layer.

Stretching factor,

$$\eta = y\sqrt{\frac{U_\infty}{\nu x}} == 0.04 \times \sqrt{\frac{15}{2.8 \times 10^{-3} \times 2}} = 2.07$$

From Table 7.2, we have (corresponding to $\eta = 2.07$)

$$f' = \frac{u}{U_\infty} = 0.63$$

or

$$u = 0.63 \times U_\infty = 0.63 \times 15 = 9.45 \text{ m/s},$$

and

$$\frac{1}{2}(\eta f' - f) = 0.305$$

which gives

$$v = \frac{1}{2}\sqrt{\frac{U_\infty \nu}{x}} \times (\eta f' - f) = \sqrt{\frac{15 \times 2.8 \times 10^{-3}}{2}} \times 0.305 = 0.044 \text{ m/s}.$$

Note: It can be seen that $v \ll$ u.

7.6 Integral Momentum Equation of Laminar Boundary Layer Over a Flat Plate: von Karman Solution

Figure 7.13 shows the boundary layer flow system. The free-stream velocity outside the boundary layer is U_∞. The thickness of the boundary layer is δ. Consider a control volume at distance x from the leading edge. The control volume is infinitesimal in the x-direction while its height H is such that it is enclosing the boundary layer, i.e. $H > \delta$.

The mass entering the control volume through plane AA' for the unit width of the plate is

$$m = \int_0^H \rho u dy$$

The mass leaving the control volume through plane BB' is

$$m + \frac{\partial m}{\partial x} dx = \int_0^H \rho u dy + \frac{\partial}{\partial x}\left(\int_0^H \rho u dy\right) dx$$

As no mass can enter the control volume through the solid wall (plane AB), the mass enters the control volume through face A'B' with the free-stream velocity U_∞.

The momentum (mass × velocity) of the mass entering the control volume through the plane AA' is

$$\int_0^H \rho u dy.u = \int_0^H \rho dy.u^2$$

Similarly, the momentum of the mass leaving the control volume through the planes BB' and A'B' is

$$\int_0^H \rho u^2 dy + \frac{\partial}{\partial x}\left(\int_0^H \rho u^2 dy\right) dx$$

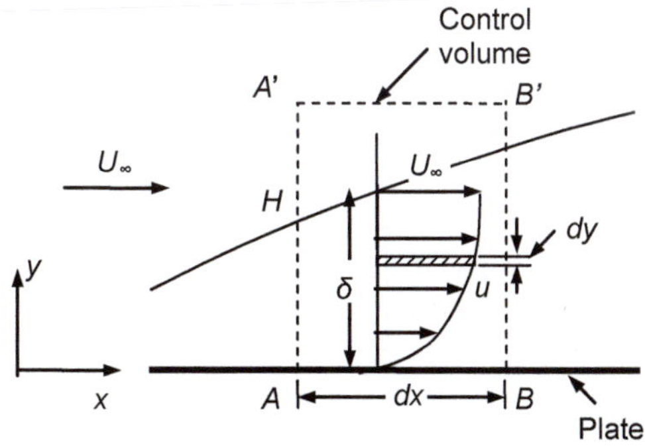

Fig. 7.13 Control volume for integral momentum analysis of laminar boundary layer over a flat plate

and

$$\left[\frac{\partial}{\partial x}\left(\int_0^H \rho u dy\right)dx\right]U_\infty,$$

respectively.

If we neglect gravity forces, the shear or drag force at the plate surface (for constant pressure condition) must equal the net momentum change for the control volume. Thus

$$\tau_w dx = \int_0^H \rho u^2 dy - \left\{\int_0^H \rho u^2 dy + \frac{\partial}{\partial x}\left(\int_0^H \rho u^2 dy\right)dx - \left[\frac{\partial}{\partial x}\left(\int_0^H \rho u dy\right)dx\right]U_\infty\right\}$$

or

$$\tau_w = U_\infty \frac{\partial}{\partial x}\left[\int_0^H \rho u dy\right] - \frac{\partial}{\partial x}\left[\int_0^H \rho u^2 dy\right].$$

Since velocity of flow is constant for $y > \delta$, i.e. from δ to H, the integrand will be zero for $y = \delta$ to H. Hence, the upper limit of the integrand can be changed to δ, which yields

$$\tau_w = U_\infty \frac{\partial}{\partial x}\left[\int_0^\delta \rho u dy\right] - \frac{\partial}{\partial x}\left[\int_0^\delta \rho u^2 dy\right]$$

or

$$\tau_w = \frac{\partial}{\partial x}\left[\int_0^\delta \rho (U_\infty - u) u dy\right]$$

or

$$\tau_w = \rho U_\infty^2 \frac{\partial}{\partial x}\left[\int_0^\delta \left(1 - \frac{u}{U_\infty}\right)\frac{u}{U_\infty} dy\right]$$

As the velocity component normal to the plate v is neglected, the equation can be written as

$$\tau_w = \rho U_\infty^2 \frac{d}{dx}\left[\int_0^\delta \left(1 - \frac{u}{U_\infty}\right)\frac{u}{U_\infty} dy\right] \tag{7.39}$$

The above equation is known as *von Karman momentum integral equation* for the hydrodynamic boundary layer. It can be used to obtain the expression for the boundary layer thickness if the velocity profile is known.

One can represent the unknown velocity profile equation as a polynomial,

$$\frac{u}{U_\infty} = C_1 + C_2\left(\frac{y}{\delta}\right) + C_3\left(\frac{y}{\delta}\right)^2 + C_4\left(\frac{y}{\delta}\right)^3 + C_5\left(\frac{y}{\delta}\right)^4 + \ldots\ldots\ldots \tag{7.40}$$

The selection of the number of terms depends on the degree of accuracy desired. The constants are determined by application of the conditions at the wall (y = 0) and at the outer limit of the boundary layer ($y = \delta$).

At the plate surface (y = 0), u = 0. The first derivative of u with respect to y at y = 0 determines the viscous shear stress at the wall. The momentum equation of the boundary layer for the zero pressure gradient is

$$u\frac{\partial u}{\partial x} + v\frac{\partial u}{\partial y} = \nu\frac{\partial^2 u}{\partial y^2} \tag{7.20}$$

Thus, at the plate surface, where $u = v = 0$,

$$\left(\frac{\partial^2 u}{\partial y^2}\right)_{y=0} = 0$$

Differentiation of the equation of motion and its evaluation at the plate surface (y = 0) will show that all the higher derivatives of u are also zero at the plate surface.

At the outer limit of the boundary layer ($y = \delta$), $u = U_\infty$. For a smooth transition from the boundary layer to the potential region, all the derivatives of u with respect to y will vanish.

These conditions at y = 0 and $y = \delta$ are summarized below:

$u/U_\infty = 1$ at $y = \delta$
$u/U_\infty = 0$ at $y = 0$
$\partial u/\partial y = 0$ at $y = \delta$
$\partial^2 u/\partial y^2 = 0$ at $y = 0$
$\partial^2 u/\partial y^2 = 0$ at $y = \delta$
$\partial^3 u/\partial y^3 = 0$ at $y = 0$
$\partial^3 u/\partial y^3 = 0$ at $y = \delta$

First degree | Second degree | Third degree

(Note: The first derivative of u with respect to y at y = 0 determines the viscous shear stress there. Hence, $\partial u/\partial y$ at y = 0 may not be specified.)

The number of boundary conditions is selected depending on the requirement of the degree of polynomials. The use of the polynomial as an approximation of the velocity profile means that all the boundary conditions cannot be satisfied. Appropriate way is to satisfy as many conditions at the plate surface as at the outer limit of the boundary layer alternating between one and the other as we use higher degrees of polynomial (Chapman 1960).

Consider as an example, a third degree polynomial

$$\frac{u}{U_\infty} = C_1 + C_2\left(\frac{y}{\delta}\right) + C_3\left(\frac{y}{\delta}\right)^2 + C_4\left(\frac{y}{\delta}\right)^3 \tag{7.41}$$

The four constants C_1 to C_4 can be determined from the four boundary conditions which the velocity function must satisfy. These conditions are

$$\begin{aligned} u/U_\infty &= 1 \text{ at } y = \delta \\ u/U_\infty &= 0 \text{ at } y = 0 \\ \partial u/\partial y &= 0 \text{ at } y = \delta \\ \partial^2 u/\partial y^2 &= 0 \text{ at } y = 0 \text{ for constant pressure condition.} \end{aligned}$$

Applying the boundary conditions, we get

$$\begin{aligned} C_1 &= 0 \\ C_2 + C_3 + C_4 &= 1 \\ C_2 + 2C_3 + 3C_4 &= 0 \\ 2C_3 &= 0 \end{aligned}$$

The values of the constants from the above equations are

$$C_1 = 0, C_2 = 3/2, C_3 = 0, \text{ and } C_4 = -1/2$$

This gives the equation of the velocity profile as

$$\frac{u}{U_\infty} = \frac{3}{2}\left(\frac{y}{\delta}\right) - \frac{1}{2}\left(\frac{y}{\delta}\right)^3 \tag{7.42}$$

Substitution in Eq. (7.39) gives

$$\tau_w = \rho U_\infty^2 \frac{d}{dx}\left\{\int_0^\delta \left[1 - \frac{3}{2}\left(\frac{y}{\delta}\right) + \frac{1}{2}\left(\frac{y}{\delta}\right)^3\right] \times \left[\frac{3}{2}\left(\frac{y}{\delta}\right) - \frac{1}{2}\left(\frac{y}{\delta}\right)^3\right] dy\right\}$$

or

$$\begin{aligned} \tau_w &= \rho U_\infty^2 \frac{d}{dx}\left\{\int_0^\delta \left[\frac{3}{2}\left(\frac{y}{\delta}\right) - \frac{9}{4}\left(\frac{y}{\delta}\right)^2 - \frac{1}{2}\left(\frac{y}{\delta}\right)^3 + \frac{3}{2}\left(\frac{y}{\delta}\right)^4 - \frac{1}{4}\left(\frac{y}{\delta}\right)^6\right] dy\right\} \\ &= \rho U_\infty^2 \frac{d}{dx}\left(\frac{39}{280}\delta\right) \\ &= \frac{39}{280}\rho U_\infty^2 \frac{d\delta}{dx} \end{aligned} \tag{i}$$

The shear stress at the wall is also given by

$$\begin{aligned}\tau_w &= \mu\left(\frac{\partial u}{\partial y}\right)_{y=0} \\ &= \mu\left\{\frac{\partial}{\partial y}\left[\frac{3}{2}\left(\frac{y}{\delta}\right) - \frac{1}{2}\left(\frac{y}{\delta}\right)^3\right]U_\infty\right\}_{y=0} \\ &= \mu U_\infty\left[\frac{3}{2\delta} - \frac{3}{2}\frac{y^2}{\delta^3}\right]_{y=0} \\ &= \frac{3}{2}\frac{\mu U_\infty}{\delta}\end{aligned} \tag{ii}$$

Equating Eqs. (i) and (ii), we get

$$\frac{39}{280}\rho U_\infty^2\frac{d\delta}{dx} = \frac{3}{2}\frac{\mu U_\infty}{\delta}$$

or

$$\delta\frac{d\delta}{dx} = \frac{140}{13}\frac{\mu}{\rho U_\infty}$$

or

$$\frac{\delta^2}{2} = \frac{140}{13}\frac{\mu}{\rho U_\infty}x + C$$

The thickness of the boundary layer at the leading edge of the plate is zero, i.e. at $x = 0$, $\delta = 0$. This gives C = 0. Hence,

$$\delta^2 = \frac{280}{13}\frac{\mu}{\rho U_\infty}x \tag{7.43a}$$

or

$$\delta = 4.64\sqrt{\frac{\mu x}{\rho U_\infty}}$$

or

$$\frac{\delta}{x} = 4.64\sqrt{\frac{\mu}{\rho U_\infty x}}$$

or

$$\frac{\delta}{x} = \frac{4.64}{\sqrt{\text{Re}_x}} \tag{7.43b}$$

Substituting the value of δ in Eq. (ii) of the wall shear stress, we get

$$\begin{aligned}
\tau_w &= \frac{3}{2}\frac{\mu U_\infty}{\delta} \\
&= \frac{3}{2}\mu U_\infty \frac{\sqrt{\mathrm{Re}_x}}{4.64x} \\
&= \frac{3}{2} \times \frac{1}{4.64}\rho U_\infty^2 \sqrt{\frac{\rho U_\infty x}{\mu}}\left(\frac{\mu}{\rho U_\infty x}\right) \\
&= \frac{1}{2}\rho U_\infty^2 \frac{0.6466}{\sqrt{\frac{\rho U_\infty x}{\mu}}} \\
&= \frac{1}{2}\rho U_\infty^2 \frac{0.6466}{\sqrt{\mathrm{Re}_x}}
\end{aligned} \tag{7.44}$$

The local skin friction coefficient is

$$C_{fx} = \frac{\tau_w}{\frac{1}{2}\rho U_\infty^2}$$

or

$$C_{fx} = \frac{0.6466}{\sqrt{\mathrm{Re}_x}} \tag{7.45}$$

The average value of the skin friction coefficient, following the procedure presented in earlier section, is

$$\overline{C_f} = \frac{1.2932}{\sqrt{\mathrm{Re}_L}} \tag{7.46}$$

where Re_L is the Reynolds number based on the total length L of the plate in the direction of flow.

Example 7.3

(a) Derive the von Karman momentum equation for flow past a flat plate in the form

$$\tau_w = \rho U_\infty^2 \frac{d}{dx}\left[\int_0^\delta \left(1 - \frac{u}{U_\infty}\right)\frac{u}{U_\infty}dy\right]$$

(b) Using the above equation and assuming the velocity distribution given by equation

$$\frac{u}{U_\infty} = 2\frac{y}{\delta} - \left(\frac{y}{\delta}\right)^2,$$

determine the boundary layer thickness, wall shear stress and the skin friction coefficient for the laminar flow over a flat plate.

Solution

(a) Refer Sect. 7.5.
(b) Substitution in Eq. (7.39) gives

$$\tau_w = \rho U_\infty^2 \frac{d}{dx}\left\{\int_0^\delta \left[1 - 2\frac{y}{\delta} + \left(\frac{y}{\delta}\right)^2\right]\left[2\frac{y}{\delta} - \left(\frac{y}{\delta}\right)^2\right] dy\right\}$$

$$= \rho U_\infty^2 \frac{d}{dx}\left\{\int_0^\delta \left[2\frac{y}{\delta} - \left(\frac{y}{\delta}\right)^2 - 4\left(\frac{y}{\delta}\right)^2 - \left(\frac{y}{\delta}\right)^4 + 4\left(\frac{y}{\delta}\right)^3\right] dy\right\}$$

$$= \rho U_\infty^2 \frac{d}{dx}\left[\frac{y^2}{\delta} - \frac{5}{3}\frac{y^3}{\delta^2} + \frac{y^4}{\delta^3} - \frac{1}{5}\frac{y^5}{\delta^4}\right]_0^\delta$$

$$= \rho U_\infty^2 \frac{d\delta}{dx}\left[1 - \frac{5}{3} + 1 - \frac{1}{5}\right] = 0.1333\rho U_\infty^2 \frac{d\delta}{dx} \tag{i}$$

At the wall,

$$\tau_w = \mu\left(\frac{\partial u}{\partial y}\right)_{y=o}$$

$$= \mu\frac{\partial}{\partial y}\left[U_\infty \frac{2y}{\delta} - U_\infty\left(\frac{y}{\delta}\right)^2\right]_{y=0}$$

$$= \mu U_\infty\left[\frac{2}{\delta} - 2\left(\frac{y}{\delta^2}\right)\right]_{y=0}$$

$$= \frac{2\mu U_\infty}{\delta} \tag{ii}$$

Equating the above two values from Eqs. (i) and (ii),

$$0.1333\rho U_\infty^2 \frac{d\delta}{dx} = \frac{2\mu U_\infty}{\delta}$$

or

$$\delta\frac{d\delta}{dx} = \frac{15\mu}{\rho U_\infty}$$

or

$$\frac{\delta^2}{2} = \frac{15\mu}{\rho U_\infty}x + C.$$

For the flow past a flat plate, the thickness of the boundary layer at the leading edge is zero, i.e. $\delta = 0$ at $x = 0$. This gives $C = 0$. Therefore,

$$\frac{\delta^2}{2} = \frac{15\mu}{\rho U_\infty} x$$

or

$$\frac{\delta}{x} = 5.477\sqrt{\frac{\mu}{\rho U_\infty x}}$$

or

$$\frac{\delta}{x} = \frac{5.477}{\sqrt{\text{Re}_x}}$$

where Re_x is the Reynolds number based on the distance x form the leading edge of the plate.

From Eq. (ii) of the example,

$$\begin{aligned}\tau_w &= \frac{2\mu U_\infty}{\delta} \\ &= 2\mu U_\infty \frac{\sqrt{\text{Re}_x}}{5.477x} \\ &= \frac{2\rho U_\infty^2}{5.477} \times \sqrt{\frac{x\rho U_\infty}{\mu}} \times \frac{\mu}{\rho U_\infty} \times \frac{1}{x} \\ &= \frac{1}{2}\rho U_\infty^2 \times \frac{0.7303}{\sqrt{\text{Re}_x}}.\end{aligned}$$

The local skin friction coefficient

$$C_{fx} = \frac{\tau_w}{\frac{1}{2}\rho U_\infty^2} = \frac{0.7303}{\sqrt{\text{Re}_x}}$$

Average value of the skin friction coefficient is

$$\begin{aligned}\overline{C_f} &= \frac{1}{L}\int_0^L C_{fx} dx \\ &= \frac{1}{L}\int_0^L \left[0.7303\sqrt{\frac{\mu}{\rho U_\infty x}} \times dx\right]\end{aligned}$$

or

$$\overline{C_f} = \frac{1.4606}{\sqrt{\text{Re}_L}}$$

where Re_L is the Reynolds number based on the plate length L from the leading edge of the plate.

Example 7.4 Assuming a fourth-degree polynomial for the velocity profile of a laminar boundary layer of a fluid flowing along a flat plate, show that the equation is

$$\frac{u}{U_\infty} = 2\frac{y}{\delta} - 2\left(\frac{y}{\delta}\right)^3 + \left(\frac{y}{\delta}\right)^4$$

Using the momentum integral equation, also determine the boundary layer thickness, wall shear stress and skin friction coefficient. Also determine the relationship between boundary layer thickness and the displacement thickness.

Solution

The fourth-degree polynomial equation is

$$\frac{u}{U_\infty} = C_1 + C_2\left(\frac{y}{\delta}\right) + C_3\left(\frac{y}{\delta}\right)^2 + C_4\left(\frac{y}{\delta}\right)^3 + C_5\left(\frac{y}{\delta}\right)^4$$

The constants C_1 to C_5 can be determined from the following boundary conditions which the velocity function must satisfy:

$$\begin{aligned} u/U_\infty &= 0 \text{ at } y = 0 \\ u/U_\infty &= 1 \text{ at } y = \delta \\ \partial^2 u/\partial y^2 &= 0 \text{ at } y = 0 \\ \partial u/\partial y &= 0 \text{ at } y = \delta \\ \partial^2 u/\partial y^2 &= 0 \text{ at } y = \delta \end{aligned}$$

Applying the boundary conditions, we get

$$\begin{aligned} C_1 &= 0 \\ C_1 + C_2 + C_3 + C_4 + C_5 &= 1 \\ C_2 + 2C_3 + 3C_4 + 4C_5 &= 0 \\ 2C_3 + 6C_4 + 12C_5 &= 0 \\ C_3 &= 0 \end{aligned}$$

The values of the constants from the above equations are

$$C_1 = 0, C_2 = 2, C_3 = 0, C_4 = -2, \text{ and } C_5 = 1$$

This gives the equation of the velocity profile as

$$\frac{u}{U_\infty} = 2\frac{y}{\delta} - 2\left(\frac{y}{\delta}\right)^3 + \left(\frac{y}{\delta}\right)^4$$

Substitution in Eq. (7.39) gives

$$\tau_w = \rho U_\infty^2 \frac{d}{dx}\left\{\int_0^\delta \left[1-2\frac{y}{\delta}+2\left(\frac{y}{\delta}\right)^3-\left(\frac{y}{\delta}\right)^4\right]\left[2\frac{y}{\delta}-2\left(\frac{y}{\delta}\right)^3+\left(\frac{y}{\delta}\right)^4\right]dy\right\}$$

or

$$\tau_w = \rho U_\infty^2 \frac{d}{dx}\left\{\int_0^\delta \left[2\left(\frac{y}{\delta}\right)-4\left(\frac{y}{\delta}\right)^2-2\left(\frac{y}{\delta}\right)^3+9\left(\frac{y}{\delta}\right)^4-4\left(\frac{y}{\delta}\right)^5-4\left(\frac{y}{\delta}\right)^6+4\left(\frac{y}{\delta}\right)^7-\left(\frac{y}{\delta}\right)^8\right]dy\right\}$$

or

$$\begin{aligned}\tau_w &= \rho U_\infty^2 \frac{d}{dx}\left\{\left[\left(\frac{y}{\delta}\right)^2-\frac{4}{3}\left(\frac{y}{\delta}\right)^3-\frac{1}{2}\left(\frac{y}{\delta}\right)^4+\frac{9}{5}\left(\frac{y}{\delta}\right)^5-\frac{2}{3}\left(\frac{y}{\delta}\right)^6-\frac{4}{7}\left(\frac{y}{\delta}\right)^7+\frac{1}{2}\left(\frac{y}{\delta}\right)^8-\frac{1}{9}\left(\frac{y}{\delta}\right)^9\right]_{y=\delta}\delta\right\}\\ &= \rho U_\infty^2 \frac{d}{dx}(0.11746\delta)\\ &= 0.11746\rho U_\infty^2 \frac{d\delta}{dx}.\end{aligned} \tag{i}$$

The shear stress at the wall is also given by

$$\begin{aligned}\tau_w &= \mu\left(\frac{\partial u}{\partial y}\right)_{y=0}\\ &= \mu\frac{\partial}{\partial y}\left\{\left[2\frac{y}{\delta}-2\left(\frac{y}{\delta}\right)^3+\left(\frac{y}{\delta}\right)^4\right]U_\infty\right\}_{y=0}\\ &= 2\frac{\mu U_\infty}{\delta}.\end{aligned} \tag{ii}$$

Equating Eqs. (i) and (ii), we get

$$0.11746\rho U_\infty^2 \frac{d\delta}{dx} = 2\frac{\mu U_\infty}{\delta}$$

or

$$\delta\frac{d\delta}{dx} = 17.03\frac{\mu}{\rho U_\infty}$$

or

$$\frac{\delta^2}{2} = 17.03\frac{\mu}{\rho U_\infty}x + C.$$

The thickness of the boundary layer at the leading edge of the plate is zero, i.e. at $x = 0$, $\delta = 0$. This gives C = 0. Hence,

$$\delta^2 = 34.06 \frac{\mu}{\rho U_\infty} x$$

or

$$\delta = 5.84 \sqrt{\frac{\mu x}{\rho U_\infty}}$$

or

$$\frac{\delta}{x} = 5.84 \sqrt{\frac{\mu}{\rho U_\infty x}} = \frac{5.84}{\sqrt{\mathrm{Re}_x}}.$$

Substituting the value of δ in Eq. (ii) of the wall shear stress, we get

$$\begin{aligned} \tau_w &= 2 \frac{\mu U_\infty}{\delta} \\ &= 2\mu U_\infty \frac{\sqrt{\mathrm{Re}_x}}{5.84x} \\ &= 2 \times \frac{1}{5.84} \rho U_\infty^2 \sqrt{\frac{\rho U_\infty x}{\mu}} \left(\frac{\mu}{\rho U_\infty x} \right) \\ &= \frac{1}{2} \rho U_\infty^2 \frac{0.685}{\sqrt{\frac{\rho U_\infty x}{\mu}}} = \frac{1}{2} \rho U_\infty^2 \frac{0.685}{\sqrt{\mathrm{Re}_x}} \end{aligned}$$

The local skin friction coefficient is

$$C_{fx} = \frac{\tau_w}{\frac{1}{2} \rho U_\infty^2}$$

or

$$C_{fx} = \frac{0.685}{\sqrt{\mathrm{Re}_x}}.$$

The average value of the skin friction coefficient, following the procedure presented in earlier section, is

$$\overline{C_f} = \frac{1.37}{\sqrt{\mathrm{Re}_L}}$$

where Re_L is the Reynolds number based on the total length L of the plate in the direction of flow.

Velocity displacement thickness,

$$\delta_{vd} = \int_0^{\infty} \left(1 - \frac{u}{U_\infty}\right) dy$$

$$= \int_0^{\delta} \left(1 - 2\frac{y}{\delta} + 2\left(\frac{y}{\delta}\right)^3 - \left(\frac{y}{\delta}\right)^4\right) dy$$

$$= \left[y - \frac{y^2}{\delta} + \frac{1}{2}\frac{y^4}{\delta^3} - \frac{1}{5}\frac{y^5}{\delta^4}\right]_0^{\delta}$$

$$= \frac{3}{10}\delta$$

or

$$\frac{\delta_{vd}}{x} = \frac{3}{10} \times \frac{5.84}{\sqrt{\mathrm{Re}_x}} = \frac{1.752}{\sqrt{\mathrm{Re}_x}}.$$

Example 7.5 If a six degree polynomial is assumed for the velocity profile of a laminar boundary layer of a fluid flowing along a flat plate, show that the equation is

$$\frac{u}{U_\infty} = 2\frac{y}{\delta} - 5\left(\frac{y}{\delta}\right)^4 + 6\left(\frac{y}{\delta}\right)^5 - 2\left(\frac{y}{\delta}\right)^6.$$

Also determine the boundary layer thickness, wall shear stress and skin friction coefficient.

Solution

The six degree polynomial equation is

$$\frac{u}{U_\infty} = C_1 + C_2\left(\frac{y}{\delta}\right) + C_3\left(\frac{y}{\delta}\right)^2 + C_4\left(\frac{y}{\delta}\right)^3 + C_5\left(\frac{y}{\delta}\right)^4 + C_6\left(\frac{y}{\delta}\right)^5 + C_7\left(\frac{y}{\delta}\right)^6$$

The constants C_1 to C_7 can be determined from the following boundary conditions which the velocity function must satisfy.

At the wall (y = 0), u = 0 and v = 0. Hence,

$$u/U_\infty = 0 \text{ at } y = 0 \quad \text{(i)}$$

$$u/U_\infty = 1 \text{ at } y = \delta \quad \text{(ii)}$$

At $y = \delta$, all the derivatives of u with respect to y must vanish, i.e.

$$\partial u/\partial y = 0 \text{ at } y = \delta \quad \text{(iii)}$$

$$\partial^2 u/\partial y^2 = 0 \text{ at } y = \delta \tag{iv}$$

$$\partial^3 u/\partial y^3 = 0 \text{ at } y = \delta \tag{v}$$

From the momentum equation, at the wall ($y = 0$, $u = 0$ and $v = 0$),

$$\partial^2 u/\partial y^2 = 0 \text{ at } y = 0 \tag{vi}$$

At the wall, all the higher derivatives are also zero, i.e.

$$\partial^3 u/\partial y^3 = 0 \text{ at } y = 0 \tag{vii}$$

Applying the boundary conditions, we get

$$\begin{aligned}
&C_1 = 0\\
&C_1 + C_2 + C_3 + C_4 + C_5 + C_6 + C_7 = 1\\
&C_2 + 2C_3 + 3C_4 + 4C_5 + 5C_6 + 6C_7 = 0\\
&2C_3 + 6C_4 + 12C_5 + 20C_6 + 30C_7 = 0\\
&6C_4 + 24C_5 + 60C_6 + 120C_7 = 0\\
&C_3 = 0\\
&C_4 = 0
\end{aligned}$$

The values of the constants from the above equations are

$$C_1 = 0, C_2 = 2, C_3 = 0, C_4 = 0, C_5 = -5, C_6 = 6, C_7 = -2$$

This gives the equation of the velocity profile as

$$\frac{u}{U_\infty} = 2\frac{y}{\delta} - 5\left(\frac{y}{\delta}\right)^4 + 6\left(\frac{y}{\delta}\right)^5 - 2\left(\frac{y}{\delta}\right)^6$$

Substitution in Eq. (7.39) gives

$$\tau_w = \rho U_\infty^2 \frac{d}{dx}\left\{\int_0^\delta \left[1 - 2\frac{y}{\delta} + 5\left(\frac{y}{\delta}\right)^4 - 6\left(\frac{y}{\delta}\right)^5 + 2\left(\frac{y}{\delta}\right)^6\right]\left[2\frac{y}{\delta} - 5\left(\frac{y}{\delta}\right)^4 + 6\left(\frac{y}{\delta}\right)^5 - 2\left(\frac{y}{\delta}\right)^6\right] dy\right\}$$

which gives

$$\begin{aligned}
\tau_w &= \rho U_\infty^2 \frac{d}{dx}(0.109335\delta)\\
&= 0.109335\rho U_\infty^2 \frac{d\delta}{dx}
\end{aligned} \tag{viii}$$

The shear stress at the wall is also given by

$$\begin{aligned}\tau_w &= \mu\left(\frac{\partial u}{\partial y}\right)_{y=0}\\ &= \mu\frac{\partial}{\partial y}\left\{\left[2\frac{y}{\delta}-5\left(\frac{y}{\delta}\right)^4+6\left(\frac{y}{\delta}\right)^5-2\left(\frac{y}{\delta}\right)^6\right]U_\infty\right\}_{y=0}\\ &= 2\frac{\mu U_\infty}{\delta}. \end{aligned} \tag{ix}$$

Equating Eqs. (viii) and (ix), we get

$$0.109335\rho U_\infty^2\frac{d\delta}{dx} = 2\frac{\mu U_\infty}{\delta}$$

or

$$\delta\frac{d\delta}{dx} = 18.2924\frac{\mu}{\rho U_\infty}$$

or

$$\frac{\delta^2}{2} = 18.2924\frac{\mu}{\rho U_\infty}x + C$$

The thickness of the boundary layer at the leading edge of the plate is zero, i.e. at $x = 0$, $\delta = 0$. This gives C = 0. Hence,

$$\delta^2 = 36.58\frac{\mu}{\rho U_\infty}x$$

or

$$\frac{\delta}{x} = 6.05\sqrt{\frac{\mu}{\rho U_\infty x}} = \frac{6.05}{\sqrt{\mathrm{Re}_x}}.$$

Substituting the value of δ in equation of the wall shear stress, we get

$$\begin{aligned}\tau_w &= 2\frac{\mu U_\infty}{\delta}\\ &= 2\mu U_\infty\frac{\sqrt{\mathrm{Re}_x}}{6.05x}\\ &= 2\times\frac{1}{6.05}\rho U_\infty^2\sqrt{\frac{\rho U_\infty x}{\mu}}\left(\frac{\mu}{\rho U_\infty x}\right)\\ &= \frac{1}{2}\rho U_\infty^2\frac{0.6612}{\sqrt{\frac{\rho U_\infty x}{\mu}}} = \frac{1}{2}\rho U_\infty^2\frac{0.6612}{\sqrt{\mathrm{Re}_x}}\end{aligned}$$

The local skin friction coefficient is

$$C_{fx} = \frac{\tau_w}{\frac{1}{2}\rho U_\infty^2}$$

or

$$C_{fx} = \frac{0.6612}{\sqrt{\mathrm{Re}_x}}.$$

The average value of the skin friction coefficient, following the procedure presented in earlier sections, is

$$\overline{C_f} = \frac{1.322}{\sqrt{\mathrm{Re}_L}}$$

where $\mathrm{Re_L}$ is the Reynolds number based on the total length L of the plate in the direction of flow.

Example 7.6 The velocity distribution in the boundary layer over a flat plate is given by

$$\frac{u}{U_\infty} = \sin\left(\frac{\pi y}{2\delta}\right)$$

Using the momentum integral equation, determine the boundary layer thickness, wall shear stress and skin friction coefficient.

Solution

Substitution in Eq. (7.39) gives

$$\tau_w = \rho U_\infty^2 \frac{d}{dx}\left\{\int_0^\delta \left[1 - \sin\left(\frac{\pi y}{2\delta}\right)\right]\left[\sin\left(\frac{\pi y}{2\delta}\right)\right] dy\right\}$$

or

$$\begin{aligned}
\tau_w &= \rho U_\infty^2 \frac{d}{dx}\left\{\int_0^\delta \sin\left(\frac{\pi y}{2\delta}\right) dy - \int_0^\delta \sin^2\left(\frac{\pi y}{2\delta}\right) dy\right\} \\
&= \rho U_\infty^2 \frac{d}{dx}\left\{\int_0^\delta \sin\left(\frac{\pi y}{2\delta}\right) dy - \frac{1}{2}\int_0^\delta \left[1 - \cos\left(\frac{\pi y}{\delta}\right)\right] dy\right\} \\
&= \rho U_\infty^2 \frac{d}{dx}\left\{\left[-\frac{2\delta}{\pi}\cos\left(\frac{\pi y}{2\delta}\right)\right]_0^\delta - \frac{1}{2}\left[y - \frac{\delta}{\pi}\sin\left(\frac{\pi y}{\delta}\right)\right]_0^\delta\right\} \\
&= \rho U_\infty^2 \frac{d}{dx}\left[\frac{2\delta}{\pi} - \frac{\delta}{2}\right] = 0.1366\rho U_\infty^2 \frac{d\delta}{dx} \qquad \text{(i)}
\end{aligned}$$

The shear stress at the wall is also given by

$$\begin{aligned}\tau_w &= \mu\left(\frac{\partial u}{\partial y}\right)_{y=0} \\ &= \mu\frac{\partial}{\partial y}\left[\sin\left(\frac{\pi y}{2\delta}\right)U_\infty\right]_{y=0} = \mu U_\infty\left[\frac{\pi}{2\delta}\cos\left(\frac{\pi y}{2\delta}\right)\right]_{y=0} \\ &= 1.571\frac{\mu U_\infty}{\delta} \end{aligned} \tag{ii}$$

Equating Eqs. (i) and (ii), we get

$$0.1366\rho U_\infty^2\frac{d\delta}{dx} = 1.571\frac{\mu U_\infty}{\delta}$$

or

$$\delta\frac{d\delta}{dx} = 11.5\frac{\mu}{\rho U_\infty}$$

or

$$\frac{\delta^2}{2} = 11.5\frac{\mu}{\rho U_\infty}x + C$$

The thickness of the boundary layer at the leading edge of the plate is zero, i.e. at $x = 0$, $\delta = 0$. This gives C = 0 and we have

$$\frac{\delta}{x} = 4.796\sqrt{\frac{\mu}{\rho U_\infty x}} = \frac{4.796}{\sqrt{\mathrm{Re}_x}}.$$

Substituting the value of δ in Eq. (ii) of the wall shear stress, we get

$$\begin{aligned}\tau_w &= 1.571\frac{\mu U_\infty}{\delta} \\ &= 1.571\mu U_\infty\frac{\sqrt{\mathrm{Re}_x}}{4.796x} \\ &= 1.571\times\frac{1}{4.796}\rho U_\infty^2\sqrt{\frac{\rho U_\infty x}{\mu}}\left(\frac{\mu}{\rho U_\infty x}\right) \\ &= \frac{1}{2}\rho U_\infty^2\frac{0.655}{\sqrt{\frac{\rho U_\infty x}{\mu}}} = \frac{1}{2}\rho U_\infty^2\frac{0.655}{\sqrt{\mathrm{Re}_x}}\end{aligned}$$

The local skin friction coefficient is

$$C_{fx} = \frac{\tau_w}{\frac{1}{2}\rho U_\infty^2} = \frac{0.655}{\sqrt{\mathrm{Re}_x}}.$$

The average value of the skin friction coefficient, following the procedure presented in earlier sections, is

$$\overline{C_f} = \frac{1.31}{\sqrt{\mathrm{Re}_L}}$$

where $\mathrm{Re_L}$ is the Reynolds number based on the total length L of the plate in the direction of flow.

Example 7.7 If the velocity distribution through the laminar boundary layer over a flat plate is assumed to be a straight line, determine the boundary layer thickness, wall shear stress and skin friction coefficient using the momentum integral equation.

Solution

The equation of the velocity profile is given as

$$\frac{u}{U_\infty} = \frac{y}{\delta}$$

(i) The wall shear stress,

$$\begin{aligned}\tau_w &= \rho U_\infty^2 \frac{d}{dx}\left[\int_0^\delta \left(1 - \frac{u}{U_\infty}\right)\frac{u}{U_\infty} dy\right] \\ &= \rho U_\infty^2 \frac{d}{dx}\left[\int_0^\delta \left(1 - \frac{y}{\delta}\right)\frac{y}{\delta} dy\right] \\ &= \rho U_\infty^2 \frac{d}{dx}\left[\frac{y^2}{2\delta} - \frac{y^3}{3\delta^2}\right]_0^\delta \\ &= 0.167\rho U_\infty^2 \frac{d\delta}{dx} \end{aligned} \tag{i}$$

The shear stress at the wall is also given by

$$\begin{aligned}\tau_w &= \mu\left(\frac{\partial u}{\partial y}\right)_{y=0} \\ &= \mu\frac{\partial}{\partial y}\left[\left(\frac{y}{\delta}\right)U_\infty\right]_{y=0} \\ &= \frac{\mu U_\infty}{\delta}\end{aligned} \tag{ii}$$

Equating Eqs. (i) and (ii), we get

$$0.167\rho U_\infty^2 \frac{d\delta}{dx} = \frac{\mu U_\infty}{\delta}$$

or

$$\delta \frac{d\delta}{dx} = \frac{1}{0.167} \frac{\mu}{\rho U_\infty}$$

or

$$\frac{\delta^2}{2} = 5.99 \frac{\mu}{\rho U_\infty} x + C$$

The thickness of the boundary layer at the leading edge of the plate is zero, i.e. at $x = 0$, $\delta = 0$. This gives C = 0 and we get

$$\frac{\delta}{x} = 3.46 \sqrt{\frac{\mu}{\rho U_\infty x}} = \frac{3.46}{\sqrt{\mathrm{Re}_x}}$$

Substituting the value of δ in Eq. (ii) of the wall shear stress, we get

$$\begin{aligned} \tau_w &= \frac{\mu U_\infty}{\delta} \\ &= \mu U_\infty \frac{\sqrt{\mathrm{Re}_x}}{3.46x} \\ &= \frac{1}{2} \rho U_\infty^2 \frac{0.578}{\sqrt{\frac{\rho U_\infty x}{\mu}}} = \frac{1}{2} \rho U_\infty^2 \frac{0.578}{\sqrt{\mathrm{Re}_x}} \end{aligned}$$

The local skin friction coefficient is

$$C_{fx} = \frac{\tau_w}{\frac{1}{2} \rho U_\infty^2} = \frac{0.578}{\sqrt{\mathrm{Re}_x}}$$

The average value of the skin friction coefficient, following the procedure presented in earlier sections, is

$$\overline{C_f} = \frac{1.156}{\sqrt{\mathrm{Re}_L}}$$

where Re_L is the Reynolds number based on the total length L of the plate.

The results for different velocity distributions are tabulated in Table 7.3.

Example 7.8 For the velocity profile equations at S. No. 1-4 and 6 of Table 7.3, determine the velocity displacement thickness relations.

Table 7.3 Effect of the velocity profile in the boundary layer on the boundary layer thickness and friction factor

Velocity profile	Boundary conditions		$(\delta/x)\sqrt{\mathrm{Re}_x}$	$\delta_{vd}\sqrt{\mathrm{Re}_x}$	$C_f\sqrt{\mathrm{Re}_x}$
	y = 0	y = δ			
$\frac{u}{U_\infty}=\frac{y}{\delta}$	$u = 0$	$u = U_\infty$	3.46	1.73	1.156
$\frac{u}{U_\infty}=2\frac{y}{\delta}-\left(\frac{y}{\delta}\right)^2$	$u = 0$	$u = U_\infty$	5.477	1.826	1.4606
$\frac{u}{U_\infty}=\frac{3}{2}\frac{y}{\delta}-\frac{1}{2}\left(\frac{y}{\delta}\right)^3$	$u = 0$ $\partial^2 u/\partial y^2 = 0$	$u = U_\infty$ $\partial u/\partial y = 0$	4.64	1.74	1.293
$\frac{u}{U_\infty}=\sin\left(\frac{\pi y}{2\delta}\right)$			4.796	1.743	1.31
$\frac{u}{U_\infty}=2\frac{y}{\delta}-2\left(\frac{y}{\delta}\right)^3+\left(\frac{y}{\delta}\right)^4$	$u = 0$ $\partial^2 u/\partial y^2 = 0$	$u = U_\infty$ $\partial u/\partial y = 0$ $\partial^2 u/\partial y^2 = 0$	5.84	1.752	1.370
$\frac{u}{U_\infty}=2\frac{y}{\delta}-5\left(\frac{y}{\delta}\right)^4+6\left(\frac{y}{\delta}\right)^5-2\left(\frac{y}{\delta}\right)^6$	$u = 0$ $\partial^2 u/\partial y^2 = 0$ $\partial^3 u/\partial y^3 = 0$	$u = U_\infty$ $\partial u/\partial y = 0$ $\partial^2 u/\partial y^2 = 0$ $\partial^3 u/\partial y^3 = 0$	6.05	1.729	1.322
Blasius solution	$u = 0$	$u = 0.99\ U_\infty$	5.0	1.729	1.3282

Solution

The velocity displacement thickness, when the limit ∞ is replaced by δ, is given by

$$\delta_{vd} = \int_0^\delta \left(1 - \frac{u}{U_\infty}\right) dy$$

(i) Substituting the equation $\frac{u}{U_\infty} = \frac{y}{\delta}$ of the velocity profile, we have

$$\delta_{vd} = \int_0^\delta \left(1 - \frac{y}{\delta}\right) dy = \left(y - \frac{y^2}{2\delta}\right)_0^\delta = \frac{\delta}{2}.$$

Substitution of value of δ in terms of Re_x gives

$$\frac{\delta_{vd}}{x} = \frac{1}{2}\left(\frac{3.46}{\sqrt{\mathrm{Re}_x}}\right) = \frac{1.73}{\sqrt{\mathrm{Re}_x}}.$$

(ii) Substituting the equation $\frac{u}{U_\infty} = 2\frac{y}{\delta} - \left(\frac{y}{\delta}\right)^2$ of the velocity profile, we have

$$\delta_{vd} = \int_0^\delta \left[1 - 2\left(\frac{y}{\delta}\right) + \left(\frac{y}{\delta}\right)^2\right] dy = \left(y - \frac{y^2}{\delta} + \frac{y^3}{3\delta^2}\right)_0^\delta = \frac{\delta}{3}.$$

Substitution of value of δ in terms of $\mathrm{Re_x}$ gives

$$\frac{\delta_{vd}}{x} = \frac{1}{3}\left(\frac{5.477}{\sqrt{\mathrm{Re}_x}}\right) = \frac{1.826}{\sqrt{\mathrm{Re}_x}}.$$

(iii) Substituting the equation $\frac{u}{U_\infty} = \frac{3}{2}\frac{y}{\delta} - \frac{1}{2}\left(\frac{y}{\delta}\right)^3$ of the velocity profile, we have

$$\delta_{vd} = \int_0^\delta \left[1 - \frac{3}{2}\left(\frac{y}{\delta}\right) + \frac{1}{2}\left(\frac{y}{\delta}\right)^3\right]dy = \left(y - \frac{3y^2}{4\delta} + \frac{y^4}{8\delta^3}\right)_0^\delta = \frac{3\delta}{8}.$$

Substitution of value of δ in terms of $\mathrm{Re_x}$ gives

$$\frac{\delta_{vd}}{x} = \frac{3}{8}\left(\frac{4.64}{\sqrt{\mathrm{Re}_x}}\right) = \frac{1.74}{\sqrt{\mathrm{Re}_x}}.$$

(iv) Substituting the equation $\frac{u}{U_\infty} = \sin\left(\frac{\pi y}{2\delta}\right)$ of the velocity profile, we have

$$\delta_{vd} = \int_0^\delta \left[1 - \sin\left(\frac{\pi y}{2\delta}\right)\right]dy = \left[y + \cos\left(\frac{\pi y}{2\delta}\right)\frac{2\delta}{\pi}\right]_0^\delta = 0.3634\delta.$$

Substitution of value of δ in terms of $\mathrm{Re_x}$ gives

$$\frac{\delta_{vd}}{x} = 0.3634\left(\frac{4.796}{\sqrt{\mathrm{Re}_x}}\right) = \frac{1.743}{\sqrt{\mathrm{Re}_x}}.$$

(v) Substituting the equation $\frac{u}{U_\infty} = 2\frac{y}{\delta} - 5\left(\frac{y}{\delta}\right)^4 + 6\left(\frac{y}{\delta}\right)^5 - 2\left(\frac{y}{\delta}\right)^6$ of the velocity profile, we have

$$\delta_{vd} = \int_0^\delta \left[1 - 2\frac{y}{\delta} + 5\left(\frac{y}{\delta}\right)^4 - 6\left(\frac{y}{\delta}\right)^5 + 2\left(\frac{y}{\delta}\right)^6\right]dy$$

$$= \left(y - \frac{y^2}{\delta} + \frac{y^5}{\delta^4} - \frac{y^6}{\delta^5} + \frac{2y^7}{7\delta^6}\right)_0^\delta = \frac{2\delta}{7}.$$

Substitution of value of δ in terms of $\mathrm{Re_x}$ gives

$$\frac{\delta_{vd}}{x} = \frac{2}{7}\left(\frac{6.05}{\sqrt{\mathrm{Re}_x}}\right) = \frac{1.729}{\sqrt{\mathrm{Re}_x}}.$$

7.7 Energy Equation of Laminar Boundary Layer Over a Flat Plate

The energy equation for the laminar boundary layer system can be developed by making energy balance for the elemental control volume shown in Fig. 7.14 The following assumptions are being made.

1. The fluid is incompressible and flow is steady.
2. The viscosity, thermal conductivity and specific heat are constant.
3. There is negligible heat conduction in the direction of flow, i.e. in the x-direction.

(a) **Rate of heat flow by conduction:**

(i) Heat inflow by conduction through the bottom face

$$= -k(\Delta x.1)\frac{\partial T}{\partial y}$$

(ii) Heat outflow by conduction through the top face

$$= -k\Delta x\frac{\partial T}{\partial y} - k\Delta x\frac{\partial}{\partial y}\left(\frac{\partial T}{\partial y}\right).\Delta y$$

So the net heat flow by conduction into the element

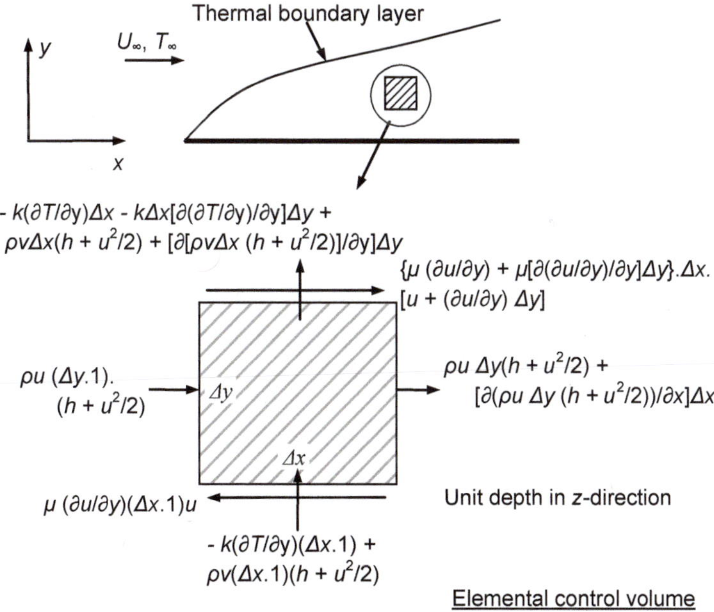

Fig. 7.14 Energy balance for an elemental control volume

$$= -k\Delta x\frac{\partial T}{\partial y} - \left[-k\Delta x\frac{\partial T}{\partial y} - k\Delta x\frac{\partial}{\partial y}\left(\frac{\partial T}{\partial y}\right).\Delta y\right]$$
$$= \left[k\Delta x\frac{\partial}{\partial y}\left(\frac{\partial T}{\partial y}\right).\Delta y\right]$$
$$= \left[k\frac{\partial^2 T}{\partial y^2}\Delta x.\Delta y\right]$$

(b) **Rate of energy flow with the mass, i.e. the energy convected:**

Energy convected into the control volume through the bottom face
= (mass flow rate) × (enthalpy + kinetic energy of the fluid)

$$= \rho v(\Delta x.1)\left(h + \frac{u^2}{2}\right)$$

where the fluid velocity $\sqrt{(u^2+v^2)} \approx u$, since $v \ll u$.

Similarly, the energy convected into the control volume through the left face

$$= \rho u(\Delta y.1)\left(h + \frac{u^2}{2}\right)$$

Hence, total inflow of energy by convection

$$= \rho v(\Delta x.1)\left(h + \frac{u^2}{2}\right) + \rho u(\Delta y.1)\left(h + \frac{u^2}{2}\right)$$

Total rate of outflow of energy by convection

$$= \rho v(\Delta x.1)\left(h + \frac{u^2}{2}\right) + \frac{\partial}{\partial y}\left[\rho v(\Delta x.1)\left(h + \frac{u^2}{2}\right)\right]\Delta y + \rho u(\Delta y.1)\left(h + \frac{u^2}{2}\right)$$
$$+ \frac{\partial}{\partial x}\left[\rho u(\Delta y.1)\left(h + \frac{u^2}{2}\right)\right]\Delta x$$

Net inflow of energy due to convection, the difference of the in- and outflow, is

$$= -\frac{\partial}{\partial y}\left[\rho v(\Delta x.1)\left(h + \frac{u^2}{2}\right)\right]\Delta y - \frac{\partial}{\partial x}\left[\rho u(\Delta y.1)\left(h + \frac{u^2}{2}\right)\right]\Delta x$$
$$= -\left\{\frac{\partial}{\partial y}\left[v\left(h + \frac{u^2}{2}\right)\right] + \frac{\partial}{\partial x}\left[u\left(h + \frac{u^2}{2}\right)\right]\right\}\rho\Delta x\Delta y$$
$$= -\left\{\frac{\partial v}{\partial y}\left(h + \frac{u^2}{2}\right) + v\frac{\partial}{\partial y}\left(h + \frac{u^2}{2}\right) + \frac{\partial u}{\partial x}\left(h + \frac{u^2}{2}\right) + u\frac{\partial}{\partial x}\left(h + \frac{u^2}{2}\right)\right\}\rho\Delta x\Delta y$$
$$= -\left\{\left(h + \frac{u^2}{2}\right)\left(\frac{\partial u}{\partial x} + \frac{\partial v}{\partial y}\right) + u\frac{\partial}{\partial x}\left(h + \frac{u^2}{2}\right) + v\frac{\partial}{\partial y}\left(h + \frac{u^2}{2}\right)\right\}\rho\Delta x\Delta y$$

From the continuity equation, $(\partial u/\partial x + \partial v/\partial y) = 0$. Knowing that enthalpy $h = c_p T$, the above equation transforms to

$$-\left[u\frac{\partial}{\partial x}\left(c_pT+\frac{u^2}{2}\right)+v\frac{\partial}{\partial y}\left(c_pT+\frac{u^2}{2}\right)\right]\rho\Delta x\Delta y$$

$$=-\left[\rho uc_p\frac{\partial T}{\partial x}+\rho u.u\frac{\partial u}{\partial x}+\rho vc_p\frac{\partial T}{\partial y}+\rho v.u\frac{\partial u}{\partial y}\right]\Delta x\Delta y$$

$$=-\left[\rho uc_p\frac{\partial T}{\partial x}+\rho vc_p\frac{\partial T}{\partial y}+u\left(\rho u\frac{\partial u}{\partial x}+\rho v\frac{\partial u}{\partial y}\right)\right]\Delta x\Delta y$$

Using Eq. (7.20) of the hydrodynamic boundary layer, the above equation transforms to

$$=-\left[\rho uc_p\frac{\partial T}{\partial x}+\rho vc_p\frac{\partial T}{\partial y}+u\left(\mu\frac{\partial^2 u}{\partial y^2}\right)\right]\Delta x\Delta y$$

(c) **The viscous work quantities:**

These quantities are indicated in Fig. 7.14. The viscous work is a product of viscous shear force and the distance this force moves in unit time, i.e. the velocity.
Viscous work at the lower face is

$$=\tau(\Delta x.1).u=\mu\frac{\partial u}{\partial y}\Delta x.u$$

Viscous work at the upper face is

$$=\left\{\mu\frac{\partial u}{\partial y}+\left[\mu\frac{\partial}{\partial y}\left(\frac{\partial u}{\partial y}\right)\Delta y\right]\right\}\times\Delta x\times\left[u+\left(\frac{\partial u}{\partial y}\right)\Delta y\right]$$

Net energy delivered to the element is

$$=\left\{\mu\frac{\partial u}{\partial y}+\left[\mu\frac{\partial}{\partial y}\left(\frac{\partial u}{\partial y}\right)\Delta y\right]\right\}\times\Delta x\times\left[u+\left(\frac{\partial u}{\partial y}\right)\Delta y\right]-\mu\frac{\partial u}{\partial y}\Delta x.u$$

Neglecting the second-order differential, the simplification of the above equation gives

$$\left[\mu\frac{\partial^2 u}{\partial y^2}u+\mu\left(\frac{\partial u}{\partial y}\right)^2\right]\Delta x\Delta y$$

The energy balance on the element gives

$$\left\{k\frac{\partial^2 T}{\partial y^2}-\left[\rho uc_p\frac{\partial T}{\partial x}+\rho vc_p\frac{\partial T}{\partial y}+u\left(\mu\frac{\partial^2 u}{\partial y^2}\right)\right]+\left[\mu\frac{\partial^2 u}{\partial y^2}u+\mu\left(\frac{\partial u}{\partial y}\right)^2\right]\right\}\Delta x\Delta y=0$$

or

$$k\frac{\partial^2 T}{\partial y^2} - \left(\rho u c_p \frac{\partial T}{\partial x} + \rho v c_p \frac{\partial T}{\partial y}\right) + \mu\left(\frac{\partial u}{\partial y}\right)^2 = 0$$

or

$$\left(u\frac{\partial T}{\partial x} + v\frac{\partial T}{\partial y}\right) = \frac{k}{\rho c_p}\frac{\partial^2 T}{\partial y^2} + \frac{\mu}{\rho c_p}\left(\frac{\partial u}{\partial y}\right)^2$$

or

$$\left(u\frac{\partial T}{\partial x} + v\frac{\partial T}{\partial y}\right) = \alpha\frac{\partial^2 T}{\partial y^2} + \frac{\mu}{\rho c_p}\left(\frac{\partial u}{\partial y}\right)^2 \tag{7.47}$$

This is the *energy equation of the laminar boundary layer* over a flat plate. The terms on the left-hand side of the equation represent net transport of the energy into the control volume. The first term on the right-hand side of the equation is the net heat conducted out of the control volume and the second term is the viscous work done on the element.

For the low-velocity incompressible flow, the magnitude of the viscous energy term is small and can be neglected.[2] This gives

---

[2]Applying order of magnitude analysis ($u \sim U_\infty$, $y \sim \delta$) to the two terms on the right-hand side of Eq. (7.47), we get

$$\alpha\frac{\partial^2 T}{\partial y^2} = \alpha\frac{T}{\delta^2}$$

and

$$\frac{\mu}{\rho c_p}\left(\frac{\partial u}{\partial y}\right)^2 = \frac{\mu}{\rho c_p}\left(\frac{U_\infty}{\delta}\right)^2$$

The ratio of these quantities is

$$\frac{\mu}{\alpha \rho c_p}\frac{U_\infty^2}{T}$$

Introducing Prandtl number $\mathrm{Pr} = \frac{\mu c_p}{k}$(a non-dimensional group of fluid properties) and $\alpha = \frac{k}{\rho c_p}$, the above ratio term transforms to

$$\mathrm{Pr}\frac{U_\infty^2}{c_p T}$$

which can be shown to be a very small term. For example, consider flow of air at $U_\infty = 100$ m/s at $T = 300$ K and putting $c_p = 1005$ J/(kg K) and Pr = 0.7, we have

$$\mathrm{Pr}\frac{U_\infty^2}{c_p T} = 0.0232 \ll 1.$$

Thus $\frac{\mu}{\rho c_p}\left(\frac{\partial u}{\partial y}\right)^2 \ll \alpha\frac{\partial^2 T}{\partial y^2}$ and can be neglected.

$$u\frac{\partial T}{\partial x} + v\frac{\partial T}{\partial y} = \alpha\frac{\partial^2 T}{\partial y^2} \tag{7.48}$$

which is similar to the momentum equation derived earlier. The solutions of the energy and momentum equations have the same form when $\alpha = \upsilon$. The ratio of α and υ is the Prandtl number Pr. The value of the Prandtl number has an important influence on the convective heat transfer since it decides the relative magnitude of the thickness of the hydraulic and thermal boundary layers[3]:

when $\Pr = 1,\ \delta = \delta_t$
when $\Pr > 1,\ \delta > \delta_t$
and when $\Pr < 1,\ \delta < \delta_t$.

7.7.1 Pohlhausen's Solution

The procedure is identical to that used for the momentum equation as the two equations are similar mathematically.

Let us assume that a dimensionless temperature can be defined as

$$\frac{T_w - T}{T_w - T_\infty} = \theta(\eta)$$

where $\eta = y\sqrt{\frac{U_\infty}{\nu x}}$.

The velocity components u and v, and the stream function ψ have been evaluated earlier as

$$u = U_\infty f',$$

$$v = \frac{1}{2}\sqrt{\frac{\nu U_\infty}{x}}(f'\eta - f) = \frac{1}{2}\sqrt{\frac{\nu U_\infty}{x}}\left(f'.y\sqrt{\frac{U_\infty}{\nu x}} - f\right)$$

$$= \frac{1}{2}\frac{U_\infty y}{x}f' - \frac{1}{2}\sqrt{\frac{\nu U_\infty}{x}}f,$$

$$\psi = \sqrt{\nu U_\infty x}f$$

where $f = \int g(\eta)\,d\eta$ and $(\partial f/\partial \eta) = f'$.

[3]Multiplying numerator and denominator of the Prandtl number equation by density ρ, we get

$$\Pr = \frac{\rho c_p}{k}.\frac{\mu}{\rho} = \frac{\nu}{\alpha} = \frac{\text{kinematic viscosity}}{\text{thermal diffusivity}} \tag{7.49}$$

Kinematic viscosity is diffusivity for momentum or velocity and thermal diffusivity refers to the diffusivity of heat or temperature. If $\Pr = 1$, the hydrodynamic and thermal boundary layers develop together at the same rate. For high Prandtl number fluids, the hydrodynamic boundary layer develops rapidly and for a fluid with $\Pr < 1$, the opposite holds true.

Now the derivatives $\partial T/\partial x$, $\partial T/\partial y$ and $\partial^2 T/\partial y^2$ of the energy equation can be determined as shown below:

$$\begin{aligned}\frac{\partial T}{\partial x} &= -(T_w - T_\infty)\frac{\partial \theta}{\partial x} \\ &= -(T_w - T_\infty)\frac{\partial \theta}{\partial \eta}\frac{\partial \eta}{\partial x} \\ &= -(T_w - T_\infty)\frac{\partial \theta}{\partial \eta}\frac{\partial}{\partial x}\left(y\sqrt{\frac{U_\infty}{vx}}\right) \\ &= -(T_w - T_\infty) \times \left(-\frac{y}{2x}\sqrt{\frac{U_\infty}{vx}}\right)\theta'\end{aligned}$$

where $(\partial\theta/\partial\eta) = \theta'$.

$$\begin{aligned}\frac{\partial T}{\partial y} &= -(T_w - T_\infty)\frac{\partial \theta}{\partial y} \\ &= -(T_w - T_\infty)\frac{\partial \theta}{\partial \eta}\frac{\partial \eta}{\partial y} \\ &= -(T_w - T_\infty)\theta'\sqrt{\frac{U_\infty}{vx}}\end{aligned}$$

$$\begin{aligned}\frac{\partial^2 T}{\partial y^2} &= \frac{\partial}{\partial y}\left(\frac{\partial T}{\partial y}\right) \\ &= -(T_w - T_\infty)\frac{\partial}{\partial y}\left(\theta'\sqrt{\frac{U_\infty}{vx}}\right) \\ &= -(T_w - T_\infty)\frac{\partial}{\partial \eta}\left(\theta'\sqrt{\frac{U_\infty}{vx}}\right) \times \frac{\partial \eta}{\partial y} \\ &= -(T_w - T_\infty)\sqrt{\frac{U_\infty}{vx}}\frac{\partial}{\partial \eta}\left(\frac{\partial \theta}{\partial \eta}\right) \times \sqrt{\frac{U_\infty}{vx}} \\ &= -(T_w - T_\infty)\left(\frac{U_\infty}{vx}\right)\left(\frac{\partial^2 \theta}{\partial \eta^2}\right) \\ &= -(T_w - T_\infty)\left(\frac{U_\infty}{vx}\right)\theta''\end{aligned}$$

where $(\partial^2\theta/\partial\eta^2) = \theta''$.

Substitution of various terms in the energy equation

$$u\frac{\partial T}{\partial x} + v\frac{\partial T}{\partial y} = \alpha\frac{\partial^2 T}{\partial y^2},$$

we get

$$U_\infty f'\left[-(T_w - T_\infty) \times \left(-\tfrac{y}{2x}\sqrt{\tfrac{U_\infty}{\nu x}}\right)\theta'\right] + \left[\tfrac{1}{2}\tfrac{U_\infty y}{x} f' - \tfrac{1}{2}\sqrt{\tfrac{\nu U_\infty}{x}} f\right] \times \left(-(T_w - T_\infty)\theta'\sqrt{\tfrac{U_\infty}{\nu x}}\right) = -\alpha(T_w - T_\infty)\left(\tfrac{U_\infty}{\nu x}\right)\theta''$$

By simplification,

$$\theta'' + \frac{1}{2}(\upsilon/\alpha)f\theta' = 0 \tag{7.50a}$$

$$\theta'' + \frac{1}{2}\mathrm{Pr}\, f\theta' = 0 \tag{7.50b}$$

The boundary conditions are

$$\begin{array}{ll} T = T_w \text{ at } y = 0 & \theta(\eta) = 0 \text{ at } \eta = 0 \\ T = T_\infty \text{ at } y \to \infty & \theta(\eta) = 1 \text{ at } \eta \to \infty \end{array}$$

Equation (7.50b) can be written as

$$\frac{d\theta'}{\theta'} + \frac{1}{2}\mathrm{Pr}\, f\theta' = 0$$

which is an ordinary differential equation and the Pohlhausen solution is

$$\theta' = C_1 \exp\left(-\frac{1}{2}\mathrm{Pr}\int_0^\eta f d\eta\right) \tag{7.51a}$$

and

$$\theta(\eta) = C_1 \int_0^\eta \left[\exp\left(-\frac{1}{2}\mathrm{Pr}\int_0^\eta f d\eta\right)\right] d\eta + C_2 \tag{7.51b}$$

The boundary condition $\theta(\eta) = 0$ at $\eta = 0$ gives $C_2 = 0$. For the boundary condition $\theta(\eta) = 1$ at $\eta \to \infty$, we get

$$C_1 = \frac{1}{\int_0^\infty \left[\exp\left(-\frac{1}{2}\mathrm{Pr}\int_0^\eta f d\eta\right)\right] d\eta}$$

Therefore,

$$\theta(\eta) = \frac{\int_0^\eta \left[\exp\left(-\frac{1}{2}\mathrm{Pr}\int_0^\eta f d\eta\right)\right] d\eta}{\int_0^\infty \left[\exp\left(-\frac{1}{2}\mathrm{Pr}\int_0^\eta f d\eta\right)\right] d\eta} \tag{7.52}$$

Since the function f has already been evaluated in the solution of the momentum equation, the above equation can be solved for known values of the Prandtl number.

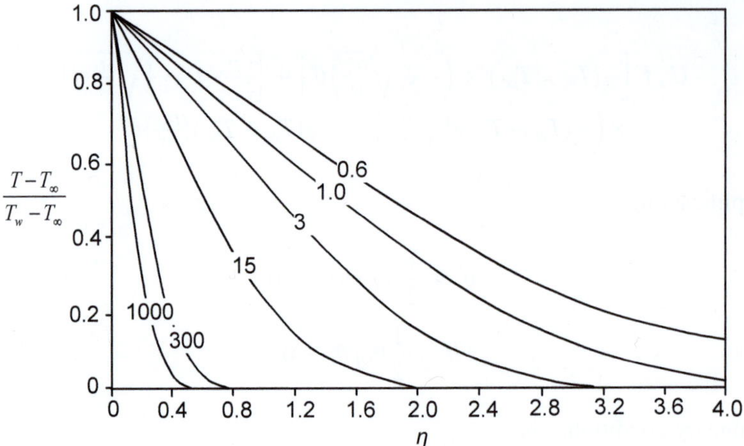

Fig. 7.15 Effect of Prandtl number on temperature distribution

The dimensionless temperature distribution for different values of Prandtl number is presented in Fig. 7.15 as function of η (Pohlhausen 1921).

The dimensionless slope of the temperature distribution at the wall (where $\eta = 0$), from Eq. (7.51a), is

$$(\theta')_{\eta=0} = \left(\frac{d\theta}{d\eta}\right)_{\eta=0} = C_1. \tag{7.53}$$

Pohlhausen showed that for the moderate values of the Prandtl number (0.6 < Pr < 15),

$$(\theta')_{\eta=0} = \left(\frac{d\theta}{d\eta}\right)_{\eta=0} = 0.332\mathrm{Pr}^{1/3}. \tag{7.54}$$

Knowing the dimensionless temperature distribution at the wall, the local heat transfer coefficient h_x at the wall can be evaluated from

$$h_x = \frac{q_w}{T_w - T_\infty}$$

where the heat transfer at the wall q_w, from the Fourier's law, is

$$q_w = -k\left(\frac{\partial T}{\partial y}\right)_{y=0}$$

Hence,

$$h_x = \frac{-k\left(\frac{\partial T}{\partial y}\right)_{y=0}}{T_w - T_\infty} \tag{7.55}$$

For $\frac{T_w - T}{T_w - T_\infty} = \theta(\eta)$,

$$\left(\frac{\partial T}{\partial y}\right)_{y=0} = -(T_w - T_\infty)\sqrt{\frac{U_\infty}{\nu x}}(\theta')_{\eta=0}$$
$$= -(T_w - T_\infty)\left(\sqrt{\frac{U_\infty}{\nu x}}\right)0.332\mathrm{Pr}^{1/3}.$$

Substitution of the value of $(\partial T/\partial y)_{y=0}$ in the equation of h_x gives

$$h_x = \frac{-k\left[-(T_w - T_\infty)\left(\sqrt{\frac{U_\infty}{\nu x}}\right)0.332\mathrm{Pr}^{1/3}\right]}{T_w - T_\infty}$$
$$= 0.332\mathrm{Pr}^{1/3}\frac{k}{x}\left(\sqrt{\frac{U_\infty x}{\nu}}\right)$$

or

$$\frac{h_x x}{k} = 0.332\mathrm{Pr}^{1/3}\left(\sqrt{\frac{U_\infty x}{\nu}}\right) \tag{7.56a}$$

The group of the terms $(h_x x/k)$ on the left-hand side of the equation is dimensionless and is known as the local *Nusselt number* Nu_x. Thus

$$\mathrm{Nu}_x = 0.332\mathrm{Pr}^{1/3}\mathrm{Re}_x^{1/2} \tag{7.56b}$$

The average heat transfer coefficient over the plate length L can be evaluated as

$$\bar{h} = \frac{1}{L}\int_0^L h_x dx$$
$$= 0.332\mathrm{Pr}^{1/3}\frac{1}{L}\int_0^L \frac{k}{x}\left(\sqrt{\frac{U_\infty x}{\nu}}\right)dx$$
$$= 0.332\mathrm{Pr}^{1/3}\frac{1}{L}k\sqrt{\frac{U_\infty}{\nu}}\int_0^L \frac{1}{\sqrt{x}}dx$$
$$= 0.332\mathrm{Pr}^{1/3}\frac{1}{L}k\sqrt{\frac{U_\infty}{\nu}}2L^{1/2}$$
$$= 0.664\mathrm{Pr}^{1/3}\frac{1}{L}k\sqrt{\frac{U_\infty L}{\nu}}$$
$$= 2(h_x)_{x=L}$$

The equation of $\bar{h}$ can also be put in the non-dimensional form as

$$\frac{\bar{h}L}{k} = 0.664\text{Pr}^{1/3}\sqrt{\frac{U_\infty L}{\nu}}$$

This gives the equation of the average Nusselt number over the plate length L as

$$\overline{\text{Nu}} = 0.664\text{Pr}^{1/3}\text{Re}_L^{1/2} \tag{7.57}$$

Equation (7.75b) can be written as

$$\frac{\text{Nu}_x}{\text{Re}_x\,\text{Pr}} = 0.332\text{Pr}^{-2/3}\text{Re}_x^{-1/2}$$

The group $[\text{Nu}_x/(\text{Re}_x\ \text{Pr})]$ is also a non-dimensional and is termed as *Stanton number* St_x. Hence,

$$\text{St}_x = 0.332\text{Pr}^{-2/3}\text{Re}_x^{-1/2}$$

or

$$\text{St}_x\text{Pr}^{2/3} = \frac{1}{2}\left(\frac{0.664}{\sqrt{\text{Re}_x}}\right). \tag{7.58}$$

Using Eq. (7.37), we have

$$\text{St}_x\text{Pr}^{2/3} = \frac{C_{fx}}{2}. \tag{7.59}$$

This establishes a relation between the Stanton number and the friction factor.

The Stanton number is used sometimes as an alternative for Nusselt number when presenting heat transfer data. Substitution of $\text{Nu} = hL/k$, $\text{Pr} = \mu c_p/k$ and $\text{Re} = \rho UL/\mu$ gives

$$\text{St} = \frac{h}{c_p\rho U} \tag{7.60}$$

7.7.2 von Karman Integral Technique (Integral Analysis of Energy Equation for the Laminar Boundary Layer)

Figure 7.16 shows the hydrodynamic and thermal boundary layers developed over a flat plate placed parallel to the fluid stream at free-stream velocity U_∞ and temperature T_∞. The wall temperature is T_w.

As in the case of the hydrodynamic boundary layer, let the profile of the temperature distribution through the thermal boundary layer is represented by a cubic parabola, i.e.

$$T = T_w + Ay + By^2 + Cy^3.$$

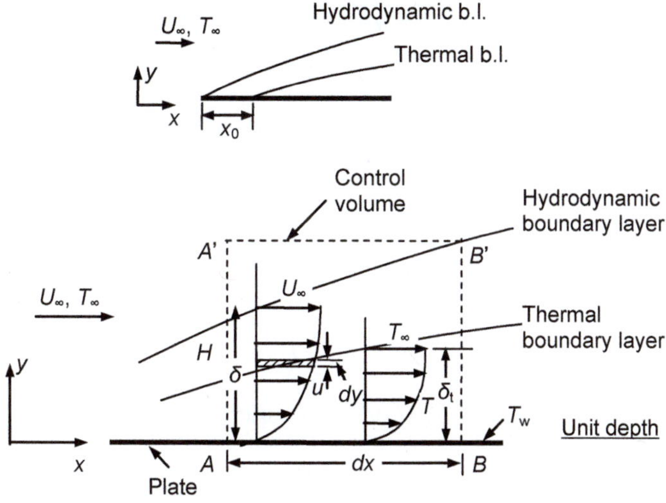

Fig. 7.16 Elemental control volume for integral energy analysis of laminar boundary layer

In terms of the temperature difference between the fluid and wall, $\theta = T - T_{\mathrm{w}}$, the expression of temperature distribution becomes

$$\theta = Ay + By^2 + Cy^3. \tag{7.61}$$

The constants A, B and C of the equation are found by applying the boundary conditions as follows.

If the thickness of the boundary layer is δ_{t}, then the following boundary conditions must be satisfied:

$$\theta = \theta_{\infty} \quad \text{at } y = \delta_{\mathrm{t}} \tag{i}$$

$$\partial\theta/\partial y = 0 \quad \text{at } y = \delta_{\mathrm{t}} \tag{ii}$$

At the wall ($y = 0$),

$$T = T_{\mathrm{w}}, \text{ i.e. } \theta = 0 \tag{iii}$$

Again at the wall, u and v are zero. Inserting this condition in the energy equation, we get

$$\left(\partial^2 T/\partial y^2\right) = 0 = \left(\partial^2\theta/\partial y^2\right) \tag{iv}$$

Applying the boundary conditions (i) to (iv) to Eq. (7.61), we obtain

$$B = 0, A = 3\theta_{\infty}/(2\delta_{\mathrm{t}}) \text{ and } C = -\theta_{\infty}/\left(2\delta_{\mathrm{t}}^3\right)$$

Substitution of the values of the constants in Eq. (7.61) gives

$$\frac{\theta}{\theta_\infty} = \frac{T - T_w}{T_\infty - T_w} = \frac{3}{2}\frac{y}{\delta_t} - \frac{1}{2}\left(\frac{y}{\delta_t}\right)^3 \tag{7.62}$$

At the wall, i.e. at $y = 0$, the fluid is at rest and hence is not having any velocity component perpendicular to the wall. Thus, the heat transfer at the wall is due to the conduction only.

Knowing the dimensionless temperature distribution at the wall, the local heat transfer coefficient h_x at the wall can be evaluated from

$$h_x = \frac{q_w}{T_w - T_\infty}.$$

where the heat transfer at the wall q_w, from the Fourier's law, is

$$q_w = -k\left(\frac{\partial T}{\partial y}\right)_{y=0}$$

Hence,

$$h_x = \frac{-k\left(\frac{\partial T}{\partial y}\right)_{y=0}}{T_w - T_\infty}$$

where

$$\left(\frac{\partial T}{\partial y}\right)_{y=0} = \left(\frac{\partial \theta}{\partial y}\right)_{y=o} = \theta_\infty\left\{\frac{\partial}{\partial y}\left[\frac{3}{2}\frac{y}{\delta_t} - \frac{1}{2}\left(\frac{y}{\delta_t}\right)^3\right]\right\}_{y=0} = \frac{3}{2\delta_t}\theta_\infty \tag{7.63}$$

This gives

$$\begin{aligned} h_x &= -\frac{3}{2}\frac{k\theta_\infty}{(T_w - T_\infty)\delta_t} \\ &= -\frac{3}{2}\frac{k(T_\infty - T_w)}{(T_w - T_\infty)\delta_t} \end{aligned}$$

or

$$h_x = \frac{3}{2}\left(\frac{k}{\delta_t}\right) \tag{7.64}$$

Thus, the heat transfer coefficient at the wall can be determined by finding the thickness δ_t of the thermal boundary layer, which can be obtained by the integral analysis of the thermal boundary layer presented below.

Let us consider the control volume in Fig. 7.16 bounded by the planes AA', AB, BB' and A'B' whose length in direction x is dx. Its height H is greater than δ_t and δ.

Energy balance for the control volume gives

$$\text{Heat transfer at the wall by conduction} + \text{Energy convected into the control volume} \\ + \text{Net viscous work} = \text{Energy convected out} \tag{a}$$

Heat transfer at the wall by conduction over the plate length dx is

$$dq_w = -k(dx.1)\left(\frac{\partial T}{\partial y}\right)_{y=0}$$

The energy convected into the control volume through plane AA' for the unit width of plate

$$= \text{mass low rate} \times (\text{enthalpy} + \text{kinetic energy})$$

$$= \int_0^H \rho u(dy.1)c_p T$$

$$= \rho c_p \int_0^H uT dy$$

where the kinetic energy term has been assumed to be negligible compared to the enthalpy of the fluid and the thermophysical properties of the fluid have been assumed to be constant.

The energy convected out of the control volume through plane BB'

$$= \rho c_p \int_0^H uT dy + \frac{d}{dx}\left(\rho c_p \int_0^H uT dy\right) dx$$

The mass flow rate through plane A'B' is difference of the mass flow through planes BB' and AA', i.e.

$$= \rho \int_0^H u dy + \frac{d}{dx}\left(\rho \int_0^H u dy\right) dx - \rho \int_0^H u dy = \frac{d}{dx}\left(\rho \int_0^H u dy\right) dx$$

This mass is at temperature T_∞, since $H > \delta_t$, and carries with it an energy equal to

$$\rho c_p T_\infty \frac{d}{dx}\left(\int_0^H u dy\right) dx$$

The viscous shear force is the product of the shear stress and the area $(dx.1)$, i.e.

$$\tau(dx.1) = \mu \frac{\partial u}{\partial y} dx$$

and the distance through which this force acts per unit time in respect to the elemental strip of thickness dy is

$$\left(\frac{\partial u}{\partial y}\right)dy$$

So the net viscous energy delivered to the control volume is

$$= \int_0^H \left(\mu \frac{du}{dy}dx\right) \times \frac{du}{dy}dy$$

$$= \mu \int_0^H \left(\frac{du}{dy}\right)^2 dxdy$$

Substituting various terms in Eq. (a) gives

$$-k(dx.1)\left(\frac{\partial T}{\partial y}\right)_{y=0} + \rho c_p \int_0^H uTdy + \rho c_p T_\infty \frac{d}{dx}\left(\int_0^H udy\right)dx + \mu \int_0^H \left(\frac{du}{dy}\right)^2 dxdy$$
$$= \rho c_p \int_0^H uTdy + \frac{d}{dx}\left(\rho c_p \int_0^H uTdy\right)dx$$

or

$$\frac{d}{dx}\left(\int_0^H (T_\infty - T)udy\right) + \frac{\mu}{\rho c_p}\left[\int_0^H \left(\frac{du}{dy}\right)^2 dy\right] = \frac{k}{\rho c_p}\left(\frac{\partial T}{\partial y}\right)_{y=0}$$

or

$$\frac{d}{dx}\left(\int_0^H (T_\infty - T)udy\right) + \frac{\mu}{\rho c_p}\left[\int_0^H \left(\frac{du}{dy}\right)^2 dy\right] = \alpha\left(\frac{\partial T}{\partial y}\right)_{y=0} \tag{7.65}$$

This is the *integral energy equation of the boundary layer*. For low velocity flow, the viscous work term can be neglected and the integral energy equation reduces to

$$\frac{d}{dx}\int_0^H (T_\infty - T)udy = \alpha\left(\frac{\partial T}{\partial y}\right)_{y=o} \tag{7.66}$$

Left-hand side of the equation can be rewritten as

$$\frac{d}{dx}\int_0^H (T_\infty - T)udy = \frac{d}{dx}\int_0^H (\theta_\infty - \theta)udy$$

$$= \theta_\infty U_\infty \frac{d}{dx}\left[\int_0^H \left(1 - \frac{\theta}{\theta_\infty}\right)\frac{u}{U_\infty}dy\right]$$

Substituting values of θ/θ_∞ and u/U_∞ from temperature and velocity distribution equations, i.e. from Eqs. (7.62) and (7.42), we get

$$\frac{d}{dx}\int_0^H (T_\infty - T)udy = \theta_\infty U_\infty \frac{d}{dx}\int_0^H \left[1 - \frac{3}{2}\frac{y}{\delta_t} + \frac{1}{2}\left(\frac{y}{\delta_t}\right)^3\right] \times \left[\frac{3}{2}\frac{y}{\delta} - \frac{1}{2}\left(\frac{y}{\delta}\right)^3\right]dy$$

$$= \theta_\infty U_\infty \frac{d}{dx}\int_0^H \left[\frac{3}{2}\frac{y}{\delta} - \frac{9}{4}\left(\frac{y}{\delta}\right)\left(\frac{y}{\delta_t}\right) + \frac{3}{4}\left(\frac{y}{\delta}\right)\left(\frac{y}{\delta_t}\right)^3 - \frac{1}{2}\left(\frac{y}{\delta}\right)^3 + \frac{3}{4}\left(\frac{y}{\delta_t}\right)\left(\frac{y}{\delta}\right)^3 - \frac{1}{4}\left(\frac{y}{\delta}\right)^3\left(\frac{y}{\delta_t}\right)^3\right]dy$$

For most of the fluids, $\delta_t < \delta$. Hence, integration is to be carried out up to $y = \delta_t$ (since for $y > \delta_t$ the integrand will be zero). Thus

$$\frac{d}{dx}\int_0^{\delta_t} (T_\infty - T)udy = \theta_\infty U_\infty \frac{d}{dx}\left[\frac{3}{2}\frac{y^2}{2\delta} - \frac{9}{4}\left(\frac{y^3}{3\delta\delta_t}\right) + \frac{3}{4}\left(\frac{y^5}{5\delta\delta_t^3}\right) - \frac{1}{2}\left(\frac{y^4}{4\delta^3}\right) + \frac{3}{4}\left(\frac{y^5}{5\delta^3\delta_t}\right) - \frac{1}{4}\left(\frac{y^7}{7\delta^3\delta_t^3}\right)^3\right]_{y=\delta_t}$$

$$= \theta_\infty U_\infty \frac{d}{dx}\left[\frac{3}{20}\left(\frac{\delta_t^2}{\delta}\right) - \frac{3}{280}\left(\frac{\delta_t^4}{\delta^3}\right)\right]$$

Representing the thickness ratio of the thermal and hydrodynamic layers δ_t/δ by ξ,

$$\frac{d}{dx}\int_0^{\delta_t} (T_\infty - T)udy = \theta_\infty U_\infty \frac{d}{dx}\left[\frac{3}{20}\delta\xi^2 - \frac{3}{280}\delta\xi^4\right] \tag{7.67a}$$

For the assumption $\delta_t < \delta$, i.e. $\xi < 1$, $(3/280)\delta\xi^4 \ll (3/20)\delta\xi^2$, hence may be neglected and the above equation transforms to

$$\frac{d}{dx}\int_0^{\delta_t} (T_\infty - T)udy = \frac{3}{20}\theta_\infty U_\infty \frac{d}{dx}(\delta\xi^2)$$

$$= \frac{3}{20}\theta_\infty U_\infty \left(2\delta\xi\frac{d\xi}{dx} + \xi^2\frac{d\delta}{dx}\right) \tag{7.67b}$$

From Eq. (7.63),

$$\alpha\left(\frac{\partial T}{\partial y}\right)_{y=0} = 3\alpha\frac{\theta_\infty}{2\delta_t}$$

Using (7.67b) and above equation, Eq. (7.66) becomes

$$\frac{3}{20}\theta_\infty U_\infty\left(2\delta\xi\frac{d\xi}{dx}+\xi^2\frac{d\delta}{dx}\right)=3\alpha\frac{\theta_\infty}{2\delta_t}$$

Using $\delta_t = \xi\,\delta$ and rearranging the terms,

$$\alpha=\frac{1}{10}U_\infty\left(2\delta^2\xi^2\frac{d\xi}{dx}+\delta\xi^3\frac{d\delta}{dx}\right)$$

From Eq. (7.43a) of hydrodynamic boundary layer,

$$\delta^2=\frac{280}{13}\frac{vx}{U_\infty}$$

and

$$\delta\frac{d\delta}{dx}=\frac{140}{13}\frac{v}{U_\infty}$$

Hence,

$$\alpha=\frac{1}{10}U_\infty\left(2\xi^2\left(\frac{280}{13}\frac{vx}{U_\infty}\right)\frac{d\xi}{dx}+\xi^3\frac{140}{13}\frac{v}{U_\infty}\right)$$

or

$$\alpha=\frac{14}{13}v\left(4\xi^2x\frac{d\xi}{dx}+\xi^3\right)$$

or

$$\xi^3+4x\xi^2\frac{d\xi}{dx}=\frac{13}{14}\frac{\alpha}{v}$$

or

$$\xi^3+\frac{4x}{3}\frac{d\xi^3}{dx}=\frac{13}{14}\frac{\alpha}{v} \tag{7.68}$$

The above equation is a linear differential equation of first order in ξ^3. Its solution is

$$\xi^3=Cx^{-3/4}+\frac{13}{14}\frac{\alpha}{v}$$

The constant of integration can be determined from the condition that at $x = x_0$ (heating of the plate has been assumed to start at $x = x_0$), the thickness of the thermal boundary layer $\delta_t = 0$, i.e. $\delta_t/\delta = \xi = 0$. This gives

$$C = -\frac{13}{14}\frac{\alpha}{\nu}(x_0)^{3/4}$$

Hence,

$$\xi^3 = -\frac{13}{14}\frac{\alpha}{\nu}(x_0)^{3/4}x^{-3/4} + \frac{13}{14}\frac{\alpha}{\nu}$$

or

$$\xi = \left\{\frac{13}{14}\frac{\alpha}{\nu}\left[1 - \left(\frac{x_0}{x}\right)^{3/4}\right]\right\}^{1/3}$$

or

$$\xi = \frac{1}{1.025}\text{Pr}^{-1/3}\left[1 - \left(\frac{x_0}{x}\right)^{3/4}\right]^{1/3} \tag{7.69a}$$

where the ratio (α/υ) is the Prandtl number Pr.

If the entire length of the plate is heated then $x_0 = 0$ and Eq. (7.69a) gives

$$\xi = \frac{1}{1.025}\text{Pr}^{-1/3} \tag{7.69b}$$

In the above analysis, the thermal boundary layer has been assumed to be thinner than the hydrodynamic boundary layer. This assumption is true provided the Prandtl number is equal to or greater than 1. Most of the liquids except molten metals have Pr > 1. The Prandtl number for most of the gases and vapours lies between 0.65 and 1. Returning to Eq. (7.67a) where based on the assumption $\xi < 1$, we dropped a term involving ξ^4, a brief consideration will show that error introduced is small even if ξ is slightly greater than unity. In fact, the thicknesses of the velocity and thermal boundary layers for gases are practically equal.

Knowing the value of the boundary layer thickness, we can obtain the value of the convective heat transfer coefficient as follows.

From Eq. (7.64), we have

$$h_x = \frac{3}{2}\left(\frac{k}{\delta_t}\right)$$

Substituting the value of the thermal boundary layer thickness δ_t, we get

$$h_x = \frac{3}{2}k\frac{1}{\delta}\left\{1.025 \times \text{Pr}^{1/3}\left[1 - \left(\frac{x_0}{x}\right)^{3/4}\right]^{-1/3}\right\}$$

Substitution of the value of hydrodynamic boundary layer thickness δ from Eq. (7.43a) gives

$$h_x = \frac{3}{2}k\sqrt{\frac{13}{280} \times \frac{U_\infty}{\nu x}}\left\{1.025 \times \mathrm{Pr}^{1/3}\left[1 - \left(\frac{x_0}{x}\right)^{3/4}\right]^{-1/3}\right\}$$

or

$$h_x = 0.331k\mathrm{Pr}^{1/3}\sqrt{\frac{U_\infty}{\nu x}} \times \left[1 - \left(\frac{x_0}{x}\right)^{3/4}\right]^{-1/3} \tag{7.70}$$

or

$$\frac{h_x x}{k} = 0.331\mathrm{Pr}^{1/3}\sqrt{\frac{U_\infty x}{\nu}} \times \left[1 - \left(\frac{x_0}{x}\right)^{3/4}\right]^{-1/3}$$

or

$$\mathrm{Nu}_x = 0.331\mathrm{Pr}^{1/3}\mathrm{Re}_x^{1/2} \times \left[1 - \left(\frac{x_0}{x}\right)^{3/4}\right]^{-1/3} \tag{7.71a}$$

Note that the exact solution to this problem yields an answer with numerical constant of 0.332.

For the plate heated for entire length,

$$h_x = 0.331k\mathrm{Pr}^{1/3}\sqrt{\frac{U_\infty}{\nu x}}$$

and

$$\mathrm{Nu}_x = 0.331\mathrm{Pr}^{1/3}\mathrm{Re}_x^{1/2} \tag{7.71b}$$

Following the procedure outlined earlier, the average values of the heat transfer coefficient and the Nusselt number for plate length L are

$$\bar{h} = 0.662k\mathrm{Pr}^{1/3}\sqrt{\frac{U_\infty}{\nu L}}$$

$$\overline{\mathrm{Nu}} = \frac{\bar{h}L}{k} = 0.662\mathrm{Pr}^{1/3}\mathrm{Re}_L^{1/2} \tag{7.72}$$

The above analysis is based on the constant physical properties of the fluid. In fact, the temperature varies from T_w at the wall to the free-stream temperature T_∞. Therefore, it is necessary that the physical properties be evaluated at the arithmetic mean temperature T_f [= $(T_w + T_\infty)/2$], defined as mean or film temperature, when temperature differences are not large.

Example 7.9 Air at atmospheric pressure and a temperature of 50°C flows at 1.5 m/s over an isothermal plate (t_s = 100°C) 1 m long with a 0.5 m long unheated starting length. Determine the local heat transfer coefficient (a) at the trailing edge with unheated starting length and (b) at the trailing edge without unheated starting length.

Solution

Thermophysical properties of air at film temperature of 75°C from Table A5 are

$$\rho = 1.0052\,\mathrm{kg/m^3}, \mu = 2.0658 \times 10^{-5}\,\mathrm{kg/(m\,s)}, k = 0.02990\,\mathrm{W/(m\,K)} \text{ and } \mathrm{Pr} = 0.697.$$

(a) At the trailing edge with unheated starting length
The Reynolds number at x = 1.5 m is

$$\mathrm{Re}_x = \frac{\rho U_\infty x}{\mu} = \frac{1.0052 \times 1.5 \times 1.5}{2.0658 \times 10^{-5}} = 1.09 \times 10^5.$$

The flow is laminar over the entire plate. From Eq. (7.71a),

$$\mathrm{Nu}_x = \frac{h_x x}{k} = 0.331\mathrm{Pr}^{1/3}\mathrm{Re}_x^{1/2} \times \left[1 - \left(\frac{x_0}{x}\right)^{3/4}\right]^{-1/3}$$

Hence, the local heat transfer coefficient at x = 1.5 m with unheated starting length x_o = 0.5 m is

$$\begin{aligned} h_x &= \frac{k}{x} \times 0.331\mathrm{Pr}^{1/3}\mathrm{Re}_x^{1/2} \times \left[1 - \left(\frac{x_0}{x}\right)^{3/4}\right]^{-1/3} \\ &= \frac{0.02990}{1.5} \times 0.331 \times 0.697^{1/3} \times (1.09 \times 10^5)^{1/2} \times \left[1 - \left(\frac{0.5}{1.5}\right)^{3/4}\right]^{-1/3} \\ &= 2.34\,\mathrm{W/(m^2\,K)}. \end{aligned}$$

(b) At the trailing edge without unheated starting length
The Reynolds number at x = 1.0 m is

$$\mathrm{Re}_x = \frac{\rho U_\infty x}{\mu} = \frac{1.0052 \times 1.5 \times 1.0}{2.0658 \times 10^{-5}} = 72989.$$

The flow is laminar over the entire plate. Nusselt number relation for this case (x_o = 0) is

$$\mathrm{Nu}_x = \frac{h_x x}{k} = 0.331\mathrm{Pr}^{1/3}\mathrm{Re}_x^{1/2}$$

Hence, the local heat transfer coefficient at x = 1.0 m is

$$\begin{aligned} h_x &= \frac{k}{x} \times 0.331\mathrm{Pr}^{1/3}\mathrm{Re}_x^{1/2} \\ &= \frac{0.02990}{1.0} \times 0.331 \times 0.697^{1/3} \times (72989)^{1/2} \\ &= 2.37\,\mathrm{W/(m^2\,K)}. \end{aligned}$$

Example 7.10 Air at 30°C and at a pressure of 1 atm is flowing over a flat plate at a velocity of 2.0 m/s. If the plate is 1 m wide and is at a temperature of 90°C, estimate the following:

(i) The critical length, x_{cr};
(ii) The hydrodynamic and thermal boundary layer thicknesses at $x = x_{cr}$;
(iii) The local and average heat transfer coefficients at $x = x_{cr}$;
(iv) The rate of heat transfer by convection form the plate to the air for the plate length of $x = x_{cr}$

Solution

The properties of air at the mean film temperature $t_{fm} = (30 + 90)/2 = 60°C$ are

$$\rho = 1.059\,\text{kg/m}^3, k = 0.0287\,\text{W/(m K)}, \mu = 2.0 \times 10^{-5}\,\text{kg/(m s)}, \text{Pr} = 0.701, c_p = 1008\,\text{J/(kg K)}.$$

(i) Critical length corresponding to $\text{Re}_{cr} = 5 \times 10^5$,

$$x_{cr} = \frac{\text{Re}_{cr}\mu}{\rho U} = \frac{5 \times 10^5 \times 2 \times 10^{-5}}{1.059 \times 2.0} = 4.72\,\text{m}.$$

(ii) Hydrodynamic boundary layer thickness at $x = x_{cr}$,

$$\delta = \frac{5}{\sqrt{\text{Re}_{cr}}} x_{cr} = \frac{5}{\sqrt{5 \times 10^5}} \times 4.72 = 0.0334\,\text{m}.$$

Ratio of thermal and hydrodynamic boundary layer thicknesses,

$$\frac{\delta_t}{\delta} = \frac{\text{Pr}^{-1/3}}{1.025},$$

which gives,

$$\delta_t = \frac{\text{Pr}^{-1/3}}{1.025}\delta = \frac{(0.701)^{-1/3}}{1.025} \times 0.0334 = 0.0366\,\text{m}.$$

(iii) The local Nusselt number equation at x is

$$\text{Nu}_x = \frac{h_x x}{k} = 0.332\text{Re}_x^{1/2}\text{Pr}^{1/3}$$

Hence, local heat transfer coefficient is

$$h_x = (0.332\text{Re}_x^{1/2}\text{Pr}^{1/3})\frac{k}{x} = [0.332 \times (5 \times 10^5)^{1/2} \times (0.701)^{1/3}] \times \frac{0.0287}{4.72}$$
$$= 1.269\,\text{W}/(\text{m}^2\,\text{K}).$$

Average heat transfer coefficient,

$$\bar{h} = 2h_x = 2 \times 1.269 = 2.538\,\text{W}/(\text{m}^2\,\text{K}).$$

(iv) Rate of heat transfer q from one side of the plate for length x_{cr},

$$q = \bar{h}A(t_w - t_\infty) = 2.538 \times (4.72 \times 1) \times (90 - 30) = 718.76\,\text{W}.$$

Example 7.11 Calculate the mass flow rate through the boundary layer from $x = 0$ to $x = x_{cr}$ for the flow of above example.

Solution

At any x-position, the mass flow in the boundary layer is given by

$$\left(\int_0^\delta \rho.u.dy\right) \text{for unit width of the plate.}$$

Assuming $u = U_\infty\left[\frac{3}{2}\frac{y}{\delta} - \frac{1}{2}\left(\frac{y}{\delta}\right)^3\right]$, we get

$$\int_0^\delta \rho.u.dy = \int_0^\delta \rho.U_\infty\left[\frac{3}{2}\frac{y}{\delta} - \frac{1}{2}\left(\frac{y}{\delta}\right)^3\right]dy = \frac{5}{8}\rho.U_\infty\delta$$

Thus, the mass flow rate through the boundary layer,

$$m = \frac{5}{8}\rho.U_\infty\delta = \frac{5}{8} \times 1.059 \times 2.0 \times 0.0334 = 0.0442\,\text{kg/s}.$$

Example 7.12 Air at 20°C and at a pressure of 1 atm is flowing over a flat plate at a velocity of 1.5 m/s. The plate length is 0.3 m and is 1 m wide. If the plate is maintained at 80°C, determine the drag force on the plate.

Solution

At the mean temperature of 50°C, air properties are

$$\rho = 1.095\,\text{kg/m}^3, k = 0.02799\,\text{W/(m K)}, \mu = 1.95 \times 10^{-5}\text{kg/(m s)}, \text{Pr} = 0.703, c_p = 1.0072\,\text{J/(kg K)}.$$

Flow Reynolds number is

$$\text{Re}_L = \frac{\rho U_\infty L}{\mu} = 25269$$

Flow is laminar. The skin friction coefficient,

$$C_{fx} = \frac{0.6641}{\sqrt{\text{Re}_x}} = \frac{0.6641}{\sqrt{25269}} = 0.00417$$

Average skin friction coefficient,

$$\overline{C_f} = 2 \times 0.00417 = 0.00835$$

Shear stress,

$$\tau_w = \overline{C_f}\left(\frac{\rho U_\infty^2}{2}\right) = 0.00835 \times \left(\frac{1.095 \times 1.5^2}{2}\right) = 0.0103\ \text{N/m}^2.$$

Drag,

$$D = \tau_w A = 0.0103 \times (0.3 \times 1) = 3.1 \times 10^{-3}\ \text{N}$$

Alternatively, the skin friction coefficient can be determined using the Reynolds–Colburn analogy.

$$St_x \text{Pr}^{2/3} = \frac{C_{fx}}{2} \tag{7.59}$$

or

$$C_{fx} = 2St_x\text{Pr}^{2/3} = 2\frac{\text{Nu}_x}{\text{Re}_x\,\text{Pr}}\text{Pr}^{2/3} = 2 \times \frac{0.332\text{Re}_x^{1/2}\text{Pr}^{1/3}}{\text{Re}_x\,\text{Pr}}\text{Pr}^{2/3}$$
$$= \frac{0.664}{\sqrt{\text{Re}_x}},$$

which is the same equation as used above.

Example 7.13 If the velocity and temperature distributions through the laminar boundary layer of a flat plate are linear, show that

$$\frac{\delta}{\delta_t} = (\mathrm{Pr})^{-1/3}$$

and

$$\mathrm{Nu}_x = 0.288(\mathrm{Re}_x)^{1/2}\mathrm{Pr}^{1/3}$$

The plate is maintained at a uniform temperature.

Solution

The velocity and temperature profile equations are

$$\frac{u}{U_\infty} = \frac{y}{\delta}$$

and

$$\frac{T - T_w}{T_\infty - T_w} = \frac{\theta}{\theta_\infty} = \frac{y}{\delta_t} \tag{i}$$

For the linear velocity profile, we have found that

$$\delta\frac{d\delta}{dx} = \frac{1}{0.167}\frac{\mu}{\rho U_\infty} = \frac{6v}{U_\infty} \tag{ii}$$

and

$$\frac{\delta}{x} = \frac{3.46}{\sqrt{\mathrm{Re}_x}} = \sqrt{\frac{12}{\mathrm{Re}_x}} = \sqrt{\frac{12v}{U_\infty x}} \tag{iii}$$

We proceed from the von Karman energy equation, which is

$$U_\infty\theta_\infty\frac{d}{dx}\int_0^{\delta_t}\left(1 - \frac{\theta}{\theta_\infty}\right)\frac{u}{U_\infty}dy = \alpha\left(\frac{\partial T}{\partial y}\right)_{y=0} \tag{iv}$$

Substitution of the values of θ/θ_∞ and u/U_∞ from Eq. (i) gives

$$U_\infty\theta_\infty\frac{d}{dx}\int_0^{\delta_t}\left(1 - \frac{y}{\delta_t}\right)\frac{y}{\delta}dy = \alpha\left[\frac{\partial}{\partial y}\left(\theta_\infty\frac{y}{\delta_t} + T_w\right)\right]_{y=0}$$

or

$$U_\infty\theta_\infty\frac{d}{dx}\left(\frac{y^2}{2\delta} - \frac{y^3}{3\delta\delta_t}\right)_0^{\delta_t} = \alpha\frac{\theta_\infty}{\delta_t}$$

or

$$U_\infty \frac{d}{dx}\left(\frac{\delta_t^2}{2\delta} - \frac{\delta_t^2}{3\delta}\right) = \frac{\alpha}{\delta_t}$$

or

$$U_\infty \frac{d}{dx}\left(\frac{\delta_t^2}{6\delta}\right) = \frac{\alpha}{\delta_t}$$

or

$$U_\infty \frac{d}{dx}\left[\frac{\delta}{6}\left(\frac{\delta_t}{\delta}\right)^2\right] = \frac{\alpha}{\delta_t}$$

or

$$\frac{U_\infty}{6}\frac{d}{dx}\left[\delta\xi^2\right] = \frac{\alpha}{\delta_t}$$

or

$$\frac{U_\infty}{6}\left[\xi^2\frac{d\delta}{dx} + 2\delta\xi\frac{d\xi}{dx}\right] = \frac{\alpha}{\delta_t}$$

or

$$\delta_t\left[\xi^2\frac{d\delta}{dx} + 2\delta\xi\frac{d\xi}{dx}\right] = \frac{6\alpha}{U_\infty}$$

or

$$\delta\xi\left[\xi^2\frac{d\delta}{dx} + 2\delta\xi\frac{d\xi}{dx}\right] = \frac{6\alpha}{U_\infty}$$

or

$$\xi^3\left(\delta\frac{d\delta}{dx}\right) + 2\delta^2\xi^2\frac{d\xi}{dx} = \frac{6\alpha}{U_\infty}$$

Putting the value of $\delta(d\delta/dx)$ from Eq. (ii), we get

$$\xi^3\left(\frac{6\nu}{U_\infty}\right) + 2\delta^2\xi^2\frac{d\xi}{dx} = \frac{6\alpha}{U_\infty}$$

Putting value of δ from Eq. (iii),

$$\xi^3\left(\frac{6v}{U_\infty}\right)+2\left(\frac{12v}{U_\infty x}x^2\right)\xi^2\frac{d\xi}{dx}=\frac{6\alpha}{U_\infty}$$

or

$$\xi^3+4\xi^2 x\frac{d\xi}{dx}=\frac{\alpha}{v}=\frac{1}{\text{Pr}}$$

which is a linear differential equation of first order in ξ^2. Its general solution is

$$\xi^3=Cx^{-3/4}+\frac{1}{\text{Pr}}$$

The constant C is determined from the boundary condition of $\xi = (\delta_t/\delta) = 0$ at $x = 0$ (assuming the heating of the plate from the leading edge of the plate). This gives $C = 0$. Hence,

$$\xi^3=\frac{1}{\text{Pr}}$$

or

$$\xi=\frac{\delta_t}{\delta}=\text{Pr}^{-1/3}$$

The local heat transfer coefficient h_x at the wall is given by

$$h_x=\frac{q_w}{T_w-T_\infty}$$

where the conduction heat transfer at the wall q_w, from the Fourier's law, is

$$q_w=-k\left(\frac{dT}{dy}\right)_{y=0}$$

Hence,

$$h_x=\frac{-k\left(\frac{dT}{dy}\right)_{y=0}}{T_w-T_\infty}$$

The dimensionless temperature distribution at the wall is given as

$$\frac{T-T_w}{T_\infty-T_w}=\frac{\theta}{\theta_\infty}=\frac{y}{\delta_t}$$

This gives the temperature gradient as

$$\left(\frac{dT}{dy}\right)_{y=0} = \frac{d}{dy}\left(\frac{T - T_w}{T_\infty - T_w}\right)_{y=0} \times (T_\infty - T_w) = \frac{d}{dy}\left(\frac{\theta}{\theta_\infty}\right)_{y=0} \times \theta_\infty = \left[\frac{d}{dy}\left(\frac{y}{\delta_t}\right)\right]_{y=0} \times \theta_\infty$$
$$= \frac{\theta_\infty}{\delta_t}$$

Hence,

$$h_x = \frac{-k\left(\frac{\theta_\infty}{\delta_t}\right)}{T_w - T_\infty} = \frac{k}{\delta_t}$$

or

$$h_x = \frac{k}{\delta_t} = \frac{k}{\delta \mathrm{Pr}^{-1/3}} = k\mathrm{Pr}^{1/3}\left(\frac{1}{\delta}\right)$$

Substituting the value of hydrodynamic layer thickness δ,

$$h_x = k\mathrm{Pr}^{1/3}\left(\sqrt{\frac{U_\infty x}{12\nu}}\right) \times \frac{1}{x}$$

or

$$\frac{h_x x}{k} = \frac{1}{\sqrt{12}}\mathrm{Pr}^{1/3}(\mathrm{Re}_x)^{1/2}$$

or

$$\mathrm{Nu}_x = 0.288\mathrm{Pr}^{1/3}(\mathrm{Re}_x)^{1/2}$$

Hence, average value of the Nusselt number is

$$\overline{Nu} = 0.576\mathrm{Pr}^{1/3}(\mathrm{Re}_L)^{1/2}$$

7.8 Turbulent Boundary Layer Over a Flat Surface

In Sect. 7.2, we discussed the transition of the flow from laminar to turbulent. The turbulent boundary layer over a flat plate basically consists of a laminar sublayer, a buffer zone and a turbulent layer, Fig. 7.17a. In the laminar sublayer, the molecular diffusion processes are dominant and the turbulent fluctuations are negligible with the result that the turbulent shear stress is much less than the laminar shear stress, in the buffer zone the molecular diffusion and eddy transport effects are of the same order, and in the turbulent region, the eddy transport effects are dominant and the turbulent shear stress dominates the laminar shear stress. The entire velocity field cannot be represented by a single equation.

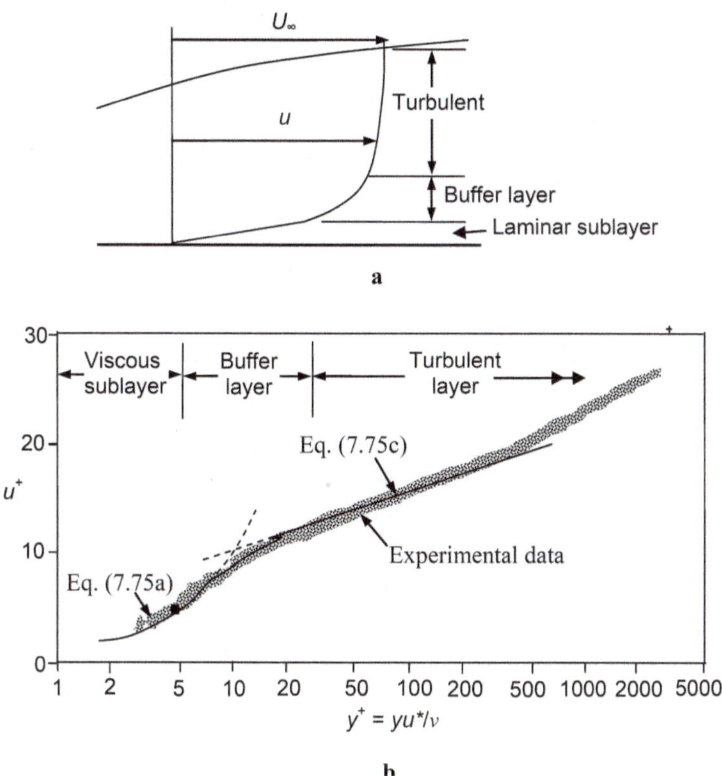

Fig. 7.17 **a** Turbulent boundary layer over a flat plate, **b** turbulent boundary layer velocity profiles in wall coordinates

Figure 7.17b is a semilogarithmic diagram of experimental data of the velocity profile in the neighbourhood of the wall using non-dimensional parameters defined as dimensionless velocity and distance from the plate surface, respectively. Mathematically, they are defined as

$$u^+ = \frac{\bar{u}}{\sqrt{\tau_w/\rho}} \tag{7.73}$$

$$y^+ = \frac{\sqrt{\tau_w/\rho}}{\nu} y \tag{7.74}$$

The term $\surd(\tau_w/\rho)$ frequently appears in turbulent flow analysis and has the dimensions of velocity. It is called the friction velocity and symbol used is u^*. The non-dimensional parameters u^+ and y^+ arise from dimensional analysis.

von Karman proposed the following set of equations termed as *universal velocity profile*, which matches well with the experimental data.

(i) **Laminar sublayer**: Adjacent to the wall, the velocity profile is linear.

$$u^+ = y^+ \quad \text{for} \quad 0 < y^+ < 5 \tag{7.75a}$$

(ii) **Transition or buffer region:**

$$u^+ = 5\ln(y^+) - 3.05 \quad \text{for} \quad 5 < y^+ < 30 \tag{7.75b}$$

(iii) **Turbulent core:** In the turbulent layer ($y^+ > 30$), the following logarithmic equation fit the experimental data:

$$u^+ = 2.5\ln(y^+) + 5.5 \quad \text{for} \quad 30 < y^+ < 500 \tag{7.75c}$$

The logarithmic equation is generally known as *law of wall* (Kays and Crawford 1980). A number of other equations have been proposed. They are presented in a tabular form in Bejan (1995).

The equations of the universal profile are too complex mathematically to use with the momentum integral equation. A suitable velocity profile for turbulent boundary layers over smooth plates is the empirical power law profile of the form $(u/U_\infty) = (y/\delta)^{1/n}$. Experimental investigations have shown that an exponent of 1/7 can be used for $5 \times 10^5 < \text{Re} < 10^7$. Thus

$$\frac{u}{U_\infty} = \left(\frac{y}{\delta}\right)^{1/7} \quad \text{for} \quad 5 \times 10^5 < \text{Re} < 10^7 \tag{7.76}$$

This is known as (1/7)th power law.

The differentiation of the above relation with respect to y yields

$$\frac{\partial u}{\partial y} = \frac{1}{7}\left(\frac{1}{\delta}\right)^{1/7}\left(\frac{1}{y}\right)^{6/7} U_\infty$$

This leads to infinite value of the wall shear stress when $y \to 0$ knowing that $\tau_w = \mu\,(\partial u/\partial y)_{y=0}$, which is not an acceptable value. In fact, the turbulent boundary layer is basically composed of two distinct regions. There is a thin layer next to the wall in which the flow is laminar. This layer is known as *laminar sublayer*. In the laminar sublayer, a linear velocity distribution is assumed. This is a reasonable assumption because the laminar sublayer is very thin. Beyond the laminar sublayer, the fully turbulent region exists in which the 1/7th power law is applicable. Since the laminar sublayer is very thin, the integral momentum analysis can be carried out using Eq. (7.76) for the velocity distribution. To evaluate the wall shear stress, an experimentally determined relation is required. One such relation is due to Blasius for the turbulent flow on flat plates:

$$C_{fx} = \frac{\tau_w}{\frac{1}{2}\rho U_\infty^2} = 0.0456\left(\frac{\nu}{U_\infty \delta}\right)^{1/4} \quad \text{for} \quad 5 \times 10^5 < \text{Re} < 10^7 \tag{7.77}$$

or

$$\tau_w = 0.0228\left(\frac{\nu}{U_\infty \delta}\right)^{1/4} \rho U_\infty^2 \tag{7.78}$$

where U_∞ is the free-stream velocity and δ is the thickness of the turbulent boundary layer. The term on the right $(\nu/U_\infty\delta)$ is known as the *thickness Reynolds number*.

The substitution of the values of u/U_∞ and τ_w from Eqs. (7.76) and (7.78), respectively, in the von Karman momentum integral equation, Eq. (7.39), yields

$$\rho U_\infty^2 \frac{d}{dx}\left\{\int_0^\delta \left[1-\left(\frac{y}{\delta}\right)^{1/7}\right]\left(\frac{y}{\delta}\right)^{1/7} dy\right\} = 0.0228\left(\frac{\nu}{U_\infty\delta}\right)^{1/4}\rho U_\infty^2$$

$$\frac{d}{dx}\left\{\int_0^\delta \left[1-\left(\frac{y}{\delta}\right)^{1/7}\right]\left(\frac{y}{\delta}\right)^{1/7} dy\right\} = 0.0228\left(\frac{\nu}{U_\infty\delta}\right)^{1/4}$$

Integration gives

$$\frac{d}{dx}\left(\frac{7}{72}\delta\right) = 0.0228\left(\frac{\nu}{U_\infty\delta}\right)^{1/4}$$

or

$$\delta^{1/4}\frac{d\delta}{dx} = \frac{72}{7}\times 0.0228\left(\frac{\nu}{U_\infty}\right)^{1/4}$$

or

$$\delta^{5/4} = 0.293\left(\frac{\nu}{U_\infty}\right)^{1/4} x + C \tag{7.79}$$

The constant of integration is found from the assumed boundary conditions as presented below.

Case A

The boundary layer is fully turbulent from the leading edge of the plate, i.e. $\delta = 0$ at $x = 0$. This gives C = 0 and

$$\delta^{5/4} = 0.293\left(\frac{\nu}{U_\infty}\right)^{1/4} x \tag{7.80a}$$

or

$$\frac{\delta}{x} = 0.375\left(\frac{\nu}{U_\infty x}\right)^{1/5}$$

or

$$\frac{\delta}{x} = 0.375(\mathrm{Re}_x)^{-1/5} \tag{7.80b}$$

The comparison of the above equation with Eq. (7.35) for the laminar boundary layer shows that the turbulent boundary layer grows at a faster rate ($\delta \propto x^{4/5}$) than a laminar boundary layer ($\delta \propto x^{1/2}$).

Substitution of the value of δ from Eq. (7.80b) in Eq. (7.77) gives

$$\begin{aligned} C_{fx} &= 0.0456\left(\frac{\nu}{U_\infty}\right)^{1/4}\left[\left(\frac{U_\infty x}{\nu}\right)^{1/5}\left(\frac{1}{x}\right)\right]^{1/4}\left(\frac{1}{0.375}\right)^{1/4} \\ &= \frac{0.0583}{\mathrm{Re}_x^{1/5}} \end{aligned} \tag{7.81}$$

The average value of the skin friction coefficient,

$$\begin{aligned} \overline{C_f} &= \frac{1}{L}\int_0^L C_{fx}dx \\ &= \frac{1}{L}\int_0^L 0.0583\left(\frac{\nu}{U_\infty}\right)^{1/5} x^{-1/5}dx, \end{aligned}$$

which gives

$$\overline{C_f} = \frac{0.0729}{\left(\frac{U_\infty L}{\nu}\right)^{1/5}}$$

or

$$\overline{C_f} = \frac{0.0729}{\mathrm{Re}_L^{1/5}} \tag{7.82}$$

The result has been found to be in a good agreement with the experimental results in the range $5 \times 10^5 < \mathrm{Re} < 10^7$.

Case B

The above analysis is based on the assumption that the boundary layer is fully turbulent from the leading edge of the plate. This is not true. The boundary layer is laminar up to a certain distance from the leading edge and the transition, as discussed earlier, to turbulent takes place at $x = x_{cr}$ corresponding to $\mathrm{Re}_c = 5 \times 10^5$. Thus

$$\delta = \delta_{\mathrm{lam}} \quad \text{at } x_{cr} = 5 \times 10^5(\nu/U_\infty)$$

From the exact solution of the Blasius, Eq. (7.35),

$$\delta_{lam} = 5.0\left(\frac{x_{cr}}{\mathrm{Re}_c^{1/2}}\right) = 5.0\left[\frac{5\times 10^5\left(\frac{\nu}{U_\infty}\right)}{(5\times 10^5)^{1/2}}\right] = 3535\left(\frac{\nu}{U_\infty}\right), \tag{7.83}$$

which must equal the boundary layer thickness at the start of the turbulent regime. Using Eqs. (7.83) and (7.79) and substituting value of x_{cr}, we have

$$\left(3535\frac{\nu}{U_\infty}\right)^{5/4} = 0.293\left(\frac{\nu}{U_\infty}\right)^{1/4}\left[5\times 10^5\left(\frac{\nu}{U_\infty}\right)\right] + C$$

Simplification gives

$$C = -119242\left(\frac{\nu}{U_\infty}\right)^{5/4}$$

Substituting value of constant C in Eq. (7.79), we get

$$\delta^{5/4} = 0.293\left(\frac{\nu}{U_\infty}\right)^{1/4} x - 119242\left(\frac{\nu}{U_\infty}\right)^{5/4}$$

or

$$\delta = 0.375\left(\frac{\nu}{U_\infty}\right)^{1/5} x^{4/5} - 11511\left(\frac{\nu}{U_\infty}\right)$$

or

$$\frac{\delta}{x} = 0.375\left(\frac{\nu}{U_\infty x}\right)^{1/5} - 11511\left(\frac{\nu}{U_\infty x}\right)$$

or

$$\frac{\delta}{x} = \frac{0.375}{\mathrm{Re}_x^{1/5}} - \frac{11511}{\mathrm{Re}_x} \tag{7.84}$$

This equation is valid for $5\times 10^5 < \mathrm{Re} < 10^7$.

Experiment results indicate that between $10^7 < \mathrm{Re} < 10^9$, the velocity distribution deviates from the 1/7th power law. Schlichting presented the following semi-empirical relation for this range.

$$\overline{C_f} = \frac{0.455}{(\log_{10}\mathrm{Re}_L)^{2.58}} \quad \text{for } 10^7 < \mathrm{Re} < 10^9 \tag{7.85a}$$

Equation (7.85a) was derived assuming the flow to be turbulent over its entire length. In fact, starting from the leading edge the flow is laminar for some distance. Considering this fact, the following relation has been suggested for $10^7 < \text{Re} < 10^9$:

$$\overline{C_f} = \frac{0.455}{(\log_{10}\text{Re}_L)^{2.58}} - \frac{A}{\text{Re}_L} \tag{7.85b}$$

where the value of A depends on the value of the critical Reynolds number at which the transition to turbulent flow takes place. For $\text{Re}_c = 5 \times 10^5$, $A \approx 1700$.

It is to note that the above-presented correlations are valid for a smooth flat plate.

Example 7.14 Determine the relationship between the boundary layer thickness and displacement thickness for a boundary layer which is (i) laminar throughout, (ii) turbulent throughout. Assume that in the laminar boundary layer, the flow obeys the law $\tau = \mu\, du/dy$, which leads to the velocity profile $(U_\infty - u) = \text{C}\,(\delta - y)^2$ where U_∞ is the free-stream velocity and u is the velocity at distance y from the plate and C is a constant. The velocity distribution in the turbulent flow is given by $(u/U_\infty) = (y/\delta)^{1/7}$.

Solution

(i) **Laminar flow**

The equation given for the velocity profile is

$$(U_\infty - u) = C(\delta - y)^2, \tag{i}$$

which can be written as

$$1 - \frac{u}{U_\infty} = \frac{C(\delta - y)^2}{U_\infty} \tag{ii}$$

Velocity displacement thickness

$$\delta_{vd} = \int_0^\infty \left(1 - \frac{u}{U_\infty}\right) dy$$

Using Eq. (ii),

$$\delta_{vd} = \int_0^\delta \left(\frac{C}{U_\infty}\right)(\delta - y)^2 dy$$

$$\text{(the limit has been changed to } \delta\text{)}$$

or

$$\delta_{vd} = \left(\frac{C}{U_\infty}\right)\left[-\frac{(\delta - y)^3}{3}\right]_0^\delta$$
$$= \left(\frac{C\delta^3}{3U_\infty}\right)$$

At the plate surface, the flow velocity is zero, i.e. $u = 0$ at $y = 0$. Hence, from Eq. (i),

$$U_\infty = C\delta^2$$

or

$$C = \tfrac{U_\infty}{\delta^2}$$
$$\delta_{vd} = \left(\tfrac{U_\infty \delta^3}{3U_\infty \delta^2}\right) = \tfrac{\delta}{3}.$$

(ii) **Turbulent flow**

The velocity distribution is given as $(u/U_\infty) = (y/\delta)^{1/7}$. Substitution in the equation of velocity displacement equation gives

$$\delta_{vd} = \int_0^\delta \left(1 - \frac{u}{U_\infty}\right) dy$$

$$\text{(the limit has been changed to } \delta\text{)}$$

$$\delta_{vd} = \int_0^\delta \left[1 - \left(\frac{y}{\delta}\right)^{1/7}\right] dy$$
$$= \left[y - \left(\frac{7y^{8/7}}{8\delta^{1/7}}\right)\right]_0^\delta = \frac{\delta}{8}.$$

7.9 Laminar Flow in Tubes

Fluid flow in tubes has been discussed earlier. The velocity distribution in terms of the tube radius R and the velocity at the centre can be derived as follows.

Consider the fluid element shown in Fig. 7.18. The pressure forces acting on the elemental area are balanced by the viscous shear forces.

Pressure force on the left face of the element $= \pi r^2 p$

Pressure force on the right face of the element $= \pi r^2 p + \frac{d}{dx}(\pi r^2 p)dx$

Viscous shear force on the element is

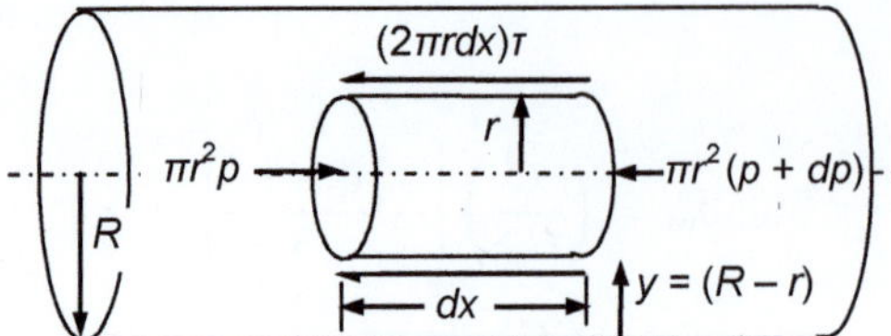

Fig. 7.18 Pressure and viscous shear forces on fluid element

$$\tau A = \tau(2\pi r dx)$$

Force balance equation is

$$\pi r^2 p - \left[\pi r^2 p + \frac{d}{dx}\left(\pi r^2 p\right)dx\right] = (2\pi r dx)\tau$$

or

$$\frac{dp}{dx} = -\frac{2\tau}{r}$$

Putting $\tau = -\mu \, du/dr$, where the negative sign indicates decrease in the velocity with the increase in r.

$$\frac{dp}{dx} = \frac{2\mu}{r}\frac{du}{dr}$$

or

$$\frac{du}{dr} = \frac{1}{2\mu}\frac{dp}{dx}r$$

or

$$u = \frac{1}{4\mu}\left(\frac{dp}{dx}\right)r^2 + C \tag{7.86}$$

The fluid velocity is zero at the wall, i.e.

$$u = 0 \,\text{at}\, r = R$$

This boundary condition gives

$$C = -\frac{1}{4\mu}\left(\frac{dp}{dx}\right)R^2$$

Hence, from Eq. (7.86),

$$\begin{aligned} u &= \frac{1}{4\mu}\left(\frac{dp}{dx}\right)r^2 - \frac{1}{4\mu}\left(\frac{dp}{dx}\right)R^2 \\ &= \frac{1}{4\mu}\left(\frac{dp}{dx}\right)(r^2 - R^2) \end{aligned} \tag{7.87}$$

The velocity at the centreline of the tube is maximum and is obtained from the above equation by putting $r = 0$,

$$U_{\max} = (u)_{r=0} = \frac{1}{4\mu}\left(\frac{dp}{dx}\right)(-R^2) \tag{7.88}$$

Dividing Eq. (7.87) by Eq. (7.88), we get

$$\frac{u}{U_{\max}} = \left(1 - \frac{r^2}{R^2}\right) \tag{7.89}$$

Thus, the velocity profile of the fully developed laminar flow is of parabolic form.

The mean velocity of the flow is calculated by dividing the volumetric flow rate of the fluid by the cross-sectional area of the tube,

$$U_{\mathrm{m}} = \frac{V}{A} = \frac{1}{A}\int_0^A u.dA \tag{i}$$

The volume flow through the annulus of radial width dr at radius r from centreline (Fig. 7.19) is

$$dV = (2\pi r dr)u$$

Hence, the volume flow rate,

$$V = \int dV = \int_0^R 2\pi ur.dr \tag{ii}$$

Using Eq. (7.89),

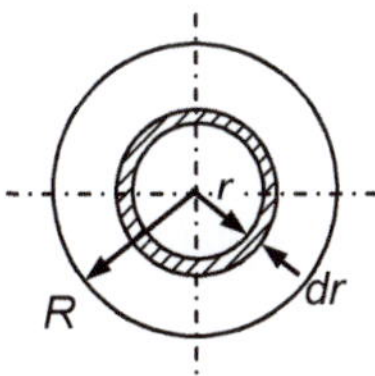

Fig. 7.19 Coordinate system for Eq. (7.89)

$$V = \int_0^R 2\pi\left(1 - \frac{r^2}{R^2}\right) U_{\max} r.dr$$

$$= \frac{1}{2}\pi R^2 U_{\max} = 0.5 A U_{\max} \tag{iii}$$

This gives

$$U_m = \frac{V}{A} = 0.5 U_{\max}$$

or

$$\left(\frac{U_m}{U_{\max}}\right)_{\text{laminar}} = 0.5 \tag{7.90}$$

i.e. for the fully developed laminar flow through a tube, the ratio of the mean velocity and maximum velocity at the centreline of the tube is 0.5.

Substitution of the value of $U_{\max}$ from Eq. (7.88) in Eq. (iii) gives

$$V = \pi R^2 \left[\frac{R^2}{8\mu}\left(-\frac{dp}{dx}\right)\right]$$

If pressure drop in pipe length L is $(p_1 - p_2)$, the equation yields

$$V = \frac{\pi R^4}{8\mu}\left(\frac{p_1 - p_2}{L}\right), \tag{7.91}$$

which is known as *Hagen–Poiseuille equation.*

The local pipe flow friction factor (the *Fanning friction factor*) f is defined in the terms of the mean velocity U_m as

$$-\frac{dp}{dx} = \frac{4f}{D}\left(\frac{\rho U_m^2}{2}\right)$$

The value of the friction factor f for the fully developed steady laminar flow can be determined by substituting value of dp/dx from above equation in Eq. (7.88). This substitution gives

$$U_{\max} = \frac{R^2}{4\mu}\left[\frac{4f}{D}\left(\frac{\rho U_m^2}{2}\right)\right]$$

Further substituting $U_{\max} = 2\,U_m$, we obtain

$$f = \frac{16}{\rho U_m D/\mu}$$

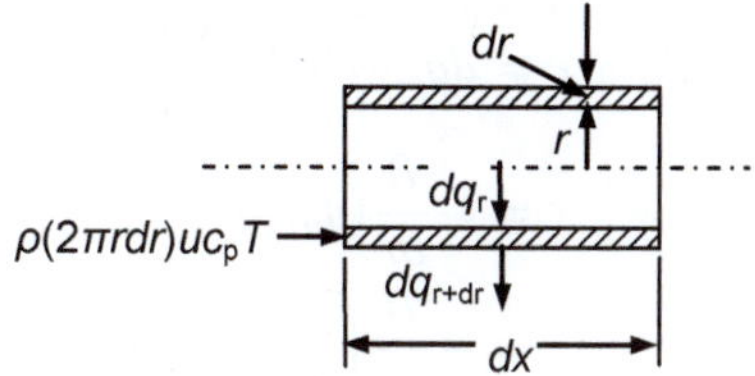

Fig. 7.20 Energy conduction and convection for the annular element

or

$$f = \frac{16}{\mathrm{Re}}, \tag{7.92}$$

which is equation of *Fanning friction factor* for the fully developed constant property laminar flow in smooth circular cross-section tube. The local friction factor in the entrance length of circular tube with laminar flow will be discussed in Chap. 8.

The temperature distribution may be obtained by making an energy balance of the energy conducted and convected for the annular element in Fig. 7.20.

The following is assumed:

(i) A constant heat flux at the tube wall, i.e.

$$\frac{dq_w}{dx} = 0$$

This condition can be approached if the tube is heated by passing electric current through it.

(ii) The conduction in the axial direction is negligible.
(iii) The energy transport in the axial direction is entirely by convection.

Radial heat flow by conduction into the annular element is

$$dq_r = -k(2\pi r dx).\frac{\partial T}{\partial r}$$

and heat outflow by conduction from the annular element

$$dq_{r+dr} = dq_r + \frac{\partial}{\partial r}(dq_r)dr$$

Net heat conducted is

$$\begin{aligned} dq_r - dq_{r+dr} &= dq_r - \left[dq_r + \frac{\partial}{\partial r}(dq_r)dr\right] \\ &= -\frac{\partial}{\partial r}(dq_r)dr \\ &= -\frac{\partial}{\partial r}\left(-k(2\pi rdx).\frac{\partial T}{\partial r}\right)dr \\ &= 2\pi kdxdr\frac{\partial}{\partial r}\left(r.\frac{\partial T}{\partial r}\right) \end{aligned}$$

Heat convected into the annular element in axial direction is

$$\begin{aligned} dq_c &= \text{mass flow rate} \times c_p T \\ &= \rho u(2\pi rdr).c_p T \end{aligned}$$

Heat convected out of the annular element in axial direction is

$$dq_c + \frac{\partial}{\partial x}(dq_c)dx$$

Net heat convected out of the element is

$$\begin{aligned} dq_c + \frac{\partial}{\partial x}(dq_c)dx - dq_c &= \frac{\partial}{\partial x}(dq_c)dx \\ &= \frac{\partial}{\partial x}\left[\rho u(2\pi rdr).c_p T\right]dx \\ &= \rho u(2\pi rdr).c_p\frac{\partial T}{\partial x}dx \end{aligned}$$

The energy balance over the element gives

$$2\pi kdxdr\frac{\partial}{\partial r}\left(r.\frac{\partial T}{\partial r}\right) = \rho u(2\pi rdr).c_p\frac{\partial T}{\partial x}dx$$

or

$$\frac{\partial}{\partial r}\left(r.\frac{\partial T}{\partial r}\right) = \frac{1}{\alpha}\frac{\partial T}{\partial x}ur \tag{7.93}$$

where $\rho c_p/k = (1/\alpha)$.

For the assumed condition of constant heat flux, the average fluid temperature will rise linearly in the axial direction, i.e. $T \propto x$ and

$$\frac{\partial T}{\partial x} = \text{constant.} \tag{i}$$

The temperature profile is such that the minimum temperature of the fluid for heating of the fluid and the maximum temperature of the fluid for cooling of the fluid must be at the axis of the tube for uniform heat flux at the wall, i.e.

$$\frac{\partial T}{\partial r} = 0 \quad \text{at } r = 0. \tag{ii}$$

Inserting the value of u from the velocity distribution given by Eq. (7.89) into Eq. (7.93), we get

$$\frac{\partial}{\partial r}\left(r.\frac{\partial T}{\partial r}\right) = \frac{1}{\alpha}\left(\frac{\partial T}{\partial x}\right) rU_{\max}\left[1 - \left(\frac{r}{R}\right)^2\right]$$

Integrating with respect to r (keeping $\partial T/\partial x$ = constant),

$$r.\frac{\partial T}{\partial r} = \frac{1}{\alpha}\left(\frac{\partial T}{\partial x}\right) U_{\max}\left[\frac{r^2}{2} - \frac{r^4}{4R^2}\right] + C_1$$

The boundary condition, Eq. (ii), gives $C_1 = 0$. Hence,

$$\frac{\partial T}{\partial r} = \frac{1}{\alpha}\left(\frac{\partial T}{\partial x}\right) U_{\max}\left[\frac{r}{2} - \frac{r^3}{4R^2}\right] \tag{7.94}$$

Again integrating it, we get

$$T = \frac{1}{\alpha}\left(\frac{\partial T}{\partial x}\right) U_{\max}\left[\frac{r^2}{4} - \frac{r^4}{16R^2}\right] + C_2$$

Let the temperature at the centre of the tube, i.e. at $r = 0$, is T_c. This condition gives $C_2 = T_c$ and the equation of the temperature distribution may finally be written as

$$T - T_c = \frac{1}{\alpha}\left(\frac{\partial T}{\partial x}\right) U_{\max}\left(\frac{R^2}{4}\right)\left[\left(\frac{r}{R}\right)^2 - \frac{1}{4}\left(\frac{r}{R}\right)^4\right] \tag{7.95}$$

Convection heat transfer coefficient

For the tube flow, the convection heat transfer coefficient is defined by

$$h = \frac{q_w}{T_w - T_b} \tag{i}$$

where T_b is known as the bulk temperature of the fluid. The bulk temperature is basically the temperature of the fluid that the fluid would assume if it is thoroughly mixed. It is calculated from

$$\begin{aligned} T_b &= \frac{\text{Total energy flow through the tube}}{(\text{mass flow} \times \text{specific heat})} \\ &= \frac{\int_0^R \rho u(2\pi r dr).c_p T}{\int_0^R \rho u(2\pi r dr).c_p} \\ &= \frac{\int_0^R (rdr)uT}{\int_0^R u(rdr)} \end{aligned}$$

Substitution of the values of u and T from Eqs. (7.89) and (7.95), respectively, gives

$$T_b = \frac{\int_0^R (rdr)U_{max}\left(1 - \frac{r^2}{R^2}\right)\left\{\left\{\frac{1}{\alpha}\left(\frac{\partial T}{\partial x}\right)U_{max}\left(\frac{R^2}{4}\right)\left[\left(\frac{r}{R}\right)^2 - \frac{1}{4}\left(\frac{r}{R}\right)^4\right]\right\} + T_c\right\}}{\int_0^R (rdr)U_{max}\left(1 - \frac{r^2}{R^2}\right)}$$

or

$$T_b = \frac{\frac{1}{\alpha}\left(\frac{\partial T}{\partial x}\right)U_{max}\left(\frac{R^2}{4}\right)\int\limits_0^R r\left(1 - \frac{r^2}{R^2}\right)\left[\left(\frac{r}{R}\right)^2 - \frac{1}{4}\left(\frac{r}{R}\right)^4\right]dr}{\int_0^R \left(1 - \frac{r^2}{R^2}\right)rdr} + T_c$$

or

$$T_b = \frac{7}{96}\left(\frac{\partial T}{\partial x}\right)\left(\frac{U_{max}R^2}{\alpha}\right) + T_c$$

Putting $r = R$ in Eq. (7.95), we get

$$T_w = \frac{3}{16}\left(\frac{\partial T}{\partial x}\right)\left(\frac{U_{max}R^2}{\alpha}\right) + T_c$$

Substituting values of T_w and T_b in Eq. (i) and knowing that q_w is also given by $k(\partial T/\partial r)_{r=R}$, we get

$$h = \frac{k\left(\frac{\partial T}{\partial r}\right)_{r=R}}{\left[\frac{11}{96}\left(\frac{\partial T}{\partial x}\right)\frac{U_{max}R^2}{\alpha}\right]}$$

From Eq. (7.94),

$$\frac{\partial T}{\partial r} = \frac{1}{\alpha}\left(\frac{\partial T}{\partial x}\right)U_{max}\left[\frac{r}{2} - \frac{r^3}{4R^2}\right]$$

Hence, $(\partial T/\partial r)_{r=R} = (\partial T/\partial x)U_{max}R/4\alpha$. This gives

$$h = \frac{k\left(\frac{\partial T}{\partial x}\right)\frac{U_{max}R}{4\alpha}}{\left[\frac{11}{96}\left(\frac{\partial T}{\partial x}\right)\frac{U_{max}R^2}{\alpha}\right]} = \frac{24\,k}{11\,R} = \frac{48\,k}{11\,d}$$

or

$$\frac{hd}{k} = 4.364$$

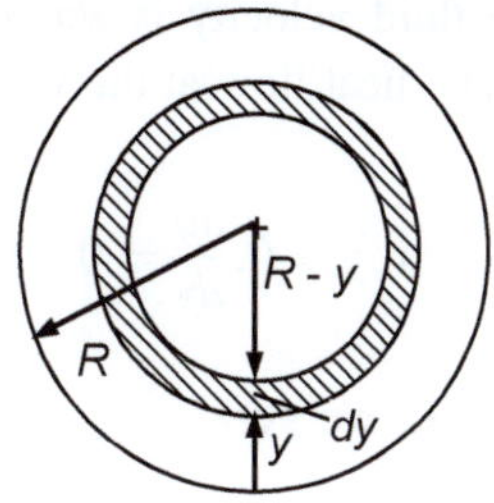

Fig. 7.21 Coordinate system for Eq. (i)

or

$$\mathrm{Nu} = 4.364 \tag{7.96}$$

for the uniform heat flux condition.

Boundary layer condition of constant wall temperature is also of interest. This condition is encountered in tubes heated on the outside by a condensing vapour. It has been reported that for the constant wall temperature condition,

$$\mathrm{Nu} = 3.658. \tag{7.97}$$

Approximate analysis

In the laminar flow, the velocity profile is given by, for the coordinate system defined in Fig. 7.21,

$$\frac{u}{U_{max}} = 2\left(\frac{y}{R}\right) - \left(\frac{y}{R}\right)^2 \tag{i}$$

where u is the velocity at a distance y from the wall and $U_{\max}$ is the velocity at the axis of the tube.

Let the temperature profile be approximated by a cubic parabola

$$T = T_w + ay + by^2 + cy^3$$

Let $T - T_w = \theta$, then

$$\theta = ay + by^2 + cy^3 \tag{ii}$$

The radial heat conduction at any radius $(R - y)$ is

$$q = -k[2\pi(R - y)dx]\frac{d\theta}{dy}$$

Differentiating with respect to y, we get

$$\frac{dq}{dy} = -2\pi k dx\left[-\frac{d\theta}{dy} + (R - y)\frac{d^2\theta}{dy^2}\right] \tag{iii}$$

In the vicinity of the wall, the fluid velocity is zero and, therefore, there is negligible convection in the axial direction. The heat flow at the wall is due to conduction only and is independent of y, i.e.

$$\text{at } y = 0, \frac{dq}{dy} = 0$$

Applying this condition to Eq. (iii), we get

$$\left(\frac{d^2\theta}{dy^2}\right)_{y=0} = \frac{1}{R}\left(\frac{d\theta}{dy}\right)_{y=0} \tag{iv}$$

From the assumed temperature profile, Eq. (ii),

$$\left(\frac{d\theta}{dy}\right)_{y=0} = a \tag{v}$$

and

$$\left(\frac{d^2\theta}{dy^2}\right)_{y=0} = 2b. \tag{vi}$$

Substitution of these values in Eq. (iv) gives

$$a = 2Rb. \tag{vii}$$

At the axis of the tube ($y = R$),

$$\theta = \theta_c$$

and

$$\left(\frac{d\theta}{dy}\right)_{y=R} = 0.$$

Applying these conditions to the assumed temperature profile, we get

$$\theta_c = aR + bR^2 + cR^3, \tag{viii}$$

$$\left(\frac{d\theta}{dy}\right)_{y=R} = 0 = a + 2bR + 3cR^2. \tag{ix}$$

Solution of Eqs. (vii) to (ix) gives

$$a = \frac{6\theta_c}{5R}, \quad b = \frac{3\theta_c}{5R^2}, \quad c = -\frac{4\theta_c}{5R^3}.$$

Substitution in Eq. (ii) gives

$$\frac{T - T_w}{\theta_c} = \frac{\theta}{\theta_c} = \frac{6}{5}\left(\frac{y}{R}\right) + \frac{3}{5}\left(\frac{y}{R}\right)^2 - \frac{4}{5}\left(\frac{y}{R}\right)^3. \tag{x}$$

Differentiating the above equation with respect to y at $y = 0$, we get

$$\left(\frac{\partial \theta}{\partial y}\right)_{y=0} = \frac{6\theta_c}{5R}$$

and the rate of heat flow at the wall

$$q_w = -k\left(\frac{\partial T}{\partial r}\right)_{y=0} = -k\left(\frac{\partial \theta}{\partial y}\right)_{y=0} = -k\frac{6\theta_c}{5R}.$$

Hence, the convection heat transfer coefficient is

$$h = \frac{q_w}{T_w - T_b} = -k\frac{6\theta_c}{5R}\frac{1}{T_w - T_b}$$

The bulk temperature, refer Fig. 7.20,

$$T_b = \frac{\text{Total energy flow through the tube}}{\text{mass flow} \times \text{specific heat}}$$

$$= \frac{\int_0^R \rho 2\pi (R - y)dy.u.c_p T}{\int_0^R \rho 2\pi (R - y)dy.uc_p} = \frac{\int_0^R (R - y)dy.u.T}{\int_0^R (R - y)dy.u}$$

Substituting value of u from Eq. (i) and value of T from Eq. (x), we have

$$T_b = \frac{\int_0^R (R - y)dy.U_{max}\left[2\left(\frac{y}{R}\right) - \left(\frac{y}{R}\right)^2\right]\left\{\left[\frac{6}{5}\left(\frac{y}{R}\right) + \frac{3}{5}\left(\frac{y}{R}\right)^2 - \frac{4}{5}\left(\frac{y}{R}\right)^3\right]\theta_c + T_w\right\}}{\int_0^R (R - y)dy.U_{max}\left[2\left(\frac{y}{R}\right) - \left(\frac{y}{R}\right)^2\right]}$$

Simplification gives

$$T_b = \frac{102}{175}\theta_c + T_w$$

or

$$T_b - T_w = \frac{102}{175}\theta_c.$$

Thus, the heat transfer coefficient is

$$h = \left(\frac{6k\theta_c}{5R}\right) / \left(\frac{102\theta_c}{175}\right) = \frac{4.12k}{D}$$

$$\frac{hD}{k} = 4.12.$$

Thus, the heat transfer correlation is

$$\text{Nu} = 4.12,$$

which is in reasonable agreement with the value of the exact solution.

The above-calculated values of the Nusselt number for the laminar flow in the tubes are valid for fully developed flow. The estimate of the heat transfer coefficient in the entrance region of the tube, where the flow is developing, is complicated and the following must be noted:

1. The temperature gradient at the wall is infinite at the beginning of the heated section and Nu = ∞ (Rogers and Mayhew 1967). In the development region, starting from the beginning, the value of the heat transfer coefficient (and hence the Nusselt number) decreases and approaches the value for the fully developed flow, refer Fig. 7.4.
2. The value of the Nusselt number in the entrance region also depends on the boundary conditions along the tube wall, i.e. whether at the beginning of the heated section the velocity profile is fully developed or developing simultaneously with the temperature profile.
3. The temperature profile may be distorted due to the natural convection currents at low Reynolds numbers especially in large diameter tubes.

Further, it is to note that the viscosity of the fluid depends on temperature and hence the velocity profile is not parabolic as discussed in Sect. 8.6. Since the temperature profile is a function of the velocity profile, the temperature profile will also change. Thus, the above-calculated values of the Nusselt number are valid for isothermal flow only, i.e. for the limiting case of heat transfer by convection as temperature difference ($T_w - T_b$) approaches zero (Rogers and Mayhew 1967).

For the details and treatment for the above-stated conditions, the readers may refer Kays and Crawford (1980), and Kays and Perkins (1985).

Example 7.15 A fluid flows at a steady mass flow rate m inside a tube of uniform section. Express the local axial rate of change of bulk temperature of the fluid, dT_b/dx, in terms of the mass flow rate m, tube perimeter P, the specific heat of the fluid c_p and the local surface heat flux q_x.

Solution

Considering the infinitesimal control volume, refer Fig. 7.22, the heat flow from the first law of thermodynamics is

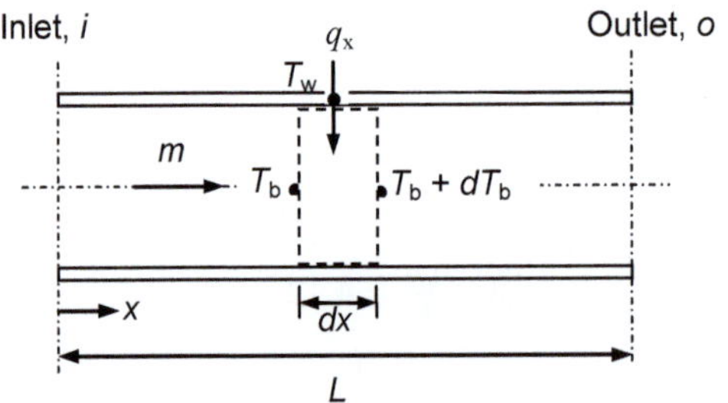

Fig. 7.22 Example 7.15

$$dq = mc_p dT_b$$

where $dq = q_x(Pdx)$. This gives

$$q_x(Pdx) = mc_p dT_b$$

or

$$\frac{dT_b}{dx} = \frac{q_x P}{mc_p}$$

From the heat transfer consideration,

$$q_x = h_x(T_w - T_b)$$

This gives

$$\frac{dT_b}{dx} = \left(\frac{P}{mc_p}\right) h_x(T_w - T_b)$$

Example 7.16 Using the differential equation of the previous example, obtain an expression for change of the bulk temperature with distance x for the cases of (i) constant surface heat flux, $q_x = q_s$, (ii) uniform tube surface temperature, T_w = constant.

Solution

(i) **Uniform heat Flux**

The first differential equation is

$$\frac{dT_b}{dx} = \frac{q_x P}{mc_p}$$

Integrating for the fixed $q_x = q_s$, P, m and c_p,

$$\int_{T_{bi}}^{T_b} dT_b = \frac{q_s P}{mc_p} \int_0^x dx$$

or

$$T_b = T_{bi} + \frac{q_s P}{mc_p} x$$

This implies that the bulk temperature increases linearly with x as depicted in Fig. 7.23a, where the flow has been assumed to be thermally developed.

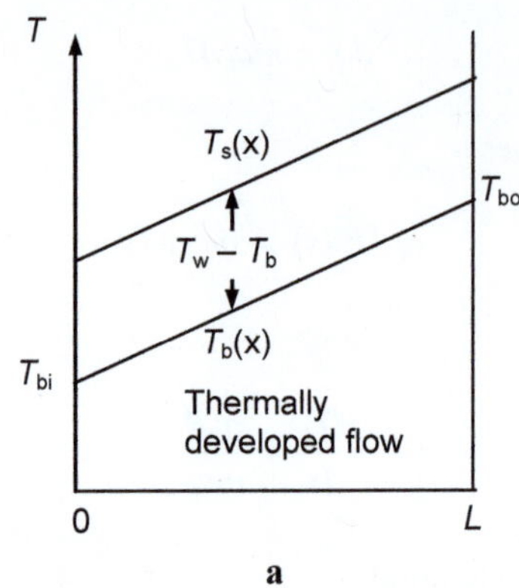

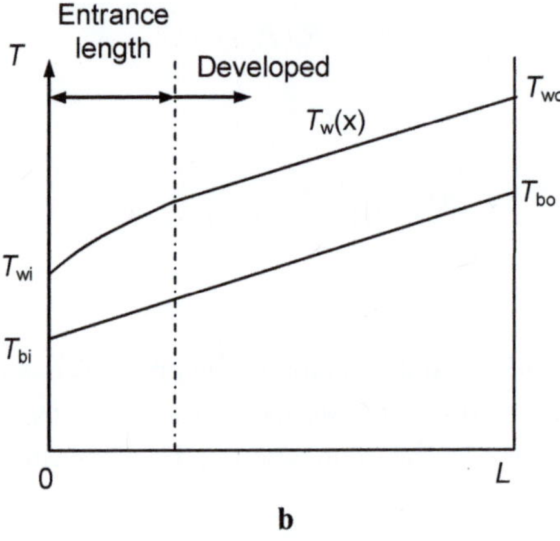

Fig. 7.23 Temperature variation for constant heat rate: **a** thermally developed flow, **b** effect of entrance length

It is to be noted that, in the entrance region, ($T_w - T_b$) is not a constant as shown in Fig. 7.23b.

(ii) **Uniform Surface Temperature**

The second differential equation is

$$\frac{dT_b}{dx} = \left(\frac{P}{mc_p}\right)h_x(T_w - T_b).$$

which gives

$$\int_{T_{bi}}^{T_b} \frac{dT_b}{T_w - T_b} = \frac{P}{mc_p}\int_0^x h_x dx.$$

From the definition of the average heat transfer coefficient,

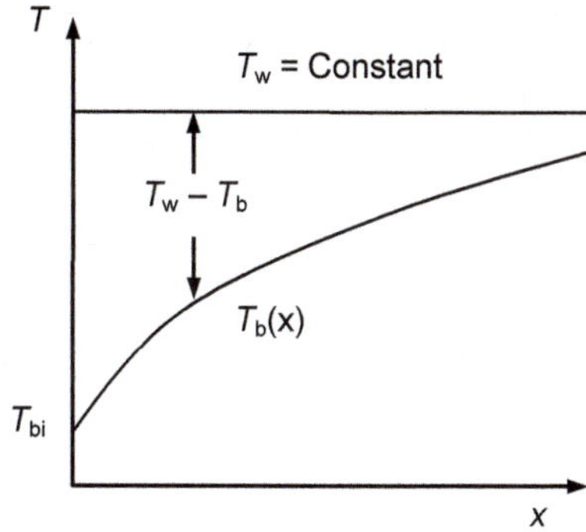

Fig. 7.24 Temperature variation of a fluid flowing through a tube with uniform wall temperature

$$\bar{h}(0,x) = \frac{1}{x}\int_0^x h_x dx$$

This gives

$$\ln\left(\frac{T_w - T_b}{T_w - T_{bi}}\right) = -\frac{P}{mc_p} x\bar{h}(0,x)$$

or

$$T_b = T_w - (T_w - T_{bi})\exp\left(-\frac{P}{mc_p} x\bar{h}(0,x)\right)$$

The function is plotted in Fig. 7.24.

Example 7.17 Determine the total heat transfer for the tube length L for the cases considered in the previous example.

Solution

(i) **Uniform Heat Flux**
Heat transfer rate,

$$\begin{aligned} q &= \text{heat flux} \times \text{area} \\ &= q_s PL \end{aligned}$$

When $(T_w - T_b)$ is constant,

$$q_s = h(T_w - T_b)$$

Thus,

$$q = hPL(T_w - T_b),$$

which is the desired equation.

(ii) **Uniform Surface Temperature**

From the previous example, Case (ii), we have

$$\ln\left(\frac{T_w - T_{bo}}{T_w - T_{bi}}\right) = -\frac{P}{mc_p}L\bar{h} \tag{i}$$

for $x = L$, where $T_b = T_{bo}$. $\bar{h} = \bar{h}(0, L)$ is the average heat transfer coefficient over the entire length L of the heat exchanger.

From the first law of thermodynamics, heat transfer rate q equals the heat transferred to the fluid, i.e.

$$q = mc_p(T_{bo} - T_{bi})$$

or

$$mc_p = \frac{q}{(T_{bo} - T_{bi})}$$

Substitution in Eq. (i) gives

$$q = -\bar{h}PL \times \frac{(T_{bo} - T_{bi})}{\ln\left(\frac{T_w - T_{bo}}{T_w - T_{bi}}\right)}$$

or

$$q = \bar{h}PL \times \frac{(T_w - T_{bo}) - (T_w - T_{bi})}{\ln\left(\frac{T_w - T_{bo}}{T_w - T_{bi}}\right)}$$

or

$$q = \bar{h}PL(LMTD)$$

where $LMTD = \frac{(T_w - T_{bo}) - (T_w - T_{bi})}{\ln\left(\frac{T_w - T_{bo}}{T_w - T_{bi}}\right)}$ is termed as log mean temperature difference (also refer Chap. 14 for the $LMTD$).

7.10 Turbulent Flow in Tubes

The fully developed velocity profile in the case of turbulent flow in tubes is shown in Fig. 7.6. The velocity curve changes abruptly near the wall and takes a somewhat blunter profile in the middle of the tube.

In the laminar sublayer, viscous forces dominate and the fluid moves in streamline pattern parallel to the wall of the tube. In the turbulent core, chunks of fluid move in a totally chaotic pattern (termed as eddying motion, shown as curved arrows in the turbulent flow region). This causes intense mixing of the fluid. The fluid in the buffer layer shows behaviour that is intermediate between that of the fluid in the laminar sublayer and turbulent core. The fluid flow pattern in turbulent flow is determined by the inertial than by the viscous forces.

The equations of the universal velocity profile presented for the turbulent flow over a flat plate also apply to the fully developed turbulent flow through a smooth tube when u^+ and y^+ are defined as

$$u^+ = \frac{\bar{u}}{\sqrt{\tau_w/\rho}} \tag{7.98a}$$

$$y^+ = \frac{\sqrt{\tau_w/\rho}}{\nu}.y \tag{7.98b}$$

where y is the distance from the wall ($y = R - r$, R is the tube radius) and $\bar{u}$ is the mean component of the velocity.

Prandtl developed the following form of velocity distribution, known as universal velocity defect law, applicable only in the turbulent core away from the wall.

$$\frac{(U_{max} - u)}{u*} = 2.5 \ln\left(\frac{R}{y}\right) \tag{7.99}$$

The defect law shows that the velocity defect (and hence the slope of the velocity profile) is a function of the distance R/y only and does not depend on the viscosity of the fluid.

As explained earlier for the turbulent flow over a flat plate, a power law instead of the universal velocity distribution equation is used for the turbulent flow in tubes. Using the Blasius empirical formula for the friction factor, Prandtl developed the following power law of the velocity distribution.

$$\frac{u}{U_{max}} = \left(\frac{y}{R}\right)^{1/n} = \left(1 - \frac{r}{R}\right)^{1/n} \tag{7.100}$$

According to the measurements by Nikuradse, the value of exponent n increases from 6 at Re = 4000 to 10 at Re = 3.2×10^6. It can be estimated from the following relation:

$$n = -1.7 + 1.8 \log_{10} \text{Re} \quad \text{for} \quad \text{Re} > 2 \times 10^4$$

However, the exponent is often taken as 7 and hence the velocity profile is termed as 'a one-seventh power profile'.

From Eq. (7.100), it can be shown that the ratio of the mean and maximum velocities U_m and U_{max}, respectively, is (see Example 7.18)

$$\frac{U_m}{U_{max}} = \frac{2n^2}{(n+1)(2n+1)} \tag{7.101}$$

Since n varies with Re, the ratio u_m/U_{max} is a function of the Reynolds number and with the increase in the Reynolds number, the velocity profile becomes flatter over most of the duct cross-section.

It is to note that the power law profile is not applicable close to the wall ($y/R < 0.04$). Since the velocity is quite low in the region near the wall, the error in the estimate of integral quantities such as mass, momentum and energy flux is relatively small. The law gives infinite velocity gradient at the wall and, hence, it cannot be used for the calculation of wall shear stress. Though the law fits the experimental data well near the centreline of the tube, it fails to give zero slope of the velocity profile there. Despite these shortcomings, the power law gives reasonably good results in many calculations.

Example 7.18 The power law equation of velocity distribution in a smooth pipe under turbulent flow condition is

$$\frac{u}{U_{max}} = \left(\frac{y}{R}\right)^{1/n} = \left(1 - \frac{r}{R}\right)^{1/n}$$

where u is the local time-averaged velocity, U_{max} is the time-averaged velocity at the centreline, R is the radius of the pipe and $y = R - r$ is the distance measured from the pipe wall.

Determine the ratio of the mean to the maximum velocity. Calculate the value of the ratio for $n = 7$.

Solution

From the definition of the mean velocity,

$$\begin{aligned} U_m &= \left(\frac{1}{\pi R^2}\right) \int_0^R U_{max}\left(1 - \frac{r}{R}\right)^{1/n} 2\pi r dr \\ &= \left(\frac{2U_{max}}{R^2}\right) \int_0^R \left(1 - \frac{r}{R}\right)^{1/n} r dr \\ &= \left(\frac{2U_{max}}{R^2}\right) \left[\frac{R^2}{1/n+2}\left(1 - \frac{r}{R}\right)^{1/n+2} - \frac{R^2}{1/n+1}\left(1 - \frac{r}{R}\right)^{1/n+1}\right]_0^R \\ &= -2U_{max}\left[\frac{n}{1+2n} - \frac{n}{1+n}\right] \end{aligned}$$

or

$$\frac{U_m}{U_{max}} = \frac{2n^2}{(n+1)(2n+1)}$$

For $n = 7$,

$$\frac{U_{\mathrm{m}}}{U_{max}} \approx 0.8$$

Note: For laminar flow, $\frac{U_{\mathrm{m}}}{U_{max}} = 0.5$. Thus, the velocity profile in the turbulent flow is much flatter over most of the duct cross-section.

7.11 Momentum and Heat Exchange in Turbulent Flow (Eddy Viscosity and Eddy Thermal Diffusivity)

In the laminar flow, the fluid particles follow well-defined streamlines. Heat and momentum are transferred by molecular diffusion (microscopic scale). There is no macroscopic mixing. In the turbulent flow, in addition to the heat and momentum transfer by molecular diffusion, there is momentum and heat exchange on the macroscopic scale due to eddy mixing. The momentum transfer takes place due to the faster moving fluid mass moving into the slow moving fluid and vice versa. Similarly, the high-temperature fluid mass mixes with the lower temperature fluid and vice versa due to the eddying motion causing transport of heat.

Since the momentum and heat exchange takes place both due to the microscopic and macroscopic means, their rates would be considerably greater than the laminar flow. This is the reason for greater rate of heat transfer in turbulent flow than the laminar flow.

Consider an arbitrary plane 1-1 parallel to the plate surface as shown in Fig. 7.25. There is exchange of mass across the surface 1-1 due to the eddying motion. The molecular transport of momentum and heat is like the laminar flow. So let us consider the macroscopic transfer of momentum and heat.

If the turbulent flow velocities at any point are averaged over a period θ longer than the period of fluctuation, then we can define mean velocities in the x- and y-directions as

$$\bar{u} = \frac{1}{\theta}\int_0^\theta u d\tau \tag{i}$$

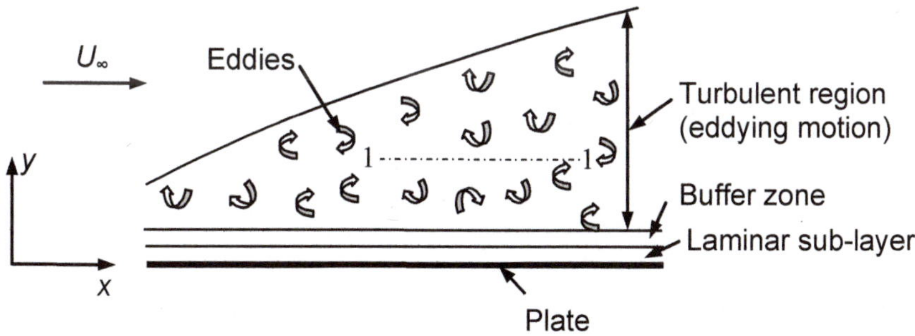

Fig. 7.25 Boundary layer over a flat plate with turbulent flow

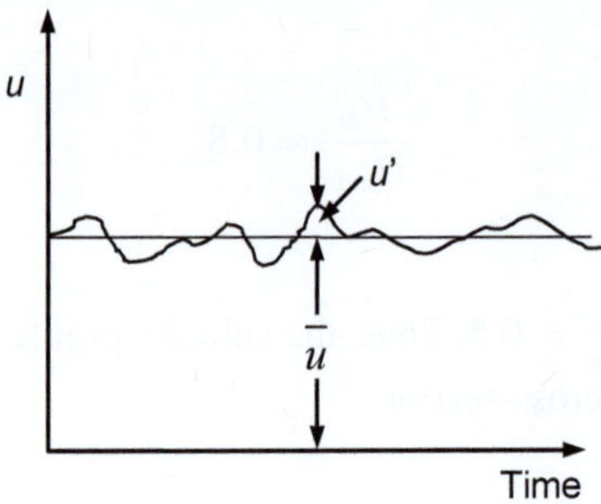

Fig. 7.26 Turbulent fluctuations with time

$$\bar{v} = \frac{1}{\theta}\int_0^\theta v d\tau \tag{ii}$$

where u and v denote the instantaneous velocities in x- and y-directions and are function of time.

The instantaneous velocities can be expressed as the sum of the mean and fluctuating components (refer Fig. 7.26)

$$\begin{aligned} u &= \bar{u} + u' \\ v &= \bar{v} + v' \end{aligned} \tag{iii}$$

where u' and v' are the fluctuating components.

For study flow, the mean components are constants. The mean values of u' and v' must be zero over the time θ, i.e.

$$\bar{u}' = \frac{1}{\theta}\int_0^\theta u' d\tau = 0 \tag{iv}$$

$$\bar{v}' = \frac{1}{\theta}\int_0^\theta v' d\tau = 0 \tag{v}$$

Due to the eddy motion, let the upward flow of mass through the unit area of plane 1-1 during time $d\theta$ is $\rho v d\theta$. The momentum in the x-direction carried with this mass is $(\rho v d\theta)$ u. The shear stress acting along 1-1 due to the eddy motion only must equal the average rate of momentum transfer through the unit area of the plane 1-1.

$$\tau_t = \frac{1}{\theta}\int_0^\theta \rho(uv) d\theta \tag{vi}$$

Substitution of the values of u and v gives

$$\tau_t = \frac{\rho}{\theta}\int_0^\theta (\bar{u}+u')(\bar{v}+v')d\theta = \frac{\rho}{\theta}\int_0^\theta (\bar{u}\bar{v}+\bar{u}v'+u'\bar{v}+u'v')d\theta \quad \text{(vii)}$$

Time constants of the fluctuating components are zero, hence

$$\frac{\rho}{\theta}\int_0^\theta \bar{u}v'd\theta = \frac{\rho}{\theta}\bar{u}\int_0^\theta v'd\theta = 0 \quad \text{(viii)}$$

$$\frac{\rho}{\theta}\int_0^\theta \bar{v}u'd\theta = \frac{\rho}{\theta}\bar{v}\int_0^\theta u'd\theta = 0 \quad \text{(ix)}$$

Therefore,

$$\tau_t = \frac{\rho}{\theta}\int_0^\theta (\bar{u}\bar{v}+u'v')d\theta \quad \text{(x)}$$

In the case of one-dimensional flow over a plate, the thickness of the boundary layer increases very slowly, hence $\bar{v}$ is negligible. For the fully developed flow through a tube, $\bar{v}$ is exactly zero. Introducing $\bar{v}=0$ in Eq. (x), we obtain

$$\tau_t = \frac{\rho}{\theta}\int_0^\theta (u'v')d\theta = \rho\left(\overline{u'v'}\right) \quad \text{(xi)}$$

where $\left(\overline{u'v'}\right)$ is the time average of product $u'v'$. The term $\rho\left(\overline{u'v'}\right)$ is referred to as *Reynolds stress*. The time averages of u' and v' are zero but the time average of their products is not zero. Let us ascertain the sign of the product $u'v'$.

Consider an instantaneous upward flow through plane 1-1, for which v' is positive. Since the average axial velocity of mass below plane 1-1 is lower than the average axial velocity of the mass above plane 1-1, this fluid mass arriving at 1-1 produces negative u' at 1-1. Hence, the product $u'v'$ will be negative. Similarly, it can be shown that for a fluid mass moving downwards through the plane 1-1, v' is negative and u' will be positive hence the product $u'v'$ will be again negative (a negative v' is accompanied with a positive u' and vice versa). Thus, on the whole the product $\overline{u'v'}$ is negative, which means there is a net flow of the momentum towards the wall resulting in a shear stress. If we define this shear stress as

$$\tau_t = \rho\varepsilon_M \frac{d\bar{u}}{dy} \quad (7.102)$$

where ε_M is called *eddy diffusivity* or *eddy viscosity*. From Eqs. (xi) and (7.102), we get

$$\tau_t = -\rho\left(\overline{u'v'}\right) = \rho\varepsilon_M \frac{d\bar{u}}{dy} \tag{7.103}$$

The negative sign has been introduced to give a positive value of ε_M like the kinematic viscosity v because $\frac{d\bar{u}}{dy}$ is always positive.

The total shear stress in a turbulent flow is the sum of the viscous shear stress τ_l and eddy or turbulent shear stress τ_t, i.e.

$$\tau = \rho v \frac{d\bar{u}}{dy} + \rho\varepsilon_M \frac{d\bar{u}}{dy} = \rho(v + \varepsilon_M)\frac{d\bar{u}}{dy} \tag{7.104}$$

It must be noted that the laminar shear stress τ_l is true shear stress, whereas the apparent turbulent shear stress τ_t is simply a concept introduced to take into account the momentum transfer by turbulent fluctuations.

The eddy viscosity is a function of the turbulence and it changes its magnitude with the distance from the wall. The eddy viscosity ε_M is analogous to the kinematic viscosity v but unlike v it is not a property of the fluid.

The turbulent boundary layer is basically composed of two distinct regions:

(i) The laminar sublayer: It is a layer just adjacent to the wall, wherein the flow is laminar. In this layer, ε_M is zero and Eq. (7.104) reduces to the familiar equation of viscous shear.

(ii) The fully turbulent region: where ε_M is much larger than v and the viscous shear stress may be neglected.

In the buffer zone (a zone between the laminar and fully turbulent zones), the eddy and kinematic viscosities are of the same order of magnitude.

Let us now consider the heat exchange. The rate of heat exchange across plane 1-1 due to the molecular transport is

$$q = -k\frac{d\bar{t}}{dy} = -\rho c_p \alpha \frac{d\bar{t}}{dy} \tag{xii}$$

where α is the thermal diffusivity of the fluid and $\bar{t}$ is the average temperature.

The instantaneous temperature t at 1-1 fluctuates due to the arrival of hot eddies from below and cold eddies from above when the wall temperature is higher than the free-stream temperature. Hence,

$$t = \bar{t} + t' \tag{xiii}$$

The mass flow rate per unit area across surface 1-1 during time interval $d\theta$ is $\rho v d\theta$ and transport of heat with this mass is $(\rho v d\theta)c_p t$. Hence, the heat transport in time interval θ is

$$q = \frac{1}{\theta}\int_0^\theta \rho v c_p t.d\theta \tag{xiv}$$

Substitution of the values of v and t gives

$$q = \frac{\rho c_p}{\theta}\int_0^{\theta}(\bar{v}+v')(\bar{t}+t')d\theta = \frac{\rho c_p}{\theta}\int_0^{\theta}(\bar{v}\bar{t}+\bar{v}t'+\bar{t}v'+v't')d\theta \tag{xv}$$

For steady flow,

$$\frac{1}{\theta}\int_0^{\theta}\bar{v}t'd\theta = \frac{1}{\theta}\bar{v}\int_0^{\theta}t'd\theta = 0$$

$$\frac{1}{\theta}\int_0^{\theta}\bar{t}v'd\theta = \frac{1}{\theta}\bar{t}\int_0^{\theta}v'd\theta = 0$$

Further, as explained earlier, $\bar{v}$ is negligible for flow over a flat plate and $\bar{v}$ is exactly zero for fully developed flow through a tube. Hence, we are left with

$$q = \frac{\rho c_p}{\theta}\int_0^{\theta}(v't')d\theta = \rho c_p\left(\overline{v't'}\right) \tag{xvi}$$

Turbulent heat transfer is analogous to the turbulent momentum transfer. In analogy to Eq. (7.102), we introduce *eddy thermal diffusivity* or *eddy heat diffusivity* defined by

$$q_t = -\rho c_p \varepsilon_H \frac{d\bar{t}}{dy} \tag{7.105}$$

Combining with Eq. (xvi), we get

$$q_t = \rho c_p\left(\overline{v't'}\right) = -\rho c_p \varepsilon_H \frac{d\bar{t}}{dy} \tag{7.106}$$

The negative sign has been introduced to give a positive value of ε_H.

To ascertain the signs of $\left(\overline{v't'}\right)$ and $\frac{d\bar{t}}{dy}$, consider an instantaneous upward flow through plane 1-1, for which v' is positive. For wall temperature higher than the free-stream temperature, t' is positive because it arrives from the region below plane 1-1. Hence, the product $v't'$ is positive. Similarly, it can be shown that for a fluid mass moving downwards through the plane 1-1, v' is negative and t' will be negative hence the product $v't'$ will be again positive (a positive v' is accompanied with a positive t' and vice versa). Thus, on the whole the product $\left(\overline{v't'}\right)$ is positive. For the assumed condition of the plate hotter than the fluid, $\frac{d\bar{t}}{dy}$ must be negative. Hence, for a positive value of ε_H, introduction of the negative sign in Eq. (7.105) is justified.

The total heat flow in a turbulent flow is the sum of the molecular and eddy or transport, i.e.

$$q = -\rho c_p \alpha \frac{d\bar{t}}{dy} - \rho c_p \varepsilon_H \frac{d\bar{t}}{dy} = -\rho c_p (\alpha + \varepsilon_H) \frac{d\bar{t}}{dy} \tag{7.107}$$

The eddy thermal diffusivity is also a function of the turbulence and it changes its magnitude with the distance from the wall.

The turbulent diffusivity is zero in the laminar sublayer. It is of the same order of magnitude as the molecular diffusivity in the buffer zone and is much greater than the molecular diffusivity in the turbulent core. However, this is true only for fluids having $\Pr \geq 1$, but not for the liquid metals ($\Pr \leq 0.01$).

The ratio of ε_M and ε_H is referred to as the *turbulent Prandtl number* $\Pr_t$ analogous to the laminar or molecule Prandtl number, i.e.

$$\Pr_t = \frac{\varepsilon_M}{\varepsilon_H} \tag{7.108}$$

The ratio $\varepsilon_M/\varepsilon_H$ is required for the prediction of the heat transfer coefficient from the measurement of the velocity field. However, the turbulent Prandtl number is difficult to measure accurately. Some investigators report value of $\varepsilon_M/\varepsilon_H$ from 0.6 to 1.7 for flow of air in tubes. They found it high near the wall, decreasing and becoming nearly a constant away from the wall. For liquid metals, $\Pr_t$ lies in the range of 1.0 to 2.0 throughout the boundary layer (Kays and Crawford 1980). At very high molecular Prandtl number, say $\Pr \gg 1$, indirect evidence suggests that the value of $\Pr_t$ is not very different from 1.00.

The values of ε_M and ε_H depend on the speed of adaptability of an eddy (Rogers and Mayhew 1967). A mass moving across a plane in turbulent region tends to acquire the momentum and temperature of the new layer into which it moves, but on immediate return of the mass to its original layer it may not be able to assume the momentum and temperature of the new layer completely. This behaviour may not be the same for the momentum and temperature. Reichardt (in Rogers and Mayhew 1967) concluded from his investigations that ε_M and ε_H are not necessarily equal, although they are nearly equal for the turbulent flow through tubes and past flat plates.

Reynolds suggested that there exists a complete similarity between forced convection fluid friction and heat transfer because the same mechanism of turbulent exchange causes the transfer of momentum and heat. This is known as Reynolds analogy and is presented in the next section. However, the analogy is a considerable simplification of a very complex process, but a reasonable approximation for fluids having the molecular Prandtl number value of nearly one.

7.12 Reynolds Analogy for Flow Past a Flat Surface

When a fluid flows past a solid surface, a thin layer of fluid is formed on the surface in which the viscous effects dominate. This film is called a laminar sublayer. The heat flow through this layer is only by conduction. In the turbulent region beyond the laminar sublayer, molecular conduction and shear effects are small compared to the turbulent exchange of heat and momentum.

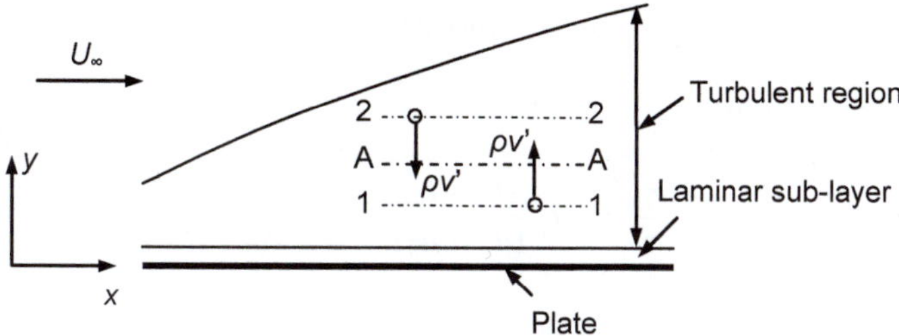

Fig. 7.27 Exchange of mass in turbulent region due to the eddying motion

The molecular shear stress and the heat flow per unit area in the laminar sublayer are given by

$$\tau_l = \mu \frac{du}{dy} \tag{i}$$

$$q_l = -k \frac{dt}{dy} \tag{ii}$$

Combining Eqs. (i) and (ii), we get

$$\frac{q_l}{\tau_l} = -\frac{k}{\mu} \frac{dt}{du} \tag{iii}$$

The relationship is also valid at the wall, hence

$$\frac{q_l}{\tau_l} = \frac{q_w}{\tau_w} = -\frac{k}{\mu} \frac{dt}{du} \tag{iv}$$

In the turbulent flow, beyond the laminar sublayer, let us consider a plane A-A parallel to the wall. Due to the effect of turbulence, a mass $\rho v'$ from the lower plane 1-1 is transferred to the upper plane 2-2 across the plane A-A, refer Fig. 7.27. In the steady-state condition, the same amount of the fluid must flow from upper to the lower plane. The interaction causes momentum and heat transfer.

The rate of momentum transfer across the plane A-A is

$$= \rho v'(u_1 - u_2) \tag{v}$$

The rate of change of momentum is the shear stress. In the turbulent region, it is called turbulent shear stress τ_t. Thus

$$\tau_t = \rho v'(u_1 - u_2) \tag{vi}$$

It is to be noted that $\rho v'$ is the mass flow rate per unit area. The heat flow with the mass transport is

$$q_t = -\rho v' c_p (t_2 - t_1) \tag{vii}$$

From Eqs. (vi) and (vii),

$$\frac{q_t}{\tau_t} = -c_p \frac{(t_2 - t_1)}{(u_1 - u_2)} = -c_p \frac{\Delta t}{\Delta u} \tag{viii}$$

In the differential form, we can write

$$\frac{q_t}{\tau_t} = -c_p \frac{dt}{du} \tag{ix}$$

For Pr = $\mu c_p/k$ = 1, k/μ in Eq. (iv) can be replaced by c_p. This gives

$$\frac{q_l}{\tau_l} = \frac{q_w}{\tau_w} = \frac{q_t}{\tau_t} \quad \text{for} \quad \text{Pr} = 1. \tag{7.109}$$

This is known as *Reynolds analogy*. The analogy establishes, 'for the ratio of heat flow to the shear stress, the law holds in laminar as well as turbulent flow'.

Since Eq. (7.109) is applicable for both the laminar and turbulent regions, the integration of this equation carried out over the limits (u = 0, t = t_w) at the wall and (u = U_∞, t = t_∞) at the outer surface of the boundary layer, we obtain

$$\frac{q_w}{c_p \tau_w} \int_0^{U_\infty} du = -\int_{t_w}^{t_\infty} dt$$

or

$$\frac{q_w}{c_p \tau_w} U_\infty = t_w - t_\infty$$

or

$$\frac{q_w}{(t_w - t_\infty)} = \tau_w \frac{c_p}{U_\infty} \tag{x}$$

By definition, $q/\Delta t$ is the heat transfer coefficient, h, hence

$$h = \tau_w \frac{c_p}{U_\infty} \tag{xi}$$

For the flow over a flat plate, the wall shear stress is given by

$$\tau_w = \frac{1}{2} \rho U_\infty^2 C_{fx} \tag{xii}$$

where C_{fx} is the local skin friction coefficient.

Inserting the value of τ_w in Eq. (xi),

$$h_x = \frac{1}{2}\rho U_\infty^2 C_{fx} \frac{c_p}{U_\infty}$$

or

$$\frac{h_x}{\rho c_p U_\infty} = \frac{1}{2} C_{fx} \tag{xiii}$$

The dimensionless group on the left-hand side is the Stanton number St. Equation (xiii) is, thus

$$\mathrm{St_x} = \frac{1}{2} C_{fx} \tag{7.110}$$

The Stanton number can also be written as

$$\mathrm{St} = \frac{h}{\rho c_p U_\infty} = \frac{hl}{k} \times \frac{\mu}{\rho U_\infty l} \times \frac{k}{\mu c_p} = \frac{\mathrm{Nu}}{\mathrm{Re\,Pr}} \tag{7.111}$$

Hence,

$$\mathrm{St_x} = \frac{\mathrm{Nu_x}}{\mathrm{Re_x\,Pr}} = \frac{C_{fx}}{2} \tag{7.112}$$

The above equation has been derived from the Reynolds analogy and hence it is valid only for the fluids having Pr = 1. It establishes the similarity of nature of heat and momentum transfers. Thus, the heat transfer coefficient can be determined by measurements of the friction factor under the conditions when no heat transfer is involved.

7.12.1 Reynolds–Colburn Analogy

For the laminar boundary layer over a flat plate, we obtained

$$\mathrm{Nu_x} = 0.332(\mathrm{Re_x})^{1/2}\mathrm{Pr}^{1/3} \tag{7.56b}$$

and

$$C_{\mathrm{fx}} = \frac{0.6641}{\sqrt{\mathrm{Re_x}}} \tag{7.37}$$

Dividing Eq. (7.56b) by Eq. (7.37), we get

$$\frac{\mathrm{Nu}_x}{C_{fx}} = \frac{1}{2}\mathrm{Re}_x\mathrm{Pr}^{1/3}$$

or

$$\frac{\mathrm{Nu}_x}{\mathrm{Re}_x\,\mathrm{Pr}}\mathrm{Pr}^{2/3} = \frac{C_{fx}}{2}$$

or

$$\mathrm{St}_x\mathrm{Pr}^{2/3} = \frac{C_{fx}}{2} \tag{7.113}$$

The relation has been proposed by Colburn and is known as *Reynolds–Colburn analogy*. This analogy has been found to be quite accurate in the range $0.5 < \mathrm{Pr} < 50$ provided the drag forces are wholly viscous in nature (i.e. when form or pressure drag is absent). It is to be noted that for Pr = 1, the Reynolds and Reynolds–Colburn analogies are the same.

Note: some references define a factor j as

$$j = \mathrm{St}\mathrm{Pr}^{2/3} \tag{7.114a}$$

Therefore, we can also write Eq. (7.113) as

$$j_x = \frac{C_{fx}}{2} \tag{7.114b}$$

Example 7.19 Air at atmospheric pressure and 20°C is flowing parallel to a flat plate at a velocity of 30 m/s on one side of the plate. The plate is 0.375 m long and is kept at a constant temperature of 180°C. For the unit width of the plate, the drag force on the plate is found to be 0.3 N. Determine heat transfer rate from the one side of the plate.

Solution

The properties of air at the mean film temperature t_{fm} = (20 + 180)/2 = 100°C are

$$\rho = 0.9452\,\mathrm{kg/m^3}, k = 0.0317\,\mathrm{W/(m\,K)}, \mu = 2.172\times10^{-5}\,\mathrm{kg/(m\,s)}, \mathrm{Pr} = 0.693, c_\mathrm{p} = 1011.3\,\mathrm{J/(kg\,K)}.$$

Skin friction coefficient,

$$C_f = \frac{F_D}{A\left(\frac{1}{2}\rho U_\infty^2\right)} = \frac{0.3}{0.375\times1\times\left(\frac{1}{2}\times0.9452\times30^2\right)} = 0.00188.$$

Colburn's analogy gives

$$\mathrm{St} = \frac{\bar{h}}{\rho U_\infty c_p} = \mathrm{Pr}^{-2/3}\frac{\overline{C_f}}{2}$$

or

$$\begin{aligned}\bar{h} &= \rho U_\infty c_p \mathrm{Pr}^{-2/3}\frac{\bar{C}_f}{2}\\ &= 0.9452 \times 30 \times 1011.3 \times 0.693^{-2/3} \times \frac{0.00188}{2}\\ &= 34.42\ \mathrm{W/(m^2\,K)}.\end{aligned}$$

Rate of heat transfer q from one side of the plate

$$q = \bar{h}A(t_w - t_\infty) = 34.42 \times (0.375 \times 1) \times (180 - 20) = 2065.2\,\mathrm{W}.$$

7.12.2 Application of Colburn Analogy to Turbulent Heat Transfer from a Flat Plate

The analogy has also been applied to predict the turbulent boundary layer heat transfer and has yielded results which are in good agreement with the experimental values.

A friction factor correlation, derived from experimental data for turbulent flow over flat plate, is

$$C_{fx} = 0.0592(\mathrm{Re}_x)^{-1/5} \quad \text{for} \quad 5 \times 10^5 < \mathrm{Re} < 10^7 \tag{7.115}$$

Applying the Colburn's analogy

$$\mathrm{St}_x \mathrm{Pr}^{2/3} = \frac{C_{fx}}{2} \tag{i}$$

or

$$\frac{\mathrm{Nu}_x}{\mathrm{Re}_x\,\mathrm{Pr}}\mathrm{Pr}^{2/3} = 0.0296\mathrm{Re}_x^{-1/5}$$

or

$$\mathrm{Nu}_x = 0.0296\mathrm{Re}_x^{4/5}\mathrm{Pr}^{1/3} \quad \text{for} \quad 5 \times 10^5 < \mathrm{Re} < 10^7 \tag{7.116}$$

The average heat transfer coefficient over the plate length L,

$$\bar{h} = \frac{1}{L}\int_0^L h_x dx,$$

which yields

$$\bar{h} = 0.037\mathrm{Pr}^{1/3}\frac{1}{L}k\left(\frac{U_{\infty}L}{\nu}\right)^{0.8} \tag{ii}$$

This gives the equation of the average Nusselt number over the plate length L as

$$\overline{\mathrm{Nu}} = \frac{\bar{h}L}{k} = 0.037\mathrm{Re}_L^{0.8}\mathrm{Pr}^{1/3} \quad \text{for} \quad 5\times10^5 < \mathrm{Re} < 10^7 \tag{7.117}$$

To determine the average heat transfer coefficient over the entire laminar turbulent boundary layer, the following equation is used:

$$\bar{h} = \frac{1}{L}\left(\int_0^{x_c} h_x dx + \int_{x_c}^{L} h_x dx\right) \tag{iii}$$

Using Eqs. (7.56b) and (7.116) for the laminar and turbulent boundary layers, respectively, we get

$$\bar{h} = \frac{1}{L}\left[\int_0^{x_c} 0.332\frac{k}{x}\mathrm{Re}_x^{1/2}\mathrm{Pr}^{1/3}dx + \int_{x_c}^{L} 0.0296\left(\frac{k}{x}\right)\mathrm{Re}_x^{4/5}\mathrm{Pr}^{1/3}dx\right]$$

Taking x_c corresponding to $\mathrm{Re}_c = 5\times10^5$, the integral yields (refer Example 7.20)

$$\overline{\mathrm{Nu}} = \frac{\bar{h}L}{k} = \left(0.037\mathrm{Re}_L^{0.8} - 871\right)\mathrm{Pr}^{1/3} \tag{7.118}$$

Example 7.20 For fluid flow over a flat plate, develop the equations of the average heat transfer coefficient if the local heat transfer coefficient is given by

$$\mathrm{Nu}_x = \frac{h_x x}{k} = 0.332(\mathrm{Re}_x)^{1/2}\mathrm{Pr}^{1/3} \quad \text{for} \quad \mathrm{Re} \le 5\times10^5$$

$$\mathrm{Nu}_x = \frac{h_x x}{k} = 0.0296\mathrm{Re}_x^{0.8}\mathrm{Pr}^{1/3} \quad \text{for} \quad 5\times10^5 < \mathrm{Re} \le 10^7$$

Solution

(a) **Laminar Region**
Refer Sect. 7.7.1.

(b) **Turbulent region**
The average heat transfer coefficient over the plate length L can be evaluated as

$$\begin{aligned}
\bar{h} &= \frac{1}{L}\int_0^L h_x dx \\
&= \frac{1}{L}\int_0^L 0.0296\mathrm{Pr}^{1/3}\frac{k}{x}\left(\frac{U_\infty x}{\nu}\right)^{0.8} dx \\
&= 0.0296\mathrm{Pr}^{1/3}\frac{1}{L}k\left(\frac{U_\infty}{\nu}\right)^{0.8}\int_0^L x^{-0.2}dx \\
&= 0.0296\mathrm{Pr}^{1/3}\frac{1}{L}k\left(\frac{U_\infty}{\nu}\right)^{0.8}\frac{L^{0.8}}{0.8} \\
&= 0.037\frac{k}{L}\left(\frac{U_\infty L}{\nu}\right)^{0.8}\mathrm{Pr}^{1/3}
\end{aligned}$$

The equation can be put in the non-dimensional form as

$$\frac{\bar{h}L}{k} = 0.037\left(\frac{U_\infty L}{\nu}\right)^{0.8}\mathrm{Pr}^{1/3}$$

$$\overline{\mathrm{Nu}} = 0.037\mathrm{Re}_L^{0.8}\mathrm{Pr}^{1/3}$$

(c) **Average heat transfer coefficient over the entire laminar turbulent boundary layer**

$$\bar{h} = \frac{1}{L}\left(\int_0^{x_c} h_x dx + \int_{x_c}^{L} h_x dx\right)$$

Using given equations for the laminar and turbulent boundary layers, respectively, we get

$$\begin{aligned}
\bar{h} &= \frac{1}{L}\left[\int_0^{x_c} 0.332\frac{k}{x}\mathrm{Re}_x^{1/2}\mathrm{Pr}^{1/3}dx + \int_{x_c}^{L} 0.0296\frac{k}{x}\mathrm{Re}_x^{4/5}\mathrm{Pr}^{1/3}dx\right] \\
&= \frac{1}{L}k\mathrm{Pr}^{1/3}\left[0.332\int_0^{x_c}\frac{1}{x}\left(\frac{U_\infty x}{\nu}\right)^{1/2}dx + 0.0296\int_{x_c}^{L}\frac{1}{x}\left(\frac{U_\infty x}{\nu}\right)^{0.8}dx\right] \\
&= \frac{1}{L}k\mathrm{Pr}^{1/3}\left[0.332\left(\frac{U_\infty}{\nu}\right)^{1/2}\int_0^{x_c}x^{-1/2}dx + 0.0296\left(\frac{U_\infty}{\nu}\right)^{0.8}\int_{x_c}^{L}x^{-0.2}dx\right] \\
&= \frac{1}{L}k\mathrm{Pr}^{1/3}\left\{0.664\left(\frac{U_\infty x_c}{\nu}\right)^{1/2} + 0.037\left[\left(\frac{U_\infty L}{\nu}\right)^{0.8} - \left(\frac{U_\infty x_c}{\nu}\right)^{0.8}\right]\right\} \\
&= \frac{1}{L}k\mathrm{Pr}^{1/3}\left\{0.664(\mathrm{Re}_c)^{1/2} + 0.037\left[(\mathrm{Re}_L)^{0.8} - (\mathrm{Re}_c)^{0.8}\right]\right\}
\end{aligned}$$

Putting $\mathrm{Re_c} = 5 \times 10^5$ and simplifying, we get

$$\bar{h} = \frac{k}{L}\mathrm{Pr}^{1/3}\left(0.037\mathrm{Re}_L^{0.8} - 871\right)$$

or

$$\overline{\mathrm{Nu}} = \frac{\bar{h}L}{k} = \left(0.037\mathrm{Re}_L^{0.8} - 871\right)\mathrm{Pr}^{1/3}$$

Example 7.21 Air, at mean bulk temperature of 20°C and 1 atm pressure, flows over a 0.5-m-long and 1-m-wide flat plate parallel to its surface at a velocity of 5 m/s. The plate is at a uniform temperature of 80°C. All other parameters remaining the same, how will the heat transfer rate change if the air pressure is increased to 10 atm?

Solution

The properties of air at the mean film temperature t_{fm} = (20 + 80)/2 = 50°C are

$$\rho = 1.0949\,\mathrm{kg/m^3}, k = 0.02799\,\mathrm{W/(m\,K)}, \mu = 1.9512 \times 10^{-5}\,\mathrm{kg/(m\,s)}, \mathrm{Pr} = 0.703.$$

(a) ***p* = 1 atm**

Flow Reynolds number is

$$\mathrm{Re}_L = \frac{\rho U_\infty L}{\mu} = \frac{1.0949 \times 5 \times 0.5}{1.9512 \times 10^{-5}} = 1.4 \times 10^5 < 5 \times 10^5.$$

The flow is laminar. The average heat transfer coefficient is

$$\bar{h} = \frac{k}{L} \times 0.664\mathrm{Re}_L^{1/2}\mathrm{Pr}^{1/3} = \frac{0.02799}{0.5} \times 0.664 \times \left(1.4 \times 10^5\right)^{1/2} 0.703^{1/3}$$
$$= 12.37\ \mathrm{W/(m^2\,K)}.$$

Heat transfer rate,

$$q = \bar{h} \times A(t_s - t_\infty) = 12.37 \times 0.5 \times 1 \times (80 - 20) = 371.1\ \mathrm{W}.$$

(b) ***p* = 10 atm**

The air density will be 10 times when pressure will increase to 10 atm from 1 atm. Now, the flow Reynolds number is

$$\mathrm{Re}_L = \frac{\rho U_\infty L}{\mu} = \frac{10.949 \times 5 \times 0.5}{1.9512 \times 10^{-5}} = 1.4 \times 10^6.$$

Now the mixed boundary condition exists over the plate. The average heat transfer coefficient is

$$\begin{aligned}\bar{h} &= \frac{k}{L} \times (0.037\mathrm{Re}_L^{4/5} - 871)\mathrm{Pr}^{1/3} \\ &= \frac{0.02799}{0.5} \times [0.037 \times (1.4 \times 10^6)^{4/5} - 871] \times 0.703^{1/3} \\ &= 108.74\ \mathrm{W/(m^2\,K)}.\end{aligned}$$

Heat transfer rate,

$$q = \bar{h} \times A(t_s - t_\infty) = 108.74 \times 0.5 \times 1 \times (80 - 20) = 3262.2\ \mathrm{W}.$$

Example 7.22 Air at 20°C flows at a velocity of 20 m/s past a flat plate 1.5 m long and 0.8 m wide. The surface of the plate is maintained at 280°C. Determine the heat transferred from both sides of the plate. The following thermophysical properties of the air at 150°C may be used:

$$\rho = 0.835\ \mathrm{kg/m^3}, \mu = 24 \times 10^{-6}\ \mathrm{kg/(m\,s)}, c_\mathrm{p} = 1.015\ \mathrm{kJ/(kg\,K)}, k = 3.56 \times 10^{-2}\mathrm{W/(m\,K)}, \mathrm{Pr} = 0.7.$$

Solution

For the given data,

$$\mathrm{Re}_L = \frac{\rho U_\infty L}{\mu} = \frac{0.835 \times 20 \times 1.5}{24 \times 10^{-6}} = 1.044 \times 10^6 > 5 \times 10^5$$

Plate length corresponding to $\mathrm{Re}_c = 5 \times 10^5$ is

$$x_c = \frac{\mathrm{Re}_c \mu}{\rho U_\infty} = \frac{5 \times 10^5 \times 24 \times 10^{-6}}{0.835 \times 20} = 0.7186\ \mathrm{m}.$$

The flow is laminar up to 0.7186 m, i.e. up to $\mathrm{Re} = 5 \times 10^5$. The Nusselt number for this part is

$$\begin{aligned}\mathrm{Nu} &= 0.664\mathrm{Re}_c^{0.5}\mathrm{Pr}^{0.33} \\ &= 0.664(5 \times 10^5)^{0.5}(0.7)^{0.33} = 417.4\end{aligned}$$

and the average heat transfer coefficient is

$$\bar{h} = \frac{\mathrm{Nu} \times k}{x_{cr}} = \frac{417.4 \times 3.56 \times 10^{-2}}{0.7186} = 20.68\ \mathrm{W/(m^2\,K)}.$$

Heat transfer from the laminar region, for the both sides of the plate,

$$\begin{aligned} q_{\text{laminar}} &= \bar{h}A_{\text{laminar}}(t_w - t_\infty) \\ &= 20.68 \times 2 \times 0.7186 \times 0.8 \times (280 - 20) = 6182\,\text{W}. \end{aligned}$$

Heat transfer from the turbulent region (x_c to L)

$$q_{\text{turbulent}} = \bar{h}A_{\text{turbulent}}(t_w - t_\infty)$$

where

$$\begin{aligned} \bar{h} &= 0.037(\text{Re}_L^{0.8} - \text{Re}_c^{0.8})\text{Pr}^{0.33}\left(\frac{k}{L - x_c}\right) \\ &= 0.037 \times [(1.044 \times 10^6)^{0.8} - (5 \times 10^5)^{0.8}] \times (0.7)^{0.33} \times \left(\frac{3.56 \times 10^{-2}}{1.5 - 0.7186}\right) \\ &= 43.56\,\text{W}/(\text{m}^2\,\text{K}). \end{aligned}$$

Hence,

$$q_{\text{turbulent}} = 43.56 \times 2 \times (1.5 - 0.7186) \times 0.8 \times (280 - 20) = 14159\,\text{W}.$$

The total heat transferred is

$$q = q_{\text{laminar}} + q_{\text{turbulent}} = 6182 + 14159 = 20341\,\text{W}.$$

Alternative solution
Use Eq. (7.118) to calculate the average heat transfer coefficient for the entire length.

$$\begin{aligned} \bar{h} &= \left(\frac{k}{L}\right)(0.037\text{Re}_L^{0.8} - 871)\text{Pr}^{0.33} \\ &= \left(\frac{3.56 \times 10^{-2}}{1.5}\right)[0.037(1.044 \times 10^6)^{0.8} - 871] \times (0.7)^{0.33} = 32.6 \end{aligned}$$

and

$$q = \bar{h}A(t_w - t_\infty) = 32.6 \times 2 \times 1.5 \times 0.8 \times (280 - 20) = 20342\,\text{W}.$$

Note: The heat transferred for the turbulent part can also be calculated by first considering the flow to be turbulent throughout the plate length and then subtracting the heat transfer from $x = 0$ to $x = x_c$.

Example 7.23 Air at 30°C flows over a flat plate 1.5 m long and 1 m wide. The plate surface is maintained at 70°C. If the desired rate of heat dissipation from the plate is 3 kW, determine the flow velocity of the air.

Solution

At the mean temperature of (70 + 30)/2 = 50°C, the properties of air are

$$\rho = 1.095\,\text{kg/m}^3, k = 0.02799\,\text{W/(m K)}, \mu = 1.95 \times 10^{-5}\,\text{kg/(m s)}, \text{Pr} = 0.703, c_p = 1007.2\,\text{J/(kg K)}.$$

From heat transfer consideration,

$$q = \bar{h}A(t_w - t_\infty)$$

Thus, for q = 3000 W, desired heat transfer coefficient,

$$\bar{h} = \frac{q}{A(t_w - t_\infty)} = \frac{3000}{1.5 \times 1 \times (70 - 30)} = 50\,\text{W/(m}^2\,\text{K)}.$$

Assuming flow to be laminar, the average heat transfer coefficient is

$$\bar{h} = \text{Nu}\frac{k}{L} = 0.664\text{Re}_L^{1/2}\text{Pr}^{1/3}\frac{k}{L}$$

or

$$50 = 0.664 \times (\text{Re}_L)^{1/2} \times (0.703)^{1/3} \times \frac{0.02799}{1.5}.$$

Simplification gives

$$\text{Re}_L = 20.59 \times 10^6 > 5 \times 10^5.$$

The flow is combination of the laminar and turbulent flows.

For the combination of the laminar and turbulent flows, the average heat transfer coefficient,

$$\bar{h} = \text{Nu}\frac{k}{L} = \left(0.037\text{Re}_L^{0.8} - 871\right)\text{Pr}^{1/3}\frac{k}{L}$$

or

$$50 = \left[0.037 \times (\text{Re})^{0.8} - 871\right] \times (0.703)^{1/3} \times \frac{0.02799}{1.5}$$

or

$$\text{Re} = 1.89 \times 10^6$$

$$U_\infty = \frac{\mathrm{Re}_L \mu}{\rho L} = \frac{1.89 \times 10^6 \times 1.95 \times 10^{-5}}{1.095 \times 1.5} = 22.43\,\mathrm{m/s}.$$

7.13 Prandtl–Taylor Modification of Reynolds Analogy for Turbulent Flow Over Flat Plates

Simple Reynolds analogy presented is applicable to fluids having Pr = 1, i.e. for $k/\mu = c_p$. In the region with high turbulence, $\varepsilon_M \gg \upsilon$ and $\varepsilon_H \gg \alpha$ except for the liquid metals hence $(\upsilon + \varepsilon_M) = (\alpha + \varepsilon_H)$ even if $\upsilon \neq \alpha$. The flow is laminar in the sublayer adjacent to the wall hence ε_M and ε_H are zero. Therefore, when Pr $\neq$ 1, the integration cannot be carried out in one step from the wall to the free stream as done in the derivation in Sect. 7.12.

Prandtl and Taylor divided the flow region into two parts: the sublayer in which the flow is laminar and turbulent diffusivities are zero, and the region outside the sublayer in which ε_M and ε_H are large enough for the assumption of $(\upsilon + \varepsilon_M) = (\alpha + \varepsilon_H)$ to be valid even if Pr $\neq$ 1.

Using Eqs. (7.104) and (7.107) of shear and heat flow, we have

$$\tau = \rho(v + \varepsilon_M)\frac{du}{dy}$$

$$q = -\rho c_p(\alpha + \varepsilon_H)\frac{dt}{dy}$$

In the turbulent region, we can write

$$q_t = -\rho c_p \varepsilon_H \frac{dt}{dy} \tag{i}$$

$$\tau_t = \rho \varepsilon_M \frac{du}{dy} \tag{ii}$$

As explained earlier $\varepsilon_M \approx \varepsilon_H$ hence in the turbulent region

$$\frac{q_t}{\tau_t c_p} = -\frac{dt}{du} \tag{iii}$$

This is valid at $y = \delta_b$ also, hence

$$\frac{q_t}{\tau_t c_p} = \frac{q_b}{\tau_b c_p} = -\frac{dt}{du} \tag{iv}$$

where the subscript 'b' refers to the interface of the laminar sublayer and turbulent region.

We rewrite the above equation as

$$dt = -\frac{q_b}{\tau_b c_p} du \tag{v}$$

Integrating this equation from $y = \delta_b$ to $y = H$, where H is greater than both δ_b and δ_t, we get

$$T_b - T_\infty = \frac{q_b}{\tau_b c_p}(U_\infty - u_b) \quad \text{(vi)}$$

The velocity and temperature distributions in the laminar sublayer are assumed to be linear because this layer is so thin that even a third degree polynomial will give an almost straight line. For this assumption,

$$\frac{q_w}{\tau_w} = \frac{q_b}{\tau_b} \quad \text{(vii)}$$

Using this relation, we get

$$T_b - T_\infty = \frac{q_w}{\tau_w c_p}(U_\infty - u_b) \quad \text{(viii)}$$

Since the velocity distribution in the laminar sublayer is linear, we have

$$\tau_w = \mu \frac{du}{dy} = \mu \frac{u_b}{\delta_b} \quad \text{(ix)}$$

Heat flow through the laminar sublayer is by conduction only. Since the sublayer is very thin, we can write

$$q_w = -k\frac{dt}{dy} = k\frac{T_w - T_b}{\delta_b} \quad \text{(x)}$$

From Eqs. (ix) and (x), we obtain

$$\frac{q_w}{\tau_w} = \frac{k}{\mu}\frac{T_w - T_b}{u_b}$$

or

$$T_b = -\frac{q_w}{\tau_w}\frac{\mu}{k}u_b + T_w \quad \text{(xi)}$$

Substitution of the value of T_b in Eq. (viii) gives

$$T_w - \frac{q_w}{\tau_w}\frac{\mu}{k}u_b - T_\infty = \frac{q_w}{\tau_w c_p}(U_\infty - u_b) \quad \text{(xii)}$$

Rearranging the terms, we obtain

$$
\begin{aligned}
T_w - T_\infty &= \frac{q_w}{\tau_w}\frac{\mu}{k}u_b + \frac{q_w}{\tau_w c_p}(U_\infty - u_b) \\
&= \frac{q_w}{\tau_w c_p}\left[\frac{\mu c_p}{k}u_b + (U_\infty - u_b)\right] \\
&= \frac{q_w}{\tau_w c_p}U_\infty\left[(\mathrm{Pr}-1)\frac{u_b}{U_\infty} + 1\right]
\end{aligned}
\tag{xiii}
$$

This gives

$$
\frac{q_w}{T_w - T_\infty} = \frac{\tau_w c_p}{U_\infty}\left[1 + (\mathrm{Pr}-1)\frac{u_b}{U_\infty}\right]^{-1} \tag{7.119}
$$

Introducing the heat transfer coefficient for $\frac{q_w}{T_w - T_b}$ and $\tau_w = (1/2)\rho U_\infty^2 C_{fx}$, we get

$$
h = \frac{C_{fx}}{2}c_p\rho U_\infty\left[1 + (\mathrm{Pr}-1)\frac{u_b}{U_\infty}\right]^{-1}
$$

or

$$
\frac{hx}{k} = \frac{C_{fx}}{2}\left(\frac{\mu c_p}{k}\right)\left(\frac{\rho U_\infty x}{\mu}\right)\left[1 + (\mathrm{Pr}-1)\frac{u_b}{U_\infty}\right]^{-1}
$$

or

$$
\mathrm{Nu}_x = \frac{C_{fx}}{2}\left[\frac{\mathrm{Re}_x\,\mathrm{Pr}}{1 + \frac{u_b}{U_\infty}(\mathrm{Pr}-1)}\right] \tag{7.120}
$$

Deduction of the velocity ratio (u_b/U_∞)
From skin friction correlation of Blasius,

$$
\tau_w = 0.0228\rho U_\infty^2\left(\frac{\nu}{U_\infty\delta}\right)^{1/4}
$$

Substitution in Eq. (ix) gives

$$
\frac{\delta_b}{\delta} = \frac{1}{0.0228}\frac{u_b}{U_\infty}\frac{\mu}{\rho U_\infty\delta}\left(\frac{U_\infty\delta}{\nu}\right)^{1/4} \tag{xiv}
$$

At the junction of the laminar sublayer and turbulent region, 1/7th power law applies. Therefore,

$$
\frac{\delta_b}{\delta} = \left(\frac{u_b}{U_\infty}\right)^7 \tag{xv}
$$

Equating these equations, we get

$$\left(\frac{u_b}{U_\infty}\right)^6 = \frac{1}{0.0228}\left(\frac{\nu}{U_\infty \delta}\right)^{3/4} \tag{xvi}$$

From Eq. (7.80b),

$$\frac{\delta}{x} = \frac{0.375}{\mathrm{Re}_x^{1/5}} = \frac{0.375}{\left(\frac{U_\infty x}{\nu}\right)^{1/5}}$$

Hence,

$$\begin{aligned}\left(\frac{u_b}{U_\infty}\right) &= \left(\frac{1}{0.0228}\right)^{1/6}\left(\frac{\nu}{U_\infty}\right)^{1/8}\left(\frac{1}{0.375}\right)^{1/8}\left(\frac{U_\infty}{\nu}\right)^{1/40}\frac{1}{(x)^{1/10}} \\ &= 2.12\left(\frac{\nu}{U_\infty x}\right)^{0.1} = 2.12(\mathrm{Re}_x)^{-0.1}\end{aligned} \tag{7.121}$$

Introducing the value of u_b/U_∞, we obtain a relation known as *Prandtl–Taylor relation*, which is applicable for turbulent flow over flat plate:

$$\mathrm{Nu}_x = \frac{C_{fx}}{2}\left[\frac{\mathrm{Re}_x \,\mathrm{Pr}}{1 + 2.12\mathrm{Re}_x^{-0.1}(\mathrm{Pr} - 1)}\right] \tag{7.122}$$

For Pr = 1, this equation reduces to simple Reynolds analogy $\mathrm{St_x} = C_{\mathrm{fx}}/2$.

7.13.1 von Karman Analogy for Flat Plates

von Karman developed the following expression of local Nusselt number for turbulent flow over flat plate using the law of wall.

$$\mathrm{St}_x = \frac{\mathrm{Nu}_x}{\mathrm{Re}_x\,\mathrm{Pr}} = \frac{\frac{C_{fx}}{2}}{1 + 5\left(\frac{C_{fx}}{2}\right)^{1/2}\left[\mathrm{Pr} - 1 + \ln\left(\frac{5\,\mathrm{Pr} + 1}{6}\right)\right]} \tag{7.123}$$

For Pr = 1, this equation also reduces to the simple Reynolds analogy. Substituting $C_{\mathrm{fx}} = 0.0592\mathrm{Re_x}^{-0.2}$ from Eq. (7.115) and rearranging the terms, we obtain

$$\mathrm{Nu}_x = \frac{0.0296\mathrm{Re}_x^{0.8}\,\mathrm{Pr}}{1 + 0.86\mathrm{Re}_x^{-0.1}\left[\mathrm{Pr} - 1 + \ln\left(\frac{5\,\mathrm{Pr} + 1}{6}\right)\right]} \quad \text{for} \quad 5 \times 10^5 < \mathrm{Re} < 10^7 \tag{7.124}$$

From the equation, it can be seen that the local Nusselt number decreases from the leading edge, but much less rapidly than for the laminar boundary layer. The average value of the Nusselt number can be determined by the numerical integration of the above equation.

7.14 Reynolds Analogy for Turbulent Flow in Tubes

The following equations for the heat flow and shear stress were developed earlier:

$$q = -\rho c_p(\alpha + \varepsilon_H)\frac{dt}{dy} \tag{7.107}$$

$$\tau = \rho(\nu + \varepsilon_M)\frac{du}{dy} \tag{7.104}$$

According to the Reynolds analogy, as given earlier, the heat and momentum are transferred by analogous processes. It means that both q and τ vary with distance from the wall in the same manner. For the turbulent flow in the tubes, the local shear stress decreases linearly in the radial direction. Hence, referring to Fig. 7.21, we can write

$$\frac{\tau}{\tau_w} = \frac{r}{R} = 1 - \frac{y}{R} \tag{i}$$

and for the heat transfer,

$$\frac{q}{q_w} = \frac{r}{R} = 1 - \frac{y}{R} \tag{ii}$$

Using these equations, Eqs. (7.107) and (7.104) yield

$$q_w\left(1 - \frac{y}{R}\right) = -\rho c_p(\alpha + \varepsilon_H)\frac{dt}{dy} \tag{iii}$$

$$\tau_w\left(1 - \frac{y}{R}\right) = \rho(\nu + \varepsilon_M)\frac{du}{dy} \tag{iv}$$

For $\varepsilon_M = \varepsilon_H$ and Pr = 1 (i.e. $\alpha = \nu$), $(\alpha + \varepsilon_H) = (\nu + \varepsilon_M)$. Dividing Eq. (iii) by Eq. (iv), we obtain

$$\frac{q_w}{\tau_w c_p} = -\frac{dt}{du} \tag{v}$$

Integrating the equation between the tube inner surface ($T = T_w$, $u = 0$) and the bulk values of fluid (T_b, U_m), we get

$$\frac{q_w}{\tau_w c_p} U_m = (T_w - T_m), \tag{vi}$$

which can be rewritten as

$$\frac{q_w}{(T_w - T_m)} = \frac{\tau_w c_p}{U_m}. \tag{vii}$$

Knowing that left-hand side is the heat transfer coefficient h and $\tau_w = (½\ \rho U_m^2)f$, we can transform the above equation to

$$h = \left(\frac{f}{2}\rho c_p U_m\right)$$

or

$$\frac{h}{\rho c_p U_m} = \left(\frac{f}{2}\right)$$

or

$$St = \left(\frac{f}{2}\right) \quad \text{for} \quad \varepsilon_M = \varepsilon_H \quad \text{and} \quad Pr = 1. \tag{7.125}$$

The analogy agrees well with experimental data for the gases whose Prandtl number is nearly unity. For fluids not having Pr = 1, the Prandtl number dependence is taken into account by introducing factor $Pr^{2/3}$ to give

$$StPr^{2/3} = \left(\frac{f}{2}\right). \tag{7.126}$$

However, the above equation is not valid for very high or low values of Pr.

The friction factor in Eq. (7.126) can be calculated from any appropriate equation. If Blasius relation $f = 0.0791\ Re^{-0.25}$ is used

$$\frac{Nu}{Re\,Pr}Pr^{2/3} = \left(\frac{0.0791Re^{-0.25}}{2}\right)$$

or

$$Nu = 0.0396Re^{0.75}Pr^{1/3}. \tag{7.127}$$

Example 7.24 It was found during a test in which water flowed with a velocity of 2.5 m/s through a tube of 25 mm inside diameter and 6.0 m long that the head lost due to the friction was 1.53 m of water. Estimate the surface heat transfer coefficient based on the Reynolds analogy. For water ρ = 998 kg/m^3, c_p = 4.187 kJ/(kg K), $\mu = 1.0 \times 10^{-3}$ kg/(m s) and Pr = 7.02.

Solution

The pressure loss in m of water head is given by

$$h_f = \frac{4fLU_m^2}{2Dg} \tag{i}$$

Given: h_f = 1.53 m, L = 6m, D = 0.025m and U_m = 2.5 m/s. Substitution in Eq. (i) gives

$$f = 5 \times 10^{-3}.$$

For water, the Prandtl number is 7.02 (given) hence the Colburn's analogy is to be applied. This gives

$$\mathrm{St} = \left(\frac{f}{2}\right) \times \frac{1}{\mathrm{Pr}^{2/3}}$$

or

$$\frac{h}{\rho U_m c_p} = \left(\frac{f}{2}\right) \times \frac{1}{\mathrm{Pr}^{2/3}}$$

or

$$h = \rho U_m c_p \left(\frac{f}{2}\right) \times \frac{1}{\mathrm{Pr}^{2/3}} = \frac{998 \times 2.5 \times 4187 \times 5 \times 10^{-3}}{2 \times 7.02^{2/3}} = 7124\ \mathrm{W/(m^2\,K)}.$$

Example 7.25 From the data given in above example, calculate the heat transfer coefficient using the following tube correlation:

$$\mathrm{Nu} = 0.0396\mathrm{Re}^{0.75}\mathrm{Pr}^{1/3}$$

Thermal conductivity k = 60 × 10^{-2} W/(m K).

Solution

Reynolds number,

$$\mathrm{Re} = \frac{\rho U_m D}{\mu} = \frac{998 \times 2.5 \times 0.025}{1.0 \times 10^{-3}} = 62375.$$

Hence,

$$\mathrm{Nu} = 0.0396 \times (62375)^{0.75} \times (7.02)^{1/3} = 299.3,$$

and heat transfer coefficient,

$$h = \mathrm{Nu}\left(\frac{k}{D}\right) = 299.3 \times \frac{60 \times 10^{-2}}{0.025} = 7183.2\ \mathrm{W/(m^2\,K)}.$$

Example 7.26 Derive an expression for the ratio of heat transfer to power required to maintain the flow in terms of the mean fluid velocity U_m and the mean temperature difference θ_m for fully developed flow through a tube. Assume that the simple Reynolds analogy is applicable. What deductions you can make from this expression?

Solution

For a tubular heat exchanger, the heat transfer rate is

$$q = h(\pi DL)(t_w - t_b) = h(\pi DL)\theta_m$$

The pumping power required is

$$\begin{aligned} P &= \text{shear force} \times \text{mean velocity} \\ &= (\tau_w \times \text{wetted area}) \times U_m \\ &= \tau_w \times (\pi DL) \times U_m \end{aligned}$$

Hence, the ratio of the heat transfer and pumping power is

$$\frac{q}{P} = \frac{h\theta_m}{\tau_w U_m}$$

Putting $\tau_w = f\rho U_m^2/2$ and $h = (f/2)\rho U_m c_p$ from the relation $St = h/(\rho U_m c_p) = (f/2)$, we get

$$\frac{q}{P} = \frac{c_p \theta_m}{U_m^2}.$$

From this relation, it can be deduced that for the given rate of heat transfer q, the pumping power can be reduced by reducing the mean velocity of the flow. However, the reduction in the velocity will cause a reduction in the heat transfer coefficient leading to an increased requirement of the surface area. Hence, in the design of the tubular heat exchanger, a compromise is made between the two.

7.14.1 Prandtl–Taylor Modification of Reynolds Analogy for Turbulent Flow in Tubes

It is similar to that presented for flow over a flat plate. For the limits between the laminar sublayer and centreline of the tube, i.e. $y = \delta_b$ (where $u = u_b$, and $\theta = T_w - T_b$) and $y = R$ (where $u = U_{max}$ and $\theta = \theta_c$), we obtain

$$\frac{q_w}{\theta_c} = \frac{\tau_w c_p}{U_{max}}\left[1 + (\Pr - 1)\frac{u_b}{U_{max}}\right]^{-1}$$

or

$$\frac{q_w}{\tau_w c_p} = \frac{\theta_c}{U_{max}}\left[1 + (\text{Pr} - 1)\frac{u_b}{U_{max}}\right]^{-1} \tag{7.128}$$

where $\theta_b/u_b = \theta_c/U_{max}$. Since $\theta_c/U_{max} \approx \theta_m/U_m$, we get

$$\frac{q_w}{\tau_w c_p} = \frac{\theta_m}{U_m}\left[1 + (\text{Pr} - 1)\frac{u_b}{U_{max}}\right]^{-1} \tag{7.129}$$

Transformation to non-dimensional form, as carried out earlier, gives

$$\text{Nu} = \frac{f}{2}\frac{\text{Re Pr}}{\left[1 + \frac{u_b}{U_{max}}(\text{Pr} - 1)\right]} \tag{7.130}$$

where the Nusselt number and Reynolds number are based on the tube diameter.

For the fully developed turbulent flow, $u_b/U_{max} = 1.99/\text{Re}^{1/8}$. Hence, the equation transforms to

$$\text{Nu}_d = \left(\frac{f}{2}\right)\frac{\text{Re Pr}}{1 + 1.99\text{Re}^{-1/8}(\text{Pr} - 1)}$$

or

$$\text{Nu}_d = \frac{0.0396\text{Re}^{0.75}\,\text{Pr}}{1 + 1.99\text{Re}^{-1/8}(\text{Pr} - 1)} \tag{7.131}$$

It can also be written as

$$\frac{\text{Nu}_d}{\text{Re Pr}} = \frac{f/2}{1 + 1.99\text{Re}^{-1/8}(\text{Pr} - 1)}$$

or

$$\text{St} = \frac{f/2}{1 + 1.99\text{Re}^{-1/8}(\text{Pr} - 1)} \tag{7.132}$$

This form is preferred because St, which also equals $h/(c_p \rho U_m)$, can be transformed directly from tube dimensions and temperature measurements.[4]

[4]From energy balance on a heated or cooled fluid in a tube of length L, we can write

$$h(\pi dL)(T_w - T_m) = \rho U_m\left(\frac{\pi}{4}d^2\right)c_p(T_o - T_i)$$

This gives

$$\frac{h}{\rho U_m c_p} = St = \frac{d}{4L}\frac{(T_o - T_i)}{(T_w - T_m)}$$

von Karman also considered the buffer zone and the result of this analogy is

$$\mathrm{St} = \frac{f/2}{1+5\sqrt{\frac{f}{2}}\{\mathrm{Pr}-1+\ln\left[1+\frac{5}{6}(\mathrm{Pr}-1)\right]\}} \tag{7.133}$$

The von Karman result differs from observed behaviour at very low and very high Prandtl numbers. For Pr = 1, both the Prandtl–Taylor and von Karman analogies transform to simple Reynolds analogy.

Introducing $f = 0.0791\ \mathrm{Re}^{-0.25}$ in von Karman relation, we obtain

$$\mathrm{Nu} = \frac{0.0396\mathrm{Re}^{0.75}\,\mathrm{Pr}}{1+\mathrm{Re}^{-1/8}\{\mathrm{Pr}-1+\ln\left[1+\frac{5}{6}(\mathrm{Pr}-1)\right]\}} \tag{7.134}$$

The empirical relations for the tubes are presented in the next chapter.

7.14.2 Friction Drag: Flow Over a Flat Plate Parallel to the Flow

The total force on a body in the direction of the flow is termed as *drag*. The total drag on a non-lifting body is termed as profile drag, which is the sum of the *skin friction drag* and *pressure drag*.

For the flow over a flat plate oriented parallel to the flow direction (zero angle of attack), the pressure drag is zero hence the total drag F_D is equal to the skin friction drag. Thus

$$F_D = \int_A \tau_w dA \tag{7.135}$$

where A is the plate surface area in contact with the fluid.

The drag coefficient is defined as

$$C_D = \frac{F_D}{\frac{1}{2}\rho U_\infty^2 A} = \frac{\int_A \tau_w dA}{\frac{1}{2}\rho U_\infty^2 A} \tag{7.136}$$

7.14.2.1 Laminar Flow

For the laminar flow over a flat plate, the skin friction coefficient is given by

$$C_{fx} = \frac{\tau_w}{\frac{1}{2}\rho U_\infty^2} = \frac{0.664}{\sqrt{\mathrm{Re}_x}} \tag{7.37}$$

For a flat plate of length L and width W, the drag force is

$$F_D = \int_A \tau_w dA = \int_A \frac{1}{2}\rho U_\infty^2 C_{fx} dA = \int_0^L \frac{1}{2}\rho U_\infty^2 C_{fx} W dx$$

$$= \frac{1}{2}\rho U_\infty^2 W \int_0^L C_{fx} dx \tag{i}$$

Substituting the value of C_{fx} from Eq. (7.37) and putting $Re_x = U_\infty x/\upsilon$, we have

$$F_D = \frac{1}{2}\rho U_\infty^2 W \int_0^L 0.664\left(\frac{U_\infty x}{\nu}\right)^{-1/2} dx \tag{ii}$$

Average value of C_D for length 0 to L is

$$\begin{aligned} C_D &= \frac{F_D}{\frac{1}{2}\rho U_\infty^2 A} = \frac{\frac{1}{2}\rho U_\infty^2 W \int_0^L 0.664\left(\frac{U_\infty x}{\nu}\right)^{-1/2} dx}{\frac{1}{2}\rho U_\infty^2 WL} \\ &= \frac{1}{L}\int_0^L 0.664\left(\frac{U_\infty x}{\nu}\right)^{-1/2} dx = \frac{1.328}{\left(\frac{U_\infty L}{\nu}\right)^{1/2}} \\ &= \frac{1.328}{Re_L^{1/2}} \end{aligned} \tag{7.137}$$

which equals the mean value of the skin friction coefficient. This is expected because here the drag is due to the skin friction only.

7.14.2.2 Turbulent Flow

(a) Assuming the flow to be turbulent from the leading edge of the plate, the skin friction coefficient from Eq. (7.81) is

$$C_{fx} = \frac{\tau_w}{\frac{1}{2}\rho U_\infty^2} = \frac{0.0583}{Re_x^{1/5}} \tag{7.81}$$

Average value of C_D for length 0 to L, following the procedure of the previous section, is

$$C_D = \frac{F_D}{\frac{1}{2}\rho U_\infty^2 A} = \frac{\frac{1}{2}\rho U_\infty^2 W \int_0^L 0.0583\left(\frac{U_\infty x}{\nu}\right)^{-1/5} dx}{\frac{1}{2}\rho U_\infty^2 WL}$$

or

$$C_D = \frac{0.0729}{Re_L^{1/5}}, \tag{7.138}$$

which is valid for $5 \times 10^5 < Re < 10^7$.

The empirical relation given by Schlichting for $5 \times 10^5 < Re < 10^9$ is

$$C_D = \frac{0.455}{(\log_{10} Re_L)^{2.58}}, \tag{7.139}$$

which agrees well with the experimental results.

(b) For the boundary layer which is initially laminar, the drag coefficient must take account of the laminar flow over the initial length. The drag coefficient for length L comprising laminar flow up to length x_c can be determined by subtracting drag coefficient value for 0 to x_c from $(C_D)_{turbulent}$ from 0 to L and then adding $(C_D)_{laminar}$ for length 0 to x_c. Thus

$$C_D = \frac{0.0729}{\mathrm{Re}_L^{1/5}} - \frac{0.0729}{\mathrm{Re}_c^{1/5}} \times \frac{x_c}{L} + \frac{1.328}{\mathrm{Re}_c^{1/2}} \times \frac{x_c}{L}$$

Substituting $x_c = \mathrm{Re}_c(\nu/U_\infty)$, $\mathrm{Re}_c = 5 \times 10^5$ and rearranging, we obtain

$$C_D = \frac{0.0729}{\mathrm{Re}_L^{1/5}} - \frac{1703}{\mathrm{Re}_L} \quad \text{for} \quad 5 \times 10^5 < \mathrm{Re} < 10^7. \tag{7.140}$$

Following the same procedure, we can obtain from Eq. (7.139),

$$C_D = \frac{0.455}{(\log_{10}\mathrm{Re}_L)^{2.58}} - \frac{1614}{\mathrm{Re}_L} \quad \text{for} \quad 5 \times 10^5 < \mathrm{Re} < 10^9. \tag{7.141}$$

Example 7.27 Air at atmospheric pressure and 20°C is flowing parallel to a flat plate at a velocity of 30 m/s. The plate is 1 m long and is kept at a constant temperature of 180°C. For the unit width of the plate, determine

(a) The drag force on the plate for the laminar portion of the flow.
(b) Heat transfer rate from the one side of the plate for the laminar portion of the flow.
(c) The drag force on the plate.
(d) The heat transfer rate from one side of the complete length of the plate.

What would be the heat transfer rate if the flow is turbulent from the leading edge of the plate?

Solution

The properties of air at the mean film temperature t_{fm} = (20 + 180)/2 = 100°C are

$$\rho = 0.9452\,\mathrm{kg/m^3}, k = 0.0317\,\mathrm{W/(m\,K)}, \mu = 2.172 \times 10^{-5}\,\mathrm{kg/(m\,s)}, \mathrm{Pr} = 0.693, c_p = 1011.3\,\mathrm{J/(kg\,K)}.$$

Laminar regime
Critical length, x_c:

$$x_c = \frac{\mathrm{Re}_c \mu}{\rho U_\infty} = \frac{5 \times 10^5 \times 2.172 \times 10^{-5}}{0.9452 \times 30} = 0.384\,\mathrm{m}.$$

(a) **Drag**

Skin friction coefficient,

$$C_f = \frac{1.328}{\sqrt{\mathrm{Re}_{cr}}} = \frac{1.328}{\sqrt{5 \times 10^5}} = 0.00188$$

Drag force,

$$F_D = C_f A\left(\frac{1}{2}\rho U_\infty^2\right) = 0.00188 \times (0.384 \times 1) \times \left(\frac{1}{2} \times 0.9452 \times 30^2\right) = 0.31\ \mathrm{N}.$$

(b) **Heat transfer rate**

The average heat transfer coefficients,

$$\begin{aligned} \bar{h} &= \mathrm{Nu}\frac{k}{x_c} = 0.664\mathrm{Re}_{cr}^{1/2}\mathrm{Pr}^{1/3}\frac{k}{x_c} \\ &= 0.664 \times (5 \times 10^5)^{1/2} \times (0.693)^{1/3} \times \frac{0.0317}{0.384} = 34.3\ \mathrm{W/(m^2\,K)}. \end{aligned}$$

Rate of heat transfer q from one side of the plate for length x_c,

$$q = \bar{h}A(t_w - t_\infty) = 34.3 \times (0.384 \times 1) \times (180 - 20) = 2107.6\ \mathrm{W}.$$

Laminar and turbulent regime
Flow Reynolds number

$$\mathrm{Re}_L = \frac{\rho U_\infty L}{\mu} = \frac{0.9452 \times 30 \times 1}{2.172 \times 10^{-5}} = 1.305 \times 10^6.$$

(c) **Drag**

Skin friction coefficient,

$$C_f = \frac{0.0729}{\mathrm{Re}_L^{1/5}} - \frac{1703}{\mathrm{Re}_L} = \frac{0.0729}{(1.305 \times 10^6)^{1/5}} - \frac{1703}{1.305 \times 10^6} = 0.00306. \tag{7.140}$$

Drag force,

$$F_D = C_f A\left(\frac{1}{2}\rho U_\infty^2\right) = 0.00306 \times (1 \times 1) \times \left(\frac{1}{2} \times 0.9452 \times 30^2\right) = 1.3\ \mathrm{N}.$$

(d) **Heat transfer rate**

The average heat transfer coefficient,

$$\bar{h} = \mathrm{Nu}\frac{k}{x_{cr}} = \left(0.037\mathrm{Re}_L^{0.8} - 871\right)\mathrm{Pr}^{1/3}\frac{k}{L}$$

Substitution of the values of various parameters gives

$$\begin{aligned}\bar{h} &= \left[0.037 \times (1.305 \times 10^6)^{0.8} - 871\right] \times (0.693)^{1/3} \times \frac{0.0317}{1.0} \\ &= 56.6\ \mathrm{W/(m^2\,K)}.\end{aligned}$$

Rate of heat transfer q from one side of the plate for length $L = 1$ m,

$$q = \bar{h}A(t_w - t_\infty) = 56.6 \times (1 \times 1) \times (180 - 20) = 9056\ \mathrm{W}.$$

Turbulent flow from the leading edge of the plate

$$\mathrm{Nu} = 0.037\mathrm{Re}_L^{0.8} \times \mathrm{Pr}^{1/3}$$

or

$$\bar{h} = \mathrm{Nu}\frac{k}{L} = 0.037 \times (1.305 \times 10^6)^{0.8} \times (0.693)^{1/3} \times \frac{0.0317}{1.0} = 81.03\ \mathrm{W/(m^2\,K)}.$$

Rate of heat transfer q from one side of the plate for length $L = 1$ m,

$$q = \bar{h}A(t_w - t_\infty) = 81.03 \times (1 \times 1) \times (180 - 20) = 12964.8\ \mathrm{W}.$$

7.15 Natural or Free Convection

Natural convection occurs whenever a heated or cooled surface is placed in a volume of fluid which is stagnant before heat is added or removed.

Consider a heated vertical plate at constant temperature T_w shown in Fig. 7.28. The density of the fluid in contact with the wall (the surface of the heated plate) decreases due to the rise in the temperature. This causes the fluid near the wall to move upwards and create free convection flow. Thus, the heat is transferred away from the wall. As the heated fluid rises, cold fluid moves near the wall and a current is set in motion. The gravity field of the earth is the pump in the considered example.

Free convection under the influence of the gravitational force is frequently encountered in engineering applications and has been extensively studied.

Flows can be caused by other body forces such as centrifugal, coriolis forces, etc. Electric or magnetic forces may arise at supersonic speeds and influence the flow. Here, only the flow under the influence of gravitational force will be considered.

A boundary layer is formed in free convection also as the fluid moves due to the buoyancy effect. Fluid relatively far from the surface will enter the layer and is heated. Thus, the

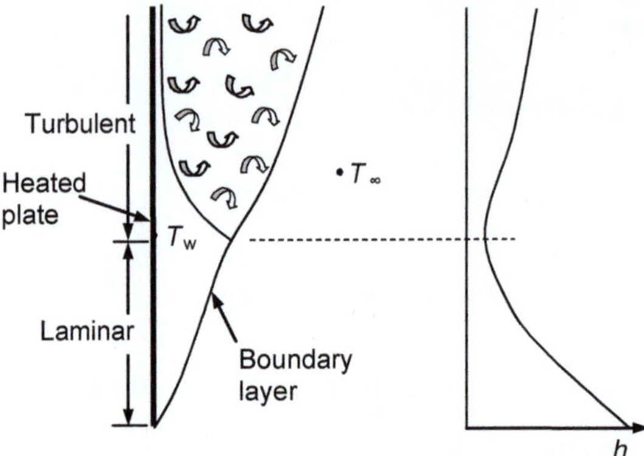

Fig. 7.28 Boundary layer over a heated vertical plate and variation of heat transfer coefficient

boundary layer, which has zero thickness at the lower edge, increases in the upward direction as shown in Fig. 7.28. Over the lower section of the wall, the flow is laminar, and eddying or turbulent along the upper section in the figure. The condition of the motion is mainly governed by the difference in the temperature of the wall and the fluid, the shape of the body, the dimension (known as the characteristic dimension) and the thermophysical properties of the fluid. In the initial section, the flow is always laminar. The heat transfer rate depends significantly on the type of flow.

It must be noted that because of the low flow velocities encountered in the free convection, the boundary layers are thicker than those in the forced convection.

Analysis is being presented for a heated vertical plate, which is simple to deal with. Experimental measurements are relied upon to obtain relations for other configurations. Some of such relations are presented in Chap. 9.

7.16 Integral Momentum and Energy Equation of Free Convection on a Vertical Plate

The integrated boundary layer equations for momentum and heat transfer can be used to calculate the heat transfer in free convection.

Since the free convection is produced by the buoyancy effect, the Archimedes' principle applies to a volume of fluid within the heated layer, refer Fig. 7.29. The buoyancy force is $(V\rho_\infty g - V\rho g)$ where ρ is the density of the heated fluid element.

The coefficient of thermal expansion is defined as

$$\beta = \frac{\rho_\infty - \rho}{\rho(\Delta T)} \tag{7.142}$$

Using this, the buoyancy force becomes $V\rho\beta g(\Delta T)$, i.e. it is a function of βg and ΔT.

Let us place the origin of the coordinates at the lower edge of the plate; distances along and perpendicular to the plate are x and y, respectively. We assume that the plate is very wide in z-direction.

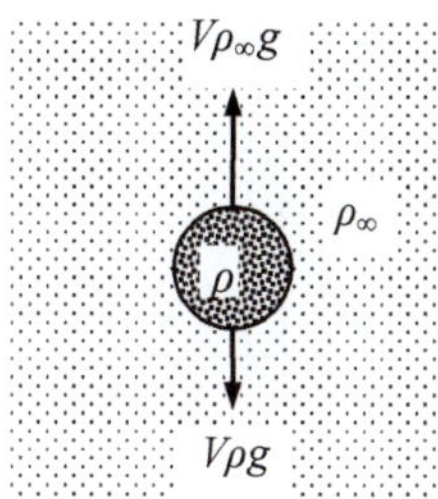

Fig. 7.29 Forces on a fluid element under free convection

To simplify the analysis, the following assumptions are being made:

1. Steady-state process.
2. Plate temperature, designated as θ_w (= $T_w - T_\infty$), is constant along its length.
3. The forces of inertia are negligible compared to those of gravity and viscosity.
4. Transfer of heat by convection and conduction along the direction of the moving fluid (i.e. in the x-direction) are ignored.
5. The pressure gradient is zero.

The temperature of the fluid, designated as θ (= $T - T_\infty$), decreases from the value of plate surface temperature θ_w to the temperature $\theta = 0$ of the fluid outside the boundary layer. The velocity, which is zero at the plate surface due to the no-slip condition, increases to some maximum value at some distance from the plate surface and then again reduces to zero at the edge of the boundary layer.

The equation of the temperature distribution is approximated by a parabola

$$\theta = \theta_w\left(1 - \frac{y}{\delta}\right)^2 \tag{7.143}$$

The equation satisfies the boundary conditions of $\theta = \theta_w$ at $y = 0$, $\theta = 0$ at $y = \delta$ and $\frac{\partial\theta}{\partial y} = 0$ at $y = \delta$.

The equation of the velocity profile is expressed as (refer Example 7.28)

$$u = u^*\frac{y}{\delta}\left(1 - \frac{y}{\delta}\right)^2 \tag{7.144}$$

where u^* is an arbitrary function with dimensions of velocity and it is function of x. This equation gives U_{max} = (4/27) u^* at $y = \delta/3$. The velocity and temperature distribution are plotted in Figs. 7.30 and 7.31.

The integral momentum equation for the free convection system can be written as follows (refer Fig. 7.32).

The mass flow through plane AA' for unit width of the plate is, considering an elemental vertical strip of thickness dy (not shown in the figure),

$$\int_0^H \rho u dy \tag{i}$$

The momentum flow through this plane is

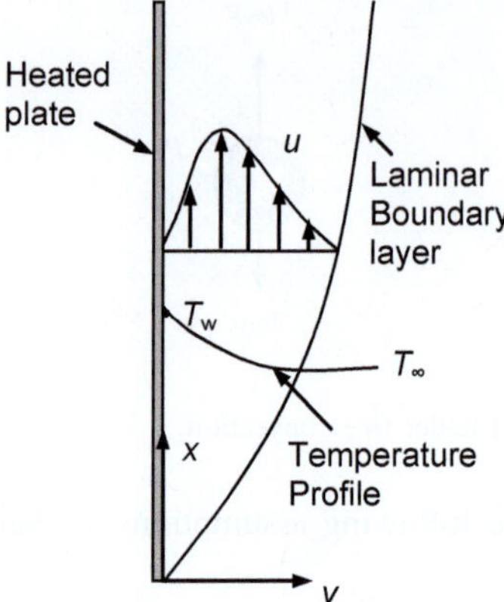

Fig. 7.30 Laminar natural convection boundary layer over a vertical wall: coordinate system, and expected velocity and temperature profiles

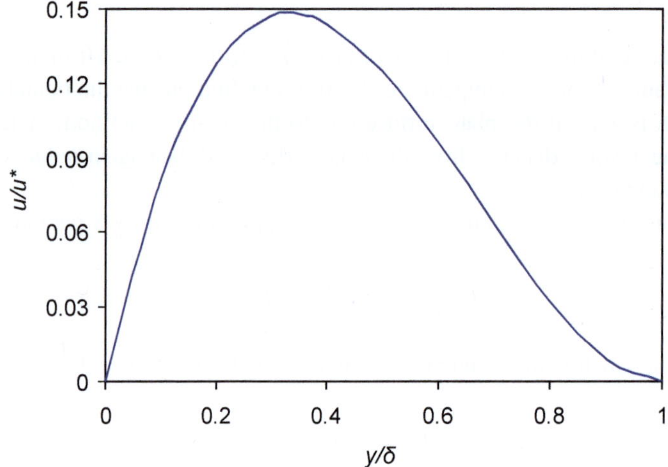

Fig. 7.31 Free convection velocity profile given by Eq. (7.144)

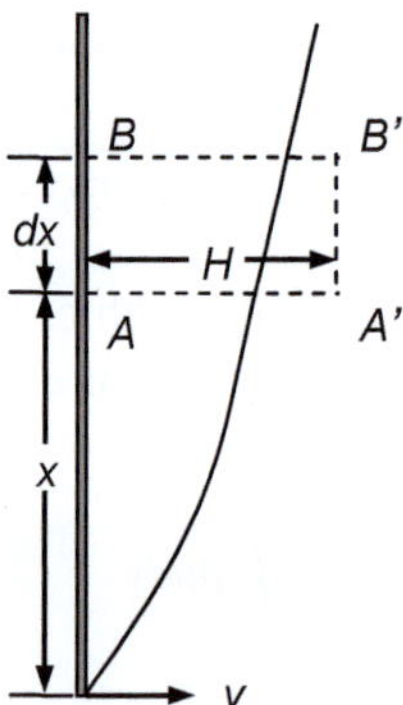

Fig. 7.32 Integral momentum equation control volume

$$\int_0^H \rho u^2 dy \tag{ii}$$

The momentum flow through plane BB' is

$$\int_0^H \rho u^2 dy + \frac{d}{dx}\left(\int_0^H \rho u^2 dy\right) dx \tag{iii}$$

The net momentum flow out of the control volume is difference of the above two equations and is, therefore,

$$\frac{d}{dx}\left(\int_0^H \rho u^2 dy\right) dx \tag{iv}$$

The shear force at the wall is

$$-\tau_w . dx = -\mu\left(\frac{\partial u}{\partial y}\right)_{y=0} dx \tag{v}$$

The buoyancy force is

$$dx \times \int_0^H \rho g\beta(T - T_\infty)dy = \left(\int_0^H \rho g\beta\theta dy\right) dx \tag{vi}$$

Combining the balancing forces given by the above equations, the integral momentum equation is obtained

$$\frac{d}{dx}\left(\int_0^\delta \rho u^2 dy\right) = -\mu\left(\frac{\partial u}{\partial y}\right)_{y=0} + \int_0^\delta \rho g\beta\theta dy \tag{vii}$$

where the upper limit of the integral has been changed to δ (which is the boundary layer thickness) because the integrand is zero for $y > \delta$ (the boundary thickness has been assumed to be same for the temperature and velocity to simplify the computational work).

The energy equation remains unchanged, Eq. (7.66),

$$\frac{d}{dx}\int_0^\delta u\theta dy = -\alpha\left(\frac{d\theta}{dy}\right)_{y=0} \tag{viii}$$

Equations (vii) and (viii) can be solved by introducing the temperature and velocity profile equations given by Eqs. (7.143) and (7.144), respectively,

$$\int_0^\delta u^2 dy = \frac{(u^*)^2\delta}{105} \quad \text{(ix)}$$

$$\int_0^\delta \theta dy = \frac{\theta_w \delta}{3} \quad \text{(x)}$$

$$\int_0^\delta \theta u dy = \frac{u^* \theta_w \delta}{30} \quad \text{(xi)}$$

The boundary layer equations now become

$$\frac{1}{105}\frac{d}{dx}\left[(u^*)^2\delta\right] = \frac{1}{3}g\beta\theta_w\delta - \nu\frac{u^*}{\delta} \quad \text{(xii)}$$

$$\frac{1}{30}\theta_w\frac{d}{dx}[u^*\delta] = 2\alpha\frac{\theta_w}{\delta} \quad \text{(xiii)}$$

Let u^* and δ are power function of x, then

$$u^* = C_1 x^a \quad \text{(xiv)}$$

$$\delta = C_2 x^b \quad \text{(xv)}$$

Introducing these expressions in Eqs. (xii) and (xiii), we get

$$\frac{2a+b}{105}C_1^2 C_2 x^{2a+b-1} = g\beta\theta_w\frac{C_2}{3}x^b - \frac{C_1}{C_2}\nu x^{a-b} \quad \text{(xvi)}$$

$$\frac{a+b}{30}C_1 C_2 x^{a+b-1} = \frac{2\alpha}{C_2}x^{-b} \quad \text{(xvii)}$$

For these equations to be dimensionally homogeneous, the exponents of x must have the same value in each term. Therefore,

$$2a+b-1 = b = a-b$$
$$a+b-1 = -b$$

Solution of these equations gives

$$a = \frac{1}{2}, b = \frac{1}{4}.$$

Introducing the values of a and b in Eqs. (xvi) and (xvii), we get

$$\frac{5C_1^2 C_2}{420} = g\beta\theta_w\frac{C_2}{3} - \frac{C_1}{C_2}\nu$$

and

$$\frac{C_1 C_2}{40} = 2\frac{\alpha}{C_2}$$

Solving these equations for C_1 and C_2, we get

$$C_1 = 5.17\nu\left(\frac{20}{21} + \frac{\nu}{\alpha}\right)^{-1/2}\left(\frac{g\beta\theta_w}{\nu^2}\right)^{1/2} \tag{xviii}$$

$$C_2 = 3.93\left(\frac{20}{21} + \frac{\nu}{\alpha}\right)^{1/4}\left(\frac{g\beta\theta_w}{\nu^2}\right)^{-1/4}\left(\frac{\nu}{\alpha}\right)^{-1/2} \tag{xix}$$

The maximum velocity is

$$U_{\max} = \frac{4}{27}u^* = 0.766\nu\left(0.952 + \frac{\nu}{\alpha}\right)^{-1/2}\left(\frac{g\beta\theta_w}{\nu^2}\right)^{1/2} x^{1/2} \tag{xx}$$

and boundary layer thickness from Eq. (xv) is

$$\begin{aligned}\delta &= 3.93\left(\frac{20}{21} + \frac{\nu}{\alpha}\right)^{1/4}\left(\frac{g\beta\theta_w}{\nu^2}\right)^{-1/4}\left(\frac{\nu}{\alpha}\right)^{-1/2} x^{1/4} \\ &= 3.93(0.952 + \mathrm{Pr})^{1/4}\left(\frac{g\beta\theta_w x^3}{\nu^2}\right)^{-1/4}\mathrm{Pr}^{-1/2} x\end{aligned}$$

or

$$\frac{\delta}{x} = 3.93 Gr_x^{-1/4}\mathrm{Pr}^{-1/2}(0.952 + \mathrm{Pr})^{1/4} \tag{7.145}$$

where $\mathrm{Gr_x} = \left(\frac{g\beta\theta_w x^3}{\nu^2}\right)$ is a dimensionless number known as *Grashof number*.

Heat flow by convection at the plate surface is given by

$$q = -k\left(\frac{d\theta}{dy}\right)_{y=0}$$

Using Eq. (7.143),

$$q = 2k\left(\frac{\theta_w}{\delta}\right) \tag{xxi}$$

The heat flow by convection is given by

$$q = h\theta_{\mathrm{w}}$$

Equating the two, the local heat transfer coefficient can be written as

$$h_x = 2\frac{k}{\delta}$$

or

$$\text{Nu}_x = \frac{h_x x}{k} = 2\frac{x}{\delta} \tag{xxii}$$

Substituting the value of x/δ from Eq. (7.145), we obtain the relation of local heat transfer coefficient as

$$\text{Nu}_x = 0.508\text{Gr}_x^{1/4}\text{Pr}^{1/2}(0.952 + \text{Pr})^{-1/4} \tag{7.146}$$

Since the local heat transfer coefficient varies as $x^{-1/4}$, the average heat transfer coefficient is

$$\bar{h} = \frac{1}{L}\int_0^L h_x dx = \frac{4}{3}h_L \tag{7.147}$$

i.e. the average value of the heat transfer coefficient is 4/3 times the local value at $x = L$. For air, Pr $\approx$ 0.7 and

$$\text{Nu}_x = 0.375\text{Gr}_x^{1/4}. \tag{7.148}$$

The numerical value of 0.375 in the equation is in good agreement with the value of 0.36 obtained in an exact solution of Pohlhusen.

In this analysis, a new dimensionless number Gr has been introduced. This number plays an important role in all free convection problems similar to that played by the Reynolds number in forced convection. The product of Grashof and Prandtl number (Gr Pr) provides a criterion for transition from laminar to turbulent boundary layer flow in free convection. For air, the critical Grashof number is about 4×10^8 for a vertical plate. For different configurations, the critical Grashof number ranges from 10^8 to 10^9.

The Grashof number Gr can be interpreted physically as a dimensionless number representing the ratio of the buoyancy to viscous forces in free convection, i.e.

$$\text{Gr} = \left(\frac{g\beta\theta_w x^3}{\nu^2}\right) = \frac{\text{Buoyancy force}}{\text{Viscous force}} \tag{7.149}$$

The analysis has been presented here for the free convection heat transfer from a vertical isothermal plate because it is the simplest case that can be treated mathematically. For the majority of the configurations, the correlations have been developed from the results of the experimental studies. The experimental approach is preferred because, in most of the cases, it is difficult to predict temperature and velocity profiles analytically. Analytical solutions for free convection turbulent flow are practically impossible. Hence, empirical data (experimental results) are presented using the method of dimensional analysis in the form of correlation

$$\text{Nu}_x = C(\text{Gr}_x\text{Pr}_x)^n \tag{7.150}$$

where the constant C and the exponent n are experimentally determined. Correlations presented by different researchers are listed in Chap. 9.

Example 7.28 (i) Derive the differential equation of motion for the free convection boundary layer on a vertical flat plate. (ii) Show that the equation of the velocity profile in a free convection flow over a vertical plate can be expressed as

$$\frac{u}{u*} = \frac{y}{\delta}\left(1 - \frac{y}{\delta}\right)^2$$

where u^* is a fictitious velocity, which is a function of the distance measured along the height of the plate.

Solution

(i) Following the method presented for the differential equation for the forced flow laminar boundary layer on a flat plate, we proceed as follows.

Momentum of mass entering the bottom face of the control volume for unit depth is, refer Fig. 7.33,

$$\rho u dy.u = \rho u^2 dy.$$

Momentum leaving the top face is

$$\rho u^2 dy + \frac{\partial}{\partial x}(\rho u^2 dy)dx.$$

Momentum in x-direction of the mass $\rho v(\Delta x.1)$, which enters the left face with velocity u in x-direction, is

$$\rho v dx.u.$$

Momentum in x-direction of this mass leaving the right face is

$$\rho uv dx + \frac{\partial}{\partial y}(\rho uv dx)dy.$$

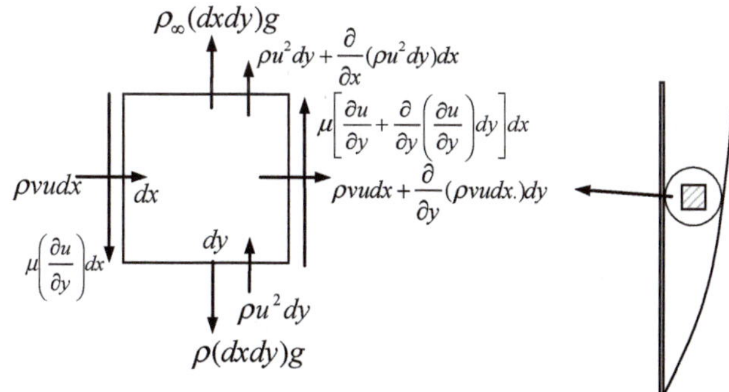

Fig. 7.33 Example 7.28

Viscous shear force on the left face, for unit width of the plate, is

$$-\mu \frac{\partial u}{\partial y} dx$$

and the shear on the right face is

$$\mu \left[\frac{\partial u}{\partial y} + \frac{\partial}{\partial y} \left(\frac{\partial u}{\partial y} \right) dy \right] dx$$

The buoyancy force acting in the upward direction is

$$(\rho_\infty - \rho) dx.dy.g.$$

The density difference $(\rho_\infty - \rho)$ may be expressed in the terms of the volumetric coefficient of expansion. This transforms the term representing the buoyancy force to

$$g\rho\beta(T - T_\infty)(dx.dy)$$

Equating the sum of the shear and buoyancy forces to the net momentum in the x-direction, we get

$$\begin{aligned}
\mu \frac{\partial^2 u}{\partial y^2} dx.dy + g\rho\beta(T - T_\infty)(dx.dy) &= \frac{\partial}{\partial x}(\rho u^2 dy)dx + \frac{\partial}{\partial y}(\rho uv dx)dy \\
&= 2\rho u \frac{\partial u}{\partial x} dxdy + \rho u \frac{\partial v}{\partial y} dxdy + \rho v \frac{\partial u}{\partial y} dxdy \\
&= \rho \left(u \frac{\partial u}{\partial x} + v \frac{\partial u}{\partial y} \right) dxdy + \rho \left(\frac{\partial u}{\partial x} + \frac{\partial v}{\partial y} \right) u.dxdy
\end{aligned}$$

From the continuity equation, $\left(\frac{\partial u}{\partial x} + \frac{\partial v}{\partial y} \right) = 0$. Using this relation and simplifying, we get

$$\rho \left(u \frac{\partial u}{\partial x} + v \frac{\partial u}{\partial y} \right) = g\rho\beta(T - T_\infty) + \mu \frac{\partial^2 u}{\partial y^2}, \quad (1)$$

which is the desired differential equation.

(ii) The boundary conditions for the velocity profile are

$$u = 0 \text{ at } y = 0, \quad \text{(i)}$$

$$u = 0 \text{ at } y = \delta, \quad \text{(ii)}$$

$$\frac{\partial u}{\partial y} = 0 \text{ at } y = \delta \quad \text{(iii)}$$

and from Eq. (1), at $y = 0$

$$\frac{\partial^2 u}{\partial y^2} = -g\beta \frac{(T_w - T_\infty)}{\nu} \tag{iv}$$

Let the equation of the velocity profile is a cubic-polynomial function of y to satisfy the four conditions. Thus

$$\frac{u}{u^*} = a + by + cy^2 + dy^3$$

where u^* is a fictitious velocity, which is function of x. This is based on the assumption that the velocity profile is having geometrically similar shapes at various x locations along the plate.

Applying the four boundary conditions, we obtain

$$\frac{u}{u^*} = \left[\frac{g\beta\delta^2(T_w - T_\infty)}{4u^*\nu}\right] \cdot \frac{y}{\delta} \cdot \left(1 - \frac{y}{\delta}\right)^2$$

The term in the first bracket on the right-hand side of the equation can be incorporated in u^*. Then the resulting expression for the velocity profile is

$$\frac{u}{u^*} = \frac{y}{\delta}\left(1 - \frac{y}{\delta}\right)^2$$

Example 7.29 Show that the coefficient of cubical expansion can be expressed as

$$\beta = \frac{\rho_\infty - \rho}{\rho dT}$$

where ρ and ρ_∞ are the densities of the heated fluid mass element and surrounding fluid, respectively. The heated fluid is at a temperature dT above the surrounding fluid.

Also show that $\beta = 1/T$ for an ideal gas having equation of state $pV = RT$.

Solution

(i) Consider a small element of fluid mass m and volume V. The mass of the unheated fluid can be expressed as

$$m = \rho_\infty V \tag{i}$$

Heating of the elemental mass increases the volume of the element to $V + dV$ and decreases the density to ρ. Hence,

$$m = \rho(V + dV) \tag{ii}$$

Equating the above two equations, we get

$$\rho_\infty V = \rho(V + dV) \tag{iii}$$

The coefficient of cubical expansion β is defined as the change in the volume per unit temperature rise at constant pressure, i.e.

$$\beta = \frac{dV}{V} \cdot \frac{1}{dT} \tag{iv}$$

Substituting the value of dV in terms of β in Eq. (iii), we get

$$\begin{aligned} \rho_\infty V &= \rho(V + \beta V dT) \\ \rho_\infty &= \rho(1 + \beta dT). \end{aligned}$$

Hence,

$$\beta = \frac{\rho_\infty - \rho}{\rho dT}$$

(ii) From the prefect gas equation, for p = constant,

$$pdV = RdT$$

Dividing both sides by V,

$$\frac{dV}{V} = \frac{R}{pV} dT$$

Using $\frac{R}{pV} = \frac{1}{T}$ from the prefect gas equation, we obtain

$$\frac{dV}{V} = \frac{dT}{T}$$

Substitution of the values in Eq. (iv) gives

$$\beta = \frac{dT}{T} \cdot \frac{1}{dT} = \frac{1}{T}$$

Example 7.30 For a flow through a circular pipe, the velocity profile is parabolic and is given by $u(r) = U_{max}(1 - r^2/R^2)$ and the non-dimensional temperature profile is given by $(t - t_w)/(t_o - t_w) = (1 - r^2/R^2)$, when U_{max} is the maximum velocity, R is the radius, t_w is the wall temperature and t_o is a constant (independent of x and r). Show that the Nusselt number of the flow is 6.

Solution

For the tube flow, the convection heat transfer coefficient is defined by

$$h = \frac{q_w}{t_b - t_w} \tag{i}$$

where T_b is the bulk mean temperature of the fluid and $q_w = -k\left(\frac{\partial t}{\partial r}\right)_{r=R}$. The bulk temperature equation is

$$t_b = \frac{\int_0^R r.dr.u.T}{\int_0^R r.dr.u.}$$

From the given equations of temperature and velocity profiles,

$$\begin{aligned} q_w &= -k\left\{\frac{\partial}{\partial r}\left[(t_0 - t_w)\left(1 - \frac{r^2}{R^2}\right)\right]\right\}_{r=R} \\ &= k(t_0 - t_w)\frac{2}{R} \end{aligned}$$

and

$$\begin{aligned} t_b &= \frac{\int_0^R r.dr.U_{\max}\left(1 - \frac{r^2}{R^2}\right)\left[(t_0 - t_w)\left(1 - \frac{r^2}{R^2}\right) + t_w\right]}{\int_0^R r.drU_{\max}\left(1 - \frac{r^2}{R^2}\right)} \\ &= \frac{\int_0^R r.dr\left(1 - \frac{r^2}{R^2}\right)\left[(t_0 - t_w)\left(1 - \frac{r^2}{R^2}\right)\right]}{\int_0^R r.dr\left(1 - \frac{r^2}{R^2}\right)} + t_w \\ &= \frac{\int_0^R \left[\left(r + \frac{r^5}{R^4} - 2\frac{r^3}{R^2}\right)\right]dr}{\int_0^R \left(r - \frac{r^3}{R^2}\right).dr}(t_0 - t_w) + t_w \end{aligned}$$

Integration gives

$$t_b - t_w = \frac{2}{3}(t_0 - t_w)$$

Substitution in Eq. (i) gives

$$h = \frac{2k}{R}(t_o - t_w)\frac{3}{2(t_o - t_w)}$$

or

$$\frac{h(2R)}{k} = 6$$

i.e. Nu = 6.

7.17 Liquid Metal Heat Transfer for Laminar Flow Over a Flat Plate

Liquid metals[5] have been employed because of their capability of removing large energy quantities, which is due to their high thermal conductivity. They remain in liquid state at higher temperatures than water. However, the liquid metal requires careful handling since they are corrosive and show a violent reaction when come into contact with water.

Let us consider flow of liquid metal over a flat plate. Since the Prandtl number of the liquid metal is quite low, ranging from 0.005 to 0.03, the thermal boundary layer thickness is substantially larger than the velocity boundary layer thickness as shown in Fig. 7.34.

We can assume the velocity to be uniform over the whole of the thermal boundary layer because the velocity profile is having very blunt shape as shown in Fig. 7.34, i.e.

$$u = U_\infty \tag{i}$$

It is an idealized flow often referred as *slug flow*.

Since the boundary conditions for the temperature profile are the same as those in Sect. 7.7.2, we have

$$\frac{\theta}{\theta_\infty} = \frac{T - T_w}{T_\infty - T_w} = \frac{3}{2}\frac{y}{\delta_t} - \frac{1}{2}\left(\frac{y}{\delta_t}\right)^3 \tag{7.62}$$

For low velocity flow, the integral energy equation [refer Eq. (7.66)] is

$$\frac{d}{dx}\int_0^{\delta_t}(\theta_\infty - \theta)u dy = \alpha\left(\frac{\partial\theta}{\partial y}\right)_{y=0}$$

Using Eqs. (i) and (7.62), we obtain

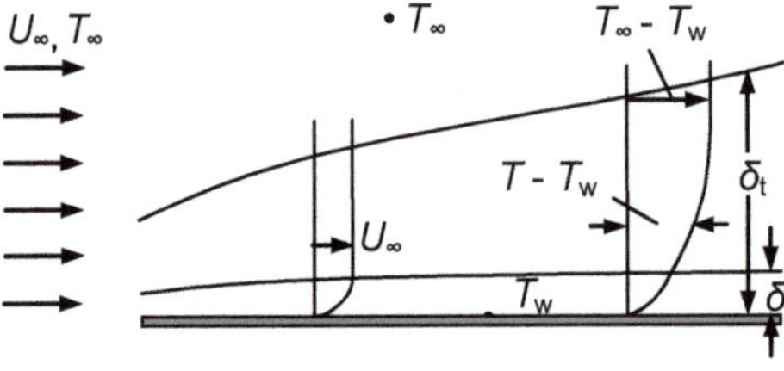

Fig. 7.34 Boundary layer regimes for flow of liquid metal over a flat plate

[5]Bismuth (Bi), lead (Pb), lithium (Li), mercury (Hg), potassium (K), sodium (Na), 22% Na + 78% K, 56% Na + 44% K, 44.5% Pb + 55.5% Bi, etc.

$$U_{\infty}\theta_{\infty}\frac{d}{dx}\int_{0}^{\delta_t}\left[1-\frac{3}{2}\frac{y}{\delta_t}+\frac{1}{2}\left(\frac{y}{\delta_t}\right)^3\right]dy=\frac{3\alpha\theta_{\infty}}{2\delta_t}$$

The integral yields

$$2\delta_t d\delta_t=\frac{8\alpha}{U_{\infty}}dx$$

The solution of this differential equation is

$$\delta_t=\sqrt{\frac{8\alpha x}{U_{\infty}}} \tag{7.151}$$

The local heat transfer coefficient,

$$\begin{aligned} h_x &= \frac{-k\left(\frac{\partial T}{\partial y}\right)_{y=0}}{T_w-T_{\infty}} \\ &= k\left\{\frac{\partial}{\partial y}\left[\frac{3}{2}\frac{y}{\delta_t}-\frac{1}{2}\left(\frac{y}{\delta_t}\right)^3\right]\right\}_{y=0} \\ &= \frac{3k}{2\delta_t}. \end{aligned}$$

Substitution of the value of δ_t from Eq. (7.151) gives

$$h_x=\frac{3k}{2}\sqrt{\frac{U_{\infty}}{8\alpha x}}=0.53\frac{k}{x}\sqrt{\frac{\rho U_{\infty}x}{\mu}\times\frac{\mu c_p}{k}}$$

or

$$\frac{h_x x}{k}=0.53\sqrt{\frac{\rho U_{\infty}x}{\mu}\times\frac{\mu c_p}{k}}$$

or

$$\mathrm{Nu}_x=0.53(\mathrm{Re}_x\,\mathrm{Pr})^{1/2}=0.53\mathrm{Pe}^{1/2} \tag{7.152}$$

The product (Re_x Pr) is a non-dimensional number called the *Peclet number* Pe. It is an important parameter in liquid metal convection. Therefore, the empirical relations are usually expressed in terms of this number.

Equation (7.151) can be transformed as follows.

$$\frac{\delta_t}{x}=\left(\frac{8}{\mathrm{Re}_x\,\mathrm{Pr}}\right)^{1/2}$$

The expression for the boundary layer thickness is

$$\frac{\delta}{x} = \frac{4.64}{\mathrm{Re}_x^{1/2}} \tag{7.154}$$

The ratio of the thermal and hydrodynamic boundary layers is

$$\frac{\delta_t}{\delta} = \xi = \frac{0.609}{\mathrm{Pr}^{1/2}} \tag{7.155}$$

For Pr = 0.01, the ratio $\frac{\delta_t}{\delta} = 6$. Thus, the approximation of the slug flow appears to be reasonable.

Notes:

1. The solution presented here is also based on the assumptions mentioned earlier, but, for the liquid metals, the heat flow by conduction in the axial direction (direction of flow) may not always be negligible and the solution must be modified. If Pe < 100, the axial conduction may cause a reduction in the effective heat transfer coefficient (Kays and Crawford 1980).
   An exact solution for flow over a plate with uniform temperature gives (Rogers and Mayhew 1967)

$$\begin{aligned} \mathrm{Nu_x} &= 0.487\,\mathrm{Pe}^{1/2} \quad \text{when Pr} = 0.03 \\ \mathrm{Nu_x} &= 0.516\,\mathrm{Pe}^{1/2} \quad \text{when Pr} = 0.01 \\ \mathrm{Nu_x} &= 0.526\,\mathrm{Pe}^{1/2} \quad \text{when Pr} = 0.006 \end{aligned}$$

2. The approximate solution presented above is also valid in the turbulent flow region where the 1/7th power velocity profile is closer to the slug flow model. This is due to fact that the heat transfer is mainly governed by the thermal conductivity of the liquid metal and the effect of eddy mixing is small.

7.18 Summary

In convection, there is an observable bulk motion of the fluid and hence can occur only in the case of liquids, gases and multiphase mixtures. The convective heat transfer can be classified into two categories: (i) natural or free convection and (ii) forced convection.

In natural convection, the fluid motion is entirely because of the density differences in the fluid caused due to the local heating in the gravity field. The fluid surrounding the heat source becomes less dense due to the heating and hence moves upwards (known as buoyancy effect). The surrounding colder fluid moves to replace it. This colder fluid is then heated and the process continues, forming the convection currents. If the motion of the fluid is caused by some external means such as a fan, blower, pump or wind, the mode of heat transfer is termed as the forced convection.

The convection is a combined problem of heat and fluid flow. The analytical solution of a convection problem involves application of the equation of motion, energy and continuity, and the Fourier's law of heat conduction. The resulting differential equations that govern the convection are complicated and their exact solution can be given only in a few simple problems of steady flow.

There are three basic methods of determining the rate of heat transfer between fluid and a solid by convection. The fluid is at rest in the immediate vicinity of the solid surface due to the viscous effect. Therefore, the heat flow at the wall is by conduction. The first method makes use of this observation and the heat transfer rate can be calculated from the Fourier's law.

The second method is based on analogy between the mechanisms of transfer of fluid momentum to the wall and the transfer of heat by convection. Using the analogy, the rate of heat transfer can be predicted from the measurement of shear stress between the fluid and the wall.

The third method is to experimentally determine the heat transfer coefficient, defined by the Newton's equation.

In this chapter, analytical treatment of some simple cases of forced and natural convection has been presented followed by the discussion of analogy between fluid friction and heat transfer.

When fluid with free-stream velocity U_∞ and temperature T_∞ flows past a heated plate of sufficient length, hydrodynamic and thermal boundary layers develop. Initially, the boundary layer development is laminar and the boundary layer thickness increases continuously with the increase in distance x along the plate surface starting from zero at the leading edge of the plate. At some critical distance x_c from the leading edge, transition from laminar to turbulent layer with a thin laminar sublayer takes place at critical Reynolds number $Re_c = 5 \times 10^5$.

In case of fluid flow through a tube, the velocity distribution is uniform at the inlet cross-section if the tube inlet is rounded and the fluid comes from a large space. Because of the friction, the velocity diminishes at the wall and increases at the centre of the tube. The flow does not become fully developed at once but at a certain distance from the tube inlet. Thus, the boundary layer gradually builds up until it reaches the centre of the tube. The velocity distribution curve, now, acquires a stable form and does not vary down the tube and the flow is said to be fully developed. The distance for the velocity profile to be fully developed is termed as hydrodynamic development length or entrance length. The development of the thermal boundary layer in a fluid which is heated or cooled in a duct is qualitatively similar to that of the hydrodynamic boundary layer.

When the flow through a straight tube is laminar, the velocity profile of the fully developed laminar flow is of parabolic form. If the flow Reynolds number exceeds a certain critical value ($Re_c \approx 2300$), the flow becomes turbulent. In turbulent flow, there are three distinct regions in the flow: a laminar sublayer in the immediate vicinity of the wall, a buffer layer and a prominent turbulent core. In the laminar sublayer, viscous forces dominate and the fluid moves in streamline pattern parallel to the wall of the tube. In the turbulent core, chunks of fluid move in a totally chaotic pattern (termed as eddying motion). This causes intense mixing of the fluid. The fluid in the buffer layer shows behaviour that is intermediate between that of the fluid in the laminar sublayer and turbulent core. The velocity changes abruptly near the wall and takes a somewhat blunter profile in the middle of the tube. With the increase in the Reynolds number, the velocity profile becomes flatter over most of the duct cross-section. Using the Blasius empirical formula for the friction factor, Prandtl developed the power law of the velocity distribution in the turbulent flow which is not applicable close to the wall ($y/R < 0.04$).

Boundary layer thickness parameters velocity displacement thickness δ_{vd} and momentum displacement thickness δ_{md} have been defined in Sect. 7.4.1 and their equations have been developed. The displacement thickness is a measure of the displacement of the free stream due to the formation of the boundary layer over the plate while the momentum thickness is a measure of the momentum flux displacement caused by the boundary layer. For the thermal boundary layer, enthalpy thickness and conduction thickness have been defined in Sect. 7.4.2.

In Sect. 7.5, the momentum equation of the laminar boundary layer over a flat plate has been derived by making a force-and-momentum balance on an elemental control volume located in the boundary layer. Numerical solution of the momentum equation by Blasius provided the equations of the thickness of the boundary layer and skin friction coefficient as function of the Reynolds number.

von Karman presented integral momentum equation of laminar boundary layer over a flat plate by equating the shear force at the plate surface (for constant pressure condition) to the net momentum change over a control volume infinitesimal in the x-direction and enclosing the boundary layer in y-direction. This integral equation for the hydrodynamic boundary layer has been used to obtain the expression for the boundary layer thickness and friction factor by assuming unknown velocity profile equation as a polynomial or other forms.

In Sect. 7.7, energy equation of the laminar boundary layer over a flat plate has been developed by making energy balance of the net transport of the energy into the elemental control volume, the net heat conducted out of the control volume and the viscous work done on the element for the control volume. For the low-velocity incompressible flow, the magnitude of the viscous energy term is small and can be neglected. This gives energy equation, which is similar to the momentum equation. Pohlhausen presented solution of the energy equation adopting a procedure identical to that used for the solution of momentum equation by Blasius. Solution provided the result in the form of Nusselt number relation for determination of heat transfer coefficient. Using the skin friction coefficient relation from the solution of momentum equation, a relation between the Stanton number and friction factor has been established.

von Karman carried out integral analysis of energy equation for the laminar boundary layer over the flat plate as presented in Sect. 7.7.2 to determine heat transfer coefficient and arrived at the Nusselt number correlation, which is quite close to the relation obtained by Pohlhausen.

The turbulent boundary layer over a flat plate consists of a laminar sublayer, a buffer zone and a turbulent layer. In the laminar sublayer, the molecular diffusion processes are dominant and the turbulent fluctuations are negligible with the result that the turbulent shear stress is much less than the laminar shear stress, in the buffer zone, the molecular diffusion and eddy transport effects are of the same order, and in the turbulent region, the eddy transport effects are dominant and the turbulent shear stress dominates the laminar shear stress. The entire velocity field cannot be represented by a single equation. von Karman proposed a set of equations termed as universal velocity profile, which matches well with the experimental data. The equations of the universal profile are too complex mathematically to use with the momentum integral equation. A suitable velocity profile for turbulent boundary layers over smooth plates is the empirical power law profile $(u/U_\infty) = (y/\delta)^{1/7}$ for $5 \times 10^5 < Re < 10^7$. This is known as 1/7th power law. In the laminar sublayer, a linear velocity distribution is assumed. To evaluate the wall shear stress τ_w, an experimentally determined relation due to Blasius for the turbulent flow on flat plates has been used. The substitution of the values of u/U_∞ and τ_w in the von Karman momentum integral equation yields the skin friction coefficient.

For laminar flow in tubes, the balancing of pressure forces acting on a fluid element in the flow by the viscous shear forces yields the equation of Fanning friction factor for the fully developed constant property laminar flow in smooth circular cross-section tube. The temperature distribution has been obtained by making an energy balance of the energy conducted and convected for an annular element, which has been utilized to determine the value of the Nusselt number for uniform heat flux condition valid for fully developed laminar flow.

In Sect. 7.11, for the momentum and heat exchange in turbulent flow, eddy viscosity and eddy thermal diffusivity have been defined. The eddy viscosity ε_M is a function of the turbulence and it changes its magnitude with the distance from the wall. It is analogous to the kinematic viscosity v but unlike v it is not a property of the fluid. In the fully turbulent region, ε_M is much larger than v and the viscous shear stress may be neglected. Turbulent heat transfer is analogous to the turbulent momentum transfer. In analogy to eddy viscosity, eddy thermal diffusivity ε_H has been introduced. The eddy thermal diffusivity is also a function of the turbulence and it also changes its magnitude with the distance from the wall. The turbulent diffusivity is zero in the laminar sublayer. It is of the same order of magnitude as the molecular diffusivity in the buffer zone and is much greater than the molecular diffusivity in the turbulent core. However, this is true only for fluids having $\text{Pr} \geq 1$, but not for the liquid metals ($\text{Pr} \leq 0.01$). The ratio of ε_M and ε_H is referred to as the turbulent Prandtl number Pr_t analogous to the laminar or molecule Prandtl number.

Reynolds suggested that there exists a similarity between forced convection fluid friction and heat transfer because the same mechanism of turbulent exchange causes the transfer of momentum and heat. This is known as Reynolds analogy and is presented as $\text{St}_x = C_{fx}/2$ for flow past a flat plate in Sect. 7.12. However, the analogy is a considerable simplification of a very complex process, but is a reasonable approximation for fluids having the molecular Prandtl number value of nearly one. Using the analogy, heat transfer coefficient can be determined by measurements of the friction factor under the conditions when no heat transfer is involved. Reynolds–Colburn analogy ($\text{St}_x\text{Pr}^{2/3} = C_{fx}/2$) has been presented, which is found to be quite accurate in the range $0.5 < \text{Pr} < 50$ provided the drag forces are wholly viscous in nature (i.e. when form or pressure drag is absent).

Colburn analogy has been applied to predict the turbulent boundary layer heat transfer from a flat plate and has yielded results which are in good agreement with the experimental values, refer Sect. 7.12.2.

In Sect. 7.13, Prandtl–Taylor modification of Reynolds analogy and von Karman analogy is also presented for flow over flat plates when $\text{Pr} \neq 1$.

Reynolds and Reynolds–Colburn analogies for turbulent flow in tubes are similar to that for flat plate, i.e. $\text{St} = f/2$ for $\text{Pr} = 1$ and $\text{St Pr}^{2/3} = f/2$ for fluids not having $\text{Pr} = 1$, respectively.

Analytical solution of natural or free convection laminar flow on a heated vertical plate is presented in Sect. 7.15. Relation of Nusselt number has been presented considering the integral momentum and energy equations. In this analysis, a new dimensionless number Gr (termed as Grashof number) has been introduced. This number plays an important role in all free convection problems similar to that played by the Reynolds number in forced convection. The product of Grashof and Prandtl number (GrPr) provides a criterion for transition from laminar to turbulent boundary layer flow in free convection. Experimental measurements are relied upon to obtain relations for other configurations. Some of such relations are presented in Chap. 9.

Liquid metals have been employed because of their high thermal conductivity. However, the liquid metals require careful handling since they are corrosive and show a violent reaction when come into contact with water. Analytical treatment for liquid metal heat transfer has been presented for laminar flow over a flat plate in Sect. 7.17 and Nusselt number correlation has been presented as function of a non-dimensional number Pe (called the Peclet number), which is an important parameter in liquid metal convection.

Review Questions

7.1 Discuss development of hydrodynamic and thermal boundary layers for flow of fluid (U_∞, T_∞) over a thin flat plate held parallel to flow and in tubes. Further discuss the effect of boundary layer development on the heat transfer coefficient. Support the discussion with appropriate figures.

7.2 Write a short note on various methods of determining convection heat transfer coefficient.

7.3 Derive the equation of continuity.

7.4 Define (i) boundary layer thickness, (ii) velocity and momentum displacement thicknesses and (iii) enthalpy and conduction thicknesses.

7.5 Starting from the fundamentals, develop the momentum equation of laminar boundary layer over a flat plate. State the assumptions made.

7.6 Develop the integral momentum equation of laminar boundary layer over a flat plate.

7.7 Prove that the Blasius solution of laminar boundary layer flow over a flat plate gives the following relation of local skin friction coefficient

$$C_{fx} = \frac{0.6641}{\mathrm{Re}_x^{1/2}}.$$

Using the above derived relation, show that the average skin friction coefficient is given by

$$\overline{C_f} = \frac{1.3282}{\mathrm{Re}_L^{1/2}}.$$

7.8 Starting from fundamentals, develop the integral momentum equation. Following the von Karman integral technique prove that

$$\overline{C_f} = \frac{1.2932}{\mathrm{Re}_L^{1/2}}$$

when a third degree polynomial is considered for the velocity distribution equation.

7.9 Show that, for laminar flow over a flat plate, the velocity (hydrodynamic) boundary layer thickness at a distance x from the leading edge is given by

$$\frac{\delta}{x} = \frac{4.64}{\sqrt{\mathrm{Re}_x}}$$

when the equation of the velocity profile is

$$\frac{u}{U_\infty} = \frac{3}{2}\left(\frac{y}{\delta}\right) - \frac{1}{2}\left(\frac{y}{\delta}\right)^3.$$

7.10 Develop the energy equation of laminar boundary layer over a flat plate. Show that for the low-velocity incompressible fluids, the energy equation reduces to

$$u\frac{\partial t}{\partial x} + v\frac{\partial t}{\partial y} = \alpha\frac{\partial^2 t}{\partial y^2}.$$

7.11 Develop integral energy equation of the boundary layer for low-velocity laminar flow over a flat plate placed parallel to the fluid stream.

7.12 Show that the Pohlhausen solution of the energy equation of the laminar boundary layer over a flat plate gives the following relation of the local Nusselt number.

$$\mathrm{Nu}_x = 0.332\mathrm{Re}_x^{1/2}\mathrm{Pr}^{1/3}$$

Deduce the equation of average Nusselt number over plate length L. Also deduce the following relation.

$$\mathrm{St}_x\mathrm{Pr}^{2/3} = \frac{C_{fx}}{2}.$$

Comment on the result.

7.13 Assuming a linear velocity distribution ($u/U_\infty = y/\delta$), for the laminar boundary layer of a fluid flowing past a flat plate, it can be shown that the boundary layer thickness is given by

$$\frac{\delta}{x} = \left(\frac{12}{\mathrm{Re}_x}\right)^{0.5}.$$

Using this result and the assumption that the temperature distribution is also linear across the boundary layer thickness, show that for a plate of uniform temperature the ratio of the thickness of the thermal and hydrodynamic boundary layers δ_t and δ, respectively, is given by

$$\frac{\delta_t}{\delta} = (\mathrm{Pr})^{-1/3}.$$

Also show that the local Nusselt number at a distance x along the plate from the leading edge of the plate is given by

$$\mathrm{Nu}_x = 12^{-0.5}\mathrm{Re}_x^{1/2}\mathrm{Pr}^{1/3}$$

7.14 Using von Karman integral technique derive the following average Nusselt number relation for laminar flow over a flat plate.

$$\mathrm{Nu}_{av} = 0.662\mathrm{Re}_L^{1/2}\mathrm{Pr}^{1/3}$$

when third degree polynomial equations are considered for both the velocity and temperature distributions.

7.15 Define bulk temperature of a fluid flowing in a tube.

7.16 Show that for the laminar flow through a smooth tube, the Fanning friction factor is given by

$$f = \frac{16}{\text{Re}}.$$

7.17 Prove that for the laminar flow in smooth tubes, the Nusselt number under uniform heat flux condition is 4.364.

7.18 State the Reynolds analogy. Prove the following relation for the turbulent flow regime over a flat plate for a fluid with Pr = 1.

$$\frac{q_l}{\tau_l} = \frac{q_w}{\tau_w} = \frac{q_t}{\tau_t}$$

where subscripts l, w and t refer to laminar, wall and turbulent.

7.19 Derive the equation of Reynolds–Colburn analogy.

7.20 Draw a schematic diagram of a free convection boundary layer development over a vertical plate for the cases:
   (i) Plate temperature higher than the surrounding fluid temperature
   (ii) Plate temperature lower than the surrounding fluid temperature
   Also show the velocity and temperature distributions.

7.21 Using the integral momentum and energy equation of free convection on a vertical heated plate show that the local heat transfer coefficient is given by

$$\text{Nu}_x = 0.508\text{Pr}^{1/2}(0.952 + \text{Pr})^{-1/4}(\text{Gr}_x)^{1/4}.$$

Problems

7.1 Water at 20°C flows over a flat plate with a velocity of 0.1 m/s. The plate is heated to a surface temperature of 40°C. At a particular location in the thermal boundary layer, the temperature profile is represented by

$$\frac{t - t_w}{t_\infty - t_w} = \frac{3}{2}\left(\frac{y}{\delta_t}\right) - \frac{1}{2}\left(\frac{y}{\delta_t}\right)^3$$

Determine the heat flux and heat transfer coefficient if the thickness of the thermal boundary layer is 10 mm at the location.

[**Ans.** $\frac{q}{A} = -k\left(\frac{\partial t}{\partial y}\right)_{y=o} = -k(t_\infty - t_w)\left\{\frac{\partial}{\partial y}\left[\frac{3}{2}\left(\frac{y}{\delta_t}\right) - \frac{1}{2}\left(\frac{y}{\delta_t}\right)^3\right]\right\}_{y=o} = \frac{3}{2}k\frac{(t_w - t_\infty)}{\delta_t}$; k = 0.617 W/(m K) for water at film temperature of 30°C; substitution gives $\frac{q}{A} =$ 1851 W/m^2; heat transfer coefficient $h = \frac{q/A}{t_w - t_\infty} = 92.55$ W/(m^2 K).]

7.2 Air flows over a flat plate parallel to its surface. All other parameters remaining the same, how will the heat transfer coefficient change if the velocity and air pressure are doubled? The flow is in the laminar regime.

[**Ans.** The air density will also be doubled when pressure is doubled; in the laminar regime, $h = [0.664 \quad (\text{Re})^{0.5}(\text{Pr})^{0.33}]k/L = [0.664(\rho U_\infty L/\mu)^{0.5}(\text{Pr})^{0.33}]k/L$, i.e. $h \propto (\rho U_\infty)^{0.5}$; hence, the new heat transfer coefficient $\propto (2 \times 2)^{0.5} = 2$ times.]

7.3 Air at 30°C flows at a velocity of 10 m/s past a flat plate 3 m long and 1 m wide. The plate surface can be assumed to be at a uniform temperature of 170°C. Determine the heat transfer rate from one side of the plate.
What would be the heat transfer rate if the air blows perpendicular to the length of the plate?
[**Ans.** Thermophysical properties of air at the film temperature of 100°C are $\rho = 0.9452$ kg/m^3, $\mu = 2.172 \times 10^{-5}$ kg/(m s), $k = 0.0317$ W/(m K), Pr = 0.693. (i) $\text{Re}_L = \frac{U_\infty L\rho}{\mu} = 1.305 \times 10^6 > 5 \times 10^5$; $h = (k/L)\ [0.037\ \text{Re}_L^{0.8} - 871]\ \text{Pr}^{0.33} =$ 18.87 W/(m^2 K); $q = hA(t_w - t_\infty) = 18.87 \times 3 \times 1 \times (170–30) = 7925$ W; (ii) $\text{Re}_L = \frac{U_\infty W\rho}{\mu} = 4.35 \times 10^5 < 5 \times 10^5$; $h = (k/W)\ 0.664\ \text{Re}_L^{0.5}\ \text{Pr}^{0.33} = 12.28$ W/(m^2 K); $q = hA(t_w - t_\infty) = 12.28 \times 3 \times 1 \times (170–30) = 5157.6$ W. In the first case, the flow becomes turbulent after x_c (= $\text{Re}_c \times \mu/\rho U_\infty$) = 1.149 m and hence the heat transfer rate is high.]

7.4 Air at 20°C flows parallel to a plate at 2 m/s. Plate length is 1 m. If the plate surface is maintained at 80°C, determine the heat loss from the trailing 0.5 m length of the plate.
[**Ans.** Thermophysical properties of air at the film temperature of 50°C are $\rho = 1.095$ kg/m^3, $k = 0.02799$ W/(m K), $\mu = 1.95 \times 10^{-5}$ kg/(m s), Pr = 0.703, $c_p = 1.0072$ kJ/(kg K); for first 0.5 m: $\text{Re}_L = \frac{\rho U_\infty L}{\mu} = 56154$; flow is laminar; $\bar{h} = 0.664\text{Re}_L^{1/2}\text{Pr}^{1/3}\frac{k}{L} = 7.83\ \text{W}/(\text{m}^2\,\text{K})$; $q = \bar{h}A(t_w - t_\infty) = 234.9\ \text{W/m width}$. For 1.0 m length of plate: $\text{Re}_L = \frac{\rho U_\infty L}{\mu} = 112308$; flow is laminar; $\bar{h} = 0.664\text{Re}_L^{1/2}\text{Pr}^{1/3}\frac{k}{L} = 5.54\ \text{W}/(\text{m}^2\,\text{K})$;$q = \bar{h}A(t_w - t_\infty) = 332.3\ \text{W/m width}$. Thus, heat transfer rate from trailing 0.5 m length of the plate = 332.3 – 234.9 = 97.4 W/m width.]

7.5 Air at atmospheric pressure and 200°C flows over a flat plate with a velocity of 5 m/s. The plate is 15 mm wide and is maintained at 120°C. Calculate the thickness of velocity and thermal boundary layers and local heat transfer coefficient at a distance of 0.5 m from the leading edge. Assuming that the flow is on one side of the plate, calculate the heat transfer rate. Given: $\rho = 0.815$ kg/m^3, $k = 0.0364$ W/(m K), $\mu = 2.45 \times 10^{-5}$ kg/(m s) and Pr = 0.7.
[**Ans.** $\text{Re}_x = \frac{\rho U_\infty x}{\mu} = 83163 < \text{Re}_{cr}$;$\delta = \frac{5}{\sqrt{\text{Re}_x}}x = 0.0087$ m; $\delta_t = \frac{\text{Pr}^{-1/3}}{1.025}\delta$ gives $\delta_t = 0.0096$ m; $h_x = 0.332\text{Re}_x^{1/2}\text{Pr}^{1/3}\frac{k}{x} = 6.19\ \text{W}/(\text{m}^2\,\text{K})$; $\bar{h} = 2h_x$;$q = \bar{h}A(t_w - t_\infty)$ $= 7.42$W.]

7.6 Assuming the boundary layer to be turbulent throughout, determine the ratio of drag on the front and rear half of the plate over which a fluid is flowing parallel to its surface.
[**Ans.** Drag force, $F_D = \tau_w A = \overline{C}_f\left(\frac{1}{2}\rho U_\infty^2\right)A$, where $\overline{C}_f = \frac{0.0729}{\text{Re}_L^{1/5}}$, and $A = WL$, i.e. $F_D = \frac{0.0729}{\text{Re}_L^{1/5}}\left(\frac{1}{2}\rho U_\infty^2\right)WL \propto \frac{L}{L^{1/5}} = CL^{0.8}$, where C is constant. For first half-length $L/2$, $F_{D1} \propto C(L/2)^{0.8}$=$0.574CL^{0.8}$; for second half-length, $F_{D2} = (1 - 0.574)\ CL^{0.8} =$ $0.426\ CL^{0.8}$; $\frac{F_{D1}}{F_{D2}} = \frac{0.574CL^{0.8}}{0.426CL^{0.8}} = 1.347$.]

7.7 Air at 20°C and at a pressure of 1 atm is flowing over a flat plate at a velocity of 1.5 m/s. Determine the boundary layer thickness at a distance of 0.3 m from the leading edge of the plate.
Calculate the mass which enters the boundary layer between distance of 0.3 m and 0.5 m from the leading edge of the plate. Assume plate width to be unity.
[Ans. The properties of air at 20°C are ρ = 1.21 kg/m^3, μ = 1.81 × 10^{-5} kg/(m s); $\text{Re}_{x=0.3} = \frac{\rho U_\infty x}{\mu} = 30083; \delta_{x=0.3} = \frac{5.0}{\sqrt{\text{Re}_x}} x = 0.00865\,\text{m}$; Similarly, $\text{Re}_{x=0.5} = 50138$, $\delta_{x=0.5} = 0.01116\,\text{m}$; for $u = U_\infty\left[\frac{3}{2}\frac{y}{\delta} - \frac{1}{2}\left(\frac{y}{\delta}\right)^3\right]$, $\int_0^\delta \rho.u.dy = \int_0^\delta \rho.U_\infty\left[\frac{3}{2}\frac{y}{\delta} - \frac{1}{2}\left(\frac{y}{\delta}\right)^3\right]dy = \frac{5}{8}\rho.U_\infty\delta$; mass flow rate, $\Delta m = \frac{5}{8}\rho.U_\infty(\delta_{x=0.5} - \delta_{x=0.3}) = 0.00285\,\text{kg/s}$.]

7.8 For the data of Q. 7.7, calculate heat transfer for the 0.3 m length of the plate if the plate is maintained at 80°C.
[Ans. At mean temperature of 50°C, air properties: ρ = 1.095 kg/m^3, k = 0.02799 W/(m K), μ = 1.95 × 10^{-5} kg/(m s), Pr = 0.703, c_p = 1.0072 kJ/(kg K); $\text{Re}_L = \frac{\rho U_\infty L}{\mu} = 25269$; flow is laminar; $\bar{h} = 0.664\text{Re}_L^{1/2}\text{Pr}^{1/3}\frac{k}{L} = 8.76$ W/(m^2 K); $q = \bar{h}A(t_w - t_\infty) = 157.7\text{W}$.]

7.9 Air at 10°C and 100 m/s flows parallel to a flat square plate (1 m × 1 m). The flow is turbulent from the leading edge. Determine (a) thickness of hydrodynamic boundary layer at the trailing end of the plate, (b) heat flow rate. The plate is maintained at 40°C.
[Ans. At mean temperature of 25°C, air properties are ρ = 1.1868 kg/m^3, k = 0.02608 W/(m K), μ = 1.8363 × 10^{-5} kg/(m s), Pr = 0.709, c_p = 1005.7 J/(kg K); $\text{Re}_L = \frac{\rho U_\infty L}{\mu} = 6.46 \times 10^6$; flow is turbulent; $\delta = \frac{0.375}{\text{Re}_L^{1/5}} L = 0.0163\,\text{m}$; $\bar{h} = 0.037\text{Re}_L^{0.8}\text{Pr}^{1/3}\frac{k}{L} = 241.5$ W/(m^2 K); $q = \bar{h}A(t_w - t_\infty) = 7245\text{W}$.]

7.10 A square plate (1 m × 1 m) is placed parallel to an air stream with flow velocity of 2 m/s. The air is at 20°C and the plate is maintained at 80°C. Using the Colburn's analogy, determine the heat transfer coefficient.
[Ans. At mean temperature of (80 + 20)/2 = 50°C, air properties are ρ = 1.095 kg/m^3, k = 0.02799 W/(m K), μ = 1.95 × 10^{-5} kg/(m s), Pr = 0.703, c_p = 1.0072 kJ/(kg K); $\text{Re}_L = \frac{\rho U_\infty L}{\mu} = 1.12 \times 10^5 < 5 \times 10^5$; for laminar flow, $\overline{C_f} = \frac{1.328}{\sqrt{\text{Re}_L}} = 0.00397$; Colburn's analogy: $\text{St}_x = \frac{h}{\rho U_\infty c_p} = \text{Pr}^{-2/3}\frac{\overline{C_f}}{2}$ gives $h = \rho U_\infty c_p \text{Pr}^{-2/3}\frac{\overline{C_f}}{2} = 5.53$ W/(m^2 K).]

7.11 Air at 20°C and at a pressure of 1 atm is flowing over a flat plate at a velocity of 30 m/s. If the plate is 0.8 m in length and 1 m wide, and is at a temperature of 80°C, estimate the heat transfer rate.
[Ans. At mean temperature of (80 + 20)/2 = 50°C, air properties are ρ = 1.095 kg/m^3, k = 0.02799 W/(m K), μ = 1.95 × 10^{-5} kg/(m s), Pr = 0.703, c_p = 1.0072 kJ/(kg K); $\text{Re}_L = \frac{\rho U_\infty L}{\mu} = 1.35 \times 10^6 > 5 \times 10^5$; the flow is turbulent, hence $\bar{h} = \text{Nu}\frac{k}{L} = (0.037\text{Re}_L^{0.8} - 871)\text{Pr}^{1/3}\frac{k}{L} = 65.2$ W/(m^2 K); $q = \bar{h}A(t_w - t_\infty) = 3129.6\text{W}$.]

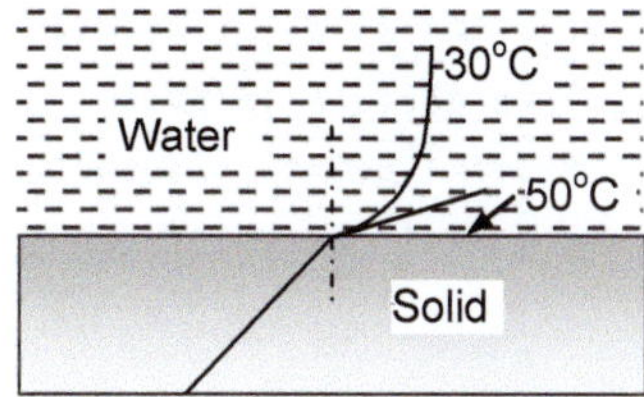

Fig. 7.35 Problem 7.13

7.12 A thin plate 1.5 m long is placed parallel to the flow of air at 15°C. The plate is maintained at 35°C. The air flows at a velocity of 25 m/s. Determine (a) the local heat transfer coefficient at the middle of the plate and (b) the average heat transfer coefficient.

[**Ans.** At mean temperature of 25°C, air properties: ρ = 1.1868 kg/m^3, k = 0.02608 W/(m K), μ = 1.8363 × 10^{-5} kg/(m s), Pr = 0.709, c_p = 1005.7 J/(kg K). (a) x = 0.75 m: $\mathrm{Re}_x = \frac{\rho U_\infty x}{\mu} = 1.21 \times 10^6 > 5 \times 10^5$; flow is turbulent; $h_x = \mathrm{Nu_x}\frac{k}{x} = 0.0296\mathrm{Re}_x^{0.8}\mathrm{Pr}^{1/3}\frac{k}{x} = 67.45$ W/(m^2 K); (b) $x = L = 1.5$ m: $\mathrm{Re}_L = \frac{\rho U_\infty L}{\mu} = 2.42 \times 10^6; \bar{h} = (0.037\mathrm{Re}_L^{0.8} - 871)\mathrm{Pr}^{1/3}\frac{k}{L} = 59.9$ W/(m^2 K).]

7.13 The surface of a heated solid [k = 15 W/(m K)] is at a uniform temperature of 50°C when it is being cooled by water [k = 0.62 W/(m K)] at 30°C, refer Fig. 7.35. If the temperature gradient ($\partial t/\partial y$) in the water at the solid–water interface is 2 × 10^4 K/m, determine the temperature gradient in the solid at the solid–water interface and the heat transfer coefficient.

[**Ans.** At the interface, the heat balance gives $[k\ (\partial t/\partial y)]_{\mathrm{water}} = [k\ (\partial t/\partial y)]_{\mathrm{solid}}$; substitution gives $(\partial t/\partial y)_{\mathrm{solid}}$ = 826.7 K/m. Heat transfer coefficient, $h = [k(\partial t/\partial y)]_{\mathrm{water}}/(t_{\mathrm{surface}} - t_{\mathrm{water}})$ = 620 W/(m^2 K).]

References

Bejan A (1995) Convective heat transfer. Wiley, New York

Blasius H (1908) Grenzschichten in flüssigkeiten mit kleiner reibung. Z Angew Math Phy 56:1–37. English version: Blasius H (1950) The boundary layers in fluids with little friction. NACA-TM-1256 Chapman AJ (1960) Heat transfer. Macmillan, New York

Kays WM, Crawford ME (1980) Convective heat and mass transfer. Tata McGraw-Hill Publishing Co., Ltd, New Delhi

Kays WM, Perkins HC (1985) Forced convection, internal flow in ducts. In: Rohsenow WM, Hartnett JP, Ganic EN (eds) Handbook of heat transfer fundamentals, Chapter 7. McGraw-Hill, New York

Mills AF (1995) Heat and mass transfer. Richard D, Irwin, Chicago

Pohlhausen E (1921) Der W¨armetausch zwischen festen K¨orpern und Fl¨ussigkeiten mit kleiner Reibung und kleiner W¨armeleitung. Z Angew Math Mech 1(2):115–121

Potter MC, Wiggert DC, Ramadan B (2012) Mechanics of fluids. Cengage Learning, USA

Rogers GFC, Mayhew YR (1967) Engineering thermodynamics work and heat transfer. English Language Book Co. and Longmans, Green & Co, London

Schlichting H (1979) Boundary-layer theory, 7th edn. McGraw-Hill, New York

8 Empirical Relations for Forced Convection Heat Transfer

8.1 Introduction

Analytical solutions, using the boundary layer equations, of some simple convection heat transfer problems, especially the convection with laminar flow, have been presented in Chap. 7. However, there are a large number of convection problems for which the analytical solutions have not met the success especially the problems involving turbulent flow or flows where detachment occurs, for example around cylinders, spheres or other curved bodies, the heat transfer coefficient is very difficult to calculate. Thus the direct measurement of the heat transfer coefficient (experimental study) is still the main approach for the solution of the most of the heat transfer problems. The experimental results are presented in the form of generalized correlations using the method of dimensional analysis. Such correlations are termed as empirical relations because they rely on the observations and experiments not on theory. Some of such correlations have been presented here, which are frequently used. The corresponding friction factor correlations are also presented.

8.2 Dimensional Analysis[1]

The dimensional analysis deals only with the dimensions of the variables to produce a relationship. Such relation consists of one or more dimensionless groups, which combine the variables involved in the physical phenomenon under study, and some unknown constants and exponents. These constants and exponents are determined experimentally. Thus the dimensional analysis combined with the experiments provides the empirical or semi-empirical relations.

It is to note that the dimensional analysis cannot produce any numerical results nor it can give any explanation of the physical nature of the process. Furthermore, the variables affecting the phenomenon must be known to apply the technique of the dimensional analysis.

The basic principle of the dimensional analysis is that the form of an equation is determined only by its dimensions as illustrated by an example that follows.

Consider the equation of mass flow rate of a fluid through a duct. The flow rate is given by

[1]For greater details of the technique of the dimensional analysis, the readers can refer texts on fluid mechanics.

R. Karwa, *Heat and Mass Transfer*,
https://doi.org/10.1007/978-981-15-3988-6_8

$$\dot{m} = \rho A U_m \tag{8.1}$$

where

$\dot{m}$	mass flow rate
ρ	density of the fluid
A	area of cross-section of the duct
U_m	mean fluid velocity.

Writing the fundamental dimensions[2] of all the variable in Eq. (8.1), we have

$$\frac{M}{T} = \frac{M}{L^3} \times L^2 \times \frac{L}{T}$$
$$= \frac{M}{T}$$

The resulting dimensions of the quantities on the right side of the equation equal to those on the left, that is, dimensional homogeneity exists. The three independent variables ρ, A and U_m give answer in *M/T* only when their relationship is in a particular way, i.e. $\rho A U_m$.

In order to establish the relationship between the variables, either *Rayleigh method* or the *Buckingham pi theorem* can be used.

According to the Buckingham pi theorem, the equation of a phenomenon is expressed as

$$\phi(\pi_1, \pi_2, \pi_3, \ldots \pi_{m-n}) = 0 \tag{8.2}$$

where π_1, π_2, etc. indicate dimensionless groups. The number of the π-terms or dimensionless groups equals the number of variables m minus the number of fundamental dimensions n.

The dimensions of the physical quantities of interest are listed in Table 8.1 in two different systems of units.

8.3 Dimensional Analysis Applied to Forced Convection

In forced convection, the fluid is forced by some external agency to flow past the heated or cold solid surface. The velocity of the fluid influences the heat transfer rate. In addition to this, it has been found from the experience that the heat transfer also depends on the viscosity, thermal conductivity and density of the fluid, temperature difference and the characteristic dimension. Hence, we may write

$$h = f(\mu, \rho, k, c_p, \Delta T, d, U) \tag{8.3}$$

where U is the velocity and d is a characteristic dimension.

[2]The *fundamental dimensions* are quantities such as length L, mass M, time T and temperature θ, which are directly measured. The derived dimensions are those which are measured in the terms of the fundamental dimensions. For example, area is a derived quantity with dimensions L^2. In heat transfer, heat Q is also sometimes considered as fundamental dimension.

Table 8.1 Dimensions of different physical quantities

Variable	Symbol (units)	Dimensions[a]	
		M-L-T-θ-Q system	M-L-T-θ system
Viscosity of fluid	μ, kg/(m s)	$ML^{-1}T^{-1}$	$ML^{-1}T^{-1}$
Density of fluid	ρ, kg/m^3	ML^{-3}	ML^{-3}
Thermal conductivity of fluid	k, W/(m K)	$QL^{-1}T^{-1}\theta^{-1}$	$MLT^{-3}\theta^{-1}$
Specific heat	c_p, J/(kg K)	$QM^{-1}\theta^{-1}$	$L^2T^{-2}\theta^{-1}$
Temperature difference	ΔT, K	θ	θ
Characteristic dimension	l, m	L	L
Velocity	U, m/s	LT^{-1}	LT^{-1}
Convection heat transfer coefficient	h, W/(m^2 K)	$QL^{-2}T^{-1}\theta^{-1}$	$MT^{-3}\theta^{-1}$

[a]T is time and θ is temperature. Q can also be considered as fundamental unit

8.3.1 Rayleigh's Method

$$h = C\mu^a \rho^b k^c c_p^d \Delta T^e d^f U^g \tag{8.4}$$

Substituting the dimensions of the different variables in Eq. (8.4), we have

$$\frac{Q}{L^2T\theta} = C\left(\frac{M}{LT}\right)^a \left(\frac{M}{L^3}\right)^b \left(\frac{Q}{LT\theta}\right)^c \left(\frac{Q}{M\theta}\right)^d (\theta)^e (L)^f \left(\frac{L}{T}\right)^g$$

Equating the indices of the fundamental dimensions on both sides, we obtain

M: $0 = a + b - d$
L:$-2 = -a - 3b - c + f + g$
T: $-1 = -a - c - g$
θ: $-1 = -c - d + e$
Q: $1 = c + d$

There are seven unknowns and five equations. Therefore the values of five of them, say a, b, c, e, and f may be determined in terms of the other two unknowns d and g. The solution gives

$$a = d - g, b = g, c = 1 - d, e = 0, \text{and}\, f = g - 1.$$

By substitution of these values in Eq. (8.4), we get

$$\begin{aligned} h &= C\mu^{d-g}\rho^g k^{1-d} c_p^d \Delta T^0 (d)^{g-1} U^g \\ &= C\left(\frac{k}{d}\right)\left(\frac{\rho U d}{\mu}\right)^g \left(\frac{\mu c_p}{k}\right)^d \end{aligned}$$

or

$$\frac{hd}{k} = C\left(\frac{\rho U d}{\mu}\right)^{g}\left(\frac{\mu c_p}{k}\right)^{d}$$

The dimensionless group $\frac{hd}{k}$ is the Nusselt number Nu, $\left(\frac{\rho U d}{\mu}\right)$ is the Reynolds number Re and $\left(\frac{\mu c_p}{k}\right)$ is the Prandtl number Pr hence

$$\mathrm{Nu} = f(\mathrm{Re}, \mathrm{Pr})$$

and the generalized correlation can be written as

$$\mathrm{Nu} = \psi(\mathrm{Re})\phi(\mathrm{Pr}) \tag{8.5}$$

The form of functions ψ(Re) and ϕ(Pr) may be specified for different conditions of heat transfer by convection on the basis of theoretical analysis or experimental investigations.

8.3.2 Buckingham's Pi-Method

$$f(\mu, \rho, k, c_p, \Delta T, d, U, h) = 0$$

There are eight variables and five fundamental units hence we expect (8–5), i.e., 3 π-terms. Taking μ, k, c, ΔT and d as repeated variables, the π-terms can be established as follows.

$$\pi_1 = \mu^a k^b U^c \Delta T^d d^e h \tag{i}$$

or

$$1 = \left(\frac{M}{LT}\right)^{a}\left(\frac{Q}{LT\theta}\right)^{b}\left(\frac{L}{T}\right)^{c}(\theta)^{d}(L)^{e}\frac{Q}{L^2T\theta}$$

Equating the indices of the fundamental dimensions on both sides, we obtain

M: $0 = a$
L: $0 = -a - b + c + e - 2$
T: $0 = -a - b - c - 1$
θ: $0 = -b + d - 1$
Q: $0 = b + 1$

Solution gives

$$a = 0, b = -1, c = 0, d = 0, \text{and } e = 1.$$

Substitution in Eq. (i) gives

$$\pi_1 = k^{-1} dh = \frac{hd}{k} = \mathrm{Nu} \tag{ii}$$

Following the above approach, we have

$$\pi_2 = \mu^a k^b U^c \Delta T^d d^e \rho \tag{iii}$$

or

$$1 = \left(\frac{M}{LT}\right)^a \left(\frac{Q}{LT\theta}\right)^b \left(\frac{L}{T}\right)^c (\theta)^d (L)^e \frac{M}{L^3}$$

Equating the indices,

M: $0 = a + 1$
L: $0 = -a - b + c + e - 3$
T: $0 = -a - b - c$
θ: $0 = -b + d$
Q: $0 = b$

Solution of above equations gives

$$a = -1, b = 0, c = 1, d = 0, \text{and } e = 1.$$

This gives from Eq. (iii),

$$\pi_2 = \mu^{-1} U d \rho = \frac{\rho U d}{\mu} = \mathrm{Re} \tag{iv}$$

Similarly,

$$\pi_3 = \mu^a k^b U^c \Delta T^d d^e c_p \tag{v}$$

or

$$1 = \left(\frac{M}{LT}\right)^a \left(\frac{Q}{LT\theta}\right)^b \left(\frac{L}{T}\right)^c (\theta)^d (L)^e \left(\frac{Q}{M\theta}\right)$$

Equating the indices, we have

M: $0 = a - 1$
L: $0 = -a - b + c + e$
T: $0 = -a - b - c$
θ: $0 = -b + d - 1$
Q: $0 = b + 1$

Solution of the above equations gives

$$a = 1, b = -1, c = 0, d = 0, \text{ and } e = 0.$$

Substitution in Eq. (v) gives

$$\pi_3 = \mu k^{-1} c_p = \frac{\mu c_p}{k} = \text{Pr} \tag{vi}$$

Thus the functional relation is

$$f(\text{Nu}, \text{Re}, \text{Pr}) = 0$$

The generalized correlation can be written as

$$\text{Nu} = \psi(\text{Re})\phi(\text{Pr}), \tag{8.5}$$

which is the same as obtained by the Rayleigh's method.

It is to note that different dimensionless groups can often be derived for a given problem. No one set is more correct than any other. However, the dimensionless groups derived here are universally accepted as the most convenient for analysis of a forced convection problem.

Note: The above functional relation has been obtained using *M-L-T-θ-Q* method, i.e. the heat has been considered as the fundamental unit. However, the friction in the forced flow causes conversion of the kinetic energy into heat and heat cannot be treated as the fundamental unit. Heat Q is expressed in the terms of fundamental units M, L, T and θ as ML^2/T^2. For the dimensional analysis with M, L, T and θ as fundamental units, refer Example 8.1.

The above analysis shows that an experimental study now needs only to investigate the variation of three dimensionless groups rather than eight variables originally specified.

Example 8.1 Using M, L, T and θ system of fundamental units, develop the functional relation for forced convection heat transfer.

Solution

From the dimensional analysis presented above, we have seen that ΔT does not appear in the final functional relation for forced convection hence we can write

$$f(\mu, \rho, k, c_p, d, U, h) = 0$$

There are seven variables and four fundamental units hence we expect (7–4), i.e., 3 π-terms. Taking μ, ρ, k and d as repeated variables, the π-terms can be established as follows.

$$\pi_1 = \mu^a \rho^b k^c d^d h$$

or

$$1 = \left(\frac{M}{LT}\right)^a \left(\frac{M}{L^3}\right)^b \left(\frac{ML}{\theta T^3}\right)^c (L)^d \frac{M}{\theta T^3}$$

Equating the indices of the fundamental dimensions on both sides, we obtain

M: $0 = a + b + c + 1$
L: $0 = -a - 3b + c + d$
T: $0 = -a - 3c - 3$
θ: $0 = -c - 1$

Solution gives

$$a = 0, b = 0, c = -1 \text{ and } d = 1.$$

Substitution gives

$$\pi_1 = k^{-1}dh = \frac{hd}{k} = \text{Nu}$$

Following the above approach, we have

$$\pi_2 = \mu^a \rho^b k^c d^d c_p$$

or

$$1 = \left(\frac{M}{LT}\right)^a \left(\frac{M}{L^3}\right)^b \left(\frac{ML}{\theta T^3}\right)^c (L)^d \frac{L^2}{\theta T^2}$$

Equating the indices,

M: $0 = a + \text{b} + \text{c}$
L: $0 = -a - 3b + c + d + 2$
T: $0 = -a - 3c - 2$
θ: $0 = -c - 1$

Solution of above equations gives

$$a = 1, b = 0, c = -1 \text{ and } d = 0$$

This gives

$$\pi_2 = \mu k^{-1} c_p = \frac{\mu c_p}{k} = \text{Pr}$$

Similarly,

$$\pi_3 = \mu^a \rho^b k^c d^e U$$

or

$$1 = \left(\frac{M}{LT}\right)^a \left(\frac{M}{L^3}\right)^b \left(\frac{ML}{\theta T^3}\right)^c (L)^d \left(\frac{L}{T}\right)$$

Equating the indices, we have

M: $0 = a + b + c$
L: $0 = -a - 3b + c + d + 1$
T: $0 = -a - 3c - 1$
θ: $0 = -c$

Solution of the above equations gives

$$a = -1, b = 1, c = 0, \text{and } d = 1.$$

This gives

$$\pi_3 = \mu^{-1}\rho dU = \frac{\rho U d}{\mu} = \text{Re}$$

Thus the functional relation is

$$f(\text{Nu}, \text{Re}, \text{Pr}) = 0$$

The generalized correlation can be written as

$$\text{Nu} = \psi(\text{Re})\phi(\text{Pr}),$$

which is the same as obtained earlier.

Note: At low flow velocities, the free convection effect may be present. In this case ($\beta g\Delta T$) group of terms is to be considered, refer Sects. 9.2 and 9.3. With this consideration, we obtain fourth dimensionless group as

$$\pi_4 = \mu^a \rho^b k^c d^d (\beta g \Delta T)$$

or

$$1 = \left(\frac{M}{LT}\right)^a \left(\frac{M}{L^3}\right)^b \left(\frac{ML}{\theta T^3}\right)^c (L)^d \left(\frac{L}{T^2}\right)$$

Equating the indices, we have

M: $0 = a + b + c$
L: $0 = -a - 3b + c + d + 1$
T: $0 = -a - 3c - 2$
θ: $0 = -c$

Solution of the above equations gives

$$a = -2, b = 2, c = 0 \text{ and } d = 3.$$

This gives

$$\pi_4 = \frac{\rho^2 d^3 (\beta g \Delta T)}{\mu^2} = \text{Gr}$$

The functional relation is now

$$f(\text{Nu}, \text{Re}, \text{Gr}, \text{Pr}) = 0$$

or

$$\text{Nu} = \psi(\text{Re}, \text{Gr}, \text{Pr}) \tag{8.6}$$

Rayleigh's Method

$$h = C\mu^a \rho^b k^c c_p^d d^e U^f \tag{i}$$

Substituting the dimensions of the different variables in Eq. (i), we have

$$\frac{M}{T^3\theta} = C\left(\frac{M}{LT}\right)^a \left(\frac{M}{L^3}\right)^b \left(\frac{ML}{\theta T^3}\right)^c \left(\frac{L^2}{\theta T^2}\right)^d (L)^e \left(\frac{L}{T}\right)^f$$

Equating the indices of the fundamental dimensions on both sides, we obtain

M: $1 = a + b + c$
L: $0 = -a - 3b + c + 2d + e + f$
T: $-3 = -a - 3c - 2d - f$
θ: $-1 = -c - d$

There are four equations but six unknowns. Therefore, the values of four of them, say a, c, e, and f may be determined in terms of the other two unknown b and d. The solution gives

$$a = d - b, c = 1 - d, e = b - 1, \text{and } f = b.$$

Substitution of these values in Eq. (i) gives

$$\begin{aligned} h &= C\mu^{d-b} \rho^b k^{1-d} c_p^d d^{b-1} U^b \\ &= C\left(\frac{k}{d}\right)\left(\frac{\rho U d}{\mu}\right)^b \left(\frac{\mu c_p}{k}\right)^d \end{aligned}$$

or

$$\frac{hd}{k} = C\left(\frac{\rho U d}{\mu}\right)^b \left(\frac{\mu c_p}{k}\right)^d$$

$$\mathrm{Nu} = f(\mathrm{Re}, \mathrm{Pr})$$

and the generalized correlation can be written as

$$\mathrm{Nu} = \psi(\mathrm{Re})\phi(\mathrm{Pr}),$$

which is the desired relation.

Note: If ΔT is considered, a fourth dimensionless group will be obtained. In this case

$$\pi_4 = \mu^a \rho^b k^c d^d \Delta T$$

or

$$1 = \left(\frac{M}{LT}\right)^a \left(\frac{M}{L^3}\right)^b \left(\frac{ML}{\theta T^3}\right)^c (L)^d (\theta)$$

Equating the indices, we have

M: $0 = a + b + c$
L: $0 = -a - 3b + c + d$
T: $0 = -a - 3c$
θ: $0 = -c + 1$

Solution of the above equations gives

$$a = -3, b = 2, c = 1, \text{and } d = 2.$$

This gives

$$\pi_4 = \frac{\rho^2 k d^2 \Delta T}{\mu^3}$$

Using dimensionless group π_2, π_3 and π_4, we can obtain a dimensionless group as

$$\frac{\pi_3^2}{\pi_2 \pi_4} = \frac{\rho^2 U^2 d^2}{\mu^2} \times \frac{k}{\mu c_p} \times \frac{\mu^3}{\rho^2 k d^2 \Delta T} = \frac{U^2}{c_p \Delta T}$$

The dimensionless group $\frac{U^2}{c_p \Delta T}$ is known as *Eckert number.*

Note 2: In the case of developing flow in ducts, the heat transfer coefficient has been found to be function of the duct length also. A fourth dimensionless group will be obtained in this case.

$$\pi_4 = \mu^a \rho^b k^c d^d L$$

or

$$1 = \left(\frac{M}{LT}\right)^a \left(\frac{M}{L^3}\right)^b \left(\frac{ML}{\theta T^3}\right)^c (L)^d (L)$$

Equating the indices, we have

M: $0 = a + b + c$
L: $0 = -a - 3b + c + d + 1$
T: $0 = -a - 3c$
θ: $0 = -c$

Solution of the above equations gives

$$a = 0, b = 0, c = 0, \text{and } d = -1.$$

This gives

$$\pi_4 = \frac{L}{d}$$

Thus the functional relation is

$$f\left(\text{Nu}, \text{Re}, \text{Pr}, \frac{L}{D}\right) = 0$$

Example 8.2 Using M, L, and T system of fundamental units, develop the functional relation for Darcy friction factor for rough pipes.

Solution

It has been found from the experience that the Darcy friction factor for rough tubes depends on the average velocity of the flow U_m, pipe diameter D, fluid density ρ, fluid viscosity μ and pipe wall roughness height e. Hence, we may write

$$f = f(U_m, D, \rho, \mu, e) \tag{i}$$

We can express Eq. (i) in the following form:

$$f = CU_m^a D^b \rho^c \mu^d e^e \tag{ii}$$

Substituting the dimensions of the different variables in Eq. (ii), we have

$$M^0L^0T^0 = C\left(\frac{L}{T}\right)^a (L)^b \left(\frac{M}{L^3}\right)^c \left(\frac{M}{LT}\right)^d (L)^e$$

Equating the indices of the fundamental dimensions on both sides, we obtain

M: $0 = c + d$
L: $0 = a + b - 3c - d + e$
T: $0 = -a - d.$

There are five unknowns and three equations. Therefore the values of three of them, say a, b and c may be determined in terms of the other two unknowns d and e. The solution gives

$$a = -d, b = -d - e, c = -d.$$

By substitution of these values in Eq. (ii), we get

$$f = CU_m^{-d} D^{-d-e} \rho^{-d} \mu^d e^e$$
$$= C\left(\frac{U_m D\rho}{\mu}\right)^{-d} \left(\frac{e}{D}\right)^e$$

The dimensionless group $\frac{U_m D\rho}{\mu}$ is the Reynolds number Re and $\frac{e}{D}$ is termed as relative roughness height. Hence,

$$f = f\left(\mathrm{Re}, \frac{e}{D}\right)$$

and the generalized correlation can be written as

$$f = \psi(\mathrm{Re})\phi\left(\frac{e}{D}\right) \quad \text{(iii)}$$

The form of functions $\psi(\mathrm{Re})$ and $\phi(e/D)$ may be specified on the basis of theoretical analysis or experimental investigations.

For smooth pipes or tubes $e = 0$ and we can readily write that

$$f = f(\mathrm{Re}) \quad \text{(iv)}$$

that is, the friction factor is function of the Reynolds number only.

8.3.3 Physical Significance of Dimensionless Numbers

There is advantage of presenting the experimental data in the terms of the nondimensional numbers (Rogers and Mayhew 1967). This method of presentation allows empirical data to be applied to a wide range of physical conditions by the application of the principle of dynamic similarity (refer Rogers and Mayhew 1967 for details), and requires few experiments only to cover a wide range of fluid properties. Considerable saving is achieved by the experimentation on a small-scale setup. The results presented in nondimensional form can be used in any consistent system of units (metric or British).

The *Reynolds number* is a measure of the relative strength of the inertial and viscous forces. A high value of the Reynolds number indicates that the inertial forces dominant and hence at such Reynolds number values the flow may become turbulent. Thus by calculating the Reynolds number, we can find whether the flow is laminar or turbulent in forced convection. The critical Reynolds number Re_c at which the flow becomes turbulent depends on the geometric configuration. For example, for flow in tubes the critical Reynolds number is about 2300, and for flow over a flat plate the same is 5×10^5.

The *Nusselt number* can be expressed as

$$\mathrm{Nu} = \frac{hd}{k} = \frac{\text{Convective heat transfer}}{\text{Conduction heat transfer}} \tag{8.7}$$

Thus the Nusselt number is a ratio of the convective to conductive heat transfer across the fluid boundary and is a measure of the rate of heat transfer by convection. A large value of Nu means a very efficient convection. A Nusselt number of the order of unity would indicate that the convection and conduction are of the same magnitude.

The *Stanton Number* is a ratio of heat transferred to a fluid to the thermal capacity of the fluid:

$$\mathrm{St} = \frac{\mathrm{Nu}}{\mathrm{Re}\,\mathrm{Pr}} = \frac{h}{c\rho U} \tag{8.8}$$

It expresses the ratio of the heat extracted from a fluid to the heat passing with it. It is used some times as an alternative for Nusselt number when presenting heat transfer data. The Reynolds analogy relates Stanton number to the friction factor. Thus the heat transfer coefficient can be determined by measurements of the friction factor under the conditions when no heat transfer is involved.

Prandtl Number Pr ($\mu c_p/k = \nu/\alpha$) involves three properties of a fluid and thus itself is a property of the fluid. It can be expressed as a ratio of the kinematic viscosity and thermal diffusivity of the fluid:

$$\mathrm{Pr} = \frac{\mu c_p}{k} = \left(\frac{\mu}{\rho}\right)\left(\frac{\rho c_p}{k}\right) = \frac{\nu}{\alpha} \tag{8.9}$$

Thus it expresses the relative magnitude of diffusion of the momentum and heat in the fluid.

The product (Re $\times$ Pr) is another dimensionless group called the *Peclet number* Pe. It is an important parameter for convection heat transfer in liquid metals.

For a gas, dividing the *Eckert number* $\frac{U^2}{c_p \Delta T}$ by $(\gamma - 1)$, we get $\frac{U^2}{\gamma RT}$, which is square of the Mach number of a perfect gas. At high-speed flows, the Mach number has considerable effect on the heat transfer. At such speeds, a large amount of kinetic energy of the gas is dissipated due to the viscous effect in the boundary layer. At low velocities, the dissipation of the kinetic energy is not significant and hence the influence of the Eckert number is small.

8.4 Experimental Determination of Forced Convection Heat Transfer Coefficient

8.4.1 Uniform Temperature Condition

The first version of the experimental setup is shown in Fig. 8.1a. The fluid flows inside a tube, which is heated by steam condensing outside the tube. The wall temperature remains constant along the tube length due to the condensing steam on the outer surface of the tube. The heat transferred to the fluid is calculated from the measured fluid inlet and outlet temperatures t_i and t_o, respectively, using the following relation.

$$q = mc_p(t_o - t_i) \tag{8.10}$$

where m is the mass flow rate of the fluid flowing through the tube and is measured with the help of suitable flow measuring device. The outlet temperature of the fluid is to be measured after proper mixing. Hence, a mixing section is provided.

8.4.2 Uniform Heat Flux Condition

In this scheme, Fig. 8.1b, the fluid flows in an electrically heated tube. Alternatively, the fluid may flow in the annulus while the inner tube is heated by an electric heater placed centrally in the tube. The tube or annular duct is properly insulated. In the steady state, the power input to the electric heater should equal to the heat transferred to the fluid.

The heat transfer coefficient is calculated from,

$$h = \frac{q}{A(t_w - t_m)} \tag{8.11}$$

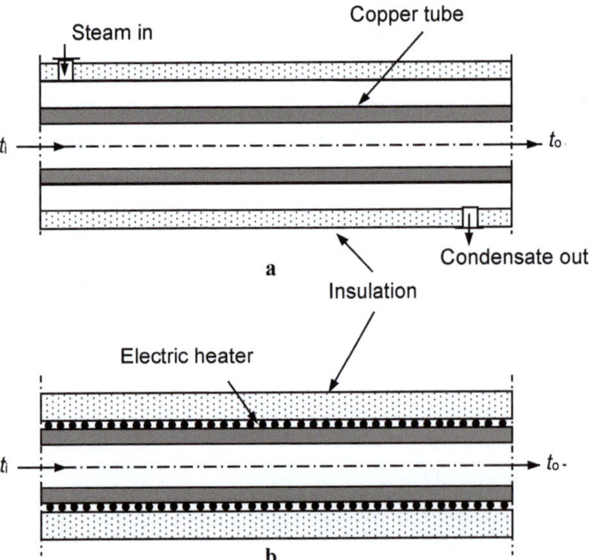

Fig. 8.1 Schematic of experimental setup

where

t_w mean wall temperature
t_m mean fluid temperature = $(t_i + t_o)/2$
A area of the heat transferring surface
= $\pi D_i L$ for tube flow
= $\pi D_o L$ for annular flow

The results of the experiment are presented in the terms of dimensionless parameters, Nusselt number Nu as function of the Reynolds and Prandtl numbers.

Example 8.3 An experimental test facility was fabricated to investigate the heat transfer characteristics for flow of air in an annular duct. The test section is 5 m long and the outer diameter of the inner tube is 50 mm, which is electrically heated. The inner diameter of the outer tube is 100 mm. For a particular test run, the following observations were made.

Air flow rate, m = 0.05 kg/s
Air inlet temperature, t_i = 30°C
Air outlet temperature, t_o = 50°C
Mean inner tube surface temperature, t_s = 80°C (outer tube surface is insulated)
Calculate the Nusselt number and the flow Reynolds number for the test run.

Solution

At t_m = 40°C, k = 0.0273 W/(m K), c_p = 1006 J/(kg K) and $\mu = 1.905 \times 10^{-5}$ kg/(m s).
The heat flow rate,

$$q = mc_p(t_o - t_i) = 0.05 \times 1006 \times (50 - 30) = 1006 \text{ W}.$$

The heat transfer coefficient,

$$h = \frac{q}{A(t_s - t_m)} = \frac{q}{\pi d_o L[t_s - (t_O + t_i)/2]} = \frac{1006}{\pi \times 0.05 \times 5 \times [80 - (50 + 30)/2]} = 32.02 \text{ W}/(\text{m}^2\,\text{K}).$$

The Nusselt number,

$$\text{Nu} = \frac{hD_h}{k} = \frac{h(D_i - d_o)}{k} = \frac{32.02 \times (0.1 - 0.05)}{0.02723} = 58.8.$$

where D_h (= $4A_c/P$) is termed as hydraulic diameter (refer Sect. 8.7).
The Reynolds number,

$$\text{Re} = \frac{GD_h}{\mu} = \left[\frac{m}{\pi/4(D_i^2 - d_o^2)}\right] \times \frac{D_h}{\mu} = \left[\frac{4 \times 0.05}{\pi(0.1^2 - 0.05^2)}\right] \times \frac{(0.1 - 0.05)}{1.905 \times 10^{-5}} = 22279.$$

8.5 Friction Factor and Heat Transfer Coefficient Correlations for Circular Ducts

Fluid flow and heat transfer characteristics of a circular cross-section duct has been investigated in a greater detail as this geometry is having a wide spread application.

8.5.1 Laminar Flow in Circular Tubes

8.5.1.1 Friction Factor Correlations

(a) **Hydrodynamically Fully Developed Laminar Flow**
The fully developed laminar flow of a constant property fluid in a smooth circular tube has been studied analytically in Chap. 7.

For laminar fully developed, constant property flow in a circular tube, the Fanning friction factor f is given by, Eq. (7.92),

$$f = \frac{16}{\text{Re}} \tag{8.12}$$

where the Reynolds number Re is based on the mean velocity of the flow U_m:

$$\text{Re} = \frac{\rho U_m D}{\mu}$$

It can be seen from Eq. (8.12) that the friction factor is inversely proportional to the Reynolds number and thus decreases with the increase in the Reynolds number.

(b) **Hydrodynamically Developing Laminar Flow**
The development length L_{hy} in laminar flow, the distance required for the friction factor to decrease to within 5% of its fully developed value, is approximately given by Kays and Crawford (1980) and Mills (1995)

$$\frac{L_{hy}}{D} \approx 0.05\text{Re} \tag{7.4}$$

At Re = 20, $L_{hy} = D$ only. At Re = 2300, the equation gives $L_{hy} = 115D$. However, the entrance effects are seen to be appreciable for a length of about 50 diameters from the entrance.

Friction factor in the hydrodynamic entrance region is higher than that for the fully developed case. In the entrance region, the pressure drop is a combined effect of surface shear and increase in the total fluid momentum flux associated with the development of the velocity profile. These effects are incorporated in the friction coefficient defined as apparent mean friction factor $\bar{f}_{app}$, refer Fig. 8.2. The apparent mean friction factor decreases with increase in x and asymptotically approaches the fully developed value of 16/Re as Re/(x/D) → 0. Knowing $\bar{f}_{app}$, the pressure drop is evaluated from:

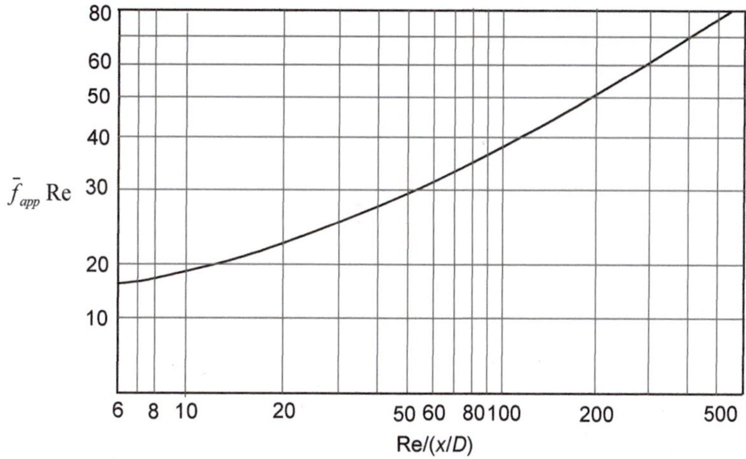

Fig. 8.2 Friction coefficients in the hydrodynamic entry length of a circular tube with laminar flow. (from Langhaar 1942)

$$\Delta p = \frac{4\bar{f}_{\text{app}}\rho U_m^2 L}{2D} \tag{8.13}$$

8.5.1.2 Heat Transfer Coefficient Correlations

(a) **Hydrodynamically and Thermally Fully Developed Laminar Flow**

For the laminar flow with fully developed velocity and temperature profiles, the Nusselt number is a function of the type of heating boundary condition. The analytical results presented in Chap. 7 for different conditions are:

(i) **Constant Heat Rate (Uniform Heat Flux)**

$$\text{Nu} = 4.364 \tag{7.96}$$

(ii) **Constant Surface Temperature**

$$\text{Nu} = 3.658 \tag{7.97}$$

It is to be noted that the above results have been obtained neglecting the frictional heating and buoyancy effects.

(b) **Thermal Entry Length (Temperature Profile Developing), Fully Developed Velocity Profile**

The thermal) entrance length L_{th} in laminar flow, the distance required for the Nusselt number to decrease to within 5% of its fully developed value, is (Mills 1995)

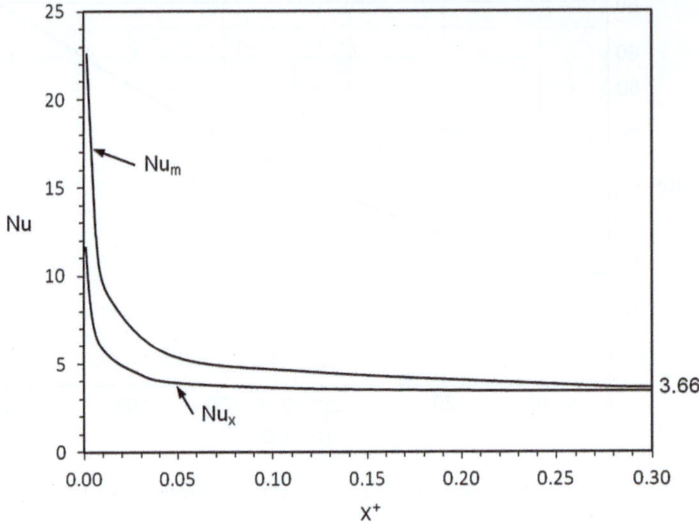

Fig. 8.3 Constant temperature thermal entry length Nusselt numbers

$$\frac{L_{th}}{D} \approx 0.05\mathrm{Re}\,\mathrm{Pr} \tag{8.14}$$

For the constant surface temperature (known as Graetz problem), the local Nusselt number $\mathrm{Nu_x}$ and the mean Nusselt number $\mathrm{Nu_m}$ in the terms of non-dimensional tube length $x^+ = (x/R)/(\mathrm{RePr})$ are given in Fig. 8.3. Dimensionless group $\mathrm{RePr}(D/x)$ is known as Graetz number Gz.

For the constant heat rate (uniform heat flux), the values of the local Nusselt number are given in Fig. 8.4. For tabulated data referring to Figs. 8.3 and 8.4, readers can refer Kays and Perkins (1973).

Hausen (1943 in Holman 1992) has presented the following empirical relation of average Nusselt number over the entire length of the tube for fully developed laminar flow in tubes at constant wall temperature:

$$\mathrm{Nu} = 3.66 + \frac{0.0668\left(\frac{d}{L} \times \mathrm{Re}\,\mathrm{Pr}\right)}{1 + 0.04\left(\frac{d}{L} \times \mathrm{Re}\,\mathrm{Pr}\right)^{2/3}} \tag{8.15}$$

It is to note that the Nusselt number approaches a constant value of 3.66 when the tube is sufficiently long and then the temperature profile is fully developed.

(c) **Thermally and Hydrodynamically Developing Flow**

If the velocity and temperature (thermal) profiles develop simultaneously, the resulting Nusselt numbers in the entry region are always higher than the preceding case. This case is having greater importance from the application point of view. Constant surface temperature and constant heat rate results are presented in Fig. 8.5a, b, respectively, in the form of mean Nusselt number $\mathrm{Nu_m}$ or local Nusselt number $\mathrm{Nu_x}$ as a function of x^+.

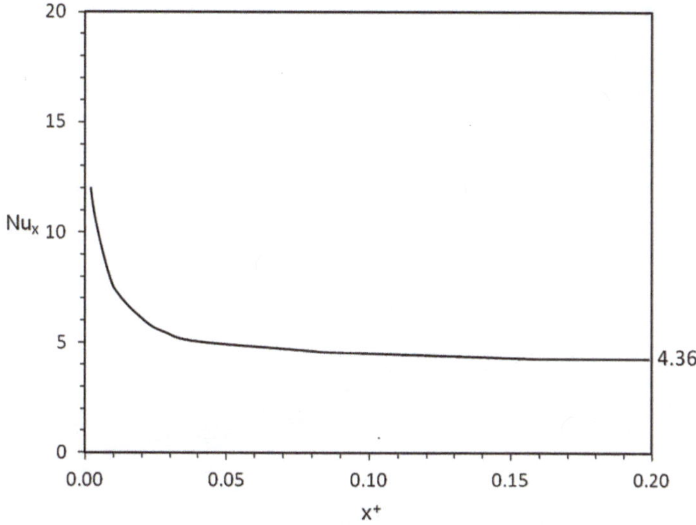

Fig. 8.4 Constant heat rate thermal entry length Nusselt numbers

For constant surface temperature condition, Sieder and Tate (1936) equation may be used for both liquids and gases in the laminar region where the thermal and velocity profiles are developing simultaneously (Bejan and Kraus 2003):

$$\mathrm{Nu} = 1.86\left(\frac{d}{L} \times \mathrm{Re}\,\mathrm{Pr}\right)^{1/3}\left(\frac{\mu_b}{\mu_w}\right)^{0.14} \tag{8.16}$$

$$\text{for } \frac{d}{L} \times \mathrm{Re}\,\mathrm{Pr} > 10, \quad 0.0044 \le \left(\frac{\mu_b}{\mu_w}\right) \le 9.75, \qquad 0.48 \le \mathrm{Pr} \le 16700.$$

where μ_b is the viscosity at the mean bulk temperature and μ_w is the viscosity at the wall temperature. Factor $\left(\frac{\mu_b}{\mu_w}\right)$ is termed as viscosity correction factor which has been discussed in Sect. 8.6. The equation must be used only for entrance region; it cannot be used for extremely long tubes because, for such tubes, it will yield a zero heat transfer coefficient.

Example 8.4 Water at 10°C enters a 10 mm inside diameter and 1.2 m long tube. Water outlet temperature is 40°C. If the tube surface temperature is maintained at 100°C by condensing steam, determine the water flow rate and heat transfer rate.

Solution

Water properties at the mean bulk temperature t_m [= $(t_i + t_o)/2$ = 25°C] from Table A4:

$$\rho = 997\,\mathrm{kg/m^3},\ c = 4181\,\mathrm{J/(kg\ K)},\ \mu = 890 \times 10^{-6}\,\mathrm{N\ s/m^2},\ k = 0.609\,\mathrm{W/(m\ K)}\ \text{and}\ \mathrm{Pr} = 6.13.$$

At T_w = 100°C, $\mu_w = 282 \times 10^{-6}$ N s/m^2.

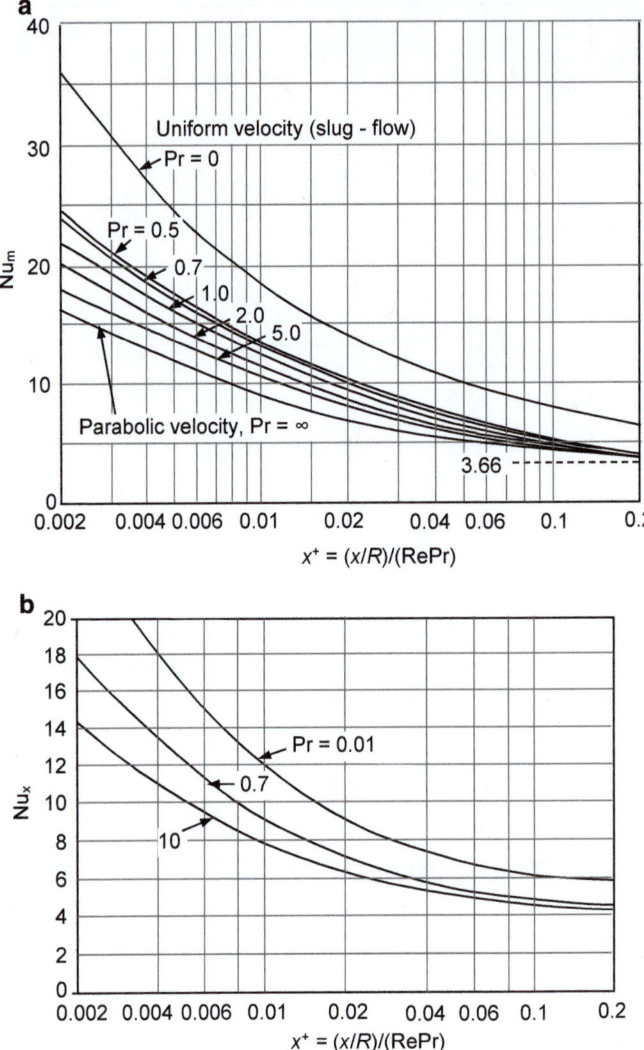

Fig. 8.5 **a** Variation of mean Nusselt number in combined thermal and hydrodynamic region of a tube with constant surface temperature (Pr = 0.5−5.0) (adapted from Kays WM, Perkins HC, Forced convection, internal flow in ducts. In: Rohsenow WM, Hartnett JP (eds) Handbook of heat transfer, Chap. 7. McGraw-Hill, New York. Copyright 1973. The material is reproduced with permission of McGraw-Hill Education.) **b** Variation of local Nusselt number in combined thermal and hydrodynamic region of a tube with constant heat rate per unit of length (Kays WM, Perkins HC, Forced convection, internal flow in ducts. In: Rohsenow WM, Hartnett JP (eds) Handbook of heat transfer, Chap. 7. McGraw-Hill, New York. Copyright 1973. The material is reproduced with permission of McGraw-Hill Education.)

For isothermal case, heat transfer consideration gives, refer Example 7.17,

$$q = \bar{h}PL \times \frac{(t_w - t_o) - (t_w - t_i)}{\ln\left(\frac{t_w - t_o}{t_w - t_i}\right)}. \tag{i}$$

From first law,

$$q = mc(t_o - t_i).$$

Equating above equations,

$$mc(t_o - t_i) = \bar{h}PL \times \frac{(t_w - t_o) - (t_w - t_i)}{\ln\left(\frac{t_w - t_o}{t_w - t_i}\right)}.$$

Simplification gives

$$\frac{t_w - t_o}{t_w - t_i} = \exp\left(-\frac{P}{mc}L\bar{h}\right),$$

which is the same as derived in Example 7.16. The equation can be rewritten as

$$\ln\left(\frac{t_w - t_i}{t_w - t_o}\right) = \frac{P}{mc}L\bar{h}$$

or

$$\ln\left(\frac{t_w - t_i}{t_w - t_o}\right) = \frac{\pi d}{mc}L\bar{h} \tag{ii}$$

where heat transfer coefficient,

$$\bar{h} = \frac{k}{d}\mathrm{Nu}$$

or

$$\bar{h} = \frac{k}{d}1.86\left(\frac{d}{L} \times \mathrm{Re\,Pr}\right)^{1/3}\left(\frac{\mu}{\mu_w}\right)^{0.14}$$

using Sieder and Tate equation for the thermal and velocity profiles developing simultaneously and assuming laminar flow.

The Reynolds number based on mass velocity of water through the tube:

$$\mathrm{Re} = \frac{md}{(\pi/4)d^2\mu} = \frac{4m}{\pi d\mu}$$

Hence,

$$\begin{aligned}\bar{h} &= \frac{k}{d} \times 1.86 \times \left(\frac{d}{L} \times \frac{4m}{\pi d\mu} \times \mathrm{Pr}\right)^{1/3}\left(\frac{\mu}{\mu_w}\right)^{0.14} \\ &= \frac{k}{d} \times 1.86 \times \left(\frac{4m}{\pi\mu L} \times \mathrm{Pr}\right)^{1/3}\left(\frac{\mu}{\mu_w}\right)^{0.14}\end{aligned}$$

Substitution in Eq. (ii) gives

$$\ln\left(\frac{t_w - t_i}{t_w - t_o}\right) = \frac{\pi d}{mc} \times L \times \frac{k}{d} 1.86 \times \left(\frac{4m}{\pi \mu L} \times \Pr\right)^{1/3} \left(\frac{\mu}{\mu_w}\right)^{0.14}$$

or

$$m^{2/3} = \frac{1}{\ln\left(\frac{t_w - t_i}{t_w - t_o}\right)} \frac{\pi k L}{c} \times 1.86 \times \left(\frac{4\Pr}{\pi \mu L}\right)^{1/3} \left(\frac{\mu}{\mu_w}\right)^{0.14}$$

Substituting values of various terms, we have

$$\begin{aligned} m^{2/3} &= \frac{1}{\ln\left(\frac{100-10}{100-40}\right)} \times \frac{\pi \times 0.609 \times 1.2}{4181} \times 1.86 \times \left(\frac{4 \times 6.13}{\pi \times 890 \times 10^{-6} \times 1.2}\right)^{1/3} \left(\frac{890}{282}\right)^{0.14} \\ &= 0.0574 \end{aligned}$$

or $m = 0.0138\,\text{kg/s}$.

This gives:

$$\text{Re} = \frac{4m}{\pi d \mu} = \frac{4 \times 0.0138}{\pi \times 10/1000 \times 890 \times 10^{-6}} = 1974.$$

Flow is laminar.

Hydrodynamic development length L_{hy},

$$L_{hy} = 0.05 \text{Re} d = 0.05 \times 1974 \times 0.01 = 0.987\,\text{m}.$$

Thermal development length L_{th},

$$L_{th} = 0.05 \text{Re} \Pr d = 0.05 \times 1974 \times 6.13 \times 0.01 = 6.05\,\text{m}.$$

Thus the thermal and velocity profiles are developing simultaneously.

The rate of heat transfer is

$$q = mc(t_o - t_i) = 0.0138 \times 4181 \times (40 - 10) = 1730.9\,\text{W}.$$

Example 8.5 One kg/s of water at 35°C flows through a 25 mm diameter 3 m long tube whose surface is maintained at uniform temperature of 100°C. Determine the outlet temperature of the water.

Solution

Water properties at the mean bulk temperature t_m [= $(t_i + t_o)/2$] of 50°C (assumed) from Table A4:

$$\rho = 988.1\ \text{kg/m}^3,\ c = 4182\ \text{J/(kg K)},\ \mu = 544 \times 10^{-6}\ \text{N s/m}^2,\ k = 0.644\ \text{W/(m K) and } \Pr = 3.55.$$

From equation derived in Example 7.16,

$$\frac{t_w - t_o}{t_w - t_i} = \exp\left(-\frac{P}{mc}L\bar{h}\right) \tag{i}$$

The Reynolds number based on mass velocity of water through the tube:

$$\mathrm{Re} = \frac{md}{(\pi/4)d^2\mu} = \frac{4m}{\pi d\mu} = \frac{4 \times 1}{\pi \times 0.025 \times 544 \times 10^{-6}} = 93620.$$

Flow is turbulent. Dittus and Boelter equation may be used for calculation of heat transfer coefficient.

$$\begin{aligned} h &= \frac{k}{d} \times 0.024\mathrm{Re}^{0.8}\,\mathrm{Pr}^{0.4} \\ &= \frac{0.644}{0.025} \times 0.024 \times 93620^{0.8} \times 3.55^{0.4} = 9735\ \mathrm{W/(m^2\ K)}. \end{aligned}$$

Substitution of values of various parameters in Eq. (i) gives

$$\frac{100 - t_o}{100 - 35} = \exp\left(-\frac{\pi \times 0.025}{1 \times 4282} \times 3 \times 9735\right)$$

or

$$t_o = 62^\circ\mathrm{C}.$$

Mean bulk temperature of water is t_m [= $(t_i + t_o)/2$] = 48.5°C, which is nearly equal to the assumed value of 50°C.

Example 8.6 The wall of a tube 1.6 m long and 20 mm diameter is held at constant temperature by condensing steam outside the tube. Water enters the tube at 30°C and leaves at 50°C at the rate of 1 kg/min. Determine the average heat transfer coefficient and the wall temperature.

Solution

The thermophysical properties of water at the mean bulk temperature t_{fm} = (30 + 50)/2 = 40°C are:

$$\rho = 992.2\,\mathrm{kg/m^3},\ k = 0.631\,\mathrm{W/(m\,K)},\ \mu = 6.5 \times 10^{-4}\,\mathrm{kg/(m\,s)},\ \mathrm{Pr} = 4.3,\ c_p = 4179\,\mathrm{J/(kg\,K)}.$$

Reynolds number,

$$\mathrm{Re} = \frac{\rho U_m d}{\mu} = \frac{md}{(\pi/4)d^2\mu} = \frac{4m}{\pi d\mu} = \frac{4 \times (1/60)}{\pi \times 0.02 \times 6.5 \times 10^{-4}} = 1632$$

Flow is laminar.

Hydrodynamic development length L_{hy},

$$L_{hy} = 0.05\text{Re}d = 0.05 \times 1632 \times 0.02 = 1.632\,\text{m} > 1.6\,\text{m}.$$

Thermal development length L_{th},

$$L_{th} = 0.05\text{Re}\,\text{Pr}\,d = 0.05 \times 1632 \times 4.3 \times 0.02 = 7.02\,\text{m} > 1.6\ \text{m}.$$

The flow is simultaneously developing. Hence, the average Nusselt number can be read from Fig. 8.5a.

$$x^+ = \frac{L}{R} \times \frac{1}{\text{Re}\,\text{Pr}} = \frac{1.6}{0.01} \times \frac{1}{1632 \times 4.3} = 0.0228.$$

Corresponding to $x^+ = 0.0228$, Nu = 7.5. Thus the heat transfer coefficient is

$$h = \text{Nu} \times \frac{k}{d} = 7.5 \times \frac{0.631}{0.02} = 236.6\,\text{W/(m}^2\ \text{K)}.$$

We have

$$\frac{t_w - t_o}{t_w - t_i} = \exp\left(-\frac{P}{mc}L\bar{h}\right) \tag{i}$$

or

$$\frac{t_w - 50}{t_w - 30} = \exp\left(-\frac{\pi \times 0.02 \times 1.6}{(1/60) \times 4179} \times 236.6\right)$$

Solution of above equation gives $t_w = 99.14°\text{C}$.

Viscosity of water at wall temperature of 99.14°C, $\mu_w \approx 2.85 \times 10^{-4}$ kg/(m s). Thus there is significant difference between viscosity values of the fluid at mean bulk and wall temperatures and the viscosity correction factor must be used, refer Eq. (8.16). The revised heat transfer coefficient is

$$h = \text{Nu} \times \frac{k}{d}\left(\frac{\mu_b}{\mu_w}\right)^{0.14} = 7.5 \times \frac{0.631}{0.02} \times \left(\frac{6.5}{2.85}\right)^{0.14} = 265.5\text{W/(m}^2\ \text{K)}.$$

With this value of heat transfer coefficient, revised t_w from Eq. (i) is obtained as

$$\frac{t_w - 50}{t_w - 30} = \exp\left(-\frac{\pi \times 0,02 \times 1.6}{(1/60) \times 4179} \times 265.5\right)$$

which gives $t_w = 92.83°\text{C}$. Further iteration is not required.

Alternatively, Sieder-Tate relation, Eq. (8.16) may be used, which is applicable for $\frac{d}{L} \times \text{Re}\,\text{Pr} > 10$.

$$\frac{d}{L} \times \mathrm{Re}\,\mathrm{Pr} = \frac{0.02}{1.6} \times 1632 \times 4.3 = 87.7 > 10.$$

Nusselt number, using the viscosity correction factor,

$$\mathrm{Nu} = 1.86\left(\frac{d}{L} \times \mathrm{Re}\,\mathrm{Pr}\right)^{0.33}\left(\frac{\mu_b}{\mu_w}\right)^{0.14} = 1.86 \times (87.7)^{0.33}\left(\frac{6.5}{2.85}\right)^{0.14} = 9.14.$$

Heat transfer coefficient,

$$h = \mathrm{Nu} \times \frac{k}{d} = 9.14 \times \frac{0.631}{0.02} = 288.4\,\mathrm{W/(m^2K)}$$

which is about 8.5% higher.

Example 8.7 If 1.8 m long unheated entrance section is provided before the heated tube of the previous example, determine the average Nusselt number.

Solution

From previous example, the hydrodynamic development length L_{hy} was 1.632 m. With an unheated entrance section of 1.8 m length, the flow is hydrodynamically developed when it enters the heated section. Flow condition is now that of thermally developing and hydrodynamically fully developed with constant wall temperature. Hausen's equation applies:

$$\begin{aligned}\mathrm{Nu} &= 3.66 + \frac{0.0668\left(\frac{d}{L} \times \mathrm{Re}\,\mathrm{Pr}\right)}{1 + 0.04\left(\frac{d}{L} \times \mathrm{Re}\,\mathrm{Pr}\right)^{2/3}} \\ &= 3.66 + \frac{0.0668 \times 87.7}{1 + 0.04 \times (87.7)^{2/3}} = 6.94\end{aligned} \tag{8.15}$$

as compared to Nu = 9.09 in the previous example. It can be seen that a simultaneously developing flow provides a greater heat transfer coefficient.

8.5.2 Turbulent Flow in Circular Tubes

The turbulent duct flows can also be divided into four categories: fully developed, hydrodynamically developing, hydrodynamically developed but thermally developing, and both hydrodynamically and thermally developing (i.e., simultaneously developing).

It is to be noted that for the turbulent duct flow, the hydrodynamic and thermal entrance lengths are much shorter than the corresponding lengths in the laminar duct flow. Hence, the results of fully developed turbulent flow friction factor and heat transfer are frequently used in design calculations neglecting the effect of the entrance regions. However, for low Prandtl number fluids, and for the heat exchangers short in length, the entrance region effects must be considered.

8.5.2.1 Friction Factor Correlations

(a) **Fully Developed Turbulent Flow in Smooth Ducts**

Several experimental friction factor correlations have been developed for fully developed turbulent flow in smooth ducts. Bhatti and Shah (1987) have compiled these correlations. Some of these correlations are presented here in Table 8.2.

The Prandtl-Karman-Nikuradse (PKN) correlation is regarded as the most accurate. This correlation is based on the universal velocity distribution law with the coefficients slightly modified to fit the highly accurate experimental data of Nikuradse. The drawback of the PKN correlation is that the friction factor f appears on both sides of the correlation.

There have been some attempts to develop a single correlation covering laminar, transition and turbulent flow regimes. Bhatti and Shah (1987) have developed the following correlation covering the three regimes:

$$f = A + \frac{B}{\mathrm{Re}^{1/m}} \tag{8.17}$$

where

$$\begin{array}{ll} A = 0, B = 16, m = 1 & \text{for Re} \le 2100 \\ A = 0.0054, B = 2.3 \times 10^{-8}, m = -2/3 & \text{for } 2100 < \text{Re} \le 4000 \\ A = 1.28 \times 10^{-3}, B = 0.1143, m = 3.2154 & \text{for Re} > 4000. \end{array}$$

(b) **Turbulent Entry length**

The average friction coefficient (termed as apparent Fanning friction factor) f_{app} based on the static pressure drop in the entrance of a circular tube with turbulent flow is presented in Fig. 8.6. The total pressure drop can be calculated from:

$$\Delta p = \bar{f}_{app}\left(\frac{x}{D}\right)(2\rho U^2) \tag{8.18}$$

Table 8.2 Fully developed turbulent flow friction factor correlations for smooth circular duct[a]

Investigators	Correlation	Reynolds number range	Remarks
Blasius	$f = 0.0791\mathrm{Re}^{-0.25}$	$4 \times 10^3 - 10^5$	Within +2.6% and −1.3% of PKN
Bhatti and Shah	$f = 0.00128 + 0.1143\mathrm{Re}^{-0.311}$	$4 \times 10^3 - 10^7$	Within +1.2% and −2% of PKN
	$f = 0.0366\mathrm{Re}^{-0.1818}$	$4 \times 10^4 - 10^7$	Within +2.4% and −3% of PKN
Prandtl, Karman and Nikuradse (PKN)	$\frac{1}{\sqrt{f}} = 1.7372\ln(\mathrm{Re}\sqrt{f}) - 0.3946$	$4 \times 10^3 - 10^7$	Its predictions agree with the highly accurate experimental data within ±2%.
Colebrook	$\frac{1}{\sqrt{f}} = 1.5635\ln\left(\frac{\mathrm{Re}}{7}\right)$	$4 \times 10^3 - 10^7$	Mathematical approximation of PKN yielding numerical values within ±1% of PKN.
Techo et al.	$\frac{1}{\sqrt{f}} = 1.7372\ln\frac{\mathrm{Re}}{1.964\ln \mathrm{Re} - 3.8215}$	$10^4 - 10^7$	Explicit form of PKN agrees within ±0.1%

[a]Bhatti MS, Shah RK, Turbulent and transition flow convective heat transfer. In: Kakac S, Shah RK, Aung W (eds) Handbook of single-phase convective heat transfer, Chap. 4, Wiley, New York, Copyright 1987. Reproduced (abridged) with the permission of John Wiley and Sons Ltd

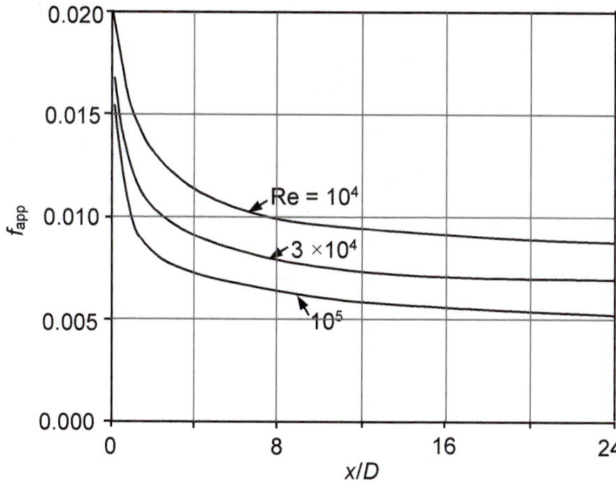

Fig. 8.6 Turbulent flow apparent friction factor f_{app} in hydrodynamic entrance region of a smooth circular duct (Bhatti MS, Shah RK, Turbulent and transition flow convective heat transfer. In: Kakac S, Shah RK, Aung W (eds) Handbook of single-phase convective heat transfer, Chap. 4, Wiley, New York, Copyright 1987. Reproduced (partially) with the permission of John Wiley and Sons Ltd.)

Bhatti and Shah (1987) recommend the following equation for the hydrodynamic entrance length in case of turbulent flow:

$$\frac{L}{D} = 1.359\mathrm{Re}^{0.25} \tag{8.19}$$

Comparison of Eq. (8.19) with the equation for the laminar flow shows that the hydrodynamic entrance length is much shorter for the turbulent flow and its dependence on the Reynolds number is weaker. For example, the laminar flow entrance length at Re = 2100 is 115*D*, whereas it is only 13.6*D* for turbulent flow at Re = 10^4. In many pipe flow of practical interest, the entrance effects are not significant beyond a pipe length of 10 diameters. When the pipe length is several times of the length of entrance region, the flow is assumed to be fully developed for the entire length of the pipe. This simplified approach gives reasonable results for long pipes. For short pipes, the simplified approach underpredicts the friction factor.

It is to note that the hydrodynamic and thermal entry lengths are almost similar in turbulent flow and are independent of the Prandtl number.

8.5.2.2 Heat Transfer Coefficient Correlations

(a) **Fully Developed Velocity and Temperature Profiles**

Two boundary conditions of interest are uniform heat flux or constant heat rate along tube length (UHF) termed as **H boundary condition** and a uniform wall temperature (UWT) termed as **T boundary condition**. The constant heat rate Nusselt number NU_H is always greater than the constant surface temperature Nusselt number Nu_T, but with exception of very low Prandtl number fluids, the difference in N_H and Nu_T is much smaller than for the laminar flow, refer Fig. 8.7. At Pr = 0.7, it is only a few percent. The difference becomes quite negligible for Pr > 1.0. The reasons for the significant effect at low values of Pr lie in the effect of Pr on the distribution of the thermal resistance. At very low Prandtl numbers

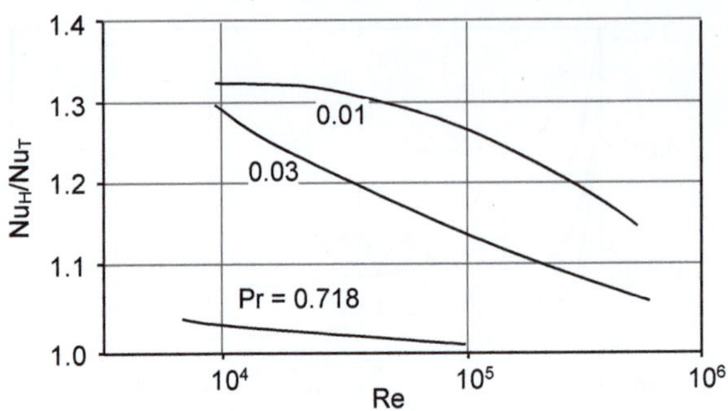

Fig. 8.7 Ratio Nu_H/Nu_T for fully developed turbulent flow (from Sleicher 1955)

where the heat transfer mechanism is mainly molecular diffusion, the thermal resistance extends over the entire cross-section of flow and different boundary conditions yield different temperature profiles (Kays and Crawford 1980). At high Pr values, the resistance is primarily very close to the wall yielding a quite square temperature profile regardless of the boundary condition. And hence the high Prandtl number fluid is relatively insensitive to a variation of surface temperature and heat flux in the direction of flow.

The correlations presented in this section are the constant heat rate Nu_H correlations and can be used to obtain Nu_T by applying the correction factor, Nu_H/Nu_T from Fig. 8.7.

A large number of correlations, both theoretical and empirical (based on experimental data), have been developed for fully developed turbulent flow in smooth tubes. An exhaustive collection of these correlations is given in a tabular form in Bhatti and Shah (1987). In Table 8.3 are presented some of these correlations for gases and liquids (Pr > 0.5) along with their comments.

Churchill (1977) presented the following correlation covering laminar, transition, and turbulent flow regimes:

$$(\mathrm{Nu})^{10} = (\mathrm{Nu}_l)^{10} + \left[\frac{\exp[(2200-\mathrm{Re})/365]}{(\mathrm{Nu}_l)^2} + \left(\frac{1}{\mathrm{Nu}_o + \frac{0.079(f/2)^{1/2}\mathrm{Re\,Pr}}{(1+\mathrm{Pr}^{4/5})^{5/6}}}\right)^2\right]^{-5} \tag{8.20}$$

valid for $0 < \mathrm{Pr} \leq 10^6$ and $10 < \mathrm{Re} \leq 10^6$

where

Nu_o 6.3 for UHF and 4.8 for UWT,
Nu_l 4.364 for UHF and 3.657 for UWT.

For $\mathrm{Re} \leq 2100$, the equation yields laminar flow values of 3.657 and 4.364 corresponding to the T and H boundary conditions (UWT and UHF), respectively. For $2100 \leq \mathrm{Re} \leq 10^4$, predictions are within +17.1% and –11.9% of Gnielinski correlation for UWT and +13.7% and –10.5% of Gnielinski for UHF (Bhatti and Shah 1987).

Note: Correlations of Gnielinski and Churchill cover transitional region.

Table 8.3 Fully developed turbulent flow Nusselt number in a smooth circular duct for Pr > 0.5[a]

Investigators	Correlations	Remarks
Dittus and Boelter (Winterton 1998)	$Nu = 0.024\ Re^{0.8}\ Pr^{0.4}$ for heating of fluid $Nu = 0.026\ Re^{0.8}\ Pr^{0.3}$ for cooling of fluid	Compared to the Gnielinski correlation, prediction of heating correlations are (i) 13.5% to 17% higher for air (ii) 15% lower to 7% higher for water (iii) 10% lower to 21% higher for oil and prediction of cooling correlations are (i) 29% to 33% higher for air (ii) 26% lower to 3% higher for water (iii) 39% to 18% lower for oil
Petukhov, Kirillov and Popov	$Nu = \frac{Re\,Pr(f/2)}{C + 12.7\sqrt{(f/2)}.(Pr^{2/3} - 1)}$ where $C = 1.07 + 900/Re - [0.63/(1 + 10Pr)]$ $Nu = \frac{Re\,Pr(f/2)}{1.07 + 12.7\sqrt{(f/2)}.(Pr^{2/3} - 1)}$ $0.5 \le Pr \le 10^6$ and $4000 \le Re \le 5 \times 10^6$	The first correlation agrees with the most reliable experimental data on heat and mass transfer to ±5% accuracy. The second correlation is a simplified version of the first.
Gnielinski[b]	$Nu = \frac{(f/2)(Re - 1000)\,Pr}{1 + 12.7\sqrt{(f/2)}.(Pr^{2/3} - 1)}$ for $0.5 \le Pr \le 2000$ and $2300 \le Re \le 5 \times 10^6$ $Nu = 0.0214\left(Re^{0.8} - 100\right)Pr^{0.4}$ for $0.5 \le Pr \le 1.5$ and $10^4 \le Re \le 5 \times 10^6$ $Nu = 0.012\left(Re^{0.87} - 280\right)Pr^{0.4}$ for $1.5 \le Pr \le 500$ and $3 \times 10^3 \le Re \le 10^6$	The first correlation is a modification of the second Petukhov et al. correlation extending it to $2300 \le Re \le 5 \times 10^6$ range. For $0.5 \le Pr \le 2000$ and $2300 \le Re \le 5 \times 10^6$, it is in an overall best accord with the experimental data. The second correlation agrees with the first within +4% and −6%. The third correlation agrees with the first within −10%.

[a]Bhatti MS, Shah RK, Turbulent and transition flow convective heat transfer. In: Kakac S, Shah RK, Aung W (eds) Handbook of single-phase convective heat transfer, Chapter 4, Wiley, New York, Copyright 1987. Reproduced (abridged) with the permission of John Wiley & Sons Ltd.
The friction factor f needed in some of the formulas of the Table 8.3 may be calculated from the PKN, Colebrook or Techo et al. correlations given in Table 8.2
[b]1. The Gnielinski's correlation agrees with the Petukhov et al. correlation within −2% and +7.8%. Hence, it has been selected as the common basis for comparison of all the correlations by Bhatti and Shah (1987) as presented in this table
[c] For developing region, Gnielinski modified the relation by introducing the correction factor $[1 + (D/x)^{2/3}]$, which takes care of the entrance effect, and factor $(Pr_b/Pr_w)^{0.11}$ to consider the effect of temperature dependent fluid properties. In case of liquids, the thermal conductivity and specific heat are relatively independent of the temperature and the Prandtl number ratio can be approximated by $(\mu_b/\mu_w)^{0.11}$ and the Gnielinski's correlation for developing liquid flow takes the form (Tam and Ghajar 2006)

$$Nu = \frac{(f/2)(Re - 1000)\,Pr}{1 + 12.7\sqrt{(f/2)}.(Pr^{2/3} - 1)}\left[1 + \left(\frac{D}{L}\right)^{2/3}\right]\left(\frac{\mu_b}{\mu_w}\right)^{0.11}$$

for $0.6 \le Pr \le 10^5$ and $2300 \le Re \le 10^6$
where $f = (1.58\ \ln Re - 3.28)^{-2}$

(b) **Turbulent Entry Length**

When the velocity and temperature profiles develop simultaneously, the entry length is strongly affected by the shape at the inlet. Bhatti and Shah (1987) provide the following correlation for average Nu with $L/D > 3$ for $Pr \approx 0.7$.

Table 8.4 Constants for the gas flow simultaneous entry length correlation for different inlet configurations

Inlet configuration	C	n
Long straight adiabatic pipe as calming section	0.9756	0.760
Square edged inlet	2.4254	0.676
180° circular bend	0.9759	0.700
90° circular bend	1.0517	0.629
90° sharp bend	2.0152	0.614

$$\frac{\mathrm{Nu}_{av}}{\mathrm{Nu}_{\infty}} = 1 + \frac{C}{(L/D)^{n}} \tag{8.21}$$

where Nu_{∞} is the fully developed value of the Nusselt number Nu_{H} or Nu_{T}, and C and n depend on the inlet configuration as given in Table 8.4.

If $L/D > 60$, entry region correction is not required.

Example 8.8 Air flows through an 80 mm diameter tube at a rate of 1 kg/min. Measurements indicate that the average wall temperature of the test section of 1 m length is 400 K and the mean temperature of the air is 300 K. Estimate the heat transfer rate from the tube wall to the air. What will be the rise in the temperature of the air through the test section?

Solution

The flow Reynolds number,

$$\mathrm{Re} = \frac{\rho U d}{\mu} = \left(\frac{\dot{m}}{A_c}\right)\frac{d}{\mu}$$

where

$A_c =$ flow area $= (\pi/4)d^2 = (\pi/4)(0.08)^2 = 5.026 \times 10^{-3}\ \mathrm{m}^2$
$\dot{m} =$ mass flow rate $= 1/60$ kg/s
$\mu =$ viscosity of the air at the mean temperature of 300 K $= 1.84 \times 10^{-5}$ kg/(m s)

Substitution gives

$$\mathrm{Re} = \left(\frac{1}{60}\right)\left(\frac{1}{5.026 \times 10^{-3}}\right)\frac{0.08}{1.84 \times 10^{-5}} = 14418.$$

For Re = 14418, the Nusselt number can be calculated from the Dittus Boelter correlation.

$$\begin{aligned}\mathrm{Nu} &= 0.024\mathrm{Re}^{0.8}\,\mathrm{Pr}^{0.4}\\ &= 0.024 \times (14418)^{0.8}(0.708)^{0.4} = 44.4.\end{aligned}$$

Hence, the convective heat transfer coefficient is

$$h = \frac{\mathrm{Nu}k}{d} = \frac{44.4 \times 0.0263}{0.08} = 14.6\ \mathrm{W/(m^2\ K)},$$

where the thermal conductivity of air is 0.0263 W/(m K) at 300 K.

Heat flow rate is

$$q = hA\Delta T$$

where

$A =$ heat transfer area $= \pi dL = \pi \times 0.08 \times 1 = 0.251\ \mathrm{m^2}$
$\Delta T =$ $T_{\mathrm{wall}} - T_{\mathrm{air}} = 400\text{–}300 = 100$ K.

This gives

$$q = 14.6 \times 0.251 \times 100 = 366.5\ \mathrm{W}.$$

Rise in the temperature of the air through the test section,

$$(\Delta T)_{air} = \frac{q}{mc_p} = \frac{366.5}{(1/60) \times 1005.7} = 21.86°\mathrm{C}$$

where the specific heat of the air at 300 K is about 1005.7 J/(kg K).

Use of Gnielinski simplified correlation gives (for $0.5 \le \mathrm{Pr} \le 1.5$ and $10^4 \le \mathrm{Re} \le 5 \times 10^6$)

$$\mathrm{Nu} = 0.0214(\mathrm{Re}^{0.8} - 100)\,\mathrm{Pr}^{0.4} = 0.0214 \times [(14418)^{0.8} - 100] \times (0.708)^{0.4} = 37.72$$

which is lower by about 18% than given by Dittus Boelter correlation.

Example 8.9 Air flows through a 1 m long 80 mm diameter tube at a rate of 1 kg/min. Measurements indicate that the average wall temperature is 150°C. The air enters the tube at 40°C. Estimate the heat transfer rate from the tube wall to the air and the rise in the temperature of the air. Assume fully developed flow.

Solution

The thermophysical properties are to be taken at mean fluid temperature, but the outlet temperature of the air is not known. For trial, we assume mean bulk temperature as 50°C. The thermophysical properties of air (refer Table A5, Appendix) at 50°C are:

$\rho = 1.095\ \mathrm{kg/m^3}$, $k = 0.02799$ W/(m K), $\mu = 1.95 \times 10^{-5}$ kg/(m s), Pr = 0.703, $c_p =$ 1007.2 J/(kg K).

The flow Reynolds number,

$$\mathrm{Re} = \frac{\rho U d}{\mu} = \left(\frac{\dot{m}}{A_c}\right)\frac{d}{\mu}$$

where

A_c flow area = $(\pi/4)d^2 = (\pi/4)(0.08)^2 = 5.026 \times 10^{-3}$ m^2
$\dot{m}$ mass flow rate = 1/60 kg/s.

This gives

$$\mathrm{Re} = \left(\frac{1}{60}\right)\left(\frac{1}{5.026 \times 10^{-3}}\right)\frac{0.08}{1.95 \times 10^{-5}} = 13604 > 10000.$$

Flow is turbulent. The Nusselt number can be calculated from the Dittus-Boelter correlation:

$$\mathrm{Nu} = 0.024\mathrm{Re}^{0.8}\,\mathrm{Pr}^{0.4} = 0.024 \times (13604)^{0.8} \times (0.703)^{0.4} = 42.3.$$

Hence, the convective heat transfer coefficient is

$$h = \frac{\mathrm{Nu}k}{d} = \frac{42.3 \times 0.02799}{0.08} = 14.8\,\mathrm{W}/\left(\mathrm{m}^2\,\mathrm{K}\right).$$

Heat flow rate is

$$q = hA\Delta T$$

where

A = heat transfer area = $\pi dL = \pi \times 0.08 \times 1 = 0.251$ m^2
ΔT = $t_{wall} - t_m$= 150–50 = 100°C.

Substitution gives

$$q = hA\Delta T = 14.8 \times 0.251 \times 100 = 371.5\ \mathrm{W}.$$

Rise in the temperature of the air,

$$(\Delta T)_{air} = \frac{q}{mc_p} = \frac{371.5}{(1/60) \times 1007.2} = 22.13^{\circ}\mathrm{C}.$$

This gives the mean bulk temperature,

$$t_m = t_i + \Delta T/2 = 40 + 22.13/2 = 51.07^{\circ}\mathrm{C}.$$

The calculated mean temperature of 51.07°C is nearly equal to the trial value of 50°C, which will have only marginal effect on the values of the thermophysical properties of the air. Hence, further iteration is not required.

Example 8.10 Air at mean bulk temperature of 25°C and 2 atm flows through a 2 m long tube of 25 mm diameter with an average velocity of 4 m/s. The tube wall is held at 75°C. Determine the heat transfer rate per unit area.

Solution

The thermophysical properties of air (refer Table A5, Appendix) at 25°C are:

At 1 atm, ρ = 1.1868 kg/m^3, k = 0.02608 W/(m K), μ = 1.8363 × 10^{-5} kg/(m s), Pr = 0.709, c_p = 1005.7 J/(kg K). At 2 atm, ρ = 2 × 1.1868 = 2.3736 kg/m^3.

The flow Reynolds number,

$$\text{Re} = \frac{\rho U d}{\mu} = \frac{2.3736 \times 4 \times 0.025}{1.8363 \times 10^{-5}} = 12926.$$

The velocity or thermal entry length,

$$\frac{L}{D} = 1.359\text{Re}^{0.25} \tag{8.19}$$

or

$$L = (1.359\text{Re}^{0.25})D = 1.359 \times (12926)^{0.25} \times 0.025 = 0.36 \text{ m},$$

which is very small as compared to the length of the tube (L/D = 80 > 60). Hence, the flow can be regarded as fully developed. Dittus-Boelter relation can be used to determine average Nusselt number.

$$\text{Nu} = 0.024\text{Re}^{0.8}\,\text{Pr}^{0.4} = 0.024 \times (12926)^{0.8} \times (0.709)^{0.4} = 40.7.$$

Heat transfer coefficient,

$$h = \frac{\text{Nu}k}{d} = \frac{40.7 \times 0.02608}{0.025} = 42.46 \text{ W/(m}^2\text{ K)}.$$

Heat transfer rate,

$$q = hA(t_w - t_b) = 42.46 \times 1 \times (75 - 25) = 2123 \text{ W/m}^2.$$

Example 8.11 Repeat the above problem if tube diameter D is 50 mm and the fluid is entering the tube through a square edged inlet.

Solution

The thermophysical properties of air (refer Table A5, Appendix) at 25°C are:

At 1 atm, ρ = 1.1868 kg/m^3, k = 0.02608 W/(m K), μ = 1.8363 × 10^{-5} kg/(m s), Pr = 0.709, c_p = 1005.7 J/(kg K). At 2 atm, ρ = 2 × 1.1868 = 2.3736 kg/m^3.

The flow Reynolds number,

$$\text{Re} = \frac{\rho U d}{\mu} = \frac{2.3736 \times 4 \times 0.05}{1.8363 \times 10^{-5}} = 25852.$$

The tube length to diameter ratio,

$$\frac{L}{D} = \frac{2000}{50} = 40.$$

Since $L/D < 60$, entry region correction is required.
For fully developed flow,

$$\text{Nu} = 0.024\text{Re}^{0.8}\,\text{Pr}^{0.4} = 0.024 \times (25852)^{0.8} \times (0.709)^{0.4} = 70.9.$$

The correlation for average Nu considering entrance region effect is (for $L/D > 3$ and $\text{Pr} = 0.7$)

$$\frac{\text{Nu}_{av}}{\text{Nu}_{\infty}} = 1 + \frac{C}{(L/D)^n} \tag{8.21}$$

where Nu_{∞} is the fully developed value of the Nusselt number, and C and n are 2.4254 and 0.676, respectively, from Table 8.4.

Thus,

$$\text{Nu}_{av} = \left[1 + \frac{2.4254}{(40)^{0.676}}\right] \times 70.9 = 85.1$$

Heat transfer coefficient,

$$h = \frac{\text{Nu}k}{d} = \frac{85.1 \times 0.02608}{0.05} = 44.39\ \text{W/(m}^2\ \text{K)}.$$

Heat transfer rate,

$$q = hA(t_w - t_b) = 44.39 \times 1 \times (75 - 25) = 2219.5\ \text{W/m}^2.$$

8.6 Effects of Temperature Varying Properties

The friction factor and heat transfer correlations given in the previous sections apply strictly to isothermal flows, where the fluid temperature is the same at all points of the stream. When the heat transfer takes place, the temperature of the fluid varies both over the cross-section and along the tube length. Since the transport properties of most fluids vary with temperature and hence will vary over the cross-section of the tube.

For the gases, the specific heat varies only slightly with the temperature, but the viscosity and thermal conductivity increase as about 0.8 power of the absolute temperature (in the moderate temperature range).The density of the gases varies inversely with the absolute temperature. However, the Prandtl number ($\text{Pr} = \mu c_p/k$) does not vary significantly with temperature.

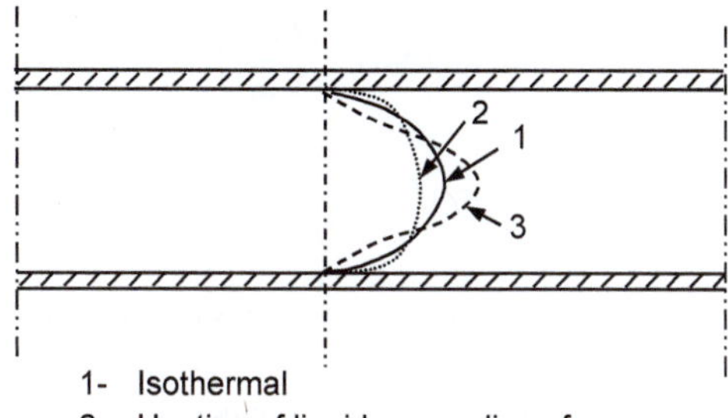

Fig. 8.8 Distortion of the velocity profile in laminar flow through a tube due to the heating or cooling of a fluid

For the liquids, the viscosity decreases markedly with the increase in temperature, while the specific heat, density and thermal conductivity are practically independent of the temperature. The Prandtl number of the liquids varies with temperature approximately the same way as the viscosity. Thus for the liquids only the temperature dependence of the viscosity is of major importance.

Figure 8.8 shows the distortion of the laminar velocity profile due to the heating or cooling of fluids. The viscosity of liquids ordinarily drops with the increasing temperature while the viscosity of the gases ordinarily drops with the decreasing temperature. Hence, if a liquid is heated or a gas is cooled, the viscosity at the wall is lower than the core. Therefore, the flow velocity is higher at the wall and lower in the core than in the isothermal conditions, see curve 2. In the case of cooling of liquids and heating of gases, the flow velocity is lower at the wall and higher in the core, curve 3.

Heat transfer from a surface to a fluid under the laminar flow condition takes place by conduction only, and thus it depends on the temperature gradient near the wall. The temperature distribution depends on (i) the velocity distribution, (ii) the thermal conductivity of the fluid, and (iii) the extent to which the fluid has been heated or cooled in traversing the passage. It has been found that the heat transfer coefficient increase when a liquid is heated or a gas is cooled and vice versa.

In the case of the turbulent flow, the effect of the variation of viscosity at the wall is mainly confined to the laminar sub-layer. The laminar sub-layer tends to be thinner for a decrease in the viscosity at the wall due to the heating of liquids or cooling of gases. The reverse is true when the liquid is being cooled or the gas is being heated.

Correction of Constant Property Results

One of the methods to take account of the effects of the variation of the fluid properties with the temperature is known as the property ratio method (Keys and Perkins 1973). In this method, all the properties are evaluated at the bulk temperature, and then the variable properties effects are lumped into a function of a ratio of the bulk to wall temperatures or viscosities as outlined below.

For liquids, viscosity correction is responsible for most of the effects, and hence the following equation is suggested.

$$\mathrm{Nu} = (\mathrm{Nu})_{c.p.}\left(\frac{\mu_w}{\mu_b}\right)^n \tag{8.22}$$

$$f = (f)_{c.p.}\left(\frac{\mu_w}{\mu_b}\right)^m \tag{8.23}$$

The subscript c.p. refers to the appropriate constant property solution or small temperature difference experimental results. The viscosity μ_w is the viscosity at the wall temperature, while μ_b is the viscosity at the bulk temperature of the fluid. The exponents m and n are functions of the geometry and type of flow.

For gases, the viscosity, thermal conductivity and density are all functions of the absolute temperature of the gas, and hence the equation suggested is

$$\mathrm{Nu} = (\mathrm{Nu})_{c.p.}\left(\frac{T_w}{T_b}\right)^n \tag{8.24}$$

$$f = (f)_{c.p.}\left(\frac{T_w}{T_b}\right)^m \tag{8.25}$$

For the laminar flow of liquids in tubes,

$$n = -0.11 \text{ to } -0.14$$
$$m = +0.5 \qquad \text{for } \mu_w/\mu_b > 1.0, \text{ i.e. cooling}$$
$$m = +0.58 \qquad \text{for } \mu_w/\mu_b < 1.0, \text{ i.e. heating}$$

(the choice of the value of exponent n depends on the viscosity-temperature relation of the particular fluid. However, the difference is small).

For the turbulent flow of liquids in tubes, the suggested values of the exponents m and n are given in Table 8.5.

The recommended values for laminar flow of a gas in a circular tube are

$$n = 0.0 \text{ and } m = 1.00 \quad 1 < T_w/T_b < 3 (\text{for heating})$$

Table 8.5 Viscosity ratio exponents for fully developed turbulent flow of liquids in a circular tube[a]

Pr	$\mu_w/\mu_b > 1.0$, i.e. cooling		$\mu_w/\mu_b < 1.0$, i.e. heating	
	m	n	m	n
1	0.12	−0.19	0.092	−0.20
3	0.087	−0.21	0.063	−0.27
10	0.052	−0.22	0.028	−0.36
30	0.029	−0.21	0.000	−0.39
100	0.007	−0.20	−0.04	−0.42
1000	−0.018	−0.20	−0.12	−0.46

[a]Kays WM, Perkins HC, Forced convection, internal flow in ducts. In: Rohsenow WM, Hartnett JP (eds) Handbook of heat transfer, Chap. 7. McGraw-Hill, New York. Copyright 1973. The material is reproduced with permission of McGraw-Hill Education

and

$$n = 0.0 \text{ and } m = 0.81 \quad 0.5 < T_w/T_b < 1 \text{(for cooling)}$$

The values for the turbulent flow of a gas are

$$\begin{array}{lll} T_w/T_b > 1.0 & n = -0.55; & m = 0.1 \\ T_w/T_b < 1.0 & n = 0.0; & m = -0.1 \end{array}$$

Information on the flow through noncircular tubes with variable properties is not complete and hence it is recommended that the circular tube results be used.

8.7 Heat Transfer and Friction in Concentric Circular Tube Annuli and Parallel Plate Duct

Hydraulic or Equivalent Diameter Concept

It has been found that nearly the same turbulence intensity and the friction factor prevail in circular and other duct geometries (such as annular, rectangular, square, triangular and irregular passages) if the ratio of flow-passage area to the wetted parameter is kept constant. This ratio is called the hydraulic radius R_h. For a circular passage, this becomes

$$R_h = \frac{(\pi/4)D^2}{\pi D} = \frac{D}{4}$$

It is often convenient to use the term equivalent diameter D_h to signify the diameter of a circular passage, which would have the same hydraulic radius as the passage geometry. Thus

$$D_h = \frac{4A}{P}$$

where A is the flow passage area and P is the wetted parameter. The hydraulic diameter for a circular passage is equal to D. This definition has been used throughout this book.

It must be noted that the use of hydraulic diameter for ducts with sharp corners (e.g., triangular ducts) may lead to error of the order of 35% in turbulent flow friction and heat transfer coefficients determined from the circular duct correlations (Bhatti and Shah 1987).

8.7.1 Laminar Flow

Lundgren et al. (1964) determined the pressure drop due to flow development in the entrance region of ducts of arbitrary cross-section. Some data of fully developed Fanning friction factor for laminar flow in concentric annular duct from their results are presented in Table 8.6 along with that for a circular tube. Natarajan and Lakshmanan (in Ebadian and Dong 1998) have presented a simple equation for $f\text{Re}$ [$f\text{Re} = 24(r^*)^{0.035}$ for $0 < r^* \leq 1$], which agrees well with the data of Table 8.6.

For the friction coefficient in the hydrodynamic entry length for the concentric tube annuli including parallel plate duct, refer Heaton et al. (1964).

Table 8.6 Fully developed friction factors for laminar flow in circular tube annuli

$r^* = r_i/r_o$	$f\text{Re}$
0.0 (circular tube)	16
0.01	20.03
0.05	21.57
0.10	22.34
0.20	23.09
0.40	23.68
0.60	23.90
0.80	23.98
1.00 (parallel plate duct)	24.00

Table 8.7 Fully developed Nusselt numbers for laminar flow in circular tube annuli; constant heat rate[a]

$r^* = r_i/r_o$	Nu_{ii}	Nu_{oo}	θ_i^*	θ_o^*
0.0	∞	4.364	∞	0
0.05	17.81	4.792	2.18	0.0294
0.10	11.91	4.834	1.383	0.0562
0.20	8.499	4.883	0.905	0.1041
0.40	6.583	4.979	0.603	0.1823
0.60	5.912	5.099	0.473	0.2455
0.80	5.58	5.24	0.401	0.299
1.00	5.385	5.385	0.346	0.346

[a]Kays WM, Perkins HC, Forced convection, internal flow in ducts. In: Rohsenow WM, Hartnett JP (eds) Handbook of heat transfer, Chap. 7. McGraw-Hill, New York. Copyright 1973. The material is reproduced with permission of McGraw-Hill Education

In the laminar flow, the fully developed Nusselt number is independent of the Reynolds number and Prandtl number. For the concentric tube flow, there are two Nusselt numbers of interest, one for inner surface (Nu_{ii} when inner surface alone is heated) and other for the outer surface (Nu_{oo} when outer surface alone is heated). Their values for constant heat rate condition are given in Table 8.7.

For the case of constant heat rate, it is possible to determine Nu_{i} and Nu_{o} (the Nusselt number on the inner and outer surfaces, respectively) for any heat flux ratio (q_o''/q_i'') on the two surfaces, in terms of Nu_{ii} and Nu_{oo}, and a pair of influence coefficients θ_i^* and θ_o^*(refer Table 8.7) using the following equations (Kays and Perkins 1973).

$$\text{Nu}_{\text{i}} = \frac{\text{Nu}_{ii}}{1 - (q_o''/q_i'')\theta_i^*} \tag{8.26a}$$

$$\text{Nu}_{\text{o}} = \frac{\text{Nu}_{oo}}{1 - (q_i''/q_o'')\theta_o^*} \tag{8.26b}$$

where

$$\mathrm{Nu_i} = \frac{h_i d_h}{k}$$

$$\mathrm{Nu_o} = \frac{h_o d_h}{k}$$

$$q_i'' = h_i(t_i - t_b)$$

$$q_o'' = h_o(t_o - t_b)$$

For the thermal entry length data of Nusselt number, readers can refer Kays and Crawford (1980).

For the case of constant surface temperature on one surface (with the other surface insulated), the Nusselt number has also been computed by Lundberg et al. (in Kays and Perkins, 1973). The data are presented in Table 8.8.

In the case of parallel plates ($r^* = 1.0$), Nu = 7.541 when both surfaces are at the same constant temperature, while it is 8.235 for identical heat rates on the two surfaces of the duct.

For the thermal entry length local Nusselt number data for constant and equal wall temperatures and heat rate refer Kays and Perkins (1973). Edwards et al. (in Mills 1995) presented the average Nusselt number relation for the thermal entrance region flow between isothermal plates as

$$\mathrm{Nu} = 7.54 + \frac{0.03(D_h/L)\mathrm{Re\,Pr}}{1 + 0.016[(D_h/L)\mathrm{Re\,Pr}]^{2/3}} \quad \text{for Re} < 2800 \tag{8.27}$$

where the hydraulic diameter is twice the spacing of the plates. Critical Reynolds number of 2800 for transition is appropriate for flow between parallel plates (Mills 1995).

For combined thermal and hydrodynamic entry length Nusselt number data relating to the circular tube annulus family (including parallel plates duct) with constant heat rate, refer Kays and Crawford (1980).

Example 8.12 Water at a mean bulk temperature of 20°C flows through the annular region formed by two concentric circular tubes. The outer tube has inner diameter of 25 mm, while the inner tube has outer diameter of 12.5 mm. The mass flow rate is 0.015 kg/s. The outer

Table 8.8 Fully developed Nusselt numbers for laminar flow in circular tube annuli; constant surface temperature on one surface, the other insulated[a]

$r^* = r_i/r_o$	Nu_i	Nu_o
0.0	∞	3.66
0.05	17.46	4.06
0.10	11.56	4.11
0.25	7.37	4.23
0.50	5.74	4.43
1.00	4.86	4.86

[a]Kays WM, Perkins HC, Forced convection , internal flow in ducts. In: Rohsenow WM, Hartnett JP (eds) Handbook of heat transfer, Chap. 7. McGraw-Hill, New York. Copyright 1973. The material is reproduced with permission of McGraw-Hill Education

surface is insulated and the inner surface is kept at a constant temperature of 50°C. Determine the friction factor and average heat transfer coefficient from the outer surface of the inner tube for the fully developed temperature and velocity profiles.

Solution

At mean bulk temperature of 20°C, the water properties are:

ρ = 998.2 kg/m^3, k = 0.601 W/(m K), μ = 10.02 × 10^{-4} kg/(m s), Pr = 7.0, c_p = 4183 J/(kg K).

Hydraulic diameter,

$$D_h = \frac{4A}{P} = \frac{4(\pi/4)(D_i^2 - d_o^2)}{\pi(D_i + d_o)} = (D_i - d_o) = 25 - 12.5 = 12.5 \text{ mm}.$$

Flow Reynolds number,

$$\begin{aligned} \text{Re} &= \frac{\rho U D_h}{\mu} = \frac{m}{(\pi/4)(D_i^2 - d_o^2)} \times \frac{(D_i - d_o)}{\mu} \\ &= \frac{4m}{\pi(D_i + d_o)} \times \frac{1}{\mu} = \frac{4 \times 0.015 \times 1000}{\pi \times (25 + 12.5)} \times \frac{1}{10.02 \times 10^{-4}} = 508. \end{aligned}$$

The flow is laminar. Nusselt number Nu_{ii} for (D_i/d_o) = 0.5, from Table 8.8 is 5.74. Thus the heat transfer coefficient is

$$h_i = \frac{\text{Nu}k}{D_h} = \frac{5.74 \times 0.601}{12.5/1000} = 276 \text{ W/(m}^2 \text{ K)}.$$

Friction factor, from Table 8.6, is

$$f = \frac{23.78}{\text{Re}} = \frac{23.78}{508} = 0.0468.$$

Example 8.13 Water at a mean bulk temperature of 20°C and 0.015 kg/s enters the annular region formed by two concentric circular tubes and is heated to 60°C. The outer tube has inner diameter of 25 mm, while the inner tube has outer diameter of 12.5 mm. The mass flow rate is 0.015 kg/s. The outer surface is insulated and the inner surface is subjected to uniform heat rate of 1000 W/m. Determine the tube length required to achieve desired outlet temperature and inner tube surface temperature at outlet.

Solution

(a) **Tube length required to achieve desired outlet temperature**

At mean bulk temperature of (20 + 60)/2 = 40°C, the water properties are:

$$\rho = 992.2 \text{ kg/m}^3,\ c = 4179 \text{ J/(kg K)}$$

The heat rate for length L is

$$q = q'L = 1000L \text{ W}$$

where q' is heat rate per m.

From first law equation,

$$q = mc(t_o - t_i) = 0.015 \times 4179 \times (60 - 20)$$

Equating the above equations, we have

$$L = \frac{0.015 \times 4179 \times (60 - 20)}{1000} = 2.5\,\text{m}.$$

(b) **Surface temperature at the outlet,**

At 60°C, the water properties are: $k = 0.654$ W/(m K), $\mu = 463 \times 10^{-6}$ kg/(m s), Pr = 3.

Hydraulic diameter,

$$D_h = \frac{4A}{P} = \frac{4(\pi/4)(D_i^2 - d_o^2)}{\pi(D_i + d_o)} = (D_i - d_o) = 25 - 12.5 = 12.5 \text{ mm}.$$

Flow Reynolds number,

$$\text{Re} = \frac{\rho U D_h}{\mu} = \frac{m}{(\pi/4)(D_i^2 - d_o^2)} \times \frac{(D_i - d_o)}{\mu}$$
$$= \frac{4m}{\pi(D_i + d_o)} \times \frac{1}{\mu} = \frac{4 \times 0.015 \times 1000}{\pi \times (25 + 12.5)} \times \frac{1}{463 \times 10^{-6}} = 1100.$$

Hence, the flow is laminar, and for $(D_i/d_o) = 0.5$, it follows from Table 8.7 that the fully developed Nusselt number $\text{Nu}_i = \text{Nu}_{ii} = 6.25$ for outer tube insulated, refer Eq. (8.26a).

Thus the heat transfer coefficient is

$$h_i = \frac{\text{Nu}k}{D_h} = \frac{6.25 \times 0.654}{12.5/1000} = 327 \text{ W/(m}^2\text{K)}.$$

At the outlet,

$$\frac{q}{L} = q' = h_i(\pi d_o)(t_{so} - t_o)$$

or

$$t_{so} = t_o + \frac{q'}{h_i(\pi d_o)} = 60 + \frac{1000}{327 \times \pi \times 12.5/1000} = 137.9^\circ\text{C}.$$

Since tube surface temperature at the outlet t_{so} > 100°C, local boiling at the tube surface will occur if water is not pressurized, which will increase the local heat transfer coefficient and hence reduce the wall temperature.

The length to hydraulic diameter ratio is 2.5/0.0125 = 200. Hydrodynamic development length is (using tube equation)

$$\frac{L_{hy}}{D_h} \approx 0.05\text{Re} = 0.05 \times 1100 = 55,$$

and the thermal entry length is

$$\frac{L_{th}}{D} \approx 0.05\text{Re Pr} = 0.05 \times 1100 \times 3 = 165.$$

At the outlet, the flow is fully developed.

Example 8.14 Air at 25°C and a pressure of 1 atm flows through a parallel plate duct with an average velocity of 0.3 m/s. The duct depth is 5 mm. For fully developed velocity and temperature profiles, what is heat transfer coefficient?

(a) If both the plates are held at a constant temperature,
(b) If identical heat rates are imposed on the two surfaces of the duct.

Solution

The air properties at 25°C are:

$$\rho = 1.1868\,\text{kg/m}^3,\ k = 0.02608\ \text{W/(m K)and}\ \mu = 1.8363 \times 10^{-5}\,\text{kg/(m s)}.$$

The hydraulic diameter of the parallel plate duct is

$$D_h = 2H = 0.01\text{m}.$$

Flow Reynolds number,

$$\text{Re} = \frac{\rho U D_h}{\mu} = \frac{1.1868 \times 0.3 \times 0.01}{1.8363 \times 10^{-5}} = 194.$$

Flow is laminar. In the case of parallel plates, Nu = 7.541 when both surfaces are at the same constant temperature, while it is 8.235 for identical heat rates on the two surfaces of the duct.

(a) Heat transfer coefficient (both sides at uniform constant temperature),

$$h = \frac{\text{Nu}k}{D} = \frac{7.541 \times 0.02608}{0.01} = 19.67.$$

(b) Heat transfer coefficient (identical heat rates on the two surfaces of the duct),

$$h = \frac{\text{Nu}k}{D} = \frac{8.241 \times 0.02608}{0.01} = 21.49.$$

Example 8.15 Air at 15°C and a pressure of 1 atm flows through a 200 mm long parallel plate duct whose one side is insulated and the other side is maintained at a constant temperature of 70°C. The duct depth is 5 mm. The pressure drop is recorded to be 1 N/m^2. Determine the outlet temperature of the air.

Solution

The air properties at 25°C (assumed mean bulk temperature) and 1 atm are:

$$\rho = 1.1868\ \text{kg/m}^3,\ c = 1005.7\ \text{J/(kg K)},\ \mu = 1.8363 \times 10^{-5}\ \text{kg/(m s)}\ \text{and}\ k = 0.02608\ \text{W/(m K)}.$$

The hydraulic diameter of the parallel plate duct is

$$D_h = 2H = 0.01\ \text{m}.$$

Flow Reynolds number,

$$\text{Re} = \frac{\rho U_m D_h}{\mu} = \frac{1.1868 \times 0.01}{1.8363 \times 10^{-5}} \times U_m = 646.3U_m.$$

Assuming flow to be laminar, the friction factor is given by (Table 8.10)

$$f = \frac{24}{\text{Re}} = \frac{24}{646.3U_m} = \frac{0.0371}{U_m}$$

Pressure drop is given by

$$\Delta p = \frac{4fL\rho U_m^2}{2D_h}$$

or

$$1 = \frac{4 \times 0.0371 \times 0.2 \times 1.1868 \times U_m^2}{U_m \times 2 \times 0.01}$$

or

$$U_m = 0.57\ \text{m/s}.$$

Hence,

$$\text{Re} = 646.3 \times 0.57 = 368.$$

The flow is laminar. From Table 8.10, the Nusselt number for given boundary conditions is 4.861.

The heat transfer coefficient is

$$h = \frac{\text{Nu}k}{D_h} = \frac{4.861 \times 0.02608}{0.01} = 12.68\ \text{W}/\left(\text{m}^2\,\text{K}\right).$$

For isothermal surface, the air outlet temperature is given by

$$\frac{t_w - t_o}{t_w - t_i} = \exp\left(-\frac{P}{mc}L\bar{h}\right)$$

or

$$t_o = t_w - (t_w - t_i)\exp\left(-\frac{P}{mc}L\bar{h}\right)$$

The mass flow rate is

$$m = \rho A_c U_m = \rho(wa)U_m = 1.1868 \times w \times 0.005 \times 0.57 = 0.00338w$$

where a is the depth and w (>> a) is width of the duct.

Substitution gives

$$t_o = 70 - (70 - 15)\exp\left(-\frac{w}{0.00338w \times 1005.7} \times 0.2 \times 12.68\right) = 43.9°\text{C}$$

where $P = w$.

Mean bulk temperature of air is (15 + 43.9)/2 = 29.45°C against assumed temperature of 25°C for the estimate of thermophysical properties. For greater accuracy, retrial may be done.

8.7.2 Turbulent Flow

Brighton and Jones (in Kays and Perkins 1973), from their data ($0.0625 < r^* < 0.562$), suggest that the friction factor in circular tube annuli is independent of the radius ratio r^*. The friction factors, as determined with water, were found by them to be 6–8% higher than the generally accepted values of smooth circular tube for 4000 < Re < 17000. The same for air flow were found to be about 1–10% higher than circular tube flow values.

Kays and Perkins (1973) recommend the following approximate relation, which gives the friction factor in circular annuli about 10% higher than pipe flow values and having little dependence on the radius ratio:

$$f = 0.085\text{Re}^{-0.25}. \qquad \text{for } 6000 < \text{Re} < 300,000. \tag{8.28}$$

However, Bhatti and Shah (1987) recommend the following correlation, which considers the dependence of the friction factor on the radius ratio r^* (= r_i/r_o),

$$f = (1 + 0.0925r^*)f_c \tag{8.29}$$

where f_c is the friction factor for a circular duct using hydraulic diameter of the annular duct. They report that the predicted friction factors from Eq. (8.29) are within ± 5% of the experimental results.

The smooth circular duct friction factor values in Eq. (8.29) may be calculated from the following correlation (Bhatti and Shah 1987):

$$f_c = 0.00128 + 0.1143\text{Re}^{-0.311}. \quad \text{for } 4000 \leq Re \leq 10^7. \tag{8.30}$$

Dalle Donne and Meerwald (1973) suggest the following relation for the fully developed isothermal friction factor:

$$f = 0.0014 + 0.125\text{Re}^{-0.32}. \quad \text{for } 8000 < \text{Re} < 2 \times 10^5 \tag{8.31}$$

For fully developed flow in parallel plate duct, Beavers et al. (in Ebadian and Dong 1998) obtained following friction factor result from very accurate experimental data:

$$f = 0.1268\text{Re}^{-0.3}. \quad \text{for } 5000 < \text{Re} < 1.2 \times 10^6 \tag{8.32}$$

Dean (in Ebadian and Dong 1998) developed the following equation based on comprehensive survey of the available data:

$$f = 0.0868\text{Re}^{-0.25}. \quad \text{for } 1.2 \times 10^4 < \text{Re} < 1.2 \times 10^6 \tag{8.33}$$

In the range of $5000 < \text{Re} < 1.2 \times 10^4$, Bhatti and Shah (in Ebadian and Dong 1998) recommend Eq. (8.32); otherwise use of Eq. (8.33) is recommended to obtain the friction factor for the fully developed turbulent flow in parallel plate duct using hydraulic diameter as twice the spacing between two plates.

Hart and Lawther (in Kays and Perkins 1973) suggest that annular flow develops more quickly than pipe flow. Olson and Sparrow's data appear to show that 30 diameter is sufficient for a close approach to fully developed flow (Kays and Perkins 1973). However, Lee (1968) typically gives the thermal development length of only 20 diameters for annular duct of $r^* = 0.67$ at $\text{Re} = 10^4$ and the development length has been shown to increase with the increase in the Reynolds number.

Kays and Leung (1963) presented results of their solution of turbulent flow energy equation of fully developed constant heat rate condition over a wide range of annulus radius ratio, Reynolds number and Prandtl numbers. The solutions are presented for one surface heated and other insulated. Typical results for Pr = 0.7 are given in Table 8.9. For detailed

Table 8.9 Fully developed constant heat rate turbulent flow Nusselt number (Pr = 0.7)

	Re = 10^4		Re = 3×10^4		Re = 10^5		Re = 10^6	
r^*	Nu_{ii}	Nu_{oo}	Nu_{ii}	Nu_{oo}	Nu_{ii}	Nu_{oo}	Nu_{ii}	Nu_{oo}
0.1	48.5	29.8	98	66	235	167	1510	1100
0.2	38.6	29.4	79.8	64.3	196	165	1270	1070
0.5	30.9	28.3	66.0	62.0	166	158	1080	1040
0.8	28.5	28	62.3	61	157	156	1050	1020
Parallel plates duct								
1.0	27.8		61.2		155		1030	

results, refer Kays and Leung. However, Bhatti and Shah, and Sparrow and Lin (in Ebadian and Dong 1998) concluded that the Nusselt number for parallel ducts can be determined using the circular duct correlations with hydraulic diameter in Nusselt number and Reynolds number as twice the spacing between the plates. Mills (1995) has extended the reasoning for the same. In the turbulent flow, the viscous sublayer around the perimeter of the duct is very thin, and the velocity and temperature are nearly uniform across the core. Since the viscous sublayer is the major resistance to momentum and heat transfer, the precise shape of the core fluid is not critical.

Recently, Gnielinski (2015) has suggested the following correlation:

$$\mathrm{Nu} = \frac{(f_{\mathrm{ann}}/2)(\mathrm{Re}-1000)\,\mathrm{Pr}}{1+12.7\sqrt{(f_{\mathrm{ann}}/2)}(\mathrm{Pr}^{2/3}-1)}\left[1+\left(\frac{D_h}{L}\right)^{2/3}\right]F_{\mathrm{ann}}K \tag{8.34}$$

$$\text{for } 0.1 \le \mathrm{Pr} \le 1000,\ (D_{\mathrm{h}}/L) \le 1, \text{and } \mathrm{Re} > 4000$$

where the hydraulic diameter D_{h} of the annulus is d_{o}–d_{i}.

The length of the annular tube is L, d_{o} is the inner diameter of the outer tube and d_{i} is the outer diameter of the inner tube.

The friction factor in Eq. (8.34) depends on the diameter or radius ratio $r^*(= d_{\mathrm{i}}/d_{\mathrm{o}})$. Gnielinski suggests

$$4f_{\mathrm{ann}} = (1.8\log_{10}\mathrm{Re}^* - 1.5)^{-2} \tag{8.35}$$

where

$$\mathrm{Re}^* = \mathrm{Re}\frac{[1+(r^*)^2]\ln r^* + [1-(r^*)^2]}{[1-r^*]^2\ln r^*} \tag{8.36}$$

A factor F_{ann} allows for the different ways of heating or cooling the flow in the two passages of the annulus. By a comparison of large volume of experimental data, the value has been given by Gnielinski (2015) as

$$F_{\mathrm{ann}} = 0.75(r^*)^{-0.17} \tag{8.37a}$$

for the boundary condition of "hear transfer at the inner wall with the outer wall insulated", and

$$F_{\text{ann}} = 0.9 - 0.15(r^*)^{0.6} \tag{8.37b}$$

for the boundary condition of "heat transfer at the outer wall with the inner wall insulated," but no experimental data have been found for heat transfer from both the walls in the annular flow (Gnielinski 2015).

The variation of fluid properties with temperature is taken into account by a factor K. As no special studies of annular ducts are known hence correction factors for circular tubes may be adopted (Gnielinski 2015).

For gases, K is given as (Gnielinski 2015)

$$K = \left(\frac{T_b}{T_w}\right)^n \tag{8.38a}$$

where T_b is the bulk temperature of the gas and T_w is the wall temperature in kelvin. $n = 0.45$ in the temperature range of $0.5 < (T_b/T_w) < 1.0$. For carbon dioxide and steam $n = 0.15$ in the temperature range.

For liquids (Gnielinski 2015),

$$K = \left(\frac{\Pr_b}{\Pr_w}\right)^{0.11} \tag{8.38b}$$

where $\Pr_b$ is the Prandtl number at absolute bulk temperature and $\Pr_w$ is at the absolute wall temperature. For hydrodynamically developing flow, hydrodynamically developed and thermally developing flow, and simultaneously developing flow in parallel plate ducts, refer Ebadian and Dong (1998).

8.8 Heat Transfer and Friction in Rectangular Duct

8.8.1 Laminar Flow

Friction Factor Correlation

The empirical equation suggested by Shah and London (in Ebadian and Dong 1998) is

$$f\text{Re} = 24(1 - 1.3553\alpha^* + 1.9467\alpha^{*2} - 1.7012\alpha^{*3} + 0.9564\alpha^{*4} - 0.2537\alpha^{*5}) \tag{8.39}$$

where α^* is aspect ratio (ratio of duct height to width).

Heat Transfer Coefficient Correlations

The fully developed Nusselt numbers Nu_T for the case of uniform temperature at four walls are approximated by the following equation (Shah and London in Ebadian and Dong 1998):

$$\text{Nu}_T = 7.541(1 - 2.610\alpha^* + 4.970\alpha^{*2} - 5.119\alpha^{*3} + 2.702\alpha^{*4} + 0.548\alpha^{*5}) \tag{8.40}$$

Table 8.10 Fully developed Nusselt number and friction factor for laminar flow in ducts of different cross-sections

Cross-section shape ($L/D_h > 100$)		Boundary conditions[a]		$fRe = C_f\ Re/4$
		Nu_T	Nu_H	
Circle		3.657	4.354	16.0
Rectangle (2b × 2a)	$\alpha^* = 2b/2a = 1$	2.976	3.608	14.227
	$\alpha^* = 0.5$	3.391	4.123	15.548
	$\alpha^* = 0.25$	4.439	5.331	18.233
	$\alpha^* = 0.125$	5.597	6.490	20.585
Parallel plates	$\alpha^* = 0$	7.541	8.235	24.0
Parallel plates, one insulated	$\alpha^* = 0$	4.861	5.385	24.0

[a]Nu_T: T_w = constant irrespective of x, y, z

Nu_H: Heat rate q_w = constant along the length of a tube, but uniform temperature around the periphery)

For rectangular ducts with a uniform heat flux at four walls constant along the length of the duct, but uniform temperature around the periphery Nu_H can be calculated from the following equation (Shah and London in Ebadian and Dong 1998):

$$Nu_H = 8.235(1 - 2.0421\alpha^* + 3.0853\alpha^{*2} - 2.4765\alpha^{*3} + 1.0578\alpha^{*4} - 0.1861\alpha^{*5}) \quad (8.41)$$

Some typical results for the friction factor and constant heat rate and surface temperature Nusselt numbers are listed in Table 8.10 along with circular and parallel plate ducts for comparison.

For Nusselt number with uniform temperature or heat rate at one or more walls, and for the hydrodynamically developing flow, thermally developing flow and simultaneously flow, refer Ebadian and Dong (1998).

8.8.2 Turbulent Flow

The entrance configuration (abrupt or smooth) exerts a marked influence on the value of critical Reynolds number Re_c for flow in smooth rectangular duct. Results of some experimental measurements are presented in Bhatti and Shah (1987), which shows that Re_c varies from 3400 ($\alpha^* = 0$, parallel plate duct) to 4300 ($\alpha^* = 1.0$, square duct) with smooth entrance while these values are 3100 and 2200, respectively, for the abrupt entrance.

A unique feature of the fully developed turbulent flow in rectangular ducts is the presence of secondary flow (flow normal to the axis of the duct). Though the secondary flow is small in magnitude (1% of the axial mean velocity), it exerts a significant effect on the turbulence fluid flow characteristics of these ducts and increases the friction factor by approximately 10% (Bhatti and Shah 1987).

For fluids with Prandtl number greater than 0.5, the resistance to heat transfer is basically confined to the laminar sublayer and the temperature distribution over most of the turbulent core region is relatively flat. Under these circumstances, the heat transfer from the surface is reasonably independent of the duct shape if the surface temperature around the duct periphery is uniform (Kays and Crawford 1980). Hence, the friction factor and Nusselt number for the fully developed flow in rectangular ducts can be estimated from the circular duct correlations. If laminar equivalent diameter D_l is used in the place of D_h, circular duct correlations provide values within ± 5% of the experimental results (Bhatti and Shah 1987). The laminar equivalent diameter D_l is defined (Bhatti and Shah 1987) as

$$\frac{D_l}{D_h} = \frac{2}{3} + \frac{11}{24}\alpha^*(2 - \alpha^*) \tag{8.42}$$

where

$$D_h = \frac{4A}{P} = \frac{4ab}{(a+b)} \tag{8.43}$$

and $\alpha^* = (2b/2a)$.

Bhatti and Shah (1987) have presented the following friction factor correlation for flow in rectangular cross-section smooth duct ($0 \leq \alpha^* \leq 1$) in the range $5000 \leq \text{Re} \leq 10^7$:

$$f = (1.0875 - 0.1125a^*)f_o \tag{8.44}$$

where

$$\begin{aligned} f_o &= \text{friction factor for the circular duct using } D_h \\ &= 0.0054 + 2.3 \times 10^{-8}\text{Re}^{1.5} \qquad \text{for } 2300 \leq \text{Re} \leq 3550 \\ \text{and} &= 1.28 \times 10^{-3} + 0.1143\text{Re}^{-0.311} \qquad \text{for } 3550\text{Re} \leq 10^7. \end{aligned}$$

The prediction of Eq. (8.44) are at par with those determined by substituting D_h by D_l (Bhatti and Shah 1987), i.e. an uncertainty of ± 5% in the predicted friction factors from the above correlation may be considered.

For hydrodynamically developing, thermally developing and simultaneously developing flows in rectangular ducts, refer Ebadian and Dong (1998) and Bhatti and Shah (1987).

With uniform heating at four walls of a rectangular duct, circular duct Nusselt number correlations provide results with ± 9% accuracy for $0.5 \leq \text{Pr} \leq 10$ and $10^4 \leq \text{Re} \leq 10^6$ while for the duct with equal heating at two long walls, circular duct correlations provide result with accuracy of ± 10% for $0.5 \leq \text{Pr} \leq 10$ and $10^4 \leq \text{Re} \leq 10^5$ (Bhatti and Shah 1987). In the case of heating at one long wall only (asymmetrical heating), the circular duct correlations provide values which may be up to 20% higher than the actual experimental values for $0.7 \leq \text{Pr} \leq 2.5$ and $10^4 \leq \text{Re} \leq 10^6$ and this is true in all probability for $0 \leq \alpha^* \leq 1$ (Bhatti and Shah 1987). This conclusion is in line with the results of the studies of Sparrow et al. (1966) for Pr = 0.7. Asymmetric heating condition is encountered in solar air heaters. The Nusselt number correlations appropriate for solar air heater ducts have been presented in Chap. 16.

8.9 Correlations for External Forced Flow Over a Flat Plate

8.9.1 Laminar Flow

Skin Friction Coefficient
The skin friction coefficient,

$$C_{fx} = \frac{0.6641}{\sqrt{\mathrm{Re}_x}} \tag{7.37}$$

The average value of the skin friction coefficient for the plate length L is

$$\overline{C_f} = \frac{1.3282}{\sqrt{\mathrm{Re}_L}} \tag{7.38}$$

where $\mathrm{Re_L}$ is based on the total length L of the plate in direction of flow.

Heat Transfer Coefficient Correlations
The local Nusselt number equation is

$$\begin{aligned} &\mathrm{Nu}_x = 0.332\,\mathrm{Pr}^{1/3}(\mathrm{Re}_x)^{1/2} \\ &\qquad \mathrm{Re_x} < 5 \times 10^5 \\ &\qquad 0.6 < \mathrm{Pr} < 50 \end{aligned} \tag{7.56b}$$

The average Nusselt number over the plate length L is

$$\mathrm{Nu}_{av} = 0.664\,\mathrm{Pr}^{1/3}(\mathrm{Re}_L)^{1/2} \tag{7.57}$$

The Stanton number $\mathrm{St} = \frac{h}{c_p \rho U}$ is used some times as an alternative for Nusselt number when presenting heat transfer data. The above relations can be written in terms of the Stanton number as

$$\mathrm{St}_x\,\mathrm{Pr}^{2/3} = \frac{0.332}{\sqrt{\mathrm{Re}_x}}. \tag{7.58}$$

Using Eq. (7.37), we have

$$j_H = \mathrm{St}_x\,\mathrm{Pr}^{2/3} = \frac{C_{fx}}{2}. \tag{7.59}$$

Equation (7.59) is known as *Reynolds-Colburn analogy*. Factor j_{H} is known as Colburn j factor for heat transfer.

8.9.2 Turbulent Flow

Skin Friction Coefficient

Case A. Boundary Layer Fully Turbulent from the Leading Edge of the Plate: The local value of the skin friction coefficient,

$$C_{fx} = \frac{0.0583}{\mathrm{Re}_x^{1/5}} \tag{7.81}$$

The average value of the skin friction coefficient,

$$\overline{C_f} = \frac{0.0729}{\mathrm{Re}_L^{1/5}} \tag{7.82}$$

The above result has been found to be in a good agreement with the experimental results in the range $5 \times 10^5 < \mathrm{Re} < 10^7$.

Case B. The boundary layer is laminar up to a certain distance from the leading edge and the transition, as discussed earlier, to turbulent flow takes place at $x = x_c$ corresponding to critical Reynolds number $\mathrm{Re}_c = 5 \times 10^5$.

Mixed average value of the skin friction coefficient, for $\mathrm{Re} < 10^7$,

$$\overline{C_f} = \frac{0.074}{\mathrm{Re}_L^{1/5}} - \frac{1742}{\mathrm{Re}_L} \tag{8.45}$$

Another relation suggested for $10^7 < \mathrm{Re} < 10^9$ is

$$\overline{C_f} = \frac{0.455}{(\log_{10}\mathrm{Re}_L)^{2.58}} - \frac{A}{\mathrm{Re}_L} \tag{7.85b}$$

where the value of A depends on the value of the critical Reynolds number at which the transition to turbulent flow takes place. For $\mathrm{Re}_c = 5 \times 10^5$, $A = 1700$.

Heat transfer coefficient correlations

The local value of the Nusselt number

$$\mathrm{Nu}_x = 0.0296\mathrm{Re}_x^{4/5}\,\mathrm{Pr}^{1/3} \quad \text{for } 5 \times 10^5 < \mathrm{Re} < 10^7 \tag{7.116}$$

Average Nusselt number equation over the plate length L (turbulent flow from the leading edge ($x = 0$)

$$\overline{Nu} = \frac{\bar{h}L}{k} = 0.037\mathrm{Re}_L^{0.8}\mathrm{Pr}^{1/3} \tag{7.117}$$

Mixed average Nusselt number over the entire laminar-turbulent boundary layer when the boundary layer is laminar up to a certain distance from the leading edge and the transition at $x = x_c$ corresponds to $\mathrm{Re}_c = 5 \times 10^5$,

$$\overline{Nu} = \frac{\bar{h}L}{k} = \left(0.037\mathrm{Re}_L^{0.8} - 871\right)\mathrm{Pr}^{1/3} \tag{7.118}$$

Example 8.16 Air at a temperature of 300°C flows with a velocity of 3 m/s over a flat plate 0.5 m long. Estimate the cooling rate per unit width of the plate needed to maintain it at a surface temperature of 20°C. The local film coefficient may be calculated from the following correlation.

$$\mathrm{Nu}_x = 0.332(\mathrm{Re}_x)^{0.5}(\mathrm{Pr})^{0.33} \quad \text{for Re} \le 5 \times 10^5$$
$$\mathrm{Nu}_x = 0.0296(\mathrm{Re}_x)^{0.8}(\mathrm{Pr})^{0.33} \text{for Re} > 5 \times 10^5$$

Solution

The mean film temperature is

$$t_{fm} = \frac{300 + 20}{2} = 160°\mathrm{C}.$$

At 160°C, the thermo-physical properties of the air are:

$\rho = 0.817\,\mathrm{kg/m^3}$, $k = 3.59 \times 10^{-2}\,\mathrm{W/(m\,K)}$, $\mu = 2.42 \times 10^{-5}\,\mathrm{kg/(m\ s)}$ and $\mathrm{Pr} = 0.6848$.

The flow Reynolds number at $L = 0.5$ m is

$$\mathrm{Re}_L = \frac{\rho U_\infty L}{\mu} = \frac{0.817 \times 3 \times 0.5}{2.42 \times 10^{-5}} = 50640.$$

Since the Reynolds number at the end of the plate is less than 5×10^5, the flow is laminar throughout. The applicable Nusselt number correlation is

$$\mathrm{Nu}_x = 0.332(\mathrm{Re}_x)^{0.5}(\mathrm{Pr})^{0.33}$$

From which,

$$h_x = \frac{k}{x} \times 0.332 \times \left(\frac{\rho x U_\infty}{\mu}\right)^{0.5}(\mathrm{Pr})^{0.33}$$

or

$$h_x = \frac{3.59 \times 10^{-2}}{x} \times 0.332 \times \left(\frac{0.817 \times x \times 3}{2.42 \times 10^{-5}}\right)^{0.5}(0.6848)^{0.33} = 3.348x^{-0.5}$$

Hence, the average convective heat transfer coefficient is

$$h_{av} = \frac{1}{L}\int_0^L h_x \mathrm{d}x$$

$$= \frac{1}{L}\int_0^L 3.348 \times x^{-0.5} \mathrm{d}x$$

$$= \frac{3.348}{0.5} \times \left[\frac{x^{0.5}}{0.5}\right]_0^{0.5} = 9.47\ \mathrm{W/(m^2\,K)}.$$

Heat flow rate is

$$q = hA\Delta T$$

where

A heat transfer area = 0.5 × 1 = 0. 5 m^2
ΔT $t_w - t_{air}$ = 300–20 = 280°C.

This gives

$$q = hA\Delta T = 9.47 \times 0.5 \times 280 = 1325.8\ \mathrm{W}.$$

Example 8.17 Air at atmospheric pressure and 15 m/s flows over a 4 mm × 4 mm chip located 100 mm from leading edge on a flat electronic circuit board and removes 20 mW of heat. Determine temperature of the chip surface if the free stream air temperature is 25°C.

If flow is turbulent throughout, determine temperature of the chip surface.

Solution

Thermophysical properties of air at film temperature t_{fm} = 50°C (assumed) from Table A5 are:

$$\rho = 1.0949\,\mathrm{kg/m^3},\ \mu = 1.9512 \times 10^{-5}\,\mathrm{kg/(m\ s)},\ k = 0.02799\ \mathrm{W/(m\ K)}\ \text{and}\ \mathrm{Pr} = 0.703.$$

Local flow Reynolds number,

$$\mathrm{Re}_x = \frac{\rho U_\infty x}{\mu} = \frac{1.0949 \times 15 \times 0.1}{1.9512 \times 10^{-5}} = 84171.$$

The flow is in the laminar regime. Hence, the local value of the Nusselt number is given by

$$\mathrm{Nu}_x = 0.332\mathrm{Re}_x^{1/2}\,\mathrm{Pr}^{1/3},$$

which gives

$$h_x = \frac{k}{x}\mathrm{Nu}_x = \frac{k}{x} \times 0.332\mathrm{Re}_x^{1/2}\,\mathrm{Pr}^{1/3}$$
$$= \frac{0.02799}{0.1} \times 0.332 \times (84171)^{1/2}(0.703)^{1/3}$$
$$= 23.97\,\mathrm{W/(m^2\,K)}.$$

Since the chip is very small, $h_{av} \approx h_x$ and Newton's equation $q = h_x A_{chip}(t_s - t_\infty)$ gives

$$t_s = t_\infty + \frac{q}{h_x A_{chip}} = 25 + \frac{20 \times 10^{-3}}{23.97 \times 4 \times 4 \times 10^{-6}} = 77.15°\mathrm{C}.$$

i.e., the chip surface temperature is 77.15°C. The film temperature $t_{fm} = (t_s + t_\infty)/2 = 51.1°$ C, which is nearly equal to the assumed film temperature hence retrial is not required.

If flow is turbulent,

$$h_x = \frac{k}{x}\mathrm{Nu}_x = \frac{k}{x} \times 0.0296\mathrm{Re}_x^{4/5}\,\mathrm{Pr}^{1/3}$$
$$= \frac{0.02799}{0.1} \times 0.0296 \times (84171)^{4/5}(0.703)^{1/3}$$
$$= 64.18\,\mathrm{W/(m^2\,K)}.$$

Since the chip is very small, $h_{av} \approx h_x$ and Newton's equation $q = h_x A_{chip}(t_s - t_\infty)$ gives

$$t_s = t_\infty + \frac{q}{h_x A_{chip}} = 25 + \frac{20 \times 10^{-3}}{64.18 \times 4 \times 4 \times 10^{-6}} = 44.48°\mathrm{C}$$

i.e., the chip surface temperature is 44.48°C. The film temperature $t_{fm} = (t_s + t_\infty)/2 = 34.74$C° against assumed temperature of 50°C. Retrial is required.

Thermophysical properties of air at film temperature $t_{fm} \approx 35$ °C from Table A5 are:

$$\rho = 1.15\,\mathrm{kg/m^3},\ \mu = 1.8823 \times 10^{-5}\,\mathrm{kg/(m\,s)},\ k = 0.02684\,\mathrm{W/(m\,K)}\ \text{and}\ \mathrm{Pr} = 0.7066.$$

Local flow Reynolds number,

$$\mathrm{Re}_x = \frac{\rho U_\infty x}{\mu} = \frac{1.15 \times 15 \times 0.1}{1.8823 \times 10^{-5}} = 91643.$$

Local heat transfer coefficient is

$$h_x = \frac{k}{x}\mathrm{Nu}_x = \frac{k}{x} \times 0.0296\mathrm{Re}_x^{4/5}\,\mathrm{Pr}^{1/3}$$
$$= \frac{0.02684}{0.1} \times 0.0296 \times (91643)^{4/5}(0.7066)^{1/3}$$
$$= 66.0\,\mathrm{W/(m^2\,K)}.$$

Since the chip is very small, $h_{av} \approx h_x$ and Newton's equation $q = h_x A_{chip}(t_s - t_\infty)$ gives

$$t_s = t_\infty + \frac{q}{h_x A_{\text{chip}}} = 25 + \frac{20 \times 10^{-3}}{66.0 \times 4 \times 4 \times 10^{-6}} = 43.94°\text{C}.$$

Further retrial is not required.

Example 8.18 Air at atmospheric pressure and a temperature of 25°C is flowing parallel to a 0.8 m long flat plate at a velocity of 10 m/s. If the plate is subjected to uniform heat flux of 1500 W/m^2 and flow is turbulent throughout, determine (a) plate surface temperature at $x = L$ and (b) average temperature of the plate surface.

Solution

The convection heat transfer equation gives

$$q'' = h_x[t_s(x) - t_\infty]$$

or

$$t_s(x) - t_\infty = \frac{q''}{h_x}$$

(a) **Plate surface temperature at $x = L$ is**

$$t_s(L) - t_\infty = \frac{q''}{h_L}$$

For turbulent flow,

$$\text{Nu}_x = 0.0296(\text{Re}_x)^{0.8}\,\text{Pr}^{1/3}$$

Hence,

$$h_L = \frac{k}{L} \times 0.0296\left(\frac{U_\infty L}{\nu}\right)^{0.8} \text{Pr}^{1/3}$$

For air at 50°C and atmospheric pressure,

$$\nu = 17.82 \times 10^{-6}\text{m}^2/\text{s},\ k = 0.02799\ \text{W}/(\text{m K})\ \text{and Pr} = 0.703.$$

This gives

$$h_L = \frac{0.02799}{0.8} \times 0.0296 \times \left(\frac{10 \times 0.8}{17.82 \times 10^{-6}}\right)^{0.8} \times 0.703^{1/3} = 30.62\ \text{W}/(\text{m}^2\,\text{K}).$$

and

$$t_s(L) = \frac{q''}{h_L} + t_\infty = \frac{1500}{30.62} + 25 = 74°\text{C}.$$

The mean film temperature $t_m = (74 + 25)/2 = 49.5°\text{C}$, nearly equal to the assumed temperature.

(b) **Mean plate surface temperature**

$$\bar{t}_s - t_\infty = \frac{1}{L}\int_0^L [t_s(x) - t_\infty]\mathrm{d}x$$

$$= \frac{1}{L}\int_0^L \frac{q''}{h_x}\mathrm{d}x$$

$$= \frac{q''}{L}\int_0^L \frac{x}{k\mathrm{Nu}_x}\mathrm{d}x.$$

Substituting $\mathrm{Nu}_x = 0.0296(\mathrm{Re}_x)^{0.8}\,\mathrm{Pr}^{1/3} = 0.0296(U_\infty x/v)^{0.8}\,\mathrm{Pr}^{1/3}$, we have

$$\bar{t}_s - t_\infty = \frac{q''}{L}\int_o^L \frac{x}{k \times 0.0296(U_\infty x/v)^{0.8}\,\mathrm{Pr}^{1/3}}\mathrm{d}x$$

$$= \frac{q''}{L} \times \frac{1}{k \times 0.0296(U_\infty/v)^{0.8}\,\mathrm{Pr}^{1/3}}\int_o^L x^{0.2}\mathrm{d}x$$

$$= \frac{q''}{L} \times \frac{1}{k \times 0.0296(U_\infty/v)^{0.8}\,\mathrm{Pr}^{1/3}} \times \frac{L^{1.2}}{1.2}$$

$$= \frac{1500}{0.8} \times \frac{1}{0.02799 \times 0.0296 \times (10/17.82 \times 10^{-6})^{0.8} \times 0.703^{1/3}} \times \frac{0.8^{1.2}}{1.2}$$

$$= 40.83°\text{C}.$$

Mean plate surface temperature

$$\bar{t}_s = 40.83 + t_\infty = 40.83 + 25 = 65.83°\text{C}.$$

8.10 Forced Convection Laminar and Turbulent Flows Around Submerged Bodies

We shall discuss in this section the convection under forced flow condition from a solid body immersed in a stream of fluid. Though the submerged bodies may be of various shapes, we shall consider the flow around cylinder and sphere in cross flow.

8.10.1 Cylinder in Cross Flow

A cylinder in cross flow has been studied extensively from fluid mechanics point of view. Here a brief account of the same is being presented. Readers can refer Schlichting (1960) for the details. Figure 8.9 shows a cylinder in cross flow. Starting from the stagnation point S ($\phi = 0^\circ$), a boundary layer builds up on the front portion of the surface of the cylinder and its thickness increases with angle ϕ. If the body is heated, thermal boundary layers also develops in a similar way.

For an ideal or non-viscous flow (known as potential flow) the pressure decreases and the velocity increases from the stagnation point and reaches its maximum at $\phi = 90^\circ$. On the back side of the cylinder, the pressure increases again and at $\phi = 180^\circ$ attains the same value as the forward stagnation point. Figure for this case has been included here. In this case the separation of the flow from the cylinder surface does not take place and the drag (the force resisting the flow) on the cylinder is zero.

In a fluid with friction, the flow separates on both sides of the surface of the cylinder at about $80^\circ - 100^\circ$ and on the back side of the cylinder a dead fluid mass is filled with vortices on the downstream side. The fluid particles outside the boundary layer are able to move against the pressure increase on the back side of the cylinder by changing their kinetic energy into pressure energy. But the fluid particles in the boundary layer do not possess so much kinetic energy. They can therefore move only a small distance into the region of increasing pressure before their kinetic energy is consumed. Then they reverse their flow direction. In this way the flow separates from the surface. At the separation point, the velocity gradient at the surface becomes zero $[(\partial u/\partial y)_{y=0} = 0]$.

Because of this separation, the pressure distribution along the back side of the cylinder is changed also. The extent of the penetration in the region of increasing pressure depends on the kinetic energy of the fluid particles, which is greater in the case of turbulent flow boundary layer than in laminar one. As a consequence, the turbulent boundary layer (at Re greater than critical Reynolds number Re_c) separates at about $\phi = 110^\circ$ and the laminar at $\phi = 82^\circ$. The pressure on the rear side of the cylinder is lower than on the front side. This causes a resisting force in the flow direction, which is known as *form resistance*. To it is

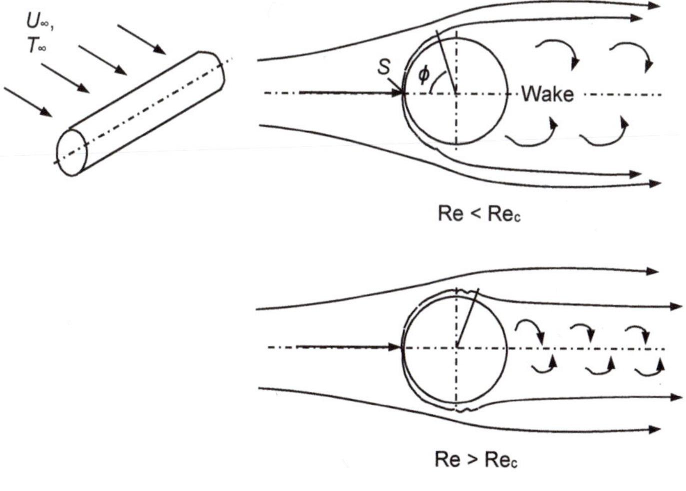

Fig. 8.9 Cylinder in cross flow

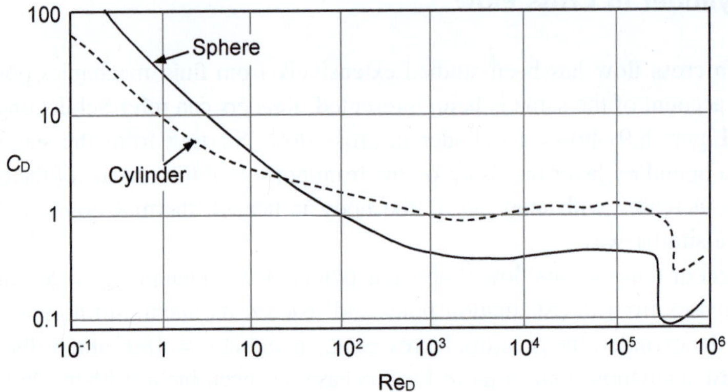

Fig. 8.10 Drag coefficient C_D versus Reynolds number for cylinder and sphere in cross flow. Schlichting H and Gersten K, Boundary layer theory, 8th Edition, Chap. 1, Figs. 1.12 and 1.19, Copyright 2000. With permission of Springer

added the frictional resistance caused by the skin friction or viscous shear acting tangential on the cylinder surface. Both parts are included in a dimensionless quantity known as the *drag coefficient,*[3] which is defined as

$$C_D = \frac{F_D/A}{\frac{1}{2}\rho U_\infty^2} \tag{8.46}$$

where

F_D drag
A frontal area (area normal to the flow) = LD for a cylinder

It is to note that the total drag on a body is the sum of two components: the pressure or form drag (drag force due to pressure) and the skin friction or viscous drag (drag force due to viscous stresses).

The drag coefficient C_D is plotted in Fig. 8.10 against the Reynolds number where the Reynolds number is based on the free stream velocity and diameter of the cylinder. In general, the drag depends on the properties of the fluid, its free stream velocity and the shape of the body. At very low values of the Reynolds number ($\leq$ 1), no separation occurs and the drag is caused only by the shear stresses (viscous friction). From Re > 1, a dead zone of fluid builds up behind the cylinder, which increases with the Reynolds number. With further increase in the Reynolds number from 100, vortices shed alternatively from upper and lower sides from the cylinder surface (known as von Karman vortex street). These vortices are carried away by the flow in a regular pattern. At Re = 4×10^5, a sudden drop in the drag coefficient is observed due to the fact that the boundary layer becomes turbulent before it separates. At Re > 4×10^5, no vortex wake forms but only irregular vortices exist in the dead fluid.

[3]If the fluid is flowing in the horizontal direction, the sum of the horizontal components of these forces constitutes the drag, and the sum of their vertical components constitutes the lift. In the study of heat transfer, we are interested only in the drag, which is the resistance to the flow to be overcome in pumping of the fluid.

At small velocities, the drag is mostly due to the frictional resistance. At Re = 10, friction and form drag are of the same order. For Re > 1000 the form resistance predominates. Each of these flow regimes influences the heat transfer coefficient in a distinct way.

The heat transfer on the back side occurs in the wake. At low Re, the heat transfer coefficient on the front side is much greater than that on the back. With the increase in the Reynolds number, the heat flow from the back side increases at a greater rate than on the front side and at Re = 5×10^4, it is as high as on the front side. The reason for this behaviour is that the vortices, while separating alternately from the upper and lower sides of the cylinder sweep the surface of the rear half of the cylinder with an intensity that increases with the Reynolds number. When the Reynolds number is > 4×10^5, the boundary layer becomes turbulent and the distribution of the heat transfer coefficient along the surface changes. The heat transfer coefficient is reported to be very high at an angle of $\phi = 100^\circ$, which indicates the transition in the flow within the boundary layer.

Hilpert (1933) for air, and Knudsen and Katz (1958) for other fluids extended the following relation for heat transfer from cylinders:

$$\mathrm{Nu} = \frac{hD}{k} = C\mathrm{Re}^n\,\mathrm{Pr}^{1/3} \tag{8.47}$$

where Re = $U_\infty D/\upsilon$.

The values of the constant C and exponent n are given in Table 8.11. All the fluid properties are to be taken at film temperature.

Douglas and Churchill (1956) presented the following correlation for heat transfer from a single cylinder for air:

$$\mathrm{Nu} = 0.46\mathrm{Re}^{0.5} + 0.00128\mathrm{Re} \quad \text{for } 500 < \mathrm{Re} < 10^6 \tag{8.48a}$$

For liquid, they recommended the following equation:

$$\mathrm{Nu} = (0.506\mathrm{Re}^{0.5} + 0.000141\mathrm{Re})\mathrm{Pr}^{1/3} \tag{8.48b}$$

For Re < 500, Hsu (1963) suggested the following equation for gases.

$$\mathrm{Nu} = 0.43 + 0.48\mathrm{Re}^{1/2} \tag{8.49}$$

Table 8.11 Constant C and n in Eq. (8.47)[a]

Re	C	n
0.4–4	0.989	0.330
4–40	0.911	0.385
40–4000	0.683	0.466
4000–40000	0.193	0.618
40000–400000	0.027	0.805

[a]Holman JP, adapted for SI units by White PRS, Heat Transfer, McGraw-Hill Book Co, New York, Copyright 1992. The material is reproduced with permission of McGraw-Hill Education (Asia)

where the fluid properties are to be taken at the film temperature and Re is defined as in Eq. (8.47).

Isachenko et al. (1977) presented following correlations based on the extensive studies on circumferential mean rate of heat transfer from the surface of a cylinder for flow of air, water and transformer oil by Zukauskas and co-researchers (1959–1972):

$$\mathrm{Nu} = 0.5\mathrm{Re}^{0.5}\,\mathrm{Pr}^{0.38}(\mathrm{Pr}/\mathrm{Pr}_w)^{0.25} \quad \text{for } 5 < \mathrm{Re} < 10^3 \tag{8.50a}$$

$$\mathrm{Nu} = 0.25\mathrm{Re}^{0.6}\,\mathrm{Pr}^{0.38}(\mathrm{Pr}/\mathrm{Pr}_w)^{0.25} \quad \text{for } 10^3 < \mathrm{Re} < 2\times10^5 \tag{8.50b}$$

$$\mathrm{Nu} = 0.023\mathrm{Re}^{0.8}\,\mathrm{Pr}^{0.37}(\mathrm{Pr}/\mathrm{Pr}_w)^{0.25} \quad \text{for } 2\times10^5 < \mathrm{Re} < 2\times10^6 \tag{8.50c}$$

where the fluid properties are to be taken at the film temperature. If the flow approaching the cylinder is artificially turbulized, the heat transfer coefficient will be larger than given by above correlation.

However, the correlations presented in Mikheyev (1964) based on the studies of Zukauskas are:

$$\mathrm{Nu} = 0.59\mathrm{Re}^{0.47}\,\mathrm{Pr}^{0.38}(\mathrm{Pr}/\mathrm{Pr}_w)^{0.25} \quad \text{for } 10 < \mathrm{Re} < 10^3 \tag{8.51a}$$

$$\mathrm{Nu} = 0.21\mathrm{Re}^{0.62}\,\mathrm{Pr}^{0.38}(\mathrm{Pr}/\mathrm{Pr}_w)^{0.25} \quad \text{for } 10^3 < \mathrm{Re} < 2\times10^5 \tag{8.51b}$$

Churchill and Bernstein (1977) have presented the following correlation for isothermal cylinder based on the data of various researchers:

$$\mathrm{Nu} = 0.3 + \frac{0.62\mathrm{Re}_D^{1/2}\,\mathrm{Pr}^{1/3}}{\left[1+\left(\frac{0.4}{\mathrm{Pr}}\right)^{2/3}\right]^{1/4}}\left[1+\left(\frac{\mathrm{Re}_D}{282000}\right)^{5/8}\right]^{4/5} \quad \text{for } \mathrm{Re_D\,Pr} > 0.2 \tag{8.52}$$

All properties in the equation are to be evaluated at the film temperature t_{fm}. For the Reynolds number range of 7×10^4–4×10^5, the equation predicts Nusselt number values that may be significantly lower than those from the direct measurement. The equation can be used for uniform heat flux condition if the perimeter-average of the temperature difference between the surface and the free stream is used.

8.10.2 Flow Around a Sphere

The flow regimes around a sphere are similar to those encountered in the case of a cylinder in cross flow and the heat transfer coefficient is governed by these regimes. Starting from $\phi = 0^\circ$, a boundary layer develops on the front portion of the surface of the sphere. At very low Reynolds number (≤ 1), a laminar flow exists around the sphere and the flow is nearly symmetric. At large Reynolds numbers, the boundary layer separates from the sides of the sphere creating a stagnant region filled with vortices in the back. On the sides of the sphere the boundary layer may become turbulent at large Reynolds numbers. This influences the position of the separation.

The drag coefficient for a sphere versus the Reynolds number is shown in Fig. 8.10. The curve can be divided in four regions. In the first region (Re $\le$ 1), the Stoke's law is applicable and the drag is

$$F_D = 3\pi D\mu U_\infty \tag{8.53}$$

and

$$C_D = \frac{F_D}{[(\pi/4)D^2] \times \frac{1}{2}\rho U_\infty^2} \tag{8.54}$$

In the second region (Re = 1–1000), the boundary layer separation begins. The coefficient of drag continues to decrease in this region. In the third region (Re = 10^3 –2 × 10^5), the drag coefficient almost remains constant. The boundary layer is entirely laminar before separation. The fourth region is characterized by sudden drop in C_D to almost half at Re = 2.5 × 10^5. This is due to the change in the flow pattern in the boundary layer from laminar to turbulent just before the separation point. The Reynolds number where the drop in C_D occurs is termed as the critical Reynolds number. Actually the transition occurs within a range of Re = 10^5 – 10^6 because it also depends on the free stream turbulence. After the transition, the coefficient of drag increases slightly.

Whitaker (1972) proposed the following correlation for the average Nusselt number between an isothermal spherical surface at T_w and free stream fluid (U_∞, T_∞):

$$\begin{aligned} \mathrm{Nu} = 2 + (0.4\mathrm{Re}_D^{0.5} + 0.06\mathrm{Re}_D^{2/3})\,\mathrm{Pr}^{0.4}(\mu_\infty/\mu_w)^{0.25} \\ \text{for} \quad 3.5 < \mathrm{Re}_D < 7.6 \times 10^4, \\ 0.71 < \mathrm{Pr} < 380, \\ \text{and} \quad 1 < \mu_\infty/\mu_w < 3.2 \end{aligned} \tag{8.55}$$

where all properties except μ_w are to be evaluated at the free stream temperature t_∞. The viscosity μ_w refers to the wall temperature t_w. No-flow limit for this equation is Nu = 2, which agrees with pure radial conduction between sphere surface and the motionless infinite medium that surrounds it (see Example 8.13).

Moving liquid drops, such as atomized fuel drops in internal combustion engines, liquid drops in spray drying chambers, or water droplets falling in the cooling tower are also examples of spheres in motion in a fluid experiencing convection in currents. In such cases, of course, mass transfer also takes place simultaneously with the heat transfer.

For the falling drops, Ranz and Marshall (1952) proposed the following equation for the convective heat transfer for a system of water drops:

$$\mathrm{Nu} = 2 + 0.6\mathrm{Re}^{0.5}\,\mathrm{Pr}^{1/3}[25(D/x)^{0.7}] \tag{8.56}$$

where x is the distance measured from start of fall of the droplet. Properties are to be evaluated at the free stream temperature. The free fall velocity is given by

$$U_\infty^2 = 2gx \tag{8.57}$$

When x is unknown, average value during the fall can be estimated by omitting the bracketed factor in Eq. (8.56), which accounts for the droplet oscillation.

8.10.3 Flow Across Tube Banks

The flow across tube banks, the tube array or bundle is frequently encountered in heat exchangers. The arrangements of the tubes frequently used are shown in Fig. 8.11. The arrangement is characterized by the relative longitudinal and transverse pitches, p/D and p_t/D of two consecutive tubes, respectively.

For both the arrangements, the flow across the first transverse row, row 1 in the figure, of the tubes is practically the same as that for a single tube. The nature of the flow across the tubes of remaining rows depends on the way the tubes are arranged. However, in most applications, the flow is turbulent after a few rows.

In the case of in-line arrangement, all the tubes in the second and subsequent rows are in the vortex region of the upstream row. Thus, a dead zone with relatively weak circulation of flow is formed in the spaces between the tubes because the fluid flows mainly through the longitudinal passages between the tubes. The points of maximum heat transfer coefficient on the tube circumference are located at the spots where the impact of the main stream occurs. The front part of the tubes is not directly exposed to the flow; therefore the heat transfer coefficient is low here.

In the staggered arrangement of the tubes, the flow past the tubes deep inside the array differs little from the nature of flow across the first row of tubes. The maximum intensity of the heat flux is at the front portion of the tubes in all rows.

In general, the flow turbulence increases as the fluid passes through a few initial rows of the tubes. After a sufficient number of rows, the flow assumes a stable character.

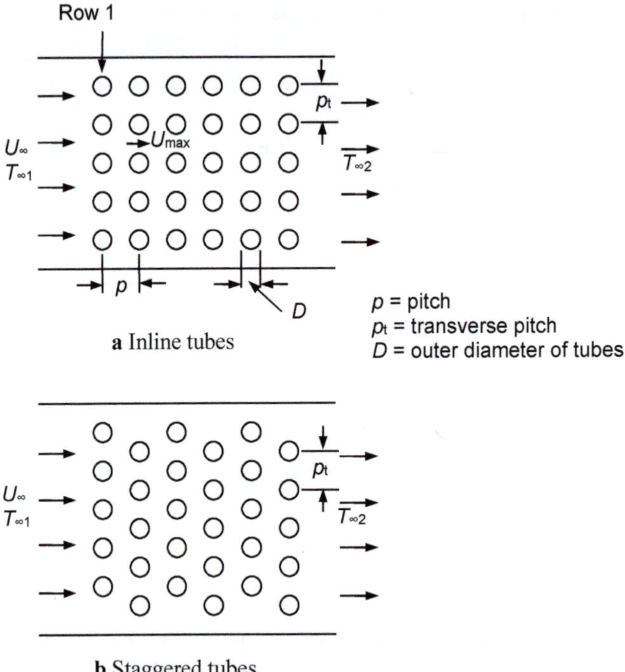

Fig. 8.11 Flow across tube banks

For the selection of the arrangement of tubes in a bank, the pressure loss (the resistance to the flow) and the rate of clogging is also considered.

Zukauskas (1987) reviewed the works published on cross-flow through tube bundles and he recommends the following correlations for the average Nusselt number for number of rows $n \geq 16$.

In-line Tube Bundles

$$\mathrm{Nu} = 0.9\mathrm{Re}_D^{0.4}\,\mathrm{Pr}^{0.36}(\mathrm{Pr}/\mathrm{Pr}_w)^{0.25} \quad \text{for } 1 \leq \mathrm{Re} \leq 100 \tag{8.58a}$$

$$\mathrm{Nu} = 0.52\mathrm{Re}_D^{0.5}\,\mathrm{Pr}^{0.36}(\mathrm{Pr}/\mathrm{Pr}_w)^{0.25} \quad \text{for } 100 < \mathrm{Re} \leq 1000 \tag{8.58b}$$

$$\mathrm{Nu} = 0.27\mathrm{Re}_D^{0.63}\,\mathrm{Pr}^{0.36}(\mathrm{Pr}/\mathrm{Pr}_w)^{0.25} \quad \text{for } 10^3 < \mathrm{Re} \leq 2 \times 10^5 \tag{8.58c}$$

$$\mathrm{Nu} = 0.033\mathrm{Re}_D^{0.8}\,\mathrm{Pr}^{0.4}(\mathrm{Pr}/\mathrm{Pr}_w)^{0.25} \quad \text{for } 2 \times 10^5 < \mathrm{Re} \leq 2 \times 10^6 \tag{8.58d}$$

Staggered Arrangement

$$\mathrm{Nu} = 1.04\mathrm{Re}_D^{0.4}\,\mathrm{Pr}^{0.36}(\mathrm{Pr}/\mathrm{Pr}_w)^{0.25} \quad \text{for } 1 \leq \mathrm{Re} \leq 500 \tag{8.59a}$$

$$\mathrm{Nu} = 0.71\mathrm{Re}_D^{0.5}\,\mathrm{Pr}^{0.36}(\mathrm{Pr}/\mathrm{Pr}_w)^{0.25} \quad \text{for } 500 < \mathrm{Re} \leq 10^3 \tag{8.59b}$$

$$\mathrm{Nu} = 0.35\mathrm{Re}_D^{0.6}\,\mathrm{Pr}^{0.36}(\mathrm{Pr}/\mathrm{Pr}_w)^{0.25}(p_t/p)^{0.2} \quad \text{for } 10^3 < \mathrm{Re} \leq 2 \times 10^5 \tag{8.59c}$$

$$\mathrm{Nu} = 0.031\mathrm{Re}_D^{0.8}\,\mathrm{Pr}^{0.4}(\mathrm{Pr}/\mathrm{Pr}_w)^{0.25}(p_t/p)^{0.2} \quad \text{for } 2 \times 10^5 < \mathrm{Re} \leq 2 \times 10^6 \tag{8.59d}$$

In Eqs. (8.58a), (8.58b), (8.58c), (8.58d) and (8.59a), (8.59b), (8.59c), (8.59d), the thermo-physical properties except $\mathrm{Pr_w}$ are evaluated at the mean bulk temperature t_m [= $(t_{\infty 1} + t_{\infty 2})/2$] of the fluid flowing around the tubes in the bundle.

The Reynolds number is based on the average velocity of the fluid through the narrowest cross-section of the flow area:

$$\mathrm{Re}_D = \frac{\rho U_{max} D}{\mu} \tag{8.60}$$

where $U_{\max} = U_\infty p_t/(p_t - D)$, refer Fig. 8.11.

The factor $(p_t/p)^{0.2}$ takes account of the effect of the change in the longitudinal and transverse pitches. However, this effect is not so evident in the case of the in-line arrangement and in staggered arrangement when $\mathrm{Re} < 10^3$.

In general, with the increase in the number of rows, the turbulence and hence the heat transfer coefficient increases. However, this effect of the number of rows is negligible for $n > 16$. For $n < 16$, the heat transfer coefficient is lower due to the lower level of turbulence. For $n < 16$, the Nusselt number values from Eqs. (8.58a), (8.58b), (8.58c), (8.58d) and (8.59a), (8.59b), (8.59c), (8.59d) must be multiplied by a factor C_n (Fig. 8.12), which is a correction factor to take account of the effect of the number of rows n less than 16.

The uncertainty in the predicted values from the above correlations is reported to be within ± 15%.

Example 8.19 Compare the predicted values of the Nusselt number from different correlations for flow of fluid (Pr = 1) across a cylinder ($L >> D$) at $\mathrm{Re} = 1 \times 10^2$, 1×10^3, 1×10^4 and 2×10^5. Neglect the viscosity correction factor ($\mathrm{Pr_f/Pr_w}$).

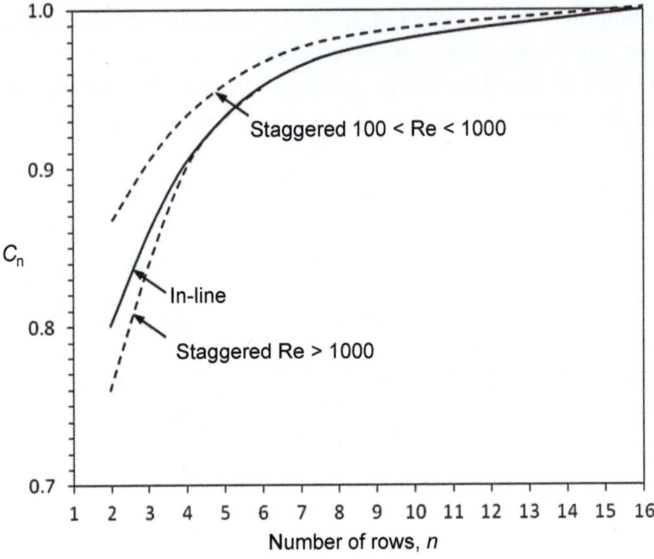

Fig. 8.12 Correction factor C_n as function of number of rows n

Table 8.12 Example 8.19

Re	Hilpert	Douglas and Churchill	Mikheyev		Isachenko et al.		Churchill and Bernstein
1×10^2	5.84	4.73	5.14		5.0		5.89
1×10^3	17.08	15.83	15.17	15.21	15.81	15.77	18.3
1×10^4	57.22	58.8		63.42		62.8	61.37
2×10^5	499.7	461.7		406.3		378.9	399.6

Solution

The results are presented in Table 8.12.

Example 8.20 An electrical heater is embedded in a long 30 mm diameter cylinder. Either water at a velocity of 1 m/s or air at atmospheric pressure and a velocity of 10 m/s flowing across the cylinder is to be used to maintain the cylinder surface temperature at 75°C. Both fluids are available at 25°C. Calculate the power of the heater required. Comment on the result.

Solution

Power required per unit length of the cylinder,

$$\frac{q}{L} = h(\pi d)(t_s - t_\infty)$$

For a cylinder in crossflow,

$$h = \frac{k}{d} Nu = \frac{k}{d} C\mathrm{Re}^n \,\mathrm{Pr}^{1/3} \tag{8.47}$$

Water as coolant

Thermophysical properties of water at film temperature t_{fm} [= (75 + 25)/2 = 50°C] from Table A4 are:

$$\rho = 988.1\,\mathrm{kg/m^3},\ \mu = 544 \times 10^{-6}\,\mathrm{kg/(m\ s)},\ k = 0.644\ \mathrm{W/(m\ K)}\ \text{and}\ \mathrm{Pr} = 3.55.$$

Flow Reynolds number,

$$\begin{aligned}\mathrm{Re} &= \frac{\rho U d}{\mu} \\ &= \frac{988.1 \times 1.0 \times 0.03}{544 \times 10^{-6}} = 54491.\end{aligned}$$

From Table 8.11, $C = 0.027$ and $n = 0.805$. Hence, heat transfer coefficient,

$$\begin{aligned}h &= \frac{k}{d} C\mathrm{Re}^n \,\mathrm{Pr}^{1/3} = \frac{0.644}{0.03} \times 0.027 \times 54491^{0.805} \times 3.55^{1/3} \\ &= 5745\,\mathrm{W/(m^2\,K)},\end{aligned}$$

and power required per unit length of the cylinder,

$$\frac{q}{L} = h(\pi d)(T_s - T_\infty) = 5745 \times (\pi \times 0.03) \times (75 - 25) = 27073\,\mathrm{W/m}.$$

Air as coolant

Thermophysical properties of air at film temperature t_{fm} [= (75 + 25)/2 = 50°C] from Table A5 are:

$$\rho = 1.0949\,\mathrm{kg/m^3},\ \mu = 1.9512 \times 10^{-5}\,\mathrm{kg/(m\ s)},\ k = 0.02799\ \mathrm{W/(m\ K)}\ \text{and}\ \mathrm{Pr} = 0.703.$$

Flow Reynolds number,

$$\begin{aligned}\mathrm{Re} &= \frac{\rho U d}{\mu} \\ &= \frac{1.0949 \times 10 \times 0.03}{1.9512 \times 10^{-5}} = 16834.\end{aligned}$$

From Table 8.11, $C = 0.193$ and $n = 0.618$. Hence, heat transfer coefficient,

$$h = \frac{k}{d} C\mathrm{Re}^n \,\mathrm{Pr}^{1/3} = \frac{0.02799}{0.03} \times 0.193 \times 16834^{0.618} \times 0.703^{1/3} = 65.5\,\mathrm{W/(m^2\,K)}.$$

and power required per unit length of the cylinder,

$$\frac{q}{L} = h(\pi d)(t_s - t_\infty) = 65.5 \times (\pi \times 0.03) \times (75 - 25) = 308.7\ \mathrm{W/m}.$$

Note that despite 10 times velocity of air, $h_{\text{water}} = 87.7\ h_{\text{air}}$. Thus water is much more effective heat transfer fluid than air.

Note: In general, Nu = C Re^{m}Pr^{n} gives

$$h = \frac{k}{L}\mathrm{CRe}^m\mathrm{Pr}^n = C\frac{k}{L}\left(\frac{UL}{\nu}\right)^m \mathrm{Pr}^n = C\frac{U^m}{L^{1-m}}\left(\frac{k\,\mathrm{Pr}^n}{\nu^m}\right) \propto \frac{k\,\mathrm{Pr}^n}{\nu^m},$$

i.e., for given fluid velocity U and characteristic dimension L, the heat transfer coefficient is high for fluids of large k and Pr, and small ν.

Example 8.21 A long cylindrical fin [k = 40 W/(m K)] of 10 mm diameter is installed on a surface (30 mm × 30 mm), refer Fig. 8.13. The surface temperature (fin base temperature) is 100°C. If air at 25°C and 5 m/s flows across the fin, determine the heat transfer rate from the fin and the surface.

Solution

Thermophysical properties of air at film temperature (100 + 25)/2 = 62.5°C from Table A5 are:

$$\rho = 1.05\,\mathrm{kg/m^3},\ \mu = 2.0085 \times 10^{-5}\,\mathrm{kg/(m\ s)},\ k = 0.02895\ \mathrm{W/(m\ K)}\ \text{and}\ \mathrm{Pr} = 0.7.$$

(a) **Heat transfer rate from the fin**

Reynolds number,

$$\mathrm{Re}_D = \frac{\rho U_\infty D}{\mu} = \frac{1.05 \times 5 \times 0.01}{2.0085 \times 10^{-5}} = 2614.$$

Nusselt number relation is

$$\mathrm{Nu} = C\mathrm{Re}_D^n\,\mathrm{Pr}^{1/3}$$

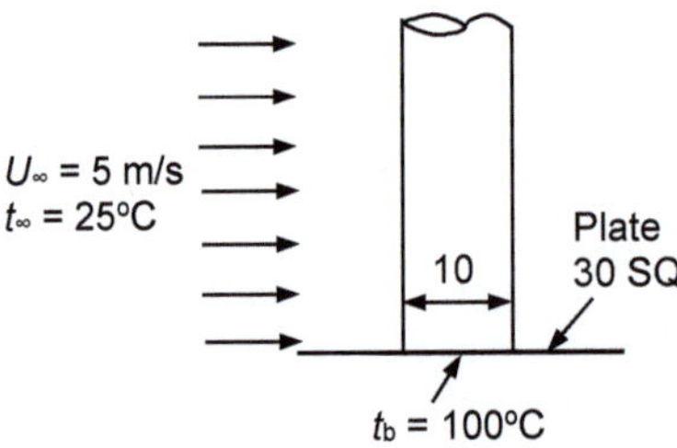

Fig. 8.13 Example 8.21

where $C = 0.683$ and $n = 0.466$ for Re = 2614 from Table 8.11. This gives

$$\begin{aligned} h &= \frac{k}{D}\mathrm{Nu} = \frac{k}{D} \times 0.683\mathrm{Re}_D^{0.466}\,\mathrm{Pr}^{1/3} \\ &= \frac{0.02895}{0.01} \times 0.683 \times (2614)^{0.466}(0.7)^{1/3} \\ &= 68.7\ \mathrm{W}/\left(\mathrm{m}^2\,\mathrm{K}\right). \end{aligned}$$

Heat rejection from the fin is given by

$$\begin{aligned} q_{\text{fin}} &= \sqrt{hPkA_c}(t_b - t_\infty) \\ &= \sqrt{h(\pi D)k(\pi/4)D^2}(t_b - t_\infty) \end{aligned}$$

Substitution of values of various terms gives

$$\begin{aligned} q_{\text{fin}} &= \sqrt{68.7 \times \pi \times 0.01 \times 40 \times (\pi/4) \times 0.01^2} \times (100 - 25) \\ &= 6.18\ \mathrm{W}. \end{aligned}$$

(b) **Heat transfer rate from surface**

Flow is parallel to the surface.

Flow Reynolds number,

$$\mathrm{Re}_L = \frac{\rho U_\infty L}{\mu} = \frac{1.05 \times 5 \times 0.03}{2.0085 \times 10^{-5}} = 7842.$$

The flow is laminar (assuming no appreciable effect of the fin on the flow structure). Relevant correlation of average Nusselt number is

$$\mathrm{Nu} = 0.664\mathrm{Re}_L^{0.5}\,\mathrm{Pr}^{1/3}$$

This gives

$$\begin{aligned} h_{av} &= \frac{k}{D}\mathrm{Nu} = \frac{k}{D} \times 0.664\mathrm{Re}_L^{0.5}\,\mathrm{Pr}^{1/3} \\ &= \frac{0.02895}{0.03} \times 0.664 \times (7842)^{0.5}(0.7)^{1/3} \\ &= 50.4\,\mathrm{W}/\left(\mathrm{m}^2\,\mathrm{K}\right). \end{aligned}$$

Heat rejection from the surface is

$$\begin{aligned} q_s &= h_{av}A_s(t_s - t_\infty) = h_{av}\left[W \times W - (\pi/4)D^2\right](t_s - t_\infty) \\ &= 50.4 \times \left[0.03 \times 0.03 - (\pi/4) \times 0.01^2\right] \times (100 - 25) \\ &= 3.11\ \mathrm{W}. \end{aligned}$$

Total heat transfer rate from surface and fin,

$$q = 6.18 + 3.11 = 9.29\,\text{W}.$$

Heat transfer rate without fin is

$$\begin{aligned} q_s &= h_{av}A_s(t_s - t_\infty) = h_{av}(W \times W)(t_s - t_\infty) \\ &= 50.4 \times (0.03 \times 0.03) \times (100 - 25) = 3.4\ \text{W}. \end{aligned}$$

Example 8.22 50 W/m of heat is to be dissipated from a 1 mm diameter electric wire using air flowing across the wire at atmospheric pressure and 20°C so as to maintain surface temperature of the wire below 80°C. Determine velocity to accomplish the task. Neglect radiation heat exchange.

Solution

For air at film temperature (80 + 20)/2 = 50°C and atmospheric pressure,

$$\nu = \mu/\rho = 17.82 \times 10^{-6}\text{m}^2/\text{s},\ k = 0.02799\ \text{W}/(\text{m K})\ \text{and Pr} = 0.703.$$

From convective heat transfer equation,

$$q/L = h\pi D(t_s - t_\infty)$$

where the heat transfer coefficient from Eq. (8.47) is

$$h = \frac{k}{D}\text{Nu} = \frac{k}{D}C\text{Re}_D^n\,\text{Pr}^{1/3} = \frac{k}{D}C\left(\frac{U_\infty D}{\nu}\right)^n \text{Pr}^{1/3}$$

Assuming Re_D lies in the range 40–4000, Table 8.11 gives $C = 0.683$ and $n = 0.466$. Substitution gives

$$h = 0.683\frac{k}{D}\left(\frac{U_\infty D}{\nu}\right)^{0.466} \text{Pr}^{1/3}$$

and

$$\begin{aligned} q/L &= \left[0.683\frac{k}{D}\left(\frac{U_\infty D}{\nu}\right)^{0.466} \text{Pr}^{1/3}\right]\pi D(t_s - t_\infty) \\ &= \left[0.683k\left(\frac{U_\infty D}{\nu}\right)^{0.466} \text{Pr}^{1/3}\right]\pi(t_s - t_\infty) \end{aligned}$$

Substitution of the values of various parameters gives

$$50 = \left[0.683 \times 0.02799 \times \left(\frac{U_\infty \times 0.001}{17.82 \times 10^{-6}}\right)^{0.466} \times 0.703^{1/3}\right] \times \pi \times (80 - 20).$$

Simplification gives $U_\infty = 6.48$ m/s.
Reynolds number for this value of U_∞ is

$$\mathrm{Re}_D = \frac{U_\infty D}{\nu} = \frac{6.48 \times 0.001}{17.82 \times 10^{-6}} = 364.$$

Re_D lies in the range 40–4000. Hence, the assumption of the Re range was correct.

Example 8.23 A long 40 mm diameter 18–8 steel rod, initially at a uniform temperature of t_i = 30°C, is suddenly exposed to cross-flow of air at t_∞ = 300°C and U_∞ = 50 m/s. Find time for the surface of the rod to reach 170°C.

Solution

For 18–8 steel, ρ = 7820 kg/m^3, c = 460 J/(kg K) and k = 21 W/(m K).
For air at film temperature [(30 +170)/2 + 300]/2 = 200°C,

$$\rho = 0.7474\,\mathrm{kg/m^3},\ \mu = 2.57 \times 10^{-5}\,\mathrm{kg/(m\ s)},\ k = 0.03859\ \mathrm{W/(m\ K)}\ \text{and}\ \mathrm{Pr} = 0.681.$$

Biot number,

$$\mathrm{Bi} = \frac{hD/4}{k} = \frac{h \times 0.04/4}{21} \tag{i}$$

Reynolds number,

$$\mathrm{Re}_D = \frac{\rho U_\infty D}{\mu} = \frac{0.7474 \times 50 \times 0.04}{2.57 \times 10^{-5}} = 58163.$$

The heat transfer coefficient in Eq. (i) can be calculated from Eq. (8.47)

$$h = \frac{k}{D}\mathrm{Nu} = \frac{k}{D}C\mathrm{Re}^n\,\mathrm{Pr}^{1/3}$$

where C = 0.027 and n = 0.805 for Re = 40000 to 400000 from Table 8.11.
Substitution gives

$$h = \frac{0.03859}{0.04} \times 0.027 \times 58163^{0.805} \times 0.681^{1/3} = 156.9\ \mathrm{W/\left(m^2\,K\right)}.$$

This gives

$$\mathrm{Bi} = \frac{156.9 \times 0.04/4}{21} = 0.0747 < 0.1.$$

Lumped capacity analysis is applicable. For this condition,

$$\frac{t - t_\infty}{t_i - t_\infty} = \exp(-\mathrm{BiFo})$$

where

$$\mathrm{Fo} = \frac{\alpha\tau}{L^2} = \frac{k}{\rho c}\frac{\tau}{L^2} = \frac{21}{7820 \times 460 \times (0.01)^2}\tau = 0.0584\tau$$

(for a cylinder, characteristic length $L = D/4 = 0.04/4 = 0.01$ m)

This gives

$$\frac{170 - 300}{30 - 300} = \exp(-0.0747 \times 0.0584\tau)$$

or

$$\tau = 167.5\,\mathrm{s}.$$

Example 8.24 A thin walled steel pipe, 500 mm outside diameter, carries wet steam at 160° C. Pipe is insulated from outside with 50 mm thick insulation [k_i = 0.04 W/(m K)]. If air at 5 m/s and 20°C flows across the pipe, determine the heat loss per m length of the pipe.

Solution

For the insulated pipe, the heat loss is given by

$$\frac{q}{L} = \frac{t_s - t_\infty}{\frac{1}{2\pi k_i}\ln\frac{D_2}{D_1} + \frac{1}{\pi D_2 h}} \tag{i}$$

per m length of the pipe assuming pipe wall surface temperature is equal to the wet steam temperature because the condensing steam heat transfer coefficient is very high.

Heat transfer coefficient at the outer surface can be determined as under.

For air at film temperature of 25°C (assumed),

$$\rho = 1.1868\,\mathrm{kg/m^3},\ \mu = 1.8363 \times 10^{-5}\,\mathrm{kg/(m\,s)},\ k = 0.02608\ \mathrm{W/(m\ K)}\ \text{and}\ \mathrm{Pr} = 0.709.$$

Reynolds number,

$$\mathrm{Re}_D = \frac{\rho U_\infty D}{\mu} = \frac{1.1868 \times 5 \times 0.6}{1.8363 \times 10^{-5}} = 1.94 \times 10^5.$$

Re_D lies in the range 40000–400000 for which $C = 0.027$ and $n = 0.805$ in Hilpert correlation. Hence, the heat transfer coefficient is

$$\begin{aligned} h = \frac{k}{D}\mathrm{Nu} &= \frac{k}{D} \times 0.027\mathrm{Re}^{0.805}\,\mathrm{Pr}^{1/3} \\ &= \frac{0.02608}{0.6} \times 0.027 \times \left(1.94 \times 10^5\right)^{0.805} \times 0.709^{1/3} \\ &= 18.9\,\mathrm{W}/\left(\mathrm{m}^2\,\mathrm{K}\right). \end{aligned}$$

Equation (i) gives

$$\frac{q}{L} = \frac{160 - 20}{\frac{1}{2\pi \times 0.04}\ln\frac{0.6}{0.5} + \frac{1}{\pi \times 0.6 \times 18.9}} = 185.8\,\mathrm{W/m}.$$

Insulation surface temperature,

$$\begin{aligned} t_o &= \frac{q/L}{\pi \times D_2 \times h} + t_\infty \\ &= \frac{185.8}{\pi \times 0.6 \times 18.9} + 20 = 25.2°\mathrm{C}. \end{aligned}$$

Film temperature is (20 + 25.2)/2 = 22.6°C, which is nearly equal to the assumed film temperature of 25°C.

Example 8.25 Show that for pure radial heat conduction between surface of a sphere and motionless infinite fluid medium surrounding the sphere, the Nusselt number Nu = 2.

Solution

For an isothermal sphere of radius r in infinite medium, the shape factor from Table 5.1 is

$$S = 4\pi R$$

and

$$q = kS(t_s - t_\infty) = k(4\pi R)(t_s - t_\infty) \tag{i}$$

From the heat transfer consideration,

$$q = hA(t_s - t_\infty) = h\left(4\pi R^2\right)(t_s - t_\infty) \tag{ii}$$

Equating Eqs. (i) and (ii), we get

$$h\left(4\pi R^2\right)(t_s - t_\infty) = k(4\pi R)(t_s - t_\infty)$$

or

$$\frac{hR}{k} = 1$$

or

$$\frac{hD}{k} = Nu = 2$$

Example 8.26 A 60 W bulb (t_s = 135°C,) is situated in an air stream flowing at 0.5 m/s. Determine the heat loss from the bulb if air temperature is 15°C. The bulb can be approximated to be a sphere of 50 mm diameter.

Solution

At t_∞ = 15°C, air properties are

$$\rho_\infty = 1.2323\ \text{kg/m}^3,\ k_\infty = 0.0253\ \text{W/(m K)},\ \mu_\infty = 1.787 \times 10^{-5}\ \text{Ns/m}^2 \text{ and } \text{Pr}_\infty = 0.7114.$$

At t_s = 135°C, $\mu_w = 2.317 \times 10^{-5}$ N s/m^2.
The Reynolds number,

$$\text{Re} = \frac{\rho U_\infty D}{\mu_\infty} = \frac{1.2323 \times 0.5 \times 50 \times 10^{-3}}{1.787 \times 10^{-5}} = 1724.$$

Correlation for the average Nusselt number between an isothermal spherical surface at T_w and free stream fluid (U_∞, t_∞):

$$\text{Nu} = 2 + (0.4\text{Re}_D^{0.5} + 0.06\text{Re}_D^{2/3})\,\text{Pr}^{0.4}(\mu_\infty/\mu_w)^{0.25} \tag{8.55}$$

Substitution gives,

$$\text{Nu} = 2 + [0.4(1724)^{0.5} + 0.06(1724)^{2/3}](0.7114)^{0.4}(1.787/2.317)^{0.25} = 22.64.$$

Heat transfer coefficient,

$$h = \text{Nu}\frac{k}{D} = 22.64 \times \frac{0.0253}{50 \times 10^{-3}} = 11.46\ \text{W/(m}^2\text{K)}.$$

The heat transfer rate,

$$q = hA(t_s - t_\infty)$$
$$= 11.46 \times 4 \times \pi \times 0.025^2 \times (135 - 15) = 10.8\ \text{W}.$$

Comments: Heat will also be rejected by free convection and radiation.

Example 8.27 A 10 mm diameter metal sphere [ρ = 8500 kg/m^3, c = 300 J/(kg K), k = 60 W/(m K)], initially at 500°C, is exposed to 25°C air stream flowing at 5 m/s. Determine time for the sphere to cool to 100°C if radiation is neglected.

Solution

Assuming lumped-heat-capacity analysis is applicable. From Eq. (6.2),

$$\ln\left(\frac{t-t_\infty}{t_i-t_\infty}\right)=-\left(\frac{hA_s}{c\rho V}\right)\tau,$$

which gives cooling time,

$$\tau=\left(\frac{c\rho V}{hA_s}\right)\ln\left(\frac{t_i-t_\infty}{t-t_\infty}\right)$$

For sphere, $V/A_s = D/6$. Hence,

$$\tau=\left(\frac{c\rho D}{6h}\right)\ln\left(\frac{t_i-t_\infty}{t-t_\infty}\right) \tag{i}$$

The heat transfer coefficient can be estimated from Eq. (8.55):

$$\mathrm{Nu}=2+(0.4\mathrm{Re}_D^{0.5}+0.06\mathrm{Re}_D^{2/3})\,\mathrm{Pr}^{0.4}(\mu_\infty/\mu_w)^{0.25}$$

where $\mathrm{Re}_D=\frac{\rho U_\infty D}{\mu}$.

For air at 25°C,

$$\rho=1.1868\,\mathrm{kg/m^3},\ \mu=1.8363\times10^{-5}\,\mathrm{kg/(m\ s)},\ k=0.02608\ \mathrm{W/(m\ K)}\ \text{and}\ \mathrm{Pr}=0.709.$$

and at 500°C, μ_w = 3.564 × 10^{-5} kg/(m s). This gives

$$\mathrm{Re}_D=\frac{\rho U_\infty D}{\mu}=\frac{1.1868\times5\times0.01}{1.8363\times10^{-5}}=3231$$

and

$$\mathrm{Nu}=2+(0.4\times3231^{0.5}+0.06\times3231^{2/3})0.709^{0.4}(1.8363/3.564)^{0.25}=28.5.$$

Hence,

$$h=\frac{k}{D}Nu=\frac{0.02608}{0.01}\times28.5=74.3$$

and from Eq. (i),

$$\tau = \left(\frac{c\rho D}{6h}\right)\ln\left(\frac{t_i - t_\infty}{t - t_\infty}\right)$$
$$= \left(\frac{300 \times 8500 \times 0.01}{6 \times 74.3}\right) \times \ln\left(\frac{500 - 25}{100 - 25}\right) = 105.6\,\text{s}.$$

Biot number,

$$Bi = \frac{h(D/6)}{k} = \frac{74.3 \times (0.01/6)}{60} = 0.002 < 0.1.$$

Lumped heat capacity analysis is applicable.

Example 8.28 20 mm diameter brass ball at 100°C is dropped in a water bath at 25°C. Determine the time for the ball surface temperature to reduce to 50°C. Given for brass ρ_b = 8520 kg/m^3, c_b = 380 J/(kg K), k_b= 125 W/(m K).

Solution

After some time ball will acquire constant speed termed as terminal velocity when the resistance of the water through which it is falling prevents further acceleration. At this instance, the force balance gives

$$C_D A \frac{1}{2}\rho_w U^2 = (\rho_b - \rho_w)Vg$$

or

$$C_D\left(\frac{\pi}{4}D^2\right)\frac{1}{2}\rho_w U^2 = (\rho_b - \rho_w)\left(\frac{\pi}{6}D^3\right)g$$

or

$$U = \sqrt{\frac{4(\rho_b - \rho_w)Dg}{3C_D\rho_w}}$$

At t_∞ = 25°C, ρ_w = 997.0 kg/m^3, μ_w = 890 × 10^{-6} kg/(m s), k_w = 0.609 W/(m K), and Pr = 6.13. At $t_s \approx$ 75°C, μ_s = 374 × 10^{-6} kg/(m s).

Substitution gives

$$U = \sqrt{\frac{4 \times (8520 - 997) \times 0.02 \times 9.81}{3 \times C_D \times 997}} = \frac{1.4}{\sqrt{C_D}}$$

where C_D is function of Reynolds number (Fig. 8.10)

$$\mathrm{Re}_D = \frac{\rho_w UD}{\mu_w} = \frac{997 \times U \times 0.02}{890 \times 10^{-6}} = 22404U.$$

Iterative solution gives $U \approx 2$ m/s. Hence,

$$\mathrm{Re}_D = 22404 \times 2 = 44808.$$

Correlation for the average Nusselt number between an isothermal spherical surface at t_s and free stream fluid (U, t_∞):

$$\mathrm{Nu} = 2 + \left(0.4\mathrm{Re}_D^{0.5} + 0.06\mathrm{Re}_D^{2/3}\right)\mathrm{Pr}^{0.4}(\mu_w/\mu_s)^{0.25} \tag{8.55}$$

Substitution gives,

$$\mathrm{Nu} = 2 + [0.4(44808)^{0.5} + 0.06(44808)^{2/3}](6.13)^{0.4}(890/374)^{0.25} = 413.3.$$

Heat transfer coefficient,

$$h = \mathrm{Nu}\frac{k_w}{D} = 413.3 \times \frac{0.609}{0.02} = 12585\ \mathrm{W/(m^2K)}.$$

Biot number,

$$\mathrm{Bi} = \frac{h(D/6)}{k} = \frac{12585 \times 0.02/6}{125} = 0.34 > 0.1.$$

Lumped heat capacity analysis is not applicable. Heisler charts will be used for which

$$\mathrm{Bi} = \frac{hr_o}{k} = \frac{12585 \times 0.01}{125} = 1.$$

From Fig. 6.22, for 1/Bi = 1 and $r/r_o = 1$, $\theta/\theta_o = 0.62$.

$$\frac{\theta_o}{\theta_i} = \frac{\theta}{\theta_i}\cdot\frac{\theta_o}{\theta} = \frac{50 - 25}{100 - 25} \times \frac{1}{0.62} = 0.54.$$

For 1/Bi = 1 and $\theta_o/\theta_i = 0.54$, Fo ≈ 0.28 from Fig. 6.22, From Fo = $\alpha\tau/r_o^2$, time

$$\tau = \frac{\mathrm{Fo}r_o^2}{\alpha_b} = \frac{\mathrm{Fo}r_o^2\rho_b c_b}{k_b} = \frac{0.28 \times 0.01^2 \times 8520 \times 380}{125} = 0.73\,\mathrm{s}.$$

It is to note that the terminal velocity is not reached immediately. Reduced velocity during this period implies reduced Re and h, and hence increased time τ.

Example 8.29 Air at 30°C and atmospheric pressure flows across the tubes of an in-line tube bundle consisting of 8 rows of 25 mm OD tubes in the direction of flow and 10 tubes in each row normal to the flow. The tube spacing is 37.5 mm in both parallel and normal to the flow. Calculate the heat transfer from the bundle per m of length when the tube surface temperature is 90°C. The maximum velocity of flow is 10 m/s.

Solution

Mean bulk temperature,

$$t_m = \frac{t_{\infty 1} + t_{\infty 2}}{2} = \frac{30 + t_{\infty 2}}{2}$$

where $t_{\infty 2}$ is the bulk temperature of the air at the outlet of the bundle. $t_{\infty 2}$ is not known. Therefore we assume that the mean temperature is 40°C.

At t_m = 40°C, ρ = 1.1317 kg/m^3, k = 0.0272 W/(m K), μ = 1.905 × 10^{-5} kg/(m s), c_p = 1.0066 kJ/(kg K) and Pr_∞ = 0.71. At t_w = 90°C, Pr_w = 0.695.

The Reynolds number,

$$Re = \frac{\rho U_{max} D}{\mu} = \frac{1.1317 \times 10 \times 0.025}{1.905 \times 10^{-5}} = 14852$$

Substitution gives,

$$\begin{aligned} Nu &= 0.27 C_n Re_D^{0.63} Pr^{0.36} (Pr/Pr_w)^{0.25} \quad \text{for } 10^3 < Re_D \le 2 \times 10^5 \\ &= 0.27 \times 0.97 \times (14852)^{0.63} (0.71)^{0.36} (0.71/0.695)^{0.25} = 98.9 \end{aligned} \tag{8.58c}$$

where $C_n \approx 0.97$ for n = 8 from Fig. 8.12.

The heat transfer coefficient,

$$\bar{h} = Nu \frac{k}{D} = 98.9 \times \frac{0.0272}{0.025} = 107.6 \text{ W/(m}^2 \text{ K)}.$$

The heat transfer rate,

$$\begin{aligned} \frac{q}{L} &= hA(t_s - t_\infty) = 107.6 \times (8 \times 10 \times \pi \times 0.025) \left(90 - \frac{30 + t_{\infty 2}}{2}\right) \\ &= 676.1 \times (75 - 0.5 t_{\infty 2}). \end{aligned} \tag{i}$$

From the first law of thermodynamics,

$$\frac{q}{L} = m c_p (t_{\infty 2} - t_{\infty 1})$$

where

$$m = \rho U_{max} [(p_t - D) \times 10] = 1.1317 \times 10 \times (0.0375 - 0.025) \times 10 = 1.415 \text{ kg/s}.$$

Hence,

$$\begin{aligned} \frac{q}{L} &= 1.415 \times 1006.6 \times (t_{\infty 2} - 30) \\ &= 1424.3 \times (t_{\infty 2} - 30) \end{aligned} \tag{ii}$$

Equating Eqs. (i) and (ii), we get

$$676.1 \times (75 - 0.5t_{\infty 2}) = 1424.3 \times (t_{\infty 2} - 30)$$

or

$$t_{\infty 2} = 53.02°\text{C}.$$

Revised mean temperature,

$$t_m = \frac{t_{\infty 1} + t_{\infty 2}}{2} = \frac{30 + 53.02}{2} = 41.51°\text{C},$$

which is nearly equal to the assumed value, hence retrial is not required.

Note: The effect of factor $(\text{Pr}/\text{Pr}_w)^{0.25}$ is very small in this example. This is practically true for low Prandtl number fluids just like gases when wall and bulk fluid temperatures do not differ too much.

Example 8.30 Air at 1 atm and 25°C flows at 7.5 m/s over inline tube bundle with $p = p_t = 25$ mm. The bundle contains 20 rows and 10 tubes per row, refer Fig. 8.11a. The tube diameter is 12.5 mm. If tube surface temperature is 400°C, determine the heat transfer rate per m length.

Solution

Air properties at the mean bulk temperature t_m [$= (t_{\infty i} + t_{\infty o})/2$] of the fluid flowing around the tubes in the bundle (assumed to be 125°C):

$\rho = 0.8872$ kg/m^3, $c = 1013.8$ J/(kg K), $\mu = 2.2776 \times 10^{-5}$ N s/m^2, $k = 0.0335$ W/(m K) and Pr = 0.689.

At $t_w = 400$°C, $\text{Pr}_w = 0.683$.

The rate of heat transfer per unit length of the tubes is, refer Example 7.17,

$$\frac{q}{L} = \bar{h}P \times \frac{(t_w - t_{\infty o}) - (t_w - t_{\infty i})}{\ln\left(\frac{t_w - t_{\infty o}}{t_w - t_{\infty i}}\right)} \tag{i}$$

where P is perimeter = πDN.

The Reynolds number is based on the velocity of the fluid through the narrowest cross-section of the flow area:

$$\text{Re}_D = \frac{\rho U_{max} D}{\mu}$$

where

$$U_{max} = U_\infty \times \frac{p_t}{p_t - D} = 7.5 \times \frac{25}{25 - 12.5} = 15\ \text{m/s}.$$

Hence,

$$\mathrm{Re}_D = \frac{0.8872 \times 15 \times 0.0125}{2.2776 \times 10^{-5}} = 7304.$$

From Eq. (8.58c), Nusselt number is

$$\begin{aligned}\mathrm{Nu} &= 0.27\mathrm{Re}_D^{0.63}\,\mathrm{Pr}^{0.36}(\mathrm{Pr}/\mathrm{Pr}_w)^{0.25} \quad \text{for } 10^3 < \mathrm{Re} \le 2 \times 10^5\\ &= 0.27 \times (7304)^{0.63} \times 0.689^{0.36} \times (0.689/0.683)^{0.25} = 64.3.\end{aligned}$$

Heat transfer coefficient,

$$\bar{h} = \frac{k}{D}\mathrm{Nu} = \frac{0.0335}{0.0125} \times 64.3 = 172.3\ \mathrm{W/(m^2\,K)}.$$

From equation derived in Example 7.16,

$$t_{\infty o} = t_w - (t_w - t_{\infty i})\exp\left(-\frac{P}{mc_p}L\bar{h}\right)$$

or

$$= t_w - (t_w - t_{\infty i})\exp\left(-\frac{\pi DN}{\rho U_\infty N_t p_t c_p}L\bar{h}\right)$$

$$\begin{aligned}&= 400 - (400 - 25)\exp\left(-\frac{\pi \times 0.0125 \times 200}{0.8872 \times 7.5 \times 10 \times 0.025 \times 1013.8} \times 1 \times 172.3\right)\\ &= 231.9^\circ\mathrm{C}.\end{aligned}$$

Thus the heat transfer rate from Eq. (i) is

$$\frac{q}{L} = 172.3 \times \pi \times 0.0125 \times 200 \times \frac{(400 - 231.9) - (400 - 25)}{\ln\left(\frac{400-231.9}{400-25}\right)} = 348.95\ \mathrm{kW/m}.$$

The mean bulk temperature $t_m = (t_{\infty 1} + t_{\infty 2})/2 = (25 + 231.9)/2 = 128.45^\circ\mathrm{C}$, which is nearly equal to the assumed value.

Note: The approach used in the previous example is approximate for uniform surface temperature, which can be used when temperature differences between hot and cold streams at inlet and exit [i.e., $(t_w - t_i)$ and $(t_w - t_o)$] are not significantly different. In the present problem the temperature differences are significantly different. It is to note that the approach used in this example may be used in all problems when isothermal heating or cooling surface is specified.

Example 8.31 Determine the value of heat transfer coefficient from the surface of water droplets falling in a cooling tower at a section where the velocity of the droplets relative to the air is 1 m/s and temperature is 80°C. The air temperature at the section is 30°C. The average diameter of droplets is 1.25 mm.

Solution

For the falling drops, the average value of Nusselt number is given by:

$$\mathrm{Nu} = 2 + 0.6\mathrm{Re}^{0.5}\,\mathrm{Pr}^{1/3}\left[25(D/x)^{0.7}\right] \tag{8.56}$$

At t_∞= 30°C, air properties are

$$\rho = 1.1684\,\mathrm{kg/m^3},\ \mu = 1.859\times10^{-6}\,\mathrm{kg/(m\ s)},\ k = 0.02646\ \mathrm{W/(m\ K)}\ \text{and}\ \mathrm{Pr} = 0.708.$$

The Reynolds number,

$$\mathrm{Re} = \frac{\rho U_\infty D}{\mu} = \frac{1.1684\times1\times1.25\times10^{-3}}{1.859\times10^{-5}} = 78.6$$

The distance x from Eq. (8.57) is

$$x = \frac{U_\infty^2}{2g} = \frac{1}{2\times9.81} = 0.051\ \mathrm{m}$$

Substitution gives,

$$\mathrm{Nu} = 2 + 0.6(78.6)^{0.5}(0.708)^{1/3}\left[25\times(1.25\times10^{-3}/0.051)^{0.7}\right] = 10.84.$$

Heat transfer coefficient,

$$h = Nu\frac{k}{D} = 10.84\times\frac{0.02646}{1.25\times10^{-3}} = 229.5\ \mathrm{W/(m^2K)}.$$

8.11 Heat Transfer in Liquid Metals

Figure 8.14 shows a typical variation of the Nusselt number with the Peclet number Pe (= RePr) number for flow of liquid metal through a tube with uniform heat flux (Eckert and Drake 1972). The Nusselt number is having a value of 4.36, as found in Chap. 7, for the laminar flow in tubes. The transition from the laminar to the turbulent flow can be seen to occur at Pe $\approx$ 40 (the critical value). At the start of the turbulent flow regime, the Nusselt number rises sharply to a value of about 7. At moderate value of the Peclet number beyond the critical value, the heat transfer by the turbulent mixing is small compared with the conductive transport. When the Peclet number exceeds a value of about 100, the heat exchange by turbulent mixing becomes appreciable and cannot be neglected.

Various empirical correlations for the Nusselt number are available in the literature. However, due to the difficulties in assessing the effect of free convection and longitudinal conduction, the experimental results of various investigators deviate from each other in the laminar and near-turbulent regions. $\mathrm{Nu_T}$ and $\mathrm{Nu_H}$ results differ significantly and separate equations are presented. Some of the correlations are given in Table 8.13.

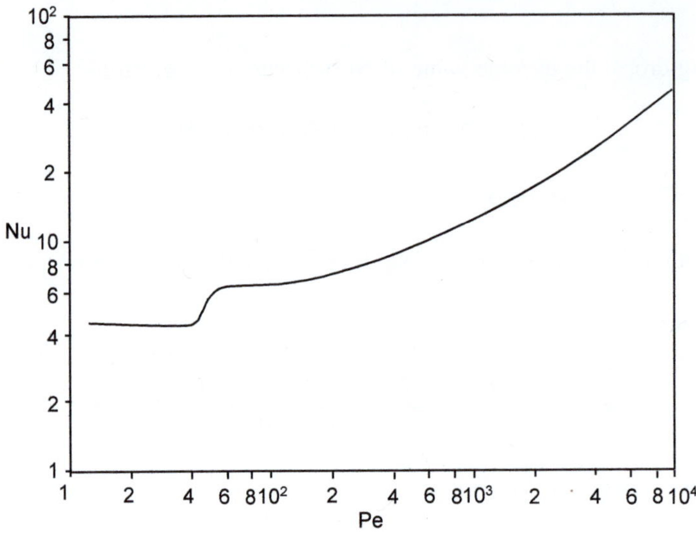

Fig. 8.14 Uniform heat flux Nusselt number for flow of liquid metal through a tube (Eckert and Drake 1972)

Table 8.13 Nusselt number for fully developed turbulent flow of liquid metals

Investigators	Correlation	Boundary condition
Tubes 1.Lyons (in Eckert and Drake 1972) (referred to as Lyons – Martinelli equation)	$Nu_d = 7 + 0.025Pe^{0.8}$ for Pe > 50	Constant heat flux
2.Sleicher and Rouse (1975)	$Nu_d = 6.3 + 0.0167Re^{0.85}Pr^{0.93}$	Constant heat flux
3.Skupinshi et al. (1965)	$Nu_d = 4.82 + 0.0185Pe^{0.827}$	Constant heat flux
4.Sleicher and Rouse (1975)	$Nu_d = 4.8 + 0.0156Re^{0.85}Pr^{0.93}$	Constant wall temperature
5.Seban and Shimazaki (1951)	$Nu_d = 5.0 + 0.025Pe^{0.8}$	Constant wall temperature
Single Cylinder Grosh and Cess (in Eckert and Drake 1972)	$Nu_d = 1.015Pe^{0.25}$ for Pe < 500	Heat transfer by vortex motion is assumed to be very small compared to the conduction effect in the back of the cylinder
Sphere Witte (1968)	$Nu_d = 2 + 0.386\ Pe^{0.5}$ for $3.56 \times 10^4 < Re < 1.525 \times 10^5$	

Although the Nusselt number tends to be low for the liquid metal, like the laminar flow, the heat transfer coefficient is very high because of the high thermal conductivity of the liquid metals.

Example 8.32 Liquid mercury at 20°C enters a metal tube of 20 mm internal diameter at the rate of 1 kg/s and is heated to 30°C. The tube wall subjected to uniform heat flux is at an average temperature of 40°C. Determine the length of the tube. Given for the mercury: $\rho = 13560$ kg/m^3, $k = 8.7$ W/(m K), $\mu = 1.5 \times 10^{-3}$ kg/(m s), Pr = 0.025, $c_p = 139$ J/(kg K).

Solution

The heat duty is

$$q_t = mc_p\Delta t = mc_p(t_o - t_i) = 1 \times 139 \times (30 - 20) = 1390 \text{ W}.$$

Flow Reynolds number,

$$\text{Re} = \frac{\rho U_m d}{\mu} = \frac{4m}{\pi d\mu} = \frac{4 \times 1}{\pi \times 0.02 \times 1.5 \times 10^{-3}} = 42441.$$

$$\text{Pe} = \text{Re}\,\text{Pr} = 42441 \times 0.025 = 1061 > 50.$$

Heat transfer coefficient from Lyons–Martinelli equation,

$$h = \text{Nu}\frac{k}{d} = (7 + 0.025\text{Pe}^{0.8})\frac{k}{d}$$
$$= [7 + 0.025 \times (1061)^{0.8}] \times \frac{8.7}{0.02} = 5909 \text{ W/m}^2\text{ K}.$$

Heat transfer rate per unit length,

$$\frac{q}{L} = h(\pi d)(t_w - t_m) = 5909 \times (\pi \times 0.02) \times (40 - 25) = 5569 \text{ W/m}.$$

Required length of the tube,

$$L = \frac{q_t}{q/L} = \frac{1390}{5569} = 0.25 \text{ m}.$$

8.12 Influence of Duct Wall Roughness in Turbulent Flow

(a) **Friction Factor and Nusselt Number Correlations**

Nikuradse (1950) carried out experiments to study the relationship of friction factor to the Reynolds number for pipes of various roughnesses. Six different degrees of relative roughness e/r ($0.002 < e/r < 0.0679$ where e = average height of roughness element and r is pipe radius) were used. Figure 8.15 shows to a logarithmic scale the relation of the friction factor and the Reynolds number for the reciprocal values r/e of the six relative roughnesses and for a smooth pipe. For a given relative roughness, three regions can be seen. At low Reynolds number (within the first region termed as *hydraulically smooth regime*), the roughness has no effect on the resistance and for all values of r/e, the curve coincides with

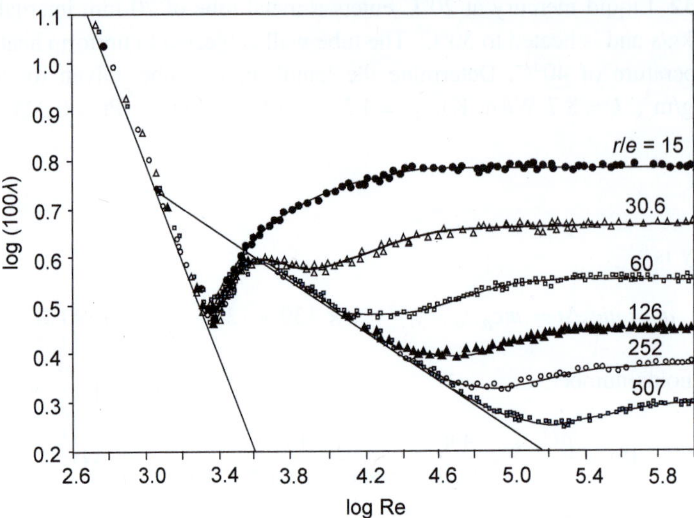

Fig. 8.15 Relation between log (100λ) and log Re; *r*/*e* is ratio of radius *r* to average projection *e* and λ is Darcy friction factor (= 4*f*) (Nikuradse 1950)

the curve for the smooth pipe and resistance factor is expressed by relation λ = 64/Re (i.e., *f* = 16/Re) for the laminar flow condition. For turbulent flow, up to about Re = 10^5, the Blasius resistance law ($\lambda = 0.361\text{Re}^{-0.25}$) holds for smooth pipes. The critical Reynolds number for all degrees of relative roughness occurs between 2160 and 2500.

Within the second range, termed as *transitional range*, the influence of the roughness becomes noticeable in an increasing degree; the resistance factor increases with an increasing Reynolds number. The resistance factor depends both on the Reynolds number and the relative roughness. In the third region, termed as *fully rough regime*, the resistance factor is independent of the Reynolds number and the curve λ = f(Re) becomes parallel to the x-axis. The dependency of the friction factor on the relative roughness height in the fully rough regime has been expressed by Nikuradse as

$$\frac{1}{\sqrt{f}} = 3.48 - 1.737\ln\left(\frac{2e}{D}\right) \tag{8.61}$$

An empirical formula correlating the entire transition regime has been given by Colebrook and White (in Bhatti and Shah 1987):

$$\frac{1}{\sqrt{f}} = 3.48 - 1.7372\ln\left(\frac{2e}{D} + \frac{9.35}{\text{Re}\sqrt{f}}\right) \tag{8.62}$$

For *e* = 0, i.e. the smooth pipes, Eq. (8.62) transforms to PKN correlation and for very high values of the Reynolds number, it transforms to Nikuradse's equation for fully rough flow regime.

The sand grain roughness used by Nikuradse in his experiment is not similar to the roughness encountered in commercial pipes. Hence, Schlichting introduced the concept of equivalent sand-grain roughness as a means of characterizing other types of roughness elements by referring to the equivalent net effect produced by Nikuradse's experiments (Kays and Crawford 1980). Moody (1944) determined the equivalent sand-grain roughness for different types of commercial pipe surfaces (refer Moody 1944).

The duct wall roughness has no effect in the case of the laminar flow. If the surface roughness height e is of the order of the magnitude of the laminar sublayer thickness δ_l, it tends to break up the laminar sublayer. This increases the wall shear stress. The ratio of the surface roughness height e and laminar sublayer thickness δ_l determine the effect of the roughness. The laminar sublayer thickness δ_l is proportional to ν/u^*, where u^* is the friction velocity ($= \sqrt{(\tau_o/\rho)}$; τ_o is shear stress $f\rho U_m^2/2$). Thus the ratio e/δ_l is proportional to eu^*/ν, which is termed as *roughness Reynolds number* e^+:

$$\begin{aligned} e^+ &= \frac{eu^*}{\nu} = \left(\frac{e}{\nu}\right) U_m \sqrt{\frac{f}{2}} = \left(\frac{e}{D}\right)\left(\frac{U_m D}{\nu}\right)\sqrt{\frac{f}{2}} \\ &= \sqrt{\frac{f}{2}} \mathrm{Re}\left(\frac{e}{D}\right) \end{aligned} \tag{8.63}$$

where D is the tube diameter, U_m mean velocity and e/D (or e/D_h for noncircular ducts) is termed as relative roughness height.

The roughness Reynolds number has been used to define the flow regimes identified by Nikuradse for flow in roughened ducts and the physical interpretation of the observed behaviour in these regimes is as follows.

1. **Hydraulically smooth ($0 \le e^+ \le 5$)**: The roughness has no effect on the friction factor because the roughness height e is so small that the roughness elements lie entirely within the laminar sublayer.
2. **Transition regime ($5 < e^+ \le 70$)**: The heights of the roughness elements in this range are of the same order of magnitude as the thickness of the laminar sublayer. Individual projections may extend through the laminar sublayer and vortices produce an additional loss of energy. As the Reynolds number increases, an increasing number of projections pass through the laminar sublayer because of the reduction in its thickness. The additional loss, then, becomes greater as the Reynolds number increases and the friction factor depends both on relative roughness height e/D and the Reynolds number Re.
3. **Fully rough regime ($e^+ > 70$)**: Finally, with the increase in the Reynolds number, the thickness of the laminar sublayer becomes so small that all projections extend through it. The energy loss due to the vortices now attains a constant value and the friction factor is independent of the Reynolds number. In this regime, the friction factor depends only upon the relative roughness height.

Moody (1944) also presented friction factor plot as shown in Fig. 8.16. The horizontal portions of the curves right to the dashed line are represented by Nikuradse's equation for fully rough regime. The dashed line corresponds to $e^+ = 70$. The downward sloping line for the laminar flow is represented by $\lambda = 4f = 64/\mathrm{Re}$. The lowermost curve for the smooth pipes with turbulent flow is represented by the PKN correlation.

It is to note that the artificial roughness creates turbulence close to the wall, which leads to the enhancement in the heat transfer coefficient also. However, the increase in the heat transfer coefficient is accompanied by a proportionately greater increase in the friction factor (Bhatti and Shah 1987).

In the case of fluids with high Prandtl number, the resistance to the heat transfer is mainly concentrated in the laminar sublayer, which is thin compared to the hydrodynamic boundary

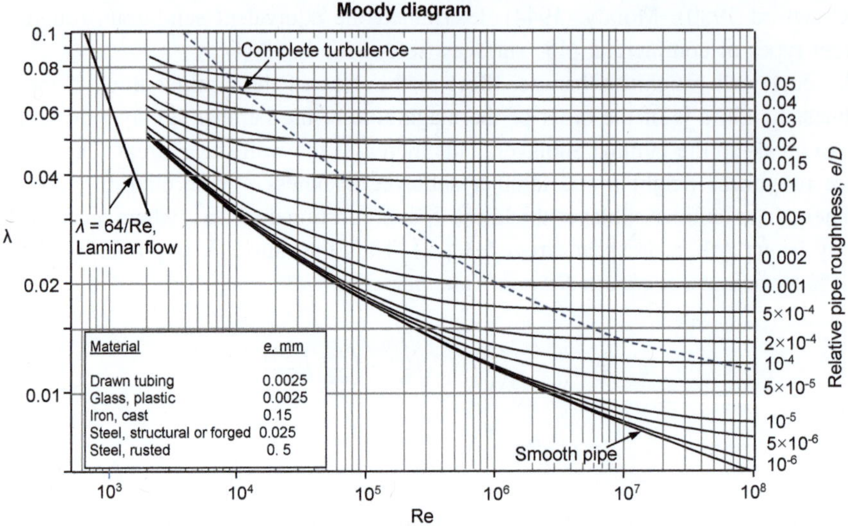

Fig. 8.16 Darcy friction factor ($\lambda = 4f$) plot for smooth and rough tubes. Moody LF (1944) Friction factors for pipe flow. Trans ASME 66: 671–684. With permission of ASME

layer. For the low Prandtl number fluids, the thermal resistance is distributed over a larger portion of the duct cross-section because the thermal boundary layer is thicker than the hydrodynamic boundary layer. Since the roughness at the wall creates turbulence near the wall, the heat transfer enhancement is greater for the fluids with high Prandtl numbers.

As pointed out above, the heat transfer coefficient is not affected as strongly as the friction coefficient. The physical explanation for this is given by Bhatti and Shah (1987) as follows.

The friction coefficient is markedly augmented by the profile drag developed by the roughness elements. As regards the heat transfer, there is no mechanism comparable to the profile drag to generate additional heat flux. Consequently, the heat transfer is affected less markedly than the friction coefficient.

Several studies on heat transfer coefficient and friction factor behaviour for different roughness types have been carried out by researchers and correlations have been proposed, refer Bhatti and Shah (1987). Some of them are presented in Table 8.14.

(b) **Roughness and Heat Transfer Functions**

The experimental results of heat transfer and fluid flow characteristics in roughened ducts have been presented either in the form of direct dependence of friction factor and Nusselt number on the system and operating parameters as presented above or in the form of interrelated roughness and heat transfer functions. This latter method makes it possible to present results in a most general form taking into account the various parameters involved including the roughness parameters.

The friction correlation has been based on the *law of wall* similarity employed by Nikuradse (1950) for sand grain roughness in pipes.

Table 8.14 Nusselt number for fully developed turbulent flow in the fully rough flow regime of a circular duct[a]

Investigators	Correlations[b]	Remarks
Dippery and Sabersky (1963)	$\mathrm{Nu} = \dfrac{\mathrm{Re\,Pr}(f/2)}{1+\sqrt{(f/2)}[5.19(e^+)^{0.2}\,\mathrm{Pr}^{0.44}-8.48]}$	This correlation is valid for $0.0024 < e/D_h \le 0.049$, $1.2 \le \mathrm{Pr} \le 5.94$, $1.4 \times 10^4 \le \mathrm{Re} \le 5\times10^5$
Gowen and Smith (1968)	$\mathrm{Nu} = \dfrac{\mathrm{Re\,Pr}\sqrt{(f/2)}}{4.5+[0.155(\mathrm{Re}\sqrt{f/2})^{0.54}+\sqrt{(2/f)}]\sqrt{\mathrm{Pr}}}$	This correlation is valid for $0.021 < e/D_h \le 0.095$, $0.7 \le \mathrm{Pr} \le 14.3$, $10^4 \le \mathrm{Re} \le 5\times10^4$
Bhatti and Shah (1987)	$\mathrm{Nu} = \dfrac{\mathrm{Re\,Pr}(f/2)}{1+\sqrt{(f/2)}[4.5(e^+)^{0.2}\,\mathrm{Pr}^{0.5}-8.48]}$	This correlation is valid for $0.002 < e/D_h < 0.05$, $0.5 < \mathrm{Pr} < 10$, $\mathrm{Re} > 10^4$. Its predictions are within ±5% of the available measurements.

[a]Bhatti MS, Shah RK, Turbulent and transition flow convective heat transfer. In: Kakac S, Shah RK, Aung W (eds) Handbook of single-phase convective heat transfer, Chap. 4, Wiley, New York, Copyright 1987. Reproduced (abridged) with the permission of John Wiley and Sons Ltd

[b]The friction factor f in these correlations may be calculated from the Nikuradse's correlation, Eq. (8.61)

The general expression for friction factor f proposed by Nikuradse is

$$f = \frac{2}{[A + B\ln(e^+) - 2.456\ln(2e/D)]^2} \tag{8.64}$$

where B = 2.5 is a constant for both smooth and rough tubes and A is a non-dimensional parameter, which is discussed below.

Nikurdse proposed a law of wall by correlating the measured velocity distribution data for these sand grain roughened tubes by a non-dimensional equation of the form

$$u^+ = u/u^* = B\ln(y/e) + A \tag{8.65}$$

Nikuradse found that the plot of parameter A as a function of $\log(e^+)$ is similar to the curve for the resistance law obtained by plotting $\lfloor 1/(2\sqrt{f}) - 2\log_{10}(r/e)\rfloor$ against e^+. From this similarity, he deduced the value of A as

$$A = \sqrt{\frac{2}{f}} + 2.5\ln\frac{2e}{D} + 3.75 \tag{8.66}$$

Thus

$$A = u^+ - 2.5\ln(y/e) = \sqrt{\frac{2}{f}} + 2.5\ln\frac{2e}{D} + 3.75$$

The non-dimensional parameter A is called by different investigators as the roughness parameter or momentum transfer roughness function or *roughness function* and has been denoted by $R(e^+)$ because Nikuradse found roughness function to be function of e^+ only. The usual relation for this function is

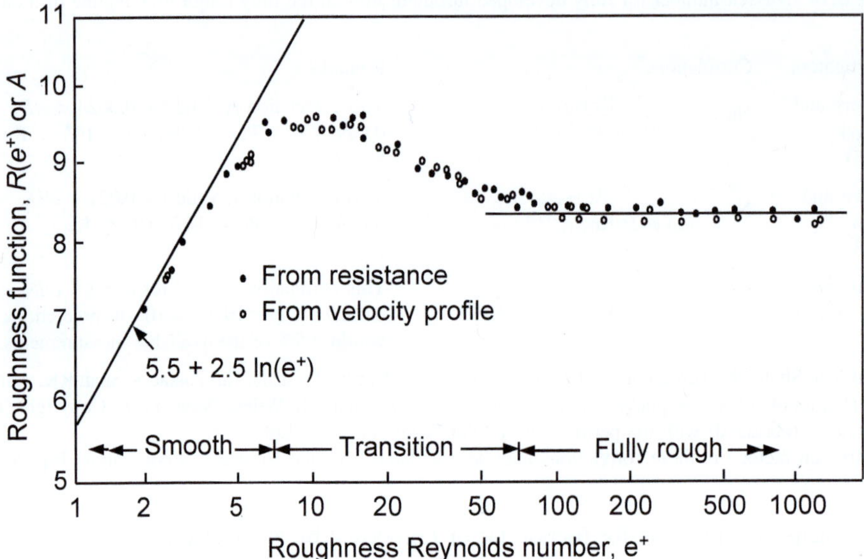

Fig. 8.17 Momentum transfer roughness function versus the roughness Reynolds number (Nikuradse 1950)

$$R(e^+) = \sqrt{\frac{2}{f}} + 2.5\ln\frac{2e}{D} + E \tag{8.67}$$

The constant E is termed as geometric parameter and is dependent on the configuration of the duct. Nikuradse reported a value of this parameter as 3.75 for pipes and the same value has been used by Dippery and Sabersky (1963) in their investigation. However, there is no general agreement on the value of constant E.

The plot of the roughness function, A or $R(e^+)$ against the roughness Reynolds number e^+ obtained by Nikuradse is shown in Fig. 8.17. The variation of the roughness function in the different flow regimes as found by Nikuradse is as follows.

(i) In the hydraulically smooth regime ($0 < e^+ \leq 5$), the measured pressure loss data were correlated by Nikuradse in the form:

$$R(e^+) = 5.5 + 2.5\ln e^+ \tag{8.68}$$

(ii) In the transition regime ($5 < e^+ \leq 70$), the roughness function is observed at first to increase with e^+ and then to attain a constant value and finally the function drops.

(iii) In the fully rough regime ($e^+ > 70$), the roughness function can be seen to be independent of the roughness Reynolds number and attains a constant value.

(iv) For the sand grain roughness, Nikuradse reported a constant value of 8.48 of the roughness function $R(e^+)$ for the fully rough region. However, a number of investigators while investigating the behaviour of wire and rib roughness in tubes and ducts (both square and rectangular) have reported different values of the roughness function for the fully rough flow. Further the investigators have shown that the criterion of hydraulically smooth or fully rough or transitional flow regimes and the values of the roughness function depend not only on the roughness Reynolds number but also on the geometric parameters of the roughness, shape of the roughness element and channel aspect ratio (ratio of width to height) in the case of flow in rectangular channels. Since the roughness function depends on e^+, roughness parameters and channel aspect ratio, it is written as R instead of $R(e^+)$.

(v) Dippery and Sabersky (1963) hypothesized that the law of wall similarity, which has been shown by Nikuradse to be valid for velocity profile in roughened tubes, also applies to the temperature profile. They developed a heat transfer similarity law for the flow in sand grain roughened tubes. They correlated turbulent data for fluids of various Prandtl numbers and tubes of three different sand grain roughness types using a heat transfer function g termed as g-function by them, which is function of e^+ and the function has also been found to have different values for different fluids. Mathematically, the g-function is defined by Dippery and Sabersky (1963) as

$$g(e^+) = \left[\frac{f}{2St} - 1\right]\sqrt{\frac{2}{f}} + R(e^+) \tag{8.69}$$

For the fully rough flow ($e^+ > 70$), Dippery and Sabersky found that the heat transfer function is a function of only roughness Reynolds number and Prandtl number. Researchers found g-function to depend also on the geometrical parameters of roughness and hence have written as g instead of $g(e^+)$.

Webb et al. (1971) used the law of wall similarity employed by Nikuradse and heat transfer-momentum analogy extended by Dippery and Sabersky to correlate data for tubes with transverse repeated rib roughness. This general strategy has been successfully used to correlate friction and heat transfer results for different types of roughness elements by a large number of researchers. For typical correlations developed using this strategy for v-discrete rib roughness elements, refer Karwa et al. (2005).

Example 8.33 A 20 mm diameter smooth surfaced tube is maintained at a constant wall temperature of 100°C. Water enters the tube at 30°C and leaves at 50°C with a velocity of 2 m/s. Determine the length of the tube necessary for the desired heating of the water. If the tube has relative roughness of 0.001, what will be the required length?

Solution

At mean bulk fluid temperature $t_m = (t_i + t_o)/2 = (30 + 50)/2 = 40$°C, the thermo-physical properties of air are

$$\rho = 992.2\ \text{kg/m}^3,\ k = 0.631\ \text{W/(m K)},\ \mu = 6.51 \times 10^{-4}\ \text{kg/(m s)}, \text{Pr} = 4.30 \text{ and } c_p = 4179\ \text{J/(kg K)}.$$

The heat duty is

$$q_t = mc_p\Delta t = [(\pi/4)d^2 U_m \rho]c_p(t_o - t_i)$$
$$= (\pi/4) \times (0.02)^2 \times 2 \times 992.2 \times 4179 \times (50 - 30) = 52105\ \text{W}.$$

Flow Reynolds number,

$$\mathrm{Re} = \frac{\rho U_m d}{\mu} = \frac{992.2 \times 2 \times 0.02}{6.51 \times 10^{-4}} = 60965 > 10000.$$

Flow is turbulent.

Smooth tube
Heat transfer coefficient from Dittus Boelter relation,

$$h = Nu\frac{k}{d} = 0.024\mathrm{Re}^{0.8}\,\mathrm{Pr}^{0.4}\frac{k}{d}$$
$$= 0.024 \times (60965)^{0.8} \times (4.3)^{0.4} \times \frac{0.631}{0.02} = 9134\ \mathrm{W/(m^2K)}.$$

Since the change in the temperature of the water is not large, a simple arithmetic mean of the end temperature differences may be used in heat transfer equation instead of the log mean temperature difference when wall temperature is constant. Hence, heat transfer rate per unit length,

$$\frac{q}{L} = h(\pi d)(\Delta t) = 9134 \times (\pi \times 0.02) \times (70 + 50)/2 = 34434\ \mathrm{W/m}.$$

Here the log mean temperature difference comes out to be 59.44°C. Required length of the tube,

$$L = \frac{q_t}{q/L} = \frac{52105}{34434} = 1.513\ \mathrm{m}.$$

Roughened tube
From Colebrook and White relation,

$$\frac{1}{\sqrt{f}} = 3.48 - 1.7372\ln\left(\frac{2e}{d} + \frac{9.35}{\mathrm{Re}\sqrt{f}}\right)$$

Staring with a trial value of $f = 0.0791\mathrm{Re}^{-0.25} = 0.005$, the solution of the above equation gives $f = 0.00585$. Alternatively, the Moody Diagram may be used to determine friction factor.

Roughness Reynolds number,

$$e^+ = \sqrt{\frac{f}{2}}\mathrm{Re}\frac{e}{d} = \sqrt{\frac{0.00585}{2}} \times 60965 \times 0.001 = 3.3.$$

Gnielinski's correlation from Table 8.3 with viscosity correction factor may be used, refer Eq. (8.20), which gives

$$\mathrm{Nu} = \frac{(f/2)(\mathrm{Re} - 1000)\,\mathrm{Pr}}{1 + 12.7\sqrt{(f/2)}.(\mathrm{Pr}^{2/3} - 1)} \times \left(\frac{\mu_w}{\mu}\right)^{-0.11}$$

Substitution gives

$$\mathrm{Nu} = \frac{(0.00585/2)(60965 - 1000) \times 4.3}{1 + 12.7\sqrt{(0.00585/2)}.(4.3^{2/3} - 1)} \times \left(\frac{2.86}{6.51}\right)^{-0.11} = 388$$

for $\mu_w = 2.86$ at 100°C.

Heat transfer coefficient,

$$h = \mathrm{Nu}\frac{k}{d} = 388 \times \frac{0.631}{0.02} = 12241 \text{ W/(m}^2\text{K)}.$$

Required length of the tube,

$$L = 1.513 \times \frac{9134}{12241} = 1.129 \text{ m}.$$

It is to note the heat transfer enhancement in roughened tube is accompanied with increase in pumping power requirement because of the increased friction.

Example 8.34 Calculate the required length if the tube relative roughness is 0.02 for the flow conditions of the previous example.

Solution

From the Moody's diagram, $f = 0.05/4 = 0.0125$ and the flow is in the fully rough region.

Alternatively, Nikuradse's correlation can be used to determine friction factor, which is

$$\frac{1}{\sqrt{f}} = 3.48 - 1.737 \ln\left(\frac{2e}{D}\right) \tag{8.61}$$

$$\frac{1}{\sqrt{f}} = 3.48 - 1.737 \ln(2 \times 0.02)$$

which gives $f = 0.01215$.

Roughness Reynolds number,

$$e^+ = \sqrt{\frac{f}{2}}\mathrm{Re}\frac{e}{d} = \sqrt{\frac{0.01215}{2}} \times 60965 \times 0.02 = 95.$$

Using Dippery and Sabersky's correlation with viscosity correction factor, Table 8.14,

$$\mathrm{Nu} = \frac{\mathrm{Re}\,\mathrm{Pr}(f/2)}{1 + \sqrt{(f/2)}.[5.19(e^+)^{0.2}\,\mathrm{Pr}^{0.44} - 8.48]} \times \left(\frac{\mu_w}{\mu}\right)^{-0.11}$$

or

$$\mathrm{Nu} = \frac{60965 \times 4.3 \times (0.01215/2)}{1 + \sqrt{(0.01215/2)} \times [5.19 \times (95)^{0.2} \times (4.3)^{0.44} - 8.48]} \times \left(\frac{2.86}{6.51}\right)^{-0.11} = 774.7.$$

Required pipe length is

$$L = 1.129 \times \frac{388}{774.7} = 0.565\ \text{m}.$$

Miscellaneous Exercises

Example 8.35 A fluid at an average bulk temperature of 40°C flows inside a 50 mm ID and 1.5 m long tube with bell mouth inlet. Because of constant heat flux condition, the tube wall is 5°C below the fluid bulk temperature. If the average bulk velocity of the fluid is 2.5 m/s, calculate the heat transfer rate from the fluid. The thermophysical properties of the fluid are given as

$$\nu = 6.0 \times 10^{-7}\text{m}^2/\text{s},\ k = 0.15\ \text{W}/(\text{m K})\ \text{and}\ \text{Pr} = 5.$$

For the entrance effect, if any, following equation may be considered.

$$h_{ent} = h\left[1 + \frac{1.4}{L/D}\right].$$

Solution

The flow Reynolds number

$$\text{Re}_L = \frac{U_\infty d}{\nu} = \frac{2.5 \times 0.05}{6 \times 10^{-7}} = 2.08 \times 10^5.$$

The flow is turbulent. Though the Prandtl number is quite high, the temperature of the wall is only 5°C above the fluid bulk mean temperature, viscosity correction is not required.

Using the Dittus-Boelter equation[4] for cooling of fluid,

$$h = \text{Nu}\frac{k}{d} = 0.026\text{Re}^{0.8}\,\text{Pr}^{0.3}\frac{k}{d}$$
$$= 0.026 \times (2.08 \times 10^5)^{0.8} \times (5)^{0.3} \times \frac{0.15}{0.05} = 2271\ \text{W/(m}^2\text{K)}.$$

L/D = 1.5/0.05 = 30 < 60 hence the corrected value of the heat transfer coefficient considering the entrance effect is

$$h_{ent} = h\left[1 + \frac{1.4}{L/D}\right]$$
$$= 2271 \times \left[1 + \frac{1.4}{30}\right] = 2377\ \text{W/(m}^2\text{K)}.$$

[4]The Dittus-Boelter equation is widely used because of its simplicity. For greater accuracy, Gnielinski or Petukhov et al. correlation (refer Table 8.3) may be considered.

Heat transfer rate,

$$q_t = hA\Delta T = 2377 \times (\pi \times 0.05 \times 1.5) \times 5 = 2800.3 \text{ W}.$$

Example 8.36 Air at 20°C and a pressure of 5.5 bar flows through a circular cross-section tube (ID = 25 mm and L = 1.6 m) with a mean velocity of 1.5 m/s. The tube is subjected to electric heating with heat flux q" of 1.5 kW/m^2. Determine the temperature rise of air and the local heat transfer coefficient at L = 1.0 m.

Solution

Assuming mean bulk temperature of 40°C, the air properties at atmospheric pressure are

$$\rho = 1.132 \text{ kg/m}^3,\ c_\text{p} = 1006.6 \text{ J/(kg K)}.$$

At 5.5 bar,

$$\rho = 1.132 \times 5.5 = 6.226 \text{ kg/m}^3.$$

From heat balance,

$$\frac{\pi}{4} D^2 \rho U c_p \Delta T = \pi DL \times q''$$

or

$$\Delta T = \frac{4Lq''}{\rho U D c_p} = \frac{4 \times 1.6 \times 1500}{6.226 \times 1.5 \times 0.025 \times 1006.6} = 40.85°\text{C}.$$

Outlet temperature, t_o = 20 + 40.85 = 60.85°C.

Mean bulk temperature, t_m = (60.85 + 20)/2 = 40.43°C, which is nearly the same as assumed. Further iteration is not required.

Other thermophysical properties are

$$\mu = 1.9 \times 10^{-5}\,\text{kg/(m s)}, \text{Pr} = 0.705 \text{ and } k = 0.0273 \text{ W/(m K)}.$$

The flow Reynolds number

$$\text{Re}_L = \frac{\rho U_m d}{\mu} = \frac{6.226 \times 1.5 \times 0.025}{1.9 \times 10^{-5}} = 12288 > 10^4.$$

Development length from Eq. (8.19),

$$\frac{L}{D} = 1.359\text{Re}^{0.25} = 1.359 \times (12288)^{0.25} = 14.31.$$

At $x = 1$ m, $x/D = 1/0.025 = 40$, which is much larger than the development length for turbulent flow. Hence, the local heat transfer coefficient at $x = 1$ m equals the heat transfer coefficient for the fully developed flow. Using Dittus-Boelter equation,

$$h = \mathrm{Nu}\frac{k}{d} = 0.024\mathrm{Re}^{0.8}\,\mathrm{Pr}^{0.4}\frac{k}{d}$$
$$= 0.024 \times (12288)^{0.8} \times (0.705)^{0.4} \times \frac{0.0273}{0.025} = 42.6\ \mathrm{W/(m^2K)}.$$

Example 8.37 Water at an average bulk temperature of 90°C is flowing through a 25 mm ID circular tube with a mean velocity of 1.5 m/s. The pipe wall temperature is 50°C along the whole length below the local value of the bulk temperature of the water because of the uniform heat flux condition. If pipe is 5 m long, determine the cooling rate. The following turbulent flow equation due to Sieder-Tate with viscosity correction factor may be considered.

$$\mathrm{Nu} = 0.027\mathrm{Re}^{0.8}\,\mathrm{Pr}^{1/3}\left(\frac{\mu_b}{\mu_w}\right)^{0.14}$$

Solution

Thermophysical properties of water at the average bulk temperature of 90°C are

$$\rho = 965.3\,\mathrm{kg/m^3}, \mathrm{Pr} = 1.97, k = 0.675\ \mathrm{W/(m\,K)}, \mu_b = 311 \times 10^{-6}\,\mathrm{kg/(m\,s)}.$$

At wall temperature of $t_w = 40$°C, $\mu_w = 651 \times 10^{-6}$ kg/(m s).
Due to a significant difference in μ_w and μ_b, viscosity correction must be considered.
Flow Reynolds number,

$$\mathrm{Re}_L = \frac{\rho U_m d}{\mu_b} = \frac{965.3 \times 1.5 \times 0.025}{311 \times 10^{-6}} = 1.16 \times 10^5.$$

Flow is turbulent. Using the given equation,

$$h = \mathrm{Nu}\frac{k}{d} = 0.027\mathrm{Re}^{0.8}\,\mathrm{Pr}^{1/3}\left(\frac{\mu_b}{\mu_w}\right)^{0.14}\frac{k}{d}$$
$$= 0.027 \times (1.16 \times 10^5)^{0.8} \times (1.97)^{1/3} \times \left(\frac{311}{651}\right)^{0.14} \times \frac{0.675}{0.025}$$
$$= 9280\ \mathrm{W/(m^2K)}.$$

Cooling rate,

$$q_t = hA\Delta t = h(\pi dL)(\Delta t) = 9280 \times (\pi \times 0.025 \times 5) \times 50 = 182.2\ \mathrm{kW}.$$

Example 8.38 Water flows through a rectangular duct of width $W = 10$ mm, height $H = 5$ mm and 3 m long. The velocity of the water is 0.25 m/s. The mean water temperature in the duct is 60°C and the duct wall temperature is 20°C. Calculate the mean heat transfer coefficient and the heat transfer rate. The following correlation may be used

$$\text{Nu} = 0.17\text{Re}^{0.33}\,\text{Pr}^{0.43}\,\text{Gr}^{0.1}\left(\frac{\text{Pr}}{\text{Pr}_w}\right)^{0.25}$$

where the subscript w refers to the wall condition.

Solution

Thermophysical properties of water at the mean temperature of 60°C are (Table A4, Appendix)

$$\rho = 983.3\ \text{kg/m}^3,\ \text{Pr} = 3.0,\ k = 0.654\ \text{W/(m K)},\ \mu = 463 \times 10^{-6}\,\text{kg/(m s)},\ \beta = 0.521 \times 10^{-3}.$$

At t = 20°C, $\text{Pr}_\text{w} = 7$.
Hydraulic diameter of the duct,

$$D_h = \frac{4WH}{2(W+H)} = \frac{4 \times 10 \times 5}{2(10+5)} = 6.66\ \text{mm}.$$

Reynolds number,

$$\text{Re} = \frac{\rho U D_h}{\mu} = \frac{983.3 \times 0.25 \times (6.66 \times 10^{-3})}{463 \text{ x } 10^{-6}} = 3536.$$

Grashof number,

$$\text{Gr} = \frac{\rho^2 \beta g \Delta T D_h^3}{\mu^2} = \frac{983.3^2 \times 0.521 \times 10^{-3} \times 9.81 \times (60 - 20) \times 0.00666^3}{(463 \times 10^{-6})^2} = 2.72 \times 10^5.$$

Heat transfer coefficient from the given Nusselt number correlation:

$$\begin{aligned} h &= \text{Nu}\frac{k}{D_h} = \frac{k}{D_h} \times 0.17\text{Re}^{0.33}\,\text{Pr}^{0.43}\,\text{Gr}^{0.1}\left(\frac{\text{Pr}}{\text{Pr}_w}\right)^{0.25} \\ &= \frac{0.654}{0.00666} \times 0.17 \times (3536)^{0.33} \times 3^{0.43} \times (2.72 \times 10^5)^{0.1} \times \left(\frac{3.0}{7.0}\right)^{0.25} \\ &= 1122.5\ \text{W/(m}^2\text{K)}. \end{aligned}$$

Heat transfer rate,

$$q_t = hA\Delta t = h(PL)(\Delta t) = 1122.5 \times (30/1000) \times 3 \times (60 - 20) = 4041\ \text{W}.$$

Example 8.39 The heat transfer coefficient for a gas flowing over a thin flat plate 3 m long and 0.3 m wide varies with the distance from the leading edge according to

$$h_x = 10x^{-1/4}\text{W/m}^2\,\text{K}.$$

Calculate the (a) average heat transfer coefficient, (b) the rate of heat transfer between the plate and the gas if the plate is at a temperature of 170°C and the gas is at 30°C, and (c) the local heat flux at 2 m from the leading edge.

Solution

$$\begin{aligned} h_{av} &= \frac{1}{L}\int_0^L h_x \text{d}x \\ &= \frac{1}{3}\int_0^3 10x^{-1/4}\text{d}x \\ &= \frac{10}{3}\left[\frac{x^{-1/4+1}}{-1/4+1}\right]_0^3 = 10.13\ \text{W/(m}^2\text{K)}. \end{aligned}$$

The rate of heat transfer between the plate and the gas,

$$\begin{aligned} q &= h_{av}A\Delta T \\ &= 10.13 \times (3 \times 0.3) \times (170 - 30) = 1276.4\ \text{W}. \end{aligned}$$

The local heat flux at $x = 2$ m is

$$\begin{aligned} q''_{x=2} &= \frac{q_{x=2}}{A} = h_{x=2}\Delta T \\ &= 10 \times (2)^{-1/4} \times (170 - 30) = 1177.3\ \text{W/m}^2. \end{aligned}$$

Example 8.40 What is pressure drop in a 20 m long smooth 20 mm ID tubing when water at 100°C flows through it? A Pitot tube measurement shows centerline velocity of 0.02 m/s.

Solution

Assuming fully developed laminar flow,

$$U_m = \frac{U_{max}}{2} = 0.01\ \text{m/s}.$$

The Reynolds number is

$$\text{Re} = \frac{\rho U_m D}{\mu} = \frac{958.3 \times 0.01 \times 0.02}{2.82 \times 10^{-4}} = 679$$

where $\rho = 958.3$ kg/m^3 and $\mu = 2.82 \times 10^{-4}$ Pa s for water at 100°C.

The flow is laminar. The hydrodynamic development length,

$$L_{hy} = 0.05\text{Re}D = 0.68\ \text{m},$$

which is very small compared to the length of the pipe. Thus the assumption of fully developed laminar flow will introduce only a small error in calculation of the friction factor.

From Eq. (8.12), the friction factor is

$$f = \frac{16}{\text{Re}} = \frac{16}{679} = 0.0235.$$

The pressure drop is

$$\Delta p = \frac{4fL\rho U_m^2}{2D} = \frac{4 \times 0.0235 \times 20 \times 958.3 \times (0.01)^2}{2 \times 0.02} = 4.5\ \text{Pa}.$$

Example 8.41 What pressure head would be if the tube diameter in the above example is halved maintaining the same volume flow?

Solution

Effect of the halving of the tube diameter on U_m, Re and f are

$$U_m = \frac{4Q}{\pi D^2} \propto \frac{1}{D^2},$$

i.e. the mean velocity U_m is four times,

$$\text{Re} = \frac{\rho U_m D}{\mu} \propto U_m D \propto \frac{1}{D^2} \times D \propto \frac{1}{D},$$

i.e., Re is two times (still laminar),

$$f = \frac{16}{\text{Re}} \propto \frac{1}{U_m D} \propto D,$$

i.e., f is half of the previous value.

In terms of the variables of this problem,

$$\Delta p \propto \left(\frac{1}{D}\right)(D)\left(\frac{1}{D^2}\right)^2 = \frac{1}{D^4},$$

which means a 16-fold increase.

Example 8.42 Air at atmospheric pressure and 25°C enters a circular tube of 30 mm diameter at mass flow rate of 0.0025 kg/s. The heat transfer coefficient is estimated to be 50 W/(m^2 K). The surface heat flux at the tube wall is function of distance and specified as ax, where a = 700 W/m^2. Determine for 2 m length of the tube, the outlet temperature of air and tube surface temperature at x = 0 and x = L.

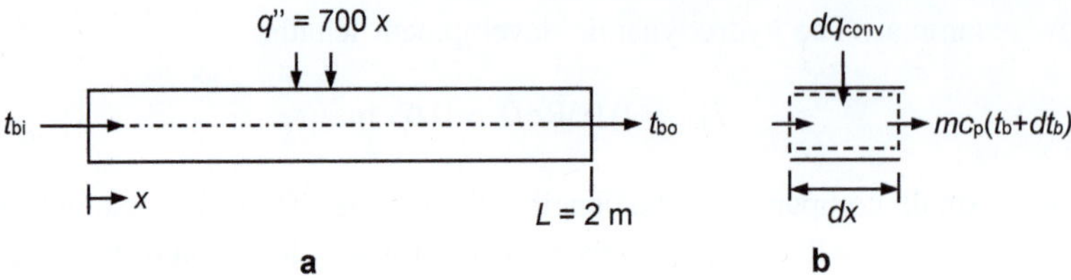

Fig. 8.18 Example 8.42

Solution

Specific heat c_p of air at the mean bulk temperature t_m [= (t_{bi} + t_{bo})/2] (assumed to be 50°C) is 1007.2 J/(kg K).

From energy balance at the control volume, Fig. 8.18b, we have

$$dq_{\text{conv}} = mc_p dt_b$$

where $dq_{\text{conv}} = q''Pdx = 700x(\pi D)dx$. Substitution gives

$$700x(\pi D)\text{d}x = mc_p \text{d}t_b$$

or

$$\text{d}t_b = \frac{700(\pi D)}{mc_p} x\text{d}x$$

Integration from x = 0 to x = L gives:

$$t_{bo} - t_{bi} = \frac{700(\pi D)}{mc_p} \int_0^L x dx$$

or

$$t_{bo} = t_{bi} + \frac{700(\pi D)}{mc_p} \left[\frac{x^2}{2}\right]_o^L = t_{bi} + \frac{700(\pi D)}{mc_p} \left(\frac{L^2}{2}\right)$$

Substitution of the values of various terms gives:

$$t_{bo} = 25 + \frac{700 \times (\pi \times 0.03)}{0.0025 \times 1007.2} \times \left(\frac{2^2}{2}\right) = 77.4°\text{C}.$$

This gives mean bulk temperature t_m [= (t_{bi} + t_{bo})/2] = 51.2°C, which is nearly equal to the assumed temperature hence retrial is not required.

At x = 0, q'' = 0, hence $t_{si} = t_{bi}$.

At x = L = 2 m, q'' = 700 L = 1400 W/m. Hence, from $q = hA(t_{so} - t_{bo})$,

$$t_{so} = t_{bo} + \frac{q}{hA} = t_{bo} + \frac{700L \times \pi D}{h(\pi D \times 1)} = 77.4 + \frac{700 \times 2}{50} = 105.4°\text{C}.$$

Example 8.43 Water at mean temperature of 60°C flows at mean velocity of 0.8 m/s through a 65 ID and 75 OD steel pipe [k = 40 W/(m K)]. Air at atmospheric pressure and 0°C flows across the pipe at a velocity of 5 m/s. Determine the heat loss from water per unit length of pipe.

Solution

Heat loss per unit length of the cylinder,

$$\frac{q}{L} = \frac{t_{\text{water}} - t_{\text{air}}}{\frac{1}{h_i \pi d_i} + \frac{1}{2\pi k_s} \ln\left(\frac{d_o}{d_i}\right) + \frac{1}{h_o \pi d_o}}$$

Thermophysical properties of water at 60°C from Table A4 are:

$$\rho = 983.3\ \text{kg/m}^3,\ \mu = 463 \times 10^{-6}\ \text{kg/(m s)},\ k = 0.654\ \text{W/(m K)}\ \text{and Pr} = 3.0.$$

The Reynolds number of water flowing through the pipe,

$$\begin{aligned} \text{Re} &= \frac{\rho U_m d_i}{\mu_w} \\ &= \frac{983.3 \times 0.8 \times 0.065}{463 \times 10^{-6}} = 110435. \end{aligned}$$

Flow is turbulent. The heat transfer coefficient can be determined from Dittus and Boelter equation (for cooling of fluid, Table 8.3):

$$\begin{aligned} h_i &= \frac{k}{d_i} \times 0.026 \text{Re}^{0.8}\, \text{Pr}^{0.3} \\ &= \frac{0.654}{0.065} \times 0.026 \times 110435^{0.8} \times 3^{0.3} = 3938\ \text{W/}\left(\text{m}^2\,\text{K}\right). \end{aligned}$$

Thermophysical properties of air at film temperature t_{fm} [= (t_s + 0)/2] assumed to be = 25°C from Table A5 are:

$$\rho = 1.1868\ \text{kg/m}^3,\ \mu = 1.8363 \times 10^{-5}\ \text{kg/(m s)},\ k = 0.02608\ \text{W/(m K)}\ \text{and Pr} = 0.709.$$

The Reynolds number of air flowing across the pipe,

$$\begin{aligned} \text{Re} &= \frac{\rho U_\infty d_o}{\mu_{air}} \\ &= \frac{1.1868 \times 5 \times 0.075}{1.8363 \times 10^{-5}} = 24236. \end{aligned}$$

For a cylinder in crossflow,

$$h = \frac{k}{d} Nu = \frac{k}{d} \mathrm{C Re}^n \, \mathrm{Pr}^{1/3} \tag{8.47}$$

From Table 8.11, $C = 0.193$ and $n = 0.618$. Hence, heat transfer coefficient,

$$\begin{aligned} h_o &= \frac{k}{d} \mathrm{C Re}^n \, \mathrm{Pr}^{1/3} \\ &= \frac{0.02608}{0.075} \times 0.193 \times 24236^{0.618} \times 0.709^{1/3} = 30.66 \ \mathrm{W}/\left(\mathrm{m}^2 \, \mathrm{K}\right). \end{aligned}$$

Hence, the heat loss per unit length of the cylinder,

$$\frac{q}{L} = \frac{60 - 0}{\frac{1}{3938 \times \pi \times 0.065} + \frac{1}{2 \times \pi \times 40} \ln\left(\frac{75}{65}\right) + \frac{1}{30.66 \times \pi \times 0.075}} = 427.84 \ \mathrm{W}.$$

Pipe outer surface temperature,

$$t_s = t_\infty + \frac{q/L}{\pi d_o h} = 0 + \frac{427.84}{\pi \times 0.075 \times 30.66} = 59.22 ^\circ \mathrm{C}.$$

Film temperature $t_{\text{-fm}}$ = (59.22 + 0)/2 = 29.6°C, which is slightly higher than the assumed values of 25°C. Retrial may be carried out for greater accuracy. However, here it will change the result marginally only.

Example 8.44 Air at 1 atm and 30°C enters a 4 m long 40 mm inside diameter smooth tube. The mass flow rate of the air is 0.05 kg/s. The tube surface temperature is maintained at 120°C by condensing steam surrounding the tube. Determine the outlet temperature of the air and pressure drop.

Solution

Air properties at the mean bulk temperature t_m [= (t_i + t_o)/2] of 50°C (assumed) from Table A5:

$$\rho = 1.0949 \ \mathrm{kg/m^3}, \ c = 1007.2 \ \mathrm{J/(kg \, K)}, \ \mu = 1.9512 \times 10^{-5} \mathrm{Ns/m^2}, \ k = 0.02799 \ \mathrm{W/(m \, K)} \text{ and } \mathrm{Pr} = 0.703.$$

For isothermal tube surface, the air outlet temperature is given by

$$\frac{t_w - t_o}{t_w - t_i} = \exp\left(-\frac{P}{mc} L\bar{h}\right) \tag{i}$$

The Reynolds number based on mass velocity of air through the tube:

$$\mathrm{Re} = \frac{md}{(\pi/4) d^2 \mu} = \frac{4m}{\pi d \mu} = \frac{4 \times 0.05}{\pi \times 0.04 \times 1.9512 \times 10^{-5}} = 81568.$$

Flow is turbulent. $L/D = 4/0.04 = 100$. Hence, the flow can be regarded as fully developed. Dittus and Boelter equation may be used for calculation of heat transfer coefficient.

$$\bar{h} = \frac{k}{d} \times 0.024\text{Re}^{0.8}\,\text{Pr}^{0.4}$$
$$= \frac{0.02799}{0.04} \times 0.024 \times 81568^{0.8} \times 0.703^{0.4} = 123.9\ \text{W}/\left(\text{m}^2\,\text{K}\right).$$

Substitution of values of various parameters in Eq. (i) gives

$$\frac{120 - t_o}{120 - 30} = \exp\left(-\frac{\pi \times 0.04}{0.05 \times 1007.2} \times 4 \times 123.9\right)$$

or

$$t_o = 93.8\,^\circ\text{C}.$$

Mean bulk temperature of water is t_m [= $(t_i + t_o)/2$] = 61.9°C. Retrial with changed air properties is required.

Air properties at the mean bulk temperature t_m of 62.5°C (assumed) from Table A5:

$$\rho = 1.05005\,\text{kg/m}^3,\ c = 1008\,\text{J}/(\text{kgK}),\ \mu = 2.0085 \times 10^{-5}\ \text{Ns/m}^2,\ k = 0.02895\ \text{W}/(\text{m K})\ \text{and Pr} = 0.7.$$

Revised Reynolds number through the tube:

$$\text{Re} = \frac{4m}{\pi d \mu} = \frac{4 \times 0.05}{\pi \times 0.04 \times 2.0085 \times 10^{-5}} = 79241.$$

Revised heat transfer coefficient.

$$\bar{h} = \frac{k}{d} \times 0.024\text{Re}^{0.8}\,\text{Pr}^{0.4}$$
$$= \frac{0.02895}{0.04} \times 0.024 \times 79241^{0.8} \times 0.7^{0.4} = 125\ \text{W}/\left(\text{m}^2\,\text{K}\right).$$

Substitution of values of various parameters in Eq. (i) gives

$$\frac{120 - t_o}{120 - 30} = \exp\left(-\frac{\pi \times 0.04}{0.05 \times 1008} \times 4 \times 125\right)$$

or

$$t_o = 94.1\,^\circ\text{C}.$$

Mean bulk temperature of water is t_m [= $(t_i + t_o)/2$] = 62.05°C. Retrial is not required.
From Blausius equation, friction factor is

$$f = 0.0791\mathrm{Re}^{-0.25} = 0.0791 \times 79241^{-0.25} = 0.0047.$$

Pressure drop,

$$\Delta p = \frac{4fL\rho U^2}{2d} = \frac{4fL\rho}{2d} \times \left(\frac{m}{\rho A_c}\right)^2$$

$$\Delta p = \frac{4 \times 0.0047 \times 4 \times 1.05005}{2 \times 0.04} \times \left[\frac{0.05}{1.05005 \times (\pi/4) \times 0.04^2}\right]^2$$

$$= 1417N/\mathrm{m}^2 = 0.014 \text{ atm.}$$

Example 8.45 Cold fluid is passing through a thin-walled tube 10 mm in diameter 1 m long which is exposed to crossflow of hot fluid at 100°C. Cold fluid flows at a rate of 0.0025 kg/s and its inlet and outlet temperatures are 30 and 60°C, respectively. Determine the outlet temperature of the cold fluid if its flow rate is increased by 50% with all other conditions remaining the same. Given that dynamic viscosity of the fluid is 0.004 N s/m^2. Flow may be assumed to be fully developed.

Solution

The fluid outlet temperature is given by

$$\frac{t_\infty - t_o}{t_\infty - t_i} = \exp\left(-\frac{P}{mc}LU\right),$$

which gives

$$\frac{U}{c} = \frac{m}{PL} \times \ln\left(\frac{t_\infty - t_i}{t_\infty - t_o}\right)$$

Substitution of values of various parameters gives the ratio of the overall heat transfer coefficient U and specific heat c as:

$$\frac{U}{c} = \frac{0.0025}{\pi \times 0.01 \times 1} \times \ln\left(\frac{100 - 30}{100 - 60}\right) = 0.0445,$$

where $U = \left(\frac{1}{h_i} + \frac{1}{h_o}\right)^{-1}$.

Flow Reynolds number is

$$\mathrm{Re} = \frac{4m}{\pi d\mu} = \frac{4 \times 0.0025}{\pi \times 0.01 \times 0.004} = 79.5.$$

The flow is laminar. For increase in mass flow rate by 50%, the Reynolds number will be 79.5 × 1.5 = 119. For fully developed laminar flow, the Nusselt number and hence the heat transfer coefficient h_i is independent of the Reynolds number. Since the hot fluid condition is unchanged, the overall heat transfer coefficient U will not change.

The fluid outlet temperature for increased mass flow rate is

$$
\begin{aligned}
t_o &= t_\infty - (t_\infty - t_i)\exp\left(-\frac{P}{mc}LU\right) \\
&= 100 - (100 - 30)\exp\left(-\frac{\pi \times 0.01}{1.5 \times 0.0025} \times 1 \times 0.0445\right) \\
&= 51.8°\text{C}.
\end{aligned}
$$

In the above analysis, it has been assumed that fluid properties c and μ are independent of temperature.

Example 8.46 Helium at 1 atm is to be heated from 400 K to 800 K while flowing at 0.005 kg/s through a tube of 20 mm diameter and 1 m length. Determine the uniform tube wall temperature required to heat the helium. Thermophysical properties of helium at the mean bulk temperature of 600 K and 1 atm are: ρ = 0.0818 kg/m^3, c = 5190 J/(kg K), $\mu = 32.2 \times 10^{-6}$ N s/m^2, k = 0.251 W/(m K) and Pr = 0.67.

Determine outlet temperature and required mass flow rate to achieve the same heat transfer rate and wall temperature if the air at 1 atm is used in place of helium.

Solution

(a) **Heating of helium**

Tube surface temperature can be determined from

$$
\frac{T_s - T_o}{T_s - T_i} = \exp\left(-\frac{P}{mc}L\bar{h}\right) \tag{i}
$$

Flow Reynolds number is

$$
\text{Re} = \frac{4m}{\pi d\mu} = \frac{4 \times 0.005}{\pi \times 0.02 \times 32.2 \times 10^{-6}} = 9885.
$$

Flow is turbulent. Assuming fully developed flow, the heat transfer coefficient from Dittus Boelter equation is

$$
\begin{aligned}
\bar{h} &= \frac{k}{d} \times 0.024\text{Re}^{0.8}\,\text{Pr}^{0.4} \\
&= \frac{0.251}{0.02} \times 0.024 \times 9885^{0.8} \times 0.67^{0.4} = 403\ \text{W}/\left(\text{m}^2\,\text{K}\right).
\end{aligned}
$$

Substituting values of various parameters in Eq. (i) gives

$$
\frac{T_s - 800}{T_s - 400} = \exp\left(-\frac{\pi \times 0.02}{0.005 \times 5190} \times 1 \times 403\right) = 0.377
$$

or

$$T_s = 1042 \text{ K}.$$

Heat transfer rate,

$$q = mc_p(T_o - T_i) = 0.005 \times 5190 \times (800 - 400) = 10.38 \text{ kW}.$$

(b) **Heating of air**
Air properties at the mean bulk temperature of 600 K (327°C) and 1 atm, from Table A5:

$$\rho = 0.5901 \text{ kg/m}^3,\ c = 1054.5 \text{ J/(kg K)},\ \mu = 3.011 \times 10^{-5} \text{ N s/m}^2,\ k = 0.04647 \text{ W/(m K) and Pr} = 0.68.$$

From first law equation,

$$q = mc_p(T_o - T_i)$$

or

$$T_o = \frac{q}{mc_p} + T_i = \frac{10380}{m \times 1054.5} + 400 = \frac{9.84}{m} + 400$$

Air outlet temperature can be determined from Eq. (i)
Flow Reynolds number is

$$\text{Re} = \frac{4m}{\pi d \mu} = \frac{4 \times m}{\pi \times 0.02 \times 3.011 \times 10^{-5}} = 2.11 \times 10^6 m.$$

Assuming the flow to be turbulent and fully developed, the heat transfer coefficient from Dittus Boelter equation is

$$\begin{aligned}\bar{h} &= \frac{k}{d} \times 0.024\text{Re}^{0.8}\,\text{Pr}^{0.4} \\ &= \frac{0.04647}{0.02} \times 0.024 \times (2.11 \times 10^6 m)^{0.8} \times 0.68^{0.4} = 5480m^{0.8}\end{aligned}$$

Substituting values of various parameters in Eq. (i) gives

$$\begin{aligned}\frac{1042 - (9.84/m + 400)}{1042 - 400} &= \exp\left(-\frac{\pi \times 0.02}{m \times 1054.5} \times 1 \times 5480m^{0.8}\right) \\ 1 - \frac{0.0153}{m} &= \exp\left(-\frac{0.3265}{m^{0.2}}\right)\end{aligned}$$

Trial and error solution gives m = 0.032 kg/s. Hence,

$$\mathrm{Re} = 2.11 \times 10^6 m = 2.11 \times 10^6 \times 0.032 = 67520.$$

$$T_o = \frac{9.84}{m} + 400 = \frac{9.84}{0.032} + 400 = 707.5\ \mathrm{K}.$$

For the same heat transfer rate, air mass flow rate is 6.4 times of that of helium.

Example 8.47 0.5 kg/s of air at 25°C is to be heated to 75°C. Available for the service are 40 mm diameter 3 m long tubes. The surface temperature of the tubes will be maintained at 100°C by condensing steam. Determine the number of tubes required for the service.

Solution

Air properties at the mean bulk temperature t_m [= $(t_i + t_o)/2$] of 50°C from Table A5 are:

$$\rho = 1.0949\ \mathrm{kg/m^3},\ c = 1007.2\ \mathrm{J/(kgK)},\ \mu = 1.9512 \times 10^5\ \mathrm{Ns/m^2},\ k = 0.02799\ \mathrm{W/(m\ K)}\ \text{and}\ \mathrm{Pr} = 0.703.$$

For isothermal tube surface, the air outlet temperature is given by

$$\frac{t_w - t_o}{t_w - t_i} = \exp\left(-\frac{P}{mc} L\bar{h}\right)$$

The equation can be rewritten as

$$\ln\left(\frac{t_w - t_i}{t_w - t_o}\right) = \frac{P}{mc} L\bar{h}$$

or

$$\ln\left(\frac{t_w - t_i}{t_w - t_o}\right) = \frac{\pi d}{mc} L\bar{h} \quad \text{(i)}$$

where heat transfer coefficient,

$$\bar{h} = \frac{k}{d}\mathrm{Nu}$$

or

$$\bar{h} = \frac{k}{d} \times 0.024\mathrm{Re}^{0.8}\,\mathrm{Pr}^{0.4}$$

using Dittus Boelter equation for fully developed turbulent flow (assumption).

The Reynolds number based on mass velocity of air through the tube:

$$\mathrm{Re} = \frac{md}{(\pi/4)d^2\mu} = \frac{4m}{\pi d\mu}$$

Hence,

$$\bar{h} = \frac{k}{d} \times 0.024 \times \left(\frac{4m}{\pi d \mu}\right)^{0.8} \text{Pr}^{0.4}$$

Substitution in Eq. (i) gives

$$\ln\left(\frac{t_w - t_i}{t_w - t_o}\right) = \frac{\pi d}{mc} \times L \times \frac{k}{d} \times 0.024 \times \left(\frac{4m}{\pi d \mu}\right)^{0.8} \text{Pr}^{0.4}$$

or

$$m^{0.2} = \frac{1}{\ln\left(\frac{t_w - t_i}{t_w - t_o}\right)} \frac{\pi k L}{c} \times 0.024 \times \left(\frac{4}{\pi d \mu}\right)^{0.8} \text{Pr}^{0.4}$$

Substituting values of various terms, we have

$$m^{0.2} = \frac{1}{\ln\left(\frac{100-25}{100-75}\right)} \times \frac{\pi \times 0.02799 \times 3}{1007.2} \times 0.024 \times \left(\frac{4}{\pi \times 0.04 \times 1.9512 \times 10^{-5}}\right)^{0.8} \times (0.703)^{0.4}$$
$$= 0.4638$$

or

$$m = 0.0215 \text{ kg/s}.$$

Reynolds number

$$\text{Re} = \frac{4m}{\pi d \mu} = \frac{4 \times 0.0215}{\pi \times 0.04 \times 1.9512 \times 10^{-5}} = 35074.$$

Length to diameter ratio $L/d = 3/0.04 = 75$. Hence, assumption of fully developed turbulent flow is correct.

For total mass flow rate of 0.5 kg/s, the required number of tubes is

$$n = \frac{\text{Total flow}}{\text{Flow per tube}} = \frac{0.5}{0.0215} = 23.25.$$

Hence, 24 tubes will be used.

Example 8.48 Water, while flowing at 15 ml/s through a rectangular cross-Section (20 mm × 5 mm) tube, is to be heated from 20°C to 50°C. Tube surface is maintained at 60°C. Determine the length of tubing.

Solution

Water properties at the mean bulk temperature t_m [= $(t_i + t_o)/2$] of 35°C from Table A4 are:

$$\rho = 994\,\text{kg/m}^3,\ c = 4178\ \text{J/(kg K)},\ \mu = 718 \times 10^{-6}\ \text{N s/m}^2,\ k = 0.624\ \text{W/(m K) and Pr} = 4.81.$$

For isothermal tube surface, the air outlet temperature is given by

$$\frac{t_w - t_o}{t_w - t_i} = \exp\left(-\frac{P}{mc}L\bar{h}\right)$$

The equation can be rewritten as

$$L = \ln\left(\frac{t_w - t_i}{t_w - t_o}\right)\frac{mc}{P\bar{h}} \tag{i}$$

Mass flow rate,

$$m = 15 \times 10^{-6} \times 994 = 0.0149\ \text{kg/s}.$$

For rectangular cross-section tubing, hydraulic diameter is

$$d_h = \frac{4WH}{2(W+H)} = \frac{4 \times 20 \times 5}{2 \times (20+5)} \times \frac{1}{1000} = 0.008\ \text{m}.$$

Reynolds number of flow,

$$\text{Re} = \frac{md_h}{(W \times H)\mu} = \frac{0.0149 \times 0.008}{(20 \times 5) \times 10^{-6} \times 718 \times 10^{-6}} = 1660.$$

Flow is laminar. Assuming flow to be fully developed, the fully developed Nusselt number Nu_T for the case of uniform temperature at four walls of a rectangular duct is approximated by the following equation:

$$\text{Nu}_T = 7.541(1 - 2.610\alpha^* + 4.970\alpha^{*2} - 5.119\alpha^{*3} + 2.702\alpha^{*4} + 0.548\alpha^{*5}) \tag{8.40}$$

where α^* is aspect ratio (ratio of duct height to width).

Hence, for $\alpha^* = 5/20 = 0.25$,

$$\text{Nu}_T = 7.541(1 - 2.610 \times 0.25 + 4.970 \times 0.25^2 - 5.119 \times 0.25^3 + 2.702 \times 0.25^4 + 0.548 \times 0.25^5) = 4.44.$$

Heat transfer coefficient,

$$\bar{h} = \frac{k}{d_h} \times \text{Nu}_T = \frac{0.624}{0.008} \times 4.44 = 346.3\ \text{W/}\left(\text{m}^2\,\text{K}\right).$$

Substitution in Eq. (i) gives

$$L = \ln\left(\frac{60-20}{60-50}\right) \times \frac{0.0149 \times 4178}{[2(20+5)/1000] \times 346.3} = 4.984 \text{ m}.$$

Length to hydraulic diameter ratio, L/D_h = 4.984/0.008 = 623. Hence, the assumption of fully developed flow is valid.

Example 8.49 A 0.4 m diameter 100 m long thin walled pipeline transports hot water. Water enters the pipe at 80°C. Water flow rate is 5 kg/s. In order to reduce the heat loss from the pipeline, it is covered with 50 mm thick insulation [k_i = 0.05 W/(m K)] and is buried 2.0 m below the Earth's surface [k_e = 1.0 W/(m K)]. If the earth's surface temperature is 0° C, determine the outlet temperature of the water and heat loss.

Solution

For a differential control volume in the water, refer Fig. 8.19b,

$$dq = mcdt_m = (t_s - t_m)/R_t$$

or

$$\frac{dt_m}{(t_s - t_m)} = \frac{1}{mcR_t}$$

where total resistance to heat flow

$$R_t = \frac{1}{\pi D_i dx h_i} + \frac{1}{2\pi k_i dx}\ln\left(\frac{D_o}{D_i}\right) + \frac{1}{k_e S}$$

where S is shape factor. From Case 8, Table 5.2 ($L >> R$, $Z > 3R_o$),

$$S = \frac{2\pi dx}{\ln\left(\frac{2Z}{R_o}\right)} = \frac{2\pi}{\ln\left(\frac{2\times 2}{0.25}\right)}dx = 2.27dx$$

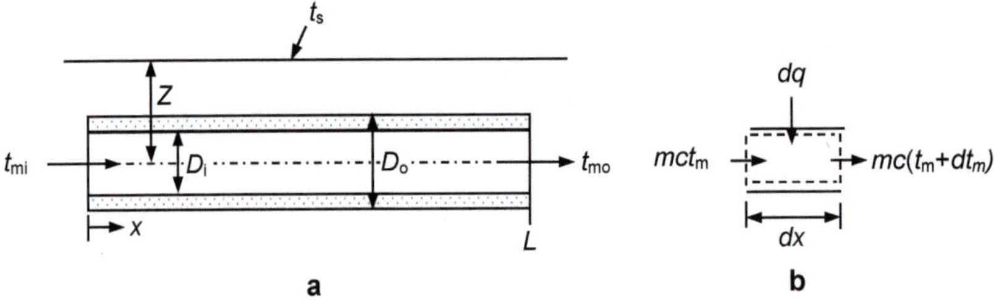

Fig. 8.19 Example 8.49

We can write

$$R_t = \left[\frac{1}{\pi D_i h_i} + \frac{1}{2\pi k_i}\ln\left(\frac{D_o}{D_i}\right) + \frac{1}{k_e \times 2.27}\right]/dx = \frac{R_t'}{dx}$$

Hence,

$$\frac{dt_m}{(t_m - t_s)} = -\frac{dx}{mcR_t'}$$

Integration for length L gives

$$\ln\left(\frac{t_s - t_{mo}}{t_s - t_{mi}}\right) = -\frac{L}{mcR_t'}$$

or

$$t_{mo} = t_s - (t_s - t_{mi})\exp\left(-\frac{L}{mcR_t'}\right) \tag{i}$$

The thermophysical properties of water at 80°C are:

$$\rho = 971.8\ \text{kg/m}^3,\ c = 4198\ \text{J/(kg K)},\ \mu = 351 \times 10^{-6}\ \text{kg/(m s)},$$
$$k = 0.670\ \text{W/(m K) and Pr} = 2.23.$$

The Reynolds number of flow in the pipe,

$$\text{Re} = \frac{\rho U_m D_i}{\mu} = \frac{m}{(\pi/4)D_i^2}\frac{D_i}{\mu} = \frac{4m}{\pi D_i \mu} = \frac{4 \times 5}{\pi \times 0.4 \times 351 \times 10^{-6}} = 45343.$$

Flow is turbulent. Length to diameter ratio of the pipe is 50/0.4 = 125. Flow can be assumed to be fully developed. From Dittus-Boelter relation, heat transfer coefficient is

$$h_i = \frac{\text{Nu}k}{D_i} = \frac{k}{D_i} \times 0.024\text{Re}^{0.8}\,\text{Pr}^{0.4} = \frac{0.67}{0.4} \times 0.024 \times 45343^{0.8} \times 2.23^{0.4}$$
$$= 294.3\ \text{W}/\left(\text{m}^2\,\text{K}\right).$$

Hence,

$$R_t' = \frac{1}{\pi D_i h_i} + \frac{1}{2\pi k_i}\ln\left(\frac{D_o}{D_i}\right) + \frac{1}{k_e \times 2.27}$$
$$= \frac{1}{\pi \times 0.4 \times 294.3} + \frac{1}{2\pi \times 0.05}\ln\left(\frac{0.5}{0.4}\right) + \frac{1}{1.0 \times 2.27}$$
$$= 1.154\ \text{K m/W}.$$

Substitution in Eq. (i) gives

$$t_{mo} = t_s - (t_s - t_{mi}) \exp\left(-\frac{L}{mcR'_t}\right)$$
$$= 0 - (0 - 80) \exp\left(-\frac{100}{5 \times 4198 \times 1.154}\right) = 79.67°\text{C}.$$

Heat loss,

$$q = mc(t_{mi} - t_{mo}) = 5 \times 4198 \times (80 - 79.67) = 6927 \text{ W}.$$

8.13 Summary

There are a large number of convection problems for which the analytical solutions for determination of heat transfer coefficient have not met the success especially the problems involving turbulent flow or flows where detachment occurs, for example around cylinders, spheres or other curved bodies. Thus the direct measurement of the heat transfer coefficient (experimental study) has been the main approach for the solution of the most of the heat transfer problems. The experimental results are presented in the form of generalized correlations using the method of dimensional analysis termed as empirical relations because they rely on the observations and experiments not on theory. Based on the dimensional analysis, the generalized correlation for forced convection is written as Nu = ψ(Re) ϕ(Pr). The form of functions ψ(Re) and ϕ(Pr) are specified for different conditions of heat transfer by convection on the basis of theoretical analysis or experimental investigations. Physical significance of the dimensionless numbers has also been explained in Sect. 8.3.3. Experimental scheme for the determination of forced convection heat transfer coefficient for uniform temperature and uniform heat flux conditions has been discussed in Sect. 8.4.

Correlations of friction factor and heat transfer have been presented for tubes, annuli, rectangular and parallel plate ducts, flat plates, submerged bodies and tube banks for different boundary conditions.

Fluid flow and heat transfer characteristics of circular cross-section smooth ducts (tubes) have been investigated in a greater detail as this geometry is having wide applications. Both laminar and turbulent flows in circular tubes have been divided into four categories: fully developed, hydrodynamically developing, hydrodynamically developed but thermally developing, and both hydrodynamically and thermally developing (i.e., simultaneously developing).

The fully developed laminar flow of a constant property fluid in a smooth circular tube has been studied analytically as presented in Chap. 7. The Fanning friction factor correlation is f = 16/Re. The hydrodynamic development length L_{hy} in laminar flow, the distance required for the friction factor to decrease to within 5% of its fully developed value, is given by $L_{\text{hy}}/D \approx 0.05\text{Re}$. Friction factor in the hydrodynamic entrance region is higher than that for the fully developed case and is defined as apparent mean friction factor $\bar{f}_{app}$, refer Fig. 8.2.

For the laminar flow with fully developed velocity and temperature profiles i.e., hydrodynamically and thermally fully developed flow, the Nusselt number is a function of the type of heating boundary condition. From the analytical results as given in Chap. 7, Nu = 4.364 for constant heat rate (uniform heat flux) and 3.658 for constant surface temperature condition.

The thermal entrance length (temperature profile developing and fully developed velocity profile) L_{th} in laminar flow required for the Nusselt number to decrease to within 5% of its fully developed value is given by $L_{hy}/D \approx 0.05\text{RePr}$. For the constant surface temperature (known as Graetz problem), the local Nusselt number Nu_x and the mean Nusselt number Nu_m in the terms of non-dimensional tube length $x^+ = (x/R)/(\text{RePr})$ are given in Fig. 8.3. Dimensionless group RePr(D/x) is known as Graetz number Gz. For the constant heat rate (uniform heat flux), the values of the local Nusselt number are given in Fig. 8.4. Empirical relation of average Nusselt number over the entire length of the tube for fully developed laminar flow in tubes at constant wall temperature is given in Eq. (8.15).

If the velocity and temperature (thermal) profiles develop simultaneously, the resulting Nusselt numbers in the entry region are always higher than the preceding case. Constant surface temperature and constant heat rate results are presented in Fig. 8.5a, b, respectively, in the form of mean Nusselt number Nu_m or local Nusselt number Nu_x as a function of x^+.

For constant surface temperature condition, Sieder and Tate equation may be used for both liquids and gases in the laminar region where the thermal and velocity profiles are developing simultaneously. The equation must be used only for entrance region.

It is to be noted that for the turbulent duct flow, the hydrodynamic and thermal entrance lengths are much shorter than the corresponding lengths in the laminar flow. Hence, the results of fully developed turbulent flow friction factor and heat transfer are frequently used in design calculations neglecting the effect of the entrance regions. However, for low Prandtl number fluids, and for the heat exchangers short in length, the entrance region effects must be considered.

Several experimental friction factor correlations have been developed for fully developed turbulent flow in smooth ducts. Some of these correlations are presented in Table 8.2, which must be critically studied for application. In comparison to the laminar flow, the hydraulic entrance length is much shorter for turbulent flow and its dependence on the Reynolds number is weaker. In many pipe flow of practical interest, the entrance effects are not significant beyond a pipe length of 10 diameters.

For fully developed velocity and temperature profiles, two boundary conditions of interest are uniform heat flux or constant heat rate along tube length (UHF) termed as H boundary condition and a uniform wall temperature (UWT) termed as T boundary condition. The constant heat rate Nusselt number NU_H is always greater than the constant surface temperature Nusselt number Nu_T, but with exception of very low Prandtl number fluids, the difference in N_H and Nu_T is much smaller than for the laminar flow. At $\text{Pr} = 0.7$, it is only a few percent.

A large number of Nusselt number correlations, both theoretical and empirical (based on experimental data), have been developed for fully developed turbulent flow in smooth tubes. Some of them are presented in Table 8.3. Remarks in third column of the table are worth consideration before selecting a correlation for use. Researchers have also presented correlations covering laminar, transition and turbulent flow regimes, which are, in general, difficult to use.

When the velocity and temperature profiles develop simultaneously in turbulent flow, entry region correction is required, which is strongly affected by the shape at the inlet. If $L/D > 60$, entry region correction is not required.

Effects of temperature varying properties on heat transfer and friction factor values have been discussed and the empirical relations have been presented in Sect. 8.6. When the heat transfer takes place, the temperature of the fluid varies both over the cross-section and along the tube length. Since the transport properties of most fluids vary with temperature and hence will vary over the cross-section of the tube. This causes distortion of the laminar velocity profile due to the heating or cooling of fluids. In the case of the turbulent flow, the effect of the variation of viscosity at the wall is mainly confined to the laminar sub-layer. The laminar sub-layer tends to be thinner for a decrease in the viscosity at the wall due to the heating of liquids or cooling of gases. The reverse is true when the liquid is being cooled or the gas is being heated.

The method to take account of the effects of the variation of the fluid properties with the temperature is to evaluate all the properties at the bulk temperature, and then the variable properties effects on Nusselt number and friction factor are lumped into functions of ratio of the wall to bulk temperatures $(T_w/T_b)^n$ and $(T_w/T_b)^m$, respectively, for gases or viscosities $(\mu_w/\mu_b)^n$ and $(\mu_w/\mu_b)^m$, respectively, for liquids. The exponents m and n are functions of the geometry and type of flow. It is to note that information on the flow through noncircular tubes with variable properties is not complete and hence it is recommended that the circular tube results be used.

It has been found that nearly the same turbulence intensity and the friction factor prevail in circular and other duct geometries (such as annular, rectangular, square, triangular and irregular passages) if the ratio of flow-passage area to the wetted parameter is kept constant. This ratio is called the hydraulic radius $R_h = A/P$. However, it has been found convenient to use the term equivalent or hydraulic diameter $D_h = 4A/P$. It has been reported that the use of hydraulic diameter for ducts with sharp corners (e.g., triangular ducts) may lead to the error of the order of 35% in turbulent flow friction and heat transfer coefficients determined from circular duct correlations.

Fully developed Fanning friction factor for laminar flow in concentric annular duct results are presented in Table 8.6 as function of radius ratio.

Fully developed laminar flow Nusselt number in concentric annular duct is independent of the Reynolds number and Prandtl number. There are two Nusselt numbers of interest, one for inner surface (when inner surface alone is heated) and other for the outer surface (when outer surface alone is heated). Their values for constant heat rate condition are given in Table 8.7. For the case of constant surface temperature on one surface (with the other surface insulated), the Nusselt number data are presented in Table 8.8.

In the case of parallel plates, Nu = 7.541 when both surfaces are at the same constant temperature, while it is 8.235 for identical heat rates on the two surfaces of the duct for laminar flow. Average Nusselt number relation for the thermal entrance region flow between isothermal parallel plates for Re < 2800 has been presented as Eq. (8.27). For parallel plates hydraulic diameter is taken as twice the spacing between the plates.

Discussion of various correlations and data of friction factor and Nusselt number for turbulent flow in circular tube annuli are presented in Sect. 8.7.2. The friction factors, as determined with water, were found to be 6–8% higher than the generally accepted values of smooth circular tube for 4000 < Re < 17000. The same for air flow were found to be about 1–10% higher than circular tube flow values. Approximate correlation of Kays and Perkins, $f = 0.085\ \mathrm{Re}^{-0.25}$ for 6000 < Re < 300,000, which gives the friction factor in circular annuli about 10% higher than pipe flow values and having little dependence on the radius ratio may be used. It is reported that the annular flow develops more quickly than pipe flow. The development length is reported to be 20–30 diameters for a close approach to the fully developed flow.

Equation (8.32), $f = 0.1268\ \mathrm{Re}^{-0.3}$ for $5000 < \mathrm{Re} < 1.2\ 10^6$, and Eq. (8.33), $f = 0.0868\ \mathrm{Re}^{-0.25}$ for $1.2 \times 10^4 < \mathrm{Re} < 1.2 \times 10^6$, are recommended to obtain the friction factor for the fully developed turbulent flow in the parallel plate duct.

Fully developed constant heat rate turbulent flow Nusselt number results for Pr = 0.7 as function of radius ratio of the annular duct are presented in Table 8.9 The solutions are presented for one surface heated and other insulated.

It is recommended that the Nusselt number for turbulent flow between parallel plates can be determined using the circular duct correlations with hydraulic diameter in Nusselt number and Reynolds number as twice the spacing between the plates.

Correlations for laminar flow friction factor and fully developed Nusselt numbers Nu_T for the case of uniform temperature at four walls and uniform heat flux at four walls constant along the length of the duct, but uniform temperature around the periphery Nu_H as function of duct aspect ratio (ratio of duct height to width) of rectangular duct have been presented in Sect. 8.8. Some typical results for the friction factor, and constant heat rate and surface temperature Nusselt numbers are listed in Table 8.10 along with circular and parallel plate ducts for comparison.

The entrance configuration (abrupt or smooth) exerts a marked influence on the value of critical Reynolds number Re_c for flow in smooth rectangular duct. Another unique feature of the fully developed turbulent flow in rectangular ducts is the presence of secondary flow (flow normal to the axis of the duct), which exerts a significant effect on the turbulence fluid flow characteristics of these ducts and increases the friction factor by approximately 10%.

With uniform heating at four walls of a rectangular duct, circular duct Nusselt number correlations provide results with ± 9% accuracy while for the duct with equal heating at two long walls, circular duct correlations provide result with accuracy of ± 10% for $0.5 \le Pr \le 10$ and $10^4 \le Re \le 10^5$. In the case of heating at one long wall only (asymmetrical heating), the circular duct correlations provide values which may be up to 20% higher than the actual experimental values for $0.7 \le Pr \le 2.5$ and $10^4 \le Re \le 10^6$.

Friction factor and Nusselt number correlations for laminar and turbulent external forced flow over a flat plate have been developed in Chap. 7 and are also presented in this chapter in Sect. 8.9.

Forced laminar and turbulent flows across cylinders and spheres have been discussed in Sect. 8.10. Plots of drag coefficient C_D versus the Reynolds number for cylinder and sphere in cross flow have been presented in Fig. 8.10. Nusselt number correlations for heat transfer from cylinders have been presented in Eqs. (8.47)–(8.52). Correlation for the average Nusselt number between an isothermal spherical surface and free stream fluid is presented in Eq. (8.55).

The flow across tube banks, the tube array or bundle is frequently encountered in heat exchangers. The in-line and staggered arrangements of the tubes are frequently used. For both the arrangements, the flow across the first transverse row of the tubes is practically the same as that for a single tube. The nature of the flow across the tubes of remaining rows depends on the way the tubes are arranged. However, in most applications, the flow is turbulent after a few rows. Correlations for the average Nusselt number for number of rows $n \ge 16$ are presented in Eqs. ((8.58a), (8.58b), (8.58c), (8.58d) and (8.59a), (8.59b), (8.59c), (8.59d) for in-line and staggered arrangements, respectively. For $n < 16$, the heat transfer coefficient is lower due to the lower level of turbulence and the Nusselt number values from Eqs. (8.58a), (8.58b), (8.58c), (8.58d) and (8.59a), (8.59b), (8.59c), (8.59d) is multiplied by a correction factor.

Typical variation of the Nusselt number with the Peclet number Pe (= RePr) for flow of liquid metal through a tube with uniform heat flux is presented in Fig. 8.14. The Nusselt number is having a value of 4.36 for the laminar flow in tubes. The transition from the laminar to the turbulent flow occurs at critical value of Pe $\approx$ 40. At moderate value of the Peclet number beyond the critical value, the heat transfer by the turbulent mixing is small compared with the conductive transport. When the Peclet number is greater than 100, the heat exchange by turbulent mixing becomes appreciable and cannot be neglected. Some of the Nusselt number correlations for fully developed turbulent flow of liquid metals are given in Table 8.13. Although the Nusselt number tends to be low for the liquid metal , the heat transfer coefficient is very high because of the high thermal conductivity of the liquid metals.

Roughness is encountered in commercial pipes and artificial roughness, which creates turbulence close to the wall, has been used to enhance heat transfer. Nikuradse carried out experiments to study the relationship of friction factor to the Reynolds number for pipes of various roughnesses. At low Reynolds number (within the first region termed as hydraulically smooth regime), the roughness has no effect on the resistance. Nikuradse noticed that the critical Reynolds number for all degrees of relative roughness occurs between 2160 and 2500. With an increasing Reynolds number, if the surface roughness height e is of the order of the magnitude of the laminar sublayer thickness δ_l, it tends to break up the laminar sublayer. This increases the wall shear stress and the ratio of the surface roughness height e and laminar sublayer thickness δ_l determines the effect of the roughness. The roughness Reynolds number $e^+ = \sqrt{(f/2)}\text{Re}(e/D)$ has been used to define three flow regimes in roughened ducts, namely (i) hydraulically smooth ($0 \le e^+ \le 5$), (ii) transition regime ($5 < e^+ \le 70$) and (iii) fully rough regime ($e^+ > 70$). Moody determined the equivalent sand-grain roughness for different types of commercial pipe surfaces and also presented plot of friction factor versus the Reynolds number as function of relative pipe roughness e/D. Moody's plot may be used to determine friction factor for flow in tubes.

Several studies on heat transfer coefficient behaviour for different roughness types have been carried out by researchers and correlations have been proposed. Some of them are presented in Table 8.14 for fully developed turbulent flow in the fully rough flow regime of a circular duct.

It is to note that in the case of fluids with high Prandtl number, the resistance of the heat transfer is mainly concentrated in the laminar sublayer, which is thin compared to the hydrodynamic boundary layer. For the low Prandtl number fluids, the thermal resistance is distributed over a larger portion of the duct cross-section because the thermal boundary layer is thicker than the hydrodynamic boundary layer. Since the roughness at the wall creates turbulence near the wall, the heat transfer enhancement due to artificial roughness is greater for the fluids with high Prandtl numbers.

The experimental results of heat transfer and fluid flow characteristics in roughened ducts have been presented either in the form of direct dependence of frition factor and Nusselt number on the system and operating parameters or in the form of interrelated roughness and heat transfer functions. This method makes it possible to present results in a most general form taking into account the various parameters involved including the roughness parameters.

Looking to the importance of this chapter in the design of heat exchangers, a significant number of numerical problems have been included in the chapter to illustrate application of various correlations presented in the chapter.

Review Questions

8.1 Explain the principle of the dimensional analysis. What are the limitations of the dimensional analysis?

8.2 What are fundamental and derived dimensions?

8.3 Express the variables in Table 8.1 in terms of M-L-T-θ and M-L-T-θ-Q systems of fundamental dimensions.

8.4 List the variables that affect the forced convection heat transfer coefficient.

8.5 Using the technique of dimensional analysis establish the following relation for forced convection heat transfer.

$$\mathrm{Nu} = \psi(\mathrm{Re})\phi(\mathrm{Pr}).$$

8.5 With the help of Buckingham pi theorem, show that for the forced convection

$$\mathrm{Nu} = f(\mathrm{Re}, \mathrm{Pr}).$$

8.6 At low flow velocities, the free convection effect may be present in forced convection heat transfer case. Using dimensional analysis, show that the following form of correlation is obtained.

$$f(\mathrm{Nu}, \mathrm{Re}, \mathrm{Pr}, \mathrm{Gr}) = 0.$$

8.7 Derive the relationship (for forced convection)

$$\mathrm{Nu} = f(\mathrm{Re}, \mathrm{Pr}, \mathrm{Ec}).$$

8.8 What is the physical interpretation of the following non-dimensional numbers?

$$\mathrm{Re}, \mathrm{Pr}, \mathrm{Nu}, \mathrm{Pe}, \mathrm{St}, \mathrm{Ec}.$$

8.9 Define hydraulic diameter.

8.10 Discuss the effect of the heating or cooling of a fluid on velocity distribution under laminar flow condition in a tube. How does this affect the heat transfer and friction factor and how this effect is taken into account?

8.11 Enlist various heat transfer-coefficient and friction-factor correlations developed for laminar and turbulent flows in circular tubes. Discuss them.

8.12 What are the effects of thermal and hydrodynamic entry lengths on the heat transfer coefficient and friction factor in both laminar and turbulent tube flows? How is this effect taken into account?

8.13 Obtain a relation between the convective heat transfer coefficient and the friction factor when Pr = 1.

Problems

8.1 What is the effect of the following on the average value of the heat transfer coefficient in the case of fully developed turbulent flow in tubes?

(a) Two fold increase in the fluid mass flow rate; all other parameters remaining the same.

(b) Two fold increase in the tube diameter; all other parameters including the flow velocity are the same.

[**Ans.** For fully developed turbulent flow in tubes, Dittus Boelter relation may be used, which gives $\text{Nu} \propto \text{Re}^{0.8} = \left(\frac{\rho U_m d}{\mu}\right)^{0.8}$; $h = \text{Nu}\frac{k}{d} \propto U_m^{0.8} d^{-0.2}$; Two fold increase in velocity will increase h $\left(h \propto U_m^{0.8}\right)$ by $2^{0.8}$ = 1.741 times. A two fold increase in the diameter will reduce h $\left(h \propto \frac{1}{d^{0.2}}\right)$ by $2^{0.2}$ i.e., 14.86%.]

8.2 Water flows through a long electrically heated smooth tube of 30 mm diameter. If velocity and temperature profiles are fully developed at a location $x = a$, determine the heat transfer coefficient and heat flux at the location. The mass flow rate of water is 0.5 kg/s. At $x = a$, water temperature is 30°C and wall temperature is 40°C.
[**Ans.** Thermo-hydraulic properties of water at bulk temperature 30°C are: ρ = 995.6 kg/m^3, k = 0.617 W/(m K), $\mu = 7.97 \times 10^{-4}$ N s/m^2, Pr = 5.4; $\text{Re} = \frac{\rho U_m d}{\mu} = \frac{4m}{\pi d \mu} =$ 26625; Turbulent flow; $h = Nu\frac{k}{d} = 0.024\text{Re}^{0.8}\,\text{Pr}^{0.4}\frac{k}{d} = 3362$ W/(m^{2}K); $\frac{q}{A} = h(t_w - t_b) = 33.62$ kW/m^2.]

8.3 One kg/s of water at 35°C flows through a 25 mm diameter tube whose surface is maintained at a uniform temperature of 100°C. Determine the required length of the tube for 65°C water outlet temperature.
[**Ans.** Water properties at the mean bulk temperature $t_m = (t_i + t_o)/2$ = 50°C from Table A4:
ρ =988.1 kg/m^3, c = 4182 J/(kg K), $\mu = 544 \times 10^{-6}$ N s/m^2, k = 0.644 W/(m K) and Pr = 3.55. $\text{Re} = \frac{md}{(\pi/4)d^2\mu} = \frac{4m}{\pi d\mu} = \frac{4\times 1}{\pi\times 0.025\times 544\times 10^{-6}} = 93620$; From Dittus and Boelter equation, $\bar{h} = \frac{k}{d} \times 0.024\text{Re}^{0.8}\,\text{Pr}^{0.4} = \frac{0.644}{0.025} \times 0.024 \times 93620^{0.8} \times 3.55^{0.4} = 9735$ W/(m^2 K); From $\frac{t_w - t_o}{t_w - t_i} = \exp\left(-\frac{P}{mc}L\bar{h}\right)$, $L = -\frac{mc}{P\bar{h}}\ln\left(\frac{t_w - t_o}{t_w - t_i}\right) = -\frac{1\times 4282}{\pi\times 0.025\times 9735}\ln\left(\frac{100-65}{100-35}\right) =$ 3.46m.]

8.4 Cold fluid is passing through a thin-walled tube 10 mm in diameter 2 m long whose surface is maintained at 100°C. The cold fluid flows at a rate of 0.05 kg/s and its inlet and outlet temperatures are 30 and 60°C, respectively. Determine the outlet temperature of the cold fluid if its flow rate is increased by 50% with all other conditions remaining the same. Given that dynamic viscosity of the fluid is 0.0004 N s/m^2. Flow may be assumed to be fully developed.
[**Ans.** $\frac{t_w - t_o}{t_w - t_i} = \exp\left(-\frac{P}{mc}L\bar{h}\right)$ gives $\frac{\bar{h}}{c} = \frac{m}{PL} \times \ln\left(\frac{t_w - t_i}{t_w - t_o}\right) = \frac{0.05}{\pi\times 0.01\times 2} \times \ln\left(\frac{100-30}{100-60}\right) = 0.445$; Reynolds number $\text{Re} = \frac{4m}{\pi d\mu} = \frac{4\times 0.05}{\pi\times 0.01\times 0.0004} = 15915$. The flow is turbulent. For increase in mass flow rate by 50%, the Reynolds number will increase by 50%. Since $h \propto \text{Re}^{0.8}$

from Dittus-Boelter relation for fully developed turbulent flow, new $\bar{h}/c \approx 0.445 \times 1.5^{0.8} = 0.616$. The fluid outlet temperature for increased mass flow rate $t_o = t_w - (t_w - t_i)\exp\left[-\frac{PL}{m}\left(\frac{\bar{h}}{c}\right)\right] \approx 100 - (100 - 30)\exp\left(-\frac{\pi \times 0.01 \times 2}{1.5 \times 0.05} \times 0.616\right) = 58.2°\text{C}$.]

8.5 Air at 1 atmospheric pressure and 40°C is heated while it passes at a velocity of 8 m/s through a tube 1 m long and 40 mm in diameter whose surface is maintained at 120°C. Determine the outlet temperature of the air.

[**Ans.** Mean bulk temperature, $t_m = \frac{t_i + t_o}{2} = \frac{40 + t_o}{2}$; Assuming a trial value of t_m = 50°C, the thermophysical properties of the air are: ρ = 1.0949 kg/m^3, k = 0.02799 W/(m K), $\mu = 1.9512 \times 10^{-5}$ kg/(m s), c_p = 1007.2 J/(kg K) and Pr = 0.703; $\text{Re} = \frac{\rho U_m D}{\mu} =$ 17957; Flow is turbulent. Dittus-Boelter equation gives $\text{Nu} = 0.024\text{Re}^{0.8}\text{Pr}^{0.4}$ = 52.77; $\bar{h} = Nu\frac{k}{D} = 36.93$ W/(m^{2}K); $m = \rho\frac{\pi}{4}D^2U_m = 0.011$ kg/s; outlet temperature $t_o = t_w - (t_w - t_i)\exp\left[-\frac{PL}{m}\left(\frac{\bar{h}}{c}\right)\right] = 120 - (120 - 40)\exp\left(-\frac{\pi \times 0.04 \times 1}{0.011} \times \frac{36.93}{1007.2}\right) = 67.38°\text{C}$; Revised mean temperature t_m = 53.69°C, retrial with this estimate of t_m may be carried out.]

8.6 Air at atmospheric pressure and 25°C mean bulk temperature flows through a rectangular duct (height H = 400 mm and width W = 800 mm) with a mean velocity of 5 m/s. The duct is at an average temperature of 40°C. Determine the heat loss per unit length of the duct.

[**Ans.** At 25°C bulk temperature, thermophysical properties of the air are: ρ = 1.1868 kg/m^3, k = 0.02608 W/(m K), $\mu = 1.8363 \times 10^{-5}$ N s/m^2, Pr = 0.709; $D_h = \frac{4WH}{2(W+H)} = 0.53$ m; $\text{Re} = \frac{\rho U_m D_h}{\mu} = 1.71 \times 10^5$;$\text{Nu} = 0.024\text{Re}^{0.8}\,\text{Pr}^{0.4} = 321.3$; $h = \text{Nu}\frac{k}{D_h} = 15.81$ W/(m^2 K); $\frac{q}{L} = h[2(W+H)](t_w - t_b) = 569.2$ W/m.]

8.7 Air at atmospheric pressure and 100°C is flowing through a 20 mm diameter tube at a velocity of 20 m/s. The wall temperature is 30°C above the air temperature all along the tube length. Calculate the heat transfer rate per unit length of the tube. Assume fully developed flow condition.

[**Ans.** At 100°C bulk temperature, thermophysical properties of the air are: ρ = 0.9452 kg/m^3, k = 0.0317 W/(m K), $\mu = 2.172 \times 10^{-5}$ N s/m^2, Pr = 0.693, c_p = 1.0113 kJ/(kg K); $\text{Re} = \frac{\rho U_m d}{\mu} = 17407$; Flow is turbulent; $h = Nu\frac{k}{d} = 0.024\text{Re}^{0.8}\,\text{Pr}^{0.4}\frac{k}{d} = 81.11$ W/(m^2 K); $\frac{q}{L} = h(\pi d)(t_w - t_b) = 152.85$ W/m.]

8.8 For flow of 0.04 m^3/s of oil at 20°C bulk temperature through a 1.5 m long tube 125 mm in diameter kept at 30°C, determine the average heat transfer coefficient. The property values are: k = 0.14 W/(m K), μ_b = 1.2 kg/(m s), μ_w = 0.6 kg/(m s), Pr = 20000, c_p = 2000 J/(kg K), ρ = 890 kg/m^3.

[**Ans.** $\text{Re}_L = \frac{\rho U_m d}{\mu} = \frac{\rho V d}{(\pi/4d^2)\mu} = \frac{4\rho V}{\pi d \mu} = 302$; Flow is laminar; $L_{hy} = 0.05\text{Re}d = 1.89\text{m} > 1.5$ m; $L_{th} = 0.05\text{Re}\,\text{Pr}\,d = 37.75$ km > 1.5 m; Flow is simultaneously developing; Sieder-Tate relation may be used; $\text{Nu} = 1.86\left(\frac{d}{L} \times \text{Re}\,\text{Pr}\right)^{0.33}\left(\frac{\mu_b}{\mu_w}\right)^{0.14} = 156$;$h = \text{Nu} \times \frac{k}{d} = 174.7$ W/(m^{2}K).]

8.9 Air at atmospheric pressure and 20°C flows across a long cylinder of 50 mm diameter at a velocity of 40 m/s. The cylinder surface temperature is maintained at 100°C. Calculate the heat transfer rate per unit length of the cylinder.
[**Ans.** At mean temperature t_m = 60°C, air properties: ρ = 1.059 kg/m^3, k = 0.02875 W/(m K), μ = 1.997 × 10^{-5} kg/(m s), Pr = 0.701, c_p = 1008 J/(kg K); $\mathrm{Re}_x = \frac{\rho U_\infty d}{\mu} = 106059$; From Table 8.11, C = 0.027, n = 0.805; $h = Nu\frac{k}{d} = 0.027\mathrm{Re}^{0.805}\,\mathrm{Pr}^{1/3}\frac{k}{d} = 153.1$ W/(m^{2}K); $\frac{q}{L} = h(\pi d)(t_w - t_b) = 1923.9\ \mathrm{W/m}$.]

8.10 Air at 25°C flows at 10 m/s parallel to the surface of a highly polished aluminium plate flat plate maintained at a uniform temperature of t_s = 75°C by a series of segmented heaters. Determine the heat removed from the section between x_1 = 0.3 m and x_2 = 0.4 m. The plate width is 0.3 m. The flow is turbulent throughout.
[**Ans.** For air at film temperature of 50°C from Table A5, ρ = 1.0949 kg/m^3, μ = 1.9512 × 10^{-5} kg/(m s), k = 0.02799 W/(m K) and Pr = 0.703; $h_x = \frac{k}{x}\mathrm{Nu_x} = \frac{k}{x} \times 0.0296\mathrm{Re}_x^{0.8}\,\mathrm{Pr}^{1/3} = \frac{k}{x} \times 0.0296\left(\frac{\rho U_\infty x}{\mu}\right)^{0.8}\mathrm{Pr}^{1/3}$; Hence, $h_{x1} = \frac{0.02799}{0.3} \times 0.0296\times\left(\frac{1.0949\times10\times0.3}{1.9512\times10^{-5}}\right)^{0.8}\times0.703^{1/3} = 37.25$ W/(m^2 K), Similarly $h_{\times2}$ = 35.17 W/(m^2 K); The average heat transfer coefficient $h_{\times1\text{-}\times2}$ for distance x_1 = 0.3 m and x_2 = 0.4 m is (37.25 + 35.17)/2 = 36.21; Heat transfer rate $q = h_{x1\text{-}\times2} \times (x_2 - x_1)\ W \times (t_s - t_\infty)$ = 54.3 W.]

8.11 Air at 1 atm, 25°C and 5 m/s is in cross flow over a long cylinder of 30 mm diameter. Determine the drag force per unit length of the cylinder.
[**Ans.** For air at 25°C from Table A5, ρ = 1.1868 kg/m^3, μ = 1.8363 × 10^{-5} kg/(m s); $\mathrm{Re}_D = \frac{\rho U_\infty D}{\mu} = \frac{1.1868\times5\times0.03}{1.8363\times10^{-5}} = 9694$; From Fig. 8.10, $C_\mathrm{D} \approx 1$; From Eq. (8.46), $F_D = C_D A\left(\frac{1}{2}\rho U_\infty^2\right)$, where $A = LD$; Substitution gives $F_D = 1\times(1\times0.03)\times\left(\frac{1}{2}\times1.1868\times5^2\right) = 0.445$ N/m.]

8.12 Water at 25°C and 10 m/s flows over a sphere of 10 mm diameter. Surface temperature of the sphere is 75°C. Determine the drag force. What will be the drag force if fluid is air?
[**Ans.** For water at film temperature of 50°C from Table A4, ρ = 988.1 kg/m^3, μ = 544 × 10^{-6} kg/(m s); $\mathrm{Re}_D = \frac{\rho U_\infty D}{\mu} == \frac{988.1\times10\times0.01}{544\times10^{-6}} = 1.8\times10^5$; From Fig. 8.10, $C_\mathrm{D} \approx 0.4$; From Eq. (8.46), $F_D = C_D A\left(\frac{1}{2}\rho U_\infty^2\right)$, where A = frontal area = (π/4)D^2; Substitution gives $F_D = 0.4\times\frac{\pi}{4}\times0.01^2\times\left(\frac{1}{2}\times988.1\times10^2\right) = 1.55$N. For air at film temperature of 50°C from Table A5, ρ = 1.0949 kg/m^3, μ = 19.512 × 10^{-6} kg/(m s); $\mathrm{Re}_D = \frac{\rho U_\infty D}{\mu} = \frac{1.0949\times10\times0.01}{19.512\times10^{-6}} = 5.6\times10^3$; From Fig. 8.10, $C_\mathrm{D} \approx 0.4$; From Eq. (8.46), $F_D = C_D A\left(\frac{1}{2}\rho U_\infty^2\right)$, where A = frontal area = (π/4)D^2; Substitution gives $F_D = 0.4\times\frac{\pi}{4}\times0.01^2\times\left(\frac{1}{2}\times1.0949\times10^2\right) = 0.0017$N. Comment: In the present case, drag force associated with water is significantly higher than for air because of higher density of water.]

8.13 Air at 1 atm and 25°C flows at 7.5 m/s over inline tube bundle with $p = p_t$ = 25 mm. The bundle contains 10 tubes per row. The tube diameter is 12.5 mm. If air outlet temperature is 225°C and tube surface temperature is 400°C, determine the number of rows of the tubes per m length.

[**Ans.** Air properties at the mean bulk temperature t_m [= ($t_{\infty i}$ + $t_{\infty o}$)/2 = 125°C] are: ρ = 0.8872 kg/m^3, c = 1013.8 J/(kg K), μ = 2.2776 × 10^{-5} kg/(m s), k = 0.0335 W/(m K) and Pr = 0.689. At t_w = 400°C, Pr$_w$ = 0.683; $U_{max} = U_\infty \times \frac{p_t}{p_t - D} = 15$ m/s; $\mathrm{Re}_D = \frac{\rho U_{max} D}{\mu} = 7304$; Assuming $N > 16$, from Eq. (8.58 c),$\mathrm{Nu} = 0.27\mathrm{Re}_D^{0.63}\,\mathrm{Pr}^{0.36}(\mathrm{Pr}/\mathrm{Pr}_w)^{0.25} = 64.3$; $\bar{h} = \frac{k}{D}\mathrm{Nu} = 172.3$ W/(m^2 K); From 7. 16, $t_{\infty o} = t_w - (t_w - t_{\infty i})\exp\left(-\frac{P}{mc_p}L\bar{h}\right)$, or $\ln\left(\frac{t_w - t_{\infty o}}{t_w - t_{\infty i}}\right) = -\frac{\pi DN}{\rho U_\infty N_t p_t c_p}L\bar{h}$, which gives $N = \ln\left(\frac{t_w - t_{\infty i}}{t_w - t_{\infty o}}\right)\frac{\rho U_\infty N_t p_t c_p}{\pi D L \bar{h}} = 190$ for L = 1 m; No. of rows = N/N_t = 190/10 = 19 > 16.]

8.14 Liquid mercury at 20°C enters a metal tube of 20 mm internal diameter at the rate of 1 kg/s and is to be heated to 30°C. The tube wall is at a constant temperature of 40°C. Determine the length of the tube. Given for the mercury: ρ = 13560 kg/m^3, k = 8.7 W/(m K), μ = 1.5 × 10^{-3} kg/(m s), Pr = 0.025, c_p = 139 J/(kg K).

[**Ans.** Reynolds number, $\mathrm{Re} = \frac{\rho U_m d}{\mu} = \frac{4m}{\pi d \mu} = \frac{4\times 1}{\pi \times 0.02 \times 1.5 \times 10^{-3}} = 42441$; $\mathrm{Pe} = \mathrm{Re\,Pr} = 42441 \times 0.025 = 1061$; For constant wall temperature, Seban and Shimazaki equation, Table 8.13, gives $\bar{h} = \mathrm{Nu}\frac{k}{d} = (5.0 + 0.025\mathrm{Pe}^{0.8})\frac{k}{d} = [5.0 + 0.025 \times (1061)^{0.8}] \times \frac{8.7}{0.02} = 5039$ W/(m^2 K); For isothermal case, $\ln\left(\frac{t_w - t_o}{t_w - t_i}\right) = -\frac{PL}{mc_p}\bar{h}$, which gives $L = \ln\left(\frac{t_w - t_i}{t_w - t_o}\right)\frac{mc_p}{P\bar{h}}$ $= \ln\left(\frac{40-20}{40-30}\right) \times \frac{1.0 \times 139}{\pi \times 0.02 \times 5039} = 0.304$m.]

References

Bejan A, Kraus AD (eds) (2003) Heat transfer handbook. Wiley, New York

Bhatti MS, Shah RK (1987) Turbulent and transition flow convective heat transfer. In: Kakac S, Shah RK, Aung W (eds) Handbook of single-phase convective heat transfer, Chap. 4, Wiley, New York

Churchill SW (1977) Comprehensive correlating equations for heat, mass and momentum transfer in fully developed flow in smooth tubes. Ind Eng Chem Fundam 16(1):109–116

Churchill SW, Bernstein M (1977) A correlating equation for forced convection from gases and liquids to a circular cylinder in crossflow. J Heat Transfer 99:300–306

Dalle Donne M, Meerwald E (1973) Heat transfer and friction coefficients for turbulent flow in smooth annuli at high temperatures. Int J Heat Mass Transfer 16:787–809

Dippery DF, Sabersky RH (1963) Heat and momentum transfer in smooth and rough tubes at various Prandtl numbers. Int J Heat Mass Transfer 6:329–353

Douglas WJM, Churchill SW (1956) Recorrelation of data for convective heat transfer between gases and single cylinders with large temperature differences. Chem Eng Prog Symp Ser 52(18):23–28

Ebadian MA, Dong ZF (1998) Forced convection, internal flow in ducts. In: Rohesenow WM, Hartnett JP, Cho YI (eds) Handbook of heat transfer. McGraw Hill Book Co, New York

Eckert ERG, Drake RM Jr (1972) Analysis of heat and mass transfer. McGraw-Hill Kogakusha, Tokya

Gnielinski V (2015) Turbulent heat transfer in annular spaces- a new comprehensive correlation. Heat Transfer Eng 36:787–789

Gowen RA, Smith JW (1968) Heat transfer from smooth and rough surfaces. Int J Heat Mass Transfer 11:1657–1673

Heaton HS, Reynolds WC, Kays WM (1964) Heat transfer in annular passages: simultaneous development of velocity and temperature fields in laminar flow. Int J Heat Mass Transfer 7:763–781

Hilpert R (1933) Heat transfer from cylinders. Forsch Geb Ingenieurwes 4:215–224

Holman JP (1992) Adapted for SI units by White PRS. Heat transfer. McGraw-Hill Book Co, New York

Hsu ST (1963) Engineering heat transfer. D. Van Nostrand, New York

Isachenko VP, Osipova VA, Sukomel AS (1977) Heat transfer, 2nd edn. MIR Publishers, Moscow

Kays WM, Crawford ME (1980) Convective heat and mass transfer. Tata McGraw-Hill Publishing Co., Ltd, New Delhi

Kays WM, Leung EY (1963) Heat transfer in annular passages-hydrodynamically developed turbulent flow with arbitrarily prescribed heat flux. Int J Heat Mass Transfer 6:537–557

Kays WM, Perkins HC (1973) Forced convection, internal flow in ducts. In: Rohsenow WM, Hartnett JP (eds) Handbook of heat transfer, Chap. 7. McGraw-Hill, New York

Knudsen JD, Katz DL (1958) Fluid dynamics and heat transfer. McGraw-Hill, New York

Langhaar HL (1942) Steady flow in the transition length of a straight tube. J Appl Mech Trans ASME 64: A55–A58

Lee Y (1968) Turbulent heat transfer from the core tube in thermal entrance regions of concentric annuli. Int J Heat Mass Transfer 11:509–522

Lundgren TS, Sparrow EM, Starr JB (1964) Pressure drop due to the entrance region in ducts of arbitrary cross-section. J Basic Eng 86(3):620–626

Mikheyev M (1964) Fundamental of heat transfer. Mir Publishers, Moscow

Mills AF (1995) Heat and mass transfer. Richard D, Irwin, Chicago

Moody LF (1944) Friction factors for pipe flow. Trans ASME 66:671–684

Nikuradse J (1950) Laws of flow in rough pipes. NACA, Technical Memorandum 1292

Rajendra Karwa, Bairwa RD, Jain BP, Karwa N (2005) Experimental study of the effect of rib angle and discretization on heat transfer and friction in an asymmetrical heated rectangular duct. J Enhanced Heat Transfer 12(4):343–355

Ranz WE, Marshall WR Jr (1952a) Evaporation from drops. Parts I. Chem Eng Progr 48:141–146

Ranz WE, Marshall WR Jr (1952b) Evaporation from drops. Parts II. Chem Eng Progr 48:173–180

Rogers GFC, Mayhew YR (1967) Engineering thermodynamics work and heat transfer. English Language Book Co and Longmans, Green and Co, London

Schlichting H (1960) Boundary layer theory (trans: Kestin J), 4th edn. McGraw-Hill, New York

Schlichting H, Gersten K (2000) Boundary-layer theory (trans: Katherine Mayes). Springer, Heidelberg

Seban RA, Shimazaki TT (1951) Heat transfer to a fluid flowing turbulently in a smooth pipe with walls at constant temperature. Trans ASME 73:803–807

Sieder EN, Tate CE (1936) Heat transfer and pressure drop of liquids in tubes. Ind Eng Chem 28:1429–1435

Skupinshi E, Tortel J, Vautrey L (1965) Determination des coefficients de convection a'un alliage sodium-potassium dans un tube circulaire. Int. J. Heat Mass Transfer 8:937–951

Sleicher CA, Rouse MW (1975) A convenient correlation for heat transfer to constant and variable property fluids in turbulent pipe flow. Int J Heat Mass Transfer 18:677–683

Sleicher Jr. CA (1955) Heat transfer in a pipe with turbulent flow and arbitrary wall-temperature distribution. Ph.D. Thesis, University of Michigen, p. 67

Sparrow EM, Lloyd JR, Hixon CW (1966) Experiments on turbulent heat transfer in an asymmetrically heated rectangular duct. J Heat Transfer 88:170–174

Tam LM, Ghajar AJ (2006) Transitional heat transfer in plain horizontal tubes. Heat Transfer Eng 27(5):23–28

Webb RL, Eckert ERG, Goldstein RJ (1971) Heat transfer and friction in tubes with repeated-rib roughness. Int J Heat Mass Transfer 14:601–617

Whitaker S (1972) Forced convection heat transfer correlations for flow in pipes, past flat plates, single cylinders, single spheres, and flow in packed beds and tube bundles. AIChE J 18(2):361–371

Winterton RHS (1998) Where did the Dittus and Boelter equation come from? Int J Heat Mass Transfer 41:809–810

Witte LC (1968) An experimental study of forced-convection heat transfer from a sphere to liquid sodium. J Heat Transfer 90:9–12

Zukauskas AA (1987) Convective heat transfer in cross-flow. In: Kakac S, Shah RK, Aung W (eds) Handbook of single-phase convective heat transfer, Chap. 6. Wiley, New York

Empirical Relations for Natural or Free Convection

9

9.1 Introduction

The development of boundary layer along a flat vertical plate under natural or free convection condition has been presented in Chap. 7. Figure 9.1 illustrates the nature of flow in free-convection near heated horizontal plates and tubes of different dimensions. In case of a large horizontal plate with heated surface facing upwards, the central portion of the plate is practically isolated due to the flow established at the edges of the plate. The central portion receives the cold fluid by flow of fluid from above. If the heated surface of the plate is facing downward, the fluid coming in contact with the heated surface cannot move directly in the upwards direction but only along the edges of the plate. Therefore, only a diminished level of convection flow is possible in this case.

The flow structure near the horizontal cylinders differs significantly from the flow along a flat vertical wall. For thin wires ($d < 0.2$–1.0 mm), the laminar flow is preserved throughout. At very small temperature differences, almost a stationary film of heated fluid may form around the wire.

9.2 Buoyancy Force in Natural Convection

In the free convection, the motion of the fluid is caused by the buoyancy force arising from the change in the density of the fluid due to the rise in the temperature.

Consider any fluid element, see Fig. 9.2, which is at a temperature ΔT above the surrounding fluid. If ρ_o is the density of the surrounding fluid and ρ is the density of the fluid element, then

$$\rho_o = \rho(1 + \beta \Delta T) \tag{i}$$

where β is the coefficient of cubical expansion.

The buoyancy force acting upwards is

$$\rho_o V g \tag{ii}$$

R. Karwa, *Heat and Mass Transfer*,
https://doi.org/10.1007/978-981-15-3988-6_9

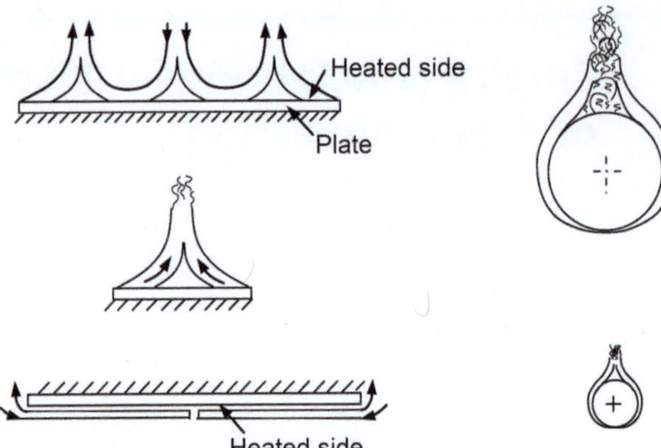

Fig. 9.1 Free convection flow

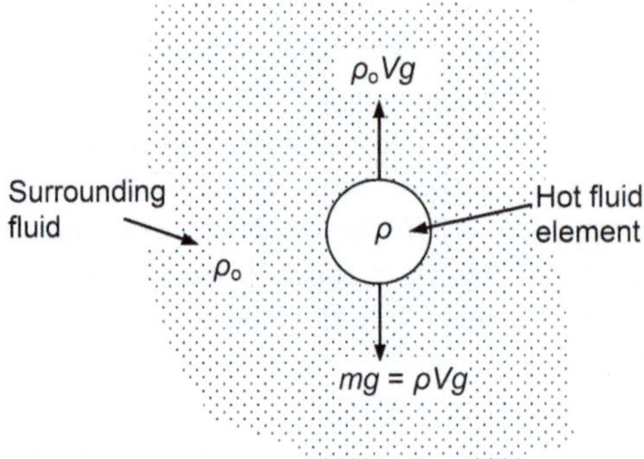

Fig. 9.2 Forces on a fluid element under free convection

The gravitational force ρVg acts downwards. The net force F acting on the fluid element is

$$\begin{aligned} F &= \rho_o Vg - \rho Vg \\ &= Vg(\rho_o - \rho) \end{aligned} \tag{iii}$$

Using Eq. (i), we obtain

$$F = \rho Vg\beta(\Delta T) \tag{9.1}$$

Thus the net force acting upwards per unit mass is $\beta g \Delta T$. This force causes the upwards motion of the fluid element.

9.3 Dimensional Analysis Applied to Natural Convection

From the experience, it has been found that the heat transfer coefficient in natural convection depends on the viscosity μ, thermal conductivity k, density ρ, temperature difference ΔT and the characteristic dimension d. Since the net buoyancy force per unit mass is $g\beta\Delta T$, the product $g\beta$ is also included as a variable. Hence we may write

$$h = f(\mu, \rho, k, c_p, \Delta T, \beta g, d) \tag{9.2a}$$

or

$$h = C\mu^a \rho^b k^c c_p^d \Delta T^e \beta g^f d^g \tag{9.2b}$$

where C is a constant of proportionality and a, b, c, etc. are arbitrary indices.

9.3.1 Rayleigh's Method

The dimensions of different physical quantities have been listed in Table 8.1. The dimensions of quantities β and g are given in Table 9.1.

Substituting the dimensions of different variables in Eq. (9.1), we have

$$\frac{Q}{L^2 T\theta} = C\left(\frac{M}{LT}\right)^a \left(\frac{M}{L^3}\right)^b \left(\frac{Q}{LT\theta}\right)^c \left(\frac{Q}{M\theta}\right)^d (\theta)^e \left(\frac{L}{\theta T^2}\right)^f (L)^g$$

Equating the indices of the fundamental dimensions on both sides,

$M \quad 0 = a + b - d$
$L \quad -2 = -a - 3b - c + f + g$
$T \quad -1 = -a - c - 2f$
$\theta \quad -1 = -c - d + e - f$
$Q \quad 1 = c + d$

There are seven unknowns and five equations. Therefore the values of five of them, say a, b, c, e, and g may be determined in terms of the other two unknowns d and f, which gives

$$a = d - 2f,\ b = 2f,\ c = 1 - d,\ e = f, \text{ and } g = -1 + 3f$$

Substituting the values in Eq. (9.2b), we get

$$h = C\mu^{d-2f} \rho^{2f} k^{1-d} c_p^d \Delta T^f (\beta g)^f d^{-1+3f}$$
$$= C\left(\frac{k}{d}\right)\left(\frac{\rho^2 \beta g \Delta T d^3}{\mu^2}\right)^f \left(\frac{\mu c_p}{k}\right)^d$$

Table 9.1 Dimensions of different physical quantities

Variable	Symbol (units)	Dimensions	
		M·L -T-θ -Q system	M-L-T-θ system
Coefficient of cubical expansion	β (1/K)	θ^{-1}	θ^{-1}
Acceleration due to gravity	g (m/s^2)	LT^{-2}	LT^{-2}

or

$$\frac{hd}{k} = f\left[\left(\frac{\rho^2 \beta g \Delta T d^3}{\mu^2}\right)^f \left(\frac{\mu c_p}{k}\right)^d\right]$$

The dimensionless group $\left(\frac{\rho^2 \beta g \Delta T d^3}{\mu^2}\right)$ is termed as the Grashof number Gr. Hence

$$\text{Nu} = f(\text{Gr}, \text{Pr})$$

The generalized correlation can be written as

$$\text{Nu} = \psi(\text{Gr})\phi(\text{Pr}) \tag{9.3}$$

The form of functions ψ(Gr) and ϕ(Pr) may be specified for different conditions of heat transfer by convection on the basis of theoretical analysis or experimental investigations.

9.3.2 Buckingham's Pi Method

$$f(\mu, \rho, k, c_p, \Delta T, \beta g, d, h) = 0$$

There are eight variables and five fundamental units hence we expect (8 – 5), i.e., 3 π-terms. Taking μ, k, βg, ΔT and d as repeated variables, the π-terms can be established as follows.

$$\pi_1 = \mu^a k^b (\beta g)^c \Delta T^d d^e h \tag{i}$$

or

$$1 = \left(\frac{M}{LT}\right)^a \left(\frac{Q}{LT\theta}\right)^b \left(\frac{L}{\theta T^2}\right)^c (\theta)^d (L)^e \frac{Q}{L^2 T\theta}$$

Equating the indices of the fundamental dimensions on both sides, we obtain

M: $0 = a$
L: $0 = -a - b + c + e - 2$
T: $0 = -a - b - 2c - 1$
θ: $0 = -b - c + d - 1$
Q: $0 = b + 1$

Solution gives

$$a = 0,\ b = -1,\ c = 0,\ d = 0, \text{ and } e = 1.$$

Substitution in Eq. (i) gives

$$\pi_1 = k^{-1}dh = \frac{hd}{k} = \text{Nu} \tag{ii}$$

Following the above procedure,

$$\pi_2 = \mu^a k^b (\beta g)^c \Delta T^d d^e \rho \tag{iii}$$

or

$$1 = \left(\frac{M}{LT}\right)^a \left(\frac{Q}{LT\theta}\right)^b \left(\frac{L}{\theta T^2}\right)^c (\theta)^d (L)^e \frac{M}{L^3}$$

Equating the indices,

M: $0 = a + 1$
L: $0 = -a - b + c + e - 3$
T: $0 = -a - b - 2c$
θ: $0 = -b - c + d$
Q: $0 = b$

Solution gives

$$a = -1,\ b = 0,\ c = 1/2,\ d = 1/2,\ \text{and}\ e = 3/2.$$

This gives

$$\pi_2 = \mu^{-1}(\beta g)^{1/2} \Delta T^{1/2} d^{3/2} \rho = \left(\frac{\rho^2 \beta g \Delta T d^3}{\mu^2}\right)^{1/2} \tag{iv}$$

Similarly,

$$\pi_3 = \mu^a k^b (\beta g)^c \Delta T^d d^e c_p \tag{v}$$

or

$$1 = \left(\frac{M}{LT}\right)^a \left(\frac{Q}{LT\theta}\right)^b \left(\frac{L}{\theta T^2}\right)^c (\theta)^d (L)^e \left(\frac{Q}{M\theta}\right)$$

Equating the indices,

M: $0 = a - 1$
L: $0 = -a - b + c + e$
T: $0 = -a - b - 2c$
θ: $0 = -b - c + d - 1$
Q: $0 = b + 1$

Solution of the above equations gives

$$a = 1,\ b = -1,\ c = 0,\ d = 0,\ \text{and}\ e = 0.$$

This gives

$$\pi_3 = \mu k^{-1} c_p = \frac{\mu c_p}{k} = \text{Pr} \tag{vi}$$

Thus the functional relation is

$$f(\text{Nu}, \text{Gr}, \text{Pr}) = 0$$

The generalized correlation can be written as Eq. (9.3):

$$\text{Nu} = \psi(\text{Gr})\phi(\text{Pr}) \tag{9.3}$$

Example 9.1 Using M, L, T and θ system of fundamental units, develop the functional relation for free convection heat transfer.

Solution

Buckingham's pi method

Analysis in previous section has shown that (βg) and (ΔT) terms appear together in the final functional relation for free convection hence we write

$$f(\mu, \rho, k, c_p, \beta g \Delta T, d, h) = 0 \tag{i}$$

There are seven variables and four fundamental units hence we expect (7 – 4), i.e., 3 π-terms. Taking μ, ρ, k, and d as repeated variables, the π-terms can be established as follows.

$$\pi_1 = \mu^a \rho^b k^c d^d (\beta g \Delta T) \tag{ii}$$

or

$$1 = \left(\frac{M}{LT}\right)^a \left(\frac{M}{L^3}\right)^b \left(\frac{ML}{\theta T^3}\right)^c (L)^d \left(\frac{1}{\theta} \times \frac{L}{T^2} \times \theta\right)$$

Equating the indices of the fundamental dimensions on both sides, we obtain

M $\quad 0 = a + b + c$

L $\quad 0 = -a - 3b + c + d + 1$

T $\quad 0 = -a - 3c - 2$

θ $\quad 0 = -c$

Solution gives

$$a = -2,\ b = 2,\ c = 0 \,\text{and}\, d = 3.$$

Substitution gives

$$\pi_1 = \mu^{-2}\rho^2 d^3 \beta g \Delta T = \frac{\rho^2(\beta g \Delta T)d^3}{\mu^2} = \text{Gr} \quad \text{(iii)}$$

Following the above procedure,

$$\pi_2 = \mu^a \rho^b k^c d^d c_p \quad \text{(iv)}$$

or

$$1 = \left(\frac{M}{LT}\right)^a \left(\frac{M}{L^3}\right)^b \left(\frac{ML}{\theta T^3}\right)^c (L)^d \left(\frac{L^2}{\theta T^2}\right),$$

which gives

$$\pi_2 = \frac{\mu c_p}{k} = \text{Pr} \quad \text{(v)}$$

Similarly,

$$\pi_3 = \mu^a \rho^b k^c d^d h, \quad \text{(vi)}$$

which gives

$$\pi_3 = \frac{hd}{k} = \text{Nu} \quad \text{(vii)}$$

Thus the functional relation is

$$f(\text{Nu}, \text{Gr}, \text{Pr}) = 0$$

The generalized correlation can be written as Eq. (9.3)

$$\text{Nu} = \psi(\text{Gr})\phi(\text{Pr}) \quad (9.3)$$

Rayleigh's Method

$$h = C\mu^a \rho^b k^c c_p^d (\beta g \Delta T)^e d^f \quad \text{(viii)}$$

or

$$\left(\frac{M}{\theta T^3}\right) = C\left(\frac{M}{LT}\right)^a \left(\frac{M}{L^3}\right)^b \left(\frac{ML}{\theta T^3}\right)^c \left(\frac{L^2}{\theta T^2}\right)^d \left(\frac{1}{\theta} \times \frac{L}{T^2} \times \theta\right)^e (L)^f$$

Equating the indices of the fundamental dimensions on both sides, we obtain

M $1 = a + b + c$

L $0 = - a - 3b + c + 2d + e + f$

T $-3 = - a - 3c - 2d - 2e$

θ $-1 = - c - d$

We have four equations but six unknowns, so we shall determine the values of a, c, e and f in the terms of b and d. Thus

$$a = d - b,\ c = 1 - d,\ e = b/2 \,\text{and}\, f = (3/2)b - 1.$$

Substitution gives

$$h = C\mu^{d-b}\rho^b k^{1-d} c_p^d (\beta g \Delta T)^{b/2} d^{3/2b-1}$$
$$= C\frac{k}{d}\left[\frac{\rho^2(\beta g\Delta T)d^3}{\mu^2}\right]^{b/2}\left(\frac{\mu c_p}{k}\right)^d$$

or

$$\frac{hd}{k} = C\left[\frac{\rho^2(\beta g\Delta T)d^3}{\mu^2}\right]^{b/2}\left(\frac{\mu c_p}{k}\right)^d$$

Thus the functional relation is

$$\mathrm{Nu} = \psi(\mathrm{Gr})\phi(\mathrm{Pr}) \tag{9.3}$$

9.3.3 Physical Interpretation of Grashof Number

As mentioned in Chap. 7, the *Grashof number* Gr can be interpreted physically as a dimensionless number representing the ratio of the buoyancy to viscous forces in free convection, i.e.

$$\mathrm{Gr} = \left(\frac{\rho^2 g\beta\Delta T d^3}{\mu^2}\right) = \frac{\text{Buoyancy force}}{\text{Viscous force}}$$

The relation Nu $= f$ (Gr, Pr) suggest that the Grashof number plays a part in the free convection analogous to the part played by the Reynolds number in forced convection. The product (Gr × Pr), called the *Rayleigh number* Ra, serves as a criterion of turbulence in free convection. However, the value of the product for the onset of turbulence depends on the geometric configuration.

In the case of low velocity flows, the free convection may play a significant role in the forced convection hence the functional relation is presented as (refer Chap. 8)

$$\mathrm{Nu} = \psi(\mathrm{Re}, \mathrm{Gr}, \mathrm{Pr}) \tag{8.6}$$

9.4 Experimental Determination of Natural Convection Heat Transfer Coefficient

A simple experimental setup for the experimental investigation of natural convection heat transfer from a long horizontal tube is shown in Fig. 9.3a. The heat of the electric heater flows in radial direction to the surface of the tube. The heat loss from the tube surface occurs both by convection (q_c) and radiation (q_r) to the surroundings. The total heat loss rate from the tube is

$$q = q_c + q_r \tag{i}$$

The convection heat transfer is given by

$$q_c = hA(T_w - T_\infty) \tag{ii}$$

and the radiation heat transfer is

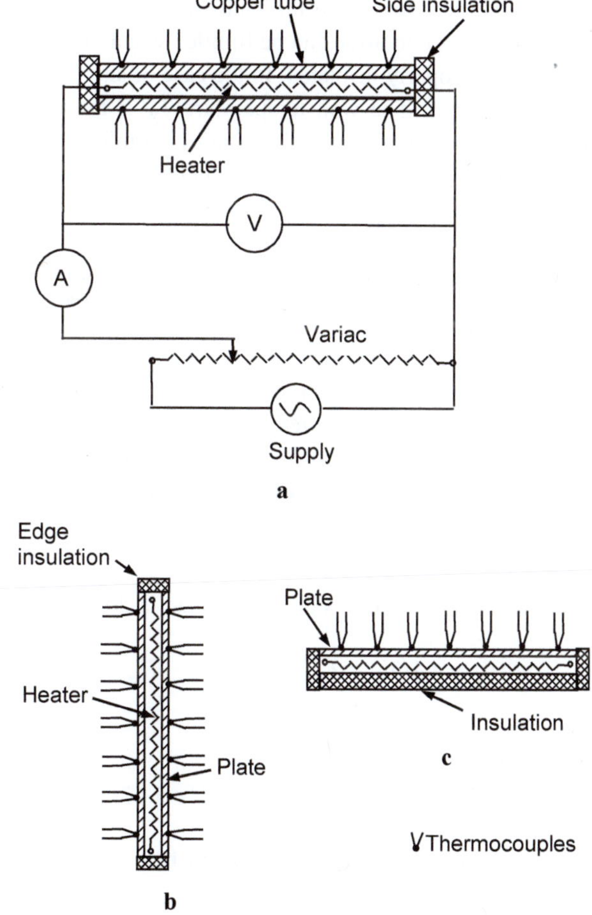

Fig. 9.3 Experimental setup for determination of natural convection heat transfer coefficient

$$q_r = \varepsilon\sigma A(T_w^4 - T_s^4) \quad \text{(iii)}$$

where

q	electric power input
h	heat transfer coefficient
A	surface area of the tube = $\pi D_o L$
D_o	tube outer diameter
L	length of the tube
T_w	tube wall temperature
T_∞	surrounding air temperature
T_s	surrounding surface temperature for radiation heat exchange
ε	emissivity of the tube surface

Substitution in Eq. (i) from Eqs. (ii) and (iii) gives

$$q = hA(T_w - T_\infty) + \varepsilon\sigma A(T_w^4 - T_s^4). \quad (9.4)$$

Equation (9.4) is solved for the heat transfer coefficient.

In the above analysis, the heat loss from the ends of the tube has not been considered assuming it to be negligible. This condition can be fulfilled when the tube is sufficiently long (say 2–3 m) and the ends are properly insulated. In the case of short tubes, guard heaters are installed at the ends of the tubes. The wall temperature T_w is average of the readings of sufficient number of thermocouples affixed axially and circumferentially on the tube wall surface. All readings are taken when the steady state condition is achieved. The experiments are conducted at different heating rates resulting in different values of wall temperature. The result is presented in the form of relation Nu = *f*(Gr Pr).

The experimental scheme presented above can be extended for determination of heat transfer coefficient from vertical and horizontal plates, refer Fig. 9.3b, c.

Example 9.2 An electrically heated 300 mm × 300 mm square metal plate is used for an experimental determination of natural convection heat transfer coefficient. The plate is placed in vertical position and is exposed to room temperature at 25°C. The plate is heated to a uniform temperature of 60°C. The electric power input is measured as 40 W. The emissivity of the surface has been estimated as 0.1. Determine the heat transfer coefficient.

Solution

In the equilibrium, the heat lost from one side of the plate is 20 W. Hence,

$$q = hA(T_w - T_\infty) + \varepsilon A\sigma(T_w^4 - T_\infty^4)$$

or

$$20 = h \times 0.3 \times 0.3 \times (333 - 298) + 0.1 \times 0.3 \times 0.3 \times 5.67 \times 10^{-8} \times (333^4 - 298^4),$$

assuming surrounding surface temperature for radiation heat exchange to be equal to the surrounding air temperature.

Solution of the equation gives

$$h = 5.635\ \text{W/m}^2\text{K}.$$

9.5 Empirical Relations for Free or Natural Convection

9.5.1 Vertical Plate and Cylinders

(i) **Uniform Temperature Condition (Isothermal Surface):**

A vertical cylinder can also be treated as a vertical plate under laminar flow conditions if the boundary layer thickness is much less than the diameter of the cylinder. This condition is satisfied when

$$\frac{D}{L} \geq \frac{C}{\text{Gr}_\text{L}^{1/4}} \tag{9.5}$$

where D is the diameter of the cylinder, and C = 35 for fluids of Pr $\leq$ 0.72 and $C \approx 25$ for fluids of Pr $\leq$ 6 [Cebeci (1974) in Popiel (2008)]. However, some references (Holman 1992; Cengel 2007. Incropera et al. 2006) use a single value of C = 35. The plate results are accurate to within 5%. The reason for deviation is the curvature effect. When the thermal boundary layer thickness in much thinner than the radius of the cylinder, the curvature can be neglected and plate relation may be used.

The vertical plane relation for laminar flow with length L as characteristic dimension (McAdams, 1954) is

$$\text{Nu}_m = 0.59\text{Ra}^{1/4} \quad \text{for } 10^4 \leq \text{Ra} \leq 10^9 \tag{9.6}$$

Various terms in the above relation are defined as

$$\text{Nu}_\text{m} = \frac{h_m L}{k} \tag{9.7}$$

$$\text{Ra} = \text{GrPr} = \frac{\beta g (T_w - T_\infty) L^3}{\nu^2} \text{Pr} \tag{9.8}$$

where h_m is average heat transfer coefficient, T_w is surface or wall temperature and T_∞ is free stream temperature. β and thermophysical properties of the fluid are taken at film temperature. The subscript 'm' refers to the mean or average value.

For turbulent flow (Ra $\geq$ 10^9), the correlation suggested (Bailey in Kays and Crawford 1980) is

$$\text{Nu}_m = 0.1\text{Ra}^{1/3} \quad \text{for } 10^9 \leq \text{Ra} \leq 10^{13} \tag{9.9a}$$

For turbulent flow over a slender vertical cylinder (Ra $\geq 4 \times 10^9$), the mean Nusselt number for air can be approximated from the following correlation of McAdams (1954) as suggested by Popiel (2008) for *L/D* from about 70 to 136.

$$\mathrm{Nu}_m = 0.13\mathrm{Ra}^{1/3} \tag{9.9b}$$

Churchill and Chu (1975b) gave the following relations of mean Nusselt number for a wide range of Ra, which is a complex relation but more accurate.

$$\mathrm{Nu_m} = \left\{0.825 + \frac{0.387\mathrm{Ra}^{1/6}}{\left[1 + (0.492/\,\mathrm{Pr})^{9/16}\right]^{8/27}}\right\}^2 \quad \text{for } 10^{-1} \leq \mathrm{Ra} \leq 10^{12} \tag{9.10}$$

Equation (9.10) is also valid for constant heat flux when factor 0.492 is replaced by 0.437 (Baehr and Stephan 2011).

In the laminar region, the following relation gives better accuracy.

$$\mathrm{Nu_m} = 0.68 + \frac{0.670\mathrm{Ra}^{1/4}}{\left[1 + (0.492/\,\mathrm{Pr})^{9/16}\right]^{4/9}} \quad \text{for } \mathrm{Ra} \leq 10^9 \tag{9.11}$$

(ii) **Uniform Heat Flux (q_w = Constant)**

The suggested equation of local Nusselt number is (Baehr and Stephan 2011)

$$\mathrm{Nu_x} = 0.616\mathrm{Ra_x}^{1/5}\left(\frac{\mathrm{Pr}}{0.8 + \mathrm{Pr}}\right)^{1/5} \quad \text{for } \mathrm{Pr} \geq 0.1;\ \mathrm{Ra_x} \leq 10^9 \tag{9.12}$$

where $\mathrm{Nu_x} = h_x x/k$; $\mathrm{Ra_x} = \mathrm{Gr_x Pr} = [\beta g q_w x^4/(\nu^2 k)].\mathrm{Pr}$.

Alternative equation for local Nusselt number in laminar region (Holman 1992) is

$$(\mathrm{Nu_x})_\mathrm{f} = \frac{hx}{k_f} = 0.60\left(\mathrm{Gr_x^*Pr_f}\right)^{1/5} \quad \text{for } 10^5 \leq \mathrm{Gr_x^*Pr_f} \leq 10^{11} \tag{9.13}$$

where $\mathrm{Gr_x^*} = \mathrm{Gr_x Nu_x} = \frac{g\beta q_w x^4}{k\nu^2}$.

The transition begins between $\mathrm{Gr_x^*Pr} = 3 \times 10^{12} - 4 \times 10^{13}$ and ends between $2 \times 10^{13} - 4 \times 10^{14}$. Fully developed turbulent flow is present beyond $\mathrm{Gr_x^*Pr} = 10^{14}$ (Holman 1992). For the turbulent region, the recommended relation is

$$(\mathrm{Nu_x})_f = 0.17(\mathrm{Gr_x^* Pr_f})^{1/4} \quad \text{for } 2 \times 10^{13} \le \mathrm{Gr_x^* Pr} \le 10^{16} \tag{9.14}$$

Subscript f in Eqs. (9.13) and (9.14) indicates that the fluid properties are to be evaluated at the local film temperature T_f [= $(T_w + T_\infty)/2$].

The average or mean heat transfer coefficient is

$$h_m = \frac{1}{L}\int_0^L h_x dx$$

Thus, for the laminar flow, Eqs. (9.12) and (9.13) yield

$$h_m = \frac{5}{4} h_{x=L} \tag{9.15}$$

It can be shown that in the case of turbulent flow condition, the local heat transfer coefficient is constant with x.

9.5.2 Inclined Plate

When a plate is inclined, the buoyancy force parallel to the surface is reduced and there is reduction in the fluid velocity along the plate causing reduction in the convection heat transfer. For $0 \le \theta \le 60°$ correlation of Churchill and Chu (1975b), Eq. (9.10), is valid for the inclined plate when the acceleration due to gravity g is replaced by its component parallel to the wall $g \cos\theta$, where θ is the angle of inclination with the vertical. This approach is satisfactory only for the top and bottom surfaces of cooled and heated plates, respectively (Incropera et al. 2012). For opposite surfaces, refer Holman (1992).

9.5.3 Horizontal Plate

For a horizontal plate, the characteristic length L is defined as

$$L = \frac{A}{P} \tag{9.16}$$

where A the heat transfer area and P is the perimeter of the plate.

When the upper side of the plate is heated or lower side is cooled, refer Fig. 9.1,

$$\mathrm{Nu}_m = 0.54\mathrm{Ra_L^{1/4}} \quad \text{for } 10^4 \le \mathrm{Ra} \le 10^7 \tag{9.17}$$

$$\mathrm{Nu}_m = 0.15\mathrm{Ra_L^{1/3}} \quad \text{for } 10^7 \le \mathrm{Ra} \le 10^{11} \tag{9.18}$$

The equation of mean Nusselt number, when upper side of the plate is cooled or lower side is heated is

$$\mathrm{Nu}_m = 0.27\mathrm{Ra}_\mathrm{L}^{1/4} \quad \text{for } 10^5 \le \mathrm{Ra} \le 10^{11} \tag{9.19}$$

(i) **Uniform Heat Flux Condition (q_w = Constant)** (Fujii and Imura 1972 in Holman 1992)

For the heated surface facing upwards,

$$\mathrm{Nu}_m = 0.13\mathrm{Ra}_\mathrm{L}^{1/3} \quad \text{for } \mathrm{Ra} \le 2 \times 10^8 \tag{9.20}$$

$$\mathrm{Nu}_m = 0.16\mathrm{Ra}_\mathrm{L}^{1/3} \text{ for } 2 \times 10^8 \le \mathrm{Ra} \le 10^{11} \tag{9.21}$$

For the heated surface facing downwards,

$$\mathrm{Nu}_m = 0.58\mathrm{Ra}_\mathrm{L}^{1/5} \quad \text{for } 10^6 \le \mathrm{Ra} \le 10^{11} \tag{9.22}$$

All properties except β, are evaluated at reference temperature $T_e = T_w - 0.25(T_w - T_\infty)$. T_w is the wall temperature related to the heat flux by $h_m = q_w/(T_w - T_\infty)$.

9.5.4 Horizontal Cylinder of Diameter D and Length L > > d

The mean Nusselt number equation for this case is given by

$$\mathrm{Nu}_m = 0.53\mathrm{Ra}_\mathrm{d}^{1/4} \quad \text{for } 10^4 \le \mathrm{Ra} \le 10^9 \tag{9.23}$$

$$\mathrm{Nu}_m = 0.13\mathrm{Ra}_\mathrm{d}^{1/3} \quad \text{for } 10^9 \le \mathrm{Ra} \le 10^{12} \tag{9.24}$$

Churchill and Chu (1975a) presented the following correlation of average or mean Nusselt number for the laminar flow ($\mathrm{Ra}_\mathrm{d} \le 10^9$)

$$\mathrm{Nu}_\mathrm{m} = 0.36 + \frac{0.518\mathrm{Ra}_\mathrm{d}^{1/4}}{\left[1 + (0.559/\mathrm{Pr})^{9/16}\right]^{4/9}} \tag{9.25}$$

where characteristic dimension for the Nusselt number and Rayleigh number is the diameter of the cylinder.

For turbulent flow ($Ra_d \geq 10^9$), the following correlation may be used.

$$Nu_m = \left\{ 0.60 + \frac{0.387 Ra_d^{1/6}}{\left[1 + (0.559/Pr)^{9/16}\right]^{8/27}} \right\}^2 \tag{9.26}$$

However, Churchill and Chu (1975a,b) recommend use of Eq. (9.26) for all Pr and Ra values.

9.5.5 Sphere of Diameter d

The characteristic dimension in the equations of the mean Nusselt number and Rayleigh number is the diameter of the sphere and the suggested equation for the average Nusselt number for $Pr \geq 0.7$ is (Churchill 1990)

$$Nu_m = 2 + \frac{0.589 Ra_d^{1/4}}{\left[1 + (0.469/Pr)^{9/16}\right]^{4/9}} \quad \text{for } Ra_d \leq 10^{11} \tag{9.27}$$

Example 9.3 For a vertical wall at 50°C exposed to still air at 20°C, what is the maximum height for laminar free convection?

Solution

At t_m = (50 + 20)/2 = 35°C, the air properties are

$$\nu = \mu/\rho = 16.36 \times 10^{-6}\, m^2/s,\ Pr = 0.7066 \text{ and } \beta = 1/(35 + 298)\, K^{-1}.$$

$$Gr_x\, Pr = \frac{g\beta(t_w - t_\infty)x^3}{\nu^2} . Pr$$

or

$$x = \left[\frac{Ra_x \nu^2}{g\beta(t_w - t_\infty)\, Pr}\right]^{1/3} \tag{i}$$

The limit for laminar flow is $Ra_x \leq 10^9$. Substitution of the values of various terms in Eq. (i) gives

$$x = \left[\frac{10^9 \times (16.36 \times 10^{-6})^2}{9.81 \times (1/303) \times (50 - 20) \times 0.7066}\right]^{1/3} = 0.731\ m.$$

Example 9.4 Determine the coefficient of heat transfer from a vertical plate of 2 m height to the surrounding still air, which at a distance from the plate is at 20°C. The mean surface temperature of the plate is measured to be 120°C.

Solution

At the mean fluid temperature t_m = (120 + 20)/2 = 70°C, the physical properties of air are:

$$k = 0.0295\,\mathrm{W/(m\ K)}, \nu = \mu/\rho = 2 \times 10^{-5}\,\mathrm{m^2/s},\ \mathrm{Pr} = 0.698 \text{ and } \beta = 1/T = 1/343\,\mathrm{K^{-1}}.$$

Hence,

$$\mathrm{Gr\,Pr} = \frac{\beta g \Delta T L^3}{\nu^2} \times \mathrm{Pr} = \frac{1/343 \times 9.81 \times (120 - 20) \times 2^3}{(2 \times 10^{-5})^2} \times 0.698 = 3.99 \times 10^{10} \geq 10^9.$$

Equation (9.9a) applies, which gives

$$\mathrm{Nu} = 0.1 \times (3.99 \times 10^{10})^{1/3} = 341.4$$

and heat transfer coefficient,

$$h = \frac{\mathrm{Nu} \times k}{L} = \frac{341.4 \times 2.95 \times 10^{-2}}{2} = 5.04\ \mathrm{W/(m^2\ K)}.$$

Example 9.5 A vertical plate is rejecting heat by free convection from one of its face which is at an average temperature of 90°C. The surrounding air is at 30°C. The height of the plate is 200 mm. Calculate the heat rejected from the plate if the width of the plate is 600 mm. The following correlation for the local Nusselt number may be used.

$$\mathrm{Nu}_x = 0.52\left(\frac{\mathrm{Pr}}{0.95 + \mathrm{Pr}}\right)^{0.25} (\mathrm{Gr\,Pr})^{0.25}.$$

Solution

From the given data,

$$t_m = \frac{t_w + t_\infty}{2} = \left(\frac{90 + 30}{2}\right) = 60^\circ\mathrm{C}.$$

The thermo-physical properties of air are:

$$\rho = 1.059\,\mathrm{kg/m^3},\ k = 0.02875\,\mathrm{W/(mK)},\ \mu = 2 \times 10^{-5}\,\mathrm{N\ s/m^2},\ \mathrm{Pr} = 0.7.$$

$$\beta = \frac{1}{T_m} = \frac{1}{(60 + 273)} = 3.003 \times 10^{-3}\,(1/\text{K}).$$

From the given correlation,

$$h_x = \frac{k}{x} \times \left[0.52 \times \left(\frac{\text{Pr}}{0.95 + \text{Pr}}\right)^{0.25} \times \left(\frac{\rho^2 \beta g \Delta T x^3}{\mu^2} \times \text{Pr}\right)^{0.25}\right].$$

Substitution of values of various terms gives

$$\begin{aligned} h_x &= \frac{0.02875}{x} \times 0.52 \times \left(\frac{0.7}{0.95 + 0.7}\right)^{0.25} \\ &\quad \times \left(\frac{1.059^2 \times 3.003 \times 10^{-3} \times 9.81 \times 60 \times x^3}{(2 \times 10^{-5})^2} \times 0.7\right)^{0.25} \\ &= 2.928 x^{-0.25}. \end{aligned}$$

The mean heat transfer coefficient

$$h = \frac{1}{L}\int_0^L h_x dx = \frac{1}{0.2}\left(2.928 \times \frac{x^{0.75}}{0.75}\right)_0^{0.2} = 5.84\ \text{W/(m}^2\ \text{K)}.$$

Heat transfer rate

$$q = hA\Delta T = 5.84 \times 0.2 \times 0.6 \times 60 = 42.05\ \text{W}.$$

Example 9.6 A vertical rectangular plate 1 m × 0.6 m has one of its surface insulted and the other face is maintained at a uniform temperature of 50°C. It is exposed to quiescent air at 20°C. Calculate the heat transfer coefficient if

(i) 1 m side of the plate is in vertical direction
(ii) 0.6 m side of the plate is in vertical direction.

Solution

For the range $10^4 < \text{Ra} < 10^9$,

$$\text{Nu} = 0.59\text{Ra}^{0.25}$$

and for the range $10^9 \le \text{Ra} < 10^{13}$,

$$\text{Nu} = 0.1\,\text{Ra}^{1/3}$$

The thermophysical properties of the air at the mean film temperature of 35°C from Table A5 are:

ρ = 1.15 kg/m^3, c_p = 1006 J/(kg K), k = 2.68 × 10^{-2} W/(m K), μ = 1.883 × 10^{-5} kg/(m s), Pr = 0.7066 and the coefficient of cubical expansion, $\beta = 1/T = 1/(35 + 273) = 1/308\ \text{K}^{-1}$.

Hence, the Rayleigh number is

$$\text{Ra} = \text{Gr Pr} = \frac{\rho^2 \beta g \Delta T L^3}{\mu^2} \times \text{Pr} = \frac{\rho^2 \beta g (t_w - t_\infty) L^3}{\mu^2} \times \text{Pr}$$
$$= \frac{1.15^2 \times 1/308 \times 9.81 \times (50 - 20) \times L^3}{(1.883 \times 10^{-5})^2} \times 0.7066 = (2.52 \times 10^9) L^3.$$

(i) **1 m side is vertical**

Characteristic dimensions, L = 1 m.

Rayleigh number,

$$\text{Ra} = 2.52 \times 10^9 \times L^3 = 2.52 \times 10^9 \times 1.0^3 = 2.52 \times 10^9 > 10^9$$

Nusselt number,

$$\text{Nu} = 0.1\text{Ra}^{1/3} = 0.1 \times (2.52 \times 10^9)^{1/3} = 136.$$

Heat transfer coefficient,

$$h = \frac{\text{Nu}k}{L} = \frac{136 \times 0.0268}{1.0} = 3.645\ \text{W/(m}^2\ \text{K)}.$$

(ii) **0.6 m side is vertical**

Characteristic dimensions, L = 0.6 m

Rayleigh number,

$$\text{Ra} = 2.52 \times 10^9 \times L^3 = 2.52 \times 10^9 \times 0.6^3 = 5.4 \times 10^8 < 10^9.$$

Nusselt number,

$$\text{Nu} = 0.59\text{Ra}^{0.25} = 0.59 \times (5.4 \times 10^8)^{0.25} = 89.94$$

Heat transfer coefficient,

$$h = \frac{\text{Nu}k}{L} = \frac{89.94 \times 0.0268}{0.6} = 4.02\ \text{W/(m}^2\ \text{K)}.$$

Example 9.7 A horizontal plate (1.0 m × 1.0 m) is maintained at 140°C by electric heating. Estimate the convective heat transfer rate to the surrounding still air at 20°C. The lower surface of the plate is insulated.

Solution

At mean film temperature of (140 + 20)/2 = 80°C, the air properties are

$$\nu = \mu/\rho = 20.87 \times 10^{-6}\text{m}^{2/\text{s}}, k = 0.0303\text{W}/(\text{mK}), \text{Pr} = 0.696, \text{and}\ \beta = 1/(50+273)\text{K}^{-1}.$$

For a horizontal plate,

$$L = \frac{A}{P} = \frac{1.0 \times 1.0}{4} = 0.25\,\text{m}.$$

The Rayleigh number,

$$\text{Ra}_L = \text{Gr}_L\,\text{Pr} = \frac{g\beta(t_w - t_\infty)L^3}{\nu^2}.\text{Pr}$$

$$= \frac{9.81 \times (1/353) \times (140 - 20) \times 0.25^3}{(20.87 \times 10^{-6})^2} \times 0.696 = 8.33 \times 10^7.$$

Equation (9.17) applies, which gives

$$\text{Nu}_L = 0.15(\text{Ra}_L)^{1/3} = 0.15 \times (8.33 \times 10^7)^{1/3} = 65.5.$$

Heat transfer coefficient,

$$h = \frac{\text{Nu}\ k}{L} = \frac{65.5 \times 0.0303}{0.25} = 7.94\ \text{W/(m}^2\ \text{K)}.$$

Heat transfer rate,

$$q = hA\Delta T = 7.94 \times (1.0 \times 1.0) \times (140 - 20) = 952.8\ \text{W}.$$

Example 9.8 If the heated surface of the plate of Example 9.7 is facing downwards, estimate the convective heat transfer rate.

Solution

For a horizontal plate with heated surface facing downwards, Eq. (9.19) applies, which gives

$$\text{Nu}_L = 0.27(\text{Ra}_L)^{1/4} = 0.27 \times (8.33 \times 10^7)^{1/4} = 25.8$$

Heat transfer coefficient,

$$h = \frac{\text{Nu}\,k}{L} = \frac{25.8 \times 0.0303}{0.25} = 3.13\ \text{W/(m}^2\ \text{K)}.$$

Heat transfer rate,

$$q = hA\Delta T = 3.13 \times (1.0 \times 1.0) \times (140 - 20) = 375.6\ \text{W}.$$

Comments: The heat rejection rate is significantly reduced, which is expected because of the difference in the flow structure of the two cases as presented in Fig. 9.1.

Example 9.9 A 400 mm square plate is inclined from vertical at an angle of 30°. The surface temperature of the plate is 330 K. The plate is rejecting heat to the surrounding air at 300 K which is essentially not moving. Determine the convective heat transfer rate from the plate.

Solution

At mean film temperature of (330 + 300)/2 = 315 K = 42°C, the air properties are

$\rho = 1.124\,\text{kg/m}^3$, $\mu = 1.91 \times 10^{-5}\,\text{kg/(m s)}$, $k = 0.0274\,\text{W/(m K)}$, $\text{Pr} = 0.705$ and $\beta = 1/315\ \text{K}^{-1}$

For an inclined plate,

$$\begin{aligned}\text{Ra}_L &= \text{Gr}_L\,\text{Pr} = \frac{(g\cos\theta)\beta(T_w - T_\infty)L^3}{\nu^2}.\text{Pr}\\ &= \frac{(9.81 \times \cos 30) \times (1/315) \times (330 - 300) \times 0.4^3}{(1.91 \times 10^{-5}/1.124)^2} \times 0.705\\ &= 1.26 \times 10^8 < 10^9.\end{aligned}$$

The flow is laminar. Equation (9.6) applies:

$$\text{Nu}_L = 0.59\,(\text{Ra}_L)^{1/4} = 0.59 \times (1.26 \times 10^8)^{1/4} = 62.5.$$

Heat transfer coefficient,

$$h = \frac{\text{Nu}k}{L} = \frac{62.5 \times 0.0274}{0.4} = 4.28\ \ \text{W/(m}^2\ \text{K)}.$$

Heat transfer rate,

$$q = hA\Delta T = 4.28 \times (0.4 \times 0.4) \times (330 - 300) = 20.54\ \text{W}.$$

Example 9.10 Determine the hourly loss of heat from a bare horizontal steam pipe whose diameter is 100 mm and the length is 5 m. The pipe wall temperature is 450 K and the temperature of the surrounding air is 300 K. Assume that the radiation heat loss is negligible.

Solution

The mean film temperature,

$$T_m = \frac{T_w + T_\infty}{2} = \frac{450 + 300}{2} = 375\ \text{K}.$$

From Table A5 of "Thermophysical Properties of Air", we have

$$\rho = 0.941\,\text{kg/m}^3,\ \text{Pr} = 0.693,\ k = 0.032\,\text{W/(m K)},\ \mu = 2.18 \times 10^{-5}\,\text{kg/(m s)}.$$

Grashof number,

$$\text{Gr} = \frac{\beta g \Delta T d^3}{\nu^2} = \frac{[1/(375)] \times 9.81 \times (450 - 300) \times 0.1^3}{(2.18 \times 10^{-5}/0.941)^2} = 7.31 \times 10^6.$$

From Eq. (9.23), Nusselt number

$$\text{Nu} = 0.53(\text{GrPr})^{0.25} = 0.53 \times (7.31 \times 10^6 \times 0.693)^{0.25} = 25.1.$$

The heat transfer coefficient at the outer surface,

$$h = \frac{\text{Nu}k}{d} = \frac{25.1 \times 0.032}{0.1} = 8.0\ \text{W/(m}^2\ \text{K)}.$$

The hourly heat loss,

$$q = hA\Delta T = h(\pi dL)(T_w - T_\infty) = 8.0 \times (\pi \times 0.1 \times 5) \times (450 - 300) \times 3600$$
$$= 6785.8\ \text{kJ}.$$

Example 9.11 If the pipe of Example 9.10 is vertical, determine the hourly loss.

Solution

Grashof number,

$$\text{Gr} = \frac{\beta g \Delta T L^3}{\nu^2} = \frac{[1/(375)] \times 9.81 \times (450 - 300) \times 5^3}{(2.18 \times 10^{-5}/0.941)^2} = 9.14 \times 10^{11}.$$

Since $\text{Gr} > 4 \times 10^9$, the flow is turbulent and $L/D = 50$, Eq. (9.9b) may be used, which gives

$$\text{Nu}_\text{m} = 0.13(\text{Ra})^{1/3} = 0.13 \times (9.14 \times 10^{11} \times 0.693)^{1/3} = 1116.$$

The heat transfer coefficient at the outer surface of the cylinder is

$$h_m = \frac{\text{Nu}_\text{m} k}{L} = \frac{1116 \times 0.032}{5} = 7.14\ \text{W/(m}^2\ \text{K)}.$$

The hourly heat loss,

$$q = h_m A(\Delta T) = h_m(\pi dL)(T_w - T_\infty) = 7.14 \times (\pi \times 0.1 \times 5) \times (450 - 300) \times 3600 \\ = 6056\ \text{kJ}.$$

Example 9.12 A thin walled vertical cylindrical tank, 0.3 m in diameter and 0.8 m in height, contains water at 70°C. The tank is placed in a larger tank with water at 10°C. Determine heat transfer rate across the tank wall.

Solution

Assuming wall temperature as mean of the water temperatures, i.e. 40°C, the temperature of the film on the inner side is (70 + 40)/2 = 55°C.

Thermophysical properties of water at the film temperature of 55°C from Table A4 are

$$\rho = 985.2\,\text{kg/m}^3,\ \mu = 501 \times 10^{-6}\,\text{N s/m}^2, k = 0.648\,\text{W/(m K)}, \text{Pr} = 3.24\ \text{and}$$
$$\beta = 0.484 \times 10^{-3}\,1/\text{K}.$$

The Grashof number based on cylinder height,

$$\text{Gr} = \frac{\rho^2 g\beta\Delta TL^3}{\mu^2} \\ = \frac{985.2^2 \times 9.81 \times 0.484 \times 10^{-3} \times (70 - 40) \times 0.8^3}{(501 \times 10^{-6})^2} = 2.82 \times 10^{11}.$$

$D/L = 0.375$ and $35/\text{Gr}^{0.25} = 0.048$. Hence, condition $\frac{D}{L} \geq \frac{35}{\text{Gr}_\text{L}^{1/4}}$ is satisfied and the cylinder can be treated as a vertical plate.

The Rayleigh number is

$$\text{Ra} = \text{GrPr} = 2.82 \times 10^{11} \times 3.24 = 9.14 \times 10^{11}.$$

From Eq. (9.10),

$$\mathrm{Nu_m} = \left\{ 0.825 + \frac{0.387\mathrm{Ra}^{1/6}}{\left[1 + (0.492/\mathrm{Pr})^{9/16}\right]^{8/27}} \right\}^2 \quad \text{for } 10^{-1} \le \mathrm{Ra} \le 10^{12}$$

$$= \left\{ 0.825 + \frac{0.387 \times (9.14 \times 10^{11})^{1/6}}{\left[1 + (0.492/3.24)^{9/16}\right]^{8/27}} \right\}^2 = 1276.9.$$

Heat transfer coefficient (inner side),

$$h_i = \frac{k}{L}\mathrm{Nu_m} = \frac{0.648}{0.8} \times 1276.9 = 1034.3\, W/(\mathrm{m^2K}).$$

The film temperature on the outer side is (40 + 10)/2 = 25°C. Properties of water at 25°C from Table A4 are:

$\rho = 997\,\mathrm{kg/m^3}$, $\mu = 890 \times 10^{-6}\,\mathrm{N\ s/m^2}$, $k = 0.609\,\mathrm{W/(m\ K)}$, $\mathrm{Pr} = 6.13$ and $\beta = 0.253 \times 10^{-3}\,1/\mathrm{K}$.

The Grashof number based on cylinder height,

$$\mathrm{Gr} = \frac{\rho^2 g\beta\Delta T L^3}{\mu^2}$$

$$= \frac{997^2 \times 9.81 \times 0.253 \times 10^{-3} \times (40 - 10) \times 0.8^3}{(890 \times 10^{-6})^2} = 4.78 \times 10^{10}.$$

D/*L* = 0.375 and $35/\mathrm{Gr}^{0.25}$ = 0.0748. Hence, condition $\frac{D}{L} \ge \frac{35}{\mathrm{Gr}_L^{1/4}}$ is satisfied and the cylinder can be treated as a vertical plate.

The Rayleigh number is

$$\mathrm{Ra} = \mathrm{GrPr} = 4.78 \times 10^{10} \times 6.13 = 2.93 \times 10^{11}.$$

From Eq. (9.10),

$$\mathrm{Nu_m} = \left\{ 0.825 + \frac{0.387\mathrm{Ra}^{1/6}}{\left[1 + (0.492/\mathrm{Pr})^{9/16}\right]^{8/27}} \right\}^2 \quad \text{for } 10^{-1} \le \mathrm{Ra} \le 10^{12}$$

$$= \left\{ 0.825 + \frac{0.387 \times (2.93 \times 10^{11})^{1/6}}{\left[1 + (0.492/6.13)^{9/16}\right]^{8/27}} \right\}^2 = 924.3.$$

Heat transfer coefficient,

$$h_o = \frac{k}{L}\text{Nu}_\text{m} = \frac{0.609}{0.8} \times 924.3 = 703.6\,\text{W}/(\text{m}^2\text{K}).$$

Overall heat transfer coefficient,

$$U = \left(\frac{1}{h_i} + \frac{1}{h_o}\right)^{-1} = \left(\frac{1}{1034.3} + \frac{1}{703.6}\right)^{-1} = 418.7\,\text{W}/(\text{m}^2\,\text{K}).$$

Heat transfer rate,

$$q = UA\Delta T = U\pi DL(t_w - t_a) = 418.7 \times \pi \times 0.3 \times 0.8 \times (70 - 10) = 18941.5\,\text{W}.$$

Wall temperature:

$$q = h_o \pi DL(t_s - t_o)$$

or

$$t_s = \frac{q}{h_0 \pi DL} + t_o = \frac{18941.5}{703.6 \times \pi \times 0.3 \times 0.8} + 10 = 45.7^\circ\text{C},$$

as against the assumed value of 40°C. Retrial with t_s = 45°C may be carried out.

Example 9.13 A horizontal steel pipeline, having inside diameter of 50 mm and outside diameter of 60 mm, carries water flowing with a velocity of 0.2 m/s. The mean water temperature is 80°C. The pipeline is lagged with asbestos. The outside diameter of the lagging is 90 mm.

Determine the loss of heat from 1 m length of the pipeline if the temperature of the still air surrounding the pipeline is 20°C. Neglect the radiation heat loss from the outer surface of the lagging.

Also determine the surface temperature of the pipeline and the lagging. For steel, k_s = 46 W/(m K) and for asbestos k_a = 0.11 W/(m K).

Solution

Thermophysical properties of water at 80°C (the mean temperature of water) are

$$\rho = 971.8\,\text{kg/m}^3,\ \text{Pr} = 2.21,\ k = 0.67\,\text{W}/(\text{m K})\ \text{and}\ \mu = 351 \times 10^{-6}\,\text{kg}/(\text{m s}).$$

(i) The coefficient of heat transfer from water to the inner surface of the pipe:

$$\text{Re} = \frac{\rho U d}{\mu} = \frac{971.8 \times 0.2 \times 0.05}{351 \times 10^{-6}} = 2.77 \times 10^4.$$

The flow is in the turbulent regime and the heat transfer coefficient can be determined from the following correlation applicable to Re > 10000.

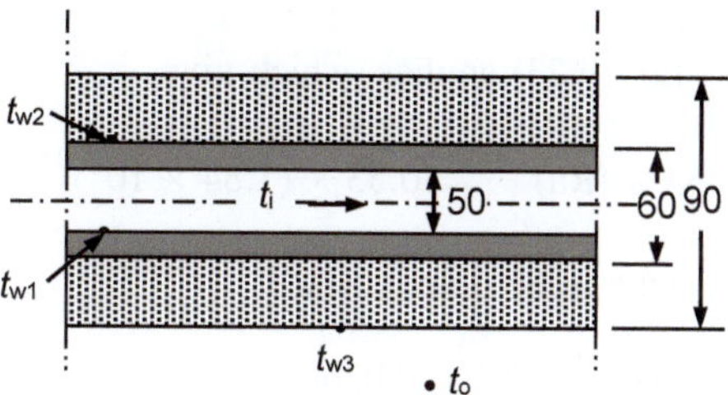

Fig. 9.4 Example 9.13

$$\begin{aligned}\text{Nu} &= 0.024\text{Re}^{0.8}\text{Pr}^{0.4} \\ &= 0.024 \times (2.77 \times 10^4)^{0.8}(2.21)^{0.4} = 118.0.\end{aligned}$$

This gives the heat transfer coefficient:

$$h_i = \frac{\text{Nu}k}{d} = \frac{118.0 \times 0.67}{0.05} = 1581 \text{ W/(m}^2 \text{ K)}.$$

(ii) The coefficient of heat transfer from water to the inside surface of the tube is very high. Further the pipe is lagged and the heat transfer coefficient from the outer surface of the lagging is likely to be low because of natural convection condition at the surface of the lagging. Hence, the pipe inside surface temperature will be nearly equal to the temperature of the water, i.e. $t_{w1} \approx t_i$.

(iii) Since the free convection heat transfer rate is a function of temperature difference ΔT, we assume the temperature of the outer surface of the lagging t_{w3} as 50°C for first trial. Refer Fig. 9.4.

With this assumption, the Grashof number is

$$\text{Gr} = \frac{\beta g \Delta T L^3}{\nu^2} = \frac{1/308 \times 9.81 \times (50 - 20) \times 0.09^3}{(1.88 \times 10^{-5}/1.15)^2} = 2.6 \times 10^6$$

where

t_m $(t_o + t_{w3})/2 = (20 + 50)/2 = 35$°C
β $1/T_m = 1/(35 + 273) = 1/308 \text{ K}^{-1}$
d outer diameter of the lagging = 0.09 m
ΔT 50–20 = 30°C
μ 1.88×10^{-5} kg/(m s) for the air at 35°C
ρ 1.15 kg/m^3 for air at 35°C

This gives

$$\text{Ra} = \text{GrPr} = 2.6 \times 10^6 \times 0.707 = 1.84 \times 10^6$$

where Pr= 0.707 for the air.

For $10^4 \leq \text{Ra} \leq 10^9$, Eq. (9.23) applies, which gives

$$\text{Nu} = 0.53 \times (\text{Ra})^{0.25} = 0.53 \times (1.84 \times 10^6)^{0.25} = 19.5.$$

The heat transfer coefficient at the outer surface,

$$h_o = \frac{\text{Nu}k}{d} = \frac{19.5 \times 0.02684}{0.09} = 5.82\,\text{W/(m}^2\,\text{K)}.$$

The heat transfer rate for the unit length of the pipeline is

$$q = \frac{t_i - t_o}{\frac{1}{2\pi r_i h_i} + \frac{1}{2\pi k_s}\ln\left(\frac{r_2}{r_1}\right) + \frac{1}{2\pi k_a}\ln\left(\frac{r_3}{r_2}\right) + \frac{1}{2\pi r_o h_o}}$$

or

$$q = \frac{80 - 20}{\frac{1}{2\pi \times 0.025 \times 1581} + \frac{1}{2\pi \times 46}\ln\left(\frac{0.03}{0.025}\right) + \frac{1}{2\pi \times 0.11}\ln\left(\frac{0.045}{0.03}\right) + \frac{1}{2\pi \times 0.045 \times 5.82}}$$
$$= 50.04\ \text{W}.$$

This gives

$$t_{w3} = t_o + \frac{q}{2\pi r_o h_o} = 20 + \frac{50.04}{2\pi \times 0.045 \times 5.82} = 50.4^\circ\text{C}.$$

Since the calculated temperature is nearly the same as the assumed one, further trial is not needed.

The inside pipe-surface temperature is

$$t_{w1} = t_i - \frac{q}{2\pi r_i h_i} = 80 - \frac{50.04}{2\pi \times 0.025 \times 1588} = 79.8^\circ\text{C},$$

which is nearly equal to the water temperature assumed earlier.

The temperature of the outside surface of the pipe,

$$t_{w2} = t_{w1} - \frac{q}{2\pi k_s}\ln\left(\frac{r_2}{r_1}\right) = 79.8 - \frac{50.04}{2\pi \times 46}\ln\left(\frac{0.03}{0.025}\right) = 79.77^\circ\text{C},$$

which indicates that the temperature drop across the wall of the pipe is negligible. This is because of the fact that the resistance offered by the metal wall to the heat flow is very small. The main resistances here are that of the insulation and outer film.

Example 9.14 If wind blows at a speed of 2 m/s across the pipe of Example 9.13, determine the increase in the heat loss.

Solution

The Reynolds number,

$$\mathrm{Re}_d = \frac{\rho U_\infty d}{\mu_f} = \frac{1.15 \times 2 \times 0.09}{1.88 \times 10^{-5}} = 11010.$$

From Table 8.11, for Re = 4000–40000, $C = 0.193$ and $n = 0.618$.
The average heat transfer coefficient,

$$\bar{h} = \frac{k}{d} \times C \times (\mathrm{Re}_d)^n \mathrm{Pr}^{1/3}$$
$$= \frac{0.02684}{0.09} \times 0.193 \times (11010)^{0.618} (0.707)^{1/3} = 16.1\ \mathrm{W/(m^2\ K)}.$$

Heat flow rate

$$q = \frac{t_i - t_o}{\frac{1}{2\pi r_i h_i} + \frac{1}{2\pi k_s}\ln\left(\frac{r_2}{r_1}\right) + \frac{1}{2\pi k_a}\ln\left(\frac{r_3}{r_2}\right) + \frac{1}{2\pi r_o h_o}}$$

or

$$q = \frac{80 - 20}{\frac{1}{2\pi \times 0.025 \times 1581} + \frac{1}{2\pi \times 46}\ln\left(\frac{0.03}{0.025}\right) + \frac{1}{2\pi \times 0.11}\ln\left(\frac{0.045}{0.03}\right) + \frac{1}{2\pi \times 0.045 \times 16.1}}$$
$$= 74.0\ \mathrm{W}.$$

Alternatively using Eq. (8.52), we have

$$\mathrm{Nu} = 0.3 + \frac{0.62\mathrm{Re}^{1/2}\,\mathrm{Pr}^{1/3}}{\left[1 + \left(\frac{0.4}{\mathrm{Pr}}\right)^{2/3}\right]^{1/4}}\left[1 + \left(\frac{\mathrm{Re}}{282000}\right)^{5/8}\right]^{4/5}$$

$$= 0.3 + \frac{0.62 \times (11010)^{1/2}(0.707)^{1/3}}{\left[1 + \left(\frac{0.4}{0.707}\right)^{2/3}\right]^{1/4}}\left[1 + \left(\frac{11010}{282000}\right)^{5/8}\right]^{4/5} = 56.46.$$

Hence,

$$\bar{h} = \frac{k}{d}\mathrm{Nu} = \frac{0.02684}{0.09} \times 56.46 = 16.84\ \mathrm{W/(m^2\ K)}$$

This gives

$$q = \frac{t_i - t_o}{\frac{1}{2\pi r_i h_i} + \frac{1}{2\pi k_s}\ln\left(\frac{r_2}{r_1}\right) + \frac{1}{2\pi k_a}\ln\left(\frac{r_3}{r_2}\right) + \frac{1}{2\pi r_o h_o}}$$

Substitution gives

$$q = \frac{80 - 20}{\frac{1}{2\pi \times 0.025 \times 1581} + \frac{1}{2\pi \times 46}\ln\left(\frac{0.03}{0.025}\right) + \frac{1}{2\pi \times 0.11}\ln\left(\frac{0.045}{0.03}\right) + \frac{1}{2\pi \times 0.045 \times 16.74}}$$
$$= 74.8\ \text{W}.$$

The heat loss due to the wind increases by about 50%.

9.6 Free Convection in Parallel Plate Channels

Array of circuit boards or fin arrays employed to enhance free convection heat transfer may be regarded as parallel vertical plates forming a channel that is open to the ambient at its ends as shown in Fig. 9.5. The heated fluid rises and exits from the top of the channel.

Boundary layers develop beginning at $x = 0$ on each wall (if both the walls of the channel are heated) and increase in thickness with height. If the ratio of the spacing S and height L is small, the boundary layers eventually merge and the flow, thereafter, is fully developed. When the ratio S/L is large (i.e. the plates are far apart), they behave like isolated vertical plates in infinite, quiescent fluid. Because of the effect of the ratio of spacing to height S/L on the flow structure, this ratio appears in the heat transfer correlations presented by researchers.

Surface thermal conditions have been classified as isothermal or isoflux and symmetrical ($T_{w1} = T_{w2}$; $q''_{w1} = q''_{w2}$) or asymmetrical ($T_{w1} \neq T_{w2}$; $q''_{w1} \neq q''_{w2}$), where q''_w is heat flux.

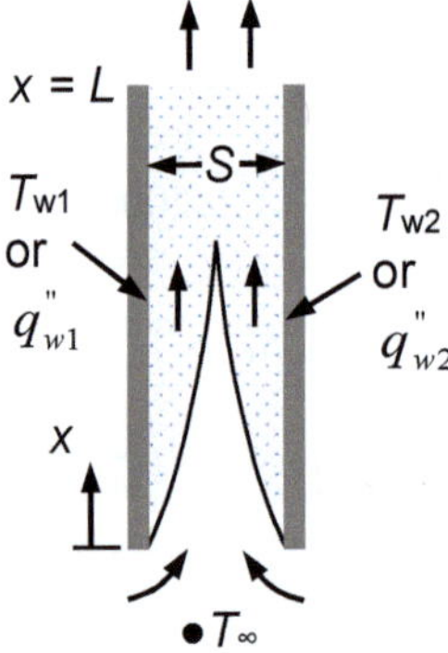

Fig. 9.5 Free convection flow between heated parallel plates

9.6.1 Vertical Channels

Bar-Cohen and Rohsenow (1984) presented correlations for symmetric and asymmetric isothermal and isoflux (uniform heat flux) conditions of vertical parallel plate channels. They have also presented relations for optimum spacing, which is the spacing to provide maximum heat transfer from an array in a given volume, i.e. spacing for the maximum volumetric heat dissipation. It is to note that as the number of plates in a given space is increased, the heat transfer coefficient h decreases with decreasing spacing and the reduction in the fluid flow rate because of the increase in the viscous force. But the increased total area A of the larger number of plates yields a maximum of the product hA. Spacing needed to maximize the heat transfer from each plate in the array has also been defined and relations with respect to the spacing for the maximum volumetric heat dissipation are given. This spacing for maximum heat transfer from each plate in the array precludes the overlap of the adjoining boundary layers and is found to be larger than the spacing for the maximum volumetric heat dissipation.

9.6.2 Inclined Channels

Experimental data for inclined channels formed by symmetric isothermal plates and isothermal-insulated plates in water (inclination 0 to 45° with vertical) for isolated plate limit of $[\mathrm{Ra}_S(S/L) > 200]$ have been correlated within ± 10% by Azevedo and Sparrow (1985) as

$$(\mathrm{Nu}_S)_\mathrm{m} = 0.645[\mathrm{Ra}_S(S/L)]^{1/4} \tag{9.28}$$

where the Rayleigh number is based on the spacing S and the fluid properties are evaluated at average temperature $T_\mathrm{m} = (T_\mathrm{w} + T_\infty)/2$.

9.7 Empirical Correlations for Enclosed Spaces

In some engineering applications, heat transfer takes place between two surfaces that are at different temperatures and a fluid is present in the space enclosed by these surfaces. Typical examples of such enclosures are double pane windows, cavities in the walls of the buildings, solar collectors, etc.

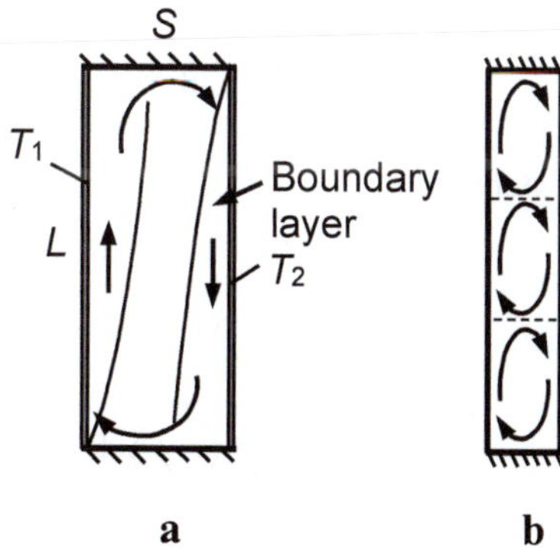

Fig. 9.6 Cellular flow in a vertical enclosure

Here the correlations are being presented only for the rectangular enclosures.

(i) **Vertical Enclosure**

Consider the rectangular enclosure shown in Fig. 9.6. The left wall is at a higher temperature than the right wall. Top and bottom surfaces of the enclosure are insulated. The heat transfers by the free convection currents in the enclosed fluid. The Rayleigh number for this case is defined as

$$\mathrm{Ra}_S = \frac{g\beta(T_1 - T_2)S^3}{\nu^2}\mathrm{Pr} \tag{9.29}$$

where S is the spacing between the heat transferring surfaces.

If the Rayleigh number is less than about 10^3, viscous forces dominate over the buoyancy forces and the fluid motion does not occur. This can happen when the spacing S is small and the viscosity is very large. In such a situation, the heat transfer takes place by conduction only through the stagnant fluid and Nu = 1.

At Ra $\approx 10^4$, the flow changes to a boundary layer type (Mills 1995). The fluid at the hot surface moves upwards along the wall and after giving heat to the cold face of the enclosure moves downwards along the wall if the spacing S is sufficiently large. A cellular flow establishes, Fig. 9.6(a), which is encountered in the thin boundary layers adjoining the vertical surfaces. The core is nearly stagnant. With small spacing, the flows affect each other which may result in internal circulation as if partitions are present, Fig. 9.6(b). At Ra $\approx 10^6$, the flow in the core becomes turbulent (Mills 1995).

Proposed empirical correlations for average Nusselt number are (MacGregor and Emery 1969)

$$\begin{aligned} \mathrm{Nu}_S &= 0.42(\mathrm{Ra}_S)^{1/4}\mathrm{Pr}^{0.012}\left(\frac{L}{S}\right)^{-0.3} \\ & 10 \le L/S \le 40 \\ & 10^4 \le \mathrm{Ra}_S \le 10^7 \\ & 1 \le \mathrm{Pr} \le 2 \times 10^4 \end{aligned} \tag{9.30}$$

$$\begin{aligned} \mathrm{Nu}_S &= 0.046(\mathrm{Ra}_S)^{1/3} \\ & 1 \le L/S \le 40 \\ & 10^6 \le \mathrm{Ra}_S \le 10^9 \\ & 1 \le \mathrm{Pr} \le 20 \end{aligned} \tag{9.31}$$

where the Nusselt number is defined as

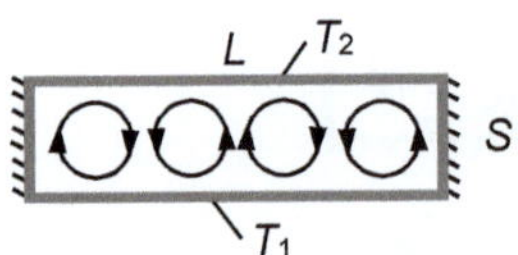

Fig. 9.7 Bernard cells

$$\mathrm{Nu}_S = \frac{hS}{k} \tag{9.32}$$

and the heat transfer rate is given by

$$q = hA(T_1 - T_2) \tag{9.33}$$

(ii) **Horizontal Enclosure**

Consider the horizontal enclosure with the lower wall heated and upper wall cold. Side walls are insulated. For Rayleigh number $\left[\mathrm{Ra}_S = \frac{g\beta(T_1 - T_2)S^3}{\alpha\nu}\right]$ less than a critical value of 1708, the viscous forces are very strong than the buoyancy forces and there is no movement of the fluid. The heat transfer takes place by conduction only and Nu = 1.

For 1708 < $\mathrm{Ra}_S \leq 5 \times 10^4$, the fluid motion consists of regularly spaced rotating cells (called as Bernard cells) as shown in Fig. 9.7. At higher Rayleigh numbers, the cells break down and turbulent motion occurs.

The correlation proposed by Globe and Dropkin (1959) for average Nusselt number is

$$(\mathrm{Nu}_S)_m = \frac{h_m S}{k} = 0.069\mathrm{Ra}_S^{1/3}\mathrm{Pr}^{0.074} \quad \text{for } 3 \times 10^5 \leq \mathrm{Ra}_S \leq 7 \times 10^9 \tag{9.34}$$

where all fluid properties are to be evaluated at the mean temperature $T_m = (T_1 + T_2)/2$.

If the upper wall is heated, then the heat transfer takes place by conduction only (Nu = 1) irrespective of the value of the Rayleigh number.

It is to note that radiation heat exchange between the hot and cold surfaces is always present when a gas is filled in the cavity.

(iii) **Tilted Enclosure (Fig. 9.8)**

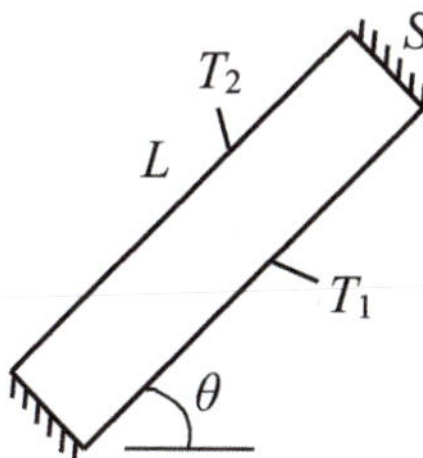

Fig. 9.8 Tilted enclosure

Table 9.2 Critical tilt angle*

L/S	1	3	6	12	>12
θ^*	25°	53°	60°	67°	70°

*Incropera Frank P, DeWitt David P, Bergman Theodore L, Lavine Adrienne S, Fundamentals of heat and mass transfer, 6th edn, John Wiley & Sons, New York, Copyright 2006. Reproduced with the permission of John Wiley & Sons Ltd

Some applications involve tilted enclosures. Typical example is that of a flat plate solar collector where the heat flows from the heated absorber plate to the glass cover placed at a distance. The tilt angle may vary from 0° (horizontal) to about 70° at high latitude applications.

In inclined enclosures, the fluid motion consists of a combination of the rolling structure of horizontal cavities and cellular structure of vertical cavities. Typically transition between the two types of motions occurs at a critical tilt angle θ^* given in Table 9.2.

For large aspect ratio duct of solar collectors (L/S > 12) and the tilt angle less than the critical value $\theta°$, the following correlation of the average Nusselt number has been proposed by Hollands et al. (1976).

$$(\mathrm{Nu_m})_S = 1 + 1.44\left(1 - \frac{1708}{\mathrm{Ra_S}\cos\theta}\right)^{+}\left[1 - \frac{1708(\sin 1.8\theta)^{1.6}}{\mathrm{Ra_S}\cos\theta}\right] + \left[\left(\frac{\mathrm{Ra_S}\cos\theta}{5830}\right)^{1/3} - 1\right]^{+}$$
$$\text{for } 0 \le \theta \le 60° \text{ and } 0 < \mathrm{Ra} \le 10^5. \tag{9.35}$$

Either of the terms in + bracket goes to zero when negative. This implies that if the Rayleigh number is less than a critical value of 1708/cos θ, there is no flow within the cavity (Incropera et al. 2012). This equation can be expected to give values of Nu with a maximum error of 5%. Equation (9.35) may also be used for θ up to 75° but error of up to 10 percent may be expected (Hollands et al. 1976).

Buchberg et al. (1976) presented the following three-region correlation for the estimate of the convective heat transfer coefficient between the absorber plate and glass cover of flat plate solar air heaters.

$$\mathrm{Nu} = 1 + 1.446\left(1 - \frac{1708}{\mathrm{Ra}'}\right)^{+} \quad \text{for } 1708 \le \mathrm{Ra}' \le 5900 \tag{9.36a}$$

(the + bracket goes to zero when negative)

$$\mathrm{Nu} = 0.229(\mathrm{Ra}')^{0.252} \quad \text{for } 5900 < \mathrm{Ra}' \le 9.23 \times 10^4 \tag{9.36b}$$

$$\mathrm{Nu} = 0.157(\mathrm{Ra}')^{0.285} \quad \text{for } 9.23 \times 10^4 < \mathrm{Ra}' \le 10^6 \tag{9.36c}$$

where Ra′ (= Ra cosθ) is Rayleigh number for the inclined air layers.

Example 9.15 A horizontal enclosure consists of two square plates (0.5 m × 0.5 m) separated by a distance of 10 mm. The lower plate is maintained at a uniform temperature of 40° C and the upper plate is at a uniform temperature of 10°C. Water at atmospheric pressure is filled in the enclosure. Calculate the heat loss rate from the lower plate if the side walls are insulated. Given $g\beta\rho^2c/(\mu k) = 1.85 \times 10^{10}\ \mathrm{m^{-3}K^{-1}}$ at the mean temperature.

Solution

At the mean temperature of (40 + 10)/2 = 25°C, k = 0.61 W/(m K) for water.

The Rayleigh number = $g\beta\rho^2 c/(\mu k) \times \Delta T \times S^3 = 1.85 \times 10^{10} \times (40 - 10) \times 0.01^3 = 5.55 \times 10^5$.

For $3 \times 10^5 \leq \mathrm{Ra_S} \leq 7 \times 10^9$, Eq. (9.34) applies:

$$\mathrm{Nu} = 0.069(\mathrm{Ra_s})^{1/3}(\mathrm{Pr})^{0.074} = 0.069 \times \left(5.55 \times 10^5\right)^{1/3} \times (6.13)^{0.074} = 6.48.$$

Mean heat transfer coefficient,

$$h = \mathrm{Nu}\left(\frac{k}{S}\right) = 6.48 \times \left(\frac{0.61}{0.01}\right) = 395.3\ \mathrm{W/(m^2\ K)}.$$

Heat loss rate,

$$q = hA(T_1 - T_2) = 395.3 \times (0.5 \times 0.5) \times (40 - 10) = 2964.7\ \mathrm{W}.$$

Example 9.16 Two horizontal surfaces of an enclosure with air between them are separated by 10 mm. Calculate the heat flow rate per m^2 of the plate surface if the upper plate is at 50°C and lower is at 20°C. The vertical sides are insulated.

Solution

In the present case, the convection currents are suppressed and the heat transfer is by conduction only. Hence,

$$q = kA\frac{T_1 - T_2}{S} = 0.027 \times 1 \times \frac{50 - 20}{0.01} = 81\ \mathrm{W/m^2},$$

where k = 0.027 W/(m K) at the mean temperature (50 + 20)/2 = 35°C.

Example 9.17 Absorber plate and cover plate combination of a flat plate solar collector can be treated as a tilted enclosure (Fig. 9.8). If T_1 = 70°C, T_2 = 30°C, L = 1.5 m, S = 50 mm and tilt θ = 45° determine heat flux due to free convection.

Solution

For large aspect ratio duct of solar collectors (L/S > 12) and the tilt angle less than the critical value θ^* given in Table 9.2, Eq. (9.35) of the average Nusselt number can be used:

$$(\mathrm{Nu_m})_S = 1 + 1.44\left(1 - \frac{1708}{\mathrm{Ra_S}\cos\theta}\right)^+ \left[1 - \frac{1708(\sin 1.8\theta)^{1.6}}{\mathrm{Ra_S}\cos\theta}\right] + \left[\left(\frac{\mathrm{Ra_S}\cos\theta}{5830}\right)^{1/3} - 1\right]^+$$

$$\text{for } 0 \leq \theta \leq 60° \text{ and } 0 < \mathrm{Ra} \leq 10^5.$$

Either of the terms in + bracket goes to zero when negative.

Thermophysical properties of air at mean temperature = (70 + 30)/2 = 50°C from Table A5 are:

$$\rho = 1.0949\,\text{kg/m}^3,\ \mu = 1.9512 \times 10^{-5}\,\text{kg/(m s)}, k = 0.02799\,\text{W/(m K)},$$
$$\alpha = 0.257 \times 10^{-4}\,\text{m}^2\text{/s and Pr} = 0.703.$$

Rayleigh number,

$$\text{Ra}_S = \frac{g\beta(T_1 - T_2)S^3}{\alpha\nu}$$
$$= \frac{9.81 \times 1/(50 + 273) \times (75 - 25) \times 0.05^3}{0.257 \times 10^{-4} \times (1.9512 \times 10^{-5}/1.0949)} = 4.14 \times 10^5.$$

Hence,

$$(\text{Nu}_\text{m})_S = 1 + 1.44\left(1 - \frac{1708}{4.14 \times 10^5 \times \cos 45}\right)^+ \left[1 - \frac{1708 \times (\sin 1.8 \times 45)^{1.6}}{4.14 \times 10^5 \times \cos 45}\right]$$
$$+ \left[\left(\frac{4.14 \times 10^5 \times \cos 45}{5830}\right)^{1/3} - 1\right]^+$$
$$= 1 + 1.44 \times 0.994 \times 0.994 + 2.689 = 4.79.$$

Heat transfer coefficient,

$$h = \frac{k}{L}(\text{Nu}_\text{m})_s = \frac{0.02799}{0.05} \times 4.79 = 2.68\,\text{W/(m}^2\text{K)}.$$

Hence, heat flux is

$$q = h(T_1 - T_2) = 2.68 \times 50 = 134\,\text{W/m}^2.$$

Here the radiation heat transfer has not been considered.

Example 9.18 A cover plate with 50 mm air gap is installed over a horizontal surface (0.4 m × 0.4 m in area) maintained at T_1= 75°C to reduce heat transfer from the surface as shown in Fig. 9.9. Determine the heat transfer rate to the still air at 25°C above the cover plate. Neglect radiation heat transfer and contact resistance between the surface and cover.

Solution

In equilibrium, the heat transfer across the air gap of the cover equals the heat rejection from the top surface of the cover, i.e.

$$h_1(T_1 - T_2) = h_2(T_2 - T_\infty) \qquad \text{(i)}$$

where T_2 is not known which is required to estimate h_1 and h_2.

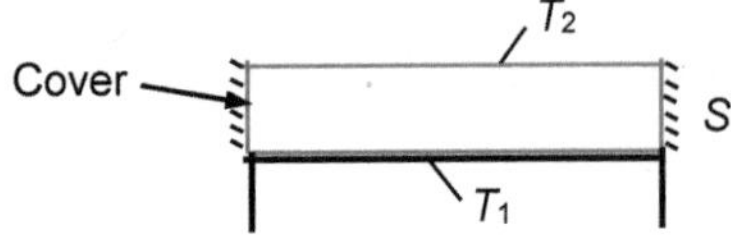

Fig. 9.9 Example 9.18

For trial, we assume T_2 = 50°C. For heat transfer through air gap of the cover, mean air temperature is $T_{m1} = (T_1 + T_2)/2 = (75 + 50)/2 = 62.5$°C. Air properties at mean temperature are:

$$\rho = 1.05\,\text{kg/m}^3,\ \mu = 2.0085 \times 10^{-5}\,\text{N s/m}^2,\ k = 0.02895\,\text{W/(m K)}\ \text{and Pr} = 0.7.$$

The Rayleigh number is

$$\text{Ra} = \frac{\beta g(T_1 - T_2)S^3}{(\mu/\rho)^2}\text{Pr} = \frac{1/(62.5+273) \times 9.81 \times (75 - 50) \times 0.05^3}{(2.0085 \times 10^{-5}/1.05)^2} \times 0.7 = 1.75 \times 10^5.$$

From Eq. (9.34),

$$(\text{Nu}_S)_1 = \frac{h_1 S}{k} = 0.069\text{Ra}_S^{1/3}\text{Pr}^{0.074}$$

or

$$\begin{aligned} h_1 &= \frac{k}{S} \times 0.069\text{Ra}_S^{1/3}\text{Pr}^{0.074} \\ &= \frac{0.02895}{0.05} \times 0.069 \times (1.75 \times 10^5)^{1/3} \times 0.7^{0.074} \\ &= 2.18\,\text{W/(m}^2\text{K)}. \end{aligned}$$

For top surface of the cover, the film temperature is $T_{m2} = (T_2 + T_\infty)/2 = (50 + 25)/2 = 37.5$°C. Air properties at the film temperature are:

$$\rho = 1.1409\,\text{kg/m}^3,\ \mu = 1.8938 \times 10^{-5}\,\text{N s/m}^2,\ k = 0.02704\,\text{W/(m K)}\ \text{and Pr} = 0.706.$$

The Rayleigh number is

$$\begin{aligned} \text{Ra} &= \frac{\beta g(T_2 - T_\infty)L^3}{(\mu/\rho)^2}\text{Pr} = \frac{1/(37.5+273) \times 9.81 \times (50 - 25) \times 0.1^3}{(1.8938 \times 10^{-5}/1.1409)^2} \times 0.706 \\ &= 2.02 \times 10^6 \end{aligned}$$

where $L = A_s/P = 0.4 \times 0.4/(4 \times 0.4) = 0.1$ m.

When the upper side of the plate is heated Eq. (9.17) gives

$$(\text{Nu}_m)_2 = h_2\frac{L}{k} = 0.54\text{Ra}_L^{1/4}$$

or

$$h_2 = \frac{k}{L} \times 0.54\mathrm{Ra}_L^{1/4}$$
$$= \frac{0.02704}{0.1} \times 0.54 \times (2.02 \times 10^6)^{1/4} = 5.5\,\mathrm{W/(m^2K)}.$$

Substitution in Eq. (i) gives

$$h_1(T_1 - T_2) = h_2(T_2 - T_\infty)$$

or

$$2.18 \times (75 - T_2) = 5.5 \times (T_2 - 25)$$

or

$$T_2 = 39.2^\circ\mathrm{C}.$$

as against the assumed value of 50°C. We repeat the above analysis with T_2 = 40°C.

For heat transfer through air gap of the cover, mean air temperature is $T_{m1} = (T_1 + T_2)/2$ = (75 + 40)/2 = 57.5°C. Air properties at mean temperature are:

$$\rho = 1.068\,\mathrm{kg/m^3},\ \mu = 1.986 \times 10^{-5}\,\mathrm{N\ s/m^2},\ k = 0.02856\,\mathrm{W/(m\ K)}\ \text{and}\ \mathrm{Pr} = 0.7012.$$

The Rayleigh number is

$$\mathrm{Ra} = \frac{\beta g(T_1 - T_2)S^3}{(\mu/\rho)^2}\mathrm{Pr} = \frac{1/(57.5 + 273) \times 9.81 \times (75 - 40) \times 0.05^3}{(1.986 \times 10^{-5}/1.068)^2} \times 0.7012$$
$$= 2.63 \times 10^5.$$

From Eq. (9.34),

$$(\mathrm{Nu}_S)_1 = \frac{h_1 S}{k} = 0.069\mathrm{Ra}_S^{1/3}\mathrm{Pr}^{0.074}$$

or

$$h_1 = \frac{k}{S} \times 0.069\mathrm{Ra}_S^{1/3}\mathrm{Pr}^{0.074}$$
$$= \frac{0.02856}{0.05} \times 0.069 \times (2.63 \times 10^5)^{1/3} \times 0.7012^{0.074}$$
$$= 2.46\,\mathrm{W/(m^2K)}.$$

For top surface of the cover, the film temperature is $T_{m2} = (T_2 + T_\infty)/2$ = (40 + 25)/2 = 32.5°C. Air properties at the film temperature are:

$$\rho = 1.1592\,\mathrm{kg/m^3},\ \mu = 1.8708 \times 10^{-5}\,\mathrm{N\ s/m^2},\ k = 0.02665\,\mathrm{W/(m\ K)}\ \text{and}\ \mathrm{Pr} = 0.7072.$$

The Rayleigh number is

$$\begin{aligned}\mathrm{Ra} &= \frac{\beta g(T_2 - T_\infty)L^3}{(\mu/\rho)^2}\mathrm{Pr} = \frac{1/(32.5+273)\times 9.81\times(50-40)\times 0.1^3}{(1.8708\times 10^{-5}/1.1592)^2}\times 0.7072 \\ &= 8.72\times 10^5.\end{aligned}$$

When the upper side of the plate is heated Eq. (9.17) gives

$$(\mathrm{Nu}_m)_2 = h_2\frac{L}{k} = 0.54\mathrm{Ra}_\mathrm{L}^{1/4}$$

or

$$\begin{aligned}h_2 &= \frac{k}{L}\times 0.54\mathrm{Ra}_\mathrm{L}^{1/4} \\ &= \frac{0.02665}{0.1}\times 0.54\times(8.72\times 10^5)^{1/4} = 4.4\,\mathrm{W/(m^2K)}.\end{aligned}$$

Substitution in Eq. (i) gives

$$h_1(T_1 - T_2) = h_2(T_2 - T_\infty)$$

or

$$2.46\times(75 - T_2) = 4.4\times(T_2 - 25)$$

or

$$T_2 = 42.9^\circ\mathrm{C}.$$

as against the second trial value of 40°C. For greater accuracy, third iteration may be carried out.

Heat transfer rate with this approximation of temperature T_2 is

$$q = h_1 A_s(T_1 - T_2) = 2.46\times 0.4\times 0.4\times(75 - 42.9) = 12.63\,\mathrm{W}.$$

9.8 Combined Free and Forced Convection (Kays and Crawford 1980; Gebhart 1961; Holman 1992; Cengel 2007)

Till now we have considered either the forced or the free convection. But in a fluid, if temperature gradient is present natural convection also occurs. When the velocity of a fluid is high (and correspondingly large value of the Reynolds number) and the Rayleigh number is small, the forced convection dominates and the effect of the free convection can be neglected. On the other hand, at low values of the Reynolds number and high value of the Rayleigh number, the buoyancy forces generate a convective velocity that alters the velocity and temperature fields in the forced convection flows. In general, the buoyancy effect is negligible in fully established turbulent flows and laminar flows with low Grashof number. Mixed convection basically occurs when moderate to large Grashof number is associated with laminar and transition flows.

The parameters which influence the heat transfer are the Reynolds number, Prandtl number, Rayleigh number, the geometry of the heat transferring surface and the orientation of the forced flow relative to the gravitational force. For a given geometry of the surface, the relative strength of the buoyancy and inertia forces is measured by Richardson number $Ri = Gr/Re^2$. In general, the effect of natural convection is negligible when $Gr/Re^2 < 0.1$. The forced convection is negligible when $Gr/Re^2 > 10$. In the range $0.1 > Gr/Re^2 > 10$, mixed flow condition occurs, and both natural and forced convection must be considered.

Sparrow and Gregg (1959) determined the conditions under which the buoyancy effect would appreciably change the heat transfer rate in forced-flow laminar boundary-layer regime over a vertical surface. The analysis considered Pr = 0.01, 1.0 and 10. They found that the effect of the buoyancy on the heat transfer coefficient will be less than 5 per cent if

$$\frac{\mathrm{Gr}_x}{\mathrm{Re}_x^2} \le 0.225 \tag{9.37}$$

where both the Grashof and Reynolds numbers are based on the distance from the leading edge of the plate. The above mentioned limits apply approximately for buoyancy force either opposed to, or in the direction of the forced flow.

Lloyd and Sparrow (1970) carried out similarity analysis to study the effect of buoyancy on forced convection over isothermal vertical surfaces. They showed that for a 5 per cent increase in the Nusselt number

$$\begin{aligned}
\frac{\mathrm{Gr}_x}{\mathrm{Re}_x^2} &= 0.24 \quad \text{for Pr} = 100 \\
\frac{\mathrm{Gr}_x}{\mathrm{Re}_x^2} &= 0.13 \quad \text{for Pr} = 10 \\
\frac{\mathrm{Gr}_x}{\mathrm{Re}_x^2} &= 0.08 \quad \text{for Pr} = 0.72 \\
\frac{\mathrm{Gr}_x}{\mathrm{Re}_x^2} &= 0.056 \quad \text{for Pr} = 0.08.
\end{aligned} \tag{9.38}$$

For flow of air on a heated horizontal plate facing upwards a 5 per cent increase in the heat transfer coefficient is reported for aiding flow by Chen et al. (1977) when

$$\frac{\mathrm{Gr}_x}{\mathrm{Re}_x^{2.5}} \approx 0.05 \tag{9.39}$$

For the turbulent flow of air in a vertical tube of $L/D \approx 5$, Eckert and Diaguila (1954) found a deviation of less than 10 per cent from pure natural convection for

$$\frac{\mathrm{Gr}_x}{\mathrm{Re}_x^{2.5}} \ge 0.007 \tag{9.40}$$

and a deviation of less than 10 per cent from the pure forced-convection was found for

$$\frac{\mathrm{Gr}_x}{\mathrm{Re}_x^{2.5}} \leq 0.0016 \tag{9.41}$$

where the region between these two limits is that of mixed convection.

Natural convection may assist or resist forced convection depending on the relative directions of buoyancy-induced currents and the forced convection motion. In *assisting flows,* the natural convection currents are in the direction of the forced flow. Therefore, natural convection assists forced convection and enhances heat transfer. In *opposing flows,* the natural convection currents are in the opposite direction to the forced flow and the natural convection resists forced convection. Therefore, the heat transfer is decreased. While in *transverse flows,* the natural convection currents are perpendicular to the forced flow. The transverse flow enhances fluid mixing and hence enhances the heat transfer.

When neither natural nor forced convection are negligible, the combined Nusselt number can be determined from the following correlation in the absence of specific relation. The given relation is based on experimental data (Cengel 2007).

$$\mathrm{Nu}_{\mathrm{combined}} = \left(\mathrm{Nu}_{\mathrm{forced}}^n \pm \mathrm{Nu}_{\mathrm{natural}}^n\right)^{1/n} \tag{9.42}$$

where $\mathrm{Nu}_{\mathrm{forced}}$ and $\mathrm{Nu}_{\mathrm{natural}}$ are determined from the correlations for pure forced and pure natural convection, respectively. The plus sign is for *assisting* and *transverse* flows and the minus sign is for the *opposing* flows. The value of the exponent n varies between 3 and 4 depending on the geometry involved. The exponent $n = 3$ generally gives good results. For transverse flows over horizontal surfaces $n = 7/2$, and for the cylinders and spheres $n = 4$ are better suited (Incropera et al. 2012).

Mori and Futagami (1967) carried out analytical and experimental study of convective heat transfer with buoyancy inside horizontal tubes for uniform heat flux condition. For air, they found that the Nusselt number began to increase when the product (Re.Gr.Pr) exceeded 10^3. They attributed the increase in the Nusselt number to the secondary flow. For very large values of (Re.Gr.Pr), the following relation for the ratio of the mixed convection Nusselt number Nu_o and pure forced-convection Nusselt number Nu_o is suggested:

$$\frac{\mathrm{Nu}}{\mathrm{Nu}_o} = 0.1634(\mathrm{ReGr\,Pr})^{0.2} \quad \text{for Pr} = 0.72 \tag{9.43}$$

where the Grashof number is defined using the tube radius while the Reynolds number has been defined using the tube diameter.

Brown and Gauvin (1965) developed a correlation of mixed-convection in horizontal tubes applicable to the laminar flow regime as

$$\mathrm{Nu} = 1.75\left[\left(\mathrm{RePr}\frac{D}{L}\right) + 0.012\left(\mathrm{RePr}\frac{D}{L}\mathrm{Gr}^{1/3}\right)^{4/3}\right]^{1/3}\left(\frac{\mu_b}{\mu_w}\right)^{0.14} \tag{9.44}$$

For the turbulent flow mixed convection in horizontal tubes, the suggested correlation due to Metais and Eckert (1964) is

$$\mathrm{Nu} = 4.69\mathrm{Re}^{0.27}\mathrm{Pr}^{0.21}\mathrm{Gr}^{0.07}\left(\frac{D}{L}\right)^{0.36} \tag{9.45}$$

Example 9.19 For a vertical isothermal plate 0.5 m high in air at 313 K, determine the plate temperature for which the natural convection effect will be less than 5 per cent at the upper end of the plate. The air flow is vertically upwards at a velocity of 2 m/s.

Solution

The condition for 5% effect of free convection on the forced convection is

$$\frac{\mathrm{Gr}_x}{\mathrm{Re}_x^2} = 0.225$$

Assuming a mean film temperature of 350 K (77°C), the fluid properties are:

$$\rho = 1.00\,\mathrm{kg/m^3},\ \mu = 2.07 \times 10^{-5}\,\mathrm{kg/(m\ s)}\ \text{and}\ \beta = 1/T = (1/350)\,\mathrm{K}^{-1}.$$

Corresponding to L = 0.5 m, the Reynolds number is

$$\mathrm{Re}_L = \frac{\rho U_\infty L}{\mu_f} = \frac{1.00 \times 2 \times 0.5}{2.07 \times 10^{-5}} = 48309.$$

Using the given condition for 5% effect of free convection on the forced convection, we have

$$\mathrm{Gr} = 0.225(\mathrm{Re})^2 = 0.225(48309)^2 = 5.25 \times 10^8.$$

The plate surface temperature corresponding to this limiting value of the Grashof number is

$$\Delta T = \frac{\mathrm{Gr}}{\rho^2 \beta g L^3/\mu^2} = \frac{5.25 \times 10^8}{(1.00)^2 \times (1/350) \times 9.81 \times 0.5^3 \times 1/(2.07 \times 10^{-5})^2} = 64.2\,^\circ\mathrm{C}.$$

The plate temperature at its upper end will be

$$t_s = \Delta t + t_{air} = 64.2 + 40 = 104.2\,^\circ\mathrm{C}.$$

This gives

$$T_m = \frac{t_s + t_{air}}{2} + 273 = 345\ \mathrm{K},$$

which is approximately equal to the assumed value of the mean temperature of 350 K. Hence, retrial is not required.

Determined plate temperature of 104.2°C is the limiting value. For a temperature less than 104.2°C, the value of the Grashof number will be less than the calculated limiting value of 5.28×10^8 and the effect of natural convection on the free convection heat transfer will be less than 5%.

Example 9.20 Air, at 1 bar and 300 K, flows through a 20 mm diameter horizontal tube at a velocity of 0.5 m/s. The tube wall is maintained at a uniform temperature of 400 K. Determine the effect of natural convection on the heat transfer coefficient if the tube length is 0.5 m.

Solution

The fluid properties at the mean bulk temperature T_b = 300 K (27°C) are:

μ = 1.845×10^{-5} kg/(m s)
ρ = 1.18 kg/m^3
k = 0.0262 W/(m K)
Pr = 0.7085
β = $1/T_b$ = (1/300) K^{-1}
μ_w = 2.285×10^{-5} kg/(m s) at the wall temperature of 400 K

For the calculation of the above values, the rise in the temperature of the air through the tube is assumed to be small.

The various parameters required are calculated as

$$\mathrm{Re}_d = \frac{\rho U_\infty d}{\mu} = \frac{1.18 \times 0.5 \times 0.02}{1.845 \times 10^{-5}} = 639$$

$$\mathrm{Gr} = \frac{\rho^2 \beta g \Delta T L^3}{\mu^2} = \frac{(1.18)^2 \times (1/300) \times 9.81 \times (400 - 300) \times 0.02^3}{(1.845 \times 10^{-5})^2} = 10.7 \times 10^4.$$

The flow is in the laminar regime hence the equation of Brown and Gauvin for horizontal tubes applies. The Nusselt number is

$$\mathrm{Nu} = 1.75\left[\left(\mathrm{RePr}\frac{D}{L}\right) + 0.012\left(\mathrm{RePr}\frac{D}{L}\mathrm{Gr}^{1/3}\right)^{4/3}\right]^{1/3}\left(\frac{\mu_b}{\mu_w}\right)^{0.14} \tag{9.44}$$

$$\mathrm{RePr}\frac{D}{L} = 639 \times 0.7085 \times \frac{0.02}{0.5} = 18.1.$$

This gives

$$\mathrm{Nu} = 1.75\left\{18.1 + 0.012\left[18.1 \times (10.7 \times 10^4)^{1/3}\right]^{4/3}\right\}^{1/3}\left(\frac{1.845}{2.285}\right)^{0.14} = 8.26.$$

The development length L_{hy} in laminar flow from Eq. (7.4) is

$$L_{hy} = (0.05\mathrm{Re}) \times D = (0.05 \times 639) \times 0.02 = 0.639 \text{ m}.$$

Flow is developing hence its effect on the mean value of the Nusselt number must be considered.

The non-dimensional tube length,

$$x^+ = \left(\frac{x}{R}\right)\frac{1}{\mathrm{Re\,Pr}} = \left(\frac{0.5}{0.01}\right) \times \frac{1}{639 \times 0.7085} = 0.11.$$

From Fig. 8.3, for $x^+ = 0.11$,

$$\mathrm{Nu_m} \approx 4.6.$$

The effect of the natural convection on the laminar forced flow Nusselt number is [(8.26 – 4.6)/4.6] × 100 = 79.6%, which is significant.

Example 9.21 A heated vertical plate, 600 mm wide and 400 mm high, at 350 K is cooled by blowing air of 300 K at 0.8 m/s. Estimate the heat rejection rate if the air is blown upwards along the plate surface.

Solution

Let us first check whether it is a case of mixed convection.

Air properties at the mean film temperature T_m = 325 K (52°C) are $\beta = \frac{1}{T_m} = \frac{1}{325}$ K^{-1}, k = 0.028 W/(m K), ρ = 1.088 kg/m^3, $\mu = 1.96 \times 10^{-5}$kg/(m s) and Pr ≈ 0.7025.

For forced flow,

$$\mathrm{Re_L} = \frac{\rho UL}{\mu} = \frac{1.088 \times 0.8 \times 0.4}{1.96 \times 10^{-5}} = 1.77 \times 10^4.$$

For the flow parallel to a plate, it is laminar.

For free or natural convection,

$$\mathrm{Gr}_L = \frac{\rho^2 \beta g \Delta T L^3}{\mu^2} = \frac{1.088^2 \times (1/325) \times 9.81 \times (350 - 300) \times 0.4^3}{(1.96 \times 10^{-5})^2} = 2.98 \times 10^8$$

$$\mathrm{Ra}_L = \mathrm{Gr}_L\,\mathrm{Pr} = 2.98 \times 10^8 \times 0.7025 = 2.09 \times 10^8 < 10^9,$$

the free convection flow is also laminar.

Richardson number Ri (= Gr/Re^2),

$$\frac{Gr_L}{Re_L^2} = \frac{2.98 \times 10^8}{(1.77 \times 10^4)^2} = 0.95,$$

i.e., the number lies in the range 0.1–10 and it is a case of mixed convection.

Forced convection Nusselt number,

$$Nu_{forced} = 0.664 Re^{1/2} \, Pr^{1/3} = 0.664(1.77 \times 10^4)^{1/2}(0.7025)^{1/3} = 78.5.$$

Free convection Nusselt number from Eq. (9.6),

$$Nu_{natural} = 0.59 Ra^{1.4} = 0.59(2.09 \times 10^8)^{1/4} = 70.94.$$

Mixed convection Nusselt number for assisting flow, from Eq. (9.42) with n = 3 for vertical plate,

$$Nu_{combined} = \left(Nu_{forced}^3 + Nu_{natural}^3\right)^{1/3} = \left[(78.5)^3 + (70.94)^3\right]^{1/3} = 94.4.$$

This gives,

$$h = Nu_{combined} \frac{k}{L} = 94.4 \times \frac{0.028}{0.4} = 6.6 \text{ W/(m}^2\text{ K)}.$$

$$q = hA(T_w - T_{air}) = 6.6 \times 0.6 \times 0.4(350 - 300) = 79.2 \text{ W}.$$

Example 9.22 Water flows across a 50 mm diameter long cylinder with surface temperature of 45°C. The free stream conditions are U_∞ = 0.05 m/s and t_∞ = 25°C. Determine the Nusselt number.

Solution

At film temperature of 35°C, thermophysical properties of water are:

$$\rho = 994\,\text{kg/m}^3,\ \mu = 718 \times 10^{-6}\,\text{N s/m}^2,\ k = 0.624\,\text{W/(m K)},\ \text{Pr} = 4.81 \text{ and}$$
$$\beta = 0.342 \times 10^{-3}\,(1/\text{K}).$$

The Grashof number,

$$Gr_D = \frac{\beta g(t_s - t_\infty)D^3}{\nu^2} = \frac{0.342 \times 10^{-3} \times 9.81 \times (45 - 25) \times 0.05^3}{(718 \times 10^{-6}/994)^2} = 16 \times 10^6.$$

Flow Reynolds number,

$$\mathrm{Re_D} = \frac{U_\infty D}{\nu} = \frac{0.05 \times 0.05}{718 \times 10^{-6}/994} = 3461$$

$$\frac{\mathrm{Gr_D}}{\mathrm{Re_D^2}} = 1.336.$$

Free convection effect will be significant.

Rayleigh number,

$$\mathrm{Ra_D} = \mathrm{Gr_D}\,\mathrm{Pr} = 16 \times 10^6 \times 4.81 = 77 \times 10^6.$$

The mean Nusselt number for natural convection from Eq. (9.23),

$$\mathrm{Nu}_{natural} = 0.53\mathrm{Ra_D^{1/4}} = 0.53 \times (77 \times 10^6)^{1/4} = 49.64.$$

From Eq. (8.47) and Table 8.11,

$$\mathrm{Nu_{forced}} = 0.683\mathrm{Re}^{0.466}\mathrm{Pr}^{1/3} = 0.683(3461)^{0.466}(4.81)^{1/3} = 51.4.$$

The combined Nusselt number from Eq. (9.42),

$$\mathrm{Nu_{combined}} = \left(\mathrm{Nu_{forced}^{\mathit{n}}} \pm \mathrm{Nu_{natural}^{\mathit{n}}}\right)^{1/n}$$

The plus sign is for *assisting* and *transverse* flows and the minus sign is for the *opposing* flows. The exponent $n = 3$ generally gives good results. For the cylinders $n = 4$ is better suited.

Hence, transverse flow combined Nusselt number is

$$\mathrm{Nu_{combined}} = \left(51.4^4 + 49.64^4\right)^{1/4} = 60.1,$$

assisting flow combined Nusselt number is

$$\mathrm{Nu_{combined}} = \left(51.4^3 + 49.64^3\right)^{1/3} = 63.67$$

and opposing flow combined Nusselt number is

$$\mathrm{Nu_{combined}} = \left(51.4^3 - 49.64^3\right)^{1/3} = 23.8.$$

9.9 Summary

In the natural or free convection, the motion of the fluid is caused by the buoyancy force arising from the change in the density of the fluid due to the rise in its temperature. Flow structure and development of boundary layer for vertical plate or cylinder, horizontal plates (with heated surface facing upwards and downwards), horizontal cylinders, sphere, parallel plate channels and enclosed spaces have been discussed.

The technique of dimensional analysis has been applied to develop functional relationship for natural convection heat transfer in terms of dimensionless numbers Gr and Pr as Nu = (Gr)ϕ(Pr). The form of functions ψ(Gr) and ϕ(Pr) are specified for different conditions of heat transfer by convection on the basis of theoretical analysis or experimental investigations. Physical interpretation of the Grashof number Gr has also been presented.

Experimental scheme for determination of the natural convection heat transfer coefficient from a long horizontal tube has been presented in Sect. 9.4, which has been extended for determination of heat transfer coefficient from vertical and horizontal plates.

In Sect. 9.5.1, empirical relations of heat transfer coefficient applicable to vertical plate for both laminar flow (Ra < 10^9) and turbulent flow (Ra $\geq$ 10^9) pertaining to uniform temperature and uniform heat flux conditions are presented. Condition for treating a vertical cylinder as a vertical plate is specified in Eq. (9.5).

For inclined plates ($0 \leq \theta \leq 60°$), correlation presented in Eq. (9.10) due to Churchill and Chu is valid when the acceleration due to gravity g is replaced by its component parallel to the wall $g \cos\theta$, where θ is the angle of inclination with the vertical. This approach is satisfactory only for the top and bottom surfaces of cooled and heated plates, respectively.

Horizontal plate correlations for both uniform temperature and uniform heat flux conditions are presented in Sect. 9.5.3, with characteristic length L defined as ratio of the heat transfer area A and perimeter P of the plate. The relations cover both the conditions of the upper side of the plate heated or lower side cooled and upper side of the plate cooled or lower side heated.

Sections 9.5.4 and 9.5.5 present heat transfer correlations for horizontal cylinder and sphere, respectively.

Array of circuit boards or fin arrays employed to enhance free convection heat transfer may be regarded as parallel vertical plates forming a channel that is open to the ambient at its ends. The heated fluid rises and exits from the top of the channel. Researchers have presented correlations for both vertical and inclined channels in terms of Rayleigh number Ra. Because of the effect of the ratio of spacing to height S/L on the flow structure, this ratio appears in the heat transfer correlations.

In some engineering applications, heat transfer takes place between two surfaces that are at different temperatures and a fluid is present in the space enclosed by these surfaces. Typical examples of such enclosures are double pane windows and cavities in the walls of the buildings (forming vertical enclosures), and solar collectors (forming horizontal or inclined enclosures), etc. Flow structures in both vertical and horizontal enclosures have been discussed and correlations for vertical, inclined and horizontal rectangular enclosure are presented in Sect. 9.7. It is to note that if the upper wall is heated in case of horizontal enclosure, then the heat transfer takes place by conduction only (Nu = 1) irrespective of the value of the Rayleigh number. The radiation heat exchange between the hot and cold surfaces is always present when a gas is filled in the cavity.

In many applications, forced and free convections may coexist. When the velocity of a fluid is high (and correspondingly large value of inertia forces and hence the Reynolds number) and the Rayleigh number is small, the forced convection dominates and the effect of the free convection can be neglected. On the other hand, at low values of the Reynolds number and high value of the Rayleigh number, the buoyancy forces generate a convective velocity that alters the velocity and temperature fields in the forced convection flows. The relative strength of the buoyancy and inertia forces is measured by Richardson number Ri = Gr/Re^2. In general, the effect of natural convection is negligible when Gr/Re^2 < 0.1. The forced convection is negligible when Gr/Re^2 > 10. In the range 0.1 > Gr/Re^2 > 10, mixed flow condition occurs, and both natural and forced convection must be considered. Some specific relations to determine combined Nusselt number for horizontal tubes are presented, refer Eqs. (9.43)–(9.45). In the absence of specific relation, Eq. (9.42) may be used to determine combined Nusselt number.

Review Questions

9.1 List the variables that affect the free convection heat transfer coefficient.

9.2 Using dimensional analysis, show that the heat transfer by natural convection can be given by a relationship of the form

$$\mathrm{Nu} = f(\mathrm{Gr,Pr})$$

where Nu is the Nusselt number, Pr is Prandtl number and Gr is the Grashof number.

9.3 With the help of Buckingham pi theorem, show that for the free convection

$$\mathrm{Nu} = f(\mathrm{Gr,Pr})$$

9.4 At low flow velocities, the free convection effect may be present in the forced convection heat transfer case. Using dimensional analysis, show that the following form of correlation is obtained

$$f(\mathrm{Nu,Re,\ Pr,Gr}) = 0$$

9.5 What is physical interpretation of the Grashof number?

9.6 What is the Rayleigh number?

9.7 Explain the physical reason for the optimum spacing for vertical channel and an array of vertical plates.

9.8 Discuss flow structures in (i) a vertical enclosure, (ii) horizontal enclosure and (iii) an inclined enclosure.

9.9 Discuss heat transfer in a horizontal enclosure when upper surface is hot.

9.10 What do you mean by mixed convection? What is the condition for the consideration of the mixed convection in the heat transfer analysis?

9.11 How do the assisting, opposing and transverse flows affect the heat transfer in mixed convection?

Problems

9.1. A 3.5 m high and 5 m wide vertical plate, maintained at 50°C, is exposed to surrounding air at 10°C. Calculate the heat transfer rate.
[**Ans.** Air properties at mean film temperature ½ (50 + 10) = 30°C:$\beta = \frac{1}{T_m} = \frac{1}{303}\ \mathrm{K}^{-1}$, k = 0.0265 W/(m K), ρ = 1.1684 kg/m^3. $\mu = 1.86 \times 10^{-5}$kg/(m s) and Pr = 0.708; Ra = $(\rho^2 g\beta \Delta T L^3)/\mu^2$.Pr= 1.55×10^{11}; Using Eq. (9.10), Nu = 604.8; $h_m = \mathrm{Nu}k/L$ = 4.58 W/(m^2 K); $q = h_m A\ (t_w - t_\infty)$ = 3206 W; Alternatively from Eq. (9.9), Nu = 0.1(Ra)$^{1/3}$ = 536.7, which is about 11% lower.]

9.2. A 20 mm diameter horizontal heated cylinder at a uniform temperature of 40°C is cooled by the surrounding air at 20°C. Calculate the heat loss rate per unit length of the cylinder.
[**Ans.** Air properties at mean film temperature ½ (40 + 20) = 30°C:$\beta = \frac{1}{T_m} = \frac{1}{303}$ K^{-1}, k = 0.0265 W/(m K), ρ = 1.1684 kg/m^3. μ = 1.86 × 10^{-5}kg/(m s) and Pr = 0.708; Ra = [($g\beta\Delta Td^3$)/v^2]Pr = 1.45 × 10^4; From Eq. (9.23), Nu$_m$ = 0.53(Ra)$^{1/4}$ = 5.82; h_m = Nuk/d = 7.71 W/(m^2 K); $q/L = h_m(\pi d)$ $(t_w - t_\infty)$ = 9.69 W/m.]

9.3. If the cylinder in the above problem is dipped in still water at 20°C, what will be the cooling rate? Given $g\beta\rho^2c/(\mu k)$ = 2.4 × 10^{10} m^{-3}K^{-1}.
[**Ans.** Ra = GrPr = $g\beta\rho^2c/(\mu k)\times\ \Delta Td^3$ = 3.84 × 10^6; From Eq. (9.23), Nu = 0.53 (Ra)$^{1/4}$ = 23.46; h_m = Nuk/d = 727.3 W/(m^2 K) for k = 0.62 W/(m K) at t_m = 30°C; $q/L = h_m$ (πd) $(t_w - t_\infty)$ = 913.9 W/m. Comments: Water cooling is very effective as compared to the air cooling.]

9.4. A vertical surface 4 m high and 1.8 m wide is subjected to uniform heat flux of 1000 W/m^2. The surface is insulated from the other side. All of the incident heat is rejected by free convection to the surrounding air at 20°C. What average temperature will the plate attain?
[**Ans.** With a trial value of h = 5 W/(m^2 K), $\Delta T = q_w/h$ = 200°C; t_w = (20 + 200) = 220°C; t_m = ½ (20 + 220) = 120°C for air properties; $\beta = \frac{1}{T_m} = \frac{1}{393}$ K^{-1}, k = 0.0331 W/(m K), ρ = 0.8988 kg/m^3. μ = 2.2565 × 10^{-5} kg/(m s) and Pr = 0.69; Gr_x^* = Gr$_x$ Nu$_x$ = $(\rho^2 g\beta q_w x^4)/(k\mu^2)$ = 3.02 × 10^{14}; From Eq. (9.14), $h_{x=L} = k/x$ (0.17) $(\text{Gr}_x^*\text{Pr})^{\mathbf{1/4}}$ = 5.34 W/(m^2 K); For turbulent heat transfer $h_x = h_m$. Since the calculated value is approximately equal to the trial value, there is no need of retrial; Revised $t_w = \Delta t + t_{air} = q_w/h + t_{air} \approx$ 210°C for $h_m \approx$ 5.3.]

9.5. Estimate the convective heat transfer rate to a cylinder of diameter 200 mm and height 1.0 m if it is placed vertically in essentially still air at a temperature of 60°C. The surface of the cylinder is maintained at 10°C.
[**Ans.** Mean film temperature = 35°C; Air properties: $v = \mu/\rho$ = 16.35 × 10^{-6} m^2/s, k = 0.0268 W/(m K), Pr = 0.7066, and β = 1/308 K^{-1};$\text{Gr}_L = \frac{g\beta(t_w - t_\infty)L^3}{v^2}$= 5.95 × 10^9; (35/Gr$^{1/4}$) = 0.126; D/L = 0.2; D/L > 35/Gr$^{1/4}$, vertical plate equation applies; Ra$_L$ = 4.2 × 10^9; From Eq. (9.9), Nu$_L$ = 0.1(Ra$_L$)$^{1/3}$ = 161.3; h = 4.32 W/(m^2 K); $q = hA\Delta T$ = 135.7 W.]

9.6. A vertical cylindrical surface, 0.4 m in diameter and 1.0 m in height, at 40°C is exposed to either water or air at 10°C. Determine relative magnitude of heat loss from two fluids by natural convection.
[**Ans. Water as coolant:** Properties of water at the mean film temperature $t_m = (t_i + t_s)/2$ = 25°C from Table A4: ρ =997 kg/m^3, μ = 890 × 10^{-6} N s/m^2, k = 0.609 W/(m K), Pr = 6.13 and β = 0.253 × 10^{-3} 1/K; $\text{Gr} = \frac{\rho^2 g\beta\Delta TL^3}{\mu^2} = \frac{997^2\times9.81\times0.253\times10^{-3}\times(40-10)\times1^3}{(890\times10^{-6})^2} =$ 9.34 × 10^{10}; D/L = 0.4; 35/Gr$^{0.25}$ = 0.0633. Hence, condition $\frac{D}{L} \geq \frac{35}{\text{Gr}_L^{1/4}}$ is satisfied and the cylinder can be treated as a vertical plate; Ra = GrPr = 9.34 × 10^{10} × 6.13 = 5.7 × 10^{11}; From Eq. (9.10),

$$\text{Nu}_m = \left\{0.825 + \frac{0.387\text{Ra}^{1/6}}{\left[1+(0.492/\text{Pr})^{9/16}\right]^{8/27}}\right\}^2 = \left\{0.825 + \frac{0.387\times(5.7\times10^{11})^{1/6}}{\left[1+(0.492/6.13)^{9/16}\right]^{8/27}}\right\}^2 = 1147.3;$$

$h_{water} = \frac{k}{L}\text{Nu}_m = \frac{0.609}{1.0} \times 1147.3 = 698.7$ W/(m^2 K).

Air as coolant: Air properties at the mean film temperature $t_m = (t_i + t_s)/2 = 25°C$ from Table A5: $\rho = 1.1868$ kg/m³, $\mu = 1.8363 \times 10^{-5}$ N s/m², $k = 0.02608$ W/(m K) and Pr = 0.709; $Gr = \frac{\rho^2 g\beta\Delta T L^3}{\mu^2} = \frac{1.1868^2 \times 9.81 \times (1/298) \times (40-10) \times 1^3}{(1.8363\times10^{-5})^2} = 4.13 \times 10^9$; $D/L = 0.4$; $35/Gr^{0.25} = 0.138$. Hence, condition $\frac{D}{L} \geq \frac{35}{Gr_L^{1/4}}$ is satisfied; Ra = GrPr = 2.93×10^9;

$$Nu_m = \left\{0.825 + \frac{0.387 Ra^{1/6}}{\left[1 + (0.492/Pr)^{9/16}\right]^{8/27}}\right\}^2 = \left\{0.825 + \frac{0.387 \times (2.93\times10^9)^{1/6}}{\left[1 + (0.492/0.709)^{9/16}\right]^{8/27}}\right\}^2 = 171.5;$$

$h_{air} = \frac{k}{L} Nu_m = \frac{0.02608}{1.0} \times 171.5 = 4.47$ W/(m² K); Relative magnitude of heat loss, $\frac{q_{water}}{q_{air}} = \frac{h_{water}\pi DL\Delta T}{h_{air}\pi DL\Delta T} = \frac{698.7}{4.47} = 156.3$; Water is much more effective heat transfer fluid than air.]

9.7. Air at atmospheric pressure is contained in an enclosure with sides made of 0.6 m 0.6 m vertical plates. The plates are separated by a distance of 20 mm. The lower and upper sides are closed by adiabatic plates. The temperatures of the left and right vertical plates are 100°C and 50°C, respectively. Calculate the free convection heat transfer rate between the plates.

[**Ans.** Fluid properties at mean temperature $(T_1 + T_2)/2 = 75°C$; $\rho = 1.0052$ kg/m³, $\mu = 2.066 \times 10^{-5}$ kg/(m s), $k = 0.0299$ W/(m K), Pr = 0.697, and $\beta = 1/348$ K⁻¹; $Ra_S = Gr_S Pr = [(\rho^2 g\beta\Delta T S^3)/\mu^2]Pr = 1.86 \times 10^4$; For $10 \leq L/S \leq 40$ and Ra = 10^4 – 10^7, Eq. (9.30) applies, which gives Nu = 1.76; $h = Nuk/L = 2.63$ W/(m² K); $q = hA(T_1 - T_2) = 47.34$ W.]

9.8. A vertical enclosure consists of two square plates (0.3 m × 0.3 m) separated by a distance of 10 mm. One of the vertical plates is maintained at a uniform temperature of 40°C while the other is at a uniform temperature of 10°C. Water at atmospheric pressure is filled in the enclosure. Calculate the heat transfer rate if the horizontal and side walls are insulated. Given $g\beta\rho^2 c/(\mu k) = 1.85 \times 10^{10}$ m⁻³K⁻¹ at the mean temperature.

[**Ans.** At the mean temperature of (40 + 10)/2 = 25°C, $k = 0.61$ W/(m K) and Pr = 6.13 for water from Table A4. The Rayleigh number = $g\beta\rho^2 c/(\mu k) \times \Delta T \times S^3 = 1.85 \times 10^{10} \times (40–10) \times 0.01^3 = 5.55 \times 10^5$; For $10 \leq L/S \leq 40$, $10^4 \leq Ra_S$ 10^7 and $1 \leq Pr \leq 2 \times 10^4$, Eq. (9.30) applies. $Nu_S = 0.42(Ra_S)^{1/4} Pr^{0.012} \left(\frac{L}{S}\right)^{-0.3} = 0.42 \times \left(5.55 \times 10^5\right)^{1/4} \times (6.13)^{0.012} \left(\frac{0.3}{0.01}\right)^{-0.3} = 4.22$; Mean heat transfer coefficient, $h = Nu\left(\frac{k}{S}\right) = 4.22 \times \left(\frac{0.61}{0.01}\right) = 257.42$ W/(m² K); Heat transfer rate, $q = hA(T_1 - T_2) = 257.42 \times (0.3 \times 0.3) \times (40 - 10) = 695$ W.]

9.9. For a vertical isothermal plate 300 mm high in air at 10°C and atmospheric pressure, determine what free stream velocity of the air will result in the forced convection effect for a plate temperature of 40°C if the flow is in vertical direction?

[**Ans.** At $T_m = 25°C = 298$ K, $\beta = 1/298$ K⁻¹, $\rho = 1.1868$ kg/m³, $\mu = 1.8363$ 10^{-5} kg/(m s); $Gr = (\rho^2 g\beta\Delta T L^3)/\mu^2 = 11.1 \times 10^7$. The forced convection is negligible when $Gr/Re^2 > 10$; $Re < 3300$; $U_\infty = Re\nu/L < 0.17$ m/s.]

References

Azevedo LFA, Sparrow EM (1985) Natural convection in open-ended inclined channels. J Heat Transfer 107 (4):893–901

Baehr Hans Dieter and Stephan Karl (2011) Heat and mass transfer. Springer-Verlag, Berlin Heidelberg

Bar-Cohen A, Rohsenow WM (1984) Thermally optimum spacing of vertical, natural convection cooled, parallel plates. J Heat Transfer 106(1):116–123

Brown CK, Gauvin WH (1965) Combined free and forced convection in aiding flows, I. II. Can. J. Chemical Engineering 43(6):306–318

Buchberg H, Catton I, Edwards DK (1976) Natural convection in enclosed spaces-a review of application to solar energy collection. J Heat Transfer 98(2):182–188

Cebeci T (1974) Laminar-free-convective-heat transfer from the outer surface of a vertical slender circular cylinder. Proc. 5th Int Heat Transfer conf., Vol. 3, Paper NC1.4, Tokyo: 15–19

Cengel Yunus A (2007) Heat transfer: a practical approach. Tata McGraw-Hill Education, India

Chen TS, Sparrow EM, Mucoglu A (1977) Mixed convection in boundary layer flow on a horizontal plate. J Heat Transfer 99:66–71

Churchill SW (1990) Free convection around immersed bodies. In: Hewitt GF (ed) Hemisphere handbook of heat exchanger design. Hemisphere Publishing, New York

Churchill SW, Chu HHS (1975a) Correlating equations for laminar and turbulent free convection from a horizontal cylinder. Int. J. Heat Mass Transfer 18:1049–1053

Churchill SW, Chu HHS (1975b) Correlating equations for laminar and turbulent free convection from a vertical plate. Int J Heat Mass Transfer 18:1323–1329

Eckert ERG, Diaguila AJ (1954) Convective heat transfer for mixed, free and forced flow through tubes. Trans ASME 76:497–504

Gebhart B (1961) Heat transfer. McGraw-Hill Book Co, New York

Globe S, Dropkin D (1959) Natural convection heat transfer in liquids confined between two horizontal plates. J Heat Transfer 81:24–28

Hollands KGT, Unny TE, Raithby GD, Konicek L (1976) Free convective heat transfer across inclined air layers. J Heat Transfer 98(2):189–193

Holman JP (1992) adapted for SI units by White PRS, Heat Transfer. McGraw-Hill Book Co, New York

Incropera Frank P, DeWitt David P, Bergman Theodore L, Lavine Adrienne S (2006) Fundamentals of heat and mass transfer, 6th edn. John Wiley & Sons, New York

Incropera Frank P, DeWitt David P, Bergman Theodore L, Lavine Adrienne S (2012) Principles of heat and mass transfer. John Wiley & Sons, New York

Kays WM, Crawford ME (1980) Convective heat and mass transfer. Tata McGraw-Hill Publishing Co., Ltd, New Delhi, India

Lloyd JR, Sparrow EM (1970) Combined forced and free convection flow on vertical surfaces. Int J Heat Mass Transfer 13:434–438

MacGregor RK, Emery AP (1969) Free convection through vertical plane layers, moderate and high Prandtl number fluids. J Heat Transfer 91:391–403

McAdams WH (1954) Heat transmission, 3rd edn. McGraw-Hill, New York

Metais B, Eckert ERG (1964) Forced, mixed and free convection regimes. J Heat Transfer. Ser. C 86:295–296

Mills AF (1995) Heat and mass transfer. Richard D, Irwin, Chicago

Mori Y, Futagami K (1967) Forced convective heat transfer in uniformly heated horizontal tubes (2nd report, theoretical study). Int J Heat Mass Transfer 10:1801–1813

Popiel CO (2008) Free convection heat transfer from vertical slender cylinders: a review. Heat Transfer Eng 29(6):521–536

Sparrow EM, Gregg JL (1959) Buoyancy effects in forced convection flow and heat transfer. J Appl Mech, Trans ASME 26, Ser E(1): 133–134

10 Laws of Thermal Radiation

10.1 Introduction

All bodies emit radiation to their surroundings through electromagnetic waves due to the conversion of the internal energy of the body into radiation. Thermal radiation or heat radiation is also a form of electromagnetic emission. Since electromagnetic waves can also travel through a vacuum hence, in contrast to the conduction and convection heat transfer, it can take place through a perfect vacuum. Thus, when no medium is present, radiation becomes the only mode of heat transfer. Common examples are the solar radiation reaching the earth and the heat dissipation from the filament of an incandescent lamp. Thus, heat is transferred between two bodies over a great distance.

In general, the radiation means propagation of the electromagnetic waves of all wavelengths. The emitted energy can range from radio waves, which have wavelengths of kilometers, to cosmic rays with wavelengths of less than 10^{-10} cm. Figure 10.1 and Table 10.1 depicts the electromagnetic spectrum that extends from very small to very large wavelengths ($\lambda = 0 - \infty$).

The most commonly used unit for the measurement of the wavelength of the electromagnetic waves is Angstrom ($\mathring{A} = 10^{-10}$ m) or micron (μm = 10^{-6} m).

At small wavelength end ($\lambda < 0.01$ μm) are gamma rays and x-rays, which are not thermally stimulated. The same is true for the radar, television and radio waves ($\lambda > 10^3$ μm). The thermal radiation ranges from around 0.1 μm to 1000 μm, which includes the visible light region between violet (0.38 μm) and red (0.78 μm). The radiation from violet to red is termed visible because the human eye can see in this wavelength range only while radiation from 0.78 μm to 1000 μm is the infrared (IR) radiation. Ultraviolet (UV) radiation occupies the region of 0.01 μm to 0.38 μm. Thermal effects are associated with thermal radiation.

We are interested in rays that are absorbed by the substances and the energy of which transforms into heat on absorption. Visible light and infrared rays possess such properties in the greatest measure. However, in most of the engineering applications, the heat emitted in the range 400–1000 μm is very small. The thermal radiation also propagates at the velocity of light (299.8×10^6 m/s in vacuum) and obeys the laws of propagation, reflection and refraction of light rays.

The radiation of thermal energy is a property of all substances, and they continuously emit energy by virtue of the molecular and atomic agitation associated with their internal energy.

R. Karwa, *Heat and Mass Transfer*,
https://doi.org/10.1007/978-981-15-3988-6_10

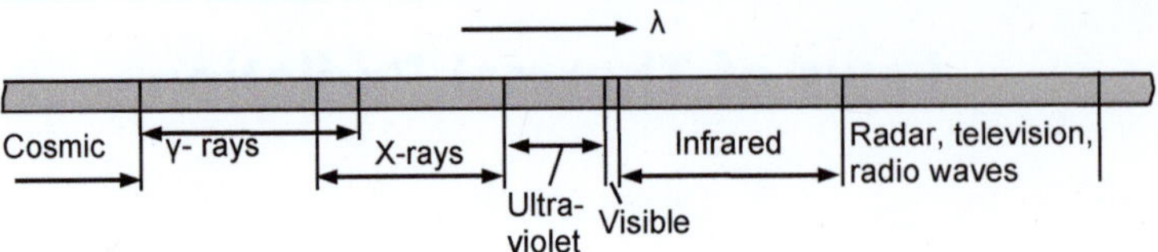

Fig. 10.1 Spectrum of electromagnetic radiation

Table 10.1 Spectrum of electromagnetic radiation

Type of rays	Wavelength λ (μm)
Cosmic rays	up to 4×10^{-7}
Gamma rays	4×10^{-7} to 1×10^{-4}
X-rays	1×10^{-5} to 2×10^{-2}
Ultraviolet rays	1×10^{-2} to 0.38
Visible (light)	0.38–0.78
Infrared rays	
Near	0.78–25
Far	25–1000
Thermal radiation	0.1–1000
Radar, television and radio	1×10^{3} to 2×10^{10}

Solids, as well as liquids and gases, are capable of radiating thermal energy and absorbing such energy. Water vapour and carbon dioxide are the main sources of the gaseous radiation in furnaces.

The true nature of electromagnetic energy is not well understood. However, the understanding of the true nature, i.e. quanta or waves, is generally not important to the engineers. In this chapter, the laws of thermal radiation are being presented.

The wavelength λ of the wave emitted in vacuum is related to the wave frequency v by the relation

$$\lambda v = c_o \tag{10.1}$$

where v is expressed in cycles/s and c_o is the velocity of light in vacuum.

The theory of radiant energy propagation has been considered from two viewpoints:

(a) **Electromagnetic Wave Theory**.
(b) **Quantum Theory**: The spectral distribution of the energy emitted from a body and the radiative properties of the gases can be explained and derived on the basis of quantum effects in which the energy is assumed to be carried by discrete particles (photons).

10.2 Reflection, Absorption and Transmission of Radiation

Radiation exchange depends on the nature of the substance, its temperature, wavelength, and the state of the emitting surface. Only thin surface layers of the solids and liquids participate in the process of radiation heat transfer; for non-conductors of heat, layer thickness is about

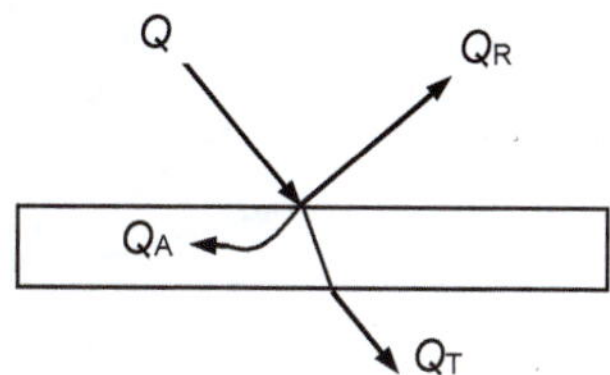

Fig. 10.2 Absorption, reflection and transmission of incident radiation from a body

1 mm, for the conductors of heat the thickness participating in the radiation is only about 1 μ. Thus, for the solids, it is a surface phenomenon. However, it also depends on the thickness of the layer and pressure in case of semi-transparent bodies (such as molten quartz, glass, etc., and gases and vapours). Therefore, emission and absorption of radiation in gases are volumetric effects.

When radiation falls on a body, a part of it may be absorbed, a part may be reflected and the remaining may pass through the body. The fraction of the incident radiation absorbed by the body is transformed into heat. If Q is the incident radiant energy, out of which Q_A is absorbed, Q_R is reflected, and Q_T is transmitted through the body, then, refer Fig. 10.2,

$$Q = Q_A + Q_R + Q_T$$

Dividing both sides of the equation by Q, we get

$$\frac{Q_A}{Q} + \frac{Q_R}{Q} + \frac{Q_T}{Q} = 1$$

The first fraction in the equation is known as *absorptivity* α, second is *reflectivity* ρ and the third fraction is *transmissivity* τ. Hence,

$$\alpha + \rho + \tau = 1 \tag{10.2}$$

Practically, most solids used in engineering applications are nontransparent or *opaque* (also termed as athermanous) to the thermal radiation, i.e. the transmissivity $\tau = 0$. However, there are some solids which are transparent to waves of a certain wavelength. For example, the ordinary glass transmits radiation very readily at wavelengths below about 2 μm but is essentially opaque to long-wavelength radiation above 3 or 4 μm. From Eq. (10.2), for the opaque solids

$$\alpha + \rho = 1 \tag{10.3}$$

or

$$\alpha = 1 - \rho$$

This means that a body with good reflectivity possesses poor absorptivity, and vice versa.

If the transmissivity τ of a body is equal to one, the absorptivity and reflectivity are equal to zero and the whole of the incident radiation would pass through the body. Such a body is termed as absolutely *transparent or diathermanous*. The only substance found to be perfectly diathermanous is crystalline pieces of rock-salt. Air has nearly zero absorptivity and

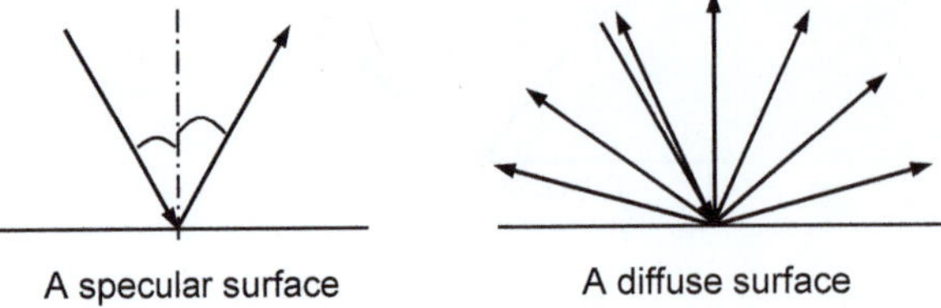

Fig. 10.3 Specular and diffuse surfaces

reflectivity. However, polyatomic gases, such as carbon dioxide, methane and water vapour are capable of absorbing heat radiation.

A body with reflectivity of unity will reflect the whole of the incident radiation and is termed a *white body*. When the reflection from a body obeys the laws of geometrical optics (angle of the reflected beam with normal equals the angle of the incident beam with the normal) the body is called *smooth or specular*. Reflections from a mirror or highly polished surface approaches the specular characteristics. However, due to the surface irregularities or roughness, the reflected radiation may be dispersed in all directions. Such a surface is known as a *diffuse surface*. If the roughness dimension (e.g., mean pit depth) for a real surface is considerably smaller than the wavelength of the incident radiation, the surface behaves as a specular reflector; if the roughness dimension is larger with respect to the wavelength, the surface behaves as diffuse one. Figure 10.3 shows the reflections from the specular and diffuse surfaces.

If the entire incident radiation is absorbed by the body, the absorptivity $\alpha = 1$. Such a body is termed as a *blackbody*. Only a few surfaces, such as carbon black, platinum black and gold black, approach the absorption capability of a blackbody. It is to be noted that the blackbody derives its name from the observation that surface appearing black to the eye is normally a good absorber of incident visible light. However, except for the visible region, this observation is not a good indicator of the absorbing capability of the surface. If a surface absorbs all incident rays, except the light rays, it does not appear black although it has a very high absorptivity. For example, white paint absorbs invisible heat rays just, as well as the black paint, although it is a poor absorber for the visible light. Similarly, the white looking ice and snow have absorptivity of 0.9 – 0.97 for longwave radiation. The absorption or reflection of heat rays depends on the state of the surface. The reflectivity of a smooth and polished surface is many times greater than that of a rough surface irrespective of the colour. Thus, the absorptivity of a surface can be increased by coating a rough surface with dark paint.

10.3 Emissivity and a Perfect Blackbody

A blackbody does not exist in nature. But the concept of a blackbody is of great importance. The ideal behaviour of the blackbody serves as a standard with which the performance of real bodies can be compared.

The radiation from a body depends on its temperature and optical property, known as emissivity ε. The emissivity of a given substance is a measure of its ability to emit radiation in comparison with a blackbody at the same temperature. A blackbody is an ideal radiator and its emissivity is assumed to be equal to one ($\varepsilon = 1$). Real bodies do not emit as much

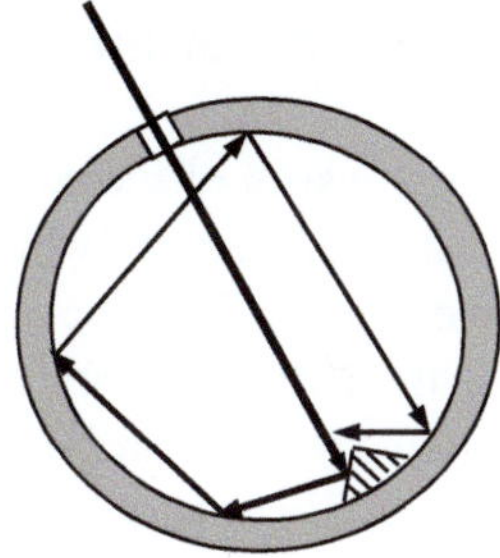

Fig. 10.4 A blackbody

energy as a blackbody and hence their emissivity is less than one. Emissivity has been further discussed in Sect. 10.8.

A model that is very close to the theoretical blackbody, known as Ferry's blackbody, is a hollow enclosure having the same temperature at any point on its inner surface and is having a very small hole as shown in Fig. 10.4. A ray entering the enclosure through the hole undergoes several reflections from the inner surface. At each incidence on the surface, a fraction of the radiation energy is absorbed. Thus, practically all the radiant energy entering the hole is absorbed inside the enclosure. The absorptivity for the model may be taken equal to unity and thus, it acts as a blackbody.

In 1879, Stefan discovered that the total emissive power of a blackbody (i.e. the heat emitted) is proportional to the fourth power of the absolute temperature of the body. Boltzmann in 1884 gave a theoretical proof of Stefan's empirical relation for blackbodies and that is why the law is known as Stefan–Boltzmann law. It is expressed as

$$q = \sigma A T^4 \tag{10.4}$$

where σ is the proportionality constant known as *Stefan–Boltzmann constant* and A is the surface area of the body.

The net heat radiated between two bodies 1 and 2 at temperatures T_1 and T_2 that see each other completely (heat exchange between them only) is proportional to the difference in T_1^4 and T_2^4. Thus

$$q_{1-2} = \sigma A (T_1^4 - T_2^4) \tag{10.5}$$

The radiation exchange between bodies that are not black is quite complex and has been dealt with in the next chapter.

It is to be noted that the conduction and convection heat transfers are proportional to the temperature difference, while the transfer of heat by thermal radiation is proportional to $(T_1^4 - T_2^4)$. Hence, the radiation heat transfer is the main mode of heat transfer at high-temperature levels such as in furnaces and combustion chambers. The radiation can be of importance in some applications where the temperature levels are not high and other modes of heat transfer are present such as in solar collectors.

10.4 Planck's Spectral Distribution of Emissive Power

The radiation emitted from a surface consists of electromagnetic waves of various wavelengths and the energy distribution varies in intensity with the wavelength. This distribution also varies with the temperature of the emitter. Thus, a change in the temperature of the body causes changes in the magnitude of the radiant energy and spectrum as well. Hence, it is important to know the law of the emissive power distribution with the wavelength at different temperatures. The energy emitted by a surface in all directions at a given wavelength is called *spectral* or *monochromatic emissive power* of the body.

Max Planck in 1900, using the quantum theory, deduced theoretically the law governing the change in emissive power of a blackbody per unit area as a function of temperature and wavelength. Mathematically, the law gives the spectral distribution of the monochromatic emissive power $E_{b\lambda}$ with the wavelength as

$$E_{b\lambda} = \frac{c_1 \lambda^{-5}}{\exp(c_2/\lambda T) - 1} \tag{10.6}$$

where

λ	wavelength, m
T	temperature of the body, K
c_1	$2\pi c_0^2 h = 3.743 \times 10^{-16} (\mathrm{J/s})\mathrm{m}^2$
c_o	velocity of light in vacuum = 2.998×10^8 m/s
h	Planck's constant = 6.6236×10^{-34} J s
c_2	$c_0 h/k = 1.4387 \times 10^{-2}\,\mathrm{m\,K}$
k	Boltzmann constant = 1.38066×10^{-23} J/K.

Figure 10.5 shows the spectral distribution of $E_{b\lambda}$ for a blackbody at some different values of absolute temperature. It can be seen that the monochromatic emissive power is zero at $\lambda = 0$. It first increases with an increase in the wavelength and reaches its maximum at a certain value of wavelength λ_{max}, then it decreases again with the increase in the wavelength and becomes zero at $\lambda = \infty$.

From the figure, it can be seen that at temperatures commonly encountered in engineering applications, a major part of the radiation is within a narrow range. Further, the energy of visible radiation is negligible compared with the infrared radiation ($\lambda = 0.74 - 40$ μm). For example, a body at 1000°C emits most of the radiation between 1 and 20 μm). On the other hand, the Sun whose surface temperature is nearly 5800 K emits 98% of its radiation between 0.1 and 3 μm.

We observe that when a body is heated from room temperature it becomes dark red, orange and finally white. This can be explained as below.

The red light becomes visible first as the temperature is raised. Higher temperatures make visible additional wavelengths of the visible light range, and at a sufficiently high-temperature, the light emitted becomes white, representing radiation composed of a mixture of all the visible wavelengths.

Equation (10.6) must be modified by including index of refraction multiplying factors for radiation into a medium where the speed of light is not close to c_o. However, for most engineering applications, the radiant emission is into the air or other gases with an index of refraction close to unity and Eq. (10.6) is applicable.

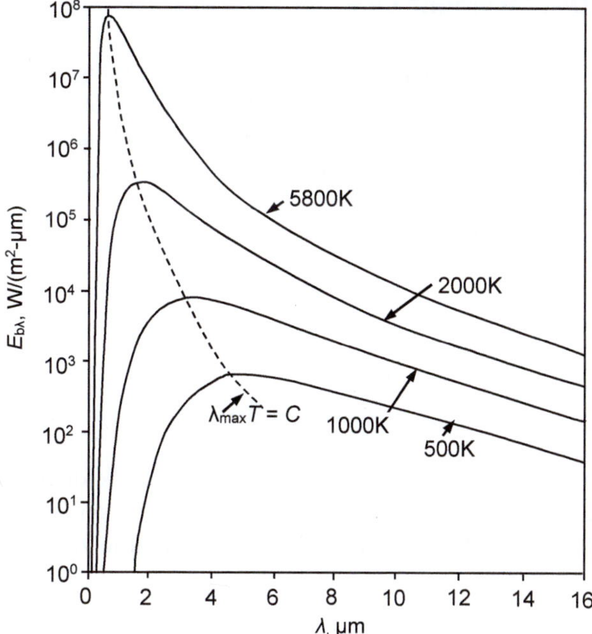

Fig. 10.5 Monochromatic emissive power of a blackbody versus wavelength at various temperatures as per the Planck's law

Note: Rayleigh and Jeans (1900)[1] made an attempt to predict theoretically the monochromatic emissive power. The law for blackbody radiation can also be derived from Planck's law (Eq. 10.6) for the limiting case of large values of λT. With this condition, we can retain only two terms of the exponential function in Planck's equation

$$e^{\frac{C_2}{\lambda T}} = 1 + \frac{1}{1!}\left(\frac{C_2}{\lambda T}\right) + \frac{1}{2!}\left(\frac{C_2}{\lambda T}\right)^2 + \cdots$$

Then Eq. (10.6) becomes

$$E_{b\lambda} = \frac{c_1 T}{c_2 \lambda^4} \tag{10.7}$$

which is the Rayleigh–Jean's law. At large wavelengths, this law describes the actual monochromatic black radiation quite well. According to this law, the radiation will increase indefinitely when the wavelength goes towards zero, which is in contradiction with the experimental observations.

[1]Readers can refer Eckert and Drake (1959) for the basic derivation of the law.

10.5 Wein's Displacement Law

With the increase in the temperature, the peak of the curve in Fig. 10.5 shifts towards the shorter wavelengths. The points corresponding to the maximum of the curves are related by the Wein's displacement law. The law states that the wavelength corresponding to the maximum value of the monochromatic emissive power is inversely proportional to the absolute temperature T, i.e.

$$\lambda_{\max} \propto \frac{1}{T}$$

or

$$\lambda_{\max} T = \text{constant} = 2897.6\,\mu\text{m K}. \tag{10.8}$$

Although the Wein's displacement law (1893) was revealed before Planck (1990), Planck made use of the quantum theory to derive his equation. Equation (10.8) can be proved or arrived at by differentiating Planck's equation and equating it to zero.

$$\frac{\mathrm{d}}{\mathrm{d}\lambda}(E_{b\lambda}) = \frac{\mathrm{d}}{\mathrm{d}\lambda}\left[\frac{c_1\lambda^{-5}}{\exp(c_2/\lambda T) - 1}\right] = 0$$

or

$$\frac{[\exp(c_2/\lambda T) - 1]c_1(-5)\lambda^{-6} - c_1\lambda^{-5}\exp(c_2/\lambda T)(c_2/T)(-1)\lambda^{-2}}{[\exp(c_2/\lambda T) - 1]^2} = 0$$

The denominator of the above equation is not zero, hence

$$-5c_1[\exp(c_2/\lambda T) - 1]\lambda^{-6} + (c_1c_2/T)\exp(c_2/\lambda T)\lambda^{-7} = 0$$

or

$$-\exp(c_2/\lambda T) + (c_2/5\lambda T)\exp(c_2/\lambda T) + 1 = 0$$

By trial and error,

$$c_2/\lambda T = 4.965$$

or

$$\lambda_{\max} T = c_2/4.965 = \frac{1.4387 \times 10^{-2}}{4.965}$$

or

$$\lambda_{\max} T = 0.0028976\text{ mK} = 2897.6\,\mu\text{m K}.$$

where $\lambda_{\max}$ denotes the wavelength at which the monochromatic emissive power $(E_b)_\lambda$ is maximum.

Substituting $\lambda T = 0.002897$, we can obtain the value of the maximum spectral monochromatic emissive power as

$$(E_{b\lambda})_{\max} = \frac{c_1\lambda^{-5}}{\exp(c_2/\lambda T) - 1} = \frac{3.743 \times 10^{-16} \times (0.002897/T)^{-5}}{\exp(4.965) - 1}$$
or
$$(E_{b\lambda})_{\max} = 1.289 \times 10^{-5} T^5 \text{ W/m}^2 \text{per meter wave length.} \quad (10.9)$$

10.6 Total Emissive Power: Stefan–Boltzmann Law

The Stefan–Boltzmann law states that the total amount of energy emitted per square meter by a blackbody, known as emissive power of the body, is proportional to the fourth power of the absolute temperature of the body, i.e.

$$E_b \propto T^4$$

or

$$E_b = \sigma T^4 \quad (10.10)$$

The value of Stefan–Boltzmann constant σ is 5.67×10^{-8} W/(m^2 K^4).

The total emissive power E_b of a blackbody is the total radiant energy emitted by the blackbody in all directions over the entire wavelength range ($\lambda = 0$ to ∞) per unit area per unit time.

From the definition of the value of the emissive power, it follows that at any temperature it is the area under the curve of $E_{b\lambda}$ pertaining to that temperature. Hence, refer Fig. 10.6,

$$E_b = \int_0^\infty E_{b\lambda} \mathrm{d}\lambda$$

Substituting the value of $(E_{b\text{-}})_\lambda$ from Eq. (10.6), we have

$$E_b = \int_0^\infty \left[\frac{c_1\lambda^{-5}}{\exp(c_2/\lambda T) - 1}\right] \mathrm{d}\lambda$$

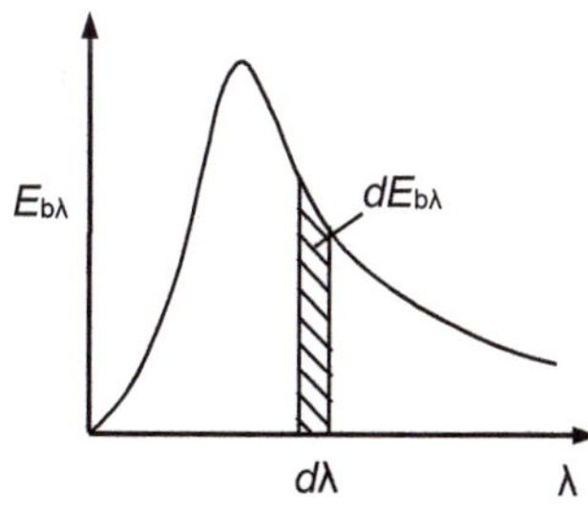

Fig. 10.6 Determination of total emissive power calculation

Let $(c_2/\lambda T) = x$, then

$$\lambda = \frac{c_2}{xT}$$

and

$$\mathrm{d}\lambda = -\frac{c_2}{x^2T}dx$$

Substitution gives

$$\begin{aligned} E_b &= -c_1 \int_0^\infty \frac{x^5T^5}{c_2^5[\exp(x) - 1]} \times \frac{c_2}{x^2T}dx \\ &= \frac{c_1T^4}{c_2^4} \int_0^\infty \left[\frac{x^3}{\exp(x) - 1}\right] dx \end{aligned}$$

Now

$$x^3[\exp(x) - 1]^{-1} = x^3[e^x - 1]^{-1} = x^3[e^{-x} + e^{-2x} + e^{-3x} + \cdots]$$

Hence,

$$E_b = \frac{c_1T^4}{c_2^4} \int_0^\infty x^3[e^{-x} + e^{-2x} + e^{-3x} + \cdots]dx$$

But

$$\int_0^\infty x^3e^{-nx}dx = \frac{3!}{n^{3+1}} = \frac{3!}{n^4}$$

Hence,

$$\begin{aligned} E_b &= \frac{c_1T^4}{c_2^4}\left[\frac{3!}{1^4} + \frac{3!}{2^4} + \cdots\right] \\ &= \frac{c_1T^4}{c_2^4}\left(\frac{6\pi^4}{90}\right) \\ &= \frac{\pi^4}{15} \times \frac{3.743 \times 10^{-16}}{(1.4387 \times 10^{-2})^4}T^4 \\ &= 5.67 \times 10^{-8}T^4 \\ &= \sigma T^4. \end{aligned} \tag{10.10}$$

It must be noted that the equation applies strictly to a blackbody.

10.7 Blackbody Radiation in a Wave Length Interval

The Stefan–Boltzmann law gives the hemispherical total emissive power of a blackbody (from $\lambda = 0$ to ∞). It is often desirable in calculations of radiation exchange to determine the fraction of the emission in a given wavelength band as shown in Fig. 10.7a. This fraction is designated by $F_{\lambda_1-\lambda_2}$ and is given by

$$F_{\lambda_1-\lambda_2} = \frac{\int_{\lambda_1}^{\lambda_2} (E_b)_\lambda \mathrm{d}\lambda}{\int_0^{\infty} (E_b)_\lambda \mathrm{d}\lambda} = \frac{\int_{\lambda_1}^{\lambda_2} (E_b)_\lambda \mathrm{d}\lambda}{\sigma T^4}$$

The integral can be expressed by the difference of two integrals each beginning at $\lambda = 0$ as shown below

$$F_{\lambda_1-\lambda_2} = \frac{1}{\sigma T^4}\left[\int_0^{\lambda_2} (E_b)_\lambda \mathrm{d}\lambda - \int_0^{\lambda_1} (E_b)_\lambda \mathrm{d}\lambda\right] = F_{0-\lambda_2} - F_{0-\lambda_1} \tag{i}$$

The fractions $F_{0-\lambda_2}$ and $F_{0-\lambda_1}$ equal the hatched area divided by the total area under the curve in Figs. 10.7b, c, respectively. The fraction of the emissive power in the wavelength interval $(F_{0-\lambda_2} - F_{0-\lambda_1})$ can be found by subtracting the value of fraction $F_{0-\lambda_1}$ from that of fraction $F_{0-\lambda_2}$.

Since $E_{b\lambda}$ depends on absolute temperature T, the fraction $F_{0-\lambda}$ is required to be tabulated for each value of T. It is possible to arrange the function in terms of single variable λT. This universal form can be found by rewriting Eq. (i) as

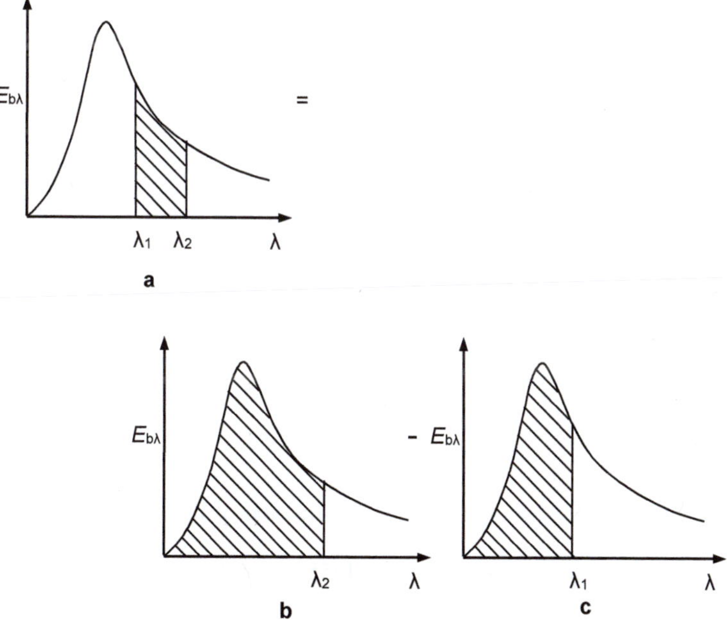

Fig. 10.7 Determination of radiation in a wavelength interval

Table 10.2 Fraction of blackbody radiation

λT (μm K)	$F_{o-\lambda T}$
1448	0.01
2898	0.25
4108	0.5
11,069	0.75
41,800	0.99

$$F_{\lambda_1-\lambda_2} = F_{\lambda_1 T-\lambda_2 T} = \frac{1}{\sigma}\left[\int_0^{\lambda_2 T} \frac{(E_b)_\lambda}{T^5} d(\lambda T) - \int_0^{\lambda_1 T} \frac{(E_b)_\lambda}{T^5} d(\lambda T)\right] \tag{10.11}$$
$$= F_{0-\lambda_2 T} - F_{0-\lambda_1 T}$$

It can be shown that $\frac{(E_b)_\lambda}{T^5}$ is only a function of (λT). Thus, the integrands in the above equation are only dependent on the variable (λT). Some typical values of $F_{o-\lambda T}$ are given in Table 10.2. It is interesting to note that exactly one fourth of the total emissive power lies in the wavelength range below the peak of the Planck's spectral distribution at any temperature. More detailed values of $F_{o-\lambda T}$ are given in Table 10.3.

Example 10.1 The surface of the Sun has an effective blackbody temperature of 5800 K. Determine the fraction of the radiant energy of the Sun lying in the following ranges.

(i) visible range ($0.35 \le \lambda \le 0.75$ μm)
(ii) ultraviolet ($0.01 \le \lambda \le 0.35$ μm)
(iii) $0 \le \lambda \le 3$ μm

At what wavelength and frequency is the maximum energy emitted? What is the maximum value of the hemispherical spectral emissive power?

Solution

(i) Visible range ($0.35 \le \lambda \le 0.75$ μm)

$$\lambda_1 T = 0.35 \times 5800 = 2030\,\mu\text{m K}$$
$$\lambda_2 T = 0.75 \times 5800 = 4350\,\mu\text{m K}.$$

Fractional emissive power, from Table 10.3,

$$F_{0-\lambda_1 T} = 0.07185$$
$$F_{0-\lambda_2 T} = 0.54055.$$

So, in the range $0.35 \le \lambda \le 0.75$ μm, the fraction of radiant energy is

$$F_{0-\lambda_2 T} - F_{0-\lambda_1 T} = 0.54055 - 0.07185 = 0.4687.$$

Table 10.3 Fraction of blackbody radiation $F_{0\text{-}\lambda T}$ as function of λT

λT (µm K)	$F_{o-\lambda T}$	λT (µm K)	$F_{o-\lambda T}$
500	0.13×10^{-7}	5200	0.65792
600	0.112×10^{-6}		
700	0.22×10^{-5}	5400	0.68031
800	0.184×10^{-4}		
900	0.912×10^{-4}	5600	0.70100
1000	0.321×10^{-3}	5800	0.72009
1100	0.926×10^{-3}		
1200	0.00216	6000	0.73777
1300	0.00436	6200	0.75408
1400	0.00783	6400	0.76917
1500	0.01285	6600	0.78315
1600	0.01977	6800	0.79607
1700	0.02860	7000	0.80806
1800	0.03941	7500	0.83435
1900	0.05215	8000	0.85624
2000	0.06672	8500	0.87455
2200	0.10093	9000	0.88997
2400	0.14027	9500	0.90303
2600	0.18312	10,000	0.91414
2800	0.22788	11,000	0.93183
3000	0.27322	12,000	0.94504
3200	0.31807	13,000	0.95508
3400	0.36170	14,000	0.96284
3600	0.40357	15,000	0.96892
3800	0.44334	16,000	0.97375
4000	0.48085	18,000	0.98080
4200	0.51596	20,000	0.98554
4400	0.54875	30,000	0.99528
4600	0.57923	40,000	0.99791
4800	0.60751	50,000	0.99889
5000	0.63371	∞	1.00000

(ii) Ultraviolet ($0.01 \leq \lambda \leq 0.35$ µm)

$$\lambda_1 T = 0.01 \times 5800 = 58\,\mu\text{m K}$$

$$\lambda_2 T = 0.35 \times 5800 = 2030\,\mu\text{m K}.$$

Fractional emissive power,

$$F_{0-\lambda_1 T} \approx 0$$

$$F_{0-\lambda_2 T} = 0.07185.$$

So, the fraction of radiant energy in the ultraviolet range ($0.01 \leq \lambda \leq 0.35$ μm) is

$$F_{0-\lambda_2 T} - F_{0-\lambda_1 T} = 0.07185 - 0 = 0.07185.$$

(iii) $0 \leq \lambda \leq 3$ μm

$$\lambda_1 T = 0 \times 5800 = 0\,\mu\text{m K}$$

$$\lambda_2 T = 3.0 \times 5800 = 17{,}400\,\mu\text{m K}.$$

Fractional emissive power,

$$F_{0-\lambda_1 T} = 0$$

$$F_{0-\lambda_2 T} \approx 0.98.$$

So, in the wavelength range $0 \leq \lambda \leq 3.0$ μm, the fraction of radiant energy is

$$F_{0-\lambda_2 T} - F_{0-\lambda_1 T} = 0.98 - 0 = 0.98,\ \text{i.e.}, 98\%.$$

(iv) $\boldsymbol{\lambda_{max}}$
From Wein's displacement law,

$$\lambda_{\max} T = 2897.6$$

or

$$\lambda_{\max} = \frac{2897.6}{T} = \frac{2897.6}{5800} = 0.5\,\mu\text{m}.$$

The corresponding frequency is

$$\frac{c_o}{\lambda} = \frac{2.998 \times 10^8}{0.5 \times 10^{-6}} = 6.0 \times 10^{14}\,\text{Hz}.$$

(v) $(E_{b\lambda})_{\max} = 1.289 \times 10^{-5} T^5 = 1.289 \times 10^{-5} \times (5800)^5 = 8.46 \times 10^{13}\,\text{W/m}^2$.
It is interesting to note that at the temperature of the Sun's surface $\lambda_{\max}$ is in the visible range and about 40% of the energy is emitted in the visible range of the spectrum. The Sun emits about 98% of its radiation in wavelength band $0 \leq \lambda \leq 3$ μm.

Example 10.2 Figure 10.8 shows a greenhouse where plants are grown in cold regions by trapping solar energy inside the greenhouse. The sheet of silica glass on the top of the greenhouse transmits 90% of the incident radiation in the wavelength band 0.35–2.7 μm and

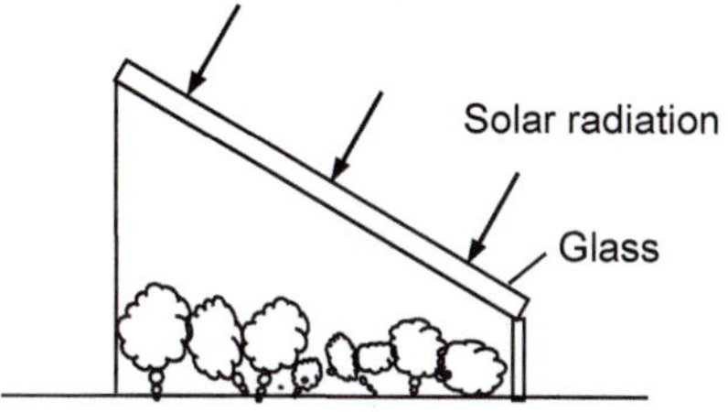

Fig. 10.8 A greenhouse

is essentially opaque to the radiation at longer and shorter wavelengths. Estimate the percent solar radiation which the glass will transmit. Consider the Sun as a blackbody at about 6000 K.

If the garden in the greenhouse behaves like a black surface and is at 40°C, what percent of this radiation will be transmitted through the glass roof?

Solution

(i) From the given data
$\lambda_1 T = 0.35 \times 6000 = 2100$ for which $F_{0\text{–}2100} = 0.08315$
and $\lambda_2 T = 2.7 \times 6000 = 16200$ for which $F_{0\text{–}16200} = 0.97460$.
Thus,

$$F_{2100-16200} = 0.97460 - 0.08315 = 0.89145.$$

Thus, the solar radiation in $0.35 \leq \lambda \leq 2.7$ is 89.15% of the total solar radiation incident and the radiation transmitted is

$$= \tau \times 0.89145 = 0.9 \times 0.89145 = 80.23\%.$$

(ii) Radiation from the garden:
$\lambda_1 T = 0.35 \times 313 = 109.6$ for which $F_{0\text{–}109.6} \approx 0$
and $\lambda_2 T = 2.7 \times 313 = 845$ for which $F_{0\text{–}845} = 0.5116 \times 10^{-4}$
Thus

$$F_{109.6-845} = 0.5116 \times 10^{-4}.$$

Thus the radiation in $0.35 \leq \lambda \leq 2.7$ is 0.005116%. The radiation transmitted through the glass is

$$= \tau \times 0.5116 \times 10^{-4} = 0.9 \times 0.5116 \times 10^{-4} = 0.0046\%.$$

Example 10.3 Determine an equivalent blackbody temperature for the solar radiation if the maximum in the spectrum wavelength occurs at about 0.5 μm.

Solution

According to the Wein's displacement law,

$$\lambda_{max} T = 2897.6\,\mu\text{m K}.$$

Hence,

$$T = \frac{2897.6}{\lambda_{max}} = \frac{0.0028976 \times 10^{-6}}{0.5} = 5795\,\text{K}.$$

Example 10.4 The average temperature of the tungsten filament of an incandescent lamp is about 2800 K. Determine the amount of energy emitted by the lamp in the visible range. Comment on the result.

Solution

The visible range of radiation extends from about 0.35 to 0.75 μm. Hence,
$\lambda_1 T = 0.35 \times 2800 = 980$ for which $F_{0\text{–}980} = 2.75 \times 10^{-4}$
and $\lambda_2 T = 0.75 \times 2800 = 2100$ for which $F_{0\text{–}2100} = 0.083825$.
Thus,

$$F_{980-2100} = 0.08355.$$

Thus, the radiation in $0.35 \leq \lambda \leq 0.75$ is about 8.36%.

Comments: The example shows that only about 8.36% of the energy is emitted in the visible range. Thus, the incandescent lamp is more efficient as a heat source than as a light source.

10.8 Real and Gray Bodies

The emissive power of a blackbody is the maximum and the emissive power of a real body is always less than that of a blackbody. The ratio of the emissive power E_λ of a real body at a particular wavelength and temperature to that of a blackbody $E_{b\lambda}$ at the same wavelength and temperature is called the *monochromatic emissivity* ε_λ of the body.

$$\varepsilon_\lambda = \frac{E_\lambda}{E_{b\lambda}} \tag{10.12a}$$

The monochromatic emissivity of a blackbody is taken as unity and that of a white body as zero. A body whose radiation spectrum is continuous and similar to that of a blackbody is shown in Fig. 10.9. Its *monochromatic emissivity* ε_λ is constant for all wavelengths, as well as temperatures. Such a body is known as *gray body*. The value of monochromatic emissivity of a gray body is $0 < \varepsilon_\lambda < 1$. For such bodies, the above equation is written as

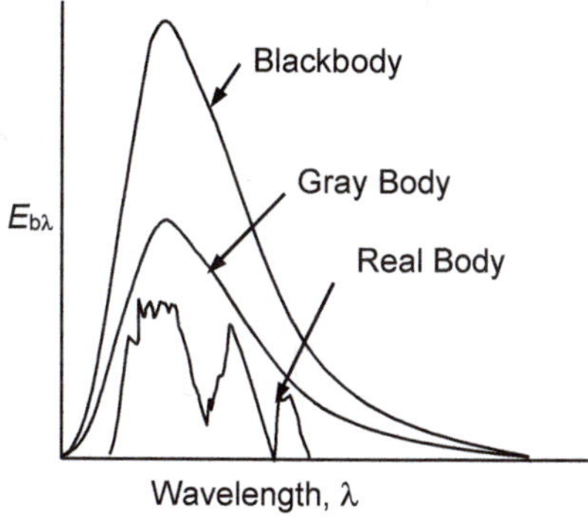

Fig. 10.9 Emissive power of a real surface versus ideal blackbody and gray body

$$\varepsilon = \frac{E}{E_b} \tag{10.12b}$$

Slate, tar-board and dark linoleum are some examples of surfaces that are good approximations of gray surfaces (Eckert and Drake 1959). These materials absorb from 85% to 92% of the incident radiation in the range of wavelengths between 0.5 and 9 μm. Unfortunately, there are not many surfaces that can be termed as gray. The monochromatic emissivity of the real bodies varies with temperature, as well as wavelength. For most of the practical bodies encountered in engineering applications an integrated average value of emissivity, as defined below, is used.

$$\varepsilon = \frac{1}{\sigma T^4}\int_0^\infty E_\lambda \mathrm{d}\lambda \tag{10.12c}$$

where $\int_0^\infty E_\lambda \mathrm{d}\lambda$ is the *total emissive power*, which is the rate of emission of radiant heat per unit area for all wavelengths and in all directions and σT^4 is the emissive power of a blackbody at the same temperature. The emissivity ε in Eq. (10.12c) is the *total hemispherical emissivity* or the *total emissivity* of a substance at a given temperature and is the ratio of the total emissive power of the substance to that of a blackbody at the same temperature.

The directional dependence of the emissive power has been discussed in the Sect. 10.10. For the time being, we define the *normal total emissive power* of a substance, which refers to the component of the total emissive power normal to the surface.

Most of the data available in the literature on emissivity of real bodies are either of normal total emissivity or total hemispherical emissivity. For engineering calculations, the information on the total emissivity of a surface is more useful. Data for some metals and non-metals are listed in the appendix.

The emissivity of polished metallic surfaces is low and it increases with the increase in the thickness of the layer of oxide on the surface. The emissivity of the non-metallic surfaces is usually very high (0.8–0.97). In the case of metals, the emissivity rises with rising temperature, but in the case of non-metallic substances, this may not be true. The average value of the ratio of the hemispherical total emissivity ε and normal emissivity ε_n of bright metal surfaces is 1.2; for other substances with smooth surfaces $\varepsilon/\varepsilon_n = 0.95$ and for rough surfaces $\varepsilon/\varepsilon_n = 0.98$.

Monochromatic absorptivity α_λ is defined as the ratio of the monochromatic absorbed radiation to the incident monochromatic irradiation G_λ. Knowing the monochromatic absorptivity, the *total absorptivity* can be determined from

$$\alpha = \frac{\int_0^\infty \alpha_\lambda(T_s) G_\lambda \mathrm{d}_\lambda}{\int_0^\infty G_\lambda \mathrm{d}\lambda} \tag{10.13a}$$

where $\int_0^\infty G_\lambda \mathrm{d}\lambda$ is the total impinging irradiation G.

If the incident radiation is coming from a blackbody at temperature T_i, whose emissive power is $E_b(T_i)$, then the total absorptivity is

$$\alpha = \frac{\int_0^\infty \alpha_\lambda(T_s) E_{b\lambda}(T_i) \mathrm{d}\lambda}{E_b(T_i)} = f(T_s, T_i) \tag{10.13b}$$

Thus, the total absorptivity depends on the nature and temperature T_s of the absorbing surface, as well as on the temperature T_i of the incident blackbody radiation.

It is to note that the monochromatic absorptivity of real solids, liquid and gaseous bodies is different at various regions of the spectrum and hence emissivity of such substances also varies. These bodies are known as *selective emitters* or absorbers. Gases show the highest selectivity.

To understand the behaviour of the selective absorbers or emitters, consider a substance which is sensitive only to the electromagnetic waves of say 5–10 μm. If the surrounding bodies emit radiation in the range 1–15 μm, this substance will absorb the radiation in the range 5–10 μm only. If the surrounding bodies emit outside the range of the sensitivity of the substance say between 20 and 100 μm, the substance will not absorb any of the incident radiation from these bodies.

For infrared radiation, the absorptivity of the non-conductors is greater than that of conductors. Absorptivity of all electric conductors, with a few exceptions, increases with increasing temperature, while that of the non-conductors decreases.

The discussion presented above for the absorptivity also applies to reflectivity of a surface since the reflectivity ρ is related to the absorptivity for the opaque solids by the relation $\rho_\lambda = 1 - \alpha_\lambda$.

10.9 Kirchhoff's Law

The law states that at any temperature the ratio of emissive power E to the absorptivity α is a constant for all bodies and equals the emissive power of a blackbody at the same temperature, i.e.

$$\frac{E_1}{\alpha_1} = \frac{E_2}{\alpha_2} = \frac{E_3}{\alpha_3} = \cdots = E_b = f(T) \tag{10.14a}$$

Since the ratio of the emissive power of a gray body to that of a blackbody at the same temperature is defined as emissivity, hence

$$\frac{E_1}{E_b} = \alpha_1 = \varepsilon_1; \frac{E_2}{E_b} = \alpha_2 = \varepsilon_2 \tag{10.14b}$$

For monochromatic radiation, the law states that the ratio of the emissive power at a certain wavelength to the absorptivity at the same wavelength is the same for all bodies and is a function of wavelength and temperature, i.e.

$$\frac{(E_\lambda)_1}{(\alpha_\lambda)_1} = \frac{(E_\lambda)_2}{(\alpha_\lambda)_2} = \frac{(E_\lambda)_2}{(\alpha_\lambda)_2} = \cdots = E_b = f(\lambda, T) \tag{10.14c}$$

Proof Consider two parallel plates as shown in Fig. 10.10. One of the plates is gray and the other is black. The gray surface emits radiation energy E, which is fully absorbed by the black surface on impingement upon it. The gray surface absorbs a portion αE_b of the radiation E_b emitted by the blackbody. The remaining radiation energy $(1 - \alpha) E_b$, reflected back by the gray surface, is absorbed by the blackbody. Thus, the net energy exchange from the gray surface is

$$q = E - \alpha E_b$$

The energy exchange also takes place when the temperatures of the two surfaces are equal, i.e. $T = T_b$. In this case, the net exchange is zero, i.e. $q = 0$. This gives

$$\frac{E}{\alpha} = E_b$$

This proves the law.

Note: For real surfaces, emissivity basically depends on direction (θ, ϕ), wavelength λ and surface temperature T_s and may be termed as *spectral-directional emittance* and denoted as $\varepsilon(\theta, \phi, \lambda, T_s)$ (Mills 1995). The fundamental statement of the Kirchhoff's law gives

$$\varepsilon(\theta, \phi, \lambda, T_s) = \alpha(\theta, \phi, \lambda, T_s)$$

A *hemispherical emittance* is a value directionally averaged over the 2π steradian hemisphere of solid angle above the surface. A total property is a value, which is averaged over the all wavelengths (from zero to infinity) and is defined as $\varepsilon(T_s)$ to indicate the

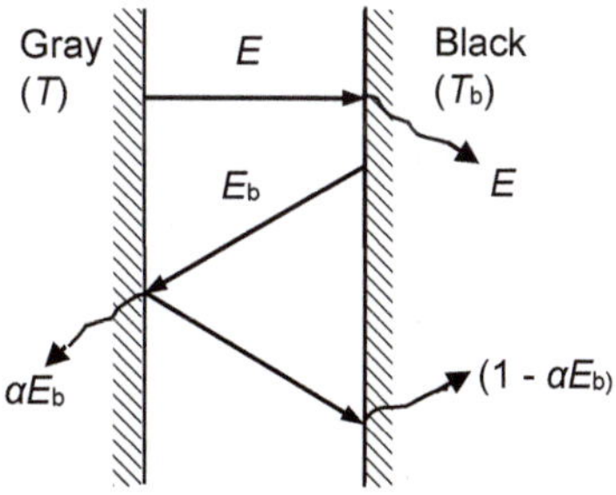

Fig. 10.10 Radiation exchange between gray and black surfaces

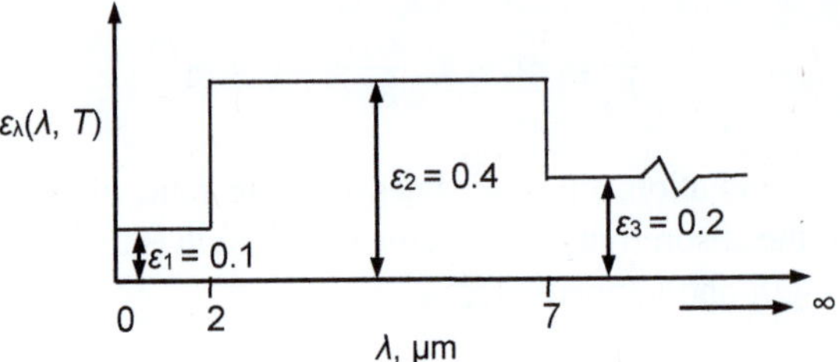

Fig. 10.11 Example 10.5

dependence on surface temperature only. However, engineers use a simpler diffuse gray surface idealization of emissivity of a surface, which is independent of temperature also; i.e. has a constant value for a given surface.

Example 10.5 The hemispherical spectral emissivity $\varepsilon_\lambda(\lambda, T)$ of a surface at temperature $T = 1000$ K can be approximated as shown in Fig. 10.11. What are the hemispherical total emissivity and the hemispherical total emissive power of the surface?

Solution

From the given figure, we have

$\varepsilon_1 = 0.1$ for $\lambda = 0$ to 2
$\varepsilon_2 = 0.4$ for $\lambda = 2$ to 7
$\varepsilon_3 = 0.2$ for $\lambda = 7$ to ∞

and it is given that $T = 1000$ K hence

$$\lambda_1 T = 2 \times 1000 = 2000 \text{ for which } F_{0-2000} = 0.06672$$

and $\lambda_2 T = 7 \times 1000 = 7000$ for which $F_{0-7000} = 0.80806$.
Thus,

$$F_{2000-7000} = 0.80806 - 0.06672 = 0.74134.$$

$$F_{7000-\infty} = 1 - 0.80806 = 0.19194.$$

Hemispherical total emissivity is given by

$$\begin{aligned}\varepsilon &= \varepsilon_1(F_{0-2000}) + \varepsilon_2(F_{2000-7000}) + \varepsilon_3(F_{7000-\infty})\\ &= 0.1 \times 0.06672 + 0.4 \times 0.74134 + 0.2 \times 0.19194 = 0.3416.\end{aligned}$$

and the total emissive power is

$$\begin{aligned}E &= \varepsilon\sigma_b T^4\\ &= 0.3416 \times 5.67 \times 10^{-8} \times 1000^4\\ &= 19368.7\,\text{W/m}^2.\end{aligned}$$

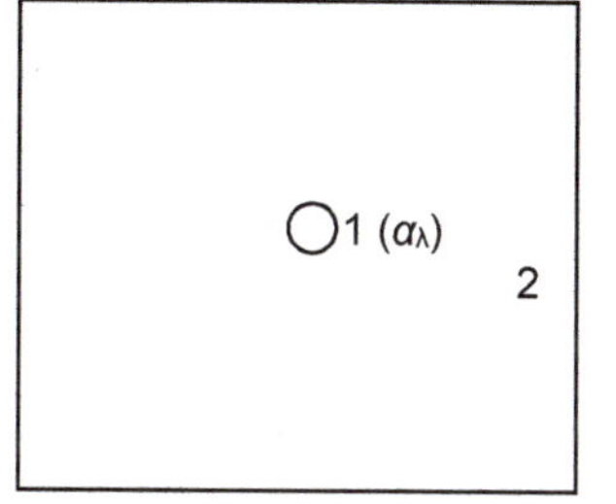

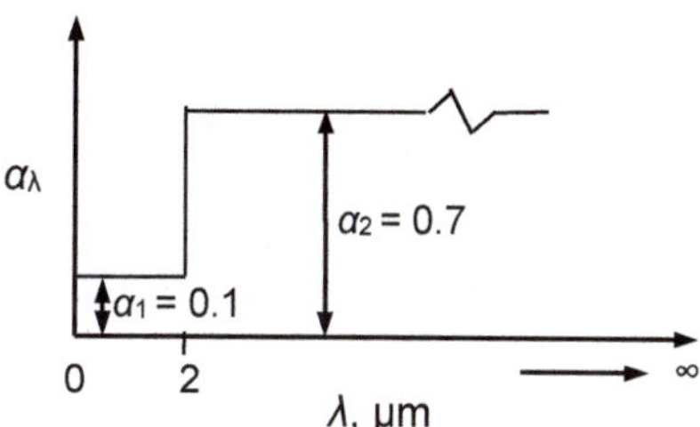

Fig. 10.12 Example 10.6

Example 10.6 A small opaque diffuse surface 1 at 1000 K with given spectral absorptivity distribution α_λ is located in a large enclosure 2 whose surface (treated as black) is at 2000 K, refer Fig. 10.12. Determine (a) total hemispherical absorptivity of the surface and (b) the total hemispherical emissivity.

Solution

(a) Since the irradiation is from the enclosure at 2000 K, the hemispherical total absorptivity of the surface, from the given spectral absorptivity of the surface, is

$$\begin{aligned}\alpha &= \alpha_1(F_{0-4000}) + \alpha_2(F_{4000-\infty}) \\ &= 0.1 \times 0.48085 + 0.7 \times (1.0 - 0.48085) = 0.4115,\end{aligned}$$

where $\lambda_1 T = 2 \times 2000 = 4000$ for which $F_{0-4000} = 0.48085$ from Table 10.3.

(b) The total hemispherical emissivity corresponds to the temperature of the surface and knowing that $\varepsilon_\lambda = \alpha_\lambda$ it is

$$\begin{aligned}\varepsilon &= \alpha_1(F_{0-2000}) + \alpha_2(F_{2000-\infty}) \\ &= 0.1 \times 0.06672 + 0.7 \times (1.0 - 0.06672) = 0.66,\end{aligned}$$

where $\lambda_1 T = 2 \times 1000 = 2000$ for which $F_{0-2000} = 0.06672$ from Table 10.3.

Example 10.7 The maximum spectral intensity of radiation from a gray surface at 1000 K is found to be 0.8×10^{10} W/m^2 per m of wavelength. Determine (i) the wavelength at which the maximum spectral intensity of radiation occurs, and (ii) the emissivity of the surface.

Solution

(i) At the point of maximum spectral emissive power,

$$\lambda_{max} T = 0.002897 \text{ mK}.$$

Hence,

$$\lambda_{max} = 0.002897/T = 0.002897/1000 = 2.897 \times 10^{-6} \text{ m}.$$

(ii) The maximum spectral intensity of a blackbody is given by

$$(E_{b\lambda})_{max} = 1.289 \times 10^{-5} T^5$$

At 1000 K,

$$(E_{b\lambda})_{max} = 1.289 \times 10^{-5} \times (1000)^5 = 1.289 \times 10^{10}\ \text{W/m}^2.$$

Emissivity of the gray surface,

$$\varepsilon = \frac{E_\lambda}{E_{b\lambda}} = \frac{0.8 \times 10^{10}}{1.289 \times 10^{10}} = 0.6206$$

Example 10.8 A body at 1100°C in black surrounding at 550°C has an emissivity of 0.4 at 1100°C and an emissivity of 0.7 at 550°C. Calculate the rate of heat flow by radiation per unit surface area.

If the body is assumed to be gray with $\varepsilon = 0.4$, what will be the heat loss?

Solution

(i) **The body is not gray**

The rate of energy emission from the body at 1100°C is

$$\begin{aligned} &= (\varepsilon)_{at1100^\circ C}\sigma T^4 \\ &= 0.4 \times 5.67 \times 10^{-8} \times (1100 + 273)^4\ \text{W per unit area.} \end{aligned}$$

The rate of energy emission from the black enclosure at 550°C is

$$\begin{aligned} &= \sigma T_2^4 \\ &= 5.67 \times 10^{-8} \times (550 + 273)^4\ \text{W per unit area.} \end{aligned}$$

The absorptivity of the body for the radiation from the enclosure will be equal to the emissivity value at the enclosure temperature, i.e. $\alpha = 0.7$. Hence,
Rate of energy absorption,

$$\begin{aligned} &= \alpha\sigma T_2^4 \\ &= 0.7 \times 5.67 \times 10^{-8} \times (550 + 273)^4\text{W per unit area.} \end{aligned}$$

The rate of heat loss is

$$\begin{aligned} \text{Rate of heat loss} &= \text{rate of emission} - \text{rate of absorption} \\ &= 5.67 \times 10^{-8} \times [0.4 \times (1100 + 273)^4 - 0.7 \times (550 + 273)^4] \\ &= 62389\ \text{W/m}^2. \end{aligned}$$

(ii) **When the body is gray**

$$\varepsilon = \alpha = 0.4.$$

The rate of heat loss,

$$= 0.4 \times 5.67 \times 10^{-8} \times [(1100 + 273)^4 - (550 + 273)^4]$$
$$= 70,193\ \text{W/m}^2.$$

Example 10.9 A small body at 300 K is placed in a large furnace whose walls are maintained at 1000 K. The total absorptivity of the body varies with the temperature of the incident radiation as follows.

Temperature:	300 K	500 K	1000 K
Absorptivity:	0.75	0.6	0.5

Determine the rate of absorption and emission of the radiation by the body.

Solution

(i) **Rate of absorption**
The rate of emission of energy by the furnace walls is

$$E = \sigma T^4$$
$$= 5.67 \times 10^{-8} \times (1000)^4 = 56,700\ \text{W/m}^2.$$

The absorptivity of a body depends on the temperature of the surface from which the radiation is coming. Here the furnace wall is at 1000 K at which $\alpha = 0.5$. Hence, the rate of absorption

$$= 0.5 \times 56,700 = 28,350\ \text{W/m}^2.$$

(ii) **Rate of emission**
The emissivity of the body at a temperature of 300 K is 0.75. Hence, the rate of emission,

$$\varepsilon\sigma T^4 = 0.75 \times 5.67 \times 10^{-8} \times (300)^4 = 344.5\ \text{W/m}^2.$$

10.10 Intensity of Radiation and Lambert's Cosine Law

The *intensity of normal radiation* I_n is defined as the rate of emission of energy from the unit surface area of a body in normal direction through unit solid angle.

Thus, from the definition of the intensity of normal radiation, the rate of emission of energy through area dA_n, refer Fig. 10.13, is

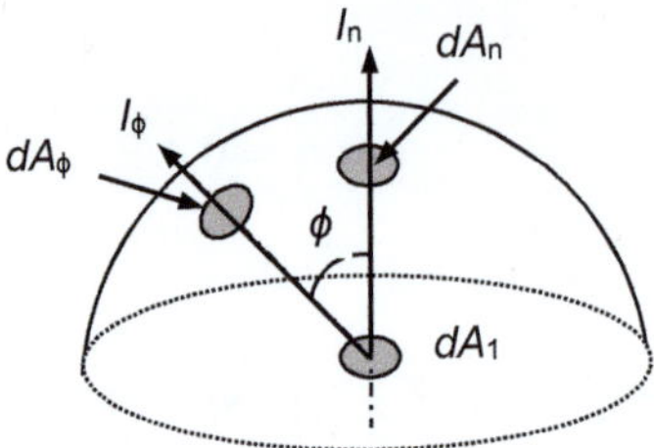

Fig. 10.13 Lambert's law

$$(dq_b)_n = I_n d\omega_n dA_1 \tag{10.15}$$

where $d\omega_n$ is the solid angle subtended by the area dA_n situated at a distance r on the normal to the small area dA_1 on the surface of the body emitting the radiation, see Fig. 10.13. The subscript b refers to the black surface.

Lambert's cosine law for diffuse-black surfaces states that the intensity of radiation in any direction ϕ with the normal I_ϕ is

$$I_\phi = I_n \cos\phi \tag{10.16}$$

Thus, the law gives the spatial distribution of intensity of radiation.

Hence, the rate of emission through area dA_ϕ at an angle ϕ with the normal and at a distance r from the emitting surface is, see Fig. 10.13,

$$(dq_b)_\phi = I_\phi d\omega_\phi dA_1 = I_n \cos\phi d\omega_\phi dA_1 \tag{i}$$

Solid angle: A solid angle is defined as a portion of the space inside a sphere enclosed by a conical surface with the vertex of the cone at the centre of the sphere. Mathematically, it is expressed as a ratio of the area of the spherical surface enclosed by the cone to the square of the radius of the sphere, refer Fig. 10.14,

$$d\omega = \frac{dA}{r^2} \tag{10.17}$$

Its unit is steradian. For a hemisphere,

$$\omega_{\text{hemisphere}} = \frac{\text{surface area}}{r^2} = \frac{2\pi r^2}{r^2} = 2\pi \text{ sr.}$$

Substitution of the value of $d\omega_\phi = dA_\phi/r^2$ gives

$$(dq_b)_\phi = (I_n \cos\phi)\left(\frac{dA_\phi}{r^2}\right)dA_1$$

The Stefan–Boltzmann law determines the energy emitted by a body in all directions. We can also calculate the total energy emitted by a body in terms of the intensity of normal radiation.

Let the elemental diffuse-black surface area dA_1 emitting radiation is located at the centre of a hemisphere of radius r. An elemental area dA_ϕ on the surface of the hemisphere at an

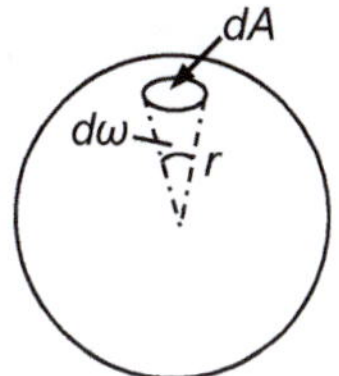

Fig. 10.14 Definition of the solid angle

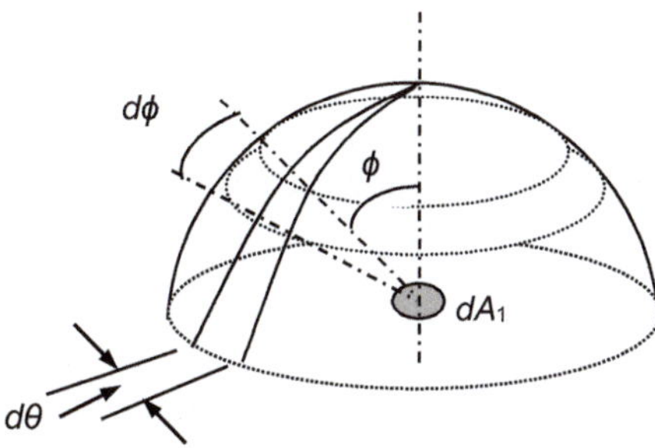

Fig. 10.15 Spherical coordinate system used in the derivation of Eq. (10.18) and elemental area

angle ϕ to the normal on the surface dA_1, subtends angles $d\phi$ and $d\theta$ at dA_1 as shown in Fig. 10.15.

The area of the element in terms of radius r and angles ϕ and θ is

$$dA_\phi = (rd\phi)(r\sin\phi)d\theta = r^2\sin\phi d\phi d\theta$$

Solid angle subtended by this elemental area at the centre of the hemisphere is

$$d\omega = \frac{dA_\phi}{r^2} = \sin\phi d\phi d\theta$$

Thus, the radiation passing through the area dA_ϕ is

$$(dq_b)_\phi = (I_n\cos\phi)(\sin\phi d\phi d\theta)dA_1$$

The total radiation through the hemisphere can be obtained by integrating between the limits $\phi = \pi/2$ and $\theta = 2\pi$

$$\begin{aligned} q_b &= I_n dA_1 \int_0^{2\pi} d\theta \int_0^{\pi/2} \sin\phi\cos\phi d\phi \\ &= 2\pi I_n dA_1 \int_0^{\pi/2} \sin\phi\cos\phi d\phi \\ &= \pi I_n dA_1 \end{aligned} \tag{10.18}$$

Equating q_b to the emissive power given by Stefan–Boltzmann law ($q_b = \sigma T^4 dA_1$), we obtain

$$I_n = \frac{\sigma T^4}{\pi} \tag{10.19}$$

This means that the emissive power in the normal direction is π times smaller than the total emissive power of the body.

It must be borne in mind that the Lambert's law is valid for diffuse radiating black surfaces only. Real surfaces are not perfectly diffuse and the experiments have shown that the law holds for such surfaces only for $\phi = 0 - 60°$. The most noticeable deviations are non-metals and metals with non-diffuse surfaces. In Fig. 10.16, the dependence of the total directional emissivity ε_ϕ on angle ϕ is shown for some of such surfaces. The *directional emissivity* is defined as

$$\varepsilon_\phi = \frac{I_\phi}{I_{b\phi}} \tag{10.20}$$

where I_ϕ refers to a real surface and $I_{b\phi}$ to that of a black surface.

The outside semicircle in the figure represents the total directional emissivity of a diffuse-black surface for which the total emissivity remains constant for all values of ϕ. For the electric insulators, the curve of the polar emissivity is almost circular up to $\phi = 60°$. Beyond $\phi = 60°$, the total emissivity gradually decreases to zero at $\phi = 90°$, refer Fig. 10.16a. For the metals, the emissivity first increases with ϕ, and thereafter decreases. At $\phi \approx 80°$, the radiation from the metal is almost twice its value at $\phi = 0°$.

The analysis of the radiation exchange involving real surfaces is quite complicated hence the practical surfaces are assumed to be gray with a constant value of the directional emissivity like that of a black surface.

Some references present the Lambert's law in the following form.

"The law states that the total emissive power E_ϕ of a flat emitting diffuse-black surface decreases as the angle ϕ with the normal to the surface increases".

Mathematically, we can put this as

$$(E_b)_\phi = (E_b)_n \cos\phi \tag{10.21}$$

where $(E_b)_n$ is the total emissive power of the radiating surface in a direction normal to the surface. Eq. (i) can be rewritten in the following form

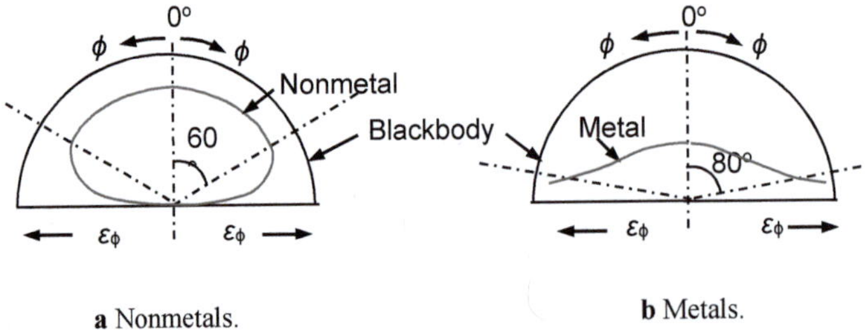

Fig. 10.16 Variation of directional emissivity with angle

$$(dq_b)_\phi = I_n d\omega_\phi (dA_1 \cos\phi) \tag{10.22}$$

where $(dA_1\cos\phi)$ is the area of the emitting surface normal to the line of view to the receiving surface dA_ϕ. The equation is interpreted to state that for a flat diffuse-black surface, the intensity of the radiation is the same in all directions based on the projected area $dA_1\cos\phi$ of the surface dA_1.

Example 10.10 A blackbody at 1100 K is radiating to space.

(a) What is spectral intensity in a direction normal to the surface at $\lambda = 5$ μm?
(b) What is directional spectral emissive power at $\phi = 40°$ away from the normal to the surface at $\lambda = 5$ μm?
(c) What is the ratio of the spectral intensity of the blackbody at $\lambda = 1$ μm to the spectral intensity of the blackbody at $\lambda = 5$ μm?
(d) How much energy is emitted by the blackbody in the range $1 \le \lambda \le 5$ μm?
(e) Calculate the wavelength such that emission from 0 to λ is equal to the emission from λ to ∞.

Solution

(a) Monochromatic emissive power at $\lambda = 5$ μm,

$$\begin{aligned} E_{b\lambda} &= \frac{c_1\lambda^{-5}}{\exp(c_2/\lambda T) - 1} \\ &= \frac{3.743 \times 10^{-16} \times (5 \times 10^{-6})^{-5}}{\exp[1.4387 \times 10^{-2}/(5 \times 10^{-6} \times 1100)] - 1} \\ &= 9.447 \times 10^9 \text{ W/(m}^2\text{-m)}. \end{aligned} \tag{10.6}$$

Spectral intensity in the normal direction,

$$(I_n)_{\lambda = 5\,\mu\text{m}} = \frac{E_{b\lambda}}{\pi} = 3.01 \times 10^9 \text{ W(m}^2\text{-m)}.$$

(b) Directional spectral emissive power at $\phi = 40°$ away from the normal to the surface at $\lambda = 5$ μm,

$$= (E_b)_\lambda \cos\phi = 9.447 \times 10^9 \times \cos 40° = 7.236 \times 10^9 \text{ W(m}^2\text{-m)}.$$

(c) Ratio of spectral intensities,

$$\begin{aligned} \frac{(E_b)_{\lambda_1}}{(E_b)_{\lambda_2}} &= \left(\frac{\lambda_1}{\lambda_2}\right)^{-5} \times \frac{\exp(c_2/\lambda_2 T) - 1}{\exp(c_2/\lambda_1 T) - 1} \\ &= \left(\frac{1}{5}\right)^{-5} \times \frac{\exp[1.4387 \times 10^{-2}/(5 \times 10^{-6} \times 1100)] - 1}{\exp[1.4387 \times 10^{-2}/(1 \times 10^{-6} \times 1100)] - 1} \\ &= 0.08275. \end{aligned}$$

(d) Energy emitted by the blackbody in the range $1 \leq \lambda \leq 5$ μm,

$$\lambda_1 T = 1.0 \times 1100 = 1100 \text{ for which } F_{0-1100} = 0.926 \times 10^{-3}$$

and

$$\lambda_2 T = 5.0 \times 1100 = 5500 \text{ for which } F_{0-5500} = 0.69065.$$

Thus,

$$F_{1100-5500} = 0.69065 - 0.926 \times 10^{-3} \approx 0.69.$$

Energy emitted by the blackbody in this range,

$$\begin{aligned} &= F_{1100-5500} \times \sigma T^4 \\ &= 0.69 \times (5.67 \times 10^{-8}) \times (1100)^4 = 57{,}280 \text{ W/m}^2. \end{aligned}$$

For $F_{0-\lambda} = 0.5$, the value of $\lambda T \approx 4109$ (μm)K from Table 10.2. Hence,

$$\lambda = 4109/T = 4109/1100 = 3.735 \text{ μm}.$$

Example 10.11 For a black surface at a temperature of 1200 K, calculate

(i) Monochromatic emissive power at $\lambda = 1$ μm
(ii) λ_{max}
(iii) Monochromatic emissive power at λ_{max}
(iv) The total rate of energy emission
(v) The intensity of radiation at an angle of 40° with the normal to the surface.

Solution

(i) From Planck's distribution law, the monochromatic emissive power is

$$\begin{aligned} E_{b\lambda} &= \frac{c_1 \lambda^{-5}}{\exp(c_2/\lambda T) - 1} \\ &= \frac{3.743 \times 10^{-16} \times (1 \times 10^{-6})^{-5}}{\exp[1.4387 \times 10^{-2}/(1 \times 10^{-6} \times 1200)] - 1} \\ &= 2.325 \times 10^{9} \text{ W(m}^2\text{-m)}. \end{aligned} \quad (10.6)$$

(ii) From the Wein's displacement law,

$$\lambda_{max} T = 2897.6$$

or $\lambda_{max} = 2897.6/T = 2897.6/1200 = 2.415$ μm.

(iii) $(E_{b\lambda})_{\max} = 1.289 \times 10^{-5} T^5 = 1.289 \times 10^{-5} \times (1200)^5 = 3.2 \times 10^{10}\,\text{W/m}^2$.

(iv) Total rate of radiation energy emission, from the Stefan Boltzmann law,

$$E_b = \sigma T^4 = 5.67 \times 10^{-8} \times (1200)^4 = 117{,}573\,\text{W/m}^2.$$

(v) Intensity of radiation at an angle of 40° with the normal to the surface,

$$I_\phi = I_n \cos\phi = \left(\frac{E_b}{\pi}\right)\cos\phi = \left(\frac{117{,}573}{\pi}\right) \times \cos 40^\circ = 28{,}669\,\text{W}/(\text{m}^2\text{ - sr}).$$

Example 10.12 Spectral directional emissivity of a diffuse surface at 1500 K is given in Fig. 10.11. Determine the fraction of emissive power over the spectral range 1.0 to 3 μm and for directions $0 \le \phi \le \pi/6$.

Solution

From the given spectral directional emissivity of the diffuse surface from Table 10.3:

$$\lambda_1 T = 1 \times 1500 = 1500 \text{ for which } F_{0-1500} = 0.01285$$
$$\lambda_2 T = 2 \times 1500 = 3000 \text{ for which } F_{0-3000} = 0.27322$$

and

$$\lambda_3 T = 3 \times 1500 = 4500 \text{ for which } F_{0-4500} = 0.56399.$$

Hemispherical total emissivity is

$$\begin{aligned}\varepsilon &= \varepsilon_1(F_{1500-3000}) + \varepsilon_2(F_{3000-4500})\\ &= 0.1 \times (0.27322 - 0.01285) + 0.4 \times (0.56399 - 0.27322)\\ &= 0.142345\end{aligned}$$

and the hemispherical total emissive power of the surface is

$$\begin{aligned}E &= \varepsilon\sigma_b T^4\\ &= 0.142345 \times 5.67 \times 10^{-8} \times 1500^4\\ &= 40{,}859\,\text{W/m}^2.\end{aligned}$$

For a diffuse surface,

$$\frac{E(0 \to \pi/6)}{E} = \frac{I_n \int_0^{2\pi} \mathrm{d}\theta \int_0^{\pi/6} \sin\phi \cos\phi \mathrm{d}\phi}{\pi I_n}$$

$$= \frac{1}{\pi}\left[2\pi \int_0^{\pi/6} \sin\phi \cos\phi \mathrm{d}\phi\right]$$

$$= \int_0^{\pi/6} 2\sin\phi \cos\phi \mathrm{d}\phi$$

$$= \left[\sin^2\phi\right]_0^{\pi/6}$$

$$= \left(\sin^2\frac{\pi}{6} - \sin^2 0\right) = 0.25.$$

Hence,

$$E(0 \to \pi/6) = 0.25E = 0.25 \times 40,859 = 10,215\,\mathrm{W/m^2}.$$

10.11 Summary

In general, the radiation means propagation of the electromagnetic waves of all wavelengths, ranging from radio waves to cosmic rays. Thermal radiation or heat radiation is also a form of electromagnetic emission. Its wavelength ranges from around 0.1 μm to 1000 μm, which includes the visible light region between violet (0.38 μm) and red (0.78 μm). Radiation from 0.78 μm to 1000 μm is termed as infrared (IR) radiation. Thermal effects are associated with thermal radiation. The thermal radiation also propagates at the velocity of light (299.8×10^6 m/s in vacuum) and obeys the laws of propagation, reflection and refraction of light rays. Solids, as well as liquids and gases, are capable of radiating thermal energy and absorbing such energy. For example, water vapour and carbon dioxide are the main sources of the gaseous radiation in furnaces. Radiation exchange depends on the nature of the substance, its temperature, wavelength and the state of the emitting surface.

When radiation falls on a body, a part of it may be absorbed, a part of it may be reflected and the remaining may pass through the body. The first fraction is known as absorptivity α, second is reflectivity ρ and the third fraction is transmissivity τ.

A body with reflectivity of unity will reflect the whole of the incident radiation and is termed a white body. When the reflection from a body obeys the laws of geometrical optics (angle of the reflected beam with normal equals the angle of the incident beam with the normal) the body is called smooth or specular. Due to the surface irregularities or roughness, the reflected radiation may be dispersed in all directions. Such a surface is known as a diffuse surface.

If the entire incident radiation is absorbed by the body, the absorptivity $\alpha = 1$. Such a body is termed as a blackbody. A blackbody does not exist in nature. But the concept of a blackbody is of great importance. The ideal behaviour of the blackbody serves as a standard with which the performance of real bodies can be compared.

The hemispherical total emissive power of a blackbody, from Stefan–Boltzmann law is proportional to the fourth power of the absolute temperature of the body.

Planck law, Eq. (10.6), gives the spectral distribution of the monochromatic emissive power of a blackbody as a function of temperature, which has also been used to determine Stefan–Boltzmann's equation of total emissive power of a blackbody. Monochromatic emissive power is zero at $\lambda = 0$. It first increases with an increase in the wavelength and reaches its maximum at a certain value of wavelength λ_{max}, then it decreases again with the increase in the wavelength and becomes zero at $\lambda = \infty$. λ_{max} is given by the Wein's displacement law, Eq. (10.8).

The fraction of the emissive power in a wavelength interval $F_{\lambda_1-\lambda_2}$ can be determined from the tabulated values of $F_{o-\lambda T}$ as function of λT in Table 10.3. The fraction calculation has been explained by some illustrative examples.

For a real body, whose radiation spectrum is not continuous, monochromatic emissivity ε_λ has been defined, which is the ratio of the emissive power E_λ of a real body at a particular wavelength and temperature to that of a blackbody $E_{b\lambda}$ at the same wavelength and temperature.

A body whose radiation spectrum is continuous and similar to that of a blackbody is known as a gray body. Its monochromatic emissivity ε_λ is constant for all wavelengths, as well as temperatures and its value is $0 < \varepsilon_\lambda < 1$. Unfortunately, there are not many surfaces that can be termed as gray but this idealization has been widely used in most of the engineering calculations, refer Chap. 11.

Kirchhoff's Law has been stated and proved. The law states that at any temperature the ratio of emissive power E to the absorptivity α is a constant for all bodies and equals the emissive power of a blackbody at the same temperature. The law establishes the equality of emissivity and absorptivity.

Concept of intensity of normal radiation I_n and Lambert's cosine law have been presented. The value of intensity of radiation in terms of total emissive power of a blackbody E_b has been determined, which along with the concept of intensity of normal radiation and Lambert's cosine law will be utilized in the next chapter to provide the base for determination of radiation exchange between two bodies.

Review Questions

10.1 What is difference between thermal radiation and other types of electromagnetic radiation?

10.2 Define absorptivity, transmissivity and reflectivity of a surface.

10.3 The Planck's law governing the change in emissive power of a blackbody with the wavelength is given by

$$E_{b\lambda} = \frac{c_1\lambda^{-5}}{\exp(c_2/\lambda T) - 1}$$

Using the above equation show that the total emissive power is given by

$$E_b = \sigma T^4.$$

where σ is the Stefan Boltzmann constant.

10.4 Distinguish between black, real and gray bodies giving suitable examples.
10.5 What is the Wein's displacement law?
10.6 State and prove the Kirchhoff's law of thermal radiation.
10.7 State the Lambert's cosine law.
10.8 Define intensity of radiation and prove that for a diffuse-blackbody, the intensity of normal radiation is $1/\pi$ times of the total emissive power of the body.
10.9 Distinguish between specular and diffuse surfaces.

Problems

10.1 A photon has an energy of 0.3×10^{-12} J. Calculate its frequency and wavelength in vacuum.
[Ans. $\nu = E/h = 0.3 \times 10^{-12}/6.6236 \times 10^{-34} = 4.53 \times 10^{20}$ Hz; $\lambda = c_o/\nu = 2.998 \times 10^8/(4.53 \times 10^{20}) = 6.62 \times 10^{-13}$ m.]

10.2 Radiant energy with an intensity of 1000 W/m^2 is incident normal to a flat surface whose absorptivity is 2.5 times the transmissivity and 2 times of the reflectivity. Determine the energy transmitted, absorbed and reflected.
[Ans. $\alpha + \rho + \tau = 1$, substitution gives $\alpha + \alpha/2.5 + \alpha/2 = 1$, hence, $\alpha = 1/1.9$; $Q_A = 1000/1.9 = 526.32$ W/m^2; $Q_T = Q_A/2.5 = 210.53$ W/m^2; $Q_R = Q_A/2 = 263.16$ W/m^2.]

10.3 The temperature of a black surface of 0.5 m^2 area is 727°C. Calculate: (a) the total amount of energy emission, (b) the intensity of normal radiation, (c) the intensity of radiation at an angle of 60° and (d) the wavelength of maximum monochromatic emissive power.
[Ans. $E_b = A\sigma T^4 = 0.5 \times 5.669 \times 10^{-8} (727 + 273)^4 = 28345$ W. $I_n = E_b/\pi = \sigma T^4/\pi = 18045$ W/(m^2 sr); $I_\phi = I_n \cos 60° = 9022.5$ W/(m^2 sr); $\lambda_{max} = 2897.6/1000 = 2.9$ μm.]

10.4 Figure 10.17 shows the variation of reflectivity with λ for an opaque surface. The irradiation G (radiation impinging) on the surface from a source is approximated as given below. Determine the energy absorbed.

$$0 \le \lambda < 0.4 \quad G = 20\ \text{W/m}^2 \qquad 0.4 \le \lambda < 1.0 \quad G = 150\ \text{W/m}^2$$
$$1.0 \le \lambda < 2.0 \quad G = 100\ \text{W/m}^2 \qquad 2.0 \le \lambda < 3.0 \quad G = 20\ \text{W/m}^2$$
$$3.0 \le \lambda < \infty \quad G = 30\ \text{W/m}^2$$

[Ans. Using the relation absorptivity = $(1 - \rho)$, we have $G_{absorbed} = (1 - 0) \times 20 + (1 - 0.5) \times 150 + (1 - 0.8) \times 100 + (1 - 0.9) \times 20 + (1 - 0) \times 30 = 147$ W/m^2.]

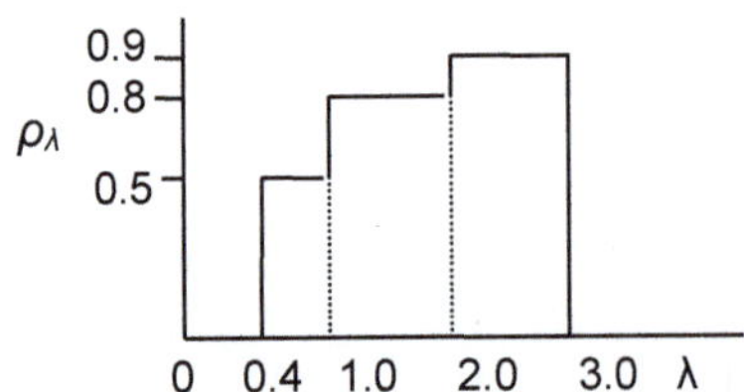

Fig. 10.17 Problem 10.4

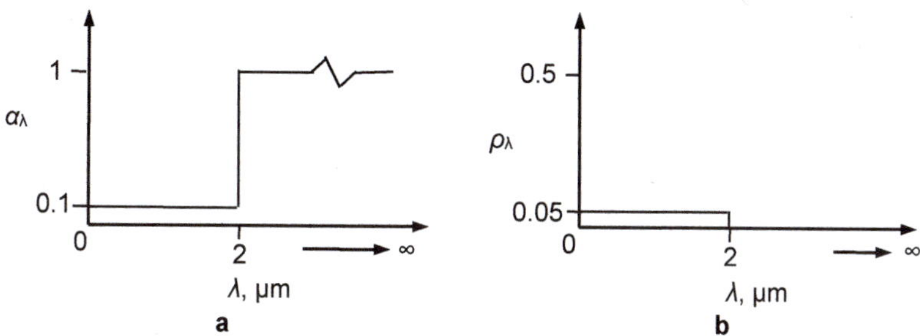

Fig. 10.18 Problem 10.5

10.5 For a spectrally selective, diffuse surface the spectral distribution of absorptivity and reflectivity are shown in Fig. 10.18. Determine spectral transmissivity, τ_λ.
[**Ans.** $\tau_\lambda = 1 - \alpha_\lambda - \rho_\lambda$. Hence, for $0 \le \lambda \le 2$ μm, $\tau_\lambda = 1 - 0.1 - 0.05 = 0.85$ and for $\lambda > 2$ μm, $\tau_\lambda = 1 - 1 - 0 = 0$, i.e. the surface is opaque for radiation of $\lambda > 2$ μm.]

10.6 A window glass transmits 90% of the radiation between wavelengths 0.3–3 μm. It is practically opaque to all other wavelengths. Determine the percentage of radiant heat flux transmitted when it is coming from a blackbody source at 2000 K.
[Ans. $\lambda_1 T = 0.3 \times 2000 = 600$ μm K, $\lambda_2 T = 3 \times 2000 = 6000$ μm K; $F_{0-\lambda_1 T} = 0.112 \times 10^{-6}$, $F_{0-\lambda_2 T} = 0.73777$ from Table 10.3, $F_{\lambda_2 T-\lambda_1 T} \approx 0.73777$; Flux transmitted = 0.90 × 0.73777 × 100 = 66.4%.]

10.7 If the incident solar radiation on the Earth is 1390 W/m^2 for a mean distance $r_{se} = 1.5 \times 10^{11}$ m, determine the radiation on planet Mercury if its orbital radius is 5.8×10^{10} m.
[Ans. The radiation incident on a surface is inversely proportional to the distance squared (the inverse square law). Hence, $G_{\text{Murcury}} = G_{\text{Earth}} \times (r_{\text{se}}/r_{\text{sm}})^2 = 1390 \times (1.5 \times 10^{11}/5.8 \times 10^{10})^2 = 9297\ \text{W/m}^2$.]

10.8 Radiation flux G_o enters into a cavity through a small opening. The absorptivity of the cavity surface is less than 1. Determine the flux after n reflections.
[Ans. On first reflection G_o reduces to $G_1 = (1 - \alpha)\, G_o$; Similarly, on the second reflection, $G_2 = (1 - \alpha)^2\, G_o$, …, after n reflection, $G_n = (1 - \alpha)^n G_o$; As n increases, $1 - (1 - \alpha)^n \to 1$, i.e. the radiation will be completely absorbed.]

10.9 A large plate ($\varepsilon_1 = 0.8$) emits 300 W/m^2. Another plate ($\varepsilon_2 = 0.4$) of the same surface area emits 150 W/m^2. If these plates are brought very close and parallel to each other, determine the net heat exchange per unit area of the plates.
[Ans. For a gray surface $E = \varepsilon E_b$. Hence, $E_1 = \varepsilon_1 E_{b1} = \varepsilon_1 \sigma T_1^4$, and $E_2 = \varepsilon_2 E_{b2} = \varepsilon_2 \sigma T_2^4$; Ratio E_1/E_2 gives $T_1 = T_2$ hence $q_{12} = 0$.]

References

Eckert ERG, Drake RM (1959) Heat and mass transfer, 2nd edn. McGraw-Hill Inc.
Mills AF (1995) Heat and mass transfer. Richard D. Irwin, Chicago.

11 Exchange of Thermal Radiation Between Surfaces Separated by Transparent Medium

11.1 Introduction

In the previous chapter, we discussed the laws of radiation, absorption, reflection, emissivity and the dependence of the emitted radiation from a diffuse surface on direction. With this background, it is possible to calculate the radiation exchange between opaque solids. Radiation exchange between solids is a complex process. It depends on the shape and size of the bodies, relative position, distance between them and their emissivity. The calculation becomes more complicated when there is an absorbing or reradiating medium between the surfaces. In this chapter, we shall consider the radiation exchange without an absorbing medium.

11.2 Radiation Heat Exchange Between Two Black Surfaces and the Shape Factor

Consider two black diffuse surfaces 1 and 2 of areas A_1 and A_2 at temperatures T_1 and T_2, respectively, as shown in Fig. 11.1. The surfaces are separated by a non-absorbing medium such as air. To determine the radiant heat exchange between these surfaces, let us first evaluate the exchange between elemental areas ΔA_1 and ΔA_2 on these surfaces at distance s apart. The normal to the elemental areas makes angles ϕ_1 and ϕ_2 with the line AB.

Projection of the area ΔA_2 normal to the line AB is ($\Delta A_2 \cos\,\phi_2$). Solid angle subtended by the area $\Delta A_2 \cos\,\phi_2$ at the center of ΔA_1 is

$$\Delta\omega_2 = \frac{\Delta A_2 \cos\phi_2}{s^2} \tag{11.1}$$

Hence, the fraction of the total heat energy emitted by ΔA_1 which is intercepted by ΔA_2, from Eq. (10.15), is

$$\begin{aligned}\Delta q_{12} &= I_{n1}\cos\phi_1(\Delta A_1)\frac{\Delta A_2\cos\phi_2}{s^2}\\ &= I_{n1}\frac{\Delta A_1 \Delta A_2\cos\phi_1\cos\phi_2}{s^2}\end{aligned} \tag{11.2}$$

R. Karwa, *Heat and Mass Transfer*,
https://doi.org/10.1007/978-981-15-3988-6_11

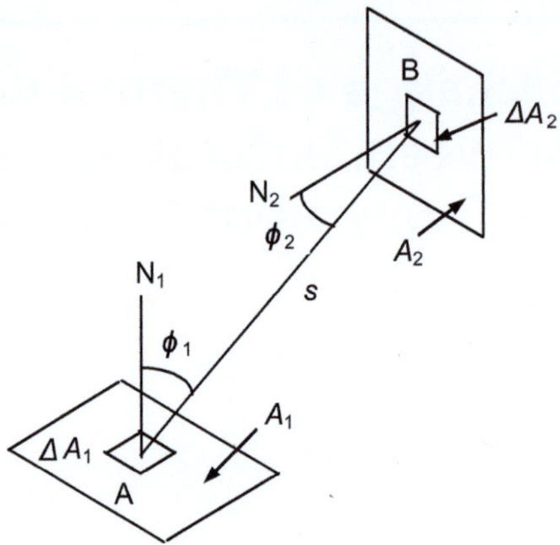

Fig. 11.1 Radiation exchange between two black surfaces

Similarly the rate at which the radiation emitted by ΔA_2 and intercepted by ΔA_1 is given by

$$\begin{aligned}\Delta q_{21} &= I_{n2}\cos\phi_2(\Delta A_2)\frac{\Delta A_1\cos\phi_1}{s^2}\\ &= I_{n2}\frac{\Delta A_1\Delta A_2\cos\phi_1\cos\phi_2}{s^2}\end{aligned} \tag{11.3}$$

Substituting $I_{n1} = E_{b1}/\pi$ and $I_{n2} = E_{b2}/\pi$, we get

$$\begin{aligned}\Delta q_{12} &= \left(\frac{E_{b1}}{\pi}\right)\frac{\Delta A_1\Delta A_2\cos\phi_1\cos\phi_2}{s^2}\\ &= \left(\frac{\sigma T_1^4}{\pi}\right)\frac{\Delta A_1\Delta A_2\cos\phi_1\cos\phi_2}{s^2}\end{aligned} \tag{11.4}$$

and

$$\begin{aligned}\Delta q_{21} &= \left(\frac{E_{b2}}{\pi}\right)\frac{\Delta A_1\Delta A_2\cos\phi_1\cos\phi_2}{s^2}\\ &= \left(\frac{\sigma T_2^4}{\pi}\right)\frac{\Delta A_1\Delta A_2\cos\phi_1\cos\phi_2}{s^2}\end{aligned} \tag{11.5}$$

For the infinitesimally small areas, Eqs. (11.4) and (11.5) transform to

$$dq_{12} = \left(\frac{\sigma T_1^4}{\pi}\right)\frac{dA_1 dA_2\cos\phi_1\cos\phi_2}{s^2} \tag{11.6}$$

and

$$dq_{21} = \left(\frac{\sigma T_2^4}{\pi}\right)\frac{dA_1 dA_2\cos\phi_1\cos\phi_2}{s^2} \tag{11.7}$$

Integration over the entire surface areas gives radiation from surface *1* incident on surface *2* as

$$q_{12} = \left(\frac{\sigma T_1^4}{\pi}\right) \int_{A_1} \int_{A_2} \frac{dA_1 dA_2 \cos\phi_1 \cos\phi_2}{s^2} \tag{11.8}$$

Similarly,

$$q_{21} = \left(\frac{\sigma T_2^4}{\pi}\right) \int_{A_1} \int_{A_2} \frac{dA_1 dA_2 \cos\phi_1 \cos\phi_2}{s^2} \tag{11.9}$$

We rewrite these equations as

$$q_{12} = A_1 F_{12} \sigma T_1^4 \tag{11.10}$$

$$q_{21} = A_2 F_{21} \sigma T_2^4 \tag{11.11}$$

where

$$\begin{aligned} F_{12} &= \frac{\text{Radiation from surface 1 incident upon surface 2}}{\text{Total radiation from surface 1}} \\ &= \frac{q_{12}}{A_1 \sigma T_1^4} \\ &= \left(\frac{1}{A_1}\right) \int_{A_1} \int_{A_2} \frac{dA_1 dA_2 \cos\phi_1 \cos\phi_2}{\pi s^2} \end{aligned} \tag{11.12}$$

Similarly,

$$\begin{aligned} F_{21} &= \frac{\text{Radiation from surface 2 incident upon surface 1}}{\text{Total radiation from surface 2}} \\ &= \frac{q_{21}}{A_2 \sigma T_2^4} \\ &= \left(\frac{1}{A_2}\right) \int_{A_1} \int_{A_2} \frac{dA_1 dA_2 \cos\phi_1 \cos\phi_2}{\pi s^2} \end{aligned} \tag{11.13}$$

The factor F_{12} is known as *radiation shape factor (configuration factor, geometrical factor, angle factor, or view factor)* of surface 1 with respect to the surface 2 and it represents the fraction of the total radiation emitted by surface 1 which has been intercepted by surface 2. Similarly factor F_{21} can be interpreted. The shape factor is merely a function of the geometry or the orientation of the two surfaces.

From Eqs. (11.12) and (11.13), we see that

$$A_1 F_{12} = \int_{A_1} \int_{A_2} \frac{dA_1 dA_2 \cos\phi_1 \cos\phi_2}{\pi s^2}$$

$$A_2 F_{21} = \int_{A_1} \int_{A_2} \frac{dA_1 dA_2 \cos\phi_1 \cos\phi_2}{\pi s^2}$$

Hence,

$$A_1F_{12} = A_2F_{21} \tag{11.14}$$

The result is known as *reciprocity relation*. It indicates that the net heat transfer by radiation can be determined by using the shape factor either ways (surface 1 to 2 or surface 2 to 1).

In the case of blackbodies, the radiation q_{12} and q_{21} incident on the surfaces 2 and 1, respectively, will be absorbed by these surfaces and the net radiation heat exchange between these surfaces from Eqs. (11.10) to (11.11) will be

$$(q_{12})_{net} = q_{12} - q_{21} = A_1F_{12}\sigma T_1^4 - A_2F_{21}\sigma T_2^4 \tag{11.15}$$

Using the reciprocity relation, we can write

$$(q_{12})_{net} = A_1F_{12}\sigma(T_1^4 - T_2^4) \tag{11.16a}$$

$$= A_2F_{21}\sigma(T_1^4 - T_2^4) \tag{11.16b}$$

Note: It is to be noted that Eqs. (11.15) and (11.16) of the net heat transfer applies only to the blackbodies. In case of the gray bodies, a part of the impinging radiation is reflected and the whole analysis becomes complicated. This has been discussed later in this chapter. The shape factor, which is a function of the geometry, is applicable in all cases.

11.3 Evaluation of the Shape Factor

The mathematical expression of the shape or configuration factor from Eq. (11.12) is

$$F_{12} = \frac{1}{A_1}\int_{A_1}\int_{A_2} \frac{dA_1 dA_2 \cos\phi_1 \cos\phi_2}{\pi s^2} \tag{11.12}$$

As an illustration for the evaluation, consider two diffuse-black flat discs oriented parallel to each other as shown in Fig. 11.2. One of the discs is having a very small surface area dA_1. The elemental area dA_2 on the other disc is a circular ring of radius r and width dr. Thus

$$dA_2 = 2\pi r dr$$

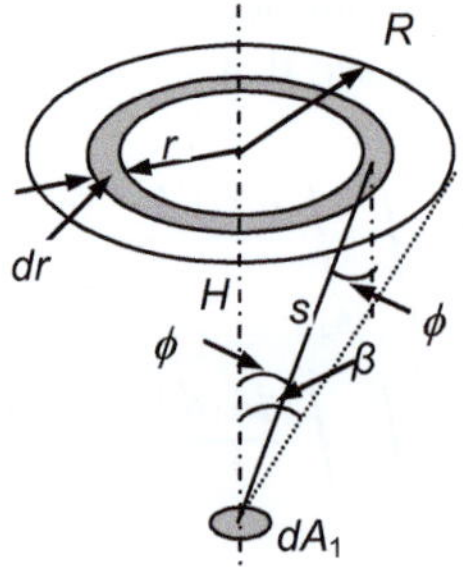

Fig. 11.2 Two diffuse-black flat discs parallel to each other

The angles subtended by the normal to the areas dA_1 and dA_2 with the line joining these areas are equal ($\phi_1 = \phi_2 = \phi$). Substitution in Eq. (11.12) gives

$$F_{12} = \frac{1}{dA_1} \int_A \frac{dA_1(2\pi r dr)\cos^2\phi}{\pi s^2}$$

From the geometry of the figure,

$$s = \sqrt{(H^2 + r^2)}$$

and

$$\cos\phi = \frac{H}{s} = \frac{H}{\sqrt{(H^2 + r^2)}}$$

Hence,

$$\begin{aligned} F_{12} &= 2H^2 \int_0^R \frac{rdr}{(H^2 + r^2)^2} \\ &= -H^2 \frac{1}{(H^2 + r^2)}\Big|_0^R \\ &= -H^2 \left[\frac{1}{(H^2 + R^2)} - \frac{1}{H^2}\right] \\ &= \frac{R^2}{(H^2 + R^2)} = \sin^2\beta \end{aligned} \tag{11.17}$$

Let us consider one more example of radiation exchange between an elemental surface area (dA_1) and a rectangular surface 2 parallel to it at distance L with one corner of the surface 2 in normal line to dA_1 as shown in Fig. 11.3.

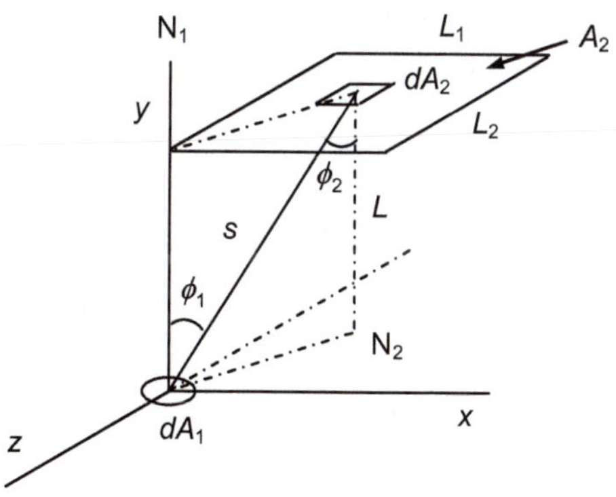

Fig. 11.3 A rectangular surface parallel to an elemental surface area

From the geometry of the figure,

$$\cos\phi_1 = \cos\phi_2 = \frac{L}{s}$$

$$s = \sqrt{(x^2 + z^2 + L^2)}$$

and

$$F_{12} = \frac{1}{dA_1}\int_{A_1}\int_{A_2} \frac{dA_1 dA_2 \cos\phi_1 \cos\phi_2}{\pi s^2}$$

$$= \int_{A_2} \frac{dA_2 \cos\phi_1 \cos\phi_2}{\pi s^2} = \int_{A_2} \left(\frac{L}{s}\right)^2 \frac{dA_2}{\pi(x^2 + z^2 + L^2)}$$

Putting $dA_2 = dxdz$

$$F_{12} = \int_0^{L_1} \int_0^{L_2} \left(\frac{L}{s}\right)^2 \frac{dxdz}{\pi(x^2 + z^2 + L^2)}$$

Evaluation of the integrals gives

$$F_{12} = \frac{1}{2\pi}\left[\frac{L_1}{\sqrt{(L_1^2 + L^2)}}\sin^{-1}\frac{L_2}{\sqrt{(L_1^2 + L_2^2 + L^2)}} + \frac{L_2}{\sqrt{(L_2^2 + L^2)}}\sin^{-1}\frac{L_1}{\sqrt{(L_1^2 + L_2^2 + L^2)}}\right] \tag{11.18}$$

Now the readers can realize that the evaluation of the integration is not an easy affair in all cases. The researchers have evaluated shape factors for some commonly encountered configurations. They are available in the form of charts; refer Fig. 11.4a–e, or equations (Table 11.1). For more information, refer Siegel and Howell (2002) and Howell (1982).

For flat surfaces, which are small in area compared with the distance between them and are uniform in temperature, an approximate solution of Eq. (11.12) can be obtained by substituting areas A_1 and A_2 together with the approximate values of ϕ_1 and ϕ_2. For the cases where such simplification is not possible, the properties of the shape factor being discussed in the next section can be useful.>

11.3.1 Salient Features of the Radiation Shape Factor

1. It the radiation coming out of a flat or convex surface 1 is intercepted by an enclosure 2, see Fig. 11.5a and b, the shape factor F_{12} is unity. In this case, the reciprocity relation gives that the shape factor F_{21} is simply the ratio of the two areas, i.e. $F_{21} = A_1/A_2$.

   The above result can also be obtained mathematically. Referring to Fig. 11.5c, $\phi_2 = 0$ and $dA_2 = 2\pi r \sin\phi_1\, r d\phi_1$. The angle ϕ_1 varies from 0 to $\pi/2$. Hence, the shape factor equation transforms to

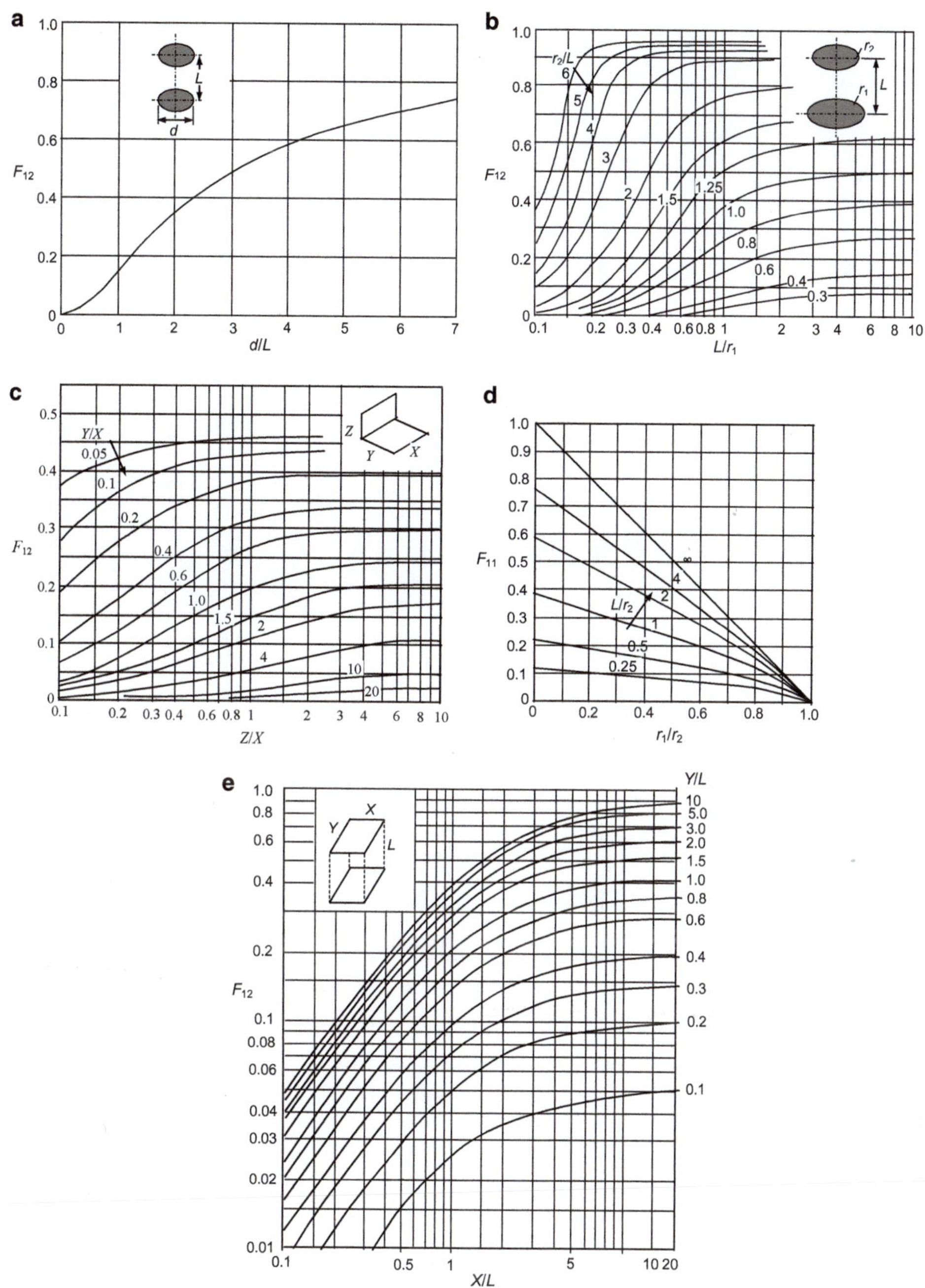

Fig. 11.4 **a** Radiation shape factor for radiation between parallel discs. **b** Radiation shape factor for radiation between two parallel concentric discs. **c** Radiation shape factor for radiation between perpendicular rectangles with a common edge. **d** Radiation shape factor for two concentric cylinders of finite length, outer cylinder to itself. **e** Radiation shape factor for radiation between parallel rectangles

Table 11.1 Shape, geometric or view factor for some configurations

Case	Configuration	Shape, Geometric or View factor
1		Two infinitely long plates of different widths joined along one of the long ages at right angles: $F_{12} = ½[1 + x - (1 + x^2)^{1/2}]$ where $x = H/L$
2		Two infinitely long plates of equal width, joined along one of the long ages: $F_{12} = 1 - \sin(\theta/2)$
3		Triangular cross-section enclosure formed by infinitely long plates of different widths: $F_{12} = (L_1 + L_2 - L_3)/2L_1$
4		Coaxial parallel discs: $F_{12} = ½\{X - [X^2 - 4(x_2/x_1)^2]^{1/2}\}$ where $x_1 = R_1/L$, $x_2 = R_2/L$, and $X = 1 + (1 + x_2^2)/x_1^2$
5		Coaxial parallel discs of equal radius: $F_{12} = 1 + (L^2/2R^2)[1 - (1 + 4R^2/L^2)^{1/2}]$ (The relation can be obtained by putting $R_1 = R_2 = R$ in the relation of Case 4)

(continued)

Table 11.1 (continued)

Case	Configuration	Shape, Geometric or View factor
6	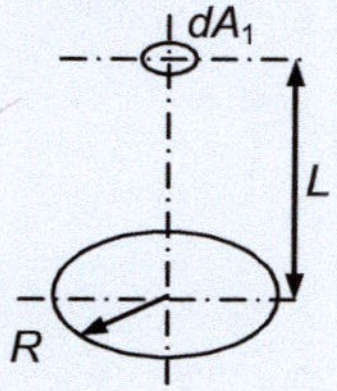 dA_1, L, R	Coaxial and parallel disc and infinitesimal area dA_1 : $F_{12} = R^2/(L^2 + R^2)$
7	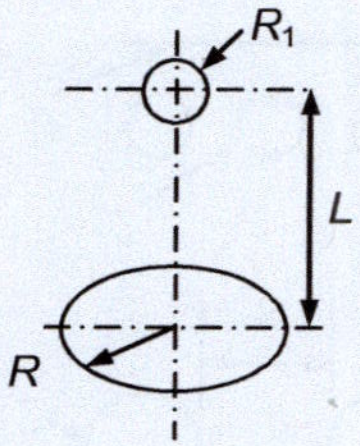 R_1, L, R	Sphere and coaxial disc: $F_{12} = ½[1 - L/(L^2 + R^2)^{1/2}]$
8	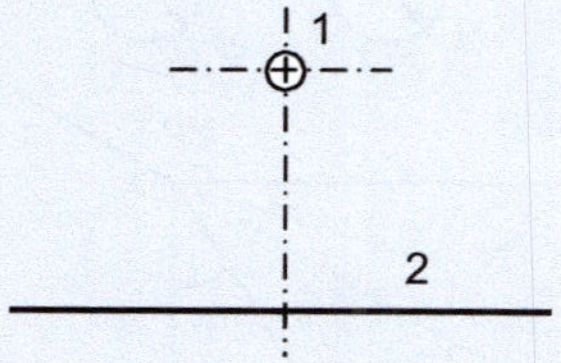 1, 2	Long cylinder parallel to a large plane area or sphere near a large plane area $F_{12} = ½$
9	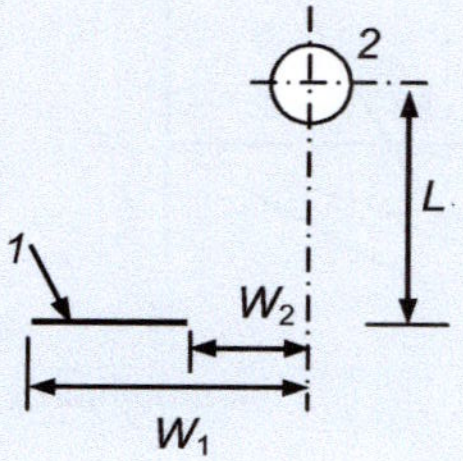 2, L, 1, W_2, W_1	Infinite cylinder of radius R parallel to an infinite plate of width $W_1 - W_2$: $F_{12} = R/(W_1 - W_2)[\tan^{-1}(W_1/L) - \tan^{-1}(W_2/L)]$

(continued)

Table 11.1 (continued)

Case	Configuration	Shape, Geometric or View factor
10	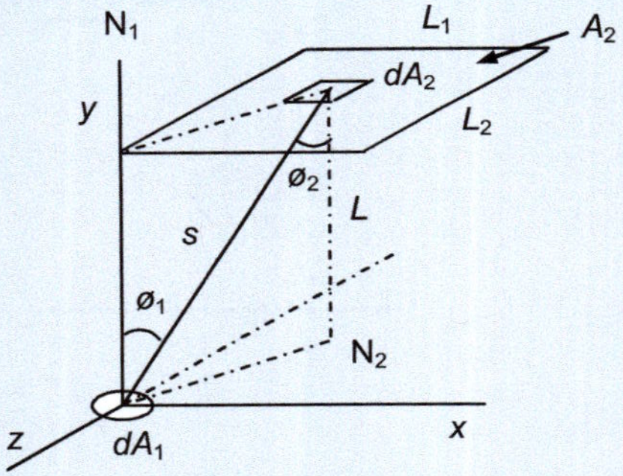	$F_{12} = (1/2\pi)\{[L_1/\sqrt{(L_1^2 + L^2)}] \sin^{-1}[L_2/\sqrt{(L_1^2 + L_2^2 + L^2)}] + [L_2/\sqrt{(L_2^2 + L^2)}] \sin^{-1}[L_1/\sqrt{(L_1^2 + L_2^2 + L^2)}]\}$
11	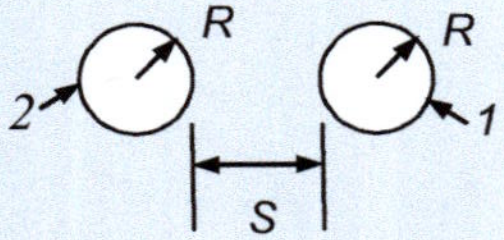	Two parallel and infinite cylinders $F_{12} = 1/\pi\ [(X^2 - 1)^{1/2} + \sin^{-1}(1/X) - X]$ where $X = 1 + S/2R$
12	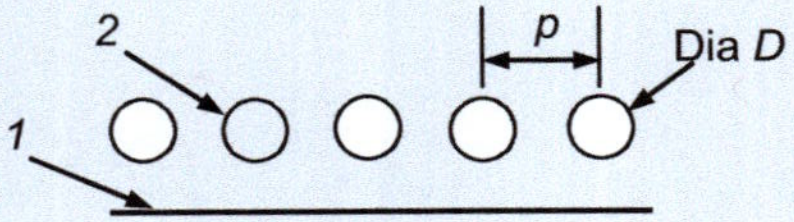	Row of infinite cylinders parallel to an infinite plate: $F_{12} = 1 - (1 - x^2)^{1/2} + x \tan^{-1}[(1 - x^2)/x^2]^{½}$ where p = pitch and $x = D/p$

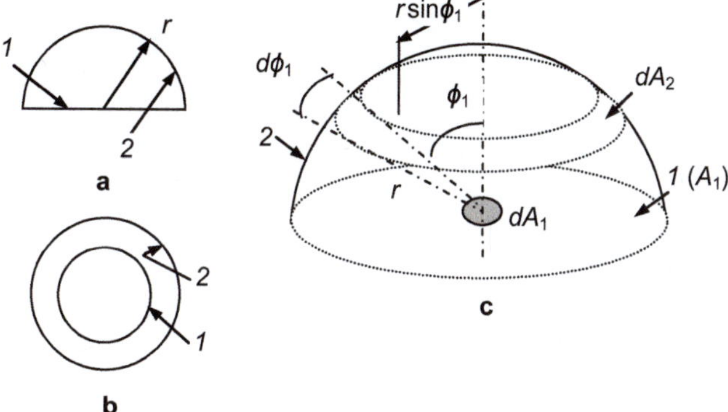

Fig. 11.5 Radiation from a flat or convex surface intercepted by an enclosure

$$F_{12} = \int_0^{\pi/2} \frac{(rd\phi_1)(2\pi r \sin\phi_1)\cos\phi_1 \cos 0}{\pi r^2} = 1 \tag{11.19}$$

2. **Subdivision of the Emitting Surface**

In Fig. 11.6, the radiating surface A_1 has been divided into two areas A_3 and A_4. Then

$$A_1F_{12} = A_3F_{32} + A_4F_{42}$$

In general, the above equation can be written as

$$A_iF_{ij} = \sum_{n=1}^{n} A_{in}F_{inj} \tag{11.20}$$

where area A_i has been divided into areas $A_{i1}, A_{i2}, \ldots A_{in}$.

3. **Subdivision of the Receiving Surface**

In Fig. 11.7, the receiving surface has been divided into areas $A_{2(1)}$ and $A_{2(2)}$. Then

$$A_1F_{12} = A_1F_{12(1)} + A_1F_{12(2)}$$

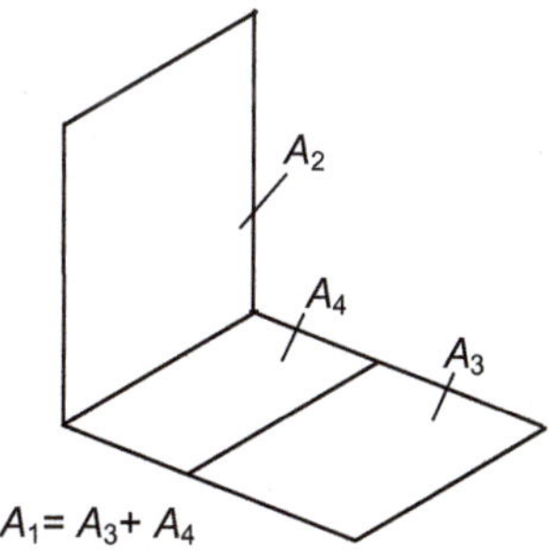

Fig. 11.6 Subdivision of emitting surface

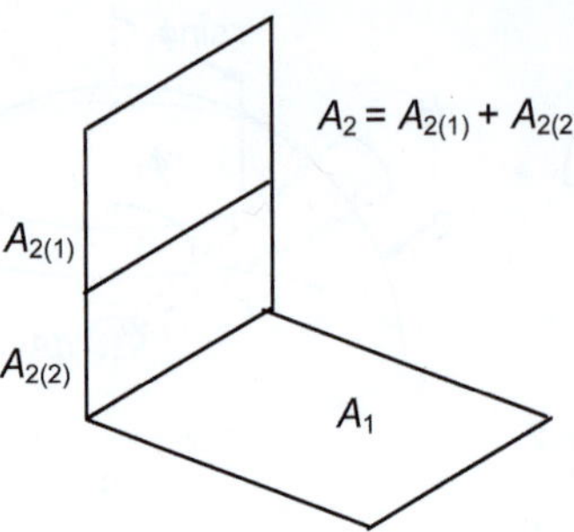

Fig. 11.7 Subdivision of the receiving surface

or

$$F_{12} = F_{12(1)} + F_{12(2)}$$

In general,

$$F_{12} = \sum_{i=1}^{n} F_{12(i)} \tag{11.21}$$

where area A_2 has been divided into areas $A_{2(1)}$, $A_{2(2)}$, ... $A_{2(n)}$. Equation expresses the *additive property* of the shape factor.

Enclosure

If a flat or convex surface 1 is completely enclosed by surface areas A_2, A_3, ... A_n, refer Fig. 11.8a, then

$$F_{12} + F_{13} + \ldots + F_{1n} = 1 \tag{11.22}$$

In case of a concave surface, a fraction of the radiant energy emitted by one part of the concave surface will be intercepted by another part of the concave surface, refer Fig. 11.8b. Thus a concave surface has shape factor with respect to itself, which can be termed as F_{11}. It follows that in this case

$$F_{11} + F_{12} + F_{13} + \ldots + F_{1n} = 1 \tag{11.23a}$$

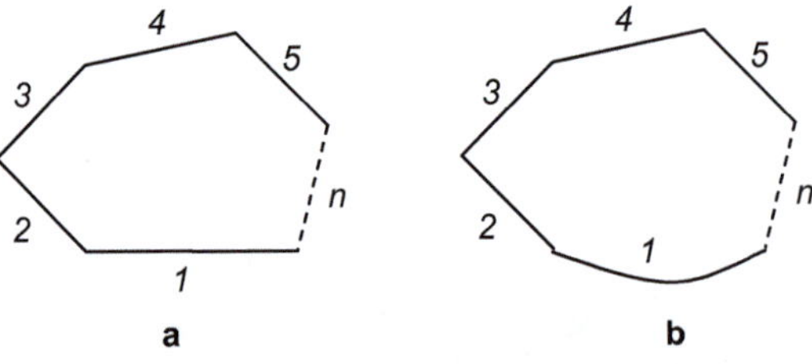

Fig. 11.8 Enclosures

or

$$\sum_{n=1}^{n} F_{1n} = 1 \quad (11.23\text{b})$$

This is known as *summation rule* and is simply based on the principle of conservation of energy. For the convex or a flat surface F_{11} is zero and Eq. (11.22) results.

11.4 Reciprocity Relation

The reciprocity relation, as presented earlier, is

$$A_1 F_{12} = A_2 F_{21} \quad (11.14)$$

It indicates that the radiation heat transfer can be determined by using the shape factor either ways. If the two surfaces have the same area ($A_1 = A_2$), the shape factor will have the same value when the surfaces 1 and 2 are interchanged.

Example 11.1 Determine the shape factors for

(a) A blackbody inside a black enclosure, Fig. 11.9a.
(b) A black sphere in a cubical box, Fig. 11.9b.
(c) A black hemisphere surface closed by a plane surface, Fig. 11.9c.
(d) A cylindrical cavity closed by a plane surface, Fig. 11.9d.

Solution The surface 1 in all cases given here is either convex or flat hence

$$F_{11} = 0.$$

The surface 2 intercepts whole of the radiation emitted by the surface 1 hence

$$F_{12} = 1.$$

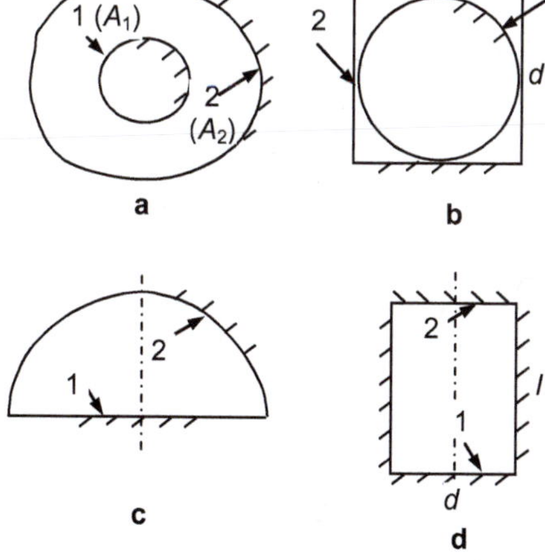

Fig. 11.9 Example 11.1

From the []reciprocity relation,

$$A_1F_{12} = A_2F_{21}$$

or

$$F_{21} = \frac{A_1}{A_2}F_{12} = \frac{A_1}{A_2}$$

From the summation rule,

$$F_{21} + F_{22} = 1$$

or

$$F_{22} = 1 - F_{21}$$

The results for different cases are given in Table 11.2.

Example 11.2 Calculate the shape factor for the ducts, whose cross-sections are shown in Fig. 11.10. Length of the duct in all cases is very large and hence the radiation loss from the ends of the ducts may be neglected. Whole the surface enclosing the surface 1 is to be considered surface 2.

Solution

For all the four cases shown in the figure, no part of the radiation leaving the surface 1 falls on the surface itself. Hence,

$$F_{11} = 0.$$

Surface 2 intercepts whole of the radiation emitted by the surface 1 hence

$$F_{12} = 1.$$

From the reciprocity relation,

$$A_1F_{12} = A_2F_{21}$$

Table 11.2 Example 11.1

Case	F_{11}	F_{12}	$F_{21} = A_1/A_2$	$F_{22} = 1 - F_{21}$
a	0	1	A_1/A_2	$1 - A_1/A_2$
b	0	1	$\frac{\pi d^2}{6d^2} = \frac{\pi}{6} = 0.5236$	$1 - 0.5236 = 0.4764$
c	0	1	$\frac{(\pi/4)d^2}{(\pi/2)d^2} = 0.5$	$1 - 0.5 = 0.5$
d	0	1	$\frac{(\pi/4)d^2}{\pi dl + (\pi/4)d^2} = \frac{d}{d+4l}$	$1 - \frac{d}{d+4l} = \frac{4l}{d+4l}$

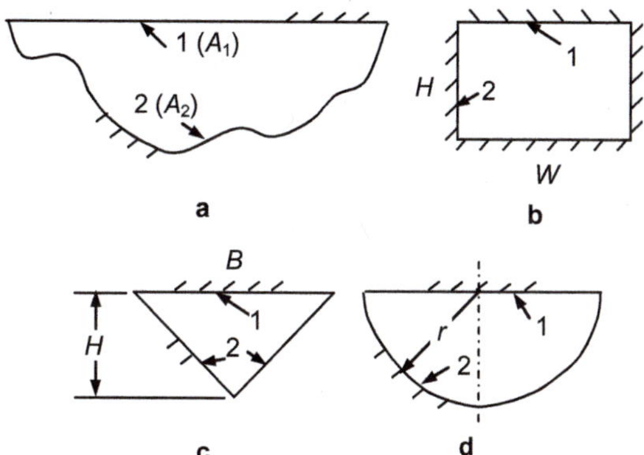

Fig. 11.10 Example 11.2

or

$$F_{21} = \frac{A_1}{A_2} F_{12} = \frac{A_1}{A_2}$$

From the summation rule,

$$F_{21} + F_{22} = 1$$

or

$$F_{22} = 1 - F_{21}$$

The results for different cases are given in Table 11.3.

Example 11.3 Determine shape factors F_{12}, F_{23}, F_{31}, etc. for the triangular cross-section enclosure formed by infinitely long plates of different widths as shown in Fig. 11.11.

Solution

Representing the sides of the enclosure by the respective areas A_1, A_2 and A_3, we can write the following:

$$A_1 F_{12} + A_1 F_{13} = A_1 \tag{i}$$

Table 11.3 Example 11.2

Case	F_{11}	F_{12}	$F_{21} = A_1/A_2$	$F_{22} = 1 - F_{21}$
a	0	1	A_1/A_2	$1 - A_1/A_2$
b	0	1	$\frac{W}{W+2H}$	$1 - \frac{W}{W+2H} = \frac{2H}{W+2H}$
c	0	1	$\frac{B}{2\sqrt{H^2+(B/2)^2}}$	$1 - \frac{B}{2\sqrt{H^2+(B/2)^2}}$
d	0	1	$\frac{2r}{\pi r} = \frac{2}{\pi}$	$1 - \frac{2}{\pi}$

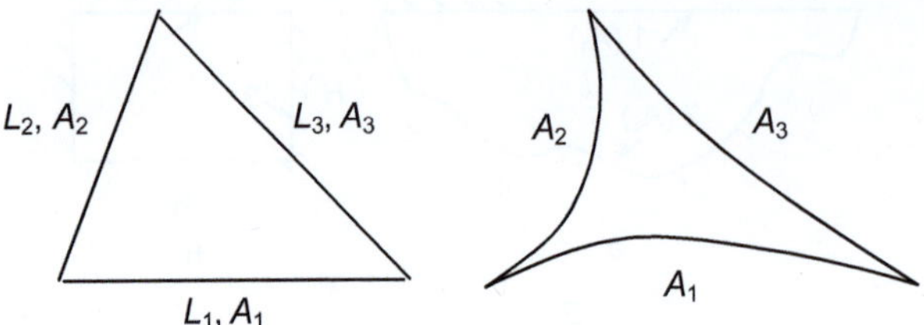

Fig. 11.11 Example 11.3

Similarly, we can write

$$A_2F_{21} + A_2F_{23} = A_2$$

$$A_3F_{31} + A_3F_{32} = A_3$$

Using the reciprocity relations, we can write these equations as

$$A_2F_{23} + A_1F_{12} = A_2 \tag{ii}$$

$$A_1F_{13} + A_2F_{23} = A_3 \tag{iii}$$

Summation of Eqs. (i)–(iii) gives

$$A_1F_{12} + A_1F_{13} + A_2F_{23} = 1/2(A_1 + A_2 + A_3) \tag{iv}$$

Subtraction of Eq. (i) from Eq. (iv) gives

$$F_{23} = \frac{A_2 + A_3 - A_1}{2A_2}$$

In terms of the widths of the plates of the enclosure, we can rewrite the above equation as

$$F_{23} = \frac{L_2 + L_3 - L_1}{2L_2}$$

Proceeding in the same manner, we can deduce the following:

$$F_{12} = \frac{L_1 + L_2 - L_3}{2L_1}, \quad F_{21} = \frac{L_2 + L_1 - L_3}{2L_2}$$

$$F_{13} = \frac{L_1 + L_3 - L_2}{2L_1}, \quad F_{31} = \frac{L_3 + L_1 - L_2}{2L_3}$$

$$F_{23} = \frac{L_2 + L_3 - L_1}{2L_2}, F_{32} = \frac{L_3 + L_2 - L_1}{2L_3}$$

The result is listed in Table 11.1. The results can be used for the approximate estimate of triangular enclosure with flat or convex surfaces of area A_1, A_2 and A_3, where length of the duct is very large compared to the widths of the sides.

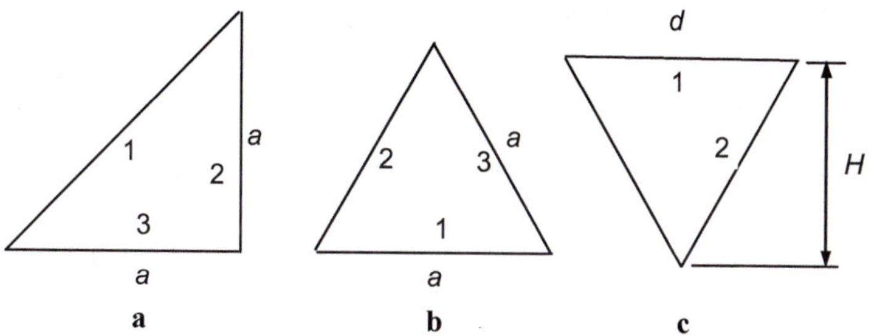

Fig. 11.12 Example 11.4

Example 11.4 Calculate the shape factor for the following:

(a) a very long duct of cross-section as shown in Fig. 11.12a
(b) a very long duct with cross-section of equilateral triangle, Fig. 11.12b
(c) A conical cavity closed by a plane surface, Fig. 11.12c.

Solution

(a) For surface 1, $F_{11} = 0$. Hence, from the summation rule,

$$F_{12} + F_{13} = 1.$$

Since $A_2 = A_3$, we have

$$F_{12} = F_{13} = 0.5.$$

Similarly, $F_{22} = 0$ and, from the summation rule,

$$F_{23} + F_{21} = 1. \quad \text{(a)}$$

From the reciprocity relation,

$$A_2 F_{21} = A_1 F_{12}.$$

or

$$F_{21} = \frac{A_1}{A_2} F_{12} = \frac{\sqrt{2}a}{a} \times 0.5 = 0.707.$$

From Eq. (a),

$$F_{23} = 1 - F_{21} = 0.293.$$

By symmetry,
$F_{33} = 0, F_{31} = 0.707$, and $F_{32} = 0.293$.

(b) $F_{11} = 0$ hence, from the summation rule,

$$F_{12} + F_{13} = 1.$$

As $A_2 = A_3$, F_{12} and F_{13} will be equal. This gives

$$F_{12} = F_{13} = 0.5.$$

By symmetry,

$$F_{22} = 0, F_{23} = F_{21} = 0.5$$

and

$$F_{33} = 0, F_{31} = F_{32} = 0.5.$$

(c) $F_{11} = 0.$
Area of the cone surface,

$$A_2 = \frac{\pi d}{2}\sqrt{H^2 + \frac{d^2}{4}}.$$

Area of the plane surface,

$$A_1 = \frac{\pi d^2}{4}.$$

From the reciprocity relation,

$$A_1 F_{12} = A_2 F_{21}.$$

Here $F_{12} = 1$. Hence,

$$F_{21} = \frac{A_1}{A_2} = \frac{d}{\sqrt{4H^2 + d^2}}$$

and

$$F_{22} = 1 - F_{21} = 1 - \frac{d}{\sqrt{4H^2 + d^2}}.$$

The result for part (a) and (b) can also be obtained using the equation presented in Table 11.1 for Case 3.

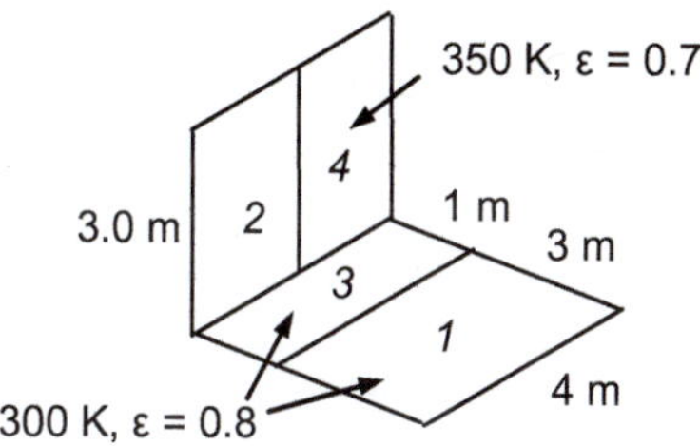

Fig. 11.13 Example 11.5

Example 11.5 Determine the shape factor F_{12} for the configuration shown in Fig. 11.13.

Solution

We have

$$A_{13}F_{(1,3)-(2,4)} = A_1F_{1-(2,4)} + A_3F_{3-(2,4)}$$

where A_{13} = 16 m^2, A_1 = 12 m^2, A_3 = 4 m^2.

This gives

$$4F_{(1,3)-(2,4)} = 3F_{1-(2,4)} + F_{3-(2,4)}.$$

From symmetry,

$$F_{12} = F_{14}.$$

Hence,

$$F_{1-(2,4)} = 2F_{12}$$

$$4F_{(1,3)-(2,4)} = 6F_{12} + F_{3-(2,4)}. \quad \text{(i)}$$

From Fig. 11.4c for Z/X = 0.75 and Y/X = 1,

$$F_{(1,3)-(2,4)} = 0.18$$

and for Z/X = 0.75 and Y/X = 0.25,

$$F_{3-(2,4)} = 0.35$$

Substitution in Eq. (i) gives

$$F_{12} = (1/6)(4 \times 0.18 - 0.35) = 0.0616.$$

Example 11.6 The ends of concentric cylinders of finite length shown in Fig. 11.14 are covered with flat annular surfaces designated as 3 and 3′. Derive the expression for F_{31}, F_{32}, F_{33}, and F_{13} in terms of F_{11}, F_{12} and areas A_1, A_2 and A_3.

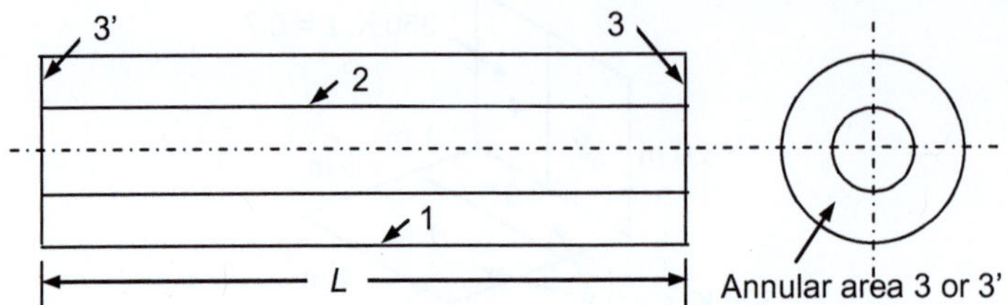

Fig. 11.14 Concentric cylinders of finite length with end covers

Solution

Surface 1

From the summation rule,

$$F_{11} + F_{12} + F_{13} + F_{13'} = 1.$$

By symmetry,

$$F_{13} = F_{13'}$$

Hence,

$$F_{13} = 1/2(1 - F_{11} - F_{12}). \quad \text{(i)}$$

Surface 2

$$F_{22} = 0.$$

Hence, from the summation rule,

$$F_{21} + F_{23} + F_{23'} = 1.$$

By symmetry,

$$F_{23} = F_{23'}.$$

Hence,

$$F_{23} = 1/2(1 - F_{21}). \quad \text{(ii)}$$

Surface 3

$$F_{33} = 0.$$

Hence, from the summation rule,

$$F_{31} + F_{32} + F_{33'} = 1.$$

or

$$F_{33'} = 1 - F_{31} - F_{32}. \tag{iii}$$

Using reciprocity relation and Eq. (i), we get

$$A_3 F_{31} = A_1 F_{13}$$

or

$$F_{31} = \frac{A_1}{A_3} F_{13} = \frac{A_1}{2A_3}(1 - F_{11} - F_{12}). \tag{iv}$$

Again from the reciprocity relations,

$$F_{21} = \frac{A_1}{A_2} F_{12}$$

$$F_{32} = \frac{A_2}{A_3} F_{23}$$

Using these relations and Eq. (ii), we get

$$F_{32} = \frac{A_2}{2A_3}(1 - F_{21}) = \frac{A_2}{2A_3}(1 - \frac{A_1}{A_2} F_{12}) \tag{v}$$

Substituting the values of F_{31} and F_{32} from Eqs. (iv) and (v), respectively, in Eq. (iii) and rearranging, we obtain

$$F_{33'} = 1 - \frac{A_1 + A_2}{2A_3} + \frac{A_1}{2A_3}(2F_{12} + F_{11}).$$

Example 11.7 Derive the equation of radiant energy exchange through the openings of the cavities (with black surface) shown in Fig. 11.15. Comment on the result.

Solution

Let T_1 be the temperature of the surface of the cavities and T_2 that of the space above the opening of the cavities. The space acts as a blackbody hence a black plane surface A_2 can replace the cavity opening.

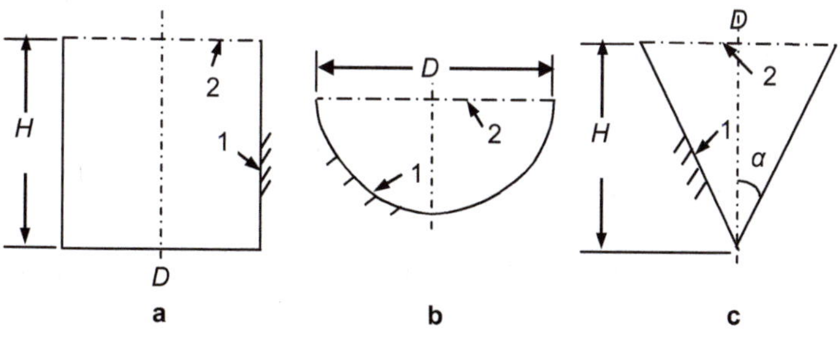

Fig. 11.15 Example 11.7

For the surface 1,

$$F_{11} + F_{12} = 1$$

or

$$F_{11} = 1 - F_{12}$$

Using the reciprocity relation, $A_1F_{12} = A_2F_{21}$, we get

$$F_{11} = 1 - F_{12} = 1 - \frac{A_2}{A_1}F_{21}. \tag{i}$$

For the surface 2,

$$F_{21} + F_{22} = 1$$

or

$$F_{21} = 1 - F_{22}$$

Surface 2 is a flat surface hence $F_{22} = 0$. Hence,

$$F_{21} = 1.$$

Substitution in Eq. (i) gives

$$F_{11} = 1 - \frac{A_2}{A_1}.$$

This is valid for all the cavities shown in Fig. 11.15.

Case (a) A Cylindrical Cavity

$$F_{11} = 1 - \frac{(\pi/4)D^2}{\pi DH + (\pi/4)D^2} = 1 - \frac{D}{D + 4H},$$

and

$$F_{12} = 1 - F_{11} = \frac{D}{D + 4H}.$$

The net heat exchange (from the cavity to the space) is

$$\begin{aligned} q_{12} &= A_1F_{12}\sigma(T_1^4 - T_2^4) \\ &= \left[\pi DH + (\pi/4)D^2\right] \times \left(\frac{D}{D + 4H}\right) \times \sigma(T_1^4 - T_2^4) \\ &= (\pi/4)D^2\sigma(T_1^4 - T_2^4) = A_2\sigma(T_1^4 - T_2^4), \end{aligned}$$

which is independent of the surface area of the cavity.

Case (b) A Hemispherical Bowl

$$F_{11} = 1 - \frac{(\pi/4)D^2}{(\pi/2)D^2} = 0.5,$$

and

$$F_{12} = 1 - F_{11} = 0.5.$$

The net heat exchange (from the cavity to the space) is

$$\begin{aligned} q_{12} &= A_1 F_{12} \sigma (T_1^4 - T_2^4) \\ &= (\pi/2)D^2 \times 0.5 \times \sigma(T_1^4 - T_2^4) \\ &= (\pi/4)D^2 \sigma(T_1^4 - T_2^4) = A_2 \sigma(T_1^4 - T_2^4). \end{aligned}$$

The result is the same as for the case (a).

Case (c) A Conical Cavity

$$F_{11} = 1 - \frac{(\pi/4)D^2}{(\pi D/2)(H^2 + D^2/4)^{1/2}} = 1 - \frac{D}{2(H^2 + D^2/4)^{1/2}},$$

and

$$F_{12} = 1 - F_{11} = \frac{D}{2(H^2 + D^2/4)^{1/2}}.$$

The net heat exchange (from the cavity to the space) is

$$\begin{aligned} q_{12} &= A_1 F_{12} \sigma (T_1^4 - T_2^4) \\ &= (\pi D/2)(H^2 + D^2/4)^{1/2} \frac{D}{2(H^2 + D^2/4)^{1/2}} \times \sigma(T_1^4 - T_2^4) \\ &= (\pi/4)D^2 \sigma(T_1^4 - T_2^4) = A_2 \sigma(T_1^4 - T_2^4). \end{aligned}$$

From the analysis of the results of the above cases, we can conclude the following:

(i) In the case of a cavity (i.e. a concave surface), since the parts of the cavity surface can see each other, the radiation escaping the cavity is less than $A_1 \sigma T_1^4$.
(ii) For a cavity with black surface, the escaping radiation is $A_2 \sigma T_1^4$, where A_2 is area of a plane surface (a surface with minimum area) covering the cavity. Thus in such cases, area of the plane surface covering the cavity can be used instead of the surface area of the cavity. This is to be noted that the conclusion is for a cavity with black surface.

Example 11.8 A small sphere of 50 mm diameter is located at the center of a hollow sphere of 200 mm inside diameter. The surface temperatures of the spheres are 600 K and 300 K, respectively. Calculate the net exchange of radiation between two spheres. Assume that the surfaces of both spheres behave as blackbody. Also determine the amount of energy radiated from the surface of the outer sphere incident on the surface of the inner sphere.

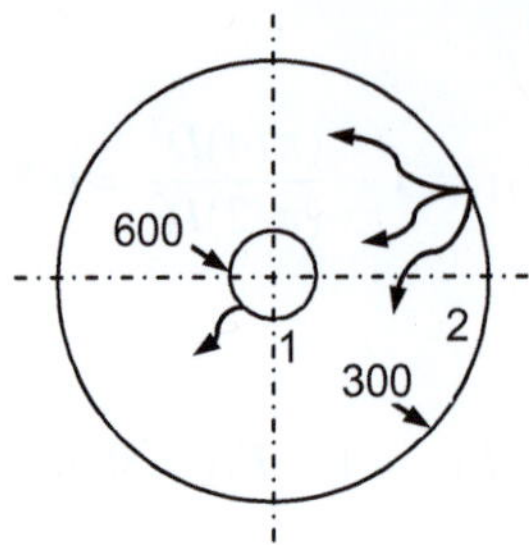

Fig. 11.16 Example 11.8

Solution

(Refer Fig. 11.16)

The configuration factor F_{12} is unity because whole of the radiation emitted by the inner sphere is falling upon the surface of the outer sphere.

By reciprocity relation

$$A_1F_{12} = A_2F_{21}$$

or

$$F_{21} = \frac{A_1}{A_2}F_{12} = \frac{A_1}{A_2} = \frac{\pi \times 50^2}{\pi \times 200^2} = 0.0625$$

i.e. only 6.25% of the radiation emitted by surface of sphere 2 is intercepted by surface of sphere 1. The remaining 93.75% of the radiation falls upon itself.

The net interchange of the heat between the two spheres is

$$\begin{aligned} q_{13} &= A_1F_{12}\sigma(T_1^4 - T_2^4) \\ &= \pi \times (0.05)^2 \times 1 \times 5.67 \times 10^{-8} \times (600^4 - 300^4) = 54.1\,\text{W}. \end{aligned}$$

Example 11.9 The Sun can be regarded as nearly a spherical radiation source emitting as a blackbody. Its diameter is approximately 1.4×10^9 m and is at a distance of 1.5×10^{11} from the Earth. On a clear day the radiation incident on the Earth's surface was measured to be 1200 W/m^2. If 250 W/m^2 of the solar radiation is estimated to be absorbed by the Earth's atmosphere, estimate the surface temperature of the Sun.

Solution

(Refer Fig. 11.17)

The radiation from the Sun impinging on the Earth is

$$q_{12} = 1200 + 250 = 1450\ \text{W/m}^2.$$

From Eqs. (11.12) and (11.16), the net radiation heat exchange is given by

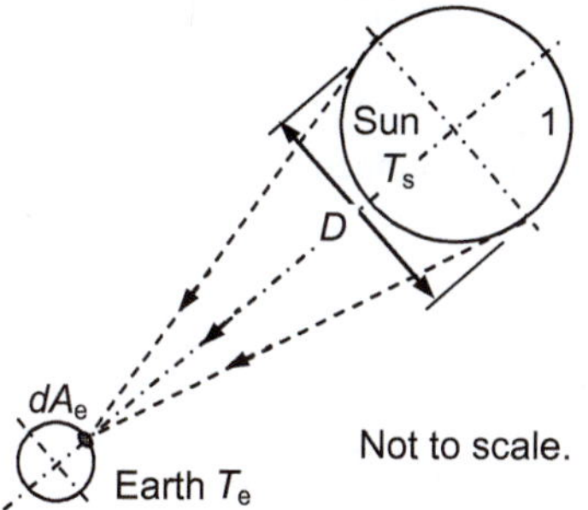

Fig. 11.17 Example 11.9

$$q_{12} = \sigma(T_s^4 - T_e^4)\frac{\cos\theta_s \cos\theta_e dA_s dA_e}{\pi s^2} \tag{i}$$

where subscript s pertains to the Sun and e to the Earth.

The distance between the Sun and the Earth is very large as compared to the diameter of the Sun hence

(i) $\cos\theta_s = \cos\theta_e \approx \cos 0^\circ = 1$, refer Fig. 11.17.
(ii) The surface of the Sun emitting radiation can be regarded as a disk of area $dA_s = (\pi/4) D_s^2$.
(iii) The radiation measurement refers to the unit area of the Earth's surface hence $dA_e = 1\ \text{m}^2$.
(iv) The temperature of the Earth's surface T_e is a very small term compared to T_s and hence can be neglected.

The above conditions transform Eq. (i) to

$$q_{12} = \sigma T_s^4 \frac{(\pi/4)D_s^2}{\pi s^2} = \sigma T_s^4 \frac{D_s^2}{4s^2}$$

or

$$1450 = 5.67 \times 10^{-8} \times T_s^4 \times \frac{(1.4 \times 10^9)^2}{4 \times (1.5 \times 10^{11})^2}$$

This gives

$$T_s = 5853.9\ \text{K}.$$

Alternative method (refer Fig. 11.18)

Total energy emitted by the Sun,

$$q_s = \varepsilon\sigma A_s T_s^4 = \sigma(\pi D_s^2)T_s^4$$

as $\varepsilon = 1$ and surface area of the Sun is $= \pi D_s^2$.

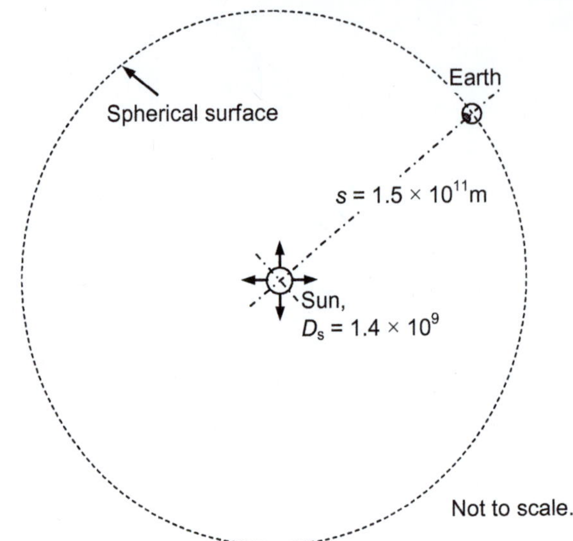

Fig. 11.18 Example 11.9 (alternative method)

The distance of the Earth from the Sun is very large compared to the diameter of the Sun hence the Sun may be regarded as a point source.

The energy received by unit area of a sphere of radius s from this point source at its center is thus

$$q = \frac{q_s}{4\pi s^2} = \frac{(\pi D_s^2)\sigma T_s^4}{4\pi s^2} = \sigma T_s^4 \left(\frac{D_s^2}{4s^2}\right)$$

which equals the energy received by the unit area at the Earth's surface. Substitution gives the above result.

Example 11.10 Obtain the configuration factor between a sphere of radius R_1 and a coaxial disc of radius R_2. What is the shape factor from the sphere to a sector of the disc as shown in Fig. 11.19a?

Solution

(i) First we consider a spherical envelope of radius a as shown in Fig. 11.19b where

$$a = \sqrt{R_2^2 + L^2}$$

The spherical segment ABC envelops the disc.

The radiation from the sphere 1 falls uniformly on the surface of the spherical envelope of radius a. Thus, the fraction falling on the spherical segment ABC is

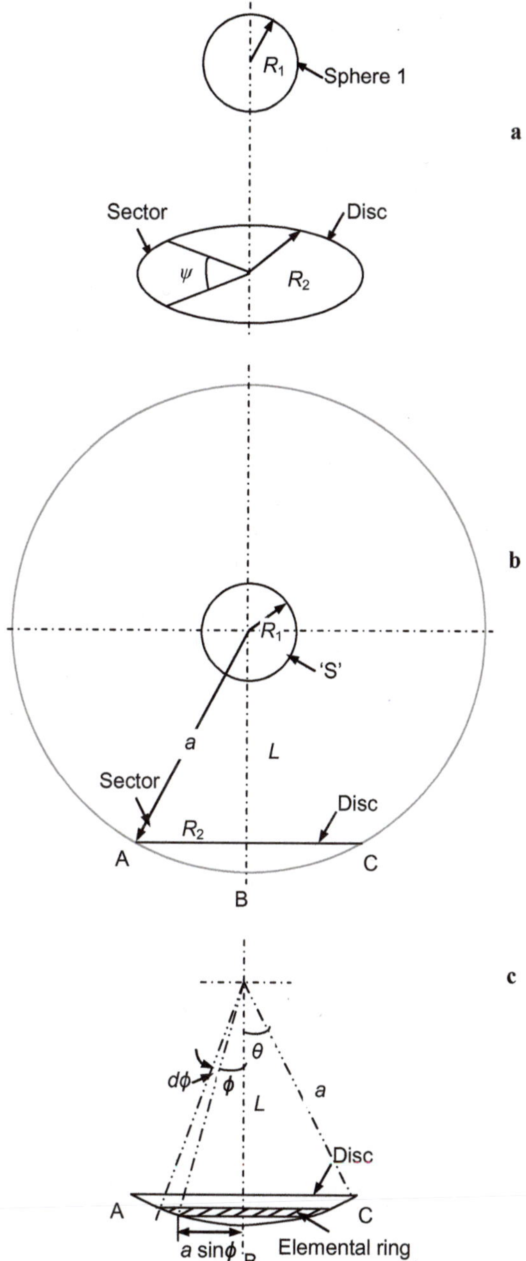

Fig. 11.19 Example 11.10

$$F_{12} = \frac{\text{Surface area of ABC}}{\text{Surface area of envelope}} \tag{i}$$

The surface area of ABC is

$$= \int_0^\theta 2\pi(a \sin\phi) a d\phi = 2\pi a^2(1 - \cos\theta)$$

where [2π (a sinϕ)$ad\phi$] is the area of the elemental strip shown in Fig. 11.19c. Substitution in Eq. (i) gives

$$F_{12} = \frac{2\pi a^2(1-\cos\theta)}{4\pi a^2} = \frac{(1-\cos\theta)}{2} = \frac{1}{2}\left(1 - \frac{L}{a}\right)$$
$$= \frac{1}{2}\left(1 - \frac{L}{\sqrt{R_2^2 + L^2}}\right)$$

Since the disc 2 intercepts the same fraction of radiation as the spherical segment, the above equation is the desired configuration factor.

(i) If we consider the sector marked in Fig. 11.19a, then the area of this sector is $\psi/2\pi$ part of the area of the disc and the configuration factor is

$$F = \frac{\psi}{2\pi}F_{12} = \frac{\psi}{4\pi}\left(1 - \frac{L}{\sqrt{R_2^2 + L^2}}\right)$$

Example 11.11 What is the configuration factor from a sphere to a portion of a ring of infinitesimal width dR_2 shown in Fig. 11.20a?

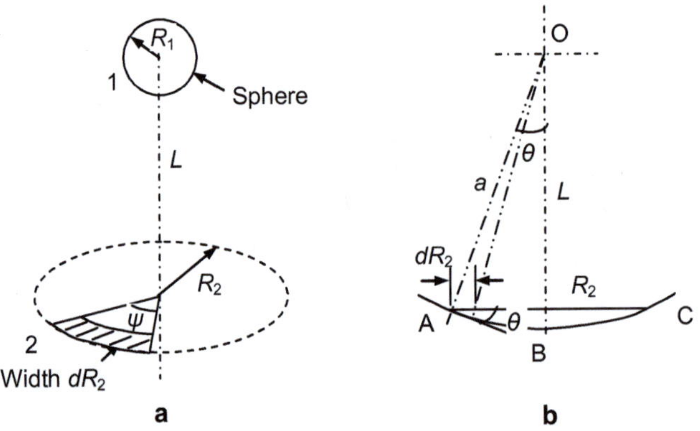

Fig. 11.20 Example 11.11

Solution

Area of the surface 2, the portion of the ring,

$$dA_2 = (\psi R_2)dR_2$$

Projection of this area normal to OA is $dA_2\cos\theta$ which lies on the surface of the spherical segment ABC in Fig. 11.20b. Thus the fraction falling on the surface 2 is

$$\begin{aligned}
F_{12} &= \frac{dA_2 \cos\theta}{\text{Surface area of spherical envelope of radius } a} \\
&= \frac{(R_2\psi)dR_2 \cos\theta}{4\pi a^2} \\
&= (R_2\psi)dR_2\left[\frac{1}{4\pi(R_2^2+L^2)}\right]\left(\frac{L}{\sqrt{R_2^2+L^2}}\right) \\
&= \frac{\psi}{4\pi}\left[\frac{R_2 L}{(R_2^2+L^2)^{3/2}}\right]dR_2.
\end{aligned}$$

Example 11.12 Determine the shape factor between a small area A_1 and a circular segment of a sphere of radius R. The area A_1 is located at the center of the sphere perpendicular to the circular segment. The segment subtends an angle 2β at the area A_1.

Solution

(Refer Fig. 11.21)

The radiation shape factor is given by

$$\begin{aligned}
F_{12} &= \frac{1}{A_1}\int\limits_{A_1}\int\limits_{A_2}\frac{\cos\phi_1\cos\phi_2 dA_1 dA_2}{\pi s^2} \\
&= \frac{1}{A_1}\int\limits_{A_1} dA_1\left(\int\limits_{A_2}\frac{\cos\phi_1\cos\phi_2 dA_2}{\pi s^2}\right) \\
&= \int\limits_{A_2}\frac{\cos\phi_1\cos\phi_2 dA_2}{\pi s^2}.
\end{aligned}$$

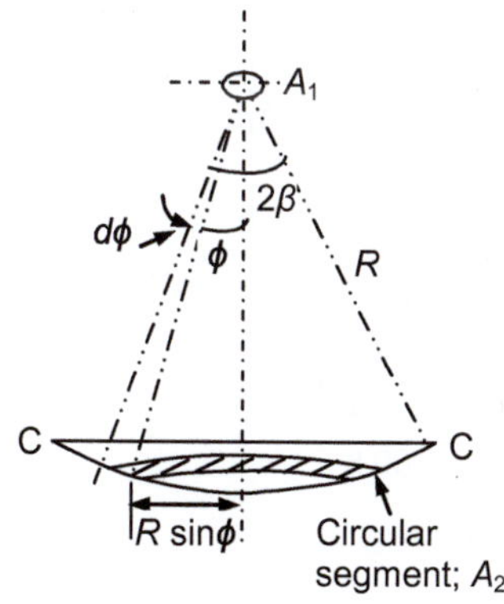

Fig. 11.21 Example 11.12

Taking an elemental area on the disc,

$$dA_2 = 2\pi R \sin \phi R d\phi$$

Also $\phi_1 = \phi$, $\phi_2 = 0$, and $s = R$. Hence,

$$\begin{aligned} F_{12} &= \int_0^\beta \frac{\cos\phi \cos 0}{\pi R^2} \times 2\pi R^2 \sin\phi d\phi \\ &= \int_0^\beta 2\sin\phi\cos\phi d\phi = \int_0^\beta \sin 2\phi d\phi = \left[\frac{-\cos 2\phi}{2}\right]_0^\beta \\ &= \left[\frac{-\cos 2\beta}{2} + \frac{1}{2}\right] \\ &= \left[\frac{1}{2} - \frac{1 - 2\sin^2\beta}{2}\right] = \sin^2\beta, \end{aligned}$$

which is the same as Eq. (11.17). Thus the shape factor can be determined by considering the projected area of the circular segment (refer plane CC in Fig. 11.21).

Example 11.13 A black 25 mm diameter sphere 1 at a temperature of 1000 K is suspended in the center of a thin 50 mm diameter partial sphere 3 having a black interior surface. The exterior surface of the sphere 3 has a hemispherical total emissivity of 0.5. The surroundings are at 300 K. A 40 mm diameter hole is cut in the outer sphere. What is the temperature of the outer sphere? What is the heat transferred from the inner sphere?

Solution

Refer Fig. 11.22.

$$\begin{aligned} F_{12} &= \frac{1}{2}\left[1 - \frac{L}{\sqrt{R^2 + L^2}}\right] \\ &= \frac{1}{2}\left[1 - \frac{15}{25}\right] = 0.2, \end{aligned}$$

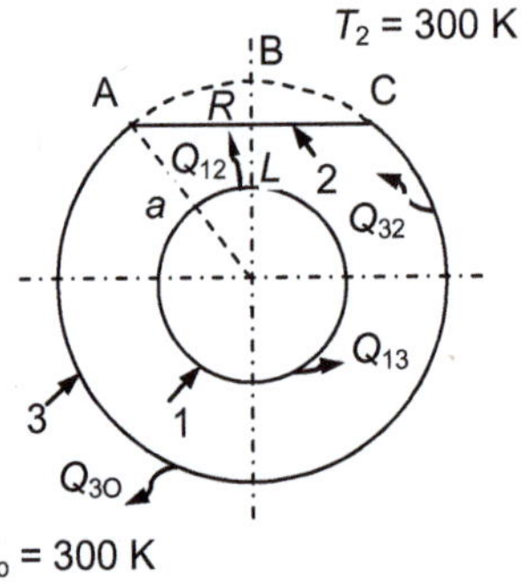

Fig. 11.22 Example 11.13

(Refer Example 11.10)

$$F_{12} + F_{13} = 1 \text{ as } F_{11} = 0,$$

or

$$F_{13} = 1 - F_{12} = 1 - 0.2 = 0.8,$$
$$F_{21} + F_{23} = 1 \text{ as } F_{22} = 0,$$

or

$$\begin{aligned} F_{23} &= 1 - F_{21} \\ &= 1 - \frac{A_1}{A_2} F_{12} \end{aligned}$$

where

$$\begin{aligned} A_1 &= 4\pi R_1^2 = 4 \times \pi \times 12.5^2 = 1963.5\,\text{mm}^2 \\ A_2 &= \pi R^2 = \pi \times 20^2 = 1256.6\,\text{mm}^2. \end{aligned}$$

Substitution gives

$$F_{23} = 1 - \frac{1963.5}{1256.6} \times 0.2 = 0.6875,$$
$$F_{32} = \frac{A_2}{A_3} F_{23}$$

where

$$\begin{aligned} A_3 &= 4\pi R_3^2 - \text{surface area of segment ABS} \\ &= 4 \times \pi \times 25^2 - \left(4 \times \pi \times 25^2\right) F_{12}, \end{aligned}$$

because

$$F_{12} = \frac{\text{Surface area of segment}}{\text{Surface area of the sphere of radius } r_3}.$$

Thus,

$$A_3 = 4 \times \pi \times 25^2 (1 - F_{12}) = 4 \times \pi \times 25^2 (1 - 0.2) = 6283.2\,\text{mm}^2.$$

This gives

$$F_{32} = \frac{1256.6}{6283.2} \times 0.6875 = 0.1375.$$

Energy balance of surface 3 gives

$$q_{13} = q_{30} + q_{32}$$

or

$$A_1 F_{13} \sigma (T_1^4 - T_3^4) = A_3 \varepsilon \sigma (T_3^4 - T_0^4) + A_3 F_{32} \sigma (T_3^4 - T_2^4)$$

or

$$A_1 F_{13} (T_1^4 - T_3^4) = A_3 \varepsilon (T_3^4 - T_0^4) + A_3 F_{32} (T_3^4 - T_2^4) \quad \text{(i)}$$

The radiation going out through the opening (hole) is lost into the surrounding at temperature 300 K. Hence, the hole can be regarded as black disc of diameter 40 mm at temperature T_2 = 300 K. Substitution of values of various parameters in Eq. (i) gives

$$1963.5 \times 0.8 \times (1000^4 - T_3^4) = 6283.2 \times 0.5 \times (T_3^4 - 300^4) + 6283.2 \times 0.1375 \times (T_3^4 - 300^4)$$

Simplification gives

$$T_3 = 732.3 \text{ K}$$

Heat transfer from the inner sphere

$$\begin{aligned} q_1 = q_{13} + q_{12} &= A_1 F_{13} \sigma (T_1^4 - T_3^4) + A_1 F_{12} \sigma (T_1^4 - T_2^4) \\ &= 1963.5 \times 10^{-6} \times 5.67 \times 10^{-8} \\ &\quad \times \left[0.8 \times (1000^4 - 732.3^4) + 0.2 \times (1000^4 - 300^4)\right] \\ &= 85.53 \text{ W}. \end{aligned}$$

Example 11.14 A hollow cylindrical heating element 150 mm long and 150 mm inside diameter with a black interior surface is to be maintained at 1000 K. The outside of the cylinder is insulated and the surroundings are in vacuum at 750 K. If both ends of the cylinder are open, estimate the energy to be supplied to the heating element.

Solution

The heat is lost by radiation through the open ends hence

$$q_3 = q_{31} + q_{32} = 2q_{31}$$

by symmetry, refer Fig. 11.23.

We have

$$F_{12} + F_{13} = 1$$

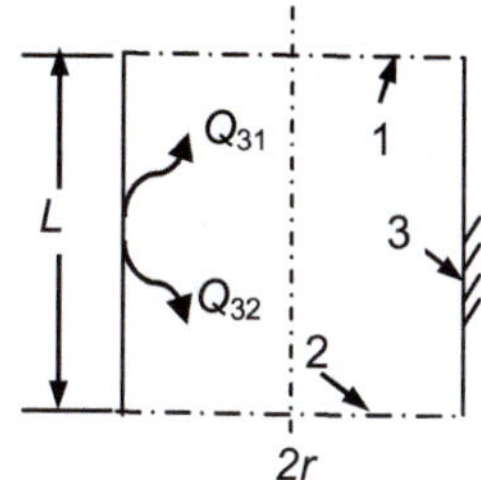

Fig. 11.23 Example 11.14

or

$$F_{13} = 1 - F_{12}$$

Further,

$$F_{31} = \frac{A_1}{A_3} F_{13} = \frac{\pi r^2}{2\pi r L}(1 - F_{12}) = \frac{r}{2L}(1 - F_{12}) \quad \text{(ii)}$$

Configuration factor (Table 11.1),

$$\begin{aligned} F_{12} &= 1 + \frac{L^2}{2r^2}\left[1 - \left(1 + \frac{4r^2}{L^2}\right)^{1/2}\right] \\ &= 1 + \frac{150^2}{2 \times 75^2}\left[1 - \left(1 + \frac{4 \times 75^2}{150^2}\right)^{1/2}\right] = 0.17157. \end{aligned}$$

Hence, from Eq. (ii),

$$F_{31} = \frac{75}{2 \times 150}(1 - 0.17157) = 0.2071$$

and

$$\begin{aligned} q_3 = 2q_{31} &= 2\left[A_3 F_{31} \sigma \left(T_3^4 - T_1^4\right)\right] \\ &= 2 \times 2 \times \pi \times 75 \times 150 \times 10^{-6} \times 0.2071 \times 5.67 \times 10^{-8} \\ &\quad \times \left(1000^4 - 750^4\right) = 1134.85\ \text{W}. \end{aligned}$$

Example 11.15 A circular cylindrical enclosure, as shown in Fig. 11.24, has black interior surfaces each maintained at uniform temperature (T_1 = 1500 K, T_2 = 500 K and T_3 = 1000 K). The outside of the entire cylinder is insulated such that the outside surface does not radiate to the surroundings. How much heat is supplied to each surface as a result of the internal radiation exchange?

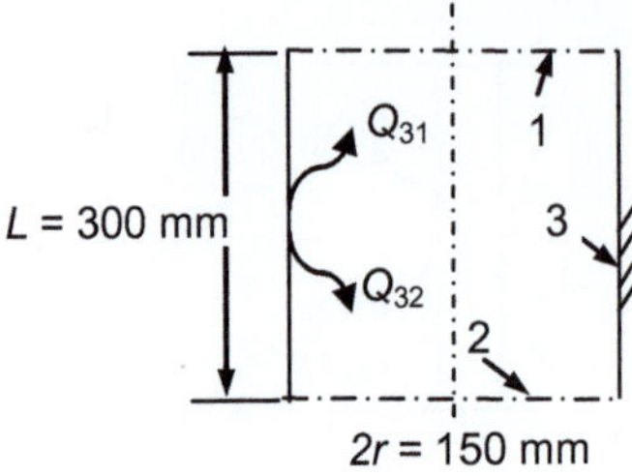

Fig. 11.24 Example 11.15

Solution

Heat transfer from various surfaces are given by (refer Fig. 11.24)

$$q_1 = q_{12} + q_{13} = \left[A_1 F_{12} \sigma (T_1^4 - T_2^4)\right] + \left[A_1 F_{13} \sigma (T_1^4 - T_3^4)\right] \quad \text{(i)}$$

$$q_2 = q_{21} + q_{23} = \left[A_2 F_{21} \sigma (T_2^4 - T_1^4)\right] + \left[A_2 F_{23} \sigma (T_2^4 - T_3^4)\right] \quad \text{(ii)}$$

$$q_3 = q_{31} + q_{32} = \left[A_3 F_{31} \sigma (T_3^4 - T_1^4)\right] + \left[A_3 F_{32} \sigma (T_3^4 - T_2^4)\right] \quad \text{(iii)}$$

where

$$A_1 = A_2 = \frac{\pi}{4}(150)^2 = 17671.5\,\text{mm}^2,$$

$$A_3 = \pi \times 150 \times 300 = 141372\,\text{mm}^2,$$

$T_1 = 1500$ K, $T_2 = 500$ K, $T_3 = 1000$ K,

$$F_{12} = 1 + \frac{L^2}{2r^2}\left[1 - \left(1 + \frac{4r^2}{L^2}\right)^{1/2}\right] = F_{21}$$

$$= 1 + \frac{300^2}{2 \times 75^2}\left[1 - \left(1 + \frac{4 \times 75^2}{300^2}\right)^{1/2}\right]$$

(Table 11.1)
or

$$F_{12} = F_{21} = 0.05573.$$

Since $F_{12} + F_{13} = 1$,

$$F_{13} = 1 - F_{12} = 1 - 0.05573 = 0.94427 = F_{23}$$

$$F_{31} = \frac{A_1}{A_3} F_{13} = 0.11803 = F_{32}.$$

Knowing the values of various parameters in Eqs. (i)–(iii), the heat transfer rates from the surfaces are

$$
\begin{aligned}
q_1 &= \left[A_1 F_{12}\sigma(T_1^4 - T_2^4)\right] + \left[A_1 F_{13}\sigma(T_1^4 - T_3^4)\right] \\
&= 17671.5 \times 10^{-6} \times 5.67 \times 10^{-8} \\
&\quad \times \left[0.05573 \times (1500^4 - 500^4) + 0.94427 \times (1500^4 - 1000^4)\right] \\
&= 4123 \text{ W.}
\end{aligned}
$$

$$
\begin{aligned}
q_2 &= \left[A_2 F_{21}\sigma(T_2^4 - T_1^4)\right] + \left[A_2 F_{23}\sigma(T_2^4 - T_3^4)\right] \\
&= 17671.5 \times 10^{-6} \times 5.67 \times 10^{-8} \\
&\quad \times \left[0.05573 \times (500^4 - 1500^4) + 0.94427 \times (500^4 - 1000^4)\right] \\
&= -1166.2 \text{ W.}
\end{aligned}
$$

$$
\begin{aligned}
q_3 &= \left[A_3 F_{31}\sigma(T_3^4 - T_1^4)\right] + \left[A_3 F_{32}\sigma(T_3^4 - T_2^4)\right] \\
&= 141372 \times 10^{-6} \times 5.67 \times 10^{-8} \\
&\quad \times \left[0.11803 \times (1000^4 - 1500^4) + 0.11803 \times (1000^4 - 500^4)\right] \\
&= -2956.6 \text{ W.}
\end{aligned}
$$

(Check: heat lost by surface 1 equals the heat gained by surfaces 2 and 3).

Example 11.16 A bead shaped thermocouple is located at the central plane along the axis of a circular pipe of radius R and length $2L$. Determine the shape or geometrical factor between the bead and the inside surface of the pipe.

Solution

We know that the shape factor equals the ratio of the radiation energy intercepted by the area receiving the radiation to the total radiation emitted from the body. Alternatively, for a spherical body, it is area of the receiving body (in the form of a segment of a sphere) at distance 'a' to the surface area of a sphere of radius a. In Fig. 11.25, we have covered the open ends of the pipe by spherical segments. The spherical thermocouple bead of surface area A_1 emits radiation in all directions. The radiation intercepted by the two spherical segments is, thus

$$
\begin{aligned}
F_{13} &= \frac{\text{Surface area of segments}}{\text{Surface area of the sphere of radius } a} \\
&= \frac{2 \times \text{Surface area of one segment}}{\text{Surface area of the sphere of radius } a}
\end{aligned}
\tag{i}
$$

Surface area of one segment can be found out as under by considering an elemental strip, see Fig. 11.25.

$$
\begin{aligned}
A_3 &= \int_0^\beta 2\pi a \sin\alpha \times a \times (d\alpha) = \int_0^\beta 2\pi a^2 \sin\alpha \times d\alpha = 2\pi a^2 \int_0^\beta \sin\alpha \times d\alpha \\
&= 2\pi a^2 [-\cos\alpha]_0^\beta = 2\pi a^2 (1 - \cos\beta) = 2\pi a^2 \left(1 - \frac{L}{a}\right).
\end{aligned}
$$

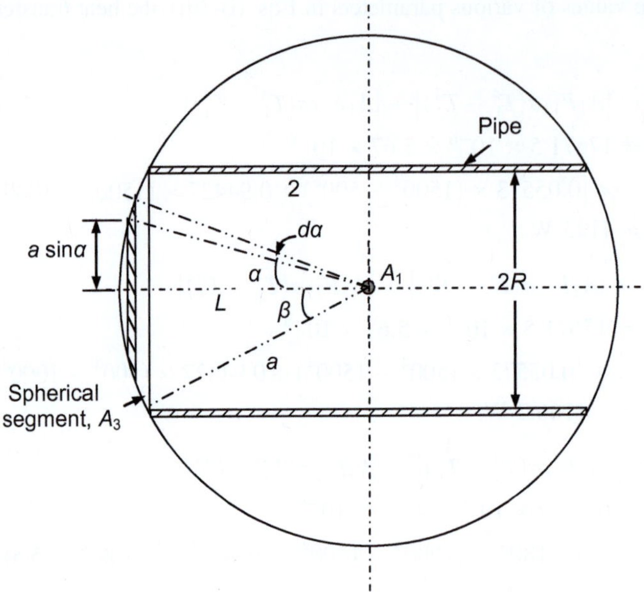

Fig. 11.25 Example 11.16

Hence,

$$F_{13} = 2 \times \frac{2\pi a^2(1 - L/a)}{4\pi a^2} = \left(1 - \frac{L}{a}\right).$$

Knowing that $F_{11} + F_{12} + F_{13} = 1$ and $F_{11} = 0$ for a spherical body, we get

$$F_{12} = 1 - F_{13} = \frac{L}{a} = \frac{L}{\sqrt{R^2 + L^2}}.$$

Example 11.17 Given that the configuration or view factor between two coaxial parallel discs of radius R distance L apart is (Table 11.1)

$$1 + \frac{L^2}{2R^2}\left[1 - \left(1 + \frac{4R^2}{L^2}\right)^{1/2}\right]$$

Show that the self view factor of the inner surface of a hollow cylinder of radius R and length L is

$$= 1 + \frac{L}{2R}\left[1 - \left(1 + \frac{4R^2}{L^2}\right)^{1/2}\right]$$

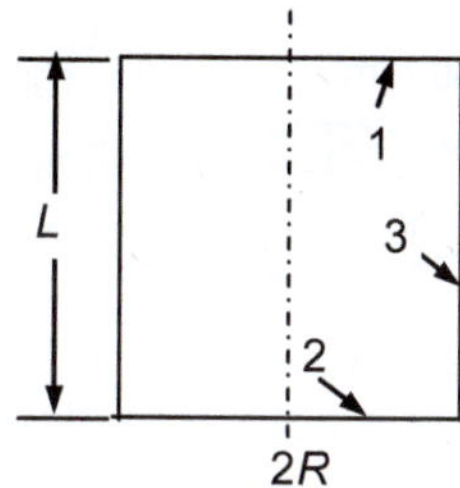

Fig. 11.26 Example 11.17

Solution

For the hollow cylinder shown in Fig. 11.26, the application of the summation rule and reciprocity relation gives for surface 1:

$$F_{12} + F_{13} = 1 \text{ as } F_{11} = 0,$$

$$F_{31} = \frac{A_1}{A_3} F_{13} = \frac{\pi R^2}{2\pi RL}(1 - F_{12}) = \frac{R}{2L}(1 - F_{12}). \tag{ii}$$

Surface 2

Geometrically surface 2 is similar to the surface 1 hence

$$F_{21} = F_{12},$$

$$F_{32} = \frac{R}{2L}(1 - F_{12}) = F_{31}.$$

Surface 3

The self view factor of the inner surface of the cylinder,

$$F_{33} = 1 - F_{31} - F_{32} = 1 - 2F_{31}$$

$$= 1 - 2\left[\frac{R}{2L}(1 - F_{12})\right] = 1 - \frac{R}{L}(1 - F_{12}).$$

Substituting the value of F_{12}, we get

$$F_{33} = 1 - \frac{R}{L}\left\{1 - 1 - \frac{L^2}{2R^2}\left[1 - \left(1 + \frac{4R^2}{L^2}\right)^{1/2}\right]\right\}$$

$$F_{33} = 1 + \frac{L}{2R}\left[1 - \left(1 + \frac{4R^2}{L^2}\right)^{1/2}\right].$$

Example 11.18 Blackbody radiation is leaving a small opening in a furnace at 1100°C. What fraction of this radiation is intercepted by the annular disc shown in Fig. 11.27? What fraction passes through the hole in the disc?

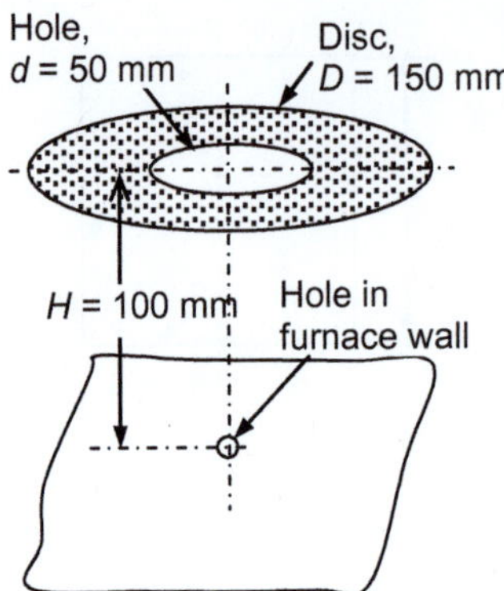

Fig. 11.27 Example 11.18

Solution

Radiation shape factor or fraction of the radiation leaving the surface 1 which reaches surface 2 is (Table 11.1 case 6)

$$F_{dA_1-A_2} = \frac{D^2}{4H^2 + D^2}$$

where area dA_1 refers to the area of the small hole in the furnace wall and A_2 is the area of the surface intercepting the radiation leaving the small hole.

Fraction of the radiation passing through the hole in the disc,

$$F_1 = \frac{d^2}{4H^2 + d^2} = \frac{50^2}{4 \times 100^2 + 50^2} = 0.0588$$

Fraction of the radiation falling on the disc of area $(\pi/4)D^2$ is

$$F_1 = \frac{D^2}{4H^2 + D^2} = \frac{150^2}{4 \times 100^2 + 150^2} = 0.36$$

Hence, fraction of the radiation intercepted by the annular disc is

$$F_{dA_1-Disc} = 0.36 - 0.0588 = 0.3012.$$

Note: The fraction F, i.e. the shape or configuration factor, does not depend on the temperatures or emissivities of the bodies.

Example 11.19 Determine the shape factor for radiation exchange between a planar point source and an infinite parallel plane.

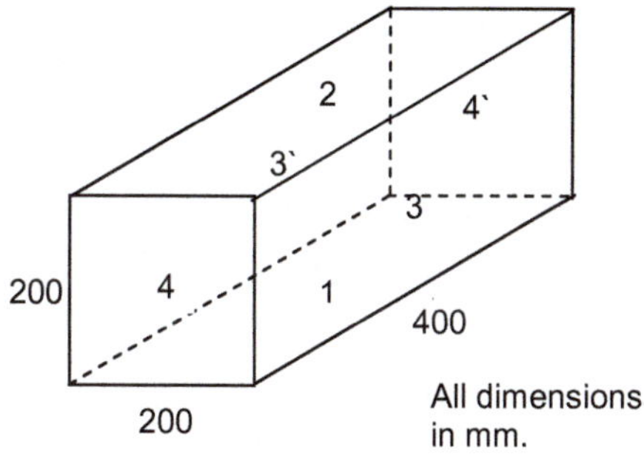

Fig. 11.28 Example 11.20

Solution

From Eq. (11.17),

$$F_{12} = \frac{R^2}{H^2 + R^2} = \frac{1}{(H/R)^2 + 1}.$$

For an infinite parallel plane, $R \to \infty$. Hence, $F_{12} \to 1$for all finite values of H.

Example 11.20 Figure 11.28 shows a box whose inner surface can be assumed to be black. The bottom surface 1 of the box is at a temperature of 600 K, the top surface 2 is at 500 K and all vertical surfaces (marked as 3, 3`, 4 and 4`) are at 550 K. Calculate the net heat transfer from the bottom to the top surface and also from the bottom to the vertical walls.

Solution

(a) **Shape Factors**
Surface 1 to surface 2:
From Fig. 11.4, with X/L = 200/200 = 1 and Y/L = 400/200 = 2.0, F_{12} = 0.28.
Applying the summation rule,

$$F_{11} + F_{12} + F_{1v} = 1$$

where subscript v refers to all vertical surfaces.

Since surface 1 is a flat surface, F_{11} = 0. This gives

$$F_{1v} = 1 - F_{12} = 1 - 0.28 = 0.72.$$

(b) **Net Heat Exchange**

$$\begin{aligned} q_{12} &= A_1 F_{12} \sigma (T_1^4 - T_2^4) \\ &= 0.2 \times 0.4 \times 0.28 \times 5.67 \times 10^{-8} \times (600^4 - 500^4) = 85.2\,\text{W}. \end{aligned}$$

$$\begin{aligned} (q_{1v})_{net} &= A_1 F_{1v} \sigma (T_1^4 - T_v^4) \\ &= 0.2 \times 0.4 \times 0.72 \times 5.67 \times 10^{-8} \times (600^4 - 550^4) = 124.4\,\text{W}. \end{aligned}$$

Note: $F_{1v\text{-}}$ can also be found as follows:

$$F_{1v} = F_{13} + F_{13'} + F_{14} + F_{14'}.$$

From Fig. 11.4, for $Z/X = 200/400 = 0.5$ and $Y/X = 200/400 = 0.5$, $F_{13} = F_{13'} = 0.245$, and for $Z/X = 200/200 = 1$ and $Y/X = 400/200 = 2$, $F_{14} = F_{14'} = 0.115$.

Hence,

$$F_{1v} = 2(0.245 + 0.115) = 0.72.$$

11.5 Radiation Exchange Between Infinite Parallel Planes

In the case of two infinite parallel diffuse gray planes, Fig. 11.29, areas A_1 and A_2 are equal and whole of the radiation leaving one surface reaches the other. Thus the shape factor is unity. The net radiant energy exchange can be determined using *ray tracing method* as under.

Let the gray surface 1 emits

$$E_1 = \varepsilon_1 \sigma T_1^4 A$$

The surface 2 absorbs

$$X_1 = \alpha_2 E_1 = \varepsilon_1 \varepsilon_2 \sigma T_1^4 A$$

(since $\alpha_2 = \varepsilon_2$).

The amount reflected by surface 2 is

$$\rho_2 E_1 = \rho_2(\varepsilon_1 \sigma T_1^4 A)$$

The radiation reflected by surface 2 falls upon the surface 1 and the surface 1 absorbs $\alpha_1 \rho_2 E_1$ and reflects back $\rho_1(\rho_2 E_1)$. Thus the amount of second incidence on surface 2 is $\rho_1 \rho_2 E_1$. The surface 2, now, absorbs

$$X_2 = \alpha_2(\rho_1 \rho_2 E_1) = \rho_1 \rho_2(\varepsilon_1 \varepsilon_2 \sigma T_1^4 A)$$

Similarly it can be shown that the radiation absorbed by surface 2 on third incidence will be

$$X_3 = (\rho_1 \rho_2)^2(\varepsilon_1 \varepsilon_2 \sigma T_1^4 A)$$

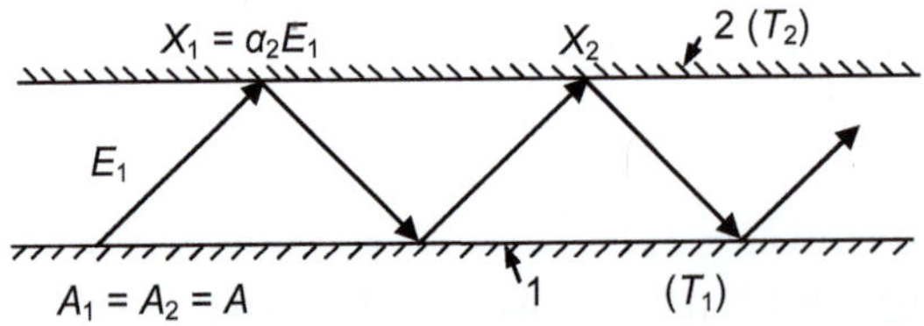

Fig. 11.29 Infinite parallel planes

and so on.

Thus the total energy absorbed by surface 2 out of the radiation of surface 1 is

$$q_2 = \varepsilon_1\varepsilon_2\sigma T_1^4 A[1+\rho_1\rho_2+(\rho_1\rho_2)^2+\ldots]$$

Similarly the amount absorbed by surface 1 out of $\varepsilon_2\sigma T_2^4 A$ radiated by surface 2 is

$$q_1 = \varepsilon_1\varepsilon_2\sigma T_2^4 A[1+\rho_1\rho_2+(\rho_1\rho_2)^2+\ldots]$$

The net radiation energy exchange is

$$(q_{12})_{net} = q_2 - q_1 = \varepsilon_1\varepsilon_2\sigma(T_1^4 - T_2^4)A[1+\rho_1\rho_2+(\rho_1\rho_2)^2+\ldots]$$

As $(\rho_1\rho_2) < 1$, the sum of the terms in the bracket is $1/(1-\rho_1\rho_2)$ and

$$(q)_{12} = \frac{\varepsilon_1\varepsilon_2\sigma(T_1^4 - T_2^4)A}{1-\rho_1\rho_2}$$

But $\rho_1 = 1 - \varepsilon_1$ and $\rho_2 = 1 - \varepsilon_2$ hence

$$\begin{aligned}(q)_{12} &= \frac{\varepsilon_1\varepsilon_2}{1-(1-\varepsilon_1)(1-\varepsilon_2)}\left[\sigma(T_1^4 - T_2^4)A\right]\\ &= \frac{\varepsilon_1\varepsilon_2}{\varepsilon_1+\varepsilon_2-\varepsilon_1\varepsilon_2}\left[\sigma(T_1^4 - T_2^4)A\right]\\ &= \frac{1}{\frac{1}{\varepsilon_1}+\frac{1}{\varepsilon_2}-1}\left[\sigma(T_1^4 - T_2^4)A\right]\\ &= f_{12}\sigma(T_1^4 - T_2^4)A,\end{aligned} \tag{11.15}$$

where $f_{12} = \left(\frac{1}{\varepsilon_1}+\frac{1}{\varepsilon_2}-1\right)^{-1}$ is known as *interchange or transfer factor.*

For black parallel surfaces $\varepsilon_1 = \varepsilon_2 = 1$ and the interchange factor is unity.

11.6 Radiation Exchange Between Infinite Long Concentric Cylinders

Consider two infinitely long concentric cylinders of surface areas A_1 and A_2, emissivities ε_1 and ε_2 and their surfaces are maintained at temperatures T_1 and T_2, respectively, as shown in Fig. 11.30.

From the reciprocity relation $A_1 F_{12} = A_2 F_{21}$. Since, the inner cylinder is completely enclosed by the outer one, entire radiation from the surface of the inner cylinder will be intercepted by the outer cylinder. Hence, $F_{12} = 1$. Therefore,

$$F_{21} = \frac{A_1}{A_2}.$$

Let the surface 1 emits radiant energy E_1. This radiation falls upon the surface 2. The radiation absorbed by the surface 2 is $\alpha_2 E_1$ and equals $\varepsilon_2 E_1$ (as from Kirchhoff's law $\alpha = \varepsilon$).

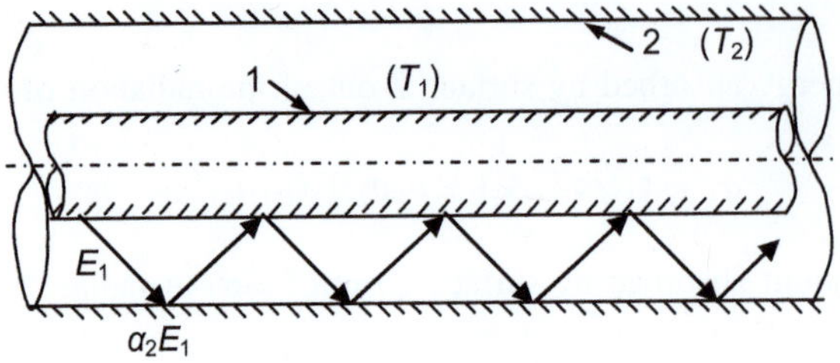

Fig. 11.30 Radiation exchange between infinite long concentric cylinders

The surface 2 reflects back $E_1 - \varepsilon_2 E_1$. A part of the reflected radiation from surface 2 is absorbed by the surface 1 and is $(E_1 - \varepsilon_2 E_1)F_{12}\,\alpha_1 = (E_1 - \varepsilon_2 E_1)\,(A_1/A_2)\varepsilon_1$. Remaining radiation, i.e. $(E_1 - \varepsilon_2 E_1) - (E_1 - \varepsilon_2 E_1)\,(A_1/A_2)\varepsilon_1 = \{(1 - \varepsilon_2)[1 - (A_1/A_2)\varepsilon_1]E_1\}$ is received back by the surface 2.

It can be shown that on the second reflection, the surface 1 will absorb

$$E_1(1-\varepsilon_2)^2\varepsilon_1\frac{A_1}{A_2}\left[1-\frac{A_1}{A_2}\varepsilon_1\right]$$

Continuation of this process will show that the total radiant energy lost by the inner cylinder per unit area is

$$\begin{aligned}
&= E_1 - E_1(1-\varepsilon_2)\varepsilon_1\frac{A_1}{A_2} - E_1(1-\varepsilon_2)^2\varepsilon_1\frac{A_1}{A_2}\left(1-\frac{A_1}{A_2}\varepsilon_1\right) - \ldots\\
&= E_1\left[1-\frac{A_1}{A_2}\varepsilon_1(1-\varepsilon_2) - (1-\varepsilon_2)^2\varepsilon_1\frac{A_1}{A_2}\left(1-\frac{A_1}{A_2}\varepsilon_1\right) - \ldots\right]\\
&= E_1\left\{1-\frac{A_1}{A_2}\varepsilon_1(1-\varepsilon_2)\left[1+(1-\varepsilon_2)\left(1-\frac{A_1}{A_2}\varepsilon_1\right).+\ldots\right]\right\}\\
&= E_1\left\{1-\frac{A_1}{A_2}\varepsilon_1(1-\varepsilon_2)\left[1-(1-\varepsilon_2)\left(1-\frac{A_1}{A_2}\varepsilon_1\right)\right]^{-1}\right\}\\
&= E_1\left[1-\frac{\frac{A_1}{A_2}\varepsilon_1(1-\varepsilon_2)}{1-(1-\varepsilon_2)\left(1-\frac{A_1}{A_2}\varepsilon_1\right)}\right]\\
&= \frac{\varepsilon_2 E_1}{\frac{A_1}{A_2}\varepsilon_1+\varepsilon_2-\frac{A_1}{A_2}\varepsilon_1\varepsilon_2}.
\end{aligned}$$

Similarly the heat lost by the outer cylinder is

$$=\frac{\varepsilon_1 E_2\frac{A_1}{A_2}}{\frac{A_1}{A_2}\varepsilon_1+\varepsilon_2-\frac{A_1}{A_2}\varepsilon_1\varepsilon_2}$$

The net heat transfer between the cylinders is

$$q_{12} = A_1\left(\frac{\varepsilon_2 E_1}{\frac{A_1}{A_2}\varepsilon_1 + \varepsilon_2 - \frac{A_1}{A_2}\varepsilon_1\varepsilon_2}\right) - A_2\left(\frac{\varepsilon_1 E_2 \frac{A_1}{A_2}}{\frac{A_1}{A_2}\varepsilon_1 + \varepsilon_2 - \frac{A_1}{A_2}\varepsilon_1\varepsilon_2}\right)$$
$$= \left(\frac{\varepsilon_2 A_1 E_1 - \varepsilon_1 A_1 E_2}{\frac{A_1}{A_2}\varepsilon_1 + \varepsilon_2 - \frac{A_1}{A_2}\varepsilon_1\varepsilon_2}\right)$$

From Stefan–Boltzmann law for gray bodies,

$$E_1 = \varepsilon_1 \sigma T_1^4$$

and

$$E_2 = \varepsilon_2 \sigma T_2^4.$$

This gives

$$q_{12} = \left(\frac{\varepsilon_1\varepsilon_2 A_1 \sigma T_1^4 - \varepsilon_1\varepsilon_2 A_1 \sigma T_2^4}{\frac{A_1}{A_2}\varepsilon_1 + \varepsilon_2 - \frac{A_1}{A_2}\varepsilon_1\varepsilon_2}\right)$$
$$= \frac{A_1\sigma(T_1^4 - T_2^4)}{\frac{1}{\varepsilon_1} + \frac{A_1}{A_2}\left(\frac{1}{\varepsilon_2} - 1\right)}$$
$$= f_{12} A_1 \sigma (T_1^4 - T_2^4)$$

where

$$f_{12} = \left[\frac{1}{\varepsilon_1} + \frac{A_1}{A_2}\left(\frac{1}{\varepsilon_2} - 1\right)\right]^{-1} \tag{11.16}$$

If $A_2 >> A_1$, i.e. a small body in a large enclosure,

$$f_{12} = \varepsilon_1$$

and

$$q_{12} = A_1 \varepsilon_1 \sigma (T_1^4 - T_2^4)$$

Some useful interchange factors are listed in Table 11.4.

11.7 Radiation from a Gray Cavity

Let the gray surface of the cavity emits, refer Fig. 11.31a,

$$E_1 = A_1 \varepsilon_1 \sigma T_1^4.$$

As the surface 1 sees itself hence

Table 11.4 Useful interchange or transfer factors[a]

S. No.	Configuration	Interchange factor
1	1 2	Infinite parallel planes: $f_{12} = f_{21} = [1/\varepsilon_1 + 1/\varepsilon_2 - 1]^{-1}$
2	2 1 2	Completely enclosed flat or convex body, very small compared to the enclosing body ($A_1 \ll A_2$): $f_{12} = \varepsilon_1$
3	1	Completely enclosed body $A_1 \approx A_2$: $f_{12} \approx [1/\varepsilon_1 + 1/\varepsilon_2 - 1]^{-1}$; $F_{12} = 1$ (f_{12} is approximately equal to that of infinite parallel planes)
4	1	Concentric spheres or infinite cylinders: $f_{12} = [1/\varepsilon_1 + (A_1/A_2)(1/\varepsilon_2 - 1)]^{-1}$

[a]For use in equation $q_{12} = A_1 f_{12} \sigma (T_1^4 - T_2^4)$.

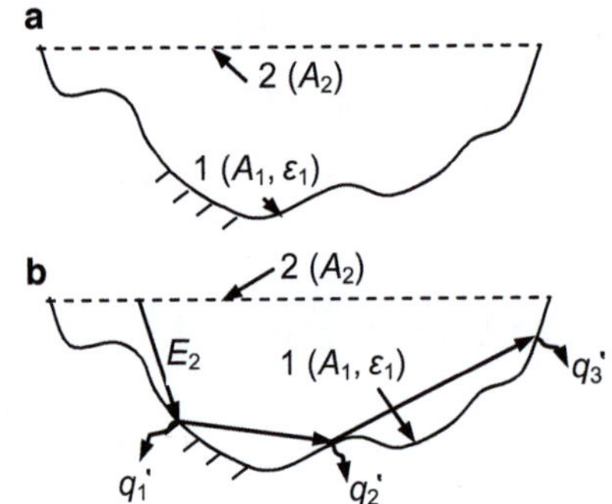

Fig. 11.31 **a** A gray cavity. **b** Radiation from surface 2

$$F_{11} \neq 0.$$

Hence, a fraction $F_{11}E_1$ of the radiant energy E_1 falls upon the surface 1 itself. If α_1 is the absorptivity of the surface, it absorbs

$$E_1' = \alpha_1 F_{11} E_1 = \alpha_1 F_{11} A_1 \varepsilon_1 \sigma T_1^4.$$

As $\alpha_1 = \varepsilon_1$, we have

$$E_1' = F_{11} A_1 \varepsilon_1^2 \sigma T_1^4.$$

Remaining $(F_{11}E_1 - E_1')$ is reflected back. A fraction F_{11} of the reflected energy, i.e. $[F_{11}(F_{11}E_1 - E_1')]$ falls back on surface 1. Out of this α_1 times is absorbed. This absorbed quantity is

$$\begin{aligned} E_1'' &= \alpha_1[F_{11}(F_{11}E_1 - E_1')] \\ &= \varepsilon_1[F_{11}(F_{11}E_1 - E_1')] \\ &= \varepsilon_1 F_{11}^2 E_1 - \varepsilon_1 F_{11} E_1' \\ &= \varepsilon_1 F_{11}^2 (A_1\varepsilon_1\sigma T_1^4) - \varepsilon_1 F_{11}(F_{11}A_1\varepsilon_1^2\sigma T_1^4) \\ &= A_1 F_{11}^2 \varepsilon_1^2 \sigma T_1^4 (1-\varepsilon_1). \end{aligned}$$

This process of absorption and reflection continues indefinitely. The net amount of energy leaving surface 1 is

$$\begin{aligned} (q_1)_{net} &= E_1 - (E_1' + E_1'' + \ldots) \\ &= A_1\varepsilon_1\sigma T_1^4 - [F_{11}A_1\varepsilon_1^2\sigma T_1^4 + A_1F_{11}^2\varepsilon_1^2\sigma T_1^4(1-\varepsilon_1) + \ldots] \\ &= A_1\varepsilon_1\sigma T_1^4\{1 - [\varepsilon_1 F_{11} + \varepsilon_1(1-\varepsilon_1)F_{11}^2 + \ldots]\} \\ &= A_1\varepsilon_1\sigma T_1^4\{1 - \varepsilon_1 F_{11}[1 + (1-\varepsilon_1)F_{11} + (1-\varepsilon_1)^2 F_{11}^2 \ldots]\}. \end{aligned}$$

As $(1 - \varepsilon_1)\, F_{11}$ is less than 1, summation of the series gives

$$(q_1)_{net} = A_1\varepsilon_1\sigma T_1^4\left[1 - \frac{\varepsilon_1 F_{11}}{1-(1-\varepsilon_1)F_{11}}\right] = A_1\varepsilon_1\sigma T_1^4\left[\frac{1-F_{11}}{1-(1-\varepsilon_1)F_{11}}\right].$$

We have

$$F_{11} + F_{12} = 1$$

or

$$F_{11} = 1 - F_{12}.$$

Substitution gives

$$\begin{aligned} (q_1)_{net} &= A_1\varepsilon_1\sigma T_1^4\left[\frac{F_{12}}{1-(1-\varepsilon_1)(1-F_{12})}\right] \\ &= A_1\sigma T_1^4\left[\frac{\varepsilon_1 F_{12}}{1-(1-\varepsilon_1)(1-F_{12})}\right] \\ &= A_1\sigma T_1^4 \frac{1}{\frac{1}{F_{12}} + \frac{1-\varepsilon_1}{\varepsilon_1}} \end{aligned}$$

Let the black surface 2 of the cover emits radiation, refer Fig. 11.31(b),

$$E_2 = A_2\sigma T_2^4.$$

If α_1 is absorptivity of the surface 1, it absorbs

$$q_1' = \alpha_1 E_2$$

on the first incidence of the radiation from surface 2.

A fraction $F_{11}\rho_1 E_2$ of the radiant energy E_2 falls upon the surface 1 out of the radiation $\rho_1 E_2$ reflected from surface 1 and the surface 1 absorbs

$$q_2' = \alpha_1 F_{11}\rho_1 E_2$$

and the remaining is reflected.

On the next incidence, the surface 1 can be shown to absorb

$$q_3' = \alpha_1 F_{11}\rho_1 (F_{11}\rho_1 E_2)$$

This process of absorption and reflection continues indefinitely.

The total radiation received by surface 1 out of the radiation E_2 is

$$\begin{aligned} q' = q_1' + q_2' + q_3' + \ldots &= \alpha_1 E_2 + \alpha_1 F_{11}\rho_1 E_2 + \alpha_1 F_{11}\rho_1 (F_{11}\rho_1 E_2) + \ldots \\ &= \varepsilon_1 E_2 + \varepsilon_1 F_{11}(1-\varepsilon_1)E_2 + \varepsilon_1(1-\varepsilon_1)^2 F_{11}^2 E_2 + \ldots \\ &= \varepsilon_1 E_2[1 + F_{11}(1-\varepsilon_1) + (1-\varepsilon_1)^2 F_{11}^2 + \ldots] \end{aligned}$$

Since $F_{11}(1 - \varepsilon_1) < 1$, summation of the series gives

$$q' = \varepsilon_1 E_2\left[\frac{1}{1-(1-\varepsilon_1)F_{11}}\right] = \varepsilon_1 A_2\sigma T_2^4\left[\frac{1}{1-(1-\varepsilon_1)F_{11}}\right].$$

From reciprocity relation,

$$A_2F_{21} = A_1F_{12}$$

or

$$A_2 = A_1F_{12} \text{ since } F_{21} = 1,$$

and

$$F_{11} = 1 - F_{12}.$$

Substitution gives

$$q' = \varepsilon_1 A_1 F_{12}\sigma T_2^4\left[\frac{1}{1-(1-\varepsilon_1)(1-F_{12})}\right].$$

Simplification gives

$$q' = \frac{A_1 \sigma T_2^4}{\frac{1}{F_{12}} + \frac{(1-\varepsilon_1)}{\varepsilon_1}}$$

Net radiation lost by surface 1 is

$$\begin{aligned} q_{12} &= (q_1)_{net} - q' \\ &= \frac{A_1 \sigma (T_1^4 - T_2^4)}{\frac{1}{F_{12}} + \frac{1-\varepsilon_1}{\varepsilon_1}} \end{aligned} \tag{11.24}$$

11.8 Small Gray Bodies

The bodies in Fig. 11.32 are very small compared to the distance between them. The radiant energy from body 1 falling upon body 2 will be partly absorbed by the body 2. The unabsorbed (i.e. the reflected) radiation from body 2 is lost in the surroundings and it is being assumed that it is not reflected back to the surface 1.

Energy emitted by body 1 is $A_1 \varepsilon_1 \sigma T_1^4$.

Energy incident upon body 2 $F_{12} A_1 \varepsilon_1 \sigma T_1^4$.

Energy absorbed body 2 is $\alpha_2 F_{12} A_1 \varepsilon_1 \sigma T_1^4$, which equals $\varepsilon_1 \varepsilon_2 F_{12} A_1 \sigma T_1^4$ as $\alpha_2 = \varepsilon_2$.

Thus the radiant energy transfer from body 1 to body 2 is

$$q_1 = \varepsilon_1 \varepsilon_2 A_1 F_{12} \sigma T_1^4$$

Similarly the radiant energy transfer from body 2 to body 1 is

$$q_2 = \varepsilon_1 \varepsilon_2 A_2 F_{21} \sigma T_2^4.$$

The net radiation energy exchange between the bodies is

$$q_{12} = q_1 - q_2 = \varepsilon_1 \varepsilon_2 A_1 F_{12} \sigma T_1^4 - \varepsilon_1 \varepsilon_2 A_2 F_{21} \sigma T_2^4$$

Applying the reciprocity relation $A_1 F_{12} = A_2 F_{21}$, we have

$$q_{12} = \varepsilon_1 \varepsilon_2 A_1 F_{12} \sigma (T_1^4 - T_2^4)$$

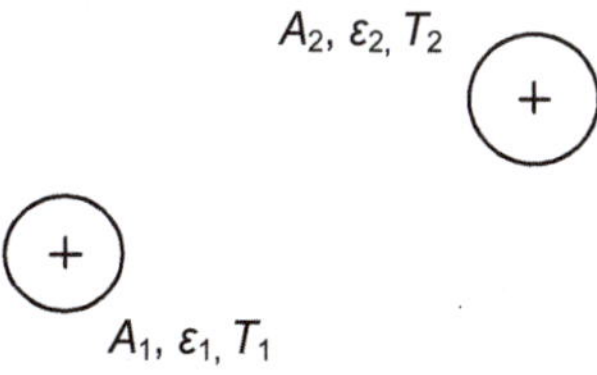

Fig. 11.32 Small gray bodies exchanging radiation

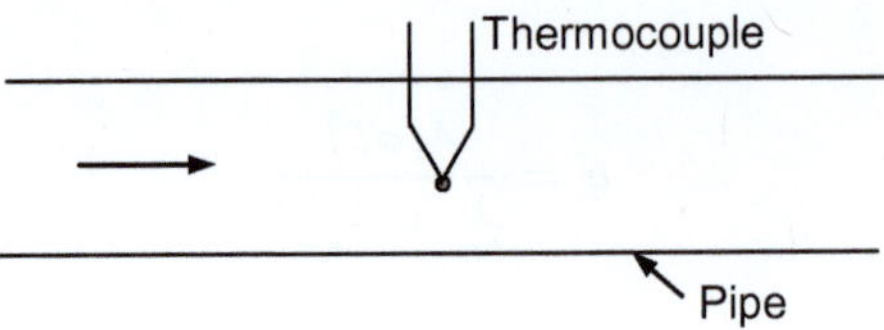

Fig. 11.33 Example 11.21

Example 11.21 A thermocouple is installed in a gas pipe as shown in Fig. 11.33 to measure temperature of the gas flowing through the pipe. The equilibrium temperature of the pipe surface is 500 K. The thermocouple indicates a temperature of 600 K. Calculate the true temperature of the gas if the emissivity of the thermocouple surface is 0.85 and the effect of the conduction along the thermocouple wires is negligible. The convection heat transfer coefficient from the gas to the thermocouple is 40 W/(m^2 K).

Solution

The thermocouple is heated due to convection from the gas. But cooling of the thermocouple takes place due to the rejection of heat by radiation to the pipe wall which is at a temperature lower than the temperature of the thermocouple. Hence, it indicates a temperature lower than the gas temperature.

Let the gas temperature is T_g and the temperature indicated by the thermocouple is T_1. Then the convective heat transfer q_c from the gas to the thermocouple is

$$q_c = hA_1(T_g - T_1) = 40 \times A_1(T_g - 600)$$

where A_1 is the surface area of the thermocouple bead.

For a small body in a large enclosure (bead area < < pipe wall area), the radiation heat transfer is

$$q_r = \varepsilon_1 A_1 \sigma(T_1^4 - T_2^4) = 0.85 \times A_1 \times 5.67 \times 10^{-8} \times (600^4 - 500^4)$$

In equilibrium, $q_c = q_r$. This gives

$$40 \times A_1(T_g - 600) = 0.85 \times A_1 \times 5.67 \times 10^{-8} \times (600^4 - 500^4)$$

or

$$T_g = 680.85\,\text{K}.$$

Thus the error is

$$\Delta T = 680.85 - 600 = 80.85\ \text{K}.$$

The error can be reduced by reducing the radiation heat exchange with the pipe surface. This can be achieved by insulating the pipe from the outside, which will increase the pipe surface temperature. Use of a radiation shield around the thermocouple bead will also reduce the heat exchange with the pipe surface and thus the error will reduce.

Example 11.22 A mercury-in-glass thermometer suspended from the roof of a room reads 20°C. The roof and wall of the room are found to be at 10°C. If the average convective heat transfer coefficient for the thermometer is estimated to be 7 W/(m^2 K), what is the true air temperature? The emissivity of the glass may be assumed as 0.8.

Solution

The thermometer is a small gray body (ε = 0.8) in a large enclosure (the room) which exchanges heat with the walls and roof of the room by radiation. The room air transfers heat to the thermometer by convection. The energy balance gives

$$q_c + q_r = 0$$

or

$$hA(T - T_{air}) + \varepsilon A\sigma(T^4 - T_w^4) = 0$$

where A is the surface area of the thermometer bulb, T is temperature indicated by the thermometer and T_w is room walls and roof temperature.

Substitution of values of various parameters gives

$$7.0 \times A(293 - T_{air}) + 0.8 \times A \times 5.67 \times 10^{-8} \times (293^4 - 283^4) = 0$$

or

$$T_{air} = 299.19\text{ K} = 260.19°\text{C}.$$

Comments: If room walls and roofs are at a higher temperature than the air temperature, the thermometer will indicate temperature higher than the room air temperature.

Example 11.23 Two parallel infinite planes are facing each other. One of the planes has an emissivity of 0.8 and is maintained at 400 K while the other has emissivity of 0.7 and is at 500 K. Calculate the radiation heat exchange between the planes. If the planes are black, what will be the heat exchange? Also calculate the interchange factor. What will be the heat transfer and interchange factor if the temperatures or emissivities of the planes are interchanged.

Solution

For infinite parallel gray planes,

$$q_{12} = \frac{A\sigma(T_1^4 - T_2^4)}{\frac{1}{\varepsilon_1} + \frac{1}{\varepsilon_2} - 1}$$
$$= \frac{1 \times 5.67 \times 10^{-8} \times (500^4 - 400^4)}{\frac{1}{0.8} + \frac{1}{0.7} - 1} = 1246.4\text{ W/m}^2.$$

If the surfaces are black,

$$q_{12} = A\sigma(T_1^4 - T_2^4)$$
$$= 1 \times 5.67 \times 10^{-8} \times (500^4 - 400^4) = 2092.2\,\text{W/m}^2.$$

The interchange factor, $f_{12} = \left[\frac{1}{\varepsilon_1} + \frac{1}{\varepsilon_2} - 1\right]^{-1} = 0.596$, which is also the ratio of heat exchange values calculated above.

From the equations of heat exchange and interchange factor, we can readily see that the heat exchange or interchange factor for parallel plane configuration will not be affected if the temperatures or emissivities of the planes are interchanged. This conclusion is not applicable when $A_1 \neq A_2$ as in the case of cylindrical or spherical configurations.

Example 11.24 A metal pipe of 50 mm diameter, painted with aluminium paint (ε = 0.3), carries a hot fluid. The surface temperature of the pipe is measured to be 400 K. In order to reduce radiation heat loss from the pipe a thin aluminium sheet (ε = 0.1) is placed around the pipe at a distance of 10 mm from the pipe wall. Estimate the percentage reduction in the radiation heat transfer from the pipe if there is no change in pipe surface temperature. The pipe is located in a large space with surrounding temperature of 300 K.

Solution

(i) Radiation heat loss from the pipe wall per meter length without covering,

$$q_{12} = A_1\varepsilon\sigma(T_1^4 - T_\infty^4)$$

where

ε emissivity of the pipe surface = 0.3
A_1 pipe surface area = $\pi DL = \pi \times 0.05 \times 1\ \text{m}^2$
T_1 pipe surface temperature = 400 K
T_∞ surrounding temperature = 300 K

Substitution gives

$$q_{12} = \pi \times 0.05 \times 1 \times 0.3 \times 5.67 \times 10^{-8} \times (400^4 - 300^4) = 46.76\,\text{W/m length.}$$

(ii) Radiation heat loss from the pipe wall per meter length when covered with aluminium sheet (ε = 0.1) of surface area A_2 (= $\pi \times 0.07 \times 1\ \text{m}^2$),

$$q_{12} = \frac{A_1\sigma(T_1^4 - T_2^4)}{\frac{1}{\varepsilon_1} + \frac{A_1}{A_2}\left(\frac{1}{\varepsilon_2} - 1\right)} \qquad \text{(i)}$$

where T_2 is aluminium sheet temperature in steady state.

The radiation heat loss from the outer surface of the aluminium sheet is

$$(q_r)_2 = A_2\varepsilon_2\sigma(T_2^4 - T_\infty^4)$$

It is to note that the sheet is thin and is made of high thermal conductivity material hence it's outer and inner surface temperatures can be assumed to be equal.
In steady state, $q_{12} = (q_r)_2$, i.e.

$$\frac{A_1\sigma(T_1^4 - T_2^4)}{\frac{1}{\varepsilon_1} + \frac{A_1}{A_2}\left(\frac{1}{\varepsilon_2} - 1\right)} = A_2\varepsilon_2\sigma(T_2^4 - T_\infty^4)$$

or

$$\frac{\pi \times 0.05 \times 1 \times (400^4 - T_2^4)}{\frac{1}{0.3} + \frac{0.05}{0.07}\left(\frac{1}{0.1} - 1\right)} = \pi \times 0.07 \times 1 \times 0.1 \times (T_2^4 - 300^4),$$

which gives $T_2 = 352.8$ K.
Radiation heat loss,

$$\begin{aligned}(q_r)_2 &= A_2\varepsilon_2\sigma(T_2^4 - T_\infty^4) = \pi \times 0.07 \times 1 \times 0.1 \times 5.67 \times 10^{-8} \times (T_2^4 - 300^4) \\ &= 9.22\,\text{W/m}^2.\end{aligned}$$

Percentage reduction in the loss is

$$\frac{46.76 - 9.22}{46.76} \times 100 = 80.3\%.$$

Example 11.25 A hemispherical cavity of radius 0.1 m has an emissivity of 0.5 and is at a temperature of 400 K. If the surrounding environment is at a temperature of 300 K, determine the net heat flux leaving the surface of the cavity. Assume the cavity to be a gray diffuse surface. If the cavity is cylindrical (D = 0.2 m and H = 0.1 m) what will be net heat flow? Compare the result with that of a black surfaced cavity.

Solution

(i) The opening of the cavity can be assumed to be a flat black surface at the surrounding temperature (refer Fig. 11.34).
From the previous section, we have

$$q_{12} = \frac{\sigma(T_1^4 - T_2^4)}{\frac{1}{A_1F_{12}} + \frac{1-\varepsilon_1}{A_1\varepsilon_1}}$$

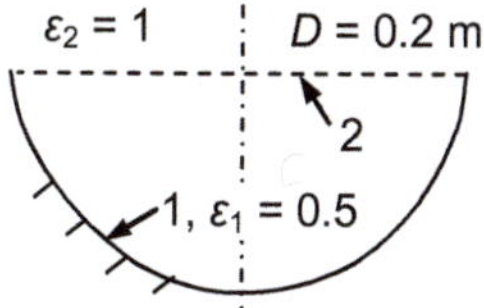

Fig. 11.34 Example 11.25

Substituting T_1 = 400 K, T_2 = 300 K, ε_1 = 0.5, $A_1 = \pi D^2/2 = \pi \times 0.2^2/2 = 0.02\ \pi$ m^2, $A_2 = \pi D^2/4 = \pi \times 0.2^2/4 = 0.01\ \pi$ m^2 and $F_{12} = \frac{A_2}{A_1} = \frac{0.1\pi}{0.2\pi} = 0.5$, we get

$$q_{12} = \frac{5.67 \times 10^{-8} \times (400^4 - 300^4)}{\frac{1}{0.02\pi \times 0.5} + \frac{1-0.5}{0.02\pi \times 0.5}} = 20.78\ \text{W}.$$

(ii) **Cylindrical cavity**
From Example 11.7, we have

$$F_{12} = \frac{D}{4H + D} = \frac{0.2}{4 \times 0.1 + 0.2} = 0.333,$$

and

$$A_1 = \pi DH + \pi D^2/4 = \pi \times 0.2 \times 0.1 + \pi \times 0.2^2/4 = 0.09425\,\text{m}^2.$$

This gives

$$q_{12} = \frac{5.67 \times 10^{-8} \times (400^4 - 300^4)}{\frac{1}{0.09425 \times 0.333} + \frac{1-0.5}{0.09425 \times 0.5}} = 23.37\,\text{W}.$$

(iii) **Black Surfaced Cavity (ε_1 = 1)**
The shape of a black surfaced cavity does not affect the heat flow. It depends on A_2.

$$(q_{12})_{black} = A_2\sigma(T_1^4 - T_2^4)$$

or

$$q_{12} = \left(\frac{\pi}{4} \times 0.2^2\right) \times 5.67 \times 10^{-8} \times (400^4 - 300^4) = 31.17\ \text{W}.$$

Example 11.26 A brick wall (ε = 0.85), 6 m × 4 m in area, faces an opening in a furnace of 0.2 m × 0.2 m in area. The wall is at a distance of 5 m from the opening as shown in Fig. 11.35. The centerline of the opening is 1 m right and 1 m below than the center point of the opening. Consider the opening as a blackbody at 1100°C and average wall temperature as 40°C. Determine the net heat transfer by radiation between the opening and wall.

Solution
Divide the wall area into four parts such that the furnace opening area (area dA) is located opposite the corner of the areas A, B, C and D. Case 10 of Table 11.1, i.e. Equation (11.18) is now applicable to each of these areas. For these areas:

A: L_1 = 3 m, L_2 = 4 m, L = 5 m;
B: L_1 = 3 m, L_2 = 2 m, L = 5 m;
C: L_1 = 1 m, L_2 = 2 m, L = 5 m;
D: L_1 = 1 m, L_2 = 4 m, L = 5 m.

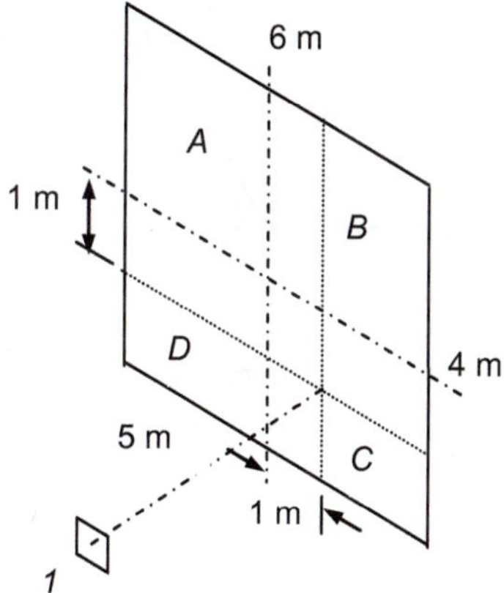

Fig. 11.35 Example 11.26

The shape factor is determined for each area from Eq. (11.18) as

$$F_{1A} = 0.0928, F_{1B} = 0.057, F_{1C} = 0.0225, F_{1D} = 0.036.$$

The total net rate of heat transfer from surface 1 is

$$\begin{aligned} q &= q_A + q_B + q_C + q_D \\ &= \varepsilon\sigma(dA) \times (F_{1A} + F_{1B} + F_{1C} + F_{1D}) \times (T_1^4 - T_{wall}^4) \\ &= 0.85 \times 5.67 \times 10^{-8} \times 0.04 \times (0.0928 + 0.057 \\ &\quad + 0.0225 + 0.036) \times (1373^4 - 313^4) \\ &= 1423.2\ \text{W}. \end{aligned}$$

11.9 Electric Network Method for Solving Radiation Problems

The problems of radiation heat exchange between blackbodies are relatively easy to handle because all of the incident radiation on a blackbody is absorbed. The main problem in such cases is the estimate of the radiation shape factor. In the case of gray bodies, a part of the incident radiation is reflected back to the other gray body. From where, again a part of the radiation is reflected back to the first body. This process of reflection and absorption basically goes on. Clearly due to this process of multiple reflection and absorption, the calculation of the radiation exchange is a difficult task. Such problems can be readily solved by reducing the actual system to an equivalent electric network as explained below.

We introduce two new terms:

(i) **Irradiation *G***: It is defined as the total radiation energy incident per unit time and unit area of a surface.

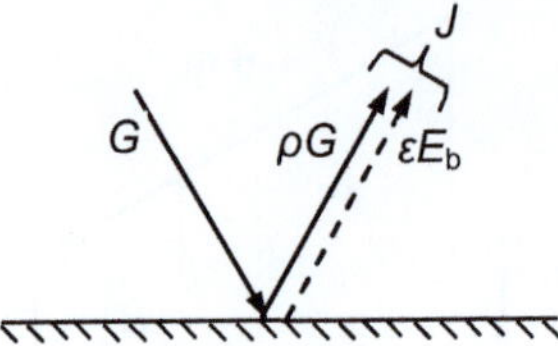

Fig. 11.36 A diffuse gray surface

(ii) **Radiosity *J***: It is the total energy which leaves a surface per unit area per unit time. It is the sum of the energy emitted and reflected when no energy is transmitted, refer Fig. 11.36. Thus radiosity is

$$J = \varepsilon E_b + \rho G \tag{i}$$

where

ε emissivity of the surface
E_b blackbody emissive power and
ρ reflectivity = $(1 - \alpha)$ (when $\tau = 0$) = $(1 - \varepsilon)$. (as $\alpha = \varepsilon$)

Substitution in Eq. (i)gives

$$J = \varepsilon E_b + (1 - \varepsilon)G$$

or

$$G = \frac{J - \varepsilon E_b}{(1 - \varepsilon)} \tag{ii}$$

For a black surface, $\rho = 0$ and $\varepsilon = 1$ hence $J = E_b$.

$$\begin{aligned} q_{net} &= A(J - G) \\ &= A\left[J - \frac{J - \varepsilon E_b}{(1 - \varepsilon)}\right] \\ &= \frac{A\varepsilon}{(1 - \varepsilon)}(E_b - J) \end{aligned}$$

or

$$q_{net} = \frac{E_b - J}{(1 - \varepsilon)/A\varepsilon} \tag{11.25}$$

The equation states that the net rate at which the radiation energy leaves a gray surface of area A is equivalent to current flow in a circuit with potential difference of $(E_b - J)$ and resistance $(1 - \varepsilon)/A\varepsilon$ as shown in Fig. 11.37a. This resistance is considered as *surface resistance* to the radiation energy transfer.

Fig. 11.37 **a** Surface resistance **b** space resistance

Now consider the radiant heat exchange between two gray surfaces. Of the total radiation J_1 leaving the surface 1, a fraction $J_1A_1F_{12}$ is received by the surface 2. Similarly of the total radiation J_2 leaving the surface 2, a fraction $J_2A_2F_{21}$ reaches surface 1. Thus the net exchange is

$$\begin{aligned} q_{12} &= J_1A_1F_{12} - J_2A_2F_{21} \\ &= A_1F_{12}(J_1 - J_2) \\ &\quad (\text{as}\, A_1F_{12} = A_2F_{21}) \end{aligned}$$

or

$$q_{12} = \frac{J_1 - J_2}{1/A_1F_{12}} \tag{11.26}$$

The equivalent electrical network is shown in Fig. 11.37(b). The term $1/A_1F_{12}$ is defined as *space resistance.*

Equations (11.25) and (11.26) provide the basis for the electric network method of solving radiation heat exchange problems. The effect of the emissivity of a diffuse and opaque gray surface is taken account by connecting potential E_b to potential J through the surface resistance $(1 - \varepsilon)/A\varepsilon$. While the shape factor effect between two radiosity potentials is accounted by the space resistance $1/A_1F_{12}$.

11.9.1 Electric Network for a System Consisting of Two Gray Surfaces

Figure 11.38 shows the network for the given system. The total resistance,

$$R_t = \frac{1 - \varepsilon_1}{A_1\varepsilon_1} + \frac{1}{A_1F_{12}} + \frac{1 - \varepsilon_2}{A_2\varepsilon_2}$$

Radiation heat exchange,

$$\begin{aligned} q_{12} &= \frac{\sigma\left(T_1^4 - T_2^4\right)}{R_t} \\ &= \frac{\sigma\left(T_1^4 - T_2^4\right)}{\frac{1-\varepsilon_1}{A_1\varepsilon_1} + \frac{1}{A_1F_{12}} + \frac{1-\varepsilon_2}{A_2\varepsilon_2}} \\ &= \frac{A_1\sigma\left(T_1^4 - T_2^4\right)}{\frac{1-\varepsilon_1}{\varepsilon_1} + \frac{1}{F_{12}} + \frac{A_1}{A_2}\left(\frac{1-\varepsilon_2}{\varepsilon_2}\right)} \end{aligned}$$

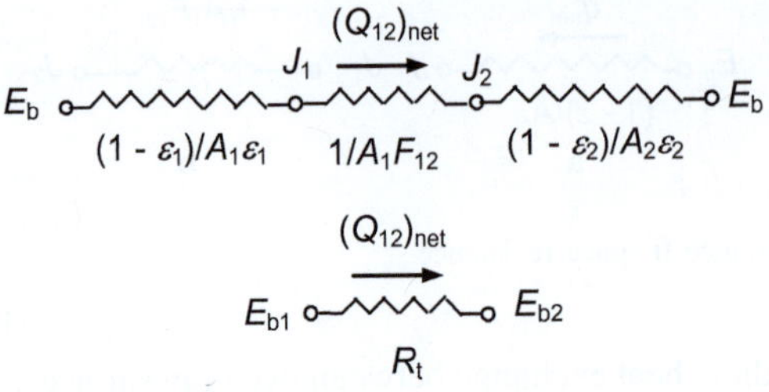

$(Q_{12})_{net}$

E_b — J_1 — J_2 — E_b

$(1 - \varepsilon_1)/A_1\varepsilon_1$ $1/A_1F_{12}$ $(1 - \varepsilon_2)/A_2\varepsilon_2$

$(Q_{12})_{net}$

E_{b1} — E_{b2}

R_t

Fig. 11.38 Electric network for a system consisting of two gray surfaces

Let,

$$q_{12} = A_1\sigma f_{12}\left(T_1^4 - T_2^4\right)$$

where f_{12} is *interchange factor.*

Comparing the above equations of q_{12}, we get

$$f_{12} = \left[\frac{1 - \varepsilon_1}{\varepsilon_1} + \frac{1}{F_{12}} + \frac{A_1}{A_2}\left(\frac{1 - \varepsilon_2}{\varepsilon_2}\right)\right]^{-1}. \tag{11.27}$$

In the case of infinite parallel planes of equal area, $A_1 = A_2$ and $F_{12} = 1$. Hence,

$$q_{12} = \frac{A_1\sigma\left(T_1^4 - T_2^4\right)}{\frac{1}{\varepsilon_1} + \frac{1}{\varepsilon_1} - 1} = \frac{A_1(E_{b1} - E_{b2})}{\frac{1}{\varepsilon_1} + \frac{1}{\varepsilon_1} - 1},$$

as found earlier.

11.9.2 System Consisting of Two Black Surfaces

For a black surface, $J = E_b$. The equivalent network for this case is shown in Fig. 11.39, and the net radiation energy transfer is

$$\begin{aligned}(q_{12})_{net} &= \frac{(E_{b1} - E_{b2})}{1/A_1F_{12}} \\ &= A_1F_{12}(E_{b1} - E_{b2})\end{aligned}$$

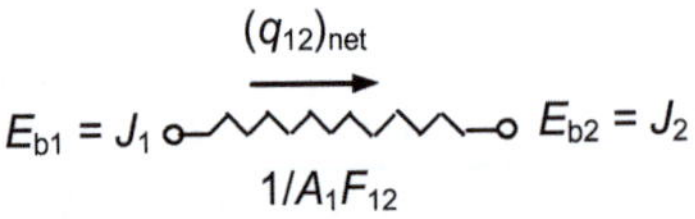

$(q_{12})_{net}$

$E_{b1} = J_1$ — $E_{b2} = J_2$

$1/A_1F_{12}$

Fig. 11.39 Network for two black surfaces

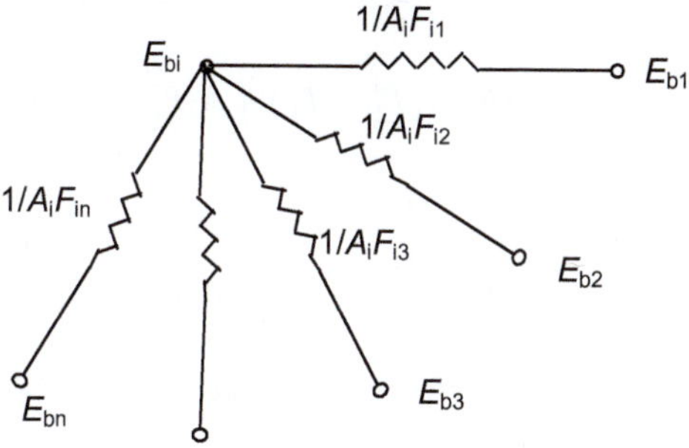

Fig. 11.40 Network for a closed system of n-black surfaces

11.9.3 Closed System of N-Black Surfaces

For a closed systems of *n*-black surfaces, we can write a general expression of net heat transfer from surface *i* as (refer Fig. 11.40)

$$\begin{aligned} q_{12} &= A_iF_{i1}\sigma(T_i^4 - T_1^4) + A_iF_{i2}\sigma(T_i^4 - T_2^4) + \ldots + A_iF_{in}\sigma(T_i^4 - T_n^4) \\ &= A_i\sigma \sum_{j=1}^{n} F_{ij}(T_i^4 - T_j^4) \end{aligned} \tag{11.28}$$

where $F_{\text{i-i}} = 0$

11.9.4 Systems Consisting of Two Black Surfaces Connected by a Single Refractory Surface

A refractory surface is also called reradiating surface. This surface does not experience a net heat gain or loss because whatever heat is absorbed by this surface, the same is reradiated to other surfaces exchanging heat.

The net radiation heat transfer from surface 1 is (Fig. 11.41)

$$q_1 = q_{12} + q_{13} = A_1F_{12}\sigma\left(T_1^4 - T_2^4\right) + A_1F_{13}\sigma\left(T_1^4 - T_3^4\right) \tag{i}$$

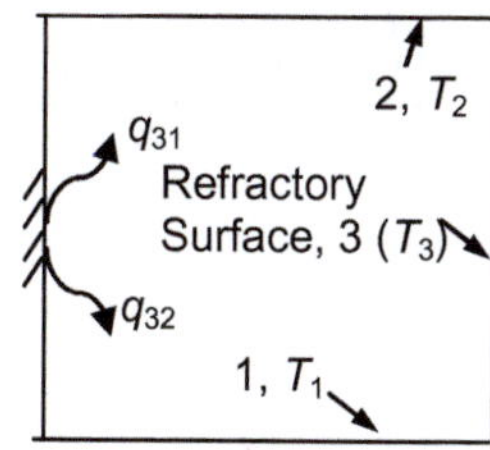

Fig. 11.41 Two black surfaces connected by a single refractory surface

Similarly,

$$q_2 = q_{21} + q_{23} = A_2F_{21}\sigma(T_2^4 - T_1^4) + A_2F_{23}\sigma(T_2^4 - T_3^4) \tag{ii}$$

and

$$q_3 = q_{31} + q_{32} = A_3F_{31}\sigma(T_3^4 - T_1^4) + A_3F_{32}\sigma(T_3^4 - T_2^4) \tag{iii}$$

Since there is no heat loss from the refractory surface, $q_3 = 0$, and from Eq. (iii)

$$0 = A_3F_{31}\sigma(T_3^4 - T_1^4) + A_3F_{32}\sigma(T_3^4 - T_2^4)$$

or

$$T_3^4 = \frac{F_{31}T_1^4 + F_{32}T_2^4}{F_{31} + F_{32}} \tag{iv}$$

Substitution of the value of T_3^4 in Eq. (i) gives

$$\begin{aligned} q_1 &= A_1F_{12}\sigma(T_1^4 - T_2^4) + A_1F_{13}\sigma\left(T_1^4 - \frac{F_{31}T_1^4 + F_{32}T_2^4}{F_{31} + F_{32}}\right) \\ &= A_1F_{12}\sigma(T_1^4 - T_2^4) + A_1F_{13}\sigma\left(\frac{F_{32}T_1^4 - F_{32}T_2^4}{F_{31} + F_{32}}\right) \\ &= \sigma(T_1^4 - T_2^4) \times \left(A_1F_{12} + \frac{A_1F_{13}F_{32}}{F_{31} + F_{32}}\right) \end{aligned} \tag{v}$$

From the reciprocity theorem,

$$A_3F_{31} = A_1F_{13},$$

and

$$A_3F_{32} = A_2F_{23},$$

which gives

$$F_{32} = \frac{A_2}{A_3}F_{23}$$

Using the above relations Eq. (v) transforms to

$$q_1 = \sigma(T_1^4 - T_2^4)\left(A_1F_{12} + \frac{1}{1/A_1F_{13} + 1/A_2F_{23}}\right) \tag{11.29a}$$

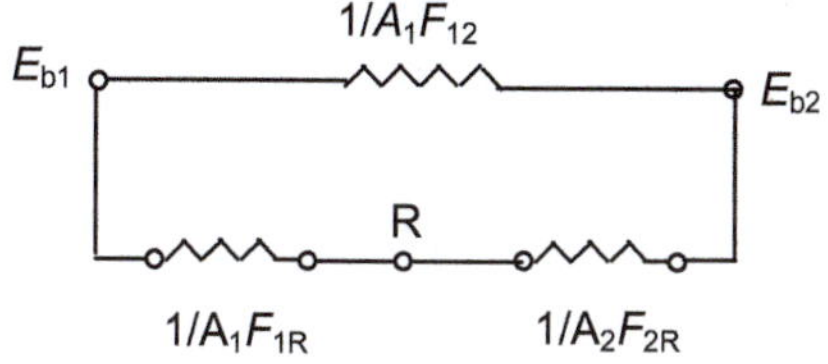

Fig. 11.42 Network for the system in Fig. 11.41

The electric network for the system is shown in Fig. 11.42 where the subscript 3 has been replaced by R. It is to be noted that unlike gray surface, the refractory surface is not connected to any potential because the net radiation energy transfer from the surface is zero. The total resistance between E_{b1} and E_{b2} is

$$R_t = \left(A_1F_{12} + \frac{1}{1/A_1F_{13} + 1/A_2F_{23}}\right)^{-1} \tag{11.29b}$$

If $A_1 = A_2$, $F_{1R} = F_{2R}$ as $F_{12} = F_{21}$. This gives

$$F_{32} = F_{R2} = \frac{A_2}{A_3}F_{2R}$$

$$F_{31} = F_{R1} = \frac{A_1}{A_3}F_{1R} = \frac{A_2}{A_3}F_{2R} = F_{32}.$$

Substitution in Eq. (iv) gives

$$T_3^4 = \frac{T_1^4 + T_2^4}{2}$$

or

$$T_3 = T_R = \sqrt{4}\frac{T_1^4 + T_2^4}{2} \tag{11.30}$$

Example 11.27 Show that the factor f_{12} for two parallel black surfaces of equal area connected by reradiating walls at constant temperature is given by

$$f_{12} = \frac{1 + F_{12}}{2}$$

Solution

The given arrangement of black surfaces is shown in Fig. 11.41 and the network is shown in Fig. 11.42. The total resistance from the network is given by

$$R_t = \left(A_1F_{12} + \frac{1}{1/A_1F_{1R} + 1/A_2F_{2R}}\right)^{-1}$$

For $A_1 = A_2 = A$,

$$R_t = \frac{1}{A}\left(F_{12} + \frac{1}{1/F_{1R} + 1/F_{2R}}\right)^{-1} \tag{i}$$

From the summation rule,

$$F_{1R} + F_{12} = 1 \text{ as } F_{11} = 0$$

and

$$F_{2R} + F_{21} = 1 \text{ as } F_{22} = 0.$$

From the above equations,

$$F_{1R} = 1 - F_{12}$$

and

$$F_{2R} = 1 - F_{21}.$$

From the reciprocity relation,

$$A_1 F_{12} = A_2 F_{21},$$

which gives

$$F_{12} = F_{21}$$

as $A_1 = A_2$. Thus

$$F_{2R} = 1 - F_{12}.$$

Substitution of the values of F_{1R} and F_{2R} in Eq. (i) gives

$$R_t = \frac{1}{A}\left(\frac{1 + F_{12}}{2}\right)^{-1}.$$

The radiation heat transfer,

$$q_{12} = \frac{E_{b1} - E_{b2}}{R_t} = A\left(\frac{1 + F_{12}}{2}\right)(E_{b1} - E_{b2}).$$

The equation of the radiation heat transfer can be written as

$$q_{12} = Af_{12}(E_{b1} - E_{b2}).$$

Comparison gives

$$f_{12} = \frac{1 + F_{12}}{2}.$$

Example 11.28 Show that the factor f_{12} for two parallel black surfaces of unequal area connected by reradiating surface is given by

$$f_{12} = \frac{A_2/A_1 - F_{12}^2}{1 - 2F_{12} + A_2/A_1}$$

Solution

The network of Fig. 11.42 applies. The total resistance from the network is

$$R_t = \left(A_1F_{12} + \frac{1}{1/A_1F_{1R} + 1/A_2F_{2R}}\right)^{-1} \quad \text{(i)}$$

We have

$$F_{1R} + F_{12} = 1 \text{ as } F_{11} = 0$$

or

$$F_{1R} = 1 - F_{12}. \quad \text{(ii)}$$

Similarly,

$$F_{2R} = 1 - F_{21}.$$

From the reciprocity relation,

$$F_{21} = \frac{A_1}{A_2}F_{12}$$

Hence,

$$F_{2R} = 1 - \frac{A_1}{A_2}F_{12} \quad \text{(iii)}$$

Substitution of the values of F_{1R} and F_{2R} from Eqs. (ii) and (iii) in Eq. (i) gives

$$R_t = \frac{1 - 2F_{12} + A_2/A_1}{A_1\left(A_2/A_1 - F_{12}^2\right)}$$

The heat flow by radiation is given by

$$q_{12} = \frac{E_{b1} - E_{b2}}{R_t}$$

Also,

$$q_{12} = A_1 f_{12}(E_{b1} - E_{b2})$$

Comparison of these relations gives

$$f_{12} = \frac{1}{R_t A_1} = \frac{A_2/A_1 - F_{12}^2}{1 - 2F_{12} + A_2/A_1},$$

which is the desired result. If $A_1 = A_2$, the result of previous example is obtained.

11.9.5 System Consisting of Two Gray Surfaces Connected by a Single Refractory Surface

Refer Fig. 11.43.

$$\begin{aligned} R_t &= \frac{1-\varepsilon_1}{A_1\varepsilon_1} + \frac{1-\varepsilon_2}{A_2\varepsilon_2} + \frac{1}{A_1F_{12} + \frac{1}{1/A_1F_{1R} + 1/A_2F_{2R}}} \\ &= \frac{1-\varepsilon_1}{A_1\varepsilon_1} + \frac{1-\varepsilon_2}{A_2\varepsilon_2} + \frac{1}{A_1\bar{F}_{12}}. \end{aligned}$$

The heat flow by radiation is given by

$$\begin{aligned} q_{12} &= \frac{E_{b1} - E_{b2}}{R_t} \\ &= \frac{E_{b1} - E_{b2}}{\frac{1-\varepsilon_1}{A_1\varepsilon_1} + \frac{1-\varepsilon_2}{A_2\varepsilon_2} + \frac{1}{A_1\bar{F}_{12}}} \\ &= (E_{b1} - E_{b2})\frac{A_1}{\frac{1-\varepsilon_1}{\varepsilon_1} + \frac{1}{\bar{F}_{12}} + \frac{A_1}{A_2}\left(\frac{1-\varepsilon_2}{\varepsilon_2}\right)} \\ &= A_1(F_g)_{12}(E_{b1} - E_{b2}) \end{aligned}$$

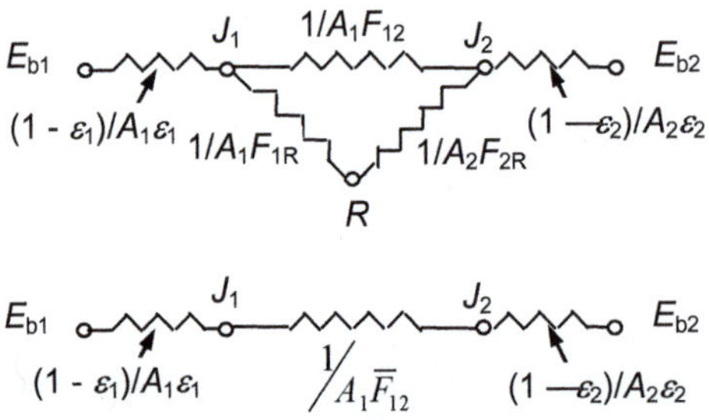

Fig. 11.43 Network for system consisting of two gray surfaces connected by a single refractory surface

where

$$(F_g)_{12} = \frac{1}{\frac{1-\varepsilon_1}{\varepsilon_1} + \frac{1}{\bar{F}_{12}} + \frac{A_1}{A_2}\left(\frac{1-\varepsilon_2}{\varepsilon_2}\right)} \tag{11.31a}$$

and

$$\bar{F}_{12} = \frac{A_2/A_1 - F_{12}^2}{1 - 2F_{12} + A_2/A_1}, \tag{11.31b}$$

from Example 11.28.

11.9.6 System Consisting of Four Gray Surfaces Which See Each Other and Nothing Else

The radiation network is shown in Fig. 11.44.

To calculate the net rate of heat flow from the each of the surfaces, we must determine the radiosities J_1, J_2, J_3 and J_4. The network can be solved by setting the sum of heat currents entering each node to zero, i.e.

$$\text{Node } J_1: \quad \frac{E_{b1} - J_1}{(1-\varepsilon_1)/A_1\varepsilon_1} + \frac{(J_2 - J_1)}{1/A_1F_{12}} + \frac{(J_3 - J_1)}{1/A_1F_{13}} + \frac{(J_4 - J_1)}{1/A_1F_{14}} = 0$$

$$\text{Node } J_2: \quad \frac{E_{b2} - J_2}{(1-\varepsilon_2)/A_2\varepsilon_2} + \frac{(J_1 - J_2)}{1/A_1F_{12}} + \frac{(J_3 - J_2)}{1/A_2F_{23}} + \frac{(J_4 - J_2)}{1/A_2F_{24}} = 0$$

$$\text{Node } J_3: \quad \frac{E_{b3} - J_3}{(1-\varepsilon_3)/A_3\varepsilon_3} + \frac{(J_1 - J_3)}{1/A_1F_{13}} + \frac{(J_2 - J_3)}{1/A_2F_{23}} + \frac{(J_4 - J_3)}{1/A_3F_{34}} = 0$$

$$\text{Node } J_4: \quad \frac{E_{b4} - J_4}{(1-\varepsilon_4)/A_4\varepsilon_4} + \frac{(J_1 - J_4)}{1/A_1F_{14}} + \frac{(J_2 - J_4)}{1/A_2F_{24}} + \frac{(J_3 - J_4)}{1/A_3F_{34}} = 0$$

The four unknowns J_1, J_2, J_3 and J_4 are determined by the solution of the above four equations. Knowing the radiosities J_1, J_2, J_3 and J_4, the net rate of heat transfer from surfaces 1 to 4 are determined from:

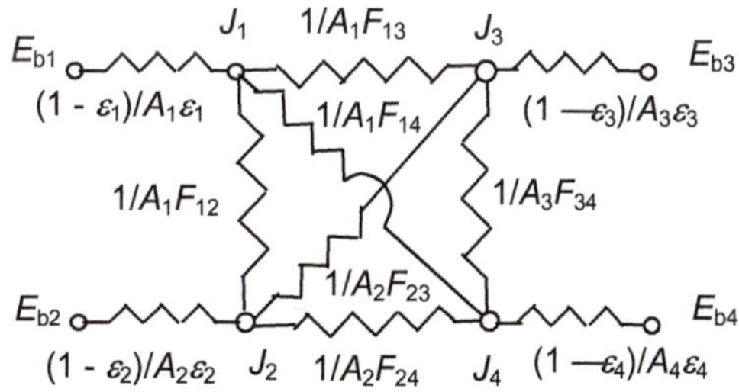

Fig. 11.44 Network for system consisting of four gray surfaces which see each other and nothing else

$$q_1 = \frac{E_{b1} - J_1}{(1-\varepsilon_1)/A_1\varepsilon_1},\ q_2 = \frac{E_{b2} - J_2}{(1-\varepsilon_2)/A_2\varepsilon_2},\ q_3 = \frac{E_{b3} - J_3}{(1-\varepsilon_3)/A_3\varepsilon_3} \text{ and } q_4 = \frac{E_{b4} - J_4}{(1-\varepsilon_4)/A_4\varepsilon_4}.$$

Example 11.29 Determine loss of heat by radiation from a steel tube ($\varepsilon = 0.8$) of 70 mm diameter and 4 m long at a temperature of 500 K if the tube is located:

(i) In a large brick room ($\varepsilon = 0.9$) with wall temperature of 300 K.
(ii) In a brick conduit (0.25 m × 0.25 m, $\varepsilon = 0.9$) at 300 K, neglect loss of the heat from ends.

Solution

(i) **Radiation exchange with large brick room**
The room surface area is very large compared to the pipe surface area, i.e. $A_2 >> A_1$. The network is shown in Fig. 11.45.
Total resistance,

$$R_t = \frac{1-\varepsilon_1}{A_1\varepsilon_1} + \frac{1}{A_1F_{12}} + \frac{1-\varepsilon_2}{A_2\varepsilon_2} = \frac{1}{A_1}\left[\frac{1-\varepsilon_1}{\varepsilon_1} + \frac{1}{F_{12}} + \frac{A_1}{A_2}\left(\frac{1-\varepsilon_2}{\varepsilon_2}\right)\right]$$

Since $F_{12} = 1$ and $\frac{A_1}{A_2}\left(\frac{1-\varepsilon_2}{\varepsilon_2}\right) \approx 0$ for $A_2 >> A_1$, we obtain

$$R_t = \frac{1}{A_1\varepsilon_1}$$

Radiation heat exchange,

$$q_{12} = \frac{\sigma(T_1^4 - T_2^4)}{R_t} = A_1\varepsilon_1\sigma(T_1^4 - T_2^4)$$
$$= (\pi \times 0.07 \times 4) \times 0.8 \times 5.67 \times 10^{-8} \times (500^4 - 300^4) = 2170.6 \text{ W}.$$

(ii) **Radiation exchange with brick conduit (Fig. 11.46)**
Total resistance,

$$R_t = \frac{1}{A_1}\left[\frac{1-\varepsilon_1}{\varepsilon_1} + \frac{1}{F_{12}} + \frac{A_1}{A_2}\left(\frac{1-\varepsilon_2}{\varepsilon_2}\right)\right]$$

Since $F_{12} = 1$, we obtain

$$R_t = \frac{1}{A_1}\left[\frac{1}{\varepsilon_1} + \frac{A_1}{A_2}\left(\frac{1-\varepsilon_2}{\varepsilon_2}\right)\right]$$

Fig. 11.45 Network Example 11.29 Part (i)

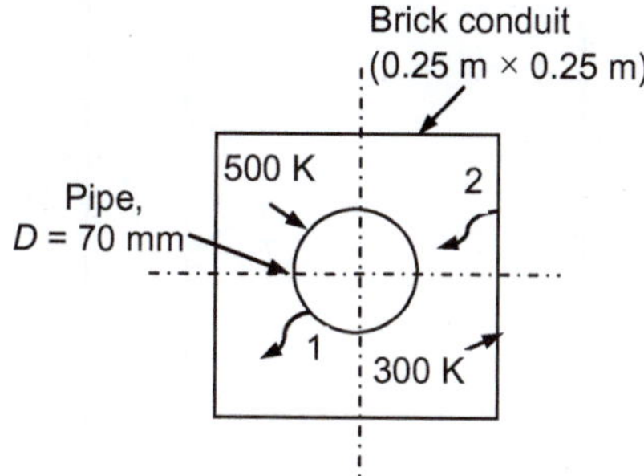

Fig. 11.46 Example 11.29 Part (ii)

Here

$$A_1 = \pi DL = \pi \times 0.07 \times 4 = 0.8796\ \text{m}^2,$$

$$A_2 = (0.25 \times 4) \times 4 = 4.0\ \text{m}^2.$$

Radiation heat exchange,

$$q_{12} = \frac{\sigma\left(T_1^4 - T_2^4\right)}{R_t} = \frac{A_1\sigma\left(T_1^4 - T_2^4\right)}{\left[\frac{1}{\varepsilon_1} + \frac{A_1}{A_2}\left(\frac{1-\varepsilon_2}{\varepsilon_2}\right)\right]}$$

$$= \frac{0.8796 \times 5.67 \times 10^{-8} \times (500^4 - 300^4)}{\left[\frac{1}{0.8} + \frac{0.8796}{4.0}\left(\frac{1-0.9}{0.9}\right)\right]} = 2128.9\ \text{W}.$$

11.10 Radiation Shields

Radiation heat transfer between two surfaces can be reduced by placing a thin opaque partition between the surfaces. This partition is known as radiation shield. The radiation shield introduces an additional resistance in the radiation path. From the analysis, which follows, it will be shown that the shields must be made of very low absorptivity and high reflectivity materials, such as thin sheets of aluminium, copper, etc.

Consider two parallel infinite black plates as shown in Fig. 11.47a. The net heat exchange between these plates is

$$q_{12} = A\sigma\left(T_1^4 - T_2^4\right)$$

Suppose a thin metal sheet 3, black on both sides, is introduced between these planes as shown in Fig. 11.47b. When the steady state is reached, the radiative heat transfer between the surfaces 1 and 3 equals the transfer between 3 and 2, i.e.

$$q_{13} = q_{32}$$

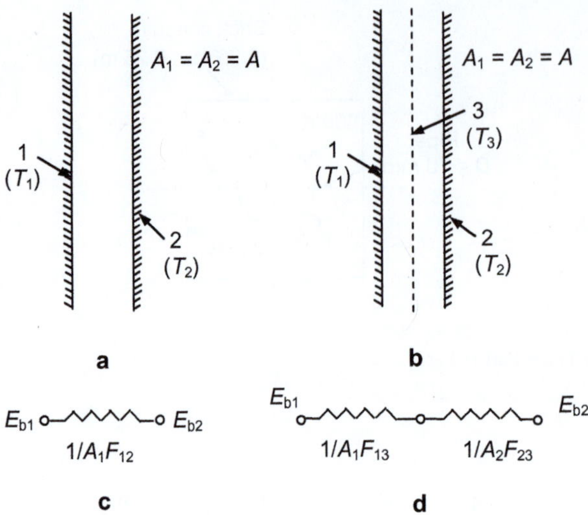

Fig. 11.47 Radiation shield and network

or

$$A\sigma\left(T_1^4 - T_3^4\right) = A\sigma\left(T_3^4 - T_2^4\right)$$

or

$$T_3^4 = \frac{1}{2}\left(T_1^4 + T_2^4\right) \tag{11.32}$$

Thus the heat transfer rate when the black surfaced radiation shield is introduced is

$$\begin{aligned} q_{13} &= A\sigma\left(T_1^4 - T_3^4\right) \\ &= A\sigma\left[T_1^4 - \frac{1}{2}\left(T_1^4 + T_2^4\right)\right] \\ &= \frac{1}{2}A\sigma\left(T_1^4 - T_2^4\right) \\ &= \frac{1}{2}q_{12}. \end{aligned}$$

The analysis shows that by inserting one shield, the radiation energy transfer is reduced to one-half of the original value. The analysis can be extended to show that with n shields the net energy exchange is

$$(q)_{\text{with}\,n-\text{shields}} = \frac{1}{n+1}(q_{12})_{\text{withoutshields}}. \tag{11.33}$$

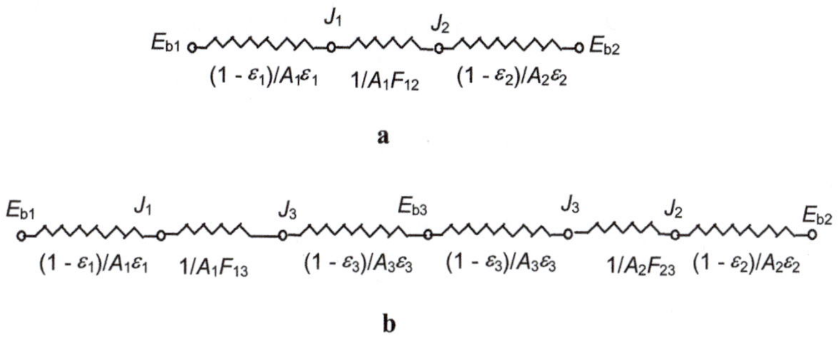

Fig. 11.48 Network for gray parallel plates **a** without shield, **b** with shield

If the surfaces 1 and 2 are not black but having emissivities ε_1 and ε_2, respectively, and the shield 3 has emissivity ε_3, the radiation energy transfer equations can be written by referring the network in Fig. 11.48.

Without shield,

$$q_{12} = \frac{A\sigma\left(T_1^4 - T_2^4\right)}{\frac{1}{\varepsilon_1} + \frac{1}{\varepsilon_2} - 1}. \tag{i}$$

With the shield,

$$q_{13} = \frac{A\sigma\left(T_1^4 - T_3^4\right)}{\frac{1}{\varepsilon_1} + \frac{1}{\varepsilon_3} - 1}, \tag{ii}$$

and

$$q_{32} = \frac{A\sigma\left(T_3^4 - T_2^4\right)}{\frac{1}{\varepsilon_3} + \frac{1}{\varepsilon_2} - 1}. \tag{iii}$$

Equating Eqs. (ii) and (iii), we get value of T_3^4 in the terms of T_1^4 and T_2^4. Substitution of the value of T_3^4 in Eq. (ii) and simplification gives

$$q_{13} = \frac{A\sigma\left(T_1^4 - T_2^4\right)}{\left(\frac{1}{\varepsilon_1} + \frac{1}{\varepsilon_3} - 1\right) + \left(\frac{1}{\varepsilon_3} + \frac{1}{\varepsilon_2} - 1\right)}. \tag{11.34}$$

The ratio of energy exchange with and without shield in this case is

$$\frac{q_{13}}{q_{12}} = \frac{\left(\frac{1}{\varepsilon_1} + \frac{1}{\varepsilon_2} - 1\right)}{\left(\frac{1}{\varepsilon_1} + \frac{1}{\varepsilon_3} - 1\right) + \left(\frac{1}{\varepsilon_3} + \frac{1}{\varepsilon_2} - 1\right)}.$$

For radiation shield with emissivity ε_3 less than 1, the ratio is less than ½ when surfaces 1 and 2 are black surfaces. Lower is the value of the emissivity of the shield greater is its effectiveness. For example, with $\varepsilon_1 = \varepsilon_2 = 1$ and $\varepsilon_3 = 0.1$, the radiation transfer can be shown to reduce to 1/20th.

Example 11.30 Two parallel square plates, each of 4 m^2, are separated by a distance of 3 mm. One of the plates is at a temperature of 500 K and its surface emissivity is 0.7, while the other plate surface is at a temperature of 300 K and has a surface emissivity of 0.6. Find the net energy exchange by radiation between the plates.

If a thin polished metal sheet of surface emissivity of 0.1 on both sides is now located centrally as radiation shield between the two plates, find out the altered net heat transfer. The convection and edge effects, if any, may be neglected. What would be the heat exchange when the emissivity of the surface of the shield is 0.9?

Solution

Case (a) Refer Fig. 11.48a.

Total resistance,

$$R_t = \frac{1-\varepsilon_1}{A_1\varepsilon_1} + \frac{1}{A_1F_{12}} + \frac{1-\varepsilon_2}{A_2\varepsilon_2} = \frac{1}{A}\left(\frac{1-\varepsilon_1}{\varepsilon_1} + \frac{1}{F_{12}} + \frac{1-\varepsilon_2}{\varepsilon_2}\right) \text{ as } A_1 = A_2$$

For parallel plates of large area, $F_{12} = 1$. Hence,

$$R_t = \frac{1}{4}\left(\frac{1-0.7}{0.7} + \frac{1}{1} + \frac{1-0.6}{0.6}\right) = 0.5238.$$

Radiation heat exchange,

$$q_{12} = \frac{\sigma\left(T_1^4 - T_2^4\right)}{R_t} = \frac{5.67 \times 10^{-8} \times (500^4 - 300^4)}{0.5238} = 5888\,\text{W}.$$

Case (b) Refer Fig. 11.48b.

(i) Total resistance,

$$R_t = \frac{1}{A}\left(\frac{1-\varepsilon_1}{\varepsilon_1} + \frac{1}{F_{13}} + \frac{1-\varepsilon_3}{\varepsilon_3} + \frac{1-\varepsilon_3}{\varepsilon_3} + \frac{1}{F_{32}} + \frac{1-\varepsilon_2}{\varepsilon_2}\right)$$

For $F_{13} = F_{32} = 1$,

$$R_t = \frac{1}{4}\left[\frac{1-0.7}{0.7} + \frac{1}{1} + \frac{1-0.1}{0.1} + \frac{1-0.1}{0.1} + \frac{1}{1} + \frac{1-0.6}{0.6}\right] = 5.2738.$$

Radiation heat exchange,

$$q_{12} = \frac{\sigma\left(T_1^4 - T_2^4\right)}{R_t} = \frac{5.67 \times 10^{-8} \times (500^4 - 300^4)}{5.2738} = 584.8\,\text{W}.$$

(ii) Total resistance,

$$R_t = \frac{1}{4}\left[\frac{1-0.7}{0.7} + \frac{1}{1} + \frac{1-0.9}{0.9} + \frac{1-0.9}{0.9} + \frac{1}{1} + \frac{1-0.6}{0.6}\right] = 0.8294.$$

Radiation heat exchange,

$$q_{12} = \frac{\sigma\left(T_1^4 - T_2^4\right)}{R_t} = \frac{5.67 \times 10^{-8} \times (500^4 - 300^4)}{0.8294} = 3718.9 \text{ W}.$$

From comparison of the results, it can be seen that a radiation shield with low emissivity (a polished surface) is very effective in reducing the heat loss.

Example 11.31 Three screens (having emissivity $\varepsilon = 0.05$) have been placed between two parallel black surfaces ($F_{12} = F_{21} = 1$) at temperatures T_1 and T_2. What is the percentage reduction in radiation heat exchange due to screens? Assume that the temperatures of the black surfaces remain the same with or without screens.

Solution

Let us consider a single screen of emissivity ε between two parallel black surfaces.

As $F_{13} = F_{32}$, the total resistance is

$$R_t = \frac{1}{A}\left[1 + 2\left(\frac{1-\varepsilon}{\varepsilon}\right) + 1\right] = \frac{1}{A}\left[2\left(\frac{1-\varepsilon}{\varepsilon}\right) + 2\right]$$

Extending the analysis to the system with three screens, the total resistance will be

$$R_t = \frac{1}{A}\left\{3 \times \left[2\left(\frac{1-\varepsilon}{\varepsilon}\right)\right] + (3+1)\right\}$$

For $\varepsilon = 0.05$,

$$R_t = \frac{118}{A}$$

Radiation heat exchange,

$$q_{\text{withshield}} = \frac{(E_{b1} - E_{b2})}{R_t} = \frac{A(E_{b1} - E_{b2})}{118} = \frac{q_{\text{withoutshield}}}{118} = 0.0085 q_{\text{withoutshield}}.$$

Thus the heat exchange is reduced by 99.15%.

Example 11.32

(a) Prove that the net radiation heat exchange between two parallel planes 1 and 2 with n radiation shields (ε_s) inserted between them is given by

$$q_{12} = \frac{\sigma\left(T_1^4 - T_2^4\right)}{\frac{1}{\varepsilon_1} + \frac{1}{\varepsilon_2} + 2\sum_{i=1}^{n}\frac{1}{\varepsilon_s} - (n+1)}$$

(b) Determine the minimum number of screens required to reduce the radiation heat transfer at least by a factor of 60 between two surfaces of emissivities 0.7 and 0.8, if the emissivity of the screens is 0.05.

Solution

(a) For a single screen, we have

$$\left(q_{12}\right)_{\text{oneshield}} = \frac{\sigma\left(T_1^4 - T_2^4\right)}{\frac{1}{\varepsilon_1} + \frac{1}{\varepsilon_2} + \frac{2}{\varepsilon_s} - 2},$$

which can be readily extended to n screens to give the desired result.

(b) Putting $\varepsilon_1 = 0.7$, $\varepsilon_2 = 0.8$, $\varepsilon_{s1} = \varepsilon_{s2} = \ldots\ldots = 0.05$, we get

$$\left(q_{12}\right)_{\text{n-shield}} = \frac{\sigma\left(T_1^4 - T_2^4\right)}{2.678 + 40n - (n+1)} \tag{a}$$

The heat transfer without the shield is

$$\left(q_{12}\right)_{\text{n=0}} = \frac{\sigma\left(T_1^4 - T_2^4\right)}{\frac{1}{0.7} + \frac{1}{0.8} - 1} = \frac{\sigma\left(T_1^4 - T_2^4\right)}{1.678} \tag{b}$$

From Eqs. (a) and (b), we get

$$\frac{\left(q_{12}\right)_{\text{n=0}}}{\left(q_{12}\right)_{\text{n- shield}}} = \frac{2.678 + 40n - (n+1)}{1.678} = 60 \text{ (given)}$$

Solving for n, we get $n = 2.538$. Thus three screens are required.

Example 11.33 Two coaxial cylinders 1 ($r_1 = 100$ mm, $\varepsilon_1 = 0.3$) and 2 ($r_2 = 200$ mm, $\varepsilon_2 = 0.7$) are maintained at temperatures of 700°C and 100°C, respectively. Determine the radiative heat transfer rate between the cylinders per m length.

Solution

Heat transfer between cylinders 1 and 2 for unit length is

$$q_{12} = \frac{A_1\sigma(T_1^4 - T_2^4)}{\frac{1}{\varepsilon_1} + \frac{A_1}{A_2}\left(\frac{1}{\varepsilon_2} - 1\right)}$$

Substitution gives

$$q_{12} = \frac{2\pi \times 0.1 \times 1 \times 5.67 \times 10^{-8} \times \left[(700+273)^4 - (100+273)^4\right]}{\frac{1}{0.3} + \frac{0.1}{0.2}\left(\frac{1}{0.7} - 1\right)}$$
$$= 8806.3\ \text{W/per unit length.}$$

Example 11.34 If a thin cylinder ($\varepsilon_1 = 0.1$) of 150 mm radius is introduced between the cylinders of above example. How much reduction in radiative heat transfer rate will result?

Solution

The total resistance to heat transfer between cylinders 1 and 2 for unit length is

$$R_t = \left(\frac{1-\varepsilon_1}{A_1\varepsilon_1}\right) + \frac{1}{A_1} + \left(\frac{1-\varepsilon_2}{A_2\varepsilon_2}\right) + \left(\frac{1-\varepsilon_2}{A_2\varepsilon_2}\right) + \frac{1}{A_2} + \left(\frac{1-\varepsilon_3}{A_3\varepsilon_3}\right)$$
$$= \frac{1}{A_1}\left[\left(\frac{1-\varepsilon_1}{\varepsilon_1}\right) + 1 + \frac{A_1}{A_2}\left(\frac{1-\varepsilon_2}{\varepsilon_2}\right) + \frac{A_1}{A_2}\left(\frac{1-\varepsilon_2}{\varepsilon_2}\right) + \frac{A_1}{A_2} + \frac{A_1}{A_3}\left(\frac{1-\varepsilon_3}{\varepsilon_3}\right)\right]$$

Substitution gives

$$R_t = \frac{1}{2\pi \times 0.1 \times 1}\left[\left(\frac{1-0.3}{0.3}\right) + 1 + 2 \times \frac{0.1}{0.15} \times \left(\frac{1-0.1}{0.1}\right) + \frac{0.1}{0.15} + \frac{0.1}{0.2} \times \left(\frac{1-0.7}{0.7}\right)\right]$$
$$= 25.8.$$

The heat transfer is given by

$$q_{12} = \frac{\sigma(T_1^4 - T_2^4)}{R_t}$$

Substitution gives

$$q_{12} = \frac{5.67 \times 10^{-8} \times \left[(700+273)^4 - (100+273)^4\right]}{25.8} = 1927\ \text{W per unit length.}$$

Percentage reduction in heat transfer

$$= \frac{8806.3 - 1927}{8806.3} = 78.12\%.$$

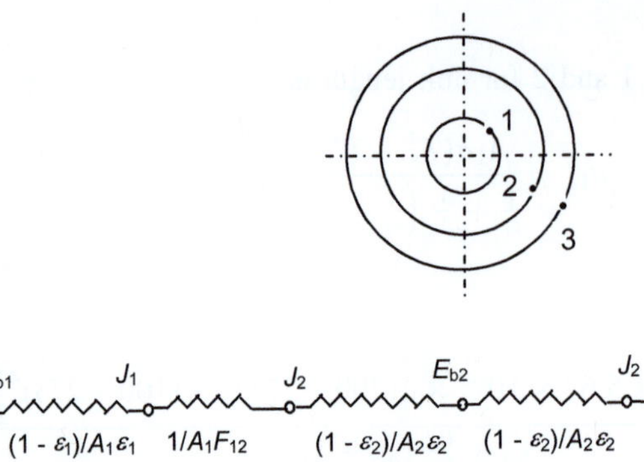

$F_{12} = 1 = F_{23}$,
$A_1 = \pi D_1 L$, $A_2 = \pi D_2 L$, $A_3 = \pi D_3 L$,
$D_1 = 0.05$ m, $\varepsilon_1 = 0.05$, $T_1 = 1200$ K,
$D_2 = 0.1$ m, $\varepsilon_2 = 0.05$, $T_2 = ?$,
$D_3 = 0.15$ m, $\varepsilon_3 = 0.05$, $T_3 = 400$ K.

Fig. 11.49 Example 11.35

Example 11.35 Three very long thin walled hollow cylinders of 50 mm, 100 mm and 150 mm in diameters are arranged coaxially. The temperatures of the surface of 50 mm and 150 mm diameter cylinders are maintained at 1200 K and 400 K, respectively. Assuming a vacuum between the annular spaces, find the steady state temperature attained by the surface of 100 mm cylinder. Also calculate the radiation heat transfer between the inner and outer most cylinders. Take $\varepsilon_1 = \varepsilon_2 = \varepsilon_3 = 0.05$.

Solution

Heat transfer between cylinders 1 and 2 for unit length is (refer Fig. 11.49)

$$q_{12} = \frac{A_1\sigma(T_1^4 - T_2^4)}{\frac{1}{\varepsilon_1} + \frac{A_1}{A_2}\left(\frac{1}{\varepsilon_2} - 1\right)}$$

Substitution gives

$$q_{12} = \frac{\pi \times 0.05 \times 1 \times 5.67 \times 10^{-8} \times (1200^4 - T_2^4)}{\frac{1}{0.05} + \frac{0.05}{0.1}\left(\frac{1}{0.05} - 1\right)}$$

$$= 3.019 \times 10^{-10}(1200^4 - T_2^4)\ \text{W per unit length.}$$

Similarly the heat transfer between cylinders 2 and 3 for unit length is

$$q_{23} = \frac{A_2\sigma(T_2^4 - T_3^4)}{\frac{1}{\varepsilon_2} + \frac{A_2}{A_3}\left(\frac{1}{\varepsilon_3} - 1\right)}$$

Substitution gives

$$q_{23} = \frac{\pi \times 0.1 \times 1 \times 5.67 \times 10^{-8} \times (T_2^4 - 400^4)}{\frac{1}{0.05} + \frac{0.1}{0.15}\left(\frac{1}{0.05} - 1\right)}$$

$$= 5.453 \times 10^{-10}(T_2^4 - 400^4)\ \text{W per unit length.}$$

In equilibrium, $q_{12} = q_{23}$. This gives

$$3.019 \times 10^{-10}(1200^4 - T_2^4) = 5.453 \times 10^{-10}(T_2^4 - 400^4)$$

or

$$T_2 = 932.3\ \text{K}$$

and

$$q_{12} = q_{23} = 3.019 \times 10^{-10}(1200^4 - 932.3^4) = 397.96\ \text{W}.$$

Alternatively, the heat transfer can be found as under.
The total resistance,

$$R_t = \frac{1-\varepsilon_1}{A_1\varepsilon_1} + \frac{1}{A_1} + \frac{1-\varepsilon_2}{A_2\varepsilon_2} + \frac{1-\varepsilon_2}{A_2\varepsilon_2} + \frac{1}{A_2} + \frac{1-\varepsilon_3}{A_3\varepsilon_3}$$

Substitution of values of various parameters gives

$$R_t = 291.78$$

and

$$q_{13} = \frac{\sigma(T_1^4 - T_2^4)}{R_t}$$

Substitution of values of various parameters gives

$$q_{13} = \frac{5.67 \times 10^{-8} \times (1200^4 - 400^4)}{291.78} = 397.98\ \text{W per unit length.}$$

Example 11.36 For the perpendicular plane surfaces shown in Fig. 11.50, calculate the heat transfer between the surfaces. Neglect radiation or reflection from the surrounding surfaces.

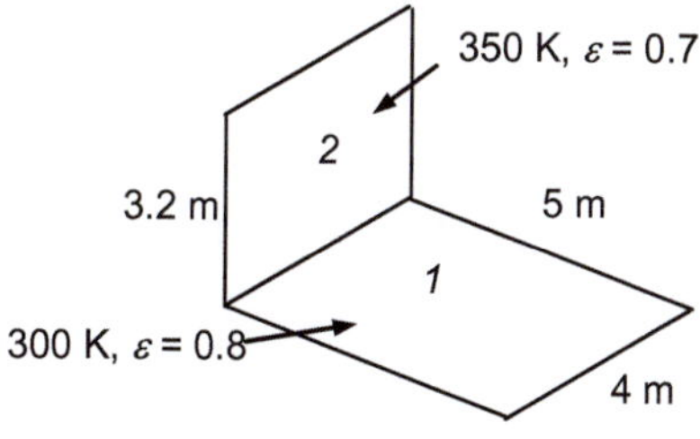

Fig. 11.50 Example 11.36

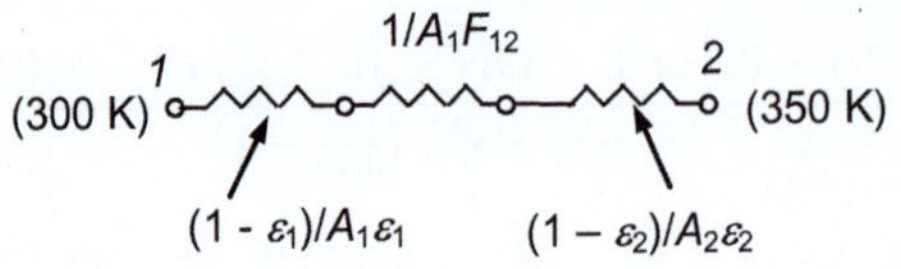

$1/A_1F_{12}$

1 (300 K) 2 (350 K)

$(1 - \varepsilon_1)/A_1\varepsilon_1$ $(1 - \varepsilon_2)/A_2\varepsilon_2$

Fig. 11.51 Network for system in Fig. 11.50

Solution

Assuming the horizontal surface to be surface *1*, we have

$Z = 3.2$ m, $X = 4$ m, $Y = 5$ m, $T_1 = 300$ K, $\varepsilon_1 = 0.8$, $T_2 = 350$ K, $\varepsilon_2 = 0.7$, $A_1 = 5 \times 4$ m^2 and $A_2 = 3.2 \times 4$ m^2.

For $Z/X = 3.2/4 = 0.8$; $Y/X = 5/4 = 1.25$, $F_{12} \approx 0.155$ from Fig. 11.4c.

The total resistance, see Fig. 11.51,

$$R_t = \frac{1 - \varepsilon_1}{A_1\varepsilon_1} + \frac{1}{A_1F_{12}} + \frac{1 - \varepsilon_2}{A_2\varepsilon_2}$$

$$R_t = \frac{1}{A_1}\left[\left(\frac{1}{\varepsilon_1} - 1\right) + \frac{1}{F_{12}} + \left(\frac{A_1}{A_2}\right)\left(\frac{1}{\varepsilon_2} - 1\right)\right]$$

Heat exchange,

$$q_{12} = \sigma\frac{\left(T_1^4 - T_2^4\right)}{R_t}$$

$$= A_1\sigma\frac{\left(T_1^4 - T_2^4\right)}{\left[\left(\frac{1}{\varepsilon_1} - 1\right) + \frac{1}{F_{12}} + \left(\frac{A_1}{A_2}\right)\left(\frac{1}{\varepsilon_2} - 1\right)\right]}$$

Substitution gives

$$q_{12} = 20 \times 5.67 \times 10^{-8} \times \frac{(300^4 - 350^4)}{\left[\left(\frac{1}{0.8} - 1\right) + \frac{1}{0.155} + \left(\frac{20}{12.8}\right)\left(\frac{1}{0.7} - 1\right)\right]} = -1062.5\,\text{W}.$$

If the surfaces are black, $\varepsilon_1 = \varepsilon_2 = 1$, hence,

$$R_t = \frac{1}{A_1F_{12}}$$

$$q_{12} = \sigma\frac{\left(T_1^4 - T_2^4\right)}{R_t}$$

$$= A_1F_{12}\sigma\left(T_1^4 - T_2^4\right)$$

or

$$q_{12} = 20 \times 0.155 \times 5.67 \times 10^{-8} \times \left(300^4 - 350^4\right) = -1213.9\ \text{W}.$$

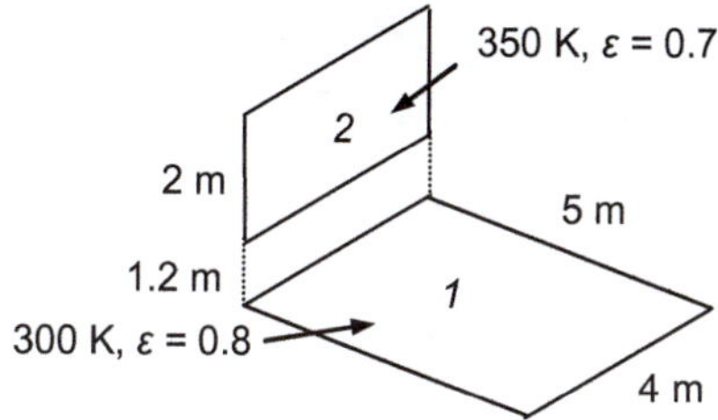

Fig. 11.52 Example 11.37

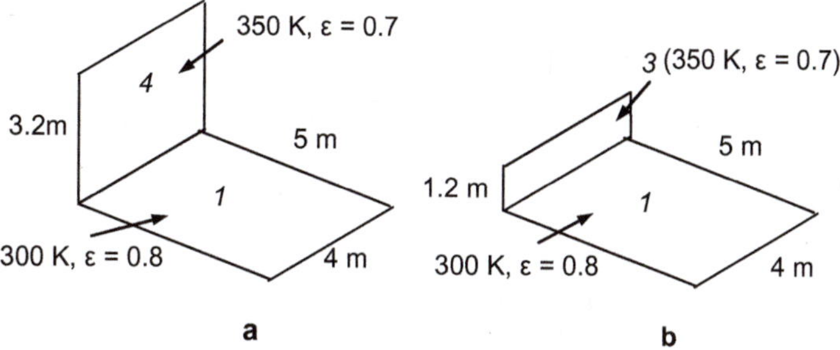

Fig. 11.53 Sub-problems of system in Fig. 11.52

Example 11.37 For the perpendicular plane surfaces shown in Fig. 11.52, calculate the heat transfer between the surfaces. Neglect radiation or reflection from the surrounding surfaces.

Solution

The given problem can be divided into the sub-problems (a) and (b) as shown in Fig. 11.53. The subtraction of value of the shape factor of the sub-problem (b) from that of (a) will give the desired shape factor, that is,

$$F_{12} = F_{14} - F_{13}$$

where

$F_{14} = 0.155$ for $Z/X = 3.2/4 = 0.8$ and $Y/X = 5/4 = 1.25$
$F_{13} = 0.085$ for $Z/X = 1.2/4 = 0.3$ and $Y/X = 5/4 = 1.25$

Hence,

$$F_{12} = 0.155 - 0.085 = 0.07$$

Total resistance is

$$R_t = \frac{1}{A_1}\left[\left(\frac{1}{\varepsilon_1} - 1\right) + \frac{1}{F_{12}} + \left(\frac{A_1}{A_2}\right)\left(\frac{1}{\varepsilon_2} - 1\right)\right]$$

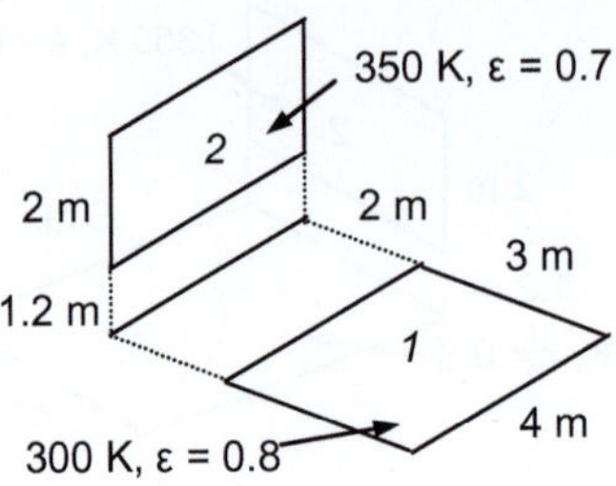

Fig. 11.54 Example 11.38

$$= \frac{1}{20}\left[\left(\frac{1}{0.8} - 1\right) + \frac{1}{0.07} + \left(\frac{20}{8}\right)\left(\frac{1}{0.7} - 1\right)\right] = 0.7803$$

This gives the heat exchange

$$q_{12} = \sigma \frac{(T_1^4 - T_2^4)}{R_t}$$

$$= 5.67 \times 10^{-8} \times \frac{(300^4 - 350^4)}{0.7803} = -501.8\,\text{W}.$$

Example 11.38 For the perpendicular plane surfaces shown in Fig. 11.54, calculate various shape factors.

Solution

Consider the configurations shown in Fig. 11.55a–c.

For the configuration of Fig. 11.55a, we have

$$A_1F_{1-(2,4)} = A_1F_{12} + A_1F_{14}$$

where $F_{1\text{-}(2,4)}$ means the shape factor considering the surface 1 and the combined area of surfaces 2 and 4. This gives

$$A_1F_{12} = A_1F_{1-(2,4)} - A_1F_{14} \tag{i}$$

From Fig. 11.55b, we have

$$A_1F_{1-(2,4)} + A_3F_{3-(2,4)} = A_{(3,1)}F_{(3,1)-(2,4)}$$

From Example 11.36,

$$F_{(3,1)-(2,4)} = 0.155$$

and from Fig. 11.4c, $F_{3\text{-}(2,4)} = 0.275$ for $Z/X = 3.2/4 = 0.8$ and $Y/X = 2/4 = 0.5$.

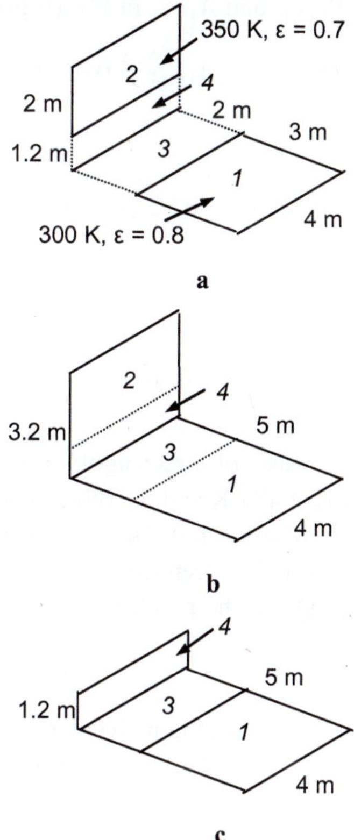

Fig. 11.55 Configurations for the solution of Example 11.38

Substitution gives

$$A_1F_{1-(2,4)} = 20 \times 0.155 - 8 \times 0.275 = 0.9.$$

For the configuration of Fig. 11.55c, we have

$$A_1F_{14} + A_3F_{34} = A_{(3,1)}F_{(3,1)-4}$$

or

$$A_1F_{14} = A_{(3,1)}F_{(3,1)-4} - A_3F_{34}$$

Again from Fig. 11.4c, $F_{34} = 0.19$ for $Z/X = 1.2/4 = 0.3$ and $Y/X = 2/4 = 0.5$ From Example 11.37, $F_{(3,1)\text{-}4} = 0.085$. Hence,

$$A_1F_{14} = 20 \times 0.085 - 8 \times 0.19 = 0.18.$$

Substitution of values of $A_1F_{1\text{-}(2,4)}$ and A_1F_{14} in Eq. (i) gives

$$A_1F_{12} = A_1F_{1-(2,4)} - A_1F_{14} = 0.9 - 0.18 = 0.72.$$

or

$$F_{12} = 0.72/A_1 = 0.72/(4 \times 3) = 0.06.$$

Note: If $A_1 = A_3$,

$$F_{12} = 2F_{(3,1)-(2,4)} - 2F_{(3,1)-4} - F_{3-(2,4)} + F_{34}.$$

Example 11.39 An enclosure consists of a rectangular parallelepiped 1 m × 2 m × 2 m. One of the 1 m × 2 m surfaces is at 475 K and the other is at 350 K. Both of these surfaces may be regarded as black. The remaining four surfaces of the enclosure act as reradiating surfaces Estimate the net heat transfer between the active surfaces of the enclosure and also estimate the equilibrium temperature of the reradiating surface.

Solution

For the given arrangement of the surfaces, the network is shown in Fig. 11.56.

From Fig. 11.4, $F_{12} = F_{21} \approx 0.1$ for $X/L = 1/2 = 0.5$ and $Y/L = 2/2 = 1.0$.

From the given data, $A_1 = A_2 = 1 \times 2 = 2\ \text{m}^2$. Hence,

$$\frac{1}{A_1F_{12}} = \frac{1}{2 \times 0.1} = 5,$$

$$\frac{1}{A_1F_{1R}} = \frac{1}{A_1(1 - F_{12})} = \frac{1}{2(1 - 0.1)} = 0.555 \quad \text{as}\ F_{12} + F_{1R} = 1,$$

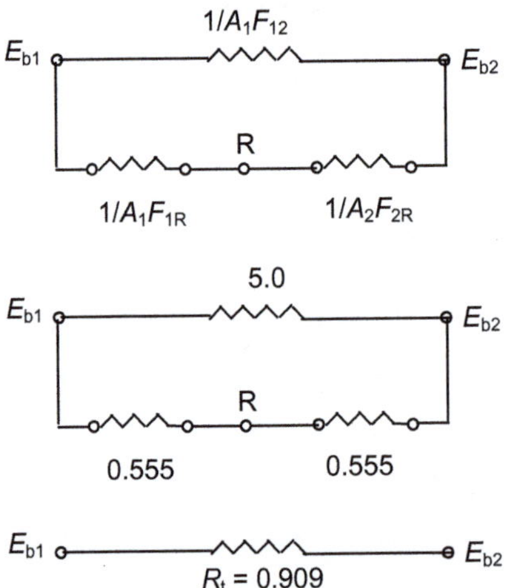

Fig. 11.56 Network for the system of Example 11.39

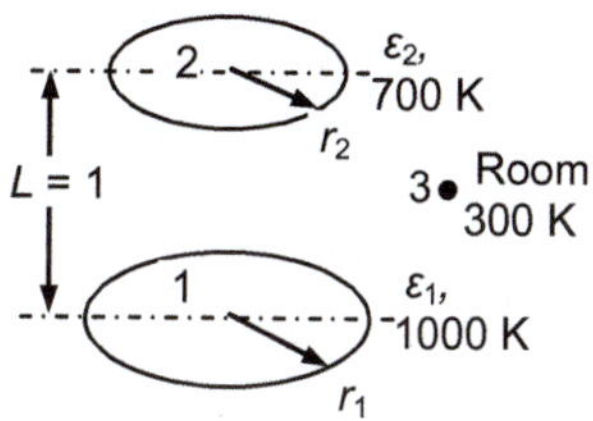

Fig. 11.57 Example 11.40

$$\frac{1}{A_2 F_{2R}} = \frac{1}{A_2(1 - F_{21})} = \frac{1}{2(1 - 0.1)} = 0.555 \quad \text{as } F_{21} + F_{2R} = 1.$$

Total resistance, refer Fig. 11.56,

$$R_t = \frac{1}{\frac{1}{5} + \frac{1}{2 \times 0.555}} = 0.909.$$

This gives the heat exchange

$$\begin{aligned} q_{12} &= \sigma \frac{(T_1^4 - T_2^4)}{R_t} \\ &= 5.67 \times 10^{-8} \times \frac{(475^4 - 350^4)}{0.909} = 2239 \text{ W} \end{aligned}$$

Here, $A_1 = A_2$, hence

$$T_R^4 = \frac{T_1^4 + T_2^4}{2},$$

which gives $T_R = 426.1$ K.

Example 11.40 Two circular discs of 0.8 m and 1.0 m radius, respectively, are placed parallel to each other (Fig. 11.57). The first disc is maintained at 700 K and the other at 1000 K. The emissivities of the surfaces of the discs are 0.2 and 0.5, respectively. The discs are located in a very large room. The walls of the room are maintained at 300 K. The discs are exchanging heat with each other and with the room but only the disc surfaces facing each other are to be considered in the present analysis as the other sides are insulated. Find the net heat transfer to each disc and the room by radiation.

Solution

The radiation network for this three-body problem (two discs and the room) is shown in Fig. 11.58.

For the given bodies, $L/r_1 = 1/1 = 1$ and $r_2/L = 0.8/1 = 0.8$. From Fig. 11.4, the shape factor $F_{12} = 0.28$.

The other shape factors are

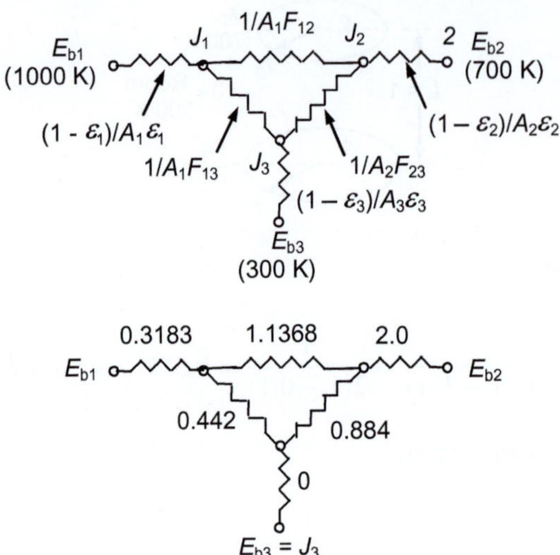

Fig. 11.58 Radiation network

$$F_{13} = 1 - F_{12} = 1 - 0.28 = 0.72,$$

$$F_{21} = \frac{A_1}{A_2} F_{12} = \frac{\pi \times 1^2}{\pi \times 0.8^2} \times 0.28 = 0.4375,$$

$$F_{23} = 1 - F_{21} = 1 - 0.4375 = 0.5625.$$

The surface resistances of the network are

$$\frac{1 - \varepsilon_1}{A_1 \varepsilon_1} = \frac{1 - 0.5}{\pi \times 1^2 \times 0.5} = 0.3183$$

$$\frac{1 - \varepsilon_2}{A_2 \varepsilon_2} = \frac{1 - 0.2}{\pi \times 0.8^2 \times 0.2} = 2.0.$$

and

$$\frac{1 - \varepsilon_3}{A_3 \varepsilon_3} \approx 0$$

because the room area A_3 is very large.

The space resistances are

$$\frac{1}{A_1 F_{12}} = 1.1368, \quad \frac{1}{A_1 F_{13}} = 0.442, \quad \frac{1}{A_2 F_{23}} = 0.8842.$$

Introducing above values of resistances, the network obtained is shown in Fig. 11.58. The network is solved by setting the sum of currents at nodes J_1 and J_2 to zero.

Node J_1: $$\frac{E_{b1} - J_1}{0.3183} + \frac{(J_2 - J_1)}{1.1368} + \frac{(E_{b3} - J_1)}{0.442} = 0$$

Node J_2: $$\frac{E_{b2} - J_2}{2.0} + \frac{(J_1 - J_2)}{1.1368} + \frac{(E_{b3} - J_2)}{0.8842} = 0$$

where

$E_{b1} = \sigma T_1^4 = 56700 \text{ W/m}^2$,
$E_{b2} = \sigma T_2^4 = 13614 \text{ W/m}^2$,
$E_{b3} = \sigma T_3^4 = 459.3 \text{ W/m}^2$.

Solution of the above simultaneous equations gives

$$J_1 = 30415 \quad \text{and} \quad J_2 = 13575.$$

The net heat lost by the disc 1,

$$q_1 = \frac{E_{b1} - J_1}{0.3183} = 82579 \text{ W},$$

and by disc 2,

$$q_2 = \frac{E_{b2} - J_2}{2.0} = 19.5 \text{ W}.$$

Heat received by the room,

$$q_3 = \frac{J_1 - E_{b3}}{0.442} + \frac{J_2 - E_{b3}}{0.8842} = 67773 + 14828 = 82601 \text{ W}.$$

The heat received by the room equals the heat lost by the discs.

Example 11.41 Two concentric parallel discs of 500 mm and 300 mm diameter are 0.5 m apart. A reradiating surface in the form of a right frustum of a cone encloses them as shown

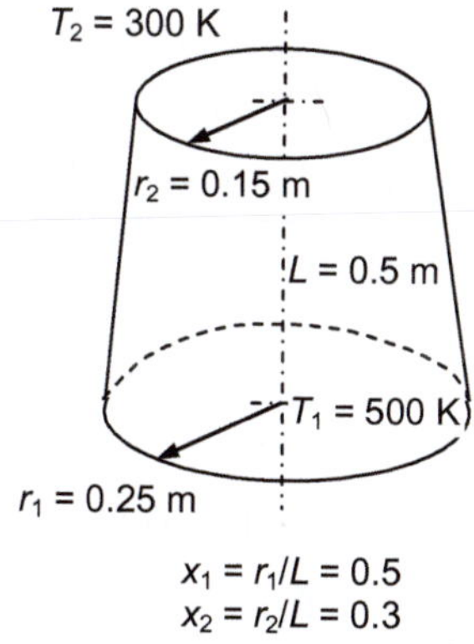

Fig. 11.59 Example 11.41

in Fig. 11.59. The larger disc is at 500 K and the smaller disc at 300 K. Determine the rate of heat transfer by radiation between the discs. The emissivity of both the discs is 0.8.

Solution

The shape factor from Table 11.1 is

$$F_{12} = \frac{1}{2}\left\{X - \left[X^2 - 4\left(\frac{x_2}{x_1}\right)^2\right]^{1/2}\right\}$$

where $x_1 = r_1/L = 0.5$, $x_2 = r_2/L = 0.3$ and $X = 1 + \left(\frac{1+x_2^2}{x_1^2}\right) = 1 + \left(\frac{1+0.3^2}{0.5^2}\right) = 5.36$.

Hence,

$$F_{12} = \frac{1}{2}\left\{5.36 - \left[5.36^2 - 4\left(\frac{0.3}{0.5}\right)^2\right]^{1/2}\right\} = 0.068.$$

Alternatively, from Fig. 11.4, $F_{12} \approx 0.07$ for $L/r_1 = 2$ and $r_2/L = 0.3$.
From Eq. (11.31),

$$q_{12} = \frac{A_1\sigma(T_1^4 - T_2^4)}{\left(\frac{1}{\varepsilon_1} - 1\right) + \frac{1}{\bar{F}_{12}} + \frac{A_1}{A_2}\left(\frac{1}{\varepsilon_2} - 1\right)}$$

where $\bar{F}_{12} = \frac{A_2/A_1 - F_{12}^2}{1 - 2F_{12} + A_2/A_1}$.
From the given data,

$$\varepsilon_1 = \varepsilon_2 = 0.8, A_1 = (\pi/4)(0.5)^2 = 0.19635\,\text{m}^2, A_2 = (\pi/4)(0.3)^2 = 0.07069\,\text{m}^2$$

Substitution gives

$$\bar{F}_{12} = 0.29034.$$

and

$$q_{12} = \frac{0.19635 \times 5.67 \times 10^{-8}(500^4 - 300^4)}{\left(\frac{1}{0.8} - 1\right) + \frac{1}{0.29034} + \frac{0.19635}{0.07069} \times \left(\frac{1}{0.8} - 1\right)} = 138\ \text{W}.$$

Alternatively, the network (Fig. 11.60) can be used to solve the problem.
We have

$$F_{1R} = 1 - F_{12} = 1 - 0.068 = 0.932$$

$$F_{21} = \frac{A_1}{A_2}F_{12} = \frac{0.5^2}{0.3^2} \times 0.068 = 0.1888,$$

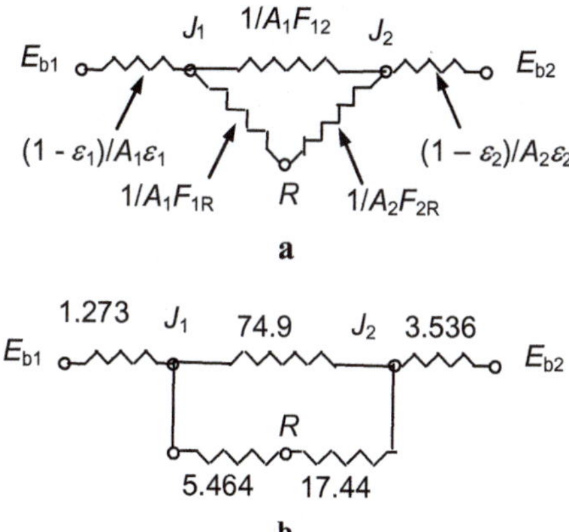

Fig. 11.60 Radiation network for the system of Fig. 11.59

$$F_{2R} = 1 - F_{21} = 0.8111$$

The various resistances of the network are

$$\frac{1-\varepsilon_1}{A_1\varepsilon_1} = \frac{0.2}{0.19635 \times 0.8} = 1.273, \quad \frac{1-\varepsilon_2}{A_2\varepsilon_2} = \frac{0.2}{0.07069 \times 0.8} = 3.536,$$

$$\frac{1}{A_1F_{12}} = \frac{1}{0.19635 \times 0.068} = 74.9, \quad \frac{1}{A_1F_{1R}} = \frac{1}{0.19635 \times 0.932} = 5.464,$$

$$\frac{1}{A_2F_{2R}} = \frac{1}{0.07069 \times 0.8111} = 17.44.$$

The network with the values of these resistances is shown in Fig. 11.60b. The total resistance comes out to be

$$R_t = 22.35.$$

$$q_{12} = \frac{\sigma(T_1^4 - T_2^4)}{R_t}$$
$$= \frac{5.67 \times 10^{-8} \times (500^4 - 300^4)}{22.35} = 138\,\text{W}.$$

Example 11.42 Two parallel square plates 1 m × 1 m are spaced 0.5 m apart. One of the plates is at 1000 K and the other is at 500 K. The emissivities of the plates are 0.4 and 0.5, respectively. The plates are located in a large space, which can be assumed to be at an effective temperature of 300 K. If the surfaces of plates not facing each other are insulated, determine the net heat transfer to each plate and to the space.

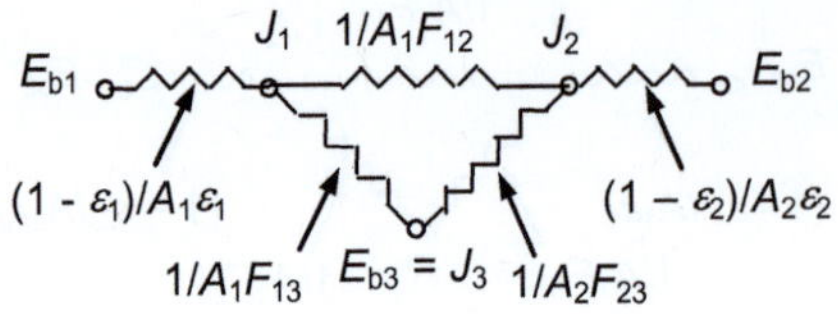

Fig. 11.61 Example 11.42

Solution

The electric network is shown in Fig. 11.61.

Since area of space A_3 is very large, the resistance $\frac{1-\varepsilon_3}{A_3\varepsilon_3}$ may be taken as zero and $E_{b3} = J_3$.

From Fig. 11.4,

$F_{12} = 0.43 = F_{21}$ for $X/L = Y/L = 2$,
$F_{13} = 1 - F_{12} = 0.57 = F_{23}$.

Various resistances in the network are

$$\frac{1-\varepsilon_1}{A_1\varepsilon_1} = \frac{1-0.4}{1\times 0.4} = 1.5, \quad \frac{1-\varepsilon_2}{A_2\varepsilon_2} = \frac{1-0.5}{1\times 0.5} = 1.0,$$

$$\frac{1}{A_1F_{12}} = \frac{1}{1\times 0.43} = 2.325, \quad \frac{1}{A_1F_{13}} = \frac{1}{A_2F_{23}} = \frac{1}{1\times 0.57} = 1.7544.$$

Radiosities J_1 and J_2 can be determined by setting the sum of currents at nodes J_1 and J_2 to zero.

Node J_1: $$\frac{E_{b1}-J_1}{1.5} + \frac{(J_2-J_1)}{2.325} + \frac{(E_{b3}-J_1)}{1.7544} = 0$$

Node J_2: $$\frac{E_{b2}-J_2}{1.0} + \frac{(J_1-J_2)}{2.325} + \frac{(E_{b3}-J_2)}{1.7544} = 0$$

where $E_{b1} = \sigma T_1^4 = 56700$ W/m^2, $E_{b2} = \sigma T_2^4 = 3543.75$ W/m^2, $E_{b3} = \sigma T_3^4 = 459.27$ W/m^2.

Substitution gives

$$-1.666J_1 + 0.43J_2 + 38061.78 = 0$$

$$0.43J_1 - 2.0J_2 + 3805.53 = 0$$

Solution of the above simultaneous equations gives $J_1 = 24699$ and $J_2 = 7213$.

The net heat lost by the plate 1,

$$q_1 = \frac{E_{b1}-J_1}{\frac{1-\varepsilon_1}{A_1\varepsilon_1}} = \frac{56700-24699}{1.5} = 21334 \text{ W},$$

and by plate 2,

$$q_2 = \frac{E_{b2} - J_2}{\frac{1-\varepsilon_2}{A_2\varepsilon_2}} = \frac{3543.75 - 7213}{1.0} = -3669.25\ \text{W}.$$

Heat received by the room,

$$q_3 = \frac{J_1 - E_{b3}}{\frac{1}{A_1F_{13}}} + \frac{J_2 - E_{b3}}{\frac{1}{A_2F_{23}}}$$
$$= \frac{24699 - 459.27}{1.7544} + \frac{7213 - 459.27}{1.7544} = 17666\ \text{W},$$

which equals $q_1 + q_2$.

The heat received by plate 2 from plate 1 is

$$q_{12} = \frac{J_1 - J_2}{\frac{1}{A_1F_{12}}} = \frac{24699 - 7213}{2.325} = 7520.86\ \text{W}.$$

The heat rejected by plate 2 to the surrounding space (room) is

$$q_{23} = \frac{J_2 - E_{b3}}{\frac{1}{A_2F_{23}}} = \frac{7213 - 459.27}{1.7544} = 3849.6\ \text{W}.$$

Example 11.43 In a muffle furnace, the floor (4.5 m by 4.5) m is constructed of refractory material (emissivity = 0.7). Two rows of oxidized tubes are placed 3 m above and parallel to the floor. But for the purpose of the analysis, these tubes can be replaced by a 4.5 m by 4.5 m plane surface having an effective emissivity of 0.9 (refer Fig. 11.62a). The average temperatures for the floor and tubes are 900°C and 270°C, respectively. Taking the geometric factor for radiation from floor to the tubes as 0.32, calculate:

(i) The net heat transfer to the tubes,
(ii) The mean temperature of the refractory walls of the furnace, assuming that these walls are insulated.

Solution

The shape factors are

$F_{12} = 0.32 = F_{21}$ (given)
$F_{13} = 1 - F_{12} = 0.68$
$F_{23} = 1 - F_{21} = 0.68$.

The surface resistances are

$$\frac{1 - \varepsilon_1}{A_1\varepsilon_1} = \frac{1 - 0.7}{(4.5 \times 4.5) \times 0.7} = 0.021$$

$$\frac{1 - \varepsilon_2}{A_2\varepsilon_2} = \frac{1 - 0.9}{(4.5 \times 4.5) \times 0.9} = 5.48 \times 10^{-3}.$$

The space resistances are

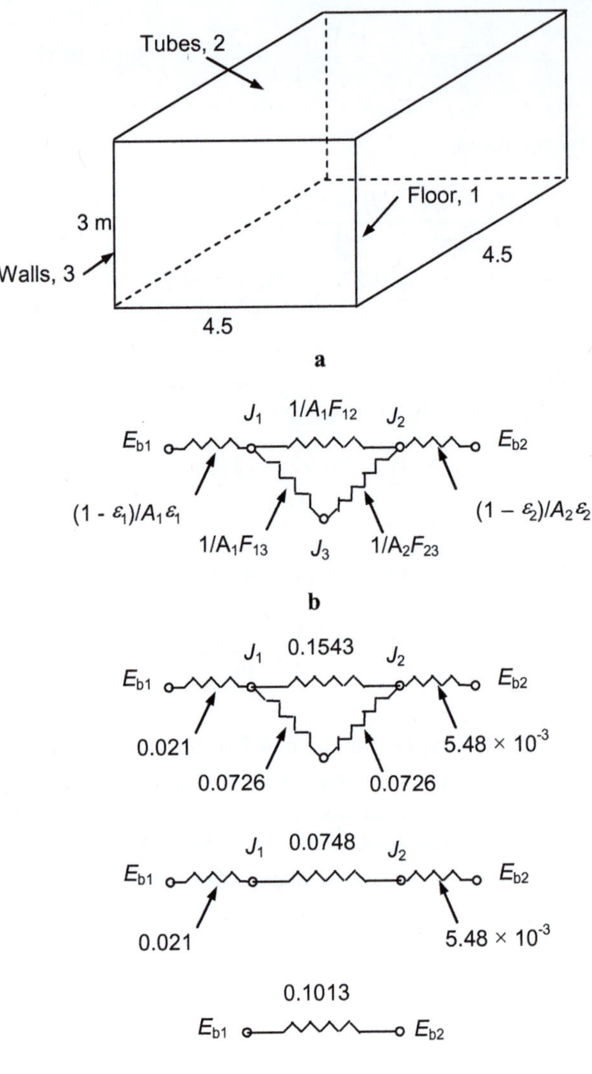

Fig. 11.62 Example 11.43

$$\frac{1}{A_1F_{12}} = \frac{1}{(4.5 \times 4.5) \times 0.32} = 0.1543$$

$$\frac{1}{A_1F_{13}} = \frac{1}{A_2F_{23}} = \frac{1}{(4.5 \times 4.5) \times 0.68} = 0.0726.$$

The network is shown in Fig. 11.62b, c. It consists of series and parallel combination of various resistances. Remember that J_3 floats in the network. The total resistance is

$$R_t = 0.1013,$$

$$q_{12} = \frac{\sigma(T_1^4 - T_2^4)}{R_t}$$

$$= \frac{5.67 \times 10^{-8} \times (1173^4 - 543^4)}{0.1013} = 1011\,\text{kW}.$$

Refractory wall temperature

The heat transfer q_{12} is also given by

$$q_{12} = \frac{E_{b1} - J_1}{(1 - \varepsilon_1)/A_1\varepsilon_1}$$

or

$$1011 \times 1000 = \frac{5.67 \times 10^{-8} \times 1173^4 - J_1}{0.021}.$$

or

$$J_1 = 86112$$

Similarly, refer Fig. 11.62,

$$q_{12} = \frac{J_1 - J_2}{0.0748}$$

or

$$1011 \times 1000 = \frac{86112 - J_2}{0.0748}$$

or

$$J_2 = 10489.$$

The value of J_3 is determined from

$$\frac{J_1 - J_3}{0.0726} = \frac{J_3 - J_2}{0.0726}$$

or

$$J_3 = \frac{J_1 + J_2}{2} = 48300,$$

and

$$E_{b3} = J_3 = \sigma T_3^4.$$

This gives

$$T_3 = \sqrt[4]{\frac{48300}{5.67 \times 10^{-8}}} = 961\ \text{K}.$$

Example 11.44 The inner sphere of a flask is of 300 mm diameter and outer sphere is of 360 mm diameter. Both the spheres are plated for which the emissivity is 0.05. The space between them is evacuated. Determine the rate at which the liquid oxygen will evaporate at -183°C when the outer sphere temperature is 20°C. The latent heat of evaporation of the liquid oxygen is 214.2 kJ/kg.

Solution

From the given data,

$A_1 = 4\pi R_1^2 = 4\pi \times 150^2 \times 10^{-6} = 0.2827\ \text{m}^2$ (where subscript 1 refers to inner sphere).
$A_2 = 4\pi R_2^2 = 4\pi \times 180^2 \times 10^{-6} = 0.4072\ \text{m}^2$
$\varepsilon_1 = \varepsilon_2 = 0.05$
$F_{12} = 1$
$T_1 = -183 + 273 = 90\ \text{K}$
$T_2 = 20 + 273 = 293\ \text{K}.$

Total resistance,

$$R_t = \frac{1-\varepsilon_1}{A_1\varepsilon_1} + \frac{1}{A_1F_{12}} + \frac{1-\varepsilon_2}{A_2\varepsilon_2}$$
$$= \frac{1-0.05}{0.2827 \times 0.05} + \frac{1}{0.2827} + \frac{1-0.05}{0.4072 \times 0.05} = 117.41.$$

Hence,

$$q_{12} = \frac{E_{b1} - E_{b2}}{R_t} = \frac{\sigma(T_1^4 - T_2^4)}{R_t}$$
$$= \frac{5.67 \times 10^{-8} \times (90^4 - 293^4)}{117.41} = 3.528\ \text{W}.$$

Rate of evaporation of liquid oxygen,

$$\dot{m} = \frac{q_{12}}{\text{Latent heat}} = \frac{3.528}{2140.2 \times 1000} = 1.647 \times 10^{-5}\ \text{kg/s} = 59.3\ \text{g/hr}.$$

Example 11.45 The Sun may be regarded as a blackbody with a surface temperature of 6000 K. The diameter of the Sun is 1.39×10^9 m and the distance between the Earth and the Sun is 1.48×10^{11} m. Calculate

(i) The solar constant
(ii) The rate of energy received by the Earth. The mean diameter of Earth is 12.8×10^6 m.

(iii) The rate of energy reaching the Earth if the transmittance of the Earth's atmosphere is 0.85.

(iv) The rate of energy received by a 1.5×1.5 m^2 solar collector whose perpendicular is inclined at 35° to the direction of the Sun. Assume diffuse radiation to be 15% of the direct radiation.

Solution

The solar constant is the amount of solar energy received by unit area of the Earth's surface placed normal to the rays of the Sun. Hence, from Example 11.9,

$$\begin{aligned}\text{Solar constant} &= \frac{\sigma T_s^4(\pi/4)D_s^2}{\pi r^2}\\ &= \frac{5.67\times10^{-8}\times6000^4\times(\pi/4)\times(1.39\times10^9)^2}{\pi(1.48\times10^{11})^2}\\ &= 1620\,\mathrm{W/m^2}.\end{aligned}$$

Energy received by the Earth is

$$\begin{aligned}&= \text{Solar constant}\times(\pi/4)D_e^2\\ &= 1620\times(\pi/4)\times(12.8\times10^6)^2 = 2.085\times10^{17}\,\mathrm{W}.\end{aligned}$$

Rate of energy reaching the Earth's surface,

$$= \tau\times1620 = 0.85\times1620 = 1377\ \mathrm{W/m^2}.$$

Area of the collector $= 1.5\times1.5\times\cos35^o = 1.843\mathrm{m}^2$ resolved perpendicular to the Sun rays.

Direct (beam) radiation falling on the collector $= 1.843\times1377 = 2539\,\mathrm{W}$.

Diffuse radiation falling on the collector $= 0.15\times2539 = 380.85\,\mathrm{W}$.

Total radiation falling on the collector $= 2539 + 380.85 = 2919.85\,\mathrm{W}$.

11.11 Radiation from a Gray Cavity (Alternative Method)

Refer to the electric network in Fig. 11.63.

Total resistance is

$$R_t = \frac{1-\varepsilon_1}{A_1\varepsilon_1} + \frac{1}{A_1F_{12}}.$$

Hence,

$$q_{12} = \frac{A_1\sigma(T_1^4 - T_2^4)}{\frac{1}{F_{12}} + \frac{1-\varepsilon_1}{\varepsilon_1}},$$

which is the same as Eq. (11.24) found in Sect. 11.6.

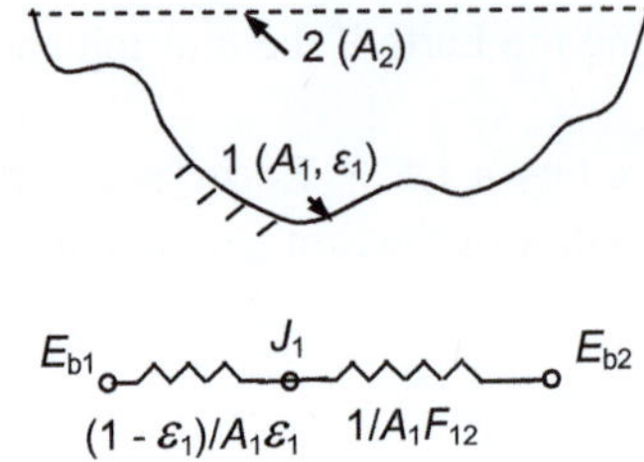

Fig. 11.63 A gray cavity

11.12 Newton's Law of Cooling and Overall Heat Transfer Coefficient

In year 1701 A.D., Newton proposed the law of cooling of bodies. The law states, 'The rate of loss of heat of a body by cooling is proportional to the excess of temperature of the body above its surrounding'. This law is valid when the temperature difference is small.

Mathematically, the law can be written as

Rate of cooling,

$$\begin{aligned} q &\propto (T - T_{\infty}) \\ &= h_o A (T - T_{\infty}) \end{aligned} \tag{i}$$

where h_o is known as heat transfer coefficient and A is the surface area of the body rejecting heat.

The heat transfer coefficient h_o accounts for both convection and radiation. Hence, we write

$$h_o = h_c + h_r$$

where h_c is convection heat transfer coefficient,

h_r is radiation heat transfer coefficient.

If temperature of a body falls by ΔT in time interval $\Delta\theta$, then from the energy consideration (rate of cooling equals the change in internal energy),

$$q = mc\left(\frac{\Delta T}{\Delta\theta}\right)$$

Substitution in Eq. (i) gives

$$mc\left(\frac{\Delta T}{\Delta\theta}\right) = -h_o A (T - T_{\infty}).$$

or

$$\frac{dT}{T - T_{\infty}} = -\left(\frac{h_o A}{mc}\right) d\theta$$

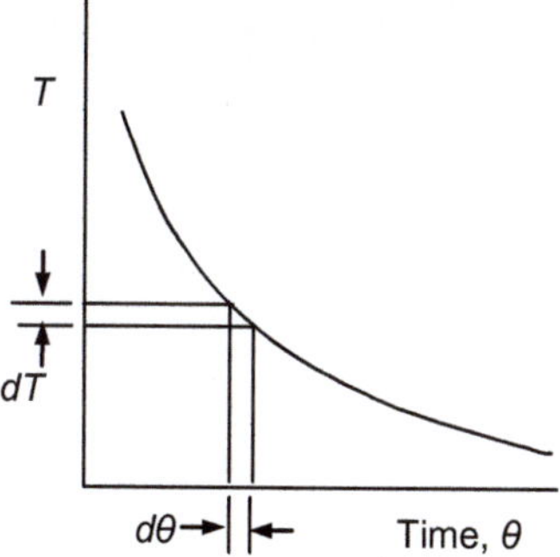

Fig. 11.64 Cooling of a body

Integration of the equation gives

$$\ln\left(\frac{T - T_\infty}{T_o - T_\infty}\right) = -\left(\frac{h_o A}{mc}\right)\tau$$

where T_o is the temperature at time $\theta = 0$, and T is the temperature at time $\theta = \tau$. The above equation can be rewritten as

$$\frac{T - T_\infty}{T_o - T_\infty} = e^{-\left(\frac{h_o A}{mc}\right)\tau} = e^{-K\tau} \tag{11.35}$$

It can be seen from the above equation that the temperature of the body falls exponentially as shown in Fig. 11.64.

The value of the radiation heat transfer coefficient can be obtained from the Stefan–Boltzmann law.

Radiation heat transfer from a blackbody at temperature T is

$$\begin{aligned} q_r &= A\sigma(T^4 - T_\infty^4) \\ &= A\sigma(T - T_\infty)(T^3 + T^2 T_\infty + T T_\infty^2 + T_\infty^3) \end{aligned}$$

Let us denote $(T + T_\infty)/2 = T_m$. Then, if $0.9 \le (T/T_\infty) \le 1.1$, it may be assumed that

$$(T^3 + T^2 T_\infty + T T_\infty^2 + T_\infty^3) = 4T_m^3$$

and hence

$$q_r = \left(4\sigma T_m^3\right)A(T - T_\infty). \tag{ii}$$

Above assumption introduces less than 1% error (Mikheyev 1968).

Hence, the radiation heat transfer coefficient,

$$h_r = 4\sigma T_m^3 \tag{11.36}$$

If a wall is exposed to liquid, then $h_r = 0$ and $h_o = h_c$.

11.12.1 Determination of Specific Heat Using Newton's Law of Cooling

For a blackbody we have seen that the overall heat transfer coefficient h_o is

$$h_o = h_c + h_r$$

or

$$h_o = h_c + 4\sigma T_m^3$$

We consider simultaneous cooling of two blackbodies at temperature T_m, of the same shape and size, then the value of the overall heat transfer coefficient for the two bodies will be equal and the rate of cooling will also be equal, i.e. $q_1 = q_2$ when $T_{m1} = T_{m2}$.

Rate of cooling equals the change in internal energy hence

$$m_1 c_1 \left(\frac{T_1 - T_2}{\theta_1}\right) = m_2 c_2 \left(\frac{T_1 - T_2}{\theta_2}\right)$$

where θ_1 and θ_2 are the time intervals for cooling from temperature T_1 to T_2, of bodies 1 and 2, respectively, (Fig. 11.65).

From above equation,

$$c_2 = m_1 c_1 \left(\frac{\theta_2}{\theta_1}\right) \times \left(\frac{1}{m_2}\right) \tag{11.37a}$$

Specific heat of body 2 can be determined from the above equation if the specific heat of body 1 is known.

For the determination of specific heat of liquids, Eq. (11.37a) is modified to take account of the effect of calorimeter. The equation becomes

$$(m_1 c_1 + m_c c_c)\left(\frac{T_1 - T_2}{\theta_1}\right) = (m_2 c_2 + m_c c_c)\left(\frac{T_1 - T_2}{\theta_2}\right)$$

or

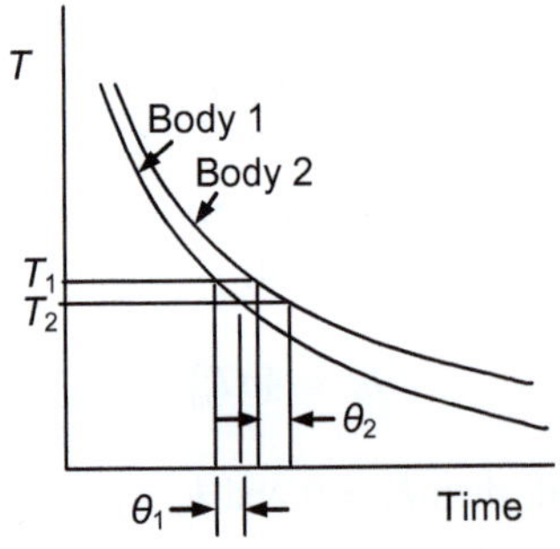

Fig. 11.65 Simultaneous cooling of two blackbodies

$$(m_2c_2 + m_c c_c) = (m_1 c_1 + m_c c_c)\left(\frac{\theta_2}{\theta_1}\right)$$

or

$$c_2 = \left[(m_1c_1 + m_c c_c)\left(\frac{\theta_2}{\theta_1}\right) - m_c c_c\right]\frac{1}{m_2}, \tag{11.37b}$$

where m_c and c_c refer to the calorimeter.

11.13 Radiation Heat Transfer Coefficient

The convection heat transfer from a body at temperature T_1 to the surrounding fluid at temperature T_∞ is

$$q_c = h_c A_1 (T_1 - T_\infty). \tag{i}$$

Radiation heat transfer exchange by a body at temperature T_1 with another body at temperature T_2 is

$$q_r = A_1 f_{12} \sigma (T_1^4 - T_2^4) \tag{ii}$$

where f_{12} is interchange factor as discussed earlier.

Equation (ii) can be written as

$$q_r = A_1\left[f_{12}\sigma(T_1^2 + T_2^2)(T_1 + T_2)\right](T_1 - T_2).$$

We can put the convection and radiation processes of heat transfer on a common basis by introducing radiation heat transfer coefficient h_r defined by

$$h_r = f_{12}\sigma(T_1^2 + T_2^2)(T_1 + T_2) \tag{11.38}$$

If the second radiation exchanging body is an enclosure which is at the fluid temperature T_∞, i.e. $T_2 = T_\infty$, we have

$$\begin{aligned} q &= q_r + q_c \\ &= (h_c + h_r)A_1(T_1 - T_\infty). \end{aligned}$$

where $h_r = \frac{\sigma(T_1^2 + T_2^2)(T_1 + T_2)}{\frac{1}{\varepsilon_1} + \frac{A_1}{A_2}\left(\frac{1}{\varepsilon_2} - 1\right)}$ for body 1 of emissivity2 of emissivity ε_2 and area A_2 ε_1 and area A_1 completely enclosed by body a large enclosure.

For a small body in a large enclosure,

$$h_r = \varepsilon_1\sigma(T_1^2 + T_2^2)(T_1 + T_2)$$

and if $0.9 \leq (T_1/T_2) \leq 1.1$, we assume that T_1 and T_2 can be replaced by $T_m = (T_1 + T_2)/2$ and then

$$h_r = 4\varepsilon_1 \sigma T_m^3$$

as explained earlier.

It should be noted that unlike convective heat transfer coefficient, the radiation heat transfer coefficient is a strong function of temperatures T_1 and T_2.

Example 11.46 A horizontal steel pipe, 3 m long and 50 mm in diameter, is located in a large room. The pipe wall is at an average temperature of 110°C. The surrounding air and walls of the room are at a temperature of 30°C. The surface emissivity of the steel pipe may be taken as 0.85. Calculate the total heat lost by the pipe. The convection heat transfer coefficient may be taken as 10 W/(m^2 K).

Solution

It is a problem of multimode heat transfer. The heat is lost by radiation as well as convection.

The convection heat loss,

$$q_c = hA(T_s - T_\infty) = 10 \times (\pi \times 0.05 \times 3) \times (110 - 30) = 377 \text{ W}.$$

The pipe (emissivity = 0.85) is a body in a large enclosure (room). Hence, the radiation heat loss is

$$q_r = \varepsilon A \sigma (T_s^4 - T_\infty^4)$$

or

$$q_r = 0.85 \times (\pi \times 0.05 \times 3) \times 5.67 \times 10^{-8} \times (383^4 - 303^4) = 297 \text{ W}.$$

The total heat loss,

$$q = q_r + q_c = 674 \text{ W}.$$

It is to be noted that in the present case, the two modes of heat transfer are equally significant. Both modes must be considered when dealing with such problems. However, with the rise in the temperature difference, the contribution of the radiative mode in the total heat transfer will increase.

Example 11.47 An electric heater 25 mm diameter and 300 mm long is used to heat a room. Calculate the electrical input to the heater when the bulk of the air in the room is at 20° C, the walls are at 15°C and the surface of the heater is at 540°C. For the convective heat transfer from the heater assume that

$$\text{Nu} = 0.4(\text{Gr})^{0.25}.$$

where all properties are at the mean film temperature and $\beta = 1/T$, and T (in K) is the bulk temperature of the air. Take the emissivity of the heater surface as 0.55 and assume that the surroundings are black.

Solution

The film temperature is

$$t_f = \frac{540 + 20}{2} = 280°\text{C}.$$

Properties of air at the film temperature are
$\rho = 0.64$ kg/m^3, $k = 0.0436$ W/(m K), $\mu = 2.86 \times 10^{-5}$ kg/(m s) and
$\beta = 1/(20 + 273) = 3.41 \times 10^{-3}$ K^{-1} (as given).
Then, Grashof number,

$$Gr = \frac{\rho^2 g\beta(\Delta T)d^3}{\mu^2} = \frac{0.64^2 \times 9.81 \times 3.41 \times 10^{-3} \times (540 - 20) \times 0.025^3}{(2.86 \times 10^{-5})^2} = 1.36 \times 10^5.$$

Hence, from the given relation,

$$\text{Nu} = 0.4(1.36 \times 10^5)^{0.25} = 7.68$$

and the heat transfer coefficient is

$$h = \frac{k}{d} \times Nu = \frac{0.0436}{0.025} \times 7.68 = 13.4\,\text{W/m}^2\text{K}.$$

Heat transfer by convection,

$$q_c = hA(\Delta T) = 13.4 \times \pi \times 0.025 \times 0.3 \times (540 - 20) = 164.2\,\text{W}.$$

Heat lost by radiation,

$$q_r = \varepsilon A\sigma(T_s^4 - T_\infty^4)$$

Substitution gives

$$q_r = 0.55 \times (\pi \times 0.025 \times 0.3) \times 5.67 \times 10^{-8} \times (813^4 - 288^4) = 316\ \text{W}.$$

Electrical input to the heater, $q = q_r + q_c = 480.2$ W.

Example 11.48 Air flows between two concentric cylindrical gray surfaces. The geometrical and thermodynamic parameters are given in Table 11.5.

At a given point, the temperature of the air is 700 K. Compare the rate of heat transfer by radiation to the lower temperature surface with the heat transfer by convection to this point. The convective heat transfer coefficient is 30 W/(m^2 K).

Table 11.5 Example 11.48

	Inner	Outer
Diameters	40 mm	100 mm
Absorptivity	0.75	0.8
Temperature	400 K	800 K

Solution

Heat transfer by radiation,

$$q_{12} = \frac{A_1 \sigma (T_1^4 - T_2^4)}{\frac{1}{\varepsilon_1} + \frac{A_1}{A_2}\left(\frac{1}{\varepsilon_2} - 1\right)}$$

Substitution of values of various parameters gives

$$q_{12} = \frac{\pi \times 0.04 \times 1 \times 5.67 \times 10^{-8} \times (800^4 - 400^4)}{\frac{1}{0.75} + \frac{40}{100}\left(\frac{1}{0.8} - 1\right)} = 1908.8\,\text{W per unit length.}$$

Heat transfer by convection,

$$q_{conv} = hA_1(T_g - T_1) = 30 \times \pi \times 0.04 \times 1 \times (700 - 400) = 1130.97\,\text{W per unit length.}$$

(Figure 11.66)
The ratio of heat transfers is

$$\frac{q_{12}}{q_{conv}} = 1.688.$$

Example 11.49 A furnace wall is 200 mm thick built up of refractory bricks having $k = 0.9$ W/(m K). The internal surface temperature is 1000°C and the surrounding temperature is 25°C. Calculate the heat loss per m^2 of the wall area if

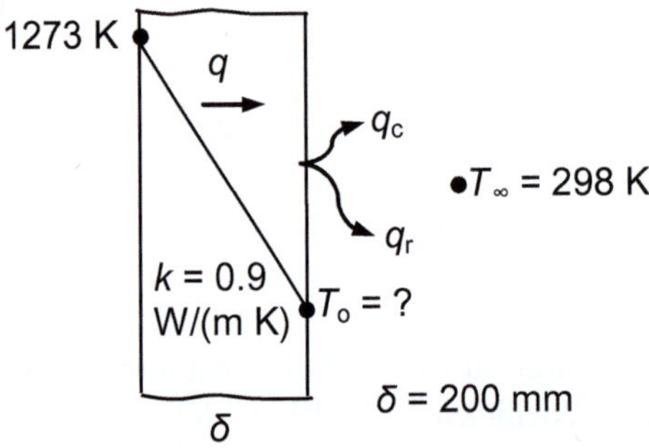

Fig. 11.66 Example 11.49

(i) the free convection heat transfer coefficient from the outer surface of the wall is 0.27 $(\Delta T)^{1/3}$ and
(ii) the radiation heat transfer coefficient h_r for the outer surface is

$$h_r = \varepsilon(T_2 + T_1)(T_2^2 + T_1^2)$$

with $\varepsilon = 0.8$.

Solution

Heat flow through the furnace wall by conduction,

$$q = kA\frac{(T_i - T_o)}{\delta} = 0.9 \times 1 \times \frac{(1273 - T_o)}{200 \times 10^{-3}} = 4.5 \times (1273 - T_o)$$

(the outer surface temperature T_o is not known)

Heat reaching the outer surface of the furnace wall is rejected by convection q_c and radiation q_r, i.e.

$$q = q_r + q_c, \tag{i}$$

where

$$\begin{aligned} q_c &= h_c A(\Delta T) \\ &= 0.27(T_o - 298)^{1/3} \times 1 \times (T_o - 298) \\ &= 0.27(T_o - 298)^{4/3}, \end{aligned}$$

and

$$\begin{aligned} q_r &= h_r A(T_o - T_\infty) \\ &= [\varepsilon\sigma(T_o + T_\infty)(T_o^2 + T_\infty^2)]A \times (T_o - T_\infty) \\ &= \varepsilon\sigma A(T_o^4 - T_\infty^4) \\ &= 0.8 \times 5.67 \times 10^{-8} \times 1 \times (T_o^4 - 298^4). \end{aligned}$$

Substitution in Eq. (i) gives

$$4.5 \times (1273 - T_o) = 0.27(T_o - 298)^{4/3} + 0.8 \times 5.67 \times 10^{-8} \times 1 \times (T_o^4 - 298^4).$$

By trial and error, $T_o = 522$ K.

Heat transfer per m^2 of the wall surface,

$$q = 4.5 \times (1273 - T_o) = 4.5 \times (1273 - 522) = 3379.5 \text{ W}.$$

Example 11.50 A hot water radiator of overall dimensions 2 m in height, 1.5 m long and 0.25 m wide is used to heat a room. The surface temperature of the radiator is 340 K and the room temperature is 290 K.

Calculate the heat transfer from the radiator if the convection heat transfer coefficient is $1.3(\Delta T)^{1/3}$. The actual surface area of the radiator is 3 times the area of its envelope. The surface of the radiator can be assumed to be nearly black.

Solution

(i) **Heat Transfer by Radiation**

As the surface of the radiator is black, the area of the radiation heat exchange is assumed to be the area of the envelope, i.e.

$$A_{\text{envelope}} = 2(LW + WH + HL) = 2(2 \times 0.25 + 0.25 \times 1.5 + 1.5 \times 2) = 7.75\,\text{m}^2.$$

where L, H, W are the length, height and width of the radiator, respectively.

Radiant heat loss,

$$\begin{aligned} q_r &= A_{envelope}\sigma(T^4_{raditor} - T^4_{room}) \\ &= 7.75 \times 5.67 \times 10^{-8} \times (340^4 - 290^4) = 2764.2\,\text{W}. \end{aligned}$$

(i) **Heat Flow by Convection**

$$\begin{aligned} q_c &= h_c A(\Delta T) \\ &= 1.3(\Delta T)^{1/3} A(\Delta T) \\ &= 1.3(\Delta T)^{4/3} A \\ &= 1.3 \times (340 - 290)^{4/3} \times (3 \times 7.75) = 5567.5\,\text{W}. \end{aligned}$$

Total heat flow from the radiator,

$$q = q_r + q_c = 8331.7\,\text{W}.$$

Example 11.51 Figure 11.67 shows cross-section of a solar collector with one glass cover. If τ is the transmissivity of the cover, α is the absorptance of the absorber plate and ρ_{d} is the reflectivity of the cover system for the diffuse radiation incident from the bottom side, evaluate the transmittance-absorptance product.

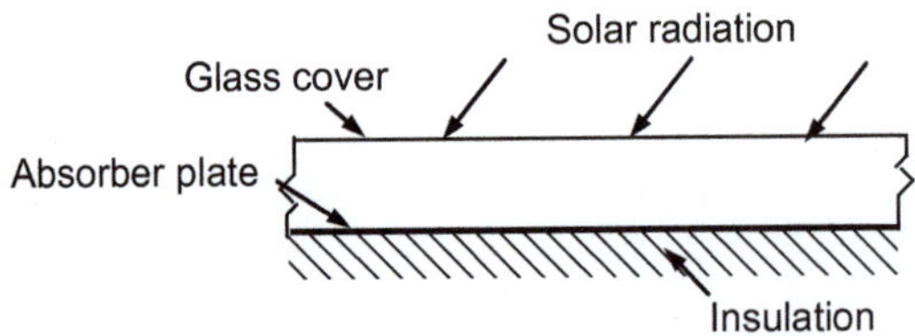

Fig. 11.67 Example 11.51

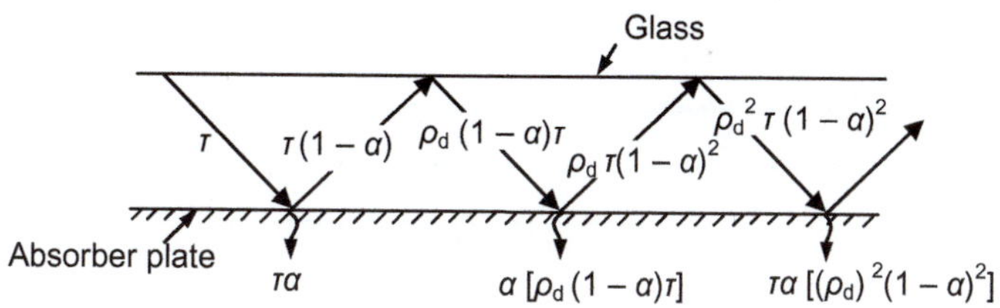

Fig. 11.68 Successive absorption and reflection of solar rays

Solution

Of the unit incident solar radiation, refer Fig. 11.68, a part τ passes through the cover and strikes the absorber plate. α times of this radiation, i.e. $\tau\alpha$ is absorbed by the plate on the first incidence and $(1 - \alpha)\tau$ is reflected back to the cover. The radiation reflected back from the cover is $\rho_d(1 - \alpha)\tau$ and again strikes the absorber plate and out of this reflected radiation $\alpha[\rho_d(1 - \alpha)\tau]$ is absorbed by the absorber plate. Similarly on the third incidence, $\tau\alpha[(\rho_{d-})^2(1 - \alpha)^2]$ will be absorbed by the plate. This reflection and absorption process continues.

The energy ultimately absorbed by the absorber plate is

$$\begin{aligned}(\tau\alpha)_e &= \tau\alpha + \tau\alpha[\rho_d(1-\alpha)] + \tau\alpha[(\rho_d)^2(1-\alpha)^2] + \ldots \\ &= \tau\alpha[1 + \rho_d(1-\alpha)] + (\rho_d)^2(1-\alpha)^2 + \ldots]\end{aligned} \tag{i}$$

where $(\tau\alpha)_e$ is the effective value of transmittance-absorptance product.

As the product $[\rho_d\,(1 - \alpha)]$ in Eq. (i) is less than 1, the summation of the terms on the right hand side gives

$$(\tau\alpha)_e = \tau\alpha\left[\frac{1}{1-(1-\alpha)\rho_d}\right].$$

Example 11.52 The following data were obtained for a single glass cover flat plate solar air heater.

Mean plate temperature	= 70°C
Ambient temperature = sky temperature[1]	= 10°C
Back insulation thickness	= 50 mm
Insulation thermal conductivity	= 0.05 W/(m K)
Coefficient of heat transfer by convection from plate to cover, h_c	= 5 W/(m^2 K)
Equivalent coefficient of heat transfer by radiation from plate to cover, h_{r1}	= 6 W/(m^2 K)
Coefficient of heat transfer by convection from cover to ambient air, h_w	= 20 W/(m^2 K)
Equivalent coefficient for radiant heat transfer from cover to sky, h_{r2}	= 5 W/(m^2 K)

Compute the total heat loss per m^2 of the collector area.

[1]The sky temperature is a function of many parameters and is generally significantly lower than the ambient temperature.

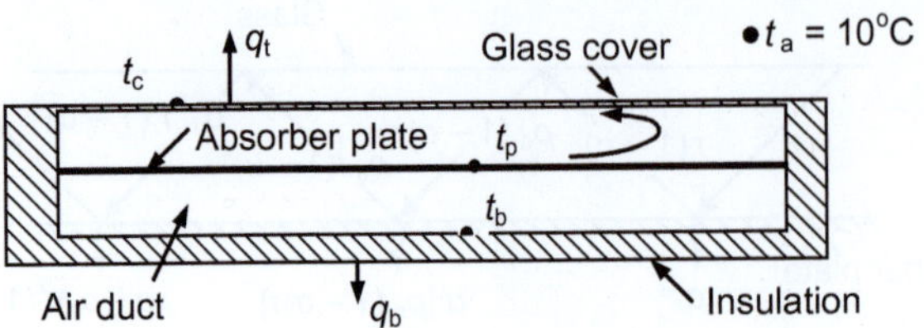

a Cross-section of a solar air heater

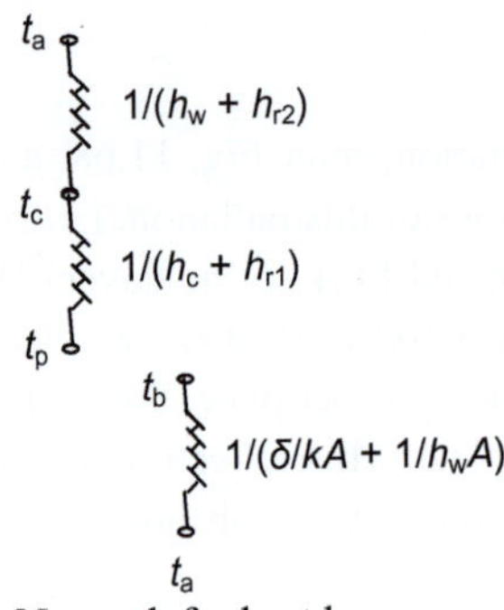

b Network for heat loss

Fig. 11.69 Example 11.52

Solution

Top Loss:
Considering convection and radiation heat flow from absorber plate to the cover,

$$q_t = (h_c + h_{r1})A_c(t_p - t_c) = (5 + 6) \times 1 \times (70 - t_c) \quad \text{(i)}$$

where t_c is the cover temperature which is not known and A_c is absorber plate area (= cover area).

Similarly, from cover to ambient, heat loss is given by

$$q_t = (h_w + h_{r2})A_c(t_c - t_a) = (20 + 5) \times 1 \times (t_c - 10) \quad \text{(ii)}$$

Equating Eqs. (i) and (ii), we get

$$t_c = 28.33^{\circ}\text{C}.$$

Figure 11.69(a) shows schematic of the solar air heater.

Hence, from Eq. (i),

$$q_t = (5 + 6) \times 1 \times (70 - 28.33) = 458.4 \text{ W}.$$

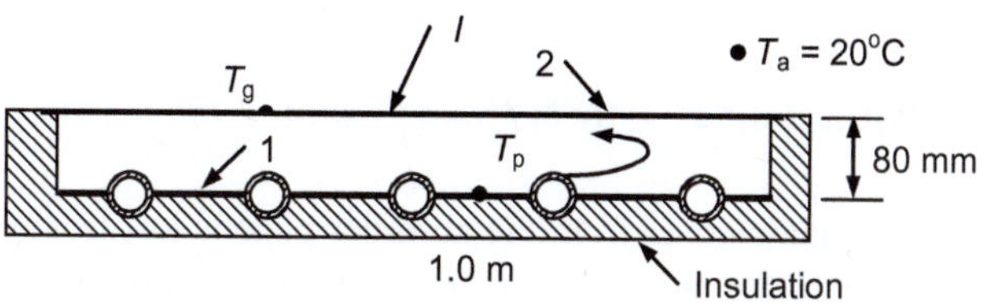

Fig. 11.70 A solar water heater

Back Loss (by conduction through the back insulation and by convection from the back):

$$q_b = \left(\frac{\Delta t}{\frac{\delta}{kA} + \frac{1}{h_w A}}\right) = \left(\frac{t_b - t_a}{\frac{\delta}{kA} + \frac{1}{h_w A}}\right) = \left(\frac{40 - 10}{\frac{0.05}{0.05 \times 1} + \frac{1}{20 \times 1}}\right) = 28.6 \text{ W},$$

assuming that the back temperature t_b equals the mean of plate temperature and ambient temperature.

$$\text{Total loss}, q_t + q_b = 458.4 + 28.6 = 487 \text{ W}.$$

Example 11.53 Figure 11.70 shows a solar water heater. The glass cover *2* is placed over a blackened absorber plate *1*. The side walls and back are adequately insulated. The glass is assumed to have emissivity of 0.88 for thermal radiation. The emissivity for the absorber plate is 0.9 for all radiations. The environment is at 20°C and the convection heat transfer coefficient from the outer surface of the glass is 25 W/(m^2 K). The solar radiation *I* incident on the solar collector is 980 W/m^2. The effective value of transmittance-absorptance product $(\tau\alpha)_e$ is 0.72. The convection heat transfer coefficient between the plate and glass is 7 W/(m^2 K). Estimate the glass and plate temperatures under stagnation condition.

Solution

The energy absorbed by the absorber plate is $(\tau\alpha)_e$ times the incident radiation *I*.

In stagnation condition when equilibrium is reached, the heat lost by convection and radiation to the glass cover equals the solar energy absorbed, i.e.

$$I(\tau\alpha)_e = \frac{\sigma(T_p^4 - T_g^4)}{\frac{1}{\varepsilon_p} + \frac{1}{\varepsilon_g} - 1} + h_c(T_p - T_g) \qquad \text{(i)}$$

where h_c is convection heat transfer coefficient.

Substitution of given values of I = 980, $(\tau\alpha)_e$= 0.72, ε_g = 0.88, ε_p = 0.9 and h_c = 7) transforms Eqs. (i) to

$$980 \times 0.72 = \frac{5.67 \times 10^{-8}(T_p^4 - T_g^4)}{\frac{1}{0.9} + \frac{1}{0.88} - 1} + 7.0 \times (T_p - T_g)$$

or

$$705.6 = 4.545 \times 10^{-8}(T_p^4 - T_g^4) + 7.0 \times (T_p - T_g) \tag{ii}$$

The heat reaching the glass cover from the absorber plate is lost from outside surface of the collector by radiation to the sky and by convection to the surrounding air. Thus

$$I(\tau\alpha)_e = \varepsilon_g \sigma (T_g^4 - T_{sky}^4) + h_w(T_g - T_a) \tag{iii}$$

T_{sky} is sky temperature and can be estimated from Swinbank's correlation,

$$T_{sky} = 0.0552 T_a^{1.5} = 276.85\,\text{K}$$

Substituting values of I, $(\tau\alpha)_e$, $\varepsilon_g = 0.88$, $h_w = 25$ W/(m^2 K), $T_a = 293$ K and $T_{sky} =$ 276.85 K in Eq. (iii), we get

$$705.6 = 0.88 \times 5.67 \times 10^{-8} \times (T_g^4 - 276.85^4) + 25 \times (T_g - 293)$$

or

$$705.6 = 4.99 \times 10^{-8} \times (T_g^4 - 276.85^4) + 25 \times (T_g - 293) \tag{iv}$$

By trial and error, Eq. (iv) gives

$$T_g = 313.6\,\text{K}.$$

Substitution of the value of T_g in Eq. (ii) gives the plate temperature:

$$T_p = 363.7\,\text{K}.$$

Note: The resistance of the glass to the conduction heat transfer has been neglected here.

Example 11.54 A 20 mm diameter horizontal pipe is laid in a large open space and carrying a hot fluid. It has a surface temperature of 250°C. Emissivity of pipe surface is 0.6. Air at a velocity of 5 m/s and 40°C is blowing across the pipe. Determine the heat loss from the pipe surface per unit length of the pipe. The effective temperature of the surface of the open space may be taken equal to the air temperature.

Solution

The heat is transferred from the pipe surface by both radiation and convection.

Radiation heat loss

The radiation heat loss from the pipe surface in large space is given by

$$q_r = \varepsilon\sigma A(T_s^4 - T_a^4) = 0.6 \times 5.67 \times 10^{-8} \times (\pi \times 0.02 \times 1) \times (523^4 - 313^4)$$
$$= 139.4\,\text{W/m}.$$

Convection heat loss

At film temperature t_{fm} = 145°C, ρ = 0.84 kg/m³, k = 0.035 W/(m K), $\mu = 2.35 \times 10^{-5}$ kg/(m s) and Pr = 0.686.

Flow Reynolds number,

$$\mathrm{Re}_L = \frac{\rho U_\infty d}{\mu} = \frac{0.84 \times 5 \times 0.02}{2.35 \times 10^{-5}} = 3574.$$

From Table 8.11, C = 0.683, n = 0.466

$$h = \mathrm{Nu}\frac{k}{d} = 0.683\mathrm{Re}^{0.466}\,\mathrm{Pr}^{1/3}\frac{k}{d}$$
$$= 0.683 \times (3574)^{0.466} \times (0.686)^{1/3} \times \frac{0.035}{0.02}$$
$$= 47.72\,\mathrm{W/m^2K}.$$

Heat loss due to convection,

$$q_c = h(\pi dL)(t_w - t_a) = 47.72 \times (\pi \times 0.02 \times 1.0) \times (250 - 40) = 629.6\,\mathrm{W/m}.$$

Total heat loss rate,

$$q = q_r + q_c = 139.4 + 629.6 = 769\,\mathrm{W/m}.$$

Example 11.55 A 50 mm internal diameter steel pipe [wall thickness 5 mm, k_s = 40 W/(m K)] is laid in a large space. Condensing steam at 200°C is flowing through the pipe giving convection heat transfer coefficient of 600 W/(m² K) at the pipe inner surface. Pipe is covered with 20 mm thick insulation [k_i = 0.05 W/(m K)]. Surrounding air temperature is 20°C and the convection heat transfer coefficient is estimated to be 20 W/(m² K). Insulation surface emissivity is 0.75. Determine the heat loss from the pipe surface per unit length of the pipe. The effective temperature of the surface of the space may be taken equal to the air temperature.

Solution

Heat flows to the outer surface of the insulation by convection and conduction and is rejected from there by both convection and radiation, refer Fig. 11.71.

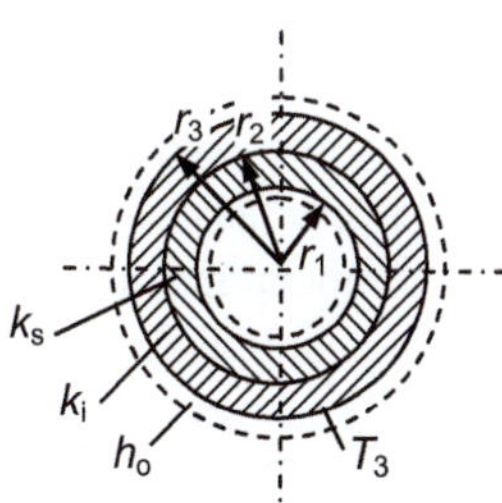

Fig. 11.71 Example 11.55

Heat flow from the steam to the outer surface of the insulation for unit length of pipe is

$$\frac{q}{L} = \frac{(T_i - T_3)}{\frac{1}{2\pi}\left[\frac{1}{h_i r_1} + \frac{1}{k_s}\ln\left(\frac{r_2}{r_1}\right) + \frac{1}{k_i}\ln\left(\frac{r_3}{r_2}\right)\right]} \tag{i}$$

Heat loss from outer surface of insulation due to convection,

$$q_c = h_o(2\pi r_3 L)(T_3 - T_\infty).$$

The radiation heat loss from the insulation surface in a large space is given by

$$q_r = \varepsilon\sigma(2\pi r_3 L) \times (T_3^4 - T_{space}^4).$$

Total heat loss rate from outer surface per unit length,

$$\begin{aligned}\frac{q}{L} &= \frac{q_c}{L} + \frac{q_r}{L} \\ &= h_o(2\pi r_3)(T_3 - T_\infty) + \varepsilon\sigma(2\pi r_3) \times (T_3^4 - T_{space}^4).\end{aligned} \tag{ii}$$

Equating Eqs. (i) and (ii),

$$\frac{(T_i - T_3)}{\frac{1}{2\pi}\left[\frac{1}{h_i r_1} + \frac{1}{k_s}\ln\left(\frac{r_2}{r_1}\right) + \frac{1}{k_i}\ln\left(\frac{r_3}{r_2}\right)\right]} = h_o(2\pi r_3)(T_3 - T_\infty) + \varepsilon\sigma(2\pi r_3) \times (T_3^4 - T_{space}^4)$$

Substituting $r_1 = 0.025$ m, $r_2 = 0.03$ m, $r_3 = 0.05$ m, $T_i = 473$ K, $T_\infty = 303$ K, $T_{space} = 303$ K, $k_s = 40$ W/(m K), $k_i = 0.05$ W/(m K), $h_i = 600$ W/(m^2 K), $h_o = 20$ W/(m^2 K) and $\varepsilon = 0.75$, we get

$$\begin{aligned}&\frac{(473 - T_3)}{\frac{1}{2\pi}\left[\frac{1}{600\times 0.025} + \frac{1}{40}\ln\left(\frac{0.03}{0.025}\right) + \frac{1}{0.05}\ln\left(\frac{0.05}{0.03}\right)\right]} \\ &= 20 \times (2\pi \times 0.05)(T_3 - 303) + 0.75 \times 5.67 \times 10^{-8} \times (2\pi \times 0.05) \times (T_3^4 - 303^4)\end{aligned}$$

Solution by trial and error gives $T_3 = 315.25$ K. Hence,

$$\frac{q}{L} = \frac{(473 - 315.25)}{\frac{1}{2\pi}\left[\frac{1}{600\times 0.025} + \frac{1}{40}\ln\left(\frac{0.03}{0.025}\right) + \frac{1}{0.05}\ln\left(\frac{0.05}{0.03}\right)\right]} = 96.35\,\text{W/m}.$$

Example 11.56 A long copper bus bar of rectangular cross-section (150 × 500 mm^2) is located in a large space whose surface may be assumed at the surrounding air temperature of 25°C. The electrical resistivity ρ of copper as a function of temperature is 1.72×10^{-8} [1 + 0.00393(t – 20)] Ωm, where t is temperature in °C. For current of 75000 A in the bus bar, determine the minimum convection heat transfer coefficient to maintain a safe operating temperature below 100°C. The emissivity of bus bar surface is 0.7.

Solution

Under steady state condition, energy balance on the bus bar for a unit length gives

$$E = q_c + q_r$$

where

E is electric energy dissipated = $I^2\rho L/A_c$, W/m
q_c is the convection heat transfer rate = $hA_s(T_s - T_\infty)$, W/m,
q_r is the radiation heat transfer rate = $\varepsilon\sigma A_s(T_s^4 - T_{sur}^4)$, W/m.
A_c area of cross-section of bus bar, m^2
A_s surface area of bus bar, m^2 and
T_s surface temperature of the bus bar, K

Substitution gives

$$\begin{aligned} 75000^2 \times \{1.72 \times 10^{-8}[1 + 0.00393(T_s - 293]\} \times 1/(0.15 \times 0.5) \\ = h \times [2(0.15 + 0.5) \times 1] \times (T_s - 298) + 0.7 \times 5.67 \times 10^{-8} \\ \times [2(0.15 + 0.5) \times 1] \times (T_s^4 - 298^4) \end{aligned}$$

(assuming that the bus bar is at a uniform temperature of T_s)

Solution of the above equation gives h = 11.32 W/(m^2 K) for T_s = 373 K. Hence, the required heat transfer coefficient must be $\geq$ 11.32 W/(m^2 K).

Example 11.57 The wall of an oven shown in Fig. 11.72a is made of an insulation material [k = 0.045 W/(m K)]. It's inner surface is subjected to radiation heat flux q_r of 125 W/m^2 and it is exposed to convection environment on both sides as shown in the figure. Determine the wall thickness for outer surface temperature $T_2 \leq$ 30°C.

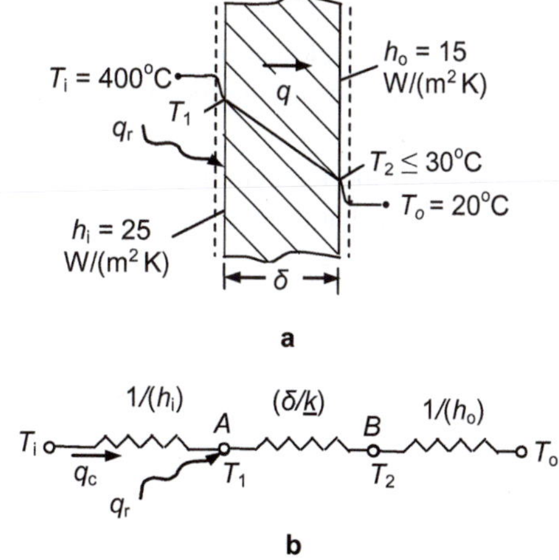

Fig. 11.72 Example 11.57

Solution

The thermal network for unit surface area is shown in Fig. 11.72b.

For nodes A and B, the energy balance equations are

$$\frac{T_i - T_1}{1/h_i} + q_r + \frac{T_2 - T_1}{\delta/k} = 0$$

$$\frac{T_1 - T_2}{\delta/k} + \frac{T_o - T_2}{1/h_o} = 0$$

Substituting various values, we have for T_2 = 30°C,

$$\frac{400 - T_1}{1/25} + 125 + \frac{30 - T_1}{\delta/0.045} = 0 \quad \text{(i)}$$

$$\frac{T_1 - 30}{\delta/0.045} + \frac{20 - 30}{1/15} = 0 \quad \text{(ii)}$$

Solution of Eqs. (i) and (ii) gives δ = 0.1107 m. Hence, the wall thickness must be greater than 0.1107 m for $T_2 < 30$°C.

Example 11.58 A 2 mm bead diameter thermocouple is installed in a 600 mm diameter duct carrying hot flue gas (may be treated as air for analysis) at about 1 atm as shown in Fig. 11.33 to measure temperature of the gas flowing through the duct. Mass flow rate of the gas is measured to be 0.5 kg/s. Emissivity of the duct outer surface is 0.8 and duct is located in a large space at surface temperature of 300 K. Temperature of the air surrounding the duct is also 300 K. The convection heat transfer coefficient h_o at the outer surface of the duct is estimated to be 20 W/(m^2 K). The thermocouple indicates a temperature of 550 K. Calculate the true temperature of the gas if the emissivity of the thermocouple surface is 0.8 and the effect of the conduction along the thermocouple wires is negligible.

Solution

The thermocouple is heated due to convection from the hot gas. But cooling of the thermocouple takes place due to the rejection of heat by radiation to the duct wall which is at a temperature lower than the temperature of the thermocouple. Hence, it indicates a temperature lower than the gas temperature.

Let the gas temperature is T_g and the temperature indicated by the thermocouple is T_t (= 550 K given). Then the convective heat transfer q_c from the air to the thermocouple is

$$q_c = h_t A_t (T_g - T_t) = h_t \times A_t (T_g - 550)$$

where A_t is the surface area of the thermocouple bead.

For a small body in a large enclosure (bead area < < duct wall area), the radiation heat transfer is

$$q_r = \varepsilon_t A_t \sigma (T_t^4 - T_s^4) = 0.8 \times A_t \times 5.67 \times 10^{-8} \times (550^4 - T_s^4)$$

In equilibrium, $q_c = q_r$. This gives

$$h_t \times (T_g - 550) = 0.8 \times 5.67 \times 10^{-8} \times (550^4 - T_s^4) \quad \text{(i)}$$

Heat transfer coefficient at the spherical thermocouple bead surface h_t can be estimated from Eq. (8.55):

$$\text{Nu} = 2 + (0.4\text{Re}_D^{0.5} + 0.06\text{Re}_D^{2/3})\text{Pr}^{0.4}(\mu_a/\mu_w)^{0.25}$$

Thermophysical properties of gas (to be treated as air) at assumed temperature of 573 K (= 300°C) from Table A5 are

$\mu_g = 2.926 \times 10^{-5}$ kg/(m s), $k_g = 0.04497$ W/(m K) and Pr = 0.680. $\mu_w = 2.841 \times 10^{-5}$ kg/(m s) at thermocouple bead temperature of 550 K (≈ 275°C).

Reynolds number,

$$\text{Re}_D = \frac{\rho U_g d}{\mu_g} = \left(\frac{d}{\mu_g}\right)\frac{m_g}{(\pi/4)D^2} = \left(\frac{0.002}{2.926 \times 10^{-5}}\right) \times \frac{0.5}{(\pi/4) \times 0.6^2} = 121$$

Hence,

$$\begin{aligned}\text{Nu} &= 2 + (0.4\text{Re}_D^{0.5} + 0.06\text{Re}_D^{2/3})\text{Pr}^{0.4}(\mu_g/\mu_w)^{0.25} \\ &= 2 + (0.4 \times 121^{0.5} + 0.06 \times 121^{2/3}) \times 0.680^{0.4}(2.926/2.841)^{0.25} \\ &= 7.1.\end{aligned}$$

Heat transfer coefficient,

$$h_t = \frac{k_g}{d}\text{Nu} = \frac{0.04497}{0.002} \times 7.1 = 159.6\,\text{W}/(\text{m}^2\text{K}).$$

Substitution in Eq. (i) gives

$$159.6 \times (T_g - 550) = 0.8 \times 5.67 \times 10^{-8} \times (550^4 - T_s^4)$$

or

$$T_g = 576 - 0.0284 \times 10^{-8} \times T_s^4 (1)$$

At the duct wall, heat input at the inner surface by convection is balanced by heat rejection at the outer surface by convection and radiation. Hence,

$$h_i \times A(T_g - T_s) = h_o \times A(T_s - T_{ao}) + \varepsilon A \times 5.67 \times 10^{-8} \times (T_s^4 - T_{sur}^4)$$

or

$$h_i(T_g - T_s) = 20 \times (T_s - 300) + 0.8 \times 5.67 \times 10^{-8} \times (T_s^4 - 300^4)(ii)$$

where T_{ao} is outside air temperature.
Reynolds number of the flow in the duct,

$$\text{Re} = \frac{\rho U_g D}{\mu_g} = \left(\frac{D}{\mu_g}\right)\frac{m_g}{(\pi/4)D^2} = \left(\frac{0.6}{2.926 \times 10^{-5}}\right)\frac{0.5}{(\pi/4) \times 0.6^2} = 36262.$$

Heat transfer coefficient at duct wall,

$$h_i = \frac{k_g}{D} \times 0.026 \times \text{Re}^{0.8}\text{Pr}^{0.3} = \frac{0.04497}{0.6} \times 0.026 \times 36262^{0.8} \times 0.680^{0.3} = 7.7$$

Substitution in Eq. (ii) gives

$$7.7 \times (T_g - T_s) = 20 \times (T_s - 300) + 0.8 \times 5.67 \times 10^{-8} \times (T_s^4 - 300^4)$$

or

$$T_g = T_s + \frac{20}{7.7} \times (T_s - 300) + \frac{0.8 \times 5.67 \times 10^{-8}}{7.7} \times (T_s^4 - 300^4)$$

or

$$T_g = 3.597T_s - 826.94 + 0.589 \times 10^{-8}T_s^4(2)$$

Equations (1) and (2) are to be solved by trial and error, which gives T_s = 360.9 K and T_g = 571.15 K.

Thus the error in reading of air temperature is $\Delta T = T_g - T_t = 571.15 - 550 = 21.15$ K. Calculated value of gas temperature T_g is nearly equal to the assumed one for thermophysical properties of the gas.

Example 11.59 A vertical copper plate [ρ = 8950 kg/m^3, c = 380 J/(kg K), k = 375 W/(m K)] at an initial uniform temperature of 300°C is suspended in a room where the ambient air and surroundings are at 25°C. Plate measures 0.25 m × 0.25 m in area and is 0.02 m in thickness. Determine the rate of cooling (K/s) when plate temperature is 275°C. Plate surface emissivity is 0.2.

Solution

At mean film temperature of ½(275 + 25) = 150°C, the thermophysical properties of air from Table A5 are

ρ = 0.8370 kg/m^3, c = 1017.1 J/(kg K), μ = 2.3769 × 10^{-5} N s/m^2, k = 0.03522 W/(m K) and Pr = 0.686.

The plate will reject heat both by natural convection and radiation. At any instant, the energy balance gives

$$-\rho(A_s\delta)c\frac{dT}{d\tau} = h(2A_s)(T_s - T_\infty) + \varepsilon(2A_s)\sigma(T_s^4 - T_{sur}^4)$$

or

$$\frac{dT}{d\tau} = \frac{-2}{\rho\delta c}\left[h(T_s - T_\infty) + \varepsilon\sigma(T_s^4 - T_{sur}^4)\right] \qquad \text{(i)}$$

where A_s is surface area and δ is thickness of the plate, and T_s is surface temperature of the plate.

The Rayleigh number,

$$\mathrm{Ra} = \mathrm{GrPr} = \frac{\beta g(T_w - T_\infty)L^3}{\nu^2}\mathrm{Pr} = \frac{(1/T_m)g(T_s - T_\infty)L^3}{(\mu/\rho)^2}\mathrm{Pr}$$

$$= \frac{(1/423)\times 9.81\times(275-25)\times 0.25^3}{(2.3769\times 10^{-5}/0.8370)^2}\times 0.686 = 1.12\times 10^8.$$

Flow is laminar. Assuming a uniform plate temperature looking to the high conductivity of plate material and small thickness, Eq. (9.11) gives

$$\mathrm{Nu_m} = 0.68 + \frac{0.670\mathrm{Ra}^{1/4}}{\left[1+(0.492/\mathrm{Pr})^{9/16}\right]^{4/9}}$$

$$= 0.68 + \frac{0.670\times(1.12\times 10^8)^{1/4}}{\left[1+(0.492/0.686)^{9/16}\right]^{4/9}} = 53.4.$$

Alternatively Eq. (9.6) may be used for Nusselt number. However, Eq. (9.11) gives better accuracy.

Heat transfer coefficient,

$$h = \frac{k}{L}\mathrm{Nu_m} = \frac{0.03522}{0.25}\times 53.4 = 7.52\,\mathrm{W/(m^2K)}.$$

Equation (i) gives

$$\frac{dT}{d\tau} = \frac{-2}{\rho\delta c}\left[h(T_s - T_\infty) + \varepsilon\sigma(T_s^4 - T_{sur}^4)\right]$$

$$= \frac{-2}{8950\times 0.02\times 380}\left[7.52\times(548-298) + 0.2\times 5.67\times 10^{-8}\times(548^4 - 298^4)\right]$$

$$= -0.083\,\mathrm{K/s}.$$

Negative sign indicates cooling of the plate.

Radiation heat transfer coefficient,

$$h_r = \varepsilon\sigma(T_s + T_{sur})(T_s^2 + T_{sur}^2)$$
$$= 0.2 \times 5.67 \times 10^{-8} \times (548 + 298) \times (548^2 + 298^2) = 3.73\,\mathrm{W/(m^2K)}.$$

Biot number,

$$\mathrm{Bi} = \frac{(h + h_r)L}{k} = \frac{(7.52 + 3.73) \times (0.02/2)}{375} = 3 \times 10^{-4}$$

Biot number Bi < < 0.1 hence assumption of uniform temperature of the plate is valid.

Example 11.60 A thin vertical plate (height $H = 0.5$ m and width $W = 1$ m) rejects 325 W heat by convection and radiation. The surrounding air temperature is 20°C. The plate is suspended in a large space with wall temperature of 20°C. If the plate surface emissivity is 0.1, determine the plate surface temperature.

Solution

Heat balance gives

$$\begin{aligned}325 &= q_c + q_r = 2A_s h(T_s - T_\infty) + 2A_s\varepsilon\sigma(T_s^4 - T_{sur}^4) \\ &= 2 \times (0.5 \times 1.0)[h(T_s - 293) + 0.1 \times 5.67 \times 10^{-8} \times (T_s^4 - 293^4)]\end{aligned} \quad \text{(i)}$$

where the heat transfer coefficient for laminar flow (assumed) is calculated from Eq. (9.6):

$$\mathrm{Nu}_m = 0.59\mathrm{Ra}^{1/4} = 0.59 \times \left[\frac{\beta g(T_s - T_\infty)H^3}{\nu^2}\mathrm{Pr}\right]^{1/4}$$

or

$$h = \frac{k}{H} \times 0.59 \times \left[\frac{\beta g(T_s - T_\infty)H^3}{\nu^2}\mathrm{Pr}\right]^{1/4} \quad \text{(ii)}$$

Assuming, for trial, film temperature to be 50°C, the air properties are

$\rho = 1.0949$ kg/m^3, $\mu = 1.9512 \times 10^{-5}$ N s/m^2, $k = 0.0.02799$ W/(m K) and Pr = 0.703.

Substitution of values of various parameters in Eq. (ii) gives

$$h = \frac{0.02799}{0.5} \times 0.59 \times \left[\frac{1/(50 + 273) \times 9.81 \times (T_s - 293) \times 0.5^3}{(1.9512 \times 10^{-5}/1.0949)^2} \times 0.703\right]^{1/4}.$$
$$= 1.778(T_s - 293)^{1/4}$$

Substitution in Eq. (i) gives

$$325 = 2 \times (0.5 \times 1.0)[1.778 \times (T_s - 293)^{1/4} \times (T_s - 293) + 0.1 \times 5.67 \times 10^{-8} \times (T_s^4 - 293^4)]$$

Solution by trial and error gives $T_s = 350.5$ K = 77.5°C. This gives film temperature $T_{fm} = (20 + 77.5)/2 = 48.75$°C, which is very close to the assumed temperature of 50°C. There is no need of retrial.

The Rayleigh number is

$$\mathrm{Ra} = \frac{\beta g(T_s - T_\infty)H^3}{\nu^2}\mathrm{Pr} = \frac{1/(48.75 + 273) \times 9.81 \times (350.5 - 293) \times 0.5^3}{(1.9512 \times 10^{-5}/1.0949)^2} \times 0.703$$
$$= 4.85 \times 10^8 < 10^9.$$

The flow is laminar as assumed.

The problem can also be solved by assuming a trial value of the heat transfer coefficient to estimate the surface temperature. For this method, refer next example.

Example 11.61 The door of a cold chamber (height H = 1.0 m and width W = 0.6 m) is provided with 25 mm thick insulation [k_i = 0.05 W/(m K)]. Temperature of the inside surface of the door is 5°C and surrounding outside air is at 27°C. The emissivity of the outside surface of the door is 0.8 and is having radiation heat exchange with the large room surface at 27°C. Determine the heat gain rate.

Solution

Heat balance for the door gives, refer Fig. 11.73,

$$q_{conduction} = q_c + q_r$$

or

$$k_i A_s \frac{(T_{so} - T_{si})}{\delta} = A_s h(T_\infty - T_{so}) + A_s \varepsilon\sigma(T_{sur}^4 - T_{so}^4)$$

Due to small temperature difference, which is the driving force in natural convection, the heat transfer coefficient is likely to have a low value. Assuming a trial value of h = 3 W/(m^2 K), we get, on substitution of values of various parameters,

$$0.05 \times \frac{(T_{so} - 278)}{0.025} = 3 \times (300 - T_{so}) + 0.8 \times 5.67 \times 10^{-8}(300^4 - T_{so}^4).$$

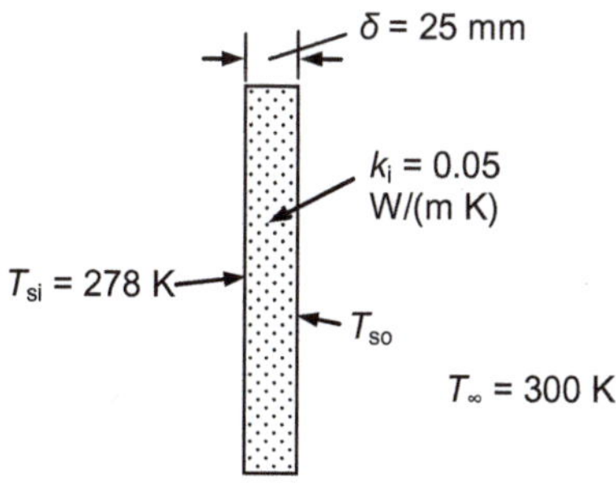

Fig. 11.73 Example 11.61

Solution of above equation by trial and error gives T_{so} = 295.5 K = 22.5°C.

For the estimate of the heat transfer coefficient, let film temperature [= (T_{so} + T_∞)/2] is 25°C. Air properties at the film temperature are

$$\rho = 1.1868\,\text{kg/m}^3, \mu = 1.8363 \times 10^{-5}\,\text{N s/m}^2,\ k = 0.02608\,\text{W/(m K)}\ \text{and}\ \text{Pr} = 0.709.$$

Using estimated value of outside surface temperature T_{so} = 295.5 K and above air properties, the Rayleigh number is

$$\text{Ra} = \frac{\beta g (T_\infty - T_{so}) H^3}{\nu^2}\text{Pr} = \frac{1/(25+273) \times 9.81 \times (300 - 295.5) \times 1^3}{(1.8363 \times 10^{-5}/1.1868)^2} \times 0.709$$
$$= 4.39 \times 10^8 < 10^9.$$

Flow is laminar. Equation (9.6) gives

$$\text{Nu}_m = 0.59\text{Ra}^{1/4} = 0.59 \times (4.39 \times 10^8)^{1/4} = 85.4$$

and

$$h = \frac{k}{H} \times \text{Nu}_m = \frac{0.02608}{1.0} \times 85.4 = 2.23\,\text{W/(m}^2\,\text{K)}.$$

With this value of the heat transfer coefficient, we have

$$0.05 \times \frac{(T_{so} - 278)}{0.025} = 2.23 \times (300 - T_{so}) + 0.8 \times 5.67 \times 10^{-8}(300^4 - T_{so}^4),$$

which gives, by trial and error, T_{so} = 295.1 K = 22.1°C. The film temperature is (22.1 + 27)/2 = 24.55°C. Retrial is not required.

The heat gain is

$$q = k_i A_s \frac{(T_{so} - T_{si})}{\delta} = 0.05 \times (0.6 \times 1.0) \times \frac{(295.1 - 278)}{0.025} = 20.52\,\text{W}.$$

This problem can also be solved by the method presented in the previous example. However, the method presented in this example is quite useful when Nusselt number correlation is not of simple form.

Example 11.62 Figure 11.74 shows an experimental setup to determine emissivity of the surface of a given specimen. Specimen is 300 mm in diameter. Surface temperature t_s is measured to be 150°C. Rate of heat dissipation (= the electric power) is 150 W. The surrounding air is at 25°C. The setup is installed in a large space whose surface temperature may be assumed to be 25°C. Determine the emissivity of the surface of the specimen.

Solution

The heat is rejected from the upper surface of the specimen by convection and radiation, which equals the electric power. Hence,

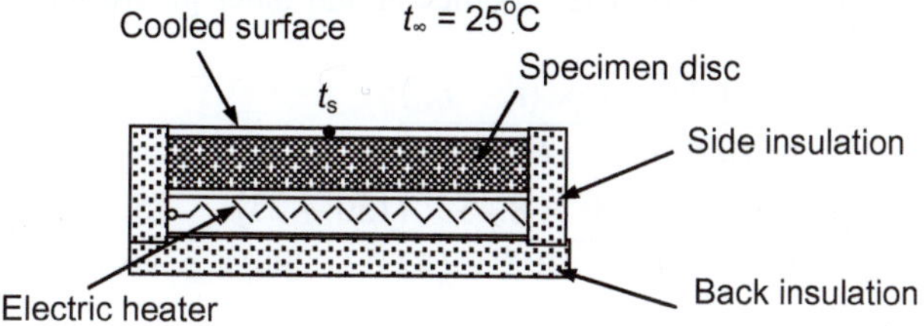

Fig. 11.74 Example 11.62

$$150 = A_s h(t_s - t_\infty) + A_s \varepsilon\sigma(T_s^4 - T_{sur}^4)$$

or

$$\begin{aligned}\varepsilon &= \frac{150 - A_s h(t_s - t_\infty)}{A_s\sigma(T_s^4 - T_{sur}^4)} \\ &= \frac{150 - (\pi/4)D^2 \times h \times (t_s - t_\infty)}{(\pi/4)D^2\sigma(T_s^4 - T_{sur}^4)}\end{aligned} \tag{i}$$

The characteristic dimension L for the horizontal plate from Eq. (9.16) is

$$L = \frac{A_s}{P} = \frac{(\pi/4)D^2}{\pi D} = \frac{D}{4} = \frac{0.3}{4} = 0.075\,\text{m}.$$

Air properties at the film temperature [= $(t_s + t_\infty)/2$] = 87.5°C are

$$\rho = 0.9752\,\text{kg/m}^3,\ \mu = 2.1189 \times 10^{-5}\,\text{N s/m}^2,\ k = 0.03080\,\text{W/(m K)},\ \text{and}\ \text{Pr} = 0.695.$$

The Rayleigh number is

$$\begin{aligned}\text{Ra} &= \frac{\beta g(t_s - t_\infty)L^3}{\nu^2}\text{Pr} = \frac{1/(87.5 + 273) \times 9.81 \times (150 - 25) \times 0.075^3}{(2.1189 \times 10^{-5}/0.9752)^2} \times 0.695 \\ &= 2.11 \times 10^6.\end{aligned}$$

When the upper side of the plate is heated,

$$\begin{aligned}\text{Nu}_m &= 0.54\text{Ra}_\text{L}^{1/4} \\ &\text{for } 10^4 \le \text{Ra} \le 10^7\end{aligned} \tag{9.17}$$

or

$$\begin{aligned}h &= \frac{k}{L}\text{Nu}_m = \frac{k}{L} \times 0.54\text{Ra}_\text{L}^{1/4} \\ &= \frac{0.03080}{0.075} \times 0.54 \times (2.11 \times 10^6)^{1/4} = 8.45\,\text{W/(m}^2\,\text{K)}.\end{aligned}$$

Substitution of value of heat transfer coefficient and other parameters in Eq. (i) gives

$$\varepsilon = \frac{150 - (\pi/4)D^2 \times h \times (t_s - t_\infty)}{(\pi/4)D^2\sigma(T_s^4 - T_{sur}^4)}$$
$$= \frac{150 - (\pi/4) \times 0.3^2 \times 8.45 \times (150 - 25)}{(\pi/4)0.3^2 \times 5.67 \times 10^{-8} \times (423^4 - 298^4)} = \frac{75.34}{96.7} = 0.78.$$

Example 11.63 A 250 × 250 × 5 mm borosilicate glass sheet [ρ =2230 kg/m^3, c = 750 J/(kg K), k = 1.1 W/(m K) and ε = 0.88] initially at a uniform temperature t_i = 150°C is placed on an insulated horizontal surface with surrounding air temperature of 25° C. The sheet is placed in a large space at 25°C. Determine the time required for the sheet surface to reduce to 50°C.

Solution

For a thin plate, lumped heat capacity analysis must be applicable. Hence,

$$\frac{t - t_\infty}{t_i - t_\infty} = \exp\left[-\left(\frac{hA_s}{c\rho V}\right)\tau\right] \tag{6.2}$$

or

$$\tau = -\ln\left(\frac{t - t_\infty}{t_i - t_\infty}\right)\left(\frac{c\rho V}{hA_s}\right) \tag{i}$$

Since the plate is rejecting heat by natural convection to the surrounding air at 25°C and by radiation to the large space at 25°C, the combined coefficient h for convection and radiation is

$$h = h_c + h_r$$

The linearized approximation of radiation coefficient based upon the average surface temperature T_{sm} = (150 + 50)/2 = 100°C = 373 K is

$$h_r = \varepsilon\sigma(T_{sm} + T_{sur})(T_{sm}^2 + T_{sur}^2)$$

or

$$= 0.88 \times 5.67 \times 10^{-8} \times (373 + 298) \times (373^2 + 298^2) = 7.63\,\text{W}/(\text{m}^2\text{K}).$$

The free convection coefficient h_c can be estimated from the correlation for the horizontal flat plate. The characteristic dimension L for horizontal plate from Eq. (9.16) is

$$L = \frac{A_s}{P} = \frac{W^2}{4W} = \frac{W}{4} = \frac{250}{4000} = 0.0625\,\text{m}.$$

Air properties at the mean film temperature [= (t_{sm} + t_∞)/2] = 62.5°C are

$\rho = 1.05\,\text{kg/m}^3$, $\mu = 2.0085 \times 10^{-5}\,\text{N s/m}^2$, $k = 0.028945\,\text{W/(m K)}$ and $\text{Pr} = 0.7$.

The Rayleigh number corresponding to the mean surface temperature is

$$\begin{aligned}\text{Ra} &= \frac{\beta g (t_{sm} - t_\infty) L^3}{\nu^2}\text{Pr} = \frac{1/(62.5 + 273) \times 9.81 \times (100 - 25) \times 0.0625^3}{(2.0085 \times 10^{-5}/1.05)^2} \times 0.7 \\ &= 1.46 \times 10^6.\end{aligned}$$

When the upper side of the plate is heated,

$$\text{Nu}_m = 0.54\text{Ra}_\text{L}^{1/4} \quad \text{for } 10^4 \le \text{Ra} \le 10^7 \tag{9.17}$$

or

$$\begin{aligned}h &= \frac{k}{L}\text{Nu}_m = \frac{k}{L} \times 0.54\text{Ra}_\text{L}^{1/4} \\ &= \frac{0.028945}{0.0625} \times 0.54 \times (1.46 \times 10^6)^{1/4} = 8.69\,\text{W/(m}^2\text{K)}.\end{aligned}$$

Substitution of the value of heat transfer coefficient and other parameters in Eq. (i) gives

$$\tau = -\ln\left(\frac{50 - 25}{150 - 25}\right) \times \frac{750 \times 2230 \times (0.25 \times 0.25 \times 0.005)}{(7.63 + 8.69) \times (0.25 \times 0.25)} = 825\,\text{s}.$$

Biot number,

$$\text{Bi} = \frac{hL}{k} = \frac{hV}{kA_s} = \frac{(7.63 + 8.69) \times 0.005}{1.1} = 0.074 < 0.1.$$

Hence, the assumption of applicability of lumped heat capacity analysis is valid.

Example 11.64 A 50 mm long horizontal metal fin [k = 15 W/(m K)] of uniform cross-section (diameter 5 mm) is rejecting heat to the surrounding air at 25°C by convection and to the surrounding surface at 25°C by radiation. The area of the fin surface is very small as compared to the surrounding surface. Determine the rate of heat transfer if the fin base temperature is 125°C. The heat transfer from the fin end may be neglected. The fin surface emissivity is 0.6.

Solution

For a fin with insulated tip, from Eq. (3.5),

$$t_L = t_\infty + (t_s - t_\infty)\frac{1}{\cosh mL}$$

where $m = \sqrt{\frac{hP}{kA_c}} = \sqrt{\frac{h\pi D}{k(\pi/4)D^2}} = \sqrt{\frac{4h}{kD}}$.

Assuming, for trial, $h = h_r + h_c = 20$ W/(m^2 K), we have

$$m = \sqrt{\frac{4 \times 20}{15 \times 0.005}} = 32.66$$

and

$$t_L = t_\infty + (t_s - t_\infty)\frac{1}{\cosh mL} = 25 + (125 - 25)\frac{1}{\cosh(32.66 \times 0.05)} = 62.6°\text{C}.$$

We assume a linear variation of temperature of the fin from base to the end to estimate the heat transfer coefficient. For the linear variation of temperature of the fin, the average temperature of the fin is

$$T_{sm} = \frac{t_s + t_L}{2} = \frac{125 + 62.6}{2} = 93.8°\text{C} = 366.8\,\text{K, say 367 K}$$

Film temperature, $T_{fm} = (t_{sm} + t_\infty)/2 = (367 + 298)/2 = 332.5$ K = 59.5°C. Air properties at film temperature are:

$$\rho = 1.059\,\text{kg/m}^3,\ \mu = 1.997 \times 10^{-5}\,\text{N s/m}^2,\ k = 0.02875\,\text{W/(m K) and Pr} = 0.7006.$$

The Rayleigh number is

$$\text{Ra} = \frac{\beta g(t_{sm} - t_\infty)D^3}{\nu^2}\text{Pr} = \frac{1/332.5 \times 9.81 \times (367 - 298) \times 0.005^3}{(1.997 \times 10^{-5}/1.059)^2} \times 0.7006 = 716$$

From Eq. (9.25),

$$\text{Nu}_\text{m} = 0.36 + \frac{0.518\text{Ra}_\text{d}^{1/4}}{\left[1 + (0.559/\text{Pr})^{9/16}\right]^{4/9}}$$

$$= 0.36 + \frac{0.518(716)^{1/4}}{\left[1 + (0.559/0.7006)^{9/16}\right]^{4/9}} = 2.38$$

Convection heat transfer coefficient,

$$h_c = \frac{k}{D}\text{Nu}_\text{m} = \frac{0.02875}{0.005} \times 2.38 = 13.68\,\text{W/(m}^2\,\text{K)}.$$

The linearized approximation of radiation coefficient based upon the average surface temperature is

$$h_r = \varepsilon\sigma(T_{sm} + T_{sur})(T_{sm}^2 + T_{sur}^2)$$

or

$$= 0.6 \times 5.67 \times 10^{-8} \times (367 + 298) \times (367^2 + 298^2) = 5.06\,\mathrm{W/(m^2\,K)}.$$

Hence, $h = h_r + h_c = 13.68 + 5.06 = 18.74$ W/(m^2 K), which is not much different from the assumed value of 20 W/(m^2 K). Retrial with $h = 19$ W/(m^2 K) may be carried out for better approximation.

Revised parameter,

$$m = \sqrt{\frac{4 \times 18.74}{15 \times 0.005}} = 31.6$$

Heat transfer rate from the fin, from Eq. (3.6),

$$\begin{aligned} q_{fin} &= \sqrt{hPkA_c}(t_s - t_\infty)\tanh mL \\ &= \sqrt{18.74 \times \pi \times 0.005 \times 15 \times \pi/4 \times 0.005^2} \times (125 - 25) \\ &\quad \times \tanh(31.6 \times 0.05) \\ &= 0.855\,\mathrm{W}. \end{aligned}$$

Note: For a better estimate, numerical method of solution may be applied.

Example 11.65 An opaque, diffuse surface with spectral reflectivity as shown in Fig. 11.75 (a) is at a temperature of 1000 K. It is subjected to a spectral irradiation, G_l given in Fig. 11.75(b). Determine net radiative heat flux to the surface.

Solution

Net radiative flux to the surface,

$$\begin{aligned} q'' &= \alpha G - \varepsilon E_b(T_s) \\ &= G_{absorbed} - \varepsilon\sigma T_s^4 \end{aligned} \tag{i}$$

We calculate ε and $G_{absorbed}$.

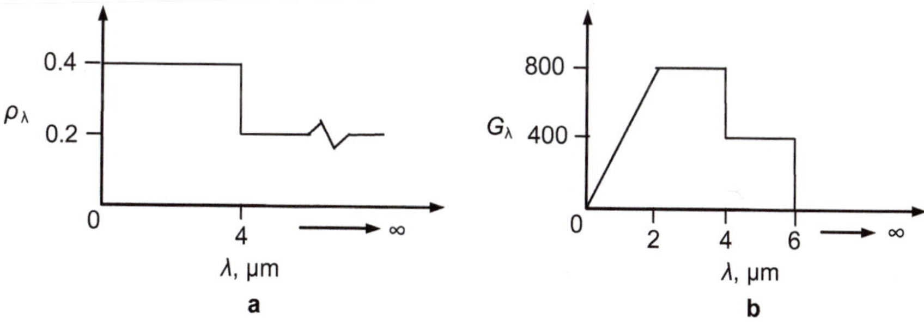

Fig. 11.75 Example 11.65

For an opaque surface, $\varepsilon_\lambda = 1 - \rho_\lambda = \alpha_\lambda$. From the given spectral reflectivity of the diffuse surface, hemispherical total emissivity is

$$\begin{aligned} \varepsilon &= \varepsilon_1 F_{0-4000} + \varepsilon_2 F_{4000-\infty} \\ &= (1-\rho_1)F_{0-4000} + (1-\rho_2)(1-F_{0-4000}) \\ &= (1-0.4)\times(0.48085) + (1-0.2)\times(1-0.48085) \\ &= 0.70383. \end{aligned}$$

where $F_{0-4000} = 0.48085$ for $\lambda_1 T = 4 \times 1000 = 4000$.

Absorbed irradiation,

$$G_{absorbed} = \alpha_{\lambda 1} \times [G_{\lambda 1} \times (\lambda_1 - 0)/2] + \alpha_{\lambda 1} \times [G_{\lambda 1} \times (\lambda_2 - \lambda_1)] + \alpha_{\lambda 2} \times [G_{\lambda 2} \times (\lambda_3 - \lambda_2)]$$

or

$$\begin{aligned} G_{absorbed} &= (1-0.4)\times[800\times(2-0)/2] + (1-0.4)\times[800\times(4-2)] + (1-0.2)\times[400\times(6-4)] \\ &= 2080\,\mathrm{W/m^2}. \end{aligned}$$

Hence, from equation (i),

$$q'' = G_{absorbed} - \varepsilon\sigma T_s^4 = 2080 - 0.70383 \times 5.67 \times 10^{-8} \times 1000^4 = -37827\,\mathrm{W/m^2}.$$

Negative sign indicates that the net radiative flux is away from the surface.

Example 11.66 What will be the temperature indicated by a thermocouple when thermocouple sheath (diameter 4 mm) is located horizontal in a large room whose wall temperature is 30°C? Surrounding quiescent air is at 22°C. Emissivity of thermocouple sheet is 0.5.

Solution

Energy balance on the thermocouple gives

$$q_c = q_r$$

or

$$hA_1(T_1 - T_a) = \varepsilon_1 A_1 \sigma (T_w^4 - T_1^4) \tag{i}$$

where A_1 is the surface area of the thermocouple sheath. Thermocouple sheath is a small body in a large enclosure (room).

We determine heat transfer coefficient h for the horizontal cylindrical sheath of thermocouple in quiescent air from Eq. (9.25).

Assuming thermocouple temperature as 26°C, the film temperature is (26 + 22)/2 = 24°C, air properties at 24°C are

$$\rho = 1.1913\,\text{kg/m}^3,\ \mu = 1.8314 \times 10^{-5}\,\text{N s/m}^2,\ k = 0.026\,\text{W/(m K)}\ \text{and}\ \text{Pr} = 0.7092.$$

Rayleigh number,

$$\begin{aligned}\text{Ra}_\text{d} &= \frac{\beta g(t_1 - t_a)d^3}{\nu^2}\text{Pr} \\ &= \frac{1/(25+273) \times 9.81 \times (26-22) \times 0.004^3}{(1.8314 \times 10^{-5}/1.1913)^2} \times 0.7092 \\ &= 25.\end{aligned}$$

Nusselt number,

$$\begin{aligned}\text{Nu}_\text{m} &= 0.36 + \frac{0.518\text{Ra}_\text{d}^{1/4}}{\left[1 + (0.559/\text{Pr})^{9/16}\right]^{4/9}} \\ &= 0.36 + \frac{0.518 \times 25^{1/4}}{\left[1 + (0.559/0.7092)^{9/16}\right]^{4/9}} = 1.24\end{aligned}$$

Heat transfer coefficient,

$$h = \frac{k}{d}\text{Nu}_\text{m} = \frac{0.026}{0.004} \times 10.24 = 8.06\,\text{W/(m}^2\,\text{K)}.$$

Substitution in Eq. (i) gives

$$8.06 \times (T_1 - 295) = 0.5 \times 5.67 \times 10^{-8} \times (303^4 - T_1^4)$$

Solution by trial and error gives T_1 = 297.2 K = 24.2°C. Retrial with T_1 = 24.2°C may be made.

The thermocouple error is because of the radiation exchange with the room wall. It will reduce with decrease in the emissivity of the sheath.

Example 11.67 A shallow pan of water is exposed to quiescent ambient air [t_∞ = 15°C and h = 7 W/(m^2 K)] and sky (T_{sky} = 235 K) at night. Determine water surface temperature. Assume water surface as diffuse gray with ε = 0.95.

Solution

Energy balance on the water surface gives

$$\varepsilon A\sigma T_s^4 - \alpha A G_{sky}^4 = hA(T_\infty - T_s)$$

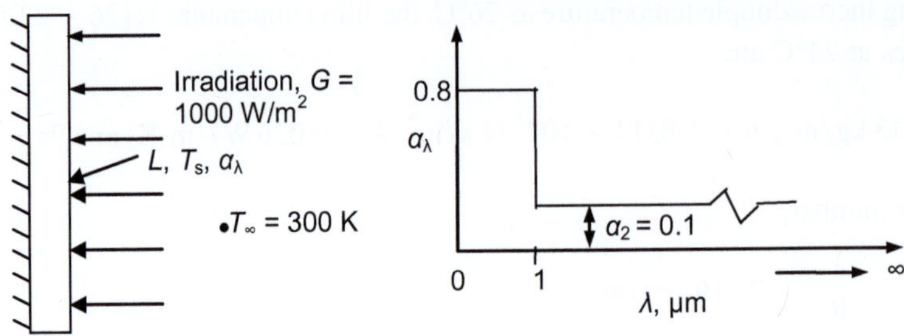

Fig. 11.76 Example 11.68

where

T_s =	water surface temperature, K
G_{sky} =	irradiation from sky = σT_{sky}^4
α =	absorptivity of water surface = ε for gray surface

Substitution gives

$$\varepsilon A\sigma T_s^4 - \varepsilon A\sigma T_{sky}^4 = hA(T_\infty - T_s)$$

or

$$\varepsilon\sigma(T_s^4 - T_{sky}^4) = h(T_\infty - T_s)$$

Substitution of the values of various parameters gives

$$0.95 \times 5.67 \times 10^{-8} \times (T_s^4 - 235^4) = 7 \times [(273 + 15) - T_s]$$

By trial and error, we get water surface temperature T_s = 270.36 K. Since water surface temperature is less than 273 K, the water surface will freeze.

Example 11.68 A vertical plate of height L = 1.8 m is suspended in still air at 300 K. One of the surfaces of the plate with diffuse coating of given spectral absorptivity distribution is exposed and subjected to an irradiation of 1000 W/m^2 from a source at 5500 K while the other side of the plate is insulated, refer Fig. 11.76. Determine the plate steady state temperature.

Solution

Heat balance for the plate surface gives

Radiation absorbed by the plate = heat radiated by the plate + heat rejected by convection

or

$$A\alpha G = A\varepsilon\sigma T_s^4 + hA(T_s - T_\infty)$$

or

$$\alpha G = \varepsilon\sigma T_s^4 + h(T_s - T_\infty) \quad \text{(i)}$$

Since the irradiation is from a source at 5500 K, the hemispherical total absorptivity of the surface, from the given spectral absorptivity distribution for the surface, is

$$\begin{aligned}\alpha &= \alpha_1(F_{0-5500}) + \alpha_2(F_{5500-\infty})\\ &= 0.8 \times 0.6906 + 0.1 \times (1.0 - 0.6906)\\ &= 0.5834\end{aligned}$$

where $\lambda_1 T = 1 \times 5500 = 5500$ for which $F_{0-5500} = 0.6906$ from Table 10.3.

Since emission from the plate is likely to at wavelength greater than 1 μm, we assume plate emissivity $\varepsilon = 0.1$. For natural convection condition, heat transfer coefficient h may be assumed to be 5 W/(m^2 K). Hence, Eq. (i) gives

$$0.5834 \times 1000 = 0.1 \times 5.67 \times 10^{-8} \times T_s^4 + 5 \times (T_s - 300).$$

By trial and error, T_s = 390 K from above equation. Using this estimate of the plate surface temperature, we can estimate the value of plate emissivity and heat transfer coefficient.

For T_s = 390 K, the total hemispherical emissivity corresponding to the temperature of the surface, knowing that $\varepsilon_\lambda = \alpha_\lambda$, is

$$\begin{aligned}\varepsilon &= \alpha_1(F_{0-390}) + \alpha_2(F_{390-\infty})\\ &= 0.8 \times 0.0 + 0.1 \times (1.0 - 0.0) = 0.1.\end{aligned}$$

where $\lambda_1 T = 1 \times 390 = 390$ for which $F_{0-390} \approx 00.0$ from Table 10.3. This estimate values of the emissivity is the same as assumed

Mean film temperature $T_m = (T_s + T_\infty)/2$ = (390 + 300)/2 = 345 K = 72°C. Considering air thermophysical properties at 75°C from Table A.5 for trial, we have

$\mu = 2.0658 \times 10^{-5}\,\text{kg/(m s)}$, $\rho = 1.0052\,\text{kg/m}^3$, $\text{Pr} = 0.697$, $k = 0.0299\,\text{W/(m K)}$ and $\beta = 1/345\,\text{K}^{-1}$

The Rayleigh number,

$$\begin{aligned}\text{Ra} &= \frac{\rho^2 \beta g (T_s - T_\infty) L^3}{\mu^2}\text{Pr}\\ &= \frac{(1.0052)^2 \times 1/345 \times 9.81 \times (390 - 300) \times 1.8^3}{(2.0658 \times 10^{-5})^2} \times 0.697\\ &= 2.46 \times 10^{10}.\end{aligned}$$

Using relation given by Eq. (9.10),

$$\mathrm{Nu_m} = \left\{0.825 + \frac{0.387\mathrm{Ra}^{1/6}}{\left[1 + (0.492/\mathrm{Pr})^{9/16}\right]^{8/27}}\right\}^2$$

for $10^{-1} \le \mathrm{Ra} \le 10^{12}$

$$= \left\{0.825 + \frac{0.387(2.46 \times 10^{10})^{1/6}}{\left[1 + (0.492/0.697)^{9/16}\right]^{8/27}}\right\}^2 = 334.76.$$

Hence, heat transfer coefficient,

$$h = \frac{\mathrm{Nu_m} k}{L} = \frac{334.76 \times 0.0299}{1.8} = 5.56\,\mathrm{W/(m^2\,K)}.$$

Substitution in Eq. (i) gives

$$0.5834 \times 1000 = 0.1 \times 5.67 \times 10^{-8} \times T_s^4 + 5.56 \times (T_s - 300).$$

By trial and error, T_s = 383 K. The new estimate of the plate surface temperature T_s will have only marginal effect on the heat transfer coefficient hence further retrial is not required.

Example 11.69 An opaque diffuse surface cylinder 1 (30 mm in diameter and 200 mm long) with given spectral absorptivity distribution is located in a large enclosure 2 whose surface is at 1000 K, refer Fig. 11.77. Determine the equilibrium temperature of the cylinder if it is subjected to cross-flow of air (air is at T_∞ = 300 K and velocity U = 3 m/s).

Solution

Heat balance for the cylinder surface gives

Radiation absorbed by the cylinder surface
= heat radiated by the cylinder + heat rejected by convection
or

$$A\alpha G = A\varepsilon\sigma T_s^4 + hA(T_s - T_\infty)$$

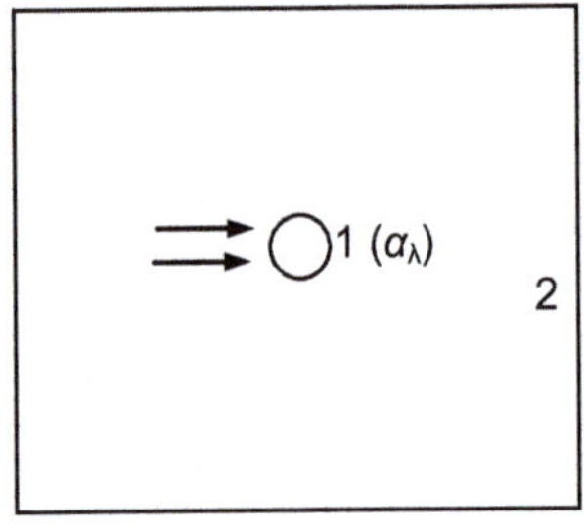

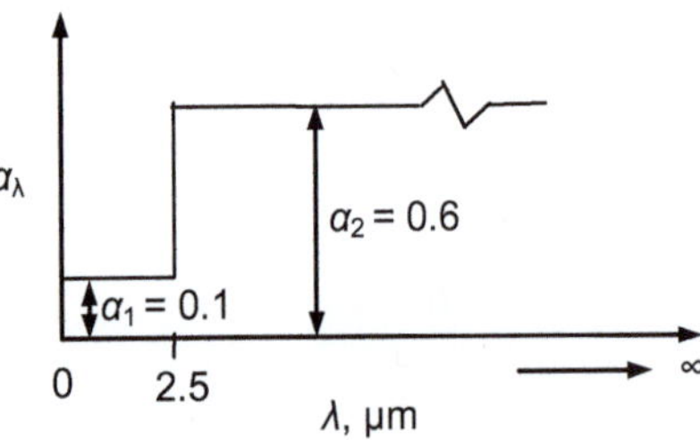

Fig. 11.77 Example 11.69

or

$$\alpha G = \varepsilon\sigma T_s^4 + h(T_s - T_\infty) \tag{i}$$

Since the irradiation is from the enclosure surface at 1000 K, the hemispherical total absorptivity of the surface, from the given spectral absorptivity of the surface, is

$$\begin{aligned}\alpha &= \alpha_1(F_{0-4000}) + \alpha_2(F_{4000-\infty})\\ &= 0.1 \times 0.1617 + 0.6 \times (1.0 - 0.1617) = 0.51915,\end{aligned}$$

where $\lambda_1 T = 2.5 \times 1000 = 2500$ for which $F_{0-2500} = 0.1617$ from Table 10.3.

Since majority of the emission from the cylinder surface is likely to at wavelength greater than 2.5 μm, we assume cylinder surface emissivity $\varepsilon = 0.6$ for first trial. For forced convection condition heat transfer coefficient h may be assumed to be 30 W/(m^2 K). Hence, Eq. (i) gives

$$0.51915 \times 5.67 \times 10^{-8} \times 1000^4 = 0.6 \times 5.67 \times 10^{-8} \times T_s^4 + 30 \times (T_s - 300).$$

By trial and error, T_s = 805 K from above equation. Using this estimate of the cylinder surface temperature, we can estimate the value of cylinder surface emissivity and heat transfer coefficient.

For T_s = 805 K, the total hemispherical emissivity corresponding to the temperature of the surface, knowing that $\varepsilon_\lambda = \alpha_\lambda$, is

$$\begin{aligned}\varepsilon &= \alpha_1(F_{0-2000}) + \alpha_2(F_{2000-\infty})\\ &= 0.1 \times 0.06886 + 0.6 \times (1.0 - 0.06886) = 0.566,\end{aligned}$$

where $\lambda_1 T = 2.5 \times 805 = 2012.5$ for which $F_{0-2012.5} = 0.06886$ from Table 10.3.

Mean film temperature $T_m = (T_s + T_\infty)/2 = (300 + 805)/2 = 552.5$ K = 279.5°C. Considering air thermophysical properties at 275°C from Table A.5 for trial, we have

$$\mu = 2.841 \times 10^{-5}\,\text{kg/(m s)},\ \rho = 0.6448\,\text{kg/m}^3,\ \text{Pr} = 0.68\ \text{and}\ k = 0.04347\,\text{W/(m K)}$$

The Reynolds number,

$$\text{Re} = \frac{\rho U d}{\mu} = \frac{0.6448 \times 3 \times 0.03}{2.841 \times 10^{-5}} = 2043.$$

Using relation given by Eq. (8.47) with values of C and n from Table 8.11, or

$$\begin{aligned}h &= 0.683\text{Re}^{0.466}\text{Pr}^{1/3}\frac{k}{D}\\ &= 0.683 \times 2043^{0.466} \times 0.68^{1/3} \times \frac{0.04347}{0.03}\\ &= 30.35.\end{aligned}$$

Using estimated emissivity and heat transfer coefficient values, the cylinder surface temperature from Eq. (i) is estimated again:

$$0.51915 \times 5.67 \times 10^{-8} \times 1000^4 = 0.566 \times 5.67 \times 10^{-8} \times T_s^4 + 30.35 \times (T_s - 300)$$

By trial and error T_s = 812 K, which is very close to the earlier estimate and hence further iteration is not required.

Example 11.70 For basically a horizontal flat roof surface (insulated from the back side), two surface coatings are available. The spectral distribution of absorptivity α_λ for the coating A is shown in Fig. 11.78a. The absorptivity of gray coating B is independent of the wavelength and is 0.6. The roof is exposed to solar irradiation during daytime. Assuming negligible irradiation to the roof surface by sky emission and negligible convection effect, suggest suitable coating for minimum temperature of the roof surface.

Solution

Energy balance on the roof surface gives

$$\varepsilon\sigma T_s^4 = \alpha G_{sun}$$

or

$$T_s = \sqrt[4]{\frac{\alpha G_{sun}}{\varepsilon\sigma}} \propto \sqrt[4]{\frac{\alpha}{\varepsilon}}, \tag{i}$$

where α is the absorptivity of the roof surface for solar radiation and ε is the emissivity of the surface for infrared radiation. T_s is the equilibrium roof surface temperature.

Coating A:
The Sun, whose surface temperature is nearly 5800 K, emits 99% of its radiation below λ = 4 μm hence absorptivity for sun irradiation in Eq. (i) is 0.5. The radiation from the heated surface is likely to be at λ > 4 hence ε = 0.1. This gives α/ε = 0.5/0.1 = 5.

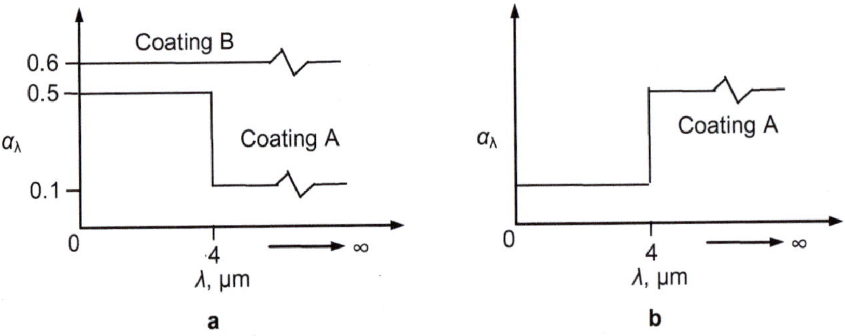

Fig. 11.78 Example 11.70

Coating B:

For coating B, $\alpha = \varepsilon = 0.6$, Hence, $\alpha/\varepsilon = 1$.

Because of lower α/ε value, the coating B is likely to give a lower roof surface temperature.

Note: The ideal surface coating for low temperature of the roof surface must have a very low absorptivity for sun radiation and high emissivity for infrared radiation, refer Fig. 11.78b.

Example 11.71 Figure 11.79 shows a very long tunnel whose upper wall is at 300 K and lower wall is insulated. The insulated wall is exposed to a row of regularly spaced cylindrical heating elements with surface temperature of 800 K. The lower surface is also exposed to forced air current at 400 K which flows parallel to the wall giving a heat transfer coefficient of 100 W/(m^2 K). Determine the temperature of the lower wall under steady state condition. Assume all surfaces to be black.

Solution

Heat balance on the lower wall gives

Radiation from the heating elements

= Radiation heat loss to the upper wall + convection heat loss to the flowing air

or

$$A_1F_{12}\sigma(T_2^4 - T_1^4) = A_1F_{13}\sigma(T_1^4 - T_3^4) + hA_1(T_1 - T_\infty)$$

or

$$F_{12}\sigma(T_2^4 - T_1^4) = F_{13}\sigma(T_1^4 - T_3^4) + h(T_1 - T_\infty) \tag{i}$$

View factor for row of infinite cylinders parallel to an infinite plate from Table 11.1 is

$$F_{12} = 1 - (1 - x^2)^{1/2} + x \tan^{-1}\left[(1 - x^2)/x^2\right]^{1/2}$$

where p = pitch and $x = D/p$. Here x = 15/30 = 0.5. Hence,

$$F_{12} = 1 - (1 - 0.5^2)^{1/2} + 0.5 \tan^{-1}[(1 - 0.5^2)/0.5^2]^{1/2} = 0.6576.$$

This gives

$$F_{13} = 1 - F_{12} = 1 - 0.6576 = 0.3424.$$

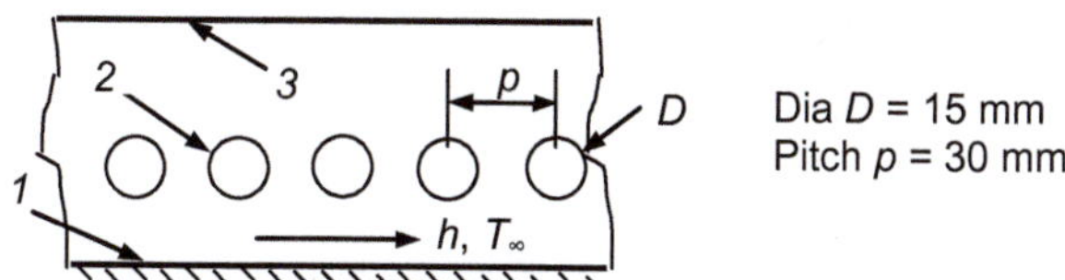

Fig. 11.79 Example 11.71

Substitution of the values of various terms in Eq. (i) gives

$$0.6576 \times 5.67 \times 10^{-8} \times (800^4 - T_1^4) \\ = 0.3424 \times 5.67 \times 10^{-8} \times (T_1^4 - 300^4) + 100 \times (T_1 - 400)$$

By trial and error, $T_1 = 514.5$ K.

Example 11.72 A vertical parallel plate enclosure consists of two square plates (1.0 m 1.0 m) separated by a distance of 30 mm. One of the plates is maintained at a uniform temperature of 40°C (ε = 0.8) and while the other is at a uniform temperature of 10°C (ε = 0.6). Air at atmospheric pressure is filled in the enclosure. Calculate the heat transfer across the space. The sides are insulated.

Solution

With the air space between plates, heat is transferred by radiation and free convection.

The radiation heat transfer between parallel plates is

$$q_{r12} = \frac{E_{b1} - E_{b2}}{\frac{1}{\varepsilon_1} + \frac{1}{\varepsilon_2} - 1} = \frac{\sigma A(T_1^4 - T_2^4)}{\frac{1}{\varepsilon_1} + \frac{1}{\varepsilon_2} - 1}$$

$$= \frac{5.67 \times 10^{-8} \times (1 \times 1) \times [(40 + 273)^4 - (10 + 273)^4]}{\frac{1}{0.8} + \frac{1}{0.6} - 1}$$

$$= 94.18\,\text{W}.$$

Air properties at mean temperature (40 + 10)/2 = 25°C are

$\rho = 1.1868\,\text{kg/m}^3, \mu = 1.8363 \times |10^{-5}\,\text{Ns/m}^2, k = 0.02608\,\text{W/(mK)}$ and $\text{Pr} = 0.709$.

Rayleigh number,

$$\text{Ra}_S = \frac{\beta g(t_1 - t_2)S^3}{\nu^2}\text{Pr}$$

$$= \frac{1/(25 + 273) \times 9.81 \times (40 - 10) \times 0.03^3}{(1.8363 \times 10^{-5}/1.1868)^2} \times 0.709$$

$$= 111380.$$

From Eq. (9.30),

$$\text{Nu}_S = \frac{hS}{k} = 0.42(\text{Ra}_S)^{1/4}\text{Pr}^{0.012}\left(\frac{L}{S}\right)^{-0.3}$$

$$= 0.42 \times (111380)^{1/4} \times 0.709^{0.012} \times \left(\frac{1}{0.03}\right)^{-0.3}$$

$$= 2.67$$

or

$$h = 2.67 \times \frac{k}{S} = 2.67 \times \frac{0.02608}{0.03} = 2.32\,\mathrm{W/(m^2\,K)}.$$

The convection heat transfer rate is given by

$$\begin{aligned} q_{c12} &= hA(t_1 - t_2) \\ &= 2.32 \times 1 \times 1 \times (40 - 10) = 69.6\,\mathrm{W}. \end{aligned}$$

Total heat transfer rate,

$$q_{12} = q_{r12} + q_{c12} = 94.18 + 69.6 = 163.78\,\mathrm{W}.$$

11.14 Summary

Considering radiation exchange between two black surfaces, the following basic integral equation of the shape factor have been derived which represent the fraction of the total radiation emitted by a surface intercepted by other surface.

$$F_{12} = \left(\frac{1}{A_1}\right) \int_{A_1} \int_{A_2} \frac{dA_1 dA_2 \cos\phi_1 \cos\phi_2}{\pi s^2} \tag{11.12}$$

and

$$F_{21} = \left(\frac{1}{A_2}\right) \int_{A_1} \int_{A_2} \frac{dA_1 dA_2 \cos\phi_1 \cos\phi_2}{\pi s^2} \tag{11.13}$$

It is to note that the shape factor is merely a function of the geometry or the orientation of the two surfaces. A number of illustrative examples on calculation of the shape factor from their basic integral equation have been given. From which, the readers must have realized that the evaluation of the integration is not an easy affair in all cases. Researchers have evaluated shape factors for some commonly encountered configurations in actual practice. They are available in the form of charts (Fig. 11.4a–e or equations (Table 11.1). Salient features of the shape factor have also been presented in Sect. 11.3.1, which may help in evaluation of the shape factor in some cases.

Radiation exchange between gray bodies is a complex process due to the process of multiple reflection and absorption. To illustrate the same, some comparatively simple cases have been discussed in Sects. 11.5 to 11.8, which clearly demonstrates that the determination of the radiation exchange between gray bodies is a difficult task.

All radiation exchange problems, between black or gray bodies, can be readily solved by electrical analogy-based method wherein the actual system is reduced to an equivalent electric network. We introduced two new terms irradiation G and radiosity J, and then defined surface resistance $(1 - \varepsilon)/A\varepsilon$ and space resistance $1/A_1F_{12}$, which provide the basis for the electric network method of solving radiation heat exchange problems. The effect of the emissivity of a diffuse and opaque gray surface is taken account by connecting potential E_b to potential J through the surface resistance $(1 - \varepsilon)/A\varepsilon$, while the shape factor effect between two radiosity potentials J_1 and J_2 is accounted by the space resistance $1/A_1F_{12}$.

Application of the electric network method has been explained taking various illustrative examples. It is to note that the solution of the network problems basically requires knowledge of electrical engineering.

Radiation heat transfer between two surfaces can be reduced by placing a thin opaque partition between the surfaces known as radiation shield. The radiation shield introduces an additional resistance in the radiation path. From the analysis, it has been shown that the shields must be made of very low absorptivity and high reflectivity materials, such as thin sheets of aluminium and copper.

Finally, Newton's law of cooling and overall heat transfer coefficient which accounts for both convection and radiation have been presented and discussed.

Review Questions

11.1 Define shape or geometrical or view factor in reference to the radiation heat exchange. Derive a general relation of the factor for radiation energy exchange between two blackbodies.

11.2 Prove the following relation for two surfaces 1 and 2 exchanging radiation energy

$$A_1F_{12} = A_2F_{21},$$

where A and F are area and shape factors, respectively.

11.3 Discuss salient features of the radiation shape factor.

11.4 Define irradiation and radiosity. Find the net rate at which the radiation energy leaves a gray surface in terms of radiosity and emissivity.

11.5 Using the definition of radiosity and irradiation develop an expression for the radiant energy exchange between two gray bodies.

11.6 Prove that the radiation heat exchange between two long concentric cylinders is given by

$$q_{12} = A_1\sigma\left(T_1^4 - T_2^4\right)/[1/\varepsilon_1 + (A_1/A_2)(1/\varepsilon_2 - 1)],$$

where ε_1 and ε_2 are the emissivities of the inner and the outer cylinders, respectively. Assume that there is no radiation heat loss from the cylinder ends.

11.7 For a system consisting of two diffuse gray surfaces 1 and 2 at different temperatures T_1 and T_2, respectively, connected by a single refractory surface, draw the radiation network (electrical analog) and write the equation of q_{12}.

11.8 A system consists of three diffuse gray surfaces which see each other and nothing else. Draw the radiation network (electrical analog). Show all the resistances on the network. Also write the nodal equations and discuss how will you proceed to find heat transfer from each surface?

11.9 Show that a thin black screen (radiation shield) introduced between two black surfaces reduces the radiation heat transfer by half. Develop general relation of reduction in heat transfer for n shields.

11.10 Discuss various methods of reducing the temperature measurement error from thermometer and thermocouples.

11.11 State Newton's law of cooling and show that the temperature of a body falls exponentially when the body is rejecting heat by convection and radiation with a small temperature difference.

11.12 Define radiation heat transfer coefficient.

11.13 How would you determine specific heat of a solid using Newton's law of cooling?

Problems

11.1. Hot flat plate of small area ΔA_1 moves a distance d from location A directly below sensor S of small area ΔA_2 at a vertical distance $H = 0.8$ m as shown in Fig. 11.80. Determine the distance d at which the sensor signal will be 50% of that directly below the sensor.

[**Ans.** The sensor signal is proportional to the radiation leaving plate (area ΔA_1) and intercepted by the sensor of area ΔA_2. Hence, at location A, $q_A = I_{n1} \frac{\Delta A_1 \Delta A_2 \cos 0 \cos 0}{H^2} = I_{n1} \frac{\Delta A_1 \Delta A_2}{H^2}$ and at location B, $q_B = I_{n1} \frac{\Delta A_1 \Delta A_2 \cos\theta \cos\theta}{s^2} = I_{n1} \frac{\Delta A_1 \Delta A_2 \cos^2\theta}{s^2}$. Ratio of the q_B and q_A is $\frac{q_B}{q_A} = \frac{\cos^2\theta}{s^2} \times H^2 = \frac{H^2/s^2}{s^2} \times H^2 = \frac{H^4}{s^4} = \frac{H^4}{(H^2+d^2)^2}$. Substituting $\frac{q_B}{q_A} = 0.5$ and $H = 0.8$, we get $d = 0.514$ m.]

11.2. Determine sh6ape or configuration factor F_{12} for a hemispherical shell and a flat surface forming an enclosure, Fig. 11.81.

[**Ans.** From reciprocity relation, $F_{12} = (A_2/A_1)F_{21}$; $F_{21} + F_{22} = 1$, but $F_{22} = 0$, so $F_{21} = 1$ hence $F_{12} = (A_2/A_1) = \pi r^2/(2\pi r^2) = 0.5$]

11.3. Two diffuse-black surfaces, a small flat disc 1 of area dA_1 and a large disc 2 of radius R are parallel to each other and directly opposed. A line joining their centers is normal to both the surfaces. The larger disc is located at a height H from the smaller one. Determine the shape factor F_{12}.

[**Ans.** $F_{12} = R^2/(R^2 + H^2)$]

11.4. A rolled steel sheet ($\varepsilon_1 = 0.4$) at 600 K is lying in a large space, which can be regarded at an average temperature of 300 K. Estimate the heat loss per m^2 of the plate area from one side of the plate by radiation only.

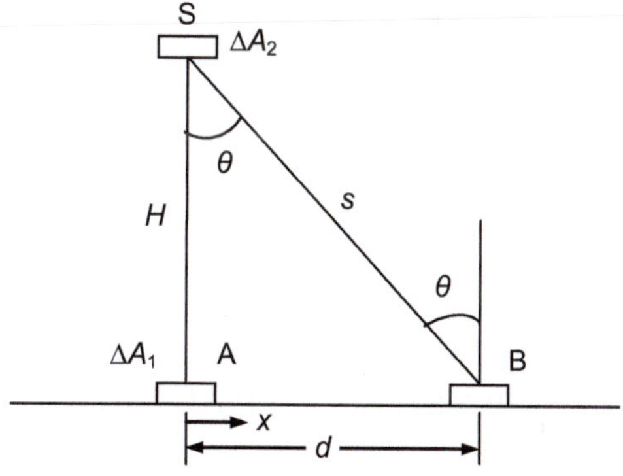

Fig. 11.80 Problem 11.1

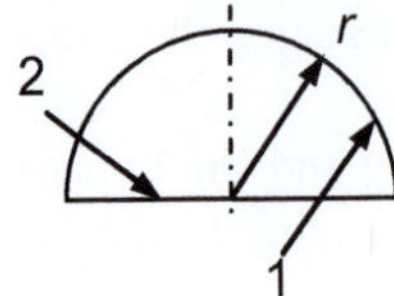

Fig. 11.81 Problem 11.2

[**Ans.** $A_1 \ll A_2$, $f_{12} = \varepsilon_1$ and $q_{12} = A_1\varepsilon_1\sigma\,(T_1^4 - T_2^4) = 2755.6$ W/m^2.]

11.5. The Sun can be regarded as nearly a spherical radiation source of diameter D_s of approximately 1.4×10^9 m and is at a distance $s = 1.5 \times 10^{11}$ m from the Earth. If the solar flux outside the Earth's atmosphere, i.e. the solar constant is 1350 W/m^2, determine the emissive power of the Sun and its surface temperature.

[**Ans.** From Example 11.9, $q_{12} = \sigma T_s^4 \frac{D_s^2}{4s^2}$ or $E = \sigma T_s^4 = q_{12} \times \frac{4s^2}{D_s^2} = 1350 \times \frac{4 \times (1.5 \times 10^{11})^2}{(1.4 \times 10^9)^2} = 6.2 \times 10^7$ W/m^2; surface temperature of the Sun, $T_s = \left(\frac{E}{\sigma}\right)^{1/4} = \left(\frac{6.2 \times 10^7}{5.67 \times 10^{-8}}\right)^{1/4} = 5750$ K.]

11.6. Prove that the interchange factor for two concentric spheres is

$$\left[\frac{1}{\varepsilon_1} + \left(\frac{R_1}{R_2}\right)^2 \left(\frac{1}{\varepsilon_2} - 1\right)\right]^{-1}$$

where ε_1 and ε_2 are the total emissivities of the two spheres. The outside radius of the inner sphere is R_1 and inside radius of the outer sphere is R_2.

[**Ans.** We have proved that $f_{12} = [1/\varepsilon_1 + (A_1/A_2)(1/\varepsilon_2 - 1)]^{-1}$ for concentric spheres. Substitution of $A_1/A_2 = (R_1/R_2)^2$ gives the result.

11.7. Two concentric spheres, 210 mm and 300 mm in diameter, with the space between them evacuated are to be used to store liquid air at –153°C in a room at 27°C. The surfaces of the spheres are flushed with aluminium ($\varepsilon = 0.03$). Find the rate of evaporation of liquid air if its latent heat of vaporization L is 200 kJ/kg. Assume that the outer sphere temperature is equal to the room temperature.

[**Ans.** $q_{12} = A_1 \times \sigma\,(T_1^4 - T_2^4)/[1/\varepsilon_1 + A_1/A_2(1/\varepsilon_2 - 1)] = \pi \times 0.21^2 \times 5.67 \times 10^{-8}$ $(120^4 - 300^4)/[1/0.03 + (0.21/0.3)^2 \times (1/0.03 - 1)] = -1.26$ W; Rate of evaporation = $(q_{12}/L) = 22.7$ g/h.]

11.8. Two parallel discs of 600 mm diameter each are spaced 300 mm apart. The temperatures of the discs are maintained at 800 K and 500 K with emissivities of 0.2 and 0.4, respectively. The discs are located in a very large space whose walls are maintained at 310 K. Determine the rate of heat loss by radiation from the inside surface of each disc.

[**Ans.** $L = R = 0.3$ m. From Table 11.1, $F_{12} = 1 + (L^2/2R^2)[1 - (1 + 4R^2/L^2)^{1/2}] = 0.382$. $E_{b1} = \sigma T_1^4 = 23224$, $E_{b2} = \sigma T_2^4 = 3543.75$, $E_{b3} = \sigma T_3^4 = 523.63$. The network is shown in Fig. 11.82.

$A_1 = A_2 = 0.2828$ m^2, $(1 - \varepsilon_1)/A_1\varepsilon_1 = 14.147$, $(1 - \varepsilon_2)/A_2\varepsilon_2 = 5.305$, $F_{13} = F_{23} = 1 - F_{12} = 0.618$, $1/A_1F_{12} = 9.258$, $1/A_1F_{13} = 1/A_2F_{23} = 5.723$.

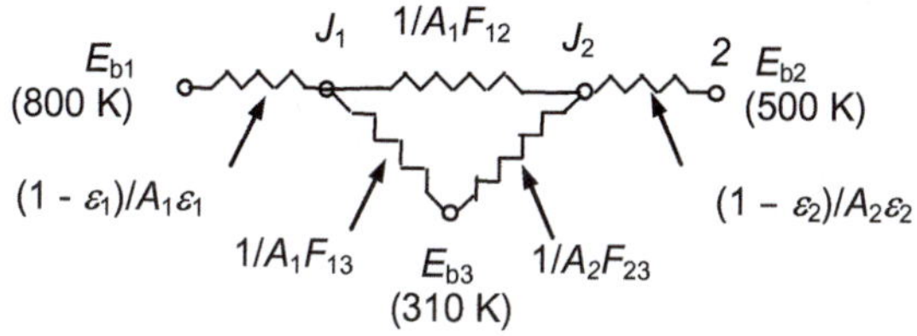

Fig. 11.82 Problem 11.8

The nodal equations are $(E_{b1} - J_1)/14.147 + (E_{b3} - J_1)/5.723 + (J_2 - J_1)/9.258 = 0$, $(E_{b2} - J_2)/5.305 + (E_{b3} - J_2)/5.723 + (J_1 - J_2)/9.258 = 0$; The equations give $J_1 = 5802.76$, $J_2 = 2941.66$. Hence, $q_1 = (E_{b1} - J_1)/14.147 = 1231.44$ W, $q_2 = (E_{b2} - J_2)/5.305 = 113.49$ W, $q_3 = [(J_1 - E_{b3}) + (J_2 - E_{b3})]/5.723 = 1344.95 \text{ W} = q_1 + q_2$.]

11.9. Following the procedure used for the estimate of F_{12} in Example 11.36, determine the heat exchange from surface 2 to 1.
[**Ans.** Changing the nomenclature of the surfaces, we have $A_1 = 3.2 \times 4 \text{ m}^2$, $A_2 = 5 \times 4 \text{ m}^2$, $\varepsilon_1 = 0.7$, $\varepsilon_2 = 0.8$, $T_1 = 350$ K, $T_2 = 300$ K. For $Z/X = 5/4 = 1.25$, $Y/X = 3.2/4 = 0.8$, $F_{12} \approx 0.24$; $R_t = [(1 - \varepsilon_1)/A_1\varepsilon_1 + 1/A_1F_{12} + (1 - \varepsilon_2)/A_2\varepsilon_2] = 0.3715$; $q_{12} = \sigma(T_1^4 - T_2^4)/R_t = 1054.1$ W, which is nearly the same as found earlier.]

11.10. Two parallel rectangular plates 1 and 2, 3 m × 2 m in size and 2 m apart, are joined on their long edge by a third plate *3* as shown in Fig. 11.83. Determine shape factors F_{12} and F_{13}.
[**Ans.**. Surface *1* to 2: $X = 3$ m, $Y = 2$ m and $L = 2$ m, for $X/L = 1.5$ and $Y/L = 1.0$, $F_{12} = 0.255$; Surface *1* to *3*: $F_{13} = 0.225$ for $Z/X = 0.66$ and $Y/X = 0.66$.]

11.11. Determine the radiation energy impinging upon a 2 m × 2 m wall 1.5 m away from a 50 mm diameter spherical body at 1000 K as shown in Fig. 11.84. Assume the wall to be a blackbody.
[**Ans.** The sphere is very small compared to the wall; hence, it can be treated as an infinitesimal disc of area $\pi/4 \times 50^2 = 1963.5 \text{ mm}^2$. Considering 1/4th of the wall *a*, $F_{12a} = 0.09$ from Eq. (11.18) for $L_1 = L_2 = 1$ m and $L = 1.5$. For the total wall area, $F_{12} = 0.09 \times 4 = 0.36$; $q_{12} = A_1F_{12}\sigma(T_1)^4 = 40.08$ W.]

11.12. Two parallel square plates, each 5 m^2, are separated by 3 mm distance. One of the plates has a temperature of 2000 K and surface emissivity of 0.7, while the other

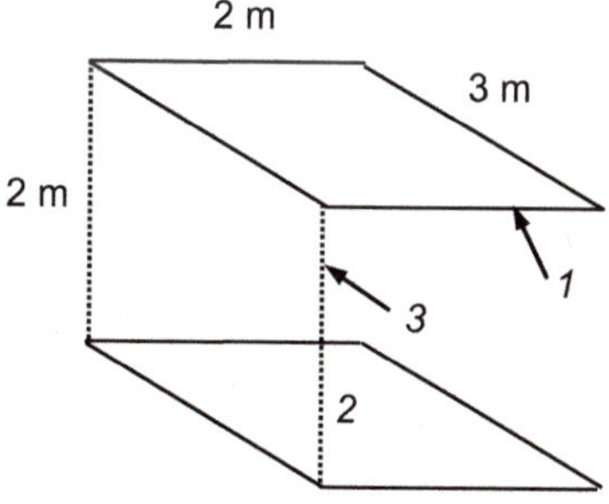

Fig. 11.83 Problem 11.10

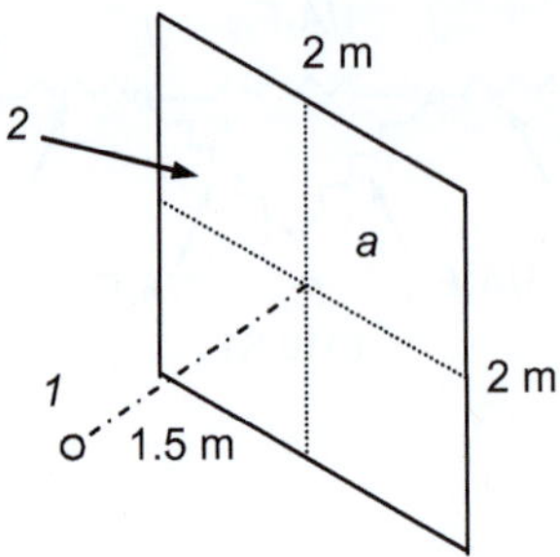

Fig. 11.84 Problem 11.11

plate surface has a temperature of 1000 K and a surface emissivity of 0.8. Find the net energy exchange by radiation between the plates.

If a thin polished metal sheet of surface emissivity of 0.15 on both sides is now located centrally between the two plates, find out the altered net heat transfer. The convection and edge effects may be neglected.

[**Ans.** The gap is very small; hence, equation of infinite parallel planes can be used. $q_{12} = f_{12}\ A_1\sigma(T_1^4 - T_2^4)$, where $f_{12} = [1/\varepsilon_1 + 1/\varepsilon_2 - 1]^{-1}$ without shield. $f_{12} = [(1/\varepsilon_1 + 1/\varepsilon_3 - 1) + (1/\varepsilon_2 + 1/\varepsilon_3 - 1)]^{-1}$ with shield. Substitution of the values gives $(q_{12})_{\text{without shield}} = 2.533 \times 10^6$ W, $(q_{12})_{\text{with shield}} = 303.5 \times 10^3$ W.]

11.13. Which curve of Fig. 11.85 represents correctly the variation of the radiation shape factor for parallel black planes of finite size?

[**Ans.** Curve *A*; As the gap between the plates increases the fraction of the radiation leaving the edge of the plate increases and F_{12} reduces. When the gap is very small, the configuration can be regarded as representing the infinite parallel plates and $F_{12} = 1$]

11.14. A thermocouple probe of emissivity 0.85 gives the temperature reading of 600 K of a gas flowing through a 0.9 m diameter duct having wall temperature of 500 K. Find out the error in the temperature measurement. The convection coefficient between the gas and thermocouple is 125 kJ/(m^2 h °C).

[**Ans.** The thermocouple bead area is very small compared to the pipe surface area, hence $f_{12} = \varepsilon_1$. The heat balance equation gives $[hA_1(T_g - T_c) = A_1\varepsilon_1\sigma\ (T_c^4 - T_s^4)]$,

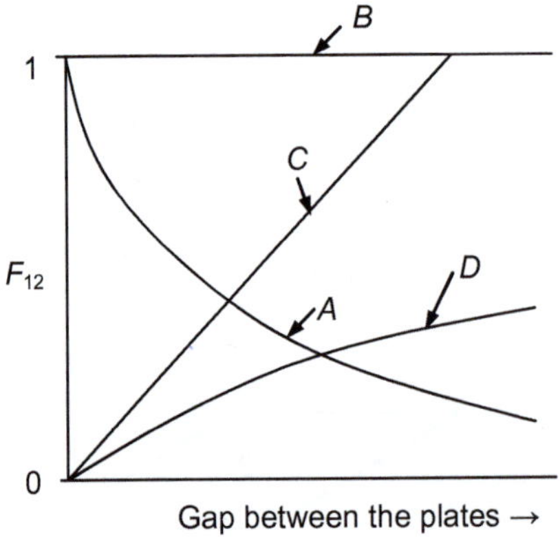

Fig. 11.85 Problem 11.13

where T_c is the thermocouple reading, T_g the true gas temperature, and T_s the pipe wall temperature. Substitution of the values gives, error = $(T_g - T_c) = 93.14$°C.]

11.15. A mercury-in-glass thermometer, hanged in a room, indicates a temperature of 25°C. The walls of the room are at a temperature of 40°C. Calculate the true temperature of the room air if the thermometer bulb emissivity is 0.9 and the heat transfer coefficient from the room air to the thermometer bulb is 8 W/(m^2 K).

[**Ans.** Use $hA_1(T_a - T_b) = A_1\varepsilon_1\sigma(T_w^4 - T_b^4)$. Error = 10.92°C and hence the actual room air temperature = 25 – 10.92 = 14.18°C.]

11.16. Air at atmospheric pressure flows over a thermocouple bead with a velocity U of 2.5 m/s. The temperature of the pipe wall is 650 K and the thermocouple indicates a temperature of 850 K. The diameter of the thermocouple bead d is 1.0 mm and emissivity ε is 0.3. Determine the true temperature of the air.

[**Ans.** Thermophysical properties of air at 850 K (for trial): $k = 0.0603$ W/(m K), $\mu = 3.756 \times 10^{-5}$ kg/(m s), $\rho = 0.4149$ kg/m^3, Pr = 0.692; Re = $\rho Ud/\mu = 27.55$; Nu = 4.28 from Eq. (8.55) neglecting (μ_∞/μ_w); $h = \text{Nu}k/d = 258$ W/(m^2 K); For $T_{bead} = 850$ K and $T_{wall} = 650$ K, equation $h(T_a - T_{bead}) = \varepsilon\sigma(T_{bead}^4 - T_{wall}^4)$ gives $T_a = 872.6$ K; Iteration may be carried out using thermophysical properties at mean air temperature $T_m = (872.6 + 850)/2 \approx 861$ K.]

11.17. Figure 11.86 shows a cylindrical cavity whose surface can be assumed to be gray ($\varepsilon = 0.6$). Find the rate of emission from the cavity to the surrounding at 293 K if the cavity surface temperature is 900°C.

[**Ans.** Assume cavity to be covered with a black surface 2 at 293 K as shown in the figure. $\varepsilon_1 = 0.6$, $T_1 = 1173$ K, $R = 0.1$ m; $H = 0.1$ m; $A_1 = \pi DH + (\pi/4) D^2 = 0.09425$ m^2; $A_2 = \pi R^2 = 0.031416$ m^2; $F_{12} = A_2/A_1 = 0.3333$; $R_t = (1 - \varepsilon_1)/(A_1\varepsilon_1) + 1/(A_1F_{12}) = 38.9$; $q_{12} = \sigma\ (T_1^4 - T_2^4)/R_t = 2748.72$ W; *Note* $(1 - \varepsilon_2)/(A_2\varepsilon_2) = 0$]

11.18. Determine view factor F_{12} for the following configurations:

(i) A sphere lying on an infinite horizontal plane, Fig. 11.87a

(ii) Long inclined planes where plate end edge indicated by B in Fig. 11.87b is 100 mm above the longitudinal axis of plane 1 passing through center C in the figure.

[**Ans.** (i) Put a parallel infinite plane 3 as shown in figure. Then $F_{11} + F_{12} + F_{13} = 1$; for spherical surface $F_{11} = 0$. Since $F_{12} = F_{13}$ (by symmetry), $F_{12} = F_{13} = 0.5$. From reciprocity relation $F_{21} = (A_1/A_2)\ F_{12}$. Since $A_2 = \infty$, $F_{21} = 0$.

(iii) $F_{11} + F_{12} + F_{13} = 1$. Since $F_{12} = F_{13}$ (by symmetry), $F_{12} = F_{13} = 0.5$. From reciprocity relation $F_{21} = (A_1/A_2)\ F_{12}$. Since $A_2 = LW\sqrt{2}$, $A_1 = 2WL$, $F_{21} = (2WL/LW\sqrt{2}) \times 0.5 = 0.707$.]

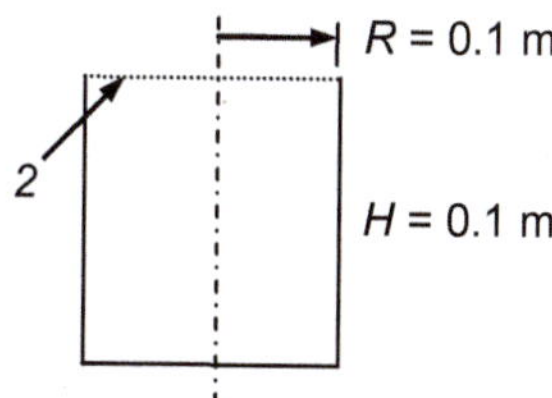

Fig. 11.86 A cylindrical cavity (Problem 11.17)

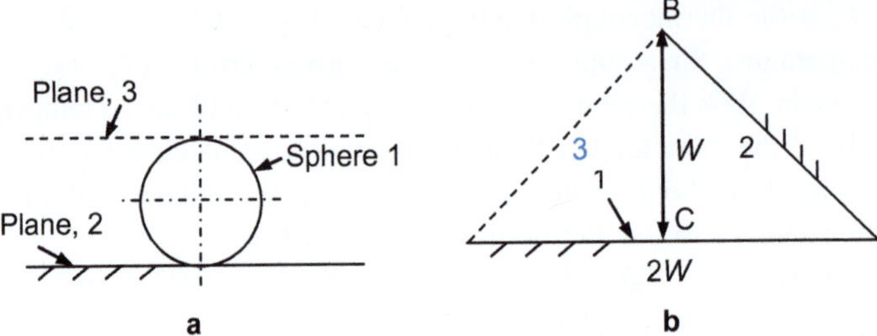

Fig. 11.87 Problem 11.18

11.19. Two infinite parallel plates, one is black (1) and other (2) is gray with emissivity 0.8, are at temperatures 1000 K and 600 K, respectively. Determine the irradiation on plate 1.
[**Ans.** $E_b = \sigma T_1^4$, radiosity of surface 2 = irradiation for surface 1 = $E_2 + \rho_2 E_b$ = $\varepsilon_2 T_2^4 + \rho_2 \sigma T_1^4 = \sigma \varepsilon_2 T_2^4 + (1 - \varepsilon_2)\ \sigma T_1^4 = \sigma[\varepsilon_2 T_2^4 + (1 - \varepsilon_2)\ T_1^4] = 5.67 \times 10^{-8} \times [0.8\ 600^4 + (1 - 0.8) \times 1000^4] = 17218$ W/m^2.]

11.20. A small object at 500 K and spectral emissivity as given in Fig. 10.11 is suspended in a large furnace (wall temperature 1300 K and total emissivity 0.8). Determine (a) hemispherical total surface emissivity and absorptivity of the object, (b) net radiative flux to the surface.
[**Ans.** (a) (i) Hemispherical total emissivity: $\lambda_1 T = 2 \times 500 = 1000$ for which $F_{0\text{-}1000} = 0.321 \times 10^{-3}$; $\lambda_2 T = 7 \times 500 = 3500$ for which $F_{0\text{-}3500} = 0.382635$; Thus, $F_{1000\text{-}3500} = 0.382635 - 0.321 \times 10^{-3} = 0.382314$; $F_{3500\text{-}\infty} = 1 - 0.382635 = 0.617365$; $\varepsilon = \varepsilon_1(F_{0-1000}) + \varepsilon_2(F_{1000-3500}) + \varepsilon_3(F_{3500-\infty}) = 0.1 \times 0.321 \times 10^{-3} + 0.4 \times 0.382314 + 0.2 \times 0.617365 = 0.27643$; (ii) Absorptivity of the object for the radiation from furnace wall at 1300 K: $\lambda_1 T = 2 \times 1300 = 2600$ for which $F_{0\text{-}2600} = 0.18312$; $\lambda_2 T = 7 \times 1300 = 9100$ for which $F_{0\text{-}9100} = 0.892582$; Thus, $F_{2600\text{-}9100} = 0.892582 - 0.18312 = 0.709462$; $F_{9100\text{-}\infty} = 1 - 0.892582 = 0.107418$; Hemispherical total absorptivity, $\alpha = \alpha_1(F_{0-2600}) + \alpha_2(F_{2600-9100}) + \alpha_3(F_{9100-\infty}) = 0.1\ 0.18312 + 0.4 \times 0.709462 + 0.2 \times 0.107418 = 0.32358$; (b) Net radiative flux to the surface, $q'' = G - J = G - \rho G - \varepsilon E_b(500) = (1 - \rho)G - \varepsilon E_b(500) = \alpha E_b(1300) - \varepsilon E_b(500) = 0.32358 \times 5.67 \times 10^{-8} \times 1300^4 - 0.27643 \times 5.67 \times 10^{-8} \times 500^4 = 51421$ W/m^2.]

11.21. A vertical copper plate [$\rho = 8950$ kg/m^3, $c = 380$ J/(kg K), $k = 375$ W/(m K)] at an initial uniform temperature of 300°C is suspended in a room where the ambient air and surroundings are at 25°C. Plate measures 0.25 m × 0.25 m in area and is 0.02 m in thickness. The rate of cooling is found to be 0.08 K/s when plate temperature was 275°C. Plate surface emissivity is 0.2. Determine the convection heat transfer coefficient.
[**Ans.** For air at mean film temperature of ½(275 + 25) = 150°C, from Table A5, $\rho = 0.8370$ kg/m^3, $c = 1017.1$ J/(kg K), $\mu = 2.3769 \times 10^{-5}$ N s/m^2 and $k = 0.03522$ W/(m K). The plate rejects sensible heat both by convection and radiation. At any instant, the energy balance gives $-\rho(A_s\delta)c\frac{dT}{d\tau} = h(2A_s)(T_s - T_\infty) + \varepsilon(2A_s)\sigma(T_s^4 - T_{sur}^4)$, where δ is thickness and A_s is area of plate. Equation gives $h = \frac{1}{(T_s - T_\infty)}\left[-\frac{dT}{d\tau}\frac{\rho\delta c}{2} - \varepsilon\sigma(T_s^4 - T_{sur}^4)\right]$; For $dT/d\tau$ = -

0.08 K/s, $h = \frac{1}{(548-298)}\left[\frac{0.08\times8950\times0.02\times380}{2} - 0.2\times5.67\times10^{-8}\times(548^4-298^4)\right] =$ 7.15 W/(m^2 K).]

11.22. Determine the heat loss from an electric bulb (T_s = 125°C) in a room if the surrounding air temperature is 25°C. The bulb can be approximated to be a sphere of 50 mm diameter. Bulb surface emissivity is 0.88 and the room surface temperature is 27°C.

[**Ans.** At film temperature t_{fm} = (125 + 25)/2 = 75°C, air properties are ρ = 1.0052 kg/m^3, k = 0.0299 W/(m K), μ = 2.0658 × 10^{-5} N s/m^2 and Pr = 0.697; Heat loss from the bulb is by convection and radiation; $q = q_c + q_r = h_c A_s(T_s - T_\infty) + \varepsilon A_s\sigma(T_s^4 - T_{sur}^4)$; $\mathrm{Ra} = \frac{\beta g(T_s - T_\infty)D^3}{\nu^2}\mathrm{Pr} = \frac{1/(75+273)\times9.81\times(125-25)\times0.05^3}{(2.0658\times10^{-5}/1.0052)^2}\times 00.697 = 5.8\times10^5$; From Eq. (9.27), $\mathrm{Nu_m} = 2 + \frac{0.589\mathrm{Ra}_d^{1/4}}{\left[1+(0.469/\mathrm{Pr})^{9/16}\right]^{4/9}} = 2 + \frac{0.589\times(50.8\times10^5)^{1/4}}{\left[1+(0.469/0.697)^{9/16}\right]^{4/9}} = 14.52$; $h_c = \mathrm{Nu_m}\frac{k}{D} = 14.52\times\frac{0.0299}{50\times10^{-3}} = 8.68$ W/(m^2 K); Hence, $q = 8.68\times\pi\times0.05^2\times(125-25) + 0.88\times\pi\times 0.05^2\times5.67\times10^{-8}\times(398^4 - 300^4) = 13.48$ W; Some heat is lost by conduction to the base.]

11.23. Determine net radiative exchange between the plates of Problem 11.19 per unit area of the plates.

[**Ans.** Radiosity of surface 2, $J_2 = E_2 + \rho_2 E_{b1} = \sigma\varepsilon_2 T_2^4 + \rho_2\sigma T_1^4 = \sigma\varepsilon_2 T_2^4 + (1-\varepsilon_2)\sigma T_1^4 = \sigma[\varepsilon_2 T_2^4 + (1-\varepsilon_2)T_1^4] = 5.67\times10^{-8}\times[0.8\times600^4 + (1-0.8)\times1000^4] = 17218$ W/m^2; Radiosity of surface 1, which is black, $J_1 = E_{b1} = T_1^4 = 5.67\times10^{-8}\times1000^4 = 56700$ W/m^2, $q_{12} = J_1 - J_2 = 39482$ W/m^2; alternatively $q_{12} = (E_{b1} - E_{b2})/(1/\varepsilon_1 - 1/\varepsilon_2 - 1) = \varepsilon_2(E_{b1} - E_{b2}) = 0.8\times5.67\times10^{-8}(1000^4 - 600^4) = 39481.3$ W/m^2]

11.24. Determine the shape factor F_{12} for the configuration shown in Fig. 11.88.

[**Ans.** $A_{(13)}F_{(1,3)-(2,4)} = A_1F_{12} + A_1F_{14} + A_3F_{32} + A_3F_{34}$; From symmetry, $A_1F_{12} = A_4F_{43}$; From reciprocity relation, $A_4F_{43} = A_3F_{34}$; Hence, $A_3F_{34} = A_1F_{12}$; This gives $A_{(13)}F_{(1,3)-(2,4)} = 2A_1F_{12} + A_1F_{14} + A_3F_{32}$; or $F_{12} = \frac{1}{2A_1}\left[A_{(13)}F_{(1,3)-(2,4)} - A_1F_{14} - A_3F_{32}\right]$; A_{13} = 9 m^2, A_1 = 3 m^2, A_3 = 6 m^2; $F_{12} = \frac{1}{2\times3}\left[9F_{(1,3)-(2,4)} - 3F_{14} - 6F_{32}\right]$; From Fig. 11.4c for Z/X = 1.0 and Y/X = 1, $F_{(1,3)-(2,4)} = 0.2$; for Z/X = 3.0 and Y/X = 3.0, $F_{14} = 0.125$ and for Z/X = 0.75 and Y/X = 0.25, $F_{32} = 0.17$; substitution gives $F_{12} = (1/6)(9\times0.2 - 3\times0.125 - 6\times0.17) = 0.0675$.]

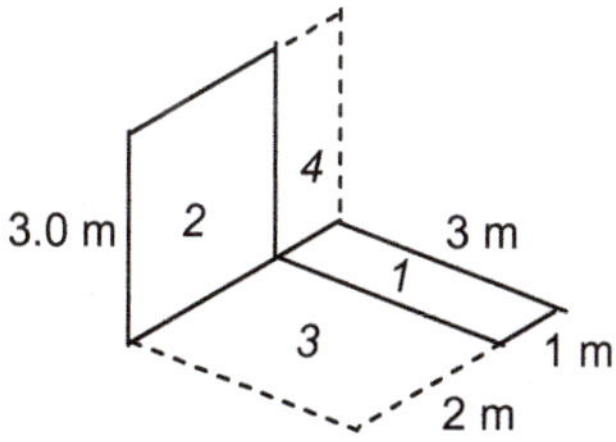

Fig. 11.88 Problem 11.24

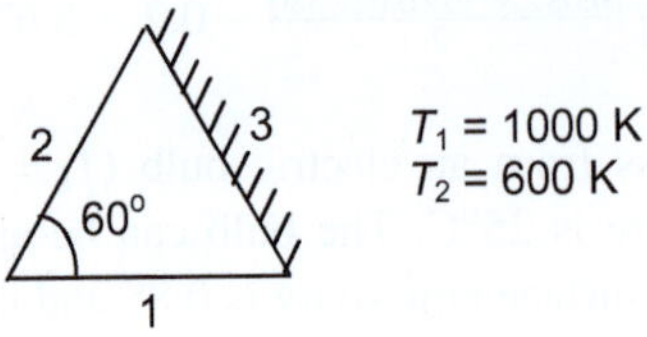

Fig. 11.89 Problem 11.25

11.25. Long, inclined black surfaces 1 and 2 in Fig. 11.89 are maintained at 1000 K and 600 K, respectively. Determine temperature of the black insulated surface 3.
[**Ans.** Since surface 3 is insulated, $q_3 = 0$, i.e. $q_{13} = q_{32}$; or $A_1F_{13}\sigma(T_1^4 - T_3^4) = A_3F_{32}\sigma(T_3^4 - T_2^4)$; By symmetry, $A_1 = A_3$ and $F_{13} = F_{32}$; Hence $T_1^4 - T_3^4 = T_3^4 - T_2^4$, or $T_3 = \left[\frac{1}{2}(T_1^4 + T_2^4)\right]^{1/4} = \left[\frac{1}{2}(1000^4 + 600^4)\right]^{1/4} =$ 866.91 K.]

11.26. A 25 mm diameter pipe is laid horizontal in a room for heating. Condensing steam is flowing through the pipe. If the temperature of the outer surface of the pipe is 105°C and temperature in the room is 20°C, find the required length of the pipe for a heating rate of 1.0 kW. Pipe surface emissivity = 0.8.
[**Ans.** At $t_{fm} = (105 + 20)/2 = 62.5$°C, air properties are: $\rho = 1.05$ kg/m^3, $\mu = 2.0 \times 10^{-5}$, and $k = 0.029$ W/(m K), $\beta = 1/T_m = 1/(62.5 + 273)$, Pr = 0.7; Gr $= \frac{g\beta\rho^2 d^3 \Delta T}{\mu^2} = 1.07 \times 10^5$, $h = 0.53(Gr_f \, \mathrm{Pr}_f)^{0.25} k/d = 100.17$ W/(m^2 K); Convection heat transfer, $q_c = hA\Delta T = 67.92$ W/m, radiation heat transfer $q_r = \varepsilon A\sigma\,(T_{pipe}^4 - T_{room}^4) = 46.48$ W/m. Required length $L = q/(q_c + q_r) = 8.74$ m.]

References

Howell JR (1982) A catalog of radiation configuration factors. McGraw-Hill Book Co, New York
Mikheyev M (1968) Fundamental of heat transfer. Mir Publishers, Moscow
Siegel R, Howell JR (2002) Thermal radiation heat transfer. Taylor & Francis, New York

Heat Transfer in Absorbing and Emitting Media (Gaseous Radiation)

12

12.1 Introduction

Monatomic and diatomic gases such as argon, helium, oxygen, nitrogen and hydrogen are transparent (diathermic) to the thermal radiation and their capacity to radiate or absorb the thermal radiation is insignificant except at extremely high temperatures. Molecules such as carbon dioxide, carbon monoxide, sulfur dioxide, water vapour and hydrocarbon gases are capable of emitting and absorbing the heat radiation. Carbon dioxide and water vapour are formed when the combustion of hydrocarbon fuels takes place and the study of these gases is of practical importance. Methane and carbon dioxide are regarded as greenhouse gases and are the cause of global warming.

The radiation exchange between a gas and a solid surface is considerably more complex than exchanges between solid surfaces. The specific features of the gaseous radiation are being discussed in this chapter. However, the discussion is basically confined to the behavior of carbon dioxide, water vapour and their mixtures.

12.2 Specific Features of Gaseous Radiation

12.2.1 Selective Emitters

Most of the solids possess continuous radiation spectra, i.e. they emit and absorb rays of wavelength zero to infinity. But the gases emit and absorb radiation in certain narrow wavelength regions called bands, Fig. 12.1. Outside these bands, these gases are practically transparent and their emissive power is zero. Thus, the gases are selective absorbers and emitters. These bands, their width and number are different for different gases. Typically for carbon dioxide, these bands are

$$\Delta\lambda_1 = 2.36\text{–}3.02\ \mu\text{m};\ \Delta\lambda_2 = 4.01\text{–}4.80\ \mu\text{m};\ \Delta\lambda_3 = 12.5\text{–}16.5\ \mu\text{m}$$

and for the water vapour, these bands are

$$\Delta\lambda_1 = 1.7\text{–}2.0\ \mu\text{m};\ \Delta\lambda_2 = 2.24\text{–}3.27\ \mu\text{m};\ \Delta\lambda_3 = 4.8\text{–}8.5\ \mu\text{m};\ \Delta\lambda_4 = 12.0\text{–}30.0\ \mu\text{m}$$

R. Karwa, *Heat and Mass Transfer*,
https://doi.org/10.1007/978-981-15-3988-6_12

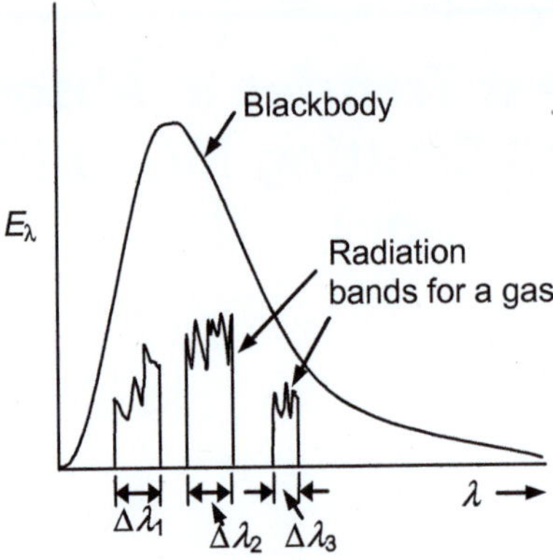

Fig. 12.1 Emissive power of a gas versus a blackbody

12.2.2 Beer's Law

In opaque solids, the emission and absorption of the heat radiation occur only in thin surface layers but in gases, the emission and absorption occur over their volume. As the radiation passes through a gas, a reduction in its intensity takes place. This reduction depends on the number of molecules encountered, i.e. the thickness of the gas volume and the partial pressure of the gas.

The absorption of the radiation in the gas layers has been expressed in a mathematical form. Let a monochromatic beam of radiation intensity $(I_\lambda)_0$ impinges on the gas volume and passes through it, refer Fig. 12.2. Then the decrease in the intensity due to the absorption in the gas layer at any plane *xx* has been found to be proportional to the thickness *dx* of the layer and the intensity of radiation $(I_\lambda)_x$ at that plane.

$$d(I_\lambda)_x \propto (I_\lambda)_x dx = -\alpha_\lambda (I_\lambda)_x dx \tag{12.1}$$

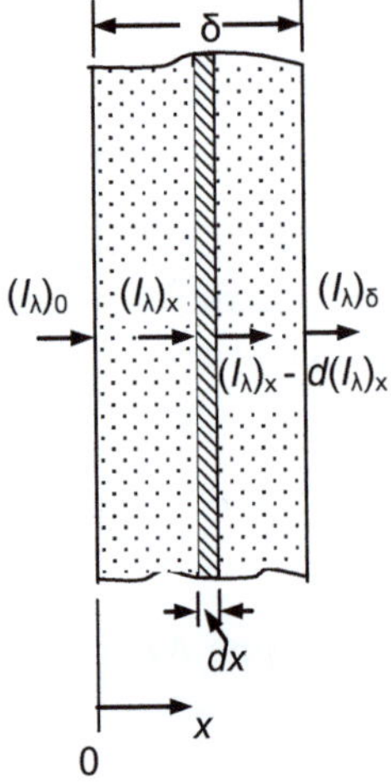

Fig. 12.2 Beer's law

where α_λ is called monochromatic absorption coefficient. The coefficient $-\alpha_\lambda$ depends on the gas temperature T_g and product (pL), where L is a characteristic dimension of the system called the mean equivalent beam length and p is the gas pressure, i.e.

$$\alpha_\lambda = f(T_g, pL) \tag{12.2}$$

Integrating Eq. (12.1), we get

$$\int_{I_{\lambda 0}}^{I_{\lambda x}} \frac{dI_{\lambda x}}{I_{\lambda x}} = \int_0^x -\alpha_\lambda dx$$

or

$$\frac{I_{\lambda x}}{I_{\lambda 0}} = e^{-\alpha_\lambda x} \tag{12.3}$$

Equation (12.3) represents exponential decay of the radiation intensity with distance as the radiation travels through the gas volume. It is known as *Beer's law*.

12.2.3 Transmissivity, Emissivity and Absorptivity

The ratio $(I_\lambda)_x/(I_\lambda)_0$, refer Fig. 12.2, is termed as monochromatic transmissivity of the gas for thickness x, i.e.

$$\tau_\lambda = \frac{I_{\lambda x}}{I_{\lambda 0}} = e^{-\alpha_\lambda x} \tag{12.4}$$

In general, the reflectivity of the gases is zero, therefore

$$\alpha_\lambda + \tau_\lambda = 1$$

We can write

$$\alpha_\lambda = 1 - \tau_\lambda = 1 - e^{-\alpha_\lambda x} \tag{12.5}$$

and, from the Kirchhoff's law, the equation for the monochromatic emissivity can be written as

$$\varepsilon_\lambda = \alpha_\lambda = 1 - e^{-\alpha_\lambda x} \tag{12.6}$$

From Eqs. (12.5) and (12.6), it can be seen that as the thickness of the gas increases, the absorptivity and emissivity increase. In the limit when gas thickness δ is very large α_λ and ε_λ approach unity and the gas volume acts as a blackbody.

The total transmissivity τ (for $\lambda = 0 - \infty$) is obtained from

$$\tau = \frac{\int_0^\infty I_{\lambda\delta} d\lambda}{\int_0^\infty I_{\lambda 0} d\lambda} \tag{12.7}$$

12.2.4 Total Emissive Power

As mentioned earlier, gases emit and absorb radiation within certain bands. The radiant energy in a band is

$$E_{\Delta\lambda} = \int_{\lambda_1}^{\lambda_2} E_\lambda d\lambda \tag{12.8}$$

and the total emissive power of the gas equals the sum of radiation of all bands,

Note: A gas may be termed as *gray gas* for which the monochromatic coefficients τ_λ, α_λ and ε_λ are independent of the wavelength λ. In this case, the gas emissivity is equal to the gas absorptivity regardless of the source of the incident radiation. That is, for a gray gas, $\alpha_g = \alpha_\lambda$, $\varepsilon_g = \varepsilon_\lambda$ and $\alpha_g = \varepsilon_g$.

12.3 Heat Exchange

The procedure for the calculation of net heat exchange was presented by Hottel (1954). In this section, a simplified case of total radiation exchange between a black enclosure and isothermal (uniform temperature) gas volume is being considered.

12.3.1 Radiation Emitted by a Gas

The geometry of most of the configurations involves a complicated analysis for the determination of the radiation heat exchange between the gas volume and the surface. Hence, Hottel considered a hemispherical volume of gas at a uniform temperature T_g exchanging radiation heat with a small black surface A_s located at its centre, as shown in Fig. 12.3, for which the analysis can be easily carried out. The radius of the hemisphere is L.

Let the gas mixture contains CO_2 (a participating gas) and N_2 and O_2 (the non-participating gases). The total pressure of the gas mixture is p. CO_2 is at a partial pressure of p_c in the mixture.

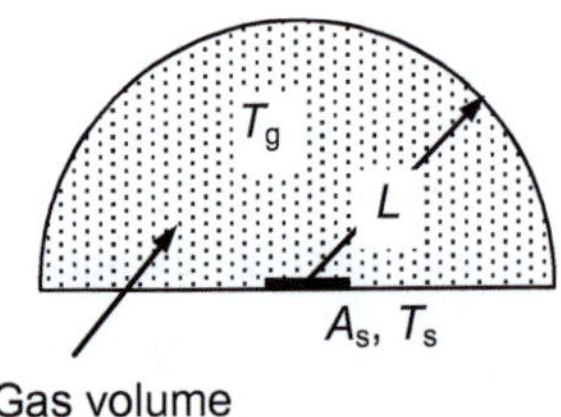

Fig. 12.3 Hemispherical gas volume

The radiation heat emitted by the gas and arriving at the surface A_s is

$$q_{gs} = \varepsilon_g A_s \sigma T_g^4 \tag{12.9}$$

Since the surface is black, it absorbs the radiation q_{gs} completely. As the gas contains only one participating gas CO_2, hence $\varepsilon_g = \varepsilon_c$, i.e. the emissivity of CO_2. ε_c is a function of the gas mixture temperature T_g, the partial pressure of CO_2 in the gas volume times the characteristic dimension ($p_c L$), and the total pressure of the mixture, that is

$$\varepsilon_c = f(T_g, p_c L, p) \tag{12.10}$$

The emissivity is determined from Fig. 12.4. When the total pressure of the gas is 1 atm, the value of emissivity obtained from the figure is the value of the emissivity ε_c. When the total pressure of the gas mixture is different from 1 atm, the emissivity value read from the figure is to be multiplied by a pressure correction factor C_c from Fig. 12.5. Thus, in general,

$$\varepsilon_c = (\varepsilon_c)_{p=1\text{atm}} (C_c)_{p \neq 1\text{atm}} \tag{12.11}$$

where $(\varepsilon_c)_{p\,=\,1\text{ atm}}$ is the emissivity value from Fig. 12.4. Note that $(C_c)_{p\,=\,1\text{ atm}} = 1$.

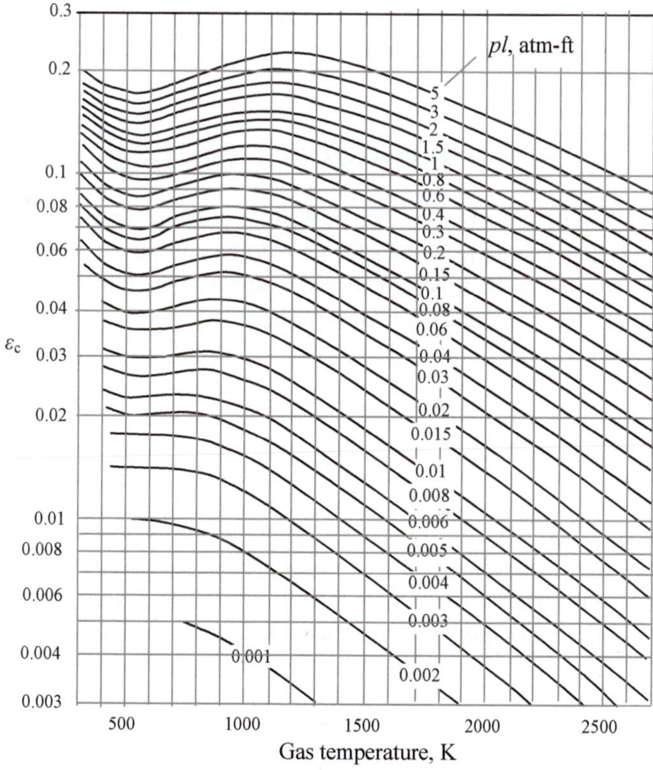

Fig. 12.4 Emissivity of CO_2 (p = 1 atm). Hottel HC in McAdams WH, Heat Transmission, 3rd edition, McGraw-Hill, NY, copyright 1954. Material is reproduced with permission of McGraw-Hill Education

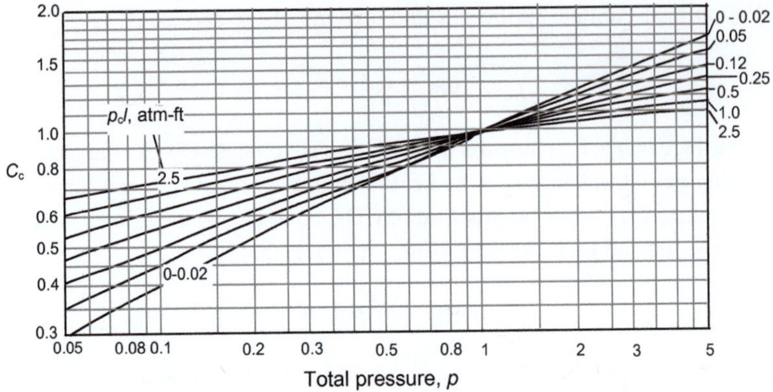

Fig. 12.5 Pressure correction factor for CO_2. Hottel HC in McAdams WH, Heat Transmission, 3rd edition, McGraw-Hill, NY, copyright 1954. Material is reproduced with permission of McGraw-Hill Education

12.3.2 Radiation Heat from Surface (Wall)

The surface (area A_s, temperature T_s) gives out $A_s\,\sigma\,T_s^4$. The fraction absorbed by the gas is

$$q_{sg} = \alpha_g(A_s\sigma T_s^4) \tag{12.12}$$

The gas absorptivity α_g equals ε_g when T_s is nearly equal to T_g. If T_s differs significantly from T_g, the estimate of α_g is done according to the following empirical rule.

Let the participating medium is CO_2 only, $\alpha_g = \alpha_c$. The absorptivity α_c is related to the ε_c by the following relation.

$$\alpha_c = \{\varepsilon_c[T_s, p_cL(T_s/T_g), p]\} \times (T_g/T_s)^{0.65} \tag{12.13}$$

where ε_c is the emissivity of CO_2 at surface or wall temperature T_s, and the parameter p_cL is replaced by $p_cL(T_s/T_g)$. The correction factor C_c is to be considered when $p \neq 1$ atm.

A similar procedure is to be followed if water vapour is the only participating medium, Eqs. (12.10) and (12.13) presented for CO_2 have been suggested in a similar way for water vapour as

$$\varepsilon_w = f(T_g, p_wL, p) \tag{12.14}$$

$$\alpha_w = \{\varepsilon_w[T_s, p_wL(T_s/T_g), p]\} \times (T_g/T_s)^{0.45} \tag{12.15}$$

The relevant charts are given as Figs. 12.6 and 12.7. Correction factor C_w is to be used when $p \neq 1$ atm. Note that when $p = 1$ atm, C_w is to be taken as unity. However, as per Fig. 12.7, $C_w = 1$ only when $(p + p_w)/2 = 0.5$. It seems that a very low concentration of water vapour has been considered in the preparation of the emissivity chart of water vapour so that $p + p_w \approx 1$.

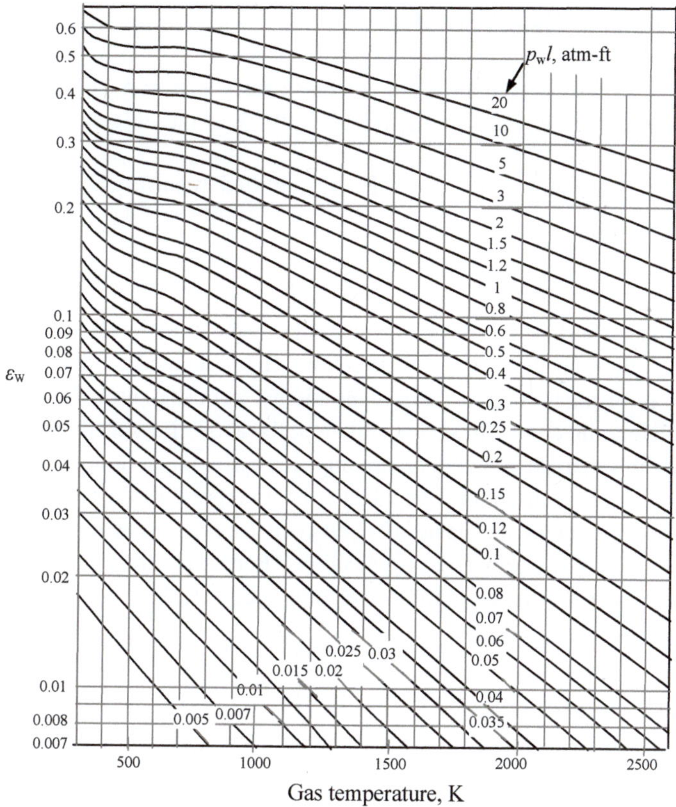

Fig. 12.6 Emissivity of water vapour. Hottel HC in McAdams WH, Heat Transmission, 3rd edition, McGraw-Hill, NY, copyright 1954. Material is reproduced with permission of McGraw-Hill Education

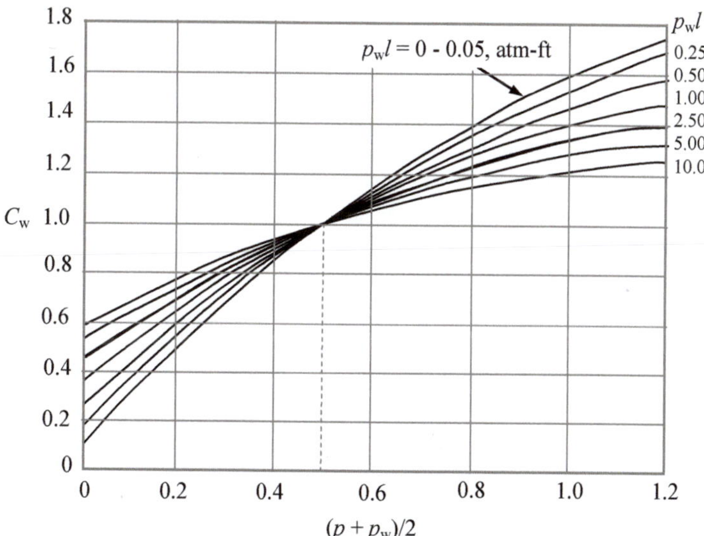

Fig. 12.7 Pressure correction factor for water vapour. Hottel HC in McAdams WH, Heat Transmission, 3rd edition, McGraw-Hill, NY, copyright 1954. Material is reproduced with permission of McGraw-Hill Education

12.3.3 Net Rate of Heat Transfer

The net rate of heat transfer from the gas to the black surface (area A_s) located at the centre of the base of the hemispheric gas volume, refer Fig. 12.3, is the difference of q_{gs} and q_{sg}

$$\begin{aligned}(q_{gs})_{net} &= \text{Energy emitted by the gas} - \text{Energy emitted by surface absorbed by the gas}\\ &= q_{gs} - q_{sg}.\end{aligned} \tag{12.16}$$

Since a small area surface has been considered here, the radiation from the surface travels equal distance L through the gas in all directions. If we consider other gas volumes and radiating surfaces of other shapes, the radiant energy from the surface will travel different distances in different directions. In such a situation, the above presented charts and equations developed for hemispherical gas volume are applicable if a mean or equivalent beam length L as given in Table 12.1 is used; the mean beam length represents the radius of an equivalent hemisphere. The table gives the value of equivalent length for several gas volume shapes frequently encountered in engineering applications. For an arbitrary volume V, surrounded by the surface of area A_s, the approximate relation suggested is

$$L = 3.6\frac{V}{A_s} \tag{12.17}$$

The net heat exchange between a gas volume and isothermal black surface can be written as, in a way similar to Eq. (12.16),

Table 12.1 Mean or equivalent beam length L for gas radiation

Shape of the gas volume	Characteristic dimension	Equivalent or mean beam length, L
Hemisphere, radiation to element on centre of the base	Radius R	R
Sphere, radiation to internal surface	Diameter D	0.65 D
Infinite cylinders, radiation to entire internal surface	Diameter D	0.95 D
Circular cylinder with height D, radiation to entire surface	Diameter D	0.6 D
Circular cylinder with height D, radiation to a spot on the centre of the base	Diameter D	0.77 D
Semi-infinite cylinder, radiation to entire base	Diameter D	0.65 D
Semi-infinite cylinder, radiation to a spot on the centre of the base	Diameter D	0.9 D
Infinite parallel planes	Spacing S	1.8S
Cube, radiation to one base	Side a	0.66a
Space between two tubes in an infinite bank of tubes with tube diameter = clearance between tubes, radiation to a single tube:	Tube diameter D	
Tube centres on equilateral triangles		3.0 D
Tube centres on squares		3.5 D

$$
\begin{aligned}
(q_{gs})_{net} &= \text{Energy emitted by the gas} - \text{Energy from the enclosure absorbed by the gas} \\
&= \varepsilon_g A_s \sigma T_g^4 - \alpha_g A_s \sigma T_s^4
\end{aligned}
\tag{12.18}
$$

12.3.4 Mixture of CO_2 and H_2O Vapour

We discussed the case when only one participating component (either carbon dioxide or water vapour) was present in the gas mixture. When the mixture contains both CO_2 and water vapour, both of them emit radiation. The total radiation is slightly less than the sum of the radiations emitted by carbon dioxide and water vapour alone because a part of the radiation of one of the gases is absorbed by the other (emission bands of these radiating gases overlap each other) before it is emitted. The combined emissivity is estimated by determining the individual emissivity values of the two separately and then applying an emissivity correction factor $\Delta\varepsilon$, i.e.

$$
\varepsilon_g = \varepsilon_c + \varepsilon_w - \Delta\varepsilon \tag{12.19}
$$

The correction factor $\Delta\varepsilon$ is plotted in Fig. 12.8. It is a function of temperature T_g, $L(p_c + p_w)$, and proportions of carbon dioxide and water vapour expressed by $p_w/(p_c + p_w)$. The value of $\Delta\varepsilon$ is the maximum when the percentages of carbon dioxide and water vapour are comparable.

Similarly, the estimate of the combined absorptivity is made

$$
\alpha_g = \alpha_c + \alpha_w - \Delta\alpha \tag{12.20}
$$

where $\Delta\alpha = \Delta\varepsilon$ and is read from Fig. 12.8 at the surface temperature T_s.

After ε_g and α_g are determined, the mixture can be treated as a single radiating gas and the net heat transfer is determined from Eq. (12.18).

The gas emissivity charts presented here (Figs. 12.4, 12.5, 12.6, 12.7 and 12.8) were developed by Hottel (1954) from experimental data. Additional charts for CO, SO_2, CH_4, etc. can be found in Hottel and Sarofim (1967).

12.3.5 Gray Enclosure

When the enclosure is not black, the net rate of heat transfer must be determined by the estimate of successive absorption and reflection between the gas and the enclosure. This analysis is cumbersome. In such cases, an appropriate estimate of the effective emissivity ε', as defined below, is used and the heat exchange is

$$
\begin{aligned}
(q_{gs})_{gray} &= \varepsilon'(q_{gs})_{black} \\
&= \varepsilon'(\varepsilon_g A_s \sigma T_g^4 - \alpha_g A_s \sigma T_s^4)
\end{aligned}
\tag{12.21}
$$

where $\varepsilon' = (1 + \varepsilon_s)/2$ when surface emissivity $\varepsilon_s > 0.8$.

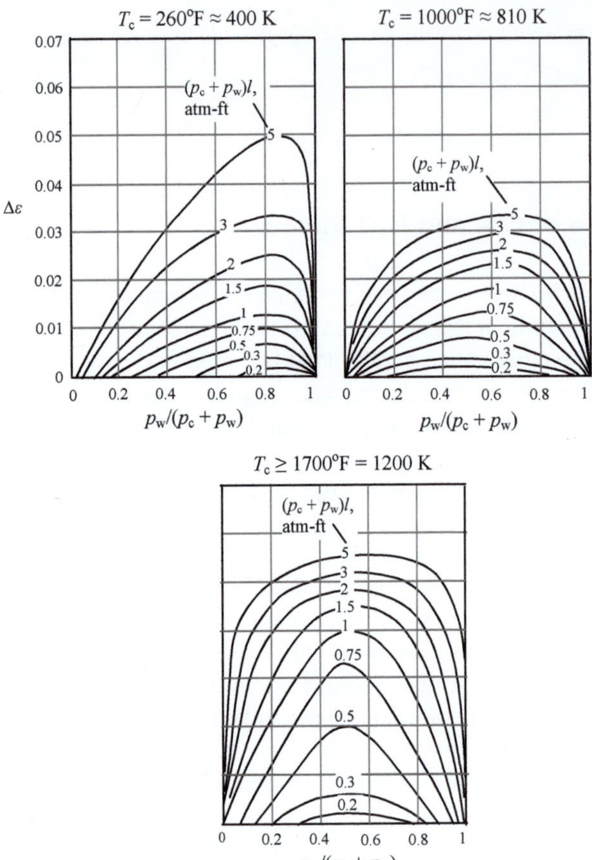

Fig. 12.8 Correction factor $\Delta\varepsilon$. Hottel HC in McAdams WH, Heat Transmission, 3rd edition, McGraw-Hill, NY, copyright 1954. Material is reproduced with permission of McGraw-Hill Education

Example 12.1 A cylinder, 200 mm in diameter and 1 m in length has closed ends. It contains a mixture of dry air and carbon dioxide. The total pressure and temperature of the mixture are 1 bar and 1000°C, respectively. The partial pressure of the carbon dioxide is 0.1 bar. Find the emissivity of the mixture of the gases in the cylinder (1 m = 3.2808 ft).

Solution

The mean beam length,

$$\begin{aligned} L &= 3.6\frac{V}{A_s} \\ &= 3.6 \times \frac{\frac{\pi}{4}D^2 L}{\pi DL + 2 \times \frac{\pi}{4}D^2} \\ &= 3.6 \times \frac{\frac{\pi}{4} \times 0.2^2 \times 1}{\pi \times 0.2 \times 1 + 2 \times \frac{\pi}{4} \times 0.2^2} = 0.1636 \text{ m}. \end{aligned} \tag{12.17}$$

Hence, for the carbon dioxide product, $p_c L \approx 0.1 \times 0.1636 = 0.01636$ atm-m. From Fig. 12.4, $\varepsilon_c \approx 0.055$ at $T_g = 1273$ K and $p_c L = 0.01636$ atm-m.

Example 12.2 A 2 m side cubic chamber is filled with a gas mixture at a total pressure of 3 atm and at a temperature of 1200 K. The partial pressure of carbon dioxide in the mixture is 0.3 atm. What is the emissivity of the gas volume if the other gases in the mixture are transparent and non-condensing? (1 m = 3.2808 ft).

Solution

The emissivity of the gas equals the emissivity of carbon dioxide in the mixture, i.e. $\varepsilon_g = \varepsilon_c$.

The mean beam length, from Table 12.1, for the cubic volume, $L = 0.66\, a = 0.66 \times 2 = 1.32$ m.

Partial pressure of CO_2, $p_c = 0.3$ atm.

Therefore, the product $p_cL = 0.3 \times 1.32 = 0.396$ atm-m.

From Fig. 12.4, corresponding to $T_g = 1200$ K and $p_cL = 0.396$ atm-m (1.3 atm-ft),

$$(\varepsilon_c)_{p=1\ \text{atm}} \approx 0.16.$$

From Fig. 12.5, corresponding to $p = 3$ atm and $p_cL = 0.396$ atm-m (1.3 atm-ft),

$$C_c = 1.1.$$

Hence,

$$\begin{aligned} \varepsilon_c &= (\varepsilon_c)_{p=1\ \text{atm}} \times (C_c)_{p\neq 1\ \text{atm}} \qquad (12.11) \\ &= 0.16 \times 1.1 = 0.176. \end{aligned}$$

Example 12.3 Calculate the net rate of heat transfer from the gas to the chamber of Example 12.2. The chamber wall is at a temperature of 800 K and the surface is black.

Solution

From Eq. (12.18), the net rate of heat transfer is

$$q = \sigma A[\varepsilon_c(T_g)T_g^4 - \alpha_c(T_s)T_s^4]. \qquad \text{(i)}$$

From Example 12.2, $\varepsilon_c(T_g) = 0.176$.

To find $\alpha_c(T_s)$, we read ε_c at $T_s = 800$ K from Fig. 12.4 at $p_cL(T_s/T_g) = 0.396$ (800/1200) = 0.264 atm-m, which gives

$$\varepsilon_c(T_s) = 0.15$$

and

$$\alpha_c(T_s) = \varepsilon_c(T_s) \times (T_g/T_s)^{0.65} = 0.15 \times (1200/800)^{0.65} = 0.195 \text{ at 1 atm}.$$

The correction factor $C_c = 1.12$ from Fig. 12.5 at $p_cL(T_s/T_g) = 0.264$ atm-m and $p = 3$ atm. This gives

$$[\alpha_c(T_s)]_{p=3\text{atm}} = 1.12 \times 0.195 = 0.218.$$

(**Note**: Some authors evaluate the pressure correction factors C_c and C_w for α_c and α_w using p_cL and p_wL, as in emissivity calculations, i.e. they presume that C_c and C_w do not change with surface temperature.)

Hence, Eq. (i) gives

$$q = \sigma A\left[\varepsilon_c(T_g)T_g^4 - \alpha_c(T_s)T_s^4\right]$$

or

$$q = 5.67 \times 10^{-8} \times (6 \times 2^2) \times \left[0.176 \times 1200^4 - 0.218 \times 800^4\right]/1000 = 375.12\,\text{kW}.$$

Example 12.4 Flue gases from a coal burning boiler furnace pass through a cylindrical pipe of 0.8 m diameter. The gases consist of 10% CO_2 by volume and rest N_2 and O_2. The temperature of the flue gases is 1000 K while that of the pipe surface is 600 K. Determine the heat exchange between the gas and pipe surface for a unit length of the pipe. The gas pressure is 1 atm and the pipe surface can be treated as black.

Solution

$$\frac{q}{A} = \sigma\left[\varepsilon_c(T_g)T_g^4 - \alpha_c(T_s)T_s^4\right] \tag{i}$$

where

$$\sigma T_g^4 = 5.67 \times 10^{-8} \times 1000^4 = 56700\,\text{W/m}^2$$

and

$$\sigma T_s^4 = 5.67 \times 10^{-8} \times 600^4 = 7348.3\,\text{W/m}^2.$$

The equivalent beam length, from Table 12.1, for an infinite cylinder, is

$$L = 0.95D = 0.95 \times 0.8 = 0.76\,\text{m}.$$

For the partial pressure of 0.1 atm of CO_2, product $p_cL = 0.1 \times 0.76 = 0.076$ atm-m. From Fig. 12.4, corresponding to $T_g = 1000$ K and $p_cL = 0.076$ atm-m,

$$\varepsilon_c(T_g) \approx 0.11.$$

For $\alpha_c(T_s)$, the required parameters are

$$T_s = 600\,K,\ p_cL(T_s/T_g) = 0.076 \times (600/1000) = 0.0456\ \text{atm-m}.$$

From Fig. 12.4,

$$\varepsilon_c(T_s) \approx 0.08.$$

The absorptivity from Eq. (12.13) is

$$\alpha_c(T_s) = \varepsilon_c(T_s) \times (T_g/T_s)^{0.65} = 0.08 \times (1000/600)^{0.65} = 0.1115.$$

Substituting values of different terms in Eq. (i) gives

$$\begin{aligned}\frac{q}{A} &= \sigma[\varepsilon_c(T_g)T_g^4 - \alpha_c(T_s)T_s^4] \\ &= 0.11 \times 56700 - 0.1115 \times 7348.3 = 5417.7\,\text{W/m}^2.\end{aligned}$$

For unit length of the pipe,

$$\frac{q}{L} = \frac{q}{A}(\pi D) = 5417.7 \times (\pi \times 0.8) = 13616\,\text{W/m}.$$

Example 12.5 If in the previous example, the total gas pressure is 0.8 atm, what will be the magnitude of heat exchange?

Solution

For the gas pressure of 0.8 atm, the partial pressure of CO_2 is 0.8 × 1.0 = 0.08 atm and the product p_cL = 0.08 × 0.76 = 0.061 atm-m.

From Fig. 12.4, corresponding to T_g = 1000 K and p_cL = 0.061 atm-m,

$$\varepsilon_c(T_g) \approx 0.1.$$

The correction factor for total pressure other than 1.0 atm, from Fig. 12.5, is

$$C_c \approx 0.95.$$

Considering this correction factor,

$$\varepsilon_c(T_g)_{p=0.8\text{atm}} = 0.95 \times 0.1 = 0.095.$$

For $\alpha_c(T_s)$, the required parameters are

$$T_s = 600\,\text{K},\ p_cL(T_s/T_g) = 0.061 \times (600/1000) = 0.0366\,\text{atm-m}.$$

From Fig. 12.4,

$$\varepsilon_c(T_s) \approx 0.075.$$

The absorptivity from Eq. (12.13), considering the correction factor $C_c \approx 0.95$ from Fig. 12.5, is

$$\alpha_c(T_s) = C_c[\varepsilon_c(T_s) \times (T_g/T_s)^{0.65}] = 0.95 \times 0.075 \times (1000/600)^{0.65} = 0.0993.$$

Substituting values of different terms in Eq. (i) of the previous example gives

$$\frac{q}{A} = \sigma[\varepsilon_c(T_g)T_g^4 - \alpha_c(T_s)T_s^4]$$
$$= 0.095 \times 56700 - 0.0993 \times 7348.3 = 4656.8\,\text{W/m}^2.$$

For unit length of the pipe,

$$\frac{q}{L} = \frac{q}{A}(\pi D) = 4656.8 \times (\pi \times 0.8) = 11704\,\text{W/m}.$$

Example 12.6 In Example 12.4, if the emissivity of the pipe surface is 0.9, what will be the heat exchange?

Solution

Using Eq. (12.21), we have

$$\frac{q_{\text{gray}}}{q_{\text{black}}} = \frac{\varepsilon_s + 1}{2} \text{ for } \varepsilon_s > 0.8$$
$$= \frac{0.9 + 1}{2} = 0.95$$

Hence, $q_{\text{gray}} = 0.95 \times q_{\text{black}}$ gives

$$\left(\frac{q}{L}\right)_{\text{gray}} = 0.95 \times 13616 = 12935.2\,\text{W/m length of the pipe.}$$

Example 12.7 The flue gas of Example 12.4 is from the burning of a hydrocarbon fuel and consists of 10% CO_2 and 8% H_2O by volume (rest of N_2 and O_2). Determine the heat exchange if the temperatures involved are the same as in Example 12.4. The gas pressure is 1 atm.

Solution

For a mixture of CO_2 and water vapour, the empirical relations for ε_g and α_g are

$$\varepsilon_g = \varepsilon_c + \varepsilon_w - \Delta\varepsilon \quad \text{(i)}$$

$$\alpha_g = \alpha_c + \alpha_w - \Delta\alpha \quad \text{(ii)}$$

Since the partial pressure of CO_2 is the same as in Example 12.4, the values of $\varepsilon_c(T_g)$ and $\alpha_c(T_s)$ will remain the same, i.e. $\varepsilon_c(T_g) \approx 0.11$ and $\alpha_c(T_s) = 0.1115$.

We now determine ε_w and α_w and correction factors $\Delta\varepsilon$ and $\Delta\alpha$.

The equivalent beam length is

$$L = 0.95D = 0.95 \times 0.8 = 0.76\,\text{m}.$$

For the partial pressure of 0.08 atm of water vapour, product $p_w L = 0.08 \times 0.76 = 0.0608$ atm-m. From Fig. 12.6, corresponding to $T_g = 1000$ K and $p_w L = 0.0608$ atm-m,

$$\varepsilon_w(T_g) \approx 0.095$$

For $\alpha_w(T_s)$, the values of the required parameters are

$$T_s = 600\,\text{K},\ p_w L(T_s/T_g) = 0.0608 \times (600/1000) = 0.0365\,\text{atm-m}.$$

From Fig. 12.6,

$$\varepsilon_w(T_s) = 0.1.$$

Hence,

$$\alpha_w(T_s) = \varepsilon_w(T_s) \times (T_g/T_s)^{0.45} = 0.1 \times (1000/600)^{0.45} = 0.1258.$$

For the correction factors $\Delta\varepsilon$ and $\Delta\alpha$, the values of the required parameters in Fig. 12.8 are

$$(p_c + p_w)L = (0.1 + 0.08) \times 0.76 = 0.1368\,\text{atm-m}.$$

and

$$\frac{p_w}{p_c + p_w} = \frac{0.08}{0.1 + 0.08} = 0.444.$$

From Fig. 12.8,

$$\Delta\varepsilon = 0.012\,\text{at}\,1000\,\text{K}(727\,°\text{C}).$$

At 327°C and $(p_c L + p_w L)(T_s/T_g) = 0.0456 + 0.0365 = 0.0821$ atm-m, $\Delta\alpha = \Delta\varepsilon = 0.002$. From Eqs. (i) and (ii),

$$\varepsilon_g = \varepsilon_c + \varepsilon_w - \Delta\varepsilon = 0.11 + 0.095 - 0.012 = 0.193 \quad \text{(i)}$$

$$\alpha_g = \alpha_c + \alpha_w - \Delta\alpha = 0.1115 + 0.1258 - 0.002 = 0.2353 \quad \text{(ii)}$$

Substituting values of different terms in the heat transfer equation gives

$$\begin{aligned}\frac{q}{A} &= \sigma[\varepsilon_g(T_g)T_g^4 - \alpha_g(T_s)T_s^4] \\ &= 0.193 \times 56700 - 0.2353 \times 7348.3 = 9214\,\text{W/m}^2.\end{aligned}$$

For unit length of the pipe,

$$\frac{q}{L} = \frac{q}{A}(\pi D) = 9214 \times (\pi \times 0.8) = 23157\,\text{W/m}.$$

In general, the emissivity (ε_c and ε_w) increases with the increase in the value of product $p_c L$ and $p_w L$ because with the increase in the partial pressure of the participating gas and the gas volume (i.e. increase of L), the number of molecules of the participating gas increases.

The emissivity can also be seen to increase in Figs. 12.5 and 12.7 with the increase in the total pressure of the mixture.

Example 12.8 Flue gases from the burning of a hydrocarbon fuel consisting of 15% CO_2 and 15% water vapour by volume (rest of N_2 and O_2) pass through a space between two large parallel black plates. The gap between plates is 0.8 m. The temperature of the flue gases T_g is 1200 K while that of the surfaces of both plates T_s is 600 K. Determine the net radiation flux to the plates. The gas pressure is 2 atm.

Solution

The equivalent beam length, from Table 12.1, is

$$L = 1.8S = 1.8 \times 0.8 = 1.44 \text{ m}.$$

For a mixture of CO_2 and water vapour, the empirical relations for ε_g and α_g are

$$\varepsilon_g = \varepsilon_c + \varepsilon_w - \Delta\varepsilon \tag{i}$$

$$\alpha_g = \alpha_c + \alpha_w - \Delta\alpha \tag{ii}$$

For the partial pressure of 0.15 × 2 = 0.3 atm of CO_2, product p_cL = 0.3 × 1.44 = 0.432 atm-m. From Fig. 12.4, corresponding to T_g = 1200 K and p_cL = 0.432 atm-m = 1.42 atm-ft, $\varepsilon_c \approx 0.175$ for total pressure p = 1 atm. Applying pressure correction factor $C_c \approx 1.05$ from Fig. 12.5, we have

$$\varepsilon_c(T_g) = C_c(\varepsilon_c)_{p=1} \approx 0.175 \times 1.05 = 0.1838.$$

Similarly for the partial pressure 0.15 × 2 = 0.3 atm of water vapour, product p_wL = 0.3 × 1.44 = 0.432 atm-m. From Fig. 12.6, corresponding to T_g = 1200 K and p_wL = 0.432 atm-m = 1.42 atm-ft, $\varepsilon_w \approx 0.23$. Applying correction factor C_w = 1.42 for $(p + p_w)/2$ = (2 + 0.3)/2 = 1.15 atm from Fig. 12.5, we have

$$\varepsilon_w(T_g) = C_w(\varepsilon_w)_{p=1} \approx 1.42 \times 0.23 = 0.3266.$$

For $\alpha_c(T_s)$, the values of the required parameters are

$$T_s = 600\,\text{K},\ p_cL(T_s/T_g) = 0.432 \times (600/1200) = 0.216 \text{ atm-m} = 0.709 \text{ atm-ft}.$$

From Fig. 12.4, $\varepsilon_c(T_s) \approx 0.13$. Pressure correction factor $C_c \approx 1.08$ from Fig. 12.5 for $p_cL(T_s/T_g)$ = 0.709 atm-ft. Hence,

$$\alpha_c(T_s) = C_c\varepsilon_c(T_s) \times (T_g/T_s)^{0.65} = 1.08 \times 0.13 \times (1200/600)^{0.65} = 0.2203.$$

For $\alpha_w(T_s)$, the values of the required parameters are

$$T_s = 600\,\text{K},\ p_wL(T_s/T_g) = 0.432 \times (600/1200) = 0.216 \text{ atm-m} = 0.709 \text{ atm-ft}.$$

From Fig. 12.6, $\varepsilon_w(T_s) \approx 0.25$. Pressure correction factor $C_w \approx 1.46$ from Fig. 12.7 for $p_w L(T_s/T_g) = 0.709$ atm-ft and $(p + p_w)/2 = (3 + 0.3)/2 = 1.15$ atm. Hence,

$$\alpha_w(T_s) = C_w \varepsilon_w(T_s) \times (T_g/T_s)^{0.45} = 1.46 \times 0.25 \times (1200/600)^{0.45} = 0.4986.$$

For the correction factor $\Delta\varepsilon$, the values of the required parameters in Fig. 12.8 are

$$(p_c + p_w)L = (0.3 + 0.3) \times 1.44 = 0.864 \text{ atm-m} = 2.84 \text{ atm-ft}$$

and

$$\frac{p_w}{p_c + p_w} = \frac{0.3}{0.3 + 0.3} = 0.5.$$

From Fig. 12.8, $\Delta\varepsilon = 0.05$ at 1200 K.

Read correction factor $\Delta\alpha$, at 600 K for $p_w/(p_c + p_w) = 0.5$ and $(p_c + p_w)L(T_s/T_g) = 0.432$ atm-m = 1.42 atm-ft. Since there is no chart for 600 K, we read $\Delta\varepsilon$ values at 400 K and 810 K, and take their average. This gives $\Delta\alpha = \Delta\varepsilon \approx 0.017$.

From Eqs. (i) and (ii),

$$\varepsilon_g = \varepsilon_c + \varepsilon_w - \Delta\varepsilon = 0.1838 + 0.3266 - 0.05 = 0.4604 \quad \text{(i)}$$

$$\alpha_g = \alpha_c + \alpha_w - \Delta\alpha = 0.2203 + 0.4986 - 0.017 = 0.7019. \quad \text{(ii)}$$

Irradiation on plate 1 is the sum of the radiation from the gas and radiation from plate 2 reaching surface of plate 1 after transmission through the gas layer, i.e.

$$\begin{aligned} G_1 &= \varepsilon_g E_g + \tau_g(T_2) E_{b2} \\ &= \varepsilon_g E_g + [1 - \alpha_g(T_2)] E_{b2} \\ &= \varepsilon_g \sigma T_g^4 + [1 - \alpha_g(T_2)] \sigma T_s^4. \end{aligned}$$

Substitution gives

$$\begin{aligned} G_1 &= 0.4604 \times 5.67 \times 10^{-8} \times 1200^4 + (1 - 0.7019) \times 5.67 \times 10^{-8} \times 600^4 \\ &= 56321.2 \text{ W/m}^2. \end{aligned}$$

Heat gained by plate 1 is the difference of irradiation G_1 on the plate and the energy emitted by the plate, i.e.

$$\begin{aligned} \frac{q_1}{A} &= G_1 - E_{b1} \\ &= 56321.2 - 7348.3 = 48972.9 \text{ W/m}^2. \end{aligned}$$

Since plate 2 is also at the same temperature, the heat gained by plate 2 is also 48972.9 W/m^2.

Example 12.9 Two infinite parallel black plates 1 and 2 are separated by a distance of 0.422 m. The gas mixture between the plates consists of 10% CO_2 and 8% H_2O by volume. The gas temperature is 1000 K while the plates are maintained at 600 K and 800 K, respectively. Calculate the heat gain by each plate. The gas pressure is 1 atm.

Solution

From the given data

$$E_g = \sigma T_g^4 = 5.67 \times 10^{-8} \times 1000^4 = 56700 \text{ W/m}^2$$
$$E_{b1} = \sigma T_1^4 = 5.67 \times 10^{-8} \times 600^4 = 7348.3 \text{ W/m}^2$$
$$E_{b2} = \sigma T_2^4 = 5.67 \times 10^{-8} \times 800^4 = 23224 \text{ W/m}^2.$$

The equivalent beam length, from Table 12.1, is

$$L = 1.8S = 1.8 \times 0.422 = 0.76 \text{ m}.$$

Since product p_cL and p_wL are the same as Example 12.7, we have

$$\varepsilon_g = 0.193$$
$$\alpha_g(T_1) = 0.2353.$$

At T_2 = 800 K,

$$p_cL(T_2/T_g) = 0.1 \times 0.76 \times (800/1000) = 0.0608 \text{ atm-m}$$
$$p_wL(T_2/T_g) = 0.08 \times 0.76 \times (800/1000) = 0.0486 \text{ atm-m}.$$

From the charts,

$$\varepsilon_c(T_2) \approx 0.095$$
$$\varepsilon_w(T_2) = 0.1.$$

For the correction factor $\Delta\alpha$ for $\alpha_g(T_s)$, the required parameters in Fig. 12.8 are (see Example 12.7)

$$\frac{p_w}{p_c + p_w} = \frac{0.08}{0.1 + 0.08} = 0.444.$$

From Fig. 12.8, $\Delta\alpha = \Delta\varepsilon \approx 0.005$.
From these values

$$\alpha_c(T_2) = \varepsilon_c(T_2) \times (T_g/T_2)^{0.65} = 0.095 \times (1000/800)^{0.65} = 0.1098$$
$$\alpha_w(T_2) = \varepsilon_w(T_2) \times (T_g/T_2)^{0.45} = 0.1 \times (1000/800)^{0.45} = 0.1156$$
$$\alpha_g(T_2) = \alpha_c(T_2) + \alpha_w(T_2) - \Delta\alpha(T_2) = 0.1098 + 0.1156 - 0.005 = 0.2204.$$

Irradiation on plate 1 is the sum of the radiation from the gas and radiation from plate 2 reaching the surface of plate 1 after transmission through the gas layer, refer Fig. 12.9, i.e.

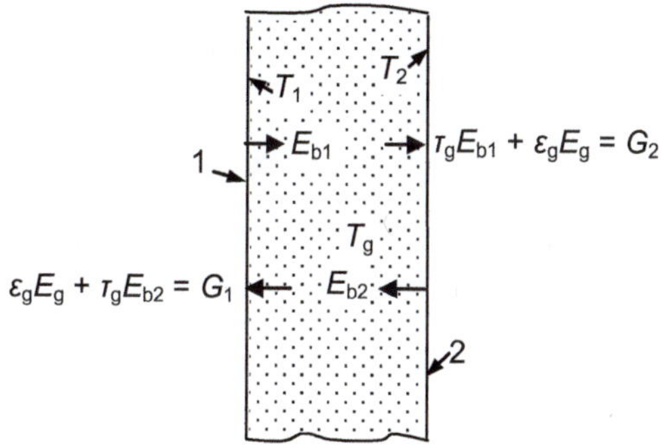

Fig. 12.9 Example 12.9

$$G_1 = \varepsilon_g E_g + \tau_g(T_2)E_{b2} = \varepsilon_g E_g + [1 - \alpha_g(T_2)]E_{b2}$$
$$= 0.193 \times 56700 + (1 - 0.2204) \times 23224 = 29048.5 \text{ W/m}^2.$$

Similarly

$$G_2 = \varepsilon_g E_g + [1 - \alpha_g(T_1)]E_{b1}$$
$$= 0.193 \times 56700 + (1 - 0.2353) \times 7348.3 = 16562.3 \text{ W/m}^2.$$

Heat gained by plate 1,

$$\frac{q_1}{A} = \text{irradiation} - \text{energy emitted by the plate 1}$$
$$= G_1 - E_{b1} = 29048.5 - 7348.3 = 21700.2 \text{ W/m}^2.$$

Similarly

$$\frac{q_2}{A} = \text{irradiation} - \text{energy emitted by the plate 2}$$
$$= G_2 - E_{b2} = 16562.3 - 23224 = -6661.7 \text{ W/m}^2.$$

Heat lost by the gas

$$\frac{q_g}{A} = \frac{q_1}{A} + \frac{q_2}{A} = 21700.2 - 6661.7 = 15038.5 \text{ W/m}^2.$$

12.4 Gray Gas Surrounded by Diffuse Gray Surfaces at Different Temperatures

The basic aim of the analysis presented in this section is to give the readers an idea of the technique to deal with the problems involving transmitting and absorbing medium.

Consider the system shown in Fig. 12.10 consisting of diffuse gray surfaces 1 and 2 at different but isothermal temperatures T_1 and T_2, respectively. The surfaces are separated by a transmitting and absorbing medium. The main simplifying assumptions are

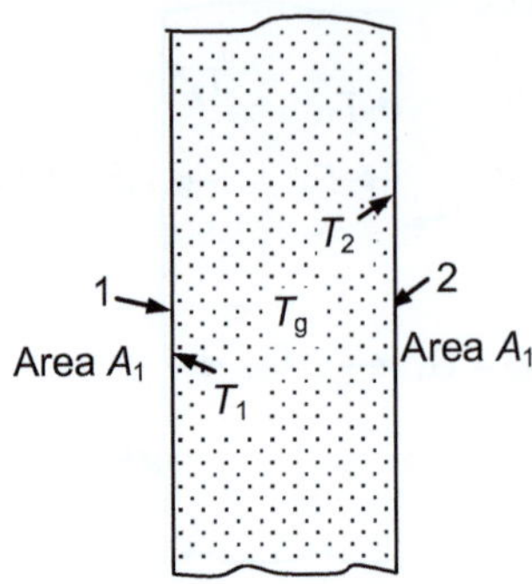

Fig. 12.10 Gray gas surrounded by diffuse gray surfaces at different temperatures

1. The medium is gas and behaves as a gray substance so that $\alpha_g = \varepsilon_g$
2. The gas is at a uniform temperature T_g.
3. The medium is non-reflecting, which is valid for gases. This may not be valid for other transmitting and absorbing media. For example, glass has reflectivity of the order of 0.1
4. Each surface is diffuse gray and isothermal.

We are interested in the calculation of the radiation exchange between the surface in the presence of the transmitting and absorbing gas.

The radiation leaving surface 1 is J_1A_1.

The radiation reaching surface 2 is $J_1A_1F_{12}\tau_g$, where τ_g is the transmissivity of the gas = 1 − α_g = 1 − ε_g for the gray gas.

Similarly, the radiation leaving surface 2 is J_2A_2 and reaching surface 1 is $J_2A_2F_{21}\tau_g$.

The net exchange between surfaces 1 and 2 is

$$J_1A_1F_{12}\tau_g - J_2A_2F_{21}\tau_g$$

Using reciprocity relation $A_1F_{12} = A_2F_{21}$, we obtain

$$A_1F_{12}\tau_g(J_1 - J_2)$$

So the resistance between J_1 and J_2 is

$$R_{12} = \frac{1}{A_1F_{12}\tau_g}$$

Now we consider the exchange process between surface 1 and the gas.

Radiation leaving the gas is

$$\varepsilon_g(\sigma T_g^4)$$

and reaching surface 1 is

$$A_1F_{1g}[\varepsilon_g(\sigma T_g^4)]$$

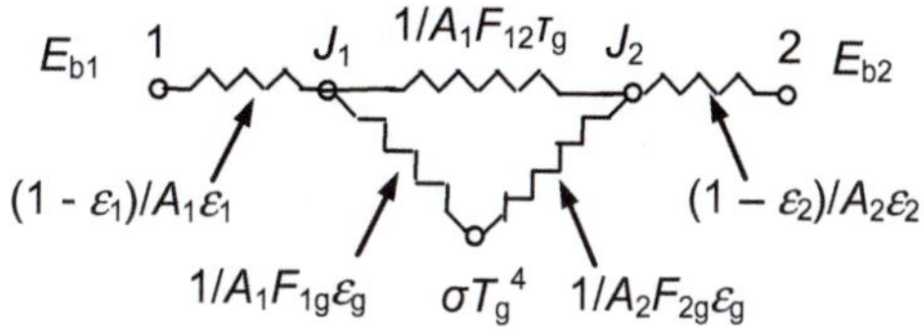

Fig. 12.11 Radiation network

The energy leaving surface 1 and absorbed by the gas is

$$J_1A_1F_{1g}\alpha_g = J_1A_1F_{1g}\varepsilon_g$$

The difference between the two is

$$J_1A_1F_{1g}\varepsilon_g - A_1F_{1g}[\varepsilon_g(\sigma T_g^4)] = A_1F_{1g}\varepsilon_g[J_1 - (\sigma T_g^4)]$$

So, the resistance is

$$R_{1g} = \frac{1}{A_1F_{1g}\varepsilon_g}$$

Following the same procedure, it can be shown that

$$R_{2g} = \frac{1}{A_2F_{2g}\varepsilon_g}$$

Figure 12.11 shows the network.

If there is no net heat change for the gas, σT_g^4 is like a reradiating body.

In the case of infinite parallel planes, F_{12}, F_{1g} and F_{2g} will be unity.

It is to note that the estimates of transmissivity and emissivity of a gas is very difficult because these properties vary both with the thickness of the gas layer and the temperature. If more than two surfaces are involved, the transmissivity can be different between different surfaces. The temperature of the gas may also vary and, in such a situation, the transmissivity and absorptivity also vary with the location. Such problems are solved by dividing the gas volume into layers.

Example 12.10 Two large parallel plates (ε_1 = 0.8 and ε_1 = 0.6) are separated by a small distance and are maintained at a temperature of T_1 = 1000 K and T_2 = 400 K, respectively. Space between the plates is filled with a gray gas (ε_g = 0.15) and rest is N_2 at a total pressure of 2 atm. Calculate the radiation heat exchange between the plates. Also, determine the equilibrium gas temperature.

Solution

Since the plates are large and separated by a small distance, $F_{12} = F_{21} = 1$, $F_{1g} = F_{2g} = 1$ and $A_1 = A_2$. The network is shown in Fig. 12.12.

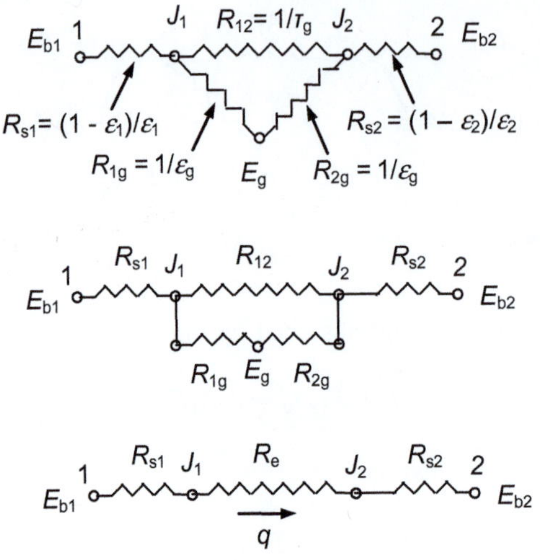

Fig. 12.12 Radiation network for the system of Example 12.10

The gas, in this problem, is behaving like a reradiating body, the resistance between J_1 and J_2 is

$$R_e = \left(\frac{1}{R_{12}} + \frac{1}{R_{1g} + R_{2g}}\right)^{-1} = \left(\tau_g + \frac{\varepsilon_g}{2}\right)^{-1}$$
$$= \left(1 - \varepsilon_g + \frac{\varepsilon_g}{2}\right)^{-1} = \left(1 - 0.15 + \frac{0.15}{2}\right)^{-1} = 1.081.$$

Surface resistances are

$$R_{s1} = \frac{1 - \varepsilon_1}{\varepsilon_1} = \frac{1 - 0.8}{0.8} = 0.25$$
$$R_{s2} = \frac{1 - \varepsilon_2}{\varepsilon_2} = \frac{1 - 0.6}{0.6} = 0.667.$$

Hence, the resistance between E_{b1} and E_{b2}, refer Fig. 12.12, is

$$R_t = R_{s1} + R_e + R_{s2} = 0.25 + 1.081 + 0.667 = 1.998.$$

Heat transfer rate between plates

$$q_{12} = \frac{E_{b1} - E_{b2}}{R_t} = \frac{5.67 \times 10^{-8}(1000^4 - 400^4)}{1.998} = 27652\ \text{W/m}^2.$$

Note: If transparent gas, such as air, is present between the plates, then

$$q_{12} = \frac{E_{b1} - E_{b2}}{\frac{1}{\varepsilon_1} + \frac{1}{\varepsilon_2} - 1} = 28825\ \text{W/m}^2.$$

Introduction of a gas having $\tau < 1$, reduces the radiation heat transfer between the surfaces. Lower the transmissivity τ of the gas, greater will be the reduction in the radiation heat transfer between the surfaces.

Since $R_{1g} = R_{2g}$, E_g is mean of J_1 and J_2, i.e.

$$E_g = \frac{J_1 + J_2}{2} = \sigma T_g^4 \tag{i}$$

where

$$J_1 = E_{b1} - q_{12}R_{s1} = 5.67 \times 10^{-8} \times 1000^4 - 27652 \times 0.25 = 49787$$
$$J_2 = E_{b2} + q_{12}R_{s2} = 5.67 \times 10^{-8} \times 400^4 + 27652 \times 0.667 = 19895.$$

This gives equilibrium gas temperature, from Eq. (i)

$$T_g = \left[\frac{1}{\sigma}\left(\frac{J_1 + J_2}{2}\right)\right]^{1/4} = \left[\frac{1}{5.67 \times 10^{-8}}\left(\frac{49787 + 19895}{2}\right)\right]^{1/4} = 885 \text{ K}.$$

12.5 Flames

Anthracite, coke and some gaseous fuels (hydrogen, blast furnace gas, generator gas, liquefied petroleum gas, etc.) have non-luminous flames. The faint bluish colour of these flames is not connected with an energy flux of any importance.

Bituminous fuels, e.g., wood, lignite and the younger coals, burn with a luminous flame of yellow glow. Oil fuels can be fired to provide a varying degree of luminosity depending on the design of the burner, extent of atomization and the percentage of excess air. Pulverized coal burners produce a flame containing incandescent solid particles and a greater degree of luminosity than the minimum obtainable with the oil burners. Stoker firing gives an incandescent fuel bed. The carbon particles (mostly soot) and flying ash are formed from the breakdown of the hydrocarbons in the flames. The soot, which is the most important product formed when burning hydrocarbons, emits in a continuous spectrum in the visible and infrared regions, and imparts the yellowish colour to the flame. Radiation of such flames may be two- to four-fold greater than that from the gases with the presence of CO_2 and H_2O molecules.

12.5.1 Luminous Flames

There are several factors that complicate the radiative heat transfer from the burning flames. There is a temperature variation within the flame due to the simultaneous production and loss of heat. The emission from the luminous flames depends on the number and distribution of the soot, which varies greatly with the amount of air, and temperature of the air and gas, as well as their mixing. The soot particles may be present in the form of spherical particles (varying in size from 500 to 3000 angstrom in diameter), agglomerated masses, or sometimes in long filaments.

Due to the uncertainties, as discussed above, the radiation properties of the soot are only known to a first approximation. The flame is considered as a blackbody and the heat transfer expression is multiplied by an empirical factor p, which takes account of the fact that the flame is not really black. Eckert and Drake (1972) suggested the following expression for the radiative heat transfer.

$$q = p[\varepsilon_w \sigma A(T_f^4 - T_w^4)] \tag{12.22}$$

where T_f is flame temperature, T_w is wall temperature, ε_w is emissivity of furnace wall and A is the area of the wall.

The factor p mainly depends on the type of the fuel and size of the furnace. For large furnaces, p is nearly unity. For furnaces of the usual size, it lies from 0.6 to 1.0. Furnaces are of different designs. Whole or part of the furnace refractory walls may be covered with tubes forming the water wall or convection tubes. The problem in the furnace design also consists in determining the absorptivity of the walls, as well as the actual surface area of the wall. However, in furnaces, the walls are usually rough and soot covered so they act practically as black surfaces. If the net radiative heat flow to a cooling surface consists of a row of tubes before a refractory wall, area A is determined by multiplying the area of the furnace wall before which the tubes are arranged by a factor (see Kern 1950).

Mikheyev (1968) has given the following approximate equation.

$$q = \varepsilon_f \varepsilon_w \sigma A(T_f^4 - T_w^4) \tag{12.23}$$

where

ε_f flame emissivity (see Table 12.2)
T_f effective flame temperature = $(T_1 T_2)^{0.5}$
T_1 theoretical temperature of combustion
T_2 temperature of combustion products at the outlet of the furnace.

12.5.2 Non-luminous Flames

The phenomenon of the radiation from these flames has been studied well. If the products of combustion from the fuel consist of carbon dioxide and water vapour, the method given in Sect. 12.3 can be used to compute the radiative heat transfer from the flame.

Table 12.2 Flame emissivity ε_f for an infinitely thick layer (Mikheyev 1968)

Kind of flame	ε_f
Non-luminous gas flame or flame of anthracite	0.4
Luminous flame of pulverized anthracite	0.45
Luminous flame of lean coal	0.6
Luminous flame of coal with large volatile content, brown coal, etc., burned in a layer or pulverized form	0.7

12.6 Summary

Gases such as carbon dioxide, carbon monoxide and sulfur dioxide, and water vapour are capable of emitting and absorbing the heat radiation. Carbon dioxide and water vapour are formed when combustion of hydrocarbon fuels takes place and the study of these gases is of practical importance hence the discussion in this chapter has been basically confined to the behavior of carbon dioxide, water vapour and their mixtures.

The radiation exchange between an absorbing and emitting gas and a solid surface is considerably more complex than exchanges between solid surfaces through a transparent medium. The specific features of the gaseous radiation are

(i) Gases emit and absorb radiation in certain narrow wavelength regions called bands. Outside these bands, these gases are practically transparent and their emissive power is zero. Thus, the gases are selective absorbers and emitters. These bands, their width and number are different for different gases.

(ii) In gases, the emission and absorption occur over their volume. As the radiation passes through a gas, reduction in its intensity takes place. There is an exponential decay of the radiation intensity with distance as the radiation travels through the gas volume, which is known as *Beer's law*, Eq. (12.3). This reduction depends on the thickness of the gas volume and the partial pressure of the gas.

(iii) As the thickness of the gas increases, the absorptivity and emissivity increase.

(iv) In general, the reflectivity of the gases is zero, therefore, $\alpha_\lambda + \tau_\lambda = 1$ where τ_λ is defined by Eq. (12.4).

(v) The total emissive power of a gas equals the sum of radiation of all bands.

The procedure for the calculation of net heat exchange was presented by Hottel. A simplified case of total radiation heat exchange between a black enclosure and isothermal (uniform temperature) gas volume consisting of CO_2 and/or water vapour is given in Sect. 12.3.

The emissivity ε_c of CO_2 is a function of the gas mixture temperature T_g, the partial pressure of CO_2 in the gas volume times the characteristic dimension (p_cL), and the total pressure of the mixture p as presented in Eq. (12.10). Characteristic dimension or equivalent beam length L as given in Table 12.1 is used. The emissivity ε_c is determined from Hottel's chart for CO_2, refer Fig. 12.4, when the total pressure of the gas mixture is 1 atm. When the total pressure of the gas mixture is different from 1 atm, the emissivity value read from Fig. 12.4 is multiplied by a pressure correction factor C_c from Fig. 12.5.

The absorptivity α_c is defined by Eq. (12.13) where the value of ε_c from Fig. 12.4 is read at surface or wall temperature T_s and the parameter p_cL is replaced by $p_cL(T_s/T_g)$. The correction factor C_c is to be considered when $p \neq 1$ atm.

A similar procedure is to be followed if the water vapour is the only participating medium. Similar to equations for CO_2, Eqs. (12.14) and (12.15) for water vapour have been presented. The relevant charts are Figs. 12.6 and 12.7. The correction factor C_w from Fig. 12.7 is to be used when $p \neq 1$ atm.

The net rate of heat transfer from the gas to the black surface (area A_s) is calculated as net of the energy emitted by the gas and energy emitted by the surface which is absorbed by the gas, refer Eq. (12.16). When the gas mixture contains both CO_2 and water vapour, the total radiation is slightly less than the sum of the radiations emitted by carbon dioxide and water vapour. The combined emissivity/absorptivity is estimated by determining the individual emissivity/absorptivity values of the two separately and then applying an emissivity/absorptivity correction factor $\Delta\varepsilon$ given in Fig. 12.8.

When the enclosure is not black, the net rate of heat transfer is determined by using effective emissivity ε' defined as $\varepsilon' = (1 + \varepsilon_s)/2$, and the heat exchange is $(q_{gs})_{gray} = \varepsilon'$ $(q_{gs})_{black}$.

In Sect. 12.4, an analysis is presented for a case when space between diffuse gray surfaces at different temperatures is filled with a gray gas that behaves as a transmitting and absorbing medium.

In the end of the chapter, a brief introduction to radiative heat transfer from flames has been presented.

Review Questions

12.1 Write a short note on radiation from gases, vapours and flames.

12.2 What is Beer's law? How would you calculate the monochromatic transmissivity and absorptivity of a gas from this law?

12.3 Define mean beam length.

12.4 Explain the reasons for the use of factors C_c, C_w and $\Delta\varepsilon$.

12.5 What do you mean by a gray gas?

Problems

12.1 Determine mean beam length for a rectangular furnace of dimensions $1 \times 1 \times 4$ m^3.
[**Ans**. $L = 3.6(V/A) = 0.8$ m.]

12.2 Determine the emissivity of water vapour at 1200 K contained in a cubical container of 1 m side. The partial pressure of water vapour is 0.1 atm and the total pressure is 1.2 atm.
[**Ans**. $L = 0.66\ a = 0.66$ m, $p_wL = 0.1 \times 0.66 = 0.066$ atm-m, $\frac{1}{2}\ (p + p_w) = \frac{1}{2}$ $(1.2 + 0.1) = 0.65$ atm. From Figs. 12.6 and 12.7, $C_w = 1.2$ and $(\varepsilon_w)_{1\ atm} = 0.081$, respectively. This gives $(\varepsilon_w)_{1.2\ atm} = 0.081 \times 1.2 = 0.097$.]

12.3 Products of combustion in a 1.5 m side cubical furnace consist of CO_2 at 0.1 atm and water vapour at 0.12 atm. The total pressure is nearly atmospheric. Estimate the effective gas emissivity. The mixture temperature is 800 K.
[**Ans**. $L = 0.66 \times 1.5 = 0.99$ m, $p_cL = 0.1 \times 0.99 \approx 0.1$ atm-m, $\varepsilon_c = 0.11$ from Fig. 12.4; $p_wL = 0.12 \times 0.99 = 0.1188$ atm-m, $\varepsilon_w = 0.17$ from Fig. 12.6; $\Delta\varepsilon = 0.012$ from Fig. 12.8 at $p_w/(p_c + p_w) = 0.545$, $L(p_c + p_w) = 0.2178$ atm-m and 800 K. This gives $\varepsilon_g = \varepsilon_c + \varepsilon_w - \Delta\varepsilon = 0.11 + 0.17 - 0.012 = 0.268$.]

12.4 Flue gases from a boiler furnace pass through a cylindrical pipe of 1 m diameter. The gas consists of 12% CO_2 by volume and rest N_2 and O_2. The temperature of the flue gases is 800 K while that of the pipe surface is 400 K. Determine the gas emissivity and absorptivity. The gas pressure is 1 atm and the pipe surface can be treated as black. What is the heat transfer rate?
[**Ans**. $L = 0.95\ \ D = 0.95$ m; $p_c = 0.12$ atm, $p_cL = 0.114$ atm-m; $\varepsilon_c = 0.115$ from Fig. 12.4; $(p_cL) \times (T_s/T_g) = 0.114 \times (400/800) = 0.0576$ atm-m, $\varepsilon_c(T_s) = 0.09$;

$\alpha_c = \varepsilon_c(T_s) \times (T_g/T_s)^{0.65} = 0.09 \times (800/400)^{0.65} = 0.1412$, $q_{gs} = \sigma \quad A_s(\varepsilon_c T_g^4 - \alpha_c T_s^4) = 7.75$ kW/m length of the pipe.]

12.5 The combustion products of a hydrocarbon fuel are $CO_2 = 15\%$, H_2O (vapour) = 7% and rest N_2 and O_2. The products are in a cylindrical region 1.5 m long and 0.7 m in diameter and assumed to be uniformly mixed at a flame (non-luminous) temperature of 1400 K. The total pressure is 1 atm. Compute the radiation heat leaving the gas.
[**Ans.** $L = 3.6\ (V/A_s) = 0.51$ m, $p_c = 0.15$ atm, $p_cL = 0.0767$ atm-m, $\varepsilon_c = 0.085$ from Fig. 12.4; $p_w = 0.07$ atm, $p_wL = 0.0358$ atm-m, $\varepsilon_w = 0.043$ from Fig. 12.6; $\Delta\varepsilon$ 0.007 from Fig. 12.8 at $p_w/(p_c + p_w) = 0.32$, $(p_c + p_w)L = 0.1125$ atm-m and $T_g >$ 1200 K. This gives $\varepsilon_g = \varepsilon_c + \varepsilon_w - \Delta\varepsilon = 0.121$; $q = \sigma\ \varepsilon_g\ AT_g^4 = 107.2$ kW.]

12.6 Flue gas at near atmospheric pressure and 600°C flows through a rectangular section duct (600 × 600 mm²). The gas contains $CO_2 = 12\%$, H_2O (vapour) = 5% and rest N_2 and O_2. The duct surface emissivity is 0.9. What is the rate of heat flow to the duct surface per m length if the duct surface temperature is 400°C?
[**Ans.** $L = 3.6\ (V/A_s) = 0.54$ m, $p_c = 0.12$ atm, $p_cL = 0.0648$ atm-m, $\varepsilon_c = 0.1$ from Fig. 12.4 at 873 K; $p_w = 0.05$ atm, $p_wL = 0.027$ atm-m, $\varepsilon_w = 0.065$ from Fig. 12.6; $\Delta\varepsilon = 0.004$ from Fig. 12.8 at $p_w/(p_c + p_w) = 0.294$, $(p_c + p_w)L = 0.0918$ atm-m and $T_g = 873$ K; This gives $\varepsilon_g = \varepsilon_c + \varepsilon_w - \Delta\varepsilon = 0.161$; $p_cL(T_s/T_g) = 0.05$ atm-m, $p_wL(T_s/T_g) = 0.021$ atm-m, $\varepsilon_c(673) = 0.085$, $\varepsilon_w(673) = 0.068$, $\alpha_c = \varepsilon_c(673) \times (T_g/T_s)^{0.65} = 0.1$, $\alpha_w = \varepsilon_w(673) \times (T_g/T_s)^{0.45} = 0.0764$, $\Delta\alpha = \Delta\varepsilon(673) \approx 0.002$, $\alpha_g = \alpha_c + \alpha_w - \alpha = 0.1744$, $\varepsilon_w' = (1 + \varepsilon_s)/2 = 0.95$, $q = A\ \varepsilon_w'\ \sigma(\varepsilon_g\ T_g^4 - \alpha_g T_w^4) = 7464$ W/m.]

References

Eckert ERG, Drake RM Jr (1972) Analysis of heat and mass transfer. McGraw-Hill Kogakusha, Tokyo

Hottel HC (1954) Radiant heat transfer. In: McAdams WH (ed) Heat transmission, 3rd edn. McGraw-Hill Book Co, New York

Hottel HC, Sarofim AF (1967) Radiative transfer. McGraw-Hill Book Co, New York

Kern DQ (1950) Process heat transfer. McGraw-Hill, New York

Mikheyev M (1968) Fundamental of heat transfer. Mir Publishers, Moscow

Heat Transfer in Condensing Vapours and Boiling Liquids

13

13.1 Part A: Heat Transfer in Condensing Vapours

13.1.1 Introduction

The phenomenon of the heat transfer in condensing vapours is quite different from the convective heat transfer in single-phase fluids because it is accompanied by change of phase of the vapour.

If saturated vapour comes in contact with a surface at a temperature lower than the saturation temperature of the vapour, condensation of the vapour on the surface takes place. There are basically two ways in which the vapour condensation may take place. We shall discuss these ways by considering an example of a vertical surface.

13.1.1.1 Different Types of Condensation

Dropwise Condensation

In the dropwise condensation, the condensate forms a large number of individual droplets of varying diameters on the condensing surface instead of a continuous liquid film. This happens when the condensate does not wet the cooling surface. Such type of condensation can be achieved artificially by applying a thin layer of oil, kerosene or some fatty acids upon the surface or by adding these substances to the vapour. Besides this, the surface must be properly polished. The dropwise condensation provides very high heat transfer coefficient as compared to the filmwise condensation being discussed below; the heat transfer coefficients can be more than 10 times larger than film condensation.

Filmwise Condensation

In this mode, the condensate forms a continuous film on the surface. The condensation of pure vapour of wetting liquids results in the formation of a continuous film of the condensate, which moves down the vertical surface due to the gravitational force.

In an industrial apparatus, dropwise condensation may occur in one section of the apparatus, while other section may experience filmwise condensation. Since the filmwise condensation is more frequent, it will be discussed in the following sections.

R. Karwa, *Heat and Mass Transfer*,
https://doi.org/10.1007/978-981-15-3988-6_13

13.1.2 Nusselt's Film Condensation Theory

When filmwise condensation takes place on a vertical surface, the film flows downwards due to the gravitational effect. If the liquid film is thin and its velocity is low, a laminar flow of the condensate happens. Nusselt presented a simple theory for the calculation of heat transfer in the case of laminar film condensation on vertical or inclined surfaces.

The following assumptions have been made for Nusselt's analysis.

(i) Steady state condition
(ii) The vapour rejects the latent heat only and there is no subcooling of the condensate
(iii) The heat is transferred across the laminar film by conduction only. Convection in the condensate film is negligible because of the laminar flow.
(iv) There is good thermal contact between the surface and the condensate.
(v) Velocity gradient at the liquid–vapour interface is zero. No viscous effect at the liquid film–vapour interface
(vi) Inertia forces are negligible because the velocity of the condensate is negligible.
(vii) Effect of surface tension has not been considered
(viii) Change in kinetic and potential energies are negligible.
(ix) There are no impurities in the vapour or condensate.

13.1.2.1 Laminar Film Condensation on a Vertical Surface

As mentioned earlier when the vapour at saturation temperature t_s comes in contact with a vertical surface held at temperature t_w, where t_w is less than t_s, condensation takes place. As the condensate flows down the vertical plate, its thickness increases as shown in Fig. 13.1 because of the condensation of the vapour at the liquid–vapour interface.

At distance x from the top (leading) edge, let the thickness of the condensate film is δx. Heat transfer from the liquid–vapour interface to the surface takes place by conduction through the condensate film. Hence,

$$q_x = k(dxdz)\frac{t_s - t_w}{\delta x} \tag{i}$$

where dz is the dimension perpendicular to the plane of the paper.

On the other hand, the quantity of heat transferred may be written as

$$q_x = h_x(dxdz)(t_s - t_w) \tag{ii}$$

where h_x is the local heat transfer coefficient.

From Eqs. (i) and (ii), we get

$$h_x = \frac{k}{\delta x} \tag{13.1}$$

Hence, to determine the heat transfer coefficient, we should determine the thickness δx of the condensate film. The thickness of the film can be found from the Nusselt theory presented here.

The element ($dx \times dy \times dz$), refer Fig. 13.1, is under equilibrium of two forces:

(i) The force of viscous friction = $d\tau$ ($dx \times dz$), and
(ii) The net of the weight and buoyancy force due to the displaced vapour,

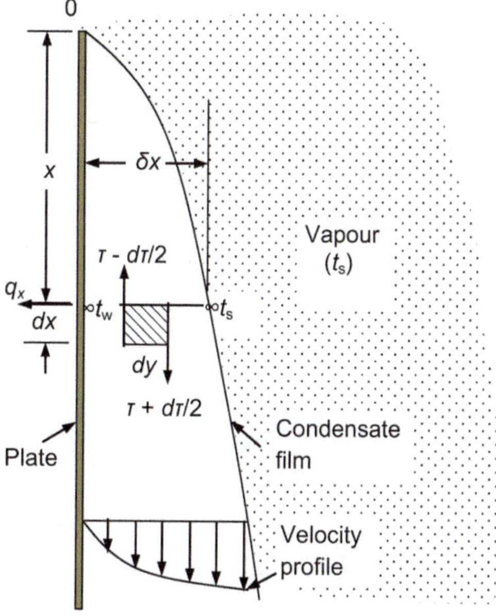

Fig. 13.1 Laminar film condensation on a vertical plate

$$= (\rho - \rho_v)(dV)g$$
$$= (\rho - \rho_v)(dx \cdot dy \cdot dz)g.$$

where ρ is the density of liquid and ρ_v is the density of vapour.
Since $\rho_v \ll \rho$, $(\rho - \rho_v) \approx \rho$, the net of the weight and buoyancy force

$$= \rho(dx \cdot dy \cdot dz)g.$$

Thus, the equilibrium of the forces gives

$$d\tau(dx \cdot dz) + \rho(dx \cdot dy \cdot dz)g = 0$$

or

$$d\tau = -\rho dyg \tag{iii}$$

According to Newton's law

$$\tau = \mu \frac{dv_x}{dy}$$

or

$$\frac{d\tau}{dy} = \mu \frac{d^2 v_x}{dy^2} \tag{iv}$$

From Eqs. (iii) and (iv),

$$\frac{d^2 v_x}{dy^2} = -\frac{\rho g}{\mu} \tag{v}$$

Integrating Eq. (v), we get

$$v_x = -\frac{\rho g}{2\mu} y^2 + C_1 y + C_2 \tag{vi}$$

The boundary conditions are

$$\text{at} \quad y = 0, \quad v_x = 0 \tag{a}$$

$$\text{at} \quad y = \delta x, \quad \frac{dv_x}{dy} = 0 \tag{b}$$

From the first boundary condition,

$$C_2 = 0.$$

Secondary boundary condition gives

$$\left(\frac{dv_x}{dy}\right)_{y=\delta x} = 0 = \left(-\frac{\rho g}{2\mu} 2y + C_1\right)_{y=\delta x} = -\frac{\rho g}{\mu}\delta x + C_1$$

or

$$C_1 = \frac{\rho g}{\mu}\delta x$$

Substitution of the values of the constants in Eq. (vi) gives

$$v_x = \left(\frac{\rho g}{\mu}\delta x\right) y - \frac{\rho g}{2\mu} y^2 \tag{vii}$$

The above equation is equation of velocity of the condensate. The mean velocity of the condensate through the film is

$$\begin{aligned} \overline{v_x} &= \frac{1}{\delta x}\int_0^{\delta x} \left[\left(\frac{\rho g}{\mu}\delta x\right) y - \frac{\rho g}{2\mu} y^2\right] dy \\ &= \frac{\rho g}{3\mu}(\delta x)^2 \end{aligned} \tag{viii}$$

The quantity of the condensate G_x flowing per hour through cross-section ($\delta x dz$) is

$$G_x = \rho(\delta x dz)\overline{v_x} = \frac{\rho^2 g(\delta x)^3}{3\mu} \cdot dz \tag{ix}$$

As we move distance dx down the plate, the condensate flow thickness increases from δx to $(\delta x + d\delta x)$ because of the condensation dG_x in distance dx. Hence,

$$\frac{dG_x}{d\delta x} = \frac{\rho^2 g(\delta x)^2}{\mu} \cdot dz$$

or

$$dG_x = \frac{\rho^2 g(\delta x)^2}{\mu} d\delta x \cdot dz \tag{x}$$

The addition dG_x in the condensate is due to the condensation. It can also be estimated from the heat transfer consideration, which gives

$$dG_x = k(dxdz)\left(\frac{t_s - t_w}{\delta x}\right)\frac{1}{h_{fg}} \tag{xi}$$

where h_{fg} is the latent heat of condensation.

Equating Eqs. (x) and (xi), we obtain

$$\frac{k}{\delta x}\left(\frac{t_s - t_w}{h_{fg}}\right)dx = \frac{\rho^2 g(\delta x)^2}{\mu} d\delta x$$

or

$$dx = \frac{\rho^2 g h_{fg}}{\mu k(t_s - t_w)} (\delta x)^3 . d\delta x$$

Integrating the above equation,

$$x = \frac{\rho^2 g h_{fg}}{\mu k(t_s - t_w)} \frac{(\delta x)^4}{4} + C$$

Since film thickness is zero at the upper edge of the plate, i.e. $\delta x = 0$ at $x = 0$, we get $C = 0$. Thus

$$x = \frac{\rho^2 g h_{fg}}{\mu k(t_s - t_w)} \frac{(\delta x)^4}{4}$$

and

$$\delta x = \left[\frac{4\mu k(t_s - t_w)x}{\rho^2 g h_{fg}}\right]^{1/4} \tag{13.2}$$

which is the equation of variation of the film thickness δx along the plate height x.

Substitution of the value of δx in Eq. (13.1) gives

$$h_x = \left[\frac{\rho^2 g h_{fg} k^3}{4\mu(t_s - t_w)x}\right]^{1/4} \tag{13.3}$$

Both the film thickness δx and heat transfer coefficient h_x vary with distance x, refer Eqs. (13.2) and (13.3). At $x = 0$, the film thickness is zero and the heat transfer coefficient h_x is maximum. With the increase in the thickness of the condensate film along the plate, the heat transfer coefficient decreases. Figure 13.2 shows the variations of δx and h_x with x. The film thickness along the plate increases as a fourth root of distance down the surface ($\delta x \propto x^{1/4}$), while the heat transfer coefficient decreases with the increase in distance ($h_x \propto x^{-1/4}$). The increase in the thickness of the film is rapid at the beginning then the increase is rather slow as shown in the figure. The figure also depicts the variation of the film coefficient.

The mean heat transfer coefficient for a vertical plate of height H is

$$\begin{aligned}\bar{h} &= \frac{1}{H}\int_0^H h_x dx \\ &= \frac{4}{3} \times \left[\frac{\rho^2 g h_{fg} k^3}{4\mu(t_s - t_w)H}\right]^{1/4}\end{aligned} \tag{xii}$$

or

$$\bar{h} = 0.943\left[\frac{\rho^2 g h_{fg} k^3}{\mu(t_s - t_w)H}\right]^{1/4} \tag{13.4}$$

Then the Nusselt number is

$$\mathrm{Nu} = \frac{\bar{h}H}{k} = 0.943\left[\frac{\rho^2 g h_{fg} H^3}{\mu(t_s - t_w)k}\right]^{1/4} \tag{13.5}$$

The heat transfer coefficient predicted by Nusselt's solution has been found to be about 20% lower than the experimentally observed values. The reasons extended for this are

(i) Effect of surface tension has not been considered in the analysis.
(ii) Even for Reynolds numbers as low as 30–40, ripples appear in the film. These ripples cause mixing action and the heat transfer coefficient is increased.

Looking to the above facts, the following correlation, arrived at by multiplying Eq. (13.5) by 1.2, is recommended by some references for a vertical plate, i.e.

$$\mathrm{Nu} = \frac{\bar{h}H}{k} = 1.13\left[\frac{\rho^2 g h_{fg} H^3}{\mu(t_s - t_w)k}\right]^{1/4} \tag{13.6}$$

For an *inclined plane*, at an angle θ with horizontal, the term g in Eq. (13.6) is replaced by $g\sin\theta$, which gives

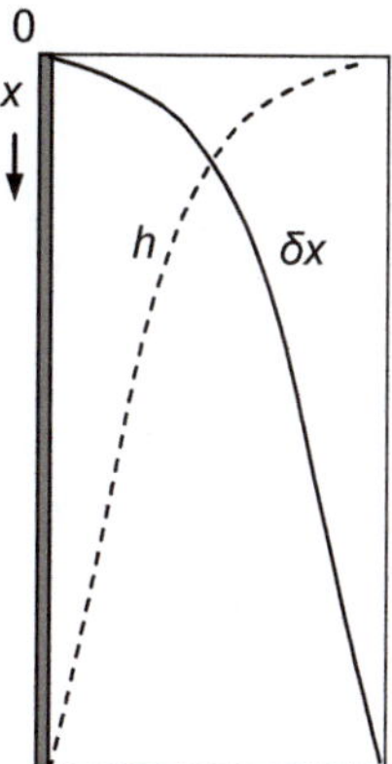

Fig. 13.2 Variation of film thickness and heat transfer coefficient along the plate

$$\mathrm{Nu} = \frac{\bar{h}H}{k} = 1.13\left[\frac{\rho^2(g\sin\theta)h_{\mathrm{fg}}H^3}{\mu(t_s - t_w)k}\right]^{1/4} \tag{13.7}$$

i.e.

$$\left(\bar{h}\right)_{\text{inclined}} = \left(\bar{h}\right)_{\text{vertical}}\sqrt{4}\sin\theta \tag{13.8}$$

This approximation gives satisfactory results for $\theta \geq 60°$.

The equations derived above for a vertical plate are also applicable to condensation on the outer surface of a vertical tube provided the tube diameter is sufficiently large compared to the film thickness.

Nusselt obtained the following equation for a *single horizontal tube* of outer diameter D_o.

$$\bar{h} = 0.725\left[\frac{\rho^2 k^3 g h_{\mathrm{fg}}}{\mu(t_s - t_w)D_o}\right]^{1/4} \tag{13.9}$$

$$\mathrm{Nu} = 0.725\left[\frac{\rho^2 g h_{\mathrm{fg}} D_o^3}{\mu(t_s - t_w)k}\right]^{1/4} \tag{13.10}$$

All the fluid properties in the heat transfer and Nusselt number correlations, except the latent heat h_{fg}, are to be taken at the mean film temperature $(t_s + t_w)/2$. The latent heat h_{fg} is taken at the saturation temperature of the condensate.

The above-presented relations are based on the assumption that the heat transfer is due to the phase change only. However, the temperature of the liquid film varies from saturation temperature t_s at the vapour–liquid interface to wall temperature t_w at the surface of the plate. This cooling of the liquid below the saturation temperature is accounted for by modification in the energy balance to include additional energy to cool the film below the saturation temperature. Use of corrected or modified latent heat of vaporization h^*_{fg} instead of h_{fg} is suggested which is defined as

$$h^*_{fg} = h_{fg} + 0.68c_1(t_s - t_w). \tag{13.11}$$

where c_l is the specific heat of the liquid and the second term on the right takes account of the cooling of the liquid below the saturation temperature.

Comparison of Horizontal and Vertical Orientation of Tubes

A comparison of the heat transfer coefficient relations for a horizontal tube of diameter D and a vertical tube of height H yields, from Eqs. (13.9) to (13.4),

$$\frac{h_{\text{horizontal}}}{h_{\text{tvertical}}} = 0.7688\left(\frac{H}{D_o}\right)^{1/4} = \left(\frac{H}{2.86D_o}\right)^{1/4} \tag{13.12}$$

Setting $h_{\text{horizontal}} = h_{\text{vertical}}$, we get $H = 2.86D_o$. This implies that for a tube of length $2.86D_o$, the heat transfer coefficient for laminar film condensation will be the same whether the tube is horizontal or vertical. For $H > 2.86D_o$, film coefficient on a horizontal tube is greater than for condensation over a vertical plate or tube. In most of the vertical condensers, the length of tube H is much greater than its outer diameter D_o (H/D_o is of the order of 50–100) and this is the reason for the preference of horizontal arrangement of tubes in condensers. However, in a condenser, there are a large number of tubes.

Figure 13.3 shows a bank of n horizontal tubes arranged in a vertical tier. For the inline arrangement the condensate trickling from the upper rows of the tubes will envelop the tubes of the lower rows. Hence, the thickness of film on these rows will increase causing a lower rate of heat transfer from these rows. The effect of thickening of the condensate film on the tubes of lower rows is less significant in the case of staggered arrangement where splashing of the condensate may occur.

In the absence of empirical relations which account for the splashing and thickening of the film, an estimate of the heat transfer can be made, by replacing D_o in Eq. (13.9) by nD_0 (Pitts and Sissom 1991). Thus, if the vapour is condensing on a bank of tubes, say n in vertical column and m in each horizontal row, the average heat transfer coefficient for the bank is given by

$$\frac{\bar{h}}{\bar{h}_1} = n^{-1/4} \tag{13.13a}$$

where n is the total number of rows and $\bar{h}_1$ is the heat transfer coefficient for the single tube

Kern (1958) proposed the following correction for the average heat transfer coefficient for n-tubes in a vertical column.

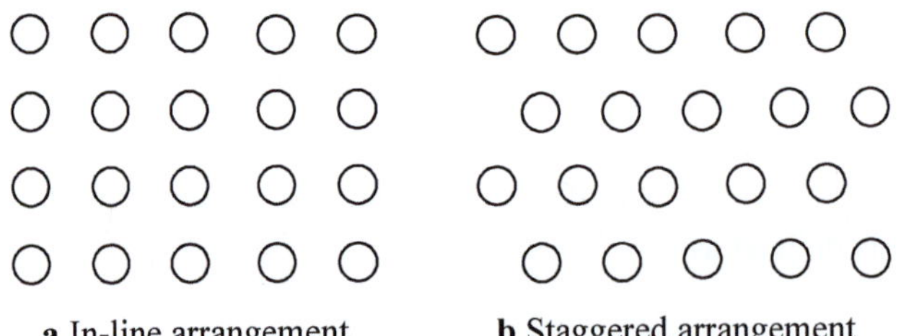

Fig. 13.3 Arrangement of tubes in a tube bundle arrangement

$$\frac{\bar{h}}{\bar{h}_1} = n^{-1/6} \quad (13.13b)$$

Example 13.1 A vertical tube, 1.2 m long and having 50 mm outer diameter is exposed to steam at 1.2 bar. If the tube surface is maintained at 85°C by flowing cooling water through it, determine the rate of heat transfer to the cooling water and the rate of condensation of steam. Assume flow to be laminar.

If the tube is held in horizontal position, estimate the condensation rate.

Solution

From steam tables, h_{fg} = 2244.2 kJ/kg, t_s = 104.8°C at p = 1.2 bar.

At the mean film temperature, $(t_s + t_w)/2$ = (85 + 104.8)/2 = 94.9°C, the fluid properties are

$$\rho = 961.5\,\text{kg/m}^3,\ \mu = 2.94 \times 10^{-4}\,\text{kg/(m s)},\ k = 0.677\,\text{W/(m K)}\ \text{and}\ c = 4213\,\text{J/(kg K)}$$

$$\begin{aligned} h^*_{fg} &= h_{fg} + 0.68c(t_s - t_w) \\ &= 2244.2 \times 10^3 + 0.68 \times 4213 \times (104.8 - 85) \\ &= 2300.9 \times 10^3\ \text{J/kg}. \end{aligned}$$

(i) **Condensation on the vertical tube surface**

The average heat transfer coefficient is given by

$$\begin{aligned} \bar{h} &= 0.943\left[\frac{\rho^2 g h^*_{fg} k^3}{\mu(t_s - t_w)H}\right]^{1/4} \\ &= 0.943\left[\frac{(961.5)^2 \times 9.81 \times 2300.9 \times 10^3 \times (0.677)^3}{2.94 \times 10^{-4} \times (104.8 - 85) \times 1.2}\right]^{1/4} \quad (13.4) \\ &= 5203.2\ \text{W/m}^2\text{K}. \end{aligned}$$

The heat flow rate,

$$q = hA(t_s - t_w) = 5203.2 \times (\pi \times 0.05 \times 1.2) \times (104.8 - 85) = 19419\,\text{W}.$$

The steam condensation rate is

$$\dot{m} = \frac{q}{h_{fg}} = \frac{19419}{2300.9 \times 10^3} = 8.44 \times 10^{-3}\ \text{kg/s}.$$

(ii) **Condensation on the horizontal tube surface**

The average heat transfer coefficient is given by

$$\bar{h} = 0.725\left[\frac{\rho^2 g h_{fg}^* k^3}{\mu(t_s - t_w)D_o}\right]^{1/4}$$

$$= 0.725 \times \left[\frac{(961.5)^2 \times 9.81 \times 2300.9 \times 10^3 \times (0.677)^3}{2.94 \times 10^{-4} \times (104.8 - 85) \times 0.05}\right]^{1/4} \quad (13.9)$$

$$= 8854.2\,\text{W/m}^2\text{K}.$$

The heat flow rate,

$$q = hA(t_s - t_w) = 8854.2 \times (\pi \times 0.05 \times 1.2) \times (104.8 - 85) = 33045.8\,\text{W}.$$

and the steam condensation rate is

$$\dot{m} = \frac{q}{h_{fg}^*} = \frac{33045.8}{2300.9 \times 10^3} = 14.36 \times 10^{-3}\,\text{kg/s}.$$

The example clearly shows the advantage of the horizontal arrangement of a single tube in the condensation process. Further, it is to note that the condensing heat transfer coefficient is much greater than the heat transfer coefficient encountered in free or forced single-phase flows.

Example 13.2 Repeat Example 13.1 if the steam is condensing on a bundle of horizontal tubes, 8 in vertical column and 10 in each horizontal row.

Solution

From Example 13.1, for the condensation on the single horizontal tube, the average heat transfer coefficient $\bar{h} = 8854.2\,\text{W}/(\text{m}^2\text{K})$.

The average heat transfer coefficient for the bank from Eq. (13.13b) is

$$\bar{h} = n^{-1/6} \times \bar{h}_1 = 8^{-1/6} \times 8854.2 = 6260.9\,\text{W}/\left(\text{m}^2\text{K}\right). \quad (13.13\text{b})$$

where n is number of tubes in a vertical column and $\bar{h}_1$ is the heat transfer coefficient for the single tube

The heat flow rate,

$$q = hA(t_s - t_w) = 6260.9 \times (80 \times \pi \times 0.05 \times 1.2) \times (104.8 - 85) = 1869.4\,\text{kW}.$$

The steam condensation rate is

$$\dot{m} = \frac{q}{h_{fg}^*} = \frac{1869.4}{2300.9} = 0.812\,\text{kg/s}.$$

13.1.2.2 Turbulent Film Flow

Nusselt's analysis is applicable to the laminar film flow. In the case of a vertical plate, the thickness of the condensate film increases downstream. For a tall vertical surface, at certain critical distance x_{cr}, the film becomes thick enough to cause a transition to the turbulent flow and heat is transferred not only by conduction but also by eddy diffusion, a characteristic of turbulence. Hence, the heat transfer rate is higher in the turbulent regime than in the laminar regime.

In order to define the point of transition from laminar to turbulent, it is necessary to define first the Reynolds number of the flow of the condensate. From the definition of the Reynolds number

$$\mathrm{Re}_x = \frac{\rho v_m d_h}{\mu} \tag{13.14}$$

at distance x from the leading (top) edge. In the equation,

v_m = mean velocity of the condensate = $\dot{m}/\rho A_c$
$\dot{m}$ = mass flow rate through the particular section of the condensate film
ρ = density of the condensate
A_c = area of flow
d_h = hydraulic diameter = $4A_c/P$
P = wetted parameter
= width W of the plate for a vertical plate
= circumference πD_o for a vertical tube.

Substitution of values of v_m and d_h in Eq. (13.14) gives

$$\mathrm{Re}_x = \frac{\dot{m}}{\rho A_c} \times \frac{\rho}{\mu} \times \frac{4A_c}{P} = \frac{4\dot{m}}{\mu P} \tag{13.15}$$

The Reynolds number at any section can be expressed in terms of the heat transfer coefficient as given below.

From the heat transfer equation,

$$q = \bar{h} A \Delta t$$

where $\bar{h}$ is the average heat transfer coefficient, A is the surface area for heat transfer and Δt is the temperature difference = $t_s - t_w$.

This equals the total latent heat of the condensate, i.e.

$$\dot{m} h_{fg} = \bar{h} A \Delta t$$

or

$$\dot{m} = \frac{\bar{h} A \Delta t}{h_{fg}}$$

Substitution of the value of $\dot{m}$ in Eq. (13.15) gives

$$\mathrm{Re}_x = \frac{4\bar{h}A\Delta t}{h_{fg}\mu P} = \frac{4\bar{h}(Wx)\Delta t}{h_{fg}\mu P}$$

(for a plate, area = Wx for height x, where W is width)
or

$$\mathrm{Re}_x = \frac{4\bar{h}(Wx)\Delta t}{h_{fg}\mu W} = \frac{4\bar{h}x\Delta t}{h_{fg}\mu}$$

(for a plate, perimeter $P = W$)
Hence for a vertical plate of height H, the Reynolds number at the lower end of the plate ($x = H$) is

$$\mathrm{Re}_H = \frac{4\bar{h}H(t_s - t_w)}{h_{fg}\mu} \tag{13.16}$$

The following empirical correlation for the turbulent flow has been suggested by Kirkbride (1934).

$$\bar{h} = 0.0077\left[\frac{\rho^2 g k^3}{\mu^2}\right]^{1/3}(\mathrm{Re}_H)^{0.4} \tag{13.17}$$

for $\mathrm{Re} \geq 1800$

or

$$\overline{\mathrm{Nu}} = \frac{\bar{h}H}{k} = 0.0077\left[\frac{\rho^2 g H^3}{\mu^2}\right]^{1/3}(\mathrm{Re}_H)^{0.4} \tag{13.18}$$

It is to be noted that the above correlation is applicable when the Reynolds number is greater than the critical value of 1800. This Reynolds number of 1800 is termed as the critical Reynolds number. In the case of single horizontal tube, the condensate is only ($\pi D_o/2$) in height and hence turbulence does not occur.

Example 13.3 Saturated steam at 0.7 bar is condensing on a vertical plate 2 m in height. The plate is maintained at 85°C. Find the average value of the heat transfer coefficient and the heat flow rate.

Solution

At 0.7 bar, h_{fg} = 2283.2 kJ/kg, t_s = 90°C from the steam tables.

Mean film temperature = $(t_s + t_w)/2$ = (85 + 90)/2 = 87.5°C.

The fluid properties at the mean film temperature are

$$\rho = 967.2\,\mathrm{kg/m^3},\ \mu = 3.205 \times 10^{-4}\,\mathrm{kg/(m\ s)}\ \text{and}\ k = 0.674\,\mathrm{W/(m\ K)}.$$

For the trial, we assume the flow to be laminar. The average heat transfer coefficient will be

$$\bar{h} = 0.943\left[\frac{\rho^2 g h_{fg} k^3}{\mu(t_s - t_w)H}\right]^{1/4}$$

$$= 0.943\left[\frac{(967.2)^2 \times 9.81 \times 2283.2 \times 10^3 \times (0.674)^3}{3.205 \times 10^{-4} \times (90 - 85) \times 2}\right]^{1/4} \quad (13.4)$$

$$= 6307.6\ \mathrm{W/(m^2K)}.$$

From this estimate of the heat transfer coefficient,

$$\mathrm{Re}_H = \frac{4\bar{h}H(t_s - t_w)}{h_{fg}\mu} = \frac{4 \times 6307.6 \times 2 \times (90 - 85)}{2283.2 \times 10^3 \times 3.205 \times 10^{-4}} = 344.8 \quad (13.16)$$

Since $30 < \mathrm{Re_H} < 1800$, flow is laminar-wavy and ripples appear in the film. Considering the ripples effect, the heat transfer coefficient is (1.13/0.943) × 6307.6 = 7558.4 W/(m² K) and the heat flow rate,

$$q = \bar{h}A(t_s - t_w) = 7558.4 \times (2 \times 1) \times (90 - 85) = 75584\ \text{W/unit width of the plate.}$$

The steam condensation rate is

$$\dot{m} = \frac{q}{h_{fg}} = \frac{75584}{2283.2 \times 10^3} = 0.0331\ \text{kgs.}$$

Example 13.4 If the plate in Example 13.3 is held at 50°C and plate height is doubled, find the average value of the heat transfer coefficient and the condensate flow rate.

Solution

At 0.7 bar, h_{fg} = 2283.2 kJ/kg, t_s = 90°C from the steam tables.

Mean film temperature = $(t_s + t_w)/2$ = (90 + 50)/2 = 70°C.

The fluid properties at the mean film temperature are

$$c = 4.191\ \mathrm{kJ/(kg\ K)},\ \rho = 977.5\ \mathrm{kg/m^3},\ \mu = 4.0 \times 10^{-4}\ \mathrm{kg/(m\ s)}\ \text{and}\ k = 0.663\ \mathrm{W/(m\ K)}.$$

Assuming the flow to be laminar, the average heat transfer coefficient will be

$$\bar{h} = 0.943\left[\frac{\rho^2 g h_{fg} k^3}{\mu(t_s - t_w)H}\right]^{1/4}$$

$$= 0.943\left[\frac{(977.5)^2 \times 9.81 \times 2283.2 \times 10^3 \times (0.663)^3}{4.0 \times 10^{-4} \times (90 - 50) \times 4}\right]^{1/4}$$

$$= 2962.8\ \mathrm{W/(m^2\ K)}.$$

From this estimate of the heat transfer coefficient,

$$\mathrm{Re}_H = \frac{4\bar{h}H(t_s - t_w)}{h_{fg}\mu} = \frac{4 \times 2962.8 \times 4 \times (90 - 50)}{2283.2 \times 10^3 \times 4.0 \times 10^{-4}} = 2076 > 1800.$$

The flow is turbulent. We should recalculate the Reynolds number and heat transfer coefficient considering the flow to be turbulent. For the turbulent flow,

$$\bar{h} = 0.0077\left[\frac{\rho^2 g k^3}{\mu^2}\right]^{1/3} (\mathrm{Re}_H)^{0.4} \tag{13.18}$$

Substituting the value of $\bar{h}$ in equation of Re_H, we get

$$\mathrm{Re}_H = \left[0.0077\left(\frac{\rho^2 g k^3}{\mu^2}\right)^{1/3} \times \frac{4H(t_s - t_w)}{h_{fg}\mu}\right]^{1/0.6}$$

or

$$\mathrm{Re}_H = \left\{0.0077\left[\frac{(977.5)^2 \times 9.81 \times (0.663)^3}{(4.0 \times 10^{-4})^2}\right]^{1/3} \times \frac{4 \times 4 \times (90 - 50)}{2283.2 \times 10^3 \times 4.0 \times 10^{-4}}\right\}^{1/0.6}$$

$$= 3727.$$

Heat transfer coefficient,

$$\begin{aligned}\bar{h} &= 0.0077\left[\frac{\rho^2 g k^3}{\mu^2}\right]^{1/3} (\mathrm{Re}_H)^{0.4}\\ &= 0.0077\left[\frac{(977.5)^2 \times 9.81 \times (0.663)^3}{(4.0 \times 10^{-4})^2}\right]^{1/3} \times (3727)^{0.4}\\ &= 5318.6\,\mathrm{W/m^2\,K}.\end{aligned}$$

The heat flow rate,

$$q = \bar{h}A(t_s - t_w) = 5318.6 \times (4 \times 1) \times (90 - 50) = 850.98\,\mathrm{kW/unit\ width\ of\ the\ plate}.$$

The steam condensation rate is

$$\dot{m} = \frac{q}{h_{fg}} = \frac{850.98 \times 10^3}{2283.2 \times 10^3} = 0.373\,\mathrm{kg/s}.$$

Since the difference in the plate surface temperature and the saturation temperature is large, we should use the modified latent heat of vaporization h^*_{fg} instead of h_{fg} to take account of subcooling of the liquid. h^*_{fg} is defined as

$$h_{\mathrm{fg}}^{*}=h_{\mathrm{fg}}+0.68c(t_s-t_w) \tag{13.11}$$

or

$$h_{\mathrm{fg}}^{*}=2283.2\times10^3+0.68\times4191\times(90-50)=2397\times10^3\ \mathrm{J/kg}.$$

The modification gives

$$\mathrm{Re}_H=\left[0.0077\left(\frac{\rho^2gk^3}{\mu^2}\right)^{1/3}\times\frac{4H(t_s-t_w)}{h_{\mathrm{fg}}^{*}\mu}\right]^{1/0.6}$$

or

$$\mathrm{Re}_H=\left\{0.0077\left[\frac{(977.5)^2\times9.81\times(0.663)^3}{(4.0\times10^{-4})^2}\right]^{1/3}\times\frac{4\times4\times(90-50)}{2397\times10^3\times4.0\times10^{-4}}\right\}^{1/0.6}$$
$$=3437.$$

Heat transfer coefficient,

$$\begin{aligned}\bar{h}&=0.0077\left[\frac{\rho^2gk^3}{\mu^2}\right]^{1/3}(\mathrm{Re}_H)^{0.4}\\&=0.0077\left[\frac{(977.5)^2\times9.81\times(0.663)^3}{(4.0\times10^{-4})^2}\right]^{1/3}\times(3437)^{0.4}\\&=5149\ \mathrm{W/m^2\,K},\end{aligned}$$

which is about 3% lower. Now it can be understood that when (t_s-t_w) is small, the modification will have a negligible effect on the estimate of $\bar{h}$. In such cases, the estimate can be made using h_{fg}.

The heat flow rate,

$$q=\bar{h}A(t_s-t_w)=5149\times(4\times1)\times(90-50)=823.8\,\mathrm{kW/unit\ width\ of\ the\ plate.}$$

The steam condensation rate is

$$\dot{m}=\frac{q}{h_{\mathrm{fg}}^{*}}=\frac{823.8\times10^3}{2397\times10^3}=0.344\,\mathrm{kg/s}.$$

Example 13.5 A horizontal tube 1 m long and 115 mm outer diameter is to be used to condense steam at the outer surface of the tube. Determine the required surface temperature of the tube for a condensation rate of 1.75 kg/min at 1 atm.

Solution

From the heat transfer rate equation

$$\dot{m} = \frac{q}{h_{\mathrm{fg}}^*} = \frac{\bar{h}A(t_s - t_w)}{h_{\mathrm{fg}}^*}$$

where $h_{fg}^* = h_{fg} + 0.68c(t_s - t_w)$, and $\bar{h} = 0.725\left[\frac{\rho^2 g h_{\mathrm{fg}}^* k^3}{\mu(t_s - t_w)D_o}\right]^{1/4}$.

Substitution gives

$$\dot{m} = 0.725\left\{\frac{\rho^2 g[h_{\mathrm{fg}} + 0.68c(t_s - t_w)]k^3}{\mu(t_s - t_w)D_o}\right\}^{1/4}\frac{A(t_s - t_w)}{h_{fg} + 0.68c(t_s - t_w)} \tag{i}$$

At 1 atm (=1.01325 bar), h_{fg} = 2256.4 kJ/kg, t_s = 100°C from the steam tables. Assuming mean film temperature of 85°C, the fluid properties are (Table A4)

$$c = 4203\,\mathrm{J/(kg\ K)},\ \rho = 969\,\mathrm{kg/m^3},\ \mu = 3.3 \times 10^{-4}\,\mathrm{kg/(m\ s)}\ \text{and}\ k = 0.673\,\mathrm{W/(m\ K)}.$$

Substitution of the values of various parameters in Eq. (i) gives

$$\frac{1.75}{60} = 0.725 \times \left\{\frac{969^2 \times 9.81 \times [2256.4 \times 10^3 + 0.68 \times 4203 \times (100 - t_w)] \times 0.673^3}{3.3 \times 10^{-4} \times (100 - t_w) \times 0.115}\right\}^{1/4} \times \frac{\pi \times 0.115 \times (100 - t_w)}{[2256.4 \times 10^3 + 0.68 \times 4203 \times (100 - t_w)]}$$

The solution of the above equation by trial and error gives $(100 - t_w) \approx 30$°C. Hence, the required surface temperature of the tube, t_w = 70°C. Mean film temperature is (70 + 100)/2 = 85°C as assumed.

Example 13.6 A horizontal pipe (25 mm outer diameter) is in contact with 40°C, 76% relative humidity air. Determine water condensation rate per unit length of the pipe if the tube surface temperature is 5°C.

Solution

Saturation pressure of water corresponding to 40°C is 0.073814 bar. Hence, water vapour pressure is 0.073814 × RH = 0.073814 × 0.76 = 0.0561 bar. The corresponding saturation temperature is about 35°C and h_{fg} = 2417.8 kJ/kg.

Water (liquid) film temperature is (35 + 5)/2 = 20°C. At 20°C, water properties are (Table A4)

$$c = 4183\,\mathrm{J/(kg\ K)},\ \rho = 998.2\ \mathrm{kg/m^3},\ \mu = 1002 \times 10^{-6}\,\mathrm{kg/(m\ s)}\ \text{and}\ k = 0.601\,\mathrm{W/(m\ K)}.$$

The condensation rate per unit length is

$$\dot{m} = \frac{q}{h^*_{fg}} = \frac{\bar{h}\pi D(t_s - t_w)}{h^*_{fg}}$$

where

$h^*_{fg} = h_{fg} + 0.68c(t_s - t_w) = 2417.8 \times 10^3 + 0.68 \times 4183 \times (35 - 5) = 2503.1 \times 10^3$ J/kg.

Heat transfer coefficient,

$$\bar{h} = 0.725\left[\frac{\rho^2 g h^*_{fg} k^3}{\mu(t_s - t_w)D_o}\right]^{1/4}$$

$$= 0.725 \times \left[\frac{998.2^2 \times 9.81 \times 2503.1 \times 10^3 \times 0.601^3}{1002 \times 10^{-6} \times (35 - 5) \times 0.025}\right]^{1/4}$$

$$= 6647.48\ \mathrm{W/(m^2\ K)}.$$

Hence, the condensation rate is

$$\dot{m} = \frac{6647.48 \times \pi \times 0.025 \times (35 - 5)}{2473.9 \times 10^3} = 6.33 \times 10^{-3}\ \mathrm{kg/s\ per\ m\ length}.$$

Example 13.7 Saturated steam at 1 atm condenses on the outer surface of a vertical tube 1 m long and having 100 mm outside diameter and 95 mm inside diameter. If the tube surface is maintained at 95°C by flowing cooling water through it, determine the rate of heat transfer to the cooling water and the rate of condensation of steam.

If water flow within the tube experiences 3°C temperature rise, determine the mean temperature of the water in the tube. Neglect pipe wall resistance.

Solution

At 1 atm (=1.01325 bar), h_{fg} = 2256.4 kJ/kg, t_s = 100°C from steam tables. At film temperature of (95 + 100)/2 = 97.5°C, the thermophysical properties of water are (Table A4)

$$c = 4212\ \mathrm{J/(kg\ K)},\ \rho = 959.9\ \mathrm{kg/m^3},\ \mu = 2.88 \times 10^{-4}\ \mathrm{kg/(m\ s)}\ \text{and}\ k = 0.6795\ \mathrm{W/(m\ K)}.$$

Though temperature difference is small, we may consider correction in h_{fg}. Corrected or modified latent heat of vaporization is

$$h^*_{fg} = h_{fg} + 0.68c(t_s - t_w)$$
$$= 2256.4 \times 10^3 + 0.68 \times 4212 \times (100 - 95)$$
$$= 2270.7 \times 10^3\ \mathrm{J/kg}.$$

The average heat transfer coefficient for the vertical tube is given by (using vertical plate equation)

$$\bar{h} = 0.943\left[\frac{\rho^2 g h_{fg}^* k^3}{\mu(t_s - t_w)H}\right]^{1/4}$$

$$= 0.943\left[\frac{(959.9)^2 \times 9.81 \times 2270.7 \times 10^3 \times (0.6795)^3}{2.88 \times 10^{-4} \times (100 - 95) \times 1.0}\right]^{1/4} \quad (13.4)$$

$$= 7711.4\,\text{W/(m}^2\text{K)}.$$

The heat flow rate,

$$q = \bar{h}A(t_s - t_w) = 7711.4 \times (\pi \times 0.1 \times 1.0) \times (100 - 95) = 12113\,\text{W}.$$

The steam condensation rate is

$$\dot{m} = \frac{q}{h_{fg}^*} = \frac{12113}{2270.7 \times 10^3} = 5.33 \times 10^{-3}\,\text{kg/s}.$$

From Eq. (13.15), the Reynolds number of flowing condensate is

$$\text{Re} = \frac{4\dot{m}}{\mu P} = \frac{4\dot{m}}{\mu\pi D} = \frac{4 \times 5.33 \times 10^{-3}}{2.88 \times 10^{-4} \times \pi \times 0.1} = 236.$$

Since $30 < \text{Re}_H < 1800$, flow is laminar-wavy and ripples appear in the film. Considering the ripples effect, the heat transfer coefficient is (1.13/0.943) 7711.4 = 9240.6 W/(m^2 K) and the heat flow rate is

$$q = \bar{h}A(t_s - t_w) = 9240.6 \times (\pi \times 0.1 \times 1.0) \times (100 - 95) = 14515\,\text{W}.$$

Assuming mean water temperature in the tube as 20°C, the thermophysical properties are

$$c = 4183\,\text{J/(kg K)},\ \rho = 998.2\,\text{kg/m}^3,\ \mu = 10.02 \times 10^{-4}\,\text{kg/(m s)},\ k = 0.601\,\text{W/(m K) and } Pr = 7.$$

From the energy balance, the flow rate of water is

$$m_w = \frac{q}{c\Delta t} = \frac{14515}{4183 \times 3} = 1.157\,\text{kg/s}.$$

The Reynolds number of flowing water in the tube is

$$\text{Re} = \frac{\rho U D_i}{\mu} = \frac{m_w}{(\pi/4)D_i^2\rho} \times \frac{\rho D_i}{\mu} = \frac{4m_w}{\mu\pi D_i} = \frac{4 \times 1.157}{10.02 \times 10^{-4} \times \pi \times 0.095} = 15476.$$

The heat transfer coefficient at the tube inner surface,

$$h_i = \frac{Nuk}{D_i} = \frac{k}{D_i} 0.024 \mathrm{Re}^{0.8} \mathrm{Pr}^{0.4} = \frac{0.601}{0.095} \times 0.024 \times 15476^{0.8} \times 7^{0.4} = 743.24\,\mathrm{W/(m^2K)}.$$

Overall heat transfer coefficient from Eq. (2.30a) neglecting wall resistance,

$$U_o = \left(\frac{A_o}{h_i A_i} + \frac{1}{h_o}\right)^{-1} = \left(\frac{D}{h_i D_i} + \frac{1}{h_o}\right)^{-1} = \left(\frac{100}{743.24 \times 95} + \frac{1}{9240.6}\right)^{-1}$$
$$= 655.96\,\mathrm{W/(m^2K)}.$$

Heat transfer equation gives

$$t_s - t_m = \frac{q}{U_o A} = \frac{q}{U_o \pi D L} = \frac{14515}{655.96 \times \pi \times 0.1 \times 1.0} = 70.44\,^\circ\mathrm{C}.$$

Hence, the mean temperature of the water is

$$t_m = t_s - 74.5 = 100 - 70.44 = 29.56^\circ\mathrm{C}.$$

Iteration may be carried out with thermophysical properties of water at the mean water temperature of 30°C.

Example 13.8 Saturated steam at 1 atm condenses on the outer surface of a horizontal tube 1 m long and having 100 mm outside diameter and 95 mm inside diameter. Determine the tube surface temperature if water flowing within the tube at the rate of 1.2 kg/s is at a mean temperature of 20°C. Neglect pipe wall resistance.

Solution

At 1 atm (= 1.01325 bar), h_{fg} = 2256.4 kJ/kg, t_s = 100°C from steam tables. The problem will require an iterative solution. For trial, let the surface temperature is 90°C. Hence, the film temperature is (90 + 100)/2 = 95°C. The thermophysical properties of water at film temperature are (Table A4):

$$c = 4208\,\mathrm{J/(kg\ K)},\ \rho = 965.3\,\mathrm{kg/m^3},\ \mu = 3.11 \times 10^{-4}\,\mathrm{kg/(m\ s)},\ k = 0.675\,\mathrm{W/(m\ K)}.$$

Though temperature difference is small, we may consider correction in h_{fg}. Corrected or modified latent heat of vaporization is

$$h_{fg}^* = h_{fg} + 0.68c(t_s - t_w)$$
$$= 2256.4 \times 10^3 + 0.68 \times 4208 \times (100 - 90)$$
$$= 2285 \times 10^3\,\mathrm{J/kg}.$$

The average heat transfer coefficient is

$$\bar{h} = 0.725\left[\frac{\rho^2 k^3 g h_{fg}}{\mu(t_s - t_w)D_o}\right]^{1/4}$$

$$= 0.725\left[\frac{(965.3)^2 \times 9.81 \times 2285 \times 10^3 \times (0.675)^3}{3.11 \times 10^{-4} \times (100 - 90) \times 0.1}\right]^{1/4}$$

$$= 8691.5\ \mathrm{W/(m^2K)}.$$

At the mean water temperature of 20°C in the tube, the thermophysical properties are

$$c = 4183\ \mathrm{J/(kg\ K)},\ \rho = 998.2\ \mathrm{kg/m^3},\ \mu = 10.02 \times 10^{-4}\ \mathrm{kg/(m\ s)},\ k = 0.601\ \mathrm{W/(m\ K)}\ \text{and}\ \mathrm{Pr} = 7.$$

The Reynolds number of flowing water in the tube is

$$\mathrm{Re} = \frac{\rho U D_i}{\mu} = \frac{m_w}{(\pi/4)D_i^2\rho} \times \frac{\rho D_i}{\mu} = \frac{4m_w}{\mu\pi D_i} = \frac{4 \times 1.2}{10.02 \times 10^{-4} \times \pi \times 0.095} = 16051.$$

The heat transfer coefficient at the tube inner surface,

$$h_i = \frac{Nuk}{D_i} = \frac{k}{D_i}0.024\mathrm{Re}^{0.8}\mathrm{Pr}^{0.4} = \frac{0.601}{0.095} \times 0.024 \times 16051^{0.8} \times 7^{0.4} = 765.25\ \mathrm{W/(m^2\ K)}.$$

Overall heat transfer coefficient, Eq. (2.30a) neglecting wall resistance,

$$U_o = \left(\frac{A_o}{h_i A_i} + \frac{1}{h_o}\right)^{-1} = \left(\frac{D}{h_i D_i} + \frac{1}{\bar{h}}\right)^{-1} = \left(\frac{100}{765.25 \times 95} + \frac{1}{8691.5}\right)^{-1} = 670.9\ \mathrm{W/(m^2\ K)}.$$

The heat transfer rate,

$$q = U_o A_o (t_w - t_m) = 670.9 \times \pi \times 0.1 \times 1.0 \times (100 - 20) = 16862\ \mathrm{W}.$$

Heat transfer equation at the inner surface of the tube gives wall temperature

$$t_w = \frac{q}{h_i A_i} + t_m = \frac{16862}{765.25 \times \pi \times 0.095 \times 1.0} + 20 = 93.8\ ^\circ\mathrm{C}.$$

For the second trial, we assume a surface temperature of 95°C. The revised film temperature is (95 + 100)/2 = 97.5°C. The thermophysical properties of water at film temperature are (Table A4)

$$c = 4212\ \mathrm{J/(kg\ K)}, \rho = 959.9\ \mathrm{kg/m^3}, \mu = 2.88 \times 10^{-4} kg/(\mathrm{m\ s})\ \text{and}\ k = 0.6795\ \mathrm{W/(m\ K)}.$$

The corrected or modified latent heat of vaporization is

$$
\begin{aligned}
h_{fg}^* &= h_{fg} + 0.68c(t_s - t_w) \\
&= 2256.4 \times 10^3 + 0.68 \times 4212 \times (100 - 95) \\
&= 2270.7 \times 10^3 \text{ J/kg.}
\end{aligned}
$$

The average heat transfer coefficient is

$$
\begin{aligned}
\bar{h} &= 0.725\left[\frac{\rho^2 k^3 g h_{fg}}{\mu(t_s - t_w)D_o}\right]^{1/4} \\
&= 0.725\left[\frac{(959.9)^2 \times (0.6795)^3 \times 9.81 \times 2270.7 \times 10^3}{2.88 \times 10^{-4} \times (100 - 95) \times 0.1}\right]^{1/4} \\
&= 10543 \text{ W/(m}^2\text{ K).}
\end{aligned}
$$

Revised overall heat transfer coefficient is

$$
U_o = \left(\frac{A_o}{h_i A_i} + \frac{1}{h_o}\right)^{-1} = \left(\frac{D}{h_i D_i} + \frac{1}{\bar{h}}\right)^{-1} = \left(\frac{100}{765.25 \times 95} + \frac{1}{10543}\right)^{-1} = 680.1 \text{ W/}\left(\text{m}^2\text{K}\right).
$$

The heat transfer rate,

$$
q = U_o A_o (t_s - t_m) = 680.1 \times \pi \times 0.1 \times 1.0 \times (100 - 20) = 17093 \text{ W.}
$$

At the inner surface, the heat transfer equation gives wall temperature:

$$
t_w = \frac{q}{h_i A_i} + t_m = \frac{17093}{765.25 \times \pi \times 0.095 \times 1.0} + 20 = 94.84^\circ\text{C.}
$$

which is nearly equal to the assumed temperature hence further trial is not required.

13.1.3 Factors Affecting Film Condensation

(i) **Velocity and direction of the vapour flow**: If the vapour flows downward, it increases the velocity of the liquid and decreases the thickness of the condensate film. This decreases the thermal resistance of the film and thus the heat transfer coefficient increases. Upward flow of the vapour increases the thickness of the condensate film and the heat transfer coefficient decreases.

(ii) **State of the surface**: If the surface is rough, the thickness of the film increases due to the greater resistance offered to the flow and the heat transfer coefficient decreases.

(iii) **Layout of the surface**: A discussed earlier, in the case of a single tube, the horizontal position is preferred. In the tube banks, the condensate flows from the upper to the lower tubes causing the condensate film to become thick on the lower tubes in the bank and heat transfer coefficient is reduced.

In the case of the vertical tubes, the heat transfer coefficient diminishes in the downward direction as the condensate film grows in thickness. The heat transfer coefficient in this case can be increased by installing condensate tapping caps as

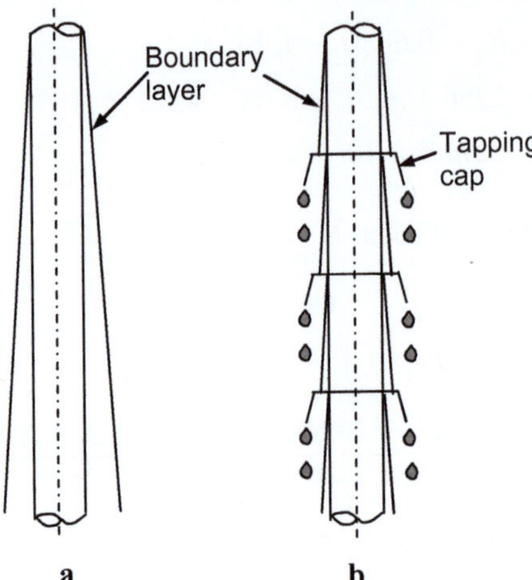

Fig. 13.4 **a** Development of boundary layer on a vertical tube, **b** boundary layer on a vertical tube with tapping caps

shown in Fig. 13.4 (Mikheyev 1968). The caps remove the condensate and thus the boundary layer redevelops after the cap. The effective thickness of the boundary layer over the tube length is reduced resulting in the increased heat transfer coefficient.

(i) **Influence of non-condensing gases**: The rate of heat transfer drops when the vapour contains non-condensable gases because the vapour has to diffuse through the insulating gas layer to vapour–liquid phase interface. It has been found that even 1% of air in the vapour reduces the heat transfer coefficient by 60% due to a larger percentage of air near the wall.
The inert gas accumulates near the liquid–vapour interface as the vapour condenses. And its pressure rises towards the phase interface. Since the total pressure of the gas and vapour is constant, the partial pressure of the vapour near the condensate film is lower than the pressure of the vapour away from the condensate film causing a reduction in the saturation temperature of the condensate below the saturation temperature which would occur if no inert gas is present. Thus the temperature difference between the phase interface and the wall is lower because of the presence of the gas causing a reduction in the heat transfer rate.

13.2 Part B: Heat Transfer in Boiling Liquids

13.2.1 Introduction

The heat transfer in boiling liquids is also quite different from the convective heat transfer in single-phase fluids. Analysis of the heat transfer to boiling liquids is rather difficult and here we shall concentrate on the qualitative discussion of the process and the factors that govern the rate of heat transfer.

Boiling heat transfer is encountered in many industrial equipments, such as boilers for vapour power cycles, evaporator for refrigeration and air-conditioning systems hence an in-depth study of this phenomenon is useful.

13.2.2 Boiling Heat Transfer[1]

13.2.2.1 Pool Boiling

When a liquid comes in contact with a heated surface (maintained at a temperature higher than the saturation temperature of the liquid), boiling occurs and vapour is formed. Boiling of a large volume of liquid by a submerged heated surface is known as *pool boiling*. In pool boiling, the fluid motion is caused by the free convection currents. Process of boiling may occur in limited space (for example, in a tube), or in a tube with forced flow of boiling liquid. In the case of forced-flow boiling, the nature of boiling is similar to the pool boiling, though some new factors appear. We shall first discuss the pool boiling.

The process of boiling takes place at a constant pressure. At a given pressure, the temperature of the bulk of the fluid t_f is somewhat higher than the saturation temperature t_s at that pressure. Typically for water boiling at atmospheric condition, $t_f - t_s = 0.4 - 0.8$°C. The difference is a function of the physical properties of the liquid and the intensity of the formation of vapour. The temperature of the liquid in contact with the heating surface equals the temperature of the heating surface. A sharp decrease in the temperature of the liquid takes place in a thin layer at the heating surface, see Fig. 13.5. Thereafter the temperature of the liquid is practically constant. The temperature excess $\Delta t = t_w - t_s$ rises with an increase in the heat flux q.

Figure 13.6 shows the typical plot of the heat flux q versus the temperature excess Δt referred to as Nukiyama boiling curve.

In region AB of low values of temperature excess, the motion of the fluid near the heating surface is due to the free-convection currents only. Both heat transfer rate and the heat transfer coefficient are low. The heat transfer coefficient can be calculated using the free-convection relations presented for single-phase liquids in Chap. 9.

At point B, nucleate boiling starts. The bubbles form on the heating surface. They grow in size, separate from the surface and move upwards through the liquid. The convection currents are intensified by the agitation of the rising vapour bubbles. The heat transfer coefficient rises rapidly in this region. If liquid superheat is not sufficient (region BC), the bubbles are dissipated in the liquid as they move upwards.

As the temperature excess is further increased, the bubbles form rapidly and rise to the surface of the liquid. The bubbles continue to grow in size as they move upwards depending on the degree of the liquid superheat. In the region indicated in the figure by CD, nucleate boiling is fully developed. Both the heat flux and heat transfer coefficient rise rapidly in this region with the increase in the temperature excess Δt. With the increase in the temperature excess along CD, the number of points where the bubbles originate increases and the bubbles also increase in size.

[1]Vaporization is changing of liquid into vapour. There are two types of vaporization: evaporation and boiling. Evaporation occurs at the surface of the liquid and takes place at temperatures lower than the boiling point at a given pressure. Boiling takes place at boiling point and is a bulk phenomenon because the bubbles are formed in the liquid.

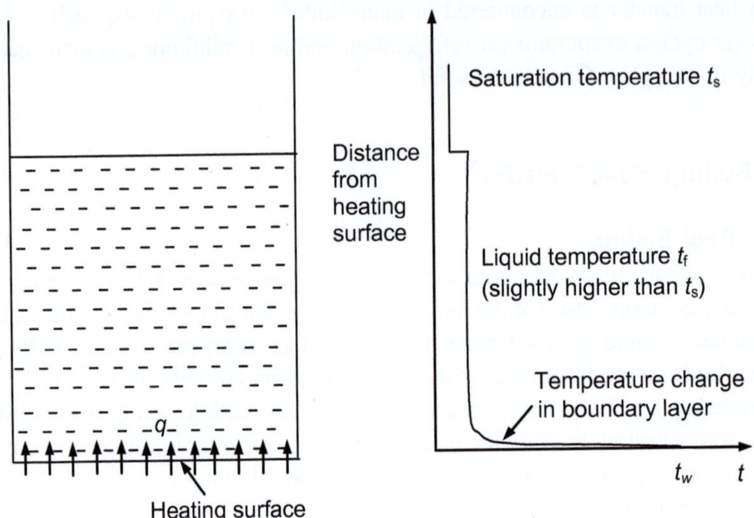

Fig. 13.5 Pool boiling

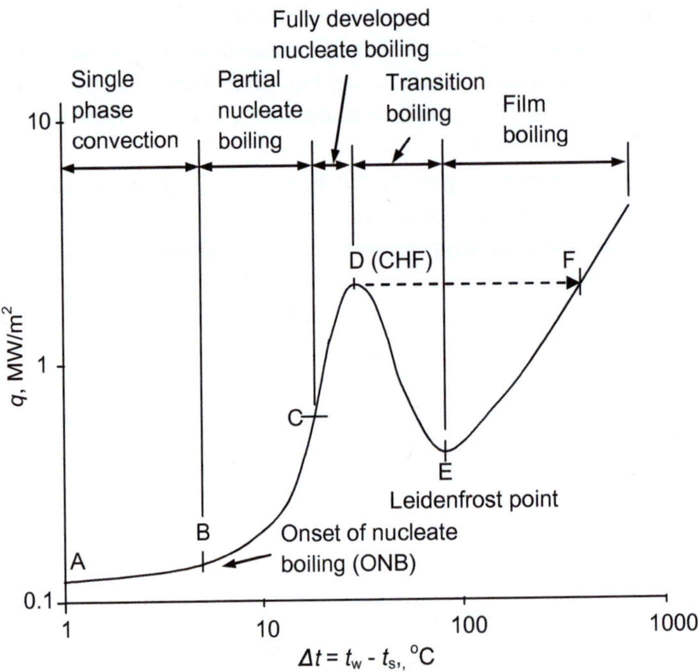

Fig. 13.6 Pool boiling curve

Beyond point D, the bubbles at the heating surface merge to form a vapour blanket or film covering the surface. The heat is now transferred by conduction through this film. The high thermal resistance of the vapour film causes a sharp drop in the heat flux and heat transfer coefficient. The region beyond D is defined as region of *film boiling*. The values of q, h, and Δt at the point D are called *critical* or *peak values*. CHF is critical heat flux. The values of these parameters are different for different liquids. For a given liquid, the critical temperature difference diminishes with the rising pressure.

The vapour film is unstable in region DE and the film breaks up frequently and separates from the heating surface in the form of large bubbles. And a new film is then formed on the surface. This process continues until point E. The point E is termed as the *Leidenfrost point*. The Leidenfrost point is the lowest temperature at which a hot body submerged in a pool of boiling water is completely blanketed from the liquid by a vapour film. This phenomenon of vapour blanketing is known as the Leidenfrost effect. There is a decrease of heat transfer and a minimum of the heat flux curve occurs at this point.

At still larger values of Δt, beyond the unstable region DE, a stable film boiling is encountered. The temperature excess Δt required for maintaining the stable film boiling is very high and thus the surface temperature t_w is also very high. The heat is transferred from the heating surface to the boiling liquid through the vapour film by conduction and radiation. The contribution of radiation heat exchange increases with increasing Δt.

It is to be noted that there is a danger of the heating surface failing or melting in the film boiling region. For example, with water boiling at atmospheric pressure Δt = 25°C at D, the temperature at point F may be of the order of 1000°C and burnout may occur. It may be noted that though Δt is only 25°C at D, the heat flux q has a very high value and hence one should try to operate as close as possible to D.

The other important observations relating to the boiling heat transfer are

(i) The vapour bubbles form on the heating surface at some individual points called starting points. Gases dissolved in the liquid and vapour trapped in the surface roughness cavities provide the starting points. The number of starting points Z depends on the degree of superheat at the heating surface, i.e. on the temperature excess Δt. With the rise in the degree of superheat, Z increases and boiling intensifies.

(ii) There is certain periodicity in the process of bubble formation, growth and separation from the starting point. The frequency f of this process depends on the separation diameter d_o of the bubble. It has been found that smaller the diameter, higher the frequency. At a given pressure, the product fd_o is approximately constant and the value of the product decreases with an increase in the pressure.
The growth, separation and the subsequent movement of the bubbles cause intense agitation of the liquid at the heating surface resulting in a sharp increase in the rate of heat transfer from the heating surface to the boiling liquid. Hence, higher the frequency the higher the heat transfer rate is.

(iii) The pressure p_v of the vapour inside a bubble is higher than the pressure p_l of the surrounding liquid due to the surface tension. For a spherical bubble, refer Fig. 13.7.

$$\text{pressure force} = \pi r^2 (p_v - p_l)$$

and

$$\text{surface tension} = 2\pi r \sigma$$

where r = radius of the bubble
σ = surface tension.
Balance of the forces gives

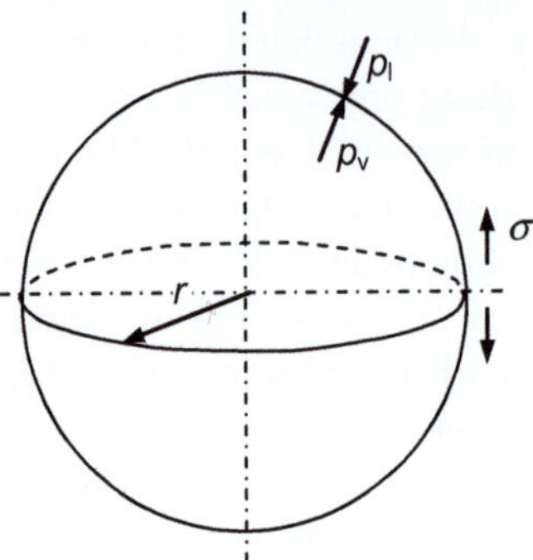

Fig. 13.7 Forces on a vapour bubble

$$\pi r^2 (p_v - p_l) = 2\pi r\sigma$$

or

$$p_v - p_l = \frac{2\sigma}{r}. \tag{13.19}$$

For a bubble in pressure equilibrium, the temperature of the vapour inside the bubble (saturation temperature of the vapour at p_v) will be higher than the surrounding liquid temperature if the liquid is not superheated (saturation temperature at p_l). In this condition, the heat will flow from the vapour to the liquid and due to the condensation of the vapour, the bubble collapses as it moves upwards through the liquid. If the liquid is superheated to such an extent that the liquid temperature is higher than the vapour temperature, the heat transfer from the liquid to the vapour takes place and the bubble grows in size as it moves through the liquid.

The conditions required for the bubble to exist and grow can be found by applying the Clapeyron equation, which gives

$$\frac{h_{fg}}{T_s v_{fg}} = \frac{dp}{dT_s} \approx \frac{p_v - p_l}{T_v - T_{sat}} \tag{13.20}$$

where T_s is the saturation temperature at liquid pressure p_l. Using Eq. (13.19), we get

$$T_v - T_s = \frac{2\sigma T_s v_{fg}}{h_{fg}} \frac{1}{r} \tag{13.21}$$

In the case of thermal equilibrium, $T_v = T_l$ and the above equation can predict the equilibrium radius of the bubble. If T_l is greater than T_v, the bubble will grow in size due to the heat transfer to the vapour from the superheated liquid. The bubble will collapse if T_v is greater than T_l because the heat will flow from the vapour to the liquid causing condensation of the vapour.

(iv) Surface conditions, roughness and material play an important role in bubble formation and growth. As far as geometry is concerned, the bubble formation and agitation depend on the surface area and not on the surface shape. The height of the liquid

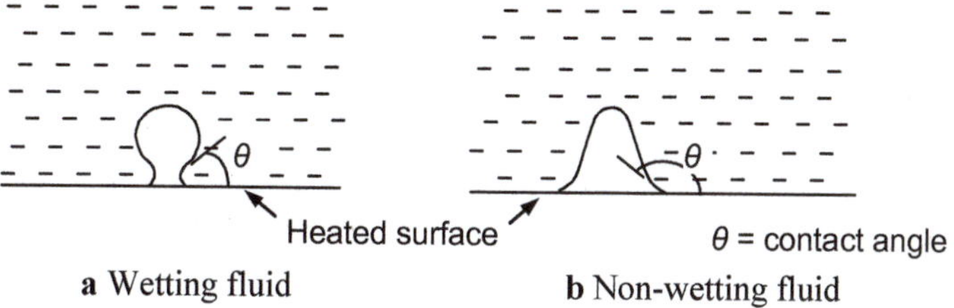

Fig. 13.8 Shape of bubbles

above the heating surface has also no effect on the rate of heat transfer. Vertical and horizontal planes and wire above a certain diameter give similar results.

(v) The boiling process and the corresponding discussion given above are true only for wetting liquids. Figure 13.8 shows the shapes of bubbles formed on a heated surface for wetting and non-wetting liquids. The bubble generated in the case of a wetting liquid can easily separate from the surface.

(vi) In the case of non-wetting liquids, the bubble spreads over the surface and forms a continuous film of the vapour. The heat transfer rate in this case is considerably lower than for the wetting fluids due to the high thermal resistance of this vapour film.

13.2.2.2 Forced-Flow Boiling

Consider the process of forced boiling occurring in limited space-say a tube. All the conditions described for pool boiling are valid but new factors appear. The arrangement of the tubes (vertical or horizontal), volume ratio of vapour to the liquid and the velocity of flow are the new important factors.

Flow Pattern in a Vertical Heated Tube with Upward Flow

First, we consider force flow of the boiling liquid through a vertical once-through tube as shown in Fig. 13.9, which is subjected to uniform heat flux q''. The liquid at the lower end of the tube (Sect. A) is at the saturation point and it goes out from the upper end of the tube in the form of superheated vapour. The mixture of liquid and vapour moves as a homogeneous emulsion with vapour as dispersed bubbles in the liquid (termed as bubbly flow) in section B when the quantity of the liquid is large compared to the vapour. The mode of boiling in this section is nucleate. With the increase in the vapour content, the vapour may flow as large slugs (Section C), which changes to annular flow in section D (liquid in the form of thin film on the wall and vapour in the centre of the tube). With further increase in the vapour content, the liquid film disappears and the liquid flows as suspended mist, refer section E. The vapour is wet at the end of the section D because of the suspended mist. The disappearance of the liquid film from the wall causes dry-out and the heat transfer coefficient reduces drastically. This causes a rise in the temperature of the tube surface and can cause tube failure. Evaporation of the mist continues during the section E as the vapour moves up and dry saturated vapour enters section F. The superheating of the vapour takes place in the section F.

Flow Pattern in Horizontal Evaporator Tube

The flow is not symmetrical in a horizontal tube. The mixture can flow as shown in Fig. 13.10a, b depending on the content of the vapour. In the lower figure, it flows in two independent streams of liquid and vapour.

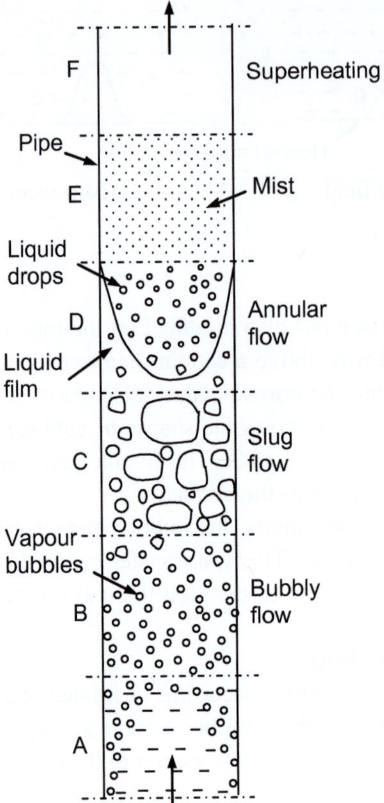

Fig. 13.9 Forced flow boiling in a vertical pipe

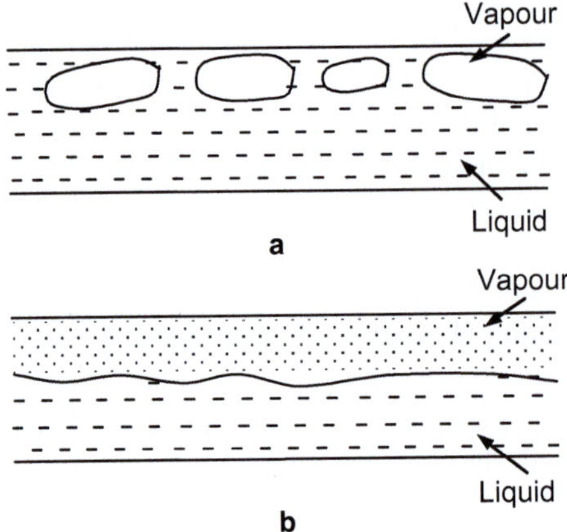

Fig. 13.10 Forced flow boiling in horizontal tubes

13.3 Relations for Boiling Heat Transfer in Pool Boiling

13.3.1 Nucleate Boiling

Experimental data for nucleate boiling on a horizontal surface facing upward has been correlated by Rohsenow (1952) as

$$\frac{c_{pl}(t_w - t_s)}{h_{fg}} = C_{nb}\left[\frac{q/A}{h_{fg}\mu_l}\sqrt{\frac{\sigma}{(\rho_l - \rho_v)g}}\right]^{1/3} \mathrm{Pr}_l^m \tag{13.22a}$$

where subscript l refers to liquid and all properties are to be evaluated at the saturation temperature. For water exponent $m = 1$ and is 1.7 for other fluids. Constant C_{nb} depends on the surface material and condition. For polished copper and stainless steel surface, $C_{nb} =$ 0.013. For copper scored, $C_{nb} = 0.0068$. For other liquids and surfaces, refer Mills (1995).

Using $q/A = h(t_w - t_s)$, and $\frac{\mu_l c_{pl}}{k_l} = \mathrm{Pr}_l$, the above equation gives the heat transfer coefficient equation for water as

$$h = \frac{\left[c_{pl}(t_w - t_s)\right]^2}{C_{nb}^3 h_{fg}^2\,\mathrm{Pr}_l^2}\left[\frac{(\rho_l - \rho_v)g}{\sigma}\right]^{1/2} k_l \tag{13.22b}$$

The heat transfer rate is given by

$$\frac{q}{A} = h(t_w - t_s) \tag{13.23}$$

It can be seen from the equation of heat transfer coefficient and heat flux that the heat flux is proportional to ΔT^3. The strong dependence of the heat flux on the temperature difference is due to the rapid increase in active nucleation sites with the increase in the superheat (Mills 1995). Equations (13.22a, 13.22b) is not very accurate and errors of 100% in q and 25% in ΔT are typical (Mills 1995).

13.3.1.1 The Peak Heat Flux

The peak heat flux relation due to Kutateladze (1948 in Mills 1995) and Zuber (1959 in Mills 1995) is

$$q_{max} = C_{max} h_{fg}\left[\sigma\rho_v^2(\rho_l - \rho_v)g\right]^{1/4} \tag{13.24}$$

where $C_{max} = 0.15$ for large flat heaters.

13.3.2 Simplified Relations for Boiling Heat Transfer with Water

Jakob and Hawkins (1957) have presented the following simplified correlations of heat transfer coefficient for water boiling on outside of submerged surfaces at standard atmospheric pressure.

For the horizontal orientation of the surface, the correlations are

$$h = 1042(\Delta t_x)^{1/3} \quad \text{for}\ q/A < 16\,\text{kW/m}^2 \quad \Delta t_x < 7.76 \tag{13.25a}$$

$$h = 5.56(\Delta t_x)^3 \quad \text{for}\ 16 < q/A < 240\,\text{kW/m}^2 \quad 7.32 < \Delta t_x < 14.4 \tag{13.25b}$$

where tT_x = temperature excess = $t_w - t_s$. For the vertical orientation of the surface, the correlations are

$$h = 537(\Delta t_x)^{1/7} \quad \text{for}\ q/A < 3\,\text{kW/m}^2 \quad \Delta t_x < 4.51 \tag{13.26a}$$

$$h = 7.96(\Delta t_x)^3 \quad \text{for}\ 3 < q/A < 63\,\text{kW/m}^2. \quad 4.41 < \Delta t_x < 9.43 \tag{13.26b}$$

To take account of the influence of the pressure, the following empirical relation is suggested

$$h_p = h\left(\frac{p}{p_a}\right)^{0.4}$$

where h_p is the heat transfer coefficient at pressure p and h is the heat transfer coefficient at standard atmospheric pressure p_a from Eqs. (13.25a, 13.25b) to (13.26a, 13.26b).

For the forced-flow local boiling inside vertical tubes, the following relation is recommended (Jakob 1957).

$$h = 2.54(\Delta t_x)^3 e^{p/1.551} \quad \text{for}\ 5 \le p \le 170\,\text{bar}. \tag{13.27}$$

where p is the pressure in MPa.

Example 13.9 Estimate heat flux for boiling of water at atmospheric pressure on a polished surface of stainless steel at 390 K. At 373 K, the properties of water are

$$h_{fg} = 2.27 \times 10^6\,\text{J/kg},\ k_l = 0.682\,\text{W/(m K)},\ c_{pl} = 4211\,\text{J/(kg K)},$$
$$\sigma = 58.9 \times 10^{-3}\,\text{N/m},\ \rho_l = 958.3\,\text{kg/m}^3,\ \text{Pr}_\text{l} = 1.76.$$

Solution

From Eq. (13.22b),

$$h = \frac{\left[c_{pl} \times (T_w - T_s)\right]^2}{C_{nb}^3 h_{fg}^2 \, \mathrm{Pr}_l^2} \left[\frac{(\rho_l - \rho_v)g}{\sigma}\right]^{1/2} k_l$$

$C_{\mathrm{nb}} = 0.013$ for stainless steel surface. Substituting the values of various terms and neglecting ρ_v ($\rho_v \ll \rho_{\mathrm{l}}$), we get

$$h = \frac{[4211 \times (390 - 373)]^2}{0.013^3 \times (2.27 \times 10^6)^2 \times 1.76^2} \times \left[\frac{958.3 \times 9.81}{58.9 \times 10^{-3}}\right]^{1/2} \times 0.682 = 39815\,\mathrm{W/(m^2\,K)}.$$

Heat flux,

$$\frac{q}{A} = h(T_w - T_s) = 39815 \times (390 - 373) = 676.855\,\mathrm{kW/m^2}.$$

Example 13.10 Estimate peak heat flux of boiling water on a large flat heater at $T_s =$ 373 K. At 373 K, the properties of water are given as $h_{\mathrm{fg}} = 2.27 \times 10^6$ J/kg, $\sigma = 58.9 \times 10^{-3}$ N/m, $\rho_{\mathrm{l}} = 958.3$ kg/m^3, and for vapour, $\rho_v = 0.6$ kg/m^3.

Solution

From Eq. (13.24) for a large flat heater,

$$\begin{aligned} q_{max} &= 0.15 h_{\mathrm{fg}} \left[\sigma \rho_v^2 (\rho_l - \rho_v) g\right]^{1/4} \\ &= 0.15 \times 2.27 \times 10^6 \times \left[58.9 \times 10^{-3} \times 0.6^2 \times (958.3 - 0.6) \times 9.81\right]^{1/4} \\ &= 1.279 \times 10^6\,\mathrm{W/m^2}. \end{aligned}$$

Example 13.11 An electrically heated flat stainless steel plate 150 mm^2 in area is operating at power levels corresponding to 40% of the critical heat flux for boiling water at 1 atm. Determine the power supplied to the electric heater and temperature of the plate surface. At boiling point 373 K (corresponding to 1 atm), the properties of water are $h_{\mathrm{fg}} = 2.27$ 10^6 J/kg, $k_{\mathrm{l}} = 0.682$ W/(m K), $c_{\mathrm{pl}} = 4211$ J/(kg K), $\sigma = 58.9 \times 10^{-3}$ N/m, $\rho_{\mathrm{l}} = 958.3$ kg/m^3 and $\mathrm{Pr}_{\mathrm{l}} = 1.76$, and for vapour, $\rho_v = 0.6$ kg/m^3.

Solution

From Eq. (13.24) for a flat heater, the critical heat flux is

$$\begin{aligned} q_{max} &= 0.15 h_{\mathrm{fg}} \left[\sigma \rho_v^2 (\rho_l - \rho_v) g\right]^{1/4} \\ &= 0.15 \times 2.27 \times 10^6 \times \left[58.9 \times 10^{-3} \times 0.6^2 \times (958.3 - 0.6) \times 9.81\right]^{1/4} \\ &= 1.279 \times 10^6\,\mathrm{W/m^2}. \end{aligned}$$

Hence, the heat flux is 0.4 × 1279 kW/m^2 = 511.6 kW/m^2 and the power supplied to the electric heater is

$$P_e = 150 \times 10^{-6} \times 511.6 \times 10^3 = 76.74\text{W}.$$

The heat transfer coefficient is

$$h = \frac{P_e}{A\Delta T} = \frac{76.74}{150 \times 10^{-6} \times \Delta T} = \frac{511.6 \times 10^3}{\Delta T}\ \text{W/(m}^2\text{ K)}$$

where $\Delta T = T_w - T_s$.

From Eq. (13.22b) for boiling of water,

$$h = \frac{(c_{\text{pl}} \times \Delta T)^2}{C_{\text{nb}}^3 h_{\text{fg}}^2 \text{Pr}_l^2}\left[\frac{(\rho_l - \rho_v)g}{\sigma}\right]^{1/2} k_l$$

C_{nb} = 0.013 for stainless steel surface. Substituting the values of various terms, we get

$$\frac{511.6 \times 10^3}{\Delta T} = \frac{(4211 \times \Delta T)^2}{0.013^3 \times (2.27 \times 10^6)^2 \times 1.76^2} \times \left[\frac{(958.3 - 0.6) \times 9.81}{58.9 \times 10^{-3}}\right]^{1/2} \times 0.682,$$

which gives $\Delta T = 15.49$ K. Hence, plate surface temperature

$$T_w = T_s + \Delta T = 373 + 15.49 = 388.49\ \text{K}.$$

Example 13.12 Estimate the heat transfer per unit area from a horizontal flat surface submerged in water at atmospheric pressure if the surface temperature is 105°C.

Solution

$$\Delta t = 105 - 100 = 5^\circ\text{C}.$$

Let q/A < 16 kW/m^2, then the heat transfer coefficient,

$$h = 1042(\Delta t)^{1/3} == 1042(5)^{1/3} = 1782\ \text{W}/\left(\text{m}^2\text{K}\right).$$

Heat flux,

$$\frac{q}{A} = h\Delta t = 1782 \times 5 = 8910\ \text{W/m}^2 < 16\ \text{kW/m}^2.$$

Example 13.13 Figure 13.11 shows an experimental setup to study pool boiling in water. Heat flows by conduction through the copper bar [k = 350 W/(m K)]. The temperature measured at x_1 = 10 mm is 130 and 145°C at x_2 = 20 mm. Determine the constant C_{nb} in the Rohsenow correlation.

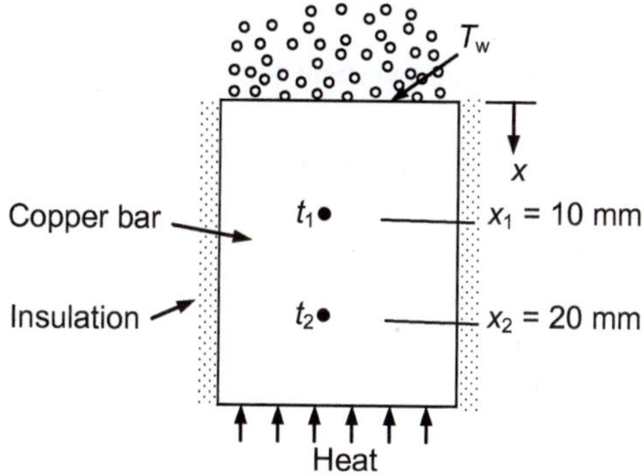

Fig. 13.11 Example 13.13

Solution

The temperature distribution in the bar is linear hence temperature at x = 0, i.e. at the boiling surface is

$$t_w = t_1 - \frac{\Delta t}{\Delta x} x_1 = t_1 - \frac{t_2 - t_1}{x_2 - x_1} x_1 = 130 - \frac{145 - 130}{10} \times 10 = 115\ ^\circ\mathrm{C}.$$

Applying Fourier's conduction law between $x1$ and $x2$, we have heat flux

$$\frac{q}{A} = k\frac{\Delta t}{\Delta x} = k\frac{t_2 - t_1}{x_2 - x_1} = 350 \times \frac{145 - 130}{10/1000} = 5.25 \times 10^5\ \mathrm{W/m^2}.$$

At boiling point 100°C (corresponding to 1 atm), the properties of water are h_{fg} = 2.27 10^6 J/kg, $\mu_l = 282 \times 10^{-6}$ N s/m^2, c_{pl} = 4211 J/(kg K), $\sigma = 58.9 \times 10^{-3}$ N/m, ρ_l = 958.3 kg/m^3 and Pr$_l$ = 1.76, and for vapour, ρ_v = 0.6 kg/m^3.

From Eq. (13.22a), for water, we have

$$C_{nb} = \frac{c_{pl}(t_w - t_s)}{h_{fg}} \left\{ \frac{q/A}{h_{fg}\mu_l} \sqrt{\frac{\sigma}{(\rho_l - \rho_v)g}} \right\}^{-1/3} \mathrm{Pr}_l^{-1}$$

Substitution gives

$$C_{nb} = \frac{4211 \times (115 - 100)}{2.27 \times 10^6} \left\{ \frac{5.25 \times 10^5}{2.27 \times 10^6 \times 282 \times 10^{-6}} \sqrt{\frac{58.9 \times 10^{-3}}{(958.3 - 0.6) \times 9.81}} \right\}^{-1/3} \times 1.76^{-1}$$

$$= 0.0124.$$

Example 13.14 For forced-convection local boiling inside a 75 mm diameter vertical tube, estimate the boiling heat transfer rate per unit length of the tube when Δt_x = 15°C and water pressure is 3 bar.

Solution

For the forced-flow local boiling inside vertical tubes, Eq. (13.27) applies which gives

$$h = 2.54(\Delta t)^3 e^{p/1.551}$$

where p is the pressure in MPa.

Substituting the given data, we have

$$h = 2.54 \times (15)^3 \times e^{0.3/1.551} = 10402\,\mathrm{W/(m^2\,K)}$$

and heat rate is

$$q = hA\Delta t = 10402 \times \pi \times 0.075 \times 1 \times 15 = 36764\,\mathrm{W/m\ length.}$$

Review Questions

13.1 Differentiate between filmwise and dropwise condensations.

13.2 Stating the assumptions made, deduce an expression of film thickness and heat transfer coefficient at a distance x from the top edge of a vertical plate for laminar film condensation.

13.3 What is the effect of direction of vapour flow, layout of the surface and non-condensing gases in vapours on the film coefficient?

13.4 Discuss various regimes in pool boiling. Also, explain the significance of critical heat flux.

13.5 Discuss forced-flow boiling in a vertical tube. Support the answer with a suitable sketch.

13.6 Discuss forced-flow boiling in a horizontal tube.

Problems

13.1 Saturated steam condenses on the outside of a 50 mm diameter vertical tube 0.5 m in height. If the saturation temperature of the steam is 302 K and the tube outer surface is maintained at 299 K, calculate (i) the average heat transfer coefficient, (ii) the condensation rate and (iii) film thickness at the lower end of the tube. Given h_{fg} = 2.44 10^6 J/kg, k_f = 0.615 W/(m K), ρ_f = 995.6 kg/m^3, ν_f = 0.854 × 10^{-6} m^2/s, c_f = 4.178 kJ/(kg K).

[**Ans.** Assuming laminar flow, $\bar{h} = 0.943\left[\frac{\rho^2 g h_{fg} k^3}{\mu(t_s - t_w)H}\right]^{0.25} = 7648\,\mathrm{W/(m^2\,K)}$;

$\dot{m} = \frac{Q}{h_{fg}} = \frac{\bar{h}A(\Delta t)}{h_{fg}} = 7.38 \times 10^{-4}\,\mathrm{kg/s}$; $\delta = \left[\frac{4\mu k(t_s - t_w)H}{\rho^2 g h_{fg}}\right]^{1/4} = 0.107\,\mathrm{mm}$;

$\mathrm{Re}_H = \frac{4\dot{m}}{\mu P} = \frac{4\dot{m}}{\mu(\pi D)} = 22.1$, which confirms that flow is laminar.]

13.2 If the tube of Q. 13.1 is held in a horizontal position, estimate the condensation rate.

[**Ans.** $\bar{h} = 0.725\left[\frac{\rho^2 g h_{fg} k^3}{\mu(t_s - t_w)D}\right]^{1/4} = 10456\,\mathrm{W/(m^2\,K)}$; $\dot{m} = \frac{Q}{h_{fg}} = \frac{\bar{h}A(\Delta t)}{h_{fg}} = 10.1 \times 10^{-4}\,\mathrm{kg/s}$. Note: For a single horizontal tube, flow is always laminar.]

13.3 If the steam in Q.13.2 is condensing on a bundle of horizontal tubes, the number of tubes is 10 in a vertical column and 12 in each horizontal row. Estimate the average heat transfer coefficient.

[**Ans.** $\bar{h} = n^{-1/6} \times \bar{h}_1 = 10^{-1/6} \times 10456 = 7123$.]

References

Jakob M (1957) Heat transfer. Wiley, New York

Jakob M, Hawkins GA (1957) Elements of heat transfer. Wiley, New York

Kern DQ (1950) Process heat transfer. McGraw-Hill, New York

Kern DQ (1958) Mathematical development of tube loading in horizontal condensers. J AICHE 4(2):157–160

Kirkbride CG (1934) Heat transfer by condensing vapors on vertical tubes. Trans AICHE 30:170–186

Mikheyev M (1968) Fundamental of heat transfer. Mir Publishers, Moscow

Mills AF (1995) Heat and mass transfer. Richard D, Irwin, Chicago

Pitts Donald R, Sissom Leighton E (1991) Heat transfer. McGraw-Hill, New York

Rohsenow WM (1952) A method of correlating heat transfer data for surface boiling of liquids. Trans ASME 74:969–976

14 Heat Exchangers

14.1 Part A: Heat Exchangers Fundamentals

14.1.1 Introduction

In general, a heat exchanger is an equipment to accomplish the task of transfer of heat between two or more entities at different temperatures. In the present chapter, the heat exchanging entities are two fluid streams. Many types of heat exchangers are employed, which vary both in application and design.

The heat exchangers may be classified on the basis of the configuration of the fluid flow paths through the heat exchanger. The common flow path configurations are illustrated diagrammatically in Fig. 14.1.

In the *parallel flow* or *co-current heat exchangers*, Fig. 14.1a, the two fluid streams enter together at one end and leave together at the other end, whereas in the *counterflow* or *countercurrent* exchangers, the two fluid streams move in opposite directions as shown in Fig. 14.1b.

In *single-pass cross-flow* exchangers, Fig. 14.1c, the fluids flow through the heat exchanger matrix at right angles to each other. An example of such type of exchanger is the automobile radiator where the air is forced across the radiator tubes carrying water to be cooled. In *multi-pass cross-flow* exchanger, one of the fluids shuttles back and forth across the flow path of the other fluid as shown in Fig. 14.1d.

Based on the principle of operation, the heat exchangers are known as *recuperator, regenerator* and *direct-contact type*. In a recuperator, the cold and hot streams flow simultaneously through the heat exchanger and the heat is transferred across a surface separating the fluids. Steam boilers, surface condensers and air pre-heaters are some examples of recuperator.

In a regenerator, one and the same heating surface is alternatively exposed to the hot and cold fluids. The heat of the hot fluid is taken away and accumulated in the walls or matrix of the exchanger which is then transferred to the cold fluid. Thus, there is a continuous flow in the recuperator, but a periodic in the regenerator.

Recuperates and regenerators are also classified as *surface heat exchangers* because the fluids are separated by a solid surface.

R. Karwa, *Heat and Mass Transfer*,
https://doi.org/10.1007/978-981-15-3988-6_14

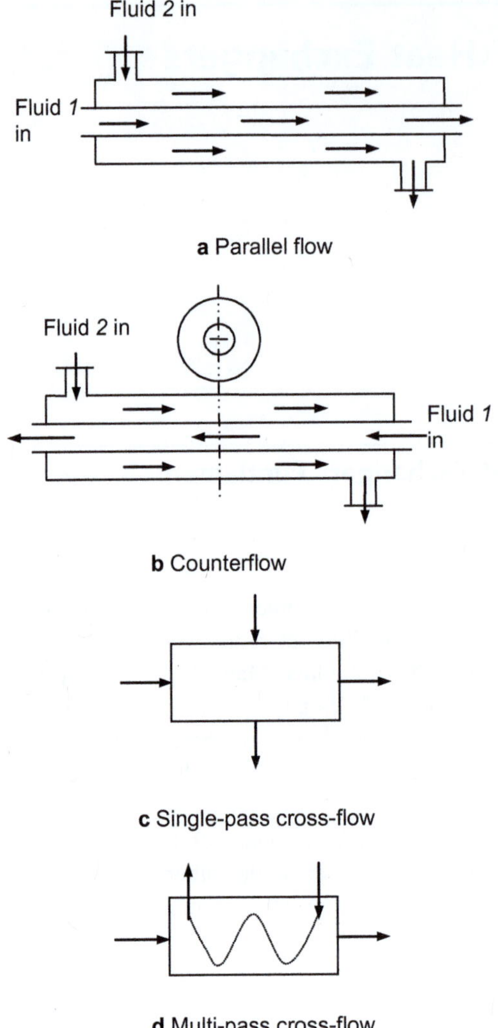

Fig. 14.1 Flow path configurations

In a direct-contact type exchanger, the heat is transferred through the direct contact and mixing of the hot and cold fluid streams. Thus, the heat transfer is also accompanied by the mass transfer.

The heat exchangers are employed in varied installations such as power plants, chemical industries, air conditioning and refrigeration systems, and automotive power plants. Hence, they are frequently known by their applications, e.g. boiler, condenser, economizer, evaporator, air pre-heater, cooling tower, radiator and cooler.

Double pipe heat exchanger, refer to Fig. 14.1a, b, consists of concentric pipes with one fluid in the inner pipe and other in the annular space surrounding the inner pipe. It is a heat exchanger which is well suited to some applications. Due to its construction, it can carry high-pressure fluids which would have required a large wall thickness in case of a shell-and-tube heat exchanger.

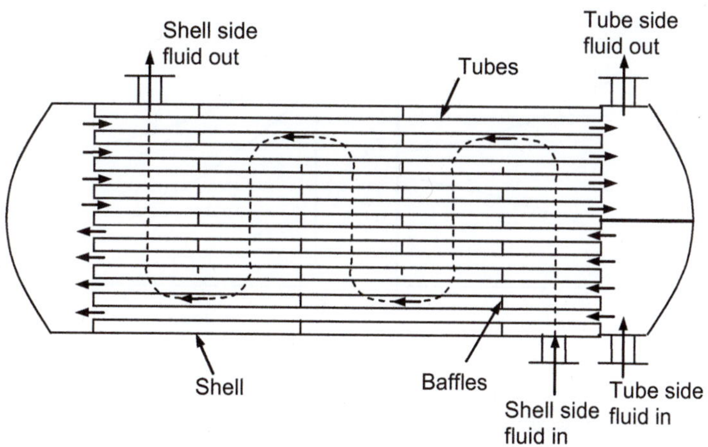

Fig. 14.2 Schematic of 1–2 shell-and-tube heat exchanger

The *shell-and-tube heat exchangers* consist of a large number of tubes mounted in cylindrical shells with their axes parallel to that of the shell. Many variations of this type of exchanger are in use. In Fig. 14.2, one of the designs known as 1–2 shell-and-tube heat exchanger is shown where the shell fluid passes once while the tube side fluid passes twice along the length of the exchanger. For other designs of the shell-and-tube heat exchangers, refer to Kern (1950).

While dealing with fluids, which provide a low heat transfer coefficient, such as air or gases, finned tubes are frequently employed in the exchangers.

14.1.2 Heat Transfer Equation for Double Pipe (Concentric Tube) Heat Exchanger

In a double pipe heat exchanger, the fluids may have either parallel or counterflow arrangement. The temperatures of both the fluids change as they pass through the exchanger. Figure 14.3a, b shows the temperature profiles of the hot and cold fluids versus the heat transfer area of the exchanger for the parallel and counterflow arrangements, respectively. Since the temperature difference between the two fluids varies from the inlet to the outlet, the heat transfer equation is based on an effective or mean temperature difference for the whole length (or heat transfer area) of the exchanger (known as *log mean temperature difference*) and is written as

$$q = U_i A_i \Delta t_m = U_o A_o \Delta t_m \tag{14.1}$$

where

U overall heat transfer coefficient,
A area transferring heat,
Δt_m log mean temperature difference.

The subscripts '*i*' and '*o*' stand for the inner and outer surface area of the pipe, respectively.

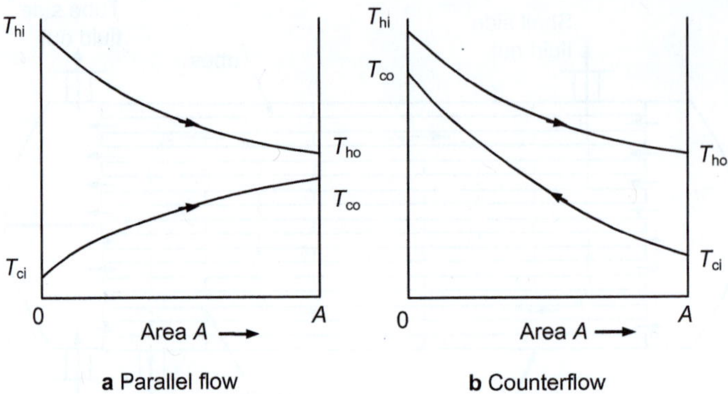

Fig. 14.3 Fluid temperature variations

The equations of overall heat transfer coefficients were developed in Chap. 2 and are reproduced here for a ready reference.

$$U_i = \left[\frac{1}{h_i} + \frac{A_i}{2\pi kL}\ln\left(\frac{R_o}{R_i}\right) + \frac{1}{h_o}\left(\frac{A_i}{A_o}\right)\right]^{-1} \tag{14.2}$$

$$U_o = \left[\frac{1}{h_i}\left(\frac{A_o}{A_i}\right) + \frac{A_o}{2\pi kL}\ln\left(\frac{R_o}{R_i}\right) + \frac{1}{h_o}\right]^{-1} \tag{14.3}$$

where $A_i = \pi d_i L$ and $A_o = \pi d_o L$ as applied to a double pipe heat exchanger. d_i and d_o are inner and outer diameters of the inner pipe. Expressions of log mean temperature difference for parallel and counterflow arrangements are derived in the next section.

14.1.3 Log Mean Temperature Difference (*LMTD*)

14.1.3.1 Parallel Flow Arrangement

Refer to Fig. 14.4. For the heat exchanger elemental area δA, the heat transfer rate is

$$\delta Q = U\delta A(T_1' - t_1') \tag{i}$$

Due to the heat transfer, the temperature of the hot fluid 1 decreases by δt_1 and the temperature of the cold fluid 2 increases by δt_2. Hence,

$$\delta Q = -m_1 c_{p1}\delta t_1 = m_2 c_{p2}\delta t_2 \tag{ii}$$

From which,

$$\delta t_1 = -\frac{\delta Q}{m_1 c_{p1}}$$

$$\delta t_2 = \frac{\delta Q}{m_2 c_{p2}}$$

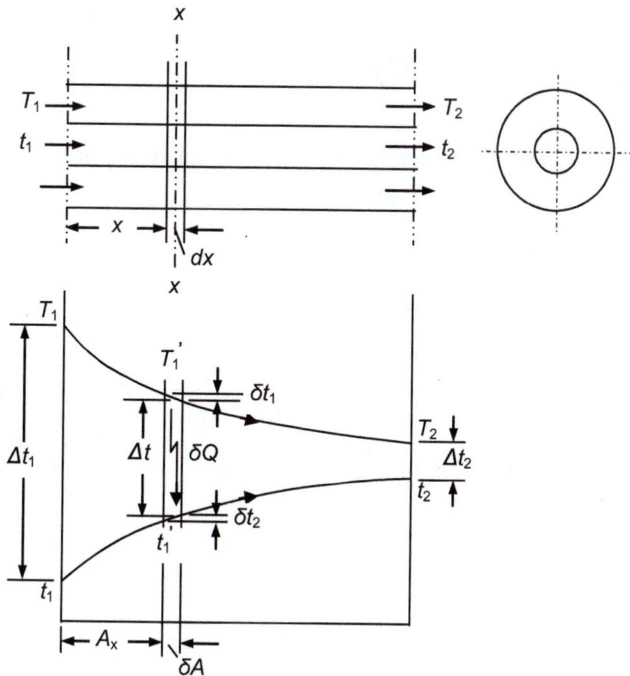

Fig. 14.4 Parallel flow heat exchanger

The difference between δt_1 and δt_2 is

$$\delta t_1 - \delta t_2 = -\left(\frac{1}{m_1 c_{p1}} + \frac{1}{m_2 c_{p2}}\right)\delta Q = -C\delta Q$$

where

$$C = \left(\frac{1}{m_1 c_{p1}} + \frac{1}{m_2 c_{p2}}\right) \tag{14.4}$$

Substituting the value of δQ from Eq. (i) gives

$$\delta t_1 - \delta t_2 = -CU(T_1' - t_1')\delta A$$

In the limit,

$$d(T_1' - t_1') = -CU(T_1' - t_1')dA$$

or

$$d(\Delta t) = -CU(\Delta t)dA$$

or

$$\frac{d(\Delta t)}{\Delta t} = -CUdA \tag{iii}$$

If C and U are constants, the integration of the above equation for the limits between the inlet to section xx (refer to Fig. 14.4) gives

$$\int_{\Delta t_1}^{\Delta t_x} \frac{d(\Delta t)}{\Delta t} = -\int_0^{A_x} CUdA$$

or

$$\ln\left(\frac{\Delta t_x}{\Delta t_1}\right) = -CUA_x$$

or

$$\Delta t_x = \Delta t_1 e^{-CUA_x} \tag{14.5}$$

Similarly, the integration of Eq. (iii) for the limits at inlet to outlet (refer to Fig. 14.4) gives

$$\int_{\Delta t_1}^{\Delta t_2} \frac{d(\Delta t)}{\Delta t} = -\int_0^{A} CUdA$$

or

$$\frac{\Delta t_2}{\Delta t_1} = e^{-CUA} \tag{14.6}$$

Equation (14.6) can be written as

$$-CUA = \ln\left(\frac{\Delta t_2}{\Delta t_1}\right) \tag{14.7}$$

From the above equations, it can be seen that the temperature difference Δt between hot and cold fluids varies in accordance with the exponential law along the heating surface of the exchanger.

The mean temperature difference for the whole surface can be defined as

$$\Delta t_m = \frac{1}{A}\int_o^{A} \Delta t_x dA_x$$

Using Eq. (14.5), we have

$$\Delta t_m = \frac{1}{A}\int_o^A \Delta t_1 e^{-CUA_x} dA_x = \frac{\Delta t_1}{-CUA}\left(e^{-CUA} - 1\right)$$

Substitution of values of terms e^{-UCA} and CUA from Eqs. (14.6) and (14.7) gives

$$\Delta t_m = \Delta t_1 \frac{1}{\ln\left(\frac{\Delta t_2}{\Delta t_1}\right)}\left(\frac{\Delta t_2}{\Delta t_1} - 1\right)$$

or

$$\Delta t_m = \frac{\Delta t_2 - \Delta t_1}{\ln\left(\frac{\Delta t_2}{\Delta t_1}\right)} = \frac{(T_1 - t_1) - (T_2 - t_2)}{\ln\left(\frac{T_1 - t_1}{T_2 - t_2}\right)} \tag{14.8}$$

This is known as *log mean temperature difference* (*LMTD*) because of the logarithmic term in the expression. The subscripts 1 and 2 in the equation can be interchanged without any effect on the value of *LMTD* Δt_m.

14.1.3.2 Counterflow Arrangement

For the elemental area δA, the heat transfer rate is

$$\delta Q = U\delta A(T_1' - t_1') \tag{i}$$

In the axial direction, the temperatures of the hot fluid 1 and cold fluid 2 change by δt_1 and δt_2, respectively, refer to Fig. 14.5. Hence,

$$\delta Q = -m_1 c_{p1}\delta t_1 = -m_2 c_{p2}\delta t_2$$

From which,

$$\delta t_1 = -\frac{\delta Q}{m_1 c_{p1}}$$

$$\delta t_2 = -\frac{\delta Q}{m_2 c_{p2}}$$

The difference between δt_1 and δt_2 is

$$\delta t_1 - \delta t_2 = -\left(\frac{1}{m_1 c_{p1}} - \frac{1}{m_2 c_{p2}}\right)\delta Q = -C\delta Q \tag{ii}$$

where

$$C = \left(\frac{1}{m_1 c_{p1}} - \frac{1}{m_2 c_{p2}}\right) \tag{14.9}$$

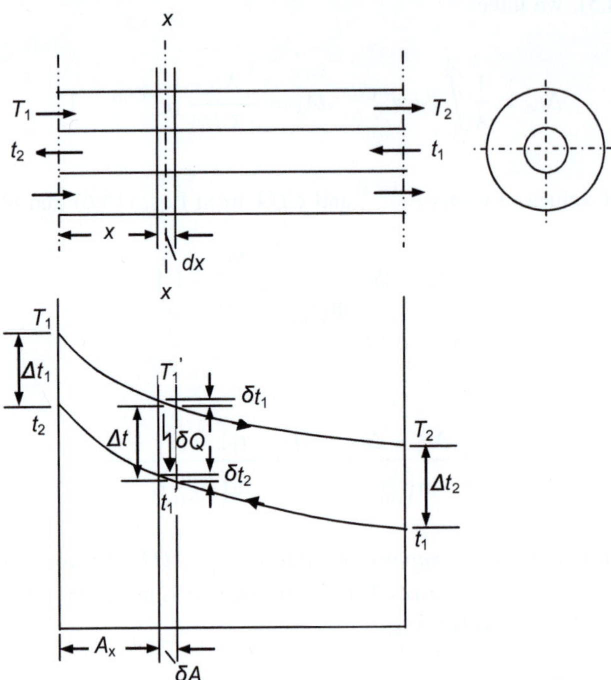

Fig. 14.5 Counterflow heat exchanger

Equation (ii) is the same as that for the parallel flow heat exchanger hence the remaining mathematical treatment will also be the same and we obtain

$$\Delta t_m = \frac{\Delta t_2 - \Delta t_1}{\ln\left(\frac{\Delta t_2}{\Delta t_1}\right)} \tag{14.10}$$

where the temperature differences in this case are defined as, refer to Fig. 14.5,

$$\Delta t_1 = (T_1 - t_2)$$

and

$$\Delta t_2 = (T_2 - t_1)$$

Hence,

$$\Delta t_m = \frac{\Delta t_2 - \Delta t_1}{\ln\left(\frac{\Delta t_2}{\Delta t_1}\right)} = \frac{(T_1 - t_2) - (T_2 - t_1)}{\ln\left(\frac{T_1 - t_2}{T_2 - t_1}\right)} \tag{14.11}$$

If the heat capacity mc_p of the two streams in a counterflow heat exchanger are equal, then $C = 0$, and from Eq. (14.5) we get

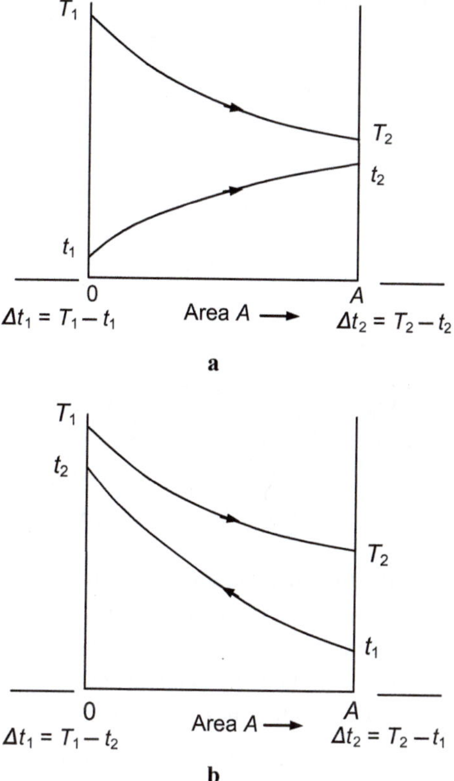

Fig. 14.6 **a** Parallel flow, **b** Counterflow

$$\Delta t_x = \Delta t_1 e^{-0} = \Delta t_1 \tag{14.12}$$

i.e. the temperature difference is constant over the entire surface.

If the temperature differences between the two fluids are denoted by Δt_1 and Δt_2 as temperature differences at the two ends of the exchanger as depicted in Fig. 14.6, then the *LMTD* equations for the parallel- and counterflow arrangements, Eqs. (14.8) and (14.11), acquire the same form:

$$\Delta t_m = \frac{\Delta t_2 - \Delta t_1}{\ln\left(\frac{\Delta t_2}{\Delta t_1}\right)}$$

Note: Arithmetic mean temperature difference is given by

$$\Delta t_{\text{am}} = \frac{\Delta t_1 + \Delta t_2}{2}$$

The ratio of the arithmetic and log mean temperature differences is

Table 14.1 $\Delta t_{am}/\Delta t_m$ as function of $\Delta t_1/\Delta t_2$

$\Delta t_1/\Delta t_2$	$\Delta t_{am}/\Delta t_m$
1.0	1.0
1.5	1.014
2.0	1.040
2.5	1.069
3.0	1.099
3.5	1.127
4.0	1.155

$$\frac{\Delta t_{am}}{\Delta t_m} = \frac{1}{2}\left(\frac{\Delta t_1 + \Delta t_2}{\Delta t_1 - \Delta t_2}\right)\ln\left(\frac{\Delta t_1}{\Delta t_2}\right)$$
$$= \frac{1}{2}\left(\frac{\Delta t_1/\Delta t_2 + 1}{\Delta t_1/\Delta t_2 - 1}\right)\ln\left(\frac{\Delta t_1}{\Delta t_2}\right)$$

The values of the ratio of the two mean temperatures are tabulated in Table 14.1 against the ratio $\Delta t_1/\Delta t_2$. It can be seen that the arithmetic mean is always greater than the log mean. The difference between the two increases with the increase in the value of the ratio $\Delta t_1/\Delta t_2$.

In cases where the temperatures of the hot and cold streams change slightly along the heat transferring surface, the mean temperature may be taken as the arithmetic mean.

Example 14.1 For what value of ratio $\Delta t_1/\Delta t_2$ is the arithmetic mean temperature difference Δt_{am} 5% larger than the log mean temperature difference Δt_m?

Solution

The ratio of the two mean temperature differences is given by

$$\frac{\Delta t_{am}}{\Delta t_m} = \frac{1}{2}\left(\frac{\Delta t_1/\Delta t_2 + 1}{\Delta t_1/\Delta t_2 - 1}\right)\ln\left(\frac{\Delta t_1}{\Delta t_2}\right)$$

olving by trial and error for $\Delta t_{am}/\Delta t_m = 1.05$, we obtain

$$\Delta t_1/\Delta t_2 = 2.2$$

It means that the arithmetic temperature difference gives results to within 5% of log mean temperature difference when the ratio of the end temperature differences $\Delta t_1/\Delta t_2$ is less than 2.2.

Example 14.2 A cold fluid is heated from 40°C to 120°C by condensing steam at 180°C. Calculate the *LMTD*.

Solution

Refer to Fig. 14.7.

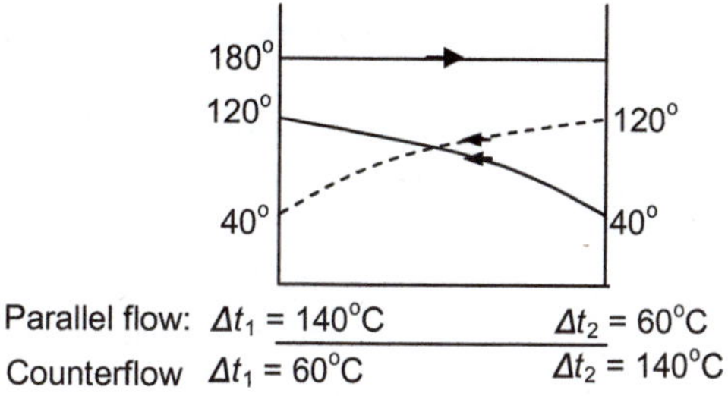

Fig. 14.7 Example 14.2

(i) **Parallel flow**

$$\Delta t_m = \frac{\Delta t_1 - \Delta t_2}{\ln \frac{\Delta t_1}{\Delta t_2}} = \frac{140 - 60}{\ln \frac{140}{60}} = 94.42$$

(ii) **Counterflow**

$$\Delta t_m = \frac{\Delta t_1 - \Delta t_2}{\ln \frac{\Delta t_1}{\Delta t_2}} = \frac{60 - 140}{\ln \frac{60}{140}} = 94.42$$

The results are identical. When one of the fluids passes through a heat exchanger isothermally (condensing vapour or boiling liquid), the flow arrangement does not affect the value of *LMTD*.

Example 14.3 In order to recover heat from a waste stream at 400°C, it is proposed to use a counterflow heat exchanger, wherein it will be cooled to 120°C while heating a cold stream from 30°C to 200°C.

Another proposal is to cool waste stream to 50°C while heating the cold stream to the same temperature by increasing the flow rate of the cold stream.

If the overall heat transfer coefficient is the same in both the cases, comment on the proposed schemes.

Solution

I case:

$$\Delta t_m = \frac{\Delta t_1 - \Delta t_2}{\ln \frac{\Delta t_1}{\Delta t_2}} = \frac{200 - 90}{\ln \frac{200}{90}} = 137.76^\circ\text{C}$$

II case:

$$\Delta t_m = \frac{\Delta t_1 - \Delta t_2}{\ln \frac{\Delta t_1}{\Delta t_2}} = \frac{200 - 20}{\ln \frac{200}{20}} = 78.17^\circ\text{C}$$

The log mean temperature difference of the second scheme is lower by 43.3%, which will require 43.3% greater area while the heat recovery is increased only by 25%. It can be shown that a further 3.6% enhancement in the heat recovery (cooling the heat stream to 40°C) will increase the heat transfer area requirement by 54%.

The log mean temperature difference reduces rapidly if the temperature of one of the streams approaches the temperature of the other stream.

Example 14.4 In a counterflow heat exchanger, hot water inlet and outlet temperatures are 90°C and 40°C, respectively. Cold water inlet and outlet temperatures are 10°C and 60°C, respectively. Mass flow rate of hot water is 10 kg/min. Calculate the *LMTD* and heat transfer rate.

Solution

Log mean temperature difference is (Fig. 14.8)

$$\Delta t_m = \frac{\Delta t_1 - \Delta t_2}{\ln \frac{\Delta t_1}{\Delta t_2}}$$

$$\Delta t_m = \frac{\frac{\partial}{\partial \Delta t_1}(\Delta t_1 - \Delta t_2)}{\frac{\partial}{\partial \Delta t_1}\left(\ln \frac{\Delta t_1}{\Delta t_2}\right)} = \frac{1}{\frac{1}{\frac{\Delta t_1}{\Delta t_2}} \times \frac{1}{\Delta t_2}} = \frac{\Delta t_2}{\Delta t_2} \times \Delta t_1 = \Delta t_1$$

From the first law of thermodynamics, heat flow rate is

$$q = mc_p\Delta t = (10/60) \times 4.1868 \times (90 - 40) = 34.9 \text{ kW}$$

Alternatively

The heat balance equation gives

$$(mc_p\Delta t)_{\text{hot}} = (mc_p\Delta t)_{\text{cold}}$$

Since the temperature changes for both hot and cold streams are the same, we get

$$(mc_p)_{\text{hot}} = (mc_p)_{\text{cold}}$$

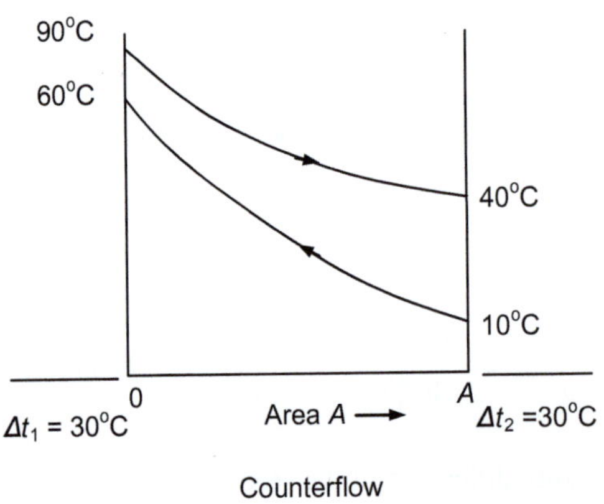

Fig. 14.8 Example 14.4

Hence,

$$C = \frac{1}{(mc_p)_{\text{hot}}} - \frac{1}{(mc_p)_{\text{cold}}} = 0$$

This gives

$$\Delta t_x = \Delta t_1 e^{-CUA} = \Delta t_1 e^{-0} = \Delta t_1 = 30°\text{C}$$

Example 14.5 Exhaust gases [c_p = 1.12 kJ/(kg K)] are to be cooled from 700 K to 400 K. The cooling is to be affected by water [c_p = 4.18 kJ/(kg K)] available at 30°C. The flow rates for the exhaust gases and water are 1000 kg/hr and 1500 kg/hr, respectively. If the overall heat transfer coefficient is estimated to be 400 kJ/(m^2 hr K), calculate the heat transfer area required for (i) counterflow arrangement (ii) parallel flow arrangement in the exchanger.

Solution

Heat given by exhaust gases is

$$\begin{aligned} q &= m_e c_{pe}(T_1 - T_2) \\ &= 1000 \times 1.12 \times (700 - 400) = 336 \times 1000 \text{ kWh} \end{aligned}$$

The unknown exit temperature of the water can be determined from the heat balance, i.e. by equating the heat gain by the water to the heat lost by the exhaust gases. Hence,

$$m_w c_{pw}(t_2 - t_1) = 1500 \times 4.18 \times (t_2 - 30) = 336 \times 1000$$

or

$$t_2 = 83.6°\text{C}$$

(i) **Counterflow arrangement** (Fig. 14.9)
Log mean temperature difference

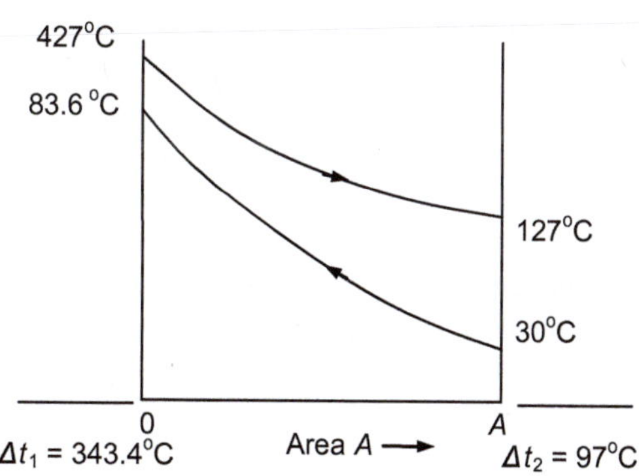

Fig. 14.9 Example 14.5, counterflow

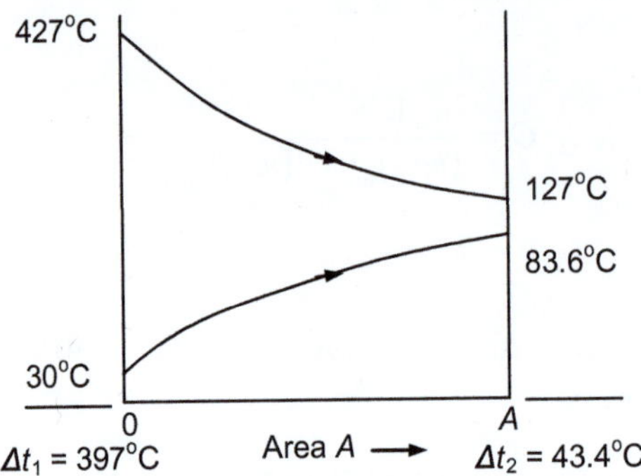

Fig. 14.10 Example 14.5, parallel flow

$$\Delta t_m = \frac{\Delta t_2 - \Delta t_1}{\ln\left(\frac{\Delta t_2}{\Delta t_1}\right)} = \frac{343.4 - 97}{\ln\left(\frac{343.4}{97}\right)} = 194.9°\text{C}$$

The heat exchange equation gives the heat transfer area as

$$A = \frac{q}{U\Delta t} = \frac{336 \times 1000}{400 \times 194.9} = 4.31 \text{ m}^2$$

(ii) **Parallel flow heat exchanger**

Log mean temperature difference, refer to Fig. 14.10, is

$$\Delta t_m = \frac{\Delta t_2 - \Delta t_1}{\ln\left(\frac{\Delta t_2}{\Delta t_1}\right)} = \frac{397 - 43.4}{\ln\left(\frac{397}{43.4}\right)} = 159.75°\text{C}$$

The heat exchange equation gives the heat transfer area as

$$A = \frac{q}{U\Delta t} = \frac{336 \times 1000}{400 \times 159.75} = 5.26 \text{ m}^2$$

From the above analysis, it can be seen that the *LMTD* is higher for the counterflow arrangement and hence a smaller heat transfer area is required for the same heat transfer rate. Hence, if the conditions allow, a counterflow design is always preferred.

Example 14.6 A heat exchanger is to cool liquid metal from 800°C to 500°C. The air used for the cooling enters the exchanger at 300°C. The flow rate of air is 10 kg/s and that of the liquid metal is 15 kg/s. Overall heat transfer coefficient is estimated to be 300 W/(m^2 K). Determine the surface area required for both counter- and parallel-flow arrangements. Average specific heat c_p of the air is 1008 J/(kg K) and is 950 J/(kg K) for the liquid metal.

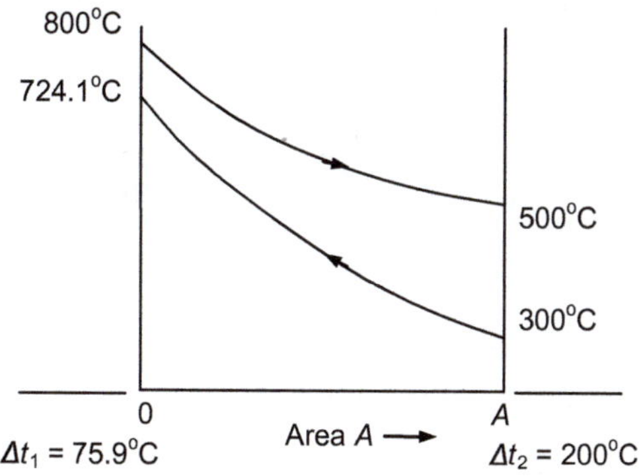

Fig. 14.11 Example 14.6

Solution

Heat balance gives

$$\begin{aligned} q &= \left(mc_p\right)_{\text{lm}}(T_1 - T_2) = \left(mc_p\right)_{\text{air}}(t_1 - t_2) \\ &= 15 \times 950 \times (800 - 500) = 10 \times 1008 \times (t_2 - 300) \end{aligned}$$

or

$$t_2 = 724.1^\circ\text{C}$$

Counterflow

Log mean temperature difference (refer to Fig. 14.11) is

$$\Delta t_m = \frac{\Delta t_2 - \Delta t_1}{\ln\left(\frac{\Delta t_2}{\Delta t_1}\right)} = \frac{200 - 75.9}{\ln\left(\frac{200}{75.9}\right)} = 128.08^\circ\text{C}$$

The heat exchange area required is

$$A = \frac{q}{U\Delta t} = \frac{\left(mc_p\right)_{\text{lm}}(T_1 - T_2)}{U\Delta t} = \frac{15 \times 950 \times (800 - 500)}{300 \times 128.08} = 111.26\ \text{m}^2$$

Parallel flow

Since t_2 cannot be greater than T_2, parallel flow arrangement is not possible, refer to Fig. 14.12.

Note: When *LMTD* approach is used, it is advisable to sketch the temperature-area diagram to avoid error in calculation of Δt_1 and Δt_2.

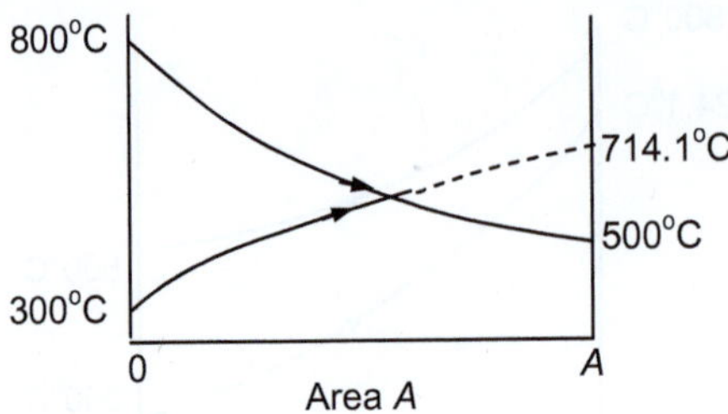

Fig. 14.12 Parallel flow arrangement with temperature cross

14.1.4 *LMTD* for Other Flow Arrangements

Besides the counterflow and parallel flow arrangements, there are many possible flow arrangements. Figure 14.13 shows the variation of temperatures of the hot and cold fluids at the outlet of cross-flow heat exchangers with both fluids unmixed (in part a) and both fluids mixed (in part b). In shell-and-tube heat exchangers with baffles (refer to Fig. 14.2), the shell fluid crosses the tubes many times (multi-pass). For these arrangements, the effective temperature difference between the fluids is a much more complicated function of the temperatures than the *LMTD* equations presented for the counter- and parallel-flow arrangements. For simplicity, the same format of the heat exchange equation is retained by introducing a correction factor F_T, i.e.

$$q = UA\Delta t_m F_T \tag{14.13}$$

The temperature difference Δt_m in Eq. (14.13) is the *LMTD* for the counterflow arrangement. The correction factor F_T, which is a function of two dimensionless parameters P and R, can be read from Fig. 14.14a–d, refer to Standards of Tubular Exchanger Manufacturers Association for details. The parameters P and R are defined as

$$P = \frac{t_2 - t_1}{T_1 - t_1} \tag{14.14a}$$

$$R = \frac{T_1 - T_2}{t_2 - t_1} \tag{14.14b}$$

A study of the correction factor charts shows that the factor F_T approaches 1 when either P or R approaches zero. Physically the limit $P \to 0$ corresponds to a heat exchanger in which the fluid represented by temperature t_1 and t_2 undergoes a phase change (boiling or condensation) so that $t_1 = t_2$. Similarly, $R \to 0$ corresponds to the phase change of the fluid stream represented by temperatures T_1 and T_2. When $F_T = 1$, the heat exchanger performance equals to that of a counterflow exchanger. It means that for the condensing vapour or boiling liquids, all flow arrangements will have the same heat transfer performance and equal to that of counterflow exchanger.

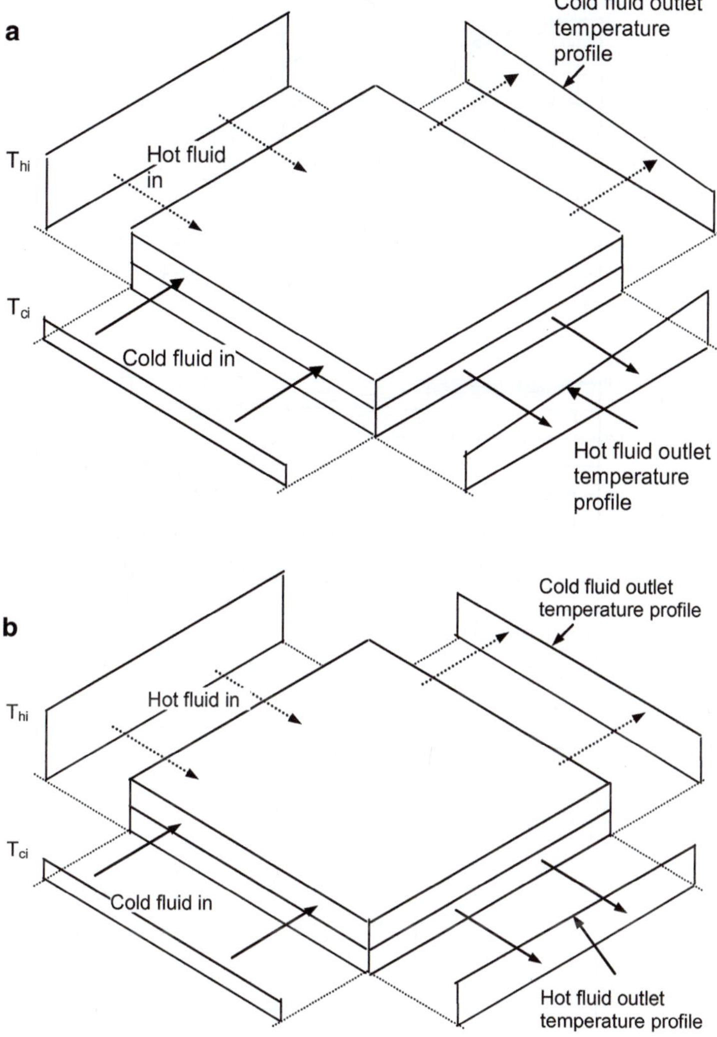

Fig. 14.13 **a** The variations of fluid temperatures across a cross-flow heat exchanger with both fluids unmixed, **b** The variation of fluid temperatures across a cross-flow heat exchanger with both fluids mixed

Using the heat balance equation, it can be shown that the parameter R is a ratio of the heat capacities C_t and C_T:

$$R = \frac{T_1 - T_2}{t_2 - t_1} = \frac{(mc_p)_t}{(mc_p)_T} = \frac{C_t}{C_T} \tag{14.15}$$

Example 14.7 Hot lubricating oil ($c_p = 2.09$ kJ/(kg K), flow rate of 0.1 kg/s) enters a cross-flow heat exchanger with both fluids unmixed at 100°C. It is to be cooled to 70°C. The cooling water enters the exchanger at 50°C. The water flow rate is 0.05 kg/s. What area of the exchanger will be required if the overall heat transfer coefficient is 250 W/(m^2 K)?

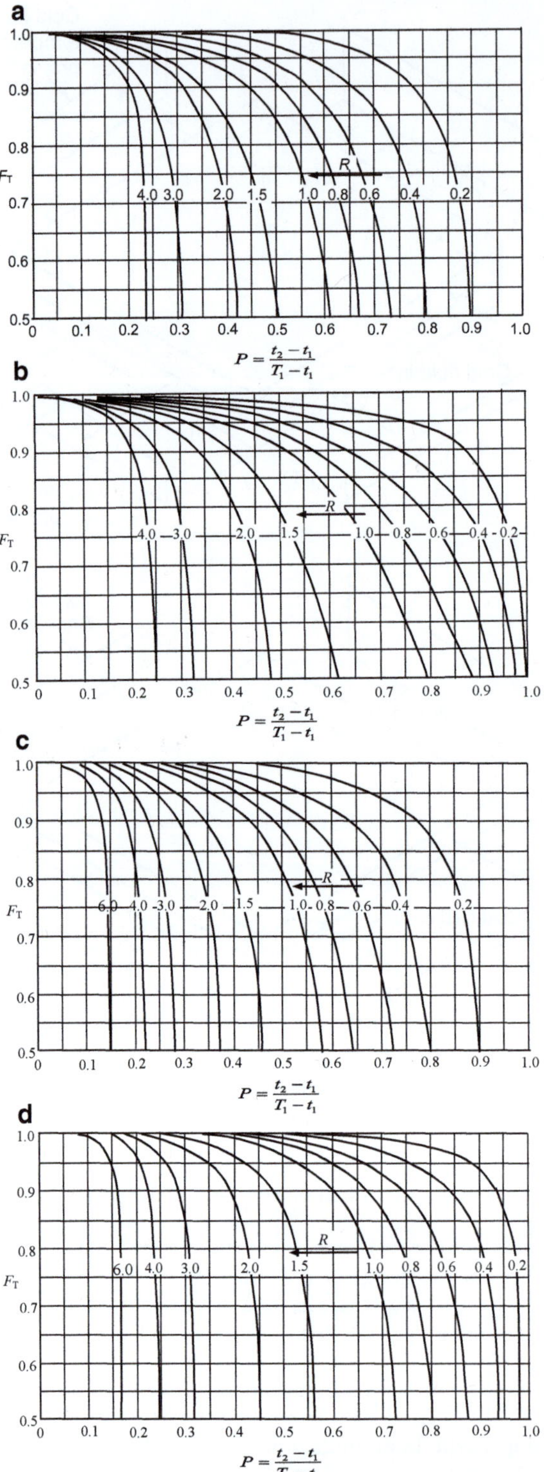

Fig. 14.14 **a** *LMTD* correction factor for cross-flow heat exchanger, one fluid mixed other unmixed, **b** *LMTD* correction factor for cross-flow heat exchanger, both fluids unmixed, **c** *LMTD* correction factor for heat exchanger, with one shell pass and even number of tube passes, **d** *LMTD* correction factor for heat exchanger, with two shell passes and 4 or multiple of 4 tube passes

Solution

(i) From the heat balance,

$$\dot{m}_w c_{pw}(t_{wo} - t_{wi}) = \dot{m}_e c_{pe}(t_{ei} - t_{eo})$$

where subscripts '*w*' and '*e*' refer to water and oil, respectively. The above equation gives

$$t_{wo} = t_{wi} + \frac{\dot{m}_e c_{pe}}{\dot{m}_w c_{pw}}(t_{ei} - t_{eo}) = 50 + \frac{0.1 \times 2.09}{0.05 \times 4.18}(100 - 70) = 80^\circ\text{C}$$

(ii) Using the nomenclature of Fig. 14.14,

$$T_1 = t_{ei} = 100^\circ\text{C}, T_2 = t_{eo} = 70^\circ\text{C}, t_1 = t_{wi} = 50^\circ\text{C}, t_2 = t_{wo} = 80^\circ\text{C}$$

Values of parameters *P* and *R* are

$$P = \frac{t_2 - t_1}{T_1 - t_1} = \frac{80 - 50}{100 - 50} = 0.6$$

$$R = \frac{T_1 - T_2}{t_2 - t_1} = \frac{100 - 70}{80 - 50} = 1.0$$

For these values of parameters *P* and *R*, the correction factor $F_T \approx 0.83$ from Fig. 14.14b.

Since $\Delta t_1 = \Delta t_2$, the log mean temperature difference for counterflow arrangement is

$$\Delta t_m = \Delta t_1 = 20^\circ\text{C}$$

Hence, the required heat transfer area is

$$A = \frac{q}{U \Delta t_m F_T} = \frac{6.27 \times 1000}{250 \times 20 \times 0.83} = 1.51 \text{ m}^2$$

where $q = \dot{m}_w c_{pw}(t_{wo} - t_{wi}) = 0.05 \times 4.18 \times (80 - 50) = 6.27\,\text{kJ/s}$.

For the same heat duty, the area required for a counterflow heat exchanger is

$$A = \frac{q}{U \Delta t_m} = \frac{6.27 \times 1000}{250 \times 20} = 1.254 \text{ m}^2$$

Example 14.8 In a heat exchanger, oil is to be cooled from 90°C to 40°C. Water [c_p = 4.1868 kJ/(kg K)] is available at 10°C, which can be heated to 50°C. If the flow rate of the water is 20 kg/min and an overall heat transfer coefficient is 200 W/(m^2 K), determine the heat exchange area. The following three alternatives are to be looked into:

(i) A double pipe counterflow heat exchanger,
(ii) a shell-and-tube heat exchanger with water making one shell pass and oil making two tube passes, and
(iii) a shell and tube heat exchanger with two shell passes and four tube passes.

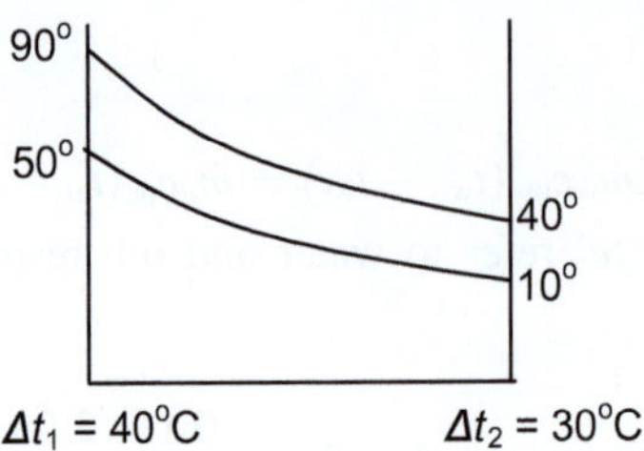

Fig. 14.15 Example 14.8

Solution

Heat transfer rate is

$$q = \dot{m}_w c_{pw}(t_{wo} - t_{wi}) = (20/60) \times 4.1868 \times (50 - 10) = 55.82\,\text{kW}$$

The log mean temperature difference (counterflow) is (Fig. 14.15)

$$\Delta t_m = \frac{\Delta t_1 - \Delta t_2}{\ln \frac{\Delta t_1}{\Delta t_2}} = \frac{40 - 30}{\ln \frac{40}{30}} = 34.76°\text{C}$$

(i) **Counterflow heat exchanger:**
The required heat transfer area is

$$A = \frac{q}{U \Delta t_m} = \frac{55.82 \times 1000}{200 \times 34.76} = 8.03\ \text{m}^2$$

(ii) **Shell-and-tube heat exchanger with one shell pass and two tube passes, Fig.** 14.14c,
Correction factor F_T:
Values of parameters P and R are

$$P = \frac{t_2 - t_1}{T_1 - t_1} = \frac{40 - 90}{10 - 90} = 0.625$$

$$R = \frac{T_1 - T_2}{t_2 - t_1} = \frac{10 - 50}{40 - 90} = 0.8$$

For these values of P and R, $F_T = 0.6$.
The required heat transfer area is

$$A = \frac{q}{U \Delta t_m F_T} = \frac{55.82 \times 1000}{200 \times 34.76 \times 0.6} = 13.38\ \text{m}^2$$

(iii) **Shell-and-tube heat exchanger with two shell pass and four tube passes**
The correction factor for this case from Fig. 14.14d is

$$F_T = 0.93$$

The required heat transfer area is

$$A = \frac{q}{U\Delta t_m F_T} = \frac{55.82 \times 1000}{200 \times 34.76 \times 0.93} = 8.63\ \text{m}^2$$

Comments: Arrangement for the shell-and-tube heat exchanger must be selected to give a high value of the correction factor F_T and the design point must lie on the flat part of the F_T curve so that the correction factor F_T is least affected by the variations in values of P and R due to any variations in the fluid temperatures. In general, the shell-and-tube heat exchanger is less effective than the concentric double pipe counterflow heat exchanger.

Example 14.9 A hot fluid [m_h = 1 kg/s, c_p = 2 kJ/(kg K)] enters a counterflow heat exchanger at 150°C. The cooling water [c_p = 4.18 kJ/(kg K)] enters the exchanger at 20°C. If the flow rate of the water is 2 kg/s, area of the heat transferring surface is 20 m^2, and overall heat transfer coefficient is 250 W/(m^2 K), determine the outlet temperature of the hot fluid.

Solution

The heat balance gives

$$\dot{m}_h c_{\text{ph}}(t_{\text{hi}} - t_{\text{ho}}) = \dot{m}_c c_{\text{pc}}(t_{\text{co}} - t_{\text{ci}})$$

or

$$1 \times 2000 \times (150 - t_{\text{ho}}) = 2 \times 4180 \times (t_{\text{co}} - 20)$$

or

$$t_{\text{co}} = 55.89 - t_{\text{ho}}/4.18 \tag{i}$$

The heat transfer equation gives

$$q = UA\Delta t_m = UA\frac{\Delta t_1 - \Delta t_2}{\ln\frac{\Delta t_1}{\Delta t_2}} = 250 \times 20 \times \frac{(150 - t_{\text{co}}) - (t_{\text{ho}} - 20)}{\ln\left(\frac{150 - t_{\text{co}}}{t_{\text{ho}} - 20}\right)} \tag{ii}$$

where $q = \dot{m}_h c_{\text{ph}}(t_{\text{hi}} - t_{\text{ho}}) = 1 \times 2000 \times (150 - t_{\text{ho}})$. Substituting value of q in Eq. (ii), we have

$$1 \times 2000 \times (150 - t_{\text{ho}}) = 250 \times 20 \times \frac{(150 - t_{\text{co}}) - (t_{\text{ho}} - 20)}{\ln\left(\frac{150 - t_{\text{co}}}{t_{\text{ho}} - 20}\right)} \tag{iii}$$

We have written the applicable expressions. The problem can be solved only by following an iterative method because of the involvement of the logarithmic term in Eq. (iii). Thus, when U, A, C_c and C_h, and inlet temperatures are specified, the calculation of q, t_{ho} and t_{co} from the *LMTD* approach requires a trial and error procedure. This procedure will be more tedious for the cross-flow and shell-and-tube exchangers. A direct approach to the solution of such problems is being discussed in the next section.

14.1.5 Effectiveness-NTU Method

14.1.5.1 Effectiveness-NTU Method for Counterflow Heat Exchanger

In Example 14.9, we have seen the difficulty in the use of the *LMTD* method when all the four temperatures (T_1, T_2, t_1 and t_2) are not specified or cannot be easily determined from the simple heat balance. In such cases, the analysis can be easily carried out using the method presented in this section. The method, known as effectiveness-NTU method, is based on the effectiveness of a heat exchanger in transferring the heat. The effectiveness ε of a heat exchanger is defined as the ratio of the actual heat transfer q to the maximum possible q_{max}. Hence,

$$\varepsilon = \frac{q}{q_{max}} \tag{14.16}$$

The actual exchange is the heat lost by the hot stream or gained by the cold stream, i.e.

$$q = (mc_p)_{hot}(T_1 - T_2) = (mc_p)_{cold}(t_2 - t_1) \tag{i}$$

The maximum possible heat exchange q_{max} is attained when the fluid with minimum value of heat capacity $(mc_p)_{min}$ undergoes the maximum temperature change $T_1 - t_1$ (Fig. 14.16). Thus

$$q_{max} = (mc_p)_{min}(T_1 - t_1) = C_{min}(T_1 - t_1) \tag{ii}$$

It is to note that the fluid with C-$_{max}$ cannot undergo the maximum temperature change. If the fluid with C_{max} undergoes the maximum temperature change, then the fluid with C_{min} will be required to undergo a temperature change greater than $T_1 - t_1$ to satisfy the heat balance equation and this is impossible.

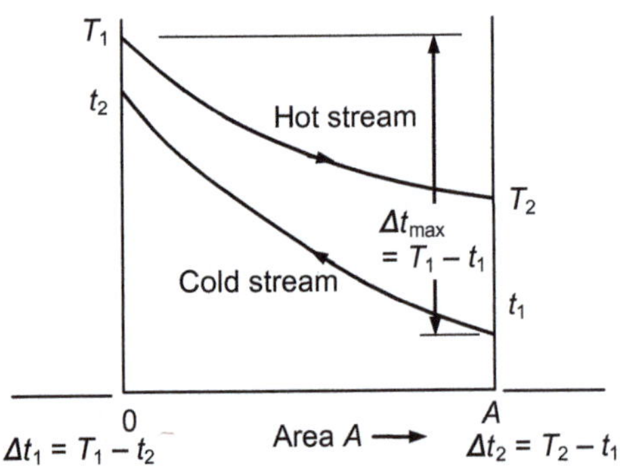

Fig. 14.16 Temperature variations in counterflow exchanger

Using Eqs. (i) and (ii) in Eq. (14.16), we have

$$\varepsilon = \frac{(mc_p)_{hot}(T_1 - T_2)}{C_{min}(T_1 - t_1)} = \frac{(mc_p)_{cold}(t_2 - t_1)}{C_{min}(T_1 - t_1)}$$

or

$$\varepsilon = \frac{C_{hot}(T_1 - T_2)}{C_{min}(T_1 - t_1)} = \frac{C_{cold}(t_2 - t_1)}{C_{min}(T_1 - t_1)} \tag{14.17}$$

where $mc_p = C$.

Let C_{hot} be C_{min} and C_{cold} be C_{max}, then

$$\varepsilon = \frac{T_1 - T_2}{T_1 - t_1} = \frac{C_{max}(t_2 - t_1)}{C_{min}(T_1 - t_1)} \tag{14.18}$$

Solving for T_2 and t_2, we get

$$T_2 = T_1 - \varepsilon(T_1 - t_1) \tag{14.19a}$$

$$t_2 = \frac{C_{min}}{C_{max}}(T_1 - t_1)\varepsilon + t_1 \tag{14.19b}$$

For the counterflow heat exchanger, from Eq. (14.6),

$$\frac{\Delta t_2}{\Delta t_1} = \frac{T_2 - t_1}{T_1 - t_2} = e^{-UA\left(\frac{1}{C_1} - \frac{1}{C_2}\right)} = e^{-\left[\frac{UA}{C_{min}}\left(1 - \frac{C_{min}}{C_{max}}\right)\right]} \tag{14.20}$$

where $C_1 = (mc_p)_{hot}$, $C_2 = (mc_p)_{cold}$ and U is the overall heat transfer coefficient.

Substitution of the values of T_2 and t_2 from Eqs. (14.19a, 14.19b) in Eq. (14.20) gives

$$\frac{T_1 - \varepsilon(T_1 - t_1) - t_1}{T_1 - \frac{C_{min}}{C_{max}}(T_1 - t_1)\varepsilon - t_1} = e^{-\frac{UA}{C_{min}}\left(1 - \frac{C_{min}}{C_{max}}\right)}$$

or

$$\frac{(T_1 - t_1)(1 - \varepsilon)}{(T_1 - t_1)\left(1 - \frac{C_{min}}{C_{max}}\varepsilon\right)} = e^{-\frac{UA}{C_{min}}\left(1 - \frac{C_{min}}{C_{max}}\right)}$$

Simplification gives

$$\varepsilon = \frac{1 - e^{-\left[\frac{UA}{C_{min}}\left(1 - \frac{C_{min}}{C_{max}}\right)\right]}}{1 - \frac{C_{min}}{C_{max}} e^{-\left[\frac{UA}{C_{min}}\left(1 - \frac{C_{min}}{C_{max}}\right)\right]}} \tag{14.21}$$

The group of terms UA/C_{min} is called *number of transfer units NTU* and is indicative of the size of the heat exchanger. It is a non-dimensional group.

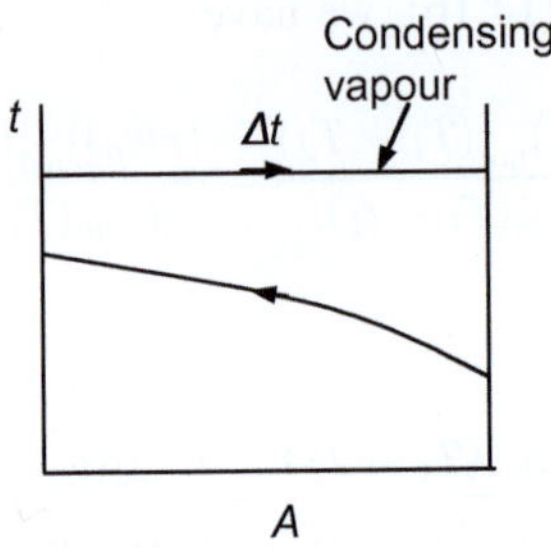

Fig. 14.17 Temperature variation in a condenser

Introducing $UA/C_{\min} = NTU$ and capacity ratio $C_{\min}/C_{\max} = C^*$ in Eq. (14.21), we obtain

$$\varepsilon = \frac{1 - e^{-NTU(1-C^*)}}{1 - C^* e^{-NTU(1-C^*)}} \tag{14.22}$$

The expression can be rewritten for NTU as

$$NTU = \frac{1}{1 - C^*} \ln\left(\frac{1 - \varepsilon C^*}{1 - \varepsilon}\right) \tag{14.23}$$

Case (A): In the case of boilers and condensers, one of the fluids undergoes only a phase change. The fluid undergoing the phase change remains at constant temperature as shown in Fig. 14.17 hence $\Delta t = 0$ for this fluid.

Equation $q = mc_p\, \Delta t$ gives

$$mc_p = \frac{q}{\Delta t} = \infty$$

or

$$C_{max} = \infty$$

Hence,

$$C^* = \frac{C_{min}}{C_{max}} = 0$$

Then Eq. (14.22) gives

$$\varepsilon = 1 - e^{-NTU} \tag{14.24}$$

Case B: If the hot and cold streams in a heat exchanger have approximately equal thermal capacity $C_h = C_c$, the heat exchanger is said to have *balanced flow*. A typical example is regenerator of a gas turbine regenerative cycle, wherein the heat of the turbine exhaust is utilized to preheat the combustion air. Though the mass flow rate and the specific heat values of the incoming air are slightly less than that of the exhaust, an assumption of balanced flow is reasonable. Hence, for balanced flow

$$\frac{C_{min}}{C_{max}} \approx 1$$

For this case, the value of effectiveness ε from Eq. (14.22) becomes non-determinant so its value is found by expanding the exponential terms as below

$$\varepsilon = \frac{1 - e^{-NTU(1-C^*)}}{1 - C^* e^{-NTU(1-C^*)}}$$

Let -*NTU* = *x*, then

$$\begin{aligned}
\varepsilon &= \frac{1 - e^{x(1-C^*)}}{1 - C^* e^{x(1-C^*)}} \\
&= \frac{e^{x(1-C^*)} - 1}{C^* e^{x(1-C^*)} - 1} \\
&= \frac{1 + x(1 - C^*) + x^2(1 - C^*)^2/2 + \ldots. - 1}{C^*\left[1 + x(1 - C^*) + x^2(1 - C^*)^2/2 + \ldots.\right] - 1} \\
&= \frac{x(1 - C^*) + x^2(1 - C^*)^2/2 + \ldots.}{-(1 - C^*) + C^* x(1 - C^*) + C^* x^2(1 - C^*)^2/2 + \ldots.} \\
&= \frac{(1 - C^*)[x + x^2(1 - C^*)/2 + \ldots.]}{(1 - C^*)[-1 + C^* x + C^* x^2(1 - C^*)/2 + \ldots\ldots..]} \\
&= \frac{x + x^2(1 - C^*)/2 + \ldots.}{-1 + C^* x + C^* x^2(1 - C^*)/2 + \ldots...}
\end{aligned}$$

In the limit $C^* = 1$, hence

$$\varepsilon = \frac{x}{-1 + x} = \frac{-NTU}{-1 - NTU}$$

or

$$\varepsilon = \frac{NTU}{1 + NTU} \tag{14.25}$$

Example 14.10 Solve Example 14.9 using the effectiveness-*NTU* method.

Solution

From the given data,

$$C_h = m_h c_h = 1 \times 2000 = 2000$$

$$C_c = m_c c_c = 2 \times 4180 = 8360$$

Thus, $C_{min} = C_h = 2000$ and $C_{max} = C_c = 8360$

$$C^* = \frac{C_{min}}{C_{max}} = \frac{2000}{8360} = 0.2392$$

$$NTU = \frac{UA}{C_{min}} = \frac{250 \times 20}{2000} = 2.5$$

The effectiveness

$$\varepsilon = \frac{1 - e^{-NTU(1-C^*)}}{1 - C^* e^{-NTU(1-C^*)}} = \frac{1 - e^{-2.5(1-0.2392)}}{1 - 0.2392 \times e^{-2.5(1-0.2392)}} = 0.8822$$

Considering the hot steam,

$$\varepsilon = \frac{T_1 - T_2}{T_1 - t_1}$$

or

$$0.8822 = \frac{150 - T_2}{150 - 20}$$

or

$$T_2 = 35.31°\text{C}$$

Now the readers can understand the usefulness of the ε-*NTU* method over the *LMTD* method when all temperatures (T_1, T_2, t_1 and t_2) are not known or cannot be determined using the heat balance equation.

Example 14.11 In Sect. 14.1.5.1, it was stated that the fluid with $C = C_{\max}$ cannot undergo the maximum temperature change $T_1 - t_1$. Verify the statement using the data of the above example.

Solution

From the data of the previous example,

$$C_{min} = C_h = 2000$$

Hence,

$$q_{max} = C_{min}(T_1 - t_1) = 2000 \times (150 - 20) = 260 \times 10^3 \text{ W}$$

Now let us see if the fluid with $C_{max} = C_c = 8360$ undergoes the maximum temperature change, then

$$q'_{max} = C_{max}(T_1 - t_1) = 8360 \times (150 - 20) = 1086.8 \times 10^3 \text{ W}$$

From the first law consideration,

$$q'_{max} = C_{min}(T_1 - T_2)$$

or

$$1086.8 \times 10^3 = 2000 \times (T_1 - T_2)$$

or

$$(T_1 - T_2) = \frac{1086.8 \times 10^3}{2000} = 543.4°\text{C}$$

which is greater than $(T_1 - t_1)$ and hence is impossible.

14.1.5.2 Effectiveness-NTU Method for Parallel Flow Heat Exchanger

Following the method presented in Sect. 14.1.5.1, we can develop the equation of effectiveness for the parallel flow heat exchanger.

The definition of effectiveness is the same, i.e.

$$\begin{aligned}\varepsilon &= \frac{(mc_p)_{\text{hot}}(T_1 - T_2)}{C_{min}(T_1 - t_1)} = \frac{(mc_p)_{\text{cold}}(t_2 - t_1)}{C_{min}(T_1 - t_1)} \\ &= \frac{C_{\text{hot}}(T_1 - T_2)}{C_{min}(T_1 - t_1)} = \frac{C_{\text{cold}}(t_2 - t_1)}{C_{min}(T_1 - t_1)}\end{aligned}$$

where $mc_p = C$.

Let C_{hot} be C_{min} and C_{cold} be C_{max}, then

$$\varepsilon = \frac{T_1 - T_2}{T_1 - t_1} = \frac{C_{max}(t_2 - t_1)}{C_{min}(T_1 - t_1)}$$

Solving for T_2 and t_2, we get

$$T_2 = T_1 - \varepsilon(T_1 - t_1) \tag{i}$$

$$t_2 = \frac{C_{min}}{C_{max}}(T_1 - t_1)\varepsilon + t_1 \tag{ii}$$

For the parallel flow heat exchanger from Eqs. (14.4) and (14.6), we have (refer to Fig. 14.18)

$$\frac{\Delta t_2}{\Delta t_1} = \frac{T_2 - t_2}{T_1 - t_1} = e^{-UA\left(\frac{1}{C_1} + \frac{1}{C_2}\right)} = e^{-\left[\frac{UA}{C_{min}}\left(1 + \frac{C_{min}}{C_{max}}\right)\right]} \tag{iii}$$

where $C_1 = (mc_p)_{\text{hot}}$, $C_2 = (mc_p)_{\text{cold}}$ and U is the overall heat transfer coefficient.

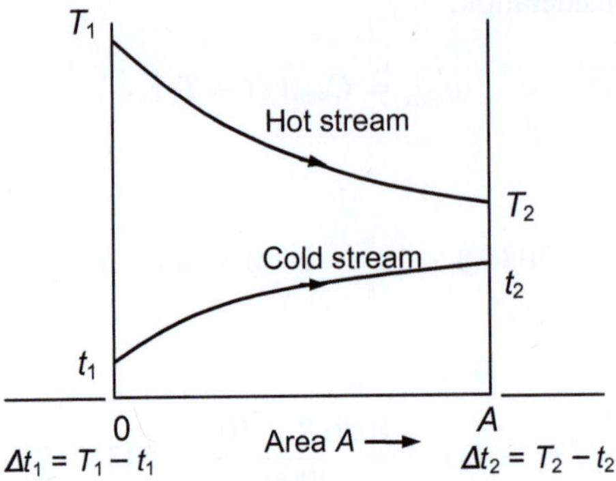

Fig. 14.18 Parallel flow

Substitution of the values of T_2 and t_2 from Eqs. (i) and (ii) in Eq. (iii), and $UA/C_{min} = NTU$ and $C_{min}/C_{max} = C^*$ gives

$$\frac{[T_1 - \varepsilon(T_1 - t_1)] - [C^*(T_1 - t_1)\varepsilon + t_1]}{T_1 - t_1} = e^{-NTU(1+C^*)}$$

or

$$\frac{(T_1 - t_1) - \varepsilon(T_1 - t_1)(1 + C^*)}{T_1 - t_1} = e^{-NTU(1+C^*)}$$

or

$$1 - \varepsilon(1 + C^*) = e^{-NTU(1+C^*)}$$

Simplification gives

$$\varepsilon = \frac{1 - e^{-[NTU(1+C^*)]}}{1 + C^*} \quad (14.26)$$

The expression can be rewritten for NTU as

$$NTU = -\frac{1}{1 + C^*}\ln[1 - \varepsilon(1 + C^*)] \quad (14.27)$$

Equations (14.26) and (14.27) have been derived by assuming that the hot stream has a smaller capacity rate, i.e. $C_h = C_{min}$ and $C_c = C_{max}$. The equations are valid irrespective of whether the hot stream or the cold stream has a smaller capacity rate.

Case (A): As explained earlier, in the case of boilers and condensers,

$$\frac{C_{min}}{C_{max}} \to 0$$

and Eq. (14.26) gives, as in the case of the counterflow,

$$\varepsilon = 1 - e^{-NTU} \tag{14.28}$$

Case B: For the '*balanced flow*', as in the case of the gas turbine generators,

$$\frac{C_{min}}{C_{max}} \approx 1$$

For this case, effectiveness ε from Eq. (14.26) is

$$\varepsilon = \frac{1}{2}\left(1 - e^{-2NTU}\right) \tag{14.29}$$

It can be seen from the above equation that the maximum value of the effectiveness is 50% in the case of parallel flow arrangement when $C_{min}/C_{max} \approx 1$. There is no such limitation in the case of counterflow arrangement. This is the main reason for the use of counterflow arrangement whenever C_{min}/C_{max} is not having a very low value. Figure 14.19a, b shows the variation of the effectiveness with *NTU* for the counter- and parallel-flow arrangements, respectively.

We have seen that the effectiveness equations for the boiler or condenser are the same both for the counterflow and parallel flow arrangements. This is expected because the temperature of the boiling fluid or condensing vapour does not vary along the heat exchanger surface hence the direction of the flow has no significance.

14.1.6 Effectiveness-NTU Relations for Other Flow Arrangements

Kays and London (1964) have given results for effectiveness-NTU analysis for various arrangements. Figure 14.20a–c shows the ε-*NTU* curves for cross-flow and shell-and-tube heat exchangers. The analytical expressions for these arrangements are somewhat tedious to use (see Table 14.2) and hence are used only when accurate analysis is required. The following observations can be made from the study of ε-*NTU* curves and from the equations in Table 14.2.

1. For a given C^*, the effectiveness increases with increase in the value of *NTU*.
2. For the given value of *NTU*, the effectiveness increases with decrease in the value of C^*.
3. For $C^* = 0$ (boilers and condensers), the value of the effectiveness is the same for all arrangements.
4. At very low values of *NTU*, the effectiveness of different arrangements does not differ significantly for all values of C^*.
5. Counterflow heat exchanger gives the highest value of the effectiveness (i.e. the heat transfer performance) for any specified value of *NTU*.
6. At high values of the *NTU*, a large increase in the heat transfer area is required for a small gain in the effectiveness.

It is to note that the effectiveness-*NTU* method offers us a tool to select the best possible exchanger for the specified conditions. Higher the effectiveness of an exchanger, better is its performance.

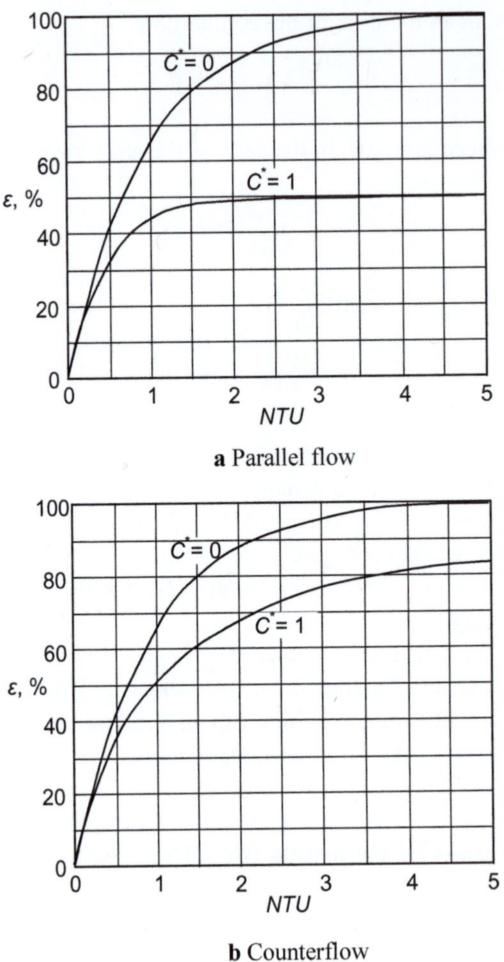

Fig. 14.19 Effectiveness-*NTU* curves

Example 14.12 Air-to-flue gas heat exchanger of a gas turbine has counterflow arrangement. The hot flue gas inlet temperature is 600 K and cold air inlet temperature is 400 K. Mass flow rates and specific heats of the two streams can be assumed to be nearly equal. If the overall heat transfer coefficient is 50 W/(m^2 K), determine the outlet temperatures of the two streams and the heat transferred per m^2 of the heat exchanger surface for 1 kg/s of the air flow rate. Given c_p = 1005 J/(kg K).

Solution

The capacity rate ratio, $C^* = C_{min}/C_{max} = 1$ hence

$$\varepsilon = \frac{NTU}{1 + NTU}$$

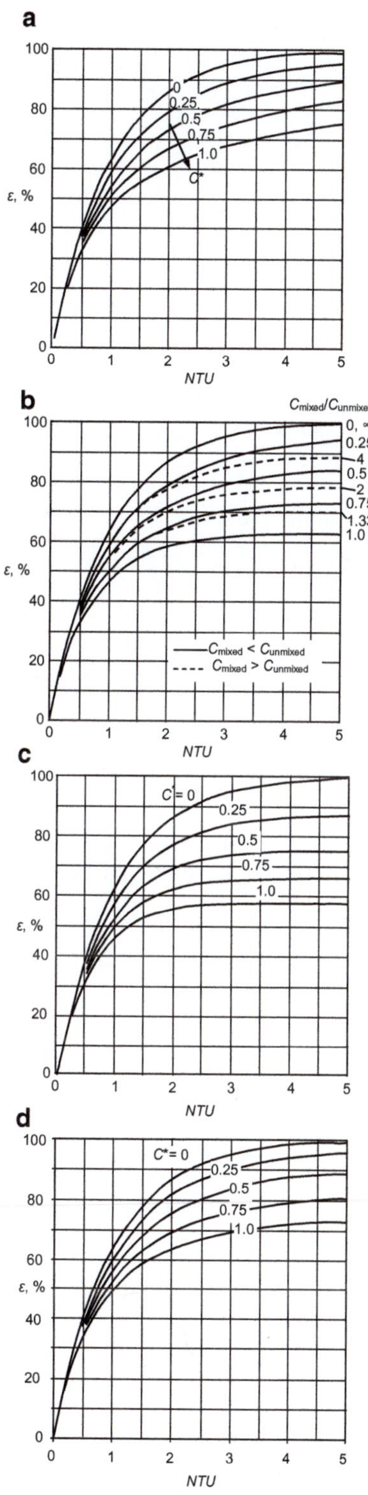

Fig. 14.20 **a** ε-*NTU* curve for cross-flow exchanger with both fluids unmixed, **b** ε-*NTU* curve for cross-flow exchanger with one fluid unmixed other mixed, **c** ε-*NTU* curve for one shell pass and 2, 4, 6, etc., tube passes (1–2 parallel counterflow) heat exchanger, **d** ε-*NTU* curve for two shell passes and 4, 8, etc., tube passes (2–4 parallel counterflow) heat exchanger

Table 14.2 Effectiveness and *NTU* expressions

Flow Arrangement	$\varepsilon(NTU, C^*)$	$NTU\ (\varepsilon, C^*)$
Counterflow	$\varepsilon = \dfrac{1 - e^{-NTU(1-C^*)}}{1 - C^* e^{-NTU(1-C^*)}}$ for $C^* < 1$ $\varepsilon = \dfrac{NTU}{1+NTU}$ for $C^* = 1$	$NTU = \frac{1}{C^*-1}\ln\left(\frac{\varepsilon-1}{\varepsilon C^*-1}\right)$ for $C^* < 1$ $NTU = \frac{\varepsilon}{1-\varepsilon}$ for $C^* = 1$
Parallel flow	$\varepsilon = \dfrac{1 - e^{-[NTU(1+C^*)]}}{1+C^*}$	$NTU = -\frac{1}{1+C^*}\ln[1 - \varepsilon(1+C^*)]$
Cross-flow: both fluids unmixed	$\varepsilon = 1 - \exp\left[\dfrac{\exp(-NTU^{0.78}C^*) - 1}{NTU^{-0.22}C^*}\right]$	–
Cross-flow: both fluids mixed	$\varepsilon = \left[\dfrac{1}{1 - e^{(-NTU)}} + \dfrac{C^*}{1 - e^{(-NTUC^*)}} - \dfrac{1}{NTU}\right]^{-1}$	–
Cross-flow, C_{max} mixed, C_{min} unmixed	$\varepsilon = \dfrac{1 - \exp\{-C^*[1 - \exp(-NTU)]\}}{C^*}$	$NTU = -\ln\left[1 + \frac{1}{C^*}\ln(1 - \varepsilon C^*)\right]$
Cross-flow, C_{min} mixed, C_{max} unmixed	$\varepsilon = 1 - \exp\left[-\dfrac{1 - \exp(-C^* NTU)}{C^*}\right]$	$NTU = -\frac{1}{C^*}\ln[1 + C^*\ln(1 - \varepsilon)]$
One shell pass, 2,4,6 tube passes	$\varepsilon = 2\left[1 + C^* + \sqrt{1+C^{*2}}\dfrac{1+\exp\left(-NTU\sqrt{1+C^{*2}}\right)}{1-\exp\left(-NTU\sqrt{1+C^{*2}}\right)}\right]^{-1}$	$NTU = \dfrac{1}{\sqrt{1+C^{*2}}} \times \ln\left[\dfrac{2 - \varepsilon\left(1+C^* - \sqrt{1+C^{*2}}\right)}{2 - \varepsilon\left(1+C^* + \sqrt{1+C^{*2}}\right)}\right]$
All exchangers with $C^* = 0$	$\varepsilon = 1 - \exp(-NTU)$	$NTU = -\ln(1 - \varepsilon)$

* $UA/C_{min} = NTU$ and capacity ratio $C_{min}/C_{max} = C^*$

where $NTU = UA/C_{min} = 50 \times 1/1005 = 0.04975$. This gives

$$\varepsilon = \frac{0.04975}{1+0.04975} = 0.0474$$

Heat balance equation is

$$q = C_h(T_1 - T_2) = C_c(t_2 - t_1) \tag{i}$$

It is given that the mass flow rates and specific heats of the air and flue gas are equal. Hence,

$$C_h = C_c \tag{ii}$$

Using this condition, Eq. (i) yields

$$(T_1 - T_2) = (t_2 - t_1)$$

or

$$T_2 + t_2 = 600 + 400 = 1000\,\text{K} \tag{iii}$$

From the equation of effectiveness,

$$\varepsilon = \frac{C_h(T_1 - T_2)}{C_{min}(T_1 - t_1)} = \frac{C_c(t_2 - t_1)}{C_{min}(T_1 - t_1)}$$

Using Eq. (ii),

$$\varepsilon = \frac{(T_1 - T_2)}{(T_1 - t_1)} = \frac{(t_2 - t_1)}{(T_1 - t_1)}$$

Substituting values of ε, T_1 and t_1, we get

$$0.0474 = \frac{600 - T_2}{600 - 400}$$

or

$$T_2 = 590.52\,\mathrm{K}$$

and

$$0.0474 = \frac{t_2 - 400}{600 - 400}$$

or

$$t_2 = 409.48\,\mathrm{K}$$

The values of T_2 and t_2 satisfy Eq. (iii).
The heat transfer rate is

$$q = m_h C_h(T_1 - T_2) = 1 \times 1005 \times 9.48 = 9527.4\,\mathrm{W}$$

Example 14.13 An oil cooler consists of a straight tube 20 mm ID and 2.5 mm thick, which is enclosed within a pipe and concentric with it. The external pipe is well insulated. Oil flows through the tube at the rate of 180 kg/hr [c_p = 2 kJ/(kg K)] and cooling water flows in the annulus in the opposite direction at a rate of 340 kg/hr [c_p = 4.2 kJ/(kg K)]. The oil enters the cooler at 170°C and leaves at 70°C while the cooling water enters at 20°C. Calculate the length of the tube required if the convection heat transfer coefficient from the oil to the tube surface is 2 kW/(m^2 K) and from the outer surface of the tube to the water is 4 kW/(m^2 K). Neglect resistance of the tube wall and the effect of fouling. Also calculate the effectiveness of the heat transfer.

Solution

(i) **Heat Balance**

Oil:

$$q = mc_p(t_{\rm hi} - t_{\rm ho}) = \frac{180}{3600} \times 2 \times (170 - 70) = 10\,\text{kW}$$

Water:

$$q = mc_p(t_{\rm co} - t_{\rm ci}) = \frac{340}{3600} \times 4.2 \times (t_{\rm co} - 20)$$

or

$$10 = \frac{340}{3600} \times 4.2 \times (t_{\rm co} - 20)$$

or

$$t_{\rm co} = 45.2^\circ\text{C}$$

(ii) ***LMTD***

$$\Delta t_m = \frac{(t_{\rm hi} - t_{\rm co}) - (t_{\rm ho} - t_{\rm ci})}{\ln\frac{(t_{\rm hi} - t_{\rm co})}{(t_{\rm ho} - t_{\rm ci})}} = \frac{(170 - 45.2) - (70 - 20)}{\ln\frac{(170-45.2)}{(70-20)}} = 81.78^\circ\text{C}$$

(iii) **Overall heat transfer coefficient referred to the outer surface of the tube**

$$U = \frac{1}{\frac{R_o}{R_i h_i} + \frac{1}{h_o}} = \frac{1}{\frac{1}{0.8 \times 2} + \frac{1}{4}} = 1.143\,\text{kW/m}^2\,\text{K}$$

neglecting pipe wall resistance.

(iv) **Heat transfer area required**

$$A = \frac{q}{U(\Delta t_m)} = \frac{10}{1.143 \times 81.78} = 0.107\ \text{m}^2$$

Required tube length,

$$L = \frac{A}{\pi d_o} = \frac{0.107 \times 1000}{\pi \times 25} = 1.36\,\text{m}$$

(v) **Effectiveness**

$$\varepsilon = \frac{t_{\rm hi} - t_{\rm ho}}{t_{\rm hi} - t_{\rm ci}} = \frac{170 - 70}{170 - 20} = 0.666$$

Example 14.14 Solve Example 14.13 Using the effectiveness-*NTU* approach.

Solution

Continuing from steps (i), (iii) and (iv) of Example 14.13,

$$C_{min} = \frac{180}{3600} \times 2 = 0.1$$

$$C_{max} = \frac{340}{3600} \times 4.2 = 0.3966$$

$$C^* = \frac{C_{min}}{C_{max}} = \frac{0.1}{0.3966} = 0.2521$$

Effectiveness

$$\varepsilon = \frac{1 - e^{-NTU(1-C^*)}}{1 - C^* e^{-NTU(1-C^*)}}$$

Let $e^{-NTU(1-C^*)} = z$, then

$$\varepsilon = \frac{1 - z}{1 - C^* z}$$

Substitution gives

$$0.666 = \frac{1 - z}{1 - 0.2521z}$$

or

$$z = 0.4014$$

This gives

$$e^{-NTU(1-C^*)} = 0.4014$$

or

$$e^{-NTU(1-0.2521)} = 0.4014$$

or

$$NTU = 1.2205$$

or

$$\frac{UA}{C_{min}} = 1.2205$$

or

$$A = \frac{1.2205 \times C_{min}}{U} = \frac{1.2205 \times 0.1}{1.143} = 0.107 \text{ m}^2$$

Alternatively, equation of NTU from Table 14.2 can be used, which is

$$NTU = -\frac{1}{1 - C^*} \ln\left[\frac{1 - \varepsilon C^*}{1 - \varepsilon}\right]$$
$$= -\frac{1}{1 - 0.2521} \ln\left[\frac{1 - 0.666 \times 0.2521}{1 - 0.666}\right] = 1.2205$$

This gives

$$A = \frac{NTU \times C_{min}}{U} = \frac{1.2205 \times 0.1}{1.143} = 0.107 \text{ m}^2$$

Example 14.15 A heat exchanger (surface area = 100 m^2) has overall heat transfer coefficient of 420 W/(m^2 K). Find the outlet temperature of hot and cold fluids for both counter and parallel flow arrangements when the inlet temperatures of the hot and cold fluids are 700°C and 100°C, respectively. The mass flow rates and specific heat of the hot and cold fluids are 1000 kg/min and 3.6 kJ/(kg K), and 1200 kg/min and 4.2 kJ/(kg K), respectively.

Solution

We have

$$C_h = (mc_p)_h = (3.6 \times 1000) \times (1000/60) = 60000$$
$$C_c = (mc_p)_c = (4.2 \times 1000) \times (1200/60) = 84000$$

Hence,
$C_h = C_{min} = 60000$, $C_c = C_{max} = 84000$ and $C_{min}/C_{max} = 0.7143$

$$NTU = UA/C_{min} = 420 \times 100/60000 = 0.7$$

(i) **Counterflow arrangement:**
The effective-*NTU* relation gives

$$\varepsilon = \frac{1 - e^{-NTU(1-C^*)}}{1 - C^* e^{-NTU(1-C^*)}} = \frac{1 - e^{-0.7(1-0.7143)}}{1 - 0.7143 \times e^{-0.7(1-0.7143)}} = 0.4366$$

We have

$$\varepsilon = \frac{t_{hi} - t_{ho}}{t_{hi} - t_{ci}}$$

which gives

$$t_{ho} = t_{hi} - \varepsilon(t_{hi} - t_{ci}) = 700 - 0.4366 \times (700 - 100) = 438.04^\circ\text{C}$$

Also

$$\varepsilon = \frac{C_c(t_{co} - t_{ci})}{C_{min}(t_{hi} - t_{ci})} = \frac{C_{max}(t_{co} - t_{ci})}{C_{min}(t_{hi} - t_{ci})} = \frac{(t_{co} - t_{ci})}{C^*(t_{hi} - t_{ci})}$$

which gives

$$t_{co} = t_{ci} + C^*\varepsilon(t_{hi} - t_{ci}) = 100 + 0.7143 \times 0.4366 \times (700 - 100) = 287.11^\circ\text{C}$$

Heat transfer rate,

$$q = (mc_p)_h(t_{hi} - t_{ho}) = 60000 \times (700 - 438.04) = 1.572 \times 10^7 \text{ W}$$

Check

$$\Delta t_m = \frac{(t_{hi} - t_{co}) - (t_{ho} - t_{ci})}{\ln\left(\frac{t_{hi} - t_{co}}{t_{ho} - t_{ci}}\right)} = \frac{(700 - 287.11) - (438.04 - 100)}{\ln\left(\frac{700 - 287.11}{438.04 - 100}\right)} = 374.22^\circ\text{C}$$

$$q = UA\Delta t_m = 420 \times 100 \times 374.22 = 1.572 \times 10^7 \text{ W}$$

(ii) **Parallel flow arrangement:**
The effective-*NTU* relation gives

$$\varepsilon = \frac{1 - e^{-NTU(1 + C^*)}}{1 + C^*} = \frac{1 - e^{-0.7(1 + 0.7143)}}{1 + 0.7143} = 0.40764$$

We have

$$\varepsilon = \frac{t_{hi} - t_{ho}}{t_{hi} - t_{ci}}$$

which gives

$$t_{ho} = t_{hi} - \varepsilon(t_{hi} - t_{ci}) = 700 - 0.40764 \times (700 - 100) = 455.42^\circ\text{C}$$

Also

$$\varepsilon = \frac{C_c(t_{co} - t_{ci})}{C_{min}(t_{hi} - t_{ci})} = \frac{C_{max}(t_{co} - t_{ci})}{C_{min}(t_{hi} - t_{ci})} = \frac{(t_{co} - t_{ci})}{C^*(t_{hi} - t_{ci})}$$

which gives

$$t_{co} = t_{ci} + C^*\varepsilon(t_{hi} - t_{ci}) = 100 + 0.7143 \times 0.40764 \times (700 - 100) = 274.71^\circ\text{C}$$

Heat transfer rate,

$$q = (mc_p)_h(t_{hi} - t_{ho}) = 60000 \times (700 - 455.42) = 1.4675 \times 10^7 \text{ W}$$

Check

$$\Delta t_m = \frac{(t_{hi} - t_{ci}) - (t_{ho} - t_{co})}{\ln\left(\frac{t_{hi} - t_{ci}}{t_{ho} - t_{co}}\right)} = \frac{(700 - 100) - (455.42 - 274.7)}{\ln\left(\frac{700-100}{455.42-274.7}\right)} = 349.41°\text{C}$$

And

$$q = UA\Delta t_m = 420 \times 100 \times 349.41 = 1.4675 \times 10^7 \text{ W}$$

Example 14.16 A double pipe heat exchanger has an effectiveness of 0.5 when the flow is countercurrent and the thermal capacity of one of the fluids is twice that of the other. Calculate the effectiveness of the heat exchanger if the direction of the flow of one of the fluids is reversed with the same mass flow rates as before.

Solution

The effectiveness-*NTU* relation for a counterflow heat exchanger is

$$\varepsilon = \frac{1 - e^{-NTU(1-C^*)}}{1 - C^* e^{-NTU(1-C^*)}}$$

where $C^* = C_{min}/C_{max}$ = ½ = 0.5 (given). Hence

$$0.5 = \frac{1 - e^{-NTU(1-0.5)}}{1 - 0.5e^{-NTU(1-0.5)}}$$

On simplification, we get

$$NTU = 0.8109$$

The effectiveness relation for parallel flow heat exchanger is

$$\varepsilon = \frac{1 - e^{-NTU(1+C^*)}}{1 + C^*}$$

Substituting the values of C^* and *NTU* from above, we obtain

$$\varepsilon = \frac{1 - e^{-0,8109(1+0.5)}}{1 + 0.5} = 0.469$$

Example 14.17 Water [c = 4.18 kJ/(kg K)] at the rate of 60 kg/min is to be heated from 35°C by oil having a specific heat of 2.0 kJ/(kg K). The oil enters the double pipe heat exchanger at 110°C. The heat transfer surface area is 15.0 m^2. The overall heat transfer coefficient is 300 W/(m^2 K). If the oil flow rate is 150 kg/min, calculate the exit temperature of the fluids.

If a parallel flow arrangement is used, calculate Q, t_{ho} and t_{co}.

Solution

The fluid capacity rates are

$$C_h = (mc_p)_h = (150/60) \times 2000 = 5000 = C_{max}$$

$$C_c = (mc_p)_c = (60/60) \times 4180 = 4180 = C_{min}$$

Hence,

$$C_{min}/C_{max} = 4180/5000 = 0.836$$

and

$$NTU = UA/C_{min} = 300 \times 15/4180 = 1.077$$

Counterflow:

The effective-*NTU* relation for counterflow arrangement gives

$$\varepsilon = \frac{1 - e^{-NTU(1-C^*)}}{1 - C^* e^{-NTU(1-C^*)}} = \frac{1 - e^{-1.077(1-0.836)}}{1 - 0.836 \times e^{-1.077(1-0.836)}} = 0.541$$

From the definition of effectiveness:

$$\varepsilon = \frac{C_c(t_{\mathrm{co}} - t_{\mathrm{ci}})}{C_{min}(t_{\mathrm{hi}} - t_{\mathrm{ci}})} = \frac{(t_{\mathrm{co}} - t_{\mathrm{ci}})}{(t_{\mathrm{hi}} - t_{\mathrm{ci}})}, \text{ as } C_c = C_{min}$$

This gives

$$t_{\mathrm{co}} = t_{\mathrm{ci}} + \varepsilon(t_{\mathrm{hi}} - t_{\mathrm{ci}}) = 35 + 0.541 \times (110 - 35) = 75.55^\circ\mathrm{C}$$

Heat duty is

$$q = (mc_p)_c(t_{\mathrm{co}} - t_{\mathrm{ci}}) = 4180 \times (75.55 - 35) = 169.5\,\mathrm{kW}$$

The exit temperature of the hot fluid can be found from the heat balance or effectiveness equation. Using the former, we have

$$q = (mc_p)_h(t_{hi} - t_{ho})$$

or

$$169.5 \times 1000 = 5000 \times (110 - t_{\mathrm{ho}})$$

or

$$t_{ho} = 76.1°C.$$

Parallel flow:

The effectiveness relation for the parallel flow arrangement is given by

$$\varepsilon = \frac{1 - e^{-NTU(1+C^*)}}{1+C^*}$$

Substituting the values of C^* = 0.836 and NTU = 1.077, we obtain

$$\varepsilon = \frac{1 - e^{-1.077(1+0.836)}}{1+0.836} = 0.4692$$

From the definition of effectiveness:

$$\varepsilon = \frac{C_c(t_{co} - t_{ci})}{C_{min}(t_{hi} - t_{ci})} = \frac{(t_{co} - t_{ci})}{(t_{hi} - t_{ci})}$$

which gives

$$t_{co} = t_{ci} + \varepsilon(t_{hi} - t_{ci}) = 35 + 0.4692 \times (110 - 35) = 70.2°C$$

From the heat balance,

$$(mc_p)_c(t_{co} - t_{ci}) = (mc_p)_h(t_{hi} - t_{ho})$$

or

$$4180 \times (70.2 - 35) = 5000 \times (110 - t_{ho})$$

or

$$t_{ho} = 80.6°C.$$

The results can be checked from the *LMTD* approach.

Comments: Comparison of the results for the counter- and parallel-flow arrangements shows the superiority of the former arrangement. The advantage of the counterflow arrangement increases with the increase in the value of C_{min}/C_{max}.

Example 14.18 Hot water at the rate of 4 kg/min and 80°C enters an automobile radiator. Air at 25°C and 60 kg/min enters the radiator and flows across the tubes of the radiator. If the overall heat transfer coefficient is 200 W/(m² K), determine the heat transfer area required to achieve water outlet temperature of 50°C.

Solution

At mean water temperature of (80 + 50)/2 = 65°C, specific heat of water is 4188 J/(kg K) from Table A4.

Heat balance equation gives

$$m_h c_h (t_{\mathrm{hi}} - t_{\mathrm{ho}}) = m_c c_c (t_{\mathrm{co}} - t_{\mathrm{ci}})$$

from which

$$t_{\mathrm{co}} = \frac{m_h c_h (t_{\mathrm{hi}} - t_{\mathrm{ho}})}{m_c c_c} + t_{\mathrm{ci}} = \frac{4 \times 4188 \times (80 - 50)}{60 \times 1006} + 25 = 33.33^{\circ}\mathrm{C}$$

for trial value of c_c = 1006 J/(kg K). Mean temperature of air is (25 + 33.33)/2 = 29.2°C and $c \approx$ 1006 from Table A5. This gives

$$C_c = m_c c_c = (60/60) \times 1006 = 1006$$

$$C_h = m_h c_h = (4/60) \times 4188 = 279.2 = C_{min}$$

$$C^* = \frac{C_{min}}{C_{max}} = \frac{279.2}{1006} = 0.278$$

$$\varepsilon = \frac{q}{q_{max}} = \frac{m_h c_h (t_{\mathrm{hi}} - t_{\mathrm{ho}})}{m_h c_h (t_{\mathrm{hi}} - t_{\mathrm{ci}})} = \frac{t_{\mathrm{hi}} - t_{\mathrm{ho}}}{t_{\mathrm{hi}} - t_{\mathrm{ci}}} = \frac{80 - 50}{80 - 25} = 0.545$$

Automobile radiator is a cross-flow heat exchanger with both fluids unmixed. From Fig. 14.20a, $NTU \approx 0.85$ for C^* = 0.278 and ε = 0.545. Hence, the required heat transfer surface area

$$A = \frac{NTU C_{min}}{U} = \frac{0.85 \times 279.2}{200} = 1.187\,\mathrm{m}^2$$

Example 14.19 A fluid *1* [c_p = 2.0 kJ/(kg K)], entering a cross-flow heat exchanger at 200°C, is used to heat fluid *2* [c_p = 4.0 kJ/(kg K)] from 15°C to 55°C. The flow rate of fluid 2 is 20 kg/min. For an exchanger surface area of 7 m^2, what flow rate of fluid *1* is required and what is its outlet temperature? The overall heat transfer coefficient is expected to be 200 W/(m^2 K). The fluid *1* is unmixed and fluid *2* is mixed.

Solution

From the given data, it is not possible to determine the fluid having smaller mc_p. Assume that the fluid *2* is having minimum mc_p. Then

$$C_{min} = (mc_p)_2 = (20/60) \times 4000 = 1333.3$$

$$\varepsilon = \frac{(t_2 - t_1)_2}{(T_1 - t_1)} = \frac{55 - 15}{200 - 15} = 0.216$$

Table 14.3 Example 14.19

Trial no.	$C_1 = C_{unmixed}$	$C_{mixed}/C_{unmixed}$	*NTU*	Δt_1, °C	Effectiveness	
					From Fig. 14.20b	Calculated
1	1000	1.333	1.4	53.33	0.58	0.288
2	500	2.666	2.8	106.66	0.78	0.576
3	400	3.333	3.5	133.33	0.82	0.72
4	350	3.808	4.0	152.38	0.86	0.8236
5	325	4.100	4.31	164.1	0.89	0.887

and

$$NTU = UA/C_{min} = 200 \times 7/1333.3 = 1.05$$

But for the above calculated values of *NTU* and ε, it can be seen from Fig. 14.20b that the ratio $C_{mixed}/C_{unmixed}$ does not exist. Therefore, the fluid with C_{min} is fluid *1*. This selection will require trial and error as outlined below.

Let $C_{min} = (mc_p)_1 = 1000$, then

$$\frac{C_{mixed}}{C_{unmixed}} = \frac{C_2}{C_1} = \frac{1333.3}{1000} = 1.333$$

$$NTU = UA/C_{min} = UA/C_1 = 200 \times 7/1000 = 1.4$$

From Fig. 14.20b, ε = 0.58.

Energy balance gives

$$\frac{(\Delta t)_1}{(\Delta t)_2} = \frac{C_2}{C_1}$$

or

$$(\Delta t)_1 = \frac{1333.3 \times (55 - 15)}{1000} = 53.33°\text{C}$$

and

$$\varepsilon = \frac{(\Delta t)_1}{(T_1 - t_1)} = \frac{53.33}{200 - 15} = 0.288$$

The calculated value of effectiveness is significantly different from the value of 0.58 from the figure. Hence, the assumed value of C_1 must be revised. The trial with the new value of C_1 is to be continued till the two values of ε are equal. The process is shown in Table 14.3. The closing value of C_1 gives m_1 = 9.75 kg/min and outlet temperature of fluid *1* = 35.9°C.

Example 14.20 Exhaust gas (0.1 kg/s) at 800 K from a gas turbine is to be used to preheat the air at 400 K before going to the combustor. It is desired to cool the exhaust to 500 K. If an overall heat transfer coefficient of 20 W/(m^2 K) can be achieved, determine the area required for a counterflow arrangement. Assume that the specific heat of the exhaust gases is the same as that of the air and is 1.005 kJ/(kg K).

What would be the outlet temperature of the exhaust gas and air if the flow arrangement is changed to parallel in the above exchanger?

Solution

We assume a balanced flow $C_h = C_c$. Actually, the exchanger is not exactly balanced because the mass flow rate of the exhaust is higher than that of the air by the amount of the fuel added in the combustor. Further, the average temperatures of the exhaust and air are different leading to a slight difference in the value of the specific heat of the two streams.

Counterflow arrangement:

$$\varepsilon = \frac{NTU}{1+NTU} = \frac{T_1 - T_2}{T_1 - t_1} = \frac{800 - 500}{800 - 400} = 0.75$$

or

$$NTU = 3$$

Hence, the required area

$$A = \frac{NTU \times C_h}{U} = \frac{3 \times 0.1 \times 1005}{20} = 15.1\,\text{m}^2$$

Parallel flow arrangement:

$$\varepsilon = \frac{1}{2}[1 - \exp(-2NTU)] = \frac{1}{2}[1 - \exp(-2 \times 3)] = 0.4987$$

Hence,

$$\varepsilon = \frac{T_1 - T_2}{T_1 - t_1} = 0.4987$$

or $T_2 = T_1 - 0.4987 \times (T_1 - t_1) = 800 - 0.4987 \times (800 - 400) = 600.5$ K.

Note: The limiting value of the effectiveness of a parallel flow heat exchanger with the balanced flow is 0.5 and thus the exhaust in the present case cannot be cooled below 600 K when parallel flow arrangement is used.

Example 14.21 Steam is condensed in the shell of a shell-and-tube (one shell, two tube passes) heat exchanger, which consists of 160 tubes of 25 mm diameter and 2 m length. Steam pressure is 1 atm and heat transfer coefficient on the condensation side is 8000 W/(m^2 K). If the water flow rate is 12 kg/s and inlet temperature is 20°C, determine water outlet temperature.

Solution

Assuming 40°C water mean temperature [= (t_{ci} + t_{co})/2], thermophysical properties of water are (Table A4)

$$c = 4179\,\text{J/(kg K)},\ k = 0.631\,\text{W/(m K)},\ \mu = 651 \times 10^{-6}\,\text{N s/m}^2\ \text{and}\ \text{Pr} = 4.3$$

Water flow rate per tube is (flow rate/no. of tubes in single pass) = 12/80 = 0.15 kg/s (refer to Fig. 14.2). Reynolds number of the flow in the tube is

$$\mathrm{Re} = \frac{md_i}{(\pi/4)d_i^2\mu} = \frac{4m}{\pi d_i \mu} = \frac{4 \times 0.15}{\pi \times 0.025 \times 651 \times 10^{-6}} = 11735$$

Flow is turbulent. Dittus–Boelter equation gives

$$\begin{aligned} h_i &= \frac{k}{d_i} \times 0.024\mathrm{Re}^{0.8}\mathrm{Pr}^{0.4} \\ &= \frac{0.631}{0.025} \times 0.024 \times 11735^{0.8} \times 4.3^{0.4} = 1955.6\,\mathrm{W/(m^2\,K)} \end{aligned}$$

Overall heat transfer coefficient,

$$U = \frac{h_i h_o}{h_i + h_o} = \frac{1955.6 \times 8000}{1955.6 + 8000} = 1571.5\,\mathrm{W/(m^2\,K)}$$

Heat transfer surface area, $A = N\pi DL = 160 \times \pi \times 0.025 \times 2 = 25.13\,\mathrm{m^2}$.
Minimum capacity rate, $C_{min} = m_c c_c = 12 \times 4179 = 50148$ W/K.
Hence,

$$NTU = \frac{UA}{C_{min}} = \frac{1571.5 \times 25.13}{50148} = 0.79$$

For a condensing fluid, $C_{\max} = \infty$. Hence, $C^* = \frac{C_{min}}{C_{max}} = 0$ The applicable ε-NTU relation is

$$\begin{aligned} \varepsilon &= 1 - \exp(-NTU) \\ &= 1 - \exp(-0.79) = 0.546 \end{aligned}$$

From the definition of effectiveness,

$$\varepsilon = \frac{q}{q_{max}} = \frac{m_c c_c (t_{\mathrm{co}} - t_{\mathrm{ci}})}{m_c c_c (t_{\mathrm{hi}} - t_{\mathrm{ci}})} = \frac{t_{\mathrm{co}} - t_{\mathrm{ci}}}{t_{\mathrm{hi}} - t_{\mathrm{ci}}}$$

or

$$0.546 = \frac{t_{\mathrm{co}} - 20}{100 - 20}$$

or

$$t_{\mathrm{co}} = 63.68^\circ\mathrm{C}.$$

Mean water temperature is (20 + 63.68)/2 = 41.84°C against the assumed mean temperature of 40°C. Retrial is not required.

14.2 Part B: Design of Heat Exchangers

14.2.1 Introduction

The basic considerations in the design or selection of heat exchanger as an off-the-shelf item are

1. Thermohydraulic design: It involves
   a. Thermal design: The heat exchanger must fulfil the heat duty requirements.
   b. Pressure drop characteristic (hydraulic consideration): Calculation of pressure drops for both hot and cold streams, which should not exceed the allowable values.
2. Size and weight
3. Cost.

The basic steps in the thermal design of the exchanger are

(i) Choose the configuration of the exchanger. In the case of shell-and-tube exchanger, the selected configuration should provide correction factor F_T greater than 0.75.
(ii) Heat balance equation is used to determine unknown temperature or the flow rate of one of the streams.
(iii) Knowing all the temperatures, *LMTD* is calculated.
(iv) Compute the heat transfer coefficients h_o and h_{io} from appropriate heat transfer correlations and obtain clean overall heat transfer coefficient.
(v) Allowing fouling factors (resistances), a value of design overall heat transfer coefficient is calculated from which the surface area is found for the given heat duty of the exchanger.

In the sections to follow, some basic fundamentals relating to the heat exchanger design have been discussed followed by some design examples with the aim of exposing the readers to the subject of heat exchanger design. A detailed discussion on heat exchanger design is beyond the scope of this book, the interested readers may refer to the listed references on heat exchanger design, for example Kern (1950). Further, it is to note that the examples given here consider only the conduction and convection modes of heat transfer. The radiation is an important mode of heat transfer in exchangers operating at high temperatures especially those dealing with high-temperature gases.

14.2.2 Double Pipe Exchangers

A double pipe exchanger is shown in Fig. 14.21. It is a concentric-pipe heat transfer apparatus, which consists of two legs and hence is also known as hairpin. The exchanger shown in the figure consists of two sets of concentric pipes, two connecting tees, a return head, a return bend and packing glands. The inner pipe is supported within the outer pipe by the packing glands. One of the fluids enters the inner pipe through the threaded connection and flows through two legs, which are connected by the return bend. The tees have nozzles or connections to permit the entry and exit of the annulus fluid, which passes from one leg of the exchanger to the other through the return head. The return bend does not contribute to the heat exchange.

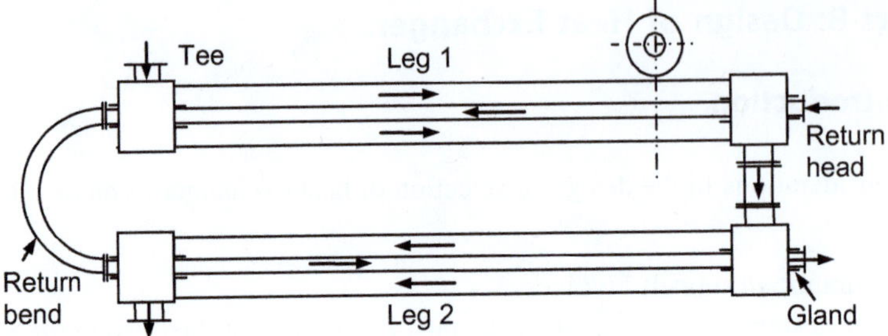

Fig. 14.21 A double pipe heat exchanger (hairpin)

Table 14.4 Double pipe exchanger fittings

Inner pipe, in	Outer pipe, in
1¼	2
1¼	2½
2	3
3	4

Table 14.4 shows some combination for the double pipe exchangers. These exchangers are usually assembled in not more than 20 ft effective leg lengths. The effective length is the distance in each leg over which the heat transfer occurs. It does not include the return bend and lengths protruding beyond the exchanger section.

The double pipe exchangers are of the greatest use where the total required heat transfer surface area is small, up to about 200 ft^2. When an industrial process requires a large amount of heat transfer surface area, a large number of hairpins are required because of the small surface area associated with a single hairpin. A single hairpin has 14 points where the leakage might occur (Kern 1950). This also makes difficult the periodic dismantling for cleaning. If the hairpins are employed in excess of 20 ft length in each leg, the inner pipe tends to sag and may touch the outer pipe causing a poor flow distribution in the annulus.

14.2.3 Clean and Design Overall Heat Transfer Coefficients

In the case of two concentric pipes, the resistances encountered to the heat transfer from one fluid to the other are the pipe fluid film resistances (at the inner and outer surfaces of the pipe) and the pipe wall conduction resistance. The total resistance to the heat exchange is, refer to Fig. 14.22a,

$$R_{\Sigma} = \frac{1}{h_i A_i} + \frac{1}{2\pi k L} \ln\left(\frac{R_o}{R_i}\right) + \frac{1}{h_o A_o} = R_i + R_k + R_o \tag{14.30}$$

Usually metal pipes are used and the pipe wall is very thin hence the pipe wall resistance is negligible. This reduces the above equation to

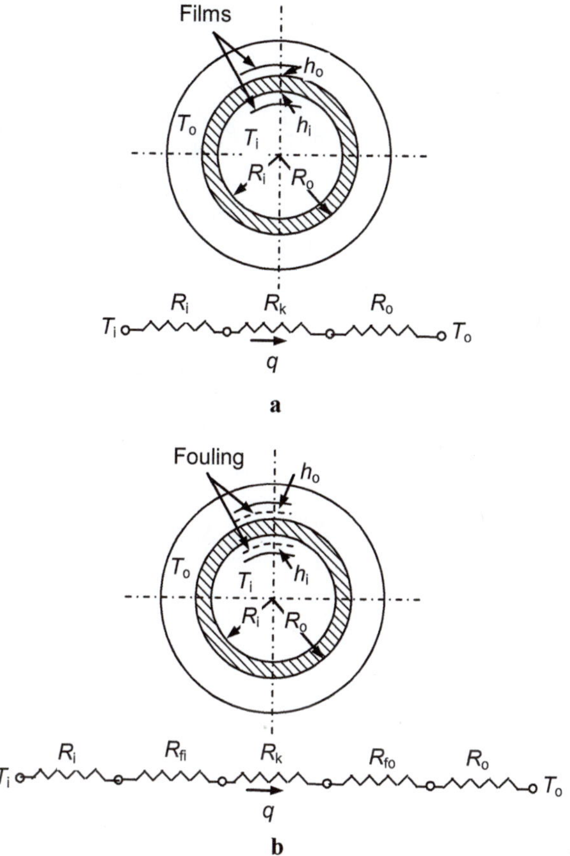

Fig. 14.22 Clean and design overall heat transfer coefficient: network

$$R_\Sigma = \frac{1}{h_i A_i} + \frac{1}{h_o A_o} \tag{14.31}$$

An overall heat transfer coefficient U_c is defined such that

$$U_c = \frac{1}{R_\Sigma A_o} \tag{14.32}$$

where the subscript c has been used to indicate that the coefficient pertains to the clean surfaces.

Then the heat exchange is

$$q = \frac{\Delta t}{R_\Sigma} = U_c A_o \Delta t \tag{14.33}$$

When heat exchange equipment is in service for some time, deposition of dirt and salts, and scaling due to the corrosive effects of the fluids on the inside and outside surfaces of the pipe takes place. These deposits on the surfaces offer additional resistances to the heat transfer. The value of the overall heat transfer coefficient considering these resistances can be written as, refer to Fig. 14.22b,

$$R'_{\Sigma} = \frac{1}{A_o}\left(\frac{A_o}{h_i A_i} + \frac{1}{h_o} + R_{\text{fi}}\frac{A_o}{A_i} + R_{\text{fo}}\right) \tag{14.34}$$

where R_{fi} and R_{fo} are the additional resistances in series with the film resistances at the inner and outer surfaces, respectively (the pipe wall resistance has been neglected). The resistances R_{fi} and R_{fo} are called dirt, scale or *fouling factors*.

We define *design overall heat transfer coefficient* as

$$U_d = \frac{1}{R'_{\Sigma} A_o} \tag{14.35}$$

and then the heat transfer rate is

$$q = \frac{\Delta t}{R'_{\Sigma}} = U_d A_o \Delta t \tag{14.36}$$

The total fouling resistance or the total fouling factor is defined as

$$R_{\text{ft}} = \frac{1}{U_d} - \frac{1}{U_c} \tag{14.37}$$

The total fouling factor can be expressed as

$$R_{\text{ft}} = R_{\text{fi}}\frac{A_o}{A_i} + R_{\text{fo}} \tag{14.38}$$

R_{fi} may not be referred to the outside diameter. This gives $R_{\text{ft}} \approx R_{\text{fi}} + R_{\text{fo}}$ (simple summation of the two fouling factors). The error introduced is negligible (Kern 1950).

Fouling refers to an undesirable accumulation of deposits on a heat transfer surface. It leads to a reduction in the thermal performance and an increase in the pressure drop in a heat exchanger. The influence of the fouling on the overall heat transfer coefficient depends on the clean overall heat transfer coefficient. For example, if U_c is low, say 10 W/(m^2 K) as in an air to air heat exchanger, a total fouling resistance of 0.0004 gives a design overall heat transfer coefficient U_d = 9.96 W/(m^2 K). Thus, the additional surface area required is very small. However, for a heat exchanger with a high value of U_c, say 1000 W/(m^2 K), R_{ft} = 0.0004 leads to U_d = 714 W/(m^2 K) and hence a 28.6% increase in the heat transfer surface area.

Recommended values of some of the fouling factors are listed in Table A7 (Appendix). Although the fouling is time dependent, only a fixed value of the factor is specified for the design. It means that the performance of the exchanger deteriorates with time and when the additional resistance due to the fouling of the heat transfer surface attains the value of the fouling factor, the exchanger is withdrawn from the service for cleaning. Thus, the cleaning schedule of an exchanger depends on the value of the design fouling factor.

Example 14.22 For water flowing inside and in cross-flow over the tubes in a shell-and-tube heat exchanger, h_i = 2500 W/(m^2 K) and h_o = 2000 W/(m^2 K). The tubes, with ID and OD of 25 mm and 30 mm, respectively, are made of mild steel [k = 45 W/(m K)].

Calculate the overall heat transfer coefficient. Determine the error in the estimate of the overall heat transfer coefficient due to the neglect of the pipe wall conduction resistance.

Solution

Considering the wall resistance, the overall heat transfer coefficient from Eqs. (14.30) and (14.32) is

$$\begin{aligned} U_{o,c} &= \left[\frac{R_o}{R_i h_i} + \frac{R_o}{k}\ln\left(\frac{R_o}{R_i}\right) + \frac{1}{h_o}\right]^{-1} \\ &= \left[\frac{15}{12.5 \times 2500} + \frac{15 \times 10^{-3}}{45}\ln\left(\frac{15}{12.5}\right) + \frac{1}{2000}\right]^{-1} \\ &= 960.8\ \text{W}/(\text{m}^2\ \text{K}) \end{aligned}$$

Neglecting wall resistance,

$$\begin{aligned} U'_{o,c} &= \left(\frac{R_o}{R_i h_i} + \frac{1}{h_o}\right)^{-1} \\ &= \left(\frac{15}{12.5 \times 2500} + \frac{1}{2000}\right)^{-1} = 1020.4\ \text{W}/(\text{m}^2\ \text{K}) \end{aligned}$$

Percent error

$$= \frac{960.8 - 1020.4}{960.8} \times 100 = -6.2\%$$

Example 14.23 Repeat Example 14.22 for water flowing inside the tubes and lubricating oil in the shell (cross-flow over the tubes); $h_i = 2500$ W/(m^2 K) and $h_o = 250$ W/(m^2 K).

Solution

Considering the wall resistance, the overall heat transfer coefficient is

$$\begin{aligned} U_{o,c} &= \left[\frac{R_o}{R_i h_i} + \frac{R_o}{k}\ln\left(\frac{R_o}{R_i}\right) + \frac{1}{h_o}\right]^{-1} \\ &= \left[\frac{15}{12.5 \times 2500} + \frac{15 \times 10^{-3}}{45}\ln\left(\frac{15}{12.5}\right) + \frac{1}{250}\right]^{-1} \\ &= 220.2\ \text{W}/\text{m}^2\,\text{K} \end{aligned}$$

Neglecting wall resistance,

$$\begin{aligned} U'_{o,c} &= \left(\frac{R_o}{R_i h_i} + \frac{1}{h_o}\right)^{-1} \\ &= \left(\frac{15}{12.5 \times 2500} + \frac{1}{250}\right)^{-1} = 223.2\ \text{W}/\text{m}^2\,\text{K} \end{aligned}$$

Percent error

$$= \frac{220.2 - 223.2}{220.2} \times 100 = -1.37\%$$

Comments: Comparison of the results of the previous two examples shows that the error in the estimate of the overall heat transfer coefficient due to the neglect of pipe wall resistance is significantly reduced when even one of the film resistances (proportional to the inverse of heat transfer coefficient) increases. In most of the heat exchangers, the pipe wall resistance can be neglected even when the film resistances are low because the fouling resistances are significantly higher than the wall resistance.

Example 14.24 For the data of Example 14.23, neglecting wall resistance, calculate the overall heat transfer coefficient when

(i) $h_i = 3500$ W/(m^2 K) and $h_o = 250$ W/(m^2 K),
(ii) $h_i = 2500$ W/(m^2 K) and $h_o = 350$ W/(m^2 K).
Comment on the result.

Solution

$$\text{(i)}\quad U'_{o,c} = \left(\frac{R_o}{R_i h_i} + \frac{1}{h_o}\right)^{-1}$$

$$= \left(\frac{15}{12.5 \times 3500} + \frac{1}{250}\right)^{-1} = 230.3\ \mathrm{W/(m^2\ K)}$$

$$\text{(ii)}\quad U'_{o,c} = \left(\frac{R_o}{R_i h_i} + \frac{1}{h_o}\right)^{-1}$$

$$= \left(\frac{15}{12.5 \times 2500} + \frac{1}{350}\right)^{-1} = 299.65\ \mathrm{W/(m^2\ K)}$$

Comment: Comparison of the results of this example with that of the previous example shows that a 40% increase in the higher heat transfer coefficient has resulted in only 3.2% increase in the overall heat transfer coefficient while a similar increase in the lower heat transfer coefficient has increased the overall heat transfer coefficient by 34.3%. From the results of Examples 14.23 and 14.24, it can be seen that the overall heat transfer coefficient is controlled by the value of the lower of the two heat transfer coefficients hence the lower of the two heat transfer coefficients is termed as *controlling film coefficient*.

Example 14.25 Consider Example 14.23 with fouling factors $R_{fi} = 0.0002$°C m^2/W and $R_{fo} = 0.0001$°C m^2/W.

Solution

Considering the fouling factor, the overall heat transfer coefficient is given by

$$U_{o,d} = \left(\frac{R_o}{R_i h_i} + R_{fi}\frac{R_o}{R_i} + \frac{R_o}{k}\ln\left(\frac{R_o}{R_i}\right) + R_{fo} + \frac{1}{h_o}\right)^{-1}$$

Substitution gives

$$U_{o,d} = \left(\frac{15}{12.5 \times 2500} + 0.0002 \times \frac{15}{12.5} + \frac{15 \times 10^{-3}}{45} \ln\left(\frac{15}{12.5}\right) + 0.0001 + \frac{1}{2000}\right)^{-1}$$
$$= 724.2 \text{ W}/(\text{m}^2 \text{ K})$$

Neglecting wall resistance,

$$U'_{o,d} = \left(\frac{R_o}{R_i h_i} + R_{\text{fi}} \frac{R_o}{R_i} + R_{\text{fo}} + \frac{1}{h_o}\right)^{-1}$$

Substitution gives

$$U'_{o,d} = \left(\frac{15}{12.5 \times 2500} + 0.0002 \times \frac{15}{12.5} + 0.0001 + \frac{1}{2000}\right)^{-1} = 757.6 \text{ W}/(\text{m}^2 \text{ K})$$

Percent error

$$= \frac{724.2 - 757.6}{724.2} \times 100 = -4.6\%$$

Example 14.26 When a heat exchanger was new, it transferred 5% more heat than it is transferring after 3 months service. Determine the overall fouling factor in terms of its clean overall heat transfer coefficient. Assume all other operating data to be the same.

Solution

The heat transfers in clean and fouled conditions of the exchanger are

$$q_{\text{clean}} = U_c A \Delta t_m$$
$$q_{\text{fouled}} = U_d A \Delta t_m$$

The given ratio of the above two heat transfer rates is 1.05. This gives

$$\frac{q_{\text{clean}}}{q_{\text{fouled}}} = 1.05 = \frac{U_c}{U_d}$$

By definition,

$$R_{\text{ft}} = \frac{1}{U_d} - \frac{1}{U_c}$$

Substituting the value of $U_d = U_c/1.05$, we get

$$R_{\text{ft}} = \frac{0.05}{U_c}$$

Example 14.27 At clean condition, oil enters a double pipe heat exchanger at 100°C and is cooled to 50°C while the cooling water is heated from 20°C to 80°C. Overall heat transfer coefficient is 400 W/(m^2 K) and the desired heat rate q = 2000 W. After operation of the exchanger for 2 years, the water outlet temperature drops to 70°C due to fouling. Determine the fouling factor after 2 years of operation. Comment on the result.

Solution

Clean condition

Log mean temperature difference is

$$LMTD = \frac{(100-80)-(50-20)}{\ln\frac{(100-80)}{(50-20)}} = 24.66^{\circ}\text{C}$$

Heat transfer area is

$$A = \frac{q}{U \times LMTD} = \frac{2000}{400 \times 24.66} = 0.203\ \text{m}^2$$

Fouled condition

From the design heat rate in clean condition, $q = C_w\,(t_{wo} - t_{wi}) = 2000$, we have C_w = 2000/(80–20) = 33.33. Similarly, C_o = 2000/(100–50) = 40.

After fouling, the heat rate is

$$q' = C_w(t'_{wo} - t_{wi}) = 33.33 \times (70-20) = 1666.5\ \text{W}$$

Oil outlet temperature for t_{wo} = 70°C is

$$t_{oo} = t_{oi} - \frac{q'}{C_0} = 100 - \frac{1666.5}{40} = 58.34^{\circ}\text{C}$$

Log mean temperature difference is

$$LMTD = \frac{(100-70)-(58.34-20)}{\ln\frac{(100-70)}{(58.34-20)}} = 34^{\circ}\text{C}$$

The new overall heat transfer coefficient is

$$U' = \frac{q'}{A \times LMTD} = \frac{1666.5}{0.203 \times 34} = 241.45\ \text{W}/(\text{m}^2\ \text{K})$$

The fouling factor is

$$R_{\text{ft}} = \frac{1}{U'} - \frac{1}{U} = \frac{1}{241.45} - \frac{1}{400} = 0.00164\ \text{m}^2\ \text{K}/\text{W}$$

which is quite high as compared to the recommended values of fouling factors for oil and water (refer to Table A7). Hence, cleaning of the heat exchanger in the discussion must be undertaken earlier than 2 years of operation.

Example 14.28 In a heat exchanger, 200 kg/min water flows through a tube of 50 mm in diameter and is heated from 40°C to 60°C. The tube wall temperature is maintained at 100°C by condensing steam. What is the length of the tube? The following equation may be used to estimate the heat transfer coefficient for tube flow.

$$\mathrm{Nu} = 0.024\mathrm{Re}^{0.8}\mathrm{Pr}^{0.4}$$

Solution

At 50°C thermophysical properties of water are

$\rho = 988.1$ kg/m^3, $c = 4.182$ kJ/(kg K), $k = 0.644$ W/(m K), $\mu = 544 \times 10^{-6}$ kg/(m s) and Pr = 3.55.

Flow velocity,

$$u = \frac{\dot{m}}{\rho A} = \frac{200}{60} \times \frac{1}{988.1} \times \frac{1}{(\pi/4) \times (0.05)^2} = 1.718\,\mathrm{m/s}$$

Reynolds number,

$$\mathrm{Re} = \frac{\rho u D}{\mu} = \frac{988.1 \times 1.718 \times 0.05}{544 \times 10^{-6}} = 1.56 \times 10^5$$

Nusselt number, from the given correlation, is

$$\mathrm{Nu} = 0.024\mathrm{Re}^{0.8}\mathrm{Pr}^{0.4} = 0.024 \times (1.56 \times 10^5)^{0.8} \times (3.55)^{0.4} = 568.6.$$

Heat transfer coefficient,

$$h = \frac{\mathrm{Nu}k}{D} = \frac{568.6 \times 0.644}{0.05} = 7323\ \mathrm{W/m^2\,K}$$

Log mean temperature difference,

$$\Delta t = \frac{\Delta t_1 - \Delta t_2}{\ln\frac{\Delta t_1}{\Delta t_2}} = \frac{(T_1 - t_2) - (T_2 - t_1)}{\ln\frac{(T_1 - t_2)}{(T_2 - t_1)}} = \frac{(100 - 60) - (100 - 40)}{\ln\frac{(100-60)}{(100-40)}} = 49.32^\circ\mathrm{C}$$

Heat transfer, from the heat transfer equation, is

$$q = hA\Delta t = 7323 \times (\pi D L) \times 49.32 \qquad \text{(i)}$$

From the first law of thermodynamics,

$$q = mc_p(t_2 - t_1) = \frac{200}{60} \times 4.182 \times (60 - 40) = 278.8\,\mathrm{kW} \qquad \text{(ii)}$$

Equating Eqs. (i) and (ii), we get

$$7323 \times (\pi DL) \times 49.32 = 278.8 \times 1000$$

or
$$L = \frac{278.8 \times 1000}{7323 \times \pi D \times 49.32} = \frac{278.8 \times 1000}{7323 \times \pi \times 0.05 \times 49.32} = 4.91\,\text{m}.$$

Example 14.29 A single-pass steam condenser is to handle 20,000 kg/hr of dry saturated steam at 0.07 bar. Available for the service are 22.5/25 mm tubes. The thermal conductivity of the tube material is 70 W/(m K). The average water velocity in each tube is limited to 1 m/s. The cooling water is available at 15°C and the outlet water temperature can be assumed to be 25°C. Steam side film coefficient is 6000 W/(m^2 K). Determine the number of tubes and length of the tube for the service.

Solution

Thermophysical properties of water at the mean temperature of 20°C are (Table A4, Appendix)

ρ = 998.2 kg/m^3, c_p = 4.183 kJ/(kg K), k = 0.601 W/(m K), μ = 10.02 × 10^{-4} kg/(m s) and Pr = 7.

For the steam at 0.07 bar, t_s = 39.0°C, i.e. $T_1 = T_2$ = 39°C and latent heat h_{fg} = 2409.1 kJ/kg from steam table.

(i) **Water-side film Coefficient**

Reynolds number,

$$\text{Re} = \frac{\rho u D_i}{\mu} = \frac{998.2 \times 1 \times 22.5 \times 10^{-3}}{10.02 \times 10^{-4}} = 22415$$

Nusselt number from the Dittus–Boelter correlation,

$$\text{Nu} = 0.024\text{Re}^{0.8}\text{Pr}^{0.4} = 0.024 \times (22415)^{0.8} \times (7)^{0.4} = 158$$

Heat transfer coefficient,

$$h = \frac{\text{Nu}k}{D} = \frac{158 \times 0.601}{22.5 \times 10^{-3}} = 4220\,\text{W/m}^2\text{K}$$

(ii) **Overall heat transfer coefficient**

$$\begin{aligned} U_{o,c} &= \left[\frac{R_o}{R_i h_i} + \frac{R_o}{k}\ln\left(\frac{R_o}{R_i}\right) + \frac{1}{h_o}\right]^{-1} \\ &= \left[\frac{25}{22.5 \times 4220} + \frac{25 \times 10^{-3}}{2 \times 70}\ln\left(\frac{25}{22.5}\right) + \frac{1}{6000}\right]^{-1} \\ &= 2228.3\,\text{W/m}^2\text{K} \end{aligned}$$

Log mean temperature difference,

$$\Delta t = \frac{\Delta t_1 - \Delta t_2}{\ln \frac{\Delta t_1}{\Delta t_2}} = \frac{(T_1 - t_2) - (T_2 - t_1)}{\ln \frac{(T_1 - t_2)}{(T_2 - t_1)}} = \frac{(39 - 15) - (39 - 25)}{\ln \left(\frac{39-15}{39-25}\right)} = 18.55\,^{\circ}\mathrm{C}$$

Heat transfer from unit length of the pipe is

$$\begin{aligned} q' &= UA\Delta t = 2228.3 \times (\pi DL) \times 18.55 \\ &= 2228.3 \times (\pi \times 25 \times 10^{-3} \times 1.0) \times 18.55 = 3246.4\ \mathrm{W/m} \end{aligned}$$

Heat to be transmitted,

$$q = \text{steam condensation rate } \times h_{fg} = (20000/3600) \times 2409.1 = 13383\,\mathrm{kW}$$

Required water flow rate from the heat balance,

$$m_w = \frac{q}{c_p(t_o - t_i)} = \frac{13383}{4.183 \times (25 - 15)} = 319.9\ \mathrm{kg/s}$$

Required number of tubes arranged in parallel to accommodate the water flow rate,

$$n = \frac{m_w}{m'_w} = \frac{319.9}{u \times (\pi/4) \times D_i^2 \times \rho} = \frac{319.9}{1 \times (\pi/4) \times 0.0225^2 \times 998.2} = 806 \approx 810$$

where m_w' is the flow rate through one tube.
Knowing the total heat duty, heat transfer rate from the unit length of the tube and number of tubes, the length of the tubes can be determined as

$$L = \frac{q}{nq'} = \frac{13383 \times 1000}{810 \times 3246.4} \approx 5.1\ \mathrm{m}$$

Example 14.30 Calculate the number and length of tubes required for a single-pass steam condenser to handle 25,000 kg of dry saturated steam per hour at 60°C. The cooling water enters the tubes at 20°C and leaves at 30°C. The tubes are of 25 mm outside and 22.5 inside diameter. The thermal conductivity of the tube material is 100 W/(m K). The average water velocity is 1.5 m/s. Assume that the steam side film coefficient is 4500 W/(m^2 K). Use the following correlation for the calculation of water-side heat transfer coefficient:

$$\mathrm{Nu} = 0.024\mathrm{Re}^{0.8}\mathrm{Pr}^{0.4}$$

Thermophysical properties of water at the mean temperature of 25°C are (Table A4)

$\rho = 997\,\mathrm{kg/m^3}$, $c_p = 4.181\,\mathrm{kJ/(kg\,K)}$, $k = 0.609\,\mathrm{W/(m\,K)}$, $\mu = 890 \times 10^{-6}\,\mathrm{kg/(m\,s)}$, Pr $= 6.13$.

Latent heat h_{fg} of the steam at 60°C (saturation temperature) is 2358.5 kJ/kg.

Solution

(i) **Water-side film Coefficient**
Flow Reynolds number,

$$\mathrm{Re} = \frac{\rho u D_i}{\mu} = \frac{997 \times 1.5 \times 22.5 \times 10^{-3}}{890 \times 10^{-6}} = 37807$$

Nusselt number from the Dittus–Boelter correlation,

$$\mathrm{Nu} = 0.024\mathrm{Re}^{0.8}\mathrm{Pr}^{0.4} = 0.024 \times (37807)^{0.8} \times (6.13)^{0.4} = 227.6$$

Water side heat transfer coefficient,

$$h = \frac{\mathrm{Nu}k}{D} = \frac{227.6 \times 0.609}{22.5 \times 10^{-3}} = 6160\ \mathrm{W/(m^2\ K)}$$

(ii) **Overall heat transfer coefficient**

$$\begin{aligned} U_{o,c} &= \left[\frac{R_o}{R_i h_i} + \frac{R_o}{k}\ln\left(\frac{R_o}{R_i}\right) + \frac{1}{h_o}\right]^{-1} \\ &= \left[\frac{12.5}{11.25 \times 6160} + \frac{12.5 \times 10^{-3}}{100}\ln\left(\frac{12.5}{11.25}\right) + \frac{1}{4500}\right]^{-1} \\ &= 2405\ \mathrm{W/(m^2\ K)} \end{aligned}$$

Log mean temperature difference,

$$\Delta t = \frac{\Delta t_1 - \Delta t_2}{\ln\frac{\Delta t_1}{\Delta t_2}} = \frac{(T_1 - t_2) - (T_2 - t_1)}{\ln\frac{(T_1 - t_2)}{(T_2 - t_1)}} = \frac{(60 - 20) - (60 - 30)}{\ln\left(\frac{60-20}{60-30}\right)} = 34.76^\circ\mathrm{C}$$

Heat transfer from unit length of the pipe is

$$\begin{aligned} q' &= UA\Delta t = 2405 \times (\pi D_0 L) \times 34.76 = 2405 \times (\pi \times 25 \times 10^{-3} \times 1.0) \times 34.76 \\ &= 6565.7\ \mathrm{W/m} \end{aligned}$$

Heat to be transferred,

$$q = \text{steam condensation rate } \times h_{fg} = (25000/3600) \times 2358.5 = 16378\,\mathrm{kW}$$

Required water flow rate from the heat balance,

$$m_w = \frac{q}{c_p(t_o - t_i)} = \frac{16378}{4.181 \times (30 - 20)} = 391.7\ \mathrm{kg/s}$$

Required number of tubes arranged in parallel to accommodate the water flow rate,

$$n = \frac{m_w}{m'_w} = \frac{391.7}{u \times (\pi/4) \times D_i^2 \times \rho} = \frac{391.7}{1.5 \times (\pi/4) \times 0.0225^2 \times 996.95} = 658.8 \approx 660$$

where m'_w is the flow rate through one tube.
Knowing the total heat duty, heat transfer rate from the unit length of the tube and number of tubes, the length of the tubes can be determined as

$$L = \frac{q}{nq'} = \frac{16378 \times 1000}{660 \times 6565.7} = 3.78 \approx 3.8\,\text{m}$$

Example 14.31 Find the length of a counterflow heat exchanger to cool 1000 kg/hr of hot water from 60°C to 30°C using cold water at 20°C. The cold water flows at a rate of 1500 kg/hr. The hot water flows through a steel pipe of 18 mm ID and 22 mm OD while the cold water flows through the annular space between the inner pipe and a 30 mm ID pipe. Neglect the pipe wall resistance.

The following correlations may be used for the estimates of the heat transfer coefficient and friction factor.

Tube flow:

$$\text{Nu} = 0.026\text{Re}^{0.8}\text{Pr}^{0.3}$$

$$f = 0.0791\text{Re}^{-0.25}$$

Annular duct:

$$\text{Nu} = 0.02\text{Re}^{0.8}\text{Pr}^{0.4}$$

$$f = 0.085\text{Re}^{-0.25}$$

The thermophysical properties of water at the mean bulk temperature may be taken from Table 14.5.

Table 14.5 Example 14.31

t, °C	ρ, kg/m^3	c, kJ/(kg K)	$k \times 10^2$, W/(m K)	$\nu \times 10^6$, m^2/s	Pr
30	995.7	4.174	61.8	0.805	5.42
40	992.2	4.174	63.5	0.659	4.31
50	988.1	4.176	64.8	0.556	3.54

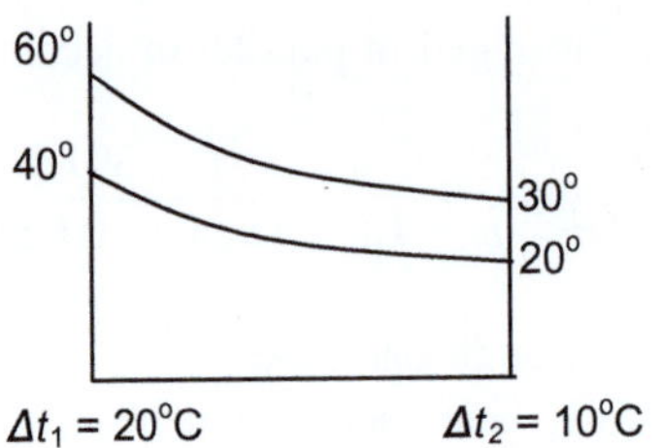

Fig. 14.23 Example 14.31

Solution

1. **Heat balance**
   Heat lost by the hot water

$$q = m_h c_h (t_{\text{hi}} - t_{\text{ho}}) = 1000 \times c_h \times (60 - 30) \qquad \text{(i)}$$

Heat gained by the cold (raw) water

$$q = m_c c_c (t_{\text{co}} - t_{\text{ci}}) = 1500 \times c_c \times (t_{\text{co}} - 20) \qquad \text{(ii)}$$

We obtain the outlet temperature of the cold water, by equating Eqs. (i) and (ii) and assuming $c_h \approx c_c$. This gives

$$t_{\text{co}} = 40^{\circ}\text{C}$$

2. **Log mean temperature difference** (refer to Fig. 14.23)

$$\Delta t = \frac{\Delta t_1 - \Delta t_2}{\ln \frac{\Delta t_1}{\Delta t_2}} = \frac{20 - 10}{\ln \frac{20}{10}} = 14.43^{\circ}\text{C}$$

3. **Flow area and equivalent diameters in double pipe exchangers**
   Tube: Flow area,

$$a_t = (\pi/4) d_i^2 = (\pi/4) \times 18^2 \times 10^{-6} = 2.544 \times 10^{-4}\ \text{m}^2$$

Annulus: Flow area,

$$a_a = (\pi/4)(D_i^2 - d_o^2) = (\pi/4) \times (30^2 - 22^2) \times 10^{-6} = 3.267 \times 10^{-4}\ \text{m}^2$$

Since the flow area of the annulus is greater than that of the inner pipe, we place the larger fluid stream in the annulus.

Hot water, tube	Cold water, annulus
4) Flow area, $a_t = 2.544 \times 10^{-4}$ m^2 $d = 18/1000 = 0.018$ m	4') Flow area, $a_a = 3.267 \times 10^{-4}$ m^2 Hydraulic diameter, $D_e = 4 \times$ flow area/perimeter where perimeter = π ($D_i + d_o$). Hence, $D_e = 4 \times 3.267 \times 10^{-4}/[\pi \times (30 + 20) \times 10^{-3}]$ = 0.008 m
5) Mass velocity, $G_t = W/a_t$ $= 1000/[3600 \times 2.544 \times 10^{-4}]$ = 1092.8 kg/(m^2 s)	5') Mass velocity, $G_a = W/a_a$ $= 1500/[3600 \times 3.267 \times 10^{-4}]$ = 1275 kg/(m^2 s)
6) $t_{av} = (60 + 30)/2 = 45$°C The temperature range and difference are moderate, the fluid properties may be evaluated at the mean temperature and the correction factor $(\mu/\mu_w)^{0.14}$ may be taken as unity. Properties at the mean temperature are $\rho = (992.2 + 988.1)/2 = 990.15$ kg/m^3 $c_p = (4.174 + 4.178)/2 = 4.176$ kJ/(kg K) $k = (0.635 + 0.648)/2 = 0.6415$ W/(m K) $v = (0.659 + 0.556) \times 10^{-6}/2$ $= 0.6075 \times 10^{-6}$ m^2/s Pr = (4.31 + 3.54)/2 = 3.925	6') $t_{av} = (20 + 40)/2 = 30$°C The temperature range and difference are moderate, the fluid properties may be evaluated at the mean temperature and the correction factor $(\mu/\mu_w)^{0.14}$ may be taken as unity. Properties at the mean temperature of the annulus fluid are $\rho = 995.7$ kg/m^3 $c_p = 4.174$ kJ/(kg K) $k = 0.618$ W/(m K) $v = 0.805 \times 10^{-6}$ m^2/s Pr = 5.42
7) Reynolds number: $\mathrm{Re} = Gd/\mu$ $= 1092.8 \times 0.018/(990.15 \times 0.6075 \times 10^{-6})$ = 32701	7') Reynolds number $\mathrm{Re} = GD_e/\mu$ $= 1275 \times 0.008/(995.7 \times 0.805 \times 10^{-6})$ = 12726
8) Heat transfer coefficient $Nu = 0.026\mathrm{Re}^{0.8}\,\mathrm{Pr}^{0.3}$ or $h = \left(0.026\mathrm{Re}^{0.8}\,\mathrm{Pr}^{0.3}\right)\frac{k}{d}$ $h_i = \left[0.026 \times (32701)^{0.8} \times (3.925)^{0.3}\right]\frac{0.6415}{0.018}$ $h_i = 5711$ $h_{io} = h_i\,(ID/OD) = 5711 \times (18/22)$ = 4673 W/(m^2 °C)	8') Heat transfer coefficient $Nu = 0.02\mathrm{Re}^{0.8}\,\mathrm{Pr}^{0.4}$ or $h = \left(0.02\mathrm{Re}^{0.8}\,\mathrm{Pr}^{0.4}\right)\frac{k}{D_e}$ $h_o = \left[0.02 \times (12726)^{0.8} \times (5.42)^{0.4}\right]\frac{0.618}{0.008}$ $h_o = 5840$ W/(m^2 °C)

Clean overall heat transfer coefficient,

$$U_c = \frac{h_0 \times h_{i0}}{h_0 + h_{i0}} = \frac{5840 \times 4673}{5840 + 4673} = 2596\ \mathrm{W/(m^2\ K)}$$

Heat duty,

$$q = mc_p(t_{\mathrm{hi}} - t_{\mathrm{ho}}) = (1000/3600) \times 4176 \times (60 - 30) = 34800\ \mathrm{W}$$

Required heat transfer area,

$$A = \frac{q}{U_c(LMTD)} = \frac{34800}{2596 \times 14.43} = 0.93\ \mathrm{m^2}$$

Required tube length,

$$L = \frac{A}{\pi d_o} = \frac{0.93}{\pi \times 0.022} = 13.46 \text{ m}$$

Example 14.32 In the previous example, calculate the required length if an overall dirt factor of 0.0003°C m^2/W is to be considered. Also calculate the pressure drop for both the streams.

Solution

Design heat transfer coefficient for combined fouling factor of 0.0003:

$$\frac{1}{U_d} = \frac{1}{U_c} + R_d = \frac{1}{2596} + 0.0003$$

or

$$U_d = 1459 \text{ W/(m}^2\text{ K)}$$

Required tube length,

$$L = \frac{13.46 \times 2596}{1459} = 23.95 \text{ m}$$

This can be fulfilled by connecting two 6.1 m long hairpins in series giving a total length of 24.4 m.

Pressure Drop

Tube side	Annulus side
$f = 0.0791\text{Re}^{-0.25}$ $= 0.0791 \times (32701)^{-0.25} = 0.00588$ $(\Delta p)_a = \frac{4fG_t^2L}{2g\rho^2 d}$ $= \frac{4 \times 0.00588 \times 1092.8^2 \times 24.4}{2 \times 9.81 \times 990.15^2 \times 0.018}$ $= 1.98$ m of water column	$f = 0.085\text{Re}^{-0.25} = 0.085 \times (12726)^{-0.25}$ $= 0.00802$ $(\Delta p)_t = \frac{4fG_a^2L}{2g\rho^2 d}$ $= \frac{4 \times 0.00802 \times 1275^2 \times 24.4}{2 \times 9.81 \times 995.7^2 \times 0.008}$ $= 8.18$ m of water column Entrance and exit losses are taken as one velocity head/hairpin (Kern 1950), i.e. $V = \frac{G_a}{\rho} = 1.28$ m/s $F_t = 2\left(\frac{V^2}{2g}\right) = 2 \times \left(\frac{1.28^2}{2 \times 9.81}\right) = 0.17$ m So total head loss is $(\Delta p)_a = 8.18 + 0.17 = 8.35$ m of water head

Example 14.33 A copper tube (20 mm ID and 25 mm OD) is used in a cross-flow heat exchanger. Water at a mean temperature of 80°C flows through the tube at a mean velocity of 0.8 m/s. The fouling factors are R_{fi} =0.0004 and R_{fo} = 0.0002 m^2 K/W. Air at 15°C flows

across the tube with velocity of 15 m/s. Thermal conductivity of copper is 350 W/(m K). Determine the overall heat transfer coefficient based upon the outer surface.

Solution

At 80°C, thermophysical properties of water are (Table A4)

$$\rho = 971.8\ \mathrm{kg/m^3},\ k = 0.67\ \mathrm{W/(m\,K)},\ \mu = 351 \times 10^{-6}\ \mathrm{kg/(m\,s)}\ \text{and}\ \mathrm{Pr} = 2.23$$

Reynolds number of the flow in the tube is

$$\mathrm{Re} = \frac{\rho U d_i}{\mu} = \frac{971.8 \times 0.8 \times 0.02}{351 \times 10^{-6}} = 44299$$

Flow is turbulent. For cooling of fluid, the heat transfer coefficient is (Table 8.3)

$$\begin{aligned} h_i &= \frac{k}{d_i} \times 0.026\mathrm{Re}^{0.8}\mathrm{Pr}^{0.3} \\ &= \frac{0.67}{0.02} \times 0.026 \times 44299^{0.8} \times 2.23^{0.3} = 5776 \end{aligned}$$

Assuming outer surface temperature t_{wo} = 60°C, the film temperature is (60 + 15)/2 = 37.5°C. Thermophysical properties of air at 27.5°C are (Table A5)

$$\rho = 1.141\ \mathrm{kg/m^3},\ k = 0.027\ \mathrm{W/(m\,K)},\ \mu = 1.894 \times 10^{-5}\mathrm{kg/(m\,s)}\ \text{and}\ \mathrm{Pr} = 0.706$$

For flow of air across the tube, the Reynolds number is

$$\mathrm{Re} = \frac{\rho U d_o}{\mu} = \frac{1.141 \times 15 \times 0.025}{1.894 \times 10^{-5}} = 22591$$

From Eq. (8.47) and Table 8.11,

$$\begin{aligned} h_o &= \frac{k}{d_o} \times 0.193 \times \mathrm{Re}^{0.618} \times \mathrm{Pr}^{1/3} \\ &= \frac{0.027}{0.025} \times 0.193 \times 22591^{0.618} \times 0.706^{1/3} = 91.1\ \mathrm{W/(m^2\,K)} \end{aligned}$$

The overall heat transfer coefficient is given by

$$\begin{aligned} U_o &= \left[\frac{R_o}{R_i h_i} + R_{\mathrm{fi}}\frac{R_o}{R_i} + \frac{R_o}{k}\ln\left(\frac{R_o}{R_i}\right) + R_{\mathrm{fo}} + \frac{1}{h_o}\right]^{-1} \\ &= \left[\frac{25}{20 \times 5776} + 0.0004 \times \frac{25}{20} + \frac{0.025/2}{350}\ln\left(\frac{25}{20}\right) + 0.0002 + \frac{1}{91.1}\right]^{-1} \\ &= [0.000216 + 0.0005 + 0.00000797 + 0.0002 + 0.01098]^{-1} = 84.0\ \mathrm{W/(m^2\,C)} \end{aligned}$$

Heat transfer equation gives

$$q = U_o A_o (t_w - t_a) = h_o A_o (t_{wo} - t_a)$$

or

$$t_{wo} = \frac{U_o}{h_o}(t_w - t_a) + t_a = \frac{84}{91.1}(80 - 15) + 15 = 74.93°\text{C}$$

Outer surface temperature t_{wo} is 74.93°C against the assumed temperature of 60°C. Retrial with t_{wo} = 75°C may be carried out.

It is to note that the significant resistance is that of the air convection and the wall resistance is negligibly small.

Example 14.34 Steam is condensed at the rate of 1 kg/s at 1 atm in the shell of a shell-and-tube (one shell pass and two tube passes) heat exchanger, which consists of 25 mm diameter tubes. Heat transfer coefficient on the condensation side is 8000 W/(m^2 K). If the mean flow velocity of water in the tube is 0.35 m/s and inlet and outlet temperature are 20°C and 60°C, respectively, determine length and number of tubes per pass.

Solution

At 1 atm, h_{fg} = 2256.4 kJ/kg and t_h = 100°C from the steam tables.

The required heat rate is

$$q = m_s h_{fg} = 1 \times 2256.4 \times 10^3 = 2256.4 \times 10^3 \text{ W}$$

At 40°C mean water temperature [= (t_{ci} + t_{co})/2], thermophysical properties of water are (Table A4)

$$\rho = 992.2\,\text{kg/m}^3. c = 4179\,\text{J/(kg K)},\ k = 0.631\,\text{W/(m K)},\ \mu = 651 \times 10^{-6}\text{N s/m}^2 \text{ and Pr} = 4.3$$

Required water flow rate, from heat balance, is

$$m_c = \frac{q}{c_c(t_{co} - t_{ci})} = \frac{2256.4 \times 10^3}{4179 \times (60 - 20)} = 13.5\,\text{kg/s}$$

Water flow rate per tube from given mean velocity U_m = 0.35 m/s is

$$m = \rho U_m (\pi/4) D^2 = 992.2 \times 0.35 \times (\pi/4) \times 0.025^2 = 0.17\,\text{kg/s}$$

Number of tubes per pass to accommodate 13.5 kg/s of water flow rate is

$$N = \frac{m_c}{m} = \frac{13.5}{0.17} = 79.4 \text{ say } 80 \text{ per pass}$$

Reynolds number of the flow in the tube is

$$\mathrm{Re} = \frac{md_i}{(\pi/4)d_i^2\mu} = \frac{4m}{\pi d_i \mu} = \frac{4 \times 0.17}{\pi \times 0.025 \times 651 \times 10^{-6}} = 13300$$

Flow is turbulent. Dittus–Boelter equation gives

$$\begin{aligned} h_i &= \frac{k}{d_i} \times 0.024\mathrm{Re}^{0.8}\mathrm{Pr}^{0.4} \\ &= \frac{0.631}{0.025} \times 0.024 \times 13300^{0.8} \times 4.3^{0.4} = 2161.6\,\mathrm{W/(m^2\,K)} \end{aligned}$$

Overall heat transfer coefficient,

$$U = \frac{h_i h_o}{h_i + h_o} = \frac{2161.6 \times 8000}{2161.6 + 8000} = 1701.8\,\mathrm{W/(m^2\,K)}$$

Effectiveness,

$$\varepsilon = \frac{q}{q_{max}} = \frac{m_c c_c(t_{\mathrm{co}} - t_{\mathrm{ci}})}{m_c c_c(t_{\mathrm{hi}} - t_{\mathrm{ci}})} = \frac{t_{\mathrm{co}} - t_{\mathrm{ci}}}{t_{\mathrm{hi}} - t_{\mathrm{ci}}} = \frac{60 - 20}{100 - 20} = 0.5$$

Hence, refer to Table 14.2,

$$NTU = -\ln(1 - \varepsilon) = -\ln(1 - 0.5) = 0.693$$

Minimum capacity rate, $C_{min} = m_c c_c = 13.5 \times 4179 = 56417\,\mathrm{W/K}$
Heat transfer surface area,

$$A = \frac{NTU \times C_{min}}{U} = \frac{0.693 \times 56147}{1701.8} = 22.86\,\mathrm{m^2}$$

From $A = 2N\pi DL$, the tube length is

$$L = \frac{A}{2N\pi D} = \frac{22.86}{160 \times \pi \times 0.025} = 1.82\,\mathrm{m}$$

Example 14.35 Steam is condensed at the rate of 1 kg/s at 1 atm in the shell of a shell-and-tube (one shell pass and two tube passes) heat exchanger, which consists of 25 mm diameter and 1.82 m long tubes. Heat transfer coefficient on the condensation side is 8000 W/(m^2 K). If the mean flow velocity of water in the tube is 0.35 m/s and inlet temperature is 20°C, determine the number of tubes per pass.

Solution

At 1 atm, h_{fg} = 2256.4 kJ/kg and t_h = 100°C from the steam tables.

The required heat rate is

$$q = m_s h_{fg} = 1 \times 2256.4 \times 10^3 = 2256.4 \times 10^3 \text{ W}$$

At assumed mean water temperature [= (t_{ci} + t_{co})/2] of 40°C, thermophysical properties of water are (Table A4)

$\rho = 992.2\,\text{kg/m}^3 . c = 4179\,\text{J/(kg K)}, k = 0.631\,\text{W/(m K)},\ \mu = 651 \times 10^{-6}\,\text{N s/m}^2$ and Pr $= 4.3$

Water flow rate per tube from given mean velocity U_m = 0.35 m/s is

$$m = \rho U_m (\pi/4) D^2 = 992.2 \times 0.35 \times (\pi/4) \times 0.025^2 = 0.17\,\text{kg/s}$$

Reynolds number of the flow in the tube is

$$\text{Re} = \frac{m d_i}{(\pi/4) d_i^2 \mu} = \frac{4m}{\pi d_i \mu} = \frac{4 \times 0.17}{\pi \times 0.025 \times 651 \times 10^{-6}} = 13300$$

Flow is turbulent. Dittus–Boelter equation gives

$$\begin{aligned} h_i &= \frac{k}{d_i} \times 0.024 \text{Re}^{0.8} \text{Pr}^{0.4} \\ &= \frac{0.631}{0.025} \times 0.024 \times 13300^{0.8} \times 4.3^{0.4} = 2161.6\ \text{W/(m}^2\,\text{K)} \end{aligned}$$

Overall heat transfer coefficient,

$$U = \frac{h_i h_o}{h_i + h_o} = \frac{2161.6 \times 8000}{2161.6 + 8000} = 1701.8\,\text{W/(m}^2\,\text{K)}$$

Heat rate for N tubes per pass,

$$q = \varepsilon q_{max} = \varepsilon m_c c_c (t_{hi} - t_{ci}) = \varepsilon (mN) c_c (t_{hi} - t_{ci})$$

or

$$2256.4 \times 10^3 = \varepsilon (0.17 \times N) \times 4179 \times (100 - 20)$$

or

$$\varepsilon N = 39.7$$

Number of transfer units,

$$NTU = \frac{UA}{C_{min}} = \frac{U(2N\pi DL)}{Nmc_c}$$
$$= \frac{1701.8 \times 2N \times \pi \times 0.025 \times 1.82}{0.17N \times 4179} = 0.685$$

Hence, refer to Table 14.2,

$$\varepsilon = 1 - \exp(-NTU)$$

or

$$\frac{39.7}{N} = 1 - \exp(-0.685) = 0.496$$

or

$$N = \frac{39.7}{0.496} = 80 \text{ per pass}$$

This gives $m_c = mN = 80 \times 0.17 = 13.6$ kg/s. From heat rate equation, $q = m_c c_c(t_{co} - t_{ci})$, we get

$$t_{co} = \frac{q}{m_c c_c} + t_{ci} = \frac{2256.4 \times 10^3}{13.6 \times 4179} + 20 = 59.7^\circ\text{C}$$

Mean water temperature is (20 + 59.7)/2 = 39.85°C. Assumed mean temperature was 40°C. Retrial is not required.

Example 14.36 2 kg/s of hot exhaust gases from a furnace, at nearly atmospheric pressure, flow across a tube bank of an *economizer* (Fig. 14.24) consisting of inline array of 16 × 16. Steel tubes have 22 mm internal and 25 mm external diameter, and are 3 m long. Upstream of the tube bank, the gas temperature is 500°C. The maximum velocity of the gas in the tube bank is 4 m/s. 10 kg/s of water to be heated enters the tubes at 30°C. Determine water and gas outlet temperatures. Treat the exhaust gas as air for thermophysical properties.

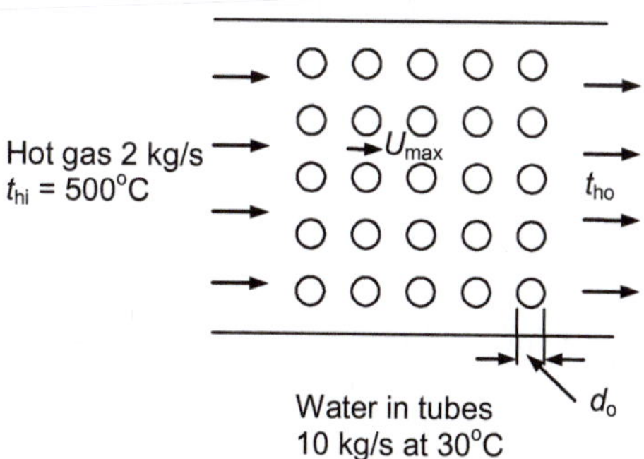

Fig. 14.24 Tube bank of the economizer (representative)

Solution

At assumed mean water temperature [= $(t_{ci} + t_{co})/2$] of 50°C, thermophysical properties of water are (Table A4)

$$\rho = 988.1\,\text{kg/m}^3.\,c = 4182\,\text{J/(kg K)},\ k = 0.644\,\text{W/(m K)},\ \mu = 544 \times 10^{-6}\,\text{N s/m}^2\ \text{and Pr} = 3.55$$

Reynolds number of flow of water in the tubes,

$$\text{Re} = \frac{md_i}{(\pi/4)d_i^2\mu} = \frac{4m}{\pi d_i \mu} = \frac{4 \times [10/(16 \times 16)]}{\pi \times 0.022 \times 544 \times 10^{-6}} = 4156$$

Assuming fully developed flow, we may use Gnielinski correlation for $1.5 \leq \text{Pr} \leq 500$ and $3 \times 10^3 \leq \text{Re} \leq 10^6$ (Table 8.3), which gives

$$\begin{aligned} h_i &= \frac{k}{d_i} \times 0.012(\text{Re}^{0.87} - 280)\text{Pr}^{0.4} \\ &= \frac{0.644}{0.022} \times 0.012 \times (4156^{0.87} - 280) \times 3.55^{0.4} \\ &= 657\,\text{W/(m}^2\,\text{K)} \end{aligned}$$

When referred to the outside diameter, $h_{io} = h_i \frac{d_i}{d_o} = 657 \times \frac{22}{25} = 578.2$

Treating exhaust gases as air and assuming a mean gas temperature of 350°C in the exchanger, the air properties are (Table A5)

$$\rho = 0.5674\,\text{kg/m}^3.\,c = 1058.9\,\text{J/(kg K)},\ k = 0.04794\,\text{W/(m K)},\ \mu = 3.9091 \times 10^{-5}\,\text{N s/m}^2\ \text{and Pr} = 0.681$$

Reynolds number for flow of gas over the tube bank,

$$\begin{aligned} \text{Re}_D &= \frac{\rho U_{max} d_o}{\mu} \\ &= \frac{0.5674 \times 4 \times 0.025}{3.9091 \times 10^{-5}} = 1451 \end{aligned} \qquad (8.60)$$

Using Eq. (8.58c) for $10^3 < \text{Re} \leq 2 \times 10^5$, we have

$$h_o = \frac{k}{d_o} 0.27\text{Re}_d^{0.63}\text{Pr}^{0.36}(\text{Pr}/\text{Pr}_w)^{0.25}$$

Neglecting factor (Pr/Pr_w) for air, we have

$$h_o = \frac{0.04794}{0.025} \times 0.27 \times 1451^{0.63} \times 0.681^{0.36} = 44.25\,\text{W/(m}^2\,\text{K)}$$

Overall heat transfer coefficient (neglecting tube wall resistance),

$$U = \frac{h_{io}h_o}{h_{io} + h_o} = \frac{578.2 \times 44.25}{578.2 + 44.25} = 41.1\,\mathrm{W/(m^2\,K)}$$

Minimum heat capacity rate, $C_{min} = 2 \times 1058.9 = 2117.8$
and maximum heat capacity rate, $C_{max} = 10 \times 4162 = 41620$.
Number of transfer units,

$$NTU = \frac{UA}{C_{min}} = \frac{U(N\pi d_o L)}{C_{min}}$$
$$= \frac{41.1 \times 256 \times \pi \times 0.025 \times 3}{2117.8} = 1.17$$

and

$$C^* = \frac{C_{min}}{C_{max}} = \frac{2117.8}{41620} = 0.05$$

From Fig. 14.20a, $\varepsilon \approx 0.68$. Hence, from the effectiveness equation,

$$\varepsilon = \frac{q}{q_{max}} = \frac{m_h c_h (t_{hi} - t_{ho})}{m_h c_h (t_{hi} - t_{ci})} = \frac{t_{hi} - t_{ho}}{t_{hi} - t_{ci}}$$

or

$$0.68 = \frac{500 - t_{ho}}{500 - 30}$$

or

$$t_{ho} = 500 - 0.68 \times (500 - 30) = 180.4°\mathrm{C}$$

From heat rate equation, $q = m_c c_c (t_{co} - t_{ci})$, which gives

$$t_{co} = \frac{q}{m_c c_c} + t_{ci} = \frac{m_h c_h (t_{hi} - t_{ho})}{m_c c_c} + t_{ci} = \frac{2117.8 \times (500 - 180.4)}{10 \times 4182} + 30 = 46.2°\mathrm{C}$$

Mean water temperature = (30 + 46.2)/2 = 38.1°C against the assumed temperature of 50°C. Mean gas temperature = (500 + 180.4)/2 = 340.2°C against the assumed temperature of 350°C. Retrial by taking thermophysical properties of water and gas at these estimated mean temperatures may be carried out.

14.3 Summary

This chapter has been presented in two parts. Part A deals with heat exchanger fundamentals while Part B deals with design of heat exchangers.

Heat exchangers have been classified on the basis of the configuration of the fluid flow paths through the heat exchanger, based on the principle of operation and by their applications. While dealing with fluids, which provide a low heat transfer coefficient, such as air or gases, finned tubes are frequently employed in the exchangers. Double pipe and shell-and-tube heat exchangers are widely used. However, all heat exchanger types, especially the compact heat exchangers, have not been presented and discussed in the present chapter. Readers should refer to specific books on heat exchangers and their design.

Heat transfer equation for double pipe (concentric tube) heat exchanger has been presented in Eq. (14.1) in terms of overall heat transfer coefficient, heat transfer area and an effective or mean temperature difference Δt_m for the whole heat transfer area of the exchanger (termed as log mean temperature difference *LMTD*).

For a double pipe heat exchanger, where the fluids may have either parallel or counterflow arrangement, expressions of log mean temperature difference Δt_m have been derived. The following important observations relating to *LMTD* must be kept in mind.

(i) If the temperature difference between two fluids is denoted by Δt_1 and Δt_2 at the two ends of the heat exchanger, then the *LMTD* equations for the parallel- and counter-flow arrangements have the same form as

$$\Delta t_m = \frac{\Delta t_2 - \Delta t_1}{\ln\left(\frac{\Delta t_2}{\Delta t_1}\right)}$$

(ii) The arithmetic mean temperature difference is always greater than the log mean and the difference between the two increases with the increase in the value of ratio $\Delta t_1/\Delta t_2$.

(iii) When one of the fluids passes through a heat exchanger isothermally (condensing vapour or boiling liquid), the flow arrangement does not affect the value of *LMTD*. It has been shown later on that this is true for all flow arrangements.

(iv) The log mean temperature difference reduces rapidly if the temperature of one of the fluid streams approaches the temperature of the other stream.

(v) In counterflow arrangement, if $\Delta t_1 = \Delta t_2$, *LMTD* equals Δt_1.

(vi) *LMTD* for the counterflow arrangement is always higher and hence such exchangers will require smaller heat transfer area for the same heat duty. If the conditions permit, a counterflow design is always preferred.

(vii) In parallel flow heat exchanger, the outlet temperature of the cold fluid can never exceed or equal the outlet temperature of the hot fluid.

Log mean temperature difference for other flow arrangements is obtained by applying a correction factor F_T to the *LMTD* of counterflow arrangement. In Sect. 14.1.4, graphs of correction factor F_T for other flow arrangements in terms of parameters P and R have been presented. Arrangement for the shell-and-tube heat exchanger must be selected to give a high value of the correction factor F_T and the design point must lie on the flat part of the F_T curve so that the correction factor is least affected by the variations in values of parameters P and R due to any variations in the fluid temperatures.

When all the four temperatures (T_1, T_2, t_1 and t_2) required for calculation of *LMTD* are not specified or cannot be easily determined from the heat balance equation, there is great difficulty in the use of the *LMTD* method. In such cases, the analysis can be easily carried out using the method of effectiveness-*NTU*. The effectiveness ε of a heat exchanger is defined as the ratio of the actual heat transfer to the maximum possible, which is attained when the fluid with minimum value of heat capacity $(mc_p)_{\min}$ undergoes the maximum temperature change $T_1 - t_1$.

Basic treatment of effectiveness-*NTU* method for counterflow and parallel flow heat exchangers is given in Sect. 14.1.5. Effectiveness-*NTU* curves for cross-flow and shell-and-tube heat exchangers are presented in Fig. 14.20a–c. The analytical expressions for these arrangements are somewhat tedious to use (refer to Table 14.2) and hence are used only when accurate analysis is required.

The following observations have been made from the ε-*NTU* curves:

7. For a given $C^*(= C_{min}/C_{max})$, the effectiveness increases with increase in the value of *NTU*.
8. For the given value of *NTU*, the effectiveness increases with decrease in the value of C^*.
9. For $C^* = 0$ (boilers and condensers), the value of the effectiveness is the same for all arrangements.
10. In case of balanced flow as in the case of the gas turbine regenerators, $C_{min}/C_{max} = 1$. The maximum value of the effectiveness is 50% for the parallel flow arrangement. There is no such limitation for the counterflow arrangement. This is the main reason for the use of counterflow arrangement, especially when C_{min}/C_{max} is not having a very low value.
11. At very low values of *NTU*, the effectiveness of different arrangements does not differ significantly for all values of C^*.
12. Counterflow heat exchanger gives the highest value of the effectiveness (i.e. the heat transfer performance) for any specified value of *NTU*. The advantage of the counterflow arrangement increases with the increase in the value of C_{min}/C_{max}.
13. At high values of *NTU*, a large increase in the heat transfer area is required for a small gain in the effectiveness for given overall heat transfer coefficient.

It is to note that the effectiveness-*NTU* method offers us a tool to select the best possible exchanger for the specified conditions. Higher the effectiveness of an exchanger, better is its performance.

As mentioned earlier, in Part B of the chapter, some basic considerations and fundamentals relating to the heat exchanger design have been discussed followed by some design examples with the aim of exposing the readers to the subject of heat exchanger design. It is to note that a detailed discussion on heat exchanger design is beyond the scope of this book, the interested readers may refer to the listed references on heat exchanger design.

The basic consideration in the design or selection of heat exchanger is the thermohydraulic design, which involves (a) thermal design to fulfil the heat duty requirements and (b) pressure drop characteristic (hydraulic consideration), i.e. calculation of pressure drops for both hot and cold streams, which should not exceed the allowable values. Other considerations, not discussed here, are size, weight and cost of the exchanger.

Overall heat transfer coefficient has been presented, vide Eq. (14.32), which considers the film resistances and wall resistance. This coefficient is termed as clean overall heat transfer coefficient because when a heat exchange equipment is in service for some time, deposition of dirt and salts, and scaling due to the corrosive effects of the fluids on the inside and outside surfaces of the pipe take place. These deposits on the surfaces offer additional resistances to the heat transfer which are termed as dirt, scale or fouling factors. Considering the fouling factor, design overall heat transfer coefficient has been defined vide Eq. (14.35).

The basic steps in thermal design of the exchanger have been outlined as

(vi) Choose the configuration of the exchanger. In the case of shell-and-tube exchanger, the selected configuration should provide *LMTD* correction factor F_T greater than 0.75.

(vii) Heat balance equation is used to determine unknown temperature or the flow rate of one of the streams.

(viii) Knowing all the temperatures, calculate *LMTD*.

(ix) Compute the heat transfer coefficients h_o and h_{io} from appropriate heat transfer correlations and obtain clean overall heat transfer coefficient, Eq. (14.32).

(x) Allowing fouling factors (resistances), the value of design overall heat transfer coefficient is calculated from which the surface area is found for the given heat duty of the exchanger, refer to Eq. (14.35).

Review Questions

14.1 Classify heat exchangers and draw diagrams to show the temperature distributions along their length or at the outlet.

14.2 Describe with the help of figures the double pipe, and shell-and-tube heat exchangers. Discuss the effect of baffles on the flow pattern in the shell-and-tube heat exchanger.

14.3 Derive the equation of overall heat transfer coefficient for tube flow. State the condition when you can neglect the pipe wall resistance to heat transfer.

14.4 Establish the expressions of *LMTD* for parallel flow and counterflow heat exchangers. State clearly the simplifying assumptions made in the derivation of the *LMTD* equation. What is the effect of these assumptions on the estimate of *LMTD*?

14.5 Define the correction factor for the *LMTD* and compare the performance of the counterflow, parallel flow and cross-flow arrangements.

14.6 Compare arithmetic mean and log mean temperature differences.

14.7 What is fouling factor? Explain with proper comments.

14.8 Define clean and design overall heat transfer coefficients.

14.9 What do you mean by controlling film coefficient? Explain giving a suitable example.

14.10 Define clearly the effectiveness of heat exchangers and develop an expression for effectiveness in terms of *NTU* and heat capacity ratio for a counterflow heat exchanger.

14.11 What advantage does the effectiveness-*NTU* method has over the *LMTD* method?

14.12 Using the effectiveness—*NTU* charts, compare the performance of parallel and counterflow heat exchangers.

14.13 Describe with suitable sketch the constructional features of a double pipe heat exchanger. What are its merits and demerits?

14.14 For a boiler or condenser, show that

$$\varepsilon = 1 - \exp(-NTU)$$

14.15 For a gas turbine heat exchanger ($C_{min}/C_{max} \approx 1$), show that

(i) for the parallel flow arrangement,

$$\varepsilon = (1/2)[1 - \exp(-2NTU)]$$

(ii) for the counterflow arrangement,

$$\varepsilon = NTU/(1 + NTU).$$

Which arrangement will you prefer?

Problems

14.1 Water [c_p = 4.1868 kJ/(kg K)] flowing at the rate of 2.0 kg/min is heated from 15°C to 45°C in a concentric, double pipe, parallel flow heat exchanger. The hot fluid is oil [c_p = 1.9 kJ/(kg K)], flowing at 2.5 kg/min, which enters the exchanger at 200°C. Determine the heat exchanger surface area required for an overall heat transfer coefficient U = 250 W/(m^2 K). Also calculate the effectiveness and NTU of the exchanger.

[**Ans.** $(T_o)_{oil}$ = 147.11°C from heat balance; $LMTD$= 139.47°C; A = 0.12 m^2; ε = 0.286; NTU= 0.38.]

14.2 A counterflow double pipe heat exchanger is to be used to cool hot oil [c_p = 2.0 kJ/(kg K)] from 200°C to 65°C. The cold stream oil [c_p = 1.8 kJ/(kg K)] at 50°C enters at a flow rate of 1.0 kg/s. If the flow rate of the hot oil is 0.7 kg/s, determine the required heat exchanger surface area [U = 300 W/(m^2 K)].

[**Ans.** t_o = 155°C from heat balance equation; $LMTD$ = 27.31°C; $A = q/(LMTD \times U)$ = 23.07 m^2.]

14.3 Determine the surface area required in a cross-flow heat exchanger with both streams unmixed to cool 300 m^3 of air per min from 50°C to 30°C. Water is available at 15°C for cooling of the air. The flow rate of water is 300 kg/min and the overall heat transfer coefficient is 150 W/(m^2 K). Given: ρ_{air} = 1.1 kg/m^3, c_p of air = 1.006 kJ/(kg K), c of water = 4.1868 kJ/(kg K).

[**Ans.** q = 110.66 kW; $(t_o)_w$ = 20.3°C; $LMTD$ = 21.52°C; P = 0.1514; R = 3.77; F_T 0.97; $A = q/(UF_T\,\Delta t_m)$ = 35.34 m^2.]

14.4 Air at 100°C [c_p = 1.006 kJ/(kg K)] flows into a cross-flow heat exchanger (both fluids unmixed) at a flow rate of 10 kg/s. Water [c_p = 4.1868 kJ/(kg K)] enters the exchanger at 15°C and at a flow rate of 5 kg/s. If the heat exchanger surface area is 100 m^2 and the overall heat transfer coefficient is 200 W/(m^2 K), determine the exit temperature of the air.

[**Ans.** $NTU = UA/C_{min}$ = 1.99; $C^* = C_{min}/C_{max}$ = 0.4806; effectiveness ≈ 0.74 (Fig. 14.20a); $(T_o)_{air}$ = 37.1°C; $(T_o)_w$ = 45.2°C from heat balance; Check: $(LMTD)_{counter}$ = 36.0°C; P = 0.355; R = 2.08; F_T = 0.88; $q = UAF_T(LMTD)_{counter}$ = 633600 W $\approx mc_p\Delta t$.]

14.5 Hot water [c_p = 4.1868 kJ/(kg K)] at 90°C is used to heat liquid ammonia [c_p = 4.8 kJ/(kg K)] from 40°C in a 1-shell pass and 2-tube passes (water in the tube). The water outlet temperature is 60°C and flow rate is 12 kg/min. If U = 100 W/(m^2 K) and A = 9.8 m^2, determine the outlet temperature of the ammonia.
[**Ans.** Assuming $C_h = C_{\min}$, $NTU = UA/C_{\min}$ = 1.17; ε = (90 – 60)/(90–40) = 0.6; from Fig. 14.20c, $C^* = C_{\min}/C_{\max} \approx 0.25$; $m_{\text{Ammonia}} = C_{\max}/c_{\text{Ammonia}}$ = 41.87 kg/min; $(t_o)_{\text{Ammonia}}$ = 47.5°C from heat balance.]

14.6 Check the result of Example 14.19 using the *LMTD* approach.
[**Ans.** P = 0.887; R = 0.2437; F_t = 0.6; $LMTD$ = 64.07°C; q = 53819 W ≈ $mc_p\Delta t = C_{\max}\,\Delta t$ = 4.1 × 325 × (55 −15).]

14.7 Check the result of Example 14.6 using ε-*NTU* approach.
[**Ans.** C_h = 14250, C_c = 10080, $C^* = C_{\min}/C_{\max}$ = 0.7074, ε = (724.1–300)/(800–300) = 0.8482, using relevant equation from Table 14.2 gives NTU = 3.311 and $A = (NTU \times C_{\min})/U$ = 111.25 m^2.]

14.8 Saturated steam at 100°C is condensing in a shell-and-tube heat exchanger with a *UA* value of 3650 W/K. Cooling water enters the tubes at 30°C. Determine cooling water flow rate required to maintain a heat rate of 200 kW.
[**Ans.** For the condensing fluid, $C_{\max} = \infty$. Hence, $C^* = C_{\min}/C_{\max} = 0$ and effectiveness relation is
$\varepsilon = 1 - \exp(-NTU)$, or $\frac{q}{q_{max}} = 1 - \exp\left(-\frac{UA}{C_{min}}\right)$, or $\frac{200\times1000}{m_c c_c(100-30)} = 1 - \exp\left(-\frac{3650}{m_c c_c}\right)$;
assuming mean temperature of water as 40°C, c_c = 4179 J/(kg K). Substitution gives $\frac{200\times1000}{m_c\times4179\times(100-30)} = 1 - \exp -\left(\frac{3650}{m_c\times4179}\right)$. Solution by trial and error gives m_c = 1.7 kg/s. From I law equation, $q = m_c \times c_c \times (t_{co} - t_{ci})$ or $200 \times 1000 = 1.7 \times 4179 \times (t_{co} - 30)$, which gives t_{co} =58.15°C. So t_{cm} = (30 + 58.15)/2 = 44.07°C. Retrial is not required as c_c will change only marginally.]

14.9 Steam is condensed in the shell of a shell-and-tube (one shell, two tube passes) heat exchanger, which consists of 160 tubes of 25 mm diameter. Steam pressure is 1 atm and heat transfer coefficient on the condensation side is 8000 W/(m^2 K). If water flow rate is 12 kg/s and it is heated from 20°C to 60°C, determine the tube length.
[**Ans.** Thermophysical properties of water at mean temperature [= $(t_{ci} + t_{co})/2$] = 40° C are (Table A4) c = 4179 J/(kg K), k = 0.631 W/(m K), $\mu = 651 \times 10^{-6}$ N s/m^2 and Pr = 4.3; water flow rate per tube is (flow rate/no. of tubes in single pass) = 12/80 = 0.15 kg/s (refer to Fig. 14.2); Reynolds number of the flow in the tube, $\text{Re} = \frac{md_i}{(\pi/4)d_i^2\mu} = \frac{4m}{\pi d_i\mu} = \frac{4\times0.15}{\pi\times0.025\times651\times10^{-6}} = 11735$; flow is turbulent. Dittus–Boelter equation gives $h_i = \frac{k}{d_i} \times 0.024\text{Re}^{0.8}\,\text{Pr}^{0.4} = \frac{0.631}{0.025} \times 0.024 \times 11735^{0.8} \times 4.3^{0.4} = 1955.6$ W/(m^2 K); $U = \frac{h_i h_o}{h_i + h_o} = \frac{1955.6\times8000}{1955.6+8000} = 1571.5$ W/(m^2 K);$\varepsilon = \frac{q}{q_{max}} = \frac{m_c c_c(t_{co}-t_{ci})}{m_c c_c(t_{hi}-t_{ci})} = \frac{t_{co}-t_{ci}}{t_{hi}-t_{ci}} = \frac{60-20}{100-20} = 0.5$; for the condensing fluid,$C_{max} = \infty$. Hence, $C^* = \frac{C_{min}}{C_{max}} = 0$ and the applicable *NTU*-ε relation gives $NTU = -\ln(1-\varepsilon) = -\ln(1-0.5) = 0.693$; $C_{min} = m_c c_c = 12 \times 4179 = 50148$ W/K; hence, $NTU = \frac{UA}{C_{min}}$ gives $A = \frac{NTU\times C_{min}}{U} = \frac{0.693\times50148}{1571.5} = 22.11$ m^2; from $A = N\pi DL$, $L = \frac{A}{N\pi D} = \frac{22.11}{160\times\pi\times0.025} = 1.76$ m.]

References

Kays WM, London AL (1964) Compact heat exchangers, 2nd edn. McGraw-Hill Book Co., New York
Kern DQ (1950) Process heat transfer. McGraw-Hill, New York
Standards of Tubular Exchanger Manufacturers Association (1999), 8th edn., New York

15 Mass Transfer

15.1 Introduction

Mass transfer can take place by molecular diffusion and convection. The molecular diffusion is mass transfer at microscopic level due to a concentration difference, i.e., from the region of high concentration to the region of low concentration. The diffusion may also take place due to pressure differences (pressure diffusion) or temperature differences (thermal diffusion). However, these forms of diffusion will not be discussed here. The molecular diffusion is analogous to heat conduction, while the mass transfer in a flowing fluid, termed as convective mass transfer, is analogous to convective heat transfer.

We can understand the diffusion process by considering some simple examples frequently encountered in our daily life. Air freshener/deodorant, perfume, flower fragrance, cigarette smoke, etc. distribute over a room by diffusion. Consider a sugar cube put in a glass of water that is not stirred. As time passes, the sugar dissolves slowly and will distribute over the water even in the absence of the convection currents. This has happened by the diffusion of sugar molecules in water from lower to the upper part of the water, i.e. from the region of high concentration to the region of low concentration.

The terms mass convection or convective mass transfer describes the process of mass transfer between a surface and a moving fluid. Just like heat convection, mass convection can be free or forced, laminar or turbulent and internal or external.

Consider the system, shown in Fig. 15.1, where two gases *B* and *C*, at the same temperature and pressure, are initially separated by a thin partition. When the partition is removed, the two gases diffuse through each other. During diffusion, the molecules of gas *B* move from the zone of high concentration of component *B* to the zone of lower concentration of the component, i.e. from left to right across the plane where the partition was present. A higher concentration means more molecules per unit volume. Similarly the molecules of the gas *C* move from right to left. Figure 15.1 also shows the concentration profile of constituent *B* at a certain instant shortly after the removal of the partition. In this example, the concentration gradient is the cause of the diffusion. It is to note that the diffusion basically happens because of the random motion of the molecules.

R. Karwa, *Heat and Mass Transfer*,
https://doi.org/10.1007/978-981-15-3988-6_15

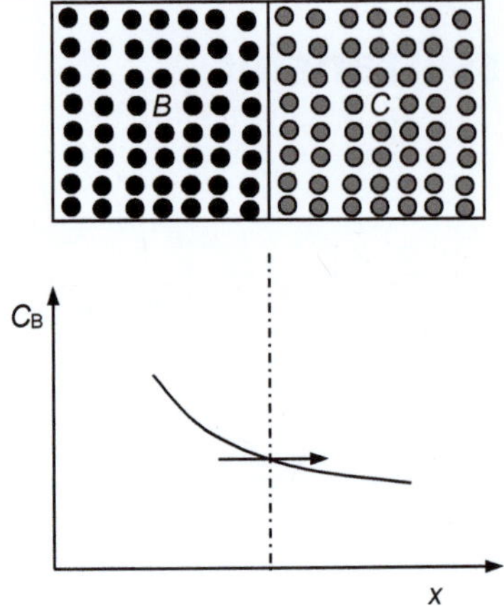

Fig. 15.1 Diffusion of gases

15.2 Fick's Law of Diffusion

The rate of mass diffusion $\dot{m}_B$ of component B in direction x is proportional to the concentration gradient of the component B in that direction, i.e.

$$\dot{m}_B = -D_{\mathrm{BC}} A \frac{dC_B}{dx} \tag{15.1}$$

where

A area through which mass is diffusing
C_{B} *mass concentration* of component B per unit volume
$\frac{dC_B}{dx}$ concentration gradient in direction x.

The proportionality factor D_{BC} is termed as diffusion coefficient or diffusivity (units m^2/s[1]) for the mixture of components B and C.

The negative sign in the equation has been introduced because the mass diffusion is in the direction of decreasing concentration.

Similarly, diffusion of constituent C is given by

$$\dot{m}_C = -D_{\mathrm{CB}} A \frac{dC_C}{dx}$$

[1]The units of mass diffusion coefficient can be determined from Eq. (15.1):

$$\begin{aligned} D &= \frac{\dot{m}_B}{A} \cdot \frac{dx}{dC_B} \\ &= \frac{kg}{s} \cdot \frac{1}{m^2} \cdot \frac{m}{(kg/m^3)} = \frac{m^2}{s} \end{aligned}$$

Physically, the diffusion coefficient implies the mass of the substance diffusing through a unit surface area in a unit time at a concentration gradient of unity. Diffusion coefficients are commonly determined experimentally. In gases, the diffusion coefficient is high because the molecules can move easily while it is lowest for the solids; the diffusion in liquids and solids is with a greater difficulty because of the influence of the molecular force fields and increased number of collisions that lead to the less freedom of movement of the molecules.

The law expressed by Eq. (15.1) is based on experimental investigation. It was first extended by A. Fick and hence is known as Fick's law. The law states that the rate of diffusion of a constituent per unit area at a location is proportional to the concentration gradient of that constituent at that location.

The Fick's law, which describes the transport of mass due to concentration gradient, is analogous to the Fourier's law of heat conduction, which gives transport of heat due to the temperature gradient,

$$\frac{Q}{A} = -k\frac{dt}{dx}$$

and to the Newton's law of viscosity for the transport of momentum across fluid layers due to the velocity gradient,

$$\tau = \mu\frac{du}{dy}$$

We know that the Prandtl number (= $\mu c_p/k$) is the connecting link between the velocity and temperature fields and when Pr = 1, the velocity and temperature distributions are similar. There is a similarity of the above equations of Fourier and Newton with Eq. (15.1). Hence, we can conclude that the velocity and concentration profiles will be similar when $v = D$ or $v/D = 1$. Similarly, when $\alpha = D$ or $\alpha/D = 1$, the temperature and concentration distributions will have the same profile.

The dimensionless ratio v/D is known as *Schmidt number* Sc, and the ratio α/D is called the *Lewis number* Le.

15.2.1 Fick's Law for Gases in Terms of Partial Pressures

For gases, the Fick's law can be expressed in terms of the partial pressures. Applying the characteristic gas equation for the gas B, we get

$$p_B = \rho_B R_B T = \rho_B\left(\frac{R_o}{M_B}\right)T$$

or

$$\rho_B = \frac{p_B M_B}{R_o T}$$

Knowing that $\rho_B = C_B$, Eq. (15.1) transforms to

$$\frac{m_B}{A} = -D_{\mathrm{BC}} \frac{d}{dx}\left(\frac{p_B M_B}{R_o T}\right)$$

For isothermal diffusion, the above equation can be written as

$$\frac{m_B}{A} = -D_{\mathrm{BC}} \left(\frac{M_B}{R_o T}\right) \frac{dp_B}{dx}. \tag{15.2}$$

The gases B and C diffuse through each other at the same time. Hence, it is expected that the diffusion coefficient for the diffusion of B into C, D_{BC}, must equal the diffusion coefficient D_{CB} for the diffusion of C into B. This can be proved.

The Fick's law for the isothermal diffusion can be written for the gases B and C as

$$N_B = \frac{m_B}{M_B} = -D_{\mathrm{BC}} \left(\frac{A}{R_o T}\right) \frac{dp_B}{dx} \tag{i}$$

$$N_C = \frac{m_C}{M_C} = -D_{\mathrm{CB}} \left(\frac{A}{R_o T}\right) \frac{dp_C}{dx} \tag{ii}$$

where N is the number of moles.

When the total pressure of the system is constant,

$$p = p_B + p_C$$

or

$$0 = \frac{dp_B}{dx} + \frac{dp_C}{dx}$$

or

$$\frac{dp_B}{dx} = -\frac{dp_C}{dx} \tag{iii}$$

In equimolar counter-diffusion condition, the steady state molar diffusion rates of components B and C are equal, i.e. each molecule of component B is replaced by a molecule of component C and vice versa. Hence,

$$N_B = -N_C$$

Substitution from Eqs. (i) and (ii) gives

$$D_{\mathrm{BC}} \frac{A}{R_o T} \frac{dp_B}{dx} = D_{\mathrm{CB}} \frac{A}{R_o T} \frac{dp_C}{dx}$$

Using Eq. (iii), we get

$$D_{BC} = D_{CB} = D \tag{15.3}$$

Thus, the diffusion coefficient for the diffusion of B into C, D_{BC}, equals the diffusion coefficient D_{CB} for the diffusion of C into B.

Equation (15.2) can be integrated to obtain mass flux of component B

$$\frac{m_B}{A} = -D\left(\frac{M_B}{R_o T}\right)\frac{p_{B2} - p_{B1}}{\Delta x} \tag{15.4}$$

15.2.2 Fick's Law on Mass Basis and Mole Basis

We can express the concentration of a species in two ways:

(a) **Mass Basis**

For volume V, the densities ρ_i and ρ of a species i and of the mixture, respectively, are given by

$$\rho_i = \frac{m_i}{V} \tag{i}$$

and

$$\rho = \frac{m}{V} = \sum \frac{m_i}{V} = \sum \rho_i \tag{ii}$$

Mass concentration can be expressed in dimensionless form by using mass fraction w as Mass fraction of species i,

$$w_i = \frac{m_i}{m} = \frac{m_i/V}{m/V} = \frac{\rho_i}{\rho} \tag{iii}$$

and

$$\sum w_i = 1 \tag{iv}$$

(b) **Mole Basis**

On a mole basis, the concentration is expressed in the terms of *molar concentration*. Again considering volume V of species i and the mixture, we have the molar concentration

$$C_i = \frac{N_i}{V} \tag{v}$$

and

$$C = \frac{N}{V} = \sum \frac{N_i}{V} = \sum C_i \tag{vi}$$

Molar concentration can be expressed in dimensionless form by using mole fraction y as Mole fraction of species i,

$$y_i = \frac{N_i}{N} = \frac{N_i/V}{N/V} = \frac{C_i}{C} \tag{vii}$$

and

$$\sum y_i = 1 \tag{viii}$$

It is to note that mole number $N = m/M$ and density $\rho = CM$, where M is molecular weight. The mass and mole fractions are related to each other:

$$w_i = \frac{\rho_i}{\rho} = \frac{C_i M_i}{CM} = y_i \frac{M_i}{M} \tag{ix}$$

Fick's law on mass basis is expressed as

$$\frac{\dot{m}_B}{A} = -\rho D_{\mathrm{BC}} \frac{d(\rho_B/\rho)}{dx} = -\rho D_{\mathrm{BC}} \frac{dw_B}{dx} \quad (\mathrm{kg/s\ m^2}) \tag{15.5}$$

On the mole basis, the law is expressed as

$$\frac{\dot{N}_B}{A} = -C D_{\mathrm{BC}} \frac{d(C_B/C)}{dx} = -C D_{\mathrm{BC}} \frac{dy_B}{dx} \quad (\mathrm{mol/s\ m^2}) \tag{15.6}$$

For constant mixture density ρ and constant molar concentration C, the above relations simplify to

$$\frac{\dot{m}_B}{A} = -D_{\mathrm{BC}} \frac{d\rho_B}{dx} \quad (\mathrm{kg/s\ m^2}) \tag{15.7}$$

and

$$\frac{\dot{N}_B}{A} = -D_{\mathrm{BC}} \frac{dC_B}{dx} \quad (\mathrm{mol/s\ m^2}). \tag{15.8}$$

15.3 Diffusion Coefficient

Diffusion coefficient is a transport property, which depends on temperature T, pressure p and nature of the component in the system. Typical values of the diffusion coefficient (binary diffusion) are given in Table 15.1.

Table 15.1 Inter-diffusion (or binary diffusion) coefficient D_{AB} at 293 K, 101 kPa

Diffusing substance A	Medium B	D_{AB}, m^2/s
H_2O (vapour)	Air	0.24×10^{-4}
CO_2	Air	0.16×10^{-4}
CO	Air	0.21×10^{-4}
CH_4	Air	0.10×10^{-4}
H_2	Air	0.63×10^{-4}
O_2	Air	0.20×10^{-4}
CO_2	Water	1.92×10^{-9}
Air	Water	2.50×10^{-9}
H_2	Water	5.50×10^{-9}
NH_3	Water	1.80×10^{-9}
Carbon	Iron	30×10^{-12} at 1000 K 150×10^{-12} at 1180 K

Assuming ideal gas behaviour, kinetic theory of gases has been used to show that the pressure and temperature dependence of the diffusion coefficient for a binary mixture of gases may be estimated from the relation

$$D_{AB} \propto p^{-1} T^{3/2}. \tag{15.9a}$$

or

$$\frac{D_{AB,1}}{D_{AB,2}} = \frac{p_2}{p_1} \left(\frac{T_1}{T_2} \right)^{3/2}. \tag{15.9b}$$

15.4 Diffusion of Vapour Through a Stationary Gas: Stefan Law

Consider isothermal evaporation of water contained in a vessel as shown in Fig. 15.2. Since the free surface of the water is exposed to the air at plane 1-1, the water evaporates from its surface and diffuses through the stagnant layer of air above the water surface. To analyze the phenomenon, some simplifying assumptions are made.

1. The system is isothermal
2. The system is in steady state condition
3. There is slight air movement over the vessel to remove the water diffusing across the plane 22.
4. Air movement over the vessel does not cause turbulence or affect the air concentration profile in the vessel.
5. Both air and water vapour behave as perfect gases.

In the steady state condition, the upward diffusion of the vapour is balanced by a downward diffusion of air so that the concentration profile remains undisturbed. Since at the surface of the water (plane 1-1), there cannot be downward movement of air, there must be a bulk mass movement upwards to balance the downward diffusion of the air. Thus, this bulk mass movement produces an additional mass flux of water vapour.

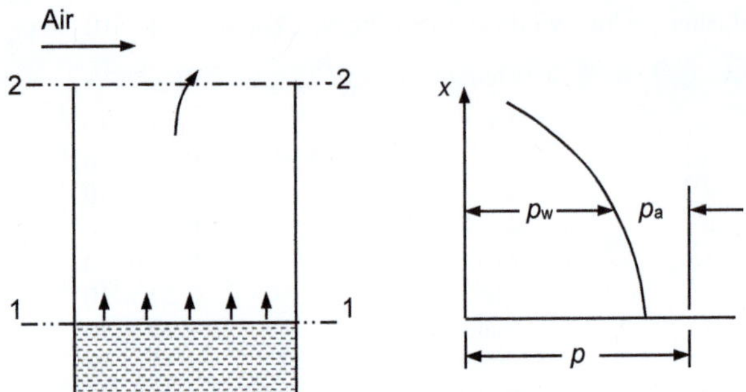

Fig. 15.2 Diffusion of vapour through a stationary gas

The downward diffusion of the air is

$$\dot{m}_a = -DA\left(\frac{M_a}{R_o T}\right)\frac{dp_a}{dx} \tag{i}$$

where A is the area of cross-section of the vessel and subscript, a, refers to the air. This diffusion must be balanced by the bulk-mass movement upwards, which is

$$-\rho_a Av = -\left(\frac{p_a M_a}{R_o T}\right)Av \tag{ii}$$

where $\rho_a = \dfrac{p_a M_a}{R_o T}$ is the density from the characteristic gas equation and v is the bulk-mass velocity in the upward direction.

From Eqs. (i) and (ii),

$$v = \frac{D}{p_a}\frac{dp_a}{dx} \tag{iii}$$

The mass diffusion of water vapour upwards is

$$\dot{m}_w = -DA\left(\frac{M_w}{R_o T}\right)\frac{dp_w}{dx} \tag{iv}$$

and the bulk transport of water vapour is

$$\rho_w Av = \left(\frac{p_w M_w}{R_o T}\right)Av = \left(\frac{p_w M_w}{R_o T}\right)A\left(\frac{D}{p_a}\frac{dp_a}{dx}\right) \tag{v}$$

Thus, the total transport of the water vapour, from Eqs. (iv) and (v), is

$$(\dot{m}_w)_{\text{total}} = -DA\left(\frac{M_w}{R_o T}\right)\frac{dp_w}{dx} + \left(\frac{p_w M_w}{R_o T}\right)A\left(\frac{D}{p_a}\frac{dp_a}{dx}\right) \tag{vi}$$

From the Dalton's law,

$$p_a + p_w = p \tag{vii}$$

Since p = constant, differentiation gives

$$\frac{dp_a}{dx} = -\frac{dp_w}{dx} \tag{viii}$$

Using Eqs. (vii) and (viii), we get the total mass flow rate of water from Eq. (vi) as

$$(\dot{m}_w)_{\text{total}} = -DA\left(\frac{M_w}{R_oT}\right)\frac{p}{p - p_w}\frac{dp_w}{dx} \tag{15.10}$$

This relation is called as *Stefan's law*. Equation (15.10) can be integrated to give

$$(\dot{m}_w)_{\text{total}} = DA\left(\frac{pM_w}{R_oT(x_2 - x_1)}\right)\ln\left(\frac{p - p_{w2}}{p - p_{w1}}\right) = \left(\frac{DApM_w}{R_oT(x_2 - x_1)}\right)\ln\left(\frac{p_{a2}}{p_{a1}}\right) \tag{15.11}$$

The Stefan law can be utilized to experimentally determine the diffusion coefficient, refer Q. 15.1.

Example 15.1 A 20 mm diameter tube is partially filled with water at 20°C. The distance of the water surface from the open end of the tube is 300 mm. Dry air at 20°C and 100 kPa is blowing over the open end of the tube so that water vapour diffusing to the open end of the tube is removed immediately. Determine the amount of water that will evaporate in 30 days.

Solution

From Eq. (15.11),

$$(\dot{m}_w)_{\text{total}} = DA\left[\frac{pM_w}{R_oT(x_2 - x_1)}\right]\ln\left(\frac{p - p_{w2}}{p - p_{w1}}\right)$$

where

D $= 0.24 \times 10^{-4}$ m²/s, Table 15.1,
A $= \pi/4\ D^2 = \pi/4 \times (0.02)^2 = 3.1416 \times 10^{-4}$ m²,
R_o = 8.314 kPa m³/(kmol K),
p = 100 kPa,
T = 293 K,
$\frac{p}{R_oT}$ $= \frac{100}{8.314 \times 293} = 0.04105\ \text{kmol/m}^3$,
M_w = 18 kg/kmol,
$x_2 - x_1$ = 0.3 m,
p_{w2} = 0 (dry air),
p_{w1} = saturation pressure at 20°C = 2.339 kPa (from steam table)

Substitution gives,

$$(\dot{m}_w)_{total} = 0.24 \times 10^{-4} \times 3.1416 \times 10^{-4} \times \left(\frac{0.04105 \times 18}{0.3}\right) \ln \left(\frac{100 - 0}{100 - 2.339}\right) = 4.395 \times 10^{-10}\ \text{kg/s}$$

$$= 4.395 \times 10^{-10} \times 30 \times 24 \times 3600 \times 1000 = 1.139\ \text{g in 30 days.}$$

15.5 Convective Mass Transfer

The convective mass transfer is defined by the following equation.

$$\dot{m}_B = h_m A \Delta C_B, \tag{15.12}$$

where h_m is the mass transfer coefficient. The equation is similar to the convective heat transfer equation defined earlier.

As in the case of convection heat transfer, when the fluid is forced to flow, it is termed as mass transfer in forced convection and when the mass transfer takes place only because of difference in the density due to pressure or temperature variations, it is termed as mass transfer in free convection. The value of the mass transfer coefficient depends on the type of flow (laminar or turbulent), the physical properties of the materials involved, geometry of the system and the concentration difference ΔC_B.

In the laminar flow, the fluid elements move in parallel paths while in the case of turbulent flow, the mass transfer takes place due to the random motion of the flowing fluid and is termed as turbulent diffusion. Even in turbulent flows, a thin layer develops close to the surface where the convective mass transfer is very low and the molecular diffusion dominates.

The mass transfer coefficient can be correlated to the diffusion coefficient. For one-dimensional steady state diffusion across this thin layer of thickness Δx, we can write, from the Fick's law,

$$\frac{\dot{m}_B}{A} = D\frac{\Delta C_B}{\Delta x} \tag{i}$$

From Eqs. (15.12) and (i), we have

$$h_m = \frac{D}{x_2 - x_1} \tag{15.13}$$

15.5.1 Convective Mass Transfer Equation in Terms of Partial Pressure Difference

We have

$$m_B = h_m A \Delta C_B = h_m A (C_{B1} - C_{B2})$$

Since $C_B = \rho_B = p_B/RT$, the mass transfer equation becomes

$$m_B = h_m A\left(\frac{p_{B1}}{RT_1} - \frac{p_{B2}}{RT_2}\right)$$

For an isothermal process, $T_1 = T_2 = T$ (say), then the mass transfer equation in the terms of partial pressures is

$$m_B = h_m A \frac{1}{RT}(p_{B1} - p_{B2}) = h_{mp} A(p_{B1} - p_{B2}) \tag{15.14}$$

where $h_{mp} = h_m \times \frac{1}{RT}$.

15.6 Dimensional Analysis Applied to Convective Mass Transfer

15.6.1 Forced

The convective mass transfer coefficient is function of diffusion coefficient D, density ρ, characteristic dimension l, viscosity μ and velocity U, i.e.

$$h_m = f(D, \rho, l, \mu, U)$$

or

$$f(D, \rho, h_m, l, \mu, U) = 0 \tag{15.15}$$

There are six variables and three fundamental units hence we expect (6–3), i.e. 3 π-terms. Taking D, ρ and l as repeated variables, the π-terms can be established as follows.

$$\pi_1 = D^a \rho^b l^c h_m \tag{i}$$

or

$$M^0 L^0 T^0 = \left(\frac{L^2}{T}\right)^a \left(\frac{M}{L^3}\right)^b (L)^c \frac{L}{T} \tag{ii}$$

Equating the indices of the fundamental dimensions on both sides, we obtain

M: $0 = b$
L: $0 = 2a - 3b + c + 1$
T: $0 = -a - 1$

Solution gives

$$a = -1,\ b = 0,\ c = 1.$$

Substitution gives

$$\pi_1 = \frac{h_m l}{D} \tag{iii}$$

The non-dimensional group $h_m l/D$ is termed as *Sherwood number* Sh.

Following the above approach, we have

$$\pi_2 = D^a \rho^b l^c \mu \tag{iv}$$

or

$$M^0 L^0 T^0 = \left(\frac{L^2}{T}\right)^a \left(\frac{M}{L^3}\right)^b (L)^c \frac{M}{LT}$$

Equating the indices,

M: $0 = b + 1$
L: $0 = 2a - 3b + c - 1$
T: $0 = -a - 1$

Solution of above equations gives

$$a = -1,\ b = -1,\ c = 0.$$

This gives

$$\pi_2 = \frac{\mu}{\rho D} = \frac{\nu}{D} = \mathrm{Sc} \tag{v}$$

The non-dimensional number ν/D, which is the dimensionless combination of the diffusivity and viscosity, is the *Schmidt number* Sc.

Similarly,

$$\pi_3 = D^a \rho^b l^c U \tag{vi}$$

or

$$M^0 L^0 T^0 = \left(\frac{L^2}{T}\right)^a \left(\frac{M}{L^3}\right)^b (L)^c \frac{L}{T}$$

Equating the indices, we have

M: $0 = b$
L: $0 = 2a - 3b + c + 1$
T: $0 = -a - 1$

Solution of the above equations gives

$$b = 0, a = -1, c = 1.$$

This gives

$$\pi_3 = \frac{Ul}{D} \quad \text{(vii)}$$

From π_2 and π_3 terms, we get

$$\pi_4 = \frac{\pi_3}{\pi_2} = \frac{Ul}{D} \times \frac{D}{\nu} = \frac{Ul}{\nu} = \text{Re} \quad \text{(viii)}$$

Thus, the functional relation is

$$\frac{h_m l}{D} = f\left(\frac{Ul}{\nu}, \frac{\mu}{\rho D}\right)$$

The generalized correlation can be written as

$$\text{Sh} = \psi(\text{Re})\phi(\text{Sc}) \quad (15.16)$$

Sherwood number is similar to the Nusselt number in convection heat transfer. The Schmidt number Sc is important in the problems where both mass transfer and convection are present. This number plays a role similar to the role of the Prandtl number in convection heat transfer.

15.6.2 Free

In this case, the convective mass transfer coefficient

$$h_m = f(D, \rho, l, \mu, g\Delta\rho)$$

or

$$f(D, \rho, h_m, l, \mu, g\Delta\rho) = 0 \quad (15.17)$$

There are six variables and three fundamental units hence we expect (6–3), i.e. 3 π-terms. Taking D, ρ and l as repeated variables, the π-terms can be established as follows.

$$\pi_1 = D^a \rho^b l^c h_m \quad \text{(ix)}$$

or

$$M^0 L^0 T^0 = \left(\frac{L^2}{T}\right)^a \left(\frac{M}{L^3}\right)^b (L)^c \frac{L}{T}$$

Equating the indices of the fundamental dimensions on both sides, we obtain

M: $0 = b$
L: $0 = 2a - 3b + c + 1$
T: $0 = -a - 1$

Solution gives

$$a = -1,\ b = 0,\ c = 1.$$

Substitution gives

$$\pi_1 = \frac{h_m l}{D} = \text{Sh} \quad \text{(x)}$$

Following the above approach, we have

$$\pi_2 = D^a \rho^b l^c \mu \quad \text{(xi)}$$

or

$$M^0 L^0 T^0 = \left(\frac{L^2}{T}\right)^a \left(\frac{M}{L^3}\right)^b (L)^c \frac{M}{LT}$$

Equating the indices,

M: $0 = b + 1$
L: $0 = 2a - 3b + c - 1$
T: $0 = -a - 1$

Solution of the above equations gives

$$a = -1,\ b = -1,\ c = 0.$$

This gives

$$\pi_2 = \frac{\mu}{\rho D} = \frac{\nu}{D} = \text{Sc} \quad \text{(xii)}$$

Similarly,

$$\pi_3 = D^a \rho^b l^c g \Delta\rho \quad \text{(xiii)}$$

or

$$M^0 L^0 T^0 = \left(\frac{L^2}{T}\right)^a \left(\frac{M}{L^3}\right)^b (L)^c \frac{M}{L^2 T^2}$$

Equating the indices, we have

M: $0 = b + 1$
L: $0 = 2a - 3b + c - 2$
T: $0 = -a - 2$

Solution of the above equations gives

$$a = -2,\ b = -1,\ c = 3.$$

This gives

$$\pi_3 = \frac{l^3(g\Delta\rho)}{D^2\rho} \tag{xiv}$$

From π_2 and π_3 terms, we get

$$\begin{aligned}\pi_4 &= \frac{\pi_3}{(\pi_2)^2} = \frac{l^3(g\Delta\rho)}{D^2\rho} \times \left(\frac{D}{\nu}\right)^2 \\ &= \frac{l^3(g\Delta\rho)}{D^2\rho} \times \left(\frac{D}{\mu/\rho}\right)^2 = \frac{\rho l^3 g(\Delta\rho)}{\mu^2} = \mathrm{Gr_m}\end{aligned} \tag{xv}$$

Thus, the functional relation is

$$\frac{h_m l}{D} = f\left(\frac{\rho l^3 g(\Delta\rho)}{\mu^2}, \frac{\mu}{\rho D}\right)$$

The dimensionless group $\frac{\rho l^3 g(\Delta\rho)}{\mu^2}$ is called *mass Grashof number* and is analogous to the Grashof number in free convection heat transfer and the generalized correlation can be written as

$$\mathrm{Sh} = \psi(\mathrm{Gr_m})\phi(\mathrm{Sc}) \tag{15.18}$$

15.7 Mass Transfer Correlations

In Chaps. 8 and 9, single phase forced and free convection heat transfer correlations in the form Nu = *f*(Re, Pr) and Nu = *f*(Gr, Pr), respectively, have been presented. These correlations are also valid for mass transfer when the Nusselt number, Prandtl number and Grashof number are replaced by Sherwood number, Schmidt number and mass Grashof number, respectively. Typical results for the forced convection are presented below.

The convective mass transfer equations for local mass transfer coefficient in case of flow past a flat plate are

$$\text{Laminar :} \qquad \mathrm{Sh}_x = \frac{h_m x}{D} = 0.332(\mathrm{Re}_x)^{0.5}(\mathrm{Sc})^{0.33} \tag{15.19a}$$

$$\text{Turbulent:} \quad \text{Sh}_x = \frac{h_m x}{D} = 0.0296(\text{Re}_x)^{0.8}(\text{Sc})^{0.33} \tag{15.19b}$$

For averaged values, the equations are

$$\text{Laminar:} \quad \overline{\text{Sh}} = \frac{\bar{h}_m L}{D} = 0.664(\text{Re}_L)^{0.5}(\text{Sc})^{0.33} \tag{15.20a}$$

$$\text{Turbulent:} \quad \overline{\text{Sh}} = \frac{\bar{h}_m L}{D} = 0.037(\text{Re}_L)^{0.8}(\text{Sc})^{0.33} \tag{15.20b}$$

For vaporization of liquid into the air inside the smooth tube where the liquid wets the surface and the air is forced through the tube, the mass transfer coefficient equations (for the fully developed region) are

$$\text{Laminar:} \quad \text{Sh} = 3.66 \quad \text{for constant wall concentration.} \tag{15.21}$$

$$\begin{aligned} \text{Turbulent:} \quad \text{Sh} &= 0.023\text{Re}^{0.83}\text{Sc}^{0.44}. \\ &\text{for} \quad 2000 < \text{Re} < 35000 \\ &\text{for} \quad 0.6 < \text{Sc} < 2.5 \end{aligned} \tag{15.22}$$

15.8 Reynolds and Colburn (or Chilton-Colburn) Analogies

These analogies for convection heat transfer have been extended to the mass transfer problems. The *Reynolds analogy* is written as

$$\begin{aligned} \frac{\text{Sh}}{\text{ReSc}} &= \frac{C_f}{2} \qquad \text{for Sc} = 1 \\ &= \frac{f}{2}.\text{(for pipe flow)} \end{aligned} \tag{15.23}$$

The group of non-dimensional terms on the left is known as *Stanton number* St_m, which is equal to (h_m/U). Thus

$$\text{St}_m = \frac{C_f}{2} = \frac{f}{2}. \tag{15.24}$$

The Reynolds-Colburn analogy for heat transfer, refer Chap. 7, for flow over the smooth plate is

$$j_H = \text{St}\,\text{Pr}^{2/3} = \frac{C_f}{2}.$$

The *Chilton-Colburn analogy* gives

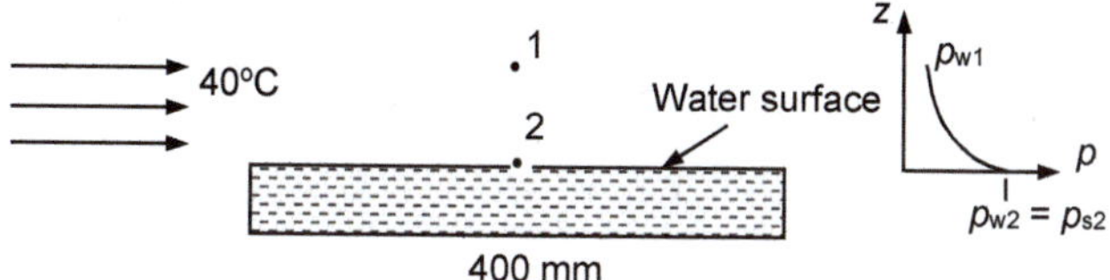

Fig. 15.3 Example 15.2

$$j_m = \text{St}_m \times \text{Sc}^{2/3} = \frac{C_f}{2} \quad \text{for } 0.5 < \text{Sc} < 3000 \tag{15.25}$$

where j_m is known as Colburn "*j*-factor" for mass transfer.

Example 15.2 Air at a velocity of 2.5 m/s is flowing along a horizontal water surface. The temperature of the air (relative humidity RH = 20%) is 40°C and the temperature of the water on the surface is 20°C. Length of the tray along the flow direction is 400 mm and it is 500 mm wide. Calculate the amount of water evaporated per hour. Atmospheric pressure = 101.326 kPa and diffusion coefficient of water vapour in air = 0.26×10^{-4} m^2/s.

Solution

In this example of mass transfer, immediately above the water surface, there is a large content of water vapour in the air, while further away from the surface there is much less. As a result of this, a concentration difference (partial pressure difference for the vapour, refer Fig. 15.3) is established near the water surface and there is macroscopic relative movement between water vapour and air.

It is to note that, when the air flows with a significantly high velocity above the surface of the water, the amount of water vapour transferred to the air will be far greater than that transferred in the same time interval in quiescent air. The mass transfer is basically by molecular diffusion only in case of quiescent air.

From the steam table, partial pressure of water vapour at 40°C, p_{s1} = 7.384 kPa. Since the air away from the water surface is unsaturated, the partial pressure of the water vapour is

$$p_{w1} = p_{s1} \times RH = 7.384 \times RH = 7.384 \times 0.2 = 1.4768 \text{ kPa}$$

and the partial pressure of water vapour at 20°C, p_{w2} = 2.339 kPa because the air near the water surface is saturated.

From air properties (Table A5, Appendix), viscosity of air at 30°C is:

$$\nu = \mu/\rho \approx 1.6 \times 10^{-5}\text{m}^2/\text{s}.$$

Flow Reynolds number is

$$\text{Re}_L = \frac{Ul}{\nu} = \frac{2.5 \times 0.4}{1.6 \times 10^{-5}} = 62500.$$

Flow is laminar.

The Schmidt number

$$\text{Sc} = \frac{\nu}{D} = \frac{1.6 \times 10^{-5}}{0.26 \times 10^{-4}} = 0.6154.$$

From the equation of convective mass transfer equation for laminar flow, we have

$$\begin{aligned}\overline{\text{Sh}} = \frac{h_m L}{D} &= 0.664(\text{Re}_L)^{0.5}(\text{Sc})^{0.33} \\ &= 0.664(62500)^{0.5}(0.6154)^{0.33} = 141.2.\end{aligned}$$

Mass transfer coefficient,

$$\begin{aligned}h_m &= \frac{\overline{\text{Sh}} \times D}{L} = \frac{141.2 \times 0.26 \times 10^{-4}}{0.4} = 9.178 \times 10^{-3}\text{m/s}. \\ m_B &= h_m A\left(\frac{p_{w1}}{RT_1} - \frac{p_{w2}}{RT_2}\right) \\ &= 9.178 \times 10^{-3} \times (0.4 \times 0.5)\left(\frac{2.339 \times 10^3}{461 \times 293} - \frac{1.4768 \times 10^3}{461 \times 313}\right) \\ &= 1.3 \times 10^{-5}\text{kg/s}.\end{aligned}$$

Example 15.3 A gas flows with uniform velocity u_m through a tube of cross-section A_c with evaporation or sublimation at the tube surface. Determine the longitudinal distribution of mean vapour density.

Solution

Considering the infinitesimal control volume, refer Fig. 15.4, the conservation of species gives

$$\rho_{\text{Am}} u_m A_c + dm_A = \left[\rho_{\text{Am}} + \left(\frac{d\rho_{\text{Am}}}{dx}\right)dx\right] u_m A_c$$

or

$$\left(\frac{d\rho_{\text{Am}}}{dx}\right) dx u_m A_c = dm_A$$

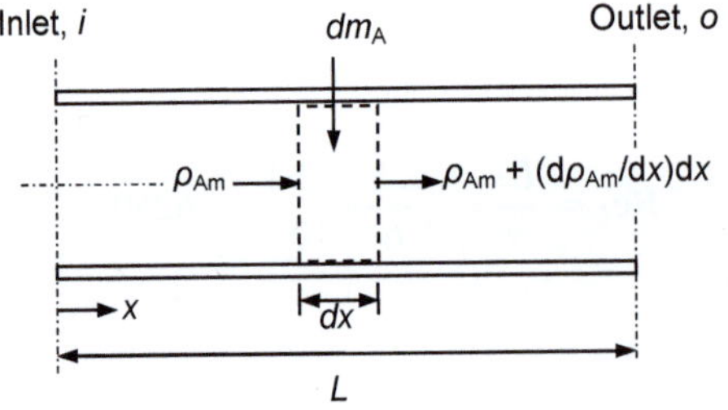

Fig. 15.4 Example 15.3

Substituting $dm_A = h_{mx}Pdx(\rho_{\mathrm{As}} - \rho_{\mathrm{Am}})$ and $u_m A_c = m/\rho$ we have

$$\left[\left(\frac{d\rho_{\mathrm{Am}}}{dx}\right)dx\right]\frac{m}{\rho} = h_{\mathrm{mx}}Pdx(\rho_{\mathrm{As}} - \rho_{\mathrm{Am}})$$

or

$$\frac{d\rho_{\mathrm{Am}}}{\rho_{\mathrm{As}} - \rho_{\mathrm{Am}}} = \frac{h_{\mathrm{mx}}\rho P}{m}dx$$

We carry out integration from inlet to outlet

$$\int_{\rho_{\mathrm{Ami}}}^{\rho_{\mathrm{Amx}}} \frac{d\rho_{\mathrm{Am}}}{\rho_{\mathrm{As}} - \rho_{\mathrm{Am}}} = \frac{\rho P}{m}\int_{o}^{x} h_{\mathrm{mx}}dx$$

or

$$\ln\left(\frac{\rho_{\mathrm{As}} - \rho_{\mathrm{Amx}}}{\rho_{\mathrm{As}} - \rho_{\mathrm{Ami}}}\right) = -\frac{\rho P}{m}\bar{h}_m x$$

or

$$\frac{\rho_{\mathrm{As}} - \rho_{\mathrm{Amx}}}{\rho_{\mathrm{As}} - \rho_{\mathrm{Ami}}} = \exp\left(-\frac{\rho P}{m}\bar{h}_m x\right)$$

where $\bar{h}_m$ is average value of the mass transfer coefficient for length x.

Example 15.4 0.002 kg/s of dry air at 25°C and 1 atm is forced through a 15 mm diameter circular tube. Inner surface of the tube is wetted. Flow is fully developed. Determine the tube length required to reach 99% of saturation. Diffusion coefficient of water vapour in air $D_{\mathrm{AB}} = 0.26 \times 10^{-4}$ m^2/s.

Solution

Thermophysical properties of air at mean film temperature = (30 + 20)/2 = 25°C from Table A5 are $\rho = 1.1868$ kg/m^3 and $\mu = 1.8363 \times 10^{-5}$ kg/(m s). For water vapour (component A) at 25°C, $\rho_A = 1/v_g = 0.0231$.

The Schmidt number,

$$\mathrm{Sc} = \frac{\nu}{D} = \frac{\mu}{\rho D} = \frac{1.8363 \times 10^{-5}}{1.1868 \times 0.26 \times 10^{-4}} = 0.595.$$

From Example 15.3, for length L

$$\ln\left(\frac{\rho_{\mathrm{As}} - \rho_{\mathrm{Amo}}}{\rho_{\mathrm{As}} - \rho_{\mathrm{Ami}}}\right) = -\frac{\rho P}{m}h_m L$$

where h_m is average value of mass transfer coefficient. We can rewrite the above equation as

$$L = -\frac{m}{P\rho h_m}\ln\left(\frac{\rho_{\text{As}} - \rho_{\text{Amo}}}{\rho_{\text{As}} - \rho_{\text{Ami}}}\right) \quad \text{(i)}$$

Here $\rho_{\text{Ami}} = 0$ for dry air and $\rho_{\text{Amo}} = 0.99\ \rho_{\text{As}}$.
Reynolds number,

$$\text{Re} = \frac{\rho U d}{\mu} = \frac{md}{(\pi/4)d^2\mu} = \frac{4m}{\pi d\mu} = \frac{4 \times 0.002}{\pi \times 0.015 \times 1.8363 \times 10^{-5}} = 9245.$$

For fully developed turbulent flow, Eq. (15.22) may be used, i.e.

$$\text{Sh} = \frac{h_m d}{D_{\text{AB}}} = 0.023\text{Re}^{0.83}\text{Sc}^{0.44}$$

or

$$\begin{aligned} h_m &= \frac{D_{\text{AB}}}{d} \times 0.023\text{Re}^{0.83}\text{Sc}^{0.44} \\ &= \frac{0.26 \times 10^{-4}}{0.015} \times 0.023 \times 9245^{0.83} \times 0.595^{0.44} = 0.0621\ \text{m/s} \end{aligned}$$

From Eq. (i),

$$L = -\frac{m}{P\rho h_m}\ln\left(\frac{\rho_{\text{As}} - \rho_{\text{Amo}}}{\rho_{\text{As}} - \rho_{\text{Ami}}}\right) \quad \text{(i)}$$

or

$$\begin{aligned} L &= -\frac{0.002}{\pi \times 0.015 \times 1.1868 \times 0.0621} \times \ln\left(\frac{\rho_{\text{As}} - 0.99\rho_{\text{As}}}{\rho_{\text{As}} - 0}\right) \\ &= -\frac{0.002}{\pi \times 0.015 \times 1.1868 \times 0.0621} \times \ln(0.01) = 2.65\ \text{m}. \end{aligned}$$

Example 15.5 Dry air at t_∞ = 30°C and U_∞ = 5 m/s is flowing across a 25 mm tube with a water-saturated fibrous coating. The tube is maintained approximately at uniform surface temperature t_s of 20°C. Determine the heat rate from the external surface of the tube considering heat and mass transfer processes. Diffusion coefficient of water vapour in the air is 0.26×10^{-4} m²/s.

Solution

Heat rate,

$$\begin{aligned} q &= q_{\text{conv}} + q_{\text{evap}} \\ &= hA_s(t_s - t_\infty) + [h_m A_s(\rho_{\text{ws}} - \rho_{\text{w}\infty})]h_{\text{fg}} \end{aligned}$$

or

$$q = \left[h(t_s - t_\infty) + h_m(\rho_{ws} - \rho_{w\infty})h_{fg}\right]A_s \tag{i}$$

Thermophysical properties of air at mean film temperature = (30 + 20)/2 = 25°C from Table A5 are:

$\rho = 1.1868\ \text{kg/m}^3,\ \mu = 1.8363 \times 10^{-5}\ \text{kg/(m s)},\ k = 0.02608\ \text{W/(m K)}$ and $\text{Pr} = 0.709.$

Water density at tube surface (at 20°C saturation temperature), $\rho_{ws} = 1/v_g = 0.0173$ kg/m^3 and h_{fg} = 2453.5 kJ/kg from steam table. For dry air, $\rho_{w\infty} = 0$.

Reynolds number,

$$\text{Re} = \frac{\rho U_\infty d}{\mu} = \frac{1.1868 \times 5 \times 0.025}{1.8363 \times 10^{-5}} = 8079.$$

From Eq. (8.47) and Table 8.11 for cross flow over a cylinder, we have

$$\text{Nu} = \frac{hd}{k} = 0.193\text{Re}^{0.618}\text{Pr}^{1/3}$$

or

$$\begin{aligned} h &= \frac{k}{d}0.193\text{Re}^{0.618}\text{Pr}^{1/3} \\ &= \frac{0.02608}{0.025} \times 0.193 \times 8079^{0.618} \times 0.709^{1/3} = 46.65\ \text{W}/\left(\text{m}^2\,\text{K}\right). \end{aligned}$$

The Schmidt number,

$$\text{Sc} = \frac{v}{D} = \frac{\mu}{\rho D} = \frac{1.8363 \times 10^{-5}}{1.1868 \times 0.26 \times 10^{-4}} = 0.595.$$

From heat mass transfer analogy,

$$\text{Sh} = \frac{h_m d}{D} = 0.193\text{Re}^{0.618}\text{Sc}^{1/3}$$

or

$$\begin{aligned} h_m &= \frac{D}{d}0.193\text{Re}^{0.618}\text{Sc}^{1/3} \\ &= \frac{0.26 \times 10^{-4}}{0.025} \times 0.193 \times 8079^{0.618} \times 0.595^{1/3} = 0.04387\ \text{W}/\left(\text{m}^2\,\text{K}\right). \end{aligned}$$

Substitution in Eq. (i) gives

or

$$q = \left[46.65 \times (20 - 30) + 0.04387(0.0173 - 0) \times 2453.5 \times 10^3\right] \times \pi \times 0.025 \times 1.0$$

or

$$q = -36.64 + 146.22 = 109.58 \text{ W per unit length.}$$

Example 15.6 A wet cloth 1.2 m high and 1 m wide at 30°C is hanging in a factory shade with still dry air at 45°C and atmospheric pressure (refer Fig. 15.5). Determine the drying rate. Diffusion coefficient of water vapour in air is 0.26×10^{-4} m²/s.

Solution

From the steam table, partial pressure of water vapour at 30°C = 4.25 kPa = p_{w1} because the water vapour near the clothe surface is saturated. Partial pressure of air near the surface of the clothe $p_{a1} = p - p_{w1}$ = 101.325 – 4.25 = 97.075 kPa, where 1 atm = 101.325 kPa. We assume perfect gas behaviour of vapour and air.

Hence, the density of air

$$\rho_{a1} = \frac{p_{a1}}{RT_1} = \frac{97.075 \times 1000}{287 \times 303} = 1.1163 \text{ kg/m}^3$$

and

$$\rho_{a2} = \frac{p_{a2}}{RT_2} = \frac{101.325 \times 1000}{287 \times 318} = 1.1102 \text{ kg/m}^3$$

Similarly the density of water

$$\rho_{w1} = \frac{p_{w1}}{RT_1} = \frac{4.25 \times 1000}{461 \times 303} = 0.03043 \text{ kg/m}^3$$

and $\rho_{w2} = 0$ since air is dry.

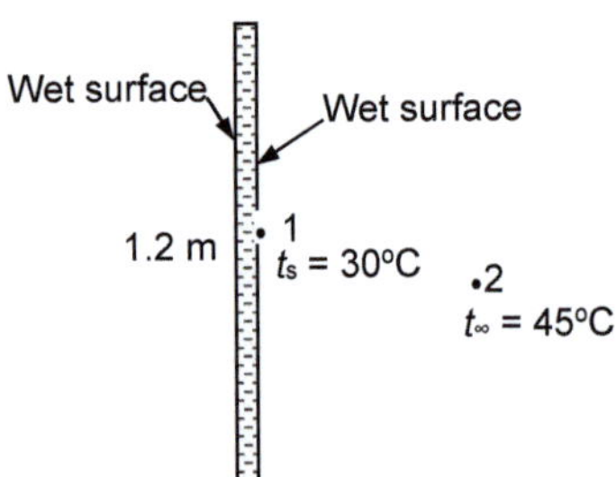

Fig. 15.5 Example 15.6

Density of air-water vapour mixture at clothe surface is

$$\rho_1 = \rho_{w1} + \rho_{a1} = 0.03043 + 1.1163 = 1.14673\,\text{kg/m}^3$$

and

$$\rho_2 = \rho_{w2} + \rho_{a2} = 0 + 1.1102 = 1.1102\,\text{kg/m}^3$$

Density in the film, ρ = (1.14673 + 1.1102)/2 = 1.128465 kg/m^3. Viscosity of air at (45 + 30)/2 = 37.5°C is 1.89375 × 10^{-5} N s/m^2.

$$\text{Gr}_\text{m} = \frac{\rho H^3 g(\Delta\rho)}{\mu^2} = \frac{1.128465 \times 1.2^3 \times 9.81 \times (1.14673 - 1.1102)}{(1.89375 \times 10^{-5})^2} = 1.948 \times 10^9.$$

The Schmidt number for $v = \mu/\rho$ = 1.66 × 10^{-5} from Table A5,

$$\text{Sc} = \frac{v}{D} = \frac{1.66 \times 10^{-5}}{0.26 \times 10^{-4}} = 0.638.$$

Analogy between heat and mass transfer applies. Hence, Eq. (9.6) can be written as

$$\text{Sh}_m = 0.59(\text{Gr}_\text{m}\text{Sc})^{1/4}$$

or

$$h_m = \frac{D}{H}\text{Sh}_m = \frac{0.26 \times 10^{-4}}{1.2} \times 0.59 \times (1.948 \times 10^9 \times 0.638)^{1/4} = 0.0024\,\text{m/s}.$$

The drying rate from both sides of the cloth is

$$\begin{aligned} m_w &= h_m A_s(\rho_{w1} - \rho_{w2}) = 0.0024 \times 2 \times 1.2 \times 1 \times (0.03043 - 0) = 17.53 \times 10^{-5}\text{kg/s} \\ &= 0.631\,\text{kg/hr}. \end{aligned}$$

Example 15.7 If the cloth in the previous example is having 1 m side as vertical, determine the drying rate.

Solution

Mass Grashof number for H = 1 m is

$$\text{Gr}_\text{m} = \frac{\rho H^3 g(\Delta\rho)}{\mu^2} = \frac{1.128465 \times 1^3 \times 9.81 \times (1.14673 - 1.1102)}{(1.89375 \times 10^{-5})^2} = 1.127 \times 10^9.$$

Mass transfer coefficient is

$$h_m = \frac{D}{H}\text{Sh}_m = \frac{0.26 \times 10^{-4}}{1.0} \times 0.59 \times (1.127 \times 10^9 \times 0.638)^{1/4} = 0.0025\,\text{m/s}.$$

The drying rate from both sides of the cloth is

$$m_w = h_m A_s(\rho_{w1} - \rho_{w2}) = 0.0025 \times 2 \times 1 \times 1.2 \times (0.03043 - 0) = 18.26 \times 10^{-5}\text{kg/s}$$
$$= 0.6574\ \text{kg/hr}.$$

This position of cloth is better because of the reduced average thickness of the boundary layer.

Example 15.8 If the cloth of Example 15.6 is lying horizontal, determine the drying rate.

Solution

We extend the analysis of the vertical surface of the previous example to the horizontal surface. The density data will be the same.

The characteristic length for the horizontal surface is

$$L = \frac{A_s}{P} = \frac{1.2 \times 1}{4.4} = 0.273.$$

$$\text{Gr}_\text{m} = \frac{\rho L^3 g(\Delta\rho)}{\mu^2} = \frac{1.128465 \times 0.273^3 \times 9.81 \times (1.14673 - 1.1102)}{(1.89375 \times 10^{-5})^2} = 2.294 \times 10^7.$$

The Schmidt number Sc = 0.638 from the previous example.

Using the analogy between heat and mass transfer, Eq. (9.18) can be written as

$$\text{Sh}_m = 0.15(\text{Gr}_\text{m}\text{Sc})^{1/3}$$

or

$$h_m = \frac{D}{L}\text{Sh}_m = \frac{0.26 \times 10^{-4}}{0.273} \times 0.15 \times (2.294 \times 10^7 \times 0.638)^{1/3} = 0.00349\ \text{m/s}$$

The drying rate is

$$m_w = h_m A_s(\rho_{w1} - \rho_{w2}) = 0.00349 \times 1.2 \times 1 \times (0.03043 - 0) = 12.74 \times 10^{-5}\ \text{kg/s}$$
$$= 0.459\ \text{kg/hr}.$$

15.9 Summary

An introduction and elementary treatment of mass transfer have been presented in this chapter. Mass transfer takes place by molecular diffusion and convection. The molecular diffusion is mass transfer at microscopic level due to a concentration difference, which is analogous to heat conduction, while the mass transfer in a flowing fluid, termed as convective mass transfer, is analogous to convective heat transfer.

Fick's law, which is based on experimental investigation, has been presented which states that the rate of diffusion of a constituent per unit area is proportional to the concentration gradient of that constituent. Diffusion coefficient has been defined, which is basically a transport property that depends on temperature T, pressure p and nature of the component in the system. Typical values of the diffusion coefficient (binary diffusion) are given in Table 15.1.

Fick's law is analogous to the Fourier's law of heat conduction and Newton's law of viscosity. The conditions for the similarity of concentration and velocity profiles, and temperature and concentration profiles have been discussed and the relevant dimensionless numbers Schmidt number Sc and Lewis number Le are presented.

For gases, the Fick's law has been expressed in terms of the partial pressures and also been expressed on the mass and mole basis.

Stefan law for diffusion of vapour through a stationary gas has been presented in Sect. 15.4, which can be utilized for experimental determination of the diffusion coefficient.

The convective mass transfer is defined as $\dot{m}_B = h_m A \Delta C_B$, where h_m is the mass transfer coefficient, which is similar to the convective heat transfer equation. Dimensional analysis has been used to determine functional relations for free and forced flow conditions, which is followed by the presentation of mass transfer correlations in Sect. 15.7. In Sect. 15.8, analogies for convection heat transfer have been extended to the mass transfer problems, which are termed as Reynolds and Colburn (or Chilton-Colburn) analogies. Applicable correlations in terms of non-dimensional terms for mass transfer Stanton number St_m and Colburn "j-factor" j_m have been presented.

Review Questions

15.1 State the Fick's law of diffusion.

15.2 Express the Fick's law for gases in terms of partial pressures for the diffusion of components B and C and vice versa.

15.3 What are the conditions for the similarity of concentration and velocity profiles, and temperature and concentration profiles?

15.4 Define and explain the physical significance of Schmidt, Sherwood and Lewis numbers.

15.5 Define mass transfer coefficient.

15.6 What is Reynolds analogy for mass transfer?

Problems

15.1 A 20 mm diameter Stefan tube is used to measure diffusion coefficient. It is partially filled with water at 20°C. The distance of the water surface from the open end of the tube is 300 mm. Dry air at 20°C and 100 kPa is blowing over the open end of the tube so that water vapour diffusing to the open end of the tube is removed immediately. If the amount of water evaporated in 30 days is 1.139 g, determine the diffusion coefficient.

[**Ans.** Refer Example 15.1, $D = 0.24 \times 10^{-4}$ m^2/s.]

15.2 Air at 25°C and 2 kg/hr is flowing through a 25 mm diameter tube having a thin water film on its inside surface. Determine the convection mass transfer coefficient. Diffusion coefficient of water vapour in air $D = 0.26 \times 10^{-4}$ m^2/s.

[**Ans.** Analogy between heat and mass transfer applies. For air at 25°C, $\mu = 1.8363 \times 10^{-5}$ from Table A5; $\mathrm{Re} = \dfrac{\rho U d}{\mu} = \dfrac{md}{(\pi/4)d^2\mu} = \dfrac{4m}{\pi d\mu} = \dfrac{4 \times 2/3600}{\pi \times 0.025 \times 1.8363 \times 10^{-5}} = 1541$; Flow is laminar. Assuming a fully developed flow, for constant wall concentration, Eq. (15.21) gives $\mathrm{Sh} = 3.66$; $h_m = \frac{D}{d}\mathrm{Sh} = \dfrac{0.26 \times 10^{-4}}{0.025} \times 3.66 = 0.0038\ \mathrm{m/s}$.]

16 Special Topic: Performance of Solar Air Heater

Nomenclature

A	Absorber plate area = WL, m^2;
B/S	Relative roughness length of discrete ribs;
D_h	Hydraulic diameter of duct = $4WH/[2(W + H)]$, m;
e	Rib height, m;
e^+	Roughness Reynolds number
e/D_h	Relative roughness height;
f	Fanning friction factor
g	Heat transfer function;
G	Mass flow rate per unit area of plate = m/A, kg/(s m^2);
h	Heat transfer coefficient, W/(m^2 K);
H	Air flow duct height (depth), m;
h_w	Wind heat transfer coefficient, W/(m^2 K);
I	Solar radiation on the collector plane, W/m;
k	Thermal conductivity of air, W/(m K);
L	Length of collector, m;
m	Mass flow rate = WLG, kg/s;
$\dot{m}$	Mass velocity = $m/(WH)$, kg/(s m^2);
Nu	Nusselt number;
p	Rib pitch, m;
P	Pumping power, W;
Pr	Prandtl number = $\mu c_p/k$;
p/e	Relative roughness pitch;
Q	Useful heat gain, W;
Q_b	Back loss, W;
Q_e	Edge loss, W;
Q_L	Heat loss, W;
Q_t	Top loss, W;
R	Roughness function;
Re	Reynolds number = $[m/(WH)]D_h/\mu$;

R. Karwa, *Heat and Mass Transfer*,
https://doi.org/10.1007/978-981-15-3988-6_16

s/l Relative roughness length of expanded metal mesh;
St Stanton number = Nu/RePr;
T_a Ambient temperature, °C, K;
T_{fm}, T_m Mean air temperature = $(T_o + T_i)/2$, °C, K;
T_i Inlet air temperature, °C, K;
T_{mpg} Mean of the plate and glass temperatures = $(T_p + T_{gi})/2$, °C, K;
T_o Outlet air temperature, °C, K;
T_p Mean plate temperature, °C, K;
T_{sky} Sky temperature, K;
U_L Overall loss coefficient, W/(m^2 K);
w Width of rib, m;
W Width of the duct, m.

Greek Symbols

α Absorptivity
α Rib angle with flow, deg
β Collector slope, deg
δ_{pg} Gap between the absorber plate and glass cover, m
ΔT Air temperature rise = $T_o - T_i$, °C, K
$\Delta\eta$ Change in thermal efficiency
$\Delta\eta_e$ Change in effective efficiency
ε Emissivity
η Thermal efficiency = Q/IA
η_e Effective efficiency
ϕ Chamfer angle, deg
μ Dynamic viscosity, Pa s
ν_{mpg} Kinematic viscosity of air at temperature T_{mpg}, m^2/s
$\tau\alpha$ Transmittance-absorptance product

Subscripts

b Duct bottom surface
g Glass
m Mean
p Plate

16.1 Introduction

Flat plate collector is the heart of a solar heat collection system designed for the delivery of heated fluid in the low to medium temperature range (5°–70°C above ambient temperature) for applications, such as water heating, space heating, drying and similar industrial applications. The flat plate collectors absorb both beam and diffuse radiation. The absorbed

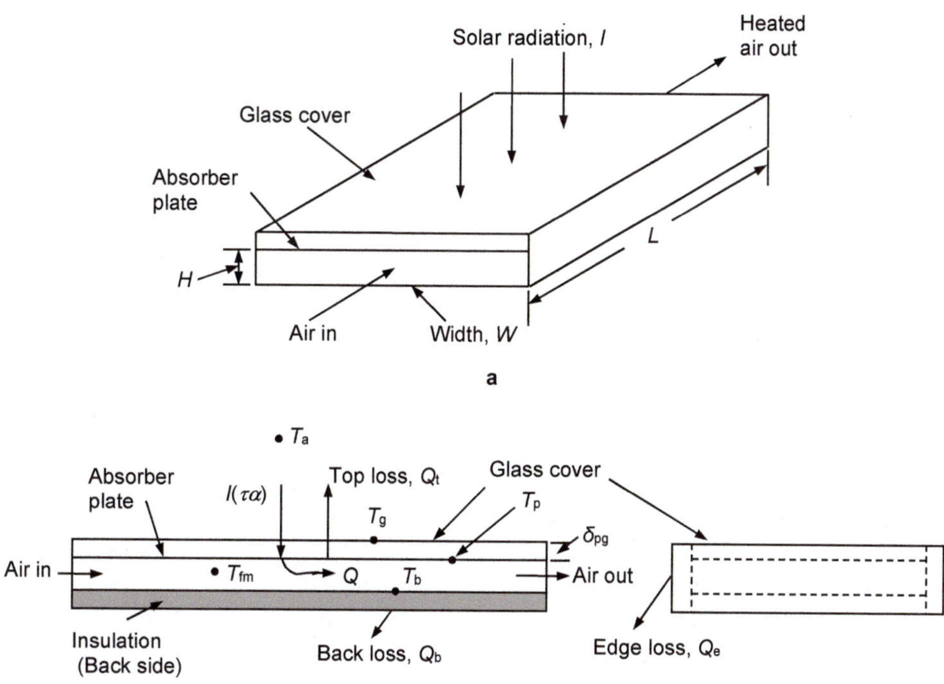

Fig. 16.1 **a** Schematic diagram of a solar air heater, **b** heat balance

radiation is converted into heat which is transferred to water or air flowing through the collector tubes or duct, respectively. Such collectors do not require tracking of the sun and little maintenance is required.

The conventional flat plate solar air heater, shown in Fig. 16.1, consists of a flat blackened absorber plate, a transparent cover (such as a glass cover) at the top and insulation at the bottom and on the sides. The air to be heated flows through the rectangular duct below the absorber plate. The glass cover transmits a major part of the solar radiation incident upon it to the absorber plate where it is converted into heat. The glass is, however, opaque to long-wavelength radiation and thus it does not allow the infrared radiation from the heated absorber plate to escape.

Karwa et al. (2002) have presented the following deductions of heat collection rate and pumping power equations for the flat plate solar air heater.

The Reynolds Number ($\dot{m}D_h/\mu$) for a solar air heater duct can be expressed in a simple form as presented below considering the fact that for a rectangular duct of high aspect ratio (typically duct width W is of the order of 1 m and height H = 5–20 mm) the hydraulic diameter $D_\mathrm{h} \approx 2H$.

$$\mathrm{Re} = \frac{\dot{m}D_h}{\mu} = \left(\frac{1}{\mu}\right)\left(\frac{AG}{WH}\right)2H = \left(\frac{1}{\mu}\right)\left(\frac{WLG}{WH}\right)2H = \frac{2GL}{\mu} \tag{16.1}$$

where A (=WL) is the absorber plate area, G is the mass flow rate of air per unit area of the plate and $\dot{m}[= WLG/(WH)]$ is the mass velocity.

Using the Dittus and Boelter correlation ($Nu = 0.024\ Re^{0.8}\ Pr^{0.4}$) for the Nusselt number,

$$h = \frac{Nuk}{D_h} = \frac{(0.024Re^{0.8}\,Pr^{0.4})k}{2H} \propto \left(\frac{GL}{\mu}\right)^{0.8}\left(\frac{1}{H}\right) \tag{16.2}$$

The useful heat gain Q can be expressed in the following form.

$$Q = hA\Delta T \propto (G)^{0.8}\left(\frac{L}{H}\right)\left(\frac{1}{L}\right)^{0.2} WL \tag{16.3}$$

Pressure loss δp and pumping power P equations for flow in the rectangular cross-section duct of a solar air heater can be written as

$$\delta p = \left(\frac{4fL}{2\rho D_h}\right)\dot{m}^2 = \left(\frac{4fL}{4\rho H}\right)\left(\frac{WLG}{WH}\right)^2 = \left(\frac{fG^2}{\rho}\right)\left(\frac{L}{H}\right)^3 \tag{16.4}$$

and

$$P = \left(\frac{m}{\rho}\right)\delta p = \left(\frac{WLG}{\rho}\right)\left(\frac{fG^2}{\rho}\right)\left(\frac{L}{H}\right)^3 = \left(\frac{fWL}{\rho^2}\right)G^3\left(\frac{L}{H}\right)^3 \tag{16.5}$$

where m (=WLG) is the mass flow rate.

Using Eq. (16.1) and Blasius equation for friction factor ($f = 0.0791\ Re^{-0.25}$), the pumping power equation can be transformed to give

$$P \propto G^{2.75}\left(\frac{L}{H}\right)^3 (WL)\left(\frac{1}{L}\right)^{0.25} \tag{16.6}$$

From Eqs. (16.3) and (16.6), it can be seen that the heat collection rate and pumping power are strong functions of the duct geometrical parameters (L, W, H) and air flow rate per unit area of the plate G. Thus, an appropriate way to evaluate the performance is to take both heat collection rate and pumping power requirement into account, i.e. to carry out a thermohydraulic performance evaluation.

16.2 Mathematical Model for Thermohydraulic Performance Prediction (Karwa et al. 2007; Karwa and Chauhan 2010)

The effect of the design and operating parameters on the performance of the solar air heater can be evaluated using a mathematical model presented here. This model calculates the useful heat gain from the iterative solution of basic heat transfer equations of top loss and equates the same with the convective heat transfer rate from the absorber plate to the air using proper heat transfer correlations for the duct of the air heaters. The model estimates the collector back loss from the iterative solution of the heat balance equation for the back surface. The edge loss can be calculated from the equation suggested by Klein (1975).

Figure 16.1b shows the longitudinal section of the air heater. The heat balance on the solar air heater gives the distribution of incident solar radiation I into useful heat gain Q and various heat losses. The useful heat gain or heat collection rate for a collector can be expressed as

$$Q = AI(\tau\alpha) - Q_L = A[I(\tau\alpha) - U_L(T_p - T_a)] \tag{16.7}$$

where $(\tau\alpha)$ is the transmittance-absorptance product of the glass cover–absorber plate combination and Q_L is heat loss from the collector. From the known values of mean absorber plate temperature T_p and the ambient temperature T_a, overall loss coefficient U_L is calculated from

$$U_L = \frac{Q_L}{A(T_p - T_a)} \tag{16.8}$$

The collected heat is transferred to the air flowing through the air heater duct. Thus

$$Q = mc_p(T_o - T_i) = GAc_p(T_o - T_i) \tag{16.9}$$

For open-loop operation $T_i = T_a$.

Estimating the Nusselt number from an appropriate correlation for the collector duct, heat transfer coefficient h between the absorber plate and air is determined.

From heat transfer consideration, the heat gain is

$$Q = hA(T_p - T_{\text{fm}}) \tag{16.10}$$

or

$$T_p = \frac{Q}{hA} + T_{\text{fm}} \tag{16.11}$$

where T_{fm} is the mean temperature of air in the solar air heater duct.

The heat loss Q_L from the collector is a sum of the losses from top Q_t, back Q_b and edge Q_e of the collector as shown in Fig. 16.1b.

16.2.1 Top Loss

The top loss Q_{t} from the collector can be calculated from the iterative solution of basic heat transfer equations given below.

Heat transfer Q_{tpg} from absorber plate at mean temperature T_p to the inner surface of the glass cover at temperature T_{gi} takes place by radiation and convection hence

$$Q_{\text{tpg}} = \frac{\sigma A(T_p^4 - T_{\text{gi}}^4)}{\frac{1}{\varepsilon_p} + \frac{1}{\varepsilon_g} - 1} + h_{\text{pg}}A(T_p - T_{\text{gi}}) \tag{16.12a}$$

The conduction heat transfer through the glass cover of thickness δ_{g} is given by

$$Q_{tg} = k_g A \frac{T_{gi} - T_{go}}{\delta_g} \tag{16.12b}$$

where k_g is the thermal conductivity of the glass and T_{go} is the temperature of the outer surface of the glass cover.

From the outer surface of the glass, the heat is rejected by radiation to the sky at temperature T_{sky} and by convection to the ambient hence

$$Q_{tgo} = \sigma A \varepsilon_g (T_{go}^4 - T_{sky}^4) + h_w A (T_{go} - T_a) \tag{16.12c}$$

where h_w is termed as wind heat transfer coefficient. It is a function of the wind velocity at the outer surface of the glass cover. In equilibrium,

$$Q_{tpg} = Q_{tg} = Q_{tgo} = Q_t \tag{16.13}$$

16.2.2 Wind Heat Transfer Coefficient

Various correlations for the estimate of the wind heat transfer coefficient h_w from the wind velocity data are available in the literature. Karwa and Chitoshiya (2013) have compiled some of these correlations and have presented a discussion on the same.

(i) McAdams (1954) proposed the following correlation:

$$h_w = 5.6214 + 3.912 V_w \quad \text{for } V_w \le 4.88\,\text{m/s} \tag{16.14a}$$

$$= 7.172 (V_w)^{0.78} \quad \text{for } V_w > 4.88\,\text{m/s} \tag{16.14b}$$

This correlation has been widely used in modelling, simulation and relevant calculations in spite of its shortcomings (Palyvos 2008).

(ii) Watmuff et al. (1977) expressed the opinion that the wind heat transfer coefficient, derived from McAdams correlation, includes radiation effect. They presented the following relation to exclude radiation and free convection contribution:

$$h_w = 2.8 + 3.0 V_w \quad \text{for } 0 < V_w < 7\,\text{m/s} \tag{16.15}$$

(iii) Kumar et al. (1997), based on indoor laboratory measurement on box-type solar cooker, proposed the following correlation for the wind heat transfer coefficient.

$$h_w = 10.03 + 4.68 V_w \tag{16.16}$$

(iv) Test et al. (1981) observed that, in the outdoor environment, the convection heat transfer coefficient is greater than the values reported from wind tunnel tests and gave the following correlation for rectangular plate exposed to varying wind directions.

$$h_w = (8.355 \pm 0.86) + (2.56 \pm 0.32) V_w \tag{16.17}$$

(v) Kumar and Mullick (2007) experimentally determined the wind heat transfer coefficient at low wind velocities ($V_w < 0.37$ m/s) from a metal plate exposed to solar radiation. Their correlation is

Table 16.1 Estimate of the wind heat transfer coefficient h_w [W/(m^2 K)] (Karwa and Chitoshiya 2013)

V_w (m/s)	Equation (16.14)	Equation (16.15)	Equation (16.16)	Equation (16.17)	Kumar and Mullick (2007)	Range*
0	5.7	2.8	10.03	7.5–9.2	h_w = 6 W/(m^2 K) at V_w = 0 m/s;	5.7–10.03
0.5	7.6	4.3	12.4	8.6–10.7		7.6–12.4
1.0	9.5	5.8	14.7	9.7–12.1		9.5–14.7
1.5	11.5	7.3	17.0	10.9–13.5		10.9–17.0
2.0	13.5	8.8	19.4	12.0–15.0		12.0–19.4

* Neglecting the values estimated from Eq. (16.15).

$$h_w = 0.8046(T_p - T_a)^{0.69} \tag{16.18}$$

The predicted values of the wind heat transfer coefficient from the different correlations presented above were found by Karwa and Chitoshiya (2013) to vary significantly as shown in Table 16.1.

Palyvos (2008) expressed the opinion that the obvious lack of generality of the existing wind convection coefficient correlations presents a challenge for future research. Field rather than laboratory measurements, as well as some sort of standardization in the choice of height above the ground and/or distance from the façade wall or roof with a suitable amendment of the velocity value on the basis of surface orientation and wind direction relative to the surface, are required.

It is to note that the uncertainty in the estimate of the wind heat transfer coefficient will have an impact on the accuracy of predicted thermal performance of a solar air heater using any mathematical model.

16.2.3 Sky Temperature

The sky is considered as a blackbody at some fictitious temperature known as sky temperature T_{sky} at which it is exchanging heat by radiation. The sky temperature is a function of many parameters. Some studies assume the sky temperature to be equal to the ambient temperature because it is difficult to make a correct estimate of it, while others estimate it using different correlations. One widely used equation due to Swinbank (1963) for clear sky is

$$T_{\text{sky}} = 0.0552(T_a)^{1.5} \tag{16.19}$$

where temperatures T_{sky} and T_a are in Kelvin.

Another approximate empirical relation is (Garg and Prakash 2000)

$$T_{\text{sky}} = T_a - 6 \tag{16.20}$$

The above relations give significantly different values of the sky temperature. Nowak (1989) comments that, in the case of large city areas, the sky temperature may be about 10°C higher than the one calculated from Swinbank's formula because of the atmospheric pollution. The sky temperature also changes with the change in the atmospheric humidity. Thus, there can be significant uncertainty in the estimate of the sky temperature, which may affect the predicted thermal performance from any mathematical model using any of the above equations.

16.2.4 Convective Heat Transfer Coefficient Between the Absorber Plate and Glass Cover

For the estimate of the convective heat transfer coefficient between the absorber plate and glass cover $h_{pg,}$ the three-region correlation of Buchberg et al. (1976) can be used, which is

$$\mathrm{Nu} = 1 + 1.446(1 - 1708/\mathrm{Ra}')^{+} \quad \text{for } 1708 \le \mathrm{Ra}' \le 5900 \quad \text{(the + bracket goes to zero when negative)} \tag{16.21a}$$

$$\mathrm{Nu} = 0.229(\mathrm{Ra}')^{0.252} \quad \text{for } 5900 < \mathrm{Ra}' \le 9.23 \times 10^4 \tag{16.21b}$$

$$\mathrm{Nu} = 0.157(\mathrm{Ra}')^{0.285} \quad \text{for } 9.23 \times 10^4 < \mathrm{Ra}' \le 10^6 \tag{16.21c}$$

where Ra′ (=Ra cosβ) is Rayleigh number for the inclined air layers. The Rayleigh number for the internal natural convection flow between parallel plates is given by

$$\mathrm{Ra} = \mathrm{Gr}\,\mathrm{Pr} = \left[\frac{g(T_p - T_{gi})\delta_{\mathrm{pg}}^3}{T_{mpg}\nu_{mpg}^2}\right]\mathrm{Pr} \tag{16.22}$$

where δ_{pg} = gap between the absorber plate and glass cover.

16.2.5 Back and Edge Losses

The back loss from the collector, refer Fig. 16.1b, can be calculated from the following equation:

$$Q_b = \frac{A(T_b - T_a)}{\frac{\delta_i}{k_i} + \frac{1}{h_w}} \tag{16.23a}$$

where δ_{i} is the insulation thickness and k_{i} is the thermal conductivity of the insulating material.

Heat transfer by radiation from the heated absorber plate to the duct bottom surface Q_{pb} is given by

$$Q_{pb} = \frac{\sigma A(T_p^4 - T_b^4)}{\frac{1}{\varepsilon_{pi}} + \frac{1}{\varepsilon_b} - 1} \tag{16.23b}$$

The heat flows from the heated bottom surface at temperature T_{b} to the surroundings through the back insulation and to the air flowing through the duct at mean temperature T_{fm}, i.e.

$$Q_{ba} = \frac{A(T_b - T_a)}{\frac{\delta_i}{k_i} + \frac{1}{h_w}} + hA(T_b - T_{fm}) \tag{16.23c}$$

The heat balance for the surface gives $Q_{pb} = Q_{ba}$. The temperature of the duct bottom surface T_b can be estimated from the iterative solution of this heat balance equation.

For the edge loss estimate, the empirical equation suggested by Klein (1975) is

$$Q_e = 0.5A_e(T_p - T_a) \tag{16.24}$$

where A_e is the area of the edge of the air heater rejecting heat to the surroundings.

The outlet air temperature is estimated from

$$T_o = T_i + \frac{Q}{mc_p} \tag{16.25}$$

The thermal efficiency η of the solar air heater is the ratio of the useful heat gain Q and the incident solar radiation I on the solar air heater plane, i.e.

$$\eta = \frac{Q}{IA} \tag{16.26}$$

Niles et al. (1978) have used the following equations to calculate the outlet air and mean plate temperatures when the solar air heater operates in open-loop mode (i.e. $T_i = T_a$):

$$T_o = T_a + \frac{I(\tau\alpha)\xi}{U_L} \tag{16.27}$$

$$T_p = T_i + \left[\frac{I(\tau\alpha)}{U_L}\right]\left(1 - \frac{G\xi c_p}{U_L}\right) \tag{16.28}$$

where $\xi = 1 - \exp[-U_L/(Gc_p)(1 + U_L/h)^{-1}] = (F_R U_L/Gc_p)$. Parameter F_R is termed as heat removal factor and is given by (Duffie and Beckman 1980)

$$F_R = \left(\frac{Gc_p}{U_L}\right)\left[1 - \exp\left(\frac{-F'U_L}{Gc_p}\right)\right] \tag{16.29}$$

where F' is termed as efficiency factor. It is given by

$$F' = \left(1 + \frac{U_L}{h}\right)^{-1} \tag{16.30}$$

Equations (16.27) and (16.28) may be used for the cross-check of the values of T_o and T_p calculated from Eqs. (16.25) and (16.11), respectively.

The mean air temperature equation in terms of F_R and F' (Duffie and Beckman 1980) is

$$T_{fm} = T_i + \frac{(Q/A)}{U_L F_R}\left(1 - \frac{F_R}{F'}\right) \tag{16.31}$$

16.2.6 Heat Transfer and Friction Factor Correlations

The results of the simulation strongly depend on the use of appropriate heat transfer and friction factor correlations for the solar air heater ducts. These correlations must take into account the effects of the asymmetric heating encountered in the solar air heaters, duct aspect ratio and developing length, and must be applicable to the laminar to early turbulent flow regimes. The geometry of interest is the parallel plate duct (a rectangular duct of high aspect ratio) since the width of the collector duct is of the order of 1 m and height of the order of 5–10 mm (Karwa et al. 2002; Holland and Shewen 1981), with one wall at constant heat rate and the other insulated. An intensive survey of the literature has been carried out by Karwa et al. (2007) for the correlations to fulfil these requirements.

For hydrodynamically developing laminar flow in parallel plate ducts, Chen (in Ebadian and Dong 1998) has obtained the following equations for the hydrodynamic length L_{hy} and apparent friction factor f_{app}, respectively:

$$\frac{L_{hy}}{D_h} = 0.011\mathrm{Re} + \frac{0.315}{1 + 0.0175\mathrm{Re}} \tag{16.32}$$

$$f_{app} = \frac{24}{\mathrm{Re}} + \left(0.64 + \frac{38}{\mathrm{Re}}\right)\left(\frac{D_h}{4L}\right) \tag{16.33}$$

From Eq. (16.32), $L_{\mathrm{hy}}/D_{\mathrm{h}}$ is less than 30 for Re $\leq$ 2550, while L/D_{h} ranges from 100 to 200 in the case of solar air heaters. Equation (16.33) takes account of the increased friction in the entrance region and the change of the momentum flux.

The thermal entrance length $L_{\mathrm{th}}/D_{\mathrm{h}}$ for the laminar flow in a flat parallel plate passage, when one wall is insulated and other subjected to uniform heat flux, is of the order of 0.1(Re Pr) (Heaton et al. 1964) for the approach of Nusselt number value within about 1% of the fully developed Nusselt number value. The appropriate Nusselt number-Reynolds number relation for the thermally developing laminar flow in a parallel plate duct has been presented by Hollands and Shewen (1981), which has been deduced from (Kays and Perkins 1973) and agrees well with the data of Heaton et al. (1964):

$$\mathrm{Nu} = 5.385 + 0.148\mathrm{Re}\left(\frac{H}{L}\right) \quad \text{for Re} < 2550 \tag{16.34}$$

The second term containing H/L takes the entrance length effect into account.

The friction factor correlation of Bhatti and Shah (1987) for the transition to turbulent flow regime in rectangular cross-section smooth duct ($0 \leq H/W \leq 1$) is

$$f = 1.0875 - 0.1125\left(\frac{H}{W}\right)f_o \tag{16.35}$$

where $f_o = 0.0054 + 2.3 \times 10^{-8}\mathrm{Re}^{1.5}$ for $2100 \leq \mathrm{Re} \leq 3550$

and $f_o = 1.28 \times 10^{-3} + 0.1143\mathrm{Re}^{-0.311}$ for $3550 < \mathrm{Re} \leq 10^7$

They report an uncertainty of ±5% in the predicted friction factors from the above correlation.

The study of the apparent friction factor plots for the entrance region of flat parallel plate duct in the turbulent flow regime presented along with those for a circular tube in Bhatti and Shah (1987) shows that the trend of variation in the friction factor in the entrance region for the parallel plates is not significantly different from that for a circular tube. Hence, the following circular tube relation (Bhatti and Shah 1987) can be used:

$$f_{app} = f + 0.0175\left(\frac{D_h}{L}\right) \tag{16.36}$$

The thermal entrance length L_{th}/D_h for the turbulent flow of air, based on the local Nusselt number approaching the fully developed value, ranges from 20 to 30 for the Reynolds number range of 8000–30000 (Lee 1968; Barrow and Lee 1964). Kays and Leung (1963) solved the fully developed turbulent-flow energy equations with constant heat rate for parallel plate duct with one side insulated. Their results are reported to be in excellent agreement with the experimental data for air (Kays and Crawford 1980). The following Nusselt number correlations, deduced by Hollands and Shewen (1981) from the data of Kays and Leung (1963) and Tan and Charters (1970) for collectors with $L/H > 125$, may be used.

$$\begin{gathered}\mathrm{Nu} = 4.4 \times 10^{-4}\mathrm{Re}^{1.2} + 9.37\mathrm{Re}^{0.471}\frac{H}{L} \\ \text{for } 2550 \le \mathrm{Re} \le 10^4 \text{ (transition flow)}\end{gathered} \tag{16.37a}$$

and

$$\begin{gathered}\mathrm{Nu} = 0.03\mathrm{Re}^{0.74} + 0.788\mathrm{Re}^{0.74}\frac{H}{L} \\ \text{for } 10^4 < \mathrm{Re} \le 10^5 \text{ (early turbulent flow)}\end{gathered} \tag{16.37b}$$

where the terms containing H/L take the entrance length effect into account.

The Nusselt number data from the correlation of Holland and Shewen for fully developed turbulent flow are in close agreement with the data of Hatton et al. (1964). They are about 10% lower than the data from the tube correlation of Petukhov et al. in (Bhatti and Shah 1987) for $\mathrm{Re} \ge 8000$ and about 15–20% lower than that of $\mathrm{Nu} = 0.023\,\mathrm{Re}^{0.8}\,\mathrm{Pr}^{0.4}$ for $\mathrm{Re} \ge 10000$. This is in agreement with the observation of Sparrow et al. (1966) in an experimental study under ideal laboratory conditions for duct with $W/H = 5$ and $\mathrm{Re} = 1.8 \times 10^4 - 1.42 \times 10^5$, and Tan and Charters (1970) for the asymmetrically heated duct of a solar air heater ($W/H = 3$; Re = 9500–22000). From the close agreement of the Nusselt number data from Eq. (16.37b) with the carefully conducted experimental results, it has been inferred by Karwa et al. (2007) that the uncertainty in the predicted Nusselt number values must be of the order of 5–6%. However, the information on the transition flow in a flat or rectangular duct is extremely sparse and a higher uncertainty in the Nusselt number values determined from Eq. (16.37a) may be possible. Holland and Shewen (1981), based on the information available in (Kays and Perkins 1973), concluded that the flow is laminar for $\mathrm{Re} < 2550$ and turbulent for $\mathrm{Re} > 10^4$. By analogy with the results for rectangular ducts with $W/H = 8$ (Kays and London 1964), they concluded that power law fits extending from the laminar result to the turbulent result would be satisfactory for the transition regime. However, the lower limit of the critical Reynolds number for a parallel plate duct is reported by Ebadian and Dong (1998) to be 2200–3400 depending on the entrance configurations and

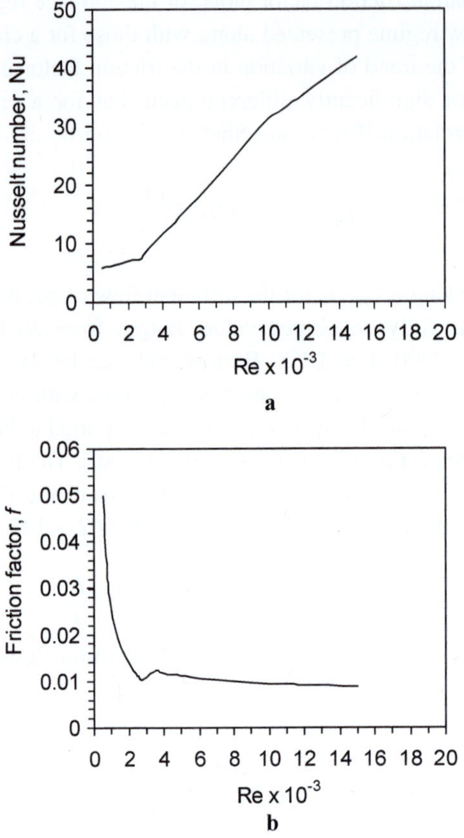

Fig. 16.2 **a** Nusselt number versus the Reynolds number, **b** Friction factor versus the Reynolds number (Karwa et al. 2007)

disturbance sources. For a high aspect ratio rectangular duct with abrupt entrance, the same is reported to be 2920–3100. Hence, the laminar flow regime may be assumed up to Re = 2800, which also limits the inconsistency of the predicted Nusselt number and friction factor values from the different correlations used here to about 5% at the laminar–transition interface.

Figure 16.2 shows the plots of the Nusselt number and friction factor from the above-presented correlations. The changes in the plots correspond to the laminar-transient and transient-turbulent regimes.

The pressure loss, from the known value of friction factor f (=f_{app} for smooth duct), and pumping power are calculated from

$$\delta p = \left(\frac{4fL}{2\rho D_h}\right)\left(\frac{m}{WH}\right)^2 \tag{16.38}$$

$$P = \left(\frac{m}{\rho}\right)\delta p \tag{16.39}$$

Cortes and Piacentini (1990) used effective thermal efficiency η_e for the collector thermohydraulic performance evaluation, which is based on the net thermal energy collection rate of a collector considering the pumping power required to overcome the friction of the solar air heater duct. Since the power lost in overcoming frictional resistance is converted into heat, the effective efficiency equation may be defined as (Karwa and Chauhan 2010)

$$\eta_e = \frac{(Q + P) - \frac{P}{C}}{IA} \tag{16.40}$$

where C is a conversion factor used for calculating equivalent thermal energy for obtaining the pumping power. It is a product of the efficiencies of the fan, electric motor, transmission and thermoelectric conversion. For example, based on the assumption of 60% efficiency of the blower–motor combination and 33% efficiency of thermoelectric conversion process referred to the consumer point, factor C will be 0.2. Since the operating cost of a collector depends on the pumping power spent, the effective efficiency based on the net energy gain is a logical criterion for the performance evaluation of the solar air heaters.

The thermophysical properties of the air are taken at the corresponding mean temperature $T_{\mathrm{m}} = T_{\mathrm{fm}}$ or T_{mpg}. The following relations of thermophysical properties, obtained by correlating data from NBS (US) (Holman 1990), can be used:

$$c_p = 1006\left(\frac{T_m}{293}\right)^{0.0155} \tag{16.41a}$$

$$k = 0.0257\left(\frac{T_m}{293}\right)^{0.86} \tag{16.41b}$$

$$\mu = 1.81 \times 10^{-5}\left(\frac{T_m}{293}\right)^{0.735} \tag{16.41c}$$

$$\rho = 1.204\left(\frac{293}{T_m}\right) \tag{16.41d}$$

$$\mathrm{Pr} = \frac{\mu c_p}{k} \tag{16.41e}$$

Equations (16.7) to (16.41a) constitute a non-linear model for the solar air heater for the computation of the useful heat gain Q, thermal efficiency η, pressure loss δp, pumping power P and effective efficiency η_{e}. The model has been solved by Karwa and Baghel (2014) following an iterative process depicted in Fig. 16.3. Karwa and Chauhan (2010) also used the model for roughened duct solar air heater with relevant heat transfer and friction factor correlations in place of smooth duct relations. For the estimate of heat collection rate, Karwa and Chauhan (2010) and Karwa and Baghel (2014) terminated the iteration when the successive values of the plate and mean air temperatures differed by less than 0.05 K. The iteration for the estimate of top loss by them has been continued till the heat loss estimates from the absorber plate to the glass cover and glass cover to the ambient, i.e. Q_{tpg} and Q_{tgo} from Eqs. (16.12a) and (16.12c), respectively, differed by less than 0.2%.

The mathematical model presented here has been validated by Karwa et al. (2007) against the data from the experimental study of a smooth duct solar air heater of Karwa et al. (2001) with reported uncertainties of ±4.65% in Nusselt number and ±4.56% in friction factor. The standard deviations of the predicted values of thermal efficiency and pumping power from the experimental values of these parameters from (Karwa et al. 2001) have been reported by Karwa et al. (2007) to be ±4.9% and ±6.2%, respectively.

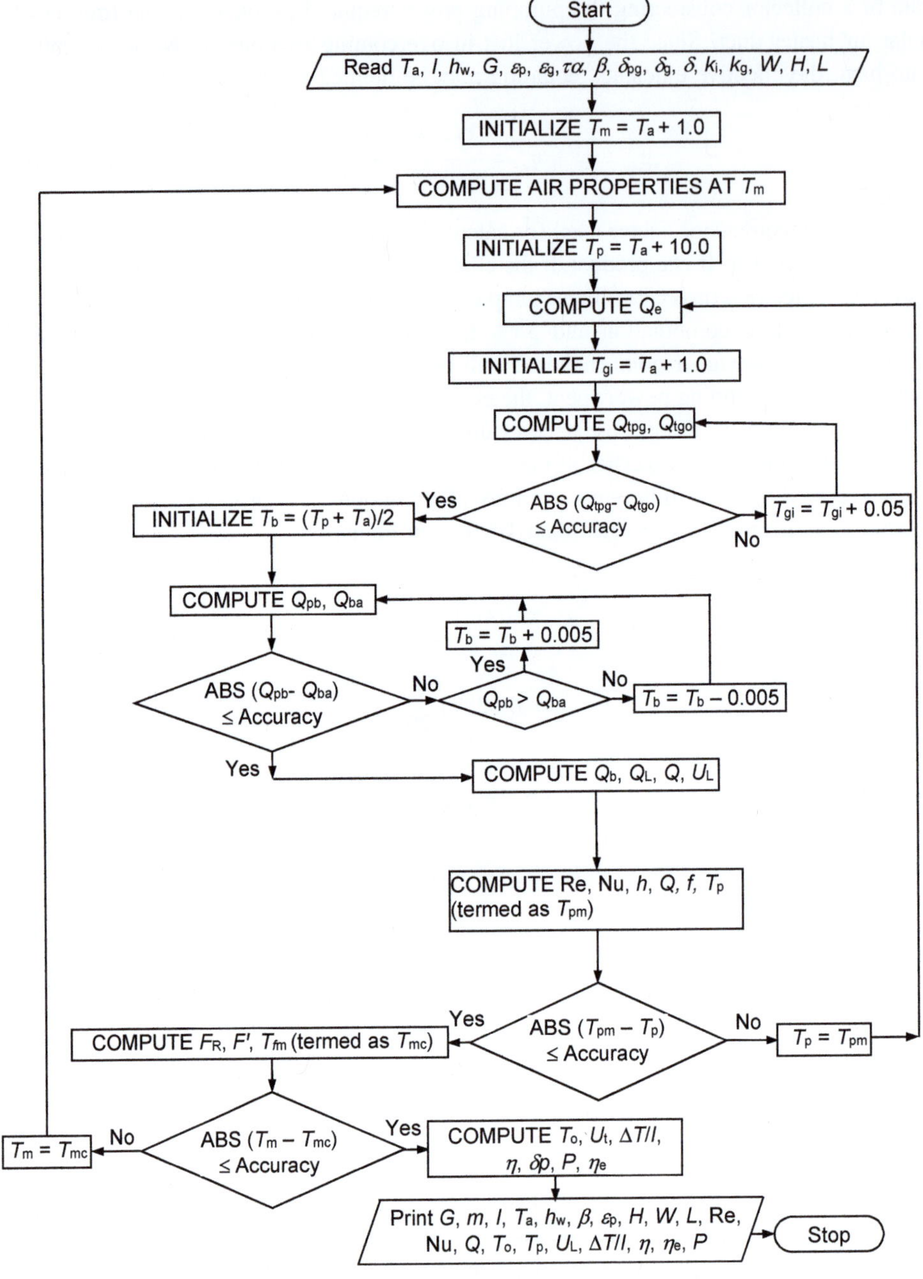

Fig. 16.3 Flow chart for iterative solution of mathematical model (Karwa and Baghel 2014)

16.3 Enhanced Performance Solar Air Heaters

16.3.1 Introduction

As compared to a solar water heater, the thermal efficiency of a smooth duct solar air heater is low because of a low value of heat transfer coefficient between the absorber plate and the flowing air through the collector duct leading to a high temperature of the absorber plate and greater heat loss. Hence, the researchers have directed their studies towards various performance improvement techniques. They are (i) the increase in the absorption of solar radiation using corrugated absorber, (ii) use of fins and corrugations to increase the area of heat transfer from the absorber, (iii) use of selective coating on the sun-facing side of the absorber plate, (iv) use of more than one glass cover over the absorber plate and (v) use of artificial roughness on the air flow side of the absorber plate to enhance heat transfer.

The main resistance to heat transfer in a solid–fluid interaction is due to the formation of the laminar sublayer on the heat transferring surface and hence the efforts for enhancing the heat transfer have been directed towards artificially destroying or disturbing this sublayer. In general, this can be achieved by active, passive or a combination of the active and passive methods.

The use of the artificial roughness elements in the form of projections is one of the most effective passive techniques of heat transfer enhancement and has been extensively used in nuclear reactors, heat exchangers, gas turbine blade cooling channels and solar collectors.

Enhancement of the heat transfer coefficient between the absorber plate surface of the solar air heater and air leads to the reduction in the absorber plate temperature which leads to the reduction in the heat loss from the collector and improvement in the thermal efficiency. Hence, the artificial roughness in the form of ribs on the air flow side of the absorber plate has been shown to be one of the most promising methods for enhancement of heat transfer coefficient and thermal performance of the solar air heater with forced flow of air. In general, any attempt to increase turbulence in the flow also increases the pumping power requirement. Since the artificial roughness on the heat-transferring surface creates turbulence near the wall and breaks the laminar sublayer at the wall, it enhances the heat transfer coefficient with a minimum pumping power penalty.

Extensive experimental studies carried out by researchers have shown that the geometry of the roughness (roughness shape, pitch, height, etc.) and the arrangement of the rib elements (orientation with respect to the flow direction) have a marked influence on the heat transfer and friction characteristics of the roughened surfaces.

Karwa et al. (2010) and Chitoshiya and Karwa (2015) have presented a parametric review of the studies carried out for the heat transfer enhancement in asymmetrically heated high aspect ratio rectangular ducts with special emphasis on the effect of the rib shapes, their arrangement with respect to flow and discretization of the ribs on the heat transfer and friction characteristic of such enhanced ducts, and on the performance of the solar air heater provided with roughness on the absorber plate. The studies referred in their review have been basically carried out for rectangular section ducts with one of the broad walls roughened and subjected to uniform heat flux while the remaining walls were insulated. These boundary conditions correspond closely to those found in flat plate solar air heaters.

16.3.2 Artificial Roughness for Heat Transfer Enhancement

16.3.2.1 Effect of Rib Shape and Pitch

The researchers have used various rib shapes: circular cross-section wires (Gupta et al. 1993 and 1997; Saini and Saini 1997; Momin et al. 2002; Muluwork et al. 1998) wedge-shaped integral ribs (Bhagoriya et al. 2002), chamfered integral rib-roughness (Karwa et al. 1999 and 2001), rib-grooved roughness (Jaurker et al. 2006), chamfered rib-grooved roughness (Layek et al. 2007) and dimple-shaped projections (Saini and Verma 2008) as shown in Fig. 16.4.

The flow detaches as it encounters the rib and reattaches at about 6–7 rib heights in the case of the square, circular and rectangular cross-section ribs, refer Fig. 16.4c–i. A typical detailed flow pattern for chamfered ribs is shown in Fig. 16.5. The laminar sublayer is completely destroyed in the reattachment region. The boundary layer redevelops after the reattachment region and hence its thickness is small. A recirculating region with frequent shading of vortices establishes in the wake of the ribs and is basically a region of low heat transfer coefficient. This region is reported to be smaller in the case of chamfered ribs. The near wall turbulence developed in the process enhances the heat transfer coefficient with a minimum pumping power penalty. Optimum relative roughness pitch (p/e) for circular and rectangular section ribs is reported to be 10 by most of the researchers. The same has been reported to be about 6–8 for the chamfered, wedge and chamfered rib-grooved surfaces (Karwa et al. 1999 and 2001; Bhagoriya et al. 2002; Jaurker et al. 2006; Layek et al. 2007).

The performance of chamfered or rib-grooved surface is reported to be superior to that of circular, square or rectangular section ribs (Karwa et al. 1999; Jaurker et al. 2006; Layek

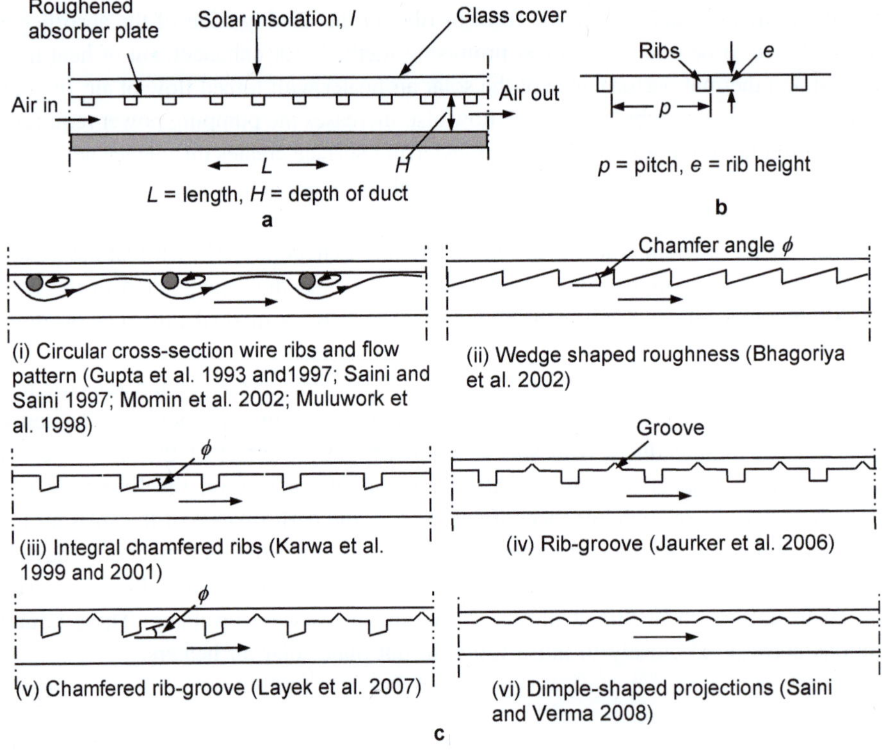

Fig. 16.4 **a** Solar air heater with roughened absorber plate, **b** Rib geometry, **c** rib shapes

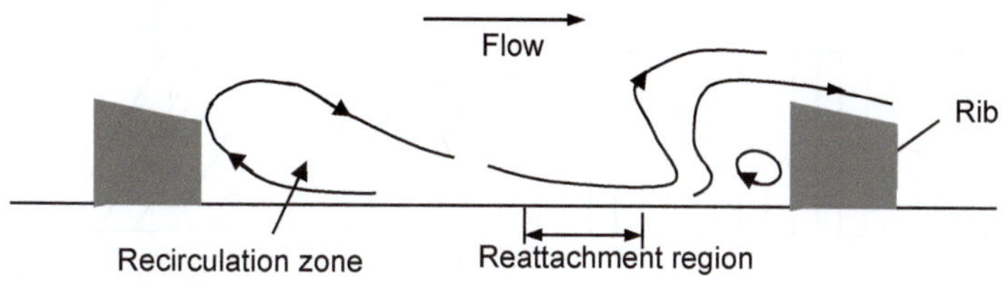

Fig. 16.5 Flow pattern for 15° chamfered ribs (Karwa et al. 2008)

et al. 2007). Positively chamfered ribs with 15° chamfer angle (Karwa et al. 1999), and 18° chamfer angle in case of chamfered and grooved (Layek et al. 2007), have been shown to provide optimum performance while the optimum angle for the wedge-shaped ribs is 10° (Bhagoriya et al. 2002). Negatively chamfered ribs are poor in performance (Karwa et al. 1999). Chamfered and wedge-shaped ribs provide better performance mainly because of the reduced recirculation zone behind the ribs (Karwa et al. 1999; Bhagoriya et al. 2002) and, in case of chamfered ribs, the vortex shedding has been shown to be vigorous. Integral ribs provide additional heat transfer enhancement due to the fin effect.

For the positively chamfered closely spaced ribs ($p/e \leq 5$), vigorous vortex shedding has been reported compared to the square or negatively chamfered ribs (Karwa et al. 2008). Unlike square ribs, reattachment effect is reported for 15° chamfered ribs even at a relative roughness pitch of 5. The recirculating region is found to reduce with increase in the rib head chamfer angle from 0° (rectangular or square section ribs) to 15° along with an increased vortex activity in the wake of the ribs for the 15° chamfered ribs at $p/e \geq 7.5$ as found from flow structure study by Karwa et al.

As compared to a smooth surface, the presence of artificial roughness has been shown to increase the Nusselt number up to 3.24 times in the transitional to early turbulent flow regime, while the friction factor increases up to 5.3 times by these researchers depending upon the relative roughness height (ratio of rib height to hydraulic diameter), shape and arrangement of the rib elements. The thermohydraulic performance (based on equal pumping power) of the roughened surface with angled discrete rib roughness has been found to improve by about 30–70% over the smooth-surfaced duct (Karwa et al. 2005).

Gupta et al. (1993), Singh et al. (2011) and Karwa et al. (2001, 2010) and Karwa and Chitoshiya (2013) carried out studies on solar air heaters with a transverse wire as roughness, chamfered or discrete rib roughness, respectively, and have shown thermal efficiency enhancement of the order of 10–40% over the smooth duct solar air heaters.

16.3.2.2 Effect of Rib Arrangement

Researchers have tried rib elements in various arrangements: inclined continuous ribs (Gupta et al. 1993 and 1997), expanded metal wire mesh (Saini and Saini 1997), inclined continuous ribs with a gap (Aharwal et al. 2008), circular cross-section wires arranged in v-shape (Momin et al. 2002), discrete ribs in v-pattern (Muluwork et al. 1998; Karwa 2003; Karwa and Chouhan 2010; Singh et al. 2011; Karwa and Chitoshiya 2013), ribs in a staggered pattern (Karmare and Tikekar 2007) and ribs in w-pattern (Lanjewar et al. 2011) as shown in Fig. 16.6.

The optimum inclination angle for v-continuous or discrete rib pattern is reported to be 60° (Momin et al. 2002; Karwa 2003). In the case of the expanded metal wire mess of Saini and Saini (1997) also, the optimum performance is found to occur for 60° inclination of the

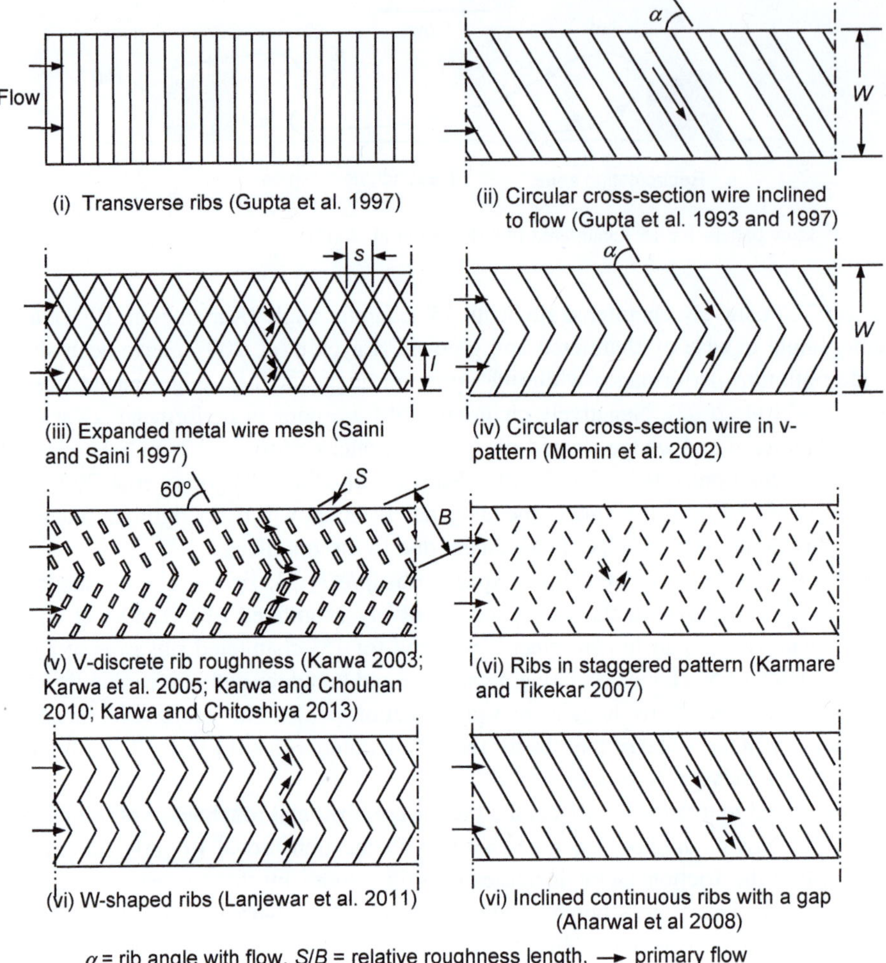

Fig. 16.6 Rib arrangements and the effect of rib orientation on the flow

elements of the mess to the flow direction though the distance between the crossing wires of the mess in this case is variable.

Karwa (2003) has presented the following flow structure in the case of inclined or v-discrete rib pattern of ribs.

The additional heat transfer enhancement observed for the inclined or the v-pattern ribs over the transverse ribs can be attributed to the secondary flow of the air induced by the rib inclination, refer Fig. 16.6; the secondary flow is depicted in the figure by inclined arrows. The heated air adjacent to the wall (the secondary flow) moves along the plate surface to the side wall in the case of the inclined ribs and towards both of the side walls in the case of v-up pattern of ribs. This exposes the absorber plate to a relatively lower temperature air of the axial or primary flow over the ribs. In the case of the v-down pattern, there are two contradictory effects. The secondary flow is towards the central axis where it interacts with the axial flow creating additional turbulence leading to the increase in the heat transfer rate while the rise in the temperature of the axial flow, just above the ribs in the central region, reduces the heat transfer rate. A similar effect of the secondary flow can be understood for w-shaped and inclined ribs with a gap.

In the case of discrete ribs, the heated air at the plate surface moving as secondary flow mixes with the primary air after only a short distance movement leading to better mixing of the heated air near the plate with the colder primary air flowing between the ribs. The uniformity of heat transfer improves with the increase in the discretization of the ribs due to the vigorous mixing of the secondary and primary flows (Karwa et al. 2005). This hypothesis has been confirmed (Karwa 2003) by the recording of the temperature variation of the heated outlet air in the span-wise (transverse to the flow) direction. Because of the secondary flow effect, the centerline temperature of the heated air has been reported to be higher in the case of the v-down arrangement of the ribs.

60° v-down discrete ribs and inclined ribs with a gap are shown to be the best of all the rib arrangements discussed here (Karwa 2003; Karwa et al. 2005; Aharwal et al. 2008).

16.4 Heat Transfer and Friction Factor Correlations for Roughened Rectangular Ducts

The results of above-presented studies on asymmetrically heated rectangular ducts are available in the form of heat transfer and friction factor correlations, which can be used for the design of solar air heaters because the boundary conditions of the ducts in these studies confirm to those encountered in solar air heaters. Some of the correlations for asymmetrically heated rectangular ducts with roughness are given in Table 16.2. These correlations have been reduced to the optimal conditions by Karwa et al. (2010) as defined by the investigators for the roughness geometries studied by them. The friction correlations of Karwa et al. (1999 and 2005) are in the terms of the roughness function R and roughness Reynolds number e^+ defined by Nikuradse (1950) as

$$R = \sqrt{\frac{2}{f}} + 2.5 \ln\left(\frac{2e}{D_h}\right) + 3.75 \tag{16.42}$$

$$e^+ = \sqrt{\frac{f}{2}} \mathrm{Re}\left(\frac{e}{D_h}\right) \tag{16.43}$$

The heat transfer function in their heat transfer correlations is defined as (Dippery and Sabersky 1963)

$$g = \left(\frac{f}{2\mathrm{St}} - 1\right)\sqrt{\frac{2}{f}} + R \tag{16.44}$$

The mathematical model presented for the smooth duct solar air heater in Sect. 16.2 can also be utilized for the performance study of solar air heaters having artificially roughened absorber plate using heat transfer and friction factor correlations of the roughness under consideration in place of correlations for smooth duct air heater in the mathematical model.

Table 16.2 Heat transfer and friction factor correlations for roughened ducts (Karwa et al. 2010)

Investigator	Roughness type	Range of investigation	Preferred geometrical parameters	Heat transfer and friction factor correlations corresponding to preferred geometrical parameters
Gupta et al. (1993 and 1997)	Inclined circular cross-section wire	e/D_h = 0.018 – 0.052 W/H = 6.8 – 11.5 p/e = 10 Re = 3×10^3 – 18×10^3	At W/H = 10, α = 60°	For $e^+ < 35$ $\text{Nu} = 2.15\times10^{-3}\,(e/D_h)^{0.001}\,(\text{Re})^{1.084}$ For $e^+ \geq 35$ $\text{Nu} = 6.65\times10^{-3}\,(e/D_h)^{-0.24}\,(\text{Re})^{0.88}$ $f = 1.54\times10^{-1}\,(e/D_h)^{0.196}\,(\text{Re})^{-0.165}$
Saini and Saini (1997)	Wire mesh (crossed)	e/D_h = 0.012–0.039 l/e = 25–71.9 s/e = 15.6–46.9 W/H = 11 Re = 1.9×10^3 –13×10^3	l/e = 46.9 s/e = 25.0	$\text{Nu} = 1.82523\times10^{-2}\,\text{Re}^{1.22}\,(e/D_h)^{0.625}$ $f = 1.9058\,\text{Re}^{-0.361}\,(10e/D_h)^{0.591}$ The average absolute percentage deviations of the predicted values are 7.98% and 4.37% for Nu and f, respectively
Karwa et al. (1999)	Integral chamfered ribs	e/D_h = 0.014 – 0.032 ϕ = (−15°)–18° Re = 3×10^3 – 20×10^3 e^+ = 5–60	p/e = 6 ϕ = 15°	$R = 15.995(e^+)^{-0.075}$ for $5 < e^+ < 20$ $R = 5.7$ for $20 < e^+ < 60$ $g = 32.255(e^+)^{-0.31}$ for $7 < e^+ < 20$ $g = 10.0(e^+)^{0.08}$ for $20 < e^+ < 60$
Bhagoria et al. (2002)	Wedge-shaped ribs	e/D_h = 0.015 – 0.033 ϕ = 8° – 15° [(60.17 $\phi^{-1.0264}$) < p/e < 12.12] Re = 3×10^3 – 18×10^3	p/e = 7.57 ϕ = 10°	$\text{Nu} = 3.874\times10^{-3}\,\text{Re}^{1.21}\,(e/D_h)^{0.426}$ $f = 4.89\,\text{Re}^{-0.18}\,(e/D_h)^{0.99}$ These correlations predict Nusselt number and friction factor values within the error limits of ± 15% and ± 12%, respectively
Momin et al. (2002)	Circular cross-section wire in v-pattern	e/D_h = 0.02–0.034 p/e = 10 α = 30° – 90° W/H = 10.15 Re = 2.5×10^3 – 18×10^3	p/e = 10 α = 60°	$\text{Nu} = 0.067\times(\text{Re})^{0.888}\times(e/D_h)^{0.424}$ $f = 6.266\times(\text{Re})^{-0.425}\times(e/D_h)^{0.565}$ Average absolute deviation for Nusselt number correlation is 3.20%, and the same is 3.50% for friction factor correlation
Karwa et al. (2005)	60° v-down discrete, B/S = 6	e/D_h = 0.047 W/H = 7.75 Re = 2.85×10^3 – 15.5×10^3	p/e = 10.6 B/S = 6	$R = 6.06(e^+)^{0.045}$ for $15 \leq e^+ \leq 75.$ $g = 15.69(e^+)^{-0.2}$ for $15 \leq e^+ < 25$ $g = 4.10(e^+)^{0.217}$ for $25 \leq e^+ \leq 75.$ The maximum deviations of the predicted values of R and g from the correlations are 1.48% and 1.86%, respectively, from the experimental data

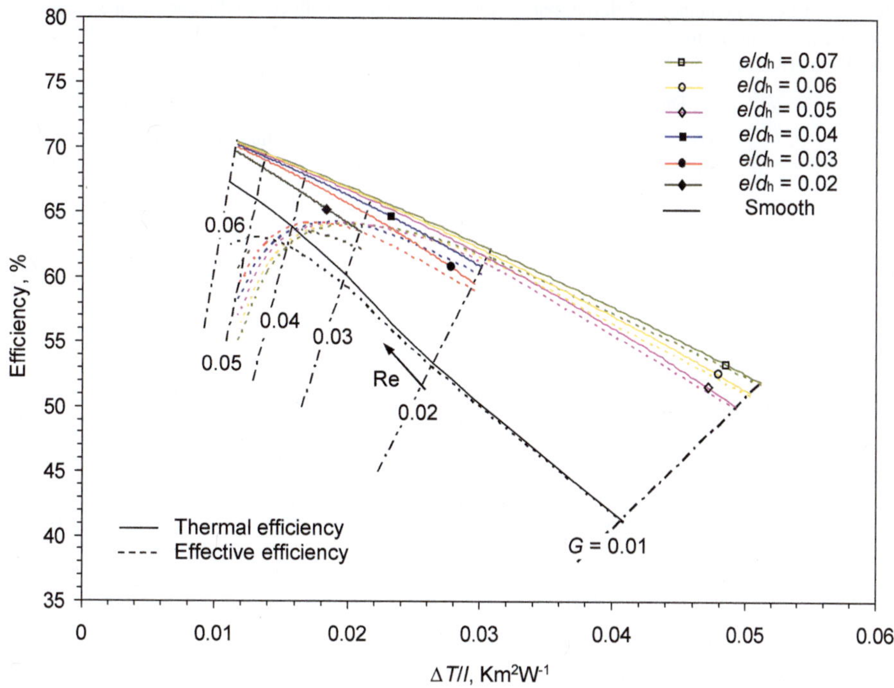

Fig. 16.7 Performance plots of solar air heaters with 60° v-down discrete rectangular rib roughness ($L = 2$ m, $H = 10$ mm, $p/e = 10$, $w/e = 2$, $\varepsilon_p = 0.95$, $\beta = 45°$, $T_a = 283$ K, $h_w = 5$ W/(m^2 K) and $I = 800$ W/m^2) (Karwa and Chauhan 2010)

Karwa and Chauhan (2010) considered 60° v-down discrete rib roughness suggested by Karwa et al. (2005), depicted in Fig. 16.6v, for the absorber plate of the roughened duct solar air heater and carried out a detailed investigation using the mathematical model presented above to study the effect of ambient, operating and design parameters on the thermal and effective efficiencies. They presented the result in the form of performance plots (thermal and effective efficiencies versus the air temperature rise parameter $\Delta T/I$) as shown in Fig. 16.7. It can be seen from the performance plots that, at the low flow rates of air per unit area of the absorber plate G, the ribs with greater relative roughness height e/D_h are beneficial; lower is the flow rate, greater is the advantage of the use of the artificial roughness on the absorber plate. At $G > 0.045$ kg/(s m^2), the smooth duct air heater is preferred from thermohydraulic consideration. The air mass flow rate per unit area of the plate, ambient temperature, solar insolation, wind heat transfer coefficient, etc. have been systematically varied by them to study their effect for a range of these parameters. Effects of change in collector length, duct height and plate emissivity on thermal and thermohydraulic performances represented by a change in thermal and effective efficiencies have also been studied by them. The results of the study are given in Table 16.3.

Table 16.3 Effect of change of different parameters on thermal and effective efficiencies; one parameter has been varied at a time from $L = 2$ m, $H = 10$ mm, $\varepsilon_p = 0.95$, $\beta = 45°$, $T_a = 283$ K, $h_w = 5$ W/(m^2 K) and $I = 800$ W/m^2 (Karwa and Chauhan 2010)

G, kg/(s m^2)	e/D_h	A. Duct height, H			B. Duct length, L ($L/H = 200$)			
		H, mm	$\Delta\eta/\eta$ (%)	$\Delta\eta_e/\eta_e$ (%)	L, m	H, mm	$\Delta\eta/\eta$ (%)	$\Delta\eta_e/\eta_e$ (%)
0.02	0.03	20 5	−11.45 6.48	−10.89 1.25	4 1	20 5	0.923 −1.04	0.95 −1.08
0.06	0.03	20 5	−4.26 2.19	10.70 −122.8	4 1	20 5	−0.59 0.2	−0.135 −0.42
0.01	0.07	20 5	−12.88 7.40	−12.74 6.18	4 1	20 5	−0.54 −1.42	−0.52 −1.43
0.06	0.07	20 5	−3.57 1.83	20.09 −194.0	4 1	20 5	−0.6 0.51	0.382 −0.69
G, kg/(s m^2)	e/D_h	C. Collector slope, β			D. Wind heat transfer coefficient, h_w			
		β, deg	$\Delta\eta/\eta$ (%)	$\Delta\eta_e/\eta_e$ (%)	h_w, W/(m^2 K)		$\Delta\eta/\eta$ (%)	$\Delta\eta_e/\eta_e$ (%)
0.02	0.03	0°	−0.40	−0.42	20		−3.84	−3.87
0.06	0.03	0°	−0.20	−0.25	20		0.07	0.08
0.01	0.07	0°	−0.64	−0.62	20		−6.88	−6.87
0.06	0.07	0°	−0.20	−0.26	20		0.17	0.20
G, kg/(s m^2)	e/D_h	E. Solar insolation, I			F. Ambient temperature, T_a			
		I, W/m^2	$\Delta\eta/\eta$ (%)	$\Delta\eta_e/\eta_e$ (%)	T_a (K)		$\Delta\eta/\eta$ (%)	$\Delta\eta_e/\eta_e$ (%)
0.02	0.03	1000 500	0.50 −2.3	0.64 −2.74	273 303		0.84 −0.91	0.90 −1.03
0.06	0.03	1000 500	0.97 −3.01	4.62 −13.98	273 303		0.16 +0.33	1.43 −2.02
0.01	0.07	1000 500	−0.01 −1.66	0.019 −1.74	273 303		1.23 −1.96	1.25 −1.99
0.06	0.07	1000 500	0.98 −3.03	6.69 −20.24	273 303		0.13 0.40	2.13 −3.42
G, kg/(s m^2)	e/D_h	G. Emissivity of plate, ε_p			H. Combined effect of emissivity, ε_p and wind heat transfer coefficient, h_w			
		ε_p	$\Delta\eta/\eta$ (%)	$\Delta\eta_e/\eta_e$ (%)	ε_p	h_w, kg/(s m^2)	$\Delta\eta/\eta$ (%)	$\Delta\eta_e/\eta_e$ (%)
0.02	0.03	0.1	11.0	10.74	0.1	20	10.21	10.27
0.06	0.03	0.1	5.40	6.34	0.1	20	6.58	7.73
0.01	0.07	0.1	15.35	15.38	0.1	20	12.75	12.79
0.06	0.07	0.1	5.13	6.53	0.1	20	6.40	8.13

16.5 Summary

Flat plate solar collectors have been designed for applications requiring delivery of heated fluid in the low to medium temperature range (5°–70°C above ambient temperature). These collectors absorb both beam and diffuse radiation. The absorbed radiation is converted into heat which is transferred to water or air flowing through the collector tubes or duct, respectively.

Heat collection rate and pumping power requirement of a solar air heater are strong functions of the duct geometrical parameters (L, W, H) and air flow rate per unit area of the absorber plate G. Thus, an appropriate way to evaluate the performance is to take both heat collection rate and pumping power requirement into account, i.e. to carry out a thermohydraulic performance evaluation.

Thermohydraulic performance evaluation and the effect of the variation of ambient, operating and design parameters on the performance of the solar air heater have been evaluated using a mathematical model presented in this chapter. The presented model is validated against the data from an experimental study.

The model calculates the useful heat gain from the iterative solution of basic heat transfer equations of top loss and equates the same with the convective heat transfer rate from the absorber plate to the air using proper heat transfer correlations for the solar air heater duct. The model estimates the collector back loss from the iterative solution of the heat balance equation for the back surface. The edge loss has been calculated from an empirical relation.

The presented equations of the mathematical model constitute a non-linear model for the computation of the useful heat gain Q, thermal efficiency η, pressure loss δp, pumping power P and effective efficiency η_e. The model presented for smooth duct solar air heater has also been utilized by researchers for performance study of roughened duct solar air heater with relevant heat transfer and friction factor correlations in place of smooth duct relations.

Uncertainties in the estimate of wind heat transfer coefficient and sky temperature affect the accuracy of predicted thermal performance of a solar air heater using any mathematical model.

Various correlations for the estimate of the wind heat transfer coefficient from the wind velocity data are available in the literature. The predicted values of the wind heat transfer coefficient from these correlations have been found to vary significantly. The obvious lack of generality of the existing wind convection coefficient correlations presents a challenge for future research. Field rather than laboratory measurements, as well as some sort of standardization in the choice of height above the ground and/or distance from the façade wall or roof with a suitable amendment of the velocity value on the basis of surface orientation and wind direction relative to the surface, are required.

The sky is considered as a blackbody at some fictitious temperature known as sky temperature at which it is exchanging heat by radiation. The sky temperature is a function of many parameters. Some studies assume the sky temperature to be equal to the ambient temperature because it is difficult to make a correct estimate of it, while others estimate it using different correlations. The presented relations give significantly different values of the sky temperature. There can be significant uncertainty in the estimate of the sky temperature because of atmospheric pollution and changes in atmospheric humidity.

The artificial roughness in the form of ribs on the air flow side of the absorber plate has been shown to be one of the most promising methods for enhancement of heat transfer coefficient and hence the thermal performance of the solar air heater with the forced flow of air. In general, any attempt to increase turbulence in the flow also increases the pumping power requirement. Since the artificial roughness on the heat-transferring surface creates turbulence near the wall and breaks the laminar sublayer at the wall, it enhances the heat transfer coefficient with a minimum pumping power penalty.

Extensive experimental studies carried out by researchers have shown that the geometry of the roughness (roughness shape, pitch, height, etc.) and the arrangement of the rib elements (orientation with respect to the flow direction) have a marked influence on the heat transfer and friction characteristics of the roughened surfaces. The mechanism of heat transfer enhancement in the case of inclined or v-discrete pattern of ribs has been explained by discussing the flow structure.

Heat transfer and friction factor correlations for some roughness geometries have been presented followed by typical results of a performance study of solar air heaters with 60° v-down discrete rectangular cross-section repeated rib roughness on the air flow side of the absorber plate. The result of performance study is presented in the form of performance plots (thermal and effective efficiencies versus the air temperature rise parameter $\Delta T/I$), which shows that at the low flow rates of air per unit area of the absorber plate G, the ribs with greater relative roughness height e/D_h are beneficial; lower is the flow rate, greater is the advantage of the use of the artificial roughness on the absorber plate. At $G > 0.045$ kg/sm^2, the smooth duct air heater is preferred from thermohydraulic consideration.

Results of variation of the air mass flow rate per unit area of the plate, ambient temperature, solar insolation, wind heat transfer coefficient, etc. have been presented for a range of these parameters. Effects of change in collector length, duct height and plate emissivity on thermal and thermohydraulic performances represented by change in thermal and effective efficiencies have also been presented.

References

Aharwal KR, Gandhi BK, Saini JS (2008) Experimental Investigation on heat transfer enhancement due to a gap in an inclined continuous rib arrangement in a rectangular duct of solar air heater. Renew Energy 33:585–596

Barrow H, Lee Y (1964) Heat transfer with unsymmetrical thermal boundary conditions. Int J Heat Mass Transf 7:580–582

Bhagoria JL, Saini JS, Solanki SC (2002) Heat transfer coefficient and friction factor correlation for rectangular solar air heater duct having transverse wedge shaped rib roughness on the absorber plate. Renew Energy 25:341–369

Bhatti MS, Shah RK (1987) Turbulent and transition flow convective heat transfer. In: Kakac S, Shah RK, Aung W (eds) Handbook of single-phase convective heat transfer. Wiley, New York Chapter 4

Buchberg H, Catton I, Edwards DK (1976) Natural convection in enclosed spaces-a review of application to solar energy collection. J Heat Transf 1976:182–188

Chitoshiya G, Karwa R (August 2015) Enhanced performance solar air heater with roughened absorber plates: a review. Int J Eng Res Technol (IJERT), NCETRASECT-2015 Conference Proceedings, pp 111–115

Cortes A, Piacentini R (1990) Improvement of the efficiency of a bare solar collector by means of turbulence promoters. Appl Energy 36:253–261

Dippery DF, Sabersky RH (1963) Heat and momentum transfer in smooth and rough tubes at various Prandtl numbers. Int J Heat Mass Transfer 6:329–353

Duffie JA, Beckman WA (1980) Solar energy thermal process. Wiley, New York

Ebadian MA, Dong ZF (1998) Forced convection, internal flow in ducts. In: Rohsenow WM, Hartnett JP, Cho YI (eds) Handbook of heat transfer. McGraw-Hill, New York Chapter 5

Garg HP, Prakash J (2000) Solar energy: fundamental and applications. Tata-McGraw-Hill, New Delhi, India

Gupta D, Solanki SC, Saini JS (1993) Heat and fluid flow in rectangular solar air heater ducts having transverse rib roughness on absorber plate. Sol Energy 51(1):31–37

Gupta D, Solanki SC, Saini JS (1997) Thermo-hydraulic performance of solar air heaters with roughened absorber plates. Sol Energy 61:33–42

Hatton AP, Quarmby A, Grundy I (1964) Further calculations on the heat transfer with turbulent flow between parallel plates. Int J Heat Mass Transfer 7:817–823

Heaton HS, Reynolds WC, Kays WM (1964) Heat transfer in annular passages. Simultaneous development of velocity and temperature fields in laminar flow. Int J Heat Mass Transfer 7:763–781

Hollands KGT, Shewen EC (1981) Optimization of flow passage geometry for air-heating, plate-type solar collectors. J Solar Energy Eng 103:323–330

Holman JP (1990) Heat transfer, 7th edn. McGraw-Hill book Co, New York

Jaurker AR, Saini JS, Gandhi BK (2006) Heat transfer and friction characteristics of rectangular solar air heater duct using rib-grooved artificial roughness. Sol Energy 80:895–907

Karmare SV, Tikekar AN (2007) Heat transfer and friction factor correlation for artificially roughened duct with metal grit ribs. Int J Heat and Mass Transfer 50:4342–4351

Karwa Rajendra (2003) Experimental studies of augmented heat transfer and friction in asymmetrically heated rectangular ducts with ribs on the heated wall in transverse, inclined, V-continuous and V-discrete pattern. Int Comm Heat and Mass Transfer 30(2):241–250

Karwa Rajendra, Bairwa RD, Jain BP, Karwa N (2005) Experimental study of the effect of rib angle and discretization on heat transfer and friction in an asymmetrical heated rectangular duct. J Enhanced Heat Transfer 12(4):343–355

Karwa Rajendra, Baghel Shweta (2014) Effect of parametric uncertainties, variations and tolerances on thermo-hydraulic performance of flat plate solar air heater. J Solar Energy, Hindawi Publishing Corporation 2014, Article ID 194764, 18 p. https://doi.org/10.1155/2014/194764

Karwa Rajendra, Chauhan Kalpana (2010) Performance evaluation of solar air heaters having v-down discrete rib roughness on the absorber plate. Energy 35:398–409

Karwa Rajendra, Chitoshiya Girish (2013) Performance study of solar air heater having v-down discrete ribs on absorber plate. Energy 55:939–955

Karwa Rajendra, Garg SN, Arya AK (2002) Thermo-hydraulic performance of a solar air heater with *n*-subcollectors in series and parallel configuration. Energy 27:807–812

Karwa Rajendra, Karwa Nitin, Misra Rohit, Agarwal PC (2007) Effect of flow maldistribution on thermal performance of a solar air heater array with subcollectors in parallel. Energy 32:1260–1270

Karwa Rajendra, Maheshwari BK, Karwa Nitin (2008) Flow visualization and local mass transfer studies for turbulent flow in a wind tunnel with chamfered ribs. J Visualization, Springer 11(3): 205–212

Karwa Rajendra, Sharma Anil, Karwa Nitin (Feb. 2010) A comparative study of different roughness geometries proposed for solar air heater ducts. Int Review of Mechanical Engineering (IREME), Special Issue on Heat Transfer 4(2):159–166

Karwa Rajendra, Solanki SC, Saini JS (1999) Heat transfer coefficient and friction factor correlations for the transitional flow regimes in rib-roughened rectangular duct. Int J Heat Mass Transfer 42:1597–1615

Karwa Rajendra, Solanki SC, Saini JS (2001) Thermo-hydraulic performance of solar air heaters having integral chamfered rib roughness on absorber plates. Energy 26:161–176

Kays WM, Crawford ME (1980) Convective heat and mass transfer. Tata McGraw-Hill Publishing Co, New Delhi, India

Kays WM, Leung EY (1963) Heat transfer in annular passages- hydrodynamically developed turbulent flow with arbitrarily prescribed heat flux. Int J Heat Mass Transfer 6:537–557

Kays WM, London AL (1964) Compact heat exchangers. McGraw-Hill, New York

Kays WM, Perkins HC (1973) Forced convection, internal flows in ducts. In: Rohsenow WM, Hartnett JP (eds) Handbook of heat transfer. McGraw-Hill, New York Chapter 7

Klein SA (1975) Calculation of flat-plate collector loss coefficients. Sol Energy 17:79–80

Kumar S, Sharma VB, Kandpal TC, Mullick SC (1997) Wind induced heat losses from outer cover of solar collectors. Renewable Energy 10(4):613–616

Kumar Suresh, Mullick SC (2007) Convective heat transfer from the top outermost cover of a flat plate solar collector at low wind speeds. J Solar Energy Society of India 17(1, 2):87–93

Lanjewar A, Bhagoria JL, Sarvaiya RM (2011) Experimental study of augmented heat transfer and friction in solar air heater with different orientation of W-rib roughness. Exp Thermal Fluid Sci 35:986–995

Layek A, Saini JS, Solanki SC (2007) Heat transfer and friction characteristics for artificially roughened duct with compound turbulators. Int J Heat and Mass Transfer 50:4845–4854

Lee Y (1968) Turbulent heat transfer from the core tube in thermal entrance regions of concentric annuli. Int J Heat Mass Transfer 11:509–522

McAdams WH (1954) Heat transmission, 3rd edn. McGraw-Hill, New York

Momin AME, Saini JS, Solanki SC (2002) Heat transfer and friction in solar air heater duct with V-shaped rib roughness on absorber plate. Int J Heat Mass Transfer 45:3383–3396

Muluwork KB, Saini JS, Solanki SC (1998) Studies on discrete rib roughened solar air heater, In: Proceeding of National Solar energy Convection, Roorkee, India, 1998, p 75–84

Nikuradse J (November 1950) Laws of flow in rough pipes. NACA Technical Memorandum 1292

Niles PW, Carnegie EJ, Pohl JG, Cherne JM (1978) Design and performance of an air collector for industrial crop dehydration. Sol Energy 20:19–23

Nowak H (1989) The sky temperature in net radiant heat loss calculations from low-sloped roofs. Infrared Phys 29(2–4):231–232

Palyvos JA (2008) A survey of wind convection coefficient correlations for building envelope energy systems' modeling. Appl Therm Eng 28(8–9):801–808

Saini RP, Saini JS (1997) Heat transfer and friction factor correlations for artificially roughened ducts with expanded metal mesh as roughness element. Int J Heat Mass Transfer 40(4):973–986

Saini RP, Verma J (2008) Heat transfer and friction factor correlations for a duct having dimple-shape artificial roughness for solar air heaters. Energy 33:1277–1287

Singh S, Chander S, Saini JS (2011) Thermal and effective efficiency based analysis of discrete V-down rib roughened solar air heaters. J Renewable and Sustainable Energy 3(023107):1–26

Sparrow EM, Lloyd JR, Hixon CW (1966) Experiments on turbulent heat transfer in an asymmetrically heated rectangular duct. J Heat Transfer May 1966: 170–174

Swinbank WC (1963) Long-wave radiation from clear skies. Q J Roy Meteor Soc 89:339

Tan HM, Charters WWS (1970) An experimental investigation of forced-convective heat transfer for fully-developed turbulent flow in rectangular duct with asymmetric heating. Sol Energy 13:121–125

Test FL, Lessmann RC, Johary A (1981) Heat transfer during wind flow over rectangular bodies in the natural environment. Trans ASME J Heat Transfer 103:262–267

Watmuff JH, Charters WWS, Proctor D (1977) Solar and wind induced external coefficients- solar collectors. Cooperation Mediterraneenne pour l'Energie Solaire, Revue Internationale d'Heliotechnique, 2nd Quarter, p. 56

Appendix A

Conversion Factors

Acceleration	$1\ \mathrm{m/s^2} = 4.252 \times 10^7\ \mathrm{ft/h^2}$
Area	$1\ \mathrm{m^2} = 10.764\ \mathrm{ft^2}$
	$1\ \mathrm{ft^2} = 0.092903\ \mathrm{m^2}$
Density	$1\ \mathrm{kg/m^3} = 0.062428\ \mathrm{lb_m/ft^3}$
Energy	1 kJ = 0.9478 Btu
	1 kcal = 4.1868 kJ
Force	$1\ \mathrm{N} = 0.22481\ \mathrm{lb_f}$
	$1\ \mathrm{lb_f} = 4.44822\ \mathrm{N}$
Heat flux	$1\ \mathrm{W/m^2} = 0.3171\ \mathrm{Btu/(h\ ft^2)} = 0.8598\ \mathrm{kcal/(h\ m^2)}$
Heat transfer coefficient	$1\ \mathrm{W/(m^2\ °C)} = 0.1761\ \mathrm{Btu/(h\ ft^2\ °F)}$
	$1\ \mathrm{Btu/(h\ ft^2\ °F)} = 5.6786\ \mathrm{W/(m^2\ K)}$
Latent heat	$1\ \mathrm{kJ/kg} = 0.4299\ \mathrm{Btu/lb_m}$
Length	1 m = 3.2808 ft
	1 ft = 0.3048 m
Mass	$1\ \mathrm{kg} = 2.2046\ \mathrm{lb_m}$
	$1\ \mathrm{lb_m} = 0.4536\ \mathrm{kg}$
	1 tonne = 1000 kg
	$1\ \text{short ton} = 2000\ \mathrm{lb_m} = 907.1847\ \mathrm{kg}$
Mass transfer coefficient	$1\ \mathrm{m/s} = 1.181 \times 10^4\ \mathrm{ft/h}$
Power	1 hp = 745.7 W
	$1\ \mathrm{hp_e} = 746\ \mathrm{W}$
	$1\ \mathrm{kW} = 737.56\ \mathrm{lb_f\,ft/s}$
Pressure	$1\ \mathrm{Pa} = 1\ \mathrm{N/m^2} = 1.4504 \times 10^{-4}\ \mathrm{psia}$
	$1\ \mathrm{psia} = 6894.8\ \mathrm{N/m^2}$
	1 atm = 14.69 psia
Specific heat	$1\ \mathrm{kJ/(kg\ °C)} = 0.23885\ \mathrm{Btu/(lb_m\ R)}$
Temperature	$T(\mathrm{K}) = t(°\mathrm{C}) + 273.15$
	$T(\mathrm{R}) = t(°\mathrm{F}) + 459.67$
Thermal conductivity	$1\ \mathrm{W/(m\ K)} = 0.57782\ \mathrm{Btu/(h\ ft\ °F)} = 2.39 \times 10^{-3}\ \mathrm{cal/(cm\ s\ °C)}$

(continued)

R. Karwa, *Heat and Mass Transfer*,
https://doi.org/10.1007/978-981-15-3988-6

Thermal resistance	1 K/W = 0.5275 °F/(h Btu)
Viscosity	1 N/(s m^2) = 1 kg/(s m) = 2419.1 lb_m/(ft h)
	1 poise = 1 g/(s cm)
Volume	1 m^3 = 1000 L
	1 US gallon = 3.7854 L
	1 m^3 = 264.17 US gallon
	1 gal (imperial) = 4.546 L
Volumetric heat generation	1 W/m^3 = 0.09665 Btu/(h ft^3)

See Table A1, A2, A3.1, Table A3.2, A4, A5, A6, A7.

Table A1 SI prefixes

Multiplier	Prefix	Symbol
10^{18}	exa	E
10^{15}	peta	P
10^{12}	tera	T
10^{9}	giga	G
10^{6}	mega	M
10^{3}	kilo	k
10^{2}	hecto	h
10	deka	da
10^{-1}	deci	d
10^{-2}	centi	c
10^{-3}	milli	m
10^{-6}	micro	μ
10^{-9}	nano	n
10^{-12}	pico	p
10^{-15}	femto	f
10^{-18}	atto	a

Table A2 Physical constants

Avogadro's number (N_A)	6.0222×10^{26} molecules/(kg mole)
Boltzmann's constant (*k*)	$(1.38032 \pm 0.00011) \times 10^{-23}$ J/(mol K)
	8.6171×10^{-5} ev/K
Planck's constant (*h*)	$(6.6237 \pm 0.0011) \times 10^{-34}$ J/s
Stefan-Boltzmann constant (σ)	5.66961×10^{-8} W/(m^2 K^4)
Speed of light in vacuum (*c*)	2.9979×10^{8} m/s
Speed of sound in dry air at 0°C (1 atm)	331.36 m/s
Universal gas constant (R_o)	(8.31436 ± 0.00038) J/(mol K)
Standard gravitational acceleration (g)	980.665 cm/s^2

(continued)

Table A2 (continued)

Standard atmospheric pressure (1 atm)	1.013246×10^5 N/m^2
	1.01325 bar
	760 mm Hg (0°C)
	10.3323 m H_20 (4°C)
	1.03323 kg$_f$/cm^2
	14.696 psi
1 cm Hg	1333 N/m^2
1 mm Hg	1 torr
1 atm	760 torr
1 bar	1×10^{-5} N/m^2
1 Newton (N)	1 kg m/s^2
1 dyne	1 g cm/s^2 = 10^{-5} N
1 joule (J)	1 kg m^2/s^2 = 10^7 erg
1 kcal	4186 J
1 ev	1.602×10^{-19} J
1 liter	1000.028 cm^3
1 Angstrom (A)	10^{-10} m
$\ln x$	2.30258 $\log_{10} x$
e	2.71828
1°	0.01745 radians
π	3.14159
Ice point (at 1 atm)	273.15 K

Note At standard conditions 1 kg-mole occupies 22.4146 ± 0.0006 m^3 volume

Table A3.1 Thermophysical properties of metals

Material	Temperature range (°C)	Density ρ (kg/m^3) × 10^{-3}	Specific heat c at 20 °C [kJ/(kg K)]	Thermal conductivity k [W/(m K)]
Aluminium	0–400	2.71	0.895	202–250
Duralumin	0–200	2.79	0.883	159–194
Brass (70%Cu, 30% Zn)	100–300	8.52	0.380	104–147
Bronze (75% Cu, 25% Sn)	0–100	8.67	0.340	26
Copper	1–600	8.95	0.380	385–350
Cast Iron	20	7.26	0.420	52
Wrought iron	0–1000	7.85	0.46	59–33
Lead	1–300	11.35	0.130	35–31
Mercury	0–300	13.6	0.14	8–14
Nickel	0–400	8.9	0.445	93–59
Silver (99.9%)	0–400	10.52	0.234	410–360
Steel (C = 0.5%)	0–1000	7.83	0.465	55–29
Steel (C = 1%)	0–1000	7.80	0.470	43–28
Cr-Steel (Cr = 1%)	0–1000	7.86	0.460	62–33
18–8 Stainless steel	0–600	7.82	0.460	16–26
Tin	1–200	7.3	0.220	66–57
Tungsten	0–800	19.3	0.133	166–76
Zinc	0–400	7.14	0.385	117–98

Table A3.2 Thermophysical properties of non-metals

Material	Temperature (°C)	Density ρ (kg/m^3)	Specific heat c [kJ/(kg K)]	Thermal conductivity k [W/(m K)]
Non-metals				
Asbestos (loose)	0–100	400–900	0.82	0.15–0.22
Brick	25	1600–1900	0.84	0.4–0.7
Calcium silicate	40	190		0.055
Clay	0–100	1460	0.88	1.3 at 20 °C
Concrete	25	1900–2300	0.88	0.81–1.40
Coal (anthracite)	25	1350	1.25	0.25
Cork	25	45–130	1.8	0.4
Cotton	25	80	1.3	0.06
Glass, window	25	2700	0.84	0.78
Glass wool	25	24–160	0.7–0.8	0.04
Granite	25	2630	0.78	2.8
Gypsum plaster/board	25	600–800	0.6-1.1	0.17–0.3
Ice	0	920	2.04	1.88
Lime stone	25	2300–2500	0.8-0.9	1.25–2.2
Magnesia 85%	40	185–270		0.05–0.07
Marble	25	2500–2700	0.8-0.85	2.0–2.8
Mica (across layers)				0.71
Mineral wool	25	160–190	1.03	0.045
Paper	25	930	1.34	0.18
Paraffin	25	900	2.9	0.24
Plaster, cement sand	25	1850		0.72
Plywood, bonded	25	545	1.21	0.12
Porcelain	25	2400		5.0
Polystyrene (expanded)	25	35–55	1.5	0.027–0.04
Polyurethane foam (rigid)	25	30–45	1.4-1.5	0.025
Rubber, hard	25	1200	2.01	0.15
Sand	25	1510	0.8	0.27
Sand stone	25	2150–2300	0.75	2.9
Sawdust	25			0.059
Wood (oak)	25	550–800	2.4	0.17–0.35
Wood (balsa)	25	140		0.055

Table A4 Thermophysical properties of water at atmospheric pressure

t (° C)	Density (kg/m^3)	Specific heat [kJ/(kg K)]	Thermal conductivity [W/(m K)]	Dynamic viscosity × 10^6 [N s/m^2 or kg/(m s)]	Prandtl number Pr	Volume expansion coefficient $\beta \times 10^3$ (1/K)
5	999.9	4.204	0.572	1501	11.1	0.016
10	999.7	4.193	0.585	1300	9.44	0.081
15	999.0	4.186	0.591	1136	8.08	0.144
20	998.2	4.183	0.601	1002	7.00	0.201
25	997.0	4.181	0.609	890	6.13	0.253
30	995.6	4.179	0.617	797	5.40	0.304
35	994.0	4.178	0.624	718	4.81	0.342
40	992.2	4.179	0.631	651	4.30	0.385
45	990.2	4.181	0.637	594	3.90	0.422
50	988.1	4.182	0.644	544	3.55	0.459
55	985.2	4.183	0.648	501	3.24	0.484
60	983.3	4.185	0.654	463	3.00	0.521
65	980.4	4.188	0.658	430	2.76	0.553
70	977.5	4.191	0.663	400	2.56	0.586
75	974.7	4.194	0.667	374	2.39	0.614
80	971.8	4.198	0.670	351	2.23	0.652
85	969.0	4.203	0.673	330	2.09	0.675
90	965.3	4.208	0.675	311	1.97	0.700
95	961.5	4.213	0.677	294	1.86	0.721
100	958.3	4.211	0.682	282	1.76	

Table A5 Thermophysical properties of air at atmospheric pressure

t, (°C)	Density (kg/m^3)	Specific heat, c_p [kJ/(kg K)]	Dynamic viscosity × 10^5 [N s/m^2 or kg/(m s)]	Thermal conductivity [W/(m K)]	Thermal diffusivity × 10^4 (m^2/s)	Pr
0	1.3005	1.0055	1.7127	0.02412	0.1860	0.715
25	1.1868	1.0057	1.8363	0.02608	0.2190	0.709
50	1.0949	1.0072	1.9512	0.02799	0.2570	0.703
75	1.0052	1.0089	2.0658	0.02990	0.2952	0.697
100	0.9452	1.0113	2.1720	0.03170	0.3340	0.693
125	0.8872	1.0138	2.2776	0.03350	0.3729	0.689
150	0.8370	1.0171	2.3769	0.03522	0.3969	0.686
175	0.7873	1.0204	2.4761	0.03693	0.4203	0.683
200	0.7474	1.0248	2.570	0.03859	0.4846	0.681
225	0.7079	1.0291	2.664	0.04025	0.5510	0.680
250	0.6762	1.0339	2.752	0.04186	0.6005	0.680
275	0.6448	1.0388	2.841	0.04347	0.6493	0.680
300	0.6174	1.0466	2.926	0.04497	0.6983	0.680
325	0.5901	1.0545	3.011	0.04647	0.7473	0.680

(continued)

Table A5 (continued)

t, (°C)	Density (kg/m^3)	Specific heat, c_p [kJ/(kg K)]	Dynamic viscosity × 10^5 [N s/m^2 or kg/(m s)]	Thermal conductivity [W/(m K)]	Thermal diffusivity × 10^4 (m^2/s)	Pr
350	0.5674	1.0589	3.091	0.04794	0.8004	0.681
375	0.5448	1.0632	3.171	0.04941	0.8535	0.682
400	0.5247	1.0690	3.248	0.05080	0.9082	0.683
425	0.5046	1.0747	3.325	0.05219	0.9628	0.684
450	0.4883	1.0799	3.400	0.05358	1.0179	0.685
475	0.4722	1.0852	3.475	0.05498	1.0730	0.686
500	0.4570	1.0913	3.564	0.05633	1.1317	0.687

The following relations for thermophysical properties of the air may be used:

$$c_p = 1006\left(\frac{T_m}{293}\right)^{0.0155} \tag{a}$$

$$k = 0.0257\left(\frac{T_m}{293}\right)^{0.86} \tag{b}$$

$$\mu = 1.81 \times 10^{-5}\left(\frac{T_m}{293}\right)^{0.735} \tag{c}$$

$$\rho = 1.204\left(\frac{293}{T_m}\right) \tag{d}$$

$$\mathrm{Pr} = \frac{\mu c_p}{k} \tag{e}$$

where temperatures are in K.

Table A6 Order of magnitude of convective heat transfer coefficients

Mode	Heat transfer coefficient, h [W/(m^2 K)]
Natural or free convection, gases	5–25
Forced convection, air (1 atm)	10–250
Forced convection, air (200 atm)	200–1000
Forced convection, organic liquids	100–1000
Forced convection, water	500–10^4
Boiling, organic liquids	500–2.5 × 10^4
Boiling, water	2500–5 × 10^4
Condensation, organic vapours	500–10^4
Condensation, water vapour	2000–5 × 10^4

Table A7 Fouling factors*, m^2 K/W

Temperature of heating medium	Up to 115 °C		115°C – 200 °C	
Temperature of water	50 °C or less		>50 °C	
Velocity	<0.9 m/s	> 0.9m/s	<0.9 m/s	>0.9 m/s
Type of water				
Seawater	0.0002	0.0002	0.0002	0.0002
Hard water	0.0005	0.0005	0.001	0.001
River water	0.00035–0.0005	0.0002–0.00035	0.0005–0.0007	0.00035–0.0005
Distilled water	0.0001	0.0001	0.0001	0.0001
Treated boiler feed water	0.0002	0.0001	0.0002	0.0002
Muddy water	0.0005	0.0005	0.001	0.001
Other Fluids				
Fuel oil			0.00035–0.0009	
Engine lubricating oil			0.0002	
Transformer oil			0.0002	
Refrigerant (liquid), including ammonia			0.0002	
Brine			0.0002–0.0005	
Methanol, ethanol, ethylene glycol solutions			0.00035	
Steam			0.0001	
Air			0.0002–0.0004	
Diesel engine exhaust			0.002	
Refrigerant vapour, including ammonia			0.0002	

*For details, refer to Standards of Tubular Exchanger Manufacturers Association (1999), 8th edn., New York

See Table A8.

Table A8 Total hemispherical emissivity of surfaces*

Surface	Temperature (°C)	Emissivity (ε)
Metals		
1. Aluminum		
Highly polished (98 % pure)	225–575	0.039–0.057
Polished	20	0.05–0.06
Commercial sheet	25	0.06–0.09
Oxidized at 600 °C	200–600	0.02–0.03
2. Brass (70% Cu, 30% Zn)		
Highly polished	250–375	0.033–0.07
Rolled natural surface	22	0.06
Dull plate	50–350	0.2–0.25
Oxidized (heated to 600 °C)	200–600*600*	0.45–0.6
Surface	Temperature (°C)	Emissivity (ε)
3. Bronze	100–300	
Polished		0.03–0.07
Commercial		0.2–0.25

(continued)

Table A8 (continued)

Surface	Temperature (°C)	Emissivity (ε)
4. Cast iron		
Polished	200	0.15–0.25
Newly turned	20	0.44
Rough	20	0.55–0.65
Rough and oxidized	40–250	0.6–0.9
5. Copper		
High polish	20	0.02
Commercial polish	80	0.03
Commercial		0.1–0.2
Heated for a long time, coated with oxide layer	25	0.78
6. Chromium	40–500	0.08–0.3
7. Gold, polished	200–600	0.018–0.035
8. Monel metal	200–600	0.4–0.45
9. Nichrome wire, oxidized	50–500	0.95–0.98
10. Nickel		
Polished	100	0.05–0.07
Oxidized	650–1250	0.35–0.8
11. Platinum: polished, pure	200–600	0.05–0.1
12. Silver, polished	30–500	0.02–0.032
13. Stainless steel (18%Cr, 8%Ni), polished	0–1000	0.07–0.17
14. Steel		
Polished	100	0.06–0.35
Oxidized at 600 °C	200–600	0.64–0.78
Oxidized and rough	40–300	0.9–0.95
Molten	1500–1550	0.42–0.53
15. Tin, polished	0–200	0.04–0.06
16. Tungsten filament	3300	0.39
17. Zinc, polished	0–400	0.02–0.03
Oxidized		0.1–0.11
Galvanized		0.2–0.3
Building and construction materials		
Asbestos	25–350	0.93–0.95
Brick		
Red and rough	20	0.9–0.95
Magnesite refractory	1000	0.38
Fireclay	1000	0.75
Lime stone	100–400	0.95–0.8
Surface	Temperature (°C)	Emissivity (ε)
Sand stone	0–300	0.83–0.90
Concrete	0–200	0.94
Marble	0–100	0.93–0.95
Plaster, rough lime	10–90	0.91
Porcelain, glazed	25	0.92
Glass, window (smooth)	0–600	0.88
Wood (oak)	0–100	0.9

(continued)

Table A8 (continued)

Surface	Temperature (°C)	Emissivity (ε)
Rubber, glossy plate	25	0.94
Paper	90	0.89
Paints		
Aluminium paint, varying age	20–100	0.27–0.67
Black shiny lacquer on iron	25	0.87
Enamel, white fused on iron	20	0.9
Flat black paint (lacquers)	35–100	0.96–0.98
Lampblack, 0.075 mm layer and thicker (rough deposit)	40–350	0.95–0.9
Others		
Water	0–100	0.95–0.96
Ice	0	0.97–0.99

*The emissivity values may vary significantly with the condition of the surface

Appendix B

Short Questions

Useful short questions for competitive examinations (including GATE and IES)

R. Karwa, *Heat and Mass Transfer*,
https://doi.org/10.1007/978-981-15-3988-6

1. For metals, the ratio of thermal conductivity and electrical conductivity equals to

(a) Prandtl number (b) Schmidt number
(c) Lewis number (d) Lorenz number

[**Ans.** (d).]

2. In the case of one-dimensional heat conduction through a plain wall without heat generation and with constant properties, $(\partial t/\partial \tau)$ is proportional to

(a) t/x (b) $\partial t/\partial x$
(c) $\partial^2 t/\partial x^2$ (d) $\partial^2 t/\partial x \partial \tau$

[**Ans.** (c).]

3. A composite slab has two layers of different materials with thermal conductivity k_1 and k_2. If each layer has the same thickness, the equivalent thermal conductivity of the slab would be

(a) $k_1\, k_2$ (b) $k_1 + k_2$
(c) $(k_1 + k_2)/\, k_1 k_2$ (d) $2k_1 k_2/(k_1 + k_2)$

[**Hint:** $\dfrac{t_1 - t_2}{\dfrac{2\delta}{kA}} = \dfrac{t_1 - t_2}{\dfrac{\delta}{k_1 A} + \dfrac{\delta}{k_2 A}}$; solve for k]

[**Ans.** (d).]

4. For steady flow of heat and no heat generation, the temperature distribution in a plane wall of constant value of thermal conductivity is

(a) linear (b) Parabolic
(c) Logarithmic (d) Cubic

[**Ans.** (a).]

5. In a composite plain wall shown in figure, the interface temperature is average of the end face temperatures. For steady state one-dimensional heat conduction the ratio of thermal conductivities is given by

(a) $k_1/k_2 = ½$
(b) $k_1/k_2 = 3/2$
(c) $k_1/k_2 = 2$
(d) $k_1/k_2 = 1$

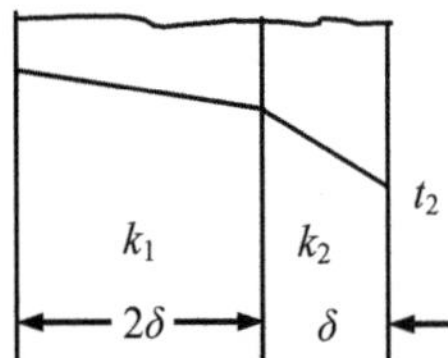

[**Hint:** $k_1 A \frac{t_1 - t_m}{2\delta} = k_2 A \frac{t_m - t_2}{\delta}$, where $t_m = \frac{t_1 + t_2}{2}$; solve for k_1/k_2]

[**Ans.** (c).]

6. The wall of a furnace is made of bricks (thickness 0.45 m and thermal conductivity 0.7 W/(m K). For the same heat loss and temperature drop, the brick wall can be replaced by a layer of insulation of conductivity 0.35 W/(m K) and of thickness

   (a) 0.9 m (b) 0.45 m,
   (c) 0.225 m (d) 0.15 m.

[**Hint:** $\frac{\delta_1}{k_1 A} = \frac{\delta_2}{k_2 A}$, solve for δ_2]

[**Ans.** (c).]

7. For the furnace wall shown in figure, the interface temperature will be

   (a) 510°C
   (b) 870°C
   (c) 1020°C
   (d) 600°C

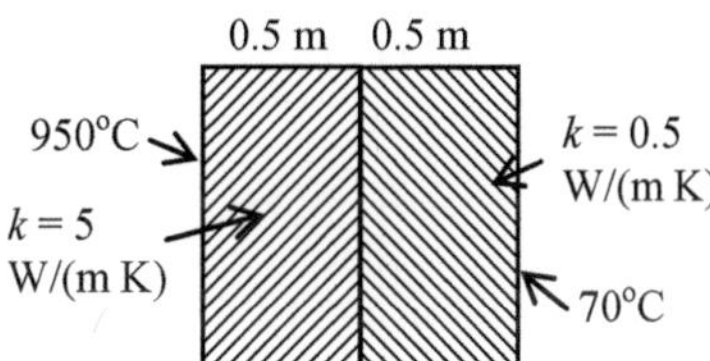

[**Hint:** $\frac{T_1 - T_2}{\frac{\delta_1}{k_1} + \frac{\delta_2}{k_2}} = \frac{T_m - T_2}{\frac{\delta_2}{k_2}}$; solve for T_m.]

[**Ans.** (b)]

8. Consider steady state heat conduction across the thickness in a plane composite wall (as shown in the figure) exposed to convection conditions on both sides. Given: $h_i = 20$ W/(m^2 K), $h_o = 50$ W/(m^2 K), $T_i = 20$°C, $T_o = -2$°C, $k_1 = 20$ W/(m K), $k_2 = 50$ W/(m K), $\delta_1 = 0.3$ m, $\delta_2 = 0.15$ m. Assuming negligible contact resistance between the wall surfaces, the temperature at the interface of the two walls will be

   (a) -0.50
   (b) 2.75
   (c) 3.75
   (d) 4.50

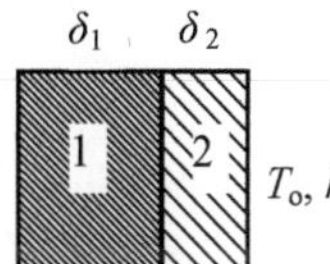

[**Hint:** $\dfrac{T_i - T_o}{\dfrac{1}{h_i} + \dfrac{\delta_1}{k_1} + \dfrac{\delta_2}{k_2} + \dfrac{1}{h_o}} = \dfrac{T_m - T_o}{\dfrac{\delta_2}{k_2} + \dfrac{1}{h_o}}$; solve for T_m.]

[**Ans.** (c)]

9. A steam pipe is covered with two layers of insulating material with better insulating material next to the pipe. If the layers of the insulating materials are interchanged, the conduction heat transfer will

(a) will decrease (b) will increase
(c) will not change (d) may increase of decrease

[**Ans.** (b).]

10. In a large plate, the steady temperature distribution is as shown in Fig. 1. If no heat is generated in the plates, the thermal conductivity k varies as (t is temperature in °C, k_o is thermal conductivity at 0°C and β is a constant)

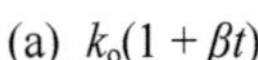

(a) $k_o(1 + \beta t)$
(b) $k_o(1 - \beta t)$
(c) $k_o - \beta t$
(d) $k_o + \beta t$

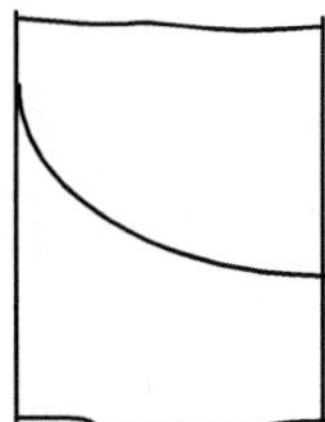

[**Ans.** (b); refer Fig. 2.34]

11. If the thermal conductivity of a material of plane wall varies with temperature as $k_0(1 + \beta t)$, then the temperature at the central plane of the wall as compared to that in the case of constant thermal conductivity will be

(a) less (b) more
(c) the same (d) depends on other factors

[**Ans.** (b); refer Fig. 2.34]

12. The temperature drop through layers 1 and 2 of a furnace wall is shown in figure. For $\delta_1 = \delta_2$ and $T_1 > T_2$, which one of the following statements is correct?

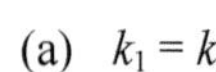

(a) $k_1 = k_2$
(b) $k_1 < k_2$
(c) $k_1 > k_2$
(d) none of the above

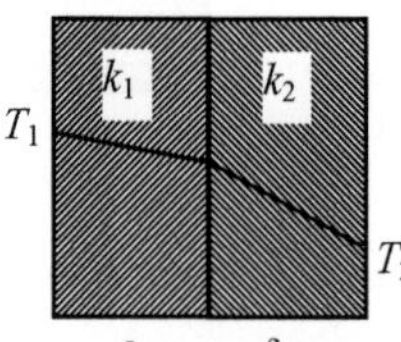

[**Hint:** Higher is the thermal conductivity, lower is the temperature drop]

[**Ans.** (c)]

13. Three plane walls of the same cross sectional area and thickness are having thermal conductivities in the ratio 1:2:3. The temperature drops in these walls will be in the ratio

(a) 1: 1: 1 (b) 1: 2:3
(c) 1/3 : 1/2 : 1 (d) 1: 1/2 : 1/3

[**Hint:** Higher is the thermal conductivity, lower is the temperature drop]

[**Ans.** (d)]

14. For a composite wall of layer thicknesses δ_1, δ_2 and δ_3 and thermal conductivities k_1, k_2 and k_3, respectively, the overall heat transfer coefficient U is given by

(a) $U = \left(\frac{k_1}{\delta_1} + \frac{k_2}{\delta_2} + \frac{k_3}{\delta_3}\right)^{-1}$ (b) $U = \frac{k_1}{\delta_1} + \frac{k_2}{\delta_2} + \frac{k_3}{\delta_3}$

(c) $U = \left(\frac{\delta_1}{k_1} + \frac{\delta_2}{k_2} + \frac{\delta_3}{k_3}\right)^{-1}$ (d) $U = \frac{\delta_1}{k_1} + \frac{\delta_2}{k_2} + \frac{\delta_3}{k_3}$

[**Ans.** (c)]

15. What will be the geometric radius of heat transfer for a hollow sphere of inner and outer radii r_1 and r_2 for the same rate of heat transfer?

(a) $\sqrt{(r_1 r_2)}$ (b) $r_1 r_2$
(c) $(r_1 + r_2)/2$ (d) $r_1 r_2/2$

[**Hint:** $k(4\pi r_1 r_2)\frac{t_1 - t_2}{r_2 - r_1} = kA_m \frac{t_1 - t_2}{r_2 - r_1}$

$= k[4\pi(r_m{}^2)]\frac{t_1 - t_2}{r_2 - r_1}$; solve for r_m]

[**Ans.** (a)]

16. If k is thermal conductivity of an insulating material and h_o is the heat transfer coefficient from the surface to air, the critical thickness of insulation for the cylinder is equal to

(a) k/h_o (b) $2k/h_o$
(c) h_o/k (d) $2h_o/k$

[**Ans.** (a)]

17. With an increase in the insulation thickness around a circular cross-section pipe, the heat loss to the surroundings due to

(a) convection and conduction decreases
(b) convection and conduction increases
(c) convection decreases and conduction increases
(d) convection increases and conduction decreases

[**Ans.** (d)]

18. Up to critical thickness of the insulation

a) Added insulation will decrease heat transfer
b) Added insulation will increase heat transfer
c) convection decreases and conduction increases
d) None of the above

[**Ans.** (b)]

19. For air, if values of thermal conductivity k and specific heat at constant pressure c_p are independent of pressure, determine the dependence of thermal diffusivity on pressure.

[**Ans.** Thermal diffusivity $\alpha = \dfrac{k}{\rho c_p}$ and from perfect gas equation $\rho = \dfrac{p}{RT} \propto p$. Hence, $\alpha \propto \dfrac{1}{p}$, i.e. thermal diffusivity α varies inversely with pressure p.]

20. Figure shows a two-dimensional body [k = 5 W/(m K)] and two isothermal surfaces A (at 50°C) and B (at 80°C). The heat enters the body at A with temperature gradient of 20 K/m. Determine temperature gradients $\partial t / \partial x$ and $\partial t / \partial y$ at surface B where the heat is leaving.

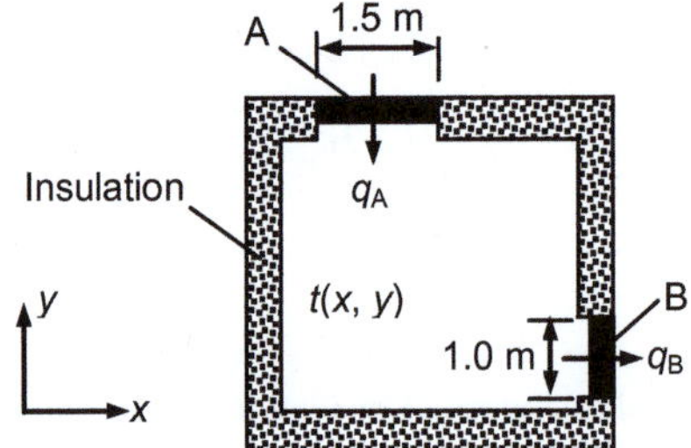

[**Ans.** Heat flux vector is always normal to an isothermal surface. Since surface B is an isothermal vertical surface, $\partial t / \partial y = 0$. From heat conservation, $q_A = q_B$, i.e. $-k\left(A\dfrac{\partial t}{\partial y}\right)_A = -k\left(A\dfrac{\partial t}{\partial x}\right)_B$; For unit depth (in direction z), $(1.5 \times 1) \times 20 = (1.0 \times 1.0) \times \left(\dfrac{\partial t}{\partial x}\right)_B$, i.e. $\left(\dfrac{\partial t}{\partial x}\right)_B = 30\,\text{K/m}$.]

21. For one dimensional axial heat flow in the conical cylinder shown in figure (with $T_1 > T_2$ and $k = k_o - \beta T$, where β is positive), the variation of heat flow rate q_x, heat flux q_x'' and temperature gradient dT/dx with increasing x is:

(i) q_x: increases/decreases/remains constant
(ii) q_x'': increases/decreases/remains constant
(iii) dT/dx: increases/decreases/remains constant

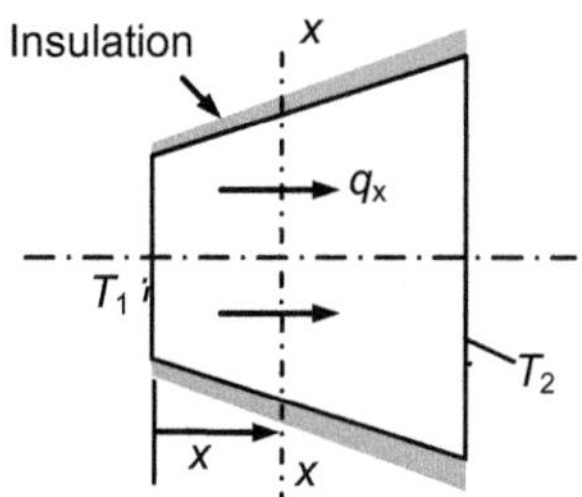

[**Ans.** (i) Sides are insulated hence q_x remains constant, (ii) since $q_x''=\dfrac{q_x}{A_x}$ and A_x increases with increasing x hence q_x'' decreases with increasing x (iii) since $\left(\dfrac{dT}{dx}\right)=-\dfrac{q_x''}{k}$, and k increases with increasing x because of decrease in temperature with increase in x, dT/dx decreases with increasing x.]

22. Temperature distribution in a solid cylinder of diameter 0.2 m is given by $T(\mathrm{r}) = a + br^2$, where $a = 50$ and $b = -1000$. The thermal conductivity of cylinder material is 40 W/(m K). If the convection coefficient at cylinder surface is 500 W/(m^2 K), find the fluid temperature.

[**Ans.** At the cylinder surface,

$$-k(2\pi r_o L)\left(\frac{dT}{dr}\right)_{r=r_o} = h(2\pi r_o L)[T(r=r_o)-T_\infty], \quad \text{or} \quad -k\left[\frac{d}{dr}(a+br^2)\right]_{r=r_o} = h[(a+br_o^2)-T_\infty];$$

simplification and substitution of values of k, a, b and h gives fluid temperature $T_\infty = 24°\mathrm{C}$.]

23. A fin of length L protrudes from a surface at temperature T_o, which is higher than the ambient temperature T_a. The heat dissipation from the free end of the fin is negligibly small. What is the temperature gradient (dt/dx) at the tip of the fin?

(a) $(T_o - T_L)/L$
(b) T_L/L
(c) $(T_L - T_a)/L$
(d) Zero

[**Ans.** (d)]

24. A fin becomes effective if the Biot number Bi is
    (a) less than one (b) more than one
    (c) equal to one (d) does not depend on Bi

[**Ans.** (a)]

25. For a fin to be effective, which one of the following is true?
    (a) $\sqrt{(hA_c/kP)} = 1$ (b) $\sqrt{(hA_c/kP)} > 1$
    (c) $\sqrt{(hA_c/kP)} < 1$ (d) $1 < \sqrt{(hA_c/kP)} < 2$

[**Ans.** (c)]

26. Effectiveness of a fin will be the maximum in an environment with
    (a) forced convection
    (b) free convection
    (c) boiling liquid
    (d) condensing vapour

[**Ans.** (b)]

27. A plane wall of thickness L has a uniform volumetric heat generation source q_g. It is exposed to local ambient temperature T_a at both faces. The surface temperature T_s of the wall under steady state condition is given by
    (a) $T_s = T_a + \frac{q_g L}{2h}$ (b) $T_s = T_a + \frac{q_g L}{2k}$
    (c) $T_s = T_a + \frac{q_g L}{h}$ (d) $T_s = T_a + \frac{q_g L}{k}$

[**Hint:** Heat generated in the wall = heat rejected by convection from both faces, i.e. $ALq_g = 2 \times hA(T_s - T_a)$]

[**Ans.** (a)]

28. Consider steady one-dimensional heat flow in a plate of 20 mm thickness with a uniform heat generation rate of 80 MW/m^3. The left and right faces are kept at constant temperature of 160°C and 120°C, respectively. The plate has a constant thermal conductivity of 200 W/(m K). The location and magnitude of the maximum temperature within the plate from its left face.

[**Hint and answer:** $\frac{d^2t}{dx^2} + \frac{q_g}{k} = 0$, $\frac{dt}{dx} = -\frac{q_g}{k}x + C_1$ and $t = -\frac{q_g}{2k}x^2 + C_1 x + C_2$.

Obtain C_1 and C_2 from condition $x = 0$, $T = 160$°C and $x = 0.02$ m, $T = 120$°C. For the location of maximum temperature, $dt/dx = 0$ gives, $x = C_1 k/q_g = 5$ mm and temperature distribution equation gives $t_{x = 5\text{mm}} = 165$°C.]

29. Temperature distribution in a one-dimensional wall of thickness 0.25 m is given as $T(x) = 100 - 1000x^2$. Thermal conductivity of the wall material is 40 W/(m K). Determine the heat generation rate in the wall.

[**Ans.** From Eq. 2.13(c), $q_g = -k\dfrac{\partial^2 T}{\partial x^2} = -k\dfrac{\partial^2}{\partial x^2}\left(100 - 1000x^2\right)$ = 80000 W/m^3; Alternatively calculate from difference of heat flow rates at two faces of the wall divided by volume of the wall, i.e. $q_g = \dfrac{1}{AL}\left[\left(-kA\dfrac{\partial T}{\partial x}\right)_{x=L} - \left(-kA\dfrac{\partial T}{\partial x}\right)_{x=0}\right]$.]

30. In transient heat conduction, the two significant dimensionless parameters are

(a) Reynolds number and Prandtl number
(b) Reynolds number and Biot number
(c) Biot number and Fourier number
(d) Fourier number and Reynolds number

[**Ans.** (c)]

31. Biot number may be expressed as

(a) ratio of buoyancy to viscous forces
(b) ratio of internal thermal resistance of a solid to the boundary layer thermal resistance
(c) ratio of gravitational and surface tension forces
(d) ratio of heat conductance rate to the rate of thermal energy storage in a solid

[**Ans.** (b)]

32. The curve for the unsteady state cooling or heating of bodies is

(a) Parabolic curve asymptotic to the time axis
(b) exponential curve asymptotic to the time axis
(c) exponential curve asymptotic both to the time and temperature axes
(d) hyperbolic curve asymptotic to the time axis

[**Ans.** (b); refer Fig 6.2]

33. A spherical thermocouple junction of diameter 0.706 mm is to be used for the measurement of temperature of a gas stream. The convective heat transfer coefficient on the bead surface is 400 W/(m^2 K). The thermophysical properties of the thermocouple material are k = 20 W/(m K), c = 400 J/(kg K) and ρ = 8500 kg/m^3. If the thermocouple initially at 30°C is placed in a hot stream of 300°C, the time taken by the bead to reach 298°C is

(a) 29.4 s (b) 2.35 s
(c) 14.7 s (d) 4.9 s

[**Hint:** Bi = hL/k = 0.0023 < 0.1, lumped heat capacity analysis applies. $\frac{T-T_\infty}{T_i-T_\infty} = \exp(-hA\tau/c\rho V)$, where $A = 4\pi r^2$ and $V = (4/3)\,\pi r^3$, solution gives $\tau = 4.9$ s]

[**Ans.** (d)]

34. A 10 mm diameter heated metal spherical ball is cooled in air. If the thermal conductivity of ball material is 20 W/(m K) and heat transfer coefficient is 10 W/(m^2 K), do you expect radial gradients to be appreciable?

[**Ans.** $\text{Bi} = \frac{h(d/6)}{k} = \frac{10\times(10/6000)}{20} = 8.33\times10^{-4}$; Since Bi << 0.1, radial temperature gradients will not be appreciable. The sphere is behaving as span-wise isothermal object.]

35. A small copper ball of 5 mm diameter at 500 K is dropped into an oil bath whose temperature is 300 K. The thermal conductivity of the copper is 400 W/(m K), its density 9000 kg/m^3 and its specific heat 385 J/(kg K). If the heat transfer coefficient is 250 W/(m^2 K) and lumped heat capacity analysis is assumed to be valid, the rate of the fall of the temperature of the ball at the beginning of the cooling will be, in K/s,

(a) 8.7 (b) 13.9
(c) 17.3 (d) 27.7

[**Hint:** $\frac{T-T_\infty}{T_i-T_\infty} = \exp(-hA\tau/c\rho V)$, $\left(\frac{dT}{d\tau}\right)_{\tau=0} = -(hA/c\rho V)(T_i - T_\infty)$, solution gives $\left(\frac{dT}{d\tau}\right)_{\tau=0}$ = 17.3 K/s]

[**Ans.** (c)]

36. The water pipe line in cold countries is laid at a certain depth from the earth surface in order to

(a) supply warm water
(b) reduce friction losses
(c) prevent water from freezing
(d) none of the above

[**Ans.** (c)]

37. It is desired to increase the heat dissipation rate from the surface of an electronic device of spherical shape of 5 mm radius exposed to air [h = 5 W/(m^2 K)] by encasing it in a spherical sheath of conductivity 0.04 W/ (m K). Determine the radius of the sheath for maximum heat flow.

[**Ans.** $r_c = r_{seath} = 2k/h_o = (2 \times 0.04/5) \times 1000 = 16$ mm.]

38. A fin is 5 mm in diameter and 100 mm long. Determine the heat flow rate from the fin. Given: thermal conductivity of the fin material k = 400 W/(m K), fin base temperature t_s = 130°C, surrounding air temperature t_a = 30°C and the convective heat transfer coefficient h = 40 W/(m^2 K).

[**Hint:** $A_c << PL; q_{fin} = \sqrt{hPkA_c}(t_s - t_\infty)\tanh mL$, where $m = \sqrt{\frac{hP}{kA_c}}$.]

[**Ans.** 5 W.]

39. A thermocouple in a thermometer well measures the temperature of a gas flowing through a pipe. For low error of measurement of temperature, the thermometer well should be made of which material of the following?

(a) copper (b) brass (c) aluminium (d) steel

[**Ans.** (d)]

40. A large concrete slab has one dimensional temperature distribution

$$T = 4 - 10\,x + 20\,x^2 + 10\,x^3$$

where T is temperature and x is distance from one face towards other face of the slab. If slab material has thermal diffusivity of 2 × 10^{-3} m^2/h, what is the rate of change of temperature at x = 1 m?

[Ans. $\frac{\partial T}{\partial x} = -10 + 40x + 30x^2$, $\frac{\partial^2 T}{\partial x^2} = 40 + 60x$. For one-dimensional transient heat conduction, $\frac{\partial^2 T}{\partial x^2} = \frac{1}{\alpha}\left(\frac{\partial T}{\partial \tau}\right)$. Hence, $\frac{\partial T}{\partial \tau} = \alpha \frac{\partial^2 T}{\partial x^2} = 2 \times 10^{-3} \times (40 + 60x)$; $\left(\frac{\partial T}{\partial \tau}\right)_{x=1} = 0.2°$ C/h.]

41. Using thermal-electrical analogy in heat transfer, match list I (electrical quantities) with List II (thermal quantities) and select the correct answer using the codes given below the lists.

List I

A. Voltage B. Current

C. Resistance D. Capacitance

List II

1. Thermal resistance
2. Thermal capacity
3. Heat flow
4. Temperature

Codes

	A	B	C	D
(a)	2	3	1	4
(b)	4	1	3	2
(c)	2	1	3	4
(d)	4	3	1	2

[**Ans.** (d)]

42. Match list I (governing equations of heat transfer) with list II (specific cases of heat transfer) and select the correct answer using the code given below the lists.

List I

A. $\frac{d^2T}{dr^2}+\frac{2}{r}\left(\frac{dT}{dr}\right)=0$ B. $\frac{\partial^2T}{\partial x^2}=\frac{1}{\alpha}\left(\frac{\partial T}{\partial \tau}\right)$

C. $\frac{d^2T}{dr^2}+\frac{1}{r}\left(\frac{dT}{dr}\right)=0$ D. $\frac{d^2\theta}{dx^2}-m^2\theta=0$

List II

1. Pin fin 1-D case
2. 1-D conduction in cylinder
3. 1-D conduction in sphere
4. Plane slab

Codes

	A	B	C	D
(a)	2	4	3	1
(b)	3	1	2	4
(c)	2	1	3	4
(d)	3	4	2	1

[**Ans.** (d)]

Fill in the blanks:

43. Thermal diffusivity controls the temperature distribution in the .... state. (unsteady/steady)

[**Ans.** Unsteady.]

44. For steady state and constant value of thermal conductivity, the temperature distribution associated with radial conduction through a cylinder has a .... curve.

[**Ans.** Logarithmic.]

45. For steady state and constant value of thermal conductivity, the temperature distribution associated with radial conduction through a spherical shell has a .... curve.

[**Ans.** Hyperbolic.]

46. In the case of steady state and one dimensional heat conduction (in the radial direction only) without heat generation, the equation of heat conduction for the cylinder is ....

[**Ans.** $\left(\frac{1}{r}\frac{dt}{dr}+\frac{d^2t}{dr^2}\right)=0$]

47. The heat flow equation through a cylindrical vessel (inner radius r_1, outer radius r_2 and length L) is desired in the same form as that for a plane wall. The equivalent area is given by ....

[Ans. $\frac{A_2 - A_1}{\ln\left(A_2/A_1\right)}$]

48. One dimensional (radial), unsteady state conduction heat transfer equation for a sphere with heat generation at the rate q_g is ....

[**Ans.** $\frac{1}{r^2}\frac{\partial}{\partial r}\left(r^2\frac{\partial t}{\partial r}\right)+\frac{q_g}{k}=\frac{1}{\alpha}\left(\frac{\partial t}{\partial \tau}\right)$]

49. If k is thermal conductivity of an insulating material and h_o is the heat transfer coefficient from the surface to air, the critical thickness of insulation for the sphere is equal to ....

[**Ans.** $2k/h_o$]

50. Fins should be used on the side where the heat transfer coefficient is .... (low/high)

[**Ans.** Low.]

51. Fins should have .... perimeter to area of cross-section ratio. (small/ Large)

[**Ans.** Large.]

52. Thermal conductivity of the fin material should be .... (low/high)

[**Ans.** High.]

53. The insulated tip temperature of a rectangular longitudinal fin having an excess (over ambient) root temperature of θ_o is ....

[**Ans.** $\frac{\theta}{\theta_o}=\frac{\cosh m(L-x)}{\cosh mL}$; for the tip $x = L$ hence $\theta_L = \frac{\theta_o}{\cosh mL}$.]

54. The overall heat transfer coefficient U for a composite plane wall of n layers is given by (the thickness of the ith layer is δ_i, thermal conductivity of the ith layer is k_i, convective heat transfer coefficients are h_i and h_o) ....

[**Ans.** $\left(\frac{1}{h_i}+\sum_{i=1}^{n}\frac{\delta_i}{k_i}+\frac{1}{h_o}\right)^{-1}$]

55. Answer the following questions:

   (i) What is equivalent diameter for an annulus of a double pipe heat exchanger for the calculation of heat transfer coefficient?
   (ii) What is equivalent diameter for a rectangular duct of width W and height H for the calculation of heat transfer coefficient?
   (iii) What is LMTD and when is it used?
   (iv) Write the equation relating overall heat transfer coefficient and individual heat transfer coefficients

56. Prandtl number is given by

   (a) $\mu c_p/k$ (b) hL/k
   (c) $k/\mu c_p$ (d) hk/L

   [Ans. (a)]

57. Stanton number is the ratio of

   (a) Reynolds number to Prandtl number
   (b) Prandtl number to Nusselt number
   (c) Nusselt number to Peclet number
   (d) Peclet number to Reynolds number

   [Ans. (c)]

58. In a heat exchanger where one of the fluids undergoes phase change,

   (a) the two fluids should flow opposite to each other
   (b) the two fluids should flow parallel to each other
   (c) the two fluids should flow normal to each other
   (d) the direction of flow of the fluids is of no significance

   [Ans. (d)]

59. For natural convection heat transfer, Nusselt number is a function of

   (a) Prandtl number Grashof number
   (b) Reynolds number and Grashof number
   (c) Reynolds number and Prandtl number
   (d) Stanton number and Peclet number

   [Ans. (a)]

60. For forced convection heat transfer, Nusselt number is a function of

(a) Prandtl number Grashof number
(b) Reynolds number and Grashof number
(c) Reynolds number Prandtl number
(d) Stanton number and Peclet number

[**Ans.** (c)]

61. Reynolds number is given by the ratio of two physical quantities. Name these quantities.
62. A cold fluid is stored in spherical vessel in order to

(a) reduce rate of heat transfer
(b) increase rate of heat transfer
(c) prevent the liquid from freezing
(d) none of the above

[**Ans.** (a)]

63. In a heat exchanger, for a given heat flow rate and also the same inlet and outlet temperatures, the heat transfer area will be minimum for

(a) counter flow arrangement
(b) parallel flow arrangement
(c) cross-flow arrangement
(d) none of the above

[**Ans.** (a)]

64. Consider the following conditions for heat transfer (thickness of thermal boundary layer is δ_t, velocity boundary layer is δ and Prandtl number is Pr)

(1) $\delta = \delta_t$, when Pr = 1
(2) $\delta > \delta_t$, when Pr > 1
(3) $\delta < \delta_t$, when Pr < 1

Which of the above conditions apply for the convective heat transfer?

(a) 1 and 2 (b) 2 and 3 (c) 1 and 3 (d) 1, 2 and 3

[**Ans.** (d)]

65. The thickness of the thermal boundary layer is equal to the hydrodynamic boundary layer when the Prandtl number is equal to

(a) 0.0 (b) 0.1 (c) 0.5 (d) 1.0

[**Ans.** (d)]

66. For flow of a fluid over a heated plate, the following fluid properties are known: $\mu = 0.001$ Pa s, $c_p = 1$ kJ/(kg K) and $k = 1$ W/(m K). The hydrodynamic boundary layer thickness at a location is 1 mm. The thermal boundary layer thickness at the same location is

(a) 0.001 mm (b) 0.01 mm
(c) 1 mm (d) 1000 mm.

[**Hint:** $\text{Pr} = \mu c_p/k = 1$ hence $\delta_t = \delta$.]

[**Ans.** (c)]

67. The velocity profile of a fluid flowing through a tube depends on

(a) the velocity of the fluid
(b) the diameter of the tube
(c) the viscosity of the fluid
(d) the Reynolds number

[**Ans.** (d)]

68. The velocity and temperature distribution in a pipe flow are given by $U(\text{r})$ and $T(\text{r})$, respectively. If u_m is the mean velocity at any section of the pipe, the bulk mean temperature at that section is ....

[**Ans.** $\dfrac{\int_0^R (2\pi r dr)U(r)T(r)}{\pi R^2 u_m}$]

69. For laminar flow over a flat plate, the local heat transfer coefficient h_x varies as $(x)^{-1/2}$, where x is the distance from the leading edge of the plate. The ratio of the average heat transfer coefficient between the leading edge and location at L on the plate to the local heat transfer coefficient $h_{\text{x}=\text{L}}$ at L is

(a) 8 (b) 4 (c) 2 (d) 1

[**Ans.** (c)]

70. Nusselt number for fully developed turbulent flow in a pipe is given by $\text{Nu} = C\,\text{Re}^\text{a}\,\text{Pr}^\text{b}$. The values of a and b are

(a) a = 0.5 and b = 0.33 for heating and cooling both
(b) a = 0.5 and b = 0.4 for heating and b = 0.3 for cooling
(c) a = 0.8 and b = 0.4 for heating and b = 0.3 for cooling
(d) a = 0.8 and b = 0.3 for heating and b = 0.4 for cooling

[**Ans.** (c)]

71. In the case of free convection over a vertical flat plate, the Nusselt number Nu varies with Grashof number Gr as

(a) $Gr^{1/4}$ and $Gr^{1/3}$ for laminar and turbulent flows, respectively
(b) $Gr^{1/3}$ and $Gr^{1/4}$ for laminar and turbulent flows, respectively
(c) $Gr^{1/3}$ and $Gr^{1/2}$ for laminar and turbulent flows, respectively
(d) $Gr^{1/2}$ and $Gr^{1/3}$ for laminar and turbulent flows, respectively

[**Ans.** (a)]

72. A fluid of thermal conductivity 1.0 W/(m K) flows as fully developed flow with Reynolds number of 1500 through a pipe of diameter 0.1 m. The heat transfer coefficient in W/(m^2 K) for uniform heat flux and uniform wall temperature conditions are, respectively

(a) 36.58 and 43.64
(b) 43.64 and 36.58
(c) 43.64 for both the cases
(d) 36.58 for both the cases

[**Hint:** Uniform heat flux, Nu = 4.364 and constant surface temperature, Nu = 3.658]

[**Ans.** (b)]

73. If $MLT\theta$ system (T being time and θ is temperature), what is the dimension of thermal conductivity?

(a) $ML^{-1}T^{1}\theta^{3}$ (b) $MLT^{-3}\theta^{-1}$
(c) $MLT^{-2}\theta^{-1}$ (d) $MLT^{-1}\theta^{-1}$

[**Ans.** (b)]

74. In the case of flow across a horizontal cylinder of diameter d, free convective heat transfer coefficient for $10^9 \leq Ra \leq 10^{12}$ will

(a) vary as $d^{0.75}$ (b) vary as $d^{0.25}$
(c) vary as $d^{0.33}$ (d) independent of d

[**Ans.** (d)]

75. A heated plate maintained at constant temperature 90°C is cooled by a coolant at 30°C. The temperature distribution in the boundary layer is given by ($20 + 50\ e^{-y}$), where temperature is in °C and distance y is in m. If the thermal conductivity of the coolant is 0.6 W/(m K), determine the heat transfer coefficient.

[**Ans.** $h = -k\,(\partial T/\partial y)_{y=0}/(T_{water} - T_{surface})$; $(\partial T/\partial y)_{y=0} = [50 \times (-1) \times e^{-y}]_{y=0} = -50$. Hence, $h = -0.6 \times (-50)/(90 - 30) = 0.5$ W/(m^2 K).]

76. A duct of rectangular cross-section 1 m × 0.5 m, carrying air at 20°C with a velocity of 10 m/s, is exposed to an ambient of 30°C. For air in the range of 20-30°C, $k = 0.025$ W/(m K), viscosity = 18×10^{-6} Pa s, density = 1.2 kg/m^3. Determine the Reynolds number for the flow.

[**Hint:** $D_h = 4WH/[2(W + H)] = 2/3$; $Re = \rho U D_h/\mu$]

[**Ans.** 4.44×10^5]

77. The average heat transfer for the surface of a thin hot vertical plate suspended in still air can be determined from the observations of the change in the plate temperature with time as it cools. Assume the plate temperature to be uniform at any instant of time and radiation heat exchange with the surroundings is negligible. The ambient temperature is 25°C, the plate has total surface area of 0.1 m^2 and a mass of 4 kg. The specific heat of the plate material is 2.5 kJ/(kg K). The convective heat transfer coefficient in W/(m^2 K), at the instant when the plate temperature is 225°C and the change in plate temperature with time $dT/d\tau = -0.02$ K/s, is

(a) 200 (b) 20 (c) 15 (d) 10.

[**Hint:** $hA\,(T - T_a) = -\,mc_p\,(dT/d\tau)$]

[**Ans.** (d)]

78. The average Nusselt number in laminar natural convection from a vertical wall at 180°C with still air at 20°C is found to be 48. If the wall temperature becomes 30°C, all other parameters including the thermophysical properties remaining the same, the average Nusselt number will be

(a) 8 (b) 16 (c) 24 (d) 32.

[**Hint:** $Nu \propto (\mathrm{Gr})^{1/4} \propto (\Delta T)^{1/4}$]

[**Ans.** (c)]

79. Determine the ratio of length to diameter for minimum surface area of a hot water container to reduce convection heat transfer.

[**Ans.** Volume of the container, $V = (\pi/4)D^2L$. Surface area, $A_s = 2 \times (\pi/4)D^2 + \pi DL$. Putting $L = 4V/\pi D^2$, $A_s = 2\times(\pi/4)D^2 + 4V/D$; $dA_s/dD = \pi D - 4V/D^2$, Equating dA_s/dD to 0 for minimum or maximum value of A_s and putting $V = (\pi/4)D^2L$, we get the necessary condition as $L/D = 1$; $d^2A_s/dD^2 = \pi + 8V/D^3$, which is a positive quantity hence the condition $L/D = 1$ refers to the minimum surface area of the container.]

80. Variation of local heat transfer coefficient h_x with distance x for laminar flow over a flat plate is given as $h_x = Cx^{-0.5}$, where C is a constant. Determine ratio of average heat transfer coefficient to the local heat transfer coefficient at distance x.

[**Ans.** $h_{av} = \frac{1}{x}\int_0^x h_x dx = \frac{C}{x}\int_0^x x^{-0.5} dx = \frac{C}{x}\left[\frac{x^{0.5}}{0.5}\right]_0^x = 2Cx^{-0.5} = 2h_x$. Hence, $h_{av}/h_x = 2$.]

81. Variation of local heat transfer coefficient h_x with distance x for free convection from a vertical heated plate is given as $h_x = Cx^{-0.25}$, where C is a constant. Determine ratio of average heat transfer coefficient to the local heat transfer coefficient at distance x.

[**Ans.** $h_{av} = \frac{1}{x}\int_0^x h_x dx = \frac{C}{x}\int_0^x x^{-0.25} dx = \frac{C}{x}\left[\frac{x^{0.75}}{0.75}\right]_0^x = \frac{4}{3}Cx^{-0.25} = \frac{4}{3}h_x$. Hence, $h_{av}/h_x = 4/3$.]

82. Local heat transfer coefficient h_x for a flat surface ($L = 3$ m) varies with distance x as $h_x = 1 + 20x - 5x^2$. Determine maximum value of the local heat transfer coefficient.

[**Ans.** $\frac{d}{dx}(h_x) = \frac{d}{dx}(1 + 20x - 5x^2) = 0$ gives $x = 2$ m; $\frac{d^2}{dx^2}(h_x) = \frac{d^2}{dx^2}(1 + 20x - 5x^2) = -10 < 0$. Hence, at $x = 2$ m maximum of h_x occurs and its value is 21W/(m^2 K).]

83. For one dimensional heat flow in the wall [$k_s = 40$ W/(m K)], determine the convection heat transfer coefficient and temperature gradient in the fluid [$k_f = 0.5$ W/(m K)] at the wall surface.

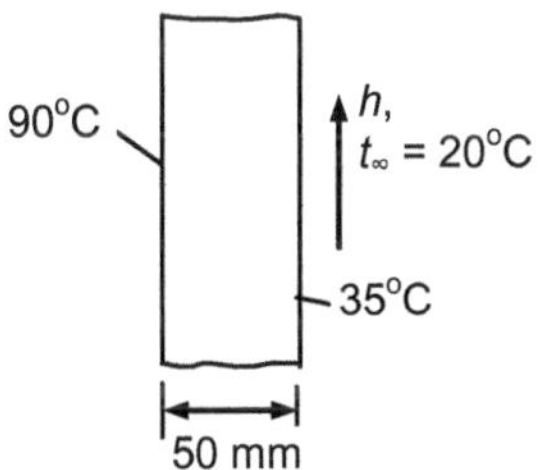

[**Ans.** $q_{\text{conduction}} = q_{\text{convection}}$ gives $k_s A\frac{t_1 - t_2}{\delta} = hA(t_2 - t_\infty)$; Substitution of values of various terms gives $h = 2933$ W/(m^2 K). Also $-k_f A\left(\frac{dt}{dx}\right)_{x=\delta} = hA(t_2 - t_\infty)$, which gives $\left(\frac{dt}{dx}\right)_{x=\delta} = -\frac{h}{k_f}(t_2 - t_\infty) = -87990\,°\text{C/m}$.]

84. For fully developed Couette flow between two parallel plates at a gap of 5 mm, determine shear stress. One of the plates is moving at 100 m/s while the other is stationary. Fluid viscosity is 2×10^{-5} Ns/m^2.

[**Ans.** For fully developed Couette flow, the velocity profile is linear. Hence, shear stress $\tau = \mu\left(\frac{du}{dy}\right) = \mu\left(\frac{U}{L}\right) = 2\times10^{-5}\times\left(\frac{100}{0.005}\right) = 0.4\,\text{N/m}^2$.]

85. Local Nusselt number correlation for turbulent flow over a flat plate is given as $Nu_x = 0.0296\ Re_x^{0.8}\ Pr^{1/3}$. Find the ratio of average heat transfer coefficient to local coefficient. Assume flow to be turbulent from leading edge of the plate.

[**Ans.** From the given correlation, $h_x = \frac{k}{x} Nu_x = 0.0296 \frac{k}{x}\left(\frac{Ux}{\nu}\right)^{0.8} Pr^{1/3} = Cx^{-0.2}$, where $C = 0.0296k\left(\frac{U}{\nu}\right)^{0.8} Pr^{1/3}$; $h_{av} = \frac{1}{x} C \int_0^x x^{-0.2} dx = \frac{C}{x}\left[\frac{x^{0.8}}{0.8}\right]_0^x = \frac{1}{0.8} Cx^{-0.2} = 1.25 h_x$.

Hence, $\frac{h_{av}}{h_x} = 1.25$.]

86. Air at 10 m/s and 25°C ($\upsilon = 0.22 \times 10^{-4}$ m^2/s) flows between infinite parallel plates, which are separated by a distance of 5 mm. Determine the distance at which the boundary layers will merge.

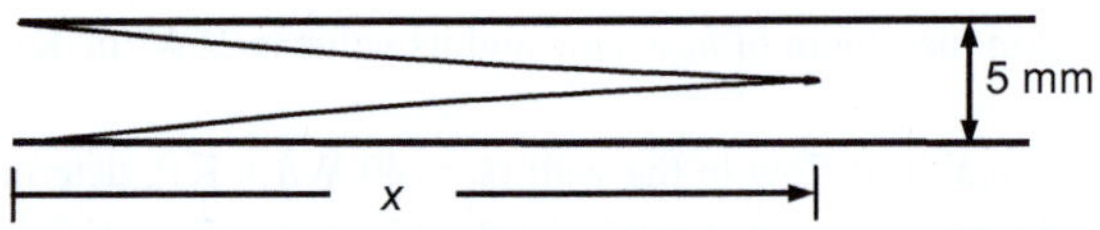

[**Ans.** Boundary layer develops on both plates as shown in figure and they merge at distance x. At x, the thickness of the boundary layer for one plate is 5/2 = 2.5 mm. Assuming the flow to be laminar, we have $\frac{\delta}{x} = \frac{5.0}{\sqrt{Re_x}}$, or $\delta = \frac{5.0x}{\sqrt{U_\infty x/\nu}} = \frac{5.0x^{1/2}}{\sqrt{U_\infty/\nu}}$, or $x = \frac{\delta^2 U_\infty}{25\nu} = \frac{0.0025^2 \times 10}{25 \times 0.22 \times 10^{-4}} = 0.114$ m;

$Re_x = \frac{U_\infty x}{\nu} = \frac{10 \times 0.114}{0.22 \times 10^{-4}} = 51818 < 5 \times 10^5$, i.e. the laminar flow assumption is correct.]

87. Air at atmospheric pressure and a temperature of 50°C flows at 1.5 m/s over an isothermal plate (t_s = 100°C) 1 m long with 0.5 m long unheated starting length. Determine the local heat transfer coefficient at the leading edge with or without unheated starting length.

[**Ans.** With or without unheated starting length, the local heat transfer coefficient at the leading edge is ∞.]

88. Air flowing through a tube at prescribed inlet temperature and mean velocity is heated by condensing steam on outer surface of the tube. What will be effect on heat transfer coefficient and pressure drop if the pressure of the air is doubled? Assume fully developed turbulent flow.

[**Ans.** For given mean velocity, the Reynolds number ($= \rho U_m d/\mu$) will be doubled because the density ($\rho \propto p$) will be doubled. For fully developed turbulent flow, heat transfer coefficient, $h = 0.024\ (k/d)\ Re^{0.8}\ Pr^{0.8}$. Since k and Pr are independent of pressure, we consider $h \propto Re^{0.8}$. Hence, the heat transfer coefficient will increase by a factor of $2^{0.8}$ = 1.74. The pressure loss is given by $\Delta p = \dfrac{4fL\rho U_m^2}{2d}$.

Here f = 0.791 $\text{Re}^{-0.25}$. Hence, $\Delta p = \dfrac{4fL\rho U_m^2}{2D} \propto f\rho \propto \text{Re}^{-0.25}\ \rho$ gives that the pressure drop will increase by factor $2^{-0.25} \times 2 = 2^{0.75}$ = 1.68.]

89. A vertical heated plate is experiencing free convection with quiescent air. Establish the dependence of average heat transfer coefficient on the plate height L if turbulent condition exists.

[**Ans.** For turbulent flow, $h_m = \dfrac{k}{L}\text{Nu}_m = \dfrac{k}{L} \times 0.1\text{Ra}_\text{L}^{1/3} = \dfrac{k}{L} \times 0.1 \times \left[\dfrac{\beta g(T_w - T_\infty)L^3}{\nu\alpha}\right]^{1/3}$

$\propto k \times \left[\dfrac{\beta g(T_w - T_\infty)}{\nu\alpha}\right]^{1/3}$. Turbulent flow heat transfer coefficient does not depend on plate height.]

90. Water flows across a 50 mm diameter long cylinder with surface temperature of 45°C. The free stream conditions are U_∞ = 0.05 m/s and t_∞ = 25°C. At 35°C, thermophysical properties of water are: ρ = 994 kg/m^3, μ = 718 × 10^{-6} N s/m^2, k = 0.624 W/(m K) and β = 0.342 × 10^{-3} (1/K). Check whether free convection effect will be significant?

[**Ans.** $\text{Gr}_\text{D} = \dfrac{\beta g(T_s - T_\infty)D^3}{\nu^2} = \dfrac{0.342\times10^{-3}\times9.81\times(45-25)\times0.05^3}{(718\times10^{-6}/994)^2} = 16\times10^6$; $\text{Re}_\text{D} = \dfrac{U_\infty D}{\nu}$

$= \dfrac{0.05\times0.05}{718\times10^{-6}/994} = 3461$; $\dfrac{\text{Gr}_\text{D}}{\text{Re}_\text{D}^2} = 1.336$, Free convection effect will be significant.]

91. The logarithmic mean temperature difference ($LMTD$) of a counterflow heat exchanger is 20 °C. The cold fluid enters at 20°C and the hot fluid enters at 100°C. Mass flow rate of the cold fluid is twice that of the hot fluid. Specific heat at constant pressure of the hot fluid is twice that of the cold fluid. The exit temperature of the cold fluid is

(a) 40°C (b) 60°C
(c) 80°C (d) cannot be determined.

[Hint: $m_hc_h = m_cc_c$ gives T_{hi} - T_{ho} = T_{co} – T_{ci}, i.e., T_{hi} – T_{co} = T_{ho} – T_{ci}. $LMTD$ = [(T_{hi} – T_{co}) – (T_{ho} – T_{ci})]/ln[(T_{hi} – T_{co})/(T_{ho} – T_{ci})] = 20°C. Since T_{hi} – T_{co} = T_{ho} – T_{ci} hence $LMTD$ equation gives T_{hi} – T_{co} = 20°C]

[Ans. (c)]

92. In a counter flow heat exchanger, hot fluid enters at 60°C and cold fluid leaves at 30°C. Mass flow rate of the hot fluid is 1 kg/s and that of the cold fluid is 2 kg/s, Specific heat of the hot fluid is 10 kJ/(kg K) and that of the cold fluid is 5 kJ/(kg K). $LMTD$ for the heat exchanger in °C is

(a) 15 (b) 30

(c) 35 (d) 45

[Hint: Here $m_h c_h = m_c c_c$ hence T_{hi} - T_{ho} = $T_{co} - T_{ci}$, i.e., $T_{hi} - T_{co} = T_{ho} - T_{ci}$ and *LMTD* in this case = $T_{hi} - T_{co}$ = 30°C.]

[Ans. (b)]

93. In a parallel flow heat exchanger operating under steady state, the heat capacity rates of the cold and hot fluids are equal. The hot fluid [c_p = 4.2 kJ/(kg K)] enters at 90°C while the cold fluid enters at 20°C. The flow rate of the hot fluid is 1 kg/s. If the overall heat transfer coefficient has been estimated as 2 kW/(m^2 K) and the surface area of the exchanger is 2.5 m^2, determine the exit temperature of the cold fluid.

[Ans. $NTU = UA/C_{min}$ = 1.19; ε = [1 – exp (-2NTU)]/2 = 0.4537. From $\varepsilon = (t_{co} - t_{ci})/(t_{hi} - t_{ci})$, t_{co} = 51.76°C.]

94. In a balanced counter flow heat exchanger, the *NTU* is equal to 1. What is the effectiveness of the heat exchanger?

(a) 1.5 (b) 0.5 (c) 0.33 (d) 0.2

[**Hint:** For a balanced flow, $C_h = C_c$; refer Eq. (14.25)]

[**Ans.** (b)]

95. For a heat exchanger clean overall heat transfer coefficient is 250 W/(m^2 K). After one year operation the fouling factors at the tube inner and outer surfaces are 0.002 and 0.001 m^2 K/W. What will be the design overall heat transfer coefficient after one year operation?

[**Ans.** From Eq. (14.37), $\frac{1}{U_d} = \frac{1}{U_c} + R_{ft} = \frac{1}{250} + 0.002 + 0.001 = 0.007$, i.e., U_d = 142.9 W/(m^2 K).]

96. Match list I (type of heat transfer) with list II (governing dimensionless parameter) and select the correct answer using the code given below the lists.

List I
A. Forced convection
B. Natural convection
C. Combined free and forced convection
D. Unsteady conduction with convection at the surface

List II
1. Re, Gr and Pr
2. Re and Pr
3. Fo and Bi

4. Pr and Gr

Codes

	A	B	C	D
(a)	2	1	4	3
(b)	3	4	1	2
(c)	2	4	1	3
(d)	3	1	4	2

[**Ans.** (c)]

97. Match List I (heat exchanger process) with List II (temperature area diagram) and select the correct answer using the codes given below the lists.

List I

A. Counter flow sensible heating
B. Parallel flow sensible heating
C. Evaporating
D. Condensing

List II

1. 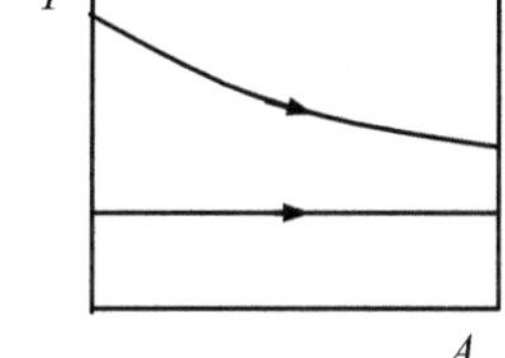

2. 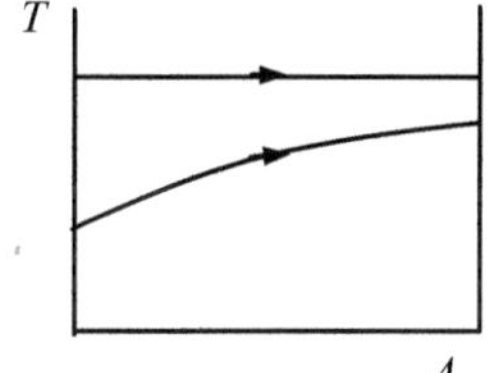

3. 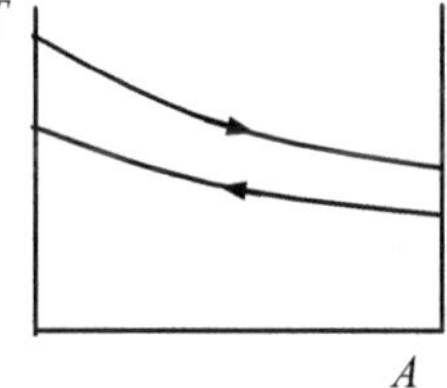

4. 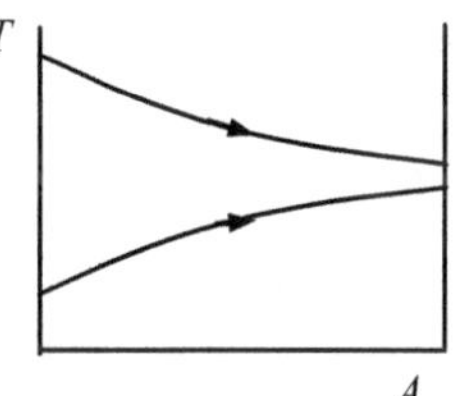

Codes:

	A	B	C	D
(a)	3	4	1	2
(b)	3	2	4	1
(c)	4	3	2	1
(d)	2	4	1	3

[Ans. (a)]

98. Match List I with List II and select the correct answer using the codes given below the lists.

List I	**List II**
A. Fin	1. UA/C_{min}

B. Heisler chart 2. $x/2\sqrt{(\alpha\tau)}$
C. Transient conduction 3. $\sqrt{(hp/kA_c)}$
D. Heat exchanger 4. hL/k

Codes:

	A	B	C	D
(a)	3	1	2	4
(b)	2	1	3	4
(c)	3	4	2	1
(d)	2	4	3	1

[**Ans.** (c)]

99. Saturated steam is allowed to condense over a vertical flat surface and the condensate flows down the surface. The local heat transfer coefficient for the condensation

(a) remains constant at all locations of the surface
(b) decreases with increasing distance from top of the surface
(c) increases with increasing thickness of the condensate film
(d) None of the above

[**Ans.** (b)]

Fill in the blanks:

100. Prandtl number is a .... of a fluid. [**Ans.** Property]
101. For flow through a pipe, the value of the critical Reynolds number is ....[**Ans.** about 2300]
102. For flow past a flat plate, the value of the critical Reynolds number is .... [**Ans.** 5×10^5]
103. According to the Reynolds analogy for Pr = 1, Stanton number is equal to ....

[**Ans.** One half of the friction factor]

104. When there is flow of fluid over a flat plate of length L and h_x is local heat transfer coefficient, the average heat transfer coefficient is given by ....

[**Ans.** $\frac{1}{L}\int_o^L h_x dx$]

105. Two vertical plate arrangements maintained at 10 °C below the saturation temperature of steam at 1 atm are available. One is a single vertical plate of height H and width W. Other arrangement consists of two vertical plates each $H/2$ in height and W in width. Which arrangement will provide the larger condensation rate? Flow is laminar.

[**Ans.** Heat transfer rate, $q = \bar{h}A\Delta T$, i.e. $q \propto \bar{h}$ for given area and temperature difference. Since for laminar flow $\bar{h} \propto H^{-1/4}$, $q \propto H^{-1/4}$, which gives $\frac{q_1}{q_2} = \left(\frac{H}{H/2}\right)^{-1/4} = 2^{-1/4} = 0.84$. Thus the two plates arrangement is 16% better.]

106. The mean condensation heat transfer coefficient for a vertical plate of height H is 5200 W/(m^2 K). What will be heat transfer coefficient if the plate is at inclination of 60° with horizontal?

[**Ans.** For an inclined plane at an angle θ with horizontal, $\left(\bar{h}\right)_{inclined} = \left(\bar{h}\right)_{vertical}\sqrt[4]{\sin\theta}$ from Eq. (13.8). Hence, $\left(\bar{h}\right)_{inclined} = 5200 \times \sqrt[4]{\sin 60} = 5016.3$ W/(m^{2}K).]

107. What will be the critical heat flux for boiling water at 1 atm pressure on Moon with respect to Earth? The gravitational field strength on the Moon is 1/6th that on the Earth.

[**Ans.** From Eq. (13.24), $q_{\max} \propto g^{1/4}$. Hence, critical heat flux at Moon is $(1/6)^{1/4} = 0.64$ times of that on the Earth.]

108. A good absorber of thermal radiation is also a good emitter. It is called

(a) Planck's law
(b) Wien's law
(c) Stefan-Boltzmann's law
(d) Kirchoff's law

[**Ans.** (d)]

109. The Sun's surface at 5800 K emits maximum radiation at a wavelength of 0.5μm. A body at 580 K will emit maximum radiation at a wavelength of nearly

(a) 0.005 μm
(b) 0.05 μm
(c) 5 μm
(d) at the same wavelength

[**Hint:** $\lambda_{\max}T$ = constant]

[**Ans.** (c)]

110. The total average emissivity of a body at a given temperature is given by the relation..........

[**Ans.** $\frac{\int_0^\infty \varepsilon_\lambda E_{b\lambda} d\lambda}{\int_0^\infty E_{b\lambda} d\lambda}$.]

111. Emission of radiation heat from a surface

(a) takes place at all temperatures
(b) takes place only above 273 K
(c) takes place only above room temperature
(d) depends on the surrounding temperature

[**Ans.** (a)]

112. Radiation heat transfer occurs at a speed of

(a) light
(b) sound
(c) 60000 km/hr
(d) none of the above

[**Ans.** (a)]

113. In case of a blackbody

(a) transmissivity is one
(b) reflectivity is one
(c) absorptivity is one
(d) none of the above

[**Ans.** (c)]

114. Planck's law is true for

(a) real bodies
(b) blackbodies only
(c) gray bodies
(d) white bodies only

[**Ans.** (b)]

115. The following figure was generated from experimental data relating spectral blackbody emissive power to wavelength at three temperatures T_1, T_2 and T_3 where $T_1 > T_2 > T_3$.

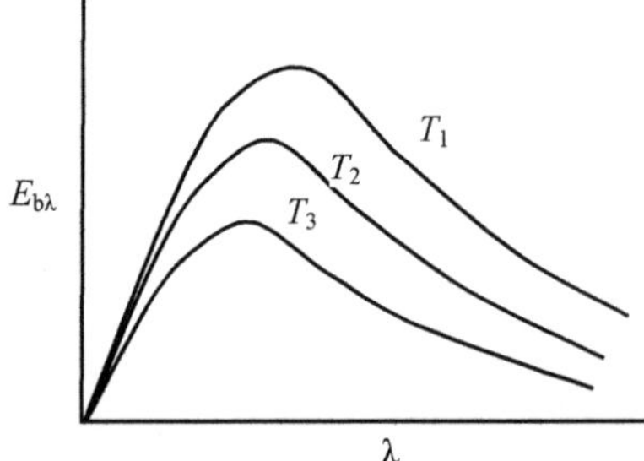

(a) Correct because the maxima in $E_{b\lambda}$ show the correct trend
(b) Correct because the Planck's law is satisfied
(c) wrong because the Stefan-Boltzmann law is not satisfied
(d) wrong because the Wein's displacement law is not satisfied

[**Ans.** (d)]

116. A gray body is that whose absorptivity

(a) changes with temperature
(b) changes with the wavelength of the incident ray
(c) changes with temperature and wavelength of the incident ray
(d) does not change with temperature and wavelength of the incident ray

[**Ans.** (d)]

117. Bodies which reflect more thermal radiation are

(a) white (b) black
(c) gray (d) rough

[**Ans.** (a)]

118. Intensity of radiation at a surface in normal direction is equal to

(a) Product of emissivity of the surface and $1/\pi$
(b) Product of emissivity of the surface and π
(c) Product of emissivity power of the surface and $1/\pi$
(d) Product of emissivity power of the surface and π

[**Ans.** (c)]

119. A …… gas does not emit thermal radiation.

[**Ans.** Monatomic or diatomic]

120. The value of shape factor for infinitely large parallel plates separated by a small distance is .....

[**Ans.** 1]

121. The equation correlating absorptivity α, reflectivity ρ, and transmissivity τ is given by $\alpha + \rho + \tau = N$. What is the value of N?

[Ans. 1]

122. Match the following.

(a) Planck's law	1. Convection
(b) Stanton number	2. Heat exchanger
(c) NTU	3. Radiation heat transfer

[Ans. a-3, b-1, c-2.]

123. Spectral distribution of E_λ for a diffuse surface is shown in figure. Find the total emissive power E.

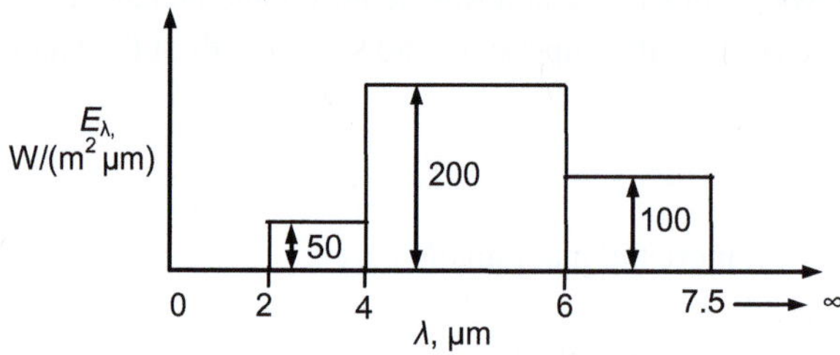

[Ans.

$$E = \int_0^2 (0)d\lambda + \int_2^4 (50)d\lambda + \int_4^6 (200)d\lambda + \int_6^{7.5} (100)d\lambda + \int_{7.5}^{\infty} (0)d\lambda$$

$$= 50\times(4-2) + 200\times(6-4) + 100\times(7.5-6) = 650 \text{ W/m}^2.]$$

124. The Sun can be regarded as nearly a spherical radiation source of diameter $D_s = 1.4 \times 10^9$ m and is at a distance $s = 1.5 \times 10^{11}$ from the Earth. Determine the solid angle subtended by the Sun about a point on the surface of a plate oriented such that its normal passes through the centre of the Sun.

[Ans. From Eq. (10.17), $d\omega = \dfrac{dA}{r^2} = \dfrac{(\pi/4)D_s^{\,2}}{s^2} = \dfrac{\pi\times(1.4\times10^9)^2}{4\times(1.5\times10^{11})^2} = 6.84\times10^{-5}$ sr.]

125. Determine fraction of emissive power leaving a surface in directions $\pi/3 \le \phi \le \pi/2$.

[Ans. For a diffuse surface,

$$\frac{E(\pi/3 \to \pi/2)}{E} = \frac{I_n \int_0^{2\pi} d\theta \int_{\pi/3}^{\pi/2} \sin\phi\cos\phi d\phi}{\pi I_n} = \frac{1}{\pi}\left[2\pi\int_{\pi/3}^{\pi/2} \sin\phi\cos\phi d\phi\right]$$

$$= \int_{\pi/3}^{\pi/2} 2\sin\phi\cos\phi d\phi = \left[\sin^2\phi\right]_{\pi/3}^{\pi/2} = \left(\sin^2\frac{\pi}{2} - \sin^2\frac{\pi}{3}\right) = 0.25.]$$

126. Diffuse flat black surface 1 of area $A_1 = 2 \times 10^{-4}$ m^2 emits diffusely with total emissive power $E = 4 \times 10^4$ W/m^2.

Determine the rate at which this emission is intercepted by the surface *2* of area $A_2 = 3 \times 10^{-4}$ m^2 located as shown in figure. What is irradiation on the surface *2*?

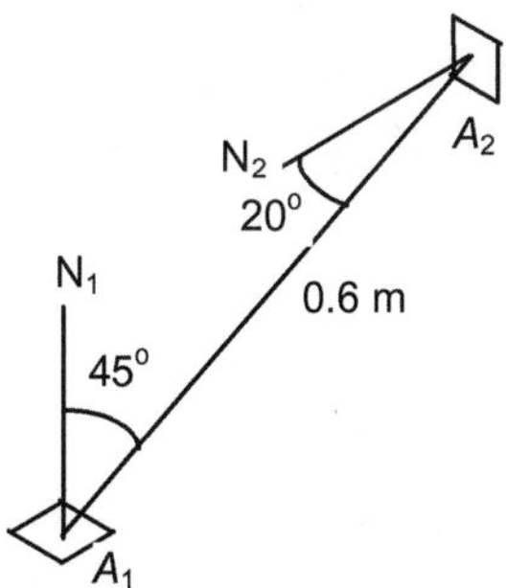

[**Ans.** Areas *1* and *2* are very small hence Eq. (11.2) gives $q_{12} = \frac{E_{b1}}{\pi} \times \frac{A_1 A_2 \cos\phi_1 \cos\phi_2}{s^2} = \frac{4\times10^4}{\pi} \times \frac{2\times10^{-4}\times3\times10^{-4}\times\cos 45^o\times\cos 20^o}{0.6^2} = 1.41\times10^{-3}$ W; Irradiation $G_2 = \frac{q_{12}}{A_2} = \frac{1.41\times10^{-3}}{3\times10^{-4}} = 2.82$ W/m^2.]

127. An enclosure of surface area 50 m^2 and emissivity 0.2 has uniform surface temperature of 600 K. How much radiation is escaping a small opening of 0.001 m^2 in the enclosure?

[**Ans.** Since the area of the aperture is very small, the radiant power emerging through it will correspond to blackbody conditions. Hence, $E = \sigma A_o T_s^4 = 5.67\times10^{-8}\times0.001\times600^4 = 7.35$ W.]

128. What will be the effect on the shape factor, if the configuration in Fig. **a** changes to that in **b**?

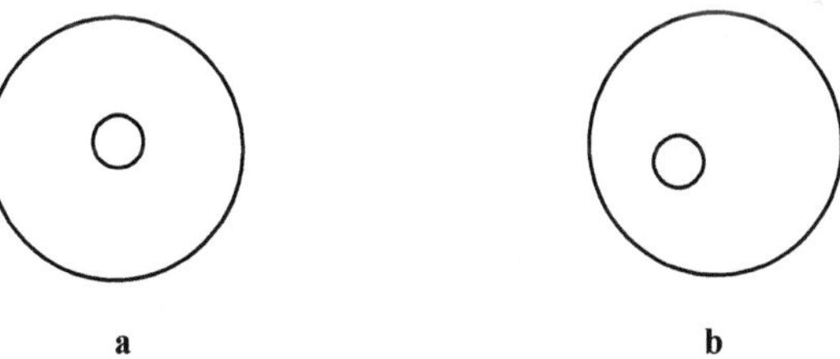

[**Ans.** No change]

129. What is equivalent emissivity for radiation heat exchange between a small body of emissivity 0.6 in a large enclosure of emissivity = 0.7)?

(a) 0.7 (b) 0.6 (c) 0.65 (d) 0.5

[**Ans.** (b)]

130. A solid cylinder (surface 2) is located at the center of a hollow sphere (surface 1). The diameter of the sphere is 1 m, while the cylinder has a diameter and length of 0.5 m each. The radiation configuration factor F_{11} is

(a) 0.375 (b) 0.625
(c) 0.75 (d) 1

[**Hint:** Relations $F_{21} = 1$, $F_{11} = 1 - F_{12}$, $A_1F_{12} = A_2F_{21}$ give $F_{11} = 1 - A_2/A_1$]

[**Ans.** (b)]

131. The shape factor of a hemispherical body 1 placed on a flat surface 2 with respect to itself is
(a) Zero (b) 0.25 (c) 0.5 (d) 1.0

[**Hint:** Relations $F_{21} = 1$, $F_{11} = 1 - F_{12}$, $A_1F_{12} = A_2F_{21}$ give $F_{11} = 1 - A_2/A_1$; $A_1 = 2\pi r^2$, $A_2 = \pi r^2$]

[**Ans.** (c)]

132. A plate having 10 cm^2 area each side is hanging in the middle of a room of 100 m^2 total surface area. The plate temperature and emissivity are 800 K and 0.6, respectively. The temperature and emissivity values for the surfaces of the room are 300 K and 0.3, respectively. The total heat loss from the two sides of the plate by radiation is

(a) 8.2 W (b) 13.66 W
(c) 27.32 W (d) 45.53 W

[**Hint:** $A_1 \ll A_2$. Hence, $q = 2A_1\varepsilon_1\sigma\,(T_1^4 - T_2^4)$]

[**Ans.** (c)]

133. Radiation heat transfer is intended between inner surfaces of two very large isothermal parallel metal plates. While the upper plate (plate 1) is a black surface and is warmer one and being maintained at 727°C, the lower plate (plate 2) is a diffuse gray surface with an emissivity of 0.7 and is kept at 227°C. Assume that the surfaces are sufficiently large to form a two-surface enclosure and steady state conditions to exist. Irradiation (in kW/m^2) for the plate 1 is

(a) 2.5 (b) 3.6
(c) 17.0 (d) 19.5

[**Hint:** Irradiation = $\varepsilon_2\sigma\, T_2^4 + \rho_2\,\sigma\, T_1^4$, where $\rho_2 = 1 - \varepsilon_2$)

[**Ans.** (d)]

134. If plate 1 in above example is also diffuse and gray surface with an emissivity of 0.8, the net radiation heat exchange (in kW/m^2) between plates 1 and 2 is

(a) 17.0 (b) 19.5
(c) 13.0 (d) 31.7

[Hint: $q_{12} = \sigma (T_1^4 - T_2^4)/(1/\varepsilon_1 + 1/\varepsilon_2 - 1)$]

[Ans. (d)]

135. Two long parallel plates of the same emissivity of 0.5 are maintained at different temperatures and have radiation heat exchange between them. If a radiation shield of emissivity 0.25 is placed in the middle, the radiation heat exchange will reduce to

(a) 1/2 (b) 1/4 (c) 3/10 (d) 3/5

[Hint: $q_{old} = \sigma (T_1^4 - T_2^4)/(1/\varepsilon_1 + 1/\varepsilon_2 - 1)$, $q_{new} = \sigma (T_1^4 - T_2^4)/(1/\varepsilon_1 + 1/\varepsilon_2 + 2/\varepsilon_s - 2)$]

[Ans. (c)]

136. The Earth receives at its surface radiation from the Sun at the rate of 1400 W/m^2. The distance of the center of the Sun from the surface of the Earth is 1.5×10^{11} m and the radius of the Sun is 7.0×10^8 m. What is approximately the surface temperature of the Sun treating the Sun as a black body?

[Hint: Refer Example 11.9**]**

[Ans. 5800 K]

137. A semi-transparent plate ($\alpha = 0.3$) is subjected to irradiation G on both sides as shown in figure. The radiosity for one side $J = 4000$ W/m^2. Air at 25°C flows parallel to the plate surface giving convection heat transfer coefficient of 20 W/(m^2 K). If plate surface temperature is at 80°C, find if the plate is behaving as a gray surface.

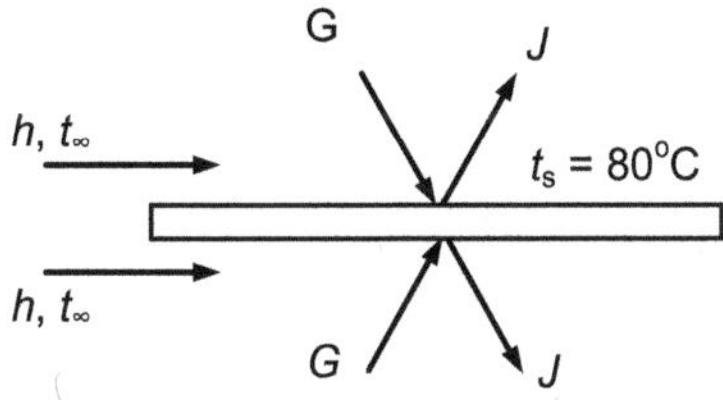

[Ans. Energy balance on the plate gives, $2G = 2J + q_c = 2J + 2hA(t_s - t_\infty)$ $= 2 \times 4000 + 2 \times 20 \times 1 \times (80 - 25) = 10200$ W/m^2. In the present case, $J = \rho G + \tau G + E = (1-\alpha)G + \varepsilon\sigma T_s^4$, or $4000 = (1 - 0.3) \times 5100 + \varepsilon \times 5.67 \times 10^{-8} \times (80 + 273)^4$. Solution gives ε = 0.49. Since $\varepsilon \neq \alpha$, surface is not gray.]

17. With an increase in the insulation thickness around a circular cross-section pipe, the heat loss to the surroundings due to

    (a) convection and conduction decreases
    (b) convection and conduction increases
    (c) convection decreases and conduction increases
    (d) convection increases and conduction decreases

    [**Ans.** (d)]

18. Up to critical thickness of the insulation

    a) Added insulation will decrease heat transfer
    b) Added insulation will increase heat transfer
    c) convection decreases and conduction increases
    d) None of the above

    [**Ans.** (b)]

19. For air, if values of thermal conductivity k and specific heat at constant pressure c_p are independent of pressure, determine the dependence of thermal diffusivity on pressure.

    [**Ans.** Thermal diffusivity $\alpha = \dfrac{k}{\rho c_p}$ and from perfect gas equation $\rho = \dfrac{p}{RT} \propto p$. Hence, $\alpha \propto \dfrac{1}{p}$, i.e. thermal diffusivity α varies inversely with pressure p.]

20. Figure shows a two-dimensional body [$k = 5$ W/(m K)] and two isothermal surfaces A (at 50°C) and B (at 80°C). The heat enters the body at A with temperature gradient of 20 K/m. Determine temperature gradients $\partial t / \partial x$ and $\partial t / \partial y$ at surface B where the heat is leaving.

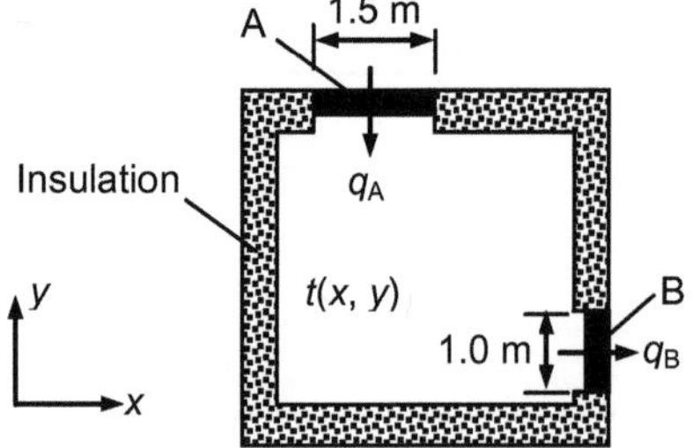

    [**Ans.** Heat flux vector is always normal to an isothermal surface. Since surface B is an isothermal vertical surface, $\partial t / \partial y = 0$. From heat conservation, $q_A = q_B$, i.e. $-k\left(A\dfrac{\partial t}{\partial y}\right)_A = -k\left(A\dfrac{\partial t}{\partial x}\right)_B$; For unit depth (in direction z), $(1.5 \times 1) \times 20 = (1.0 \times 1.0) \times \left(\dfrac{\partial t}{\partial x}\right)_B$, i.e. $\left(\dfrac{\partial t}{\partial x}\right)_B = 30$ K/m.]

140. Match List I with List II and select the correct answer using the codes given below the lists.

List I

A. Infinite parallel plates
B. Body 1 completely enclosed by body 2 but body 1 is very small
C. Radiation exchange between two small gray bodies
D. Two concentric cylinders with large lengths

List II

1. ε_1
2. $\left(\frac{1}{\varepsilon_1}+\frac{1}{\varepsilon_2}-1\right)^{-1}$
3. $\left[\frac{1}{\varepsilon_1}+\frac{A_1}{A_2}\left(\frac{1}{\varepsilon_2}-1\right)\right]^{-1}$
4. $\varepsilon_1\varepsilon_2$

Codes

	A	B	C	D
(a)	3	1	4	2
(b)	2	4	1	3
(c)	2	1	4	3
(d)	3	4	1	2

[**Ans.** (c)]

141. Match List I with List II and select the correct answer using the codes given below the lists.

List I

A. Momentum transfer
B. Mass transfer
C. Heat transfer

List II

1. Thermal diffusivity
2. Kinematic viscosity
3. Diffusion coefficient

Codes:

	A	B	C
(a)	1	3	2
(b)	2	3	1
(c)	3	2	1
(d)	1	2	3

[**Ans.** (b)]

142. Match List I with List II and select the correct answer using the codes given below the lists.

List I

A. Schmidt number
B. Thermal diffusivity
C. Lewis number
D. Sherwood number

List II

1. $\dfrac{k}{\rho c_p D}$
2. $\dfrac{h_m L}{D}$
3. $\dfrac{\nu}{D}$
4. $\dfrac{k}{\rho c}$

Codes

	A	B	C	D
(a)	4	3	2	1
(b)	4	3	1	2
(c)	3	4	2	1
(d)	3	4	1	2

[Ans. (d)]

143. **Fill in the blanks:**

(i) The velocity and temperature distributions are similar when ………. (Pr = 1)
(ii) The velocity and concentration profiles will be similar when …… (Sc = 1)
(iii) The temperature and concentration distributions will have the same profile when …….. (Le = 1)

Bibliography for Further Reading

Baehr Hans Dieter and Stephan Karl (2011) Heat and mass transfer. Springer-Verlag, Berlin Heidelberg
Bejan A (1995) Convective heat transfer. Wiley, New York
Bejan A (1993) Heat transfer. Wiley, New York
Bejan A, Kraus AD (eds) (2003) Heat transfer handbook. Wiley, New York
Bhatti MS, Shah RK (1987) Turbulent and transition flow convective heat transfer. In: Kakac S, Shah R. K, Aung W (eds) Handbook of single-phase convective heat transfer, Chapter 4. Wiley, New York
Cengel Yunus A (2007) Heat and mass transfer: a practical approach, 3rd edn. Tata McGraw-Hill Publishing Co., Ltd., New Delhi, India
Cengel Yunus A, Ghajar AJ (2011) Heat and mass transfer: fundamentals and applications. McGraw-Hill, New York
Chapman AJ (1960) Heat transfer. Macmillan, New York
Chemical Engineers' Handbook (1963) Perry RH, Chilton CH, Kirkpatrick SD (eds) 4th edn. McGraw-Hill Book Co, New York
Ebadian MA, Dong ZF (1998) Forced Convection, Internal Flow in Ducts. In: Rohsenow WM, Hartnett JP, Cho YI (eds) Handbook of heat transfer. McGraw Hill Book Co, New York
Eckert ERG, Drake RM (1959) Heat and Mass Transfer, 2nd edn. McGraw-Hill Inc.
Eckert ERG, Drake RM Jr (1972) Analysis of heat and mass transfer. McGraw-Hill Kogakusha, Tokya
Fraas Arthur P (1989) Heat Exchanger Design, 2nd edn. John Wiley & Sons
Gardner KA (1945) Efficiency of extended surfaces. Trans ASME 67:621–631
Gebhart B (1961) Heat transfer. McGraw-Hill Book Co, New York
Grober H, Erk S, Grigull U (1961) Fundamentals of heat transfer. McGraw-Hill Book Co, New York
Holman JP (1992) adapted for SI units by White PRS, Heat Transfer, McGraw-Hill Book Co, New York
Hottel HC (1954) Radiant heat transfer. In: McAdams WH (ed) Heat Transmission, 3rd edn. McGraw-Hill Book Co, New York
Hottel HC, Sarofim AF (1967) Radiative transfer. McGraw-Hill Book Co, New York
Howell John R (1982) A catalog of radiation configuration factors. McGraw-Hill Book Co, New York
Hsu ST (1963) Engineering heat transfer. D. Van Nostrand, New York
https://archive.org/details/SolutionManualFundamentalsOfHeatAndMassTransfer6thEdition
Incropera Frank P, DeWitt David P, Bergman Theodore L, Lavine Adrienne S (2012) Principles of Heat and mass Transfer. Wiley, International Students Edition
Isachenko VP, Osipova VA, Sukomel AS (1977) Heat transfer, 2nd edn. MIR Publishers, Moscow
Jakob M (1957) Heat transfer. Wiley, New York
Jakob M, Hawkins G (1957) Elements of heat transfer. Wiley, New York
Kakac S, Liu H (1998) Heat Exchangers: selection, rating, and thermal design. CRC Press, Boca Raton, Florida, p 3343
Kakac S, Yerner Y (1995) Convective heat transfer, 2nd edn. CRC Press, New York
Kays WM, Crawford ME (1980) Convective heat and mass transfer. Tata McGraw-Hill Publishing Co., Ltd, New Delhi
Kays WM, Perkins HC (1973) Forced convection, internal flow in ducts. In: Rohsenow WM, Hartnett JP (eds) Handbook of heat transfer, Section 7. McGraw-Hill, New York
Kays WM, Perkins HC (1985) Forced convection, internal flow in ducts. In: Rohsenow WM, Hartnett JP, Ganic EN (eds) Handbook of heat transfer fundamentals, Chapter 7. McGraw-Hill, New York
Kern DQ (1950) Process heat transfer. McGraw-Hill, New York
Knudsen JD, Katz DL (1958) Fluid dynamics and heat transfer. McGraw-Hill, New York
Krasnoshchekov EA, Sokomel AS (1977) Problems in heat transfer. MIR Publishers, Moscow

R. Karwa, *Heat and Mass Transfer*,
https://doi.org/10.1007/978-981-15-3988-6

Lienhard IV JH, Lienhard V JH (2000) A heat transfer textbook. Published by Lienhard V JH, Cambridge, Massachusetts, U.S.A.
Luikov AV (1980) Heat and mass transfer. Mir Publishers, Moscow
McAdams WH (1942) Heat transmission, 2nd edn. McGraw-Hill, New York
McAdams WH (1954) Heat transmission, 3rd edn. McGraw-Hill, New York
Mikheyev M (1968) Fundamental of heat transfer. Mir Publishers, Moscow
Mills AF (1995) Heat and mass transfer. Richard D, Irwin, Chicago
Mostinsky IL, Diffusion coefficient. http://www.thermopedia.com/content/696/ DOI: 10.1615/AtoZ.d.diffusion_coefficient
Pitts Donald R, Sissom Leighton E (1991) Heat transfer. McGraw-Hill, New York
Potter M, Wiggert D, Ramadan B (2012) Fluid mechanics. Cengage Learning, USA
Rogers GFC, Mayhew YR (1967) Engineering thermodynamics work and heat transfer, English Language Book Co. and Longmans, Green & Co., London
Schlichting H (1960) Boundary layer theory (trans: Kestin J), 4th edn. McGraw-Hill, New York
Siegel R, Howell JR (2002) Thermal radiation heat transfer. Taylor & Francis, New York
Standards of Tubular Exchanger Manufacturers Association (1949), 2nd edn. New York
Welty James R, Wicks Charles E, Wilson Robert E (2008) Fundamentals of momentum, heat, and mass transfer, 5th edn. Wiley
White Frank M (2011) Fluid Mechanics. McGraw-Hill, New York
Zhukauskas AA (1987) Convective heat transfer in cross-flow. In: Kakac S, Shah RK, Aung W (eds) Handbook of single-phase convective heat transfer, Chapter 6. Wiley, New York, pp 6.1–6.45

Index

R. Karwa, *Heat and Mass Transfer*,
https://doi.org/10.1007/978-981-15-3988-6

R

S

MIX
Papier aus verantwortungsvollen Quellen
Paper from responsible sources
FSC® C105338

ı have any concerns about our products,
you can contact us on
ˈuctSafety@springernature.com

ˈlisher is established outside the EU,
authorized representative is:
ıre Customer Service Center GmbH
platz 3, 69115 Heidelberg, Germany

Printed by Libri Plureos GmbH
in Hamburg, Germany